W9-AOL-871

BIOLOGY OF PLANTS

THIRD EDITION

Irises, by Vincent van Gogh (1853–90), was painted in May of 1889, when van Gogh was in a mental hospital in Saint-Rémy-de-Provence, France.

BIOLOGY OF PLANTS

THIRD EDITION

PETER H. RAVEN

Missouri Botanical Garden and
Washington University, St. Louis

RAY F. EVERT

University of Wisconsin, Madison

HELENA CURTIS

WORTH PUBLISHERS, INC.

BIOLOGY OF PLANTS, Third Edition

PRINTED IN THE UNITED STATES OF AMERICA

LIBRARY OF CONGRESS CATALOG CARD NO. 80–54233

ISBN: 0–87901–132–7

THIRD PRINTING, JULY 1982

EDITOR: GORDON BECKHORN

PRODUCTION: GEORGE TOULOUMES

DESIGN: MALCOLM GREAR DESIGNERS

COPY EDITOR: ROBYN BEM

PICTURE EDITOR: ANNE FELDMAN

TYPOGRAPHER: GRAPHIC ARTS COMPOSITION

PRINTING AND BINDING: R. R. DONNELLEY & SONS

COVER: DETAIL FROM *Irises* BY VINCENT VAN GOGH, JOAN WHITNEY PAYSON

GALLERY OF ART, WESTBROOK COLLEGE, PORTLAND, MAINE

WORTH PUBLISHERS, INC.

444 PARK AVENUE SOUTH

NEW YORK, NEW YORK 10016

PREFACE

While we are aware that most of you have only a semester, sometimes just a quarter, in which to teach botany, we nevertheless have decided against drastically shortening our book in this third edition. We believe that the reasons for this decision will be evident as you examine the book. Not only did we judge it impossible to present clearer, more accessible introductions to most topics with fewer words and fewer illustrations, but also our conversations with colleagues who have used the book did not reveal any consensus on topics that should be eliminated. Given this lack of uniformity in botany courses, together with the value of being free to choose from a larger menu, so to speak, and the small price differential between this and much more limited presentations, we felt free to indulge our own strong inclination—to present you with an introductory textbook that we think does justice to the subject.

Despite appearances, there are fewer words in this edition. But because the book is more profusely illustrated, this trimming has not resulted in fewer pages. There are many new photographs, micrographs, and drawings, and many of those that have been retained have been improved in ways suggested by classroom experience with the previous edition. Our chief concern in this revision has been to make more accessible to students an understanding of, and appreciation for, the basic facts and concepts of botany. We believe the book is written more clearly, that the drawings are thought out more carefully, prepared more skillfully, and labeled more fully, and that the photographs were selected more perceptively, than ever before.

We have also attempted to include as many of the recent advances in our understanding of plant science as feasible. For those interested in what is going on in botany today, a pretty comprehensive answer can be gleaned from this book. To mention some examples from the diversity section alone, the so-called "blue-green algae" are now treated as the cyanobacteria, one of several lines of bacteria capable of photosynthesis; the fungi and the algae are presented as coherent groups, with the water molds and slime molds in a chapter of their own. Much up-to-date material has been added on plant pathogens and the diseases they cause, as well as on the fungus-plant relationship of mycorrhizae. Newly available information about green algae has been used as the basis for a thoroughly revised treatment of the division Chlorophyta. Progymnosperms are presented as the probable ancestors of modern gymnosperms, and fossil plants in general are accorded more emphasis. Also, new biochemical evidence on such diverse subjects as alternative pathways of photosynthesis, genetic recombination and episomes, protein production in eukaryotes, and crown-gall tumors is now integrated into the text.

We have retained what we think of as the book's interlocking themes: (1) the plant body as the dynamic result of the processes of growth and development mediated by chemical interactions; (2) evolutionary relationships as the guide to understanding form and function in organisms; and (3) ecology as a way of emphasizing our dependence upon plants to sustain our lives and those of all other living organisms.

BIOLOGY OF PLANTS is still organized in such a way that you can assign topics in any preferred sequence. For the many courses in which only selected parts of the book will be assigned, we believe that students will have a clear and interesting introduction to botany, as well as an up-to-date reference to topics that may catch their imagination and into which they may wish to delve more deeply.

By introducing young people not only to what is known about botany but particularly to the many unsolved problems, we hope to enlist new talents and new enthusiasms in working toward the solutions on which all of our futures depend.

PETER H. RAVEN
RAY F. EVERT

January 1981

ACKNOWLEDGMENTS

This book could not have been written without the help, at every stage, of experts in various areas of botanical research and teaching. We had substantial assistance from L.A.S. Johnson, Royal Botanic Gardens and National Herbarium, Sydney, Australia, who reviewed the whole book and made dozens of useful suggestions for changes; from George B. Johnson, Washington University, in connection with the genetics chapters; from Gary L. Floyd, Ohio State University, Linda Graham, University of Wisconsin, and Jeremy Pickett-Heaps, University of Colorado, with the material on the algae; and from Arthur Kelman, University of Wisconsin, with the discussion of prokaryotes. We are grateful to John W. Einset of the University of California, Riverside, for his assistance in the revision of the chapter on plant hormones, and to Peter Quail and Gerald C. Gerloff of the University of Wisconsin for their assistance with the revision of Chapters 26 and 27, respectively. Charles B. Beck of the University of Michigan provided expert guidance for our treatment of the ancestors of the seed plants.

Others to whom we wish to express our thanks for reviewing parts of the manuscript or supplying information for this third edition are:

Robert M. Arnold, *Colgate University*
Harlan P. Banks, *Cornell University*
Wayne M. Becker, *University of Wisconsin*
Patricia M. Bonamo, *State University of New York at Binghamton*
Robert N. Bowman, *Colorado State University*
Winslow R. Briggs, *Stanford University*
John Burris, *Pennsylvania State University*
Mary-Dell Chilton, *Washington University*
W. Dennis Clark, *Arizona State University*
Theodore S. Cochrane, *University of Wisconsin*
James T. Colbert, *University of Wisconsin*
George E. Corson, Jr., *California State University, Chico*
Marshall R. Crosby, *Missouri Botanical Garden*
Judith G. Croxdale, *University of Wisconsin*
David L. Dilcher, *Indiana University*
John Dwyer, *St. Louis University*
Erna Eisendrath, *Washington University*
John J. Engel, *Field Museum of Natural History, Chicago*
Walter Eschrich, *University of Göttingen*
M. A. Gillott, *McGill University*
Ursula Goodenough, *Washington University*
Alan S. Heilman, *University of Tennessee*
John Heslop-Harrison, *University College of Wales*
Yolande Heslop-Harrison, *University College of Wales*
Kenneth G. Keegstra, *University of Wisconsin*
David L. Kirk, *Washington University*
Virginia M. Kline, *University of Wisconsin*
Robert R. Kowal, *University of Wisconsin*
Robert E. Magill, *Botanical Research Institute, Pretoria*

John Mitchell, *Ohio University*
Walter Mozgala, *Holyoke Community College*
Eldon H. Newcomb, *University of Wisconsin*
James W. Perry, *University of Wisconsin*
Tom L. Phillips, *University of Illinois*
Richard W. Pippen, *Western Michigan University*
Stephen E. Scheckler, *Virginia Polytechnic Institute and State University*
Daniel C. Scheirer, *Northeastern University*
Folke Skoog, *University of Wisconsin*
Champ B. Tanner, *University of Wisconsin*
John W. Thomson, *University of Wisconsin*
Kenneth G. Todar, *University of Wisconsin*
James M. Trappe, *U.S.D.A., Corvallis, Oregon*
Joseph E. Varner, *Washington University*
William F. Whittingham, *University of Wisconsin*
Clarice Yentsch, *Bigelow Laboratory for Ocean Sciences, West Boothbay Harbor, Maine*

We would like, once again, to thank Daniel I. Axelrod, University of California, Davis; Allan M. Campbell, Stanford University; Barbara G. Pickard, Washington University; and G. Ledyard Stebbins, University of California, Davis, for their contributions to past editions of BIOLOGY OF PLANTS.

We are especially indebted to Susan E. Eichhorn of the University of Wisconsin, who supervised much of the work done there for this third edition—the typing, the preparation of drawings and photographs, and the organization of innumerable details. We are also grateful to Rhonda Nass for her many superb drawings and to Damian S. Neuberger of the University of Wisconsin for his outstanding photographic work.

Others at the University of Wisconsin who contributed to the preparation of the third edition and whom we wish to thank are Gene Coffman, Patricia A. Evert, and David W. Gray, photographers; Jean Arnold, Mary E. Lauder, and Jane Liess, typists; Wendy K. Ballas and Catherine M. Ballard, general assistance; and Lucy C. Taylor, for photographic retouching.

The first author would like to express his appreciation to his wife, Tamra Engelhorn Raven, for her assistance and encouragement. The second author would like to express his gratitude to his wife and children for their patience and encouragement during the preparation of the third edition.

Helena Curtis did not play an active role in this revision, but she remains listed as a coauthor because so much of her work continues to exist in the book.

Finally, we wish to express our thanks to the many people at Worth Publishers who have made contributions to the quality and the usefulness of this textbook.

PETER H. RAVEN
RAY F. EVERT

January 1981

CONTENTS IN BRIEF

Introduction 1

SECTION 1 **The Plant Cell** 13

CHAPTER 1 Introduction to the Eukaryotic Cell 15

CHAPTER 2 The Molecular Composition of Cells 45

CHAPTER 3 The Movement of Substances Into and Out of Cells 59

SECTION 2 **Energy and the Living Cell** 71

CHAPTER 4 The Flow of Energy 73

CHAPTER 5 Respiration 84

CHAPTER 6 Photosynthesis 96

SECTION 3 **Genetics and Evolution** 115

CHAPTER 7 The Chemistry of Heredity 117

CHAPTER 8 The Genetics of Eukaryotic Organisms 132

CHAPTER 9 The Evolution of Plant Diversity 147

SECTION 4 **Diversity** 167

CHAPTER 10 The Classification of Living Things 169

CHAPTER 11 The Prokaryotes 184

CHAPTER 12 The Fungi 212

CHAPTER 13 Heterotrophic Protista: Water Molds and Slime Molds 241

CHAPTER 14 Autotrophic Protista: the Algae 251

CHAPTER 15 The Bryophytes 287

CHAPTER 16 The Vascular Plants: An Introduction 304

CHAPTER 17 The Seedless Vascular Plants 317

CHAPTER 18 The Seed Plants 338

CHAPTER 19 Evolution of the Flowering Plants 372

SECTION 5 The Angiosperm Plant Body: Structure and Development 403

CHAPTER 20 Early Development of the Plant Body 405

CHAPTER 21 Cells and Tissues of the Plant Body 416

CHAPTER 22 The Root: Primary Structure and Development 432

CHAPTER 23 The Shoot: Primary Structure and Development 445

CHAPTER 24 Secondary Growth 476

SECTION 6 Growth Regulation and Growth Responses 499

CHAPTER 25 Regulating Growth and Development: The Plant Hormones 501

CHAPTER 26 External Factors and Plant Growth 518

SECTION 7 Uptake and Transport in Plants 535

CHAPTER 27 Plant Nutrition and Soils 537

CHAPTER 28 The Movement of Water and Solutes in Plants 559

SECTION 8 Ecology 577

CHAPTER 29 The Dynamics of Ecosystems 579

CHAPTER 30 Terrestrial Biomes 595

APPENDIX A Fundamentals of Chemistry 615

APPENDIX B Metric Table 629

APPENDIX C Classification of Organisms 631

APPENDIX D Geologic Eras 637

GLOSSARY 639

ILLUSTRATION ACKNOWLEDGMENTS 665

INDEX 670

TABLE OF CONTENTS

INTRODUCTION 1

The Evolution of Plant Life 1
The Evolution of Communities 7
Plants and Humans 8
Essay: Genetic Diversity in Crop Plants 12

SECTION 1 The Plant Cell 13

CHAPTER 1 Introduction to the Eukaryotic Cell 15

Prokaryotes and Eukaryotes 16
Essay: Viewing the Microscopic World 16
Essay: Origin of the Cell Theory 18
The Plant Cell 18
Nucleus 18
Plastids 20
Mitochondria 22
Microbodies 24
Vacuoles 24
Essay: The Evolutionary Origin of Mitochondria and Chloroplasts 25
Ribosomes 26
Golgi Bodies 26
Endoplasmic Reticulum 27
Lipid Droplets and Spherosomes 28
Membrane Dynamics 28
Microtubules 29
Flagella and Cilia 30
The Cell Wall 32
Plasmodesmata 35
Cell Division 36
The Cell Cycle 36
Summary 43

CHAPTER 2 The Molecular Composition of Cells 45

Organic Compounds 45
Carbohydrates 46
Essay: Radiocarbon Dating 47

Lipids 50
Proteins 51
Nucleic Acids 56
Other Nucleotide Derivatives 56
Summary 56

CHAPTER 3 The Movement of Substances Into and Out of Cells 59

Principles of Water Movement 59
Structure of Cellular Membranes 64
Transport Across Membranes 64
Essay: Imbibition 65
Endocytosis and Exocytosis 67
Transport Via Plasmodesmata 68
Summary 68
Suggestions for Further Reading 69

SECTION 2 Energy and the Living Cell 71

CHAPTER 4 The Flow of Energy 73

The Laws of Thermodynamics 74
Essay: $E = mc^2$ 76
Enzymes and Living Systems 77
Enzymes as Catalysts 78
The Active Site 78
Cofactors in Enzyme Action 79
Enzymatic Pathways 80
Regulation of Enzyme Activity 81
The Energy Factor: ATP 81
Summary 83

CHAPTER 5 Respiration 84

Glycolysis 85
Aerobic Pathway 88
Essay: The Mechanism of Oxidative Phosphorylation 91
Essay: Bioluminescence 93
Anaerobic Pathways 94
Summary 95

CHAPTER 6 Photosynthesis 96

Overview of Photosynthesis 96
Essay: Light and Life 98
The Light Reactions 101
The Dark Reactions 106
Summary 112
Suggestions for Further Reading 112

SECTION 3 Genetics and Evolution 115

CHAPTER 7 The Chemistry of Heredity 117

The Chemistry of the Gene: DNA Versus Protein 117
The Nature of DNA 118
How Do Genes Work? 123
Regulating Gene Transcription 126
Control of Multicellular Differentiation 128
Summary 131

CHAPTER 8 The Genetics of Eukaryotic Organisms 132

Eukaryotes Versus Prokaryotes 132
Structure of Eukaryotic Chromosomes 133
Meiosis 134
Genes and Diploid Organisms 139
Mutations 143
Pleiotropy 144
Gene Interactions in Diploid Organisms 144
Summary 146

CHAPTER 9 The Evolution of Plant Diversity 147

The Behavior of Genes in Populations 149
Responses to Selection 151
The Divergence of Populations 154

Essay: Vegetative Reproduction: Some Ways and Means 154
The Evolutionary Role of Hybridization 159
Essay: Evolutionary Radiation 160
Summary 165
Suggestions for Further Reading 166

SECTION 4 Diversity 167

CHAPTER 10 The Classification of Living Things 169

The Binomial System 170
The Major Groups of Organisms 175
Formal Classification of Living Organisms 177
Sexual Reproduction and Diploidy 178
Summary 182
Suggestions for Further Reading 183

CHAPTER 11 The Prokaryotes 184

General Characteristics of Bacteria 185
Essay: Sensory Responses in Bacteria 190
Essay: Using Bacteria to Manufacture Useful Complex Molecules 194
Essay: Nature as Genetic Engineer 195
Essay: Evolution of Photosynthesis 197
Essay: Mycoplasmas 202
The Viruses 204
Essay: Viral Infections 206
Essay: Viroids 209
Summary 210
Suggestions for Further Reading 210

CHAPTER 12 The Fungi 212

Biology of the Fungi 213
The Evolution of Fungi 215
Fungal Reproduction 215
The Major Groups of Fungi 216
Essay: Phototaxis in a Fungus 218
Essay: Ergotism 222
Essay: Predaceous Fungi 224
Mycorrhizae 237
Summary 239
Suggestions for Further Reading 240

CHAPTER 13 Heterotrophic Protista: Water Molds and Slime Molds 241

Division Oomycota 242
Essay: Hormonal Control of Sexuality in a Water Mold 244
Division Chytridiomycota 245
Division Acrasiomycota 247
Division Myxomycota 247
Summary 249
Suggestions for Further Reading 250

CHAPTER 14 Autotrophic Protista: The Algae 251

Characteristics of the Algae 252
Essay: Symbiotic Algae 254
The Green Algae: Division Chlorophyta 255
Essay: Rhizoplasts: Musclelike Structures in an Alga 258
Essay: A Complex Semiterrestrial Green Alga 263
The Brown Algae: Division Phaeophyta 268
The Red Algae: Division Rhodophyta 272
Essay: The Economic Uses of Algae 275
The Diatoms and Golden-Brown Algae: Division Chrysophyta 276
The Dinoflagellates: Division Pyrrophyta 279
The Euglenoids: Division Euglenophyta 281
Essay: Coleochaete: An Advanced Green Alga 282
Essay: Mitosis in Dinoflagellates 283
Essay: Derivation of the Divisions of Algae 284
Summary 285
Suggestions for Further Reading 285

CHAPTER 15 The Bryophytes 287

Characteristics of the Bryophytes 288
The Liverworts: Class Hepaticopsida 291
Essay: Spore Discharge in Liverworts 294
The Hornworts: Class Anthocerotopsida 296
The Mosses: Class Muscopsida 296
Summary 303
Suggestions for Further Reading 303

CHAPTER 16 The Vascular Plants: An Introduction 304

Early Vascular Plants 305
Organization of the Vascular Plant Body 307
Reproductive Systems 311
Summary 314
Suggestions for Further Reading 315

CHAPTER 17 The Seedless Vascular Plants 317

Division Psilophyta 317
Division Lycophyta 319
Division Sphenophyta 327
Essay: Coal Age Plants 330
Division Pterophyta 333
Summary 337
Suggestions for Further Reading 337

CHAPTER 18 The Seed Plants 338

The Progymnosperms 339
The Gymnosperms 340
The Angiosperms 356
Summary 371
Suggestions for Further Reading 371

CHAPTER 19 Evolution of the Flowering Plants 372

Origin of the Angiosperms 372
Essay: Early Angiosperms and Their Flowers 376
Evolution of the Flower 377
Essay: Genetic Self-Incompatibility 383
Evolution of Fruits 393
Biochemical Coevolution 397
Summary 399
Suggestions for Further Reading 401

SECTION 5 The Angiosperm Plant Body: Structure and Development 403

CHAPTER 20 Early Development of the Plant Body 405

The Mature Embryo and Seed 405
Formation of the Embryo 409
Essay: Wheat 409
Requirements for Seed Germination 413
From Embryo to Adult Plant 413
Summary 415

CHAPTER 21 Cells and Tissues of the Plant Body 416

The Tissue Systems 417
Tissues and Their Component Cells 417
Summary 429

CHAPTER 22 The Root: Primary Structure and Development 432

Root Systems 432
Origin and Growth of Primary Tissues 433
Primary Structure 437
Origin of Lateral Roots 440
Aerial Roots 441
Adaptations for Food Storage 442
Summary 444

CHAPTER 23 The Shoot: Primary Structure and Development 445

Origin and Growth of the Primary Tissues of the Stem 446
Primary Structure of the Stem 447
Relation Between the Vascular Tissues of the Stem and the Leaf 453
Development of the Leaf 454
Morphology of the Leaf 457
Structure of the Leaf 459
Essay: Plants and Air Pollution 464
Leaf Abscission 466
Transition Between Vascular Systems of the Root and the Shoot 466
Development of the Flower 467
Stem and Leaf Modifications 471
Essay: Convergent Evolution 473
Summary 475

CHAPTER 24 Secondary Growth 476

The Vascular Cambium 477
Effect of Secondary Growth on the Primary Plant Body 479
The Wood: Secondary Xylem 489
Summary 495
Suggestions for Further Reading 496

SECTION 6 Growth Regulation and Growth Responses 499

CHAPTER 25 Regulating Growth and Development: The Plant Hormones 501

Auxin 502
Essay: Plants in Test Tubes 506
Cytokinins 507
Ethylene 510
Abscisic Acid 512
Gibberellins 512
Summary 517

CHAPTER 26 External Factors and Plant Growth 518

The Tropisms 518
Circadian Rhythms 520
Photoperiodism 522
Chemical Basis of Photoperiodism 524
Hormonal Control of Flowering 527
Dormancy 528
Cold and the Flowering Response 530
Touch Responses in Plants 531
In Conclusion 532
Summary 533
Suggestions for Further Reading 533

SECTION 7 Uptake and Transport in Plants 535

CHAPTER 27 Plant Nutrition and Soils 537

General Nutritional Requirements 537
Functions of Inorganic Nutrients in Plants 539

The Soil 541
Essay: The Water Cycle 544
Nutrient Cycles 545
Nitrogen and the Nitrogen Cycle 545
Essay: Carnivorous Plants 547
Essay: Lectins 550
The Phosphorus Cycle 552
Human Impact on Nutrient Cycles 553
Soils and Agriculture 554
Essay: Compost 554
Plant Nutrition Research 555
Summary 557

CHAPTER 28 The Movement of Water and Solutes in Plants 559

Movement of Water Through The Plant Body 559
The Movement of Inorganic Nutrients Throughout the Plant Body 568
Translocation: The Movement of Substances in the Phloem 569
Essay: Radioactive Tracers and Autoradiography in Plant Research 572
Summary 574
Suggestions for Further Reading 575

SECTION 8 Ecology 577

CHAPTER 29 The Dynamics of Ecosystems 579

Interactions Between Organisms 579
Cycling of Nutrients 585
Essay: Plant Defense in Solanaceae 586
Trophic Levels 587
Development of Ecosystems 589
Summary 594

CHAPTER 30 Terrestrial Biomes 595

Life on the Land 595
Essay: Alexander von Humboldt 597
Tropical Rain Forest 597
Essay: Rodent Pollination of Tropical Plants 598
Savanna 601
Desert 602
Grasslands 604
Temperate Deciduous Forest 606
Essay: The Conservation of Plants 609
Taiga 610
Tundra 611
Summary 612
Suggestions for Further Reading 612

APPENDIX A Fundamentals of Chemistry 615

Atoms 615
Bonds and Molecules 619
Water and the Hydrogen Bond 622
Water as a Solvent 623
Acids and Bases 624
Chemical Reactions 625

APPENDIX B Metric Table 629
Temperature Conversion Scale 630

APPENDIX C Classification of Organisms 631

APPENDIX D Geologic Eras 637

GLOSSARY 639

ILLUSTRATION ACKNOWLEDGMENTS 665

INDEX 670

BIOLOGY OF PLANTS

THIRD EDITION

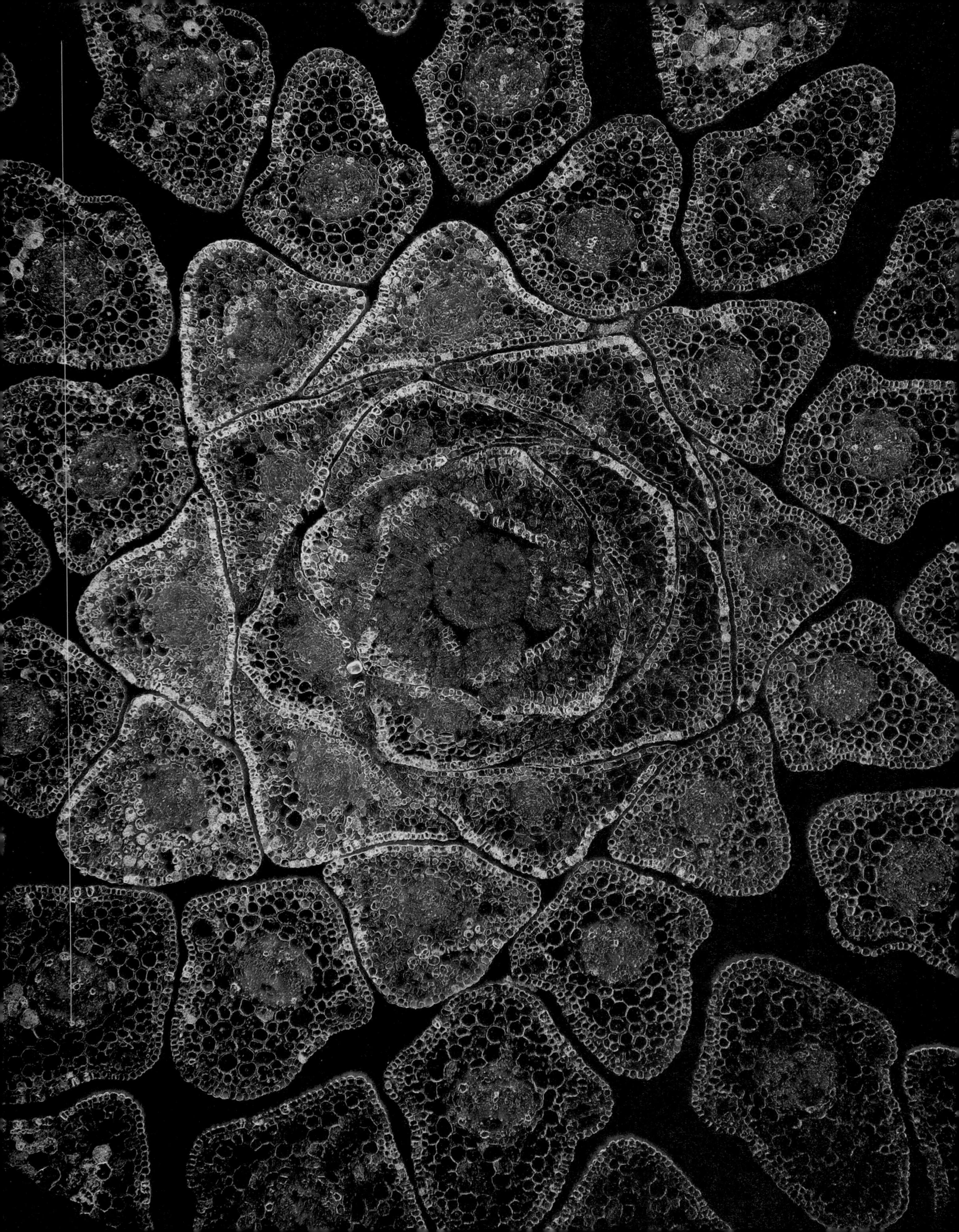

Introduction

When a particle of light strikes a molecule of chlorophyll, an electron is jolted out of the molecule and raised to a higher energy level. Within a fraction of a second, it returns to its previous energy state. All life on this planet is dependent upon the energy momentarily gained by the electron. The process by which some of the energy given up by the electron in returning to its original energy level is converted into chemical energy—energy in a form usable by living systems—is known as photosynthesis. Photosynthesis is the vital link between the physical and the biological world, or, as Nobel laureate Albert Szent-Györgyi said, more poetically: "What drives life is . . . a little current, kept up by the sunshine."

Only a few types of organisms—the green plants, the algae, and some bacteria—possess a molecule known as chlorophyll, which can, when embedded in the membranes of a living cell, carry out this energy conversion. Once the energy is trapped in chemical form, it becomes available as an energy source to all other organisms, including humans. We are totally dependent upon photosynthesis, a process for which green plants are so exquisitely adapted.

THE EVOLUTION OF PLANT LIFE

Like all other living organisms, plants have had a long evolutionary history. The planet earth itself—an accretion of dust and gases swirling in orbit around the star that is our sun—is some 4.5 billion years old. The earliest known fossils are about 3.5 billion years old and consist of several kinds of small, relatively simple cells.

As events are reconstructed, these first cells were formed by a series of chance events. The raw materials, namely, carbon, oxygen, hydrogen, and nitrogen, were present in the gases of the early atmosphere. These four elements make up about 98 percent of the material found in living organisms today.

I–1
Transverse section of the terminal bud of a spruce (Picea) *tree, as seen with fluorescence microscopy. The numerous roughly triangular structures surrounding the center of the bud are leaves, the principal photosynthetic organs of the plant.*

Through the thin atmosphere, the rays of the sun beat down on the harsh, bare surface of the young earth, bombarding it with light, heat, and ultraviolet radiation. Liquid water first appeared on earth about 3.8 billion years ago. Water vapor cooled in the upper atmosphere, fell on the crust of the earth as rain, and steamed up again, driven by the sun's heat. Violent rainstorms, accompanied by lightning, released electrical energy. Radioactive substances in the earth's crust emitted their energy, and molten rock and boiling water erupted from beneath the earth's surface. The energy in this vast crucible broke apart the simple gases of the atmosphere and reformed them into more complicated molecules.

According to present hypotheses, the compounds that were formed in the atmosphere tended to be washed out by the driving rains and to collect in the oceans, which grew larger as the earth cooled. As a consequence, the ocean became an increasingly rich mixture of organic molecules. Some organic molecules have a tendency to aggregate in groups; in the primitive ocean, these groups probably took the form of droplets, similar to the droplets formed by oil in water. Such droplets of organic molecules appear to have been the forerunners of primitive cells, the first forms of life.

These organic molecules also served, according to present theories, as the source of energy for the earliest forms of life. The primitive cells or cell-like structures were able to use these compounds, which were abundant in the "primordial soup," to satisfy their energy requirements.

Such cells are known as *heterotrophs,* a category of organisms that today includes all living things classified as animals or fungi and many of the one-celled organisms—the bacteria and the protists. *Hetero* comes from the Greek word meaning "other," and *troph* comes from *trophos,* "one that feeds." A heterotrophic organism is one that is dependent upon others—that is, upon an outside source of organic molecules—for its energy.

As the primitive heterotrophs increased in number, they began to use up the complex molecules on which their existence depended—and which had taken millions of years to accumulate. Organic molecules in free solution (not inside a cell) became more and more scarce. Competition began. Under the pressure of this competition, cells that could make efficient use of the limited energy sources now available were more likely to survive than cells that could not. In the course of time, by the long, slow process of weeding out the less fit, cells evolved that were able to make their own energy-rich molecules out of simple nonorganic materials. Such organisms are called *autotrophs,* "self-feeders." Without the evolution of autotrophs, life on earth would soon have come to an end.

The most successful of the autotrophs were those that evolved a system for making direct use of the sun's energy—the process of photosynthesis.

The earliest photosynthetic organisms, although simple in comparison to modern plants, were much more complex than the primitive heterotrophs. To capture and use the sun's energy required, first, a complex pigment system that could catch and hold the energy of a ray of light and, linked to this system, a way of fixing the energy in an organic molecule.

Thus the flow of energy in the biosphere came to assume its modern form: radiant energy channeled through photosynthetic autotrophs to all other forms of life.

Photosynthesis and the Coming of Oxygen

As photosynthetic organisms increased in number, they changed the face of the planet. This biological revolution came about because one of the most efficient strategies of photosynthesis—the one employed by nearly all living autotrophs—involves splitting the water molecule (H_2O) and releasing its oxygen. Thus, as a consequence of photosynthesis, the amount of oxygen gas (O_2) in the atmosphere increased. This had two important consequences. First, some of the oxygen molecules in the outer layer of atmosphere were converted to ozone (O_3) molecules. When there is a sufficient quantity of ozone molecules in the atmosphere, they filter the ultraviolet rays—highly destructive to living organisms—from the sunlight that reaches the earth. By about 450 million years ago, organisms, protected by the ozone layer, could survive in the surface layers of water and on the land.

Second, the increase in free oxygen opened the way to a much more efficient utilization of the energy-rich molecules formed by photosynthesis. As will be discussed in Chapter 5, respiration* yields far more energy than can be extracted by any anaerobic (oxygenless) process. Until the atmosphere became aerobic, the only cells that evolved were *prokaryotic*—simple cells that lacked nuclear envelopes and did not have their genetic material organized into complex chromosomes. The living prokaryotes are called bacteria. According to the fossil record, the increase of relatively abundant free oxygen was accompanied by the first appearance of *eukaryotic* cells—cells that have complex chromosomes, nuclear envelopes, and membrane-bound organelles. Eukaryotic organisms

*It should be noted that respiration has two meanings in biology. One is the breathing in of oxygen and breathing out of carbon dioxide; this is also the ordinary, nontechnical meaning of the word. The second meaning of respiration is the oxidation of food molecules by cells—that is, the breaking down of energy-rich, carbon-containing molecules and their use by the cell as an energy source. This process, sometimes qualified as cellular respiration, is what we are concerned with here.

(a) (b)

I–2
A modern heterotroph and a photosynthetic autotroph. (a) *A fungus,* Coprinus atramentaris, *growing on a forest floor in California.* Coprinus, *which, like other fungi, absorbs its food (often from other organisms), is heterotrophic.* (b) *Large-flowered trillium* (Trillium grandiflorum), *one of the first plants to flower in spring in the deciduous woods of eastern and midwestern North America. Like most vascular plants, trillium is rooted in the soil; photosynthesis occurs chiefly in the leaves of this autotrophic organism. The flowers are produced in well-lighted conditions before the leaves appear on the surrounding trees. The underground portions (rhizomes) of the plant live for many years and spread to produce new plants vegetatively under the thick cover of decaying leaves and other organic material on the forest floor. Trillium also reproduces by producing seeds that are dispersed by ants.*

were well established and relatively diverse by about 850 million years ago, and they may have first appeared about 1.4 billion years ago. All living systems are composed of eukaryotic cells, except for the bacteria and the viruses.

The Sea and the Shore

Early in evolutionary history, the principal photosynthetic organisms were microscopic cells floating on the surface of the sunlit waters. Energy abounded, as did carbon, hydrogen, and oxygen, but as the cellular colonies multiplied, they quickly depleted the mineral resources of the open ocean. (It is this shortage of essential minerals that is the limiting factor in any modern plans to harvest the seas.) As a consequence, life began to drift toward the shores, where the waters were rich in nitrates and minerals carried down from the mountains by rivers and streams and scraped from the coasts by the ceaseless waves.

The rocky coast presented a much more complicated environment than the open sea and, in response to these evolutionary pressures, living organisms became more complex in structure and more diversified. Some 650 million years ago, organisms evolved in which many cells, connected by strands of protoplasm, were linked together to form an integrated, multicellular body. In these primitive organisms we see the beginnings of the modern plants and animals.

On the turbulent shore, those photosynthetic organisms that were multicellular were better able to maintain their position against the action of the waves, and, in meeting the challenge of the rocky coast, new forms developed. Typically, these organisms evolved relatively strong walls for support and specialized structures to anchor their bodies to the rocky surfaces. As these multicellular organisms increased in size, they were confronted with the problem of how to supply food to the dimly lit portions of their bodies where photosynthesis was not taking place. As a consequence of these new pressures, specialized food-conducting tissues evolved that, extending down the center of their bodies, connected the photosynthesizing parts of an organism with lower, nonphotosynthesizing structures.

The Transition to Land

The body of the familiar plant can best be understood in terms of its long history and, in particular, in terms of the evolutionary pressures involved in the transition to land. The requirements of a photosynthetic organism are relatively simple: light, water, carbon dioxide for photosynthesis, oxygen for respiration, and a few minerals, or inorganic ions. On the land, light is abundant, as are oxygen and carbon dioxide—both of which circulate more freely than in the water—and the soil is generally rich in inorganic ions. The critical factor is water. Land animals, generally speaking, are mobile and are able to seek out water just as they seek out food. Fungi, though immobile, remain largely below the surface of the soil or within whatever damp organic material they feed upon. Plants utilize an alternative evolutionary strategy. Roots anchor the plant in the ground and collect the water required for maintenance of the plant body and for photosynthesis. A continuous stream of water moves into the root hairs, up through the roots and stems, and then out through the leaves. All of the above-ground portions of the plant that are ultimately concerned with photosynthesis are covered with a waxy *cuticle* that retards water loss. However, the cuticle also prevents the necessary exchange of gases between the plant and the surrounding air. The solution to this dilemma is found in specialized openings called *stomata* (singular: stoma), which open and close in response to environmental and physiological signals, thus helping the plant maintain a balance between its water losses and its oxygen and carbon dioxide requirements.

I–3
Ponderosa pine, or western yellow pine (Pinus ponderosa), *an example of a vascular plant, is found in commercial quantities in every state west of the Great Plains.*

In younger plants and those with a year-long life cycle (annuals), the stem is also a photosynthetic organ. In longer-lived plants (perennials), the stem may become thickened and woody and covered with cork, which also retards water loss. In both cases, the stem serves both to support the chief photosynthetic organs, holding them to the light, and to conduct—via the vascular system—a variety of substances between the photosynthetic and nonphotosynthetic parts of the plant body. The vascular system has two major components: the *xylem,* through which water passes upward through the plant body, and the *phloem,* through which food manufactured in the leaves and other photosynthetic parts of the plant is transported throughout the plant body. It is this efficient conducting system that has given the main group of modern plants—the vascular plants—their name.

Although vascular plants are immobile as whole individuals, they continue to grow throughout their life spans, a condition that does not hold for animals. All plant growth originates in localized regions of perpetually embryonic tissues; these regions are called *meristems.* Meristems located at the tips of all roots and shoots are called *apical meristems.* These are involved with the extension of the plant body. Thus the roots move continually and actively through the soil, and the photosynthetic regions are continually expanded and extended toward the light. This type of growth that originates from apical meristems is known as *primary growth.*

The type of growth that results in a thickening of stems, branches, and roots is known as *secondary growth,* which originates in a second kind of meristematic tissue—the *vascular cambium.*

Thus, in summary, the vascular plant is characterized by a root system that serves to anchor the plant in the ground and to collect water and inorganic ions from the soil; a stem or trunk that raises the photosynthetic parts of the plant body toward its energy source, the sun; and leaves, the highly specialized photosynthetic organs. All of these—roots, stems, and leaves—are interconnected by a complicated and efficient system for the transport of food and water. All of these characteristics are related to a photosynthetic existence on land.

I–4
Diagram of a young broad bean (Vicia faba) *plant, showing the principal organs and tissues of the modern vascular plant body. The organs—root, stem, and leaf—are composed of tissues, which are groups of cells with distinct structures and functions. Collectively, the roots make up the root system, and the stems and leaves together make up the shoot system of the plant. In the great majority of vascular plants, the shoot system is above the soil surface and the root system is below. Unlike roots, stems are divided into nodes and internodes. The node is the part of the stem at which one or more leaves are attached, and the internode is the part of the stem between two successive nodes. (In the broad bean, the first few foliage leaves are divided into two leaflets each.) Buds (embryonic shoots) commonly arise in the axils (upper angle between leaf and stem) of the leaves. Lateral, or branch, roots arise from the inner tissues of the roots. The vascular tissues—xylem and phloem—occur together and form a continuous vascular system throughout the plant body.*

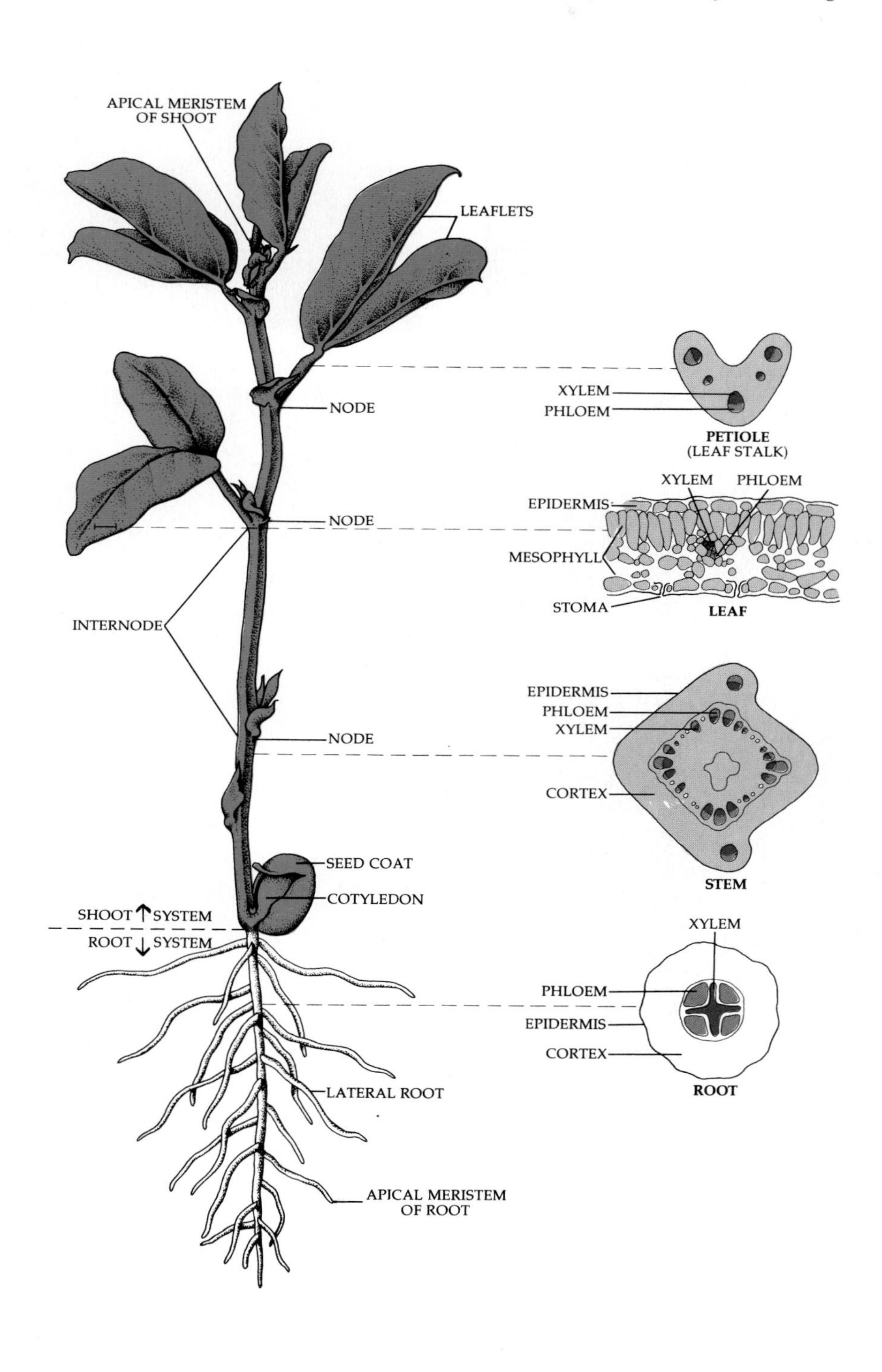

I–5
The four seasons in a deciduous forest in Illinois. In such forests, characteristic of much of the North Temperate Zone, the trees produce their leaves early in spring and begin to manufacture food; they lose them again in the autumn and enter an essentially dormant condition, in which they pass the unfavorable growing conditions of winter. Food manufactured in the leaves is carried throughout the plant and deep into the earth, reaching the farthest roots. At the same time, water and minerals are carried in a continuous stream up through the roots, stems, and leaves. Most of the water is lost from the leaf as water vapor through the same specialized pores in the leaves, the stomata, that carbon dioxide enters. Many herbs grow under the trees, and a number of these flower very early in the spring, before the leaves of the trees have reached full size and shade the forest floor. Most of the trees shed their pollen in large quantities in spring, and it is carried by the wind, sometimes reaching the flowers of other trees of the same species.

I–6
The prairies of the central United States are among the richest agricultural areas in the world, capable of retaining their productivity indefinitely if properly managed. This uncultivated prairie is dotted with wild flowers.

THE EVOLUTION OF COMMUNITIES

The invasion of the land by plants changed the face of the continents. Looking down from an airplane on one of our country's great expanses of desert or on the peaks of the Sierra Nevada, one can begin to imagine what the world looked like before the coming of the plants. Yet even in these inhospitable regions, the traveler who goes by land will find green plants of an astonishing variety punctuating the expanses of rock and sand. And in those parts of the world where the climate is more temperate and the rains are more favorable, the plant communities dominate the land and determine its character. In fact, to a large extent, they *are* the land. Rain forest, meadow, woods, prairie, tundra—each of these words brings to mind the portrait of a landscape. And the main features of the landscape are the plants—enclosing us in a dark green cathedral in our imaginary rain forest, carpeting the ground beneath our feet with wild flowers in a meadow, moving in great golden waves as far as the eye can see across our imaginary prairie. Only when we have sketched these biomes in broad strokes, that is, in terms of trees and shrubs and grasses, can we fill in the other features of our landscape—a deer, an antelope, a rabbit, a wolf.

How do vast plant communities, such as those seen on a continental scale, come into being? To some extent we can trace the evolution of the different kinds of plants and animals that populate them. Even with accumulating knowledge, however, we have only begun to glimpse the far more complex pattern of development, through time, of the whole system of organisms that make up these various communities. Such communities, along with the nonliving environment of which they are a part, are known as ecological systems, or *ecosystems*. Ecosystems will be discussed in greater detail in the final chapters of this book. For now, however, it is sufficient to regard an ecosystem as forming a sort of corporate entity, made up of transient individuals. Some of these individuals, the larger trees, live as long as several thousand years; others, the microorganisms, live only a few hours or even minutes. Yet the ecosystem as a whole tends to be remarkably stable; once in balance, it will not change for centuries. Our grandchildren will someday perhaps walk along a woodland path once followed by our great-grandparents, and where they saw a pine tree, a mulberry bush, a meadow mouse, wild blueberries, or a towhee, this child, if this woodland still exists, will see these same kinds of plants and animals and in the same numbers.

Although many of the organisms in an ecosystem are competing for resources, the system as a whole functions as an integrated unit. The death of a solitary cell floating on the ocean's surface is likely to involve the dissipation of its stored energy and the breakdown of its chemical constituents. In an ecosystem, virtually every living thing, down to the smallest bacterial cell or fungal spore, provides a food source for some other living organism. In this way, the energy captured by green plants is transferred in a highly regulated way through a number of different types of organisms before it is dissipated. Moreover, interactions among the organisms themselves, and between the organisms and the nonliving environment, produce an orderly cycling of elements such as nitrogen and phosphorus. Energy must be constantly added to the ecosystem, but the elements are cycled through the organisms, returned to the soil, decomposed by soil bacteria and fungi, and recycled. These transfers of energy and this cycling of elements involve complicated sequences of events, and in these sequences each group of organisms has its own particular and highly specific place. As a consequence, it is impossible to change a single element in an ecosystem without the risk of destroying the carefully developed balance on which its stability depends.

I–7
The clockface of biological time. Life appears relatively early in the earth's history, before 6:00 A.M. on a 24-hour scale. The first multicellular organisms do not appear until the early evening of that 24-hour day, and humans are a very late arrival—at about 30 seconds to midnight.

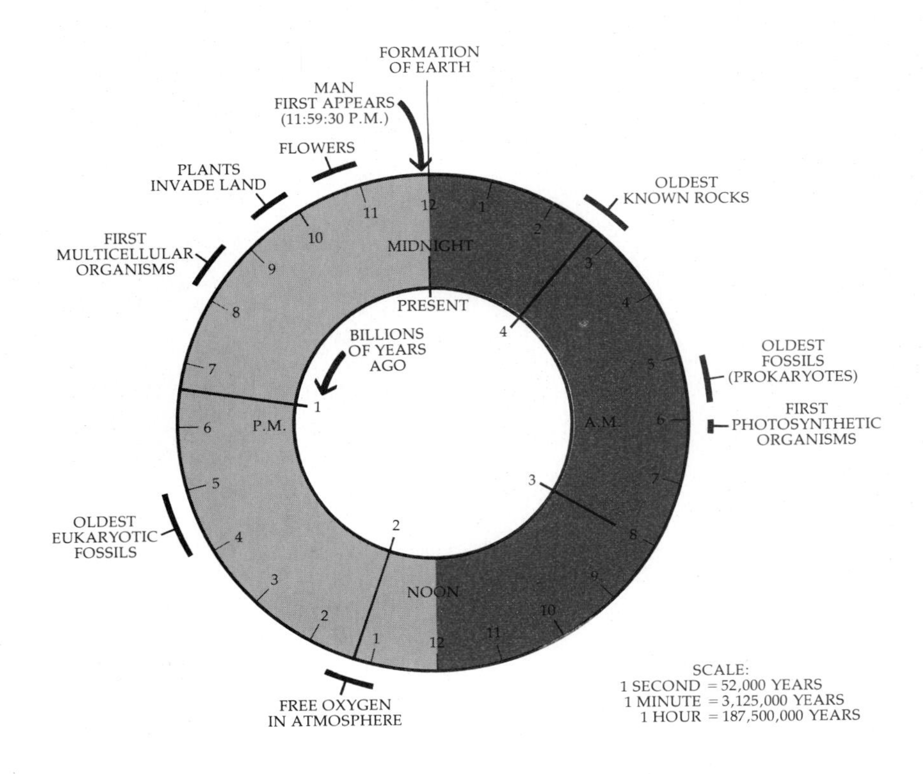

PLANTS AND HUMANS

Humans are relative newcomers to the world of living things. If the entire history of the earth were measured on a 24-hour time scale, starting at midnight, cells would appear in the warm seas at about dawn. The first multicellular organisms would not be present until well after dark, and the earliest appearance of humans (about one million years ago) would be at less than 30 seconds before the day's end. Yet humans, more than any other animal—indeed almost as much as the plants that invaded the land—have changed the surface of the planet, shaping the biosphere according to their own needs, ambitions, or follies.

At the close of the Pleistocene epoch, some 12,000 years ago, humans were already the most widely distributed land mammal, then numbering about five million. At that time, humans were hunter-gatherers, living in small nomadic bands. Thus the stage was set for the first major advance that allowed rapid expansion of the human population: the development of agriculture.

Origins of Agriculture

The reasons for the change to the agricultural way of life are not clear. One factor seems to have been changes in the climate. The most recent of the glaciations began to retreat about 18,000 years ago, withdrawing slowly for about 6000 years. As the glaciers retreated, the plains of northern Europe and of North America—once cold grasslands or steppes—gave way to forest. The great herbivores that roamed these steppes retreated northward and eventually vanished; the woolly mammoth was last seen in Siberia about 12,000 years ago. While some animals became extinct, humans adapted, as they had during the entire period of violent climatic changes that have marked their evolutionary history. With the migratory animals gone, humans shifted their attention to smaller game, such as deer, which tended to be resident in the same area throughout the year. Fishing became an important part of the economy at this time also; ponds and lakes were filling, and streams were rushing with waters from melting glaciers. Hunters made canoes and paddles and seines and other fishing equipment. The hunting and gathering of small animals, rather than large migratory herbivores, undoubtedly resulted in a less nomadic existence for the hunters and formed a prelude to the agricultural revolution.

The Transition

The earliest definite traces of agriculture are found in an area in the Near East generally known as the Fertile Crescent. Here were the raw materials required by an agricultural economy: cereals—which are grasses with seeds capable of being stored for long periods without serious deterioration—and herbivorous herd animals, which can be readily domesticated. Among the wild grasses in the Fertile Crescent were wheats and barley; they still grow wild in these foothills. The animals present included wild sheep and goats.

Although we do not know exactly when or where it first happened, the first deliberate planting of seeds can be seen as the logical end of a simple series of events. The wild cereals are weeds, ecologically speaking; that is, they grow readily on open or disturbed areas, patches of bare land where there are few other plants to compete with them. Also, early humans, judging from their campsites, were not very tidy. It is easy to envision how seeds might be spilled or discarded with garbage on open land around human habitations. And the presence of the archaeological remains of what are clearly permanent dwellings indicates that humans were staying in one place long enough to recognize and reap their accidental harvest. From this it would have been an easy step to deliberately saving seeds and tending crops.

Because humans selected the seeds that they gathered and planted, they soon produced changes in the wild strains. For instance, the stalk (rachis) of wild wheats, on which the flower clusters grow and on which the seeds eventually develop, becomes brittle when the seeds mature and breaks off, scattering the seed. Among these wild plants, occasional mutants can be found in which the rachis is not brittle. It is these plants, at a disadvantage in the wild, that are more likely to be harvested by humans and, as a consequence, are more likely to be planted. Thus, just as humans came to be dependent upon their crop plants, the plants they grew for food—such as cereals that could not seed themselves—came to be dependent upon humans.

About 11,000 years ago, new cultures appeared around the Fertile Crescent. They were characterized by implements associated with the harvesting and processing of grains, such as flint sickle blades, grinding stones, or stone mortars and pestles.

By 8100 years ago, agricultural communities were established in eastern Europe; by 7000 years ago (about 5000 B.C.), agriculture had spread to the western Mediterranean and up the Danube into central Europe, and by 4000 B.C., agriculture had spread to Britain. During this same period, agriculture originated separately in Central and South America, and perhaps slightly later in the Far East.

From the Old World sites, we have fossil imprints of cultivated wheat and barley, remains of domesticated goats, sheep, and cattle, and pottery vessels, stone bowls, and mortars. The farmers of the New World grew corn, pumpkins, squash, gourds, and cotton. The potato, the sweet potato, the peanut, and the tomato are also examples of New World crops.

I–8
Primitive methods of agriculture in Morocco and Thailand.

Today, of course, the vast majority of all the foods we eat is the result of deliberate cultivation.

The Consequences of the Agricultural Revolution

The change to agriculture had profound consequences. Populations were no longer nomadic. Thus, they could store food not only in silos and granaries but in the form of domesticated animals. In addition to food stores, other possessions could be accumulated to an extent far beyond that previously possible.

Even land could be owned and accumulated and passed on by inheritance. Thus, the world became divided into semipermanent groups of haves and have-nots, as it is today.

Because the efforts of a few could produce enough food for everyone, the communities became diversified. People became tradesmen, artisans, bankers, scholars, poets, all the rich mixture of which a modern community is composed. And these people could live much more densely than ever before. For hunting and food-gathering economies, 5 square kilometers, on the average, are required to provide enough food for one family to eat.

One immediate and direct consequence of the agricultural revolution was an increase in populations. A striking characteristic of hunting groups is that they vigorously limit their numbers. A woman on the move cannot carry more than one infant along with her household baggage, minimal though that may be. When simple means of birth control—often just abstention—are not effective, she resorts to abortion or, more probably, infanticide. In addition, there is a high natural mortality, particularly among the very young, the very old, the ill, the disabled, and women at childbirth. As a result, populations dependent upon hunting tend to remain small.

Once families became sedentary, there was no longer the same urgent need to limit the number of births, and probably there was also a decrease in the mortality rates.

The Population Explosion

About 25,000 years ago, there were perhaps 3 million people. By the close of the Pleistocene epoch, some 10,000 years ago, the human population probably numbered a little more than 5 million, spread over the entire world. By 4000 B.C., about 6000 years ago, the population had increased enormously, to more than 86 million; by the time of Christ, it is estimated that there were 133 million people. In other words, the population increased more than 25 times between 10,000 and 2000 years ago.

By 1650, the world population had reached 500 million, with many people living in urban centers, and the development of science, technology, and industrialization had begun, bringing about further profound changes in the life of humans and their relationship to nature.

By 1980, there were more than 4.4 billion people on our planet (Table I-1). This is an almost incomprehensibly large figure; moreover, the rate of increase of this enormous population is also unprecedented. The birthrate in the United States has decreased drastically since 1972, and the population could stabilize during the next century. For the world as a whole, however, the population is growing at about 1.8 percent per year. This means that about 150 people are added to the world population every minute, about 220,000 each day, and 80 million every year. If this rate of increase is sustained, in place of the 4.4 billion people living in the year 1980, there will be more than 6 billion people on earth by the year 2000.

The World Bank estimated in 1981 that some 780 million people were living in "absolute poverty," a condition of life that Robert McNamara defined as "so characterized by malnutrition, illiteracy, disease, squalid surroundings, high infant mortality, and low life expectancy as to be beneath any reasonable definition of human decency." In addition to solving this problem, we must grow enough food between now and the end of the century for about 1.7 billion addi-

Table I–1 *A 1980 Demographic Summary of Peoples of the World**

AREA	POPULATION ESTIMATE (MILLIONS)	GROWTH RATE (%)	NUMBER OF YEARS TO DOUBLE THE POPULATION	ESTIMATED POPULATION IN THE YEAR 2000
Africa	472	2.9	24	832
Asia	2563	1.8	39	3578
Europe	484	0.4	176	521
USSR	266	0.8	82	311
Latin America	360	2.6	26	595
North America**	247	0.7	98	289
Oceania	23	1.1	61	30
World	4414	1.7	41	6156

*Taken from 1980 World Population Data Sheet of the Population Reference Bureau, Inc.
**Canada and the United States. At least half the growth shown derives from immigration.

I–9
(a) *Norman Borlaug, who was awarded the Nobel Peace Prize in 1970. Borlaug was the leader of a research project sponsored by The Rockefeller Foundation under which new strains of wheat were developed in Mexico. Widely planted, these new strains have changed the status of Mexico from that of a wheat importer when the program began in 1944 to that of an exporter by 1964.*
(b) *Field workers weeding a rice plot at the International Rice Research Institute in the Philippines. The most recent strains are highly disease-resistant; however, they will grow only on irrigated land, which forms only about 30 percent of the cropland in Asia.*

(a)

(b)

tional people. If there is continued growth in affluence worldwide, and if a genuine effort is made to solve the problems of malnutrition, the demand for food in 1985 will be nearly 50 percent greater than in 1970. The International Food Policy Research Institute has estimated that the 1985–86 annual deficit of cereal grains in developing countries will range between 100 million and 200 million metric tons.

The high cost of producing fertilizer due to the increasing world energy shortage is contributing severely to the problem. Consider the fact that the more than 5 million tractors in the United States alone require 30 billion liters (8 billion gallons) of fuel, the equivalent of the energy content in the food produced. As fossil fuels become more scarce and more expensive, the costs of food will continue to increase. Land can generally be planted to energy-producing crops such as sugarcane only at the expense of lowered food production. For Americans, who spend on the average less than 20 percent of their personal income on food, the increasing cost of food already occasions serious concern. For those in developing nations, who may spend 80 to 90 percent of their income on food, it can be a death sentence. Indeed, in countries such as Bangladesh, India, and Haiti, and in regions such as the Sahel in northern Africa, people are dying in increasing numbers because of the lack of available food.

Although there are gains to be made in bringing additional land under cultivation, the most promising approach seems to lie in the development of existing crops grown on presently cultivated land, in terms not only of their yield but also of their protein content. Agricultural technology, including methods of irrigation, can also be improved considerably. These are among the most vital areas of concern to the plant scientist. The effort to stimulate agriculture by the development of new crop plants—especially grains—has been called the Green Revolution.

Enormous progress is being made. The production of wheat in Mexico has quadrupled since 1950. Between 1968 and 1972, India and Pakistan doubled their wheat production. China, the most populous nation in the world, has become agriculturally self-sustaining, largely as the result of adopting these new strains. Techniques of breeding, fertilizing, and irrigation are being applied to rice and other crops in developing countries throughout the world. A new hybrid of wheat and rye, *Triticale,* is one of the most promising products of this program (see pages 164–65). Among the important areas of current research are the improvement of photosynthetic efficiency and the fixation of atmospheric nitrogen; both will be discussed in later chapters.

Despite its acknowledged success, this massive effort has not been without its drawbacks (see "Genetic Diversity in Crop Plants") or critics. One reason is the increasing cost of fertilizer; these new grains require intensive cultivation. Indeed, it is said that they produce good yields only when they are well supplied with fertilizer. Because the large landowners are able to afford the investment in fertilizer and farming equipment that the small-scale farmers cannot, these new agricultural developments are seen as accelerating the consolidation of farmlands into a few large holdings by the very wealthy.

The problem of how to ease the burdens of hunger and poverty borne by nearly a fifth of the world's population remains very severe. Food production is already falling below population growth in Africa, and it cannot long keep pace with the rapid growth of the population in tropical Asia and Latin America.

I–10
Modern wheat production.

The Green Revolution must of course go forward, but at the same time, we must recognize that the broader solutions to these problems are social, political, and ethical. They involve not only the growth of food but its distribution, including the provision of jobs whereby the poor can earn the means to buy food. They involve not only the limiting of populations but also the raising of living standards of these populations to tolerable levels.

This introduction has ranged from the beginnings of life on this planet to the evolution of land plants and of plant communities to the development of agriculture and of modern society, with its most pressing current problem, the unprecedented growth of the human population. These broad topics are of interest to many people other than botanists. As we turn to Chapter 1, in which our attention narrows to a cell so small it cannot be seen by the unaided eye, it is well to keep in the back of our minds these broader concerns. A basic knowledge of plant biology is useful in its own right and essential in many fields of endeavor, but it is also increasingly relevant to some of society's most crucial problems and to the difficult decisions that will face us in choosing among the proposals for diminishing them. Thus this book is dedicated not only to the botanists of the future, whether teachers or researchers, but also to the informed citizens, scientists and laymen alike, in whose hands such decisions lie.

GENETIC DIVERSITY IN CROP PLANTS

Until recently, the major food plants of the world, like the wild ancestors from which they arose, were genetically quite diverse. In fact, since the beginnings of agriculture, more than 10,000 years ago, huge reserves of variability have accumulated in the important crop plants by the processes of mutation, hybridization, artificial and unintentional selection, and adaptation to a wide range of conditions. Thus, for crops such as wheat, potatoes, and corn, there are literally thousands of known strains. This genetic variability provides an important safety factor. If one strain were to prove unusually susceptible to a particular pathogen, for example, others could be found that were genetically resistant to it. From these resistant strains, new varieties could be bred.

As a result of selection for crop strains that had a dependable, high yield, however, in recent years crop plants have become more and more uniform and hence are now especially vulnerable to destruction. In 1970, for example, the southern corn leaf blight fungus, Helminthosporium maydis, *destroyed approximately 15 percent of the U.S. corn crop—a loss of approximately a billion dollars. These losses were apparently related to the appearance of a new race of the fungus that is highly destructive to some of the strains of corn that are used extensively in hybrid seed production.*

Such dangers increase as the improved strains being developed continue to replace the many distinct types that existed before. Time is growing short. Indeed, it is already becoming difficult to locate seeds of many of the strains of crop plants that have already been identified.

It is now generally agreed, by both critics and supporters of the Green Revolution, that aggressive research programs on our crop plants are needed, first to monitor the appearance of new strains of plant pathogens, and second to maintain the genetic diversity of these crops. Varieties threatened with extinction must be collected throughout the world and preserved in suitable gene banks. One such genetic reservoir has been established in Fort Collins, Colorado, by the U.S. Department of Agriculture. If such programs are not expanded and maintained, survival of the human race may once again be threatened by its own technological successes.

SECTION 1 The Plant Cell

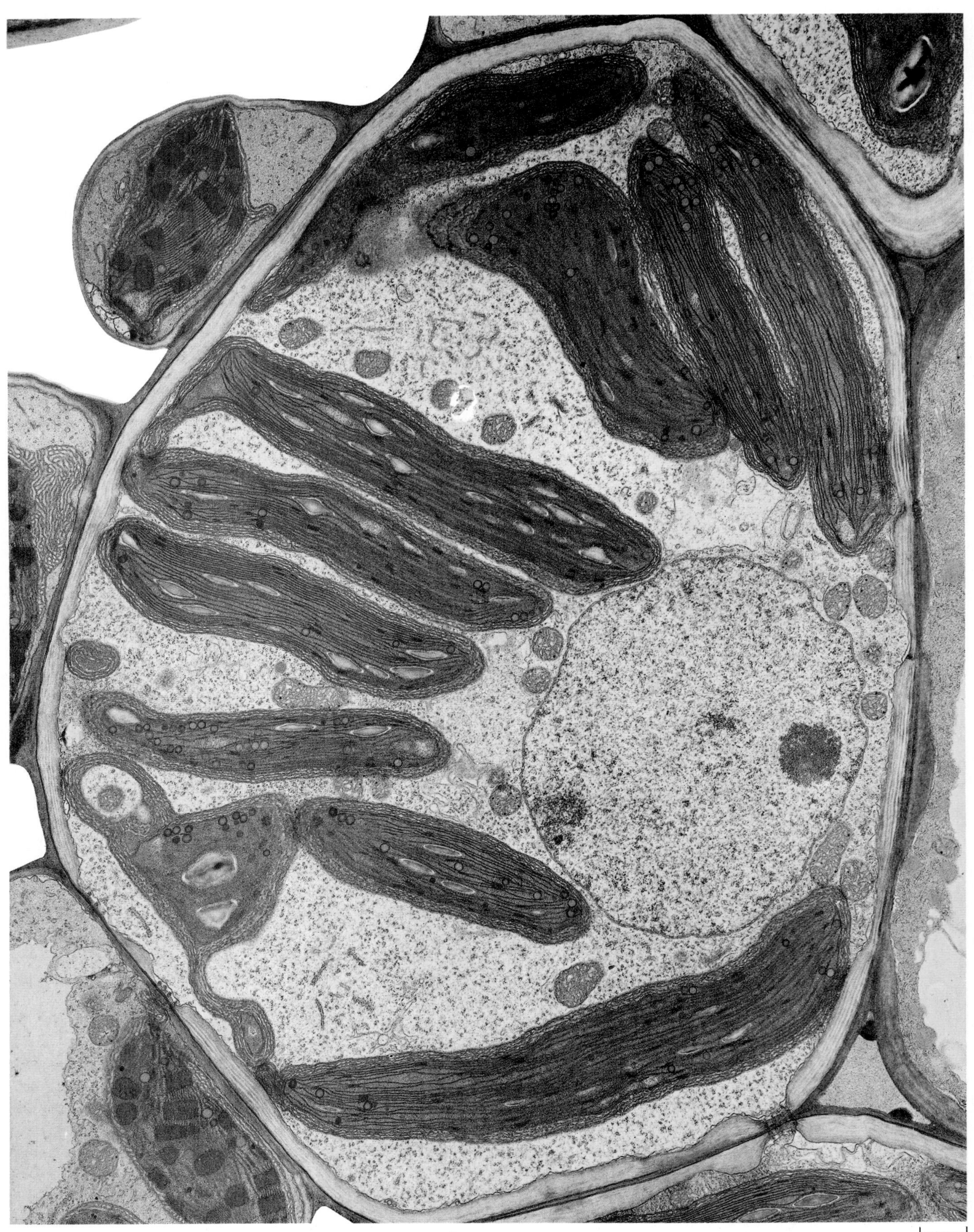

1.0 μm

CHAPTER 1

Introduction to the Eukaryotic Cell

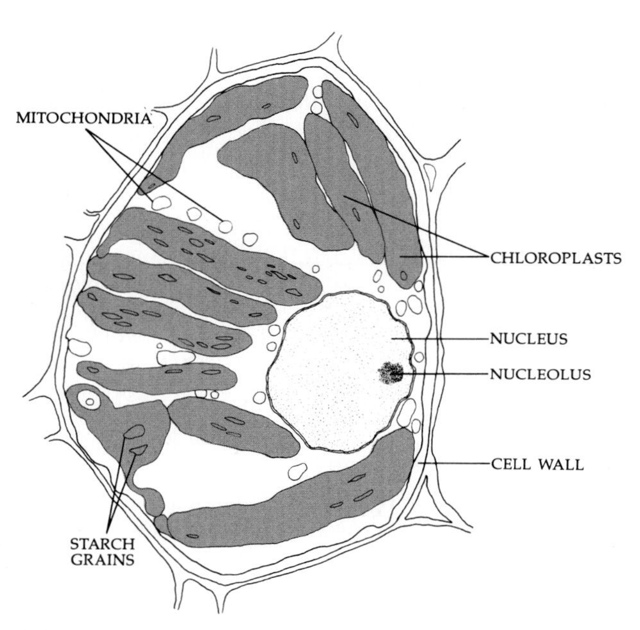

1–1
At left is a cross section of a plant cell enlarged just over 9000 times its actual size by an electron microscope. In the labeled drawing (above), some of the cell's components are identified.

This particular cell is a bundle-sheath cell from the leaf of a corn plant (Zea mays). *Cells like this completely surround the water- and food-conducting tissues (xylem and phloem) of the veins. The nucleus—site of the cell's genetic information—can be seen at the lower right. A single, well-defined nucleolus is visible within the nucleus. Several chloroplasts (organelles involved in photosynthesis) and mitochondria (organelles involved in the conversion of food to usable energy) can also be seen in this cell. Some of the chloroplasts contain oval starch grains that are lighter in appearance.*

Cells are the structural and functional units of life. The smallest living organisms are composed of single cells. The largest are made up of billions of cells, each of which still lives a partly independent existence. The realization that all organisms are composed of cells was one of the most important conceptual advances in the history of biology because it provided a unifying theme for the study of all living things. When studied at the cellular level, even the most diverse organisms are remarkably similar to one another, both in their physical organization and in their biochemical properties. The *cell theory* was formulated early in the nineteenth century, well before the presentation of Darwin's *theory of evolution,* but these two great unifying concepts are, in fact, closely related. In the similarities among cells, we catch a glimpse of a long evolutionary history that links modern organisms, including plants and ourselves, with the first cellular units that took shape on earth billions of years ago.

There are many, many different kinds of cells. Within our own bodies are more than 100 distinct cell types. In a teaspoon of pond water, one can find several different one-celled organisms, and in a whole pond, there are probably several hundred clearly different kinds. Plants are composed of cells that are superficially quite different from those of our own body, and insects have many kinds of cells not found in plants or in vertebrates. Thus, one remarkable fact about cells is their diversity.

A second, even more remarkable fact about cells is their similarity. Every living cell is a self-contained and at least partially self-sufficient unit, and each is bounded by an outer membrane—the plasma membrane, or plasmalemma (often simply called the cell membrane)—that controls the passage of materials into and out of the cell and so makes it possible for the cell to differ biochemically and structurally from its surroundings. Within this membrane is the cytoplasm which, in most cells, includes a variety of discrete bodies and various dissolved or suspended molecules. In addition, every cell contains DNA (deoxyribonucleic acid), which encodes the genetic characteristics (see Chapter 7), and this *genetic code* is very much the same from bacterium to oak tree to human.

PROKARYOTES AND EUKARYOTES

Two fundamentally distinct groups of organisms can be recognized: *prokaryotes* and *eukaryotes*. These terms are derived from the Greek word *karyon*, meaning "kernel" (nucleus). The name *prokaryote* thus means "before a nucleus." Modern prokaryotes are represented by the bacteria, including the cyanobacteria, or blue-green algae (see Chapter 11).

Prokaryotic cells differ most notably from eukaryotic cells in that their DNA is not organized into chromosomes—complex, protein-containing, thread-like bodies—nor is it surrounded by a membranous envelope (Figure 1–2). Also, prokaryotes have no specialized structures, or organelles, to perform specific functions.

Eukaryotic cells are divided into distinct compartments that perform different functions (Figure 1–1). The DNA, combined with protein, is located in chromosomes, which are in a nucleus bounded by a special membrane called the nuclear envelope. Eukaryotic cells are usually larger than prokaryotic cells.

Compartmentalization in eukaryotic cells is accomplished by means of membranes which, when seen with the aid of an electron microscope, look remarkably similar in various organisms. When suitably preserved and stained, these membranes have a three-layered appearance (Figure 1–3), consisting of two dark layers (each about 25 Å thick) separated by a lighter layer about 35 Å thick. The term *unit membrane* is commonly used to designate such a visually definable, three-layered membrane.

VIEWING THE MICROSCOPIC WORLD

Most cells can be seen only with the aid of a microscope. The units of measurement generally used for describing cells are micrometers and nanometers (see Table 1–1). Unaided, the human eye has a resolving power of about 1/10 millimeter, or 100 micrometers. This means that if you look at two lines that are less than 100 micrometers apart, they merge into a single line. Similarly, two dots less than 100 micrometers apart look like a single blurry dot. In order to separate structures closer than this, optical instruments such as microscopes must be used. The best light microscope has a resolving power of 0.2 micrometer, or about 200 nanometers, and so is about a 500-fold improvement on the naked eye. It is theoretically impossible to build a light microscope that will do better than this.

Notice that resolving power and magnification are different; using the best light microscope, if you take a picture of two lines that are less than 0.2 micrometer, or 200 nanometers, apart, you can enlarge that photograph indefinitely, but the two lines will still blur together. By using more powerful lenses, you can increase the magnification, but this will not improve the resolution.

THE TRANSMISSION ELECTRON MICROSCOPE

With the electron microscope, resolving power has been increased almost 400 times over that provided by the light microscope. This is achieved by using "illumination" consisting of electron beams instead of light rays. Under the very best conditions, electron microscopy currently affords a resolving power of about 0.5 nanometer, roughly 200,000 times greater than that of the human eye. (A hydrogen atom is about 0.1 nanometer in diameter.)

In a transmission electron microscope, a beam of electrons passing through the specimen leaves its imprint on a screen. Areas in the specimen that permit the transmission of more electrons, that is, "electron-transparent" regions, show up light in color, and "electron-dense" areas are dark.

The transmission electron microscope has one great disadvantage. Electrons have a very small mass and must travel in a vacuum; electron beams can only pass through specimens that are exceedingly thin. To prepare them for the electron microscope, specimens must therefore be killed and embedded in hard materials so that they can be sliced by special cutting instruments. This means, of course, that the high resolving powers of the electron microscope can be applied only to tissues that are no longer alive. Also, it is sometimes difficult to determine what changes may be produced in the material in the course of its preparation.

A few kinds of cells, almost all viruses, and many of the structures within cells can be seen only with the transmission electron microscope. Figure 1–1 is one of the many transmission electron micrographs found in this book.

THE SCANNING ELECTRON MICROSCOPE

In a scanning electron microscope, the electrons whose imprints are recorded come from the surface of the specimen. The electron beam is focused into a fine probe which is used to scan the specimen. As a result of the electron bombardment from the probe, the specimen emits low-energy secondary electrons. Variations in the surface of the specimen alter the number of secondary electrons emitted. Holes and fissures give off fewer secondary elec-

---

1–2
Escherichia coli, *a bacterium that is a common, usually harmless inhabitant of the human digestive tract, is an example of a modern prokaryote. Each of these rod-shaped organisms has a cell wall, a plasma membrane, and cytoplasm. The two cells in the center have just undergone cell division and have not yet separated completely. The DNA is found in the less granular area in the center of each cell. The densely granular appearance of the cytoplasm is largely due to the presence of numerous ribosomes. These small bodies are only about 20 nanometers in diameter. The scale marker at the bottom corner of the micrograph provides a reference for size. These bacteria are magnified some 29,000 times their actual size. [(29,000) (0.5 μm) = 1.5 cm = ½ inch, the length of the scale marker.]*

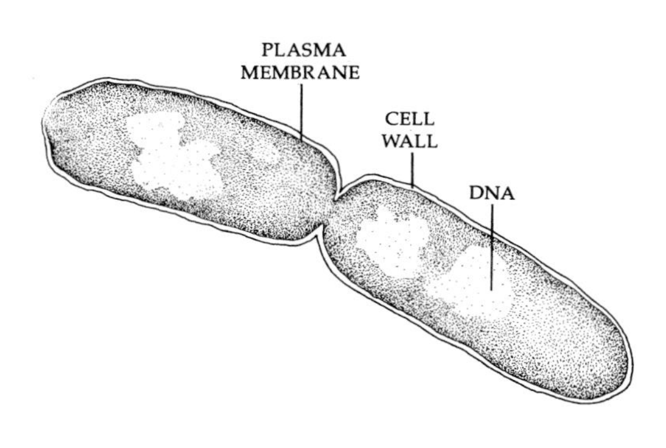

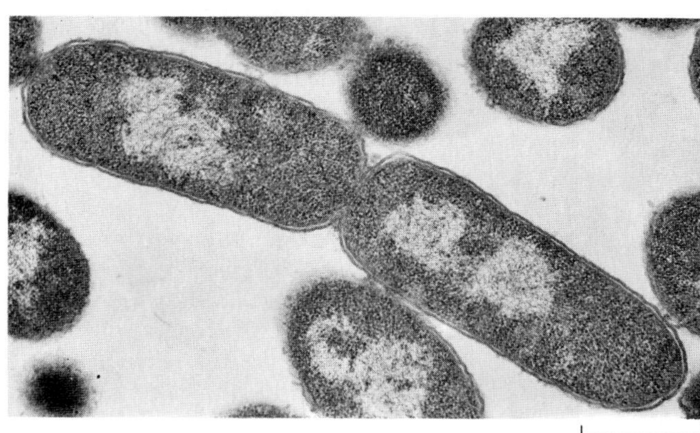

trons and thus appear dark, whereas the larger number of secondary electrons given off by high points and ridges cause these surfaces to appear light, imparting a three-dimensional image. Electrons scattered from the surface, plus any secondary electrons, are collected, amplified, and transmitted to a screen which is scanned in synchrony with the electron probe. Figure 2–4e on page 48 was prepared with a scanning electron microscope, as were others in this textbook.

Table 1–1 *Measurements Used in Microscopy*

1 centimeter (cm)	= 1/100 meter = 0.4 inch
1 millimeter (mm)	= 1/1,000 meter = 1/10 cm
1 micrometer (μm)*	= 1/1,000,000 meter = 1/10,000 cm
1 nanometer (nm)	= 1/1,000,000,000 meter
	= 1/10,000,000 cm
1 angstrom (Å)	= 1/10,000,000,000 meter
	= 1/100,000,000 cm
	or
1 meter	= 10^2 cm = 10^3 mm = 10^6 μm
	= 10^9 nm = 10^{10} Å

*Micrometers were formerly known as microns, indicated by the Greek letter μ, which corresponds to the letter *m* in our alphabet and is pronounced "mew."

1–3
Under high magnification, cellular membranes have a three-layered (dark-light-dark) appearance, as seen in the plasma membranes (see arrows) on either side of the common wall between two endodermal cells from a leaf of the fern Vittaria guineensis. *Such membranes are called unit membranes.*

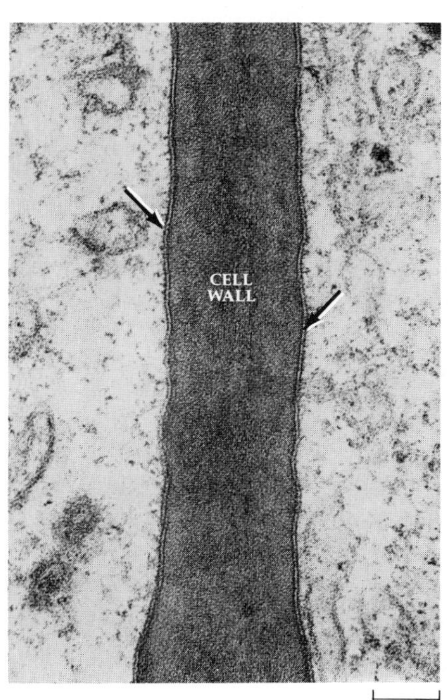

ORIGIN OF THE CELL THEORY

In the seventeenth century, Robert Hooke, using a microscope of his own construction, noted that cork and other plant tissues are made up of small cavities separated by walls. He called these cavities "cells," meaning "little rooms." The word did not take on its present meaning, however, until 150 years later.

In 1838, Matthias Schleiden, a German botanist, came to the conclusion that all plant tissues are organized in the form of cells. In the following year, zoologist Theodor Schwann extended Schleiden's observation to animal tissues and proposed a cellular basis for all life. The cell theory is of tremendous and central importance to biology because it emphasizes the basic similarity of all living systems and so brings an underlying unity to widely varied studies involving many different kinds of organisms.

In 1858, the cell theory took on broader significance when the great pathologist Rudolf Virchow generalized that cells can arise only from preexisting cells: "Where a cell exists, there must have been a preexisting cell, just as the animal arises only from an animal and the plant only from a plant. . . . Throughout the whole series of living forms, whether entire animal or plant organisms or their component parts, there rules an eternal law of continuous development."

In the broad perspective of evolution, Virchow's concept takes on an even larger significance. There is an unbroken continuity between modern cells—and the organisms that they compose—and the primitive cells that first appeared on earth at least 3.5 billion years ago.

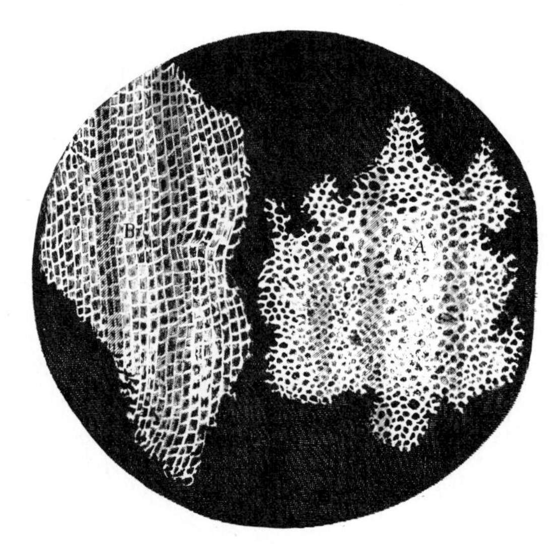

Robert Hooke's drawing of two slices of a piece of cork, reproduced from his *Micrographia* (1665). Hooke was the first to use the word "cells" to describe the tiny compartments into which living organisms are organized.

THE PLANT CELL

The plant cell typically consists of a more or less rigid *cell wall* and a *protoplast*. The term *protoplast* is derived from the word *protoplasm*, which has long been used to refer to the living contents of cells. A protoplast is the protoplasm of an individual cell—in the case of a plant cell, the unit of protoplasm inside the cell wall.

A protoplast consists of *cytoplasm* and a *nucleus*. The cytoplasm includes certain distinct entities, or organelles (such as ribosomes, microtubules, plastids, and mitochondria) and systems of membranes (the endoplasmic reticulum and Golgi bodies), all of which can be seen in detail only with the aid of an electron microscope. In addition, the cytoplasm includes an apparently unstructured *ground substance* in which these entities and membrane systems are suspended. The cytoplasm is separated from the cell wall by the *plasma membrane*, which is a unit membrane. In contrast to most animal cells, plant cells develop one or more liquid-filled cavities, or *vacuoles*, within their cytoplasm. The vacuole is bounded by a unit membrane called the *tonoplast*.

In a living plant cell, such as that of the pondweed *Elodea* (Figure 1–4), the ground substance is frequently in motion; the organelles, as well as various compounds suspended in the ground substance, can be observed being swept along in an orderly fashion in the moving currents. This movement is known as cytoplasmic streaming, or *cyclosis*, and it continues as long as the cell is alive. How cyclosis occurs remains a mystery. It undoubtedly facilitates the exchange of materials within the cell and between the cell and its environment; however, it is not known if this is a primary function.

NUCLEUS

The nucleus is often the most prominent structure within the cytoplasm of eukaryotic cells. The nucleus performs two important functions: (1) it controls the ongoing activities of the cell by determining which protein molecules are produced by the cell and when they are produced, as we shall see in Chapter 7; and (2) it stores the genetic information, passing it on to the daughter cells in the course of cell division.

In eukaryotic cells, the nucleus is bounded by a double membrane (two unit membranes) called the *nuclear envelope*. The nuclear envelope, as seen with the aid of an electron microscope, contains a large number of circular pores which are 30 to 100 nanometers in diameter (Figure 1–5); the inner and outer membranes are joined around each pore to form the margin of its opening. The pores are not merely

1–4

An Elodea *cell.* (a) *Upper surface of the cell.* (b) *Middle of the cell. The numerous disk-shaped structures are chloroplasts located in the cytoplasm along the wall. In surface view* (a), *the disk-shaped chloroplasts appear to be circular in outline. In* (b) *the chloroplasts have their broad surfaces facing the surface of the wall and appear to be elongated. Notice in* (b) *the absence of chloroplasts in the center of the cell—that is, within the vacuole.*

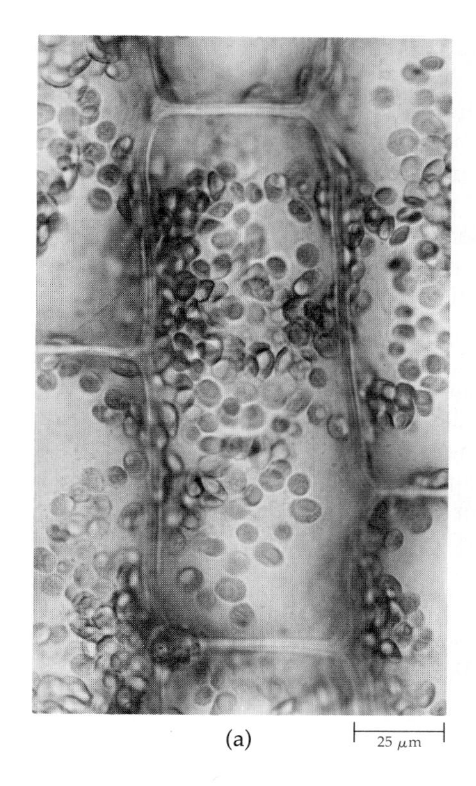

(a)

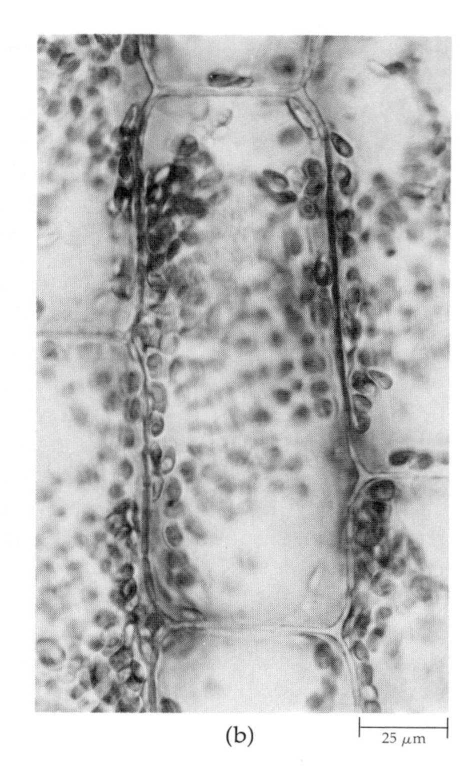

(b)

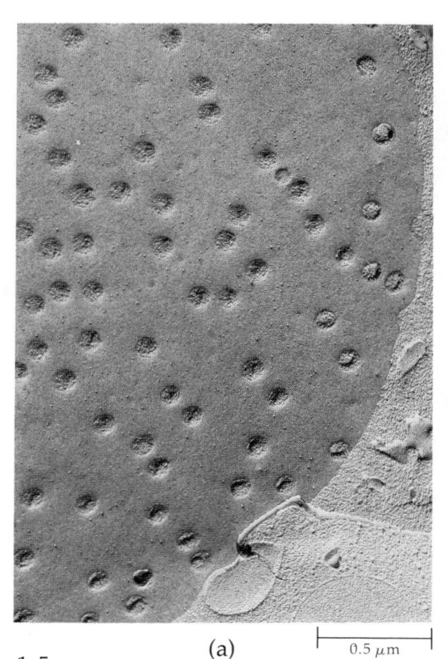

(a)

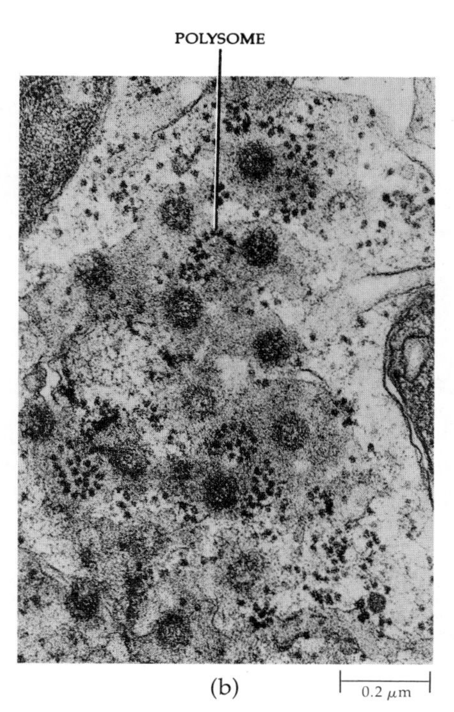

(b)

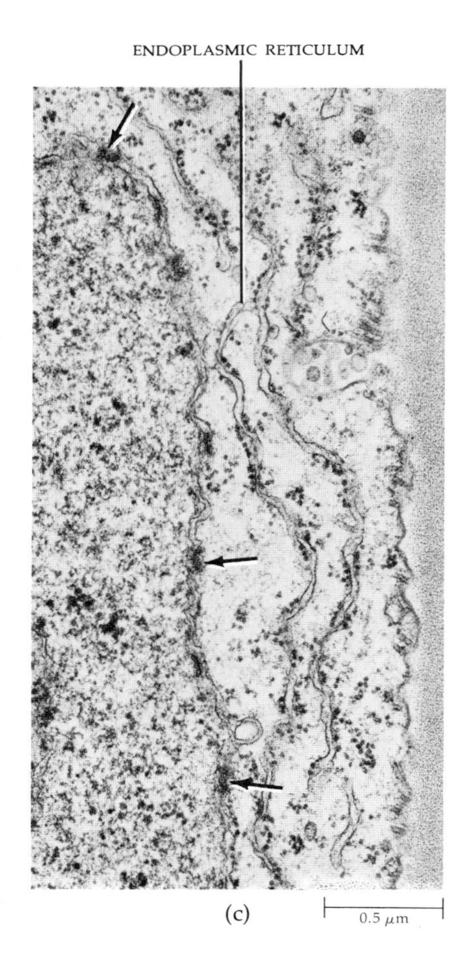

(c)

1–5

Nuclear pores as revealed (a) *in a freeze-etch preparation of the surface of the nuclear envelope of an onion* (Allium cepa) *root tip cell, and* (b) *and* (c) *in electron micrographs of nuclei in parenchyma cells of the seedless vascular plant* Selaginella kraussiana. *In* (b) *the pores are shown in surface view, whereas in* (c) *they are shown in sectional view (see arrows). Note the polysomes (coils of granules) on the surface of the nuclear envelope in* (b) *and the rough endoplasmic reticulum paralleling the nuclear envelope in* (c).

holes in the envelope; each has a complicated structure. In various places the outer membrane of the envelope may be continuous with the endoplasmic reticulum.

If the cell is treated by special staining techniques, thin strands and grains of *chromatin* can be distinguished from the *nucleoplasm,* or nuclear ground substance. Chromatin is made up of DNA combined with proteins. During the process of cell division the chromatin becomes progressively more condensed until it takes the form of *chromosomes.* As in prokaryotic organisms, the hereditary information is carried in molecules of DNA. The content of DNA per cell is much higher in eukaryotic organisms than in prokaryotes. In prokaryotes, the DNA molecules are essentially free in the cytoplasm. In most eukaryotes, they are organized into much larger units—the chromosomes.

Different organisms vary in the number of chromosomes present in their cells. *Haplopappus gracilis,* a desert annual, has 4 chromosomes; cabbage, 20; sunflowers, 34; bread wheat, 42; humans, 46; and one species of the fern *Ophioglossum,* about 1250.

The number of chromosomes usually present in the cells of higher plants and animals is known as the *diploid* number, and this number is generally constant in all of the cells of such organisms. When reproductive cells, either spores or gametes, are formed, the number of chromosomes is reduced to one-half the diploid number. Such cells are known as *haploid* cells.

Often the only structures within the nuclei that are discernible with the light microscope are the spherical structures known as *nucleoli* (singular: nucleolus). Nucleoli are present in nondividing nuclei (Figure 1–1) and are the sites of the formation of ribosomes. The number of nucleoli in a nucleus is variable. The nucleoli often fuse and appear as one large structure. Consisting mostly of protein, the nucleoli also contain about 5 percent RNA (ribonucleic acid).

PLASTIDS

Together with vacuoles and cell walls, *plastids* are characteristic components of plant cells. Each plastid is bounded by an envelope consisting of two unit membranes. Internally, the plastid is differentiated into a system of membranes and a more or less homogeneous ground substance, the *stroma*.

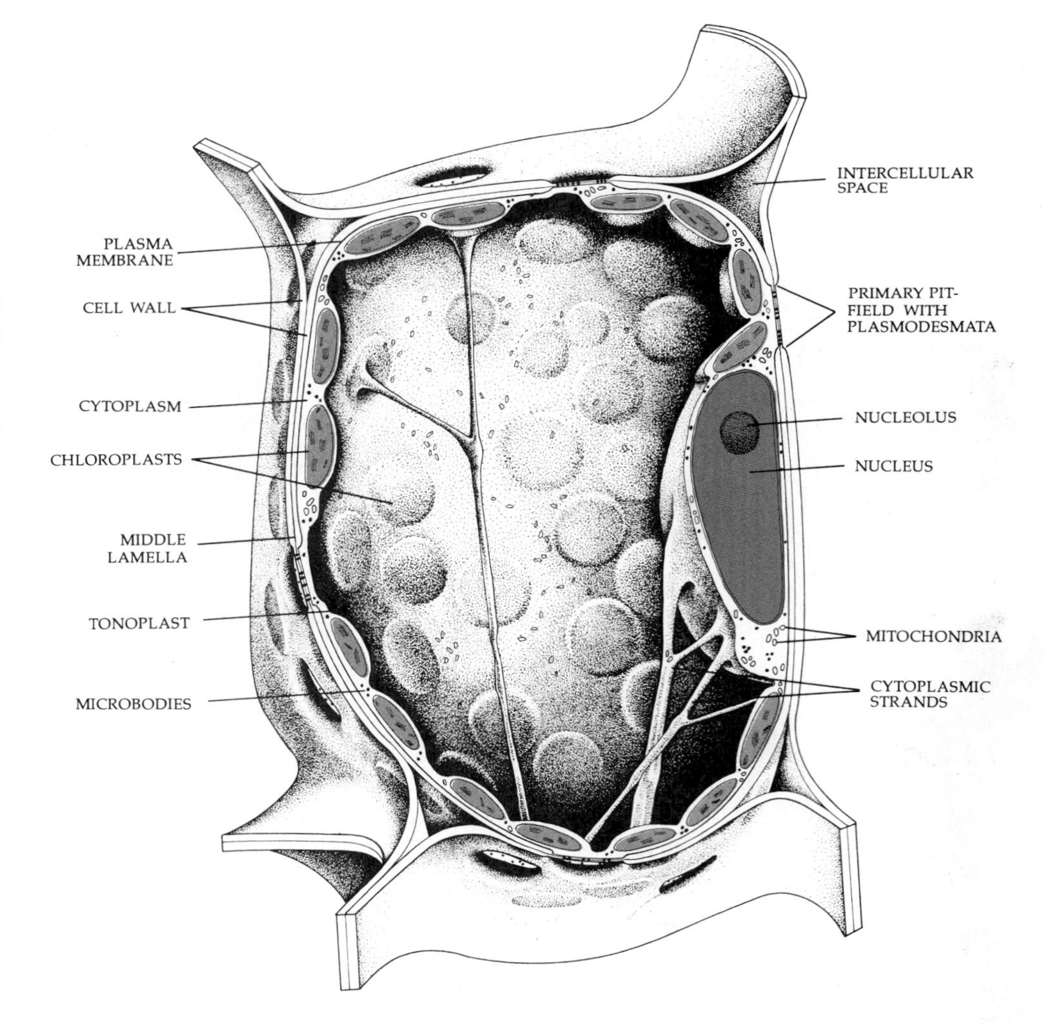

1–6
A three-dimensional diagram of a chloroplast-containing plant cell. The numerous disk-shaped chloroplasts are located in the cytoplasm along the wall and have their broad surfaces facing the surface of the wall. Most of the volume of this cell is occupied by the vacuole, which is traversed here by a few strands of cytoplasm. In this cell, the nucleus lies in the cytoplasm along the wall, although in some cells it might appear suspended by strands of cytoplasm in the center of the cell.

Mature plastids are commonly classified on the basis of the kinds of pigments they contain. The sites of photosynthesis, called *chloroplasts*, contain chlorophylls and carotenoid pigments. In higher plants, chloroplasts are usually disk-shaped and measure between 4 and 6 micrometers in diameter. A single mesophyll ("middle of the leaf") cell may contain 40 to 50 chloroplasts; a square millimeter of leaf contains some 500,000. In the cytoplasm, the chloroplasts usually are found with their broad surfaces parallel to the cell wall, as shown in Figure 1–6.

The internal structure of the chloroplast is complex (Figures 1–7 and 1–8). The stroma is traversed by an elaborate system of membranes in the form of flattened sacs called *thylakoids*. Each thylakoid is similar to the plastid envelope in that it consists of two membranes. The thylakoids are believed to constitute a single, interconnected system. Chloroplasts are generally characterized by the presence of *grana* (singular: granum)—stacks of disklike thylakoids which resemble a stack of coins. The thylakoids of the various grana are connected with each other by thylakoids (the stroma thylakoids, or intergrana thylakoids) that traverse the stroma. Chlorophylls and carotenoid pigments are found within the thylakoid membranes.

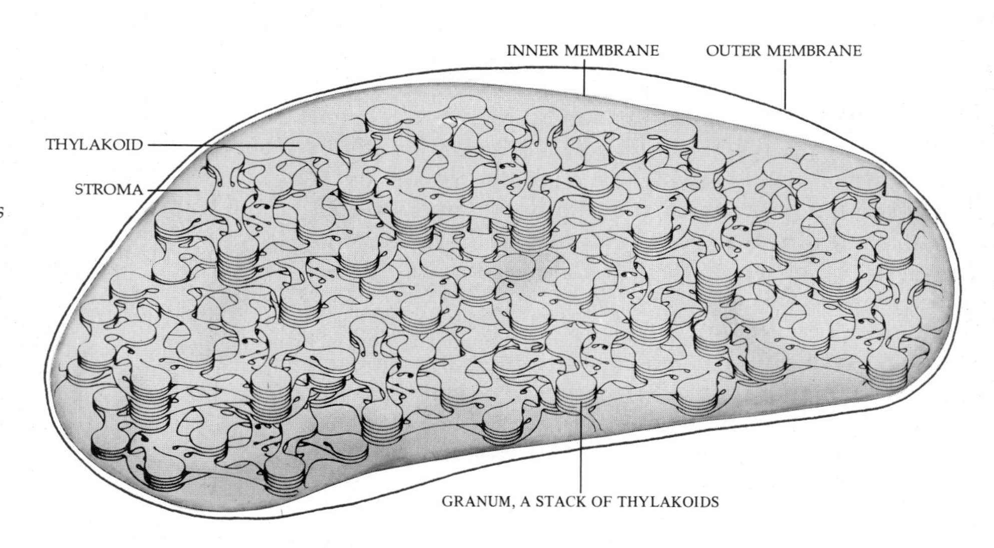

1–7
The inner structure of a chloroplast. A thylakoid is composed of a pair of membranes fused to form a closed disk. A granum is made up of stacked thylakoids. All the chlorophyll in the chloroplast is associated with these membranous structures, which are believed to form a single interconnected system.

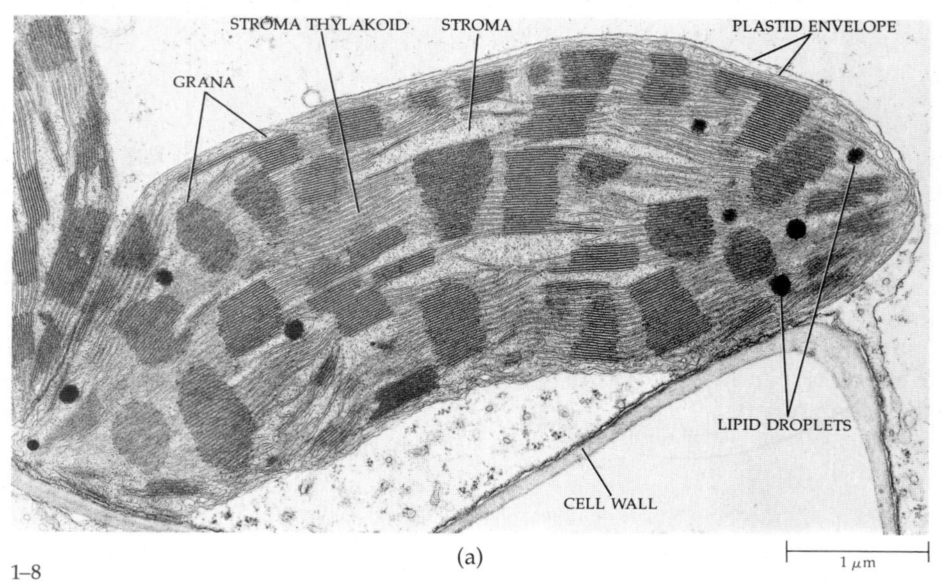

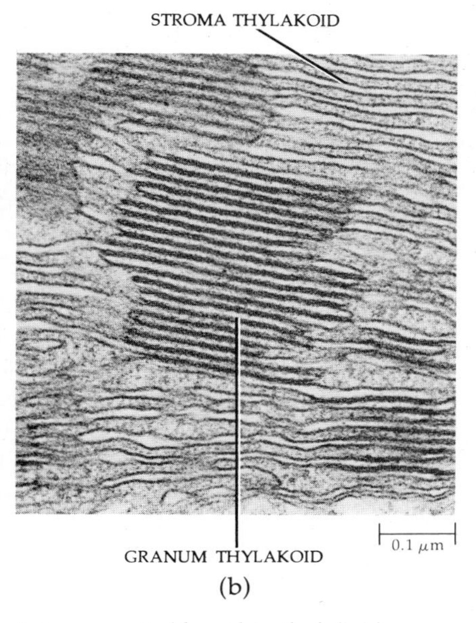

1–8
(a) *A chloroplast of a corn* (Zea mays) *leaf.* (b) *Detail showing grana composed of stacks of disklike thylakoids. The thylakoids of the various grana are interconnected by other thylakoids, commonly called stroma thylakoids.*

Chloroplasts often contain starch grains and small lipid (oil) droplets. The starch grains are temporary storage products and accumulate only when the plant is actively photosynthesizing (Figure 1–1). They may be lacking in the chloroplasts of plants that have been kept in the dark for as little as 24 hours but often reappear after the plant has been in the light for only 3 or 4 hours.

The interior surfaces of thylakoids contain numerous granules called quantasomes (Figure 1–9), which are believed to represent the basic structural units involved in the light-requiring reactions of photosynthesis. Chloroplasts also contain small ribosomes similar to those of prokaryotes, and strands of DNA can often be seen in the grana-free regions.

Two other kinds of plastids commonly found in the cells of higher plants are *chromoplasts* and *leucoplasts*. As the name implies, chromoplasts (Figure 1–10) are pigmented plastids. Of variable shape, chromoplasts synthesize and retain the carotenoid pigments, which are yellow, orange, or red in color. They may develop from previously existing green chloroplasts by a transformation in which the chlorophyll and internal membrane structure of the chloroplast disappear and masses of carotenoids accumulate, as occurs during the ripening of many fruits.

Leucoplasts (Figure 1–11) are nonpigmented plastids. Some synthesize starch, but others are thought to be capable of forming a variety of substances, including oils and proteins. Upon exposure to light, leucoplasts may develop into chloroplasts.

New plastids—chloroplasts, chromoplasts, and leucoplasts—may develop from small, colorless bodies known as *proplastids* (Figure 1–12); they also frequently arise from the simple fission (division) of preexisting, mature plastids.

1–9
A replica of a freeze-fractured thylakoid membrane showing its inner surface. The particles that are revealed on the inner surface are called quantasomes, which are thought to be involved in the light-requiring reactions of photosynthesis.

MITOCHONDRIA

Like chloroplasts, mitochondria are bounded by two unit membranes (Figures 1–13 and 1–14). The inner membrane is extensively folded into pleats or shelf-like projections known as *cristae* (singular: crista), which greatly increase the surface area available to enzymes and the reactions associated with them. Mitochondria are generally smaller than plastids, measuring about half a micrometer in diameter, and apparently there is great variability in length and shape. As a rule, mitochondria are barely visible with a light microscope, but they can be seen with the aid of an electron microscope.

Mitochondria are the sites of respiration, a process involving the release of energy from organic molecules and its conversion to molecules of ATP (adenosine triphosphate), the chief chemical energy

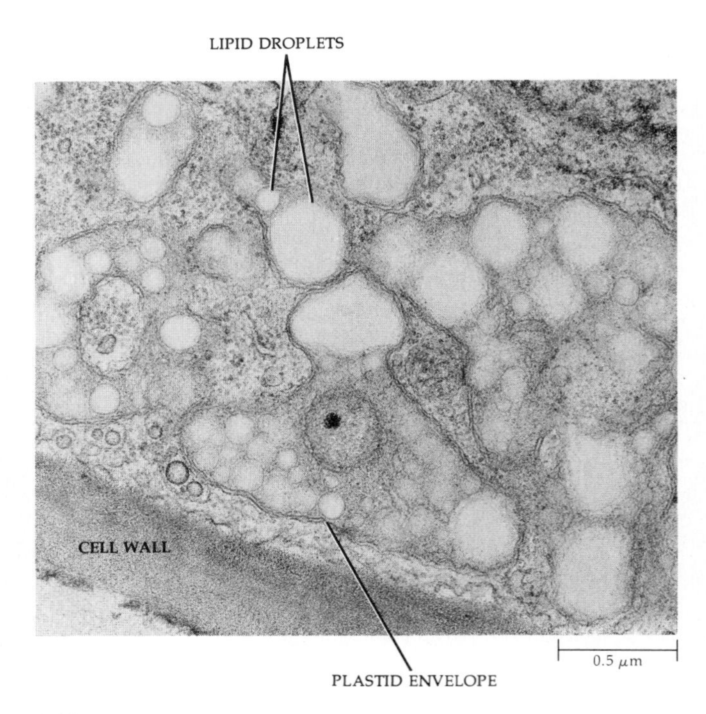

1–10
Chromoplasts from a marigold (Tagetes) *petal. Each chromoplast contains numerous lipid droplets in which the pigment responsible for the color of the petal is stored.*

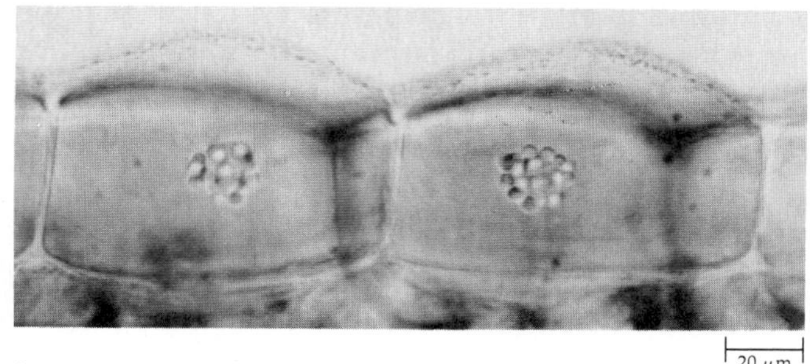

1–11
Leucoplasts clustered around the nucleus in epidermal cells of a Zebrina *leaf.*

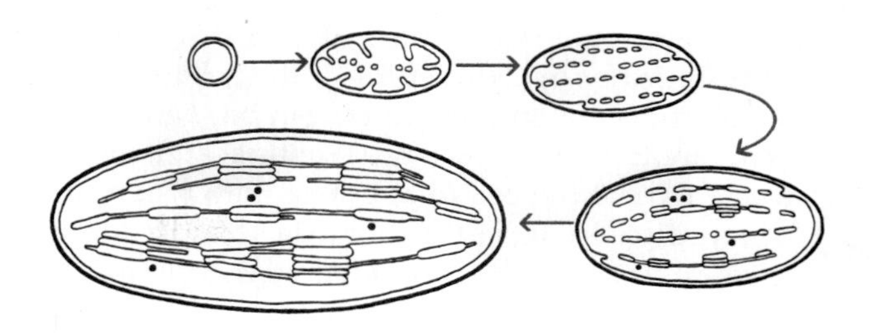

1–12
Stages in the development of a chloroplast from a proplastid. Initially the proplastid contains few internal membranes. As the proplastid differentiates, flattened vesicles develop from the inner membrane of the chloroplast envelope and eventually align themselves into grana and stroma thylakoids. At maturity, the thylakoid system appears discontinuous with the envelope.

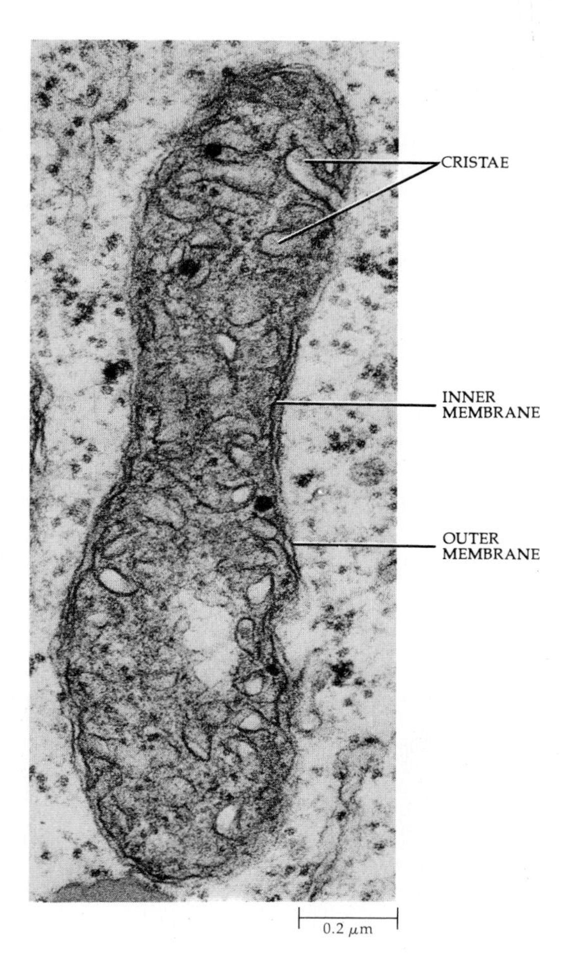

1–13
Mitochondrion in a leaf cell of the fern Regnellidium diphyllum. *The envelope of the mitochondrion consists of two separate membranes. The inner membrane folds inward to form cristae, which are embedded in a dense stroma. The many small granules in the stroma are ribosomes; the few relatively large granules are lipid droplets.*

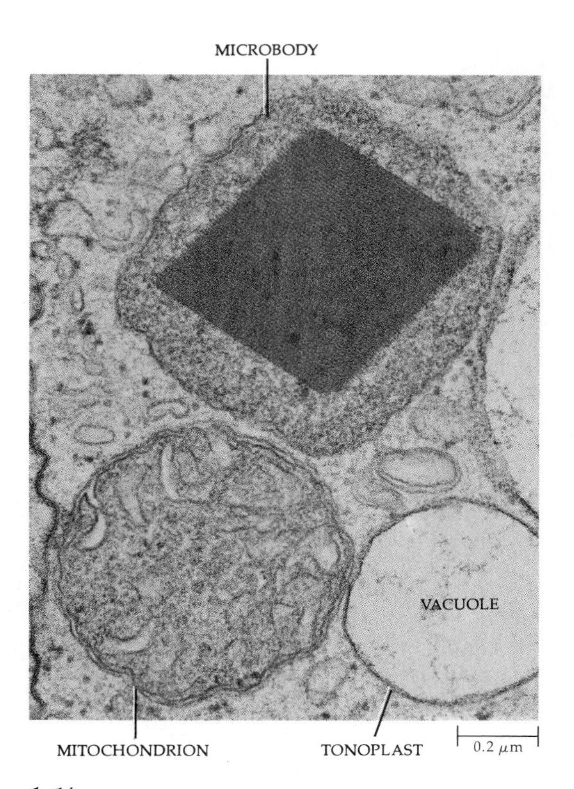

1–14
Organelles in a leaf cell of potato (Solanum tuberosum). *A microbody (above) with a large crystalline inclusion and a single bounding membrane may be contrasted with the mitochondrion bounded by a double membrane (below). A small vacuole (the relatively clear area at the lower right) is clearly separated from the cytoplasm by a single membrane, the tonoplast.*

source for all eukaryotic cells. (These processes are discussed further in Chapter 5.) With the aid of time-lapse photography, mitochondria can be seen to be in constant motion, turning and twisting and moving from one part of the cell to another; they also appear to fuse and to divide. Mitochondria tend to congregate where energy is required. In cells in which the plasma membrane is very active in transporting materials into or out of the cell, they often can be found arrayed along the membrane surface. In motile, single-celled algae, mitochondria are typically clustered at the bases of the flagella.

The greater the energy requirements of a cell, the more cristae its mitochondria are likely to contain. The cristae are surrounded by a liquid matrix which contains proteins, RNA, strands of DNA, ribosomes similar to those of prokaryotes, and various solutes (dissolved substances).

MICROBODIES

Unlike plastids and mitochondria, which are bounded by two membranes, *microbodies* are spherical organelles bounded by a single membrane. They range in diameter from 0.5 to 1.5 micrometers. Microbodies have a granular interior and sometimes contain a crystalline body composed of protein (Figure 1–14). Microbodies generally are associated with one or two segments of endoplasmic reticulum, the three-dimensional system of internal membranes within cells.

Some microbodies, called *peroxisomes*, play an important role in glycolic acid metabolism associated with photorespiration (see page 110). Others, called *glyoxysomes*, contain enzymes necessary for the conversion of fats into carbohydrates during germination in many seeds.

VACUOLES

Vacuoles are membrane-bounded regions within the cell that are filled with a liquid called *cell sap*. They are surrounded by the tonoplast, or vacuolar membrane (Figures 1–14 and 1–20).

The immature plant cell typically contains numerous small vacuoles, which increase in size and fuse into a single vacuole as the cell enlarges. In the mature cell, as much as 90 percent of the volume may be taken up by the vacuole, with the cytoplasm consisting of a thin peripheral layer closely pressed against the cell wall (Figure 1–6). Most of the increase in size of the cell results from enlargement of the vacuoles.

The principal component of the cell sap is water, with other components varying according to the type of plant and its physiological state. Vacuoles typically contain salts and sugars, and some contain dissolved proteins. The tonoplast plays an important role in the active transport of certain ions into the vacuole and their retention there. Thus, ions may accumulate in the cell sap in concentrations far in excess of those in the surrounding cytoplasm. Sometimes the concentration of a particular material in the vacuole is sufficiently great for it to form crystals. Calcium oxalate crystals, which can assume several different forms, are especially common (Figures 1–15 and 1–16). Vacuoles are usually slightly acidic. Some of them, like the vacuoles in citrus fruits, are very acidic—hence the tart, sour taste of the fruit.

1–15
Druses, compound crystals composed of calcium oxalate, in vacuoles of parenchyma cells of a Begonia *stem, as seen in* (a) *ordinary and* (b) *polarized light.*

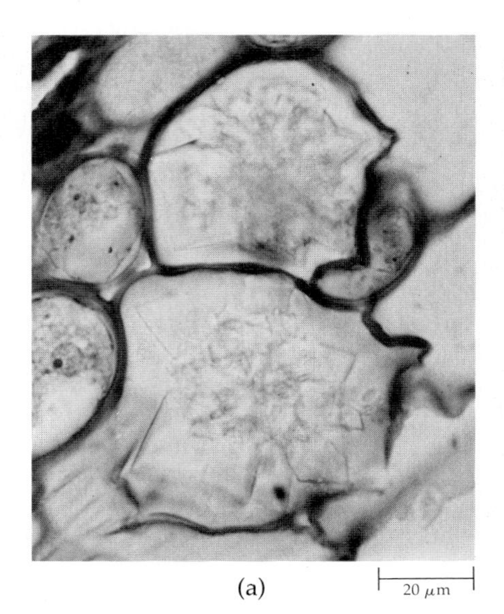

(a)

(b)

THE EVOLUTIONARY ORIGIN OF MITOCHONDRIA AND CHLOROPLASTS

Mitochondria and chloroplasts share a number of unusual properties. Both are bounded by a double membrane and have a complex internal membrane structure. Both have a capacity for growth and division that is relatively independent of the rest of the cell, and, as discussed in Chapters 5 and 6, both are involved in the production of ATP—the molecule that provides chemical energy for all cells—in similar ways.

In addition to these striking similarities, it has been found in recent years that both mitochondria and chloroplasts contain characteristic forms of DNA and also RNA, as well as ribosomes. In some of these organelles, it has been demonstrated that the DNA is present in the form of a closed circle, like that of bacteria.

The genetic role of this DNA is currently being investigated in several laboratories. Isolated chloroplasts have the capacity to synthesize RNA, which, as discussed in Chapter 7, is usually carried out only under the direction of chromosomal DNA. The ability to form chloroplasts and the pigments associated with them is largely controlled by chromosomal DNA interacting in some poorly understood fashion with chloroplast DNA. Therefore, chloroplasts cannot be formed in the absence of chloroplast DNA.

Chloroplast ribosomes resemble bacterial ribosomes in several ways. For example, the ribosomes of both prokaryotes and chloroplasts are only about two-thirds as large as the ribosomes found in the cytoplasm and endoplasmic reticulum of the eukaryotic cell. The synthesis of protein in the ribosomes of mitochondria, chloroplasts, and bacteria is inhibited by chloramphenicol, an antibiotic substance that has no effect on the ribosomes of eukaryotic cells.

On the basis of accumulating evidence, it now seems probable that mitochondria and chloroplasts originated as free-living prokaryotes that found shelter within larger heterotrophic cells. These larger cells were the forerunners of the eukaryotes. The smaller cells, which contained (and still contain) all the mechanisms necessary to trap and convert energy from their surroundings, donated these useful capacities to the larger cells. Cells with these respiratory "assistants" had a clear advantage over their contemporaries and undoubtedly soon multiplied at their expense. All modern eukaryotes contain mitochondria, and all autotrophic eukaryotes contain chloroplasts; both seem to have been acquired by independent symbiotic events. (Symbiosis is the close association between two or more different organisms that may be, but is not necessarily, beneficial to each.) The larger and more complex cells of eukaryotes seem to protect their symbiotic organelles from environmental extremes. As a consequence, eukaryotes were able to invade the land and the acidic waters, where the prokaryotic cyanobacteria, or blue-green algae, are absent but the eukaryotic green algae abound.

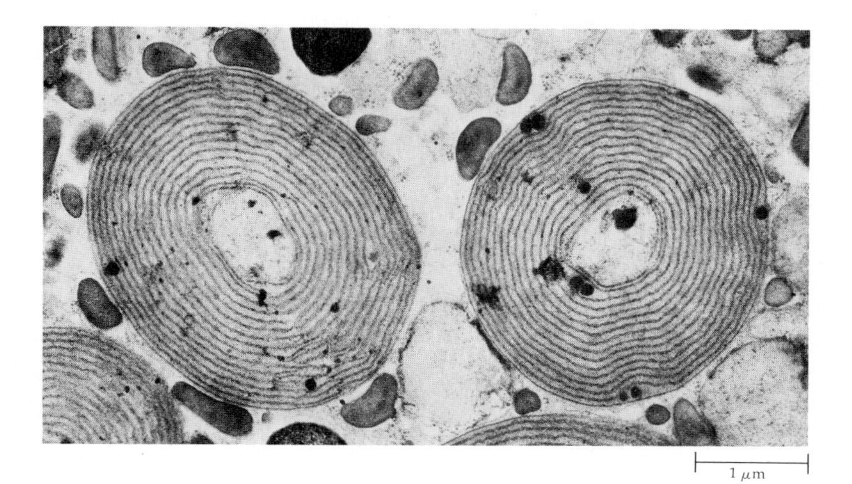

Cells of the cyanobacterium *Glaucocystis nostochinearum* contained within the larger cell of a green alga. Chloroplasts are believed to have had their evolutionary origin in an association such as this.

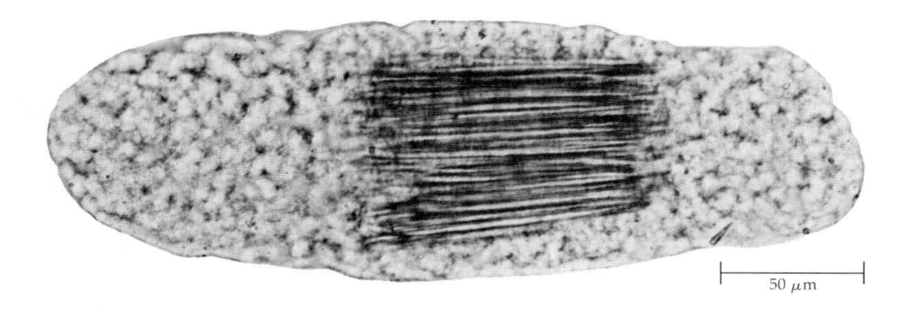

1–16
A bundle of raphides, needlelike crystals of calcium oxalate, in a vacuole of a leaf cell of the snake plant (Sansevieria). *The tonoplast is not discernible in this photomicrograph. The granular substance surrounding the crystals is cytoplasm.*

The vacuole may also be a site of pigment deposition. The blue, violet, purple, dark red, and scarlet of plant cells are usually caused by a group of pigments known as the anthocyanins (see Chapter 19). Unlike most other plant pigments, the anthocyanins are readily soluble in water, and they are dissolved within the cell sap. They are responsible for the red and blue colors of many vegetables (radishes, turnips, cabbages), fruits (grapes, plums, cherries), and a host of flowers (cornflowers, geraniums, delphiniums, roses, and peonies). Sometimes the pigments are so brilliant they mask the chlorophyll in the leaves, as in the ornamental red maple. Anthocyanins are also the pigments that are responsible for the brilliant red colors of some leaves in autumn. These pigments form in response to cold weather, at the same time the leaves stop producing chlorophyll. As the chlorophyll that is present disintegrates, the newly formed anthocyanins are unmasked. In leaves that do not form anthocyanin pigments, the breakdown of chlorophyll in autumn may unmask more stable yellow to orange carotenoid pigments already present in the chloroplasts.

The vacuole, in addition to being involved with the accumulation of ions and molecules of various types and in the water balance of the cell, may also be involved with the breakdown of macromolecules and the recycling of their components within the cell. Entire cell components, such as ribosomes, mitochondria, and plastids, may be deposited in vacuoles to be degraded. For example, a portion of the tonoplast next to a mitochondrion may invaginate into the vacuole, become distended, and pinch off into the vacuole. The resulting vesicle, with its enclosed mitochondrion, becomes suspended in the vacuole. After the mitochondrion has been degraded, or lysed, the bounding membrane disappears. Because of this digestive activity, vacuoles are thus comparable in function with organelles known as lysosomes that occur in animal cells.

RIBOSOMES

Ribosomes are small particles, only about 17 to 23 nanometers in diameter, consisting of protein and RNA. They are the sites at which amino acids are linked together to form proteins and are abundant in the cytoplasm of metabolically active cells. Ribosomes occur freely in the cytoplasm or are attached to the endoplasmic reticulum; they commonly occur at both places in the same cell. Ribosomes are also found in nuclei. As mentioned previously, plastids and mitochondria contain ribosomes similar to those in prokaryotes.

Ribosomes actively involved in protein synthesis occur in clusters or aggregates called polyribosomes, or *polysomes* (Figure 1–17). Cells that are synthesizing proteins in large quantities often contain extensive systems of polysome-bearing endoplasmic reticulum. Polysomes are often attached to the outer surface of the nuclear envelope (Figure 1–5b).

GOLGI BODIES

Golgi bodies, or *dictyosomes,* are groups of flat, disk-shaped sacs, or *cisternae* (singular: cisterna), which are often branched into a complex series of tubules at their margins (Figures 1–18 and 1–19). Commonly, numerous vesicles that are formed and pinched off the edges of the disks can be seen around the Golgi bodies. The term *Golgi apparatus* is used to refer collectively to all of the Golgi bodies of a given cell.

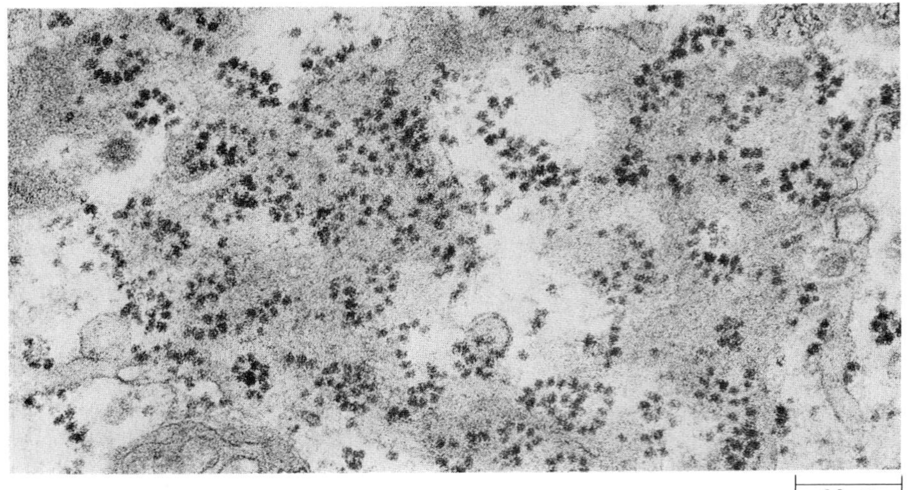

1–17
Groups of ribosomes (polysomes) on the surface of the endoplasmic reticulum. The endoplasmic reticulum is a network of membranes that is found throughout the cytoplasm of the eukaryotic cell, dividing it into compartments and providing surfaces on which chemical reactions can take place. Ribosomes are the sites at which amino acids are assembled into proteins. This electron micrograph shows a portion of a parenchyma cell from a leaf of the fern Regnellidium diphyllum.

1–18
The Golgi body consists of a group of flat, membranous sacs associated with vesicles that apparently bud off from the sacs. It serves as a "packaging center" for the cell and is concerned with secretory activities in eukaryotic cells.

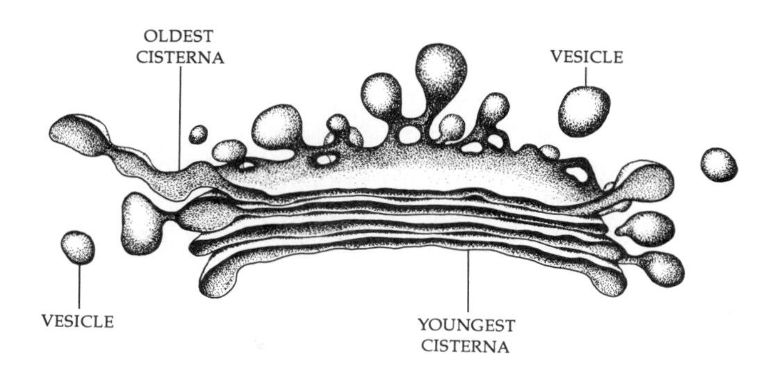

Golgi bodies are involved in secretion. Evidently the polysaccharides synthesized by many plant cells collect in the vesicles, which migrate to and fuse with the plasma membrane (Figure 1–22). The vesicles discharge their contents to the cell exterior, and their polysaccharide contents become part of the cell wall. The products secreted by the Golgi bodies are not necessarily synthesized entirely by the Golgi bodies but may be formed in some other cellular component, such as the endoplasmic reticulum, and then transferred to the Golgi bodies.

Golgi bodies seem always to be found in association with the basal bodies of flagella and cilia, as well as with centrioles. This association strongly suggests that the Golgi bodies may also produce or deliver the enzymes involved in synthesizing the structural proteins (fibrous proteins) that make up flagella, cilia, or spindle fibers.

ENDOPLASMIC RETICULUM

The *endoplasmic reticulum* is a complex three-dimensional membrane system of indefinite extent. In sectional view, the endoplasmic reticulum appears as two unit membranes with a narrow, transparent space between them (Figures 1–5c and 1–20). The membrane-bounded cavities vary considerably in size and shape in different types of cells and under different physiological conditions. In some cells, the network consists of fine tubules, 50 to 100 nanometers in diameter. In others, the cavities may be much larger, forming flattened sacs, or cisternae. Both forms of endoplasmic reticulum may be found within a given cell.

The endoplasmic reticulum appears to function as a communications system within the cell. In some electron micrographs, it can be seen to be continuous with the outer membrane of the nuclear envelope. In fact, these two structures together seem to form a single membrane system. When the nuclear envelope breaks down at cell division, its fragments are similar to portions of endoplasmic reticulum. It is easy to visualize the endoplasmic reticulum as a system for

1–19
Golgi bodies from a stem parenchyma cell of the horsetail Equisetum hyemale. *In* (a) *the cisternae of a Golgi body are shown in sectional view, whereas* (b) *shows a single cisterna in a surface view. The arrows indicate numerous secretory vesicles along the margins of the cisternae.*

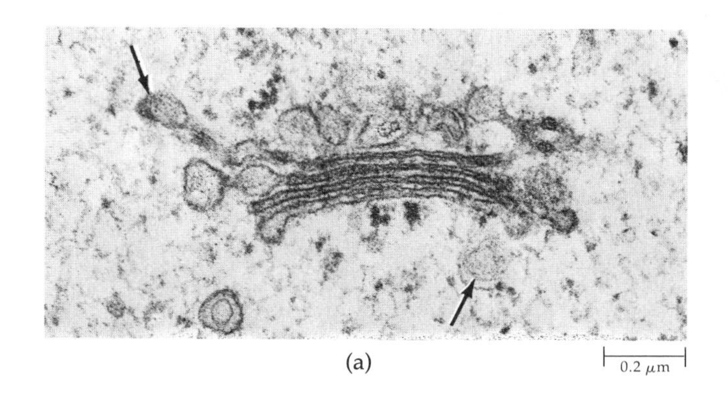

(a)

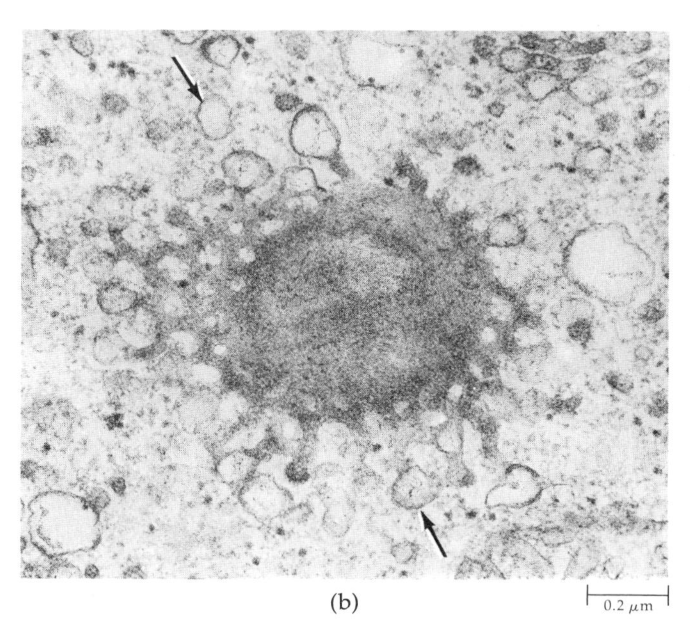

(b)

1–20
Parallel arrays of rough endoplasmic reticulum—endoplasmic reticulum studded with ribosomes—seen in sectional view in a leaf cell of the fern Vittaria guineensis. *The relatively clear areas are vacuoles.*

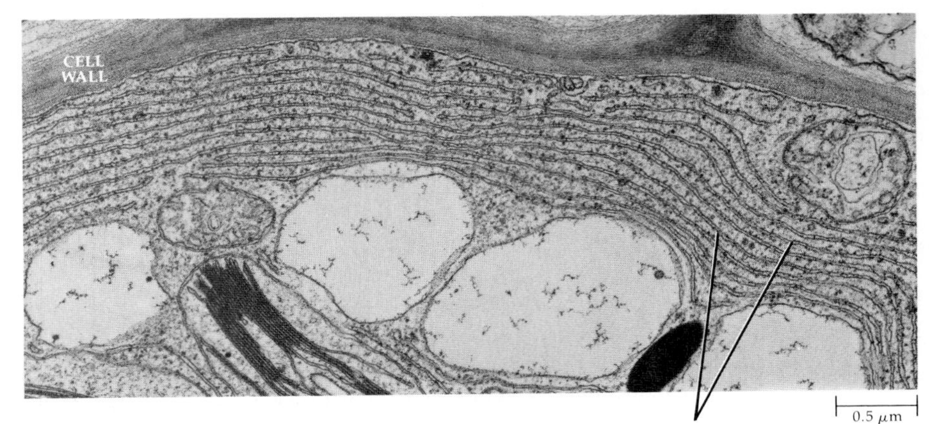

channeling materials—such as proteins and lipids—to different parts of the cell. In addition, the endoplasmic reticulum of adjacent cells is interconnected by way of cytoplasmic strands, called *plasmodesmata*, which traverse their common walls.

As mentioned previously, ribosomes are commonly attached to the endoplasmic reticulum. Such endoplasmic reticulum is called rough endoplasmic reticulum (Figures 1–5c, 1–17, and 1–20); that lacking ribosomes is called smooth endoplasmic reticulum. The rough endoplasmic reticulum is involved with protein synthesis in both plant and animal cells.

LIPID DROPLETS AND SPHEROSOMES

Lipid droplets are more or less spherical structures that impart a granular appearance to the cytoplasm of a plant cell when viewed with the light microscope. Apparently similar but usually smaller droplets are common in plastids (Figure 1–8a).

Some investigators believe that the lipid droplets originate as oil-containing vesicles cut off from endoplasmic reticulum. In many preparations these lipid droplets are bounded by a single membrane and resemble organelles, whereas in others they appear to lie freely in the cytoplasmic ground substance (Figure 1–21). Membrane-bound lipid droplets are known as *spherosomes*.

MEMBRANE DYNAMICS

The membranes and membrane systems of the cell have thus far been primarily presented as static structures, as derived from observations of electron micrographs. Membranes are, however, dynamic, mobile structures that continually change their shape and surface area. An example of the mobility of cellular membranes is provided by the concept of the *endomembrane system*. According to this concept, the

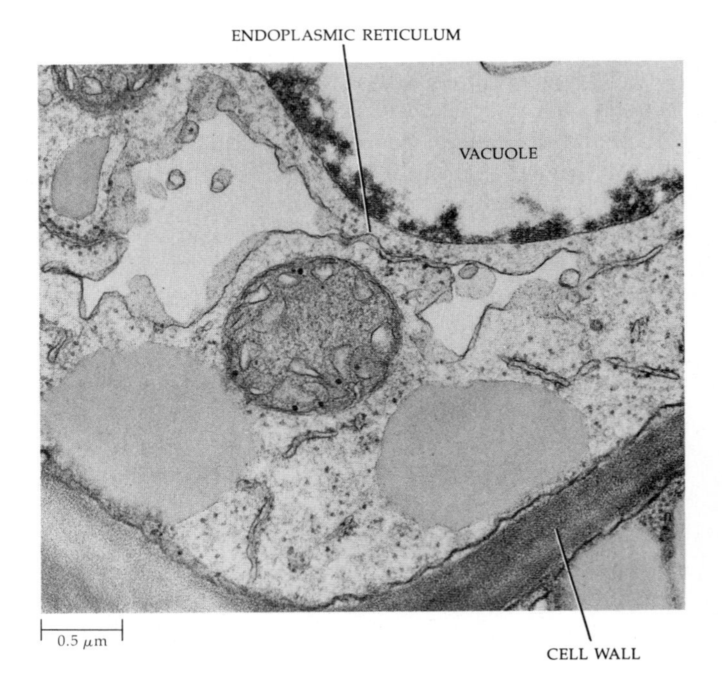

1–21
Cytoplasmic components present in a parenchyma cell from the thickened stem, or corm, of the quillwort Isoetes muricata. *Two lipid droplets can be seen below, one on either side of the mitochondrion near the center of this electron micrograph. Above, and on either side of the mitochrondrion, a cisterna of the endoplasmic reticulum appears distended. Some vacuoles are thought to arise from the endoplasmic reticulum in this way. Part of a vacuole can be seen at the top of the micrograph. The dense material lining this vacuole is tannin.*

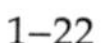

internal cytoplasmic membranes (with the possible exclusion of mitochondrial and plastid membranes) constitute a continuum, with the endoplasmic reticulum being the initial source of membranes. New cisternae are added to the Golgi bodies from the endoplasmic reticulum, while Golgi vesicles eventually contribute to growth of the plasma membrane (Figure 1–22). In addition, membranes of the endoplasmic reticulum and Golgi apparatus may give rise to vacuolar (Figure 1–21), microbody, and spherosome membranes. As mentioned previously, the endoplasmic reticulum and nuclear membrane together apparently form a continuous membrane system. Finally, it is important to note that even in tissues undergoing very little growth or cell division, a turnover of membrane components is continually taking place. Although the membranes persist, their constituent molecules are continuously replaced.

MICROTUBULES

Microtubules, which are found in virtually all cells, are long, thin structures about 24 nanometers in diameter and of varying lengths. They are polymers—large molecules composed of many identical subunits. Each subunit is a protein molecule called tubulin. Microtubules occur just inside the plasma membrane of nondividing cells and are believed to be involved in the orderly growth of the cell wall, especially through their control of the alignment of the cellulose microfibrils that are deposited by the cytoplasm when adding to the cell wall (Figure 1–23). Microtubules also serve to direct Golgi vesicles toward the developing wall and to orient other cytoplasmic components,

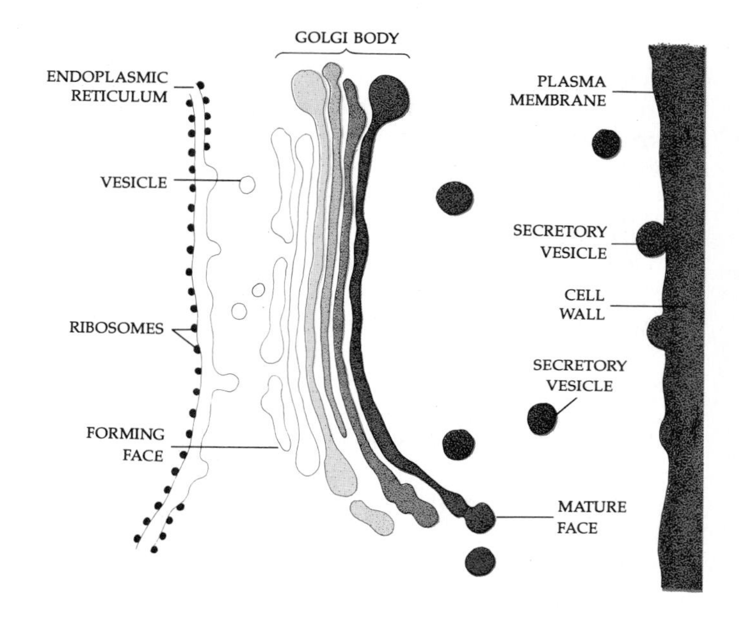

1–22
A diagrammatic representation of the endomembrane concept. This drawing illustrates the idea that new membranes are synthesized at the rough endoplasmic reticulum. Small vesicles pinch off the smooth surface of the endoplasmic reticulum and carry membranes and enclosed materials to the forming face of the Golgi body. Secretory vesicles derived from cisternae at the mature face of the Golgi body migrate to the plasma membrane and fuse with it, contributing new membrane to the plasma membrane and discharging their contents in the region of the wall.

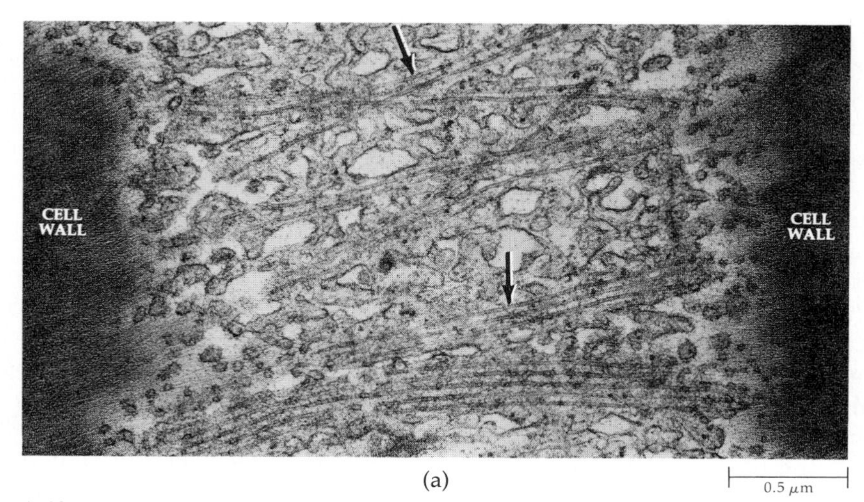

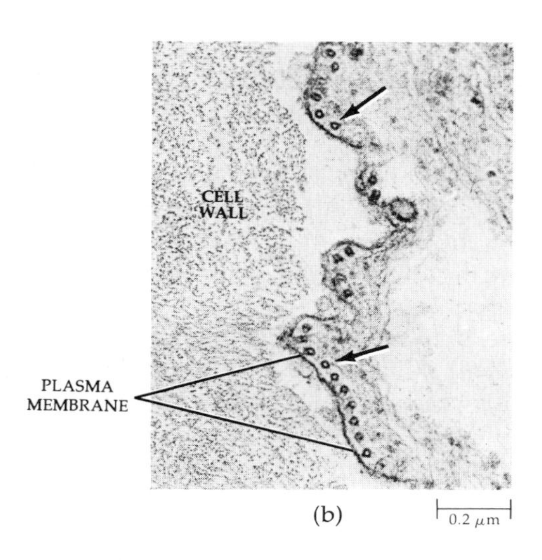

1–23
Microtubules (indicated by arrows) seen in (a) *longitudinal and* (b) *transverse views in leaf cells of* Botrychium virginianum, *a fern. A glancing section more or less parallel to and just inside the wall and plasma membrane is shown in* (a). *In* (b) *the microtubules can be seen to be separated from the wall by the plasma membrane.*

such as the nucleus, mitochondria, plastids, and lipid droplets, within the cell. Microtubules are present in the spindle fibers which form in dividing cells and apparently play a role in the formation of the cell plate (the initial partition between dividing cells). In addition, microtubules are important components of flagella and cilia and apparently are involved in the movement of these structures.

FLAGELLA AND CILIA

Flagella and *cilia* (singular: flagellum and cilium) are hairlike structures that extend from the surface of many different types of eukaryotic cells. They are relatively thin and constant in diameter (about 0.2 micrometer), but they vary in length from about 2 to 150 micrometers. By convention, the ones that are longer or are present alone or in small numbers are usually referred to as flagella, whereas the shorter ones or those occurring in greater numbers are called cilia. There is no definite distinction between these two types of organelles, however, and we shall use the term *flagellum* to refer to both.

In some algae and fungi, flagella are locomotor organs, propelling the organisms through the water. In plants, these organelles are found only in the sex cells (gametes) and then only in plants that have motile sperm, such as mosses, liverworts, ferns, and certain gymnosperms. Some flagella, called tinsel flagella, bear one or two rows of minute, lateral appendages, whereas others, called whiplash flagella, lack such processes (Figure 1–24).

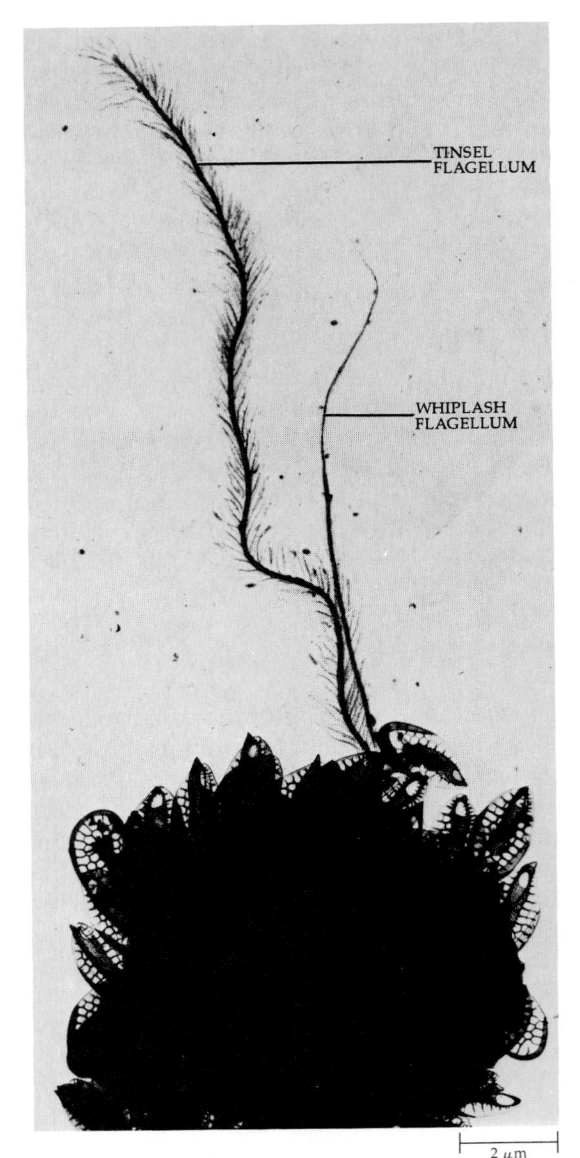

1–24
The two types of flagella—tinsel and whiplash—are represented in this single cell of the colonial organism Synura petersenii, *a golden-brown alga. This organism has a long tinsel flagellum (left) and a somewhat shorter whiplash flagellum (right).*

One of the most intriguing of the discoveries made possible by the electron microscope is the internal structure of flagella. Each flagellum has a precise internal organization (Figures 1–25 and 1–26). An outer ring of nine pairs of microtubules surrounds two additional microtubules in the center of the flagellum proper. Enzyme-containing "arms" extend from one of the microtubules of each outer pair. This basic 9-plus-2 pattern of organization is found in all flagella of eukaryotic organisms.

The movement of flagella and cilia originates from within the structures themselves; flagella and cilia are capable of movement after they are detached from cells. Some investigators believe that flagellar movement is based on a sliding microtubule mechanism, in which the outer pairs of microtubules move past one another without contracting. As the pairs of microtubules slide past one another, their movement causes localized bending of the flagellum. Sliding of the pairs of microtubules is thought to result from cycles of attachment and detachment of the "arms" on one of the microtubules to the nearer microtubule of the adjacent outer pair.

Flagella grow out of cylinder-shaped structures in the cytoplasm known as *basal bodies,* which then form the basal portion of the flagellum (Figures 1–25 and 1–26). Basal bodies have an internal structure that somewhat resembles the structure of the flagellum proper, except that the outer tubules in the basal body occur in triplets rather than in pairs, and the two central tubules are absent.

1–25
A diagram of a typical eukaryotic flagellum. In the flagellum proper, nine pairs of microtubules encircle two central microtubules. Flagella and cilia are found in nearly all major groups of organisms, and all of those found in eukaryotes have this same basic 9-plus-2 structure. The main cylinder of tubules is connected at its base to a basal body, which consists of nine sets of tubules, each a triplet. The nine outer pairs of tubules are thought to be extensions of two of the three tubules in each of the nine basal body sets. In each of the nine outer pairs, one of the tubules has enzyme-containing "arms" extending from it.

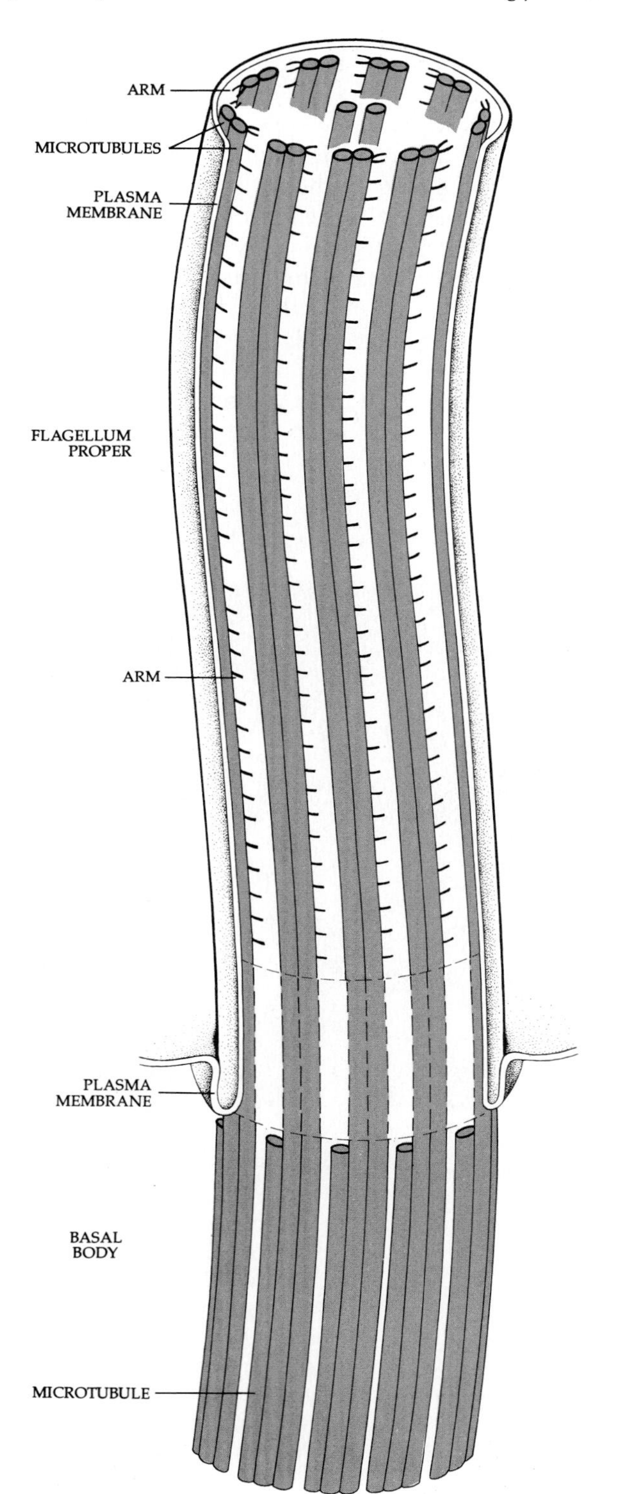

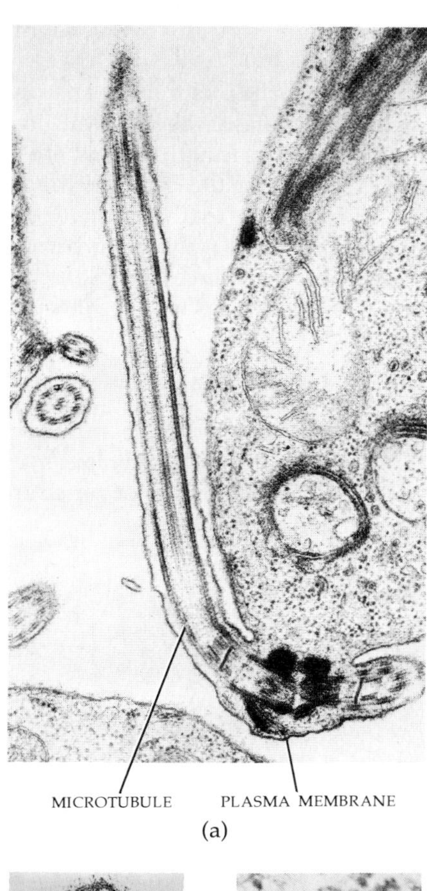

(a)

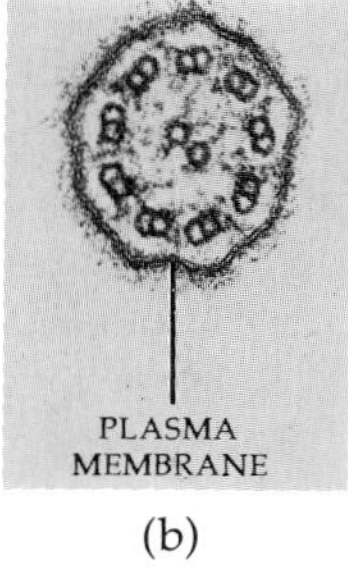

(b)

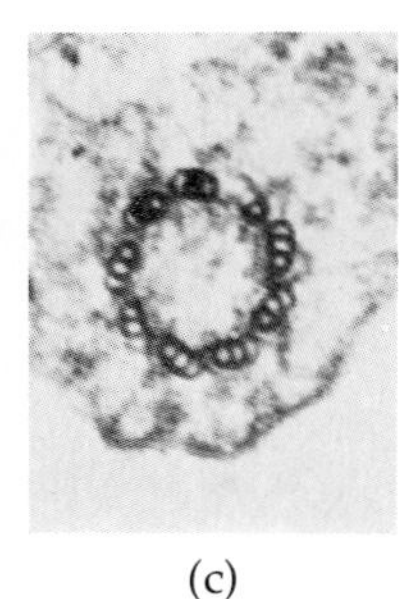

(c)

1–26
(a) *Longitudinal section of a flagellum of a gamete of the green alga* Ulvaria. *Notice that the membrane surrounding the flagellum proper is continuous with the plasma membrane.* (b) *Transverse section through the flagellum proper of the* Ulvaria *flagellum showing the typical 9-plus-2 structure.* (c) *Transverse section through the basal body of the* Ulvaria *flagellum. Notice that the basal body has a circle of nine triplet microtubules and no central microtubules.*

THE CELL WALL

The presence of a cell wall, above all other characteristics, distinguishes plant cells from animal cells. Its presence is the basis of many of the characteristics of plants as organisms. The cell wall limits the size of the protoplast and prevents it from being ruptured due to enlargement resulting from the intake of water by the vacuole.

Once regarded as merely an outer, inactive product of the protoplast, the cell wall is now recognized as having specific functions that are essential not only to the existence of the cell and the tissue in which it is found but also to the entire plant. Cell walls play important roles in the absorption, transport, and secretion of substances in plants, and they may also serve as sites of lysosomal, or digestive, activity.

Components of the Cell Wall

The most characteristic component of plant cell walls is *cellulose*, which largely determines their architecture. Cellulose is made of repeating molecules of glucose attached end to end (see Chapter 2). These long, thin cellulose molecules are united into *microfibrils* about 10 to 25 nanometers wide (Figure 1–27). Cellulose has crystalline properties (Figure 1–28) because of the orderly arrangement of cellulose molecules in certain parts, the *micelles*, of the microfibrils (Figure 1–29). The microfibrils wind together to form fine threads which may coil around one another like strands in a cable. Each "cable," or *macrofibril*, measures about 0.5 micrometer in width and may reach 4 micrometers in length. Cellulose molecules wound in this fashion are as strong as an equivalent thickness of steel.

The cellulose framework of the wall is interpenetrated by a cross-linked *matrix* of noncellulosic molecules. Some of these are polysaccharides called *hemicelluloses* and others are *pectic substances*, which are closely related chemically to hemicelluloses.

Another important constituent of the walls of many kinds of cells is *lignin* which, aside from cellulose, is the most abundant polymer found in plants. Physically, lignin is rigid, and it serves to add rigidity to the wall; it is commonly found in the walls of plant cells that have a supporting or mechanical function.

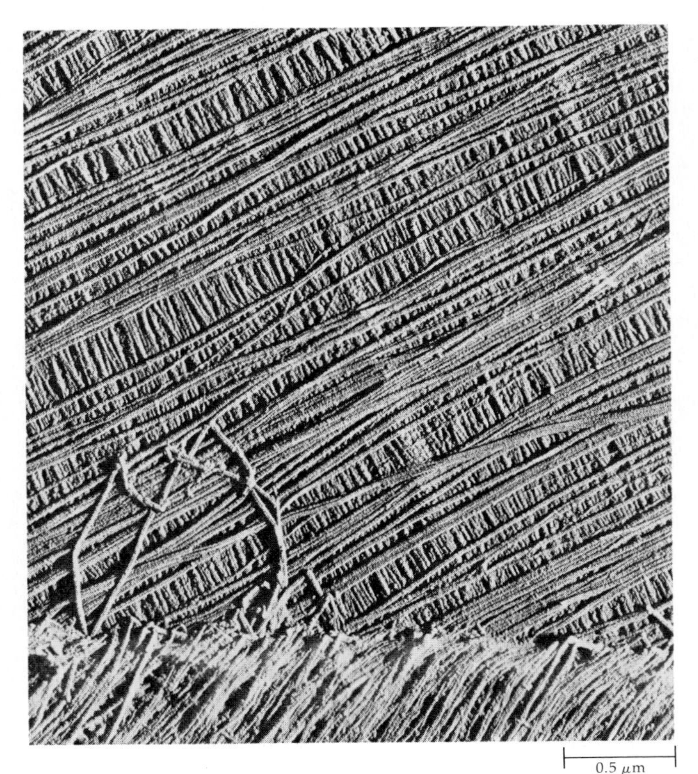

1–27
Surface of the cell wall of the algal cell Chaetomorpha, *showing cellulose microfibrils, each of which is made up of hundreds of molecules of cellulose.*

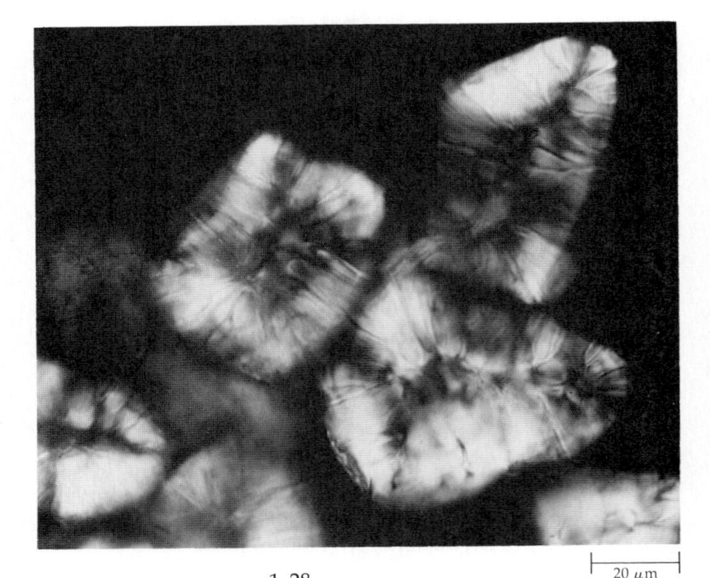

1–28
Stone cells (sclereids) from the flesh of a pear (Pyrus communis) *seen in polarized light. Clusters of such stone cells are responsible for the gritty texture of this fruit. The stone cells have very thick secondary walls traversed by numerous simple pits, which appear as lines in the walls. The walls appear bright in polarized light because of the crystalline properties of their principal component, cellulose.*

1–29
The detailed structure of a cell wall.
(a) *Portion of wall showing the middle lamella, primary wall, and three layers of secondary wall. Cellulose, the principal component of the cell wall, exists as a system of fibrils of different sizes.*
(b) *The largest fibrils, macrofibrils, can be seen with the light microscope.*
(c) *With the aid of an electron microscope, these fibrils can be resolved into microfibrils about 100 Å wide.*
(d) *Parts of the microfibrils, the micelles, are arranged in an orderly fashion and impart crystalline properties to the wall.*
(e) *A fragment of a micelle shows parts of the chainlike cellulose molecules in a lattice arrangement.*

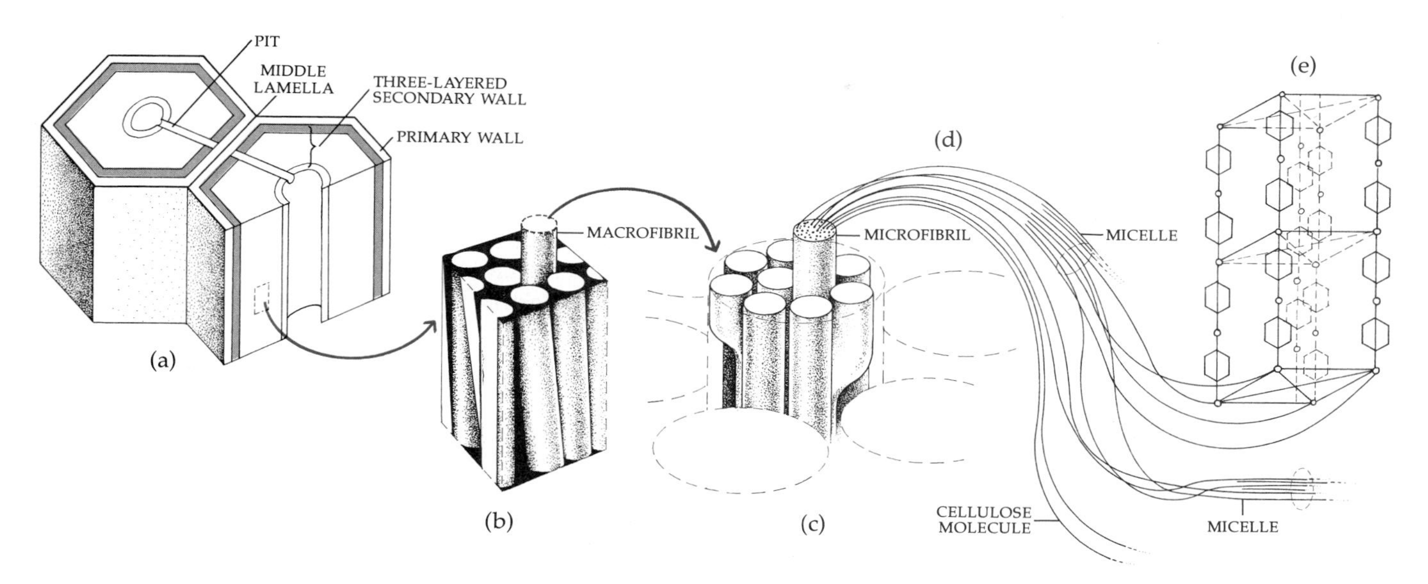

Cutin, *suberin*, and *waxes* are fatty substances commonly found in the walls of the outer, protective tissues of the plant body. Cutin, for example, is found in the walls of the epidermis, and suberin is found in those of the secondary protective tissue, cork. Both substances occur in combination with waxes and function largely to reduce water loss from the plant.

Cell Wall Layers

Plant cell walls vary greatly in thickness, depending partly on the role the cells play in the structure of the plant and partly on the age of the individual cell. Developmental studies, coupled with the use of the electron microscope, polarized light, and X-rays, indicate that there are two layers in all plant cell walls: the *middle lamella* (also called intercellular substance) and the *primary wall*. In addition, many cells deposit another wall layer, the *secondary wall*. The middle lamella occurs between the primary walls of adjacent cells; the secondary wall, if present, is laid down by the protoplast of the cell on the inner surface of the primary wall (Figure 1–30).

The Middle Lamella

The middle lamella is composed mainly of pectic substances. Frequently, it is difficult to distinguish the middle lamella from the primary wall, especially in cells that develop thick secondary walls. When a cell wall becomes lignified, this process characteristically begins in the middle lamella, and then spreads to the primary wall, and finally to the secondary wall.

The Primary Wall

The cellulosic wall layer deposited before and during the growth of the cell is called the primary wall. In addition to cellulose, hemicelluloses, and pectin, primary walls contain glycoprotein, which is a carbohydrate-protein compound. They may also become lignified. The pectic component imparts plastic properties to the wall, which make it possible for the primary wall to be stretched permanently during elongation of a root, stem, or leaf.

Actively dividing cells commonly contain only primary walls, as do most mature cells involved with such metabolic processes as photosynthesis, respiration, and secretion. These cells—that is, cells with primary walls and living protoplasts—are able to lose their specialized cellular form, divide, and differentiate into new types of cells. Therefore, principally cells with only primary walls are involved in wound healing and regeneration in the plant.

Usually, primary cell walls are not of uniform thickness throughout but have thin areas called *primary pit-fields* (Figure 1–30). Cytoplasmic strands, or plasmodesmata, which connect the living protoplasts of adjacent cells, are commonly aggregated in the primary pit-fields.

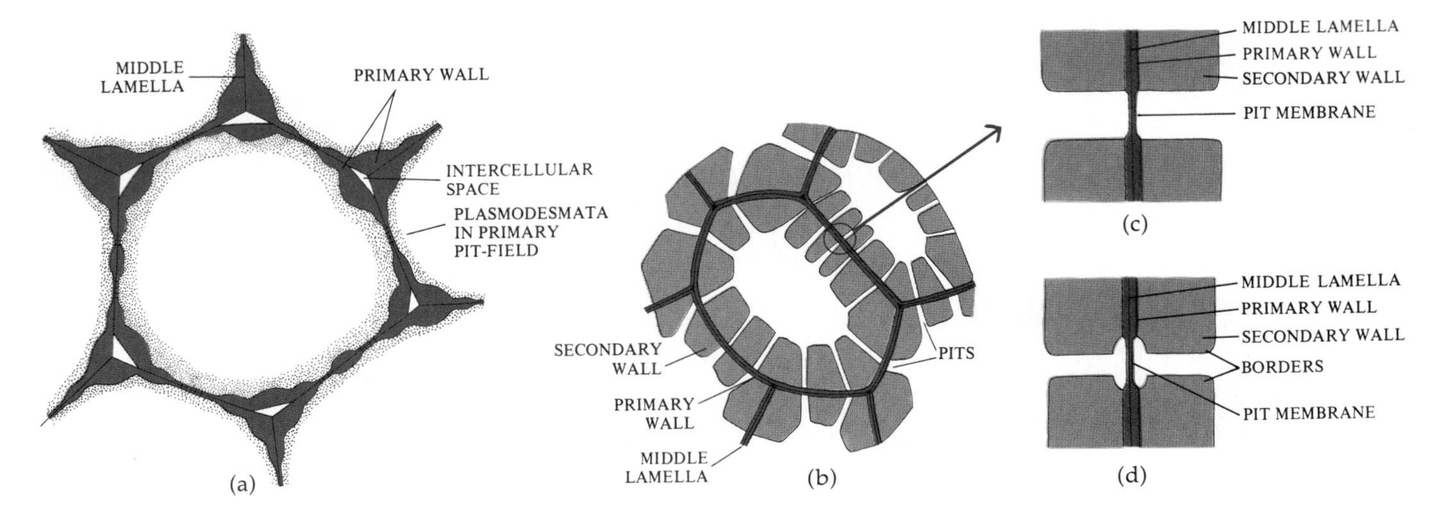

1–30
Primary pit-fields, pits, and plasmodesmata. (a) *Parenchyma cell with primary walls and primary pit-fields, thin areas in the walls. As shown here, plasmodesmata commonly traverse the wall at the primary pit-fields.* (b) *Cells with secondary walls and numerous simple pits.* (c) *A simple pit-pair.* (d) *A bordered pit-pair.*

The Secondary Wall

Although many plant cells have only a primary wall, in others a secondary wall is deposited by the protoplast inside the primary wall. This occurs mostly after the cell has stopped growing and the primary wall is no longer enlarging in area. It is partly because of this that the secondary wall is structurally distinct from the primary wall. Secondary walls are particularly important in specialized cells that have functions such as strengthening and conduction of water; in these cells, the protoplast often dies after the secondary wall has been laid down. Cellulose is more abundant in secondary walls than in primary walls and pectic substances are lacking; the secondary wall is therefore rigid and not readily stretched. Glycoproteins, which are relatively abundant in primary cell walls, are absent in secondary cell walls.

Frequently, three distinct layers—designated S_1, S_2, and S_3, for the outer, middle, and inner layer, respectively—can be distinguished in the secondary wall (Figure 1–31). The layers differ from one another in the orientation of their cellulose microfibrils. Such multiple-wall layers are found in certain cells of the secondary xylem, or wood. The laminated structure of such secondary walls—like that seen in plywood—greatly increases the strength of the wall. The cellulose microfibrils are laid down in a denser pattern, with the matrix of other polysaccharides more limited, than in the primary wall. Lignin is common in the secondary walls of cells found in wood, as discussed in Chapter 21.

When the secondary cell wall is deposited, it is not laid down over the primary pit-fields of the primary wall. Consequently, characteristic depressions, or *pits,* are formed in the secondary wall (Figure 1–30). In some instances, pits are also formed in areas where there are no primary pit-fields.

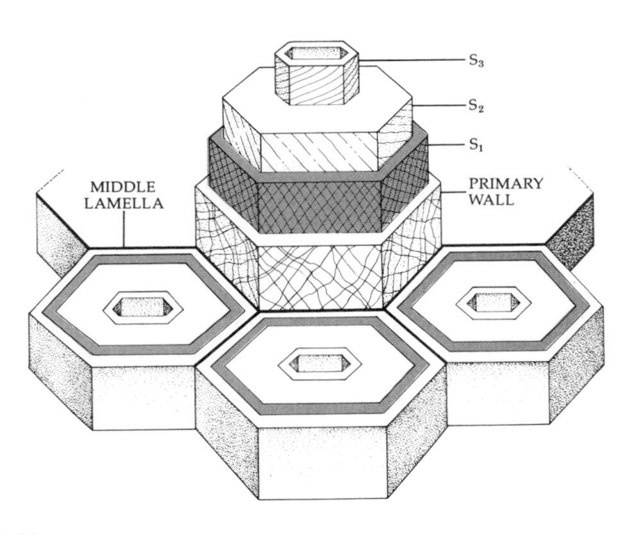

1–31
Diagram showing microfibril organization in primary walls and in three layers (S_1, S_2, S_3) *of secondary wall.*

A pit in a cell wall usually occurs opposite a pit in the wall of the cell with which it is in contact. The middle lamella and two primary walls between the two pits are called the *pit membrane*. The two opposite pits plus the membrane constitute a *pit-pair*. Two principal types of pits are found in cells that have secondary walls: *simple* and *bordered*. In bordered pits, the secondary wall arches over the *pit cavity*. In simple pits, there is no overarching.

Growth of the Cell Wall

Cell walls grow both in thickness and in surface area. Extension of the wall is a complex process under the close biochemical control of the protoplast. It requires loosening of wall structure—a phenomenon regulated by the hormone auxin (see Chapter 25)—as well as an increase in protein synthesis, respiration, and uptake of water by the cell. Most of the new microfibrils are placed on top of those previously formed (layer upon layer), although some may be inserted into the existing wall structure.

Initially, the cellulose microfibrils of the primary wall form an irregular mesh or network with a predominantly transverse orientation. As the wall increases in surface area, the orientation of the outer microfibrils becomes more nearly longitudinal, or parallel to the long axis of the cell.

Several protoplasmic components have been associated with cell wall synthesis. Most notable among these are Golgi vesicles, microtubules, and plasma membrane granules (granular bodies in the outer surface of the plasma membrane). Many studies indicate that matrix substances (hemicelluloses and pectic substances) are carried to the wall in Golgi vesicles. When they reach the plasma membrane, the vesicles fuse with the membrane and discharge their contents into the region of the developing wall. In contrast, cellulose synthesis occurs at or near the plasma membrane by an as yet still poorly understood process.

PLASMODESMATA

The protoplasts of adjacent plant cells are connected with one another by fine strands of cytoplasm known as *plasmodesmata* (singular: plasmodesma). Although such structures have long been seen with the light microscope (Figure 1–32), they were difficult to interpret. It was not until they could be observed with an electron microscope that their nature was confirmed (Figure 1–33).

Plasmodesmata may occur throughout the cell wall, or they may be aggregated in primary pit-fields or the membranes between pit-pairs. With the electron microscope, plasmodesmata appear as narrow canals (about 30 to 60 nanometers in diameter) lined by a plasma membrane and traversed by a tubule of endoplasmic reticulum known as the *desmotubule*. Many plasmodesmata are formed during cell division as strands of endoplasmic reticulum become trapped within the developing cell plate (see Figure 1–43). Plasmodesmata are also formed in walls of nondividing cells. These structures are thought to provide an effective pathway for the transport of certain substances between cells (see page 68).

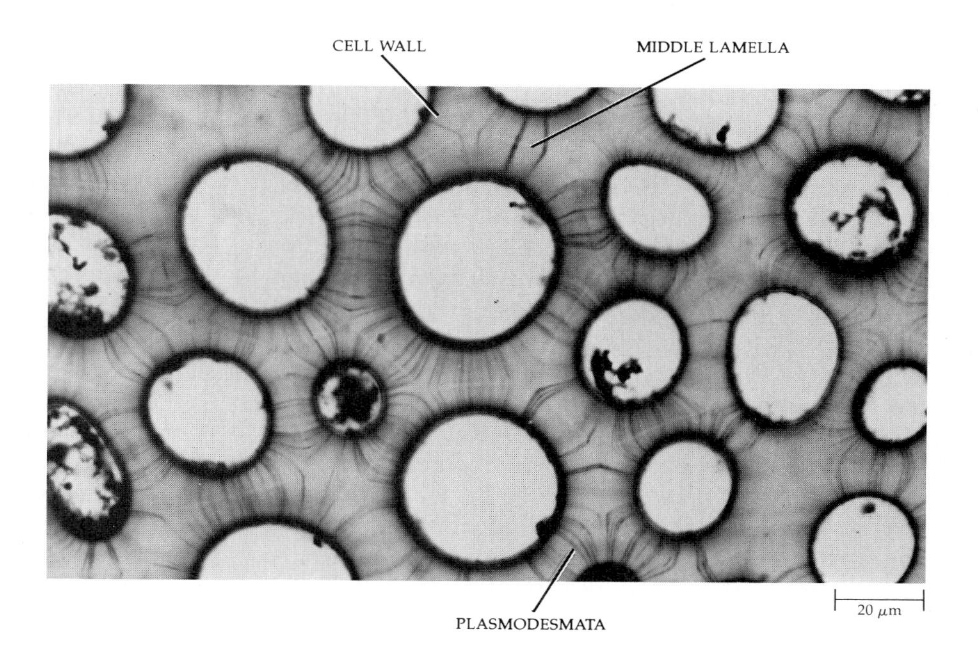

1–32
Light micrograph of plasmodesmata in the thick primary walls of persimmon (Diospyros) *endosperm, the nutritive tissue within the seed. The plasmodesmata appear as fine lines extending from cell to cell across the walls.*

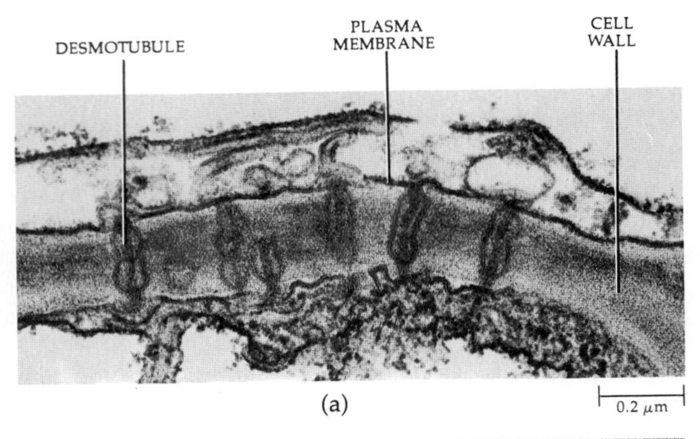

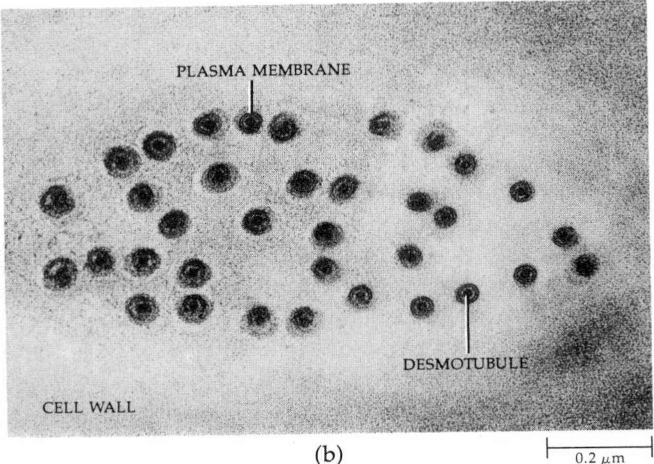

1–33
Electron micrograph of (a) *plasmodesmata, in longitudinal view, connecting the protoplasts of two corn* (Zea mays) *leaf cells.* (b) *Transverse view of plasmodesmata in primary pit-field of corn leaf cell.*

CELL DIVISION

In one-celled organisms, cells grow by absorbing substances from the environment and using these materials to synthesize new structural and functional molecules. When such a cell reaches a certain size, it divides. The two daughter cells, each about half the size of the original mother cell, then begin growing again. In a one-celled organism, cell division may occur every day or even every few hours, producing a succession of identical organisms. In many-celled organisms, cell division, together with cell enlargement, is the means by which the organism grows. In all of these instances, the new cells that are thus produced are structurally and functionally similar both to the parent cell and to one another.

Cell division in eukaryotes consists of two overlapping stages: *mitosis* and *cytokinesis*. Mitosis is the process by which a nucleus gives rise to two daughter nuclei, each morphologically and genetically equivalent to the other. Cytokinesis involves the division of the cytoplasmic portion of a cell and the separation of daughter nuclei into separate cells.

THE CELL CYCLE

Dividing cells pass through a regular sequence of events known as the cell cycle (Figure 1–34). Completion of the cycle requires varying periods of time, depending on both the type of cell and external factors, such as temperature or availability of nutrients. Commonly, the cycle is divided into *interphase* and the four phases of *mitosis*.

Interphase

Interphase, the phase between successive mitotic divisions, was once regarded as the "resting phase" in the cell cycle. Nothing could be further from the truth, however, for interphase is a period of intense cellular activity.

Interphase can be divided into three periods, which are designated G_1, S, and G_2 (Figure 1–34). The G_1 period (G stands for gap) occurs after mitosis and is primarily a time of growth of the cytoplasmic material, including the various organelles. Also during the G_1 period, according to a current hypothesis, substances are synthesized that either inhibit or stimulate the S period and the rest of the cycle, thus determining whether cell division will actually occur.

The S (synthesis) period follows the G_1 period. During the S period the genetic material (DNA) is duplicated. During the G_2 period that follows, structures directly involved with mitosis, such as the components of the spindle fibers, are formed.

Some cells pass through successive cell cycles repeatedly. This group includes the one-celled organisms and certain cells in regions of active growth (meristems). Some specialized cells lose their capacity to replicate once they are mature. A third group of cells, such as those that form wound tissue (callus), retains the capacity to divide but does so only under special circumstances.

Mitosis

Mitosis, or nuclear division, is a continuous process that is conventionally divided into four major phases: *prophase*, *metaphase*, *anaphase*, and *telophase* (Figures 1–35 through 1–38). These four phases make up the process by which the genetic material duplicated during interphase is divided equally between two daughter nuclei.

1–34
The cell cycle can be divided into mitosis and interphase, the phase between successive mitotic divisions. Cells usually spend most of their time in interphase, which can be divided into three periods. The first of these (G_1) is a period of general growth and replication of cytoplasmic organelles. During the second period (S), DNA synthesis takes place. The third (G_2) is a period during which structures directly associated with mitosis, such as the spindle fibers, are formed. After the G_2 period, mitosis begins. Mitosis is divided into four phases.

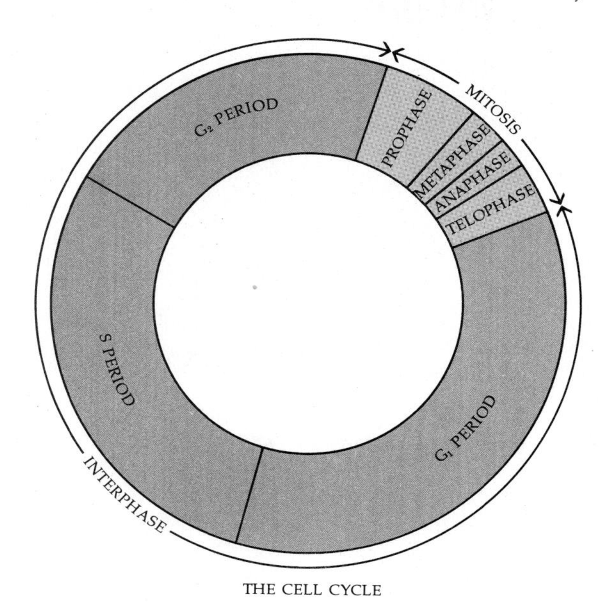

THE CELL CYCLE

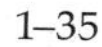

1–35
A diagram of a prophase chromosome. The chromosomal DNA was replicated during interphase, so that each chromosome now consists of two identical parts, called chromatids, which lie parallel to each other. The chromatids are joined together at their centromere.

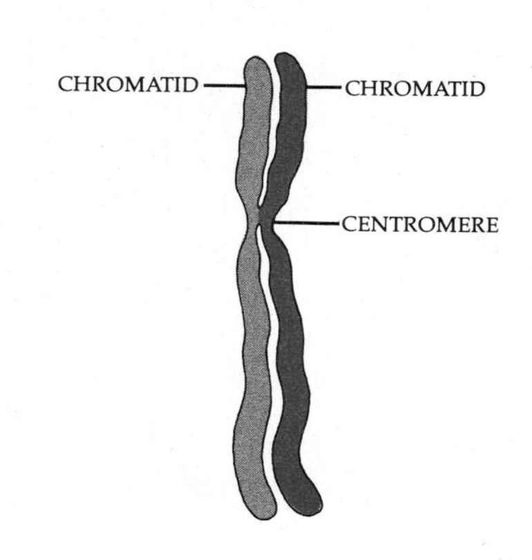

1–36
Mitosis, a diagrammatic representation with four chromosomes. During early prophase, the chromosomes become visible as long threads scattered throughout the nucleus; the chromosomes shorten and thicken until each can be seen to consist of two threads (chromatids); finally, the nucleolus and nuclear membrane disappear.

The appearance of the spindle marks the beginning of metaphase, during which the chromosomes migrate to the equatorial plane of the spindle. At full metaphase (shown here) the centromeres of the chromosomes lie in that plane.

Anaphase begins as the centromeres divide and separate, providing each of the sister chromatids, which now are called daughter chromosomes, with a centromere. As shown here, the daughter chromosomes then move to opposite poles of the spindle.

Telophase—more or less the reverse of prophase—begins when the daughter chromosomes have completed their migration.

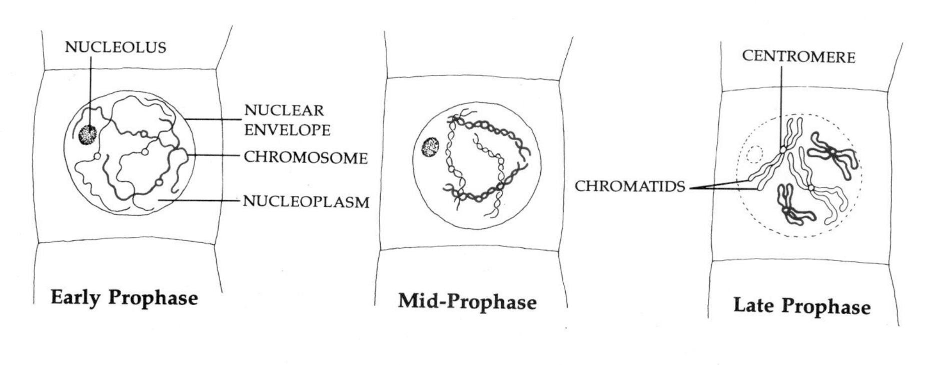

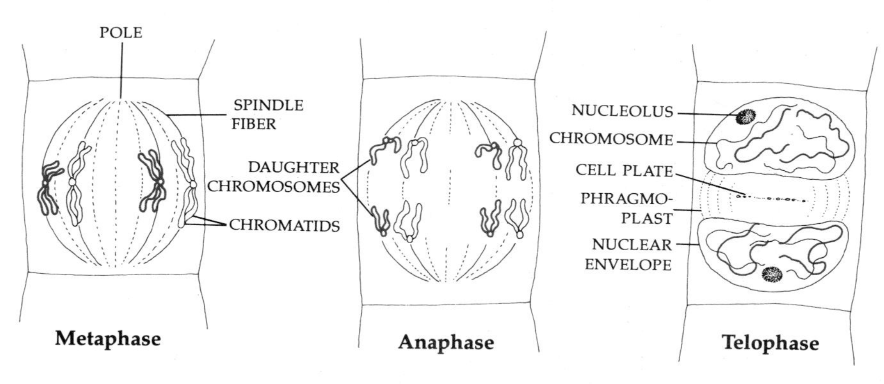

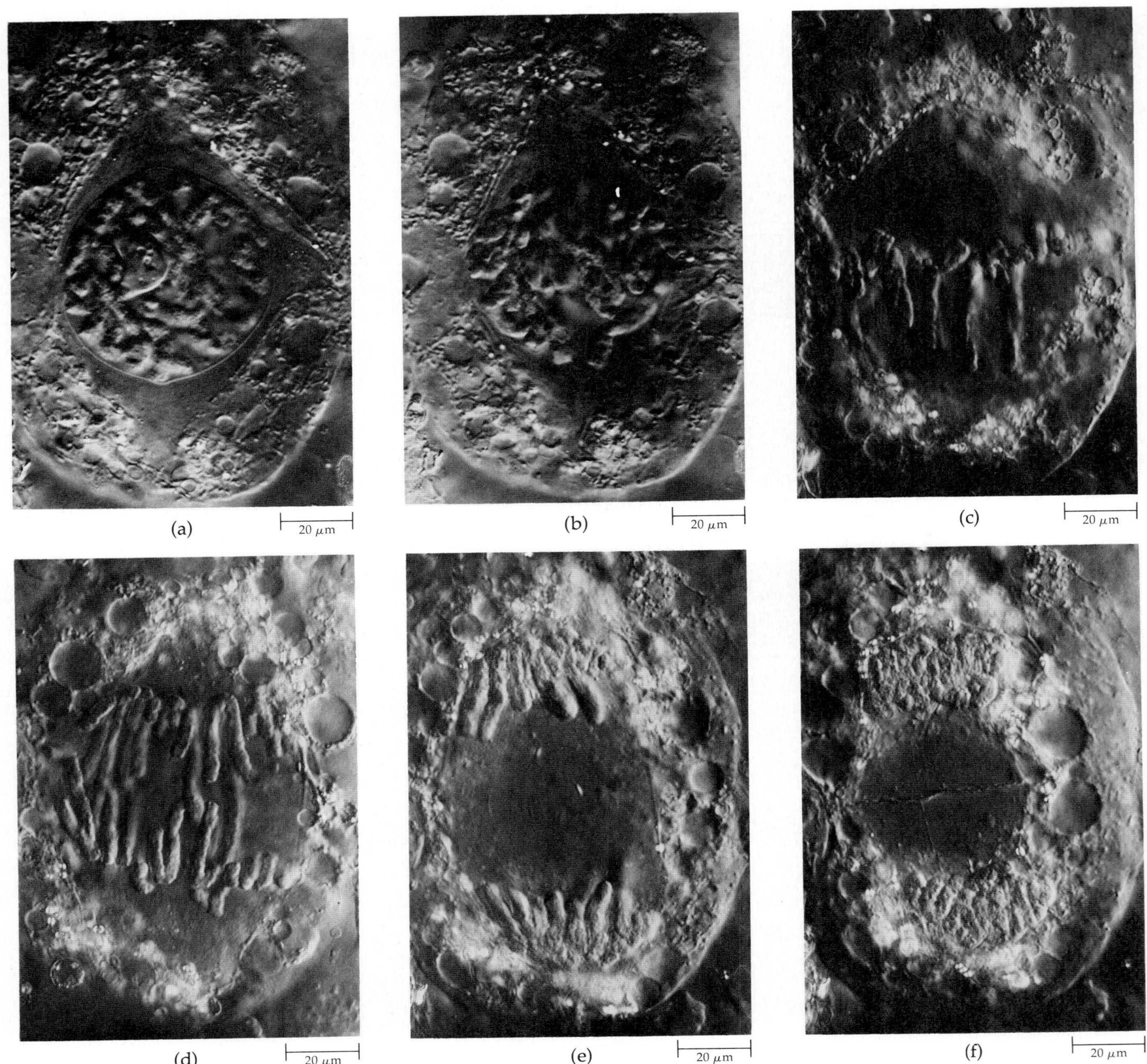

1–37
Mitosis as seen with Normarski system optics in living endosperm cells from the seeds of the blood lily (Haemanthus katherinae). (a) **Prophase:** *The chromosomes have condensed but the nuclear envelope is still present. A clear zone has developed around the nucleus.* (b) **Late prophase-early metaphase:** *The nuclear envelope has disappeared but the spindle fibers are as yet barely visible.* (c) **Metaphase:** *The chromosomes are arranged with their centromeres in the equatorial plane of the spindle. Spindle fibers are now visible, especially in the upper half of the spindle.* (d) **Anaphase:** *The sister chromatids (now called daughter chromosomes) have separated and are moving to opposite poles of the spindle.* (e) **Telophase:** *The daughter chromosomes have reached the opposite poles, and the two chromosome masses have begun the formation of two daughter nuclei. Phragmoplast fibers have begun to develop in the center of the cell.* (f) **Late telophase-cytokinesis:** *A nuclear envelope has begun to form around each of the two new nuclei. A cell plate is almost continuous across the cell, dividing it into two daughter cells with genetically equivalent nuclei.*

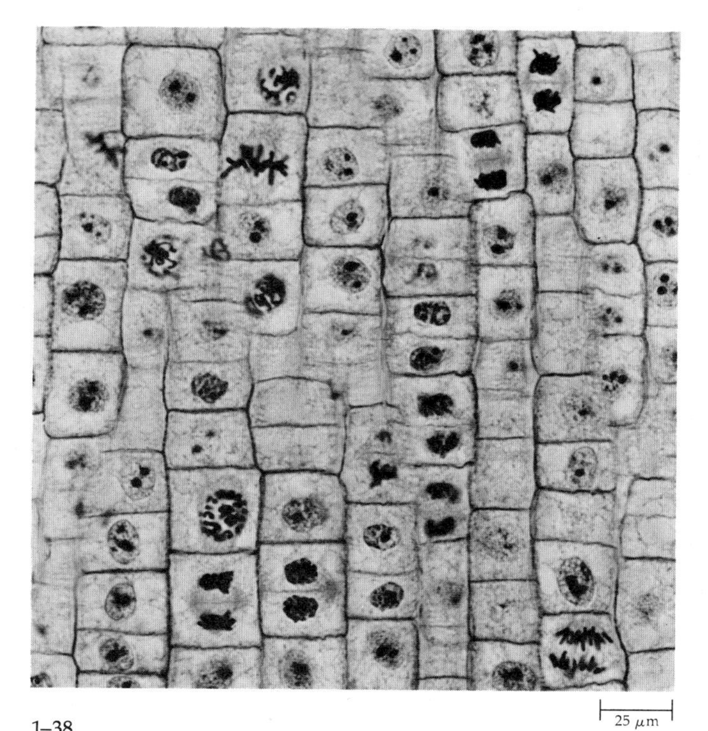

1–38
Photomicrograph of dividing cells in an onion (Allium) *root tip. By comparing the phases of mitosis illustrated in Figures 1-36 and 1-37 to these cells, you should be able to identify the mitotic phases shown here.*

Before mitosis begins, however, the nucleus must assume the proper position within the cell for division to take place. Recent evidence indicates that microtubules may play a role in the positioning of the nucleus. In many cells, microtubules form a ringlike band, called the preprophase band, that encircles the nucleus in a plane corresponding to the equatorial plane of the future mitotic spindle.

Prophase

At the beginning of prophase, the chromosomes gradually become visible as elongated threads scattered irregularly throughout the nucleus. As prophase advances, the threads shorten and thicken, and as the chromosomes become more distinct, each can be seen to be composed of not one but two threads, called *chromatids,* coiled about one another. By late prophase, after further shortening, the two duplicate chromatids of each chromosome lie side by side and nearly parallel, joined to one another in a narrowed area known as the *centromere,* or *kinetochore* (Figure 1–35). The centromere, which has a characteristic location on any one chromosome, divides the chromosome into two arms of varying lengths.

Toward the end of prophase, the nucleolus gradually becomes indistinct and finally disappears. Shortly afterward, the nuclear envelope breaks down, marking the end of prophase (Figure 1–39).

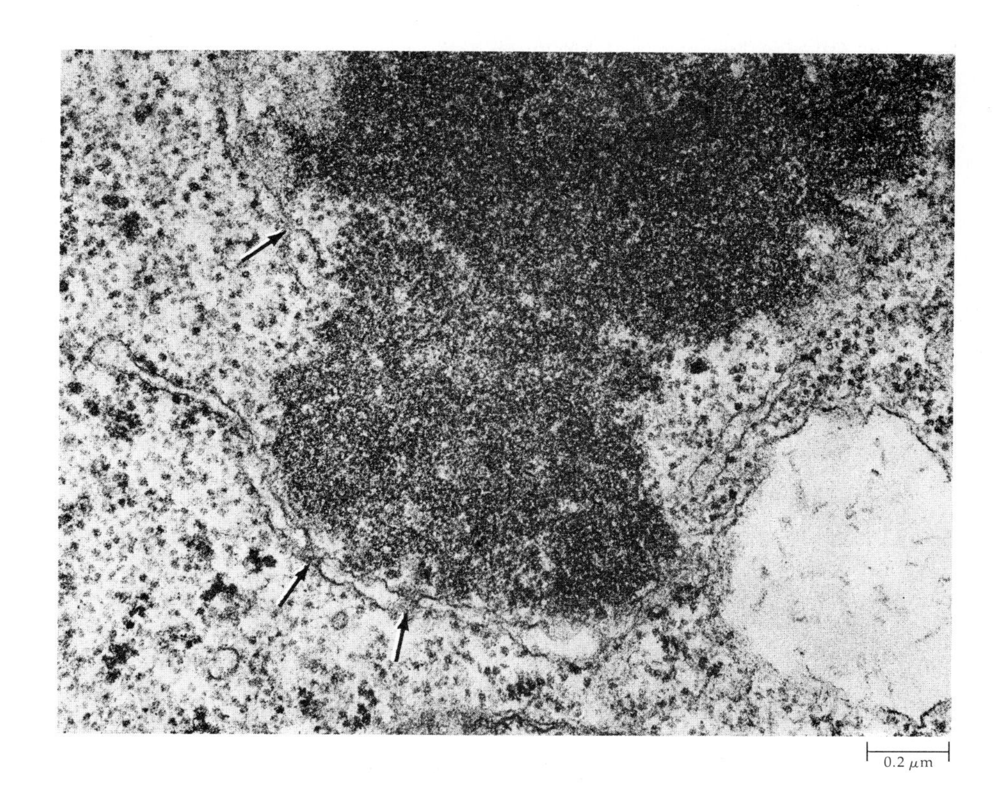

1–39
A portion of a late prophase nucleus of a parenchyma cell in the horsetail (Equisetum hyemale) *stem. The arrows point to nuclear pores in fragments of the nuclear envelope, which is in the process of breaking down.*

Metaphase

Metaphase begins with the appearance of the *spindle*, a three-dimensional structure that is widest in the middle and tapers to a point at each pole. The spindle consists of *spindle fibers*, which are bundles of microtubules, each microtubule about 20 nanometers in diameter (Figure 1–40). During metaphase, the chromosomes—each consisting of two chromatids—become arranged so that each centromere lies on the equatorial plane of the spindle. Each chromosome appears to be attached to the spindle fibers by its centromere. Some of the spindle fibers pass from one pole to the other and have no chromosome attached to them.

When the chromosomes have all moved to the equatorial plane, the cell has reached full metaphase. The chromatids are now in position to separate.

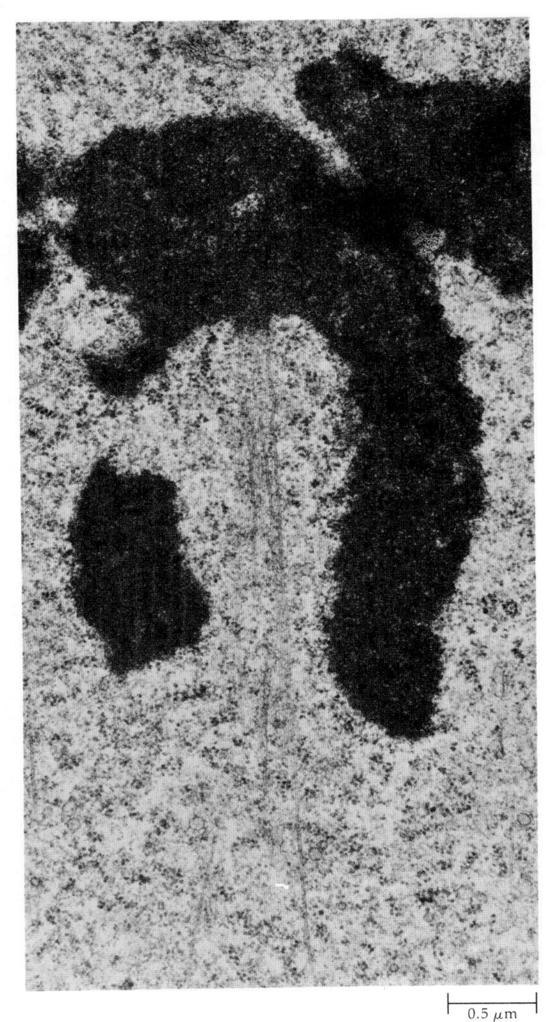

1–40
Portion of a chromosome at metaphase in a stem parenchyma cell of the horsetail (Equisetum hyemale). *Some microtubules (components of the spindle fibers) lead directly into the dense chromosomal material at a centromere. Other microtubules (not shown) extend from pole to pole.*

Anaphase

During anaphase, the chromatids of each chromosome separate from one another; they are now called daughter chromosomes. First, the centromere divides and the two daughter chromosomes move away from one another toward opposite poles. Their centromeres, which still appear to be attached to the spindle fibers, move first, and the arms drag behind. The two daughter chromosomes pull apart, with the tips of the longer arms separating last.

The spindle fibers attached to the chromosomes shorten, apparently causing the chromatids to separate and the daughter chromosomes to move apart. It has been suggested that the microtubules are continuously formed at one end of the spindle fiber and disassembled at the other. In the process, the spindle fibers tug the daughter chromosomes toward the poles by their centromeres.

By the end of anaphase, the two identical sets of daughter chromosomes have been separated and moved to opposite poles.

Telophase

During telophase, the separation of the two identical sets of chromosomes is made final as nuclear envelopes are organized around each of them (Figure 1–41). The membranes of these nuclear envelopes are derived from rough endoplasmic reticulum. The spindle apparatus disappears. Nucleoli also re-form at this time. In the course of telophase, the chromosomes become increasingly indistinct, elongating to become slender threads again. When these processes are completed and the chromosomes have once more disappeared from view, mitosis is completed and the two daughter nuclei have entered interphase.

During mitosis, the two daughter nuclei that are produced are genetically equivalent to one another and to the nucleus that divided to produce them. This is important, for the nucleus is the control center of the cell, as is described in more detail in Chapter 7. It contains coded instructions that specify the production of proteins, many of which mediate cellular processes by acting as enzymes and some of which serve directly as structural elements in the cell. This hereditary blueprint must be exactly passed on to the daughter cells, and its exact duplication is ensured in eukaryotic organisms by the organization of chromosomes and their division in the process of mitosis.

The duration of mitosis varies with the tissue and the organism involved. However, prophase is the longest phase and anaphase is the shortest. In a root tip, the relative lengths of time for each of the four phases may be as follows: prophase, 1 to 2 hours; metaphase, 5 to 15 minutes; anaphase, 2 to 10 minutes; and telophase, 10 to 30 minutes. By contrast, the duration of interphase may range from 12 to 30 hours.

Organization of the Mitotic Spindle

Near the nucleus of many eukaryotic cells, it is possible to detect a pair of structures called *centrioles*. Centrioles are similar to the basal bodies of flagella and cilia and, like them, apparently have the ability to organize some of the protein molecules into long, slender microtubules similar to those that make up the 9-plus-2 structure of flagella. During the mitotic prophase, the two members of a pair of centrioles migrate to the opposite poles of the cell and then apparently organize the formation of the spindle fibers. Two sets of fibers, each radiating from an opposite pole, form the spindle.

In motile cells, the same structures often function as centrioles and then as basal bodies in organizing flagella or cilia. Nonmotile plant cells lack centrioles or similar structures, but in these cells the organization of spindle fibers proceeds just as it does in cells that possess these structures. Recently it has been shown that the polymerization of tubulin into microtubules can be initiated by the chromosomal centromeres themselves. In special cases, where plants produce motile cells, as in the formation of flagellated sperm cells, centrioles appear in the cells, organize spindle fibers in the last mitotic division leading to the production of the sperm cells, and then function as basal bodies to organize flagella.

Cytokinesis

As noted previously, cytokinesis is the process by which the cytoplasm is divided. In most organisms, cells divide by ingrowth of the cell wall, if present, and constriction of the plasma membrane, a process that cuts through the spindle fibers. In all plants (bryophytes and vascular plants) and in a few algae, cell division occurs by the formation of a *cell plate* (Figures 1–41 and 1–42).

In early telophase, a spindle-shaped system of fibrils called the *phragmoplast* forms between the two daughter nuclei. The fibrils of the phragmoplast, like those of the mitotic spindle, are composed of microtubules. As seen with the light microscope, small droplets appear across the equatorial plane of the phragmoplast and gradually fuse, forming the cell plate. The cell plate grows outward until it reaches the wall of the dividing cell, completing the separation of the two daughter cells. With the electron microscope, the fusing droplets can be seen to be vesicles derived

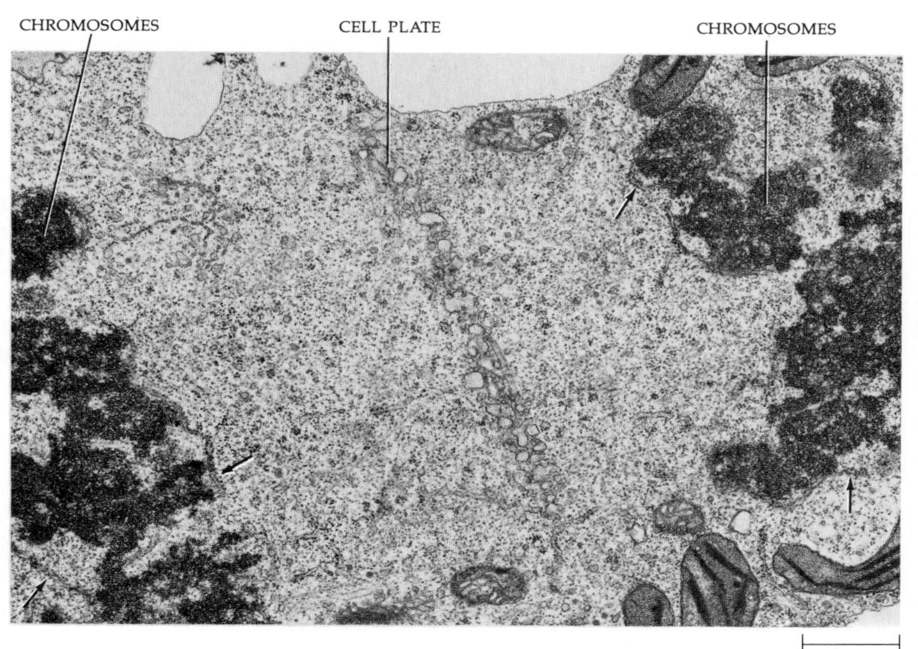

1–41
In plant cells, separation of the daughter chromosomes is followed by formation of a cell plate, which completes the separation of the dividing cells. Here numerous Golgi vesicles can be seen fusing in an early stage of cell-plate formation. The two groups of chromosomes on either side of the developing cell plate are at telophase. Arrows point to portions of the nuclear envelope reorganizing around the chromosomes.

from Golgi bodies. The vesicles presumably contain pectic substances that form the middle lamella and contribute their membranes to the formation of the plasma membrane on either side of the plate. Plasmodesmata are formed at this time, as segments of endoplasmic reticulum are "caught" between fusing vesicular contents (Figure 1–43).

Following the formation of the middle lamella, each protoplast deposits a primary wall next to the middle lamella. In addition, each daughter cell deposits a new wall layer around the entire protoplast; this new wall is continuous with the wall at the cell plate. The original wall of the parent cell stretches and ruptures as the daughter cells enlarge (Figure 1–44).

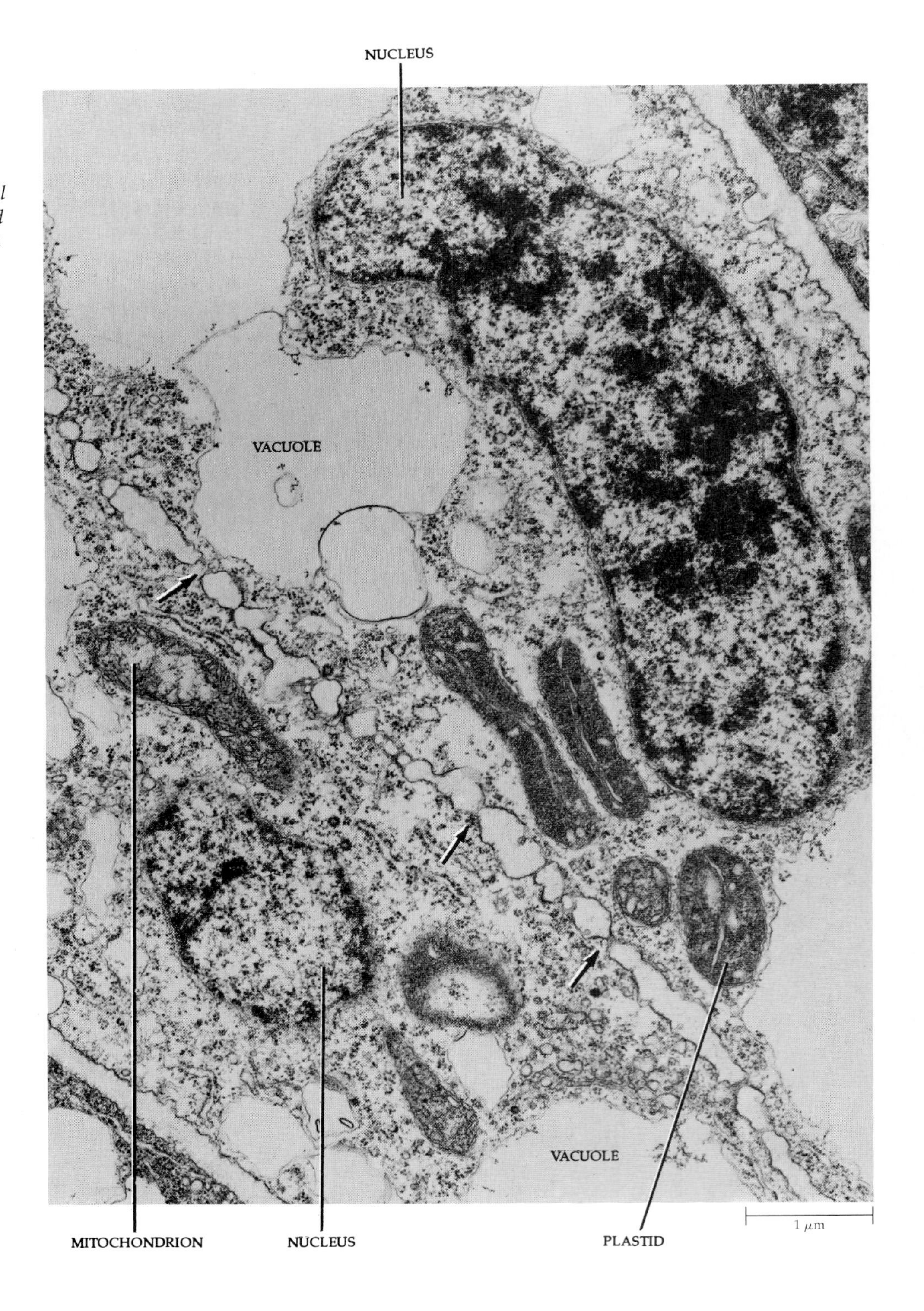

1–42
A cell from the vascular cambium of the American elm tree (Ulmus americana), *in which a cell plate can be seen at a late stage of development. The interruptions in the cell plate (see arrows) are newly formed or developing plasmodesmata. Each cell now has an intact nucleus.*

1–43
Late stage of cell-plate formation in a parenchyma cell in a bean (Phaseolus vulgaris) *root. Notice the segments of endoplasmic reticulum (see arrows) that traverse the cell plate at three places. Sites such as these will eventually become plasmodesmata.*

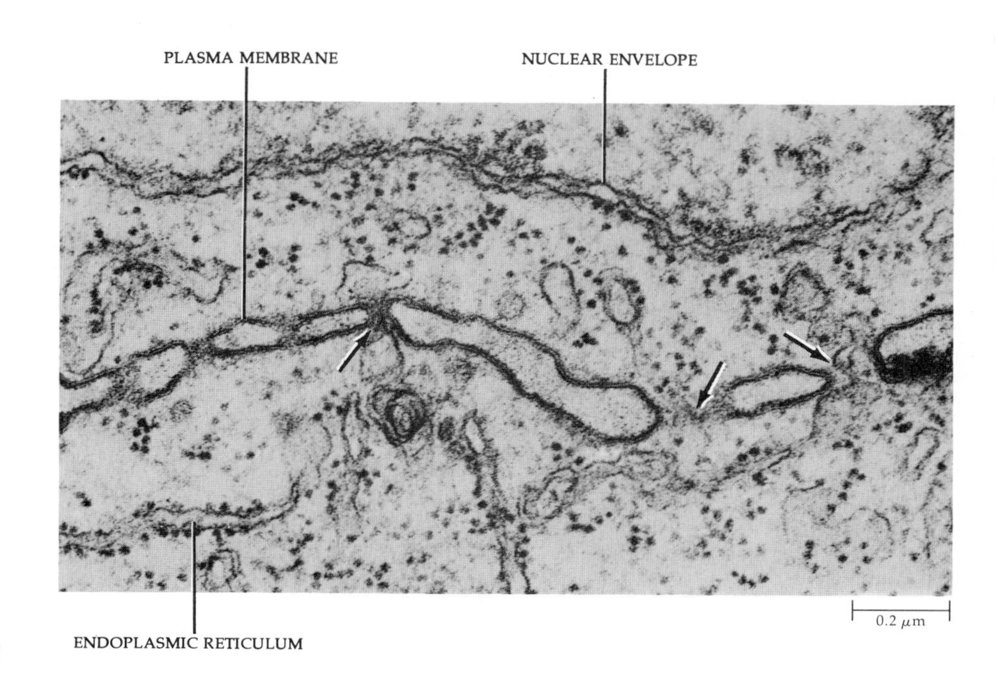

1–44
Cell-wall formation during cytokinesis. (a) *The cell plate forms at the equator of the phragmoplast between the daughter nuclei and* (b) *grows outward until it reaches the wall of the dividing cells. The phragmoplast microtubules disappear where the cell plate has been formed but are re-formed at the periphery of the developing cell plate.* (c) *Each sister cell forms its own primary wall.* (d) *With enlargement of the daughter cells, the mother cell wall is torn.*

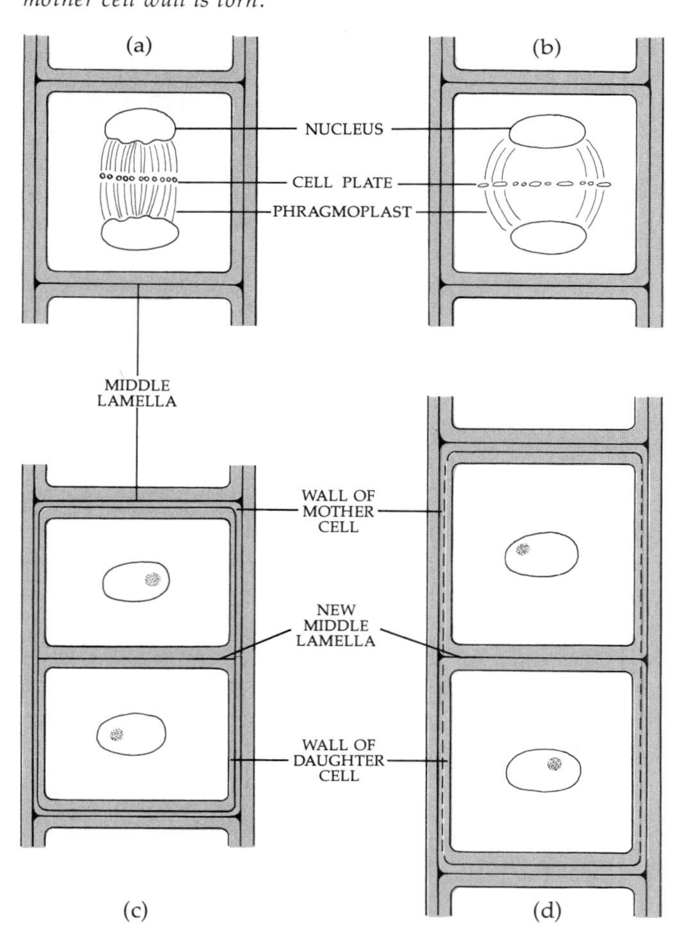

SUMMARY

All living matter is composed of cells. Cells are extremely varied, ranging in structure and function from independent, single-celled organisms to the highly specialized, interdependent cell types found in complex multicellular plants and animals. However, cells are remarkably similar in their basic structure. All cells are bounded by an outer membrane (known as the plasma membrane, or plasmalemma). Within this membrane is the cytoplasm and the hereditary information in the form of DNA. Collectively, the living contents of the cell are called protoplasm.

Cells are of two fundamentally different types: prokaryotic and eukaryotic. Prokaryotic cells lack nuclei and cellular organelles, and are represented today by bacteria, including cyanobacteria. Eukaryotic cells have true nuclei and are compartmentalized, containing distinct structures that perform different functions.

Eukaryotic cells are divided into compartments by membranes that generally have a three-layered appearance. Such membranes are called unit membranes. Membranes not only control the passage of materials into and out of the cell but also into and out of organelles.

Plant cells typically have a semirigid cell wall and a protoplast, the unit of protoplasm inside the cell wall. The protoplast includes the cytoplasm and the nucleus. The cytoplasm of plant cells is frequently in motion, a phenomenon known as cytoplasmic streaming, or cyclosis. The cytoplasm is separated from the cell wall by the plasma membrane, a unit membrane.

In addition to cell walls, plant cells are characterized by the presence of vacuoles within their cytoplasm. Vacuoles are cavities that are filled with cell sap, an aqueous solution of materials including a variety of salts, sugars, and other substances. In addition, many vacuoles are involved in the breakdown of macromolecules and the recycling of their components within the cell. The vacuole is bounded by a unit membrane called the tonoplast. Young cells commonly contain numerous small vacuoles, which increase in size and fuse into a single large vacuole as the cell matures. The enlargement of the vacuoles is the major cause of increase in the size of the cell.

The nucleus, which is the control center of the cell, is often the most prominent structure within the protoplast. It is surrounded by a nuclear envelope composed of two unit membranes. Within the envelope is the chromatin which becomes visible as distinct chromosomes during nuclear division. Chromatin and chromosomes consist of DNA and proteins.

Together with vacuoles and cell walls, plastids are characteristic components of plant cells. Each plastid is bounded by an envelope consisting of two unit membranes. Mature plastids are classified on the basis of the kinds of pigments they contain: chloroplasts contain chlorophylls and carotenoid pigments; chromoplasts contain carotenoid pigments; and leucoplasts are nonpigmented. Chloroplasts contain grana—stacks of flattened sacs of membranes called thylakoids—in which the pigments are located. The various kinds of plastids may develop from small, colorless bodies known as proplastids.

Like plastids, mitochondria are bounded by two unit membranes; the inner one is folded to form an extensive inner membrane system, increasing the surface area available to enzymes and the reactions associated with them. Mitochondria are the principal sites of respiration in eukaryotic cells.

In addition to organelles, the cytoplasm of eukaryotic cells contains two membrane systems, the Golgi bodies and the endoplasmic reticulum. Golgi bodies, or dictyosomes, are groups of flat, disk-shaped membranes, from which numerous vesicles bud off. The Golgi bodies apparently serve as collection and packaging centers for complex carbohydrates and other substances, which are transported to the surface of the cell in the vesicles. The vesicles also serve as a source of plasma membrane material.

The endoplasmic reticulum is an extensive three-dimensional system of membranes. It may have numerous connections with the nuclear envelope and often has numerous ribosomes attached to it. Ribosomes are also found free (not attached to endoplasmic reticulum) and, in eukaryotes, are present in both the cytoplasm and nucleus. Ribosomes are the sites at which amino acids are linked together to form proteins. During protein synthesis, the ribosomes occur in clusters called polyribosomes, or polysomes.

Microtubules are thin structures of variable length and are composed of subunits of the protein tubulin. They play a role in mitosis, cell-plate formation, the growth of the cell wall, and the movement of flagella.

Flagella are hairlike structures that project from the surface of some plant cells, serving as locomotor organelles. All flagella of eukaryotic cells have the same highly characteristic 9-plus-2 internal structure; that is, an outer ring of nine pairs of microtubules surrounding an inner pair of microtubules in the center of the flagellum.

The cell wall is the major distinguishing feature of the plant cell. It determines the structure of the cell, the texture of plant tissues, and many important characteristics that distinguish plants as organisms. The cell wall may be composed of three layers: the middle lamella, the primary wall, and the secondary wall. Cellulose is present in the cell walls of higher plants as rigid fibrils composed of a large number of cellulose molecules. These cellulose fibrils exist in a cross-link matrix of noncellulosic molecules, such as hemicelluloses and pectin. Lignin may also be present in all three wall layers but is especially characteristic of secondary cell walls.

Cells with secondary walls often die after the wall is laid down, leaving the cell to serve as a conduit for water or as a rigid supporting structure.

Integration of the plant body is accomplished by means of the plasmodesmata, which cross the walls and connect the protoplasts of adjacent cells.

Dividing cells pass through a cell cycle consisting of interphase and mitosis. Interphase can be divided into three periods: a G_1 period of general growth and replication of organelles; an S period, during which the genetic material, DNA, is duplicated; and a G_2 period of synthesis of the structures involved in mitosis.

When the cell is in interphase, the chromosomes are in an uncoiled state and are difficult to distinguish from the nucleoplasm. During mitosis, the chromosomal material, or chromatin, condenses, and each chromosome can be seen to consist of two parallel threads—the chromatids—held together at the centromere. The centromere divides, and the duplicate chromatids—now called daughter chromosomes—move to opposite poles. The separation of the two identical sets of chromosomes is completed when new nuclear envelopes are formed around them. In this way, the genetic material is equally distributed between the two new nuclei.

Mitosis is generally followed by cytokinesis, the division of the cytoplasm. In plants and certain algae, the cytoplasm is divided by a cell plate that begins to form during mitotic telophase. Once the cytoplasm is divided, the protoplasts lay down new cell walls.

CHAPTER 2

The Molecular Composition of Cells

2–1
Starch grains in amyloplasts of the potato tuber, photographed with polarized light. Starch is the chief storage polysaccharide in plants (see Figure 2–4d).

As discussed in the previous chapter, cells are remarkably similar in their basic structures, despite their variety. These similarities become even more striking when we examine cells at the molecular level.

The earth and its atmosphere contain 92 different types of naturally occurring atoms, or elements. (The basic chemical principles that describe the behavior of these elements are reviewed in Appendix A.) Of all these available types of atoms, only a relatively few were selected in the course of evolution to form the complex, highly organized material of the cells of living organisms. In fact, as shown in Table 2–1, about 99 percent (by weight) of living matter is composed of only six elements. In cells, these elements are found chiefly in organic compounds and, in particular, in water.

Water makes up more than half of all living tissue and more than 90 percent of most plant tissues. (See Appendix A for a review of some important properties of water.) By contrast, ions such as K^+, Na^+, and Ca^{2+} account for no more than 1 percent. Almost all the rest of the tissue, chemically speaking, is composed of organic molecules.

Table 2–1 *Atomic Compositon of Three Representative Organisms (Percentage by Fresh Weight)*

ELEMENT	HUMANS	ALFALFA	BACTERIA
C	19.37	11.34	12.14
H	9.31	8.72	9.94
N	5.14	0.83	3.04
O	62.81	77.90	73.68
P	0.63	0.71	0.60
S	0.64	0.10	0.32
C H N O P S total:	97.90	99.60	99.72

SOURCE: H.J. Morowitz, *Energy Flow in Biology* (New York: Academic Press, 1968).

ORGANIC COMPOUNDS

Organic compounds, by definition, are compounds that contain carbon. In addition to carbon, nearly all organic compounds contain hydrogen, and most of them contain oxygen as well. Nitrogen and sulfur occur less frequently than oxygen, and only a small percentage of the organic compounds contain other elements. Just as protoplasm is composed of relatively few types of substances, the atoms of these few elements are arranged to form a relatively few organic compounds that, in various combinations, make up most of the dry weight of living organisms. The four principal types of organic compounds are carbohydrates, lipids, proteins, and nucleic acids (see Table 2–2).

Table 2–2 *Some Important Classes of Organic Compounds*

ORGANIC COMPOUNDS	FUNCTIONS	COMPONENTS	COMPOSITION
Carbohydrates	Energy source, structural material, building blocks for other molecules	Simple sugars	Carbon, hydrogen, and oxygen
Lipids (fats)	Energy storage, structural material	Fatty acids, glycerol	Carbon, hydrogen, and oxygen
Proteins	Structural materials, enzymes	Amino acids	Carbon, hydrogen, oxygen, nitrogen, and sulfur
Nucleic acids	Protein synthesis	Nucleotides (nitrogenous bases, sugars, and phosphates)	Carbon, hydrogen, oxygen, nitrogen, and phosphorus

CARBOHYDRATES

Carbohydrates (sugars, starches, and various related substances) are compounds that contain carbon combined with hydrogen and oxygen. Carbohydrates are the most abundant organic compounds in nature. Large molecules that are made up of similar or identical subunits are known as *polymers;* the subunits are called *monomers.* In the case of carbohydrates, the monomers are monosaccharides, and the polymers are polysaccharides.

Monosaccharides and Disaccharides

Monosaccharides are the simplest carbohydrates and are made up of a chain of carbon atoms to which H atoms and O atoms are attached in the proportion of one carbon atom to two hydrogen atoms to one oxygen atom (CH_2O). Examples of several common monosaccharides are shown in Figure 2–2. As the figure indicates, the five-carbon and six-carbon sugars can also exist in ring form; in fact, they are normally found in this form when dissolved in water.

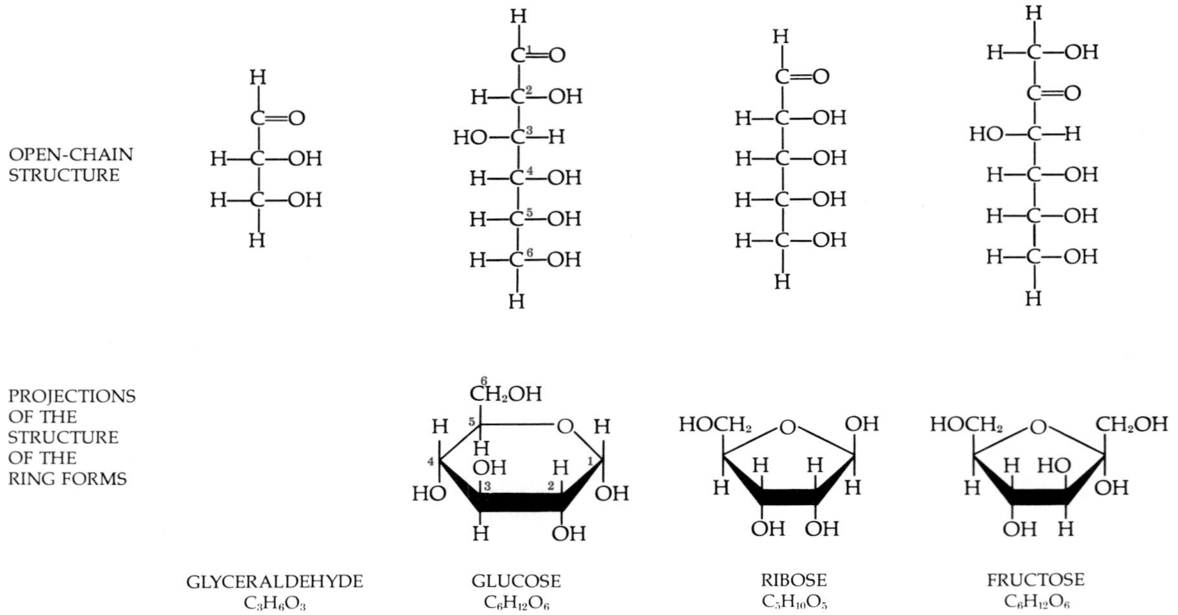

OPEN-CHAIN STRUCTURE

PROJECTIONS OF THE STRUCTURE OF THE RING FORMS

GLYCERALDEHYDE $C_3H_6O_3$

GLUCOSE $C_6H_{12}O_6$

RIBOSE $C_5H_{10}O_5$

FRUCTOSE $C_6H_{12}O_6$

2–2
Examples of biologically important monosaccharides. Five-carbon sugars (pentoses) and six-carbon sugars (hexoses) can exist in both chain and ring forms, as shown here. The position occupied by each carbon atom has a number. The numbering of the carbon atoms is shown for the two forms of the glucose molecule. By convention, carbon atoms are not shown in the ring forms; their presence is assumed at each empty corner of the ring.

Disaccharides consist of two monosaccharides. As shown in Figure 2–3, the union is accomplished by the removal of a molecule of water from the pair of monosaccharide molecules. Such joined molecules can be broken apart by *hydrolysis*—the addition of a molecule of water at each linkage—to form the monosaccharide units again. Hydrolysis is an exergonic, or "downhill," reaction; the chemical bonding energy of its products is less than that of the original molecule. Thus, energy is released in such a reaction. Conversely, linking two monosaccharides together to form a disaccharide requires the input of energy.

Glucose, a monosaccharide, is the form in which sugar is most often transported through animal systems. A combination of glucose and fructose forms sucrose, the form in which most sugar is transported in plants. Sucrose is common table sugar (cane or beet sugar). Lactose (milk sugar) is a disaccharide composed of glucose and galactose.

The ultimate source of sugar in all plant cells is photosynthesis. In the process of photosynthesis, energy from the sun is converted to the chemical bond energy needed to form a sugar molecule. When the sugar molecule is broken down again, the energy of the chemical bonds is released.

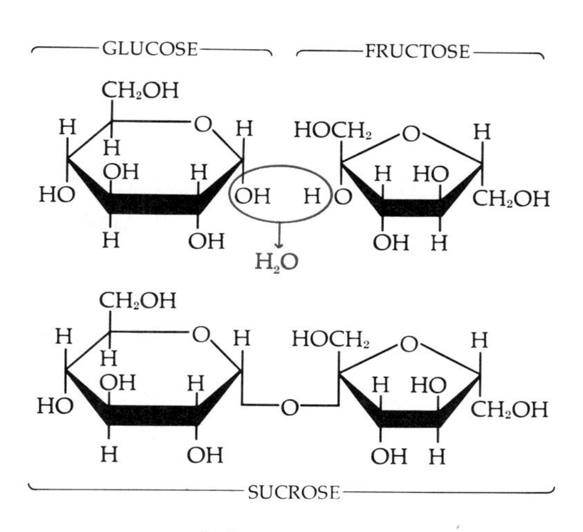

2–3
The monosaccharides glucose and fructose can combine to form the disaccharide sucrose. Sucrose is the form in which sugar is generally transported in plants. Bonds between monosaccharides are formed by the removal of a molecule of water. Formation of sucrose requires an energy input of 5.5 kcal per mole by the cell.

Polysaccharides

Polysaccharides are made up of monosaccharides linked together in long chains. Some polysaccharides are storage forms of sugar. *Starch*, which is built of many glucose molecules (glucose residues), is the chief storage polysaccharide in higher plants; and glycogen is the common storage form of sugar in fungi, bacteria, and higher animals (Figure 2–4). These carbohydrates must be hydrolyzed before they can be used as energy sources or transported through living systems.

Polysaccharides are also important structural compounds. In plants, the principal structural polysaccharide is cellulose (Figure 2–5). Although cellulose is made of the same building materials as starch, the arrangement of its long-chain molecules makes it rigid, and so its biological role is very much different. Also, because the bonds linking the glucose units in cellulose are different from those found in starch, cellulose is not readily hydrolyzed. In addition to cellulose, plant cell walls may contain a variety of other polysaccharides (Figure 2–6), pectic substances (Figure 2–7), and lignin.

Once monosaccharides are incorporated into the plant cell wall in the form of cellulose, they are no longer available to the plant as an energy source. In fact, only some bacteria, fungi, and protozoa, and a very few animals (silverfish, for example), possess enzyme systems capable of breaking down cellulose.

RADIOCARBON DATING

All organic materials contain carbon. All of this carbon previously existed in the form of carbon dioxide and made its way into the living organisms by means of photosynthesis. Most of the carbon atoms present in carbon dioxide have an atomic weight of 12 (^{12}C), but a fixed proportion of the atoms are carbon 14 (^{14}C), a radioactive isotope of carbon. Carbon 14 is produced as a result of bombardment by high-energy particles from outer space and occurs in small amounts as heavy carbon dioxide. Plant cells use carbon dioxide to make glucose and other organic molecules, accepting $^{14}CO_2$ almost as readily as $^{12}CO_2$. All animals are directly or indirectly dependent upon the products of plant metabolism for food; thus a fixed proportion of carbon atoms in the tissues of all living things is radioactive carbon 14. After death, carbon is no longer ingested and the proportions shift, with the radioactive carbon 14 decaying slowly and the carbon 12 remaining the same. Carbon 14 has a half-life of 5730 years, so a fossil of this age should contain just half the carbon 14 of a living plant. Objects, such as a fossil or even a man-made structure of wood or some other once-living material, can be dated quite accurately by measuring the ratio of ^{14}C to ^{12}C. Radiocarbon dating is particularly useful for studying archeological remains. This method of dating depends on the assumption that the proportion of ^{14}C to ^{12}C has remained constant in the atmosphere within the time span under study. Atmospheric nuclear testing has made it impossible for future archeologists to date current history in this way.

(a)

AMYLOSE

(b)

AMYLOPECTIN

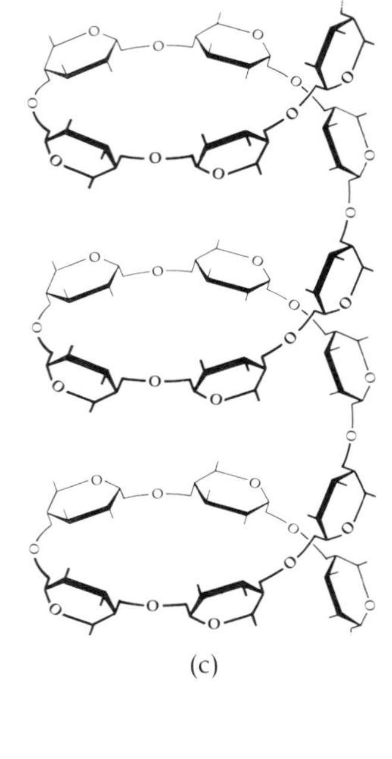

(c)

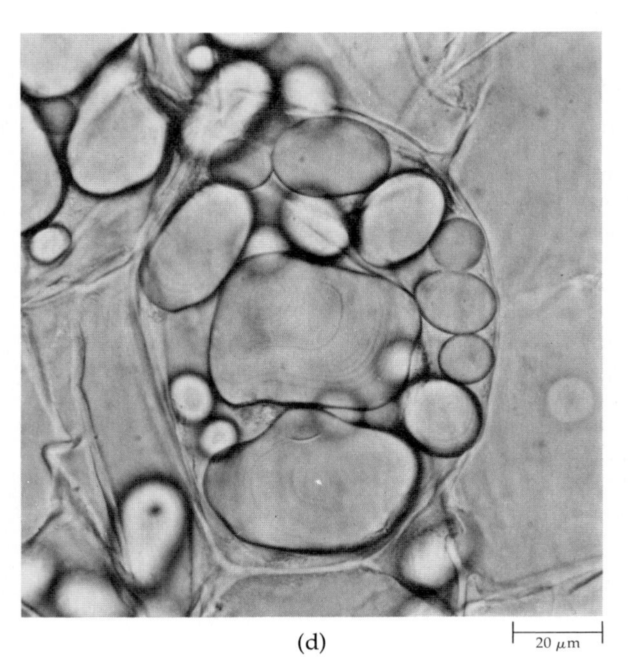

(d) 20 μm

(e) 10 μm

2–4
In plants, accumulated sugars are stored in the form of starch. Starch is composed of two different types of polysaccharides: (a) *amylose and* (b) *amylopectin. A single molecule of amylose contains 1000 or more glucose units in a long, unbranched chain which winds to form a helix* (c). *A molecule of amylopectin has a molecular weight of 1 to 6 million and branches approximately every 20 to 25 glucose units. The glucose units in starch are alpha-glucose monomers (see Figure 2–5). Perhaps because of their helical nature, starch molecules tend to cluster into grains. The photomicrograph in* (d) *shows starch grains within starch-storing leucoplasts (amyloplasts) in cells of the potato tuber.* (e) *In this scanning electron micrograph, starch grains appear as spherical and egg-shaped objects in a cell of a potato.*

The common storage form for sugar in higher animals, fungi, and bacteria is glycogen, which resembles amylopectin in its general structure, with the exception that branching occurs every 6 to 12 glucose units. The molecular weight of glycogen ranges up to 100 million.

2–5
(a) *In living systems, alpha-ring and beta-ring forms of the glucose molecule occur in equilibrium. The molecules pass through the straight-chain form to get from one ring structure to another. Whereas starch consists of alpha-glucose monomers, cellulose* (b) *consists entirely of beta-glucose monomers. The* $-OH$ *groups (in color), which project from both sides of the chain, form hydrogen bonds with neighboring* $-OH$ *groups, resulting in the formation of bundles of cross-linked parallel chains* (c). *In the starch molecule, most of the* $-OH$ *groups capable of forming hydrogen bonds face toward the exterior of the helix.*

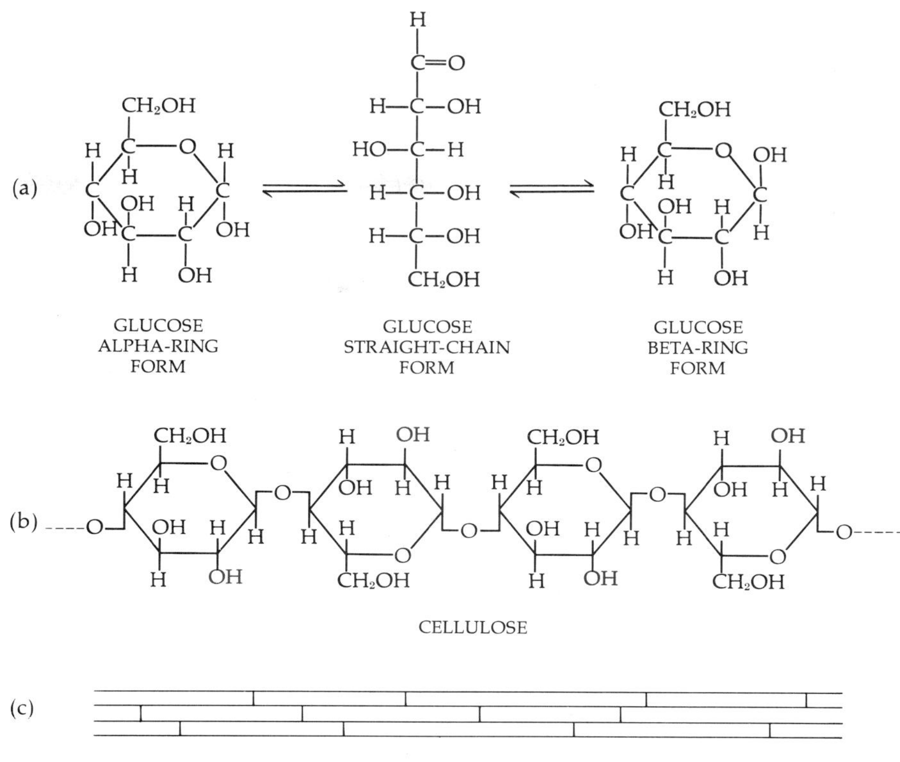

2–6
Building blocks of some polysaccharides other than cellulose that are commonly found in the cell walls of higher plants. The monosaccharides shown here are the building blocks of the polysaccharides called, respectively, galactans, mannans, xylans, and arabinans.

GALACTOSE

MANNOSE

XYLOSE

ARABINOSE

2–7
Pectic compounds are built up of residues of alpha-galacturonic acid (a), *which is a derivative of glucose. Pectic acid is the pectic compound shown in* (b). *Calcium and magnesium salts of pectic acid make up most of the middle lamella that binds adjacent plant cells. In pectin and protopectin, different proportions of the hydrogen ions (shown in color) are replaced with methyl* ($-CH_3$) *groups. Protopectin is a common constituent of the cell wall and is less soluble than pectin, which is commonly found dissolved in plant juices.*

α-GALACTURONIC ACID
(a)

PECTIC ACID
(b)

Other organisms, such as cattle (and other ruminants), termites, and cockroaches, are able to utilize cellulose as a source of energy only because of the microorganisms (which possess the necessary enzyme systems) that inhabit their digestive tracts.

Chitin is another important structural polysaccharide; it is the principal structural component of fungal cell walls and also of the relatively hard outer coverings, or exoskeletons, of insects and crustaceans. The monomer of chitin is a six-carbon sugar to which a nitrogen-containing group has been added (see Figure 12–4, page 214).

LIPIDS

Lipids are fats and fatlike substances. They have two principal distinguishing characteristics: (1) they are generally hydrophobic ("water-fearing") and thus are insoluble in water (although they are soluble in other lipids), and (2) they contain a large number of carbon-hydrogen bonds and, as a consequence, release a larger amount of energy in oxidation than other organic compounds. Fats yield, on the average, about 9.3 kcal per gram, as compared to a yield of about 3.8 kcal per gram for carbohydrates. These two characteristics of lipids determine their roles as structural materials and as energy reserves.

Fats

Fat is the principal form in which lipids are used for storage. Some plants store food energy as fats (oils), especially in seeds and fruits. Cells synthesize fats from sugars. A fat consists of three fatty acids joined to a glycerol molecule (Figure 2–8). Fatty acids are long hydrocarbon chains that carry a terminal carboxyl group, giving them the characteristics of a weak acid. The glycerol forms a link with the carboxyl group, releasing a molecule of water. The glycerol thus serves as a binder or carrier for the fatty acids. (Like polysaccharides and proteins, fats are broken down by hydrolysis.)

The physical nature of the fat is determined by the chain lengths of the fatty acids and by whether the acids are *saturated* or *unsaturated*. In saturated fatty acids, all the carbon atoms hold as many hydrogen atoms as possible. Unsaturated fatty acids contain carbon atoms joined by double bonds; such carbon atoms are able to form additional bonds with other atoms (hence the term unsaturated). Unsaturated fats, which tend to be oily liquids, are more common in plants than in animals; examples are olive oil, peanut oil, and corn oil. Animal fats, such as lard, contain saturated fatty acids and usually have higher melting temperatures than unsaturated fats.

Waxes

Waxes also contain fatty acids. They are formed by the union of long-chain alcohols (instead of glycerol) and fatty acid molecules. Plant waxes are important constituents of *cutin*, which covers the epidermis of leaves, stems, and fruits, and of *suberin*, the waterproofing material of cork cell walls (Figure 2–9).

Phospholipids

Closely related to the fats are the phospholipids—various compounds in which glycerol is attached to only two fatty acids, with the third space occupied by a molecule containing phosphorus (Figure 2–10). Phospholipids are very important in cellular structure, particularly in the cell membranes.

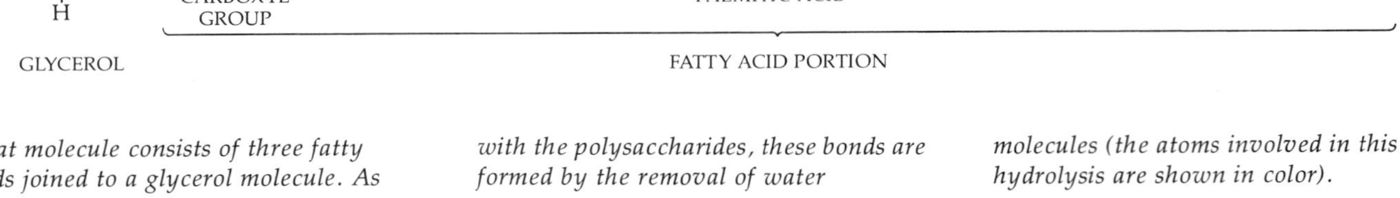

$$\text{H–C–OH} \quad \text{HO–C(=O)–}(\text{CH}_2)_{16}\text{–CH}_3 \ \text{(STEARIC ACID)}$$

$$\text{H–C–OH} \quad \text{HO–C(=O)–}(\text{CH}_2)_7\text{–CH=CH–}(\text{CH}_2)_7\text{–CH}_3 \ \text{(OLEIC ACID)}$$

$$\text{H–C–OH} \quad \text{HO–C(=O)–}(\text{CH}_2)_{14}\text{–CH}_3 \ \text{(PALMITIC ACID)}$$

GLYCEROL — CARBOXYL GROUP — FATTY ACID PORTION

2–8
A fat molecule consists of three fatty acids joined to a glycerol molecule. As with the polysaccharides, these bonds are formed by the removal of water molecules (the atoms involved in this hydrolysis are shown in color).

2–9

The upper leaf surface of Eucalyptus cloeziana *showing deposits of wax. Beneath these deposits is the cuticle, a wax-containing layer covering the outer walls of the epidermal cells. Waxes protect exposed plant surfaces from water loss.*

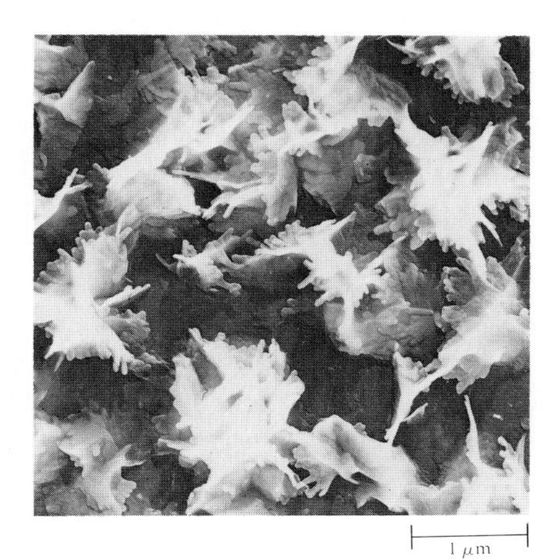

1 μm

2–10

A phospholipid molecule consists of two fatty acids linked to a glycerol molecule, as in a fat, and a phosphate group (indicated by color) linked to the third carbon of the glycerol. The phosphate group also usually contains an additional chemical group, indicated by the letter R. *The fatty acid tails are nonpolar (lacking positive and negative charges) and are therefore hydrophobic (insoluble in water); the polar head containing the phosphate and* R *groups is hydrophilic (soluble).*

POLAR HEAD — NONPOLAR TAILS

$$R{-}O{-}\overset{O}{\underset{O}{\overset{|}{\underset{\|}{P}}}}{-}O{-}CH_2$$

$$H{-}\underset{|}{\overset{|}{C}}{-}O{-}\overset{O}{\overset{\|}{C}}{-}(CH_2)_7{-}CH{=}CH{-}(CH_2)_7{-}CH_3$$

$$H{-}\underset{\underset{H}{|}}{\overset{|}{C}}{-}O{-}\overset{O}{\overset{\|}{C}}{-}(CH_2)_{16}{-}CH_3$$

GLYCEROL

2–11

The glycerol-phosphate combination is hydrophilic (soluble in water), whereas the fatty acids are hydrophobic. Consequently, when placed in water, the molecules tend to form a film (a) *on the water surface, with the polar, or hydrophilic, heads beneath the surface and the nonpolar, or hydrophobic tails protruding above the surface. In the cell, they tend to align themselves in rows* (b) *with their soluble heads pointing outward.*

SYMBOL OF PHOSPHOLIPID MOLECULE

TAIL

HEAD

FILM ON WATER

(a)

POLAR HEADS

NONPOLAR TAILS

POLAR HEADS

(b)

The phosphate end of the phospholipid molecule is hydrophilic ("water-loving") and thus is soluble in water, in contrast to the hydrophobic fatty acids. If phospholipids are added to water, they tend to form a film along its surface, with their polar heads under the water and the insoluble fatty acid chains (tails) protruding above the surface (Figure 2–11a). In the watery interior of the cell, phospholipids tend to align themselves in rows, with the insoluble fatty acids oriented toward one another and the phosphate ends directed outward (Figure 2–11b). Such configurations are important in the structure of cellular membranes. The inner layer of the unit membrane characteristic of all cells consists of a double row of phospholipid molecules.

PROTEINS

Proteins, like polysaccharides, are polymers made up of a number of similar units. In proteins, the molecular monomers are nitrogen-containing molecules known as amino acids. Only 20 different kinds of amino acids are found in proteins, but because protein molecules are large and complex—often containing several hundred amino acids—the number of different amino acid sequences and, hence, the possible variety of protein molecules are enormous. A single cell of the bacterium *Escherichia coli* can contain 600 to 800 different kinds of proteins at any one time, and the cell of a plant or animal probably has several times that number.

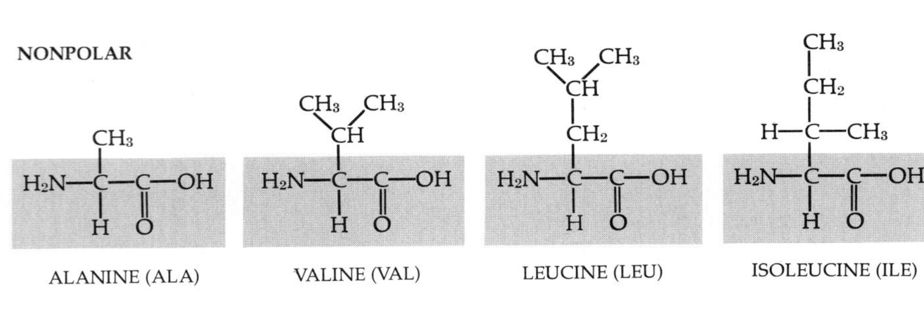

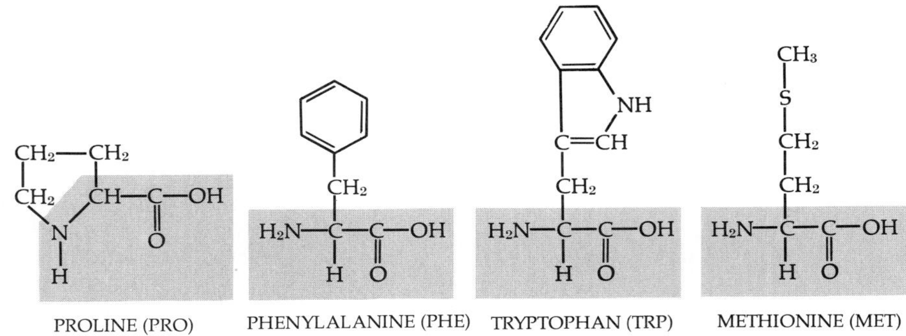

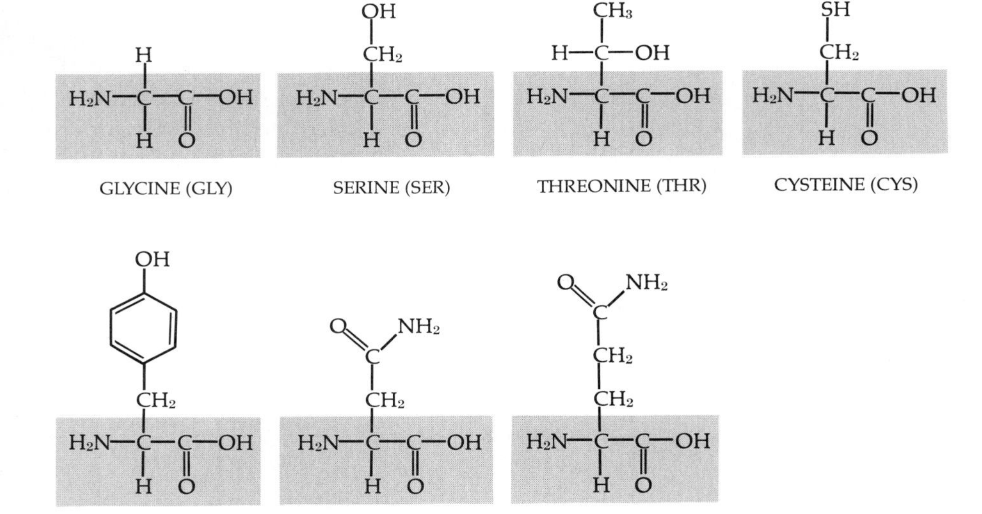

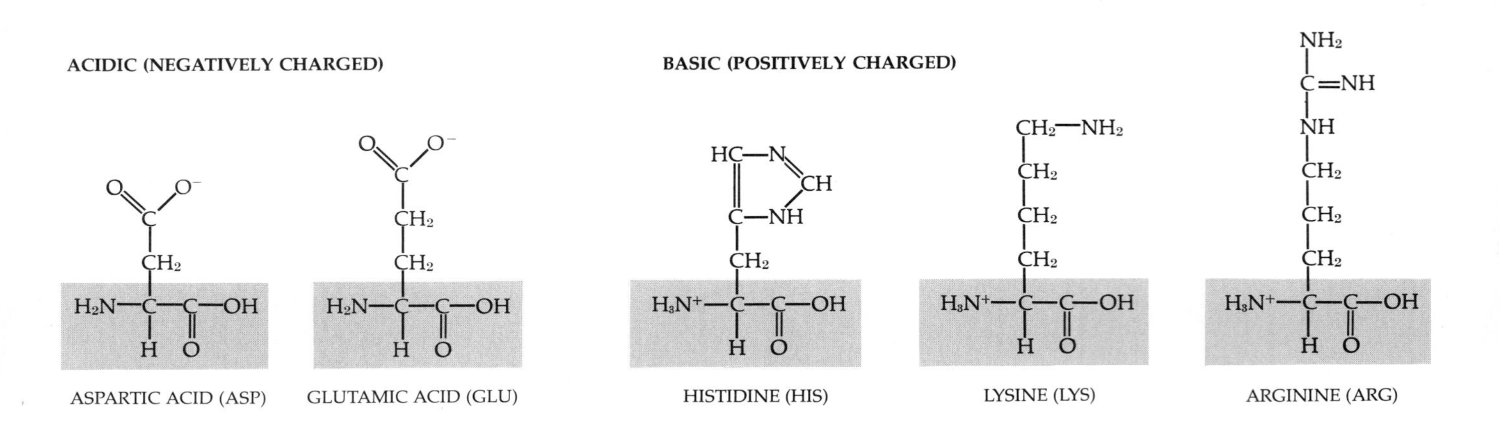

2–12
Every amino acid contains an amino group ($-NH_2$) *and a carboxyl group* ($-COOH$) *bonded to a central carbon atom. A hydrogen atom and a side group* (R) *are also bonded to the same carbon atom. The* R *represents the side group, which is different in each kind of amino acid. This basic structure is the same in all amino acids.*

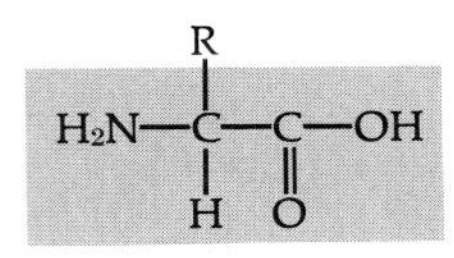

The 20 kinds of amino acids found in proteins are shown to the left and below. As you can see, their basic structures are the same; they differ only in their side groups. Because of differences in their side groups, amino acids may be nonpolar (with no difference in charge between one part of the molecule and another), polar (but with the two charges balancing one another so that the amino acid as a whole is uncharged), negatively charged (acidic), or positively charged (basic). The nonpolar molecules are insoluble in water, whereas the charged and polar molecules are soluble.

In plants, the largest concentration of proteins is found in seeds, in which as much as 40 percent of the dry weight may be protein. These proteins appear to function as storage forms of amino acids that are used by the embryo when it resumes growth upon germination of the seed.

Amino Acids

The basic structure of an amino acid is shown within the caption of Figure 2–12. It consists of an amino group and a carboxyl group bonded to a carbon atom, the so-called alpha-carbon. The "R" represents the rest of the molecule, which varies in structure from one amino acid to another. It is this R group that determines the identity of any particular amino acid. Figure 2–12 shows the complete structure of the 20 amino acids found in proteins.

The amino acids are grouped according to their electrical charges. These charges are important in determining the properties of the various amino acids and, in particular, of the proteins formed from their combination.

Polypeptides

A chain of amino acids is known as a *polypeptide*. As in the case of polysaccharides, a molecule of water is removed to create each link in the chain. The amino group of one amino acid always links to the carboxyl group of the next amino acid; this bond is known as the *peptide bond* (Figure 2–13). Amino acids joined by peptide bonds are often referred to as amino acid residues. Because of the way in which the amino acids are bonded together, there is always a free amino group at one end of the chain and a free carboxyl group at the other. The end with the amino group is called the N terminal; the one with the carboxyl group is called the C terminal.

Amino acids can also be cross-linked by disulfide bonds, which are covalent bonds formed between two sulfur atoms of two cysteine residues. A disulfide bond may link two cysteine residues in the same polypeptide chain or in two different chains (Figure 2–14).

Proteins are large polypeptides; they have molecular weights ranging from 10^4 (10,000) to more than 10^6 (1,000,000). In comparison, water has a molecular weight of 18, and glucose has a molecular weight of 180. (The average molecular weight of an amino acid residue is about 120; if you remember this figure, you can rapidly calculate the approximate number of amino acids in a protein on the basis of its total molecular weight. The way in which molecular weights are determined is described in Appendix A.)

Levels of Protein Organization

The sequence of amino acids in the polypeptide chain is referred to as the *primary structure* of the protein. The polypeptide chains often coil in a helix *(secondary structure)*, and the helix may be folded to produce a so-called globular protein molecule *(tertiary structure)*. Finally, the polypeptide chains of a protein having two or more chains may interact to form a *quaternary structure*.

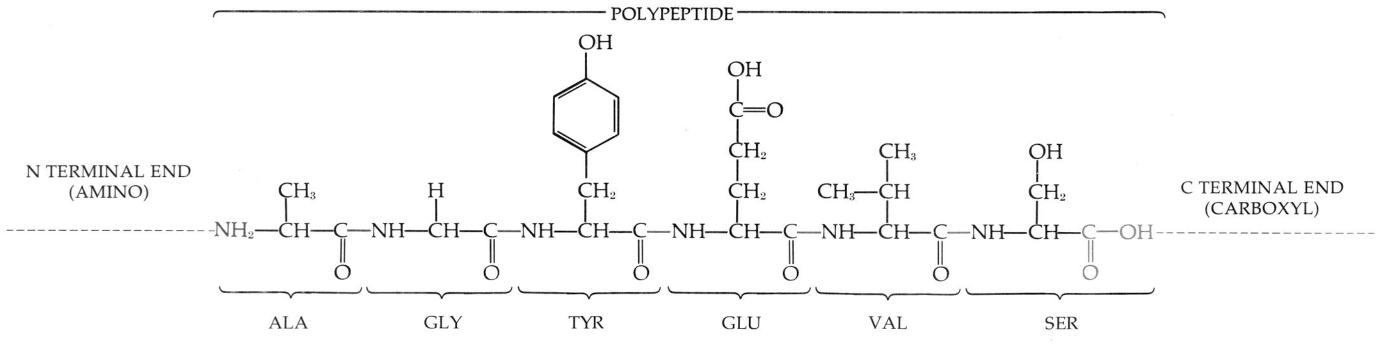

POLYPEPTIDE

N TERMINAL END (AMINO)

C TERMINAL END (CARBOXYL)

ALA GLY TYR GLU VAL SER

2–13
The links between amino acid residues are known as peptide bonds. Peptide bonds are formed by the removal of a molecule of water, as are the bonds between sugar residues. The bonds, shown here in color, always form between the carboxyl ($-COOH$) *group of one amino acid and the amino* ($-NH_2$) *group of the next. Consequently, the basic structure of a protein is a long, unbranched molecule. This linear arrangement of the amino acids is known as the primary structure of the protein. The end with the amino group is called the N terminal; the one with the carboxyl group is called the C terminal.*

Primary Structure

The primary structure of a protein is simply the linear sequence of amino acids in the polypeptide chain. Each kind of protein has a different primary structure. The primary structure of one protein, the enzyme lysozyme, is shown in Figure 2–14.

Secondary Structure

In the cell, polypeptide chains do not lie flat, as they usually appear in most diagrams; they spontaneously assume regular coiled structures. This three-dimensional coiled arrangement is the secondary structure of a protein. The most common secondary structure is the alpha helix, which resembles a spiral staircase (Figure 2–15). The alpha helix is very uniform in its geometry, with a turn of the helix occurring every 3.6 amino acids. The helical structure of polypeptide chains is maintained by hydrogen bonds across successive turns of the spiral, with the hydrogen atom in the amino group of one amino acid bonded to an oxygen atom of the carboxyl group of an amino acid in the next coil.

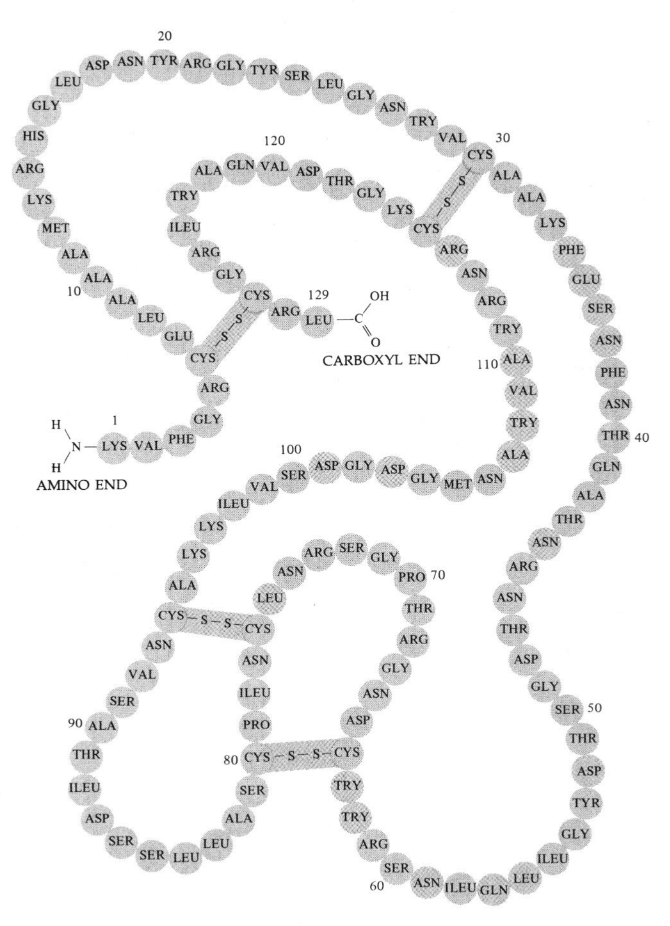

2–14
The primary structure of the enzyme lysozyme is shown in this two-dimensional model. Lysozyme is a protein containing 129 amino acid subunits, commonly called residues. These residues form a polypeptide chain that is cross-linked at four places by disulfide (–S–S–) bonds. (From The Three-Dimensional Structure of an Enzyme Molecule, *by D. C. Phillips. Copyright © 1966 by Scientific American, Inc.*

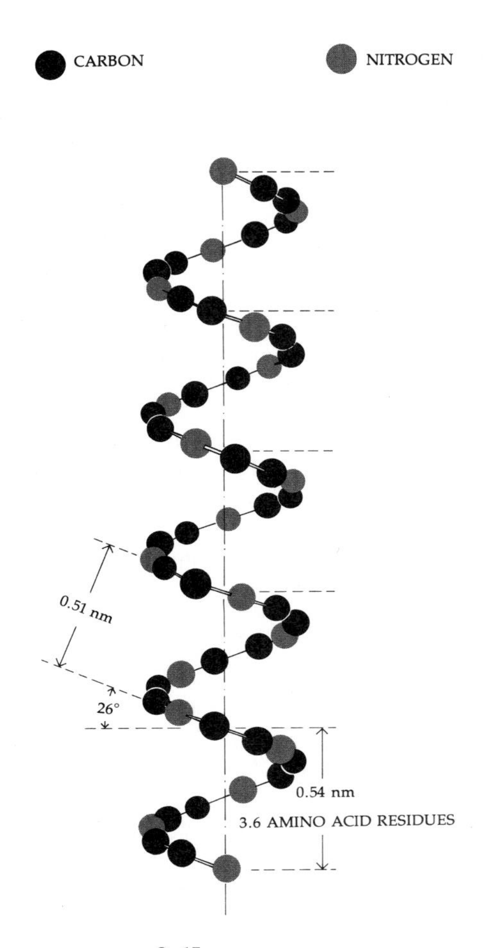

2–15
The alpha helix is the most common secondary structure of proteins. The configuration of the alpha helix is very regular in its geometry. In this figure, only the carbon and nitrogen atoms of the "backbone" of the amino acid residues are shown. One turn of the helix occurs every 3.6 residues. The helix is held in shape by hydrogen bonds that form between the amino acids, linking oxygen and hydrogen atoms at regular intervals.

Tertiary and Quaternary Structure

Polypeptide chains may also fold up to form globular structures—the tertiary structure of protein. Figure 2–16 shows the folding of the main chain in the enzyme lysozyme. Most biologically active proteins, such as enzymes and hormones, are globular. The specialized proteins of cellular membranes are globular in structure. Microtubules are composed of a large number of spherical subunits, each of which is a globular protein.

The tertiary structure of protein is determined by the primary structure. It is maintained largely by hydrogen bonds and by interactions among the various amino acid residues, based on their various charges (see Figure 2–12), and between the amino acid residues and water molecules. These bonds are relatively weak and can be broken quite easily by physical or chemical changes in the environment, such as heat or increased acidity. This breakdown is called denaturation. The coagulation of egg white upon boiling is a common example of protein denaturation. When pro-

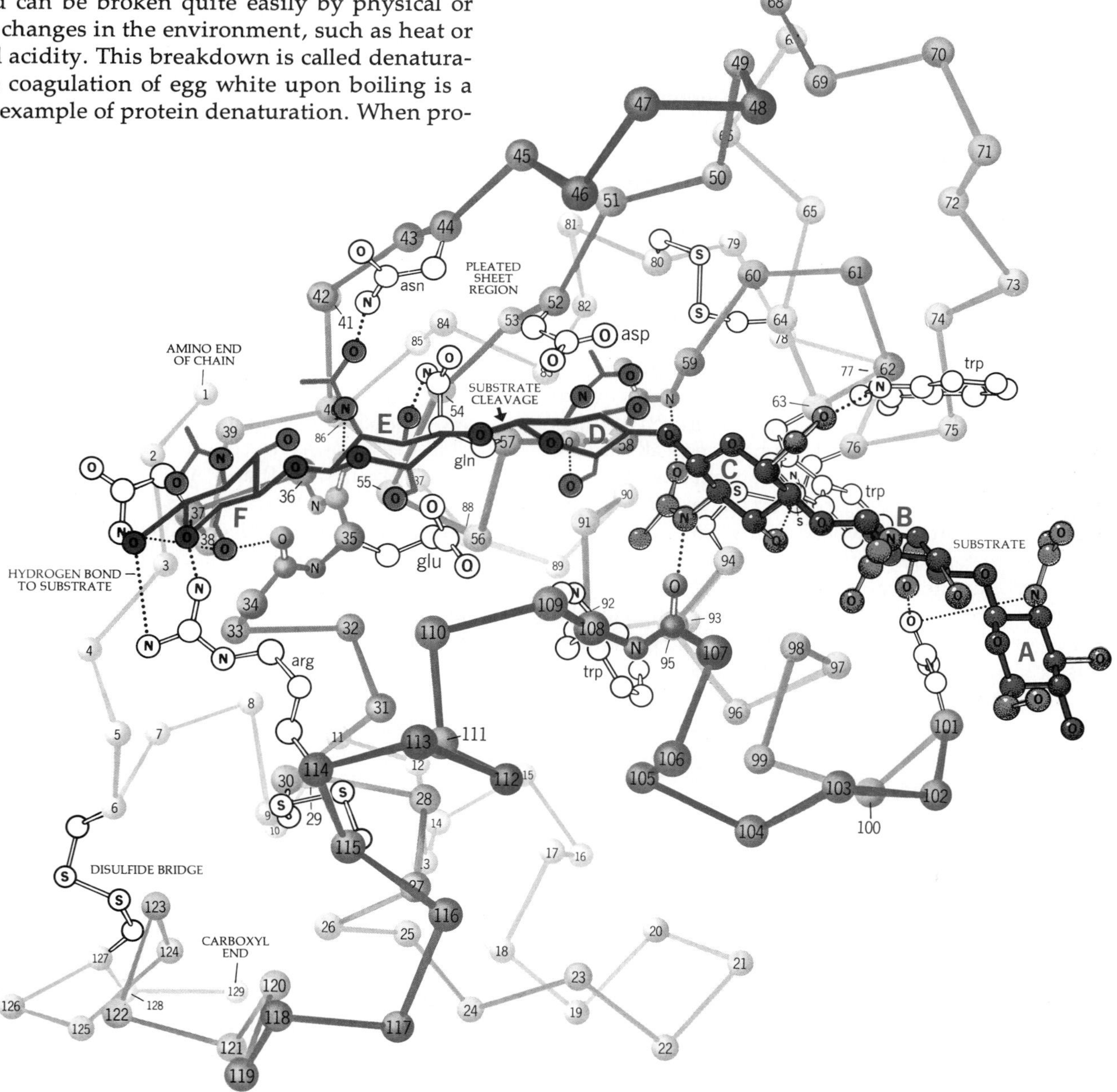

2–16
This drawing gives an idea of the complex nature of the precise folding and coiling that make up the tertiary structure of a polypeptide. (Compare this drawing with that of the relatively simple primary structure of lysozyme in Figure 2–14.) Only the main chain of the enzyme lysozyme is shown. The substrate is shown in a darker color, running horizontally across the enzyme molecule. (From The Structure and Action of Proteins, *by R. E. Dickerson and I. Geis. Menlo Park, Calif.: W. A. Benjamin, Publisher. Copyright 1969 by Dickerson and Geis.)*

teins are denatured, the polypeptide chains are unfolded, causing a loss of the biological activity of the protein. Organisms cannot live at extremely high temperatures because their enzymes and other proteins become unstable and nonfunctional due to denaturation.

The quaternary structure refers to how the polypeptide chains of a protein consisting of two or more chains are arranged in relation to each other. Most larger proteins contain two or more polypeptide chains. These chains may or may not be identical. Many, probably most, enzymes consist of two or more polypeptide chains.

Enzymes

Enzymes are large globular proteins that act as catalysts. (Different enzymes vary in their molecular weight from about 12,000 to more than 1 million.) Catalysts are substances that accelerate the rate of a chemical reaction by lowering the energy of activation (see Appendix A), but which remain unchanged in the process. Because they remain unaltered, catalysts can be used over and over again and so are typically effective in very small amounts.

In the laboratory, the rates of chemical reactions are usually accelerated by the application of heat, which increases the force and frequency of collisions among molecules. But in a cell in nature, hundreds of different reactions are going on at the same time, and heat would speed up all these reactions indiscriminately. Moreover, heat would melt the lipids, denature the proteins, and have other generally destructive effects on the cell. Because of catalytic enzymes, cells are able to carry out chemical reactions at great speeds and at relatively low temperatures. If enzymes were not present, the reactions would occur anyway, but at a rate so slow that their effects would be negligible.

Enzymes are generally named by adding the ending *-ase* to the root of the name of the substrate. Amylase catalyzes the hydrolysis of amylose (starch), and sucrase catalyzes the hydrolysis of sucrose into glucose and fructose. Over 1000 enzymes are now known, and each of them is capable of catalyzing some specific chemical reaction. The behavior of enzymes in biological reactions is further explained in Chapter 4.

NUCLEIC ACIDS

Nucleic acids are polymers of nucleotides—complex molecules made up of three subunits: (1) a phosphate group, (2) a five-carbon sugar (pentose), and (3) a nitrogenous base, so named because its ring structure contains nitrogen as well as carbon.

There are two types of nucleic acids—DNA (deoxyribonucleic acid) and RNA (ribonucleic acid). DNA is the molecule in which the genetic information is stored. RNA serves as a translator and transmitter of this genetic information. It is through RNA that the DNA dictates the structure of proteins and thus the structure and function of the cell.

Nucleotides contain two types of nitrogenous bases: *pyrimidines,* which have a single ring, and *purines,* which have two fused rings (Figure 2–17). The three pyrimidines found in nucleotides are thymine, cytosine, and uracil. DNA contains thymine and cytosine; RNA contains cytosine and uracil. The two purines in nucleotides are adenine and guanine; RNA and DNA both contain these purines. DNA and RNA also contain slightly different types of sugar; DNA contains deoxyribose, and RNA contains ribose. "Deoxy" means "minus one oxygen," and as is shown in Figure 2–17, this is the only difference between these two pentose sugars.

The way in which these molecules are put together, the significance of their structure, their role in the cell, and how this information was discovered are presented in Chapter 7.

OTHER NUCLEOTIDE DERIVATIVES

Nucleotides and compounds derived from nucleotides serve a variety of functions within the cell. Two nucleotide derivatives of prime importance are ATP (adenosine triphosphate; see Figure 4–10) and ADP (adenosine diphosphate). Almost all energy exchanges within the cell involve the transfer of phosphate groups; ATP and ADP are the most important molecules in phosphate transfer. The role of these energy-exchange molecules is discussed in Chapter 4.

Nucleotides play important roles in a variety of other energy-exchange molecules. One such compound is NAD (nicotinamide adenine dinucleotide), which contains a nucleotide of adenine plus a phosphate (see Figure 4–8). Widespread use of the purine-sugar-phosphate combination in energy transfers may have to do with the fact that these comparatively large, charged molecules do not pass through membranes and thus cannot "escape" into and out of cells or membrane-bounded organelles.

SUMMARY

Living matter is composed of only a few of the naturally occurring elements. The bulk of living matter is water. Most of the rest of living material is composed of organic compounds—carbohydrates, lipids, proteins, and nucleic acids.

2–17

The building blocks of RNA and DNA. Each nucleotide building block contains a phosphate group, a sugar, and a nitrogenous base, which can be either a purine or a pyrimidine.

The purines in both groups are the same, but one type of DNA nucleotide contains thymine, whereas its RNA counterpart contains uracil. The only other difference between the two is the presence of one more oxygen atom on the sugar (ribose) component of the RNA. As described in Chapter 7, however, the biological roles of the two are profoundly different.

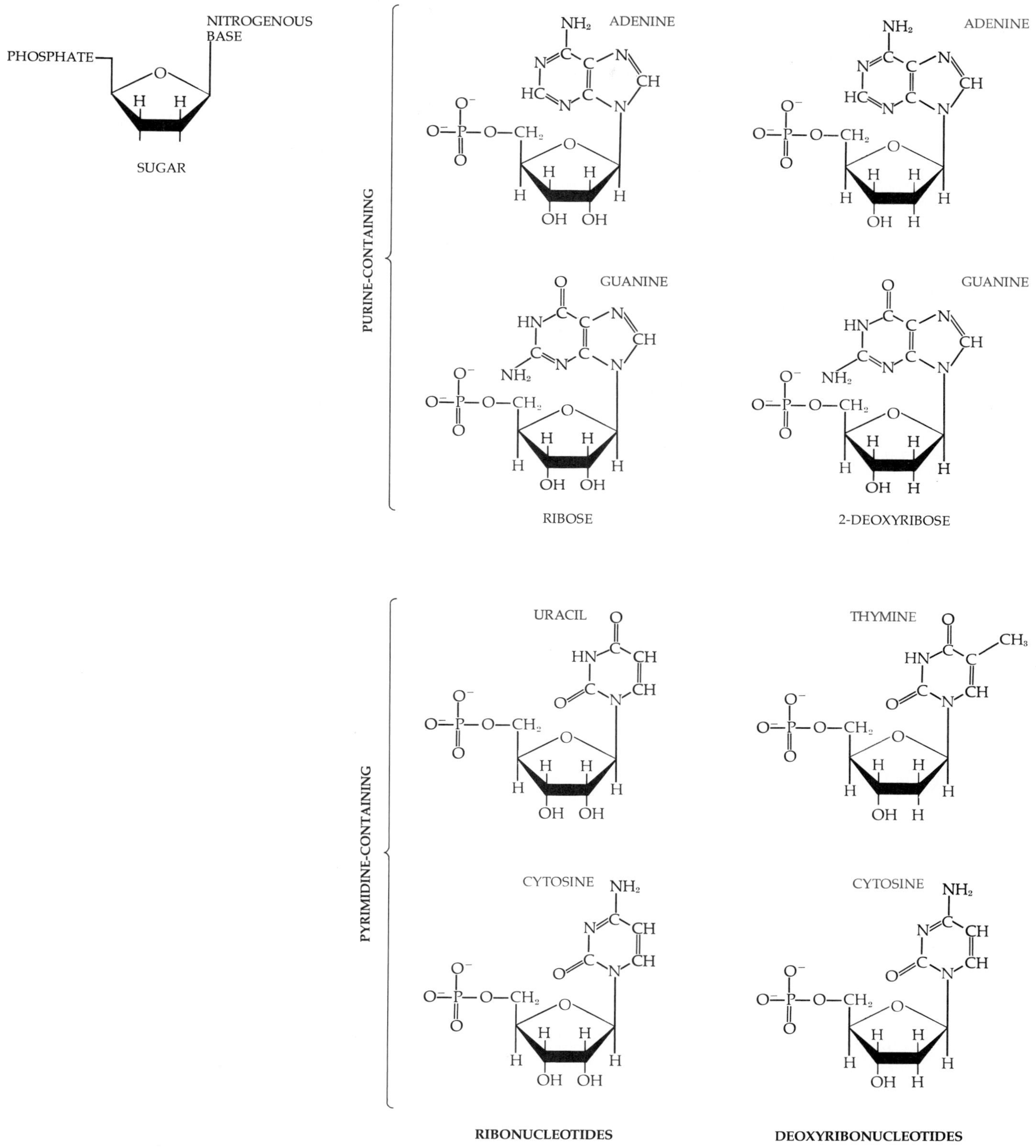

NITROGENOUS BASE

PHOSPHATE

SUGAR

PURINE-CONTAINING

ADENINE

GUANINE

RIBOSE

2-DEOXYRIBOSE

PYRIMIDINE-CONTAINING

URACIL

THYMINE

CYTOSINE

CYTOSINE

RIBONUCLEOTIDES

DEOXYRIBONUCLEOTIDES

Carbohydrates serve as a primary source of chemical energy for living systems and as important structural elements in cells. The simplest carbohydrates are the monosaccharides, such as glucose and fructose. Monosaccharides can be combined to form disaccharides, such as sucrose, and polysaccharides (chains of many submolecules of sugar), such as starch and cellulose. These molecules can usually be broken apart by the addition of a water molecule at each linkage, a chemical reaction known as hydrolysis.

Lipids are another source of energy and structural material for cells. Compounds in this group—fats, waxes, and phospholipids—are generally insoluble in water.

Proteins are very large molecules composed of long chains of amino acids known as polypeptides. Twenty different amino acids are found in proteins, and from these amino acids, enormous numbers of different protein molecules are built. The principal levels of protein organization are: (1) primary structure, the linear amino acid sequence; (2) secondary structure, the coiling or spiraling of the polypeptide chain; (3) tertiary structure, the folding of the coiled chain into globular shapes; and (4) quaternary structure, which results from specific interactions between two or more polypeptide chains.

Enzymes are globular proteins that act as catalysts in chemical reactions. Because of enzymes, cells are able to accelerate the rate of chemical reactions at relatively low temperatures.

Nucleic acids consist of nucleotides linked together in long chains. Nucleotides are composed of three subunits: a nitrogenous base, a five-carbon sugar, and a phosphate group. Nucleotides containing deoxyribose sugar form DNA; those containing ribose sugar form RNA.

Two nucleotide derivatives—ATP and ADP—are involved in most of the energy exchanges within the cell.

CHAPTER 3

The Movement of Substances Into and Out of Cells

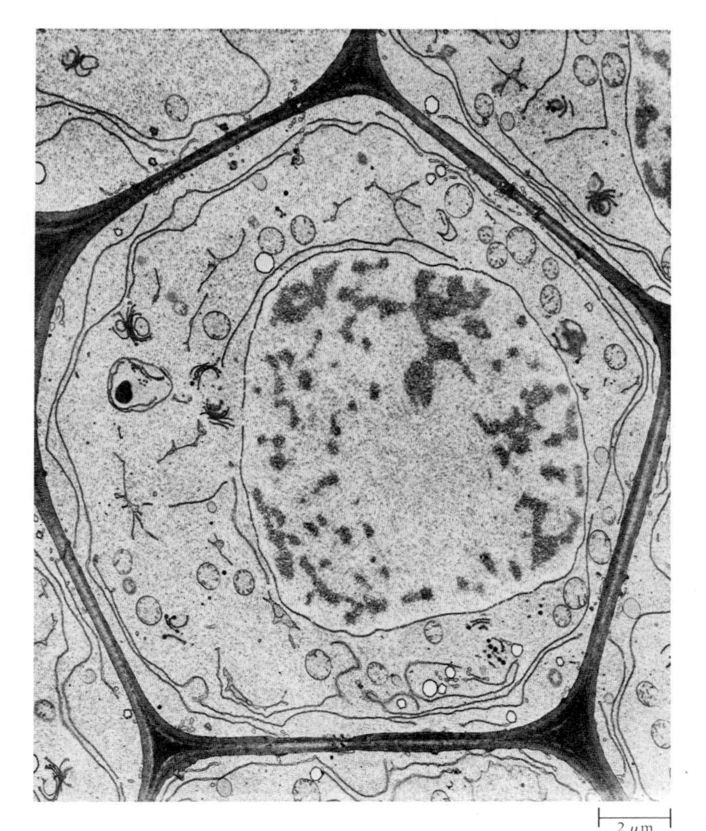

3–1
Electron micrograph of a cell from the root tip of a corn plant (Zea mays). *This cell has been fixed in potassium permanganate, a chemical that combines selectively with biological membranes; hence, the endoplasmic reticulum, Golgi bodies, and bounding membranes of various organelles are clearly defined. These membranes regulate the movement of substances into and out of the cell and control their passage from one part of the cell to another.*

All cells are separated from their surroundings by a surface membrane—the plasma membrane; eukaryotic cells are further divided internally by a variety of membranes, including those of the endoplasmic reticulum, Golgi bodies, and the bounding membranes of organelles (Figure 3–1). These cellular membranes are not impenetrable barriers, for cells are able to regulate the amount, kind, and often the direction of movement of substances that pass across these membranes. This is an essential capacity of living cells because few metabolic processes could occur at reasonable rates if they depended upon the concentration of necessary substances found in the cell's surroundings. In fact, one criterion by which we identify living systems is the difference in concentration of a variety of substances in living matter and in the surrounding, nonliving environment (Figure 3–2).

Control of the exchange of substances across membranes depends on the physical and chemical properties of the membranes and of the ions or molecules that move through them. Water is the most important of the molecules moving into and out of cells.

PRINCIPLES OF WATER MOVEMENT

The movement of water, whether in living systems or in the nonliving world, is governed by three basic principles: bulk flow, diffusion, and osmosis.

Bulk Flow

Bulk flow is the overall movement of water (or some other liquid). It occurs in response to differences in the potential energy of water, usually referred to as *water potential.*

A simple example of water that has potential energy is water behind a dam or at the top of a waterfall. As this water runs downhill, its potential energy can be converted to mechanical energy by a waterwheel or to

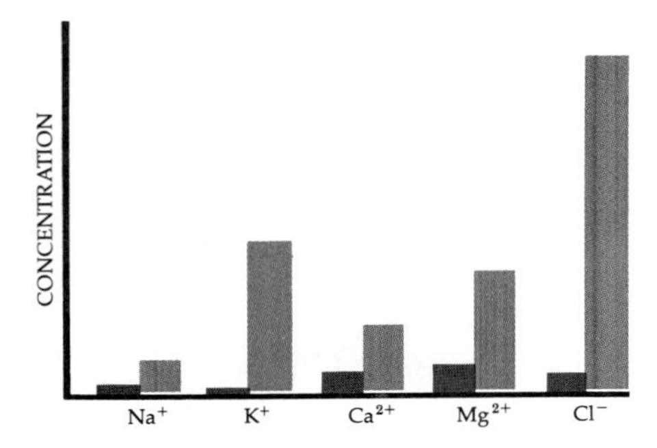

3–2
A diagram showing the concentrations of different ions in pond water (gray bars) and in the cytoplasm of the green alga Nitella *(colored bars). Such differences between cells and their surroundings indicate that cells regulate the passage of materials across their membranes.*

3–3
Water at the top of a waterfall is said to possess potential energy. As the water falls, this potential energy becomes energy of motion (kinetic energy), which can be converted into mechanical energy to do work.

mechanical and then electrical energy by a hydroelectric turbine.

Pressure is another source of water potential. If we put water into a rubber bulb and squeeze the bulb, this water, like the water at the top of a waterfall, has water potential, and it will move to an area of less water potential. Can we make the water that is running downhill run uphill by means of pressure? Yes, we can, but only so long as the water potential produced by the pressure exceeds the water potential produced by gravity. Water moves from an area where water potential is greater to an area where water potential is less, regardless of the reason for the difference.

The concept of water potential is a useful one because it enables physiologists to predict the way in which water will move under various conditions. Measurements of water potential are usually made in terms of the pressure required to stop the movement of water—that is, the hydrostatic pressure—under the particular circumstances involved. The unit used to express this pressure is usually the bar. (A bar is a metric unit of pressure equal to the average pressure of the air at sea level.)

Diffusion

Diffusion is a familiar phenomenon. If a few drops of perfume are sprinkled in one corner of a room, the scent will eventually permeate the entire room even if the air is still. If a few drops of dye are put in one end of a glass tank full of water, the dye molecules will slowly become evenly distributed throughout the tank. This process may take a day or more, depending on the size of the tank, the temperature, and the size of the dye molecules.

Why do the dye molecules move apart? If you could observe the individual dye molecules in the tank (Figure 3–4), you would see that each one of them moves individually and at random. Looking at any single molecule—at either its rate of motion or its direction of motion—gives you no clue at all as to where the molecule is located with respect to the others. So how do molecules get from one side of the tank to the other? Imagine a thin section through the tank, running from top to bottom. Dye molecules will move into and out of the section, some moving in one direction, some moving in the other. But you would see more dye molecules moving from the side of greater dye concentration. Why? Simply because there are more dye molecules at that end of the tank. Since there are more dye molecules on the left, more dye molecules will move randomly to the right, even though there is an equal probability that any one molecule of dye will move from right to left. Consequently, the overall (net) movement of dye

3–4
(a) *A diagram of the diffusion process. Diffusion is the result of the random movement of individual molecules, which produces a net movement from a more concentrated area to a less concentrated area with respect to that particular type of molecule. Notice that as one type of molecule (indicated by color) diffuses to the right, the other diffuses in the opposite direction. The result will be an even distribution of both types of molecules. Can you see why the net movement of molecules will slow down as equilibrium (even distribution) is reached?* (b) *Graphs showing the concentration gradients of dye and water.*

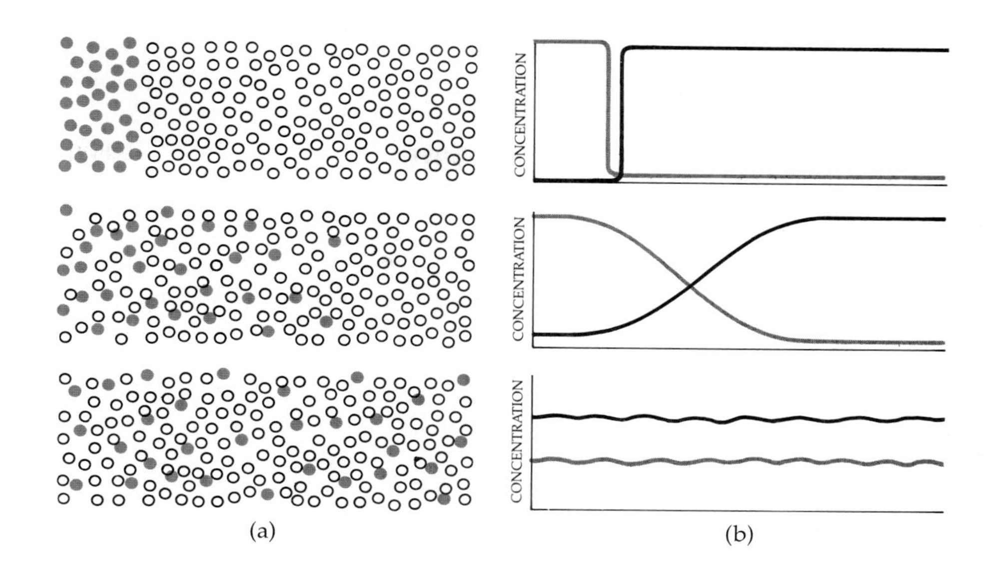

molecules will be from left to right. Similarly, if you could see the movement of the individual water molecules in the tank, you would see that net movement is from right to left.

What happens when all the molecules are evenly distributed throughout the tank? The even distribution does not affect the behavior of the molecules as individuals; they still move at random. But there are now as many molecules of dye and as many molecules of water on one side of the tank as on the other, so there is no net direction of motion. There is, however, just as much individual motion as before, provided the temperature has not changed.

Substances that are moving from a region of higher concentration to a region of lower concentration are said to be moving *along a gradient*. Diffusion only occurs along a gradient. A substance moving in the opposite direction, toward a higher concentration of its own molecules, would be moving *against a gradient*, which is analogous to being pushed uphill. The steeper the downhill gradient—that is, the larger the difference in concentration—the more rapid the net flow. Also, diffusion is more rapid in gases than in liquids and is more rapid at higher than at lower temperatures. Can you explain why?

Notice that there are two gradients in our imaginary tank; the dye molecules are moving along one of them, and the water molecules are moving in the opposite direction along the other. In both cases, the movement is along a gradient. When the molecules have reached a state of equal distribution, that is, when there are no more gradients, they continue to move, but there is no net movement in either direction. In other words, the net transfer of the molecules is zero; the system may be said to be in a state of *equilibrium*.

The concept of water potential is also useful in understanding diffusion. A high concentration of solute (dissolved substance) in one region, such as one corner of the tank, means low water concentration there and, thus, low water potential. If the pressure is equal everywhere, water molecules, as they move along the gradient, are moving from a region of higher water potential to a region of lower water potential. The region of the tank in which there is pure water has a greater water potential than the region containing water plus some dissolved substance. When equilibrium is reached, water potential is equal in all parts of the tank.

The essential characteristics of diffusion are: (1) each molecule moves independently of the others, and (2) these movements are random. The net result of diffusion is that the diffusing substance eventually becomes evenly distributed. Briefly, *diffusion* may be defined as *the dispersion of substances by a movement of their ions or molecules, which tends to equalize their concentrations throughout the system.*

Cells and Diffusion

Diffusion is essentially a slow process, except over very short distances. It is efficient only if the concentration gradient is steep and the volume is relatively small. For instance, the rapid spread of a perfume through the air is not due primarily to diffusion but, rather, to the circulation of air currents. Similarly, in many cells, the transport of materials is speeded by active streaming of the cytoplasm. Cells also hasten diffusion by their own metabolic activities. For example, oxygen is used up within the cell almost as rapidly as it enters, thereby maintaining a steep oxygen gradient from outside to inside. Carbon dioxide is produced by the cell, and so a gradient from inside to outside is maintained for carbon dioxide.

Similarly, within a cell, materials are often produced in one place and used in another. Thus, a concentration gradient can be established between two areas, and materials can diffuse along the gradient from the site of production to the area of use.

Most organic molecules are polar (hydrophilic) and so cannot freely diffuse through the lipid barrier of cellular membranes. However, carbon dioxide and oxygen, which are soluble in lipids, move freely through these same membranes. Water also moves in and out freely. Since water is not soluble in lipids, the fact that water moves so freely has led biologists to postulate the presence of pores in the membrane that permit the passage of water molecules (and some small ions).

Osmosis

While permitting the passage of water, cellular membranes block the passage of most dissolved substances. (These substances are the *solutes,* and the water is the *solvent* of the solution.) Such a membrane is known as a *differentially permeable membrane,* and the diffusion of water through it is known as *osmosis.* Osmosis involves a net flow of water from a solution that has a high water potential to a solution that has a low water potential. In the absence of other factors that influence water potential (such as pressure), the movement or diffusion of the water by osmosis will be from a region of lower solute concentration (and therefore of higher water concentration) into a region of higher solute concentration (and lower water concentration). The presence of solute decreases the water potential and so creates a gradient of water potential along which water moves.

Osmosis can result in a buildup of pressure as water molecules continue to diffuse across the membrane into regions of lower water concentration. As shown in Figure 3–5, if water is separated from a solution by a semipermeable membrane (a membrane that allows the ready passage of water but no passage of solute), water will move across the membrane and cause the solution to rise in the tube until equilibrium is reached—that is, until the water potential on both sides of the membrane is equal. If enough pressure is applied from the upper part of the tube, it is possible to prevent the movement of water into it. The pressure that must be applied to the solution to stop the movement of water is called the *osmotic pressure.* (Increasingly, botanists working with plant-water relationships use the term "osmotic potential" or "solute potential" rather than "osmotic pressure." The terms are numerically equal, but osmotic potential is negative.) The term "osmotic pressure" is also used to express the reduction of water potential caused by the solute. An increase in solute concentration increases the osmotic pressure and decreases the water potential of a solution.

The movement of water is affected not by what is dissolved in the water but only by how much solute it contains—the number of particles of solute (molecules or ions). The word *isotonic* was coined to

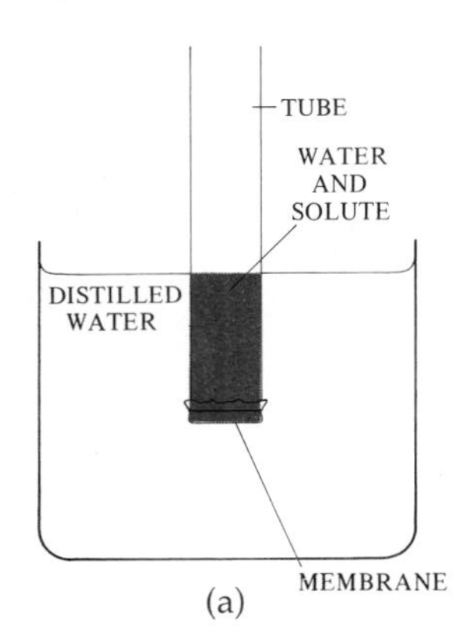

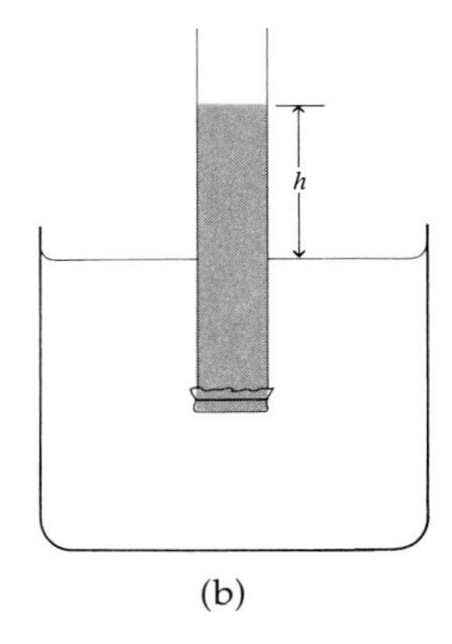

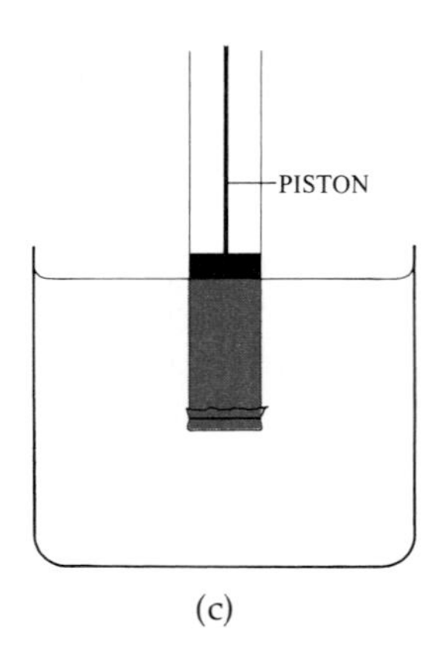

3–5
Osmosis and osmotic pressure. (a) *The tube contains a solution; the beaker contains distilled water.* (b) *The semipermeable membrane permits the passage of water but not solute. The movement of water into the solution causes the solution to rise in the tube until the osmotic pressure, resulting from the tendency of water to move into a region of lower water concentration, is counterbalanced by the height,* h, *and density of the column of solution.* (c) *The force that must be applied to the piston to oppose the rise of the solution in the tube is a measure of the osmotic pressure. It is proportional to the height and density of the solution in the tube.*

describe solutions that have equal numbers of dissolved particles and therefore exert equal osmotic pressures. There is no net movement of water across a membrane separating two solutions that are isotonic to one another, unless, of course, physical pressure is exerted on one side. In comparing solutions with different concentrations, the solution that contains less solute and therefore exerts a lower osmotic pressure is known as *hypotonic*, and the one that has more solute and exerts a higher osmotic pressure is known as *hypertonic*. (Note that *iso-* means "the same"; *hyper-* means "more"—in this case, more molecules of solute; and *hypo-* means "less"—in this case, fewer molecules of solute.)

Since solutes decrease water potential, a hypotonic solution has a higher water potential than a hypertonic solution. *In osmosis, water molecules move through a differentially permeable membrane into a hypertonic solution until the water potential is equal on both sides of the membrane.*

Osmosis and Living Organisms

The movement of water across the plasma membrane from a hypotonic to a hypertonic solution causes some crucial problems for living systems, particularly those in an aqueous environment. These problems vary according to whether the cell or organism is hypotonic, isotonic, or hypertonic in relation to its environment. For example, one-celled organisms that live in salt water are usually isotonic with the medium they inhabit, which is one way of solving the problem. Similarly, the cells of higher animals are isotonic with the blood and lymph that constitute the watery medium in which they exist.

Many types of cells live in a hypotonic environment. In all single-celled organisms that live in fresh water, such as *Euglena*, the interior of the cell is hypertonic relative to the surrounding water; consequently, water tends to move into the cell by osmosis. If too much water moves into the cell, it can dilute the cell contents to the point of interfering with function and can eventually even rupture the plasma membrane. This is prevented by a specialized organelle known as a contractile vacuole, which collects water from various parts of the cell body and pumps it out with a rhythmic contraction.

Turgor

If a plant cell is placed in a hypotonic solution, the protoplast expands and the plasma membrane stretches and exerts pressure against the cell wall. However, the plant cell will not rupture because it is restrained by the relatively tough cell wall.

Plant cells tend to concentrate relatively strong solutions of salts within their vacuoles, and they can also accumulate sugars, organic acids, and amino acids. As a result, plant cells constantly absorb water by osmosis and build up their inner hydrostatic pressure. This pressure against the cell wall keeps the cell stiff, or *turgid*. Consequently, the hydrostatic pressure in plant cells is commonly referred to as turgor pressure. *Turgor pressure* may be defined as *the pressure that develops in a plant cell as a result of osmosis (and/or imbibition)*. Equal to and opposing the turgor pressure is the inwardly directed mechanical pressure of the cell wall, called the *wall pressure*.

Turgor in the plant is especially important in the support of nonwoody plant parts. As discussed in Chapter 1, most of the growth of a plant cell is the direct result of water uptake. The hormone auxin presumably contributes to this water uptake by relaxing the cell wall and thus decreasing the resistance the wall exerts to turgor pressure.

Turgor is maintained by most plant cells because they generally exist in a hypotonic medium. However, if a turgid plant cell is placed in a hypertonic solution, water will leave the cell by osmosis, and the vacuole and protoplast will shrink, thus causing the plasma membrane to pull away from the cell wall (Figure 3–6). This phenomenon is known as *plasmolysis*. The process can be reversed if the cell is then transferred to pure water. Figure 3–7 shows *Elodea* leaf cells before and after plasmolysis. Al-

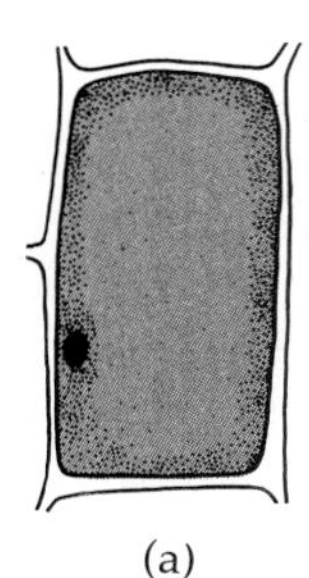

(a)

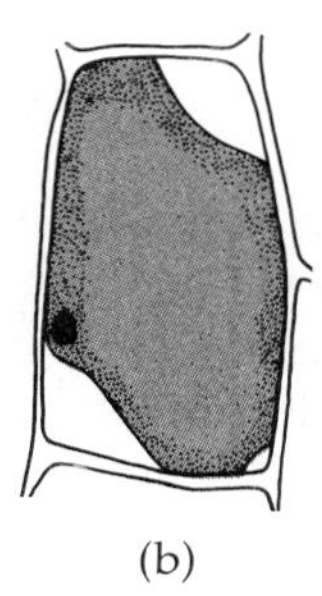

(b)

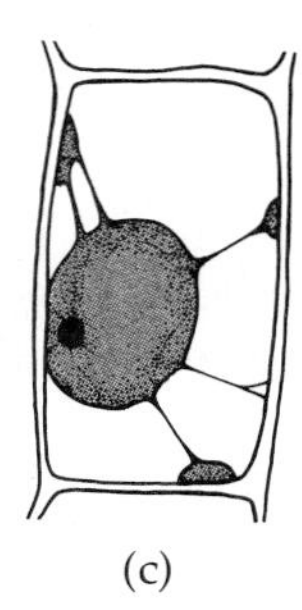

(c)

3–6
Plasmolysis in a leaf epidermal cell. (a) *Under normal conditions, the protoplasm fills the space within the cell walls.* (b) *When the cell is placed in a relatively strong sucrose solution, water passes out of the cell into the hypertonic medium and the plasma membrane contracts slightly.* (c) *When immersed in a stronger (more concentrated) solution, the cell loses even larger amounts of water and contracts still further.*

3–7
Elodea *leaf cells.* (a) *Turgid cells and* (b) *cells after being placed in a relatively strong sucrose solution. The cells in* (b) *are plasmolyzed.*

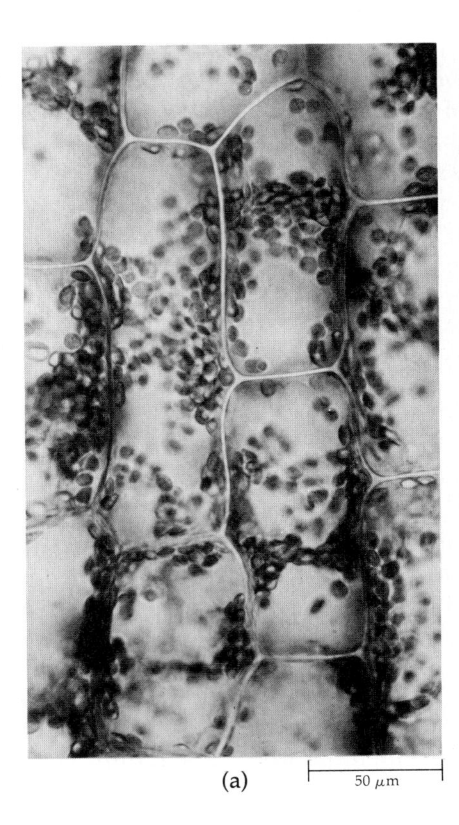

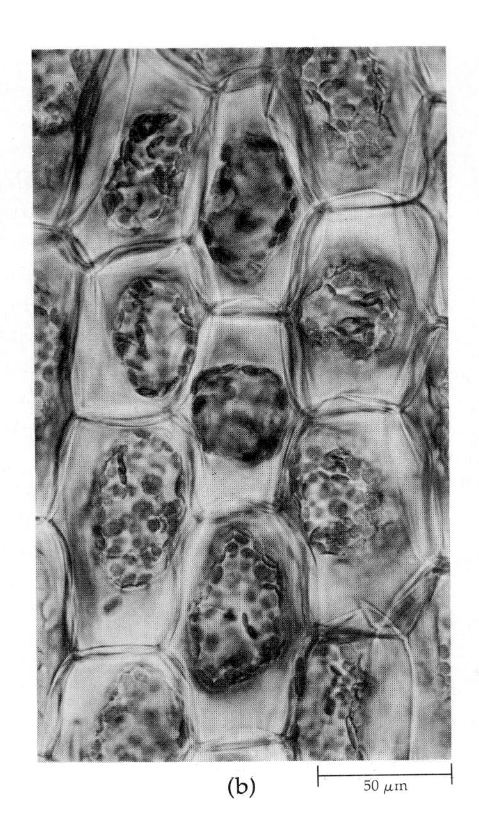

though the plasma membrane and the tonoplast—the membrane surrounding the vacuole—are permeable only to water, the cell walls allow both solutes and water to pass freely through them. The loss of turgor by plant cells may result in *wilting,* or drooping of the leaves and stems.

STRUCTURE OF CELLULAR MEMBRANES

The model of membrane structure most widely accepted today is the *fluid mosaic model.* As shown in Figure 3–8, the membrane is composed of a phospholipid bilayer in which large globular proteins are suspended. The proteins are not fixed in position but float more or less freely in and on the phospholipid bilayer, creating a mosaic of molecules. Some of the proteins span the bilayer and protrude from both surfaces; others are embedded in one or the other surface, so that the two surfaces may differ considerably in chemical composition. The portion of the protein embedded in the bilayer is hydrophobic, while that exposed on the outside is hydrophilic. The proteins that are more hydrophilic float nearer the surface of the lipid bilayer.

Most membranes are composed of 40 to 50 percent lipid (by weight) and 50 to 60 percent protein. Most membranes also contain short-chain carbohydrates attached either to the lipids or to the proteins on the outer surfaces of the membrane. In animal cells, these carbohydrates are believed to play important roles in immunological responses, cell-to-cell adhesion processes, and cell-surface transformations. Relatively little is known about the carbohydrate components of plant cellular membranes.

TRANSPORT ACROSS MEMBRANES

Molecules may move across membranes by three different processes: simple diffusion, facilitated diffusion, and active transport. Diffusion is the spontaneous and independent movement of a specific substance along a concentration gradient. Nonpolar (hydrophobic) substances that are soluble in phospholipids usually cross a membrane by simple diffusion. (The observation that hydrophobic molecules diffused readily across plasma membranes provided the first evidence of the lipid nature of the membrane.)

Water and other polar (hydrophilic) molecules and ions might be expected to be excluded by a membrane's phospholipid bilayer, and yet hydrophilic molecules and ions do cross through the membrane. How is this accomplished? In the case of water and certain other small polar molecules, there is evidence for the presence of small hydrophilic channels or pores in the membrane through which such small

IMBIBITION

Water molecules exhibit a tremendous cohesiveness because of the difference in charge between one end of a water molecule and the other (see Appendix A). Similarly, because of this difference in charge, water molecules can cling either to positively charged or to negatively charged surfaces. Many large biological molecules, such as cellulose, tend to develop charges when they are wet and so attract water molecules. The adherence of water molecules is responsible for another biologically important phenomenon called imbibition or, sometimes, hydration.

Imbibition ("drinking up") is the movement of water molecules into substances such as wood or gelatin, which swell or increase in volume as a result of adhesion to the water molecules. The pressures developed by imbibition can be astonishingly large. It is said that stone for the ancient Egyptian pyramids was quarried by driving wooden pegs into holes drilled in the rock face and then soaking the pegs with water. The swelling wood created a force that split the slab of stone. In living plants, imbibition occurs particularly in seeds, which may increase to many times their original size as a result.

(a)

(b)

(a) Milk bottle filled with marble chips and field corn and placed in water overnight. (b) By morning the bottle was broken. Can you explain why?

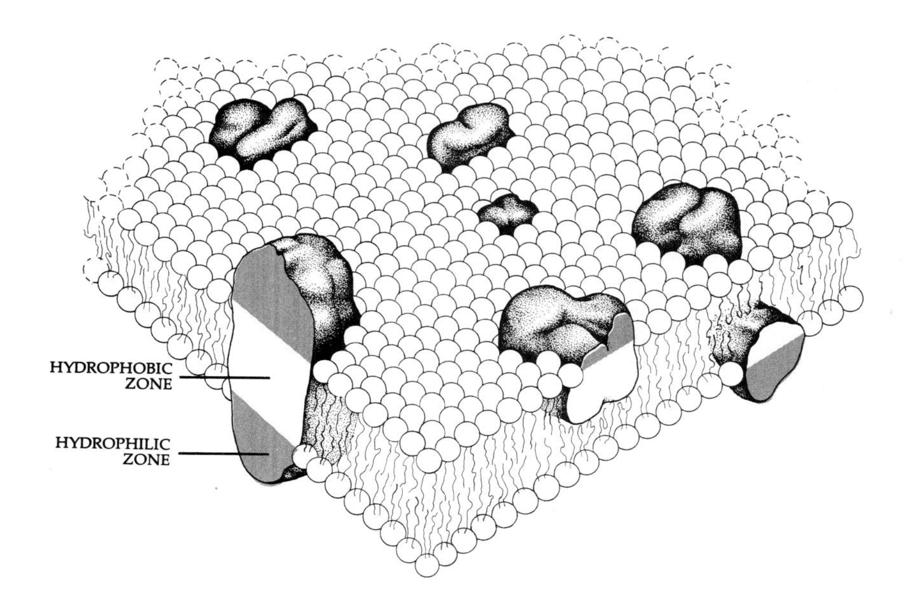

3–8
Fluid mosaic model of membrane structure. The membrane is composed of a double layer (bilayer) of phospholipid molecules—with their hydrophobic "tails" facing inward—and large globular protein molecules. The portion of the protein molecule embedded in the lipid bilayer is hydrophobic; the portion exposed on the outside is hydrophilic. The whole structure is quite fluid and, hence, the proteins can be thought of as floating in a lipid "sea."

molecules can diffuse. These may be either organized pores or momentary openings resulting from movements of molecules forming the membrane. The permeability of the membrane to small polar molecules varies inversely with the size of the molecules; this indicates that the apertures are small and that the membrane acts like a sieve. It has been estimated that the pores have a minimum diameter of 0.8 to 1.0 nanometer. Although the pores are small, their diameters are greater than the diameter of a water molecule (which is less than 0.3 nanometer)—the smallest polar molecule of biological importance.

Most substances required by cells are polar in nature and require *carrier molecules* (transport proteins) that are embedded within the membrane to ferry them across (Figure 3–9).

Two basically different categories of carrier-assisted transport have been proposed: facilitated diffusion and active transport. *Facilitated diffusion* is driven by the concentration gradient and so moves molecules from an area of higher concentration to one of lower concentration. Neither simple diffusion nor facilitated diffusion is capable of moving solutes *against* a concentration gradient. The ability to move solutes against a concentration gradient requires energy, and this process is called *active transport.*

Carrier-assisted transport, whether by facilitated diffusion or active transport, is highly selective; the transport protein may accept one molecule and exclude a nearly identical one. The carrier molecule is not permanently altered in the process. In this respect, carrier molecules are like enzymes; to emphasize this aspect of their function, transport proteins have been named *permeases*. However, they are unlike enzymes in the fact that they do not necessarily produce chemical change in the molecules with which they are temporarily bound.

Glucose is an example of a molecule whose entry into cells takes place by facilitated diffusion. There are several lines of evidence that support this conclusion. First, because glucose is rapidly broken down when it enters a cell, a steep concentration gradient is maintained across the plasma membrane. Second, if large numbers of glucose molecules are present outside the cell, the rate of entry into the cell does not increase as it would if glucose were moving by simple diffusion; instead, it reaches a peak and then levels off. This limitation on the rate of entry is attributed to the limited number of carriers. Third, some molecules that are shaped like glucose compete with glucose for transport across the membrane. Thus, although the exact carrier molecule for glucose has not been isolated or identified, it is generally agreed that it exists.

The Sodium-Potassium Pump

One of the most important and best understood active transport systems is the sodium-potassium pump. (Mechanisms that perform active transport are commonly called pumps.) Most cells maintain a differential concentration gradient of sodium ions (Na^+) and potassium ions (K^+) across the plasma membrane: Na^+ is maintained at a low concentration inside the cell, and K^+ is kept at a high concentration. High concentrations of K^+ and low concentrations of Na^+ are necessary for the cell to function properly. For example, potassium is required for a number of vital cellular processes, such as the activation of certain enzymes that are inhibited by sodium. The sodium-potassium pump apparently is important in salt tolerance in halophytes—plants that grow in saline (salty) environments. Sodium is the dominant cation of saline soil, and the sodium-potassium pump provides a means of keeping the concentration of sodium

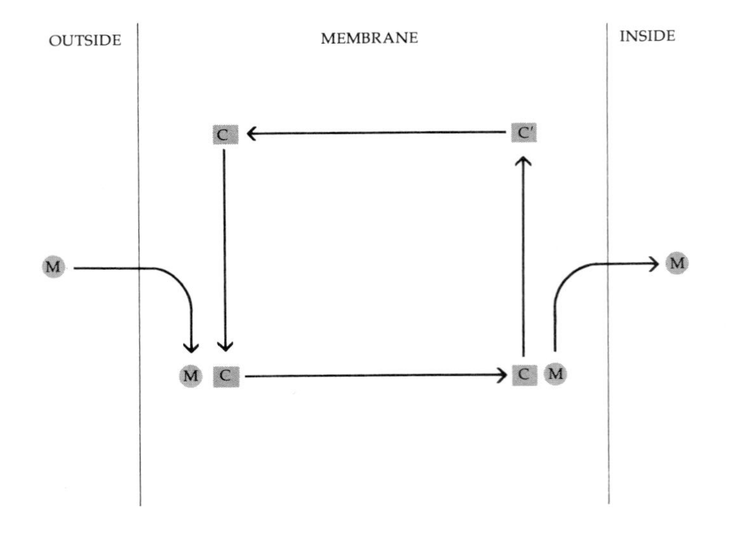

3–9
The carrier hypothesis. M *represents the penetrating molecule,* C *the carrier,* MC *the mobile complex, and* C′ *the mobile carrier precursor. The complex* MC *cannot leave the membrane but can move about within it. When the complex reaches the inside of the membrane, it is broken down, releasing the molecule and forming the carrier precursor* C′, *which is left behind. The precursor moves back across the membrane, where it is reconverted to the carrier* C. *The carrier is then ready to accept another molecule at the surface of the membrane.*

in the cytoplasm as low as possible. Moreover, in such cells, some of the sodium may be pumped across the tonoplast and into the vacuole where it is not only harmless but may help in maintaining the turgor pressure of the cells under salt stress. The energy needed to drive the sodium-potassium pump is provided by ATP manufactured in respiration. A measure of the importance of this mechanism to an organism is that more than a third of the ATP used by a resting animal is consumed by the sodium-potassium pump. Unfortunately, similar data have not been obtained for plants, but there is little doubt of the importance of the sodium-potassium pump.

The transport of Na^+ and K^+ ions is accomplished by a transport protein that is believed by some investigators to exist in two alternative forms. One has a cavity opening to the inside of the cell, into which a Na^+ ion can fit; the other has a cavity opening to the outside, into which a K^+ ion fits. As shown in Figure 3–10, Na^+ binds to the transport protein. ATP is broken down to ADP, and the released phosphate is attached to the protein (the protein is phosphorylated). This causes a change in the conformation (form) of the protein such that the bound Na^+ is brought to the outer surface of the membrane and released. The transport protein then picks up a K^+ ion, which results in a dephosphorylation of the protein, thus causing it to return to the first conformation and release the K^+ ion to the inside of the cell. This process generates a gradient of Na^+ and K^+ ions across the membrane.

Many ingenious models have been proposed to show how carrier molecules might accept and eject their passengers. One of the first models suggested that the carrier rotated, like a revolving door. A more recent model hypothesizes that the molecule has a hydrophilic core, through which the transported molecule is squeezed, propelled by changes in the conformation of the protein. It is likely that cellular membranes contain a large variety of transport proteins and that they employ a variety of different techniques for carrying out the transport process.

ENDOCYTOSIS AND EXOCYTOSIS

In *endocytosis,* materials are taken into cells by means of invagination of the plasma membrane. The invaginations form small, saclike structures that are pinched off the plasma membrane and carried, with their enclosed material, into the cytoplasm.

When the substance to be taken in is a solid, such as a bacterial cell, the process is usually called *phagocytosis,* from the Greek word *phagein,* "to eat." Many one-celled organisms, such as amoebas, feed in this way. Among the organisms discussed in this book, the true slime molds and cellular slime molds exhibit phagocytosis (Figure 3–11).

The taking in of dissolved molecules, as distinct from solid particles, is sometimes given the special name of *pinocytosis,* although it is, in principle, the same process as phagocytosis. Pinocytosis occurs not only in single-celled organisms but also in multicellular plants and animals.

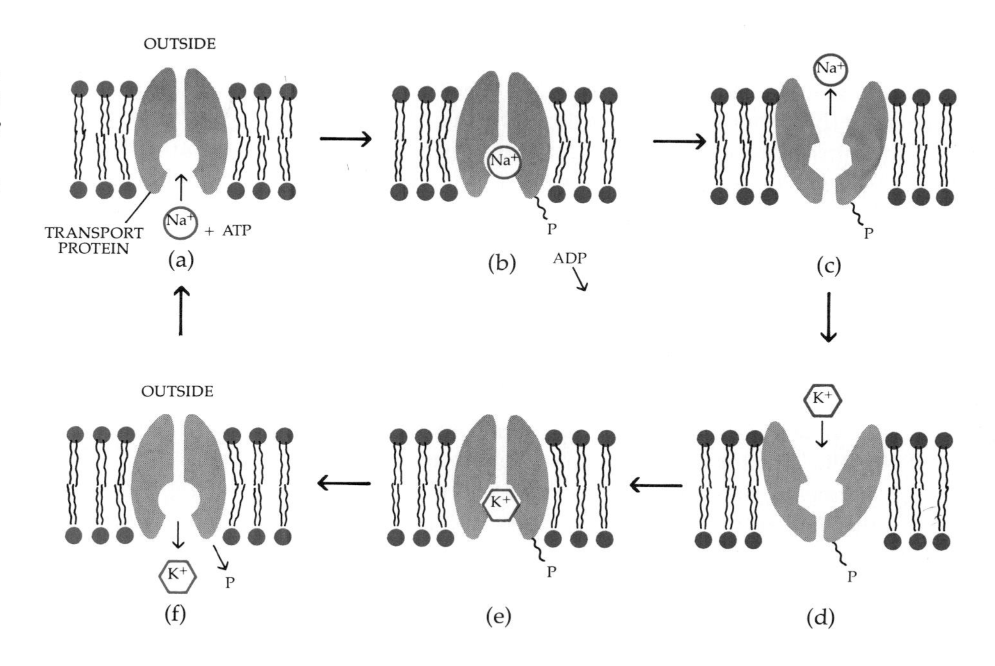

3–10
A model of the sodium-potassium pump. (a) Na^+ *is bound to the transport protein;* (b) *the protein is phosphorylated;* (c) *the protein changes form, and the* Na^+ *is released to the cell's exterior;* (d) *and* (e) K^+ *is bound to the transport protein, which in this form provides a better fit for* K^+ *than for* Na^+; (f) *the protein is dephosphorylated, inducing conversion back to the original form, and* K^+ *is released to the cell's interior.*

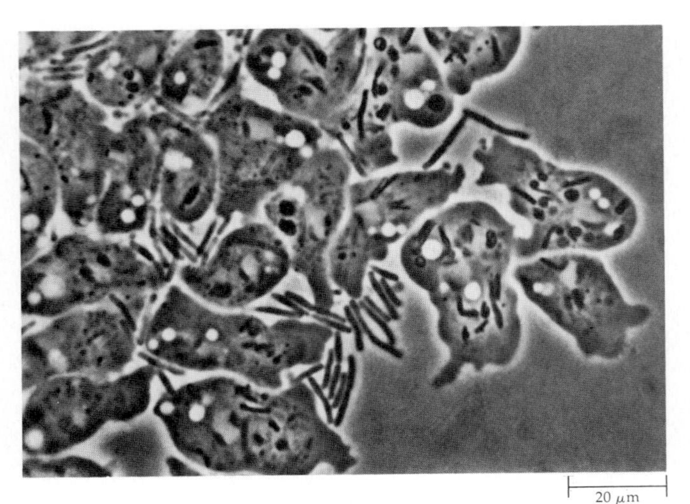

3–11
The amoeboid feeding stage of the cellular slime mold Dictyostelium aureum. *These amoebas have been feeding upon the rod-shaped bacterium* Escherichia coli *by phagocytosis. Notice the whole bacterial cells both inside and outside the amoebas.*

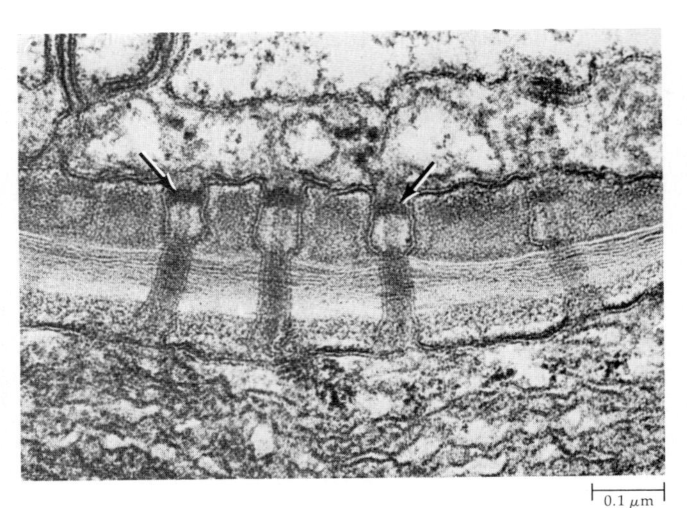

3–12
Plasmodesmata traversing the common wall between a bundle-sheath cell (below) and a mesophyll cell (above) in a corn (Zea mays) *leaf. The arrows indicate the electron-dense structures that may serve as "valves" in the plasmodesmata on the mesophyll-cell side of the wall.*

Phagocytosis and pinocytosis can also work in reverse. Many substances are exported from cells in vesicles or special vacuoles. One example cited in Chapter 1 is the role of Golgi vesicles in cell-wall formation. The vesicles, with their enclosed cell-wall precursors, move to the surface of the cell. When they reach the plasma membrane, their bounding membranes fuse with it, and their contents are expelled into the region of the developing cell wall. This reverse endocytosis is referred to as *exocytosis.*

Although phagocytosis and pinocytosis appear to be superficially different from membrane-transport systems involving carrier molecules, they are fundamentally the same; all three depend on the capacity of the membrane to "recognize" particular molecules.

TRANSPORT VIA PLASMODESMATA

As described in Chapter 1, neighboring cells of the plant body are interconnected by narrow strands of cytoplasm called plasmodesmata, which provide potential pathways for the passage of substances from cell to cell. The term *symplast* is used to refer to the interconnected protoplasts and their plasmodesmata. The movement of substances from cell to cell by means of the plasmodesmata is called *symplastic transport.*

Plasmodesmata may provide a more efficient pathway between neighboring cells than the less direct, alternative route of plasma membrane, cell wall, and plasma membrane. It is believed that cells and tissues that are far removed from direct sources of nutrients can be supplied with nutrients either by simple diffusion or bulk flow through plasmodesmata. In addition, some substances are believed to move through plasmodesmata to and from the xylem and phloem—the tissues concerned with long-distance transport in the higher plant body. The substances may be transported within the desmotubules, which are continuous with the endoplasmic reticulum of adjacent cells and/or by way of the canals surrounding the desmotubules, provided that the canals are not constricted or blocked. It has not yet been determined whether the plasmodesmata have any control over the movement of substances from cell to cell, although some investigators have found what might be "valves" in some plasmodesmata (Figure 3–12).

SUMMARY

The plasma membrane regulates the passage of materials into and out of the cell, a function that makes it possible for the cell to maintain its structural and functional integrity. This regulation depends on interaction between the membrane and the materials that pass through it.

Water is one of the principal materials passing into and out of cells. Water moves by bulk flow, diffusion, and osmosis. Water potential determines the direction the water moves; that is, the movement of water is from where the water potential is higher to where it is lower. Bulk flow is the overall movement of water molecules, as when water flows downhill or moves in response to pressure.

Diffusion involves the random movement of molecules and results in net movement along a concentration gradient. Carbon dioxide and oxygen are two important molecules that move into and out of cells by diffusion across the membrane. Diffusion is most efficient when the distance involved is small and when the concentration gradient is steep. The rate of movement of substances within cells is increased by cytoplasmic streaming.

Osmosis is the diffusion of water through a membrane that permits the passage of water but inhibits the movement of solutes; such a membrane is called a differentially permeable membrane. In the absence of other forces, the movement of water by osmosis is from a region of lower solute concentration (a hypotonic medium, one of higher water potential) to a region of higher solute concentration (a hypertonic medium, one of lower water potential). The turgor (rigidity) of plant cells is a result of osmosis.

According to the fluid mosaic model of a membrane, cellular membranes are phospholipid bilayers in which globular proteins are suspended. Some of these proteins act as carriers, ferrying molecules through the membrane. If the movement is driven by a concentration gradient, the process is known as facilitated diffusion. If the movement requires chemical energy, it is known as active transport. Active transport can move substances against a concentration gradient. One of the most important active-transport systems is the sodium-potassium pump, which maintains sodium ions at a low concentration and potassium ions at a high concentration in the cytoplasm.

Controlled movement into and out of a cell may also occur by endocytosis or by exocytosis, processes in which substances are transported in vesicles or in vacuoles. Endocytosis of solids is called phagocytosis, whereas endocytosis of dissolved molecules is called pinocytosis.

Movement of substances between plant cells may also occur by way of narrow strands of cytoplasm, the plasmodesmata, which interconnect the protoplasts of neighboring cells.

SUGGESTIONS FOR FURTHER READING

AVERS, CHARLOTTE J.: *Basic Cell Biology*, D. Van Nostrand Co., New York, 1978.

A book written for introductory courses in cell biology; designed for students with a minimal background in biology and chemistry.

BONNER, JAMES, and JOSEPH VARNER (Eds.): *Plant Biochemistry*, 3rd ed., Academic Press, Inc., New York, 1976.

A good modern account of this rapidly expanding field; a collection of articles by leading specialists that includes much structural interpretation.

BROWN, WALTER V., and ELDRIDGE M. BERTKE: *Textbook of Cytology*, 2nd ed., C. V. Mosby Co., St. Louis, 1974.

A well-illustrated account of all aspects of cellular structure but emphasizing chromosomes.

CRAM, JANE M., and DONALD J. CRAM: *The Essence of Organic Chemistry*, Addison-Wesley Publishing Co., Reading, Mass., 1978.

An elementary textbook emphasizing the features of organic chemistry most relevant to life processes.

GIESE, ARTHUR C.: *Cell Physiology*, 5th ed., W. B. Saunders Co., Philadelphia, 1979.

A description of the major problems in cell physiology; intended for the more advanced student.

GUNNING, B. E., and A. W. ROBARDS (Eds.): *Intercellular Communications in Plants: Studies on Plasmodesmata*, Springer-Verlag, New York, 1976.

An outstanding series of reviews of plasmodesmatal structure, distribution, and probable function by leading specialists in the field.

GUNNING, B. E., and M. W. STEER: *Ultrastructure and the Biology of Plant Cells*, Crane, Russak & Co., Inc., New York, 1975.

A collection of plant electron micrographs and accompanying explanations.

HALL, J. L., and D. A. BAKER: *Cell Membranes and Ion Transport*, Longman, Inc., New York, 1978.*

An introductory text that integrates information and concepts emerging from recent research on membranes of bacterial, plant, and animal cells.

*Available in paperback.

LEDBETTER, MYRON C., and KEITH R. PORTER: *Introduction to the Fine Structure of Plant Cells*, Springer-Verlag, New York, 1970.

A collection of plant electron micrographs and accompanying explanations.

LEHNINGER, ALBERT L.: *Biochemistry*, 2nd ed., Worth Publishers, Inc., New York, 1975.

The best biologically oriented biochemistry text available.

LOEWY, A. G., and P. SIEKEVITZ: *Cell Structure and Function*, 2nd ed., Holt, Rinehart and Winston, Inc., New York, 1970.*

An outstanding elementary text on cell structure and function; now somewhat out-of-date but still very useful.

McELROY, WILLIAM D.: *Cell Physiology and Biochemistry*, 3rd ed., Prentice-Hall, Inc., Englewood Cliffs, N.J., 1971.*

A brief introduction to the basic principles of cell physiology and biochemistry.

O'BRIEN, T. P., and MARGARET E. McCULLY: *Plant Structure and Development: A Pictorial and Physiological Approach*, The Macmillan Company, London, 1969.*

A beautifully illustrated atlas of cellular structure and development in the higher plants.

ROBARDS, A. W. (Ed.): *Dynamic Aspects of Plant Ultrastructure*, McGraw-Hill Book Company, New York, 1974.

An outstanding series of reviews of the cellular components discussed in this section and of the specialized conducting cells considered in Section 5, by leading specialists in each field.

THOMAS, LEWIS: *The Lives of a Cell: Notes of a Biology Watcher*, Bantam Books, Inc., New York, 1975.*

Thomas, a physician and medical researcher, reveals the extent to which science can tune our intellectual antennae, broaden our perceptions, and extend our appreciation of ourselves and of the world around us.

TROUGHTON, JOHN H., and F. B. SAMPSON: *Plants: A Scanning Electron Microscope Survey*, John Wiley & Sons, Inc., New York, 1973.*

A scanning electron microscope study of some anatomical features of plants and the relationship of these features to physiological processes.

WHITE, EMIL H.: *Chemical Background for the Biological Sciences*, 2nd ed., Prentice-Hall, Inc., Englewood Cliffs, N.J., 1970.*

A brief account of the principles of chemistry that are essential for an understanding of biological processes.

*Available in paperback.

SECTION 2 Energy and the Living Cell

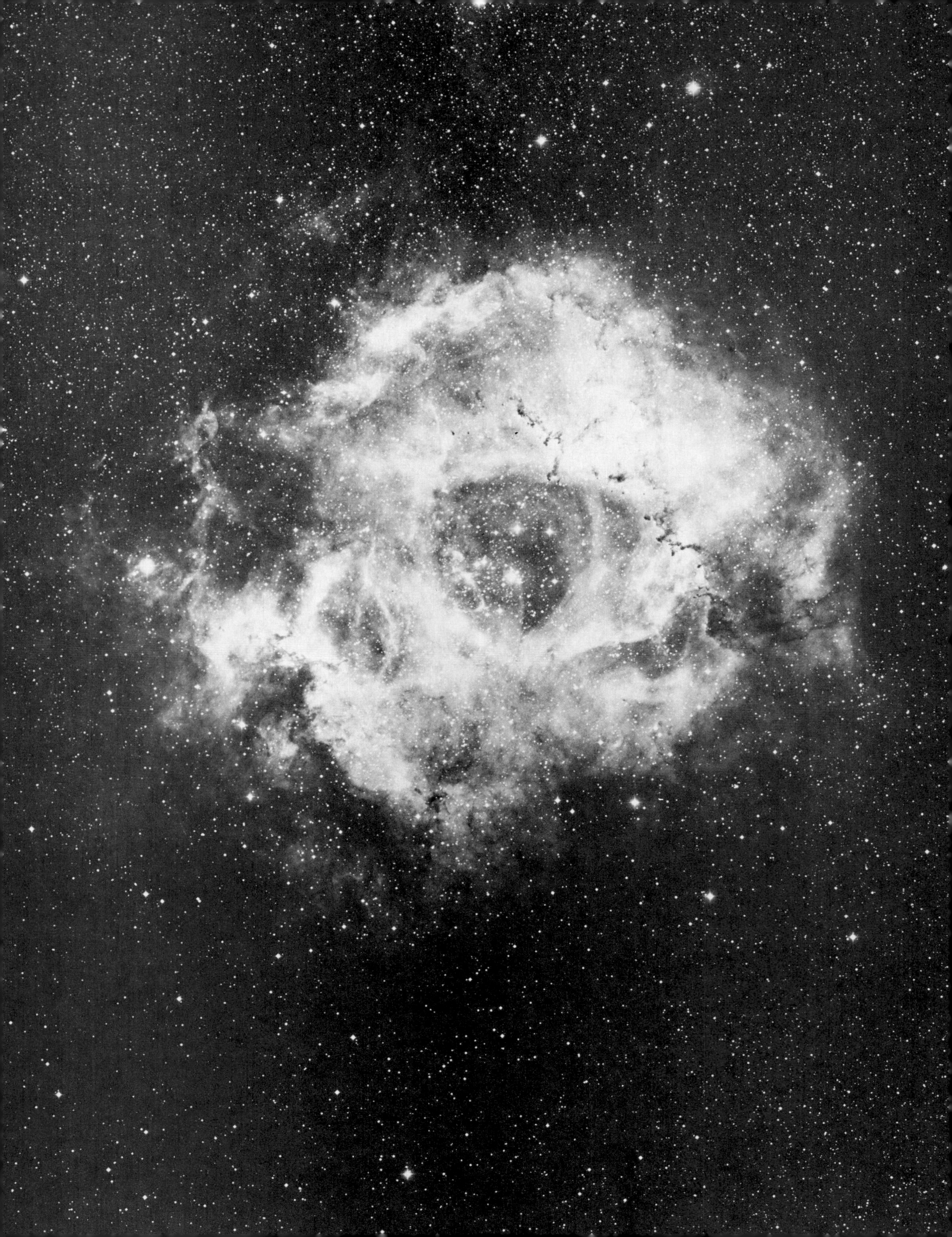

CHAPTER 4

The Flow of Energy

4–1
The birth of a star. Our own sun, like all stars, began as a swirling cloud of hydrogen gas in interstellar space.

Life here on earth depends on the flow of energy from thermonuclear reactions taking place at the heart of the sun. The amount of radiant energy delivered by the sun to the earth is 13×10^{23} (the number 13 followed by 23 zeros) calories per year. This is certainly a difficult quantity to imagine. For example, the amount of energy striking the earth every day is the equivalent of about 1 million Hiroshima-sized atomic bombs.

About a third of this solar energy is immediately reflected back into space as light (just as light is from the moon). Much of the remaining two-thirds is absorbed by the earth and converted to heat. Some of this absorbed heat energy serves to evaporate the waters of the oceans, producing the clouds that, in turn, produce rain and snow. Solar energy, in combination with other factors, is also responsible for the movements of air and water that help set patterns of climate over the surface of the earth.

Less than 1 percent of the solar energy reaching the earth becomes, through a series of operations performed by the cells of plants and other photosynthetic organisms, the energy that drives nearly all the processes of life. Living systems change energy from one form to another, transforming the radiant energy of the sun into the chemical and mechanical energy used by living organisms (see Figure 4–2).

These concepts of a vital relationship between plants and animals, and between energy and life, are relatively recent ones. They form part of the study of *thermodynamics*—the science of energy exchanges. Energy is such a common word in our energy-conscious world that it is surprising to discover that the word itself did not come into existence until about 100 years ago. Let us first consider a few key thermodynamic principles and then explore the manner in which enzymes mediate many of the reactions and processes carried out by cells.

4–2
An example of the flow of biological energy. The radiant energy of sunlight is produced by the fusion reactions taking place in the sun. Chloroplasts, present in all photosynthetic eukaryotic cells, capture this energy and use it to convert water and carbon dioxide into carbohydrates, such as glucose, starch, and other foodstuff molecules. Oxygen is released into the air as a product of these photosynthetic reactions. Mitochondria—organelles present in all eukaryotic cells—break down these carbohydrates and capture their stored energy in ATP molecules. This process—cellular respiration—consumes oxygen and produces carbon dioxide and water, completing the cycle.

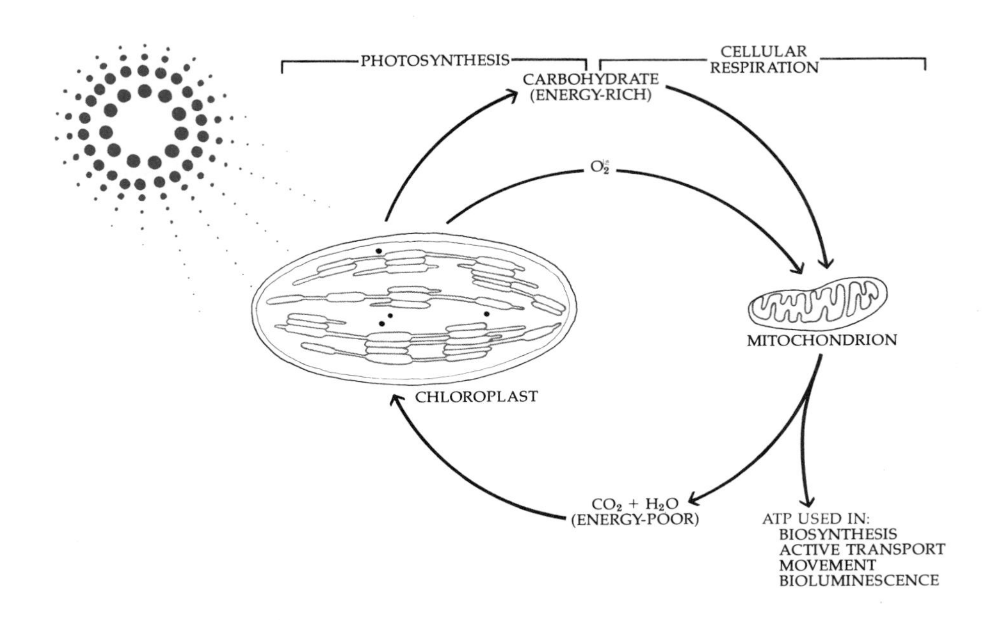

THE LAWS OF THERMODYNAMICS

Energy is an elusive concept. It is usually defined operationally, that is, by what it does rather than what it is. Until less than 200 years ago, heat—the form of energy most readily studied—was even considered to be a separate, though weightless, substance called "caloric." An object was hot or cold depending upon how much caloric it contained; when a cold object was placed next to a hot one, the caloric flowed from the hot object into the cold one; when metal was pounded with a hammer, it became warm because the caloric was forced to the surface. Even though the idea of a caloric substance proved incorrect, the concept is surprisingly useful.

The First Law

The rapid development of the steam engine in the latter part of the eighteenth century, more than any other single chain of events, changed scientific thinking about the nature of energy. Energy came to be associated with work, and heat and motion came to be seen as forms of energy. The way was opened for the formulation of the laws of thermodynamics. The *first law of thermodynamics* states, quite simply: *Energy can be changed from one form to another but cannot be created or destroyed.*

In engines, for instance, chemical energy (such as in coal or gasoline) is converted to heat, which is then partially converted to mechanical movements (kinetic energy). Some of the energy is converted back to heat by the friction of these movements, and some leaves the engine in the form of exhaust products. Unlike the heat in the engine or the boiler, the heat produced by friction and lost in the exhaust cannot produce "work"—that is, it cannot turn the gears—because it is dissipated into the environment. But it is nevertheless part of the total equation. In fact, engineers calculate that most of the energy consumed by an engine is dissipated randomly as heat; most engines work with less than 25 percent efficiency.

The notion of potential energy developed in the course of such engine-efficiency studies. A barrel of oil or a ton of coal could be assigned a certain amount of potential energy, expressed in terms of the amount of heat it would liberate when burned. The efficiency of the conversion of the potential energy to usable energy depended on the design of the energy conversion system.

Although these concepts were formulated in terms of engines running on heat energy, they apply to other systems as well. For example, a boulder pushed to the top of a hill contains energy—potential energy. Given a little push (the energy of activation), it rolls down the hill again, converting the potential energy to energy of motion and to heat produced by friction. Water can also possess potential energy. As it moves by bulk flow from the top of a waterfall or over a dam, it can turn waterwheels that turn gears, for example, to grind corn. Thus the potential energy of water, in this system, is converted to the kinetic energy of the wheels and gears and to heat, which is produced by the movement of the water itself as well as the turning wheels and gears.

Light is another form of energy, as is electricity.

Light can be changed to electrical energy, and electrical energy can be changed to light (say, by letting it flow through the tungsten wire in a light bulb).

The first law of thermodynamics states that in energy exchanges and conversions, wherever they take place and whatever they involve, the energy of the products of the reaction plus the energy released in the reaction is equal to the energy possessed by the initial reactants.

The Second Law

In a biological sense, the *second law of thermodynamics* is the more interesting one. It predicts the direction of all events involving energy exchanges; thus it has been called "time's arrow." One way of stating the second law is: *In all energy exchanges and conversions, if no energy leaves or enters the system under study, the potential energy of the final state will always be less than the potential energy of the initial state.* The second law is in keeping with everyday experience (Figure 4–3). Of its own accord a boulder will roll downhill but never uphill. Heat will flow from a hot object to a cold one and never the other way. Our cells can process glucose enzymatically to yield carbon dioxide and water but—since we cannot capture the energy of the sun as plants do—our enzymatic processes cannot produce glucose from carbon dioxide and water.

A process in which the potential energy of the final state is less than that of the initial state is one that yields energy (otherwise it would be in violation of the first law). An energy-yielding process is called an *exergonic reaction.* As the second law predicts, only exergonic reactions can take place spontaneously. (The word "spontaneously" says nothing about the rate of reaction but only whether or not it can occur.) In contrast, *endergonic reactions* are energy-requiring reactions, and in order for them to proceed, an input of energy is required that is greater than the difference in energy between product and reactants.

One important factor in determining whether or not a reaction is exergonic is ΔH, the change in heat content of the system (Δ stands for change, H for heat content). In general, the change in heat content is approximately equal to the change in potential energy. As noted in Appendix A, the energy change that occurs when glucose, for instance, is oxidized can be measured in a calorimeter and expressed in terms of ΔH. The oxidation of a mole of glucose yields 673 kilocalories.

$$C_6H_{12}O_6 + 6O_2 \longrightarrow 6CO_2 + 6H_2O$$

$$\Delta H = -673 \text{ kcal/mole}$$

Generally, a chemical reaction with a negative ΔH is an exergonic reaction.

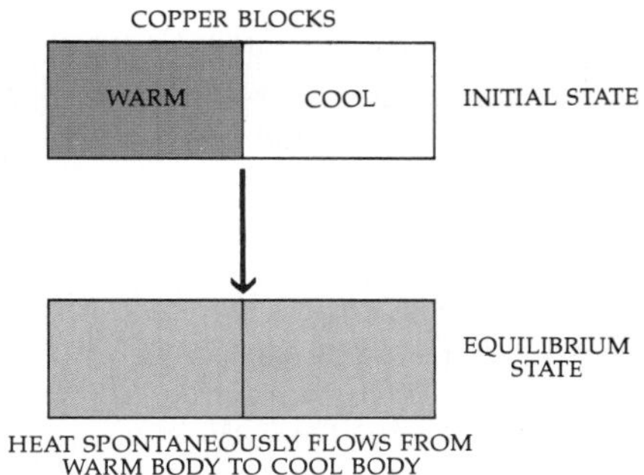

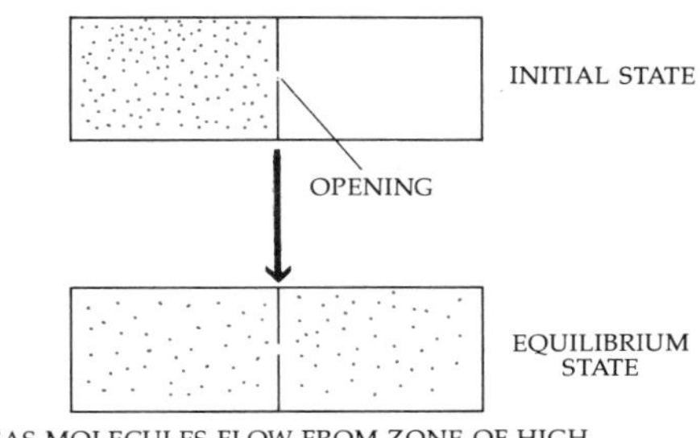

4–3
Illustrations of the second law of thermodynamics, showing the increase of disorder, or randomness, in two physical systems. Such processes do not reverse spontaneously.

Another factor besides the gain or loss of heat can determine the direction of a process. This factor is called *entropy* and is a measurement of the disorder, or randomness, of a system. Let us return to water as an example. The change from ice to liquid water and the change from liquid water to water vapor are both endothermic processes—a considerable amount of heat is removed from the surroundings as they occur. Yet, under the appropriate conditions, they proceed spontaneously. The key factor in these processes is the increase in entropy. In the case of the change of ice into liquid water, a solid is being turned into a liquid, and some of the bonds that hold the water molecules together in a crystal (ice) are being broken. As the liquid water turns to vapor, the rest of the hydrogen bonds are ruptured as the individual water molecules separate, one by one. In each case, the disorder of the system has increased.

The notion that there is more disorder associated with more numerous and smaller objects than with fewer, larger ones is in keeping with our everyday experience. If there are 20 papers on a desk, the possibilities for disorder are greater than if there are 2 or even 10. If each of the 20 papers is cut in half, the

$E = mc^2$

Protons and neutrons are arbitrarily assigned an atomic weight of 1 (see Appendix A). One would therefore expect that an element with, for example, twice as many protons and neutrons as another element would weigh twice as much. This assumption is true—almost. If the weights of nuclei are measured with great accuracy, as they can be by instruments developed by modern physics, small nuclei always have, proportionately, slightly greater weights than larger nuclei. For example, the most common isotope of carbon has a combined total of 12 protons and neutrons; and so carbon, by convention, is assigned an atomic weight of 12. The hydrogen atom, however, has an atomic weight not of 1, as would be expected, but of 1.008. Helium has two protons and two neutrons. However, it does not have a weight of 4, or of 4.032 (four times the weight of hydrogen); it weighs, in relation to carbon, 4.0026. Similarly, oxygen, with a combined total of 16 protons and neutrons, has an atomic weight, in relation to that of carbon, of 15.995. In short, when protons and neutrons are assembled into an atomic nucleus, there are slight changes in weight which reflect changes in mass.

One of the oldest and most fundamental concepts of chemistry is the law of conservation of mass—that mass is never created or destroyed. Yet under conditions of extremely high temperature, atomic nuclei fuse to make new elements. What happens to the mass "lost" in the course of this fusion? This is the question answered by Einstein's fateful equation $E = mc^2$*, where* E *stands for energy,* m *stands for mass, and* c *is a constant equal to the speed of light. Einstein's equation simply means that under certain extreme and unusual conditions, mass is turned into energy.*

Albert Einstein in 1905, the year he published his paper on the theory of relativity. He was 26 years old and working at the Swiss Patent Office in Bern as a technical expert third class.

The sun consists largely of hydrogen nuclei. At the extremely high temperatures at the core of the sun, hydrogen nuclei strike each other with enough velocity to fuse. In a series of steps, four hydrogen nuclei fuse to form one helium nucleus. In the course of these reactions, enough energy is released to keep the fusion reaction going and to emit tremendous amounts of radiant energy into space. Life on this planet depends on energy emitted by the sun in the course of this reaction. This same reaction provides—as Einstein foresaw—the energy of the hydrogen bomb.

entropy of the system—the capacity for randomness—increases. The relationship between entropy and energy is also a commonplace idea. If you were to find your room tidied up and your books in alphabetical order on the shelf, you would recognize that someone had been at work—that energy had been expended. To organize the papers on a desk similarly requires the expenditure of energy.

Now let us return to the question of the energy changes that determine the course of chemical reactions. As discussed, both the change in the heat content of the system (ΔH) and the change in entropy (which is symbolized as ΔS) contribute to the overall change in energy. This total change—which takes into account both heat and entropy—is called the <u>free-energy change</u> and is symbolized as ΔG, after the American physicist Josiah Willard Gibbs (1839–1903), who was one of the first to integrate all of these ideas.

Keeping ΔG in mind, let us examine once again the oxidation of glucose. The ΔH of this reaction is -673 kcal/mole. The ΔG is -686 kcal/mole. Thus, the entropy factor has contributed 13 kcal/mole to the free-energy change of the process. The change in heat content and in entropy both contribute to the lower energy state of the products of the reaction.

The free-energy change, ΔG, can also enable one to predict processes that occur when ΔH is zero or even positive. For instance, it confirms our earlier observations that heat will flow from a warm object to a cold one or that dye molecules will diffuse in a beaker of water. In each of these processes, the final state has more entropy—and therefore less potential energy—than the initial state.

The relationship between ΔG, ΔH, and entropy is given in the following equation:

$$\Delta G = \Delta H - T\Delta S$$

This equation states that the free-energy change is

equal to the change in heat content (a minus figure in exothermic reactions, which give off heat) minus the change in entropy, which is affected by the absolute temperature T. In exergonic reactions, ΔG is always negative, but ΔH may be zero or even positive. Since T is always positive, the greater the change in entropy, the more negative ΔG will be; that is, the more exergonic the reaction will be. Therefore it is possible to state the second law in another, simpler way: *In general, all natural processes are exergonic.*

Biology and the Second Law

The laws of thermodynamics are of crucial importance to biology, as they are to physics and chemistry. They are the organizing principles under which a number of different kinds of processes can be unified. Also, as we shall see in the following chapter, they permit a kind of biochemical bookkeeping.

The most interesting implication of the second law, as far as biology is concerned, is the relationship between entropy, on the one hand, and order and organization on the other. Living systems are continually expending large amounts of energy to maintain order. Stated in terms of chemical reactions, living systems continually expend energy to maintain a position far from equilibrium. If equilibrium were to occur, the chemical reactions in a cell would, for all practical purposes, stop and no further work could be done. At equilibrium, a cell would soon die.

The order, the organization, and the disequilibrium state found in every living cell are maintained only by a continuous uptake of energy from the environment. How this is accomplished is considered in part in Chapters 5 and 6, but first let us examine the way in which enzymes mediate many of the reactions and processes carried out by living cells.

ENZYMES AND LIVING SYSTEMS

In any living system, thousands of different chemical reactions occur, many of them simultaneously. The sum of all these reactions is referred to as *metabolism* (from the Greek *metabole,* meaning "change"). If one merely listed the individual chemical reactions, it would be difficult to understand a cell's metabolic activities. Fortunately, there are some guiding principles that lead one through the maze of cell metabolism. First, virtually all the chemical reactions that take place in a cell involve *enzymes*—the catalysts and regulators for the metabolic processes of living systems. Second, biochemists group these reactions in an ordered series of steps, called a pathway, often containing a dozen or more sequential reactions. Each pathway serves a function in the overall life of the cell or organism. Furthermore, certain pathways have many steps in common, such as those that are concerned with the synthesis of the different amino acids or the various nitrogenous bases. Some pathways converge; for example, the pathway by which fats are broken down to yield energy leads to the same pathway by which glucose is broken down to yield energy.

Many living systems have unique pathways. Plant cells expend energy building cellulose walls, an activity not engaged in by animal cells. Red blood cells specialize in the synthesis of hemoglobin molecules, which are made nowhere else in the animal body. It is not surprising that the distinctive differences among cells and organisms are reflected not only in their forms and functions but also in their biochemistry. What is surprising, however, is that much of the metabolism of even the most diverse of organisms is exceedingly similar; the differences in many of the metabolic pathways of humans, oak trees, mushrooms, and jellyfish are very slight. Some pathways are found in virtually all living systems.

The magnitude of the chemical work carried out by a cell is illustrated by the fact that, for the most part, the thousands of different molecules found within a cell are synthesized there. The total of chemical reactions involved in this synthesis is called *anabolism.*

4–4
A modern highway and the meandering old road to its right are alternative paths through these mountains. Identical chemical reactions also can occur via alternative pathways that involve vastly different rates.

Anabolic reactions usually involve an increase in atomic order (a decrease in entropy) and are almost always energy-requiring (endergonic). Cells are also constantly involved in the breakdown of larger molecules; these activities, known collectively as *catabolism,* involve a decrease in atomic order (an increase in entropy) and are usually energy-liberating (exergonic). Catabolic reactions serve two purposes: (1) they release the energy for anabolism and other work of the cell, and (2) they serve as a source of raw materials for anabolic processes. Hence, both aspects of metabolism—anabolism and catabolism—are essential for normal cellular activities.

Living systems carry out this multitude of chemical activities under extraordinarily difficult conditions. Most of their chemical reactions are carried out within living cells, and thousands of different kinds of molecules are intermingled. Temperatures cannot be high; otherwise, many of the fragile structures on which life depends would be destroyed. How is all this complex chemical work accomplished? The question can be answered in a single word: enzymes. Without enzymes, biochemical reactions would take place so slowly that, for all practical purposes, they would not occur at all, and the activities we associate with life would cease to exist.

ENZYMES AS CATALYSTS

Enzymes are the catalysts of biological reactions. They differ from other catalysts in that they are very selective in their action. Some enzymes catalyze a reaction with only a single set of reactants. In enzyme-catalyzed reactions, the reactant (or reactants) on which an enzyme operates is called its *substrate*. The selectivity an enzyme exhibits in choosing a substrate is known as its *specificity*. Enzymes otherwise resemble other catalysts in that they are not used up in the course of the reaction and so can be used over and over again.

Enzymes enormously accelerate the rate at which reactions occur. For instance, the reaction of carbon dioxide with water,

$$CO_2 + H_2O \rightleftharpoons H_2CO_3$$

can occur spontaneously, as it does in the oceans. In the human body, however, this reaction is catalyzed by an enzyme, carbonic anhydrase, which is one of the most effective enzymes known—each enzyme molecule leads to the production of 10^5 (100,000) molecules of the product, carbonic acid, per second. The catalyzed reaction is 10^7 times faster than the uncatalyzed one. In animals, this reaction is essential in the transfer of carbon dioxide from the cells, where it is produced, to the bloodstream, which transports it to the lungs.

THE ACTIVE SITE

Enzymes are complex globular proteins consisting of one or more polypeptide chains (see Chapter 2). They are folded so as to form a groove or pocket into which the reacting molecule or molecules—the substrate—fit and where the reactions occur. This portion of the enzyme is known as the *active site*. The active site is formed by the very exact folding of the polypeptide chain. The relationship between the active site and the substrate is very precise and is often compared to that of a lock and key (Figure 4–5).

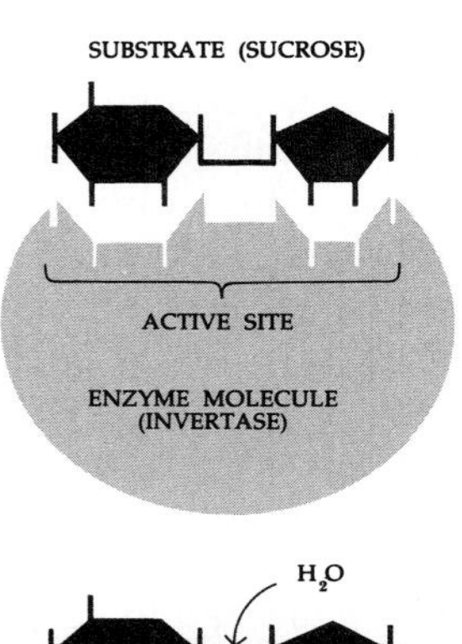

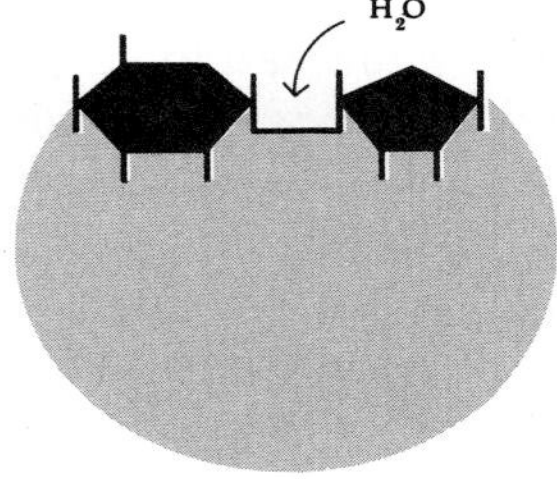

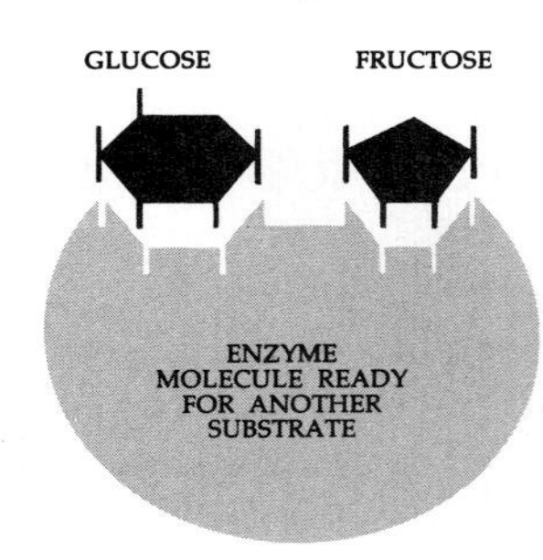

4–5
A model of the lock-and-key hypothesis of enzyme action. A molecule of sucrose is hydrolyzed to yield one molecule of glucose and one molecule of fructose. The enzyme involved in the reaction is specific for this process; the active site of the enzyme exactly fits the opposing surface of the sucrose molecule.

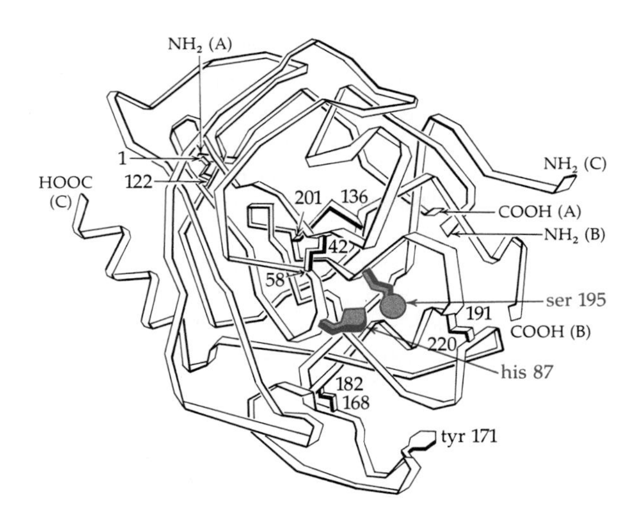

4–6
A model of an enzyme. This enzyme (the digestive enzyme chymotrypsin) is composed of three polypeptide chains. The amino ($-NH_2$) *and the carboxyl* ($-COOH$) *groups of each are labeled. The numbers represent the positions of particular amino acids in the chains, and disulfide bridges connect certain amino acids. The three-dimensional shape of the molecule is the result of a combination of disulfide bonds and of interactions among the chains and between the chains and the surrounding water molecules, based on the positive or negative charges (the polarity) of the various amino acids. As a result of this bending and twisting of the polypeptide chains, particular amino acids come together in a highly specific configuration to form the active site of the enzyme. Two amino acids known to be part of the active site are shown in color.*

The active site not only has a precise three-dimensional shape but also has exactly the correct array of charged and uncharged, or hydrophilic and hydrophobic, areas on its binding surface. If a particular portion of the substrate has a negative charge, the corresponding feature on the active site has a positive charge, and so on. Thus, the active site not only confines the substrate molecule but also orients it in the correct direction.

The amino acids involved in the active site need not be adjacent to one another on the polypeptide chains. In fact, in an enzyme with quaternary structure, they may even be on different polypeptide chains, as shown in Figure 4–6. The amino acids are brought together at the active site by the precise folding of the polypeptide chains in the molecule.

The Induced-Fit Hypothesis

During the last several years, studies of enzyme structure have suggested that the binding that takes place between enzyme and substrate alters the conformation of the enzyme, thus inducing an even closer fit between the active site and the reactants. It is believed that this induced fit may place some strain on the reacting molecules and so further facilitate the reaction (Figure 4–7).

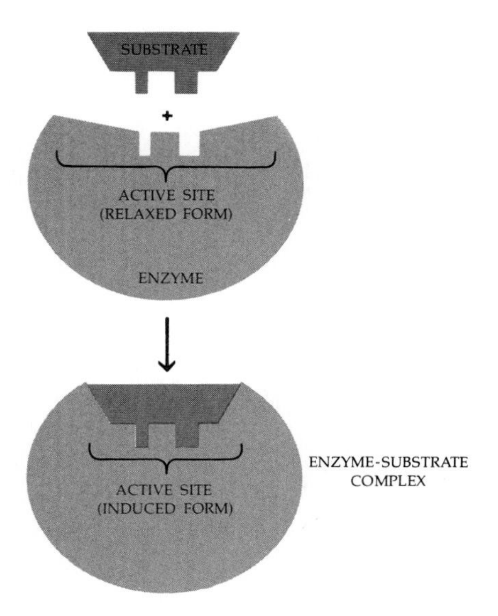

4–7
A model of the induced-fit hypothesis of enzyme action. The active site is believed to be flexible and thus adjustable in shape, or conformation, to that of the substrate molecule. This induces a particularly close fit between the active site and the substrate.

COFACTORS IN ENZYME ACTION

The catalytic activity of some enzymes appears to depend only on their structure as proteins. Many enzymes, however, require one or more nonprotein components, known as cofactors, in order to function.

Ions as Cofactors

Certain ions are cofactors for particular enzymes. For example, the magnesium ion (Mg^{2+}) is required in all enzymatic reactions involving the transfer of a phosphate group from one molecule to another. Its two positive charges hold the negatively charged phosphate group in position. Other ions, such as Na^+ and K^+, play similar roles in other reactions. In some cases, ions serve to hold the enzyme protein together.

Coenzymes and Vitamins

Nonprotein organic cofactors can also play a crucial role in enzyme-catalyzed reactions. Such cofactors are called *coenzymes*. For example, in some oxidation-reduction reactions, electrons are passed along to a molecule that serves as an electron acceptor. (Oxidation is the loss of an electron; reduction is the gain of an electron. These two processes occur simultaneously. See Appendix A.) There are several different electron acceptors in any given cell, and each is tailor-made to hold the electron at a slightly different energy level. As an example, let us look at just one, nicotinamide adenine dinucleotide (NAD), which is shown in Figure 4–8. At first glance, NAD looks complex and unfamiliar, but if you look at it more closely, you will find that you recognize most of its component parts. The two units labeled ribose are five-carbon sugars; they are linked by two phosphate groups. One of the sugars is attached to the nitrogenous base adenine. The other is attached to another nitrogenous base, nicotinamide. (A nitrogenous base plus a sugar plus a phosphate is called a *nucleotide*; a molecule that contains two of these combinations is called a dinucleotide.)

The nicotinamide ring is the active end of NAD, that is, the part that accepts electrons. Nicotinamide is a vitamin—niacin. Vitamins are compounds that are required in small quantities by many living organisms; humans and other animals cannot synthesize vitamins and so must obtain them in their diets. When nicotinamide is present, our cells can use it to make NAD. Many vitamins are coenzymes or parts of coenzymes.

Nicotinamide adenine dinucleotide, like many other coenzymes, is recycled. That is, NAD^+ is regenerated when NADH + H^+ passes its electrons to another electron acceptor. Thus, although this coenzyme is involved in many cellular reactions, the actual number of NAD molecules required is relatively small.

ENZYMATIC PATHWAYS

Enzymes work in an ordered series of steps—the pathways we referred to earlier. Consequently, a living organism carries out its chemical activities with remarkable efficiency. First, there is little accumulation of waste products, since each product tends to be used up in the next reaction along the pathway. A

4–8
Nicotinamide adenine dinucleotide (NAD) in its oxidized form, NAD^+, and in its reduced form, NADH.

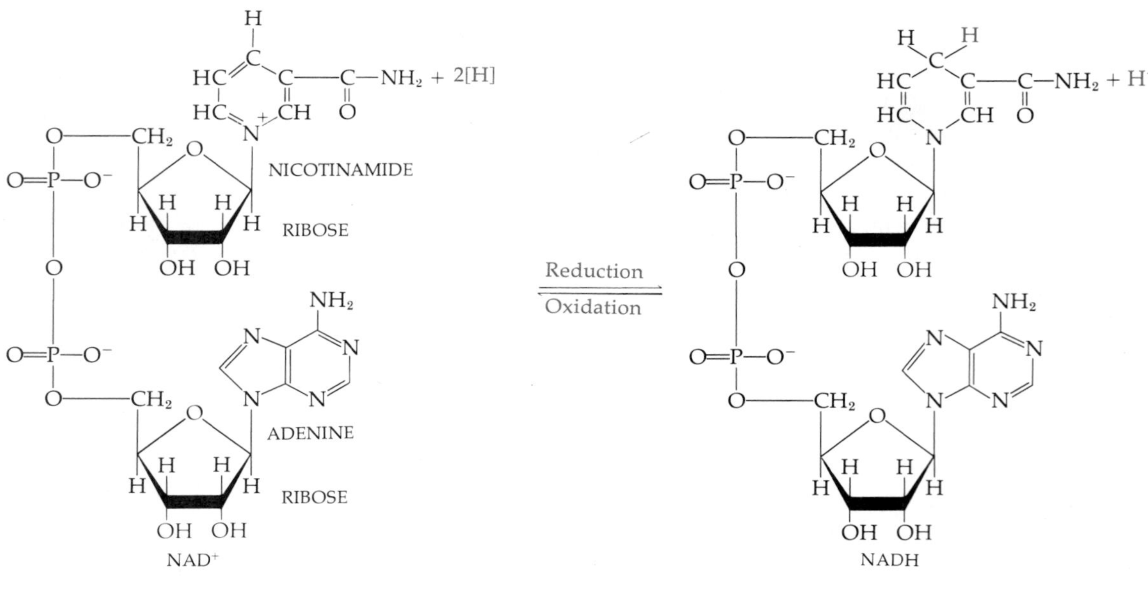

second advantage of such sequential reactions is understandable when one considers that chemical reactions can go in either direction; that is, they are reversible (see Appendix A). If each product along a series of reactions is used up by the next reaction almost as rapidly as it is formed, the tendency for reversal of the reactions will be minimized. Moreover, if the eventual end product is also used up rapidly, the whole series of reactions will move toward completion. Another advantage is that the groups of enzymes making up a common pathway can be segregated within the cell. Some are found in small vesicles, or membrane-surrounded sacs, in the cytoplasm. Others are embedded in the membranes of specialized organelles, such as mitochondria and chloroplasts.

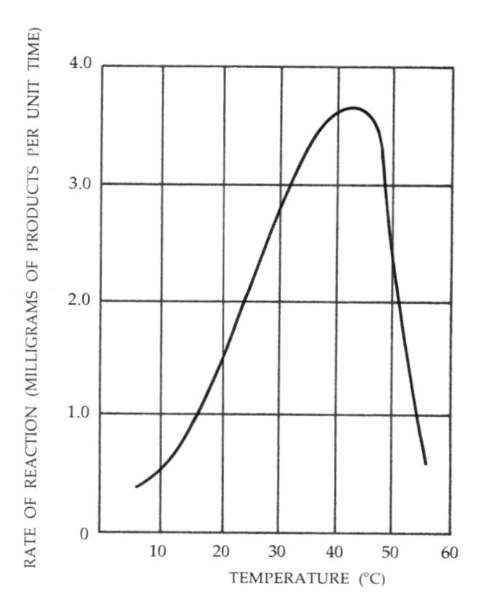

4–9
The effect of temperature on the rate of an enzyme-controlled reaction. The concentrations of enzyme and reacting molecules (substrate) were kept constant. The rate of the reaction, as in most chemical reactions, approximately doubles for every 10°C rise in temperature up to about 40°C. Above this temperature, the rate decreases, and at about 60°C, the reaction stops altogether, presumably because the enzyme is denatured.

REGULATION OF ENZYME ACTIVITY

Another remarkable feature of the metabolic activity of cells is the extent to which each cell regulates the synthesis of the products necessary for its well-being, making them in the appropriate amounts and at the rates required. At the same time, cells avoid overproduction, which would waste both energy and raw materials. The availability of reactant molecules or of cofactors is a principal factor in limiting enzyme action, and for this reason most enzymes probably work at a rate well below their maximum.

Temperature affects enzymatic reactions. An increase in temperature increases the rate of enzyme-catalyzed reactions, but only to a point. As can be seen in Figure 4–9, the rate of most enzymatic reactions approximately doubles for each 10°C rise in temperature and then drops off very quickly after about 40°C. The increase in reaction rate occurs because of the increased energy of the reactants; the decrease in the reaction rate occurs as the enzyme molecule itself begins to move and vibrate, disrupting the hydrogen bonds and other relatively fragile forces that hold it together.

The pH of the surrounding solution also affects enzyme activity. Among other factors, the conformation of an enzyme depends on the attraction and repulsion between negatively charged (acidic) and positively charged (basic) amino acids. As the pH changes, these charges change, and so the shape of the enzyme changes until it is so drastically altered that it is no longer functional. More important, probably, the charges of the active site and substrate are changed so that the binding capacity is affected. Some enzymes are frequently found at a pH that is not their optimum, suggesting that this discrepancy may not be an evolutionary oversight but a way of controlling enzyme activity.

Living systems also have more precise ways of turning enzyme activity on and off. Some enzymes are produced only in an inactive form and are activated only when they are needed, usually by another enzyme. Specific mechanisms by which enzymes may be activated or inactivated are discussed in Chapter 7.

THE ENERGY FACTOR: ATP

All of the biosynthetic (anabolic) activities of the cell (and many other activities as well) require energy. A large proportion of this energy is supplied by a single molecule, *adenosine triphosphate* (ATP), a nucleotide derivative that is the cell's chief energy currency. Glucose, glycogen, and starch are like money in the bank. ATP is like change in your pocket.

At first glance, ATP (shown in Figure 4–10 on page 82) also appears to be a complex molecule. However, as with NAD, you will find that its component parts are familiar. ATP is made up of adenine, the five-carbon sugar ribose, and three phosphate groups. These three phosphate groups, with strong negative charges, are covalently bonded to one another; this is an important feature in ATP function.

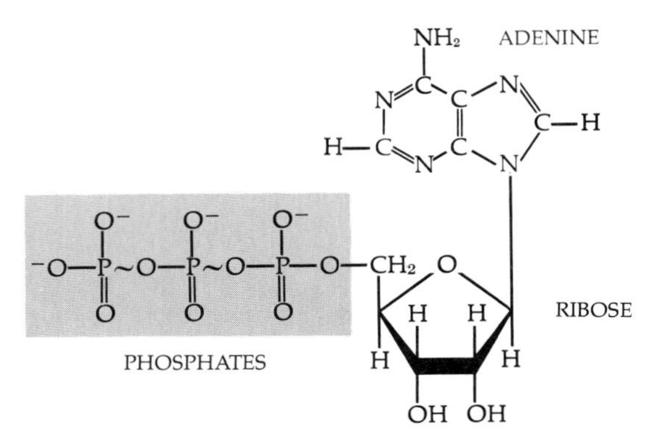

4–10
Adenosine triphosphate (ATP) is the cell's chief energy currency. The bonds between the three phosphate groups in the molecule are important in ATP function. The symbol ~ designates a high-energy bond.

To understand the role of ATP, let us briefly review the concept of the chemical bond and bond energies. A chemical bond is the total of the forces that hold the constituent atoms together in a molecule. Because a bond is a stable configuration, energy is needed to break one bond and form a new one. This energy is the energy of activation (see Appendix A). Because of enzymes, which greatly reduce the energy of activation required, the reactions essential to life can proceed at an appropriate rate. However, there is an important additional limitation on chemical reactions in living systems: the bond energies of the products must always be less than the bond energies of the reactants. Following this statement to its logical conclusion, one might think that *biosynthetic reactions* could never occur. This, of course, is not true. Cells circumvent this requirement by *coupled reactions* in which energy-requiring chemical reactions are linked to energy-releasing reactions. The molecule that participates most frequently in such coupled reactions is ATP.

Because of its internal structure, the ATP molecule is well suited to this role in living systems. Energy is released from the ATP molecule when one phosphate group is removed by hydrolysis, producing a molecule of ADP (adenosine diphosphate):

$$\text{ATP} + H_2O \longrightarrow \text{ADP} + \text{phosphate}$$

In the course of this reaction, some 7.3 kcal/mole of ATP are released—a relatively large amount of chemical energy. Removal of a second phosphate group produces AMP (adenosine monophosphate) and releases an equivalent amount of chemical energy:

$$\text{ADP} + H_2O \longrightarrow \text{AMP} + \text{phosphate}$$

To indicate that relatively large amounts of chemical energy are involved, the covalent bonds linking these two phosphates to the rest of the molecule are often called high-energy bonds (Figure 4–10). However, this name is somewhat misleading because the energy released during the reaction does not arise entirely from the bond. The difference in energy between reactants and products is due only in part to the energy in the bonds. It is also partly the result of the rearrangement of the electron orbitals of the ATP or ADP molecules. The phosphate groups each carry a negative charge and so tend to repel each other. When a phosphate group is removed, the molecule undergoes a change in electron configuration that results in a structure with less energy.

In most reactions that take place within a cell, the terminal phosphate group of ATP is not simply removed but is transferred to another molecule. This addition of a phosphate group to a molecule is known as *phosphorylation;* the enzymes that catalyze such transfers are known as kinases. This reaction transfers some of the energy in the high-energy bond to the phosphorylated compound, which, thus energized, can participate in the reaction.

Let us look at a simple example of energy exchange involving ATP in the formation of sucrose in sugarcane. Sucrose is formed from the monosaccharides glucose and fructose; under standard thermodynamic conditions, its synthesis is strongly endergonic, requiring an input of 5.5 kcal for each mole of sucrose formed:

$$\text{glucose} + \text{fructose} \rightleftharpoons \text{sucrose} + H_2O$$

However, the synthesis of sucrose in sugarcane, when coupled to the breakdown of ATP, is actually exergonic. During the series of reactions involved in formation of sucrose, a phosphate group is transferred to a glucose molecule and to a fructose molecule, thus energizing each of them; hence we have the overall equation:

$$\text{glucose} + \text{fructose} + 2\text{ATP} \longrightarrow \text{sucrose} + 2\text{ADP} + 2 \text{ phosphates} + H_2O$$

In this reaction, 5.5 kcal are utilized in the formation of sucrose, and the overall energy difference in products and reactants is about 8.5 kcal. Thus, the sugarcane plant is able to form sucrose by coupling the breakdown of two molecules of ATP to the synthesis of a covalent bond between glucose and fructose.

Where does the ATP originate? The energy released in the cell's catabolic reactions, such as in the breakdown of glucose, is used to "recharge" the ADP molecule. Of course, the energy released in these catabolic reactions is originally derived from the sun as radiant energy which is converted, during photosynthesis, into chemical energy. Some of this chemical energy appears as the high-energy bonds of ATP before being converted to chemical bond energies of other organic molecules. Thus the ATP/ADP system serves as a universal energy-exchange system, shuttling between energy-releasing and energy-requiring reactions.

SUMMARY

Life on this planet is dependent on the flow of energy from the sun. A small fraction of this energy, captured in the process of photosynthesis, is converted to the energy that drives the many other metabolic reactions associated with living systems and from which living systems derive their order and organization.

The thermodynamic relationship between plants and animals—photosynthesizers and nonphotosynthesizers—is exceedingly complicated. Stated in its simplest form: In the course of photosynthesis, the energy of the sun is used to forge high-energy carbon-carbon and carbon-hydrogen bonds; then, in the course of respiration, these bonds are once more broken down to carbon dioxide and water, and energy is released. As in machines, some useful energy is lost in each step of these energy conversions.

Living systems thus operate in accord with the laws of thermodynamics. The first law of thermodynamics states that energy can never be created or destroyed, although it can be changed from one form to another. The potential energy of the initial state (or reactants) is equal to the potential energy of the final state (or products) plus the energy released in the process or reaction. The second law of thermodynamics states that, in the course of energy conversions, the potential energy of the final state will always be less than the potential energy of the initial state. The difference in energy between the initial and final state is known as the free-energy change and is symbolized as ΔG. Exergonic (energy-yielding) reactions have a negative ΔG. Factors that determine the ΔG include ΔH, the change in heat content, and ΔS, the change in entropy, which, multiplied by the absolute temperature T, is a measure of randomness, or disorder:

$$\Delta G = \Delta H - T\Delta S$$

Metabolism is the total of all the chemical reactions that take place in cells. Reactions resulting in the breakdown or degradation of molecules are known, collectively, as catabolism. Biosynthetic reactions—the building of new molecules—are called anabolism. Metabolic reactions take place in an ordered series of steps, called pathways, each of which serves a particular function in a cell. Each step in the pathway is controlled by a particular enzyme.

Enzymes serve as catalysts; they enormously increase the rate at which reactions take place but remain unchanged in the process. Enzymes are large protein molecules folded in such a way that particular groups of amino acids form an active site. The reacting molecules, known as the substrate, fit precisely into this active site. Many enzymes require cofactors, which may be simple ions, such as Mg^{2+} or Na^{+}, or nonprotein organic molecules, such as NAD. The latter are known as coenzymes.

Enzyme-catalyzed reactions are under tight cellular control. The rate of enzymatic reactions is affected by temperature and pH.

ATP supplies the energy for most of the activities of the cell. The ATP molecule consists of a nitrogenous base, adenine; the five-carbon sugar ribose; and three phosphate groups. The phosphate groups are linked by two high-energy bonds—bonds that release a relatively large amount of energy when they are broken. ATP participates as an energy carrier in most series of reactions that occur in living systems.

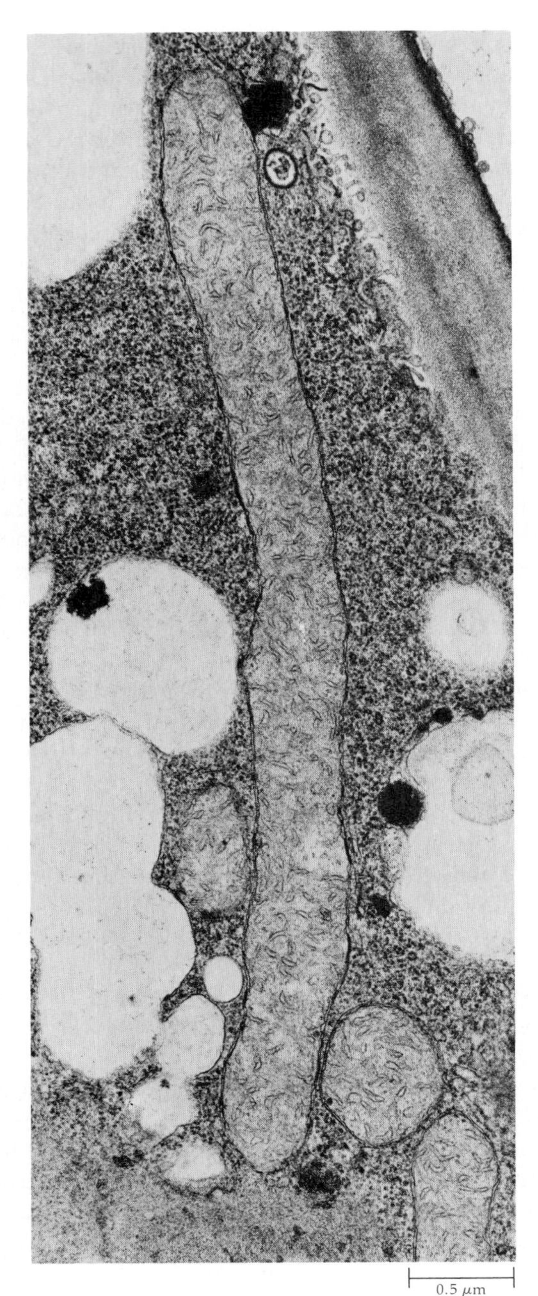

5–1
Mitochondria from a leaf cell of the fern Regnellidium diphyllum. *Mitochondria are the sites of cellular respiration by which chemical energy is transferred from carbon-containing compounds to ATP. Most of the ATP is produced on the surfaces of the cristae by enzymes that form a part of the structure of these membranes.*

CHAPTER 5

Respiration

Respiration is the means by which the energy of carbohydrates is transferred to ATP—the universal energy-carrier molecule—and is thus made available for the immediate energy requirements of the cell (Figure 5–1). In the following pages, we will describe in some detail how a cell breaks down carbohydrates and captures and stores much of the released energy in the high-energy phosphate bonds of ATP. The process is described in detail, not because we believe it should be memorized, but because it provides an excellent illustration both of chemical principles described in previous chapters and of the way in which cells perform biochemical work.

As mentioned in Chapter 2, energy-yielding carbohydrate molecules are generally found stored in plants as sucrose or starch. A necessary preliminary step to the respiratory sequence is the hydrolysis of these storage molecules to monosaccharides. Respiration itself is generally considered to begin with glucose, the building block of sucrose and starch.

Glucose can be used as a source of energy under both aerobic (with O_2 present) and anaerobic (with O_2 absent) conditions. However, maximum energy yields for oxidizable organic compounds (compounds from which electrons can be removed; see Appendix A) are generally achieved only under aerobic conditions. Consider, for example, the overall reaction for the complete oxidation of glucose:

$$C_6H_{12}O_6 + 6O_2 \longrightarrow 6CO_2 + 6H_2O + \text{energy}$$

With oxygen as the ultimate electron acceptor, this reaction is highly exergonic (energy-yielding), with a ΔG of -686 kcal/mole. This reaction represents the process called *respiration*. (When energy is extracted from organic compounds without the involvement of oxygen, the process is called *fermentation*.)

Respiration involves three distinct stages: *glycolysis*, the *Krebs cycle*, and the *electron transport chain*. In glycolysis, the six-carbon glucose molecule is broken down to a pair of three-carbon molecules of pyruvic acid or pyruvate. (Pyruvic acid dissociates, producing pyruvate and a hydrogen ion. Pyruvic acid and pyruvate exist in dynamic equilibrium, and the two terms can be used interchangeably.) In the Krebs cycle and the electron transport chain, the pyruvate molecules are further broken down to carbon dioxide and water.

As the glucose molecule is oxidized, some of its energy is extracted in a series of small, discrete steps and is stored in the high-energy bonds of ATP.

In accord with the second law of thermodynamics, some of this chemical energy is dissipated as heat energy. In birds, mammals, and to some extent, lower vertebrates, the heat generated by cellular respiration is conserved by various mechanisms so that the body temperature of the organism generally remains above that of the environment. In plants, respiration is generally so slow that any heat produced has virtually no effect on body temperature. However, during the rapid growth associated with flowering, parts of some plants, such as *Philodendron* and eastern skunk cabbage (*Symplocarpus foetidus*), have been found to have temperatures as much as 20°C above the air temperature.

GLYCOLYSIS

Glycolysis (from *glyco,* meaning "sugar," and *lysis,* meaning "splitting") occurs in a series of nine steps, each catalyzed by a specific enzyme (Figure 5–2). This series of reactions is carried out by virtually all living cells, from prokaryotes to the eukaryotic cells of plants and animals. Glycolysis takes place under anaerobic conditions and occurs in the ground substance of the cytoplasm (sometimes referred to as the cytosol). Biologically, glycolysis may be considered a primitive process, in the sense that it most likely arose before the appearance of atmospheric oxygen and before the origin of cellular organelles.

The glycolytic pathway is shown in detail in Figure 5–3. As the series of reactions is discussed, notice how the carbon skeleton of the molecule is disassembled as its atoms are rearranged step by step. *Do not try to memorize these steps; simply follow them closely.* Note especially the formation of ATP from ADP and the formation of $NADH_2$ from NAD. (To be exact, NAD and $NADH_2$ should be written as NAD^+ and $NADH + H^+$, as in Figure 4–8, but we will follow this simpler convention.) ATP and $NADH_2$ represent the cell's net energy harvest from this process.

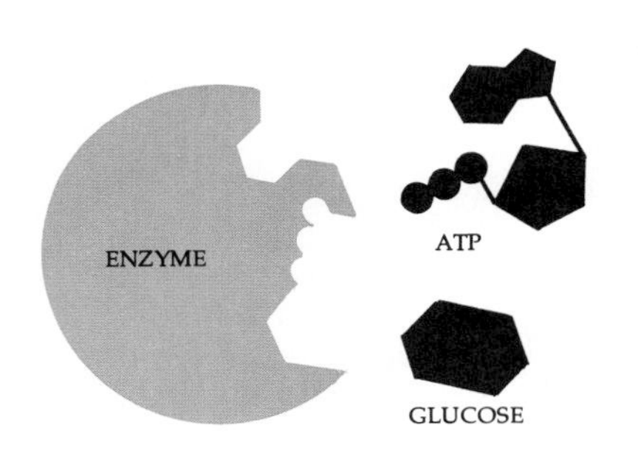

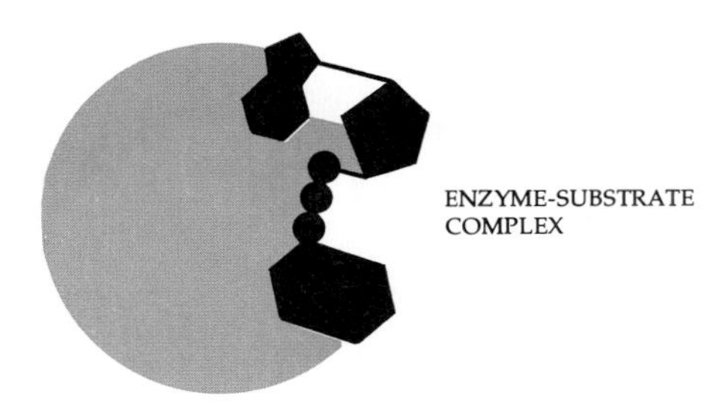

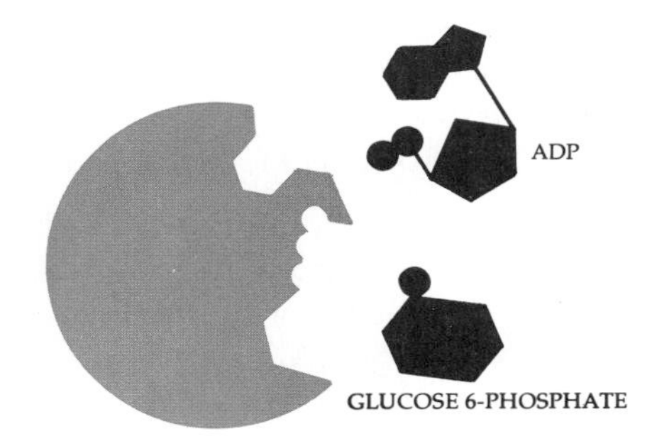

5–2
The first of the nine steps of glycolysis described on the following pages involves the transfer of a "high-energy" phosphate group from ATP to glucose. By this step, energy is put into the reaction. Like all the steps in glycolysis, this one is catalyzed by a specific enzyme.

Step 1 The first step in glycolysis requires an input of energy. This activation energy is supplied by the hydrolysis of ATP to ADP. The terminal phosphate group is transferred from an ATP molecule to the glucose molecule, producing glucose 6-phosphate. The combining of ATP with glucose to produce glucose 6-phosphate and ADP is an energy-yielding reaction. Some of the energy released from the ATP is conserved in the chemical bond linking the phosphate to the sugar molecule. This reaction is catalyzed by a specific enzyme (hexokinase), and each of the reactions that follows is similarly catalyzed by a specific enzyme.

Step 2 In this step, the molecule is rearranged, again with the help of a particular enzyme. The six-sided ring characteristic of glucose becomes a five-sided fructose ring. As shown in Figure 2–2, glucose and fructose have the same number of atoms ($C_6H_{12}O_6$) and differ only in the arrangement of their atoms. This reaction can proceed in either direction; it is pushed forward by the accumulation of glucose 6-phosphate from step 1 and the disappearance of fructose 6-phosphate as it enters step 3.

Step 3 This step, which is similar to step 1, results in the attachment of a phosphate to the first carbon of the fructose molecule, which produces fructose 1,6-bisphosphate, that is, fructose with phosphates in the 1 and 6 positions. The conversion of the glucose molecule to the higher-energy fructose 1,6-bisphosphate compound is accomplished at the expense of two molecules of ATP. Thus far, no energy has been recovered, but the overall yield will more than compensate for this initial investment.

Step 4 This is the cleavage step of glycolysis. The molecule is split into two interconvertible three-carbon molecules—glyceraldehyde 3-phosphate and dihydroxyacetone phosphate. However, because the glyceraldehyde 3-phosphate is used up in subsequent reactions, all of the dihydroxyacetone phosphate is eventually converted to glyceraldehyde 3-phosphate. Thus, all subsequent steps must be considered twice to account for the fate of each glucose molecule. With the completion of step 4, the preparatory reactions that require an input of ATP energy are complete.

Step 5 Next, glyceraldehyde 3-phosphate molecules are oxidized—that is, hydrogen atoms with their electrons are removed—and NAD is converted to $NADH_2$. This is the first reaction from which the cell gains energy. Energy from this

5–3
The steps of glycolysis.

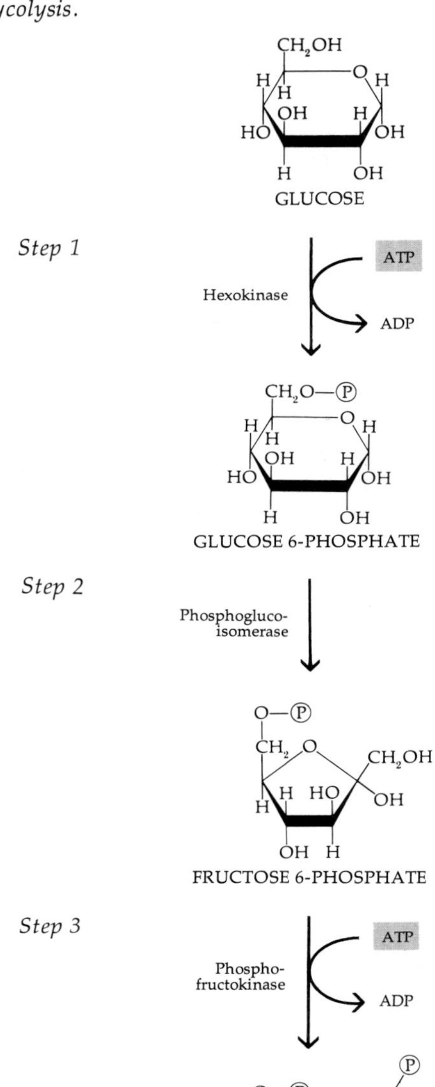

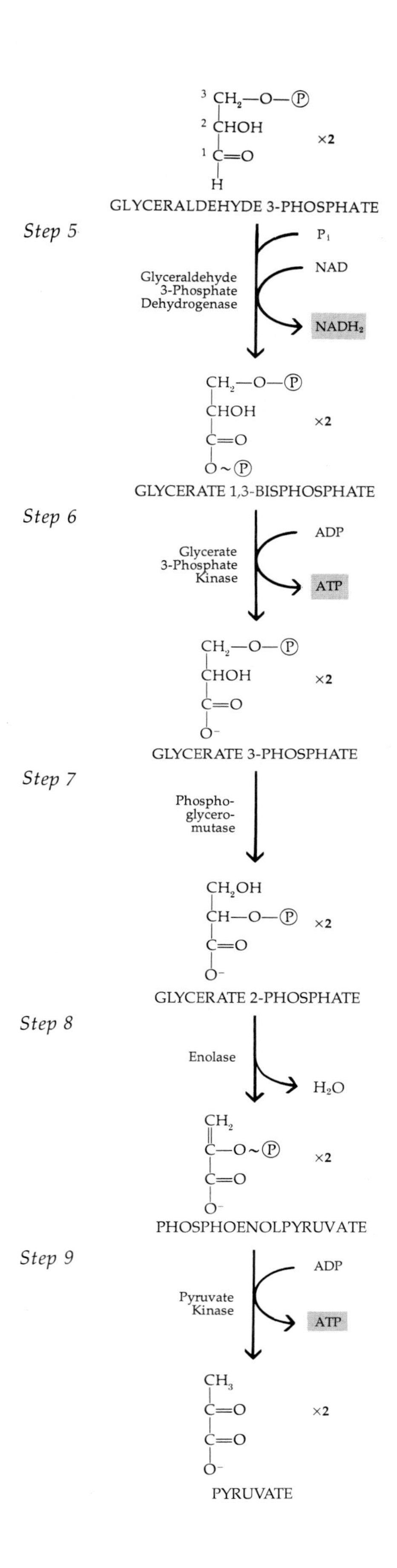

oxidation reaction is also used to attach an additional phosphate group to what is now the 1 position of each of the glyceraldehyde molecules. (The designation P_i indicates inorganic phosphate that is available as a phosphate ion in solution in the cytoplasm.) Note that a high-energy bond (~) is formed.

Step 6 The high-energy phosphate is released from the glycerate 1,3-bisphosphate molecule and is used to recharge a molecule of ADP (a total of two molecules of ATP are formed per molecule of glucose). This is a highly exergonic reaction and pulls all the previous reactions forward.

Step 7 The remaining phosphate group is transferred from the number 3 to the number 2 carbon.

Step 8 In this step, a molecule of water is removed from the three-carbon compound, and as a consequence of this internal rearrangement of the molecule, a high-energy phosphate bond is formed.

Step 9 The high-energy phosphate is transferred to a molecule of ADP, forming another molecule of ATP (again, a total of two molecules of ATP are formed per molecule of glucose). This is also a highly exergonic reaction, and so the sequence runs to completion with an accelerating force.

Summary of Glycolysis

The complete sequence of glycolysis begins with one molecule of glucose (Figure 5–4). Energy enters the sequence at steps 1 and 3 by the transfer of a phosphate group from an ATP molecule—one at each step—to the sugar molecule. The six-carbon molecule splits at step 4, and after this point the sequence yields energy. At step 5, two molecules of NAD, in being reduced to two molecules of $NADH_2$, store some of the energy from the oxidation of glyceraldehyde 3-phosphate. At both steps 6 and 9, two molecules of ADP take energy from the system, form additional phosphate bonds, and are phosphorylated to become two molecules of ATP. (Phosphorylation that occurs during glycolysis is known as *substrate phosphorylation*.)

Glycolysis (from glucose to pyruvate) can be summarized by the overall equation:

$$\underset{\text{GLUCOSE}}{C_6H_{12}O_6} + 2NAD + 2ADP + 2P_i \longrightarrow \underset{\text{PYRUVATE}}{2C_3H_4O_3} + 2NADH_2 + 2ATP$$

Thus one glucose molecule is converted to two molecules of pyruvate. The net harvest—the energy

5–4
A summary of glycolysis. Two molecules of ATP and two molecules of $NADH_2$ represent the energy yield. Most of the energy stored in the original glucose molecule is found in the two molecules of pyruvate.

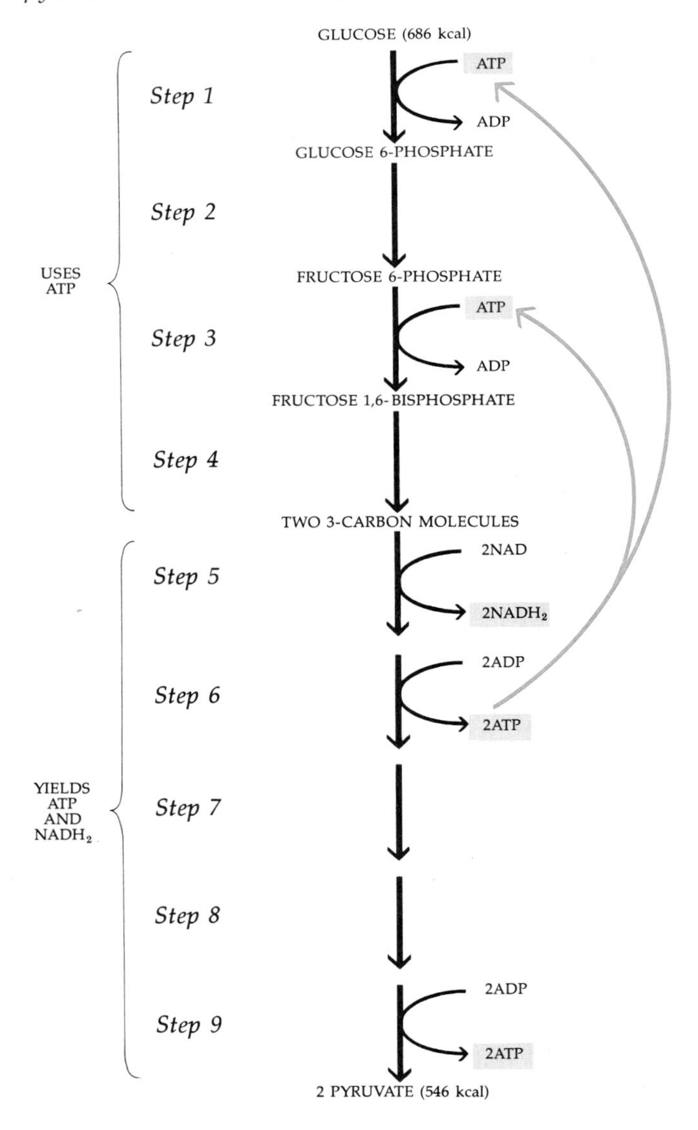

recovered—is two molecules of ATP and two molecules of $NADH_2$. The two molecules of pyruvate have a total energy content of about 546 kcal, a large portion of the 686 kcal stored in the original glucose molecule.

AEROBIC PATHWAY

Pyruvate is a key intermediate compound in cellular energy metabolism, and it can be utilized in one of several pathways. Which pathway it follows depends in part upon the conditions under which metabolism takes place and in part upon the specific organism involved. The principal environmental factor that operates to affect which pathway is followed is the availability of oxygen.

In the presence of oxygen, pyruvate is oxidized completely to carbon dioxide, and glycolysis is but the initial phase of respiration. This aerobic pathway results in the complete oxidation of glucose and a much greater ATP yield than can be achieved by glycolysis alone. These reactions take place in two stages—the Krebs cycle and the electron transport chain—both occurring within mitochondria in eukaryotic cells (Figure 5–1).

Recall that mitochondria are surrounded by two membranes, the inner one convoluted inwardly into folds called cristae. The more metabolically active a cell, the more numerous are both its mitochondria and the cristae within them. Within the inner compartment, surrounding the cristae, is a dense solution containing enzymes, coenzymes, water, phosphates, and other molecules involved in respiration. Thus, a mitochondrion is a self-contained chemical factory. The outer membrane allows most small molecules to move in or out freely, but the inner one only permits the passage of certain molecules, such as pyruvate and ATP, while it prevents the passage of others. The enzymes of the Krebs cycle are found in solution within the inner compartment. The enzymes and other components of the electron transport chain are built into the surfaces of the cristae.

The Krebs Cycle

The Krebs cycle is named in honor of Sir Hans Krebs, who is largely responsible for its elucidation. Krebs postulated this metabolic pathway in 1937 and later received the Nobel prize in recognition of his brilliant work. The Krebs cycle is also called the tricarboxylic acid (TCA) cycle because it is initiated with the formation of an organic acid (citrate) that has three carboxylic acid groups.

Before entering the Krebs cycle, pyruvate is both oxidized and decarboxylated. In the course of this exergonic reaction, a molecule of $NADH_2$ is produced from NAD. The original glucose molecule has now been oxidized to two acetyl (CH_3CO) groups; two molecules of CO_2 have been liberated; and two molecules of $NADH_2$ have been formed from NAD (Figure 5–5).

Each acetyl group is then temporarily attached to coenzyme A (CoA)—a large molecule, a portion of which is a nucleotide and a portion of which is pantothenic acid, one of the B-complex vitamins. The combination of the acetyl group and CoA is known as acetyl CoA (Figure 5–5).

5–5
The three-carbon pyruvate molecule is oxidized and decarboxylated to form the two-carbon acetyl group, which then combines with coenzyme A to form acetyl CoA. The oxidation of the pyruvate molecule is coupled to the production of $NADH_2$ from NAD. Acetyl CoA is the molecule needed to enter the Krebs cycle.

Fats and amino acids can also be converted to acetyl CoA and enter the respiratory sequence at this point. A fat molecule is first hydrolyzed to glycerol and three fatty acids. Then, beginning at the carboxyl end, two-carbon groups are successively removed from the fatty acids. A molecule such as palmitic acid (Figure 2–8), which contains 16 carbon atoms, yields eight molecules of acetyl CoA.

Upon entering the Krebs cycle (Figure 5–6), the two-carbon acetyl group is combined with a four-carbon compound (oxaloacetate) to produce a six-carbon compound (citrate). In the course of the cycle, two of the six carbons are oxidized to CO_2, and oxaloacetate is regenerated—thus literally making this series a cycle. Each turn around the cycle uses up one acetyl group and regenerates one molecule of oxaloacetate, which is then ready to begin the Krebs

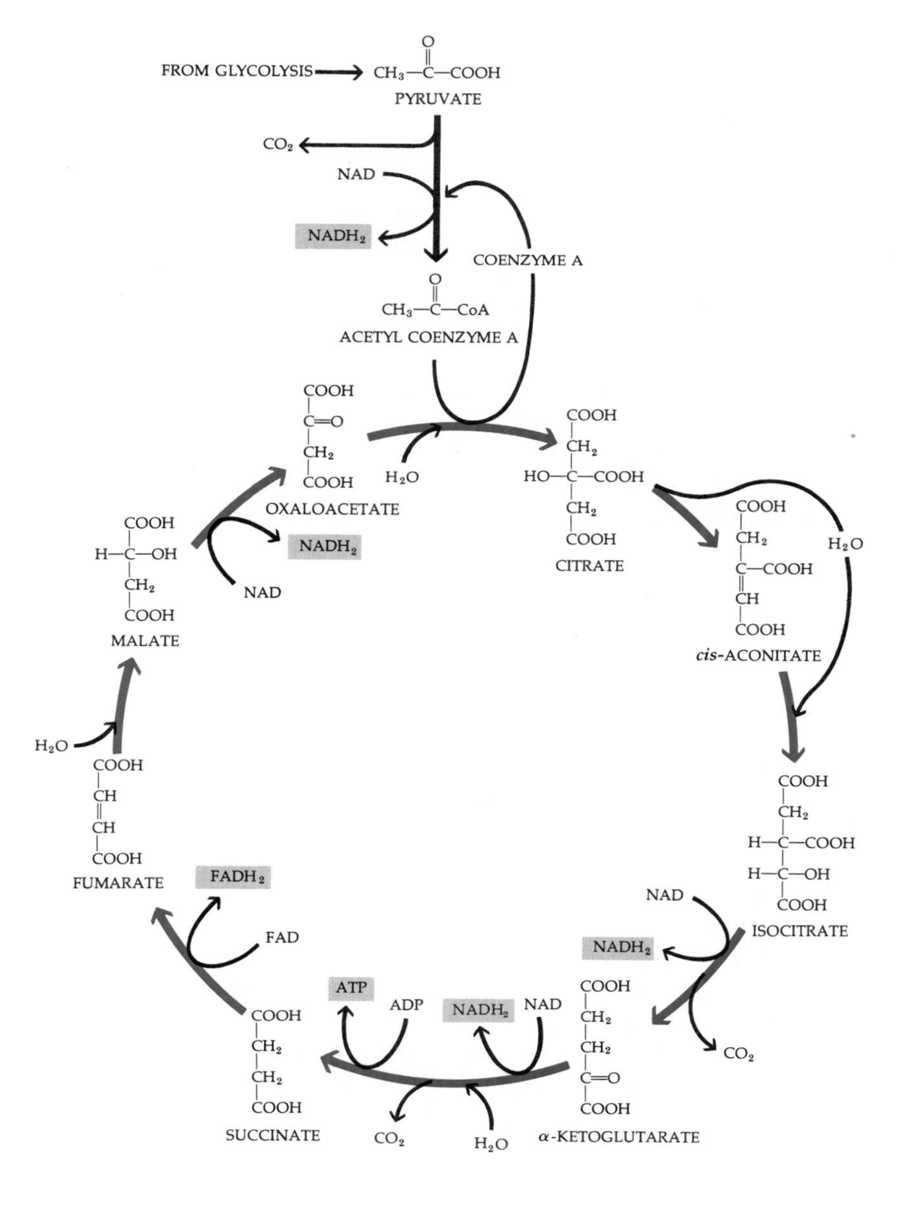

5–6
In the course of the Krebs cycle, the carbons donated by the acetyl group are oxidized to carbon dioxide, and the hydrogen atoms are passed to electron carriers. As in glycolysis, a specific enzyme is involved at each step.

5–7
Flavin adenine dinucleotide, an electron acceptor, in its oxidized form (FAD) and its reduced form ($FADH_2$). Riboflavin is a vitamin (vitamin B_2) made by all plants and many microorganisms. It is a pigment and, in its oxidized form, is bright yellow.

A related electron acceptor, flavin mononucleotide (FMN), consists of riboflavin and the first phosphate group shown here. It accepts electrons from NADH in the electron transport chain.

FAD $\overset{2e^- + 2H^+}{\rightleftharpoons}$ $FADH_2$

cycle again. In the course of these steps, some of the energy released by the oxidation of the carbon atoms is used to convert ADP to ATP (one molecule per cycle), and some is used to convert NAD to $NADH_2$ (three molecules per cycle). In addition, some of the energy is used to reduce a second electron carrier—the coenzyme flavin adenine dinucleotide (FAD) (Figure 5–7). One molecule of $FADH_2$ is formed from FAD in each turn of the cycle. Oxygen is not directly involved in the Krebs cycle; the electrons and protons removed in the oxidation of carbon are all accepted by NAD and FAD:

$$\text{oxaloacetate} + \text{acetyl CoA} + \text{ADP} + 3\text{NAD} + \text{FAD} \longrightarrow$$
$$\text{oxaloacetate} + 2CO_2 + \text{CoA} + \text{ATP} + 3NADH_2 + FADH_2$$

The Krebs cycle is summarized in Figure 5–8.

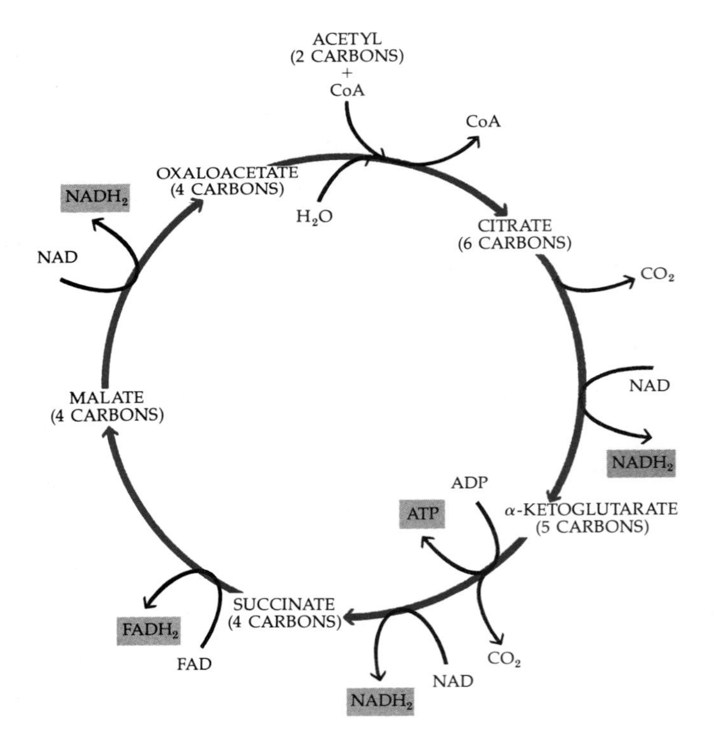

5–8
A summary of the Krebs cycle. One molecule of ATP, three molecules of $NADH_2$, and one molecule of $FADH_2$ represent the energy yield of each acetyl group passing through the cycle.

The Electron Transport Chain

The glucose molecule is now completely oxidized. Some of its energy has been used to produce ATP from ADP. Most of it, however, still remains in the electrons removed from the carbon atoms as they were oxidized. These electrons were passed to the electron carriers NAD and FAD and are at a high energy level. In the course of the electron transport chain, they are passed "downhill" to oxygen, and the energy released is used to form ATP from ADP. This process is known as *oxidative phosphorylation*.

THE MECHANISM OF OXIDATIVE PHOSPHORYLATION

One of the most intriguing puzzles now facing biochemists is the way in which the transfer of electrons from one electron carrier to the next in the electron transport chain drives the phosphorylation of ADP. *There are three current hypotheses.*

CHEMICAL COUPLING

According to this hypothesis, which is the oldest established one, oxidative phosphorylation occurs in a manner similar to the phosphorylation reactions that occur in glycolysis and in the Krebs cycle. By this model, then, energy transfer involves the formation of a high-energy, phosphorylated intermediate, followed by phosphate transfer to ADP. *However, it has not been possible to verify this hypothesis experimentally because researchers have been unable to isolate the hypothetical high-energy intermediate.*

CHEMIOSMOTIC COUPLING

This model is the most popular at the present time. It proposes that oxidative phosphorylation is driven, not by chemical coupling between electron transport and phosphorylation reactions, but by a proton gradient (a differential H^+ *concentration) produced between the two sides of the inner mitochondrial membrane during electron transport. According to this ingenious concept—first proposed in the 1960s by the British biochemist Peter Mitchell (Nobel laureate in chemistry, 1978)—protons are pumped out of the mitochondrial matrix to the outer mitochondrial compartment as electrons from* $NADH_2$ *are passed along the electron transport chain, which forms a part of the mitochondrial membrane. Each pair of electrons crosses the membrane three times as it moves from one electron carrier to the next (and finally to oxygen), transporting two protons across the membrane each time. This results in a* pH *gradient across the membrane that drives the protons back into the matrix through diffusion channels in knoblike structures (which are fancifully called "lollipops") that protrude into the matrix. The movement of the proton back across the membrane provides the energy to form* ATP *from* ADP + P_i.

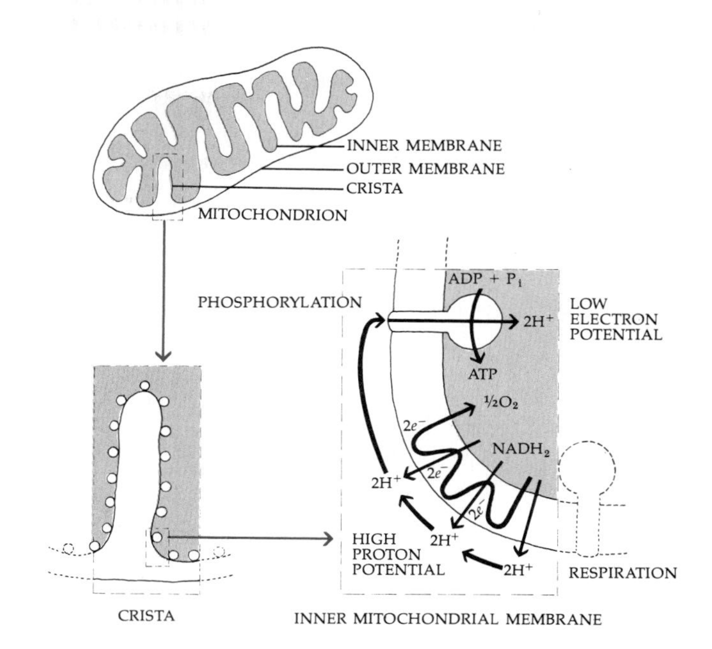

Structure and organization of a mitochondrion in relation to the chemiosmotic explanation for ATP production.

CONFORMATIONAL COUPLING

The third and most recent hypothesis, which was proposed by David E. Green of the University of Wisconsin and Paul D. Boyer of the University of California, suggests that the energy released by electron transport produces temporary changes in the shape of large molecules in the inner mitochondrial complex. The subsequent return of these molecules to their low-energy configuration releases sufficient energy to drive the phosphorylation of ADP. *Support for this model comes from electron micrographs that show changes in mitochondria undergoing active transport.*

The electron carriers of the electron transport chain of the mitochondria differ from NAD and FAD in their chemical structure. Most of them belong to the class of compounds known as cytochromes—protein molecules with an iron-containing porphyrin ring attached (Figure 5–9). Each cytochrome differs in its protein chain and in the energy level at which it holds the electrons.

At the top of the electron transport chain are the electrons held by $NADH_2$ and $FADH_2$. The yield of the Krebs cycle was two molecules of $FADH_2$ and six molecules of $NADH_2$. The oxidation of pyruvate to acetyl CoA yielded two molecules of $NADH_2$. Recall that an additional two molecules of $NADH_2$ were produced in glycolysis; in the presence of oxygen, these $NADH_2$ molecules are transported into the mitochondrion where they are transferred to the electron acceptor flavin mononucleotide (abbreviated FMN; see Figure 5–7) and then are fed into the electron transport chain.

5–9
Cytochromes are molecules in which an atom of iron is held in a nitrogen-containing (porphyrin) ring. They are involved in electron transfer. The iron combines with the electrons, each iron atom accepting an electron as it is reduced from Fe^{3+} *to* Fe^{2+}.

COOH
CH_2—CH_2
CH_3
CH_3
S
CH—CH_3
N
N
Fe
N
N
PROTEIN
CH_2—CH_2
COOH
CH_3
CH_3
CH—S
CH_3

CYTOCHROME

As electrons flow along the electron transport chain from a higher to a lower energy level, the cytochromes harness the released energy and use it to convert ADP to ATP. At the end of the chain, the electrons are accepted by oxygen and combine with protons (hydrogen ions) to produce water. Each time one pair of electrons passes from $NADH_2$ to oxygen, three molecules of ATP are formed. Each time a pair of electrons passes from $FADH_2$, which holds them at a slightly lower energy level than $NADH_2$, two molecules of ATP are formed (see Figure 5–10).

Electrons continue to flow down the electron transport chain only if ADP is available for conversion to ATP. Thus, oxidative phosphorylation is regulated by the "law of supply and demand." When the energy requirements of the cell decrease, fewer molecules of ATP are used, fewer molecules of ADP become available, and the electron flow is decreased.

5–10
A summary of respiration. Glucose is first broken down to pyruvate, with a yield of two ATP molecules and the reduction (dashed arrows) of two NAD molecules. Pyruvate is oxidized to acetyl CoA, and one molecule of NAD is reduced (note that this and subsequent reactions occur twice for each glucose molecule; this electron passage is indicated by solid arrows). In the Krebs cycle, the acetyl group is oxidized and the electron acceptors, NAD and FAD, are reduced. $NADH_2$ *and* $FADH_2$ *then transfer their electrons to the electron-transport chain which consists mostly of a series of cytochromes. As these cytochromes pass the electrons downhill, the energy released is used to form ATP from ADP.*

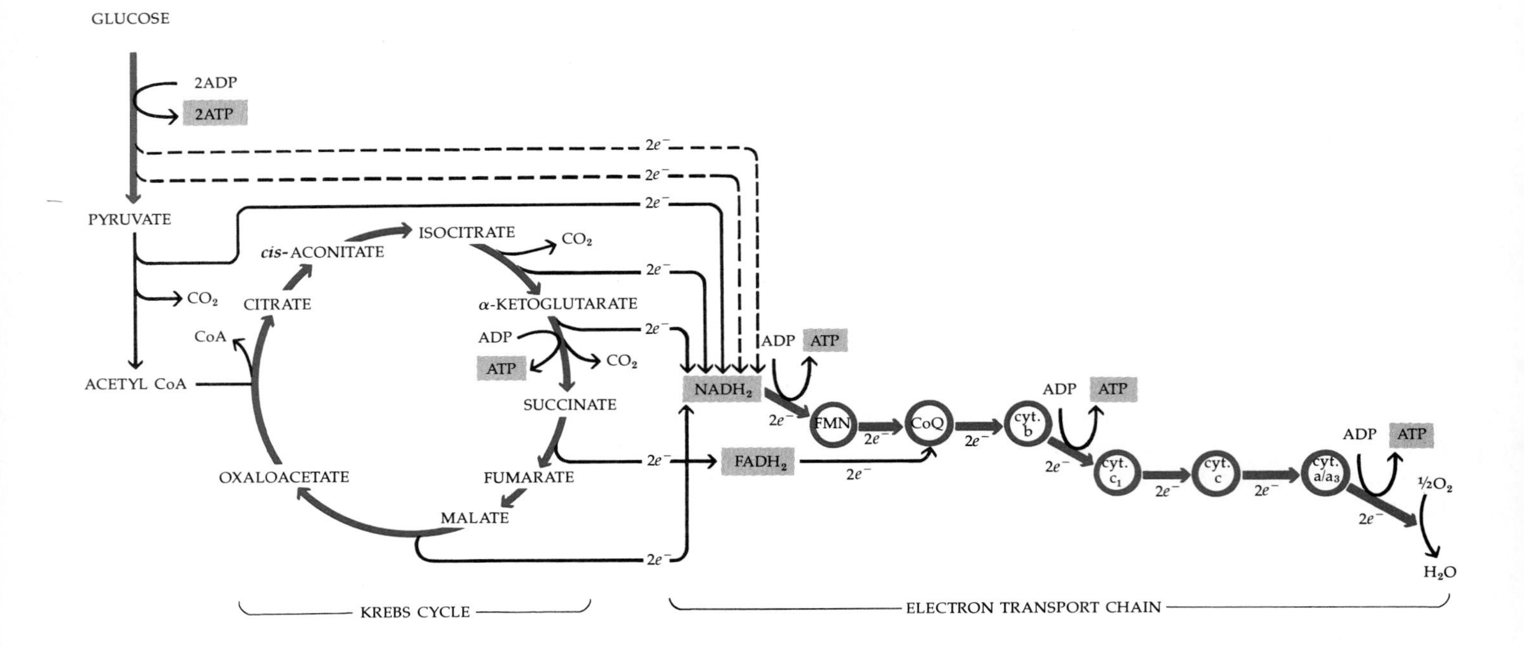

BIOLUMINESCENCE

In most organisms, the energy transferred to ATP is used for cellular work. In some, however, the chemical energy is sometimes reconverted to light energy. Bioluminescence is probably an accidental byproduct of energy exchanges in most luminescent organisms, such as the fungi (Mycena lux-coeli) *shown here, photographed by their own light. In some luminescent organisms, such as fireflies, in which the flashes of light serve as mating signals, bioluminescence serves a useful function.*

Overall Energy Harvest

We are now in a position to see how much of the energy originally present in the glucose molecule has been recovered in the form of ATP. The "balance sheet" for ATP yield given in Table 5–1 may help you keep track of the discussion that follows.

Glycolysis, in the presence of oxygen, yields two molecules of ATP directly, plus two molecules of $NADH_2$ (from which an additional six molecules of ATP are formed). The total gain, however, is not 8 ATPs, as you might calculate, but rather 6 ATPs. There is a "cost" of 2 ATPs to transport the electrons held by the two molecules of $NADH_2$ across the mitochondrial membranes.

The conversion of pyruvate to acetyl CoA yields two molecules of $NADH_2$ (inside the mitochondrion) for each molecule of glucose and so produces 6 molecules of ATP.

The Krebs cycle yields, for each molecule of glucose, two molecules of ATP, six molecules of $NADH_2$, and two molecules of $FADH_2$, for a total of 24 molecules of ATP.

As the balance sheet shows, the complete yield from a single molecule of glucose is 36 molecules of ATP. All but two of the 36 molecules of ATP have come from reactions in the mitochondrion, and all but four involve the oxidation of $NADH_2$ or $FADH_2$ along the electron transport chain.

The total difference in free energy between the reactants (glucose and oxygen) and the products (carbon dioxide and water) is 686 kilocalories. Approximately 39 percent of this, or 263 kilocalories (7.3 × 36), has been captured in the high-energy bonds of the 36 ATP molecules (Figure 5–11).

Table 5–1 *Energy Yield from One Molecule of Glucose*

Glycolysis:	2 ATP $2\ NADH_2 \longrightarrow 4$ ATP		$\longrightarrow$ 6 ATP
Pyruvate $\longrightarrow$ Acetyl CoA:	$1\ NADH_2 \longrightarrow 3$ ATP	(× 2)	$\longrightarrow$ 6 ATP
Krebs cycle:	1 ATP $3\ NADH_2 \longrightarrow 9$ ATP $1\ FADH_2 \longrightarrow 2$ ATP	(× 2)	$\longrightarrow$ 24 ATP

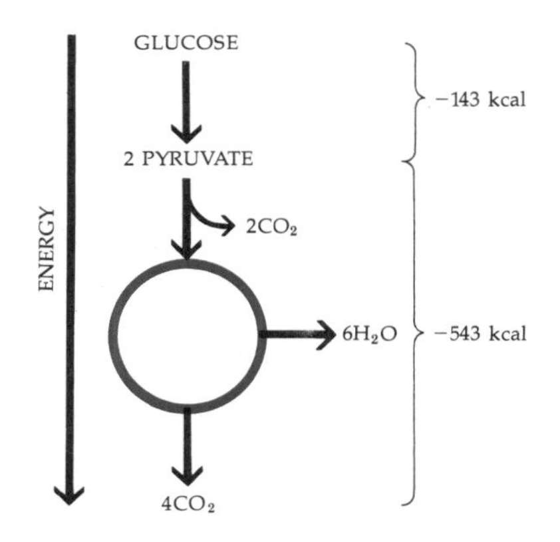

5–11
The energy changes in the oxidation of glucose. The complete respiratory sequence proceeds with an energy drop of 686 kilocalories. Of this, about 39 percent (263 kilocalories) is conserved in 36 ATP molecules. By contrast, in anaerobic respiration only two ATP molecules are produced, representing only about 2 percent of the available energy of glucose.

ANAEROBIC PATHWAYS

In eukaryotic cells (as well as most prokaryotes), pyruvate usually follows the aerobic pathway and is completely oxidized to carbon dioxide and water. However, in the absence of oxygen, pyruvate is not the end product of glycolysis because the $NADH_2$ produced during the oxidation of glyceraldehyde 3-phosphate must be reoxidized to NAD. Without this reoxidation, glycolysis would soon stop, because the cell would run out of NAD as an electron acceptor. This oxygenless, or anaerobic, process may result in the formation of lactate in many bacteria, fungi, and animal cells; it is therefore called *lactate fermentation*. For example, lactate is produced in muscle cells during bursts of unusually vigorous work, as when an athlete runs a sprint. The muscles accumulate what is known as an "oxygen debt" by producing lactate from glucose. The lactate lowers the pH of the muscles and reduces the capacity of the muscle fibers to contract, producing the sensation of muscle fatigue. The lactate diffuses into the blood and is carried to the liver where it is resynthesized to pyruvate and then to glucose or glycogen.

In yeasts and most plant cells, pyruvate is broken down to ethanol and carbon dioxide under anaerobic conditions. Under anaerobic conditions, $NADH_2$ is reoxidized by the transfer of electrons (and protons) to pyruvate. This process is preceded by the removal of carbon dioxide (decarboxylation) and results in the production of ethanol and carbon dioxide instead of lactate (Figure 5–12). Since the principal end product of glycolysis is ethanol, this process is called *alcoholic fermentation*.

Yeast cells are present as a "bloom" on the skin of grapes. When the glucose-filled juices of grapes and other fruits are extracted and stored in airtight kegs, these yeast cells turn the fruit juice to wine by converting glucose to ethanol. Yeasts, like all living things, have a limited tolerance for alcohol, and when the critical concentration (about 12 percent) is reached, the yeast cells cease to function.

Thermodynamically, lactate fermentation and alcoholic fermentation are similar. In both, the $NADH_2$ is reoxidized, and the energy yield is only the 2 ATPs harvested during glycolysis. The complete, balanced equation for the fermentation of glucose can be written as follows:

$$\text{glucose} + 2\text{ADP} + 2\text{P}_i \longrightarrow \left\{ \begin{matrix} 2\text{ lactate} \quad \textit{or} \\ 2\text{ ethanol} + 2CO_2 \end{matrix} \right\} + 2\text{ATP}$$

During alcoholic fermentation, approximately 7 percent of the total available energy of the glucose molecule—about 52 kilocalories—is released, with about 93 percent remaining in the two alcohol molecules. Only 14.6 kilocalories of the 52 kilocalories released during fermentation are trapped and stored in the two molecules of ATP. So, in terms of energy yield, anaerobic fermentation is relatively inefficient.

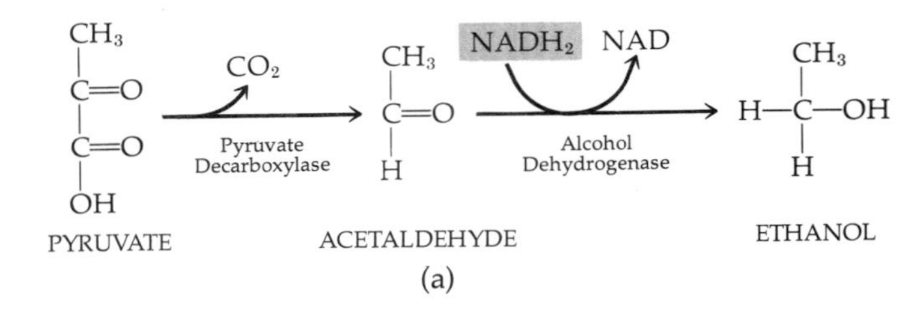

(a)

(b)

5–12
(a) *The steps by which pyruvate is converted anaerobically to ethanol. In the first step, carbon dioxide is released. In the second,* $NADH_2$ *is oxidized and acetaldehyde is reduced. Most of the energy of the glucose remains in the alcohol, which is the principal end product of the sequence. However, by regenerating NAD, these steps allow glycolysis to continue, with its small but sometimes vital yield of ATP.* (b) *The consequences of anaerobic glycolysis. Yeast cells visible on the grapes as dustlike "bloom" mix with the juice when the grapes are crushed. Storing the mixture under anaerobic conditions causes the yeast to break down the glucose in the grape juice to alcohol.*

SUMMARY

Respiration, or the complete oxidation of glucose, is the chief source of energy in most cells. As the glucose is broken down in a series of small enzymatic steps, the energy in the molecule is packaged in the form of high-energy bonds in molecules of ATP.

The first phase in the breakdown of glucose is glycolysis, in which the six-carbon glucose molecule is split into two three-carbon molecules of pyruvate; two new molecules of ATP and two of $NADH_2$ are formed. This reaction occurs in the cytoplasmic ground substance.

In the course of respiration, the three-carbon pyruvate molecules are broken down within the mitochondrion to two-carbon acetyl groups, which then enter the Krebs cycle. In the Krebs cycle, the acetyl group is broken apart in a series of reactions to yield carbon dioxide. In the course of the oxidation of each acetyl group, four electron acceptors (three NAD and one FAD) are reduced, and another molecule of ATP is formed.

The final stage of respiration is the electron transport chain, which involves a series of electron carriers and enzymes embedded in the inner membranes of the mitochondrion. Along this series of electron carriers, the high-energy electrons accepted by $NADH_2$ and $FADH_2$ during the Krebs cycle pass "downhill" to oxygen. Each time a pair of electrons passes down through the electron-transport chain, ATP molecules are formed from ADP and phosphate. In the course of the breakdown of the glucose molecule, 36 molecules of ATP are formed, most of them in the mitochondrion.

In the absence of oxygen, pyruvate can be converted either to lactate (in many bacteria, fungi, and animal cells) or to ethanol and carbon dioxide (in yeasts and most plant cells). These anaerobic processes—called fermentation—yield 2 ATPs for each glucose molecule (two pyruvate molecules).

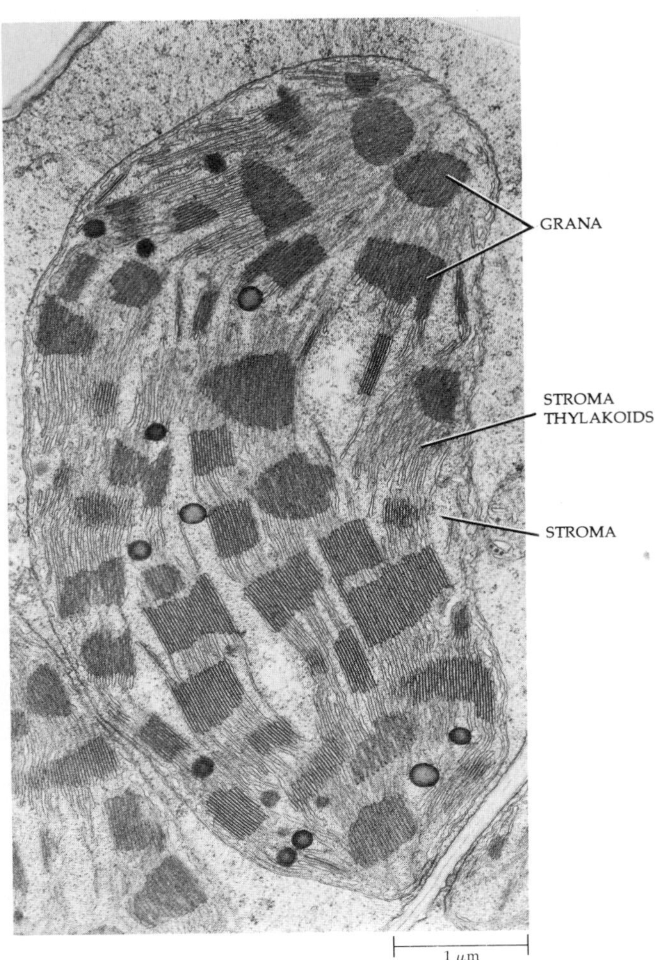

6–1
The chloroplast is the site of photosynthesis in eukaryotic organisms. The light-capturing reactions occur in thylakoids, where the chlorophylls and other pigments are found. In chloroplasts, many of the thylakoids characteristically occur in disklike stacks called grana. The series of reactions by which this energy is transferred to carbon-containing compounds occurs in the stroma, the material surrounding the thylakoids.

CHAPTER 6

Photosynthesis

In the previous chapter, we described the breakdown of carbohydrates to yield the energy required for the many kinds of activities carried out by living systems. In the pages that follow, we will complete the circle by describing the way in which energy coming from the sun in the form of light is captured and converted to chemical energy.

This process—photosynthesis—is the route by which virtually all energy enters our biosphere. Each year more than 150 billion tons of sugar are produced worldwide by the photosynthetic process. The importance of photosynthesis, however, extends far beyond the sheer weight of this product. Without this flow of energy from the sun, channeled largely through the chloroplasts of eukaryotic cells, the pace of life on this planet would swiftly diminish and then, following the inexorable second law of thermodynamics, would cease altogether.

OVERVIEW OF PHOTOSYNTHESIS

The importance of photosynthesis in the economy of nature was not recognized until comparatively recent times. Aristotle and other Greeks, observing that the life processes of animals were dependent upon the food they ate, thought that plants derived their food from the soil.

A little over 300 years ago, in one of the first carefully designed biological experiments ever reported, the Belgian physician Jan Baptista van Helmont offered the first experimental evidence that soil alone does not nourish the plant. Van Helmont grew a small willow tree in an earthenware pot, adding only water to the pot. At the end of 5 years, the willow had increased in weight by 74.4 kilograms, whereas the earth had decreased in weight by only 57 grams. On the basis of these results, van Helmont concluded that all the substance of the plant was produced from the water and none from the soil.

Toward the end of the eighteenth century, Joseph Priestley (1733–1804) reported that he had "accidently hit upon a method of restoring air that had been injured by the burning of candles." On 17 August 1771, Priestley "put a [living] sprig of mint into air in which a wax candle had burned out and found that, on the 27th of the same month, another candle could be burned in this same air." The "restorative which nature employs for this purpose," he stated, was "vegetation." Priestley extended his observations and soon showed that air "restored" by vegetation was not "at all inconvenient to a mouse." Priestley's experiments offered the first logical explanation of how air remained "pure" and able to support life despite the burning of countless fires and the breathing of many animals. When he was presented with a medal for his discovery, the citation read in part: "For these discoveries we are assured that no vegetable grows in vain . . . but cleanses and purifies our atmosphere." Today we would explain Priestley's experiments simply by saying that plants take up the CO_2 produced by combustion or exhaled by animals, and that animals inhale the O_2 released by plants.

Later, the Dutch physician Jan Ingenhousz (1730–1799) confirmed Priestley's work and showed that the air was restored only in the presence of sunlight and only by the green parts of plants. In 1796, Ingenhousz suggested that carbon dioxide is split in photosynthesis to yield carbon and oxygen, with the oxygen then released as gas. Subsequently, the proportion of carbon, hydrogen, and oxygen atoms in sugars and starches was found to be about one atom of carbon per molecule of water (CH_2O), as the word "carbohydrate" indicates. Thus, in the overall reaction for photosynthesis,

$$CO_2 + H_2O + \text{light energy} \longrightarrow (CH_2O) + O_2$$

it was generally assumed that the carbohydrate came from a combination of the carbon and water molecules and that the oxygen was released from the carbon dioxide molecule. This entirely reasonable hypothesis was widely accepted; but, as it turned out, it was quite wrong.

The investigator who upset this long-held theory was C. B. van Niel of Stanford University. Van Niel, then a graduate student, was investigating the activities of different types of photosynthetic bacteria. In their photosynthetic reactions, one particular group of such bacteria—known as the purple sulfur bacteria—reduces carbon to carbohydrates, but they do not release oxygen. The purple sulfur bacteria require hydrogen sulfide for their photosynthetic activity. In the course of photosynthesis, globules of sulfur accumulate inside the bacterial cells (Figure 6–2). In these bacteria, van Niel found that the following reaction takes place during their photosynthesis:

$$CO_2 + 2H_2S \xrightarrow{\text{light}} (CH_2O) + H_2O + 2S$$

This finding was simple and did not attract much attention until van Niel made a bold extrapolation.

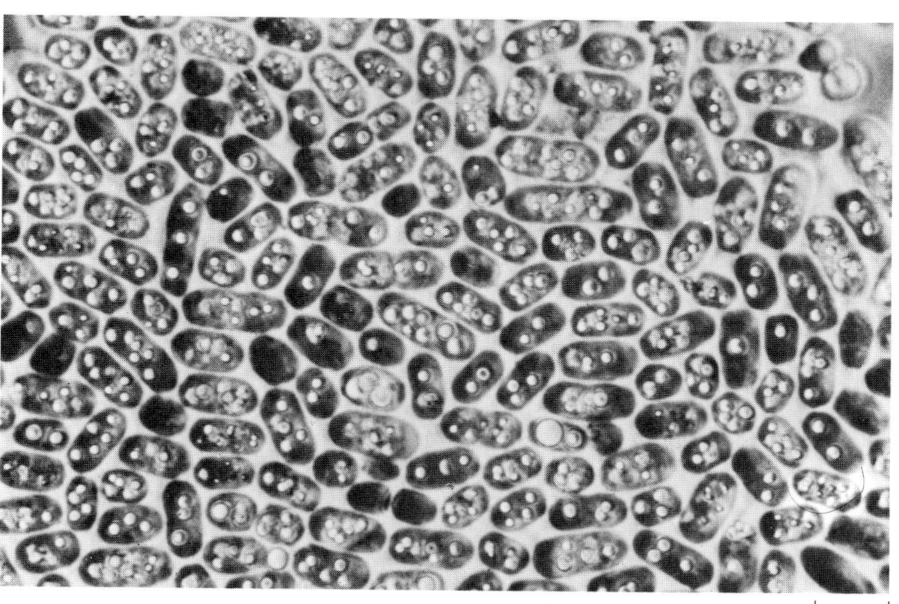

6–2
Purple sulfur bacteria. In these cells, hydrogen sulfide plays the same role as water does in the photosynthetic process of plants. The hydrogen sulfide is split, and the released sulfur accumulates in globules, visible within these cells.

LIGHT AND LIFE

Almost 300 years ago, the English physicist Sir Isaac Newton (1642–1727) separated light into a spectrum of visible colors by letting it pass through a prism. By this experiment, Newton showed that white light is actually made up of a number of different colors, ranging from violet at one end of the spectrum to red at the other. Their separation is possible because light of different colors is bent at different angles in passing through a prism.

In the nineteenth century, through the genius of James Clerk Maxwell (1831–1879), it came to be known that what we experience as light is but a very small part of a vast continuous spectrum of radiation, the electromagnetic spectrum. As Maxwell showed, all the radiations included in this spectrum travel in waves. The wavelengths—that is, the distances from one peak to the next—vary from those of X-rays, which are measured in nanometers, to those of low-frequency radio waves, which are measured in kilometers. The shorter the wavelength, the greater the energy. Within the spectrum of visible light, red light has the longest wavelength and violet has the shortest. Another feature that these radiations have in common is that, in a vacuum, they all travel at the same speed—300,000 kilometers per second.

By 1900 it had become clear, however, that the wave theory of light was not adequate. The key observation, a very simple one, was made in 1888: when a zinc plate is exposed to ultraviolet light, it acquires a positive charge. The metal, it was soon deduced, becomes positively charged because the radiation energy dislodges electrons, forcing them out of the metal atoms. Subsequently it was discovered that this photoelectric effect, as it is known, can be produced in all metals. Each metal has a critical wavelength for the effect; the radiation (visible or invisible) must be of that particular wavelength or shorter (must be more energetic) for the effect to occur. (The hypothesis that electrons orbit atoms at particular energy levels, as formulated by Bohr and others, is based on these observations.)

With some metals, such as sodium, potassium, and selenium, the critical wavelength is within the spectrum of visible light, and as a consequence, visible light striking the metal can set up a moving stream of electrons (an electric current). The electric eyes that open doors at supermarkets or airline terminals, exposure meters, and television cameras all operate on this principle of turning light energy into electrical energy.

WAVE OR PARTICLE?

Now here is the problem. The wave theory of light would lead one to predict that the brighter the light, the greater the force with which the electrons would be dislodged. Whether or not light can eject the electrons of a particular metal, however, depends on its wavelength, not on its brightness. A very weak beam of the critical wavelength is effective, whereas a stronger (brighter) beam of a longer wavelength is not. Furthermore, increasing the brightness of light of a critical intensity increases the number of electrons dislodged but not the velocity at which they are ejected from the metal. To increase the velocity, one must use a shorter wavelength of light. Nor is it necessary for energy to accumulate in the metal. With even a dim beam of a critical wavelength, an electron may be emitted the instant the light hits the metal.

To explain such phenomena, the particle theory of light was proposed by Albert Einstein in 1905. According to this theory, light is composed of particles of energy called photons, or quanta of light. The energy of a photon (a quantum of light) is not the same for all kinds of light but is, in fact, inversely proportional to the wavelength—the longer the wavelength, the lower the energy. Photons of violet light, for example, have almost twice the energy of photons of red light, the longest visible wavelength.

The wave theory of light permits physicists to describe certain aspects of its behavior mathematically, whereas the photon theory permits another set of mathematical calculations and predictions. These two models are no longer regarded as opposing one another; rather, they are complementary, in the sense that both are required for a complete description of the phenomenon we know as light.

The coexistence of these two theories further illustrates a facet of the scientific method. If a scientist defines light as a wave and measures it as such, it behaves like a wave. Similarly, if one defines it as a particle, one gets information about light as a particle. As Einstein once said, "It is the theory which determines what we can observe."

THE FITNESS OF LIGHT

Light, as Maxwell showed, is only a tiny band in a continuous spectrum. From the physicist's point of view, the difference between light and darkness—so dramatic to the human eye—is only a few nanometers of wavelength. There is no qualitative difference marking the borders of the light spectrum. Why is it that this tiny portion of the electromagnetic spectrum is responsible for vision, for phototropism (the movement of an organism toward light), for photoperiodism (the seasonal changes that take

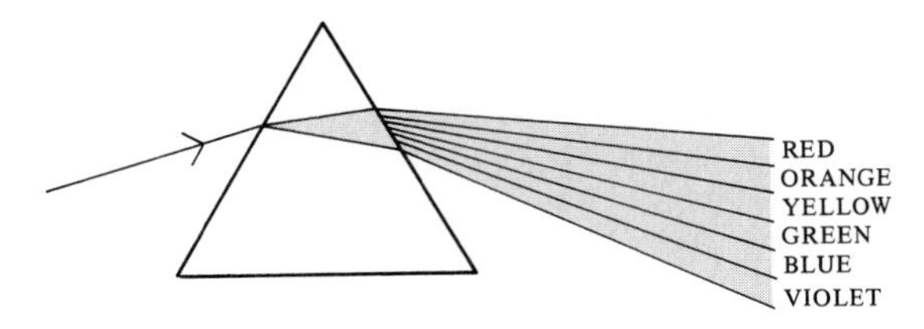

When white light passes through a prism, it is sorted into a spectrum of different colors. This separation occurs because each color has slightly different wavelengths.

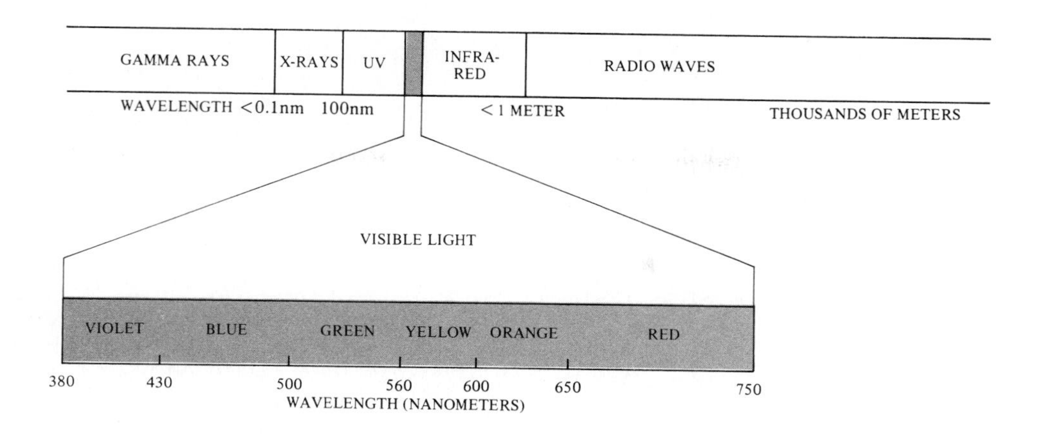

place in an organism with the changing length of day and night), and also for photosynthesis, on which all life depends? Is it an amazing coincidence that all these biological activities depend on these same wavelengths?

George Wald of Harvard University, one of the greatest living experts on the subject of light and life, says no. He thinks that if life exists elsewhere in the universe, it is probably dependent upon this same fragment of the vast spectrum of radiation. Wald bases this conjecture on two points. First, living things are composed of large, complicated molecules held in special configurations and relationships to one another by hydrogen bonds and other weak bonds. Radiation of even slightly higher energies than the energy of violet light can break these bonds and so disrupt the structure and function of the molecules. Radiation with a wavelength below 200 nanometers drives electrons out of atoms to create ions; hence, it is called ionizing radiation. The energy of light of wavelengths longer than those of the visible band is largely absorbed by water, which makes up the bulk of all living things. When such light does reach organic molecules, its lower energy causes them to increase their motion (increasing heat) but does not trigger changes in their structure. Only that radiation within the range of visible light has the property of exciting molecules—that is, of raising electrons from one energy level to another—and so of producing biological changes.

The second reason that the visible band, above all others of the electromagnetic spectrum, was "chosen" by living things is that it, above all, is what is available. Most of the radiation reaching the earth from the sun is within this range. Higher-energy wavelengths are screened out by the oxygen and ozone high in the atmosphere. Much infrared radiation is screened out by water vapor and carbon dioxide before it reaches the earth's surface.

This is an example of what has been termed "the fitness of the environment"; the suitability of the environment for life and the suitability of life for the physical world are exquisitely interrelated. If they were not, life could not exist.

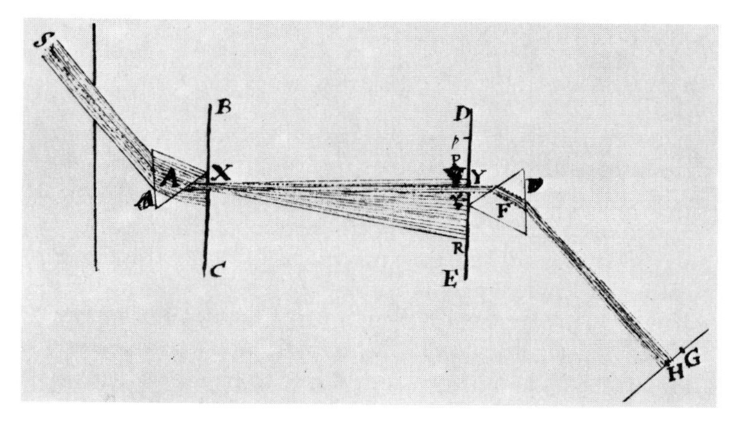

Manuscript drawing from one of Newton's papers, showing what he considered to be a "crucial experiment" in his study of the mixture of colors in "white" light. Sunlight enters a darkened room from the left, producing a spectrum of colors on passing through the first prism *A*. The hole at *Y* in the opaque board at the right permits light of a single color to pass through the second prism *F*, with the result that the beam of light is bent, but there is no further alteration in color. Newton thus demonstrated that a prism does not alter the light in producing a spectrum; rather, it merely "bends" the various colors to different degrees.

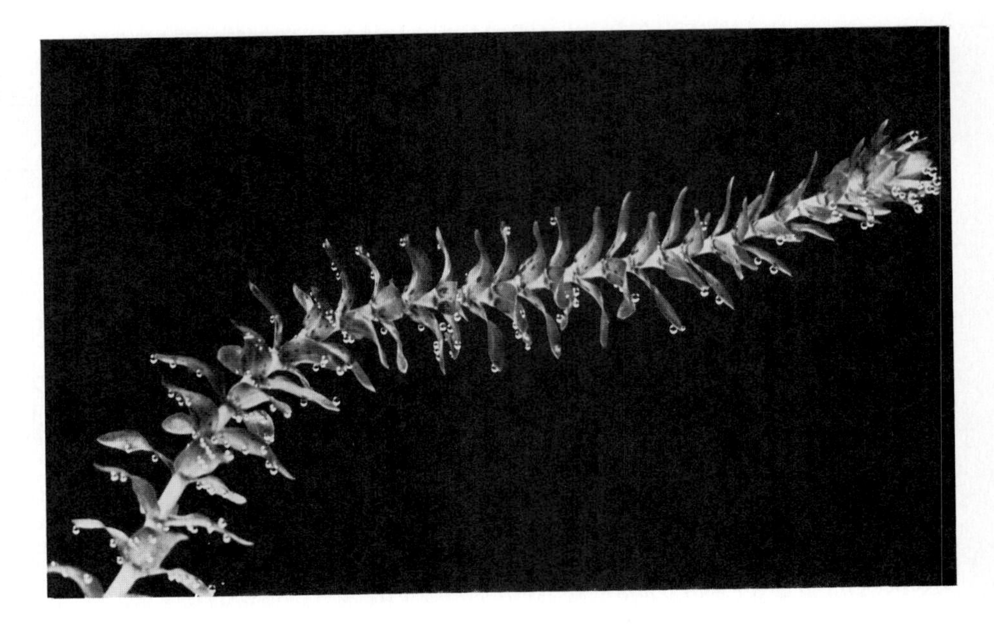

6–3
The bubbles on the leaves of this submersed pondweed, Elodea, *are bubbles of oxygen, one of the products of photosynthesis. Van Niel was the first to propose that the oxygen produced in photosynthesis came from the splitting of water rather than the breakdown of carbon dioxide.*

Van Niel proposed the following generalized equation for photosynthesis:

$$CO_2 + 2H_2A \xrightarrow{\text{light}} (CH_2O) + H_2O + 2A$$

In this equation, H_2A represents either water or some oxidizable substance, such as hydrogen sulfide or free hydrogen. In algae and green plants, H_2A is water (Figure 6–3). In short, van Niel proposed that water, and *not* carbon dioxide, was split in photosynthesis.

This brilliant speculation, first proposed in the early 1930s, was not proven until many years later, when investigators using a heavy isotope of oxygen ($^{18}O_2$) traced the oxygen from water to oxygen gas:

$$CO_2 + 2H_2{}^{18}O \xrightarrow{\text{light}} (CH_2O) + H_2O + {}^{18}O_2$$

Thus, in the case of algae and green plants, in which water serves as the electron donor, the complete, balanced equation for photosynthesis becomes:

$$6CO_2 + 12H_2O \xrightarrow{\text{light}} C_6H_{12}O_6 + 6O_2 + 6H_2O$$

About two hundred years ago, as noted above, light was discovered to be required for the process we now call photosynthesis. It is now known that photosynthesis occurs in two stages, only one of which actually requires light. Evidence for this two-stage process was first presented in 1905 by the English plant physiologist F. F. Blackman, as the result of experiments in which he measured the individual and combined effects of changes in light intensity and temperature on the rate of photosynthesis.

Blackman made the following conclusions on the basis of his experiments: (1) There was one set of light-dependent reactions which were temperature-independent. The rate of these reactions in the dim to moderate light range could be accelerated by increasing the amount of light (Figure 6–4a), but it was not accelerated by increases in temperature (Figure 6–4b). (2) There was a second set of reactions which were dependent, not on light, but on temperature. When both light and temperature were increased, the rate of photosynthesis was greatly accelerated (Figure 6–4b). Both sets of reactions seemed to be required for the process of photosynthesis. Increasing the rate of only one set of reactions increased the rate of the entire process, but only to the point at which the second set of reactions began to hold back the first (that is, the second set became rate-limiting). Thereafter, it was necessary to increase the rate of the second set of reactions in order for the first to proceed unimpeded.

Photosynthesis was thus shown to have both a light-dependent stage, the "light reactions," as well as a light-independent stage, the "dark reactions." It is important to remember that the dark reactions normally occur in the light; they require the products of the light reactions. The expression "dark reactions" merely indicates that light is not directly involved.

The dark reactions increased in rate as the temperature was increased, but only up to about 30°C, after which the rate began to decrease. From this evidence it was concluded that these reactions were controlled by enzymes, since this is the way enzymes are expected to respond to temperature (see Figure 4–9). This conclusion has since been proven to be correct.

6–4
(a) *In dim to moderate light, increasing the light intensity increases the rate of photosynthesis, but at higher intensities, a further increase in light intensity has no effect. A curve such as the one shown here indicates that some other factor—known as a rate-limiting factor—is involved in the process under study.* CO_2 *concentration is commonly the rate-limiting factor in photosynthesis.*
(b) *At a low intensity of light, an increase in temperature does not increase the rate of photosynthesis (lower curve). At a high intensity, however, an increase in temperature has a very marked effect (upper curve). Based on these data, Blackman concluded that photosynthesis included both light-dependent and light-independent reactions*

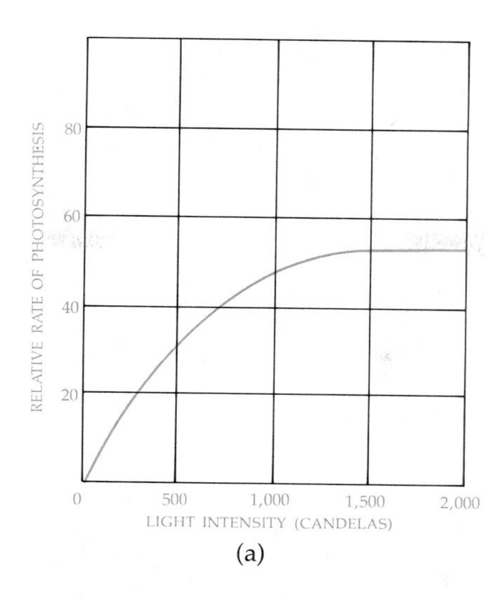

(a)

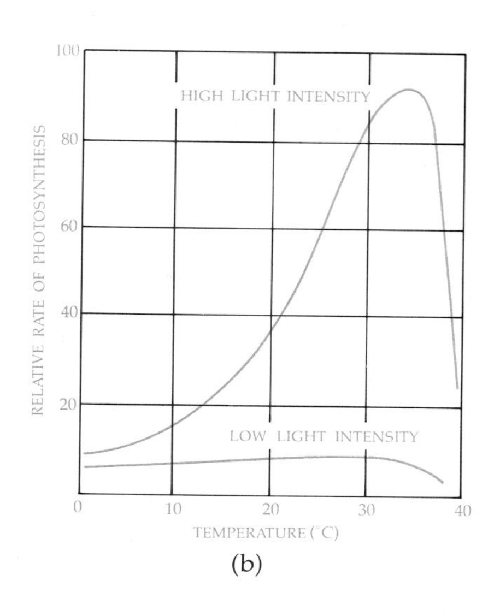

(b)

In the first stage of photosynthesis—the light reactions—light energy is used to form ATP from ADP and to reduce electron carrier molecules. In the second stage of photosynthesis—the dark reactions—the energy products of the first stage are used to reduce carbon from carbon dioxide to a simple sugar, thus converting the chemical energy of the carrier molecules to forms suitable for transport and storage and, at the same time, forming a carbon skeleton on which other organic molecules can be built. This conversion of CO_2 into organic compounds is known as *carbon fixation*.

THE LIGHT REACTIONS

The Role of Pigments

The first step in the conversion of light energy to chemical energy is the absorption of light. A *pigment* is any substance that absorbs visible light. Some pigments absorb all wavelengths of light and so appear black. Some absorb only certain wavelengths and transmit or reflect the wavelengths they do not absorb. Chlorophyll, the pigment that makes leaves green, absorbs light principally in the violet and blue wavelengths and also in the red; because it reflects green light, it appears green. The absorption pattern of a pigment is known as the *absorption spectrum* of that substance (Figure 6–5).

One of the best proofs we have that chlorophyll is the principal pigment involved in photosynthesis is the similarity between its absorption spectrum and the action spectrum for photosynthesis (Figure 6–6).

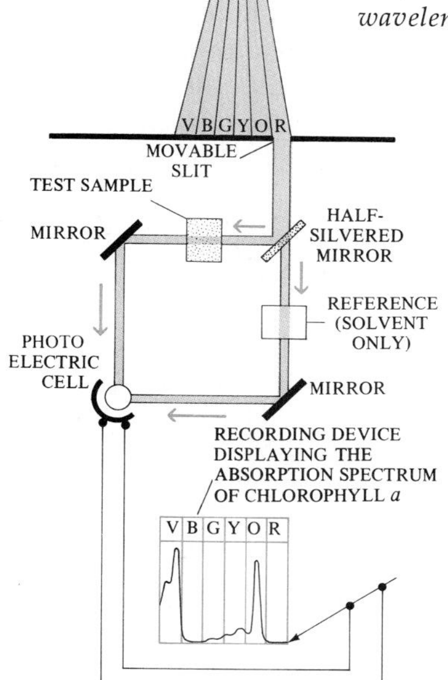

6–5
The absorption spectrum of chlorophyll a *as measured by a spectrophotometer. This device directs a beam of light of each wavelength at the object to be analyzed and records the percentage of light of each wavelength absorbed by the pigment sample as compared with a reference sample. Because the mirror is lightly (half-) silvered, half of the light is reflected and half is transmitted. The photoelectric cell is connected to an electronic device that records the percentage absorption at each wavelength.*

6–6
Results of an experiment performed in 1882 by T. W. Englemann revealed the action spectrum of photosynthesis in a filamentous alga. Like investigators working more recently, Englemann used the rate of oxygen production to measure the rate of photosynthesis. Unlike his successors, however, he lacked sensitive devices for detecting oxygen. As his oxygen indicator, he chose bacteria that are attracted by oxygen. In place of the mirror and diaphragm usually used to illuminate objects under view in his microscope, he substituted a "microspectral apparatus" which, as its name implies, produced a tiny spectrum of colors that it projected upon the slide under the microscope. Then he arranged a filament of algal cells parallel to the spread of the spectrum. The oxygen-seeking bacteria congregated mostly in the areas where the violet and red wavelengths fell upon the algal filament. As you can see, the action spectrum for photosynthesis that Englemann revealed in his experiment paralleled the absorption spectrum of chlorophyll. He concluded that photosynthesis depends on the light absorbed by chlorophyll. This is an example of the sort of experiment that scientists refer to as "elegant," for it was not only brilliant but also was simple in design and conclusive in its results.

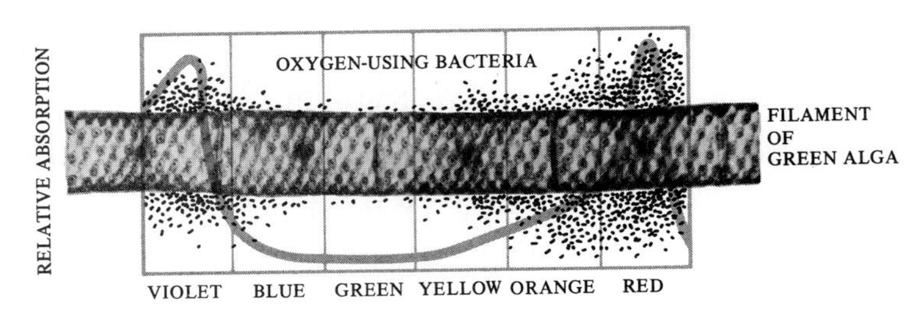

An action spectrum defines the relative effectiveness of different wavelengths of light for light-requiring processes, such as photosynthesis, flowering, and phototropism (the movement or bending of an organism toward the light). Similarity between the absorption spectrum of a pigment and the action spectrum of a process is considered as evidence that that particular pigment is responsible for that particular process (Figure 6–7).

When pigments absorb light, electrons are boosted to a higher energy level, with three possible consequences: (1) the energy may be dissipated as heat; (2) it may be reemitted almost instantaneously as light energy of longer wavelength, a phenomenon known as fluorescence (when it is reemitted as light at a later time, it is known as phosphorescence); or (3) the energy may be captured in a chemical bond, as it is in photosynthesis.

If chlorophyll molecules are isolated in a test tube and light is permitted to strike them, they fluoresce. In other words, the pigment molecules absorb light energy, so the electrons are momentarily raised to a higher energy level and then fall back again to a lower one. As they fall to a lower energy level, they release much of this energy as light. None of the light absorbed by isolated chlorophyll molecules is converted to any form of energy useful to living systems. Chlorophyll can convert light energy to chemical energy only when it is associated with certain proteins and embedded in a specialized membrane (a thylakoid).

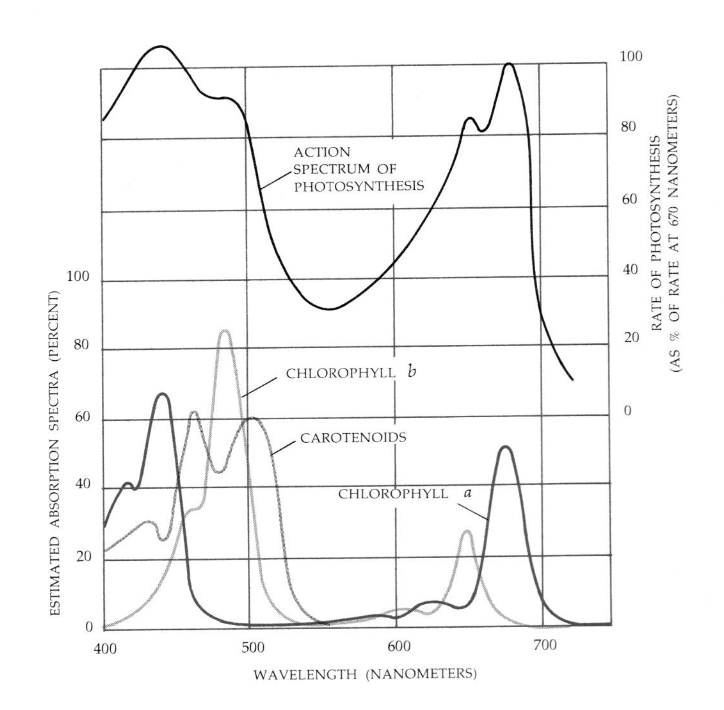

6–7
The upper curve shows the action spectrum for photosynthesis, and the lower curves show the absorption spectra for chlorophyll a, *chlorophyll* b, *and the carotenoids in the chloroplast. The action spectrum of photosynthesis indicates that chlorophyll* a, *chlorophyll* b, *and the carotenoids all absorb the light used in photosynthesis.*

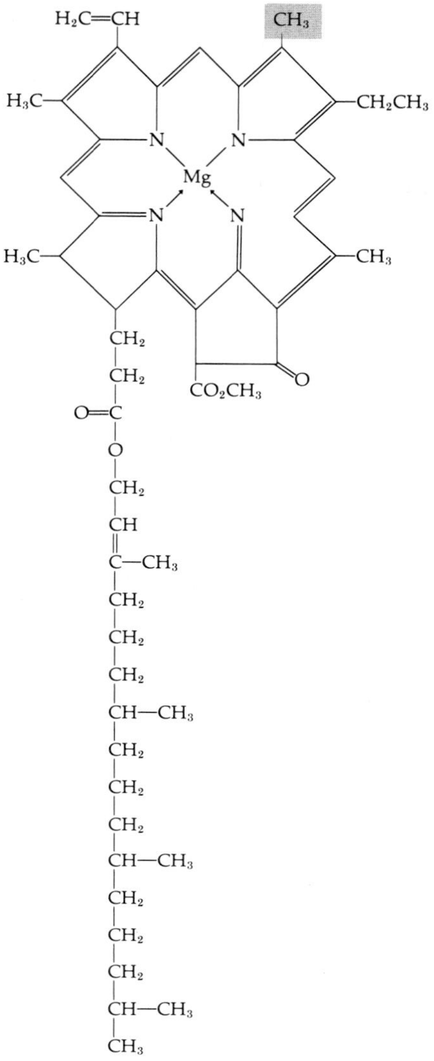

6–8
Chlorophyll a *is a large molecule with a central core of magnesium held in a porphyrin ring. Attached to the ring is a long, insoluble carbon-hydrogen chain, which serves to anchor the molecule in the internal membranes of the chloroplast. Chlorophyll* b *differs from chlorophyll* a *in having a* $-CHO$ *group in place of the* $-CH_3$ *group indicated in color here. Alternating single and double bonds, known as conjugate bonds, such as those in the porphyrin ring of chlorophylls, are common among pigments. Note the similarity between the chlorophyll* a *molecule shown here and the cytochrome molecule of Figure 5–9.*

Photosynthetic Pigments

Pigments that participate in photosynthesis include the chlorophylls, the carotenoids, and the phycobilins.

There are several different kinds of chlorophyll which differ from one another only in the details of their molecular structure. Chlorophyll *a* (Figure 6–8) occurs in all photosynthetic eukaryotes and in the prokaryotic cyanobacteria, and it is considered to be essential for the type of photosynthesis carried out by organisms of these groups.

The vascular plants, bryophytes, green algae, and euglenoid algae also contain chlorophyll *b*. Chlorophyll *b* is an accessory pigment, and like the other accessory pigments, it serves to broaden the spectrum of light absorption in photosynthesis. When a molecule of chlorophyll *b* absorbs light, the excited molecule transfers its energy to a molecule of chlorophyll *a*, which then proceeds to transform it into chemical energy during the course of photosynthesis. Because chlorophyll *b* absorbs light of different wavelengths than does chlorophyll *a* (see Figure 6–7), it extends the range of light that can be used for photosynthesis. In the leaves of green plants, chlorophyll *b* generally constitutes about one-fourth of the total chlorophyll content.

Chlorophyll *c* takes the place of chlorophyll *b* in some groups of algae, most notably the brown algae and the diatoms. The photosynthetic bacteria (other than cyanobacteria) cannot extract electrons from water and thus do not evolve oxygen. They contain either bacteriochlorophyll (in purple bacteria) or chlorobium chlorophyll (in green sulfur bacteria). Chlorophylls *b* and *c* and the photosynthetic pigments of purple bacteria and green sulfur bacteria are simply chemical variations of the basic structure shown in Figure 6–8.

Two other classes of pigments involved in the capture of light energy are the *carotenoids* and the *phycobilins*. The energy absorbed by these accessory pigments must be transferred to chlorophyll *a;* they cannot substitute for chlorophyll *a* in photosynthesis.

Carotenoids are red, orange, or yellow fat-soluble pigments found in all chloroplasts and, associated with chlorophyll *a*, in cyanobacteria. Like the chlorophylls, the carotenoid pigments of chloroplasts are embedded in the thylakoid membranes. Two groups of carotenoids—*carotenes* and *xanthophylls*—are normally present in chloroplasts (xanthophylls contain oxygen and carotenes do not). The beta-carotene found in plants is the principal source of the vitamin A required by humans and other animals (Figure 6–9). In green leaves, the color of the carotenoids is masked by the much more abundant chlorophylls.

6–9
Related carotenoids. Cleavage of the beta-carotene molecule at the point shown yields two molecules of vitamin A. Oxidation of vitamin A yields retinene, the pigment involved in vision. In the carotenoids, the conjugated bonds are located in the carbon chains. Zeaxanthin is the pigment responsible for the yellow color of corn kernels.

CH_3 CH_3 —CH=CH—C(CH_3)=CH—CH=CH—C(CH_3)=CH—CH≑CH—CH=C(CH_3)—CH=CH—CH=C(CH_3)—CH=CH— CH_3 CH_3 (CH_3, CH_3 on rings)

BETA-CAROTENE

CH_3 CH_3 —CH=CH—C(CH_3)=CH—CH=CH—C(CH_3)=CH—CH_2OH (CH_3 on ring)

VITAMIN A

CH_3 CH_3 —CH=CH—C(CH_3)=CH—CH=CH—C(CH_3)=CH—C(=O)—H (CH_3 on ring)

RETINENE

HO— CH_3 CH_3 —CH=CH—C(CH_3)=CH—CH=CH—C(CH_3)=CH—CH=CH—CH=C(CH_3)—CH=CH—CH=C(CH_3)—CH=CH— CH_3 CH_3 —OH (CH_3, H_3C on rings)

ZEAXANTHIN

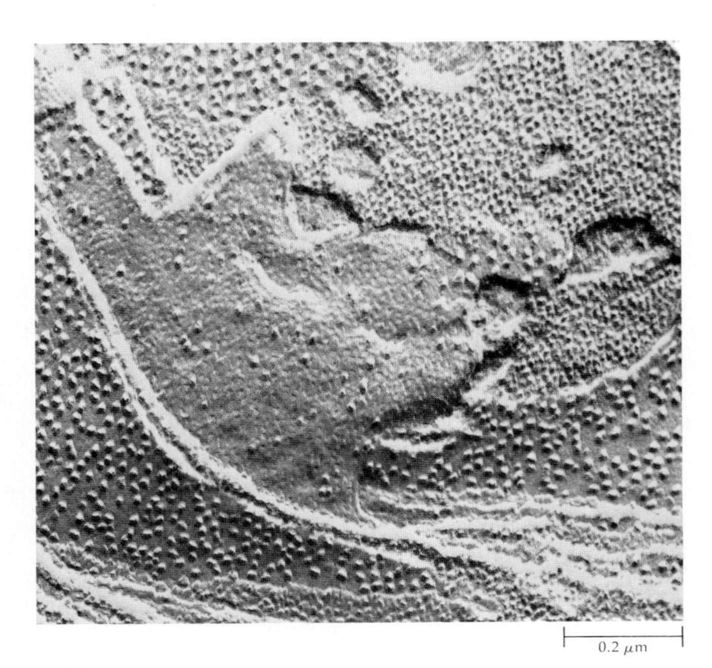

6–10
Internal surface of a thylakoid, as viewed by the freeze-fracture technique. The particles embedded within the membrane are thought to be the structural units of the photosystems involved in the light reactions.

The third major class of accessory pigments, the phycobilins, are found in the cyanobacteria and in the chloroplasts of the red algae. Unlike the carotenoids, the phycobilins are water soluble.

The Photosystems

In the chloroplast (Figure 6–1), the chlorophyll and other pigment molecules are embedded in the thylakoids in discrete units of organization called *photosystems* (Figure 6–10). Each photosystem is an assembly of about 250 to 400 pigment molecules.

All of the pigments within a photosystem are capable of absorbing *photons* (particles of light energy; see "Light and Life," page 98), but only one chlorophyll molecule per photosystem can actually use the energy in the photochemical reaction. This special chlorophyll molecule is called the *reaction center* of the photosystem, and the other pigment molecules are termed *antenna pigments*, so named because they form an antennalike network for the gathering of light.

Light energy absorbed by a pigment molecule anywhere in the network is transferred from one pigment molecule to the next until it reaches the reaction center, which is a special form of chlorophyll *a*. When this particular chlorophyll molecule absorbs the energy, one of its electrons is boosted to a higher energy level and transferred to an acceptor molecule to initiate electron flow. The chlorophyll molecule is thus oxidized and positively charged (it is minus an electron).

According to present evidence, there are two different kinds of photosystems. In Photosystem I, the reaction center is a form of chlorophyll *a* known as P_{700}. The "P" stands for pigment and the subscript "700" designates the optimal absorption peak in nanometers (nm). The reaction center of Photosystem II also contains a special form of chlorophyll *a*. Its optimal absorption peak is at 680 nm, and accordingly, it is called P_{680}.

A Model of the Light Reactions

The two photosystems probably evolved separately, with Photosystem I developing first, for as we shall soon see, Photosystem I apparently can operate independently of Photosystem II. Figure 6–11 shows the present model of how the two photosystems work together. According to this model, light energy enters Photosystem II where it is trapped by the reaction center, P_{680}, either directly or indirectly via one or more of the pigment molecules. When P_{680} is excited, its energized electron is transferred to an acceptor molecule. By an as yet poorly understood reaction, the electron-deficient P_{680} molecule is able to replace its electrons (the electrons actually move in pairs) by extracting them from a water molecule. When the electrons pass from the water molecule to the P_{680} molecule, the water molecule falls apart and is dissociated into protons and oxygen gas. This light-dependent oxidative splitting of water molecules is called *photolysis*.

The electrons then pass downhill along an electron transport chain, which includes two cytochromes, to Photosystem I. As the electrons pass along this transport chain, ATP is formed from ADP, probably via the same mechanism by which ATP is formed along the electron transport chain of the mitochondrion (see page 90). This process is called *photophosphorylation*.

In Photosystem I, light energy boosts the electrons from P_{700} to an electron acceptor from which they are passed downhill to the coenzyme NADP, resulting in the reduction of the NADP to $NADPH_2$ and oxidation of the P_{700} molecule. (NADP is similar to NAD [see Fig-

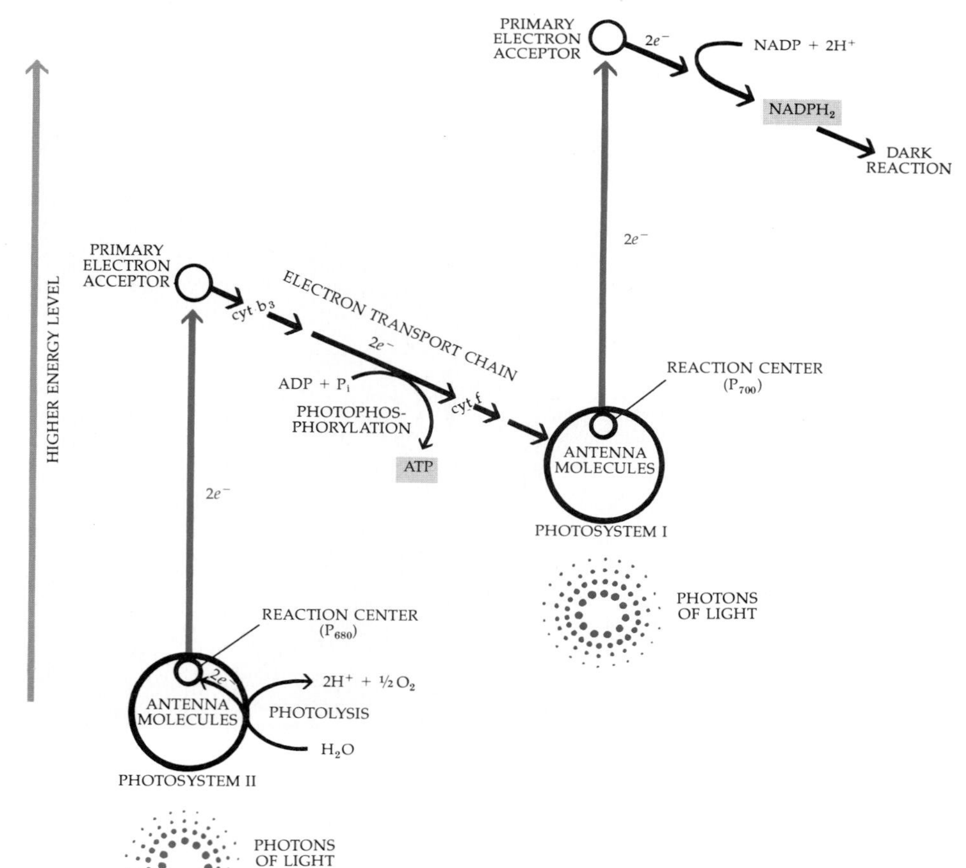

6–11
Noncyclic electron transport flow and photophosphorylation. Light energy trapped in the reaction center, P_{680}*, of Photosystem II boosts electrons to a higher energy level. These electrons are replaced by electrons pulled away from water molecules, releasing protons and oxygen gas. The electrons are passed from the primary electron acceptor along an electron transport chain to a lower energy level, the reaction center,* P_{700}*, of Photosystem I. As they pass along this electron transport chain, some of their energy is packaged in the form of ATP. Light energy absorbed by Photosystem I boosts the electrons to another primary electron acceptor. From this acceptor, they are passed via other electron carriers to NADP to form* $NADPH_2$*. Electrons removed from Photosystem I are replaced by ones from Photosystem II. ATP and* $NADPH_2$ *represent the net gain from the light reactions.*

ure 4–8, page 80] but with an additional phosphate group. Their biological roles, however, are distinctly different. $NADH_2$ transfers its electrons to other electron carriers, which continue to pass them down in discrete steps to successively lower electron levels. In the course of this electron transfer, ATP molecules are formed. $NADPH_2$ provides energy directly to biosynthetic processes of the cell that require large energy inputs.) The electrons removed from the P_{700} molecule are replaced by the electrons from Photosystem II.

Thus, in the light there is a continuous flow of electrons from water to Photosystem II to Photosystem I to NADP. The overall process involving the two photosystems is called *noncyclic* photophosphorylation, because the flow of electrons from the water to NADP is unidirectional. The energy harvest from this noncyclic electron flow (based on the passage of 12 pairs of electrons from H_2O to NADP) is 12ATP and $12NADPH_2$. To generate one molecule of $NADPH_2$, four photons of light energy must be absorbed, two by Photosystem I and two by Photosystem II.

Cyclic Photophosphorylation

It is generally accepted that Photosystem I can work independently of Photosystem II (Figure 6–12). In this process, electrons are boosted from P_{700} to an electron acceptor and from there pass downhill through a series of intermediates, including cytochromes, back into the reaction center. ATP is produced in the course of this passage, and inasmuch as this process involves a cyclic flow of electrons, it is called *cyclic* photophosphorylation. It is believed that the most primitive photosynthetic mechanisms worked in this way, and this is apparently the way in which some prokaryotes carry out photosynthesis. Eukaryotic cells are able to synthesize ATP by cyclic electron flow. However, no water is split, no O_2 is evolved, and no $NADPH_2$ is formed.

THE DARK REACTIONS

In the second stage of photosynthesis, the energy generated by the light reactions is used to reduce carbon. Carbon is available to photosynthetic cells in the form of carbon dioxide. For algae and cyanobacteria, this carbon dioxide is found dissolved in the surrounding water. In most plants, carbon dioxide reaches the photosynthetic cells through special openings, called stomata, which are found in their leaves and green stems (Figure 6–13).

ELECTRON ACCEPTOR

ELECTRON TRANSPORT CHAIN

$2e^-$

$2e^-$

cyt b_6

ADP + P_i

ATP

cyt f

REACTION CENTER (P_{700})

ANTENNA MOLECULES

PHOTOSYSTEM I

PHOTONS OF LIGHT

6–12
Cyclic photophosphorylation involves only Photosystem I. ATP is produced from ADP, but oxygen is not released and NADP is not reduced.

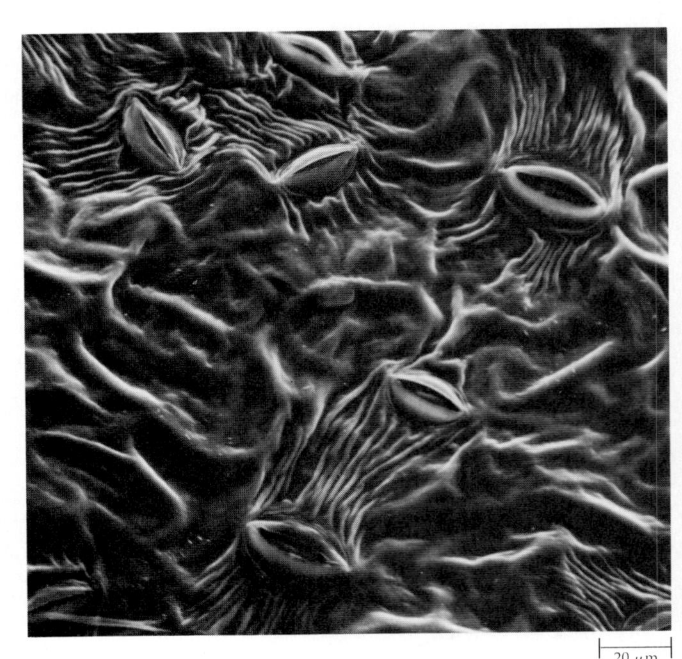

6–13
Scanning electron micrograph showing stomata on the lower surface of a cottonwood (Populus) *leaf. Carbon dioxide reaches the photosynthetic cells through the stomata.*

The Calvin Cycle: The Three-Carbon Pathway

The reduction of carbon occurs in the stroma of the chloroplast by means of a series of reactions known as the Calvin cycle (named after its discoverer, Melvin Calvin, who received the Nobel prize for his work on the elucidation of this pathway). The Calvin cycle is analogous to the Krebs cycle (see page 88) in that, by the end of each turn of the cycle, the starting compound is regenerated. The starting (and ending) compound in the Calvin cycle is a five-carbon sugar with two attached phosphate groups—ribulose 1,5-bisphosphate (RuBP). The process begins when carbon dioxide enters the cycle and is fixed to RuBP. The resultant compound then splits to form two molecules of 3-phosphoglycerate, or PGA (Figure 6–14). (Each PGA molecule contains three carbon atoms; hence the designation of the Calvin cycle as the three-carbon pathway.)

RuBP carboxylase, the enzyme catalyzing these crucial initial reactions, is quite abundant in chloroplasts, making up more than 15 percent of the total chloroplast protein. (It is said to be the most abundant protein in the world. Can you say why?) This enzyme is located on the surface of the thylakoid membranes.

The complete cycle is diagramed in Figure 6–15. As in the Krebs cycle, each step is regulated by a specific enzyme. At each full turn of the cycle, a molecule of carbon dioxide enters the cycle and is reduced, and a molecule of RuBP is regenerated. Six revolutions of the cycle, with the introduction of six atoms of carbon, are necessary to produce a six-carbon sugar,

6–14

Calvin and his collaborators briefly exposed photosynthesizing algae to radioactive carbon dioxide ($^{14}CO_2$). They found that the radioactive carbon is first incorporated into ribulose 1,5-bisphosphate (RuBP). The RuBP then immediately splits to form two molecules of 3-phosphoglycerate (PGA). The radioactive carbon atom, indicated here in color, next appears in one of the two molecules of PGA. This is the first step in the Calvin cycle.

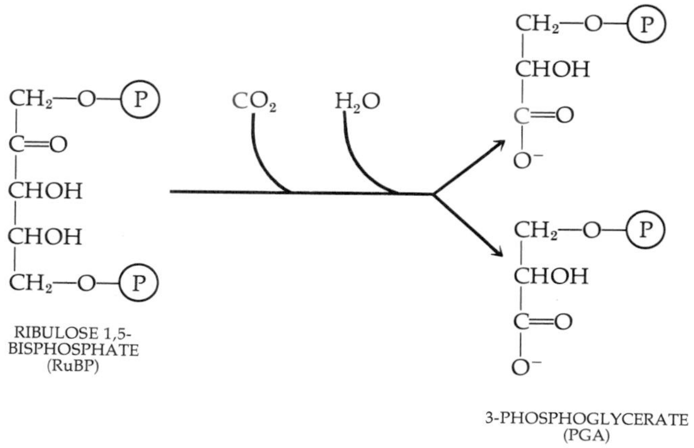

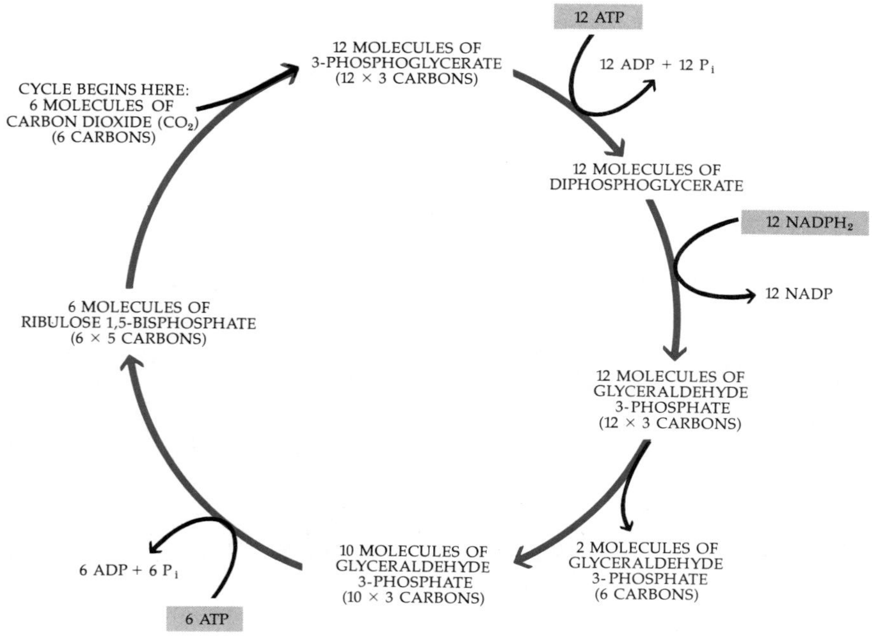

6–15

A summary of the Calvin cycle. At each full turn of the cycle, one molecule of carbon dioxide enters the cycle. Six turns are summarized here—the number required to make two molecules of glyceraldehyde 3-phosphate. Six molecules of ribulose 1,5-bisphosphate (RuBP), a five-carbon compound, are combined with six molecules of carbon dioxide, yielding twelve molecules of 3-phosphoglycerate, a three-carbon compound. These are converted to twelve molecules of glyceraldehyde 3-phosphate. Ten of these three-carbon molecules are combined and rearranged to form six five-carbon molecules of RuBP. The "extra" molecules of glyceraldehyde 3-phosphate represent the net gain from the Calvin cycle. The energy that "drives" the Calvin cycle is the ATP and $NADPH_2$ produced by the light reactions.

such as glucose. On a per-glucose basis, the net overall equation is:

$$6CO_2 + 12NADPH_2 + 18ATP \longrightarrow 1\ \text{glucose} + 12NADP + 18ADP + 18P_i + 6H_2O$$

The immediate product of the cycle is glyceraldehyde 3-phosphate. This same sugar-phosphate is formed when the fructose 1,6-bisphosphate molecule is split at the fourth step in glycolysis (see page 86). These same steps, using the energy of the phosphate bond, can be reversed to form glucose from glyceraldehyde 3-phosphate.

The Four-Carbon Photosynthetic Pathway

The Calvin cycle is not the only carbon-fixation pathway used in the dark reactions. In some plants, the first product of CO_2 fixation to be detected is not the three-carbon molecule PGA, as it is in the Calvin cycle; it is the four-carbon compound oxaloacetate (which is also an intermediate in the Krebs cycle). Plants that employ this pathway are commonly called C_4 (for four-carbon plants), as distinct from the C_3 plants that use *only* the Calvin cycle. (The C_4 pathway is also referred to as the Hatch-Slack pathway after two Australian plant physiologists who played key roles in its elucidation.)

The oxaloacetate is formed when carbon dioxide is fixed to a compound known as phosphoenolpyruvate (PEP). This reaction is catalyzed by the enzyme PEP carboxylase (Figure 6–16). The oxaloacetate is then reduced to malate or converted with the addition of an amino group to the amino acid aspartate. These steps occur in mesophyll cells. The next step is a surprise: the malate (or aspartate, depending on the species) moves from the mesophyll cells to bundle-sheath cells surrounding the vascular bundles of the leaf, where it is decarboxylated to yield CO_2 and pyruvate. The CO_2 then enters the Calvin cycle and reacts with RuBP to form PGA and other intermediates of the cycle, whereas the pyruvate returns to the mesophyll cells where it reacts with ATP to form more molecules of PEP (Figure 6–17). Hence, the anatomy of the plant imparts a spatial separation between the C_4 pathway and the Calvin cycle in the leaves of C_4 plants.

COOH–$COPO_3^{2-}$=CH_2 (PEP) + $CO_2 + H_2O$ —PEP Carboxylase→ COOH–C=O–CH_2–COOH (OXALOACETATE) → COOH–HOCH–CH_2–COOH (MALATE) or COOH–H_2NCH–CH_2–COOH (ASPARTATE)

6–16
Carbon dioxide fixation by the C_4 pathway. Carbon dioxide is incorporated into phosphoenolpyruvate (PEP) by the enzyme PEP carboxylase. The resulting oxaloacetate is either reduced to malate or converted to aspartate through the addition of an amino group ($-NH_2$). These steps will be reversed later, releasing carbon dioxide for use in the Calvin cycle.

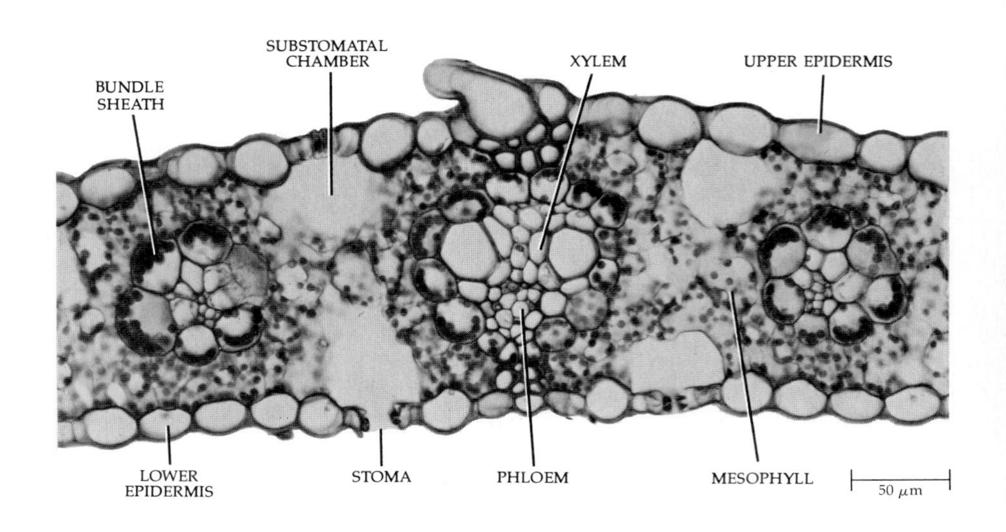

6–18
Transverse section of a portion of a corn (Zea mays) *leaf. As is typical of C_4 plants, the vascular bundles are surrounded by large, chloroplast-containing, bundle-sheath cells that are, in turn, surrounded by a layer of mesophyll cells. The C_4 pathway takes place in the mesophyll cells; the Calvin cycle occurs in the bundle-sheath cells.*

6–17
A pathway for carbon fixation in C_4 *plants.* CO_2 *is first fixed in mesophyll cells as oxaloacetate, which is rapidly converted to malate. The malate is then transported to bundle-sheath cells, where the carbon dioxide is released. The* CO_2 *thus formed enters the Calvin cycle, ultimately yielding starch and sucrose. Pyruvate returns to the mesophyll cell for regeneration of phosphoenolpyruvate (PEP). The* C_4 *leaf shown here is from a corn plant* (Zea mays).

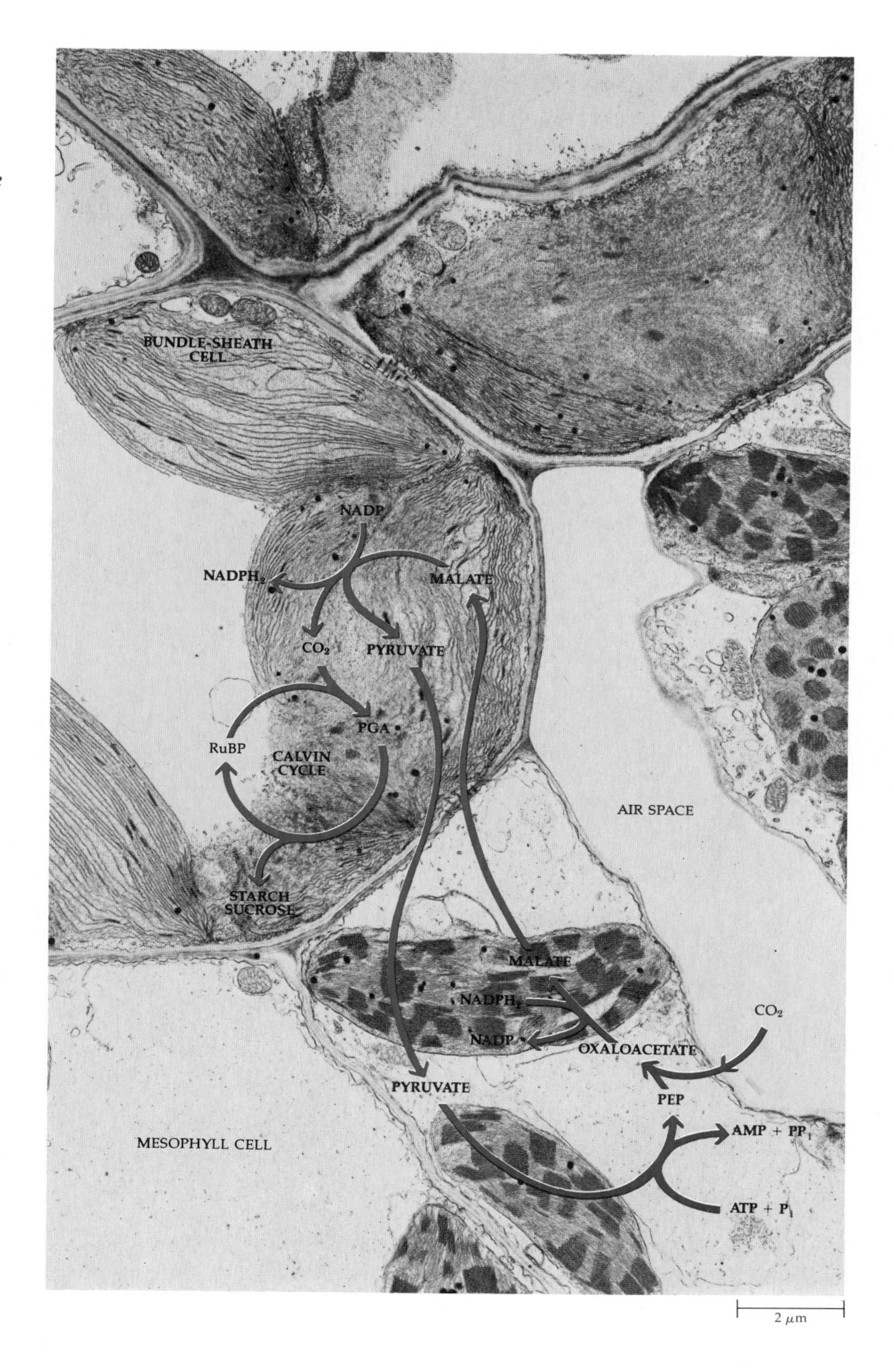

Typically, the leaves of C_4 plants are characterized by an orderly arrangement of the mesophyll cells around a layer of large bundle-sheath cells, so that together the two form concentric layers around the vascular bundle (Figure 6–18). This wreathlike arrangement has been termed Kranz anatomy. *Kranz* is the German word for "wreath." In some C_4 plants the chloroplasts of the mesophyll cells have well-developed grana, while those of the bundle-sheath cells have poorly developed grana or none at all (Figure 6–19). In addition, when photosynthesis is occurring, the bundle-sheath chloroplasts commonly form larger and more numerous starch grains than the mesophyll chloroplasts.

6–19
Electron micrograph of a portion of a chloroplast with well-developed grana in a mesophyll cell (right) and a portion of a chloroplast with poorly developed grana in a bundle-sheath cell (left) of a corn (Zea mays) *leaf. Note the plasmodesmata in the wall between these two cells.*

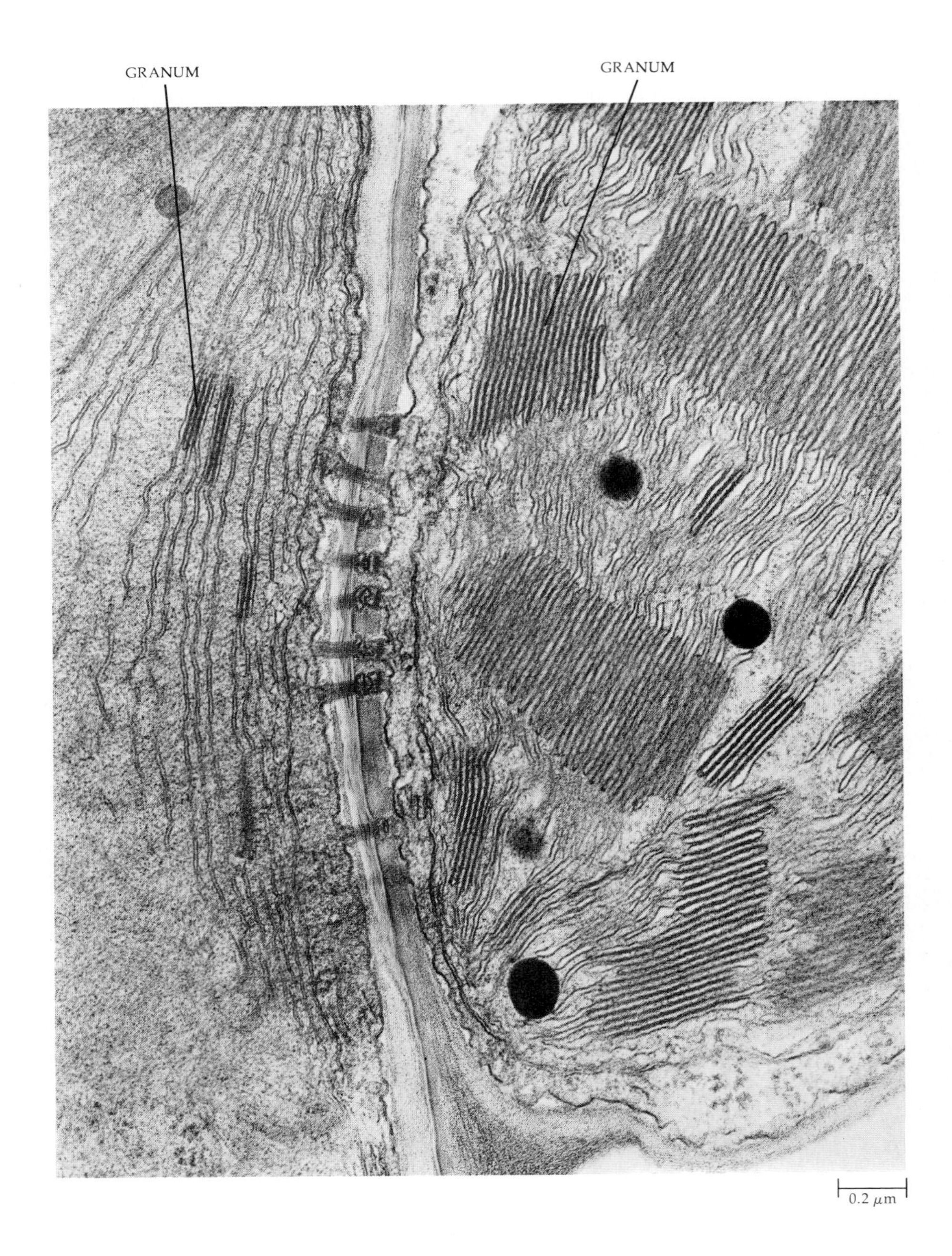

Efficiency of C_4 Plants

One might well ask why C_4 plants should have evolved such a seemingly clumsy and energetically expensive method of providing carbon dioxide to the Calvin cycle. The C_4 pathway can be better appreciated when one learns that the Calvin cycle is always accompanied by *photorespiration,* a process that *consumes* oxygen and *releases* CO_2 in the presence of light (Figure 6–20). Like the respiratory processes that take place in the mitochondria, photorespiration involves the oxidation of carbohydrates; however, unlike these other processes, photorespiration, which occurs in microbodies called peroxisomes, yields neither ATP nor $NADH_2$. Under normal atmospheric conditions, as much as 50 percent of the photosynthetically fixed carbon may be reoxidized to CO_2 during photorespiration. Thus, photorespiration lessens the efficiency of C_3 photosynthetic carbon fixation.

High CO_2 and low O_2 concentrations limit photorespiration. Consequently, C_4 plants have a distinct advantage over C_3 plants because CO_2 fixed by the C_4 pathway is essentially "pumped" from the mesophyll cells into the bundle-sheath cells, thus maintaining a high CO_2:O_2 ratio at the site of the action of RuBP carboxylase. In addition, since the Calvin cycle and its associated photorespiration is localized in the inner, bundle-sheath layer, any CO_2 liberated by

6–20
Reactions catalyzed by RuBP carboxylase. Reaction 1 is favored by high CO_2 and low oxygen concentrations. Reaction 2 has been proposed to occur in the presence of low CO_2 and high oxygen concentrations (normal atmospheric conditions).

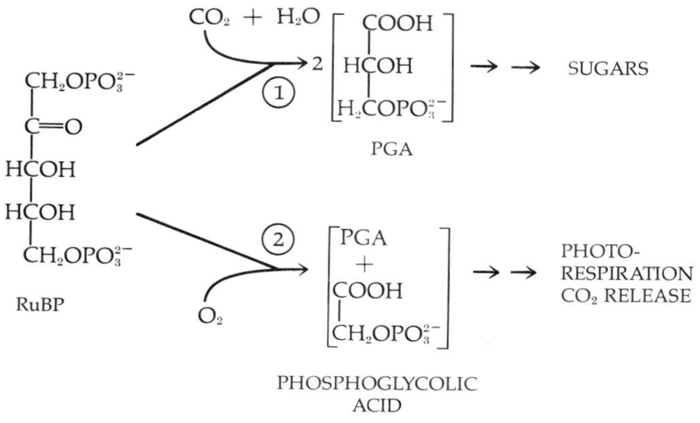

$CO_2 + H_2O$ → ① → 2 [COOH–HCOH–$H_2COPO_3^{2-}$] (PGA) → → SUGARS

RuBP ($CH_2OPO_3^{2-}$–C=O–HCOH–HCOH–$CH_2OPO_3^{2-}$)

O_2 → ② → [PGA + COOH–$CH_2OPO_3^{2-}$] (PHOSPHOGLYCOLIC ACID) → → PHOTO-RESPIRATION CO_2 RELEASE

photorespiration into the outer, mesophyll layer can be refixed by the C_4 pathway that operates there. The CO_2 liberated by photorespiration can thus be prevented from escaping from the leaf. Moreover, compared to C_3 plants, C_4 plants are superior utilizers of available CO_2; this is in part due to the enzyme PEP carboxylase having a much greater affinity for CO_2 at low concentrations than does the RuBP carboxylase of the Calvin cycle. As a result, the net photosynthetic rates of C_4 grasses, such as corn *(Zea mays)*, sugarcane *(Saccharum officinale)*, and sorghum *(Sorghum vulgare)*, can be two to three times the rates of C_3 grasses, such as wheat *(Triticum)*, rye *(Secale cereale)*, oats *(Avena sativa)*, and rice *(Oryza sativa)*.

Efforts are being made to breed for low rates of photorespiration in some C_3 crop plants as a means of increasing yield. However, the place of photorespiration in a plant's overall economy is still not properly understood, and there may be limiting factors with respect to other reactions. For example, during studies on the growth of soybeans *(Glycine max)*, which normally exhibit high rates of photorespiration, plants exposed to low levels of O_2 underwent less photorespiration and grew better than plants grown in normal air. However, these same plants failed to produce harvestable seeds when the O_2 level was below 5 percent, indicating that photorespiration, or some reaction or reactions associated with it, may be necessary for normal seed development.

It should be noted that C_4 plants evolved primarily in the tropics and are especially well adapted to high light intensities, high temperatures, and dryness. The optimal temperature range for C_4 photosynthesis is much higher than that for C_3 photosynthesis; and C_4 plants flourish even at temperatures that would eventually be lethal to many C_3 species. Because of their more efficient use of carbon dioxide, C_4 plants can attain the same photosynthetic rate as C_3 plants but with smaller stomatal openings and, hence, with considerably less water loss.

A striking illustration of differing growth patterns of C_4 plants is found in our lawns, which, in the cooler parts of the country at least, consist mainly of C_3 grasses such as Kentucky bluegrass *(Poa pratensis)* and creeping bent *(Agrostis tenuis)*. Crabgrass *(Digitaria sanguinalis)*, which all too often overwhelms these dark green, fine-leaved grasses with patches of its yellowish green, broader leaves, is a C_4 grass that grows much more rapidly in the heat of summer than the temperate C_3 grasses mentioned above.

The list of plants known to utilize the C_4 pathway has grown to over 100 genera, including both monocots and dicots, in at least 18 families. It has thus evolved independently many times. At least 19 genera are known to have both C_3 and C_4 species. An example of such a genus is *Atriplex*, the saltbushes: *A. patula* is a C_3 plant and *A. rosea* is a C_4 plant.

Crassulacean Acid Metabolism

Crassulacean acid metabolism, commonly designated CAM, is a variant of the C_4 pathway that has evolved independently in many succulent plants, including cacti (Cactaceae) and stonecrops (Crassulaceae). It also involves the initial fixation of atmospheric CO_2 into four-carbon compounds and the ultimate transfer of the CO_2 to the Calvin cycle. In CAM plants, however, there is no spatial separation of the C_4 pathway and the Calvin cycle, as there is in C_4 plants. Instead, in CAM plants, PEP carboxylase fixes CO_2 into C_4 compounds at night and then, during the daytime, the fixed CO_2 is transferred to RuBP of the Calvin cycle *within the same cell.* The C_4 compounds, largely malic acid, are stored overnight in large vacuoles. Thus we find that in CAM plants there is a temporal separation of the C_4 pathway and the Calvin cycle.

CAM plants are largely dependent upon nighttime accumulation of carbon for their photosynthesis because their stomata are closed during the day, retarding water loss. This is obviously advantageous in the conditions of high light intensity and water stress under which most CAM plants live. By accumulating C_4 compounds at night, much more CO_2 can be fixed in a 24-hour cycle than would otherwise be possible, given the pattern of stomatal opening and closing.

CAM plants have now been reported in 25 diverse families of flowering plants, mostly dicotyledons, including such familiar houseplants as the maternity plant *(Kalanchoë daigremontiana)*, wax plant *(Hoya carnosa)*, snake plant *(Sansevieria zeylanica)*, many bromeliads, and the epiphytic orchids. Some nonflowering plants have also been reported to show CAM activity, including *Welwitschia mirabilis* (see Figure 18–34, page 353), a most bizarre gymnosperm. Not all CAM plants are highly succulent; two examples of ones that are less so are the pineapple (*Ananas comosus*) and Spanish "moss" (*Tillandsia usneoides*), both members of the family Bromeliaceae.

SUMMARY

In photosynthesis, light energy is converted to chemical energy, and carbon is "fixed" into organic compounds. The generalized equation for this reaction is

$$CO_2 + 2H_2A + \text{light energy} \longrightarrow (CH_2O) + H_2O + 2A$$

in which H_2A represents water or some other substance that can be oxidized, that is, from which electrons can be removed.

The first step in photosynthesis is the absorption of light energy by pigment molecules. The pigments involved in photosynthesis in eukaryotes include the chlorophylls and the carotenoids, which are packed in the thylakoids of chloroplasts as photosynthetic units called photosystems. Light absorbed by pigment molecules boosts their electrons to a higher energy level. Because of the way the pigment molecules are arranged in the photosystems, they are able to transfer this energy to a special pigment molecule called the reaction center. Two photosystems, Photosystem I and Photosystem II, are now recognized.

Not all of the photosynthetic reactions require light. The series of reactions requiring light are referred to as the "light reactions," and those not requiring light are called the "dark reactions."

In the currently accepted model of the light reactions in photosynthesis, light energy enters Photosystem II, where it is trapped by the reaction center P_{680}. Electrons are boosted uphill from P_{680} to an electron acceptor. As the electrons are removed, they are replaced by electrons from water molecules, and oxygen is produced. Pairs of electrons then pass downhill to Photosystem I along an electron transport chain, in the course of which ATP is generated (photophosphorylation). Meanwhile, light energy absorbed in Photosystem I is passed to its reaction center, P_{700}. The energized electrons are ultimately accepted by the coenzyme molecule NADP, and the electrons removed from P_{700} are replaced by the electrons from Photosystem II. The energy yield from the light reactions is stored in the molecules of $NADPH_2$ and in the ATP formed by photophosphorylation.

In the dark reactions, which take place in the stroma of the chloroplast, the $NADPH_2$ and ATP produced in the light reactions are used to reduce carbon dioxide to organic carbon. This is accomplished by means of the Calvin cycle. In the Calvin cycle, a molecule of carbon dioxide is combined with the starting material, a five-carbon sugar called ribulose 1,5-bisphosphate (RuBP), to form a three-carbon compound, 3-phosphoglycerate (PGA). At each turn of the cycle, one carbon atom enters the cycle. Three turns of the cycle produce a three-carbon molecule, glyceraldehyde 3-phosphate, two molecules of which (in six turns of the cycle) can combine to form a glucose molecule. At each turn of the cycle, RuBP is regenerated.

Plants in which the Calvin cycle is the only carbon fixation pathway, and in which the first detectable product of CO_2 fixation is PGA, are called C_3 plants. In the so-called C_4 plants, carbon dioxide is initially fixed to a compound known as PEP (phosphoenolpyruvate) to yield oxaloacetate, a four-carbon compound. The oxaloacetate is then rapidly converted to either malate or aspartate, which transfers the CO_2 to RuBP of the Calvin cycle. The Calvin cycle in C_4 plants occurs in bundle-sheath cells, whereas the C_4 pathway takes place in the mesophyll cells. C_4 plants are more efficient utilizers of CO_2 than C_3 plants, in part because they minimize the loss of CO_2 resulting from photorespiration, an oxygen-consuming process that occurs only in the light.

A variant of the C_4 pathway, called Crassulacean acid metabolism (CAM), is found in many succulent plants. In CAM plants the fixation of CO_2 by PEP carboxylase into C_4 compounds occurs at night, when the stomata are open. The C_4 compounds are stored overnight and then, during the daytime, when the stomata are closed, the fixed CO_2 is transferred to RuBP of the Calvin cycle. The Calvin cycle and the C_4 pathway occur within the same cells in CAM plants; hence, these two pathways, which are spatially separated in C_4 plants, are temporally separated in CAM plants.

SUGGESTIONS FOR FURTHER READING

ARNON, D. I.: "The Light Reactions of Photosynthesis," *Proc. Nat. Acad. Sci. U.S.* 68: 2883–2892, 1971.

An excellent historical review of the two major aspects of photosynthesis. For the ambitious student who desires to read what scientists write about science. See also Bassham, 1971 (below).

BASSHAM, J. A.: "Photosynthetic Carbon Metabolism," *Proc. Nat. Acad. Sci. U.S.* 68: 2877–2882, 1971.

An excellent review of photosynthesis; should be read with Arnon, 1971 (above).

BECKER, WAYNE M.: *Energy and the Living Cell: An Introduction to Bioenergetics*, Harper & Row, Publishers, Inc., New York, 1977.*

An outstanding short introduction to topics in bioenergetics and cellular energy metabolism; intended for undergraduates.

BJÖRKMAN, O., and J. BERRY: "High-Efficiency Photosynthesis," *Scientific American* 229(4): 80–93, 1973.

A discussion of the superior photosynthetic performance of C_4 plants and their potential role in agriculture.

CONANT, JAMES B. (Ed.): *Harvard Case Histories in Experimental Science*, vol. 2, Harvard University Press, Cambridge, Mass., 1964.

Case No. 5, Plants and the Atmosphere, *edited by Leonard K. Nash, describes early work on photosynthesis. The narrative, often presented in the words of the investigators themselves, illuminates the historical context in which these initial discoveries were made.*

GABRIEL, MORDECAI L., and SEYMOUR FOGEL (Eds.): *Great Experiments in Biology*, Prentice-Hall, Inc., Englewood Cliffs, N.J., 1955.*

Presents many of the fundamental discoveries of biology as seen firsthand through the eyes of their discoverers. The examples are well chosen, and their value is greatly enhanced by accompanying explanatory notes and chronological tables of key developments in various areas of biology.

HALL, J. L., and D. A. BAKER: *Cell Membranes and Ion Transport*, Longman, Inc., New York, 1978.*

An introductory text that integrates information and concepts emerging from recent research on membranes of bacterial, plant, and animal cells.

HATCH, MARSHALL D., C. B. OSMOND, and R. O. SLATYER (Eds.): *Photosynthesis and Photorespiration*, Wiley-Interscience, New York, 1971.

A collection of outstanding review articles on various aspects of photosynthesis and photorespiration; dated, but of historical importance. The first author was a key figure in the development of the Hatch-Slack pathway of C_4 plants.

HINKLE, P. C., and R. E. McCARTY: "How Cells Make ATP," *Scientific American* 238(3): 104–23, 1978.

A clearly written article that provides evidence in support of the chemiosmotic theory of ATP synthesis.

KLUGE, M., and I. P. TING: *Crassulacean Acid Metabolism: Analysis of an Ecological Adaptation*, Ecological Studies: Vol. 30, Springer-Verlag, New York, 1979.

A comprehensive treatment of all aspects of Crassulacean acid metabolism, with special emphasis on CAM as an ecological adaptation. The significance of CAM in agriculture is also considered.

KREBS, H. A.: "The History of the Tricarboxylic Acid Cycle," *Perspectives in Biology and Medicine* 14:154–70, 1970.

The historical development of a major metabolic pathway, seen from the perspective of the Nobel laureate who contributed more than any other person to the elucidation of the cycle that bears his name.

LEHNINGER, ALBERT L.: *Bioenergetics: The Molecular Basis of Biological Energy Transformations*, 2nd ed., W. A. Benjamin, Inc., Menlo Park, Calif., 1971.*

A solid but readily understandable account of the flow of energy in cells, including a treatment of both mitochondria and chloroplast function.

LEVINE, R. P.: "The Mechanism of Photosynthesis," *Scientific American* 221(6):58–70, 1969.

A lucid description of the two photochemical systems and how they provide the electrons, protons, and energy-rich molecules needed to convert carbon dioxide and water into food.

McELROY, WILLIAM D.: *Cell Physiology and Biochemistry*, 3rd ed., Prentice-Hall, Inc., Englewood Cliffs, N.J., 1971.*

A good short introduction to cell physiology, especially useful in its descriptions of the exchanges and uses of energy within cells.

RABINOWITCH, EUGENE B., and GOVINDJEE: *Photosynthesis*, John Wiley & Sons, Inc., New York, 1969.*

A comprehensive and readable account of the entire field, now out of date but still a most useful overall treatment of the subject.

SAN PIETRO, ANTHONY, F. A. GREER, and T. J. ARMY (Eds.): *Harvesting the Sun: Photosynthesis in Plant Life*, Academic Press, Inc., New York, 1967.

A general account of photosynthesis, which sheds much light on the factors that affect it.

STRYER, LUBERT: *Biochemistry*, W. H. Freeman and Company, San Francisco, 1975.

A good introduction to cellular energetics.

ZELITCH, ISRAEL: *Photosynthesis, Photorespiration and Plant Productivity*, Academic Press, New York, 1971.

A modern treatment of the process of photosynthesis within cells, leaves, and plant communities.

*Available in paperback.

SECTION 3 Genetics and Evolution

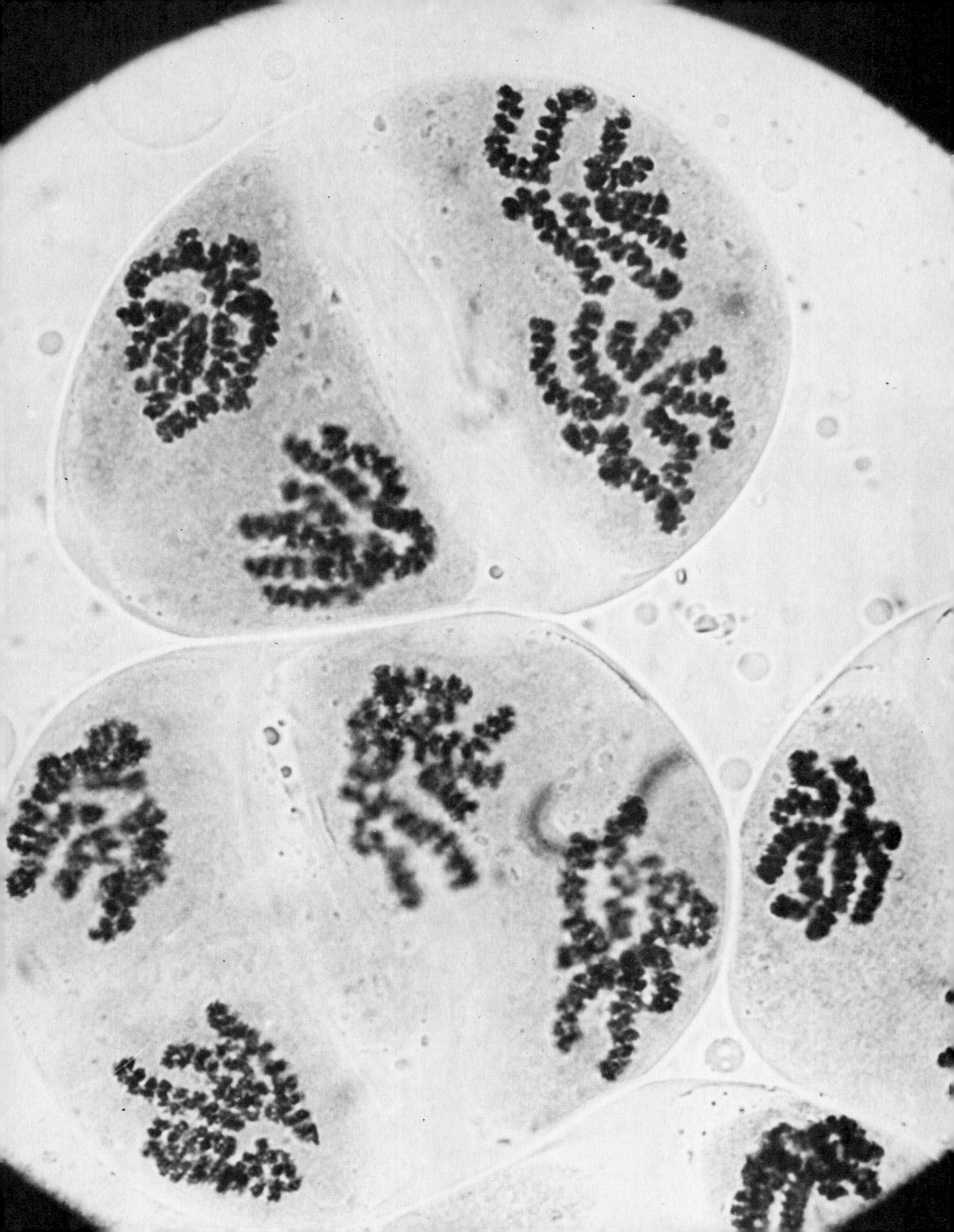

CHAPTER 7

The Chemistry of Heredity

7–1
The second anaphase of meiosis during microspore formation in the wake-robin (Trillium erectum). *Separation of the clearly visible chromosomes, which carry the genetic material—DNA—is nearly complete. Each of the newly forming nuclei will contain only half the number of chromosomes present in the original nucleus at the beginning of meiosis.*

Ever since humans first started to look at the world around them, they have puzzled and wondered about heredity. Why is it that the offspring of all living things—whether dandelions, dogs, aardvarks, or oak trees—always resemble their parents and never some other species? Why does a child have his mother's eyes or his father's chin or, even more puzzling, his grandfather's nose?

These questions were recorded as far back as the writings of the Greeks, and they were probably old even then. Such questions have always been important. Throughout history, biological inheritance has been a major factor in determining the distribution of wealth, power, land, and privilege. From the point of view of biology, self-replication is at the heart of any definition we may have of life.

Some say that the twentieth century will be remembered as the time in history when man first reached the moon, but others predict that it will be best known as the age in which man discovered the nature of DNA and thus began to unravel the secrets of heredity.

THE CHEMISTRY OF THE GENE: DNA VERSUS PROTEIN

Chromosomes in eukaryotes are complexes of DNA and protein. Once it became clear to researchers that chromosomes carried the genetic information, the problem became one of deciding which of these two components played this essential role. By the early 1950s, many investigators—particularly those who were studying proteins—believed that the genes were proteins, that the chromosomes contained master models of all the proteins required by the cell, and that enzymes and other proteins necessary for cellular life were copied from these master models. Although this was a logical hypothesis, it turned out that it was wrong.

Table 7–1 *Composition of DNA in Several Species**

	PURINES		PYRIMIDINES	
SOURCE	ADENINE	GUANINE	CYTOSINE	THYMINE
Human	30.4	19.6	19.9	30.1
Ox	29.0	21.2	21.2	28.7
Salmon sperm	29.7	20.8	20.4	29.1
Wheat germ	28.1	21.8	22.7	27.4
Escherichia coli	26.0	24.9	25.2	23.9
Sheep liver	29.3	20.7	20.8	29.2

*In moles per 100 gram-atoms, percent; after Chargaff, Erwin, *Essays on Nucleic Acids*, 1963.

At about the same time, evidence for the role of DNA as the genetic material was provided by a variety of data: (1) As revealed by specific staining techniques, DNA is present in the chromosomes of all cells, and it is mainly found in chromosomes. (2) In general, the body cells of plants and animals contain twice as much DNA as is found in their gametes. (3) As shown in Table 7–1, the proportions of purines and pyrimidines vary from one species to another. Such variation is essential in a molecule that spells out the "language of life." Do you notice anything else about the proportions of purines and pyrimidines? (4) DNA isolated from bacterial cells can act as a transforming factor, providing other bacterial cells with new genetic characteristics (Figure 7–2). (5) In the course of the infection of bacterial cells by certain bacterial viruses (bacteriophages), DNA alone enters the cell and acts to direct the formation of new viruses (Figure 7–3).

Despite all this evidence, it was not until the discovery of the structure of DNA that its genetic role came to be generally understood.

THE NATURE OF DNA

In 1951, James D. Watson went to England, where he arranged to work with Francis Crick of the Cavendish Laboratory in Cambridge. Watson and Crick were among the scientists who were convinced that DNA, rather than protein, was the genetic material. In Watson's words, DNA was "the most golden of all molecules."

Watson and Crick had two types of information with which to work as they set out to determine how the DNA molecule was put together. *First,* if DNA were the genetic material, it would have to meet at least four requirements:

1. It must carry genetic information from cell to cell and from generation to generation; furthermore, it must carry a great deal of information. (Consider how many instructions must be contained in the set of genes that directs, for example, the development of an elephant, a tree, or even a bacterium.)

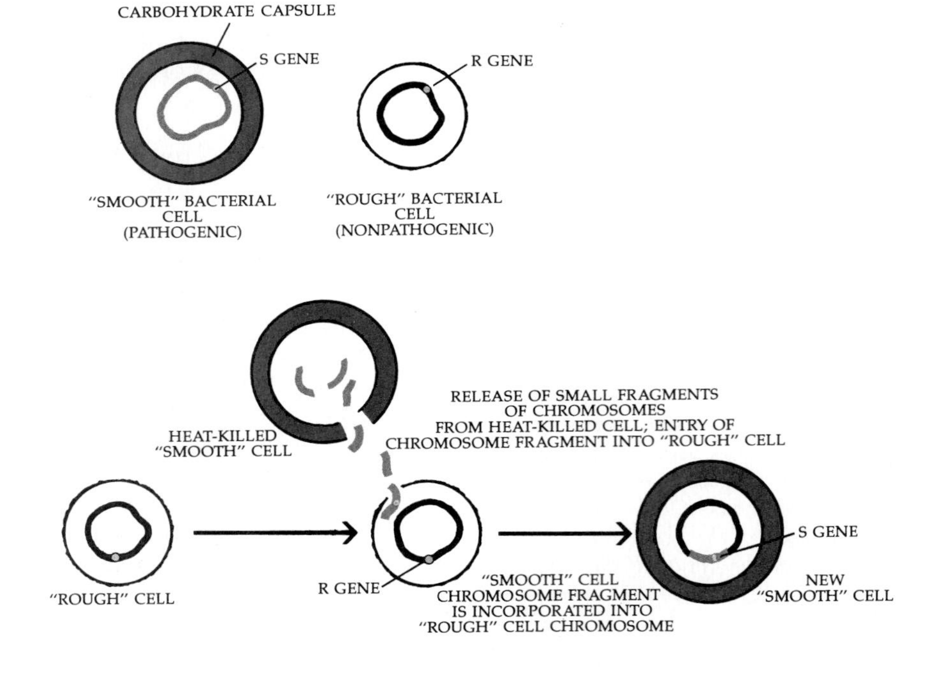

7–2
Identification of the transforming factor, a crucial experiment in establishing the role of DNA. "Smooth" pneumococcal cells—so called because, when grown on agar, they produce a polysaccharide coat that gives their colonies a glistening, smooth appearance—are pathogenic (disease-causing). "Rough" pneumococci are not pathogenic. Both of these traits are hereditary: progeny of "smooth" cells produce colonies that have the "smooth" morphology, and those of "rough" cells produce colonies that appear dull and rough. If "smooth" cells are killed and the cellular remains are added to a live culture of "rough" cells, some of the "rough" cells acquire the characteristics of "smooth" cells and produce "smooth" colonies. This phenomenon, known as transformation, was first observed in 1928. Some 16 years later, in 1944, it was proved that the transforming factor—DNA—actually changes the genetic makeup of the "rough" cells.

7–3
A summary of a series of important experiments in the history of molecular biology. Bacteriophages (or phages, from the Greek phagein, *"to eat")—viruses that infect bacterial cells—consist of protein and DNA. When virus-infected bacterial cells were grown in media containing radioactive sulfur (^{35}S) or radioactive phosphorus (^{32}P), phages were produced in which the proteins and the DNA were labeled differently. (DNA has no sulfur, and proteins have no phosphorus.) The labeled phages were then permitted to infect cells growing in a nonradioactive medium. These ingenious experiments revealed that only the DNA of the phage enters the cell and, hence, it must be the DNA that carries the hereditary information necessary for the production of new phage particles.*

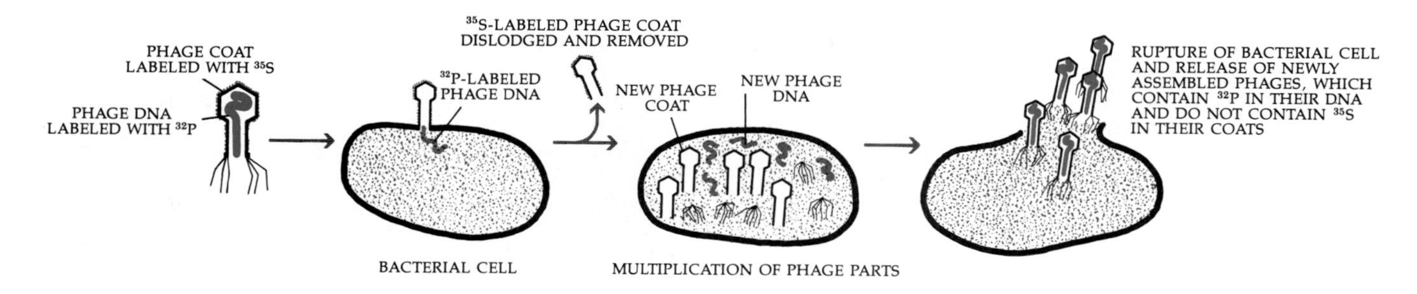

2. It must copy itself, for the chromosome does this before every cell division; moreover, it must replicate itself with great precision. (For instance, we know from accumulated data on human mutation rates that any one human gene has been copied, on average, for millions of years without a mistake.)
3. On the other hand, a gene must sometimes change, or mutate. (When a gene changes, that is, when a mistake is made, the "mistake"—rather than what was originally there—must be copied. This is a most important property, perhaps the unique attribute of living things; without the capacity to replicate "errors," there could be no evolution by natural selection.)
4. It must have some mechanism for "reading out" the stored information and translating it into action in the living individual.

Watson and Crick were well aware that the DNA molecule could be the genetic material only if it could be shown to have the size, configuration, and complexity required to code the tremendous store of information needed by living things and to make exact copies of this code.

The *second* type of information that Watson and Crick used came from previous biochemical studies of the DNA molecule. Among these data were the following:

1. The molecule was very large and also long and thin.
2. The three components (a nitrogenous base, a sugar, and a phosphate) were arranged in nucleotides, as shown in Figure 2–17 on page 57.
3. According to X-ray diffraction studies conducted by Rosalind Franklin and Maurice Wilkins at King's College in London, the long molecule of DNA was made up of regularly repeating units that appeared to be arranged in a spiral, or more accurately, a helix.
4. The data shown in Table 7–1 indicated that the ratio of nucleotides containing adenine to those containing thymine is 1 to 1, as is the ratio of nucleotides containing guanine to those containing cytosine.

Watson and Crick did not actually perform experiments in the usual sense; rather, they chose to assemble all the facts then known about DNA into a meaningful whole. They worked with the accumulated data, with measurements provided by the X-ray photographs of Franklin and Wilkins, and with a tin model that they attempted to construct in such a way that it would be consistent with all the physical and chemical data about DNA (see Figure 7–4a).

The most important question for them was whether or not the chemical structure of DNA would reflect its biological function. "In pessimistic moods," Watson recalled in a review of these investigations, "we often worried that the correct structure might be dull—that is, that it would suggest absolutely nothing." The structure of DNA turned out, in fact, to be unbelievably interesting.

The Double Helix

By piecing together the various data, Watson and Crick were able to deduce that DNA is not a single-stranded helix, as are many proteins, but is instead a huge, entwined double helix.

(a)

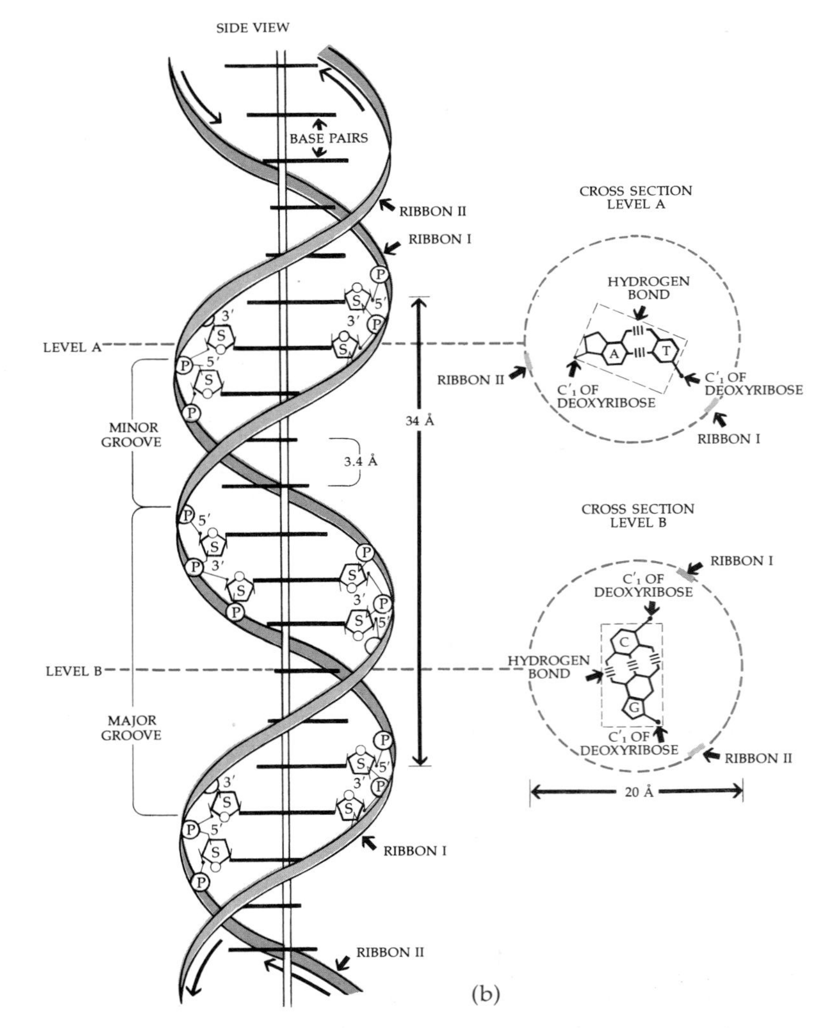

(b)

7–4
(a) *Watson (left) and Crick with their tin model of DNA.* (b) *A representation of the Watson-Crick concept of the DNA molecule. On the left, the molecule is drawn in side view with the axis indicated by the vertical rod. The backbone of the molecule consists of two polynucleotide chains which form helices coiling to the right. These chains are coiled together to form a double helix with two grooves, one shallow (minor) and one deep (major), and an overall diameter of 2 nanometers (20 angstroms).*

The two chains that make up the backbone are composed of deoxyribose sugar molecules (S), *and each is linked to a phosphate group* (P) *above and below to form a continuous chain projecting into the cylinder.*

The nitrogenous base pairs (indicated by heavy horizontal lines) are flat molecules occupying the central area in the cylinder (the dashed rectangles in the cross sections on the right). The base pairs, represented by thymine (T) *and adenine* (A) *at level A and by cytosine* (C) *and guanine* (G) *at level B, are linked by hydrogen bonds. The broken circular line in the cross sections indicates the outer edge of the double helix when the molecule is viewed end on.*

The bases are stacked above one another at intervals of 3.4 angstroms and are rotated 36° with each step. Thus, there are ten base pairs per complete turn of the helix. As a result of this rotation, the successive side views of the base pairs appear as lines of varying lengths, depending upon the viewing angle. The hydrogen bonding between the bases and the hydrophobic interactions resulting from the parallel stacking of the bases serve to stabilize the helical structure.

In the side view, the molecular composition of the backbones is shown over only small sections of the molecule because the twisting of the backbones would otherwise make the drawing confusing. Instead, the continuity of the backbones is represented by ribbons drawn as if on the surface of a cylinder to show the projection of their paths. These ribbons are separated by about 120° from each other around the circumference of the cylinder (Etkin, BioScience, *1973, as modified by Kelln and Gear,* BioScience, *1980).*

The banister of a spiral staircase forms a single helix. If one were to twist a ladder into the shape of a helix, keeping the rungs perpendicular to the vertical sides, a crude model of a double helix would be formed. The two vertical railings of the DNA molecule are made up of alternating deoxyribose sugar molecules and phosphate groups (see Figures 7–4b and 7–7).

The perpendicular rungs of the ladder are formed by the nitrogenous bases—adenine (A), thymine (T), guanine (G), and cytosine (C)—one base for each sugar-phosphate, with two bases forming each horizontal rung. The paired bases meet across the helix and are joined together by hydrogen bonds, the relatively weak chemical bonds found to be such an important part of the secondary structure of proteins.

The distance between the two sides, or railings, according to Wilkins's measurements, is 2 nanometers. Two purines in combination would extend a distance of more than 2 nanometers, and two pyrimidines would not reach all the way across this space. But a perfect fit is obtained if each purine is paired with a pyrimidine. The paired bases—the rungs of the ladder—are therefore always purine-pyrimidine combinations (see Figure 7–7). For this reason, the ratio of purines to pyrimidines in a molecule of DNA is always 1 to 1.

Watson and Crick noticed that the nucleotides along any one chain of the double helix could be assembled in any order, such as ATGCGTACATT, and so on. Because a DNA molecule may be several thousand nucleotides long, a wide variety of arrangements is possible. The number of paired bases ranges from about 5,000 for the simplest virus up to an estimated 5,000,000,000 in the 46 chromosomes of humans. The DNA from a single human cell—which, if extended in a single thread, would be about 1.5 meters long—contains an amount of information equal to some 600,000 printed pages averaging 500 words each, which is the equivalent of a library of about 1,000 books. In short, the DNA molecule does indeed have the capacity to store the necessary genetic information.

The Molecule that Copies Itself

The most exciting discovery came when Watson and Crick set out to construct the matching strand. They encountered an interesting and important restriction: not only could purines not pair with purines, and pyrimidines not pair with pyrimidines, but adenine could pair only with thymine, and guanine could pair only with cytosine. Because of the configurations of the molecules, this restriction could have been predicted from data like that in Table 7–1. Thus, one strand of DNA dictates precisely the structure of the other strand, which is *complementary* to it. When the molecule reproduces itself, it simply "unzips" down the middle, the nitrogenous bases breaking apart at the hydrogen bonds (Figure 7–5). The two strands

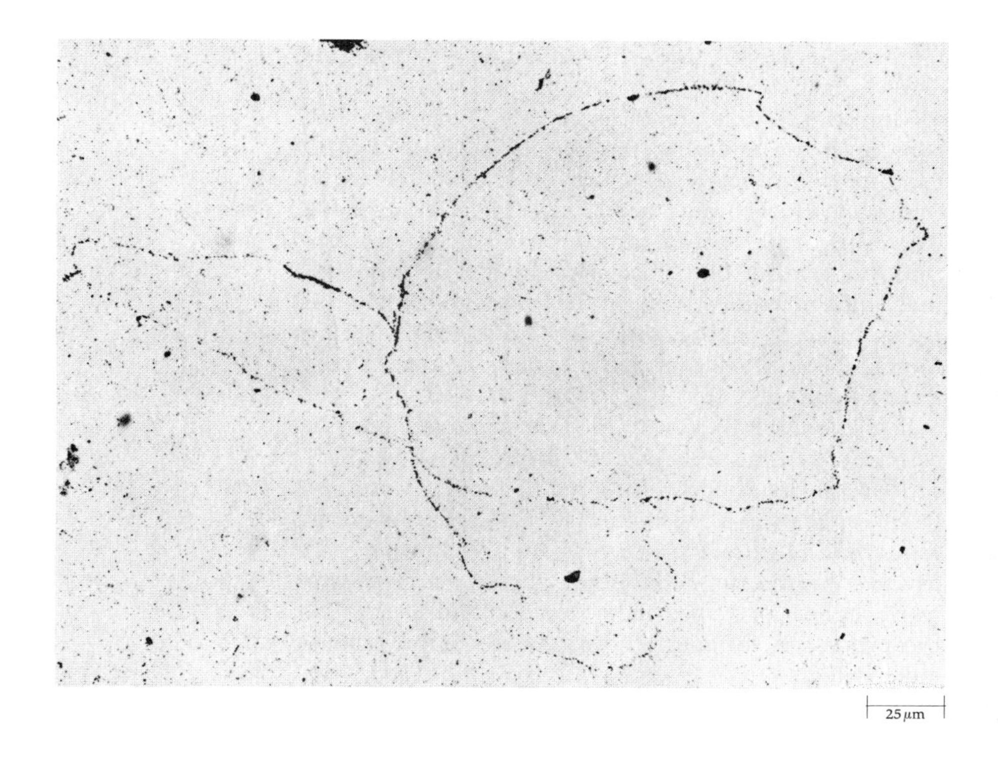

7–5
DNA replicating in Escherichia coli. *The bacterium was placed in a medium containing thymine labeled with tritium* (^{3}H) *and was allowed to incorporate this radioactive amino acid into its DNA for two generations. Then the DNA was extracted and placed on a photographic plate which recorded the position of the* ^{3}H *atoms (a technique known as autoradiography). Shown here is the circular DNA molecule of one bacterial cell undergoing the act of duplication in which the two strands of the double helix separate and have new bases synthesized along them to form complementary base pairs.*

separate, and new strands form along each old one. If a T is present on the old strand, only an A can fit into place on the new strand; a G will pair only with a C, and so on. In this way, each strand forms a complementary copy of itself, and two exact replicas of the original molecule are produced. The age-old question of how hereditary information is duplicated and passed on for generation after generation had, in principle, been answered by the discovery of the structure of DNA.

In what could be considered one of the great understatements of all time, Watson and Crick wrote in their original brief publication: "It has not escaped our notice that the specific pairing we have postulated immediately suggests a possible copying mechanism for the genetic material."

Subsequently, it has been shown that because the DNA molecule is very long, double-stranded, and helical, the replication process is complex, requiring many specific enzyme-catalyzed steps. Specific enzymes are required to separate and unwind the strands of DNA, as well as to place the newly formed nucleotides in the correct position. The way in which these enzymes function provides the key to understanding how DNA is replicated.

As the chemistry of DNA replication came to be better understood, an added complication became apparent—the ends of the two strands of the DNA double helix were different. In each strand, the phosphate group that joins two deoxyribose molecules is attached to one sugar at the 5′ position (carbon 5 of deoxyribose) and to the adjacent sugar at the 3′ position (the third carbon in the ring of deoxyribose). This configuration of sugar-phosphate-sugar attachments gives each strand a 5′ end and a 3′ end (Figure 7–6).

Moreover, while the direction of DNA synthesis runs from the 3′ end to the 5′ end in one strand, it runs in the opposite direction in the other strand. Thus, the two strands are antiparallel (Figure 7–7). Synthesis of the strand running from 3′ to 5′ is continuous, adding one nucleotide at a time, whereas that of the strand running in the opposite direction is discontinuous, being built from a series of short fragments that are eventually joined to form a continuous strand.

In 1957, Arthur Kornberg discovered an enzyme active in synthesizing a new strand of DNA. Subsequently, many additional details of the process of DNA replication have been discovered. For example, more than 14 different enzymes are now known to be involved in the steps that occur during the process of DNA replication in the bacterium *Escherichia coli,* and the process is, of course, even more complicated in eukaryotes.

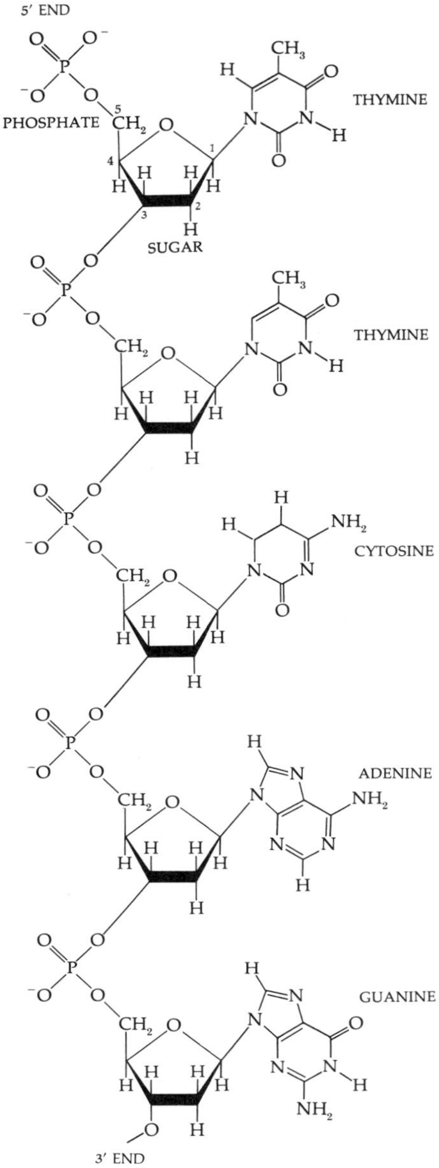

7–6
A closer look at the structure of a portion of one strand of a DNA molecule. Each nucleotide consists of a sugar (deoxyribose), a phosphate group, and a nitrogenous base (a purine or a pyrimidine). Note the repetitive sugar-phosphate-sugar-phosphate sequence that forms the backbone of the molecule. Each phosphate group is attached to the 5′ carbon of one sugar and to the 3′ carbon of the sugar in the adjacent nucleotide, so that the strand has a 5′ end and a 3′ end.

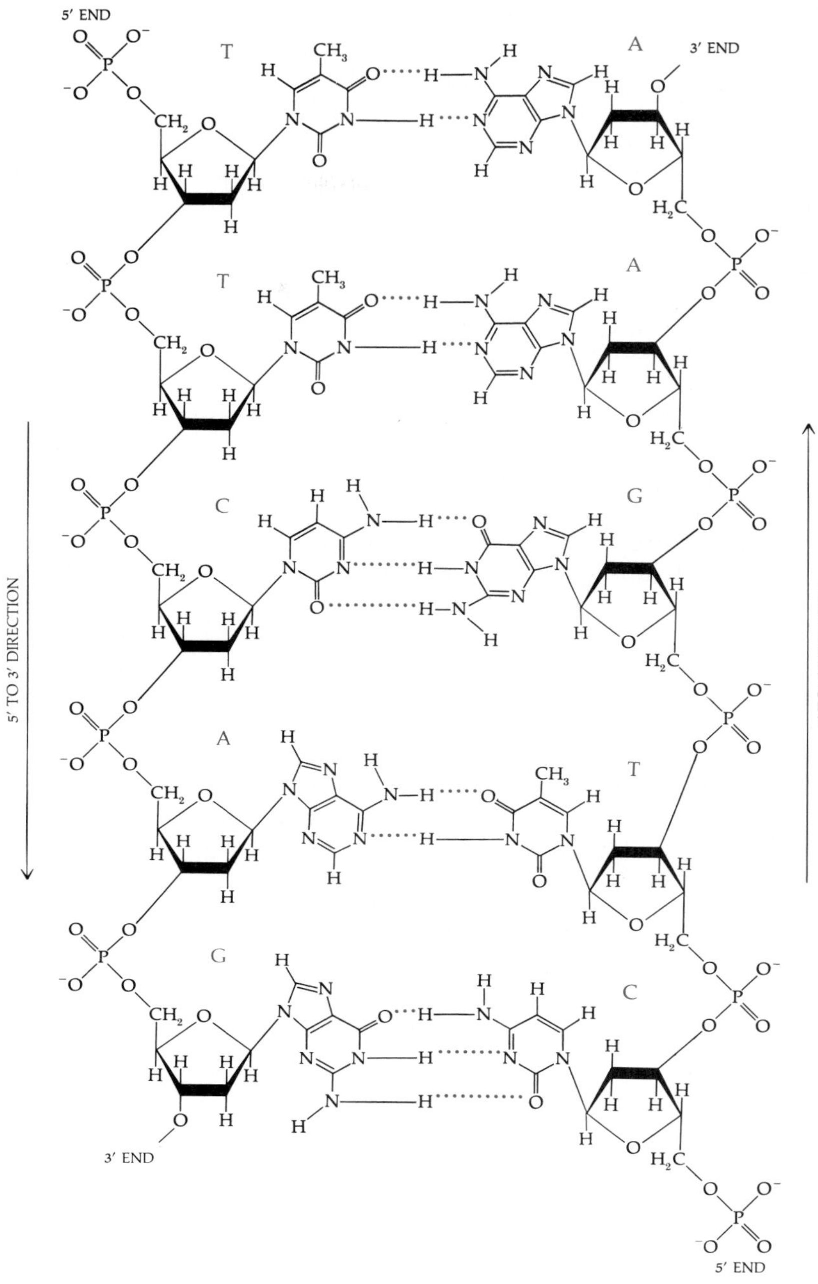

7–7
The double-stranded structure of a portion of the DNA molecule. Notice that the bridges formed by the phosphate groups between the nucleotides in the two strands run in opposite directions; that is, the strands are antiparallel. *(Compare this figure with Figure 7–4b on page 120.)*

HOW DO GENES WORK?

Watson and Crick disclosed the chemical nature of the gene and suggested the way in which it duplicated itself. Still unanswered, however, was the question of how the information stored in the DNA molecule could influence structure or function through the production of specific proteins. How, for example, does DNA change a harmless pneumococcus to a virulent one, determine the shape of a leaf or the odor of a flower, or explain why your eyes are the same color as those of your mother?

The Molecules of Heredity

What cells are and how they function depends almost entirely upon the proteins they make, and especially upon their enzymes. Therefore, the steps by which the genetic information of the DNA molecule is utilized in the formation of specific proteins is of special interest. The process comprises a number of steps that involve the sister molecule of DNA—*ribonucleic acid* (RNA). Long before the details of this step-by-step process began to be understood, it was suspected that RNA is somehow involved in protein synthesis. This idea is based on the observation that cells synthesizing large amounts of protein invariably contain large amounts of RNA.

RNA differs from DNA in a few significant respects, which were reviewed in Chapter 2: the sugar component of the molecule is ribose rather than deoxyribose, and in place of thymine, RNA has another pyrimidine—uracil (U). The RNA molecule occurs only rarely in a fully double-stranded form; its properties and activities are therefore quite different from those of DNA. There are three types of RNA: *messenger* RNA (mRNA), *transfer* RNA (tRNA), and *ribosomal* RNA (rRNA).

Messenger RNA (mRNA) is a large molecule, ranging in size from a few hundred to about 10,000 nucleotides. It is formed along one strand of the DNA helix by the same base-pair copying principle that applies to the formation of a new strand of DNA. The presence of adenine in the parent DNA strand causes the insertion of a uracil unit in the mRNA strand. Each sequence of three bases in the mRNA molecule specifies a single amino acid; such a sequence is known as a *codon*.

Transfer RNA (tRNA) has sometimes been called "the dictionary of the language of life." There are many different types of tRNA, apparently one specific type for each of the codons of the genetic code (see Figure 7–8). Each tRNA molecule consists of only about 80 nucleotides which form a long single strand that folds back on itself (Figure 7–9). As can be seen in Figure 7–11, part of the specificity of a particular tRNA molecule resides in the nature of its antiparallel triplet sequence, or *anticodon*, which determines that it will recognize only one particular triplet codon on the mRNA.

Equally important to the specificity of the tRNA molecule is its ability to bind a particular amino acid—the one coded by the anticodon. This specificity is achieved by *activating enzymes*—specific enzymes that recognize both a particular amino acid and a particular tRNA. These activating enzymes are thus the key elements in the translation of the genetic message; they determine which amino acid will be associated with which tRNA (and thus with which triplet anticodon).

The base sequences of several different tRNA molecules have been discovered. Each type of tRNA differs in its nucleotide sequence, but all have a similar number of bases and a similar shape.

In eukaryotic cells, ribosomal RNA (rRNA) is formed on the DNA of the nucleolus. A ribosome is composed of two subunits, each with its own characteristic kinds of RNA and proteins. In *Escherichia coli*, for instance, the smaller subunit of the ribosome has one type of rRNA, and the larger subunit has two other types. Although somewhat larger than those of *E. coli*, the ribosomes of eukaryotic cells are apparently similar in both structure and function.

Messenger RNA and tRNA come together at the ribosome. Presumably, the function of the ribosome is to position the mRNA, tRNA, amino acids, and the protein being formed in a precise relationship to one another as protein synthesis occurs.

Transcribing RNA from DNA

RNA is transcribed, or copied, from DNA in a complementary fashion similar to that in which DNA is replicated (Figure 7–10). In eukaryotic cells, there are different enzymes, called RNA polymerases, for the

FIRST LETTER	SECOND LETTER U		SECOND LETTER C		SECOND LETTER A		SECOND LETTER G		THIRD LETTER
U	UUU	phe	UCU		UAU	tyr	UGU	cys	U
	UUC		UCC	ser	UAC		UGC		C
	UUA	leu	UCA		UAA	stop	UGA	stop	A
	UUG		UCG		UAG	stop	UGG	trp	G
C	CUU		CCU		CAU	his	CGU		U
	CUC	leu	CCC	pro	CAC		CGC	arg	C
	CUA		CCA		CAA	gln	CGA		A
	CUG		CCG		CAG		CGG		G
A	AUU		ACU		AAU	asn	AGU	ser	U
	AUC	ile	ACC	thr	AAC		AGC		C
	AUA		ACA		AAA	lys	AGA	arg	A
	AUG	met	ACG		AAG		AGG		G
G	GUU		GCU		GAU	asp	GGU		U
	GUC	val	GCC	ala	GAC		GGC	gly	C
	GUA		GCA		GAA	glu	GGA		A
	GUG		GCG		GAG		GGG		G

7–8
The genetic code, consisting of 64 codons (triplet combinations of bases) and their corresponding amino acids (see page 52). Of the 64 codons, only 61 specify particular amino acids. The other three codons are "stop signals," which cause the chain to terminate. Since 61 triplets code 20 amino acids, there obviously must be "synonyms"; for example, leucine has as many as six. Most of the synonyms differ only in the third nucleotide. The code is shown here as it would appear in the mRNA molecule.

7–9
(a) *Two-dimensional and* (b) *three-dimensional representations of the structure of the tRNA molecule involved in the transfer of the amino acid phenylalanine in yeast. This tRNA molecule contains 76 nucleotides, 14 of which are not the typical* A, G, U, *or* C, *but rather are closely related structures. The significance of these atypical bases is that they prevent certain hydrogen bonds from forming while promoting other cross-chain associations, thus determining the final folding of the molecule. Colored dots indicate regions of hydrogen-bonded base pairing. Each portion of the tRNA molecule appears to have a unique function. The acceptor end is where the amino acid is bound. The* TΨC *loop is the same in every tRNA molecule; it probably governs the binding of tRNA to ribosomes. The* DHU *loop, in contrast, differs from one tRNA molecule to another; it may be involved in determining which amino acid is enzymatically added to the acceptor end. Finally, the anticodon loop contains the three bases complementary to the codon of mRNA, in this case* GAA.

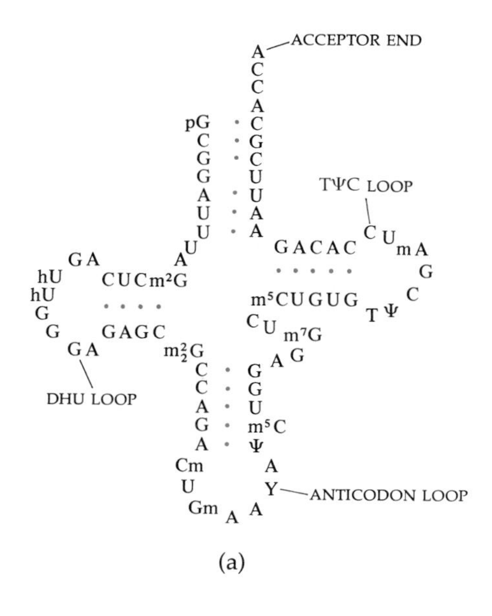

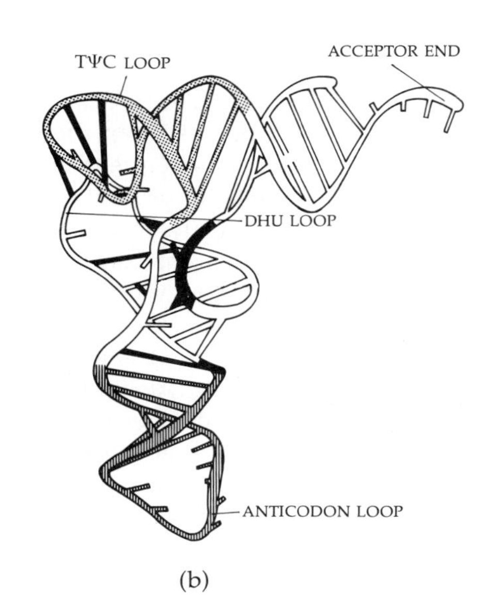

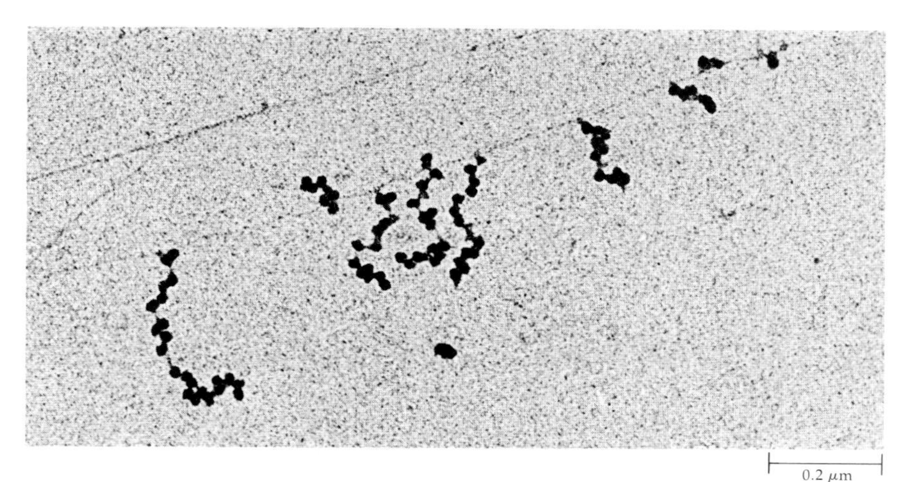

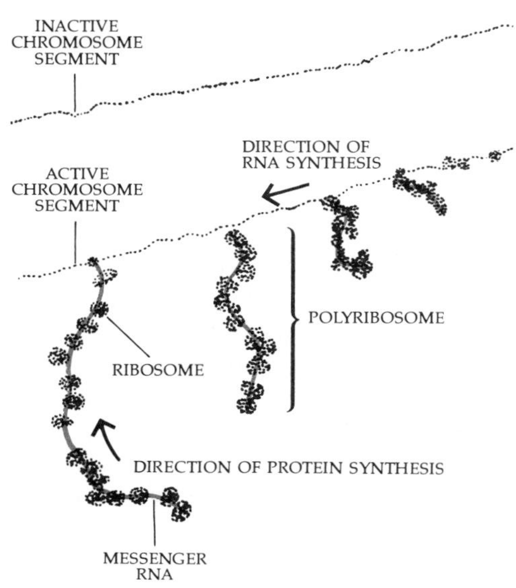

7–10
A bacterial gene in action. The micrograph (left) shows several different mRNA strands being formed simultaneously. (The strands are shown in color in the diagram at the right.) The longest strand was the first one synthesized. As each mRNA strand peels off the active chromosome segment (the DNA molecule), ribosomes attach to the mRNA strand, translating it into protein. The protein molecules are not visible in this micrograph.

transcription of tRNA, rRNA, and mRNA. The mRNA precursor molecules synthesized from the DNA template in the nucleus in eukaryotes are 10 to 20 times larger than the mRNAs found in prokaryotes. If one labels newly synthesized RNA in eukaryotes by making radioactive uracil available for a short time, the long molecules of mRNA precursor within the nucleus display radioactivity. However, only about 10 percent of this radioactivity can be detected outside the nucleus, thus demonstrating that most of the mRNA precursor formed within the nuclei of eukaryotes does not play a role in protein synthesis.

The long mRNA precursor molecules of eukaryotes are "processed" within the nucleus after they are synthesized; only small portions subsequently reach the cytoplasm and act as templates for protein synthesis. Large segments of mRNA precursor are excised from the middle portions of the molecule by processing enzymes before the mRNA passes into the cytoplasm. In eukaryotes, cytoplasmic mRNA is similar in size to the mRNA found in prokaryotes. It appears that not only the mRNAs but also the rRNAs and tRNAs are initially transcribed from eukaryotic chromosomes as large precursor molecules, which then undergo an extensive and highly specific enzymatic processing. This suggests that there are important sequences included in eukaryotic genes that do not relate to the functional RNA or to the protein ultimately formed but, rather, serve some as yet unknown function in the transcription of the gene.

In eukaryotes, therefore, the mRNA that serves as a template for protein synthesis is actually spliced together from segments of the mRNA precursor that is transcribed directly from the chromosomal DNA. The relationship in prokaryotes between one gene, one mRNA molecule, and one polypeptide chain—which used to be regarded as the "central dogma" of genetics—simply does not hold true for eukaryotes.

Translating mRNA into Protein

In prokaryotes, the sequence of codons is read from the 5′ end to the 3′ end of the molecule, as in the duplication of a DNA strand. The initial portion at the 5′ end is involved in guiding the binding of ribosomes to the mRNA molecule. The smaller ribosome subunit, which is bound to and carries a special tRNA (the one for n-formyl methionine), finds the mRNA at the 5′ leader region and scans down the mRNA until it encounters a codon specifying n-formyl methionine. The smaller ribosome unit then binds n-formyl methionine to a special site on the ribosome, forming an initiation complex.

One end of the growing protein chain (the amino, or N-terminal, end) is attached to the ribosome-mRNA complex through tRNA. The next amino acid to be added enters the complex already bound to its tRNA as a unit called amino-acyl-tRNA and binds, via the tRNA anticodon, to the next three bases (the codon) of the message. The amino group of the incoming amino acid loses a proton to the ribosome and forms a peptide bond with the carboxyl end of the peptide. As a result, the growing polypeptide chain is now bound to the newly entered amino-acyl-tRNA, while the tRNA from which an amino acid has just been removed and added to the growing peptide chain is free to dissociate and leave. Finally, the ribosome moves along the mRNA in the 3′ direction in one-codon increments, translocating the peptide chain in the process. The next amino acid can now be added, and the process is repeated (Figure 7–11).

Synthesis of polypeptide chains is terminated when a ribosome encounters one of three termination signals, or "stop" codons: UAA, UAG, or UGA. These codons do not specify amino acids, but their presence initiates a process that leads to the cleavage of the bond that joins the last amino acid in the peptide to its tRNA. This marks the termination of synthesis.

REGULATING GENE TRANSCRIPTION

The reading-out of genes into mRNA is under elaborate cellular control. Even in prokaryotes, several interlocking systems operate to regulate which genes are transcribed and to what extent. Eukaryotes appear to share many of the properties of prokaryotic gene regulation, although less is known about the details of control mechanisms in eukaryotes.

Because there has been intensive research on some prokaryotic genes, we know a great deal about how their transcription is regulated. Perhaps the best understood case is the lactose *(lac)* system of the bacterium *Escherichia coli.* Normally, these bacteria do not encounter the disaccharide lactose in their environment, so they do not synthesize the enzymes necessary to metabolize it. If lactose is added to the bacterial growth medium, the cells soon begin to produce large amounts of the enzyme β-galactosidase, which splits the disaccharide into glucose and galactose, thus permitting the lactose to be metabolized (Figure 7–12).

In cells growing on a medium that contains lactose, approximately 3000 molecules of β-galactosidase are present in every cell. This represents about 3 percent of all the protein in the cell. It is as if the lactose calls forth, or induces, the formation of the enzyme necessary to metabolize it. Many other cases of such *induction* of enzymes by energy-rich substances are

7–11
A schematic view of the processes of transcription and translation. (a) *During transcription, a strand of the DNA helix serves as a template for the synthesis of a complementary molecule of messenger RNA (mRNA).* (b) *During translation, three classes of RNA interact with a variety of enzymes and proteins to generate the formation of a new polypeptide chain. Ribosomal RNA (rRNA) is a component of the ribosomes, on which polypeptide synthesis takes place. The ribosomes contain a large* (50S) *and a small* (30S) *subunit. Transfer RNA (tRNA) interacts with amino acids and mediates their correct insertion into the growing polypeptide chain. Messenger RNA (mRNA) carries the information contained in a gene to the ribosome. The information is encoded as groups of three nucleotides, with each specifying a particular amino acid. Each codon is recognized by a complementary anticodon on a transfer RNA molecule previously associated with that particular amino acid. In this figure most of the amino acids are represented by numbered circles; the amino acid glycine (gly) has just been bound to its site on the ribosome by the corresponding transfer RNA. It will form a peptide bond with leucine (leu), thus extending the growing polypeptide chain by one amino acid. The ribosome then moves the length of the codon along the messenger RNA and so comes into position to bind the transfer RNA carrying serine (ser). (From* Ribosomes, *by M. Nomura, copyright © October 1969 by Scientific American, Inc.)*

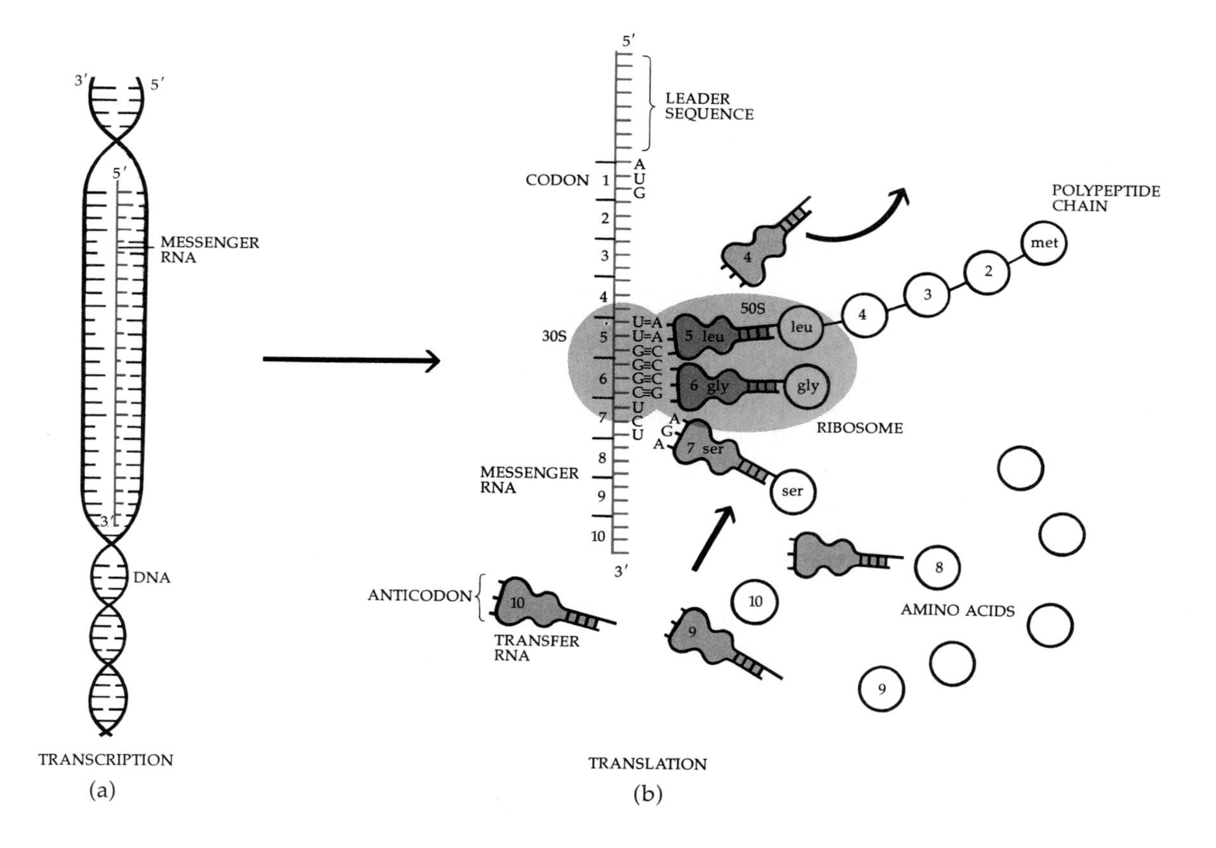

7–12
In the bacterium Escherichia coli, *the splitting of a molecule of lactose into galactose and glucose requires the enzyme β-galactosidase, which is an inducible enzyme; that is, its production is regulated by an inducer—in this case, the disaccharide lactose.*

β-Galactosidase

H_2O

GALACTOSE

GLUCOSE

known. In other cases, the enzymes involved in the synthesis of particular amino acids or other metabolites are not produced by the cell when that amino acid is present; this *repression* of enzyme synthesis by key metabolites involves the same control system as induction.

The Operon

Francois Jacob and Jacques Monod, working in France on studies for which they received the Nobel Prize in 1965, developed the concept of the operon to explain how bacterial cells regulate enzyme biosynthesis. An *operon* is a group of related genes that are aligned along a single segment of DNA. The operon is a unit of translation; all of its structural genes are coordinately produced as they are transcribed onto the same mRNA molecule. The *lac* operon region of *E. coli*, for example, contains several different kinds of genes involved in highly specific ways in the control of production of β-galactosidase and other enzymes involved in lactose metabolism. The interactions between the proteins that these genes produce are well understood at a molecular level. There are two control systems, one negative and one positive, that interact to produce large amounts of lactose-utilizing enzymes when glucose is scarce and lactose is plentiful, but the control systems shut off the splitting of lactose into galactose and glucose when glucose is abundant.

In eukaryotes, groups of related genes (operons) are not transcribed in clusters onto composite mRNA molecules, as they are in prokaryotes. Each cytoplasmic mRNA of a eukaryotic cell carries the information for only one peptide. Control systems do occur in plants, however. Mutations that result in the unrestrained production of groups of enzymes have been detected in both corn *(Zea mays)* and evening primrose *(Oenothera)*.

Regulation by Feedback Inhibition

In addition to the genetic mechanisms that change the functions of cells by altering the kinds of enzymes they contain, there are a number of physiological control systems that work by direct feedback inhibition of enzymes.

Fine adjustment control systems, such as end-product feedback inhibition, are examples of *allosteric interactions*. The basis of this kind of inhibition is that the bonds determining the tertiary structure of a protein (see page 55) are weak, and the binding of a particular molecule (an allosteric effector) at one site on the protein may so affect the weak interactions determining the protein's shape that the conformation of the protein is changed. When a protein molecule undergoes a change in shape, a second site, perhaps a site that allows the protein to function as an enzyme, may also be affected. Interactions of this sort can effectively control the levels of particular kinds of enzymatic activity in cells, thus enhancing their effective functioning. The allosteric effector is, in some instances, the end product of the pathway that begins with the reaction catalyzed by the enzyme under consideration.

CONTROL OF MULTICELLULAR DIFFERENTIATION

Differentiation is the developmental process by which a relatively unspecialized cell or tissue undergoes progressive changes to become more specialized in function or structure. This process is the means by which genes ultimately express themselves in the formation of the mature organism. One well-studied system that serves to illustrate this process involves a group of protists known as the cellular slime molds. One of these, *Dictyostelium discoideum* (Figure 7–14), illustrates the way in which influences originating outside developing cells or tissues can also affect their mature characteristics.

7–13
Cyclic AMP (adenosine monophosphate) is formed from ATP. "Cyclic" refers to the fact that the atoms of the phosphate group form a ring. This is the chemical that attracts the amoebas of the cellular slime molds, causing them to aggregate. In mammals, this same substance acts as a hormone "messenger"; it is triggered by extracellular hormones, such as adrenalin, and acts inside the cells to produce the characteristic hormonal effects.

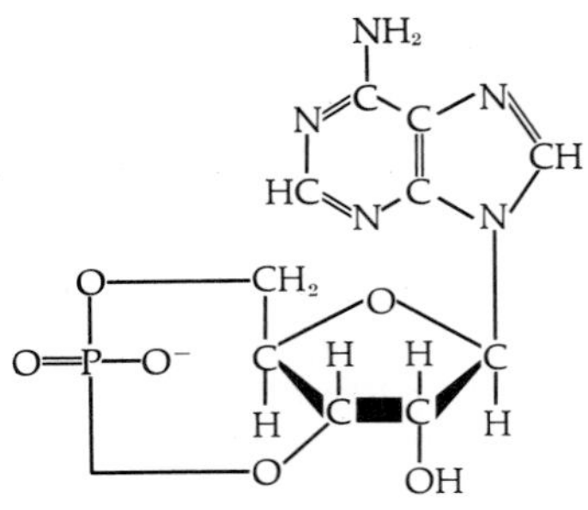

NH_2 N C C N CH HC N C N O CH_2 O O=P—O⁻ C H H C H C C H O OH

CYCLIC AMP

7–14

Life cycle of the cellular slime mold Dictyostelium discoideum. (a) *The feeding stage of the amoebas. The light gray area in the center of each cell is the nucleus, and the white areas are contractile vacuoles.* (b) *Amoebas aggregating. The direction in which the stream is moving is indicated by an arrow.* (c) *Migrating pseudoplasmodia, each formed of many amoebas. Each sluglike mass deposits a thick slime sheath which collapses behind it.* (d) *At the end of the migration, the pseudoplasmodium begins to rise vertically, differentiating into a stalk and mass of spores suspended in a droplet.*

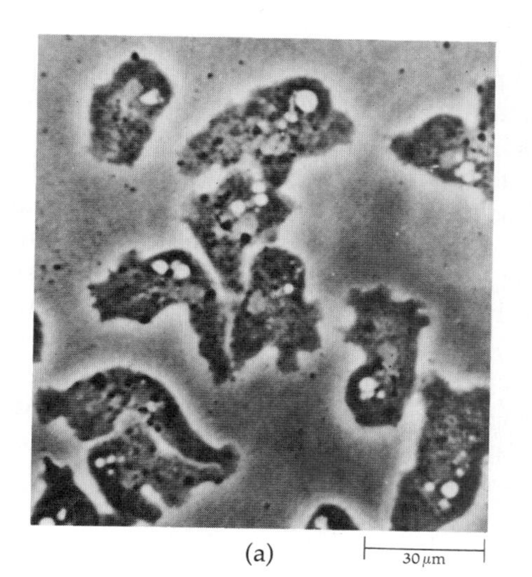

(a)

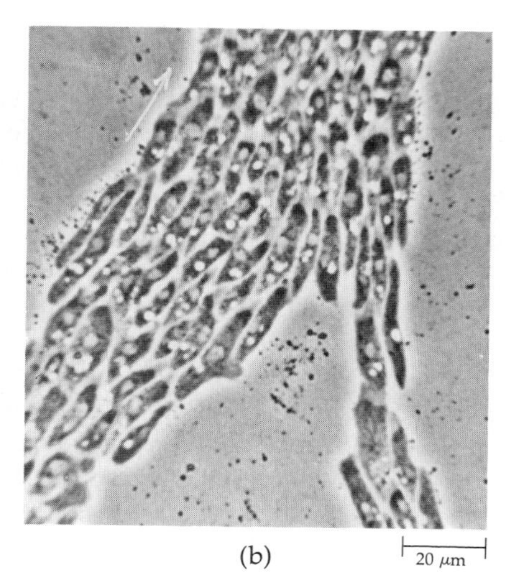

(b)

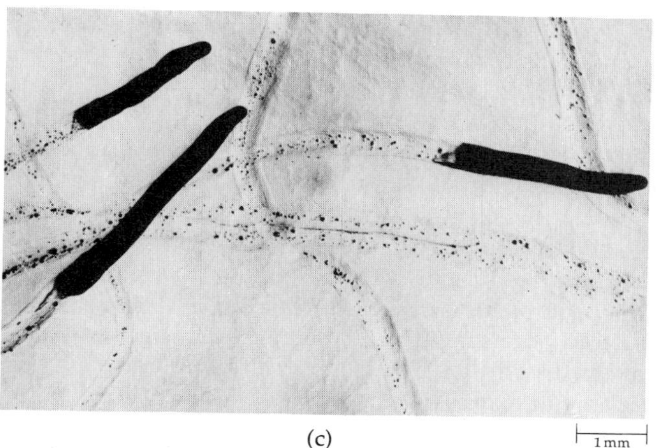

(c)

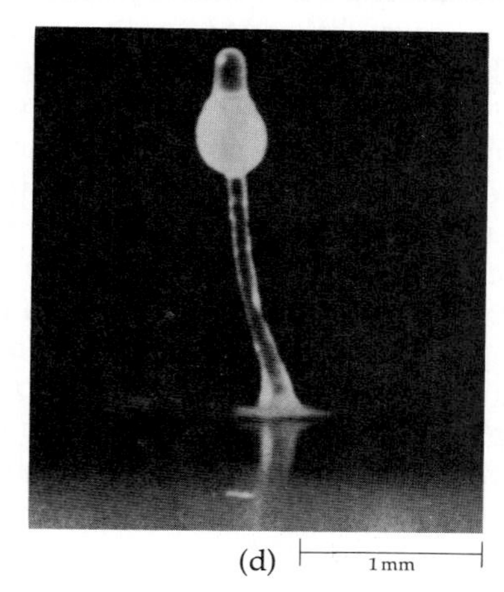

(d)

The cellular slime molds usually exist as free-living amoebalike cells, or myxamoebae, which feed on bacteria by surrounding and engulfing them (see Figure 3–11, page 68). They reproduce by cellular division and show little morphological differentiation until they exhaust the available supply of bacteria. In response to starvation, however, the cellular slime molds produce spores—but not in a simple way—and therein lies their interest to us. The individual cells first aggregate to form a sluglike motile mass called a pseudoplasmodium, which migrates to a new place before developing and releasing spores. This is a marvelous developmental advance, ensuring that new spores will not be released into old habitats in which the bacteria have been depleted.

The aggregation of myxamoebae is initiated when one or more starved cells begin to secrete cyclic adenosine monophosphate (cAMP) into the medium (Figure 7–13). The cAMP diffuses out and so establishes a concentration gradient, with the highest concentration of cAMP in the vicinity of the secreting cells. The surrounding cells respond by moving toward the higher concentrations of cAMP (Figure 7–14b). As they converge on the center of the aggregation, these cells are stimulated to emit a new "pulse" of cAMP. These higher levels of cAMP diffuse farther out into the medium, recruiting new members that migrate inward and join the aggregation. After emitting a pulse of cAMP, a period of approximately 5 minutes is required before more cAMP is released by the group of cells. At least three waves of cells are recruited in this fashion, with the characteristics of each wave determined by the concentrations of the recruiting cAMP. As the cells accumulate at the aggregation center, their plasma membranes become sticky, which causes them to adhere to each other, resulting in formation of a pseudoplasmodium surrounded by a cellulose sheath.

The eventual developmental fate of an individual cell (myxamoeba) is determined by its position in the aggregation. The first cells to aggregate usually give rise to the anterior portion of the pseudoplasmodium, whereas those cells entering the aggregation last form the base. After the formation of the sluglike structure, the pseudoplasmodium bends over and becomes horizontally oriented. In this position, the slug can migrate for several days, but only in the direction of the anterior tip (Figure 7–14c).

When migration ceases, the base of the pseudoplasmodium forms a flat disc (the basal disc), and the anterior portion develops a papilla (Figure 7–14d). The cells of the apical region (the first cells to aggregate) now become stalk cells of the developing fruiting body. These cells push down through the cell mass until the elongating stalk comes in contact with solid support. Further stalk extension occurs at the top of the stalk, which consists mainly of cellulose. What were initially posterior cells of the pseudoplasmodium move to the top of the stalk and become spores. Eventually, the basal disc and stalk cells die, and the spores are dispersed. If the spores fall on warm damp ground, they germinate. Each spore releases a single myxamoeba, and the entire cycle can now begin anew.

Thus, even in this relatively simple eukaryotic system, the phenomena of tissue recognition, differential cell migration, and localized cell death are present. Much more complex systems, with more precise control, characterize cellular differentiation in plants.

Control of Plant Development

The development of higher plants involves the organization of the complex sets of tissues that constitute the adult plant. Such a plant consists of an intricate array of many different tissues in precise physiological and morphological relation to one another. All originate from the same ultimately single cell—the fertilized egg, or zygote. As more and more complex tissues are organized, their developmental controls become an increasingly critical and difficult problem.

The process of development in the higher plants is radically different from that in animals. In animals, key developmental controls are internalized. Animal development proceeds according to a specific blueprint and in a precise sequence; the order and timing are critical to the outcome. At each stage, highly specific cell adjustments and changes in gene expression occur. The course of development in animals is not affected greatly by outside environmental perturbations that might upset the delicate balance required to produce the complex kinds of tissues that make up a mature animal.

In higher plants, on the other hand, development is directly linked to environmental influences and is unrestricted, or unlimited. Tissue-specific differentiation in plants is under the direction of hormones, whose production is sensitive to changes in the environment. The ability of plants to respond to external stimuli contributes to the development of individuals that are well suited to their particular habitats—an important factor in view of the inability of plants to seek more favorable conditions by moving from place to place.

Unlike the situation in higher animals, much of the differentiation in plants is fully reversible; this is related in part to the lack of a fixed developmental blueprint. In 1958, the plant physiologist F. C. Steward isolated small bits of secondary phloem tissue from carrot *(Daucus carota)* root and placed them in liquid growth medium in a rotating flask. Individual cells continually broke away from the growing cell mass and swirled free into the medium. Steward observed that many of the new cell clumps developed into roots. Transferred to a solid agar medium, some clumps developed shoots. If transplanted to soil, the new plantlets leafed out, flowered, and produced viable seed. This indicates that at least some cells of the differentiated phloem tissue—those that retained living protoplasts with nuclei—still contained all the genetic potential for full plant development. It also illustrates the fact that such differentiated cells are capable of expressing portions of their previously unexpressed genetic message when suitable environmental signals are provided to trigger those particular developmental patterns.

Cytoplasmic Influences on Differentiation

In many organisms, various cytoplasmic components play a direct role in cellular differentiation; these include organelles such as plastids and mitochondria, which contain their own DNA complements. If these organelles are unequally partitioned between the daughter cells during cell division, the resulting cell lines may have very different fates.

In a similar way, microchemical gradients within cells—that is, minute changes in the quantity of a substance present as one moves across the cell—play a major role in differentiation. For example, in the brown alga *Fucus* (see Figure 14–28, page 272), a gradient of stored insoluble food particles is apparently set up in the zygote by the action of gravity. This gradient in turn determines the position of the spindle apparatus in the first division of the zygote, setting up two distinctive cell lines from the very first cell division in the life of the individual. Unequal division of a cell may also be of fundamental importance in parceling out different kinds of elements present in the cytoplasm and in determining the fate of the resulting cell lines.

In the cells of higher organisms, where many substances are diffusing at different rates and in different directions, and tissues of various kinds are often packed together in close proximity, comparable but more complex effects must be common. Not only are gradients set up within individual cells and tissues that lead to an extremely subtle control of developmental processes, but also the differentiation of any one cell may be determined largely by its position in the body of the developing plant or animal. Some of the ways in which hormones and other factors interact in the development of higher plants are discussed in Section 6.

SUMMARY

The molecular key to understanding the process of heredity was provided in the 1940s with the discovery of the role of DNA. A variety of evidence strongly indicated that the genetic information resided in DNA, but the description by James Watson and Francis Crick in the early 1950s of the double-helical structure of DNA, with its complementary base pairing, provided a molecular model that has led to rapid advances in our knowledge. The intimate workings of the cell's genetic machinery can now be described in surprising detail.

DNA is replicated by synthesizing a complementary copy of each of the two strands of the double helix. A variety of enzymes act in concert to unravel DNA coils, unwind the double helix locally, and add new bases to each of the two separated strands.

The genetic information in the DNA is not expressed directly but is transferred via messenger RNA (mRNA)—long molecules that are assembled by complementary base-pairing along one strand of the DNA helix and then are carried to the cytoplasmic ribosomes. This process, called transcription, is under rigid genetic control. Each sequence of three nucleotides in the mRNA molecule is the codon for a specific amino acid.

At the ribosomes, mRNA is combined with a series of small molecules called transfer RNA (tRNA). Each tRNA has a 3-base sequence (anticodon) that is complementary to a codon of the mRNA. The mRNA-designated tRNA molecule fits its complementary 3-base anticodon onto the next codon of the mRNA base sequence and, guided by the ribosome, transfers its specific amino acid to the end of the growing peptide chain. This process is called translation.

Each of the 20 amino acids is coded by one or more 3-base sequences (codons) found in mRNA. The sequence of amino acids in a protein is determined by the sequence of codons along the mRNA molecule that directs the synthesis of that particular protein. The sequence of codons in each type of mRNA molecule depends ultimately on the sequence of bases in the DNA from which the mRNA is transcribed. Most amino acids are specified by three or four alternative codons, each with a different specific tRNA. Which codon corresponds to which amino acid has been determined, and the full code dictionary is now known.

Not all gene information represents information about protein sequence. Much of the information processed in eukaryotic nuclear mRNA consists of internal segments that are excised from the message before it reaches the cytoplasm.

The regulation of expression of simple prokaryotic gene systems, such as the *lac* system of *Escherichia coli,* is straightforward: one system activates transcription in the presence of the potential substrate—the "inducer" lactose—while another shuts down transcription in the presence of excess product—glucose. In eukaryotes, more complicated developmental sequences occur, as exemplified by those of the cellular slime mold *Dictyostelium discoideum.*

In higher plants, gene transcription programs are organized in blocks of developmental expression, each corresponding to a different pattern of cellular differentiation. The choice of a particular path of differentiation is influenced by the environment and may be reversible. In principle, any differentiated plant cell that retains a protoplast with a nucleus can be dedifferentiated and induced to produce a complete and viable plant.

CHAPTER 8

The Genetics of Eukaryotic Organisms

8–1
Gregor Mendel (1822–84), shown here standing at the right, holding a plant, carried out his studies in the garden of a monastery in Brno in what is now Czechoslovakia. His work in the field of genetics, which was not understood and therefore was largely ignored during his lifetime, was rediscovered in 1900.

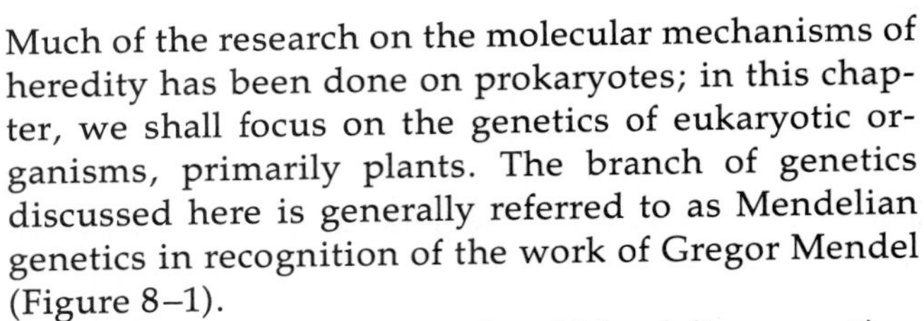

Much of the research on the molecular mechanisms of heredity has been done on prokaryotes; in this chapter, we shall focus on the genetics of eukaryotic organisms, primarily plants. The branch of genetics discussed here is generally referred to as Mendelian genetics in recognition of the work of Gregor Mendel (Figure 8–1).

In the context of our study of Mendelian genetics, we shall also examine its relationship and relevance to evolutionary theory. Charles Darwin's monumental work on evolution, *On the Origin of the Species,** was accomplished in ignorance of Mendel's studies. However, the study of evolution in the twentieth century is almost as dependent upon Mendel and his followers as it is on Darwin.

EUKARYOTES VERSUS PROKARYOTES

One of the major differences between eukaryotes and prokaryotes is that most eukaryotic organisms undergo sexual reproduction, a process not present in prokaryotes. Although some eukaryotic organisms do not reproduce sexually, it is evident in most cases that they lost the capacity to do so in the course of their evolutionary history.

Sexual reproduction involves a regular alternation between meiosis and syngamy. *Meiosis* is the process of nuclear division by which the chromosomes are reduced from the diploid ($2n$) to the haploid (n) number. During meiosis, the nucleus of a diploid cell undergoes two divisions, one of which is a reduction division. These divisions result in the production of four nuclei, each having only one-half the number of chromosomes of the original nucleus. *Syngamy* is the process by which two haploid cells fuse to form a diploid zygote. Syngamy thus results in the reestablishment of the diploid chromosome number.

*The complete title of Darwin's treatise is *On the Origin of Species by Means of Natural Selection, or the Preservation of Favoured Races in the Struggle for Life.*

All of the organisms discussed in this chapter are diploid for a major portion of their life cycle. Diploid organisms have two sets of chromosomes, one derived from each parent. Corresponding chromosomes, which form pairs during the course of meiosis, are called homologous chromosomes, or *homologues*. The interaction of products derived from genes on each of these sets of chromosomes determines the genetic characteristics of the diploid plant or animal at maturity.

STRUCTURE OF EUKARYOTIC CHROMOSOMES

Chromosomes of plants are made up of both DNA and protein. The DNA content of individual chromosomes varies widely, and this variation is reflected in great differences in chromosome length. Each eukaryotic chromosome apparently contains only a single thread or molecule of DNA, which is continuous throughout its entire length. This is certainly an impressive feat of packaging, for any one chromosome must contain a single, continuous DNA double helix that can be several centimeters in length.

Chromosomal DNA forms complexes with a variety of proteins, the majority of which are histones—proteins that are highly positive because of their high concentration of the amino acids lysine and arginine (see Figure 2–12, page 52). Most histones consist of 100 to 130 amino acid residues, with 20 to 40 lysine and arginine residues clustered at one end of the molecule. There are five major classes of histones, all of which are found in the chromosomes of most cells. Since they are so widely distributed, they clearly do not play an important role in the differential transcription of genes. However, they are important in DNA packaging; their positively charged regions interact with the negatively charged backbone of the DNA double helix with its exposed phosphate groups to produce the chromosomal complex.

The DNA of eukaryotic cells does not look at all like prokaryotic DNA. Instead of a slender thread of DNA, there is a regular array of projections, similar to beads on a string. Each "bead," or *nucleosome* (Figure 8–2a), consists of two histone tetramers (four-molecule complexes), with the DNA thread curved around them. This complex interaction of eukaryotic DNA and histones serves to package the DNA efficiently and to protect it from enzymatic degradation.

Chromosomes also contain nonhistone proteins. Their great diversity, coupled with evidence that certain classes of nonhistone proteins are restricted to particular cell types, suggests that they may play a regulatory role in gene transcription.

Some portions of chromosomes are packaged in a highly condensed state called *heterochromatin*. The rest of the chromatin, which is not condensed during interphase, is called *euchromatin*. The condensed organization of heterochromatin precludes transcription; indeed, the expression of euchromatic genes can be inhibited if they are located near a heterochromatic region.

During early meiosis or mitosis the euchromatin condenses. DNA initially aggregates into a series of tight "coils" separated by uncoiled regions. These coiled regions, or *chromomeres* (Figure 8–2b), are much larger than nucleosomes and can be detected with a light microscope.

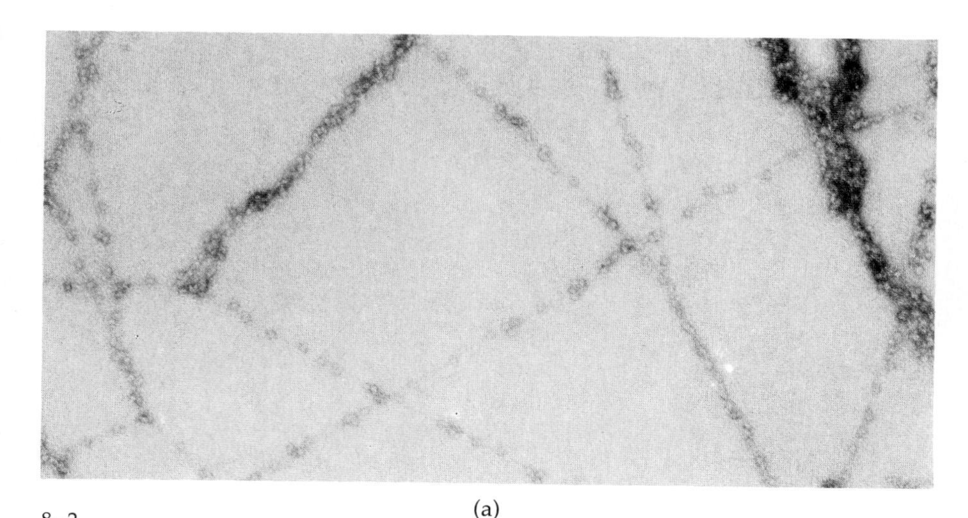

(a)

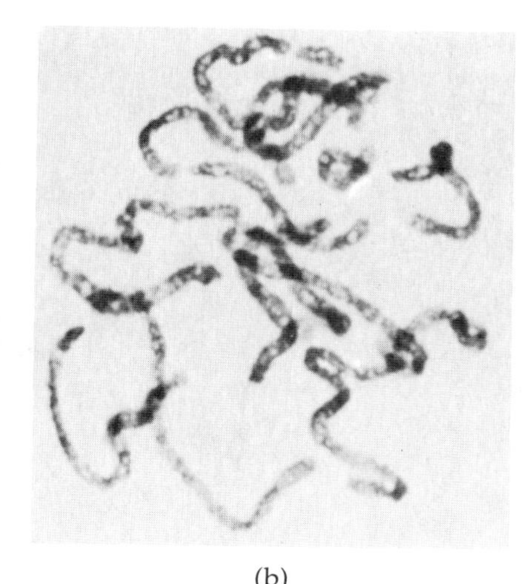

(b)

8–2
(a) *Electron micrograph of nucleosomes and connecting threads of DNA from a chicken erythrocyte (red blood cell). The nucleosomes—the structures that appear like beads on a string—are each approximately 1 micrometer in diameter.* (b) *Photomicrograph of mitotic chromosomes in a peony* (Paeonia californica). *The beadlike structures visible along the chromosomes are chromomeres, or regions of aggregated chromatin.*

MEIOSIS

Meiosis occurs only in specialized diploid cells and only at particular times in the life cycle of a given organism. Through meiosis, haploid nuclei are produced. Following cytokinesis, a single diploid cell gives rise to four haploid cells, either gametes or spores. A *gamete* is a cell that combines with another gamete to produce a diploid zygote. The zygote may then divide either meiotically to produce four unicellular haploid organisms or mitotically to form a multicellular diploid organism. The multicellular diploid organism eventually produces haploid spores or gametes by meiosis. A *spore* is a cell that can develop into an organism without uniting with another cell. Spores often divide mitotically, producing organisms that are entirely haploid and that eventually give rise to gametes by mitosis (see Figure 10–10, page 180).

First Meiotic Division

Meiosis consists of two successive nuclear divisions. Referring to Figure 8–6 on page 136 will help you keep track of the processes being described.

In prophase I (prophase of the first meiotic division), the chromosomes—present in the diploid number—first become visible as long, slender threads. As in mitosis (page 36), the chromosomes have duplicated during interphase, so that at the beginning of prophase I each chromosome actually consists of two identical chromatids attached at the centromere. However, at this early stage of meiosis each chromosome appears to be single rather than double. Before the chromatids become apparent, homologous chromosomes pair up. The pairing is very precise, beginning at one or more sites along the lengths of the chromosomes and proceeding in a zipperlike fashion, such that the same portions of the homologous chromosomes lie next to one another. Each homologue is derived from a different parent and is made up of two identical chromatids. Thus, a homologous pair consists of four chromatids. Because of this pairing, meiosis cannot occur in haploid cells in which homologues are not present. The pairing process itself is called *synapsis*, and associated pairs of homologous chromosomes are called *bivalents*.

During the course of prophase I, the paired threads become more and more tightly contracted, and the chromosomes shorten and thicken. With the aid of an electron microscope, it is possible to identify a densely staining axial core of proteins in each chromosome (Figure 8–3a). During mid-prophase, the axial cores of a pair of homologous chromosomes approach each other to within 0.1 micrometer to form what is known as a synaptonemal complex (Figure 8–3b).

In favorable material, it can now be seen that each axial core is double; that is, each bivalent is made up of four chromatids, two per chromosome. In at least one point on each bivalent, during the time when the synaptonemal complex exists, the two homologous chromatids break apart and exchange segments, thus re-forming complete chromatids but with exchanged segments. This event is known as *crossing-over*, and Figure 8–4 shows the visible evidence that such an exchange has taken place—the X-like configuration called a *chiasma* (plural: chiasmata).

As prophase I proceeds, the synaptonemal complex ceases to exist. Eventually, the nuclear envelope breaks down. The nucleolus usually disappears as RNA synthesis is temporarily suspended. Finally, the

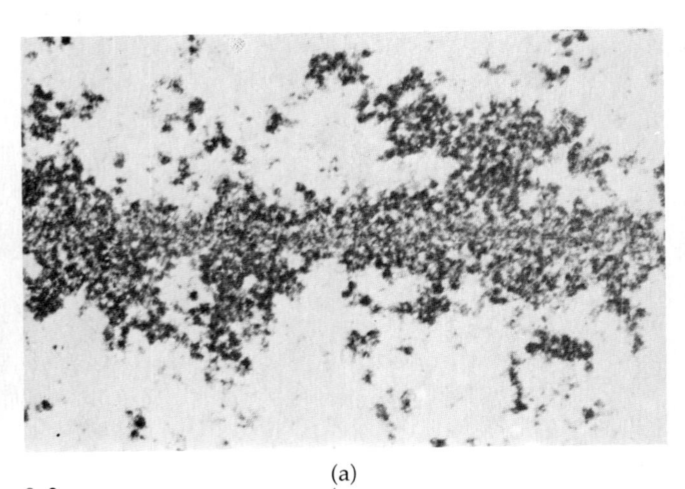

(a)

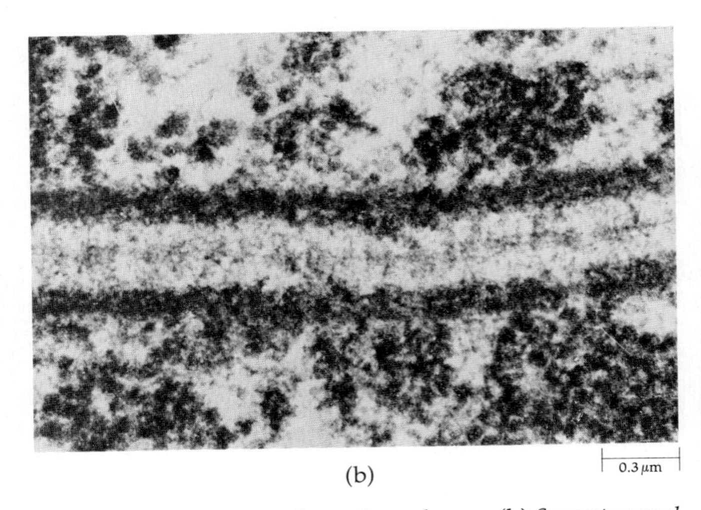

(b)

8–3
(a) Portion of a chromosome of Lilium *early in prophase I, prior to pairing. Note the dense axial core. This core, which consists mainly of proteins, may arrange the genetic material of the chromosome in preparation for pairing and genetic exchange. (b) Synaptonemal complex in a bivalent of* Lilium. *Only two of the chromatids are visible.*

homologous chromosomes appear to repulse one another. However, their chromatids are held together at the chiasmata. These chromatids tear apart very slowly. As they separate, some of the chiasmata slip toward the end of the chromosome arm. There may be one or more chiasmata in each arm of the chromosome, or only one in the entire bivalent; the appearance of the bivalent varies widely, depending on the number of chiasmata present (Figure 8–4).

In metaphase I, the spindle apparatus becomes conspicuous (Figure 8–5). The paired chromosomes, still joined together at their chiasmata, move to the equatorial plane of the cell, and their centromeres become attached to spindle fibers. Unlike mitotic metaphase, in which the centromeres line up on the equatorial plane, during meiotic metaphase I the centromeres of the paired chromosomes line up on either side of the equatorial plane.

Anaphase I begins when the homologous chromosomes separate entirely and begin to move toward the poles. (Notice again the contrast with mitosis. In mitotic anaphase, the centromeres divide and the identical chromatids separate.) In meiotic anaphase I, the centromeres do not divide and the chromatids remain together; it is the homologues that separate. Because of the exchanges of chromatid segments resulting from crossing-over, however, the chromatids are not identical, as they were at the onset of meiosis.

In telophase I, the coiling of the chromosomes relaxes and the chromosomes tend to become elongated and once again indistinct. Nuclear envelopes are reorganized out of the endoplasmic reticulum as telophase changes gradually to interphase, which may be short or absent depending on the organism involved. The nucleolus reappears. In many plants and animals, the nuclei enter promptly into prophase II of the second meiotic division.

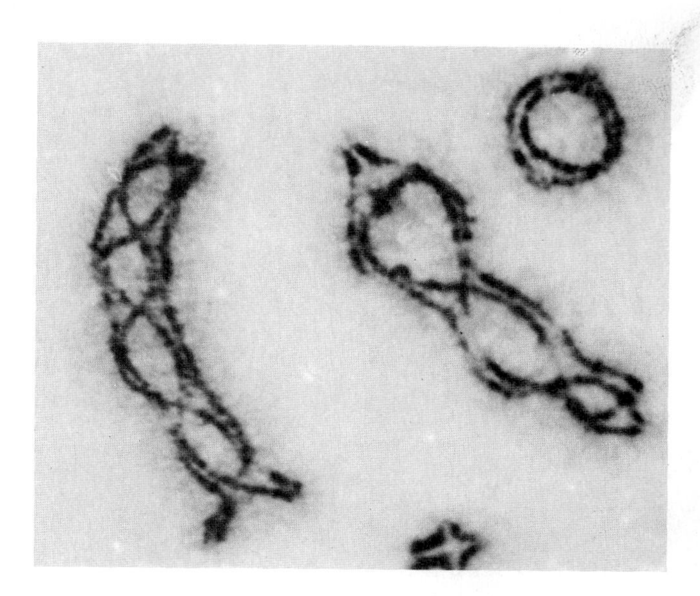

8–4
Variations in the number of chiasmata can be seen in the paired chromosomes of a grasshopper, Chorthippus parallelus.

Second Meiotic Division

At the beginning of the second meiotic division, the chromatids are still attached by their centromeres. This division proceeds just like mitosis: the nuclear envelope becomes disorganized and the nucleolus disappears at the end of prophase II. At metaphase II, the spindle apparatus again becomes obvious, and the chromosomes—each consisting of two chromatids—line up with their centromeres on the equatorial plane. At anaphase II, the centromeres divide and are pulled apart, and the newly separated chromatids, now called chromosomes, move to opposite poles (see Figure 7–1 on page 116). At telophase II, new nuclear envelopes and nucleoli are organized, and the contracted chromosomes relax as they fade into an interphase nucleus.

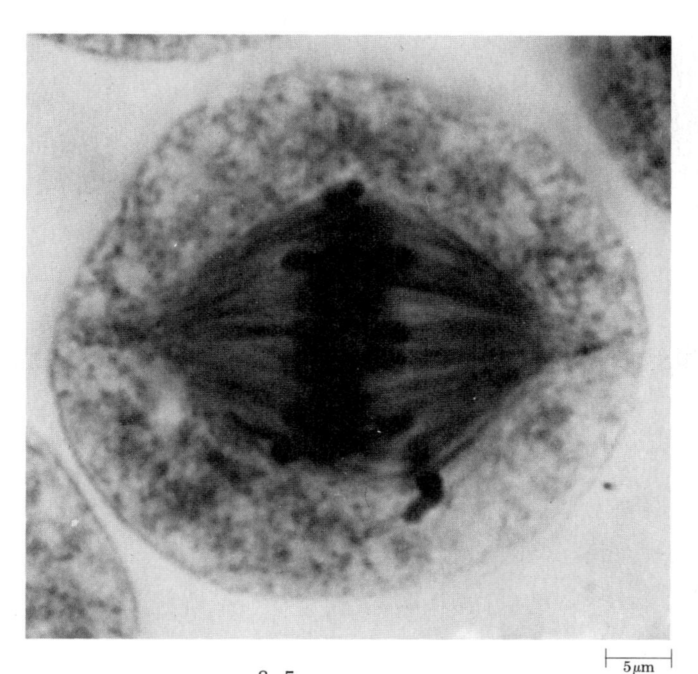

8–5
The spindle in a pollen mother cell of wheat (Triticum) *during metaphase I of meiosis.*

8–6
Meiosis, a diagrammatic representation with two pairs of chromosomes. Not all stages are shown.
Prophase I: *The chromosomes become visible as elongated threads, homologous chromosomes come together in pairs, the pairs coil round one another, and the paired chromosomes—held together at the chiasmata—become very short.*
Metaphase I: *The paired chromosomes come to lie with their centromeres evenly distributed on either side of the equatorial plane of the spindle.*
Anaphase I: *The paired chromosomes separate and move to opposite poles. The second meiotic division is essentially the same as ordinary mitosis.*
Metaphase II: *The chromosomes are lined up at the equatorial plane with their centromeres lying on the plane.*
Anaphase II: *The centromeres divide and the chromatids separate and move toward opposite poles of the spindle.*
Telophase II: *The chromosomes have completed their migration. Four new nuclei, each with the haploid number of chromosomes, are formed.*

Meiosis and microspore formation in crested wheat grass (Agropyron cristatum), n = 7, *is shown opposite.*

Early Prophase I

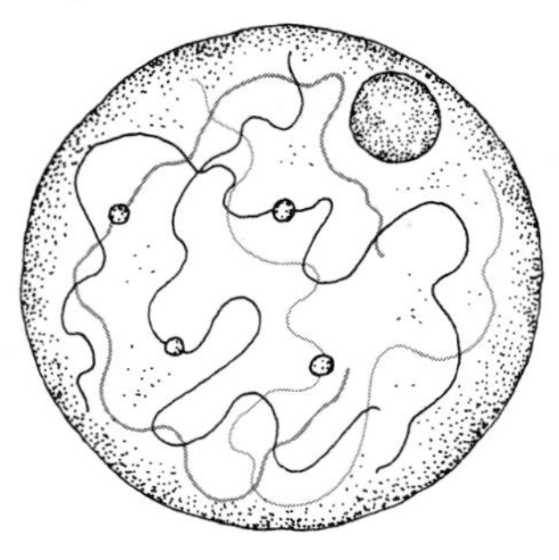

Prophase I

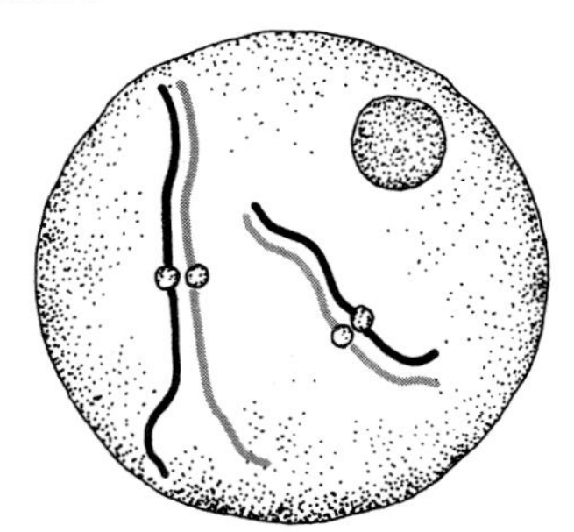

Prophase I

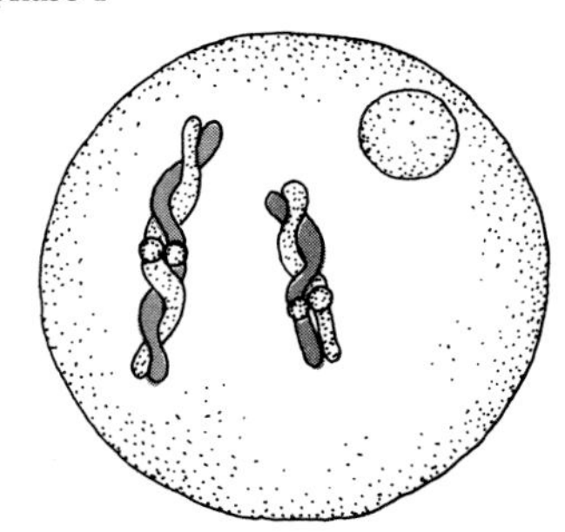

Late Prophase I

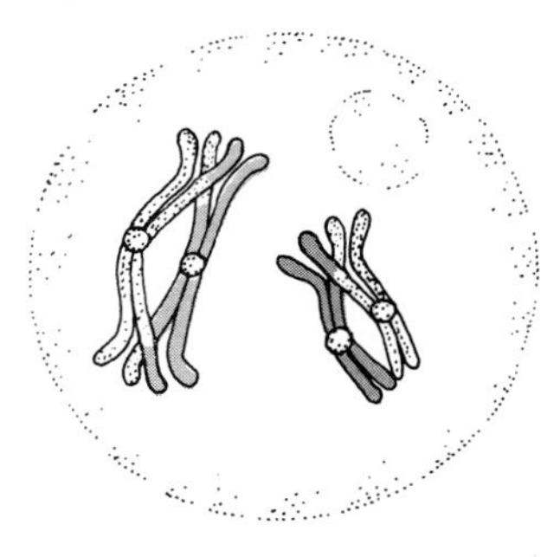

Metaphase I

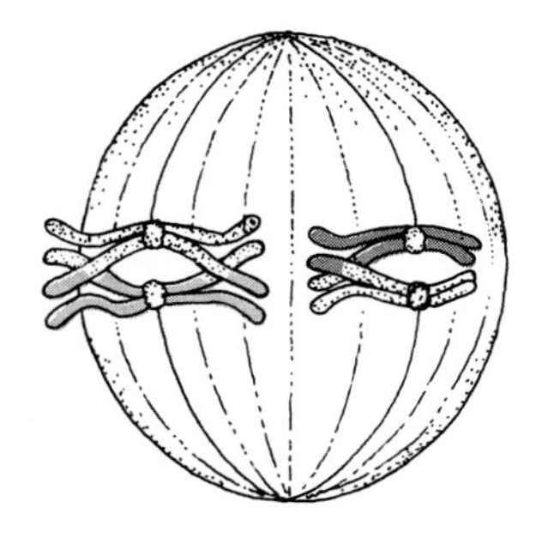

Anaphase I

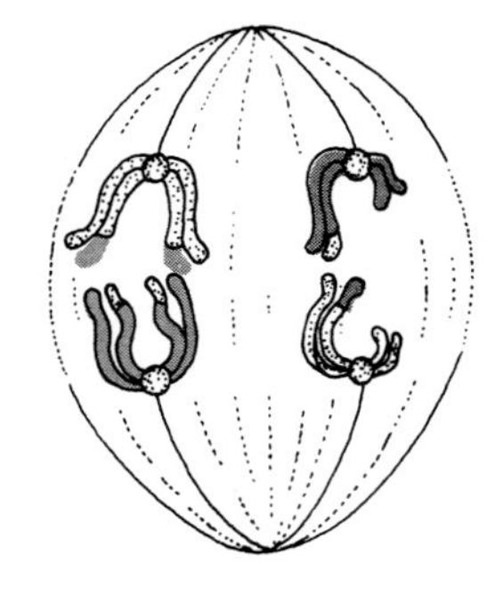

Metaphase II

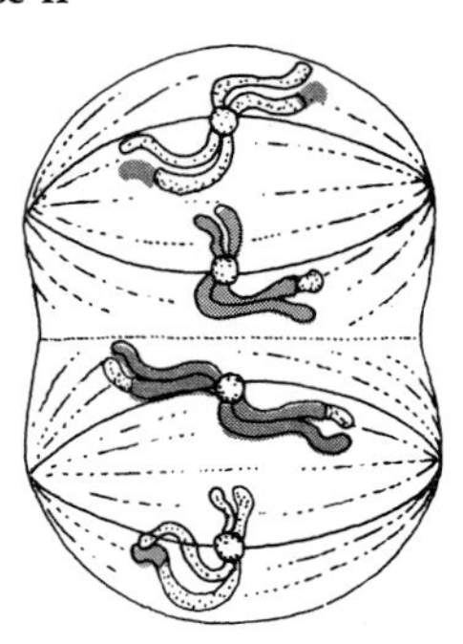

Anaphase II

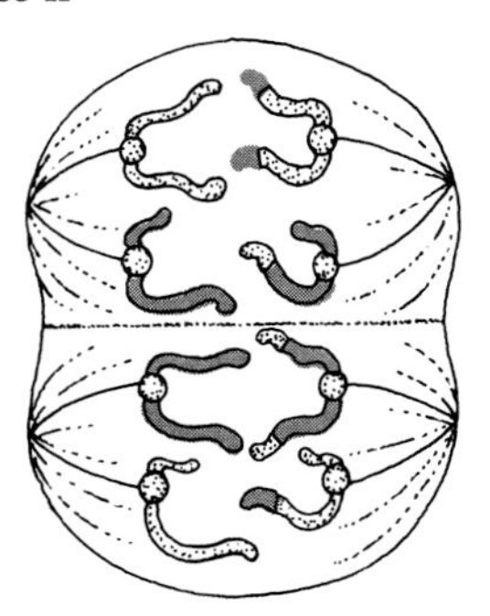

Late Telophase II

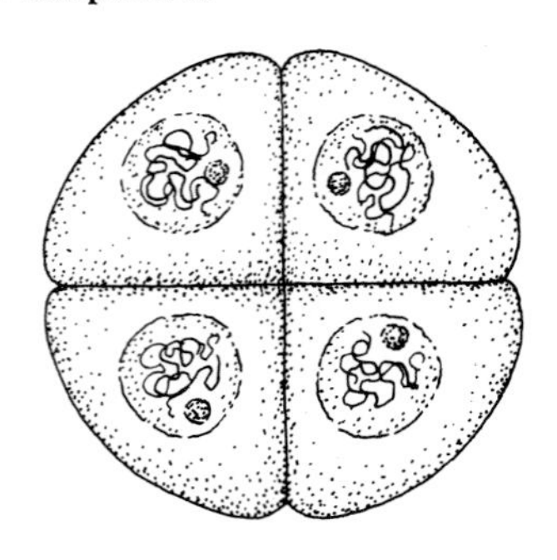

Early Prophase I. Chromosomes appear as threads. Each thread is actually double-stranded, composed of two identical chromatids.

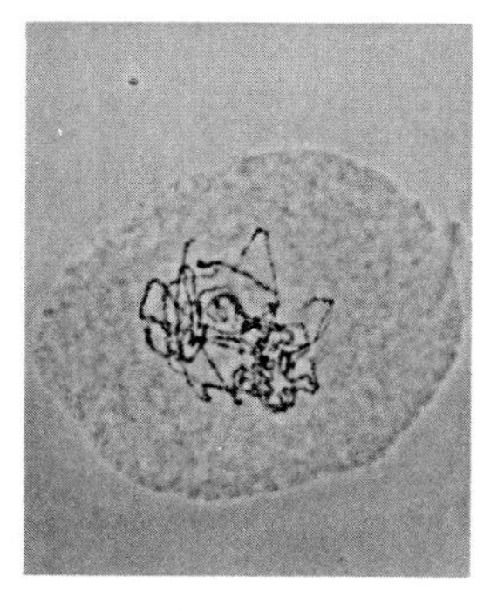

Prophase I. Homologous chromosomes pair. This is a crucial point of difference between meiosis and mitosis. Chiasmata are visible.

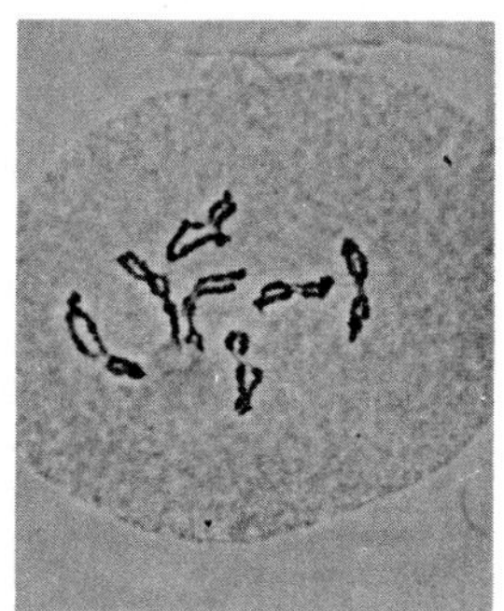

Metaphase I. Bivalents are now lined up at the equatorial plane of the cell, with their centromeres evenly distributed on either side of the plane.

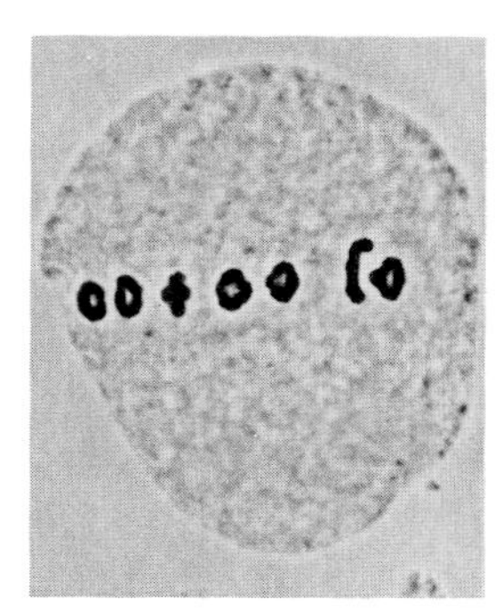

Anaphase I. The homologous chromosomes have separated from each other and are moving toward opposite poles of the cell.

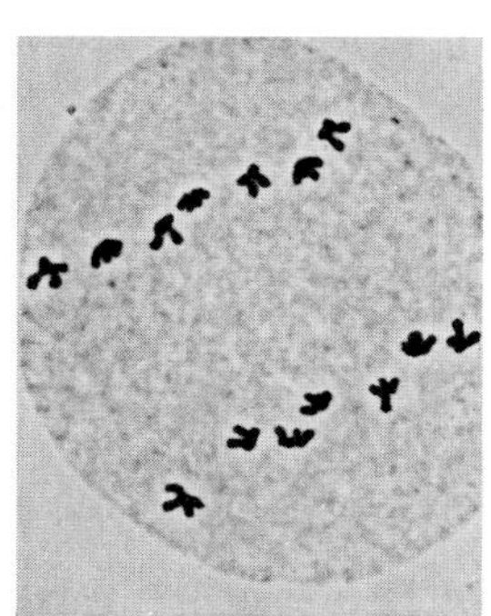

Late Telophase I. Chromosomes are regrouped at each pole, and the cell is dividing to form two haploid cells.

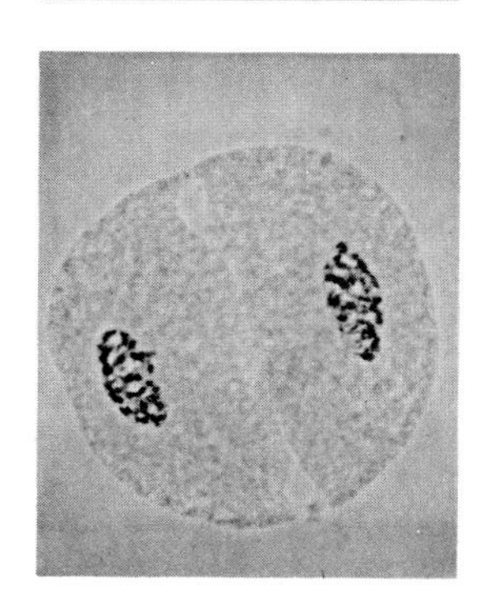

Prophase II. The chromosomes are reappearing. Each still consists of two chromatids. Because of crossing-over, the chromatids are no longer identical to each other.

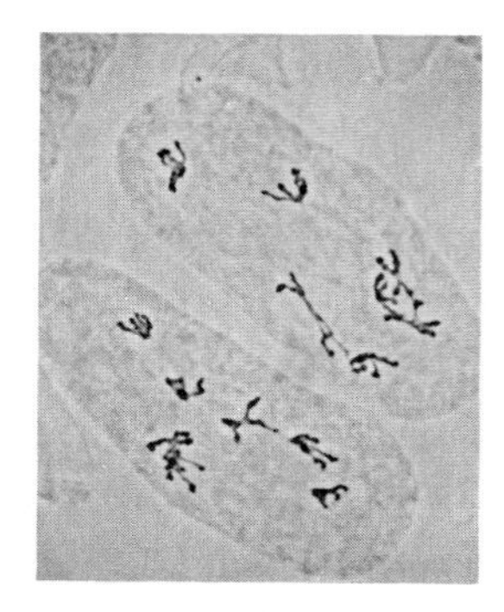

Metaphase II. The chromosomes are lined up at the equatorial plane of the cell, with their centromeres on the plane.

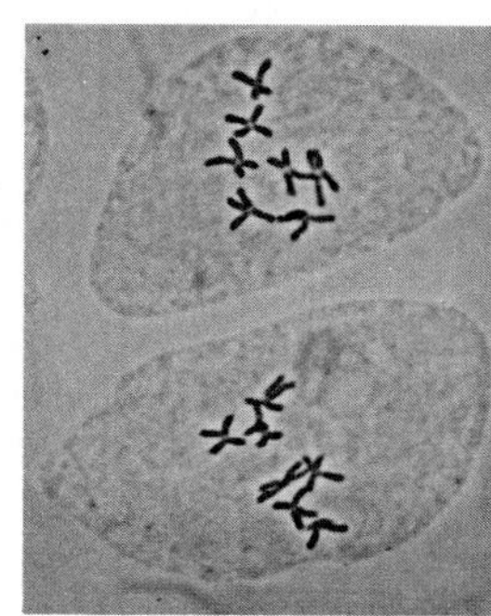

Anaphase II. The centromere of each chromosome has divided and the chromatids, now chromosomes, are moving toward opposite poles.

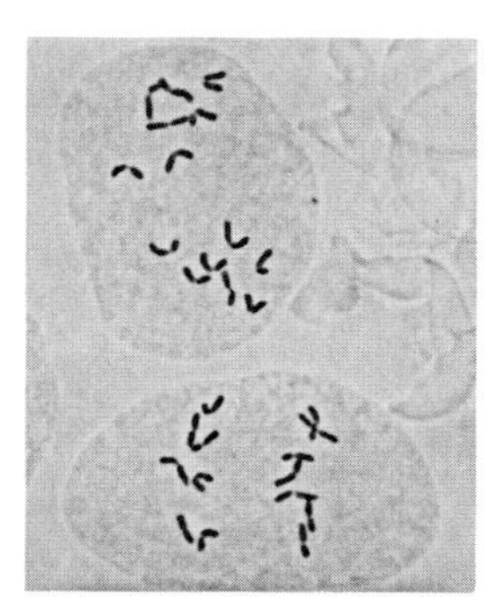

Telophase II. The chromosomes have now completely separated and new cell walls are forming.

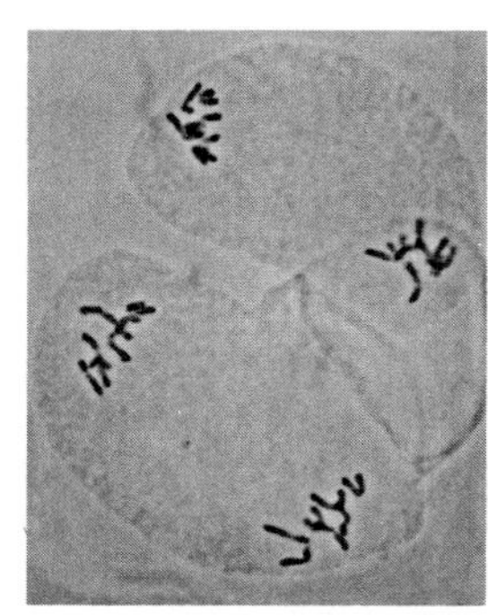

Tetrad. New plasma membranes and cell walls form. These four cells will become pollen grains.

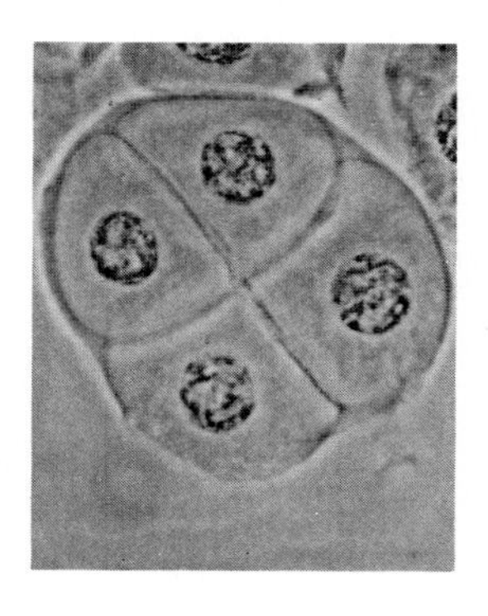

The Consequences of Meiosis

The end result of meiosis is that the genetic material present in the diploid nucleus has been *replicated* only once but has been *divided* twice. Therefore, each cell has only half as many chromosomes as the original diploid nucleus. More important are the genetic consequences of the process. At metaphase I, the orientation of the bivalents is random; that is, the chromosomes derived from one parent are randomly divided between the two new nuclei. In addition, because of crossing-over, each chromosome may contain segments derived from either parent. If the original diploid cell had two pairs of homologous chromosomes, $n = 2$, there are 4 possible ways the pairs could line up with respect to one another. If $n = 3$, there are 8 possibilities; if $n = 4$, there are 16. The general formula is 2^n. In humans, $n = 23$, so the number of possible combinations is 2^{23}, which is equal to 8,388,608.

As the number of chromosomes increases, the chance of reconstituting the same set that was present in the original diploid nucleus becomes smaller and smaller. Quite apart from this, the existence of at least one chiasma in each bivalent makes it almost impossible that any cell produced by meiosis could be genetically the same as any that fused to produce the diploid line of cells undergoing meiosis.

Meiosis differs from mitosis in three fundamental ways (see Figure 8–7):

1. Although the genetic material is replicated only one time, there are two nuclear divisions involved, producing a total of four nuclei.
2. Each of the four nuclei is haploid, containing only one-half of the number of chromosomes present in the original diploid nucleus from which it was produced.
3. The nuclei produced by meiosis contain wholly new combinations of chromosomes.

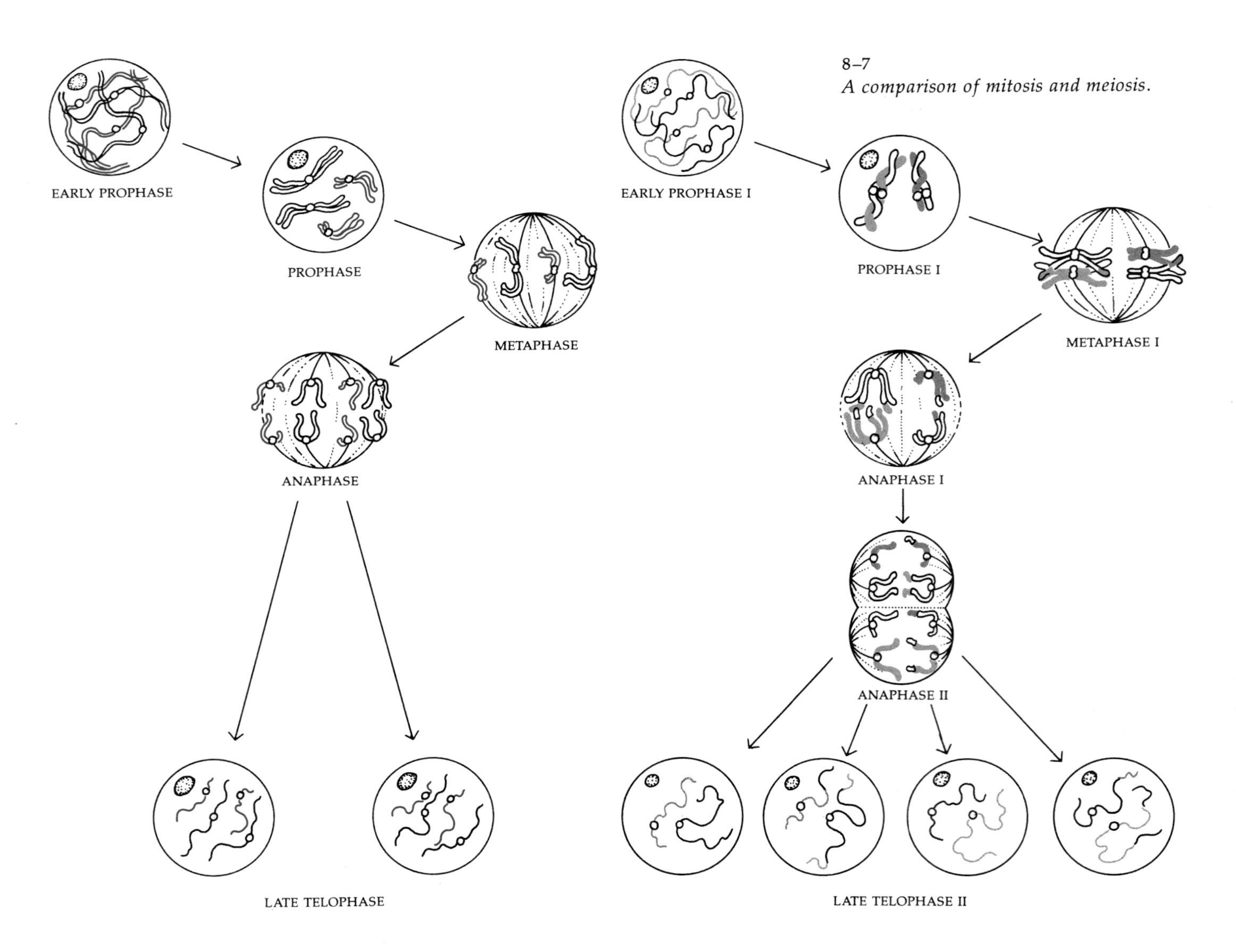

8–7
A comparison of mitosis and meiosis.

In meiosis, nuclei *different* from the original nucleus are produced, in contrast with mitosis, in which nuclei with chromosome complements *identical* to those of the original nucleus are produced. The genetic consequences of the behavior of chromosomes in meiosis are profound. Because of meiosis and syngamy, populations of diploid organisms in nature are far from uniform; they consist of individuals that differ from one another in many characteristics.

GENES AND DIPLOID ORGANISMS

The characteristics of diploid organisms are determined by the interactions between alleles. An *allele* is one of two or more alternative forms of the same gene. Alleles occupy the same site, or *locus*, on homologous chromosomes. The ways in which alleles interact to produce particular characteristics were first revealed by Gregor Mendel. For his studies, Mendel chose cultivated varieties of the garden pea *(Pisum sativum)* and considered well-defined but contrasting traits, such as differences in flower color or seed shape. He then made large numbers of experimental crosses (Figure 8–8), and perhaps most important of all, he studied the offspring not only of the first generation but also of subsequent generations and their crosses.

Table 8–1 lists seven traits of the pea plants that Mendel used in his experiments. When Mendel crossed plants with these contrasting traits, he observed that, in every case, one of the alternate traits could not be seen in the first generation (F_1). For example, the seeds of all the progeny of the cross between yellow-seeded plants and green-seeded plants were as yellow as those of the yellow-seeded parent. Mendel called the trait for yellow seeds, as well as the other traits that were seen in the F_1 generation, *dominant*. He called the traits that did not appear in the first generation *recessive*. When plants of the F_1 generation were allowed to self-pollinate, the recessive trait reappeared in the F_2 generation in ratios of approximately 3 dominant to 1 recessive (Table 8–1).

Table 8–1 *Mendel's Pea-Plant Experiment*

			F_2 GENERATION	
TRAIT	DOMINANT	RECESSIVE	DOMINANT	RECESSIVE
Seed form	Round	Wrinkled	5474	1850
Seed color	Yellow	Green	6022	2001
Flower site	Axial	Terminal	651	207
Flower color	Red	White	705	224
Pod form	Inflated	Tight	882	299
Pod color	Green	Yellow	428	152
Stem length	Tall	Dwarf	787	277

These results can be easily understood in terms of meiosis. Consider a cross between a white-flowered plant and a red-flowered plant. The allele for white flower color, which is a recessive trait, is indicated by the lowercase letter w. The contrasting allele for red flower color, which is a dominant trait, is indicated by the capital letter W. In the strains of garden pea with which Mendel worked, white-flowered individuals had the genetic constitution, or *genotype*, ww. Red-flowered individuals had the genotype WW. Individuals such as these, which have two identical alleles at a particular site, or locus, on their homologous chromosomes, are said to be *homozygous*. When plants with these contrasting traits are crossed, every individual in the F_1 generation receives a W allele from the red-flowered parent and a w allele from the white-flowered parent, and thus have the genotype Ww. Such an individual is said to be *heterozygous* for the gene for flower color.

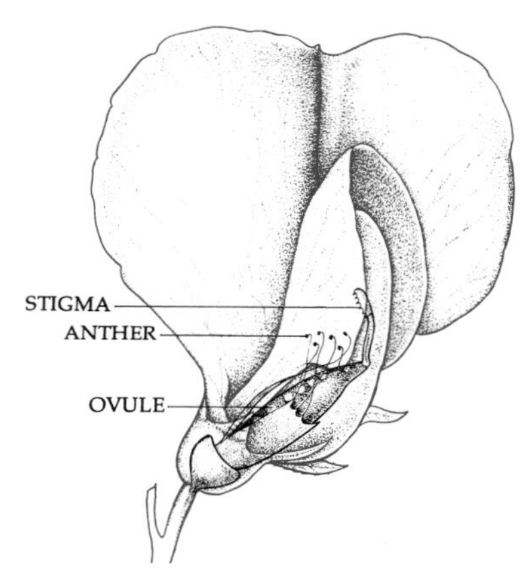

8–8
In a flower, the pollen develops in the anthers, and the egg cells develop in the ovules. Pollination occurs when pollen grains, trapped on the stigma, germinate and grow down to the ovules. In most species this involves pollen from one plant being caught on the stigma of another plant (cross-pollination).

In the garden pea flower, the stigma and anthers are completely enclosed by petals. Because the pea flower, unlike most, does not open until after pollination and fertilization have taken place, the plant normally self-pollinates. In his crossbreeding experiments, Mendel pried open the bud before the pollen matured and removed the anthers with forceps. Then he pollinated the flower artificially by dusting the stigma with pollen collected from other plants. The fertilized egg develops into an embryo within the ovule, and the ovule and embryo form the pea (the seed).

In the course of meiosis, a heterozygous individual will form two kinds of gametes, *W* and *w*, which will be present in equal proportions. As indicated in Figure 8–9a, these gametes will recombine to form one *WW* individual, one *ww* individual, and two *Ww* individuals, on average, for every four offspring produced. In terms of their appearance, or *phenotype*, the heterozygous *Ww* individuals will be red-flowered and are thus indistinguishable from the homozygous *WW* individuals. The action of the allele from the red-flowered parent is sufficient to mask the action of the allele from the white-flowered parent. This, then, is the basis for the 3 to 1 phenotypic ratios that Mendel observed.

How can you tell whether the genotype of a plant with red flowers is *WW* or *Ww*? As shown in Figure 8–9b, you can tell by crossing such a plant with a white-flowered plant and counting the progeny of the cross. Mendel performed just this kind of experiment, which is known as a *testcross*—the crossing of an individual showing a dominant trait with a second individual that is homozygous recessive for that trait. (A testcross is sometimes called a backcross because it is often performed by crossing a member of the F_1 generation with the recessive parent; however, the crossing of a hybrid with either of its parents, or with an individual that is genetically equivalent to one of its parents, would constitute what is formally known as a *backcross*.)

The Principle of Segregation

The principle established by these experiments is the *principle of segregation*, which is sometimes known as Mendel's first law. According to this principle, hereditary characteristics are determined by discrete factors (now called genes) that appear in pairs, one of each pair being inherited from each parent. During meiosis, the pairs of factors are separated, or *segregated*. Hence, each gamete produced by an offspring at maturity contains only one member of the pair. This concept of a discrete factor explained how a characteristic could persist from generation to generation without blending with other characteristics, as well as how it could seemingly disappear and then reappear in a later generation. This principle of segregation was of the utmost importance for evolutionary theory.

Incomplete Dominance

In the previous examples, the action of the dominant allele, when it was present, concealed the existence of the second, recessive allele. Dominant and recessive characteristics are not always so clear-cut, however. In cases of *incomplete dominance*, for example, a given trait appears in the heterozygote in a form intermediate between that seen in the two homozygotes, since the action of one allele does not completely mask the action of the other.

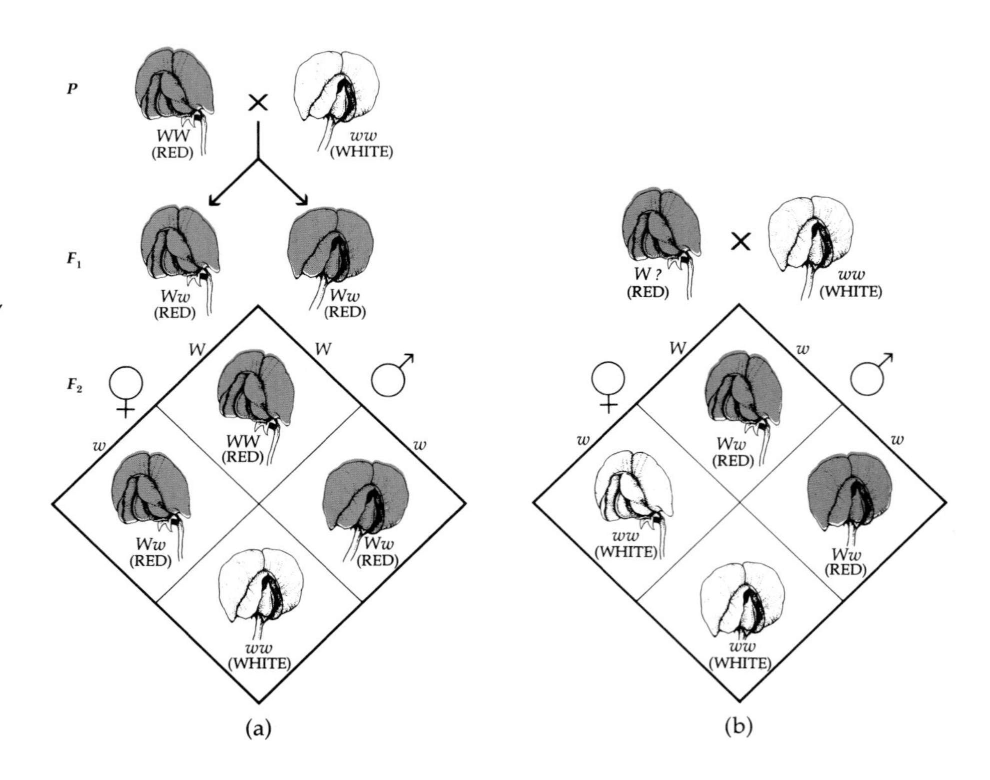

8–9
(a) *A cross between a pea plant with two dominant genes for red flowers* (WW) *and one with two recessive genes for white flowers* (ww). *The phenotype of the offspring in the* F_1 *generation is red, but note that the genotype is* Ww. *The* F_2 *generation is shown in the diagram below. The* W *allele, being dominant, determines the appearance, or phenotype, of the flower. Only when the offspring receives a* w *allele from each parent, producing a* ww *genotype, does the recessive trait (white flowers) appear. The ratio of dominant to recessive phenotypes is thus always as expected, 3 to 1.*
(b) *A testcross between a pea plant with red flowers and one with white flowers. Although the red-flowered plant is phenotypically identical to the* WW *plant shown in* (a), *results of the testcross reveal that it is heterozygous* (Ww) *for this gene; only this parent genotype could produce the genotypes observed in the progeny.*

8–10
When a snapdragon with red flowers is crossed with one with white flowers, the resulting progeny all have pink flowers (Ww). *When the pink-flowered plants are crossed, the results show that although the gene products blend in the heterozygote, the genes themselves remain discrete and segregate according to the Mendelian ratio.*

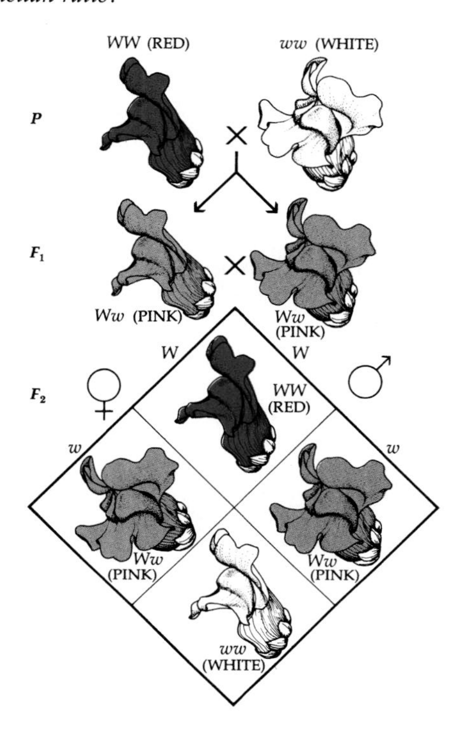

8–11
In the lima bean (Phaseolus lunatus), *the gene for spotting* (S) *affects the pigment distribution of the seed coat, but* S *is incompletely dominant. The top row of beans is homozygous* (SS) *for the spotting trait; the middle row is heterozygous* (Ss), *but shows the effect of incomplete dominance; and the bottom row is homozygous* (ss) *and thus shows no spotting.*

For instance, in the snapdragon, a cross between a red-flowered plant and a white-flowered plant produces one that has pink flowers. When the F_1 generation is crossbred, the traits sort themselves out again, the result being one red-flowered (homozygous) plant to two pink-flowered (heterozygous) plants to one white-flowered (homozygous) plant (Figure 8–10). Thus, cases of incomplete dominance also conform to Mendel's principle of segregation. Distribution of pigmentation in the seed coats of lima beans (Figure 8–11) is another example of a genetic trait that is incompletely dominant.

Independent Assortment

In inheritance patterns involving more than one gene, certain differences in the patterns depend on whether the genes are located on the same chromosome or on different chromosomes. We shall consider first the relatively simple situation in which the genes are located on different chromosomes and then the more complicated one in which they are located on the same chromosome.

Mendel himself studied hybrids that involved two pairs of contrasting characteristics. For example, he crossed strains of garden peas in which one parent plant had seeds that were round and yellow and the other had seeds that were wrinkled and green. As Table 8–1 shows, the traits for round seeds and for yellow seeds are both dominant, and the wrinkled and green traits are recessive. All the plants of the F_1 generation had seeds that were round and yellow. When the F_1 seeds were planted and the flowers were allowed to self-pollinate, 556 plants were produced. Of these, 315 showed the two dominant characteristics, round and yellow, and 32 combined the recessive traits, green and wrinkled. All the rest of the seeds were unlike either parent: 101 were wrinkled and yellow, and 108 were round and green. Totally new combinations of characteristics had appeared.

8–12
Independent assortment. In one of Mendel's experiments, he crossed a plant with round (RR) *and yellow* (YY) *peas and a plant with wrinkled* (rr) *and green* (yy) *peas. The seeds of the* F_1 *generation were round and yellow. In the* F_2 *generation, however, as shown in the diagram, the recessive traits reappeared. Furthermore, they appeared in new combinations. In a cross such as this, involving two pairs of alleles on different chromosomes, the expected phenotypic ratio in the* F_2 *generation is* 9:3:3:1.

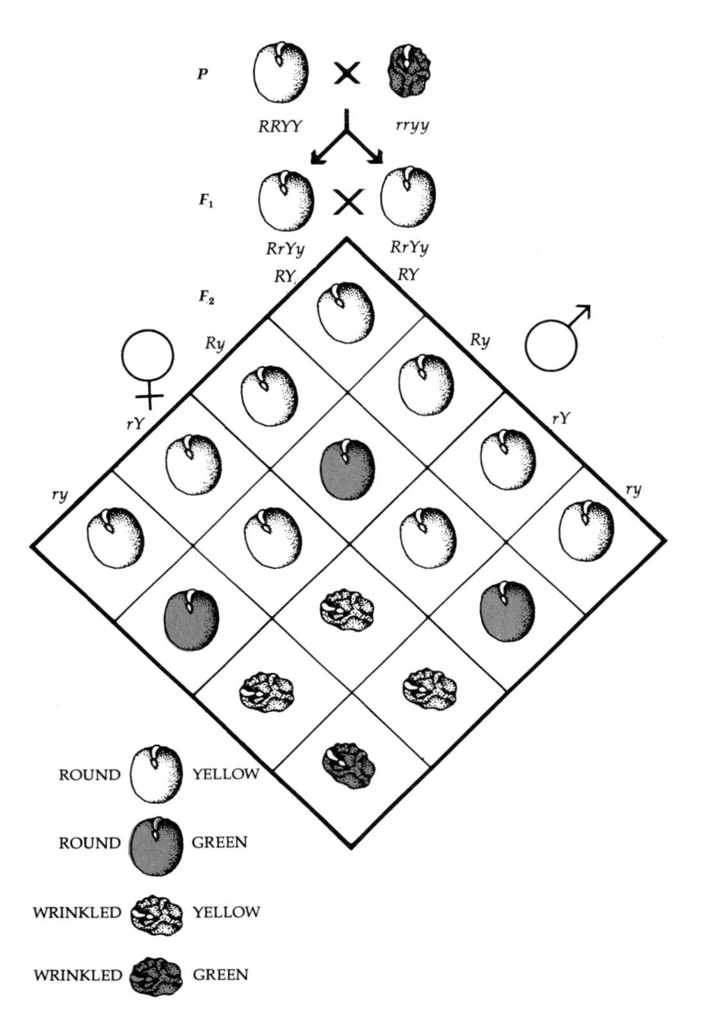

Figure 8–12 shows the basis for such results. In a cross involving two pairs of dominant and recessive alleles, with each pair on a different chromosome, the ratio of distribution of phenotypes is 9:3:3:1. The 9 represents the proportion of the F_2 progeny that shows both dominant traits; 1 is the proportion that shows both recessive traits; and 3 represents the proportions of the two possible combinations of dominant and recessive traits. In the example shown, one parent carries both dominant traits and the other parent carries both recessive traits. Suppose that each of the parents carried one recessive and one dominant trait. Would the results be the same? If you are not sure of the answer, try making a diagram of the possibilities, as was done in Figure 8–12.

From these experiments, Mendel formulated his second law, the *principle of independent assortment*. It states that the inheritance of a pair of factors for one trait is independent of the simultaneous inheritance of factors for other traits, such factors "assorting independently" as though there were no other factors present.

Linkage

When Mendel was performing his experiments, chromosomes had not yet even been visualized, much less recognized as the genetic material. Knowing that genes are located on chromosomes, one can readily see that if two different genes controlling certain characteristics are located on the same chromosome pair, they generally will not segregate independently, and a 9:3:3:1 ratio will not be obtained in the F_2 generation. Either through good fortune or by careful selection of the traits he discussed, Mendel avoided this phenomenon of *linkage*, which he probably would not have been able to understand. The pea plant has seven pairs of chromosomes, and the seven major traits studied by Mendel are located on five separate chromosomes, with those on the same chromosome so far apart as to segregate independently.

Linkage of genes was first discovered by the British geneticist William Bateson and his co-workers in 1905, while they were working with the sweet pea *(Lathyrus odoratus)*. They crossed a double recessive line with red petals and round pollen to a wild-type line with purple petals and long pollen, and they obtained the following characteristics in the F_2 generation:

284	purple	long
21	purple	round
21	red	long
55	red	round

If the two genes were on the same chromosome, there should have been only two types of progeny. If they were on two different chromosomes, then the ratio of the four classes should have been 216:72:72:24, or 9:3:3:1. Clearly, the genes of the "parental" types were being held together, as if they were *linked* to one another.

Linkage and the related phenomenon of crossing-over were clarified as a result of work with the fruit fly *Drosophila* by T. H. Morgan and his research group in the early 1900s. *Drosophila* is a particularly useful organism for use in genetic studies: colonies are easy to breed and maintain; a new generation of flies can be produced every two weeks; each female lays hundreds of eggs at a time; and the most common species has only four pairs of chromosomes. Female fruit flies have two homologous sex chromosomes, *XX*, whereas the males have one *X* chromosome and a much smaller *Y* chromosome on which many of the genes of the *X* chromosome are not present.

In Morgan's experiments, an unusual type of fruit fly with white eyes was produced. Because males always expressed this recessive trait, and male *Drosophila* are *XY*, Morgan concluded that the gene determining eye color was located on the *X* chromosome. When other recessive "sex-linked" traits, such as a yellow body, were also observed, Morgan reasoned that they must be on this same chromosome since the male contributes only one *X* chromosome.

Morgan crossed a fly showing two recessive sex-linked traits with a "wild-type" fly. As expected, all the F_1 offspring were phenotypically "wild." A testcross was then carried out between an F_1 fly and its double-recessive parent, with the expectation that the progeny would be divided between the wild type and the white-eyed, yellow-bodied type. However, flies resulted that had white eyes and wild-type bodies, as well as flies with wild-type eyes and yellow bodies.

The explanation for these results is that the two genes are "linked" on one chromosome but are sometimes exchanged between homologous chromosomes. It is now known that crossing-over—the actual breakage and rejoining of chromosomes that results in the appearance of chiasmata—occurs in prophase I of meiosis (see Figure 8–4). The farther apart two genes are located on a chromosome, the greater the chance that a crossover will occur between them. The closer together two genes are, the greater will be their tendency to assort together in meiosis and, thus, the greater is their "linkage."

MUTATIONS

The studies on independent assortment just described depend on the existence of differences between the alleles of a gene. What is the source of such differences? The first answer to this question was given by the Dutch scientist Hugo de Vries.

8–13
Hugo de Vries, shown standing next to Amorphophallus titanum, *a member of the same family as the calla lily. The plant, a native of the Sumatran jungles, has one of the most massive inflorescences, or flower clusters, of any of the angiosperms. This picture was taken in the arboretum of the Agricultural College at Wageningen, Holland, in 1932.*

In 1901, while studying the inheritance of characteristics in a type of evening primrose *(Oenothera glazioviana)*, de Vries found that although the patterns of heredity were generally well ordered and predictable, a characteristic occasionally appeared that had not been observed previously in either parental line. De Vries hypothesized that this new characteristic was the phenotypic expression of a change in a gene. Moreover, according to his hypothesis, the changed gene would then be passed along just as the other genes were. De Vries spoke of this hereditary change as a *mutation* and of the organism carrying it as a *mutant*.

Ironically, only 2 of about 2000 changes in the evening primrose observed by de Vries were actually mutations. The other 1998 were due to new genetic combinations or the presence of extra chromosomes rather than actual abrupt changes in any particular gene. However, although most of de Vries's examples proved to be invalid, his definition of a mutant and his recognition of the role that mutations play in producing variations are still essentially accurate.

It is now known that a mutation can involve as little as one change in a single nucleotide pair, or it may involve larger changes: deletions of a portion of the chromosome; repetitions of some of the genetic material; translocations, in which a piece of one chromosome becomes attached to another; or inversions, which occur when a segment of chromosome breaks loose, rotates 180°, and becomes reattached by the "wrong" ends. Although most mutations are harmful, the capacity to mutate is extremely important, for it allows the individuals within a species to vary, to adapt to changing conditions. Mutations thus provide the basis for evolutionary change. Mutations in eukaryotes occur spontaneously at a rate of about 5×10^{-6} per locus per cell division (that is, 1 mutant gene at a given locus per 200,000 cell divisions).

In a diploid plant, every gene is normally present in duplicate. When a mutation occurs in a predominantly haploid organism, such as the fungus *Neurospora* or one of the prokaryotes, it is immediately exposed to the environment. If favorable, it is selected for; if unfavorable, as is usually the case, it is quickly eliminated from the population. In diploid organisms, the situation is very much different. Each chromosome is present in duplicate, and a mutation on one of the homologues, even if it would be unfavorable in a double dose, may have much less effect or even be advantageous when present in a single dose. Thus, it may persist in the population. The mutant gene may eventually alter its function, or the selective forces on the population may change in such a way that its effects become advantageous.

PLEIOTROPY

In contrast with the examples given previously, genes normally affect more than one characteristic of an organism. Their products are proteins, which are usually involved in multiple interactions in the course of the development of the organism. When the science of genetics was relatively new, genes were described in terms of their more obvious phenotypic effects, that is, as purple-petaled, hairy, and so forth. However, it was eventually found that single genes could control whole complexes of characteristics. The capacity of a single gene to affect many characteristics is known as *pleiotropy*.

One clear-cut example was provided by Hans Gruneberg, who studied a whole complex of inherited deformities in the rat, including thickened ribs, narrowing of the tracheal passage, loss of elasticity of the lungs, hypertrophy of the heart, blocked nostrils, a blunt snout, and as a consequence, greatly increased mortality. He was able to demonstrate that all these changes were caused by a single mutation, that is, a mutation involving only one gene. This particular gene produces a protein involved in the formation of cartilage, and because cartilage is one of the most common structural substances of an animal's body, the widespread effects of a mutation of such a gene are not difficult to understand.

GENE INTERACTIONS IN DIPLOID ORGANISMS

Not only does a single gene often affect more than one phenotypic characteristic but, conversely, there are many characteristics that are controlled by the combined effects of several genes. Some examples of this process are discussed under two broadly overlapping headings—epistasis and polygenic inheritance.

Epistasis

Early in the present century, William Bateson and his group of researchers obtained some surprising results, which at first seemed impossible to explain. They crossed two pure-breeding, white-flowered varieties of sweet pea *(Lathyrus odoratus)* and found that the progeny all had purple petals. When these F_1 plants were allowed to self-pollinate, they found that of the 651 plants that flowered in the F_2 generation, 382 were purple-petaled and 269 were white-petaled. This is essentially a 9:3:3:1 ratio in which only some of the plants (9 out of 16) showed the effects of the *two* dominant genes by having flowers with purple petals. A situation of this sort is described as *epistasis*, a form of gene interaction whereby one gene modifies the phenotypic expression of another, nonallelic gene.

Epistasis is a widespread and important phenomenon. The more we know about heredity, the clearer it is that no gene ever acts in isolation; its effects are always modified by the internal environment, which is, of course, the end product of the interaction of thousands of genes. This is not surprising because genes act by the production of proteins, which can produce their effects only within the context of the cell. All such interactions between nonallelic genes in the expression of the phenotype can be considered to be examples of epistasis.

Polygenic Inheritance

Figure 8–14 is a plot of the distribution of ear length in a variety of corn plants, but a curve of this shape could just as well represent the distribution of weight among a random assortment of pinto beans or height among oak trees. Environmental factors, such as rain in a cornfield, might affect the actual measurements involved in preparing such a graph, but they rarely affect the shape of the curve.

Table 8–2 *The Genetic Control of Color in Wheat Kernels*

	Genotype	Phenotype	
Parents	$R_1R_1R_2R_2 \times r_1r_1r_2r_2$ (dark red) (white)		
F_1	$R_1r_1R_2r_2$ (medium red)		
F_2	Genotype	Phenotype	
1	$R_1R_1R_2R_2$	Dark red	15 red to 1 white
2 } 4	$R_1R_1R_2r_2$	Medium-dark red	
2	$R_1r_1R_2R_2$	Medium-dark red	
4 } 6	$R_1r_1R_2r_2$	Medium red	
1	$R_1R_1r_2r_2$	Medium red	
1	$r_1r_1R_2R_2$	Medium red	
2 } 4	$R_1r_1r_2r_2$	Light red	
2	$r_1r_1R_2r_2$	Light red	
1	$r_1r_1r_2r_2$	White	

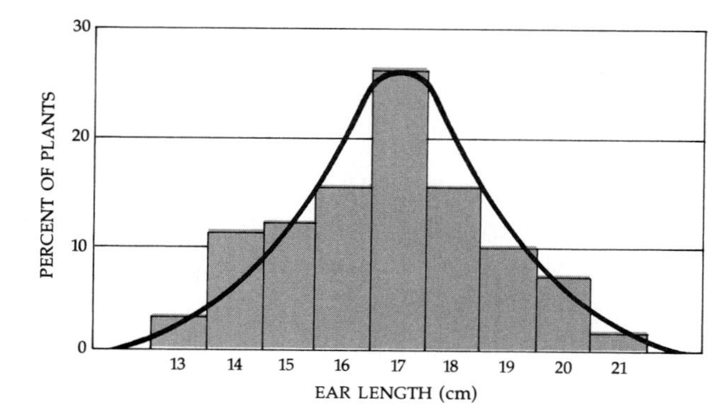

8–14
Distribution of ear length of the Black Mexican variety of corn (Zea mays). *This is an example of a characteristic that is determined by the interaction of a number of genes. Such characteristics show continuous variations; if these variations are plotted as a curve, the curve is bell-shaped, with the mean, or average, falling in the center of the curve.*

A bell-shaped curve of this sort is known as a "normal" curve. The pattern of variation is said to be *continuous*. The smooth, symmetrical shape of the curve indicates that populations cannot be divided into a series of sharply contrasting forms.

The first experiment that illustrated how many genes can interact to produce a continuous pattern of variation in plants (in this case, wheat) was carried out by the Swedish scientist H. Nilson-Ehle. Table 8–2 shows the phenotypic effects of various combinations of two genes, each with two alleles, which act together to control the intensity of color in wheat kernels. Skin color in humans follows a similar pattern of polygenic inheritance.

Quantitative characteristics are determined by the combined effects of multiple genes. Some of the effects of these genes are additive, and some may work in opposite directions. Both epistasis and pleiotropy are important in producing the phenotype, which is the result of such complex interactions that it often is very difficult or impossible to isolate the effect of any one gene on a particular characteristic.

SUMMARY

Sexual reproduction involves two events: the reduction of chromosome number through meiosis and the reestablishment of the diploid number through syngamy.

Meiosis results in the production of either gametes or spores. A gamete is a haploid cell that fuses with another gamete, producing a diploid zygote. A spore is a cell that, without fusing with another cell, develops into a mature haploid organism.

Meiosis involves two sequential nuclear divisions and results in a total of four nuclei (or cells), each of which has the haploid number of chromosomes.

In the first meiotic division, the homologous chromosomes pair lengthwise. The chromosomes are double, each consisting of two chromatids; chiasmata may form between homologous chromatids. These chiasmata are the visible evidence of crossing-over—the exchange of chromatid segments between homologous chromosomes. The bivalents line up at the equatorial plane in a random manner (but with the centromeres of the paired chromosomes on either side of the plane), so that the chromosomes that came from the female parent and those that came from the male parent are completely reassorted during anaphase I. This reassortment, together with crossing-over, ensures that each of the products of meiosis differs from the parental set of chromosomes and from each other. In this way, meiosis permits the expression of the variability that is stored in the diploid genotype.

In the second meiotic division, the chromosomes divide as in mitosis.

The genetic makeup of an organism is known as its genotype; its outward characteristics constitute its phenotype. In diploid organisms—a category that includes most familiar plants and animals—every gene is present twice. The members of these gene pairs are known as alleles. The phenotype in diploid organisms is determined by the interaction of the two alleles present at the same locus on homologous chromosomes. Both alleles may be alike (homozygous), or they may be different (heterozygous). Consequently, mutations in diploid organisms are more difficult to detect than in haploid organisms.

Although both alleles are present in the genotype, only one may be detected in the phenotype. The gene that is expressed in the phenotype is the dominant gene. The one that is concealed in the phenotype is the recessive gene. When two organisms that are each heterozygous for a given pair of alleles are crossed, the ratio of dominant to recessive in the phenotype of the offspring is 3:1. If the action of one allele is insufficient to mask the action of its alternative form (incomplete dominance), the heterozygotes are distinguishable, and the phenotypic ratio is 1:2:1.

Differences between alleles are the results of mutations, which are changes in genes or groups of genes. Most mutations are disadvantageous and so are rapidly eliminated in haploid organisms. In diploid organisms, however, they may persist in the population in a masked form, thus adding to the store of genetic variability.

Genes that are located on different chromosomes segregate independently of one another, whereas those that are located on the same chromosomes do not. Genes located on the same chromosome are said to be linked. In meiosis, genes do not always remain in the same linkage groups because of crossing-over. Genes act by producing proteins, and it is therefore not surprising that they usually affect not one but many characteristics of the organism. This multiple effect of a gene is known as pleiotropy.

A given characteristic is usually controlled by the interaction of more than one gene. Epistasis is the control of the expression of one gene by another, nonallelic gene or genes. When multiple genes affect a single trait, inheritance is said to be polygenic, and the pattern of variation for the characteristic is continuous. Continuous variation follows a distribution that can be represented graphically by a bell-shaped curve. Most quantitative characteristics of organisms, such as weight and height, are polygenic.

CHAPTER 9

The Evolution of Plant Diversity

9–1
"Afterwards, on becoming very intimate with FitzRoy (captain of the Beagle*), I heard that I had run a very narrow risk of being rejected on account of the shape of my nose! He . . . was convinced that he could judge of a man's character by the outline of his features; and he doubted whether anyone with my nose could possess sufficient energy and determination for the voyage. But I think he was afterwards well satisfied that my nose had spoken falsely" (from Darwin's book* The Voyage of the Beagle*).*

In 1831, as a young man of 22, Charles Darwin set forth on a five-year voyage as ship's naturalist on the British navy ship *H.M.S. Beagle*. The book he wrote about the journey, *The Voyage of the Beagle*, not only is a classic work of natural history but also provides some insight into the experiences that led directly to Darwin's proposal of his theory of evolution by natural selection.

At the time that Darwin's historic voyage took place, most scientists—and nonscientists as well—still believed in the theory of "special creation." According to this concept, all the different kinds of organisms had come into existence or had been created in their present form. Some scientists, such as Jean Baptiste de Lamarck (1744–1829), had challenged the idea of special creation, but their arguments were not sufficiently convincing to shake this basic belief, which was not just a scientific hypothesis but was firmly embedded in all of Western culture.

Darwin's theory was able to bring about a great intellectual revolution simply because the mass of evidence he presented was so overwhelming and so convincing that there was no longer room for reasonable scientific doubt. Particularly important in the genesis of Darwin's ideas were his experiences during a stay of some five weeks in the Galapagos Islands, an archipelago that lies in equatorial waters about 950 kilometers off the west coast of South America (Figure 9–2). Here he made two particularly important observations: First, that the plants and animals found on the islands, although distinctive, were similar to those on the nearby South American mainland. If each kind of plant and animal was created separately and was unchangeable, as was then thought, why did they not resemble the plants and animals of Africa, for example, rather than those of South America? Or, indeed, why were they not unique forms, found nowhere else on earth? Second, people familiar with the islands pointed out varia-

9–2
The Galapagos are a small cluster of volcanic islands some 950 kilometers off the coast of Ecuador. Since they came into existence more than a million years ago, they have been colonized from time to time by plant and animal voyagers accidentally swept by wind or water from the mainland. Some of these organisms have managed to survive, reproduce, and adapt themselves to these bleak islands.

The Galapagos are far enough apart that the intervening sea forms natural barriers to the easy movement of many organisms. As a consequence, slightly different species have evolved on neighboring islands. Were the living things on each island the product of a separate "special creation"? This question continued to haunt Darwin for many years after the voyage of the Beagle *and eventually led to his formulation of the theory of evolution by natural selection.*

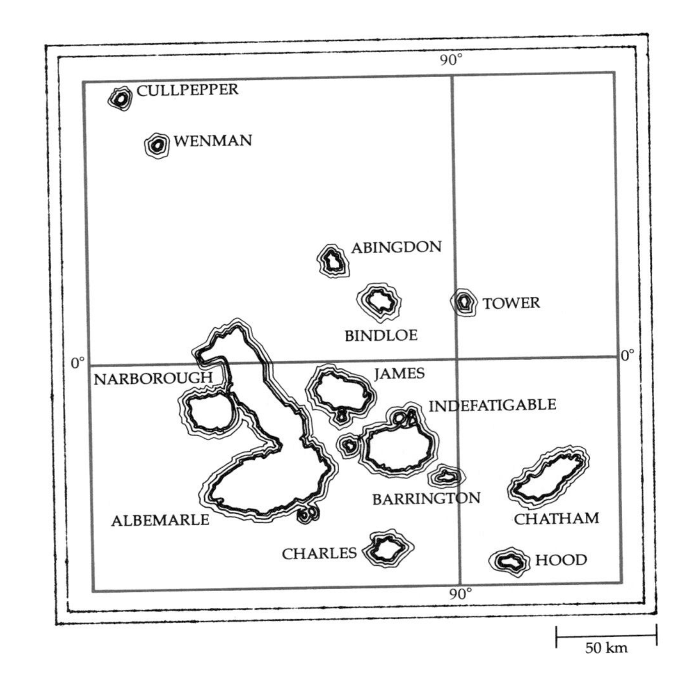

tions that occurred from island to island in such familiar forms as the giant tortoises. Sailors who took these tortoises on board and kept them as sources of fresh meat on their sea voyages were able to tell from which island a particular tortoise came simply by noting its appearance. If the Galapagos tortoises had been specially created, why did they not all look alike?

Darwin began to wonder if all the tortoises and other strange plants and animals of the Galapagos might not have been derived at different times from organisms that existed on the mainland of South America. Once they reached this remote archipelago, they might have spread slowly from island to island, changing bit by bit in relation to local conditions and eventually becoming distinct races that could be easily recognized by any observer.

In 1838, after returning from his historic voyage, Darwin read a book that had been published in 1798 by the Reverend Thomas Malthus, a clergyman who had sounded an early warning about the explosive growth of the human population. Darwin saw that Malthus's reasoning was theoretically correct, not only for the human population but also for all other populations of plants and animals. For example, a single breeding pair of elephants—which have the slowest rate of reproduction of all animals—could produce some 19 million individuals in 750 years, were all their progeny to survive. Despite this, the number of elephants on the earth remains relatively constant; where there were two individuals 750 years ago, there are, in general, only two at the present time. But what determines which two elephants, out of a possible 19 million, are the surviving progeny?

Darwin named the process by which the two particular survivors are chosen as *natural selection*. He used this expression in contrast with the process of *artificial selection*, by which breeders of domesticated plants and animals deliberately change the characters of the strains or races in which they are interested. They do this by controlling which individuals are allowed to breed and which are not. Darwin recognized that organisms are also variable in nature. Those individuals that have favorable characteristics are more apt to breed than those that lack them, and this process can be counted on, in the long run, to produce slow but steady changes in the characteristics of populations.

In artificial selection, breeders can normally concentrate their efforts on one or a few characteristics of interest (although there are certain definite limits). In natural selection, however, the entire organism must be "fit" in terms of the total environment in which it lives. Such a process might reasonably be expected to require a long period of time, and it is no coincidence that the writings of the geologist Charles Lyell, who postulated that the earth was much older than previously thought, had a profound influence on Darwin. Darwin needed such an earth as a stage upon which to view the unfolding of the diversity of living things. The process of natural selection soon became known as "survival of the fittest," which is an apt phrase but one that must be employed cautiously.

THE BEHAVIOR OF GENES IN POPULATIONS

The Hardy-Weinberg Law

In the nineteenth century, when most biologists believed in some sort of blending inheritance, it was difficult to understand why rare characteristics did not simply become so diluted that they essentially disappeared. Darwin could not solve this problem in terms of the knowledge of genetics that existed then.

After the rediscovery of Mendel's work, the question reappeared, framed in more modern terms: Why did the dominant alleles not eventually drive out the recessive ones, with a consequent loss of variability for the population as a whole? The answer, although it was not immediately obvious, lay in a proper understanding of the particulate nature of the gene; this particulate nature was simultaneously pointed out in 1908 by G. H. Hardy, an English mathematician, and G. Weinberg, a German physician.

The *Hardy-Weinberg law*, as it is now known, states that in a large population in which random mating occurs, and in the absence of forces that change the proportions of genes (discussed below), the original proportion of dominant alleles to recessive alleles will be retained from generation to generation.

For example, consider the alleles of a single gene, say, gene *A*. Let us make up an artificial population, so that half of the individuals are homozygous *AA* and the other half are homozygous *aa*. Figure 9–3 illustrates by means of a diagram that the proportion of *AA* individuals (or *aa*, or *Aa*) in the third (or fourth or fifth) generation will, on the average, be the same as it was in the second generation.

For studies in population genetics, the elements of the Hardy-Weinberg law are usually stated in algebraic terms, with the fractions used in Figure 9–3 expressed as decimals. In the case of a gene for which

9–3
(a) *If two populations, one homozygous for a dominant gene* (AA) *and one homozygous for its recessive allele* (aa), *interbreed, all of the first generation* (F_1) *resemble the dominant parents, even though the progeny are heterozygous* (Aa). *When the* F_1 *generation interbreeds, however, the Mendelian (phenotypic)* 3:1 *ratio will appear in the* F_2 *generation, which, on the average, will be* 1/4 AA, 1/2 Aa, *and* 1/4 aa.

The probability of producing AA *individuals in the third generation* (F_3) *as a result of random breeding is as follows:* (b) *If* AAs *crossbreed, the progeny will all be* AA. *The chances of this happening are* 1/4 × 1/4 *or* 1/16. (c) *If* AAs *breed with* Aas, *the chances of producing* AAs *are* 1/16 *because only half of the offspring are* AAs. (d) *If* Aas *cross with* AAs, *the chances of producing* AAs *are, again,* 1/16. (e) *If* Aas *crossbreed, the chances of producing* AAs *are* 1/16 *because only a fourth of the offspring are* AAs. *Thus, in the third generation,* AA *individuals, on the average, number* 1/16 + 1/16 + 1/16 + 1/16 *or* 1/4, *which is the same as in the second generation. In short, sexual recombination alone does not change the proportions of different alleles.*

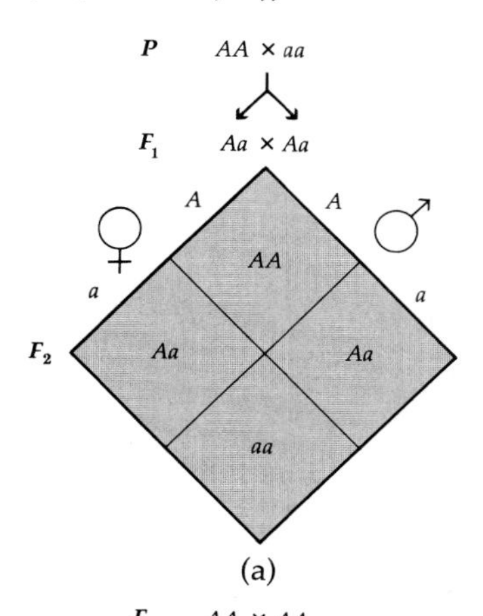

(a)

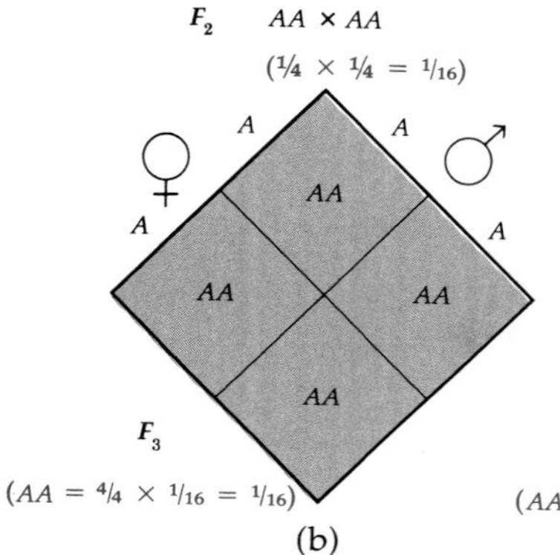

(b)

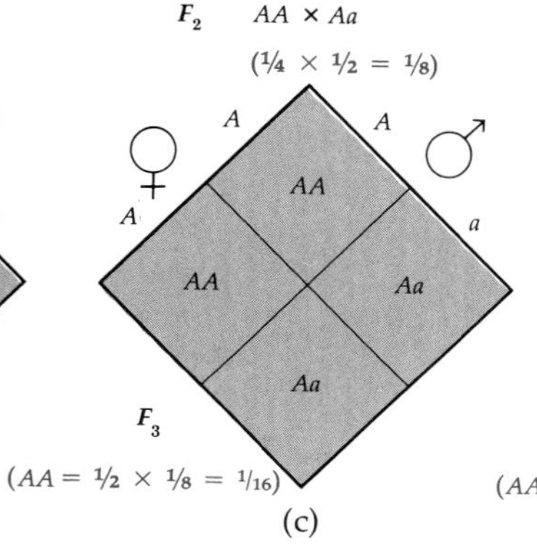

(c)

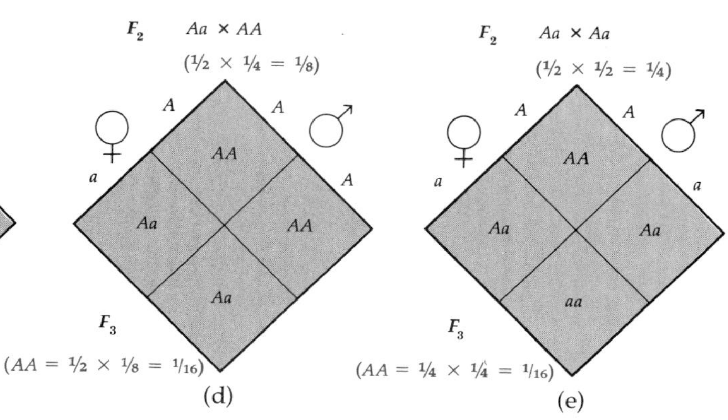

(d)

(e)

there are two alleles in the gene pool, the frequency (p) of the dominant allele plus the frequency (q) of the recessive allele must together equal the frequency of the whole; $p + q = 1$. (The frequency of an allele is simply the proportion of that allele in a gene pool in relation to all alleles of the same gene.) This is equivalent to saying that if there are only two alleles, A and a, of a particular gene and if half (0.50) of the alleles in the gene pool are A, the other half (0.50) must be a. Similarly, if 99 out of 100 (or 0.99) are A, then 1 out of 100 (or 0.01) is a.

How then do we find the relative proportions of individuals who are AA, Aa, and aa? These proportions can be calculated by multiplying the frequency of male A by that of female A [A^2]; male A times female a [Aa]; male a times female A [Aa]; and male a times female a [a^2]. These multiplications can be expressed in algebraic terms as $p^2 + 2pq + q^2$. This equals, as you may remember from the binomial theorem, $(p + q)^2$, and if $p + q = 1$, then $(p + q)^2 = 1$. So, if half (0.50) of the gene pool is A and half is a, the proportion of AA will be 0.25, the proportion of Aa will be 0.50, and the proportion of aa will be 0.25, which is exactly what Mendel said—although in slightly different terms:

$$\begin{aligned} A^2 + 2Aa + a^2 &= (0.50)^2 + 2(0.50)(0.50) + (0.50)^2 \\ &= 0.25 + 2(0.25) + 0.25 \\ &= 0.25 + 0.50 + 0.25 \\ &= 1 \end{aligned}$$

What happens in subsequent generations? As can be seen in Figure 9–4, the genotype frequencies remain constant indefinitely. Similarly, the gene frequencies remain constant at $p(A) = 0.5$ and $q(a) = 0.5$, so that the Hardy-Weinberg law can be viewed as predicting a state of "genetic equilibrium."

We know that many populations do not exist in a state of equilibrium, however, for if they did, no evolution could occur. Four factors operate to produce deviations from the Hardy-Weinberg equilibrium.

Population Size. The Hardy-Weinberg law operates only when there is a large population. In a small population, the random loss of one or more individual genotypes—by a failure to breed, for example—can lead to the elimination of one of the alleles from the population.

Migration. Further distortion is caused by migration into or out of a particular population. If individuals with particular genetic characteristics leave a population or enter it in a proportion different from that in which they are already represented, the gene and genotype frequencies will obviously change. Migration—by the dispersal of seeds, for example—is as characteristic of plants as it is of animals.

Mutation Rate. If a particular gene is mutating to an allelic form at a rate higher than that at which the reverse mutation is occurring, the population cannot remain in equilibrium for the gene in question.

Selection. The major factor causing exceptions to the Hardy-Weinberg law is selection, which provides virtually the entire basis for our current understanding of evolutionary change. Mutations are, of course, the basis for variability in individuals, but changes in populations occur as a result of selection.

Selection is simply the *nonrandom* reproduction of genotypes. In any variable population, some individuals leave more progeny than others. As a result, certain genes become more frequent in populations, and others become rarer. Although selection is sometimes thought of as a creative force, it is important to remember that it is merely a description of events that have already taken place. When the proportion of a certain gene is higher in a given generation than it was in the preceding generation, in the absence of an alternative explanation, it is said that selection has taken place. Selection does not actually *cause* the changes that occur. However, by continually eliminating certain genetic types from a population under a given set of conditions, it does *channel* the variations in characteristics caused by mutation and recombination. In this way, selection results in the production of individuals that can survive better under given conditions than those that were elimi-

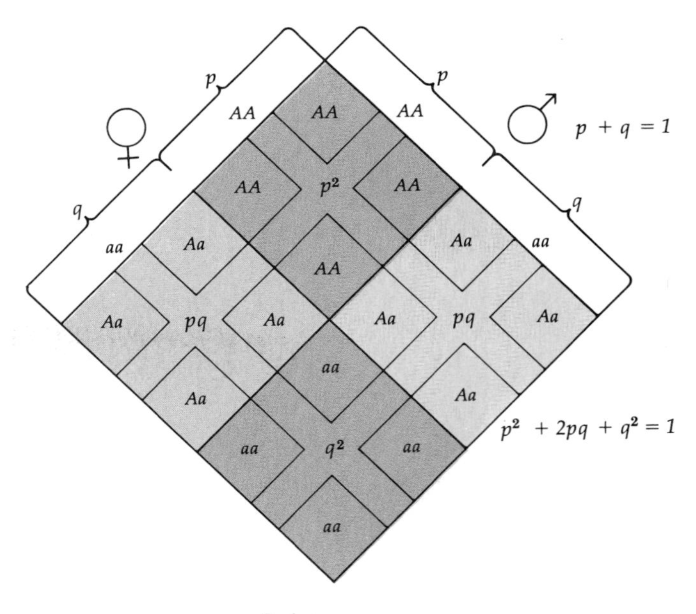

9–4
Possible combinations of gametes in a population consisting of AA, Aa, *and* aa *individuals, illustrating the principle expressed in the Hardy-Weinberg law.*

nated. In unusual or marginal environments, genotypes that were poorly represented in the main part of the range of a population may be greatly increased by the action of selection.

In the case of a deleterious recessive gene, even one that causes the death of individuals homozygous for it, the lower the frequency of the gene in the population, the less it will be acted on by selection. This is because the proportion of recessive genes that occur in homozygotes decreases precipitously as the frequency of the gene decreases. In short, the lower the frequency of a recessive gene, the less it is exposed to the action of selection in a homozygous form. As the gene frequency drops, the progress toward removal of the gene from the population also slows down.

RESPONSES TO SELECTION

Genetic Factors

The response of a population to selection is affected by the principles of genetics discussed in Chapter 8. In general, only the phenotype is being selected, and it is the relationship between the phenotype and the environment that determines the reproductive success of an organism. As nearly all characteristics of natural populations are determined by the interaction of many genes, similar individuals can have very different genetic properties. When some feature, such as tallness, is strongly selected for, there is generally an accumulation of genes that contribute to this feature and an elimination of those genes that work in the opposite direction.

But selection for a polygenic characteristic is not a simple accumulation of one set of genes, with an elimination of their alleles. Gene interactions, such as epistasis and pleiotropy, are of fundamental importance in determining the course of selection in a population. Some of the effects of a gene on another gene may be beneficial to the individual or may contribute to the characteristic being selected for. Other effects may be harmful or may work against the characteristic being selected for. As a consequence of gene interactions, the phenotypic effects of a particular gene can be measured only within a particular genetic context. As selection for particular genes changes the representation of the other genes in the population, the interactive effects of the selected genes gradually become changed. Thus, the value of the genes in determining a particular characteristic, or the fitness of the organism as a whole, also changes.

Another feature of importance in determining the nature of selective change is the necessity of producing an organism that "works." Strong selection for one feature may be prohibited because it would lead to the accumulation of so many undesirable side effects that the organism would no longer be able to survive or reproduce. The results of the breeding experiment on fruit flies shown in Figure 9–5 suggest that, as a result of selection, the population reached a new internal genetic balance that allowed the production of fertile, otherwise normal individuals with 42, instead of 36, abdominal bristles. This balance, which can be looked upon as a measure of the resistance of the population to genetic change, has been termed *genetic homeostasis* by I. Michael Lerner of the University of California. (Homeostasis in physics is a sort of dynamic equilibrium, that is, a resistance to change.) Genetic homeostasis tends to keep the population as a whole producing a high proportion of individuals that are well adapted to their environment.

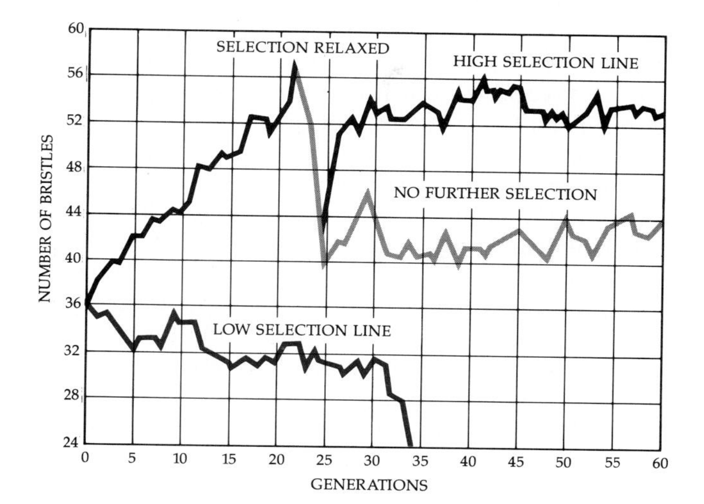

9–5
A selection experiment involving bristle number in Drosophila melanogaster. *In the parental stock, the average number of bristles on the ventral surface of the abdomen was 36. One breeding group was selected for an increase in bristle number and one for a decrease in bristle number. The high selection line reached 56 bristles in 21 generations, but soon the stock began to become sterile. Selection was discontinued after generation 21 and was resumed at generation 25. This time the previously high bristle number was regained without loss of fertility. The low selection line, which was not permitted to achieve such a new internal genetic balance, died out because of sterility. Selection for this single "neutral" gene apparently had widespread effects on the genetic constitution of these organisms.*

Phenotypic Factors

Other kinds of limits to selection are imposed by the need for an individual organism to meet conflicting environmental demands. For example, the long tail of the male peacock, with its brilliant coloration, is highly correlated with the mating success of the bird, and in this respect it is clearly advantageous for the bird to have the largest and most showy feathers possible. On the other hand, the bird must also escape from its predators, and here such tail feathers can be a disadvantage. Natural selection strikes some sort of balance between the two selection forces, and peacocks with tail feathers that are either too long and colorful or too short and drab probably do not make as great a reproductive contribution to subsequent generations as those with the tails that are "just right."

Similarly, in harsh alpine environments, plants must grow and photosynthesize rapidly if they are to store enough carbohydrates to survive the long, severe winters. Therefore, they must respond to the first sign of favorable conditions. However, if their response is so finely tuned that they respond even to a temporary thaw in mid-winter, they will be eliminated from the population. Desert plants are in a similar position; their seeds must germinate whenever sufficient water is available but must not germinate when too little is available. Premature germination is prevented in some plants by the production of germination inhibitors in their seed coats: only when enough rain has fallen to leach away these substances will the seeds germinate.

An interesting evolutionary strategy among tropical woody members of the pea family (Fabaceae) has been detected by Daniel Janzen of the University of Pennsylvania. In Central America, Janzen found two kinds of species: those with numerous small seeds and those with rather few, but much larger seeds. These plants can spend a certain amount of their food reserve each year on the production of seeds, and they can therefore produce either a large number of small seeds or a small number of large seeds. Why do some of the species follow one path while others follow the second path?

Large seeds provide a more abundant food supply for the germinating seedlings. Because each parent tree or shrub will be replaced, on the average, by only one other tree or shrub, it would seem most efficient for it to produce a small number of large seeds with an abundant stored reserve of food. But this is not always possible in the tropics, owing to the very heavy predation on legume seeds by insects, especially the seed beetles (Bruchidae). The adult bruchids lay their eggs on the fruits, and their larvae complete their development within the seeds. Bruchids are often so abundant that they virtually destroy the entire seed crop of a particular legume in a given year, except for the few seeds removed by a dispersal agent, such as a bird or mammal, before the bruchids have discovered and infested a particular fruit.

Janzen investigated bruchid infestation among large-seeded and small-seeded groups. To his surprise, he found that the small-seeded species were heavily attacked by bruchids, so that very few viable seeds remained, whereas nearly all the large-seeded species were untouched. Subsequently, he found that these large seeds, but not the small ones, contained chemical substances that apparently protected them from the beetles.

Summing up these results, it seems clear that, in an evolutionary sense, the legumes had two options. Either they could produce a very large number of small seeds, with a relatively limited food supply for the germinating seedlings but with a better chance of some of the seeds escaping the attacks of the beetles, or they could produce a smaller number of large seeds, with a good food supply for the germinating seedlings. However, this second option would succeed only if the seeds were somehow protected from predation by beetles. This is an example of the sort of evolutionary compromise often found in nature.

In the light of Janzen's reasonable explanation of the situation in tropical legumes, it is interesting that the seeds and fruits of plants on oceanic islands are often larger than those of mainland relatives of these plants (Figure 9–6). The kinds of pests that attack the plants on the mainland are often absent on the islands, and it seems likely that this provides the opportunity for the plants to produce larger fruits and seeds, which provide a better start for their seedlings. This isolation from pests, like the biochemical defenses of the large-seeded legumes of the American tropics, is a factor that allows large seed size, selected for on other grounds, to prosper more than it could otherwise.

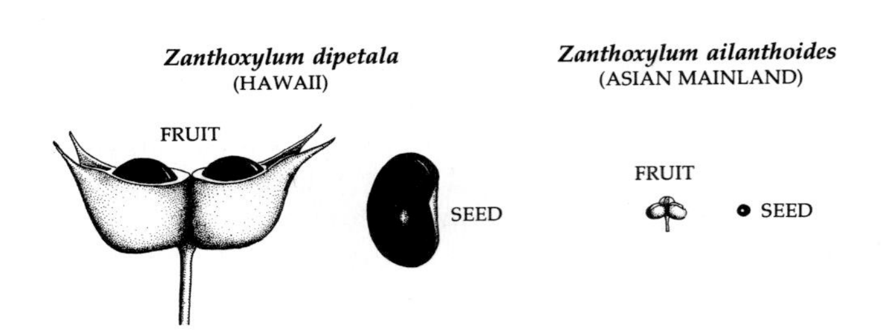

9–6
Island species, isolated from their natural enemies, are often able to produce larger fruits and seeds. Shown here are two related species, one found in Hawaii and one found on the Asian mainland.

Changes in Natural Populations

In recent years, it has become possible to study changes that have occurred in natural populations. Changes in climate and natural catastrophes have been a constant feature of the world since its beginning, and the populations of organisms present at earlier times responded to these changes in ways similar to those presently observed. The nature and rate of some of these changes have been such that we can speak about them as "evolution in action." It is not surprising that many of the relatively rapid changes in natural populations that have been observed over the last century or so have resulted from human activities—an ecologically dominant feature without parallel in the history of the world.

Some of the most spectacular changes have been observed in populations of microorganisms, which are haploid. Their mutations are expressed immediately in the phenotype, and selection for or against such changes proceeds at once. All haploid organisms have a short generation time, which is probably essential for successful changes, since they do not have the store of variability for recombination that is found in diploid organisms.

In plants, strong selective forces have been observed to produce rapid changes in natural populations. For example, plants in the grazed part of an experimental pasture in Maryland were much lower than those in the ungrazed part, and it was thought that this might be because of the direct action of grazing. This hypothesis was tested by digging up some of the plants from the grazed part and planting them elsewhere. It was assumed that if they were low merely because they were grazed, they would become tall under these changed conditions, like the plants in the ungrazed part of the pasture. Some did, but plants of white clover *(Trifolium repens)*, Kentucky bluegrass *(Poa pratensis)*, and orchard grass *(Dactylis glomerata)* remained low in stature, indicating that these populations had been modified genetically by the selective force of grazing. A similar example is illustrated in Figure 9–7.

In Wales, the tailings and dumps around a number of abandoned lead mines are rich in lead and almost bare of plants. One species of grass *(Agrostis tenuis)* colonizes this mine soil, which can contain up to 1 percent lead and 0.03 percent zinc. In an experiment, samples of plants taken from mine tailings and from a nearby pasture were grown together in both normal and lead-mine soil. In normal soil, the lead-mine plants were definitely slower-growing and smaller than the pasture plants. On the lead-mine soil, however, the mine sample was normal, but the pasture sample did not grow at all. Half of the pasture plants were dead in three months and had misshapen roots that were rarely more than 2 millimeters long. But a few of the pasture plants (3 out of a sample of 60) showed some resistance to the effects of the lead-rich soil. These were doubtless similar to the plants originally selected in the development of the lead-resistant strain of *A. tenuis*. The mine was no more than 100 years old, so the lead-resistant race had developed in a relatively short period of time.

(a)

(b)

9–7
Prunella vulgaris *is a common herb of the mint family; it is widespread in woods, meadows, and lawns in temperate regions of the world. Most populations consist of erect plants such as those shown in* (a), *which were found in an abandoned pasture. Populations found in lawns, however, always consist of prostrate plants such as those shown in* (b); *erect plants cannot survive in lawns. When lawn plants are grown in an experimental garden, some remain prostrate, whereas others grow erect. The prostrate "habit" is determined genetically in the first group and environmentally in the second group.*

Evolution in Asexual Populations

Asexual reproduction (also known as vegetative reproduction, or apomixis) results in progeny that are identical to their single parent. In plants, there is a wide variety of means of asexual reproduction, ranging from the development of an unfertilized egg cell to division of the parent organism into almost equal parts. In all such cases, however, the new plants are the product of mitosis and thus are genetically identical to the parent.

Higher plants often reproduce both sexually and asexually, thus hedging their evolutionary bets (Figure 9–8), but many species reproduce only asexually. Even among these, however, it is clear that their ancestors were capable of sexual reproduction; therefore, vegetative reproduction represents an alternative—a "choice" made in response to evolutionary pressures calling for extreme uniformity. This choice, if rigidly adhered to, severely restricts the ability of the population to adapt to varying conditions. However, different asexually reproducing populations of a given species may occur in a wide variety of different habitats.

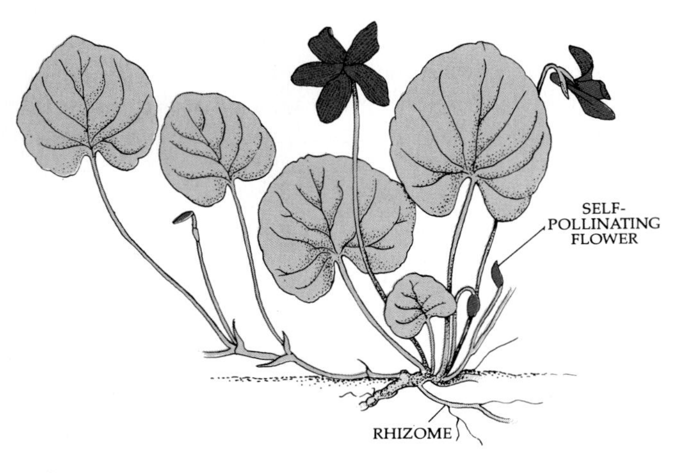

9–8
Violets reproduce both sexually and asexually. The large flowers are cross-pollinated by insects, and the seeds may be carried some distance from the parent plant. The smaller flowers, closer to the ground, are self-pollinating and never open. Seeds from these flowers drop close to the parent plant and produce plants that are genetically similar to the parent. Presumably, such plants are better able (on the average) to grow successfully near the parent. Horizontally creeping stems (either stolons or rhizomes) also can eventually produce new, genetically identical plants close to the parent.

THE DIVERGENCE OF POPULATIONS

The world is physically and biologically complex, and individual kinds of plants almost always have discontinuous distributions. Their habitats are certainly discontinuous, whether they be lakes, streams, the tops of mountains, shaded spots in the woods, or a certain kind of soil. Given this observation, the movement of genes between isolated populations is more or less limited, and the populations are correspondingly free to respond to the selective demands of their local situations in different ways. Even though normal *gene flow* (the movement of genes from one population to another as a consequence of the

VEGETATIVE REPRODUCTION: SOME WAYS AND MEANS

The forms of vegetative reproduction, or apomixis, in plants are many and varied. Some plants reproduce by means of runners, or stolons—long slender stems that grow along the surface of the soil. In the cultivated strawberry (Fragaria × ananassa), *for example, leaves, flowers, and roots are produced at every other node on the runner. Just beyond the second node, the tip of the runner turns up and becomes thickened. This thickened portion first produces adventitious roots and then a new shoot, which continues the runner.*

Rhizomes, or underground stems, are also important reproductive structures, particularly in grasses, in which they produce stems bearing leaves and flowers. During growth, rhizomes are capable of invading areas adjacent to the parent plant. Each new node can give rise to a new plant. The noxious character of many weeds results from this type of growth pattern, but many garden plants, such as irises, are propagated almost entirely from rhizomes. Corms, bulbs, and tubers—all different types of

The strawberry *(Fragaria × ananassa)* is propagated by runners, or stolons. Strawberry plants also form flowers and reproduce sexually.

migration of individuals) may often provide a source of variability for a distant population, it cannot completely counteract the effects of local selection.

The distances necessary to isolate populations effectively vary with the dispersibility of the plants in question, but they are often quite small. For several kinds of insect-pollinated plants that grow in temperate regions, a gap of 15 meters may effectively isolate two populations. Rarely will more than 1 percent of the pollen that reaches a given individual come from this far away. Even in plants in which the pollen is spread by the wind, under normal circumstances, very little falls more than 50 meters from the parent plant, and its chance of reaching a receptive stigma is correspondingly decreased beyond this distance. This is not to say, of course, that two pine trees separated by 50 meters are out of genetic contact but, rather, that the sort of progeny that each produces will be affected more by the demands of its local environment than by cross-pollination.

Any two separated populations diverge from one another because they are responding to different selective forces. If they regain contact, they may merge, or the differences that have accumulated between them may lead to a degree of genetic isolation. Genetic isolation, or reproductive isolation, can come about in a variety of ways, as will be discussed later in this chapter.

---

underground stems—are structures specialized for storage and reproduction. Potatoes are propagated artificially from tuber segments, each with one or more "eyes." It is the eyes of the "seed pieces" of potato that give rise to the new plant. Propagation of sweet potatoes generally is accomplished with whole small roots, in which new shoots arise from adventitious buds.

The roots of some plants—for example, cherry, apple, raspberry, and blackberry—produce "suckers," or sprouts, which give rise to new plants. Commercial varieties of banana do not produce seeds and are propagated by suckers that develop from buds on underground stems. When the root of a dandelion is injured, as by attempting to pull it from the ground, each root fragment gives rise to another entire plant.

In a few species, even the leaves are reproductive. One example is the house plant Kalanchoë daigremontiana, *familiar to many people as the "maternity plant," or "mother of thousands," so-called because numerous plantlets arise from meristematic tissue located in notches along the margins of the leaves. The maternity plant is propagated primarily from these small plants, which drop to the soil and take root when mature. Another example is provided by the walking fern* (Asplenium rhizophyllum), *in which young plants form where the leaf tips touch the ground.*

In certain kinds of seed plants, including many kinds of citrus, some grasses, and also dandelions, the embryos in the seeds may be produced asexually from the parent plant. Such seeds give rise to individuals that are genetically identical to their parents and therefore provide another instance of asexual reproduction.

In general, asexual reproduction allows the exact replication of individuals that are particularly well suited to a certain environment or habitat. This suitability may include characteristics that are viewed as desirable in a cultivated plant or that facilitate survival under a particular set of natural conditions.

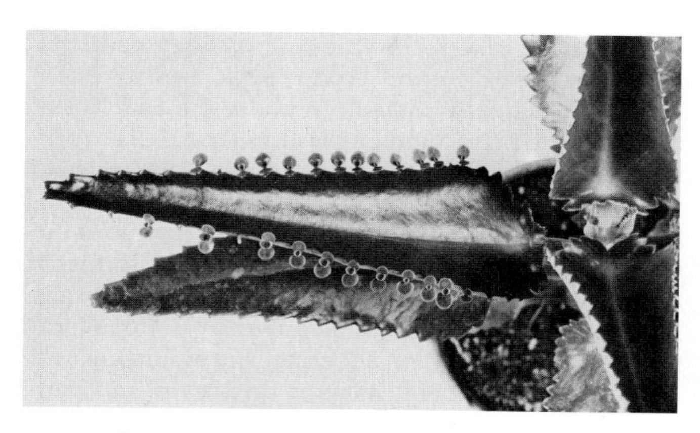

Kalanchoë daigremontiana. Notice the small plants that have arisen in the notches along the margins of the leaf.

Walking fern *(Asplenium rhizophyllum)* making its way over a rocky ledge. The fern is capable of forming large colonies of genetically identical plants.

(a)

(b)

(c)

9–9
Transplant gardens of the Carnegie Institution. (a) *Stanford, near sea level;* (b) *Mather, in the central Sierra Nevada at about 1400 meters elevation; and* (c) *Timberline, at 3050 meters elevation.*

Ecotypic and Clinal Variation

Developmental plasticity is the tendency of individual, genetically identical organisms to differ greatly among themselves because of the action of the environment. Such plasticity is much greater in plants than it is in animals, since the open system of growth characteristic of plants can more easily be modified in various ways that produce striking differences between the individuals. Environmental factors, as every gardener knows, can cause profound changes in the phenotype of various species of plants. Leaves that develop in the shade are thinner, broader, have a greater volume of airspace, thinner palisade tissue, and fewer stomata than leaves of the same plant that develop in the sun. Day length—the length of time per 24 hours that a plant is exposed to the light—can also affect leaf form. In *Kalanchoë,* for example, plants grown on short days (8 hours of light) had small succulent leaves with smooth edges, whereas plants grown on long days (16 hours) had large, thin leaves with notched edges. In view of such observations, it is not surprising that a number of workers proposed, until about 1930, that much of the variability of plants observed in nature was caused by the environment and did not have a genetic basis.

Are the differences between races of plants inhabiting different habitats genetically or environmentally controlled? The first definitive answer to this question was given by the Swedish botanist Göte Turesson, who transplanted races of a number of the plant species found in southern Sweden into his experimental gardens. In the great majority of the 31 species he investigated, the differences he observed in nature were genetically controlled; in a very few, environmental modification predominated. Differences in plant stature, time of flowering, color of leaves, and so on, were usually genetically controlled. Turesson termed such genetic races, differentiated with respect to particular habitats, *ecotypes.*

Of particular interest in the study of ecotypes is the work of Jens Clausen, David Keck, and William Hiesey, which was carried out under the auspices of the Carnegie Institution of Washington. These workers dealt experimentally with a number of species of plants from the western United States. They established transplant stations at three locations in California (Figure 9–9). Working mainly with plants that could be propagated asexually, so that genetically identical individuals could be grown at all three locations, they expanded on Turesson's experiments.

The western United States is an area of environmental contrasts, and it is not surprising that many plant species have developed sharply defined ecotypes in this region. One species studied by the Carnegie Institution group was the perennial herb *Potentilla glandulosa.* A close relative of the strawberry

(Fragaria), this herb ranges through a wide variety of climatic zones in California, and native populations occur near each of the three experimental stations mentioned above. When *P. glandulosa* plants from the different locations were grown side by side in an experimental garden, a number of ecotypic differences became apparent. There appeared to be four major ecotypes, each with morphological characteristics that are strongly correlated with particular physiological responses critical for survival.

For example, the Coast Range ecotype consists of plants that grow actively both winter and summer when cultivated at Stanford. They also survive at Mather, where they are subjected to about 5 months of cold winter. Here they become winter-dormant but store enough food in the remainder of the year to carry them through this unfavorable season. At Timberline, plants of this ecotype almost invariably die the first winter; the short growing season at this high elevation does not permit their survival. Other plants that occur in California's Coast Ranges produce ecotypes with identical physiological responses. Indeed, it is often true that different species of plants growing together are more similar to one another physiologically than they are to other populations of their respective species.

The physiological and morphological characteristics of ecotypes usually have a very complex genetic basis, involving dozens or, in some cases, perhaps even hundreds, of chromosomal loci. Sharply defined ecotypes are characteristic of regions where the breaks between adjacent habitats are themselves sharply defined. On the other hand, if the environment changes gradually, with no clear break, then populations of plants may do likewise. A gradual change of this sort is called a *cline*, and in the case of ecotypes, such changes are called *ecoclines*.

Ecoclines are frequently encountered in marine organisms, which live in a habitat where the temperature often rises or falls gradually with changes in latitude. They are also characteristic of areas such as the eastern United States, where rainfall gradients may extend over thousands of kilometers. When populations of plants are sampled along a cline, the differences are often proportional to the distance between them. On a small scale, populations of plants vary in similar ways—either gradually or abruptly, depending on the nature of the local environment.

Physiological Differentiation

Presently, one of the most active fields of botanical research concerns the physiological basis for ecotypic differentiation. For example, Scandinavian strains of goldenrod *(Solidago virgaurea)* from shaded and exposed habitats show experimental differences in the photosynthetic response to light intensity during growth. The plants from shaded environments grow rapidly under low light intensities, whereas the growth rate is markedly retarded under high light intensities. In contrast, plants from exposed habitats grow rapidly under conditions of high light intensity.

In another experiment, arctic and alpine populations of the widespread herb *Oxyria digyna* were studied, using strains from an enormous latitudinal gradient extending from Greenland and Alaska to the mountains of California and Colorado. Plants of northern populations had a higher leaf chlorophyll content, as well as higher respiration rates at all temperatures, than plants from farther south. High-elevation plants near the southern limits of the species distribution carried out photosynthesis more efficiently at high light intensities than low-elevation plants from farther north. The existence of *Oxyria* over such a wide area and range of ecological conditions is made possible, in part, by differences in metabolic potential among its constituent populations.

Reproductive Isolation

The genetic system of a population of plants responds to selection as an integrated unit. For this reason, plants from widely different populations may be unable to form *hybrids* (the offspring of genetically dissimilar parents) or, if they do, the hybrids may not be fertile. In general, the greater the difference in appearance of two populations, the less likely it is that they can hybridize, although there are important exceptions.

When two populations become reproductively isolated—unable to form fertile hybrids with one another—they have no effect on each other's subsequent evolution, at least in a genetic sense. Such isolation is therefore one of the more crucial steps in the evolutionary divergence of populations.

Reproductive isolation has often been used as a basis for defining the category *species*. However, the very different-looking, ecologically distinct species in some groups of plants—particularly long-lived groups such as trees and shrubs—often can hybridize. In contrast, species of herbaceous plants often form sterile hybrids when crossed, as may individual populations within such species. Thus, it is impossible to use a single criterion to define a species in plants. (The question of what constitutes a species is discussed more fully in Chapter 10.)

Populations of groups of annual plants presumably change much more rapidly than populations of long-lived groups. Not only do they have much shorter life cycles but each year they depend on reestablishment from seed, a factor that heightens the effects of natural selection and consequently causes them to diverge from one another more rapidly.

There are a number of factors, in addition to hybrid sterility, that tend to keep populations of plants distinct when they grow with one another. Some prevent the formation of the hybrid in the first place. For example, two kinds of plants that are capable of forming fertile hybrids may occur in different habitats. In the eastern United States, the scarlet oak *(Quercus coccinea)* can be found growing together with the black oak *(Q. velutina)* over a very wide area, yet hybrids are rare. In general, scarlet oaks are found in relatively moist, low areas with acidic soil, whereas the black oak is found in drier, well-drained habitats. Only where the habitat is badly disturbed does hybridization become more frequent, for here the distinction between the two habitats becomes blurred.

Among the other mechanisms that can prevent the formation of hybrids between species that occur together are seasonal differences in the time of flowering. If two species do not flower together, they will not hybridize in nature even when they grow side by side. The abundant variety of photoperiodic mechanisms discussed in Chapter 26 also allow differentiation of this kind. In addition, two species that occur together may differ in their pollination systems; various possibilities of this sort are discussed in Chapter 19. If two species are visited and pollinated by different kinds of insects, they will hybridize only when these insects mistakenly visit the "wrong" flower.

Clusters of Species

The evolutionary processes we have been discussing lead to the production of groups of related species in many different geographic areas. The clusters of species on the Galapagos archipelago are famous because of their role in the development of Darwin's theory of evolution. The development of such patterns is known as evolutionary radiation. (The essay on pages 160–61 illustrates this process.)

Differentiation on islands is particularly striking because, in the absence of competition, organisms seem more likely to produce bizarre forms than on the mainland. Apparently, islands provide favorable situations for the kinds of major evolutionary changes that occur when new genera and families arise. In such localities, plants and animals often change much more rapidly than on the mainland, assuming forms never encountered elsewhere. A good example is the genus *Cyanea,* a member of the family Lobeliaceae that is confined entirely to the Hawaiian Islands. There are at least 60 species in this genus, and they differ strikingly in almost every characteristic (Figure 9–10). However, clusters of species that are differentiated to a less spectacular degree are characteristic of all flourishing groups of organisms and are the inevitable result of the evolutionary patterns we have been discussing.

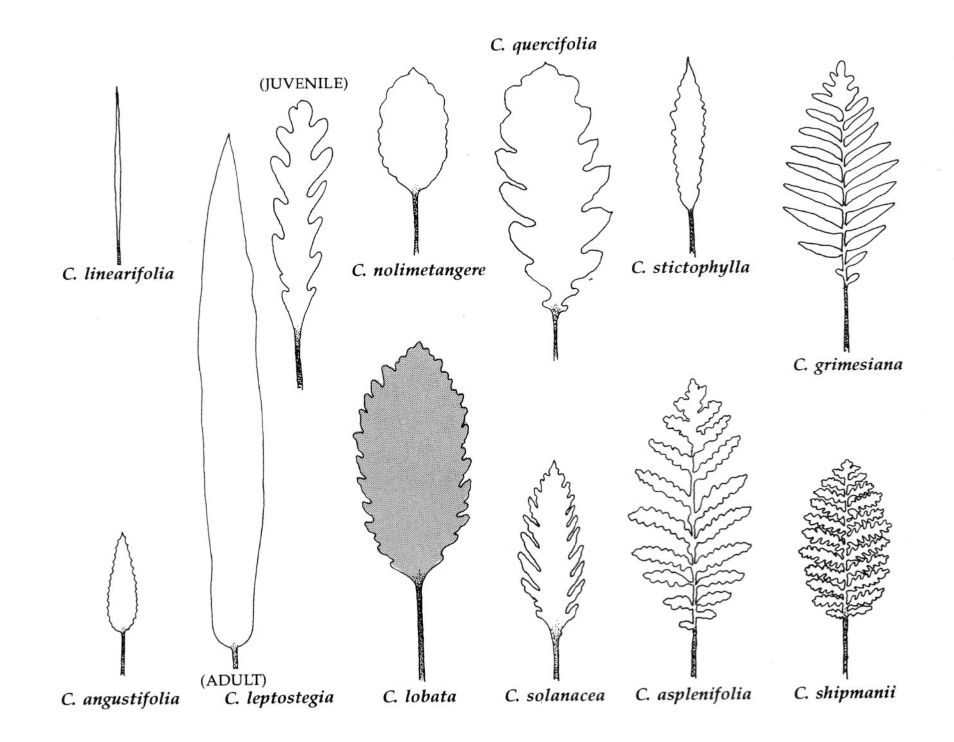

9–10
The species of Cyanea *(a genus found only in the Hawaiian islands) differ widely in leaf size and leaf shape. All of these species probably evolved from ones like* C. lobata *(center). The differences shown here are often connected with the diverse growth forms and leaf outlines of present-day species.* Cyanea linearifolia *(left), for instance, is usually found in dry, sunny locations. Plants with fern-like leaves, like those on the right, are found in shady locations, where the thin, wide-bladed leaves are more efficient in light-gathering.*

9–11
Plane trees (Platanus), *which are called sycamores in North America, offer examples of well-differentiated populations that have retained the ability to hybridize. Present-day species of this genus have been isolated from one another for at least 50 million years in widely scattered localities. One of these, the oriental plane* (P. orientalis), *is native from the eastern Mediterranean region to the Himalayas. This handsome tree has been widely cultivated in southern Europe since Roman times, but it cannot be grown in northern Europe away from the moderating influence of the sea. After the discovery of the New World,* P. occidentalis *was brought into cultivation in the colder portions of northern Europe, where it flourished. About 1670, these two very different trees hybridized to produce the intermediate and fully fertile London plane* Platanus × hybrida. *This hybrid, which is capable of growing in regions with cold winters, is much more vigorous than either parent. It is now grown as a street tree throughout the temperate regions of the world, because it also has a high tolerance for urban pollution.*

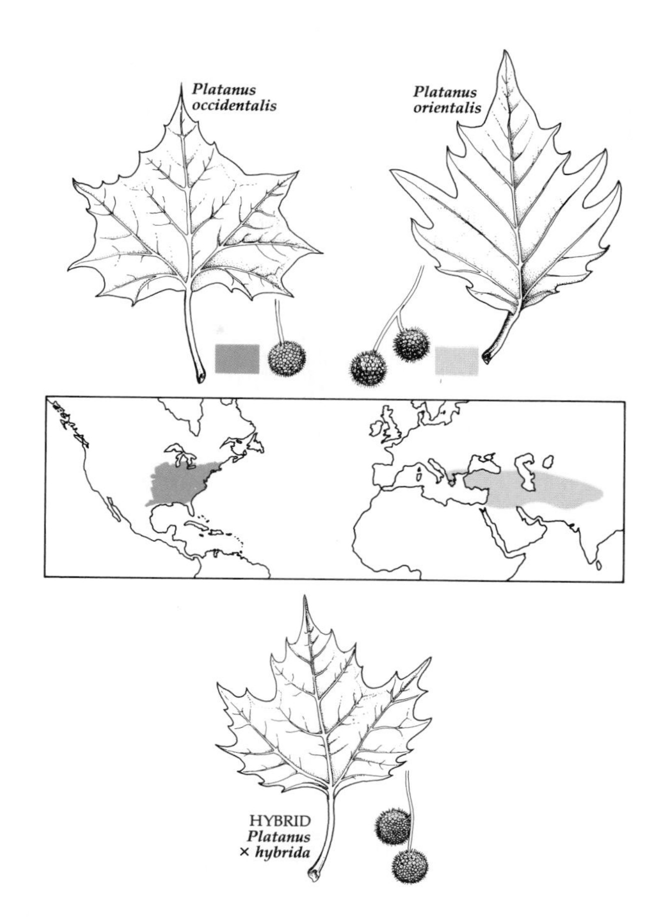

THE EVOLUTIONARY ROLE OF HYBRIDIZATION

Even if species hybridize rarely in nature, these hybrids may be important because of the way in which they recombine the parental characteristics. The environment can change repeatedly, and individuals of hybrid origin can often present genetic combinations better suited to the new environment than either parent; they may even be able to colonize some habitat where neither parent can grow. When the habitats of the parental species are next to one another and are sharply distinct (as in the case of the scarlet oak and the black oak), there may be little opportunity for the establishment of hybrids; but where the habitats intergrade or are disturbed, the situation may be very different. Here the recombination of genetic material shared by two species has a greater potential for producing well-adapted offspring than do changes within a single population. The degree of variability available when two species hybridize is much greater than that available to either of the species by itself, and the chances of producing individuals able to flourish in the new habitat are correspondingly greater (Figure 9–11).

Obviously, hybridization may weaken or diminish the distinctness of two clearly defined species by the production of a series of intermediate populations of differing characteristics. Any one of these might become stabilized in its characteristics if it were better adapted to some set of habitat conditions than either of its parents.

It is often difficult to demonstrate with certainty that a particular intermediate population has originated as a hybrid between two other species. Nevertheless, evidence is rapidly accumulating that hybridization is an important evolutionary mechanism, operating in group after group of plants. In some genera, recombination of genetic material between species and the production of new entities seem to be the chief means of developing new species and responding to a changing environment. These are primarily woody genera, such as manzanitas (*Arctostaphylos*) and mountain lilacs (Figure 9–12). Clusters of species that interact in this way tend to be most common on islands or in geologically diverse, rapidly changing areas such as California and its neighboring states. Another example, in this case involving pollination agents, is shown in Figure 9–13.

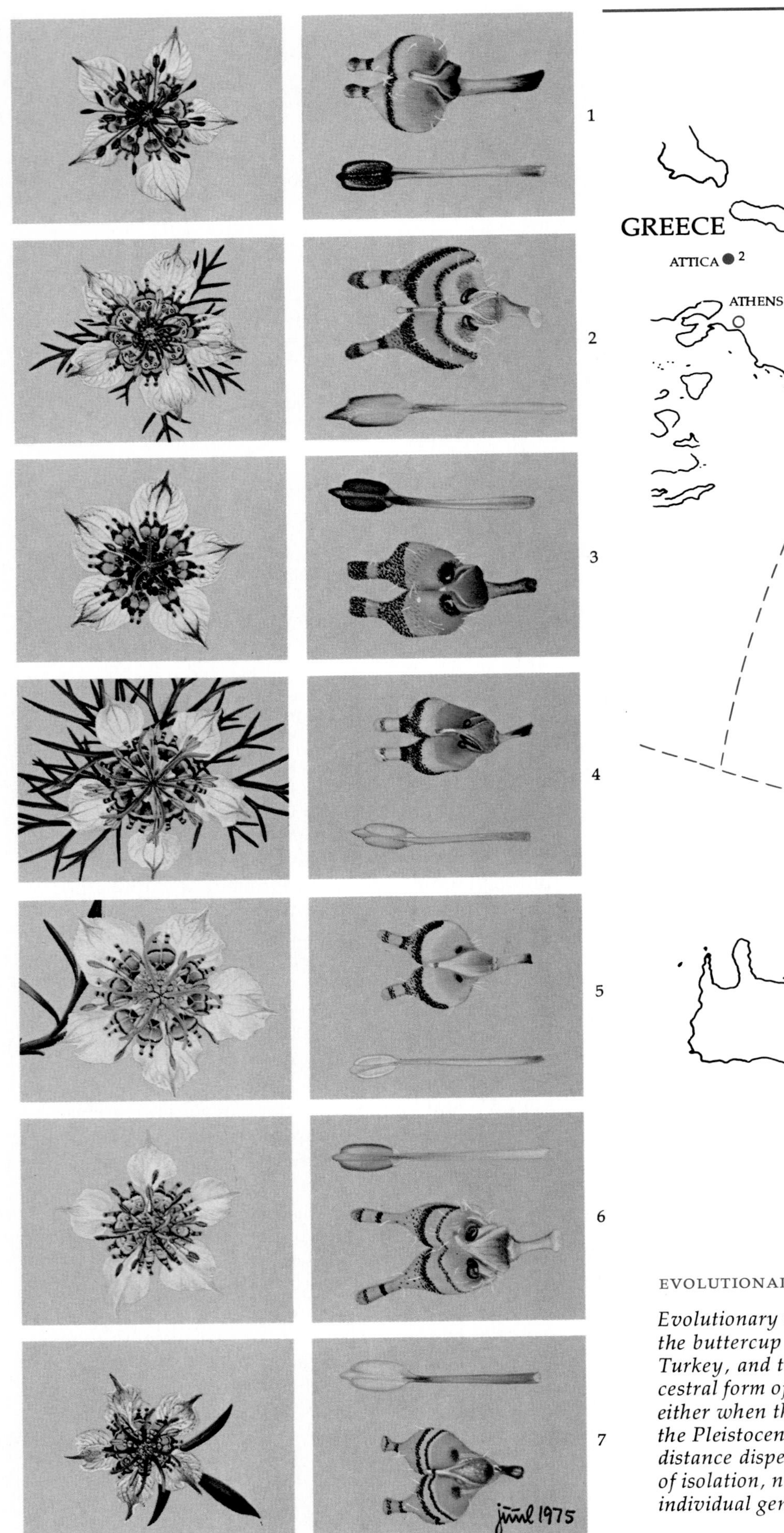

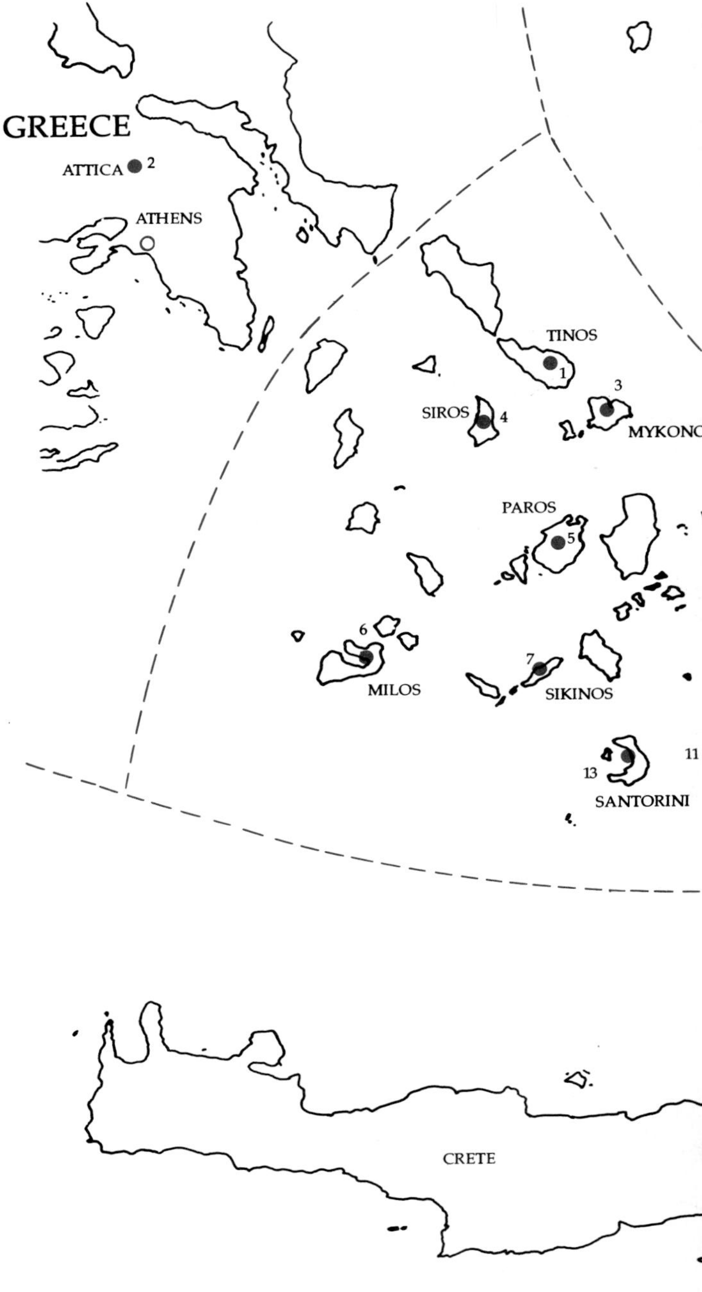

EVOLUTIONARY RADIATION

Evolutionary radiation in Nigella, *an annual herb of the buttercup family, shown on a map of Greece, Turkey, and the islands of the Aegean Sea. An ancestral form of this group spread across these islands, either when they were connected by land during the Pleistocene or subsequently by random long-distance dispersal over water. The combined effects of isolation, natural selection, and the random loss of individual genotypes in the small populations that*

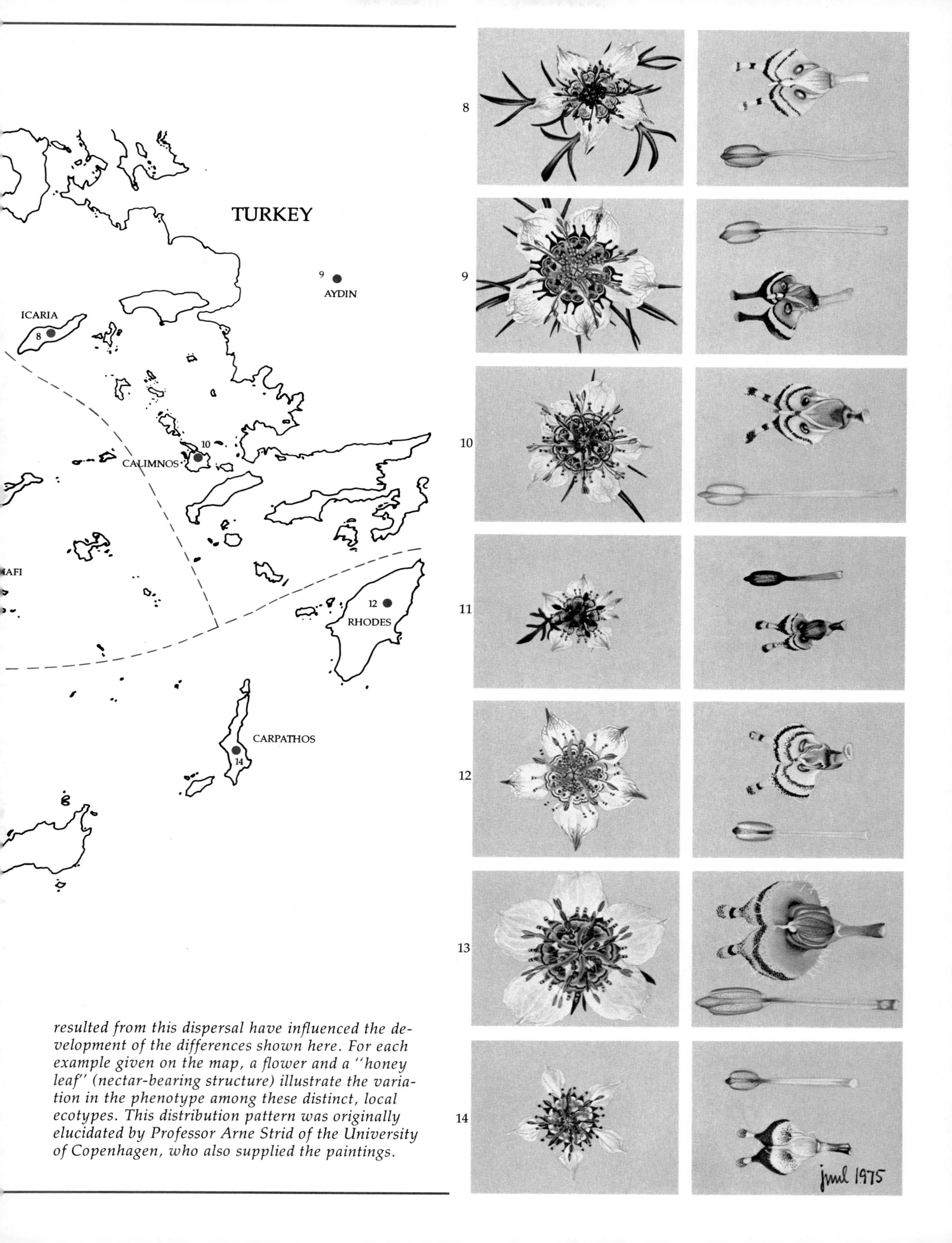

resulted from this dispersal have influenced the development of the differences shown here. For each example given on the map, a flower and a "honey leaf" (nectar-bearing structure) illustrate the variation in the phenotype among these distinct, local ecotypes. This distribution pattern was originally elucidated by Professor Arne Strid of the University of Copenhagen, who also supplied the paintings.

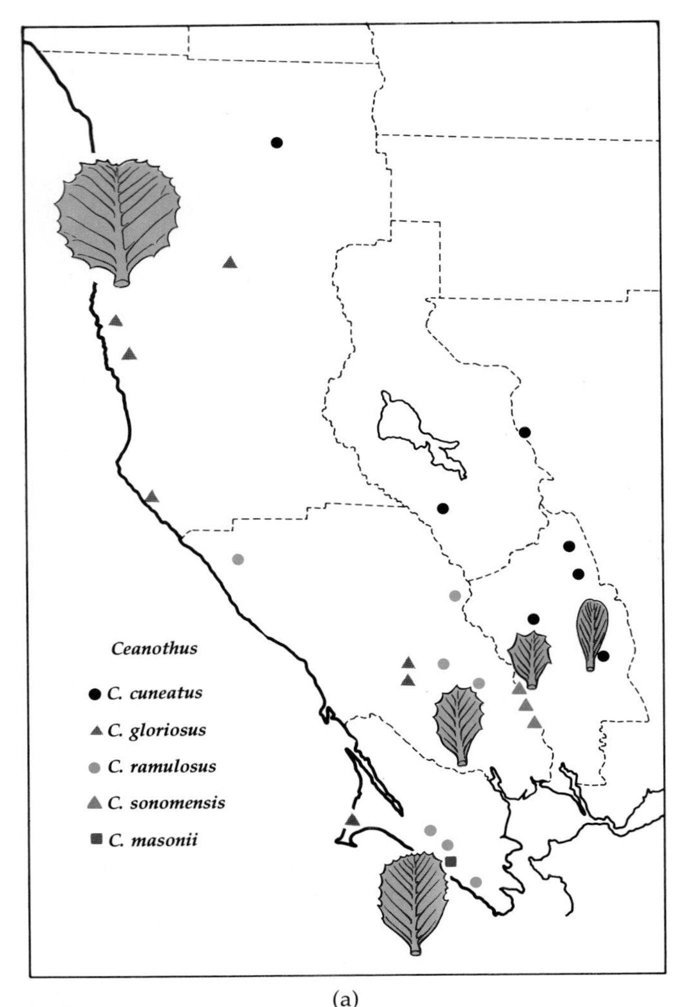

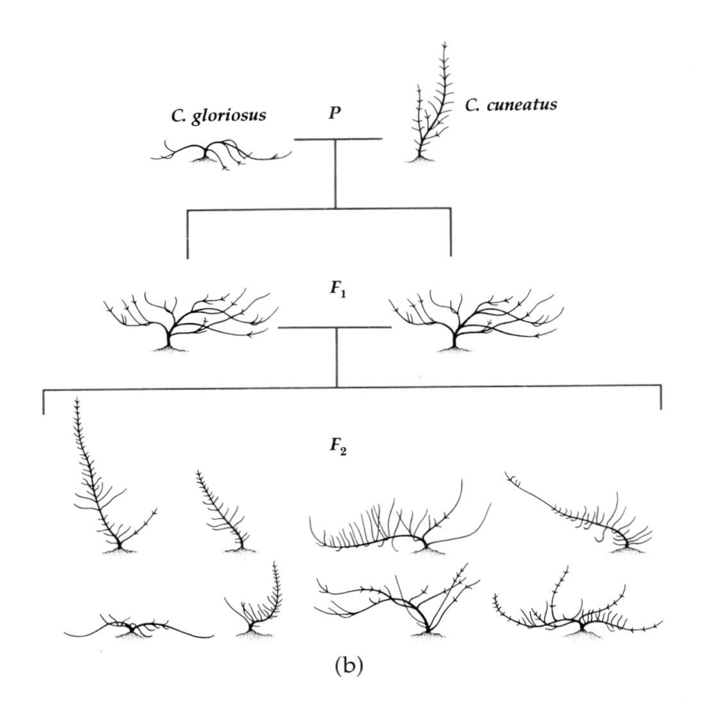

9–12
The establishment of hybrid populations is an important evolutionary mechanism in many groups of woody plants such as Ceanothus. (a) *Map showing a portion of central California (note San Francisco Bay at the bottom), a geologically complex region. Two relatively widespread and distinct species of mountain lilac, the coastal* C. gloriosus *and the interior* C. cuneatus *(which ranges far to the east out of the area depicted), have produced three series of hybrid populations; each is variable but stabilized and is able to grow better than either parent in the areas where they occur. Leaves from representative populations are shown on the map.* (b) *Segregation of the extremes in appearance in the progeny of an artificial cross between the parental species. In habit, some of the intermediates resemble the named intermediate populations shown on the map to the left.*

The establishment and stabilization of hybrid populations depends on the fertility of the hybrids involved. Even if the hybrids are sterile, however, they can still propagate themselves, either asexually by apomixis or by regaining their fertility through polyploidy.

Apomixis and Hybridization

Sterile hybrids may reproduce by a variety of vegetative means. Systems in which apomixis is prevalent, but in which occasional hybridization occurs to produce variable and novel combinations of genes, are the most flexible.

An outstanding example of such a system is the extremely variable Kentucky bluegrass *(Poa pratensis)*, which in one form or another occurs all around the Northern Hemisphere. Occasional hybridization with a whole series of related species has produced hundreds of apomictic races, each well adapted to the ecological characteristics of the region where it grows. Thus we see a flexible system in which new genotypes are produced constantly and the best are preserved by apomixis. In such a system, it makes no difference whether the well-adapted individuals are sterile or not, for their success does not depend on sexual reproduction. Furthermore, apomictically reproducing individuals are particularly successful in arctic regions. Pollination by insects is difficult under these conditions, and certain narrowly defined genotypes may be more successful under the rigorous conditions of the Arctic than more variable populations of sexually reproducing organisms.

9–13

Richard Straw of California State University, Los Angeles, has hypothesized that hybridization accounts for the origin of Penstemon spectabilis, *a flowering plant found in the mountains of southern California. One parent species,* P. grinnellii, *has broad, two-lipped, pale blue flowers that are pollinated mainly by large bees, such as carpenter bees. Another species,* P. centranthifolius, *has long, slender red flowers that are visited mostly by hummingbirds. Their suspected hybrid derivative,* P. spectabilis, *shown here, is ecologically and morphologically intermediate between the two, having rose-purple flowers. It is pollinated by a specialized family of pollen-gathering wasps, which visit neither of the parental species.*

The hundreds of species of hawthorn *(Crataegus)* and blackberry *(Rubus)* found in the eastern United States are apomictic derivatives of complexes in which occasional hybrids occur. In all these cases, the wholesale destruction of natural habitats by humans has made possible the establishment of many new genotypes that would have had no place in the primeval forests of the region.

Polyploidy

Cells or individuals with more than two sets of chromosomes are called polyploids. Polyploid cells arise at a low frequency as the result of a "mistake" of mitosis in which the chromosomes divide but the cell does not. Thus, a cell with twice the usual number of chromosomes is produced. If such cells then go through interphase and divide, they can give rise to a new individual, either sexually or asexually, that will have twice the number of chromosomes of its parents or parent. (Polyploid individuals can be produced by the use of colchicine, a drug that inhibits the formation of the spindle apparatus in mitosis through its disruptive effect on microtubule formation.)

In a polyploid, the range of variation is often narrowed considerably from that in a related diploid because each gene is present twice as often. Complete recessives for a given gene are represented by only $^1/_{16}$ of the individuals, assuming random segregation, rather than ¼ of the individuals as in a diploid. (Both cases assume a frequency for the recessive gene of 0.50.) Polyploids are often self-pollinated, even when the related diploids are largely cross-pollinated, which further reinforces their decreased variability.

Many polyploids originate in hybrids, so that polyploidy, like apomixis, provides a means for the fixation of specific desirable characteristics. Polyploid derivatives of differentiated diploid populations, which occur at low frequencies in such populations, often produce hybrid polyploids. Hybridization between hybrid polyploid individuals is often easier than that between their respective diploid parents, since the barriers to crossing may be more poorly developed. Less commonly, polyploids of hybrid origin have arisen by chromosome doubling in sterile diploid hybrids; such chromosome doubling sometimes functions to restore fertility.

One of the earliest well-documented cases of polyploidy involved just such a process in the production of polyploid hybrids between the radish *(Raphanus sativus)* and the cabbage *(Brassica oleracea).* Both of these species have 18 chromosomes in their vegetative cells and regularly form 9 pairs of chromosomes at meiotic metaphase I. Their hybrid, which was obtained with some difficulty, had 18 unpaired chromosomes in meiosis and was completely sterile. In the polyploid that appeared spontaneously among these hybrid plants, there were 36 chromosomes in the vegetative cells, and 18 pairs were formed regularly in meiosis. In other words, the hybrid had *all* the chromosomes of the radish and the cabbage in its cells, and these functioned normally to give the polyploid a relatively high fertility.

A number of polyploids originated as weeds in habitats associated with the activities of humans, and sometimes they have been spectacularly successful. One of the best known is a salt marsh grass of the genus *Spartina.* One native species, *S. maritima,* occurs in marshes along the coasts of Europe and Africa. A second species, *S. alterniflora,* was introduced into Great Britain from eastern North America in about 1800, and it spread to form large but local colonies.

In Britain, the native *S. maritima* is low in stature, whereas *S. alterniflora* is much taller, frequently growing to 0.5 meter and occasionally to 1 meter or even more in height. Near the harbor at Southampton, both the native species and the introduced species existed side by side throughout the nineteenth century, and by 1870, a sterile hybrid that reproduced vigorously by rhizomes was collected. Of the two parental species, *S. maritima* has a somatic chromosome number of 60 and *S. alterniflora* has 62; the hybrid, owing perhaps to some minor meiotic misdivision, also has 62. This sterile hybrid, which was named *Spartina* × *townsendii,* still persists. About 1890, a vigorous seed-producing polyploid was derived naturally from this hybrid, and it spread rapidly along the coasts of Great Britain and northwestern France. This polyploid, *S. anglica,* with 122 chromosomes, is often planted to bind mud flats, and such use has contributed to its rapid spread (Figure 9–14).

One of the most important polyploid species is wheat. The most commonly cultivated crop is bread wheat, *Triticum aestivum,* which has 42 chromosomes. Bread wheat was derived at least 8000 years ago, probably in central Europe, following the spontaneous hybridization of a cultivated wheat with 28 chromosomes and a wild grass of the same genus with 14 chromosomes, which grew in the fields with its cultivated relative. The hybridization that gave rise to bread wheat was probably between spontaneous polyploids produced within the populations of its two ancestors. The desirable characteristics of the new, fertile, 42-chromosome wheat were easily recognized, and it was selected for cultivation by the early farmers of Europe when it appeared in their fields. One of its parents, the 28-chromosome wheat, had itself originated following hybridization between two wild 14-chromosome species in the Near East. Wheats with 28 chromosomes are also cultivated and are the chief grain used in macaroni products because of the agglutinating properties of their proteins.

Recent studies indicate that new strains produced by artificial hybridization can improve agricultural performance. Particularly promising is *Triticale,* a group of man-made hybrids between wheat *(Triticum)* and rye *(Secale).* Some of these hybrids, combining the high yield of wheat with the rugged-

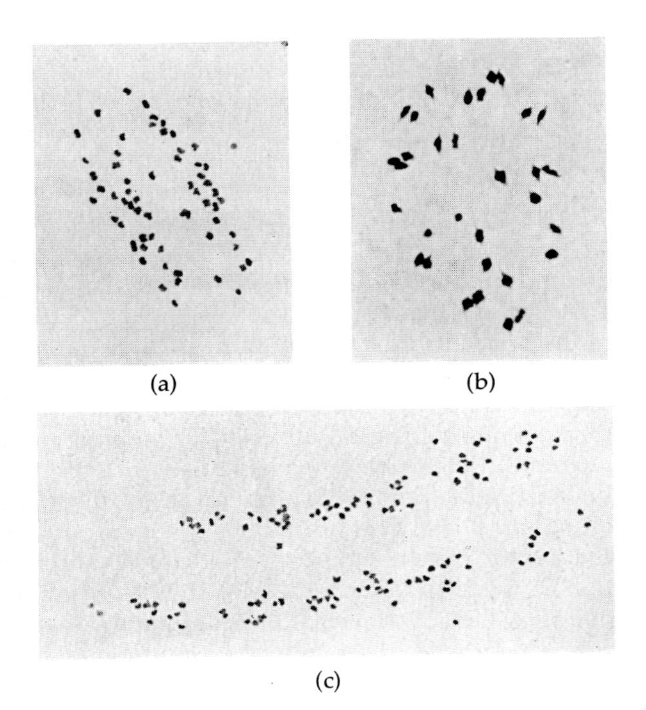

9–14
Polyploidy has been investigated extensively among grasses of the genus Spartina, *which grow in salt marsh habitats found along the coasts of North America and Europe. This salt marsh is on the coast of Great Britain, and the plant shown is a* Spartina *hybrid.* (a) Spartina maritima, *the native British species of salt marsh grass, has 60 chromosomes, visible here in a cell at meiotic anaphase I.* (b) Spartina alterniflora, *a North American species with 62 chromosomes (there are 30 bivalents and 2 unpaired chromosomes in meiotic metaphase I shown here), was first collected from the shores of Southampton Water in 1839, but the date of its introduction to Great Britain is unknown. Although the two species have never been crossed artificially, specimens of their sterile hybrid,* Spartina × townsendii, *were collected from along Southampton Water in 1870.* (c) *A vigorous polyploid,* S. anglica, *arose spontaneously from this sterile hybrid and was first collected in the early 1890s. This polyploid, which has 122 chromosomes, shown here in meiotic anaphase I, is now extending the salt marshes of Great Britain and other temperate countries.*

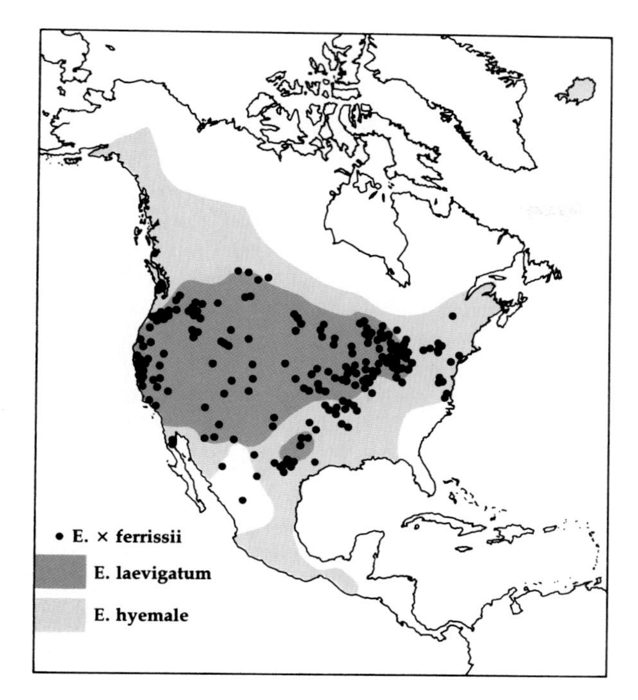

9–15
One of the most abundant and vigorous of the horsetails, genus Equisetum *(see Figure 27–2b, page 538), found in North America is* Equisetum × ferrissii, *a completely sterile hybrid of* E. hyemale *and* E. laevigatum. *Horsetails propagate readily from small fragments of underground stems, and the hybrid maintains itself over its wide range through such vegetative propagation.*

ness of rye, are more resistant to wheat rust—an economically important fungal disease. This characteristic could be particularly important in the subtropical and tropical highland regions of the world, where rust is the chief factor limiting wheat production. The most important strain of *Triticale* has 42 chromosomes and was derived by chromosome doubling following hybridization of 28-chromosome wheat with 14-chromosome rye.

In nature, polyploids are selected by the environment, and not directly by humans. Polyploidy is a major evolutionary mechanism. Events similar to those we have reviewed in the history of wheat must have taken place well over 100,000 times just to account for the present representation of polyploids in the flora of the world, which amount to about half of the total species. Among these polyploids are many of our most important crops—not only wheat but also cotton, tobacco, sugarcane, bananas, potatoes, and safflower, just to mention a few. To this list can be added many of our most attractive garden flowers, such as chrysanthemums, pansies, and daylilies.

SUMMARY

Natural selection is the process by which, under environmental pressure, organisms with more favorable characteristics leave more surviving progeny. Natural selection can occur because of genetically determined variations in natural populations.

The simple genetic situation is defined by the Hardy-Weinberg law, which states that in a large population in which there is random mating, and in the absence of forces that change the proportions of genes, the proportion of dominant to recessive alleles remains constant from generation to generation.

Four principal factors that can cause changes in the proportions of genes and deviations from the Hardy-Weinberg equilibrium are population size, differential migration, mutation rate, and selection. Of these, selection is the most consequential. It is defined as the nonrandom reproduction of genotypes, some being favored over others.

Recessive genes are relatively inaccessible to selection in diploid organisms because the less frequent they become, the higher is the proportion of genes that are hidden (masked) in heterozygotes.

Response to selection is complex for a number of reasons. Only the phenotype is accessible to selection, and it can be based in many instances on a wide variety of genotypes. Because of epistasis and pleiotropy, single genes cannot be selected in isolation; selection affects the whole genotype.

Populations adjust to particular environments and form sharply defined units, or ecotypes, if the lines between the environments are sharply drawn. If environments gradually intergrade, populations of plants may form clines with respect to the characteristics under consideration.

While populations are changing, they also diverge in those factors that allow them to intercross successfully. After a period of isolation, two populations may be more or less incompatible or may produce sterile hybrids. Related species of relatively long-lived plants, such as trees and shrubs, are less apt to be reproductively isolated than are related species of annuals and other short-lived plants.

Hybrid populations derived from two species are common in plants and predominate in most genera of woody plants. This is particularly true in environments with relatively few species, such as oceanic islands, where adjustment to environmental changes may be especially critical, and in regions with sharply defined environmental breaks and a rapidly shifting climate, such as California.

Even if the hybrids between two species are sterile, they may be propagated by vegetative reproduction (apomixis) or become fertile following doubling in chromosome number (polyploidy).

SUGGESTIONS FOR FURTHER READING

ANDERSON, EDGAR: *Plants, Man and Life*, University of California Press, Berkeley and Los Angeles, 1967.*

A fascinating account of genetic evolutionary studies of plants that reflects the personality of one of the greatest contemporary students of plants.

AYALA, FRANCISCO J., and JAMES W. VALENTINE: *Evolving: The Theory and Processes of Organic Evolution*, Benjamin-Cummings Publishing Co., Menlo Park, Calif., 1979.

A new and very readable account of modern views of evolution.

EHRLICH, PAUL R., RICHARD W. HOLM, and DENNIS R. PARNELL: *The Process of Evolution*, 2nd ed., McGraw-Hill Book Co., New York, 1974.

A concise and thoughtful treatment of modern evolutionary theory that provides a useful review of the entire field.

GOODENOUGH, URSULA: *Genetics*, 3rd ed., Holt, Rinehart & Winston, New York, 1978.

A thoroughly up-to-date account of genetics from a molecular point of view, emphasizing prokaryotes.

GOULD, STEPHEN JAY: *Ever Since Darwin: Reflections in Natural History*, W. W. Norton & Co., Inc., New York, 1979.*

A delightful collection of essays concerning the impact of Darwinism on modern science. Highly recommended.

GRANT, VERNE: *Plant Speciation*, Columbia University Press, New York, 1971.

This comprehensive treatment of plant evolution provides a thorough introduction to most aspects of the field.

JUDSON, HORACE: *The Eighth Day of Creation*, The S&S Co., Central Point, Oreg., 1979.

The story of the unraveling of modern genetics, told by the participants. Quite a different view from James Watson's, and a superb account of science in action.

KREBS, CHARLES J.: *Ecology: The Experimental Analysis of Distribution and Abundance*, 2nd ed., Harper & Row Publishers, Inc., New York, 1978.

A modern synthesis of ecology, written from an evolutionary point of view and with a strong emphasis on experimental approaches.

MOOREHEAD, ALAN: *Darwin and the Beagle*, Harper & Row Publishers, Inc., New York, 1969.

An engaging account of Darwin's work by a famous writer and historian.

PETERS, JAMES A. (Ed.): *Classic Papers in Genetics*, Prentice-Hall, Inc., Englewood Cliffs, N.J., 1960.*

Includes papers by most of the scientists responsible for the important developments in genetics—Mendel, Sutton, Morgan, Beadle and Tatum, Watson and Crick, Benzer, and so on. This book is very interesting and surprisingly readable, and the original papers give a feeling of immediacy that no other account can achieve.

SRB, ADRIAN M., KAY D. OWEN, and ROBERT S. EDGAR: *General Genetics*, 2nd ed., W. H. Freeman and Company, San Francisco, 1965.

This older book is particularly useful for someone who wishes to study classical genetics; the authors provide not only a clear and relatively simple text but also a remarkably good, long list of questions and problems following each chapter.

STEBBINS, G. LEDYARD: *Processes of Organic Evolution*, 3rd ed., Prentice-Hall, Inc., Englewood Cliffs, N.J., 1977.*

A brief review of the entire field by one of its outstanding practitioners.

STRICKBERGER, MONROE W.: *Genetics*, 2nd ed., Macmillan Publishing Co., Inc., New York, 1976.

A traditional and cohesive account of the science, this book is much broader in coverage than most texts at this level, although it does not emphasize a molecular approach.

WAGNER, R. P., B. H. JUDD, B. G. SANDERS, and R. H. RICHARDSON: *Introduction to Modern Genetics*, John Wiley & Sons, Inc., New York, 1980.

The best short genetics text available.

WATSON, JAMES D.: *Double Helix: Being a Personal Account of the Discovery of the Structure of DNA*, Atheneum Publishers, New York, 1968.*

A lively account of how to become a Nobel laureate in molecular biology.

__________: *Molecular Biology of the Gene*, 3rd ed., W. A. Benjamin, Inc., Menlo Park, Calif., 1976.

For the student who wants to go more deeply into the subject of modern genetics; this book is generally agreed to be outstanding.

*Available in paperback.

SECTION 4 Diversity

CHAPTER 10

The Classification of Living Things

10–1
A fritillary butterfly (Agraulis vanillae), *visiting an inflorescence of a tropical member of the spurge family* (Cnidoscolus aconitifolius) *in southern Mexico.*

At least 5 million different kinds of living organisms share our biosphere. Humans differ from these other organisms both in the degree of their curiosity and in their power of speech. As a consequence of these two characteristics, humans have long sought to inquire about the other creatures of the world and to exchange information about them. As knowledge about organisms grew, it became necessary to know the names that others had given to the organisms in order to learn what was known about them or to report new information about them. Most familiar organisms have been given common names, but even for the simplest of purposes, common names may be inadequate. Sometimes they are misleading, particularly when we are exchanging information with people from different parts of the world. A sycamore or a cowslip in Great Britain may or may not be the same as a plant bearing a similar name in North America. A pine in Europe or the United States is not the same as a pine in Australia. A yam in the southeastern United States is a totally different vegetable from one a few hundred kilometers away in the West Indies. When different languages are involved, the problems become hopelessly complex. For this reason, biologists refer to organisms by Latin names that are officially recognized by international organizations of bacteriologists, botanists, and zoologists.

The practice of referring to organisms by Latin names began in medieval times, when Latin was the language of scholarship. As the practice became more systematic, organisms were first grouped into *genera* (singular: *genus*) and then were identified by descriptive Latin phrase names, known as polynomials. By the end of the seventeenth century, the first word of such a polynomial came to designate the name of the group, or genus, to which the plant belonged. Thus, all willows were identified by polynomials beginning with the word *Salix;* the phrases describing different kinds of roses all began with *Rosa;* and those referring to oak trees began with *Quercus.* All of these Latin names now designate genera.

10–2
Carolus Linnaeus (1707–1778), the naturalist who devised the binomial system of classification, believed that each living thing corresponded more or less closely to some ideal model and that by classifying them, he was revealing the grand pattern of creation.

THE BINOMIAL SYSTEM

A simplification in the system of naming living things was made by the eighteenth-century Swedish professor and naturalist, Carolus Linnaeus, whose ambition was to classify all the known kinds of plants and animals according to their genera. In 1753, he published a two-volume work, *Species Plantarum,* which contained brief analytical descriptions of every known species of plant, with references to the earlier works about each one. Linnaeus used polynomial designations for all species, and in many cases he changed them so that they would be comparable with those referring to other species in the same genus. Although he regarded the polynomials as the proper names for species, he also added an important innovation to the system. In the margin of his book, next to the "proper" name of each species, Linnaeus entered a single word, which, together with the generic name, formed a convenient "shorthand" designation for the species. For example, catnip, which had previously been designated as *Nepeta floribus interrupte spicatus pedunculatis* (meaning "*Nepeta* with flowers in an interrupted pedunculate spike"), was described under *Nepeta,* but Linnaeus took a more familiar characteristic of the plant and put "cataria" (meaning "cat-associated") in the margin, making it *Nepeta cataria,* which is its name today.

The convenience of this system was obvious, and Linnaeus and subsequent authors soon began to replace "proper" names with "shorthand" ones. The binomial ("two-term") system of Linnaeus is still used today. The earliest binomial name applied to a particular species is accepted as the correct name for that species, and new names that might be applied to the same species are rejected. The rules governing the application of botanical names to plants are embodied in the *International Code of Botanical Nomenclature,* which is revised at successive International Botanical Congresses held every six years.

A species name consists of two parts—the generic name and the specific name. However, a generic name may be written alone when one is referring to the entire group of species comprising that genus. Figure 10–3, for instance, shows three species of the violet genus, *Viola.* Another example is the evening primrose genus, *Oenothera,* which includes some 120 species native to North and South America. Some species have become widespread, either in gardens or as weeds. One of the rarer species of this genus is the spectacular *Oenothera muelleri,* which has large white flowers, 10 to 12 centimeters in diameter; it has been found only on a few mountaintops in the northeastern Mexican state of Nuevo León. Another species is the weedy *Oenothera biennis,* which has much smaller yellow flowers, 3 to 6 centimeters in diameter; it is widespread throughout eastern North America and has been introduced in many other temperate regions.

(a)

(b)

(c)

10–3
Three members of the violet genus.
(a) *Long-spurred violet,* Viola rostrata.
(b) *Common violet,* V. papilionacea.
(c) *Pansy,* V. tricolor *var.* hortensis.
These photographs indicate the kinds of differences in leaf shape and margin, flower color and size, and other features that distinguish different species of a single genus, even though there is an overall similarity between all three species. The pansy shown here is an annual selected from western European progenitors for cultivation in gardens; the other species are perennials. There are about 500 species of the genus Viola; *most of them are found in temperate regions of the Northern Hemisphere.*

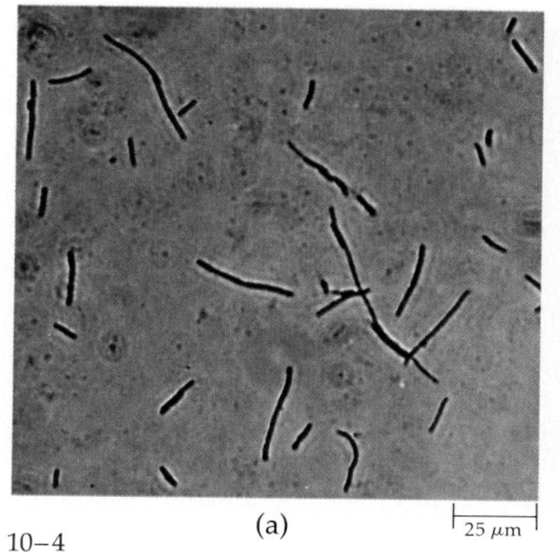

(a) 25 μm

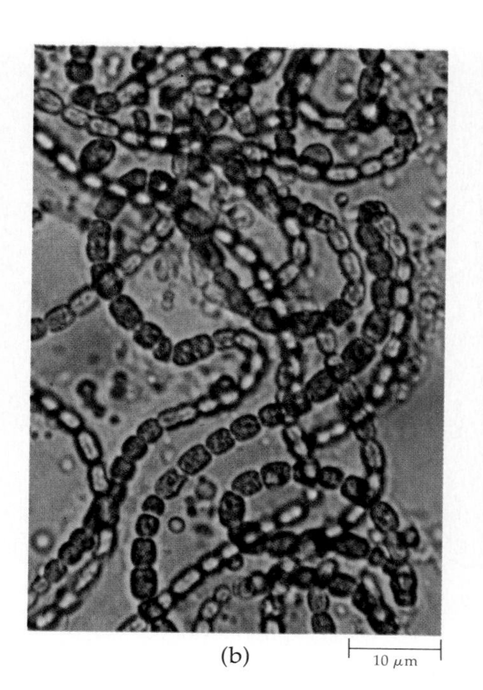

(b) 10 μm

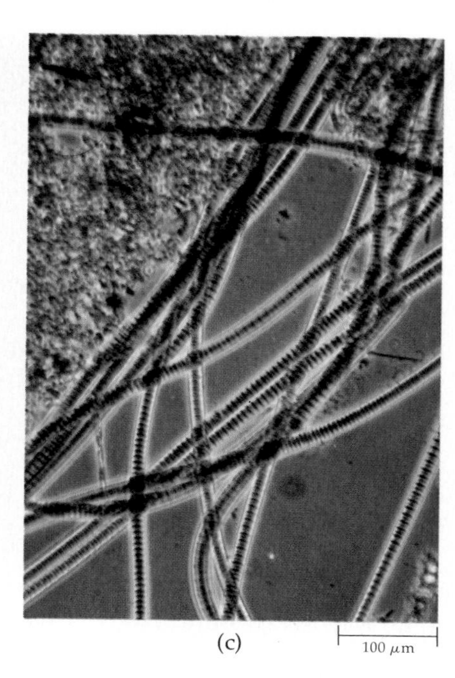

(c) 100 μm

10–4

Monera. *This kingdom is made up of the only living prokaryotes—the bacteria.* (a) Lactobacillus acidophilus, *bacteria that sour milk.* (b) *A gelatinous colony of the cyanobacterium ("blue-green alga")* Nostoc, *a member of one of the groups of photosynthetic bacteria.* (c) Oscillatoria, *another cyanobacterium, in which the growth form is filamentous.*

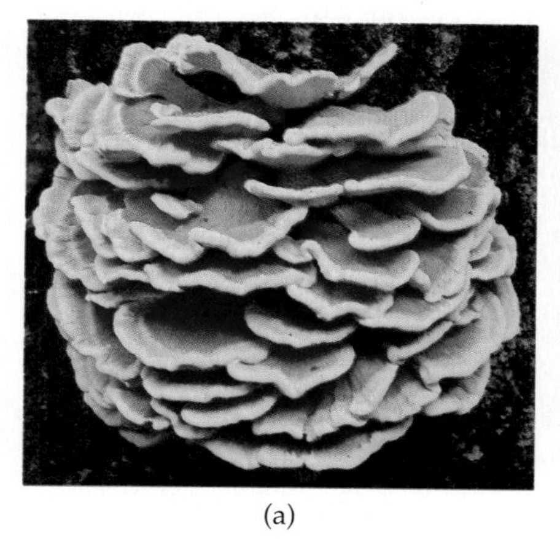

(a)

(b)

(c)

(d)

(e)

10–5

Fungi. (a) Polyporus sulphureus, *the sulfur shelf fungus.* (b) Coprinus, *an inky cap mushroom.* (c) Cantharellus aurantiacus, *a chanterelle fungus.* (d) Pleurotus ostreatus, *the oyster mushroom.* (e) *A white coral fungus (family Clavariaceae). All of these fungi belong to the division Basidiomycota.*

(a)

(b)

(c)

(d)

(e)

(f)

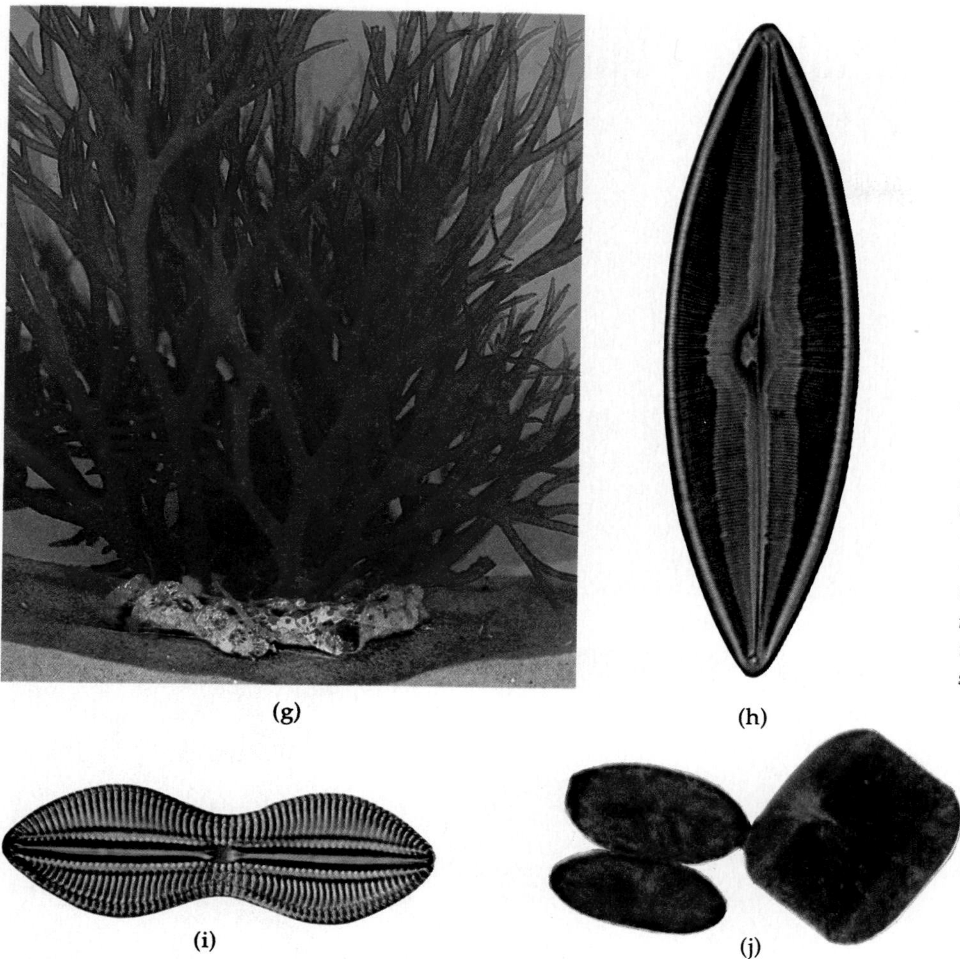

(g) (h) (i) (j)

10–6

Protista. (a) *Fruiting bodies of the plasmodial slime mold* Stemonitis *(division Myxomycota).* (b) *Plasmodium of a plasmodial slime mold,* Physarum, *growing on an agar culture medium.* (c) Chara, *a green alga (division Chlorophyta) that grows in shallow waters of temperate lakes.* (d) Postelsia palmiformis, *the "sea palm" (division Phaeophyta), growing on exposed intertidal rocks off Vancouver Island, British Columbia.* (e) Volvox, *a motile colonial green alga (division Chlorophyta).* (f) Fucus, *the rockweed (division Phaeophyta), is found along both coasts of the United States as well as along the shores of Europe.* (g) Sebdenia polydactyla, *a red alga (division Rhodophyta).* (h) *and* (i) *Pennate diatoms (division Chrysophyta), in side view, showing the intricately marked shells characteristic of this group.* (j) *Two centric diatoms, as seen from the side, with one dividing.*

(a)

(b)

10–7

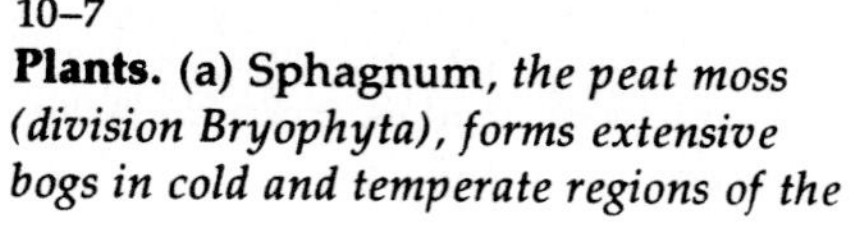

Plants. (a) Sphagnum, *the peat moss (division Bryophyta), forms extensive bogs in cold and temperate regions of the world.* (b) Marchantia *is by far the most familiar of the thallose liverworts (division Bryophyta). It is a widespread, terrestrial genus that grows on moist soil and rocks.*

10–7 (continued)

Plants. (c) Lycopodium complanatum, *a "club moss" (division Lycophyta).* (d) *Wood horsetail,* Equisetum sylvaticum *(division Sphenophyta).* (e) *Marsh ferns,* Thelypteris palustris *(division Pterophyta).* (f) *The dandelion,* Taraxacum officinale, *and* (g) *fishhook cactus,* Mammillaria microcarpa, *are dicots (class Dicotyledones, division Anthophyta).* (h) *Foxtail barley,* Hordeum jubatum, *and* (i) Cymbidium *orchids are monocots (class Monocotyledones, division Anthophyta).* (j) *Sugar pine,* Pinus lambertiana, *and incense cedar,* Calocedrus decurrens, *are both conifers (division Coniferophyta).*

A specific name is meaningless when written alone; for example, *biennis* could refer to any of the scores of species in different genera that happen to have this word as part of their name. For example, *Artemisia biennis* and *Lactuca biennis* are two very different members of the sunflower family. For this reason, the specific name is always preceded by the name or the initial letter of the genus that includes the species in question, that is, *Oenothera biennis* or *O. biennis.* Names of genera and species are printed in italics or are underlined when written or typed.

Species can be further divided into subspecies or varieties. Subspecies of one species bear an overall resemblance to one another but exhibit one or more important differences. As a result of these subdivisions, although the binomial name is still the basis of classification, the names of some plants and animals may consist of three parts. Thus the peach tree is *Prunus persica* var. *persica,* whereas the nectarine is *Prunus persica* var. *nectarina.* Names of subspecies and varieties are also written in italics or underlined, and the first-named subspecies or variety (chronologically speaking) repeats the name of the species.

What Is a Species?

Groups of populations that resemble one another relatively closely and other groups of populations less closely are called *species,* but the application of this term differs widely from one group of organisms to another. The word "species" itself has no special connotation; it simply means "kind" in Latin. The patterns of variation that occur in different groups of organisms as a result of the evolutionary processes discussed in Chapter 9 differ greatly from one another, so that the term "species" simply cannot be applied in a uniform way.

Thus the word "species" means different things for different kinds of organisms. For example, genetic recombination is unknown in some groups (for example, algae related to *Euglena*), so we should not expect the units we call species in these groups to vary in the same way as species among, for example, the oaks. However, the term "species" remains useful in that it provides a convenient way to talk about and catalogue organisms.

Other Taxonomic Groups

Linnaeus (and earlier scientists) recognized the plant, animal, and mineral kingdoms, and the *kingdom* is still the major unit used in biological classification. In the nineteenth and twentieth centuries, however, subsequent authors added a number of categories between the level of genus and the level of kingdom. Thus, genera are grouped into *families,* families into *orders,* orders into *classes,* and classes into *divisions.* "Phylum," the term that is equivalent to "division" in zoological (animal) classification, is not recognized as a taxonomic category by the *International Code of Botanical Nomenclature.*

Sample classifications of corn *(Zea mays)* and the commonly cultivated edible mushroom *(Agaricus campestris)* are given in Table 10–2 on page 179.

THE MAJOR GROUPS OF ORGANISMS

In Linnaeus's time, it was thought that there were three types of "objects": animal, vegetable, and mineral. Thus living things were thought of as either plant or animal. Animals moved, ate things, and breathed, and the size of their bodies was definitely limited. Plants did not move, eat, or breathe; they were presumed to manufacture their own food and seemed to grow indefinitely.

As new groups of organisms were discovered, they were classified either as plants or as animals. Thus the fungi and bacteria were grouped with the plants, and the protozoa were grouped with the animals. However, taxonomists (individuals specializing in the science of classification) began to have trouble with forms such as *Chlamydomonas,* a swimming green alga that moves *and* manufactures its own food. Such organisms could not be classified clearly as either plant or animal, and by the 1930s it was evident that the traditional division of living organisms into two kingdoms was little more than a historical curiosity.

Unfortunately, no completely acceptable alternative has been proposed. The relationships among organisms are complex at all levels, and it is perhaps unreasonable to expect to be able to group all organisms into a series of clearly and precisely defined kingdoms. As a result, the old division into plants and animals is still widely reflected in the organization of college and university science departments, research projects, and textbooks (including this one). All groups traditionally considered to be plants are discussed in this book.

The Prokaryotes

As new information has gradually been accumulated, it has become evident that the most fundamental division in the living world is the distinction between the prokaryotes and the eukaryotes (Figure 10–8). The living prokaryotes are bacteria (including cyanobacteria, or blue-green "algae," which were formerly treated as a separate group). The prokaryotes do not have membrane-bound cellular organelles, microtubules, or the complex 9-plus-2 structure of flagella. Their genetic material is borne on a single

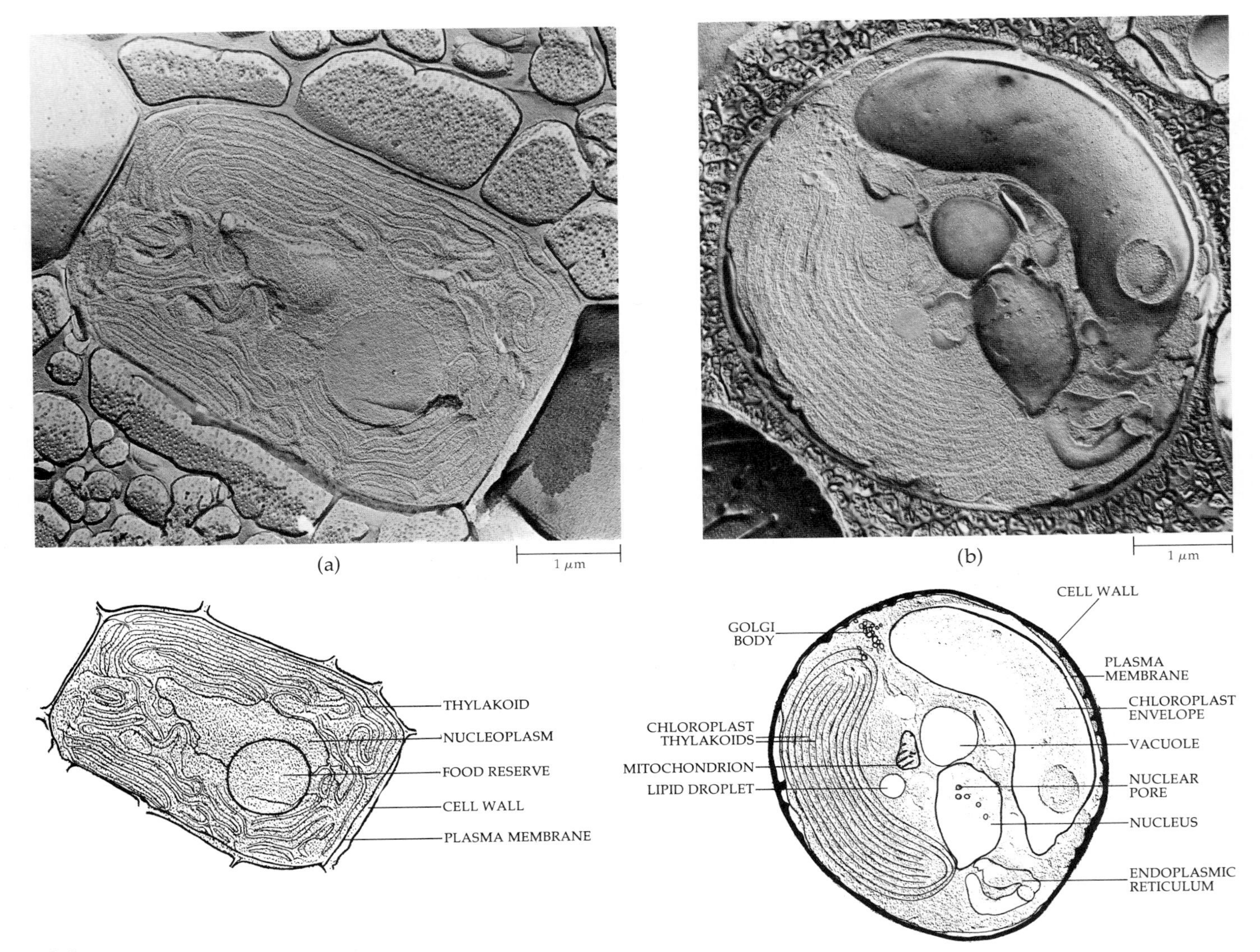

10–8
Electron micrographs of (a) Anabaena cylindrica, *a prokaryotic cyanobacterium, and* (b) Cyanidium caldarium, *a eukaryotic, single-celled green alga. Note the much greater complexity of the eukaryote. The large, lobed chloroplast of* Cyanidium, *which appears here in two parts, is biochemically similar and comparable in size to an entire cyanobacterium and is unlike the chloroplasts of most other eukaryotes. It is likely that it is in fact a symbiotic cyanobacterium that has assumed the function of a chloroplast. The three-dimensional quality of these electron micrographs are due to a technique known as freeze-etching. When this technique is used, the specimen is frozen rapidly and then can be split apart. A replica of the surface of the specimen is what is shown here.*

circular molecule of DNA that is not associated with proteins. Although several mechanisms leading to genetic recombination are known in prokaryotes, it occurs infrequently and is accomplished by means other than sexual reproduction. The cell walls of most prokaryotes contain muramic acid, and in this and a number of other biochemical peculiarities, they differ from all other organisms. These characteristics provide ample grounds for recognizing the prokaryotes as a separate kingdom—the kingdom Monera.

The Eukaryotes

All eukaryotes have a definite nucleus that is bounded by a double membrane. Within the nuclear envelope are complex chromosomes in which the

DNA is associated with proteins. These chromosomes segregate regularly by mitosis. The flagella and cilia of eukaryotes that possess them have the complex 9-plus-2 pattern of microtubules. Microtubules also occur within the cytoplasm of eukaryotes. Complex organelles, such as mitochondria, occur in the cells of all eukaryotes; and vacuoles, which are bounded by a single membrane, or tonoplast, also occur widely in eukaryotic organisms, especially plants.

Many eukaryotes also exhibit two important features that are not found in prokaryotes: integrated multicellularity and sexual reproduction. The cells of prokaryotes sometimes remain together in filamentous or even three-dimensional masses following cell division, but there are no protoplasmic connections between the individual cells and hence no overall integration of the entire filament or mass. In plant eukaryotes, the protoplasts of contiguous cells are connected by plasmodesmata, which traverse their cell walls. In animals, there are no cell walls, and the protoplasts are in more direct contact.

FORMAL CLASSIFICATION OF LIVING ORGANISMS

One practical arrangement of living organisms divides them among five kingdoms: the prokaryotic Monera and four eukaryotic groups (Figure 10–11). Of the eukaryotic groups, the Protista are believed to have given rise to the other three eukaryotic groups—the plants, the animals, and the fungi. These mostly multicellular groups differ fundamentally in their mode of nutrition. In general, plants manufacture their food, animals ingest it, and fungi absorb it. For the purposes of this classification, the viruses (see Chapter 11) are considered separately, since they differ in organization from all cellular forms of life.

The following is a synopsis of this system of classification, which is the one used in this book (see Table 10–1).

Kingdom Monera

Monera comprise prokaryotic organisms; they do not possess nuclear envelopes, plastids, mitochondria, and advanced flagella, and they exhibit solitary unicellular or colonial unicellular organization (see Figure 10–4) and lack protoplasmic connections between the cells. Absorption is the mode of nutrition in most groups, but some are photosynthetic or chemosynthetic. Reproduction is predominantly by cell division, although genetic recombination occurs in several groups. They are either motile by simple flagella or by gliding, or are nonmotile. Monera are discussed in Chapter 11.

Table 10–1 *Classification of Living Organisms Traditionally Regarded as Plants. (See Appendix C for summary descriptions of these groups.)*

PROKARYOTES	KINGDOM MONERA
	Bacteria, including cyanobacteria
EUKARYOTES	KINGDOM PROTISTA
ALGAE	Division Chlorophyta (green algae)
	Division Phaeophyta (brown algae)
	Division Rhodophyta (red algae)
	Division Chrysophyta (diatoms and golden-brown algae)
	Division Pyrrophyta (dinoflagellates)
	Division Euglenophyta (euglenoids)
HETEROTROPHIC PROTISTA	Division Oomycota (water molds)
	Division Chytridiomycota (chytrids)
	Division Acrasiomycota (cellular slime molds)
	Division Myxomycota (plasmodial slime molds)
	KINGDOM FUNGI
FUNGI	Division Zygomycota (bread molds)
	Division Ascomycota (sac fungi)
	Division Basidiomycota (club fungi)
	KINGDOM PLANTAE
BRYOPHYTES	Division Bryophyta (bryophytes)
	Class Muscopsida (mosses)
	Class Anthocerotopsida (hornworts)
	Class Hepaticopsida (liverworts)
	VASCULAR PLANTS
SEEDLESS VASCULAR PLANTS	Division Psilophyta (whisk fern)
	Division Lycophyta (lycopods)
	Division Sphenophyta (horsetails)
	Division Pterophyta (ferns)
SEED PLANTS	Division Cycadophyta (cycads)
	Division Ginkgophyta (ginkgo)
	Division Coniferophyta (conifers)
	Division Gnetophyta (gnetophytes; vessel-containing gymnosperms)
	Division Anthophyta (angiosperms; flowering plants)
	Class Dicotyledones (dicots)
	Class Monocotyledones (monocots)

Kingdom Protista

Protista are here considered to comprise all organisms traditionally regarded as protozoa (one-celled animals), as well as all eukaryotic algae (see Figure 10–6). Included in the Protista are the water molds and their relatives (division Oomycota), the chytrids (division

Chytridiomycota), the cellular slime molds (division Acrasiomycota), and the plasmodial slime molds (division Myxomycota)—four groups of heterotrophic organisms that have traditionally been placed with the fungi. The reproductive cycles of protista are varied but typically involve both cell division and sexual reproduction. They may be motile by 9-plus-2 flagella or cilia or by amoeboid movement, or they may be nonmotile. The heterotrophic protista included in this book are treated in Chapter 13, and the algae are treated in Chapter 14. Discussions of the heterotrophic protista known as protozoa are not included in this text. In sum, Protista constitute a heterogeneous assemblage of unicellular, colonial, and multicellular eukaryotes not having the distinctive characteristics of the animals, plants, or fungi.

Kingdom Animalia

The animals are multicellular organisms with wall-less eukaryotic cells lacking plastids and photosynthetic pigments. Nutrition is primarily ingestive with digestion in an internal cavity, but some forms are absorptive, and a number of groups lack an internal digestive cavity. The level of organization and tissue differentiation in higher forms far exceeds that of other kingdoms, with evolution in particular of complex sensory and neuromotor systems. The motility of the organism (or, in sessile forms, of its parts) is based on contractile fibrils. Reproduction is predominantly sexual. Animals are not discussed in this book.

Kingdom Fungi

Fungi are nonmotile filamentous eukaryotes that lack plastids and photosynthetic pigments and absorb their nutrients from either dead or living organisms (see Figure 10–5). The fungi have traditionally been grouped with plants, but there is overwhelming evidence that they are an independent evolutionary line. Aside from the low level of differentiation of their bodies, they have little in common with the algae. The cell walls of fungi include a matrix of chitin. Bodies in which spores are formed are often complex, and their reproductive cycles, which also can be quite complex, typically involve both sexual and asexual processes. Fungi are discussed in Chapter 12.

Kingdom Plantae

The plants—bryophytes (mosses and liverworts) and the nine divisions of vascular plants—include all the more specialized green organisms derived from the green algae. All are multicellular and composed of vacuolate eukaryotic cells with cellulosic cell walls. Their principal mode of nutrition is photosynthesis, although a few have become heterotrophic. In general, their evolution has occurred in relation to their successful invasion of the land. Structural differentiation has occurred, with trends toward organs of photosynthesis, anchorage, and support (see Figure 10–7). In higher forms, such organization has produced specialized photosynthetic, vascular, and covering tissues. Reproduction is primarily sexual, with cycles of alternating haploid and diploid generations, the former reduced in the more advanced members of the kingdom. The bryophytes are discussed in Chapter 15, the vascular plants in Chapters 16–19.

SEXUAL REPRODUCTION AND DIPLOIDY

All prokaryotic organisms are haploid. They have several means of achieving genetic recombination, however. As described in more detail in Chapter 11, a portion of the bacterial genetic material may be transferred from one cell to another, but there is no mechanism comparable to meiosis by which it can then be transmitted regularly along with the genetic material of the cell that it joins. The only method for regular transmission of the new genetic information found among prokaryotes is the incorporation of the new fragment into the bacterial DNA molecule, a system that is neither precise nor readily repeatable.

Sexual reproduction has a high selective advantage. As discussed, it occurs only in eukaryotic organisms and involves a regular alternation between meiosis and syngamy. One of the most significant features of sexual reproduction is that this is the mechanism that produces variability in natural populations and, to a certain extent, helps to maintain it. As such, it provides the basis for understanding the process of evolution, since living organisms must continuously adjust to variable and changing environments.

The Evolution of Diploidy

The first eukaryotic organisms were probably haploid and asexual, but once sexual reproduction was established among them, the stage was set for the evolution of diploidy. There is no direct evidence to indicate when the diploid condition evolved, but most, if not all, of the multicellular animals and algae that appear in the fossil record of some 650 to 700 million years ago were diploid. It seems likely that diploidy arose when two haploid cells combined to form a diploid zygote; such an event probably took place repeatedly. The zygote then presumably divided immediately by meiosis, thus restoring the haploid condition (Figures 10–9 and 10–10a). In organisms with this simple kind of life cycle, the zygote is the only diploid cell.

Table 10–2 *Biological Classification. Notice how much you can tell about an organism when you know its place in the system. The descriptions here do not define the various categories but tell you something about their characteristics.*

CORN

CATEGORY	NAME	DESCRIPTION
Kingdom	Plantae	Organisms that are terrestrial, have chlorophyll *a* and chlorophyll *b* contained in chloroplasts, and show structural differentiation.
Division	Anthophyta	Vascular plants with seeds and flowers; ovules enclosed in an ovary, pollination indirect; the angiosperms.
Class	Monocotyledones	Embryo with one cotyledon; flower parts usually in threes; many scattered vascular bundles in the stem.
Order	Commelinales	Monocots with fibrous leaves; reduction and fusion in flower parts.
Family	Poaceae	Hollow-stemmed monocots with reduced greenish flowers; fruit a specialized achene (caryopsis); the grasses.
Genus	*Zea*	Robust grasses with separate staminate and carpellate flower clusters; caryopsis fleshy.
Species	*Zea mays*	Corn.

EDIBLE MUSHROOM

CATEGORY	NAME	DESCRIPTION
Kingdom	Fungi	Nonmotile, multinucleate, heterotrophic, absorptive organisms in which chitin predominates in the cell walls.
Division	Basidiomycota	Dikaryotic fungi that form a basidium bearing four spores (basidiospores); the basidiomycetes.
Class	Homobasidiomycetes	Basidiomycetes that produce basidiocarps, or "fruiting bodies," and club-shaped aseptate basidia.
Order	Agaricales	Fleshy fungi with radiating gills or pores.
Family	Agaricaceae	Agaricales with gills.
Genus	*Agaricus*	Dark-spored soft fungi with a central stalk and gills free from the stalk.
Species	*Agaricus campestris*	The common edible mushroom.

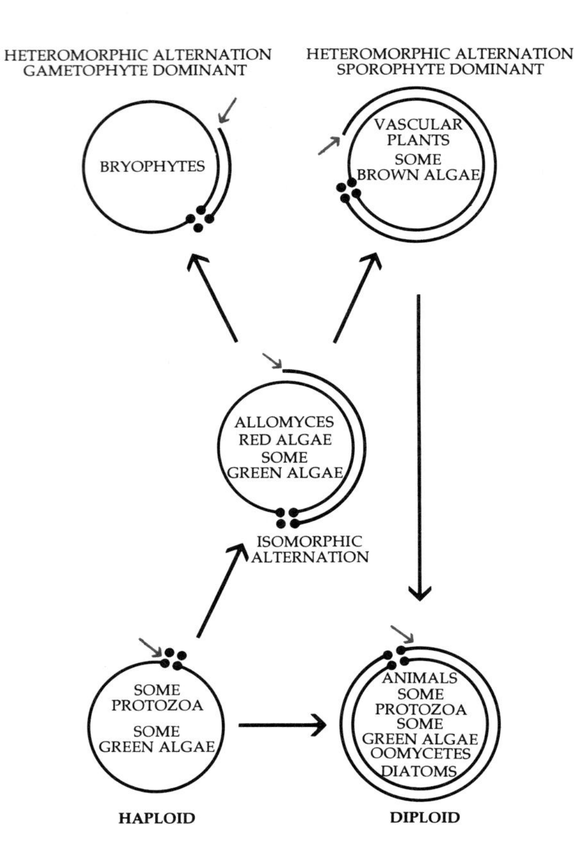

10–9
The evolution of genetic systems. Each of these circles represents a different type of life cycle. The most primitive eukaryotes were undoubtedly haploid for most of their life cycle, as indicated by the single circle (lower left). In this type of life cycle, meiosis (indicated by four black spheres) occurs right after fertilization (designated by the small arrow in color). The other life cycles differ from the haploid one in the point at which fertilization takes place, as well as in the proportion of the life cycle spent in the diploid state (indicated by the extent of the outer circle). The groups named within the circles refer to some of the modern organisms with these particular types of life cycles; not all groups are included here.

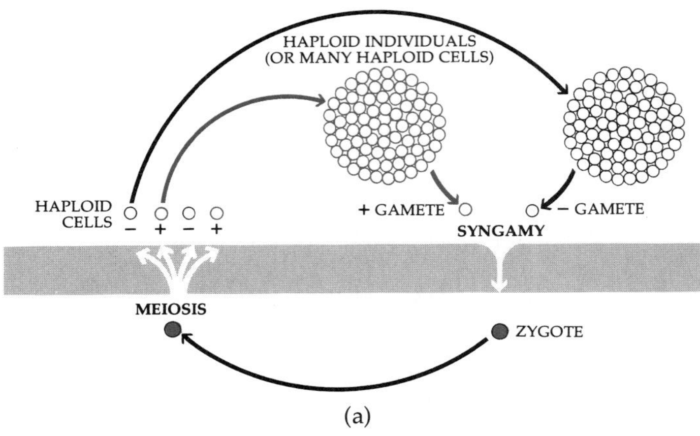

10–10
Diagrams of the principal types of life cycles. In these diagrams, the diploid phase of the cycle takes place below the broad bar, and the haploid phase occurs above it. The four white arrows signify the products of meiosis; the single white arrow represents the fertilized egg.

(a) *In zygotic meiosis, the zygote divides by meiosis to form four haploid cells that divide by mitosis to produce more haploid cells or a multicellular individual that eventually gives rise to gametes by differentiation. This type of life cycle is found in* Chlamydomonas *and a number of other algae.*

(b) *In gametic meiosis, the haploid gametes are formed by meiosis in a diploid individual and fuse to form a diploid zygote that divides to produce another diploid individual. This type of life cycle is characteristic of most animals and some protista (Oomycota). It is also the type of life cycle exhibited by the brown alga* Fucus.

(c) *In sporic meiosis, the sporophyte, or diploid individual, produces haploid spores as a result of meiosis. These spores do not function as gametes but undergo mitotic division. This gives rise to multicellular haploid individuals (gametophytes) that eventually produce gametes that fuse to form diploid zygotes. These zygotes in turn differentiate into diploid individuals. Such a life cycle, known as alternation of generations, is characteristic of the plants and many algae. A similar sort of life cycle is found in the* Allomyces *and in one closely related genus of Chytridiomycota.*

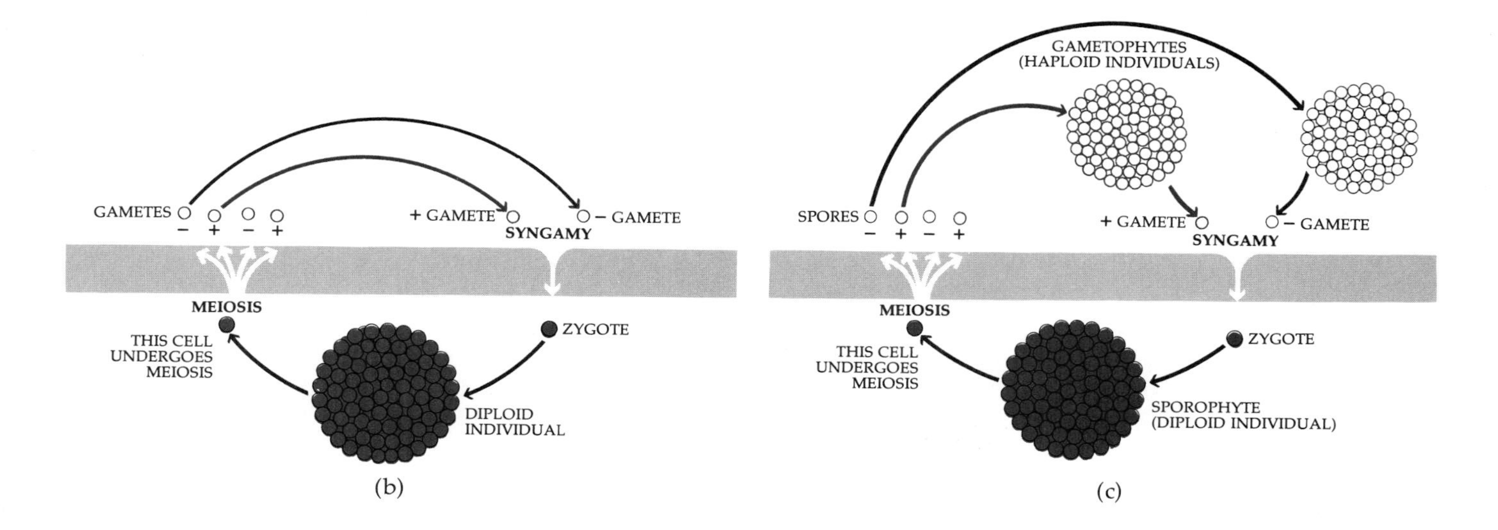

By "accident"—an accident that occurred in a number of separate evolutionary lines—the zygote divided mitotically instead of meiotically and, as a consequence, produced an organism composed of a number of diploid cells. In a few groups of organisms, the haploid stage was lost completely, as in *Euglena* and most of the amoebas. Sexuality is absent in all of these organisms. In the majority of organisms, however, meiosis was not completely suppressed; it was simply delayed. In the animals, this delayed meiosis results in the production of gametes—eggs and sperm. These gametes then fuse, which immediately restores the diploid state (see Figures 10–9 and 10–10b). Therefore, in animals, gametes are the only haploid cells.

In plants, meiosis results in the production of spores, not gametes. Spores are cells that can divide directly by mitosis to produce a multicellular haploid organism; this is in contrast to gametes, which can develop only following fusion with another gamete. Multicellular haploid organisms, which appear in alternation with diploid forms, are found in plants, in algae, and in two closely related genera of chytrids, a group of protista. Such organisms are said to show the phenomenon known as *alternation of generations* (see Figures 10–9 and 10–10c). Among the plants, the haploid, gamete-producing generation is called the *gametophyte*, and the diploid, spore-producing generation is called the *sporophyte*.

In some modern algae—most of the red algae, many of the green algae, a few of the brown algae—the haploid and diploid forms are the same in external appearance. Such types of life cycles are said to exhibit *isomorphic* alternation of generations (Figure 10–9).

In certain groups, however, mutations occurred which were expressed in only one generation although they were present in both the haploid and diploid generations. In this type of life cycle, the gametophyte and sporophyte became notably different from one another, and *heteromorphic* alternation of generations originated. Such life cycles are characteristic of plants and some brown algae (Figure 10–9). In some algae, the gametophyte and sporophyte, although markedly different, are equally large and complex. Among the bryophytes (mosses, hornworts, and liverworts), the gametophyte is dominant—it is nutritionally independent from and usually larger than the sporophyte. Among the vascular plants, the sporophyte dominates—it is much larger and more complex than the gametophyte, which is nutritionally dependent on the sporophyte in nearly all groups.

As mentioned previously, diploidy permits the storage of more genetic information and so perhaps allows a more subtle expression of the organism's genetic background in the course of development. This may be the reason that the sporophyte is the large, complex, and nutritionally independent generation in vascular plants. Among the bryophytes, the gametophyte is nutritionally independent, even though the sporophyte is structurally more complex.

In the vascular plants, which are now dominant on land, one of the clearest evolutionary trends is the increasing dominance of the sporophyte and suppression of the gametophyte. Among the flowering plants, the female gametophyte is a microscopic body that consists of only seven cells and the male gametophyte consists of only three cells. Both of these gametophytes are completely dependent upon the sporophyte.

10–11

One possible scheme of the evolutionary relationships among organisms. The solid lines indicate phylogenetic relationships and the dashed lines indicate the establishment of symbiotic relationships. All eukaryotic organisms were derived from a single line of cells containing mitochondria which in turn had their derivation in symbiotic bacteria. From this single line of cells the diverse assemblage of one-celled organisms known as the Protista developed. Those members of the Protista that developed symbiotic relationships with photosynthetic prokaryotes gave rise to the several different lines of modern algae. From particular single-celled protista, the fungi, plants, and animals evolved.

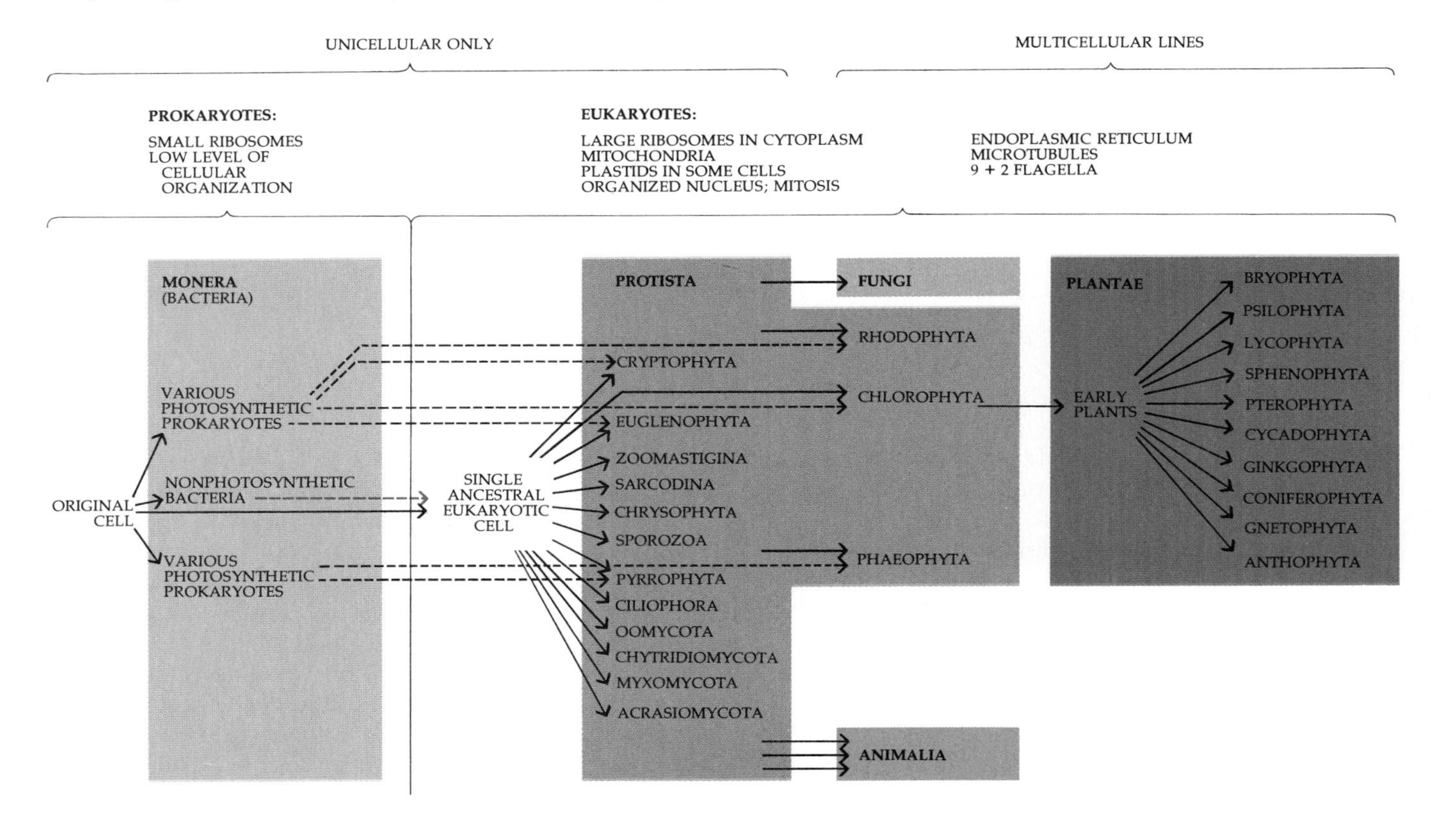

SUMMARY

Biologists have developed methods of naming and classifying living things that permit them to designate very precisely the organism with which they are working, an essential factor in scientific communication. The classification of an organism also reveals its relationship to other living things.

Organisms are designated scientifically by two words—a binomial. The first word in the binomial is the name of the genus (plural: genera), and the second word designates the species. The entire binomial is the species name. Species are sometimes subdivided into subspecies or varieties. Genera are grouped into families, families into orders, orders into classes, and classes into divisions. Divisions are grouped into kingdoms, the kingdom being the largest unit used in classification of the living world. In this text, living organisms are grouped into five kingdoms: (1) Monera, which includes all the prokaryotes (bacteria); (2) Protista, which includes the protozoa, eukaryotic algae, slime molds, and water molds; (3) Animalia, which includes multicellular organisms that are not photosynthetic; (4) Fungi, the fungi; and (5) Plantae, which includes the bryophytes and vascular plants, those photosynthesizing organisms that are more complex than the algae.

In the evolution of organisms, diploidy evolved subsequent to the process of sexual reproduction. In primitive eukaryotes and all fungi, the zygote formed by syngamy divides immediately by meiosis. From early cycles of this sort, more modern life cycles involving diploid phases were derived on a number of occasions, when the zygote divided by mitosis. If the haploid cells produced by meiosis function immediately as gametes, the result is the type of life cycle found in animals and in some groups of protista. If they divide by mitosis, as in many algae, all plants, and two genera of chytrids (Protista), they are considered spores; the diploid generation that gave rise to such spores is called the sporophyte. The haploid generation to which they give rise is called the gametophyte, which eventually produces gametes by mitosis. If the gametophyte and sporophyte in a particular life cycle are approximately equal in size and complexity, the alternation of generations is said to be isomorphic; if they differ widely in size and complexity, the alternation of generations is considered to be heteromorphic.

SUGGESTIONS FOR FURTHER READING

BELL, PETER R., and C. L. WOODCOCK: *The Diversity of Green Plants,* 2nd ed., Addison-Wesley Publishing Co., Inc., Reading, Mass., 1972.*

A concise modern survey of the diversity of green plants.

BOLD, HAROLD G., C. S. ALEXOPOULOS, and T. DELEVORYAS: *Morphology of Plants and Fungi,* 4th ed., Harper & Row, Publishers, Inc., New York, 1980.

A well-illustrated and ample treatment of the diversity of plants, algae, and fungi.

CORNER, E.J.H.: *The Life of Plants,* Mentor Books, New American Library, Inc., New York, 1968.*

A renowned botanist with a flair for poetic prose describes the evolution of plant life, telling how plants modified their structures and functions to meet the challenge of a new environment as they invaded the shore and spread across the land.

DELEVORYAS, THEODORE: *Plant Diversification,* 2nd ed., Holt, Rinehart & Winston, Inc., New York, 1977.*

This brief volume stresses evolutionary trends in presenting a readable account of the groups traditionally considered to be plants.

MARGULIS, LYNN: *Symbiosis in Cell Evolution,* W. H. Freeman and Company, San Francisco, 1980.

A fascinating discourse on the origin of eukaryotic cells by serial symbiotic events.

ROSS, H. H.: *Biological Systematics,* Addison-Wesley Publishing Co., Inc., Reading, Mass., 1974.

A nicely balanced brief overview of taxonomy, both plant and animal.

SCAGEL, ROBERT F., et al.: *Plant Diversity: An Evolutionary Approach,* Wadsworth Publishing Co., Inc., Belmont, Calif., 1969.

An exhaustive, thoroughly illustrated review of plant diversity.

*Available in paperback.

CHAPTER 11

The Prokaryotes

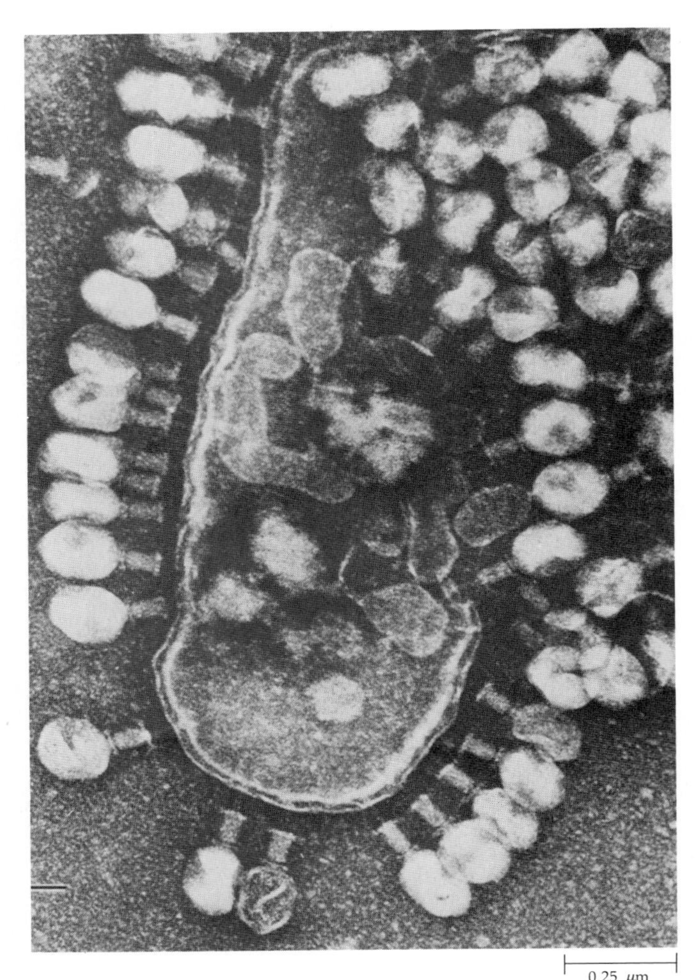

11–1
Bacterial viruses (T2) infecting a cell of Escherichia coli. *Some of the viruses have discharged their DNA into the bacterial cell, which has begun to lyse, or break down.*

The living prokaryotes, or bacteria, are the simplest of all known organisms; they bear the closest resemblance of any living organism to the earliest forms of life on earth. Collectively, they constitute the kingdom Monera. One group of bacteria, the so-called "blue-green algae," or cyanobacteria, were previously (and incorrectly) associated with the group of eukaryotic organisms known as algae, but it has now been established that they are prokaryotes. Viruses, which once were thought to be exceedingly small bacteria but actually do not fit easily into any of the traditional categories of classification, will also be considered in this chapter.

All prokaryotes lack an organized nucleus surrounded by a nuclear envelope. They do not have complex chromosomes like those of eukaryotic organisms, and they do not reproduce sexually, although some groups possess mechanisms that can lead to genetic recombination.

No prokaryote is truly multicellular. Although some bacteria, including most cyanobacteria, form filaments or masses of cells, these are connected only because their cell walls fail to separate completely following cell division or because they are held together within a common mucilaginous capsule or sheath. There are no protoplasmic connections, such as plasmodesmata, between the individual cells in such assemblages. "True" multicellularity, like sexual reproduction, occurs only in eukaryotic organisms.

Prokaryotic organisms lack cellular organelles in their cytoplasm, but they have other structures that play similar roles. Their plasma membranes often have many folds and convolutions extending into the interior of the cell. Such membranes increase the surface area to which enzymes are bound, and they facilitate the separation of different enzymatic functions. Some photosynthetic bacteria bear their photosynthetic pigments in internal membranes. In others, the chlorophyll is located in discrete spherical bodies called chromatophores.

GENERAL CHARACTERISTICS OF BACTERIA

Bacteria are the smallest and most abundant organisms in the world. Most are only about 1 micrometer in diameter, with some being only one-tenth of that size, while others range up to 10 micrometers (or rarely, even 30 micrometers or more) in length. Even though individual bacteria are so small, the total weight of all the bacteria in the world is estimated to exceed that of all the other living organisms combined. About 2500 species of bacteria are currently recognized.

As a group, bacteria are the most ancient of all organisms. They have been found as far back as 3.5 billion years in the fossil record (Figure 11–2), far beyond any known fossil eukaryotes. Until quite recently, the oldest known fossils were approximately 0.6 billion years old. Our knowledge of the origins of life has dramatically extended back in time because of developments in several seemingly unrelated fields. Radiometric dating methods, comparing the proportions of isotopes in minerals found in rocks, have made it possible to determine accurately the age of very old sediments. Sophisticated methods can now be used for the detection of organic molecules in layers of rock that have been identified as being very old. Finally, the electron microscope has been used to find very small fossils in ultrathin sections of ancient rocks. Data collected by these methods suggest that life was in existence at least 3.5 billion years ago.

Bacteria occur in all habitats and, largely because of their great metabolic diversity, can survive in many environments that support no other form of life. Some bacteria are *obligate anaerobes*, that is, they live only in the absence of air, whereas others are *facultative anaerobes* and can survive without oxygen but grow more vigorously if it is supplied. Aerobic respiration yields much more energy than anaerobic respiration.

Bacteria have been found in the icy expanses of Antarctica, the boiling waters of natural hot springs, and even in the dark depths of the ocean. The bacterium *Thermoanaerobacter ethanolicus*, living in hot springs at Yellowstone National Park, can thrive at temperatures as high as 78°C. Lars Ljungdahl of the University of Georgia has proposed that, since *T. ethanolicus* produces ethanol at much higher temperatures than yeasts, it might eventually displace the yeasts presently used to produce ethanol in certain industrial processes. Living bacteria have been found in samples of rock and ice recovered by drilling from depths of up to 430 meters in Antarctica. These bacteria, which are at least 10,000 and possibly a million years old, were lying dormant at temperatures ranging from about −7° to −14°C but immediately re-

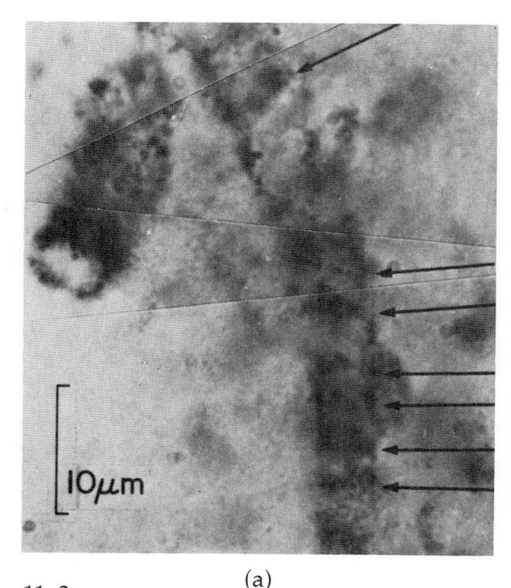

(a)

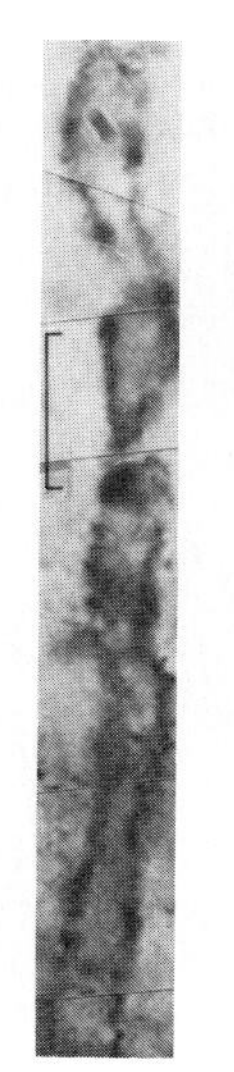

(b)

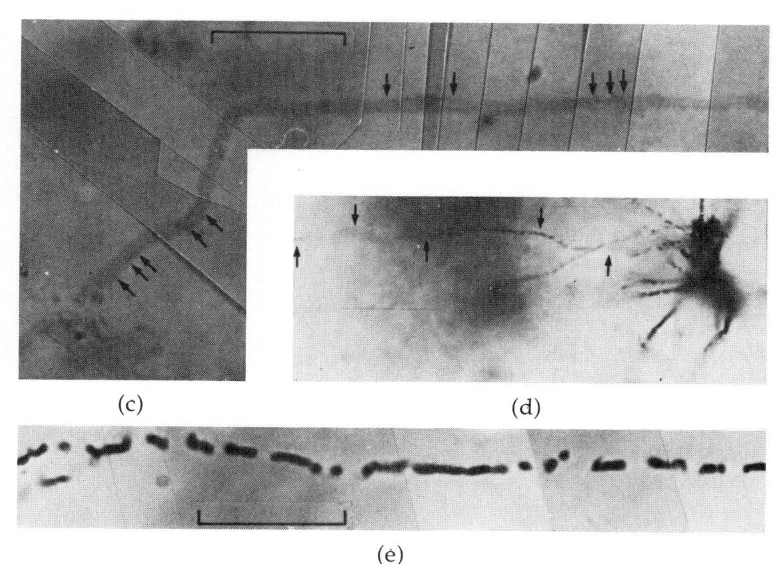

(c) (d) (e)

11–2
Chainlike bacteria from ancient rocks in Australia. Dated at 3.5 billion years of age, these are the oldest fossil organisms presently known. The discovery of these fossils in carbonaceous black chert from Western Australia was announced in June 1980 by paleobiologist J. William Schopf of the University of California, Los Angeles. (a) *A septate microbial filament (the arrows indicate cross walls);* (b) *an empty tubular sheath that at one time presumably contained a bacterial filament;* (c) *a long, very narrow bacterial filament (the arrows point to possible cell walls);* (d) *a rosettelike microbial colony (the arrows point out a single long bacterial filament); and* (e) *a single long filament, such as that indicated in part* (d), *at a high magnification. These fossils, which are about three-quarters as old as the earth itself, demonstrate that an unexpected complexity of bacterial forms had evolved even in those ancient times.*

(a) 50 μm

(c) 200 μm

(d) 10 μm

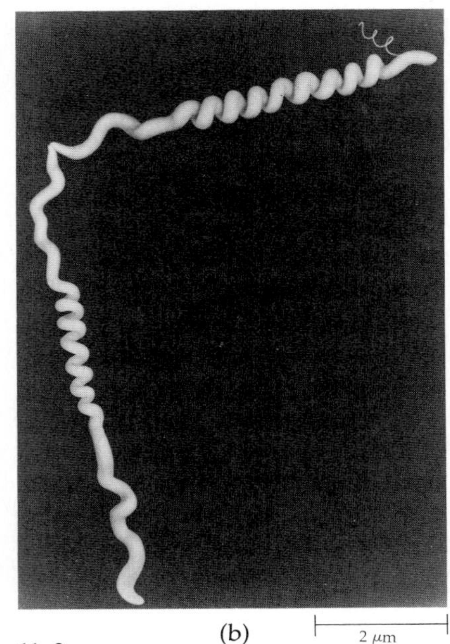

(b) 2 μm

11–3
Some of the diverse types of bacteria. (a) *The myxobacteria have a pattern of organization similar to that of the slime molds (see Chapter 13). This is a scanning electron micrograph of the "fruiting bodies" of* Chondromyces crocatus, *each consisting of as many as 1 million cells.* (b) *The spirochaetes are spiral bacteria that are up to 500 micrometers long, which is an enormous size for bacteria. Their motility is by means of a helical wave moving along the body of the cell, or by a snakelike motion.* Treponema pallidum, *shown here, is the causative agent of syphilis.* (c) *Actinomycetes are abundant in soil, where they are largely responsible for the "moldy" odor of decaying organic material.* Streptomyces fradiae, *shown here, is the commercial source of the antibiotic neomycin. Actinomycetes are the organisms responsible for nitrogen fixation in plants other than legumes. They form symbiotic associations with the roots of many genera.* (d) *Gliding bacteria consist of filamentous forms with relatively large cells in which waves of contraction cause periodic alterations in the form of the cells and are responsible for their "gliding" movements. In* Beggiatoa, *shown here, the conspicuous granules in the cells are sulfur, produced by the oxidation of hydrogen sulfide—a process utilized by the bacteria for the production of energy.*

sumed normal activity when their temperature was elevated. These findings show that bacteria are capable of surviving in a state of "suspended animation" for extremely long periods of time.

Recently, the limits of tolerance of certain bacteria have been investigated in relation to the possible existence of extraterrestrial life on various planets with known atmospheres. For example, the extreme alkalinity of Jupiter's atmosphere has been considered one of the arguments against the chance of life there; however, bacteria from the Livermore Valley of California have been found in water with an alkalinity as high as pH 11.5, and they grow and reproduce in solutions of sodium hydroxide that are equally basic. Other bacteria can tolerate not only extreme alkalinity but also the presence of high concentrations of ammonia.

Bacteria are most important, however, for the role they play in the world ecosystem. Many are autotrophic and so make major contributions to the world's carbon balance. Some of these bacteria are being grown experimentally as possible commercial sources of protein. The heterotrophic bacteria, like the fungi, are decomposers. Through the action of the decomposers, materials incorporated into the bodies of once-living organisms are released and made available for successive generations of living organisms. In a single gram of fertile agricultural soil, there may be 2.5 billion bacteria, 400,000 fungi, 50,000 algae, and 30,000 protozoa. The role of bacteria in fixing atmospheric nitrogen is of crucial ecological importance.

In addition to their ecological role, bacteria are important in a number of other ways. They are responsible for many of the most serious diseases of humans and other animals—including tuberculosis, cholera, anthrax, gonorrhea, diphtheria, and tetanus—and cause a wide range of economically significant diseases of plants. More than 200 species of bacteria are currently recognized as plant pathogens in the United States alone. Fire blight of pears and apples, for example, destroyed thousands of pear trees as it spread across the United States. By the 1930s, some 50 years after its introduction on the Eastern seaboard, this disease virtually wiped out the pear industry. Currently, the commercial production of pears in the United States is restricted to a small area in the Northwest.

Bacteria are used as the commercial source of a number of important antibiotics, such as tyrothricin, bacitracin, subtilin, and polymyxin B. Many bacteria are used experimentally and, in some instances, commercially in the production of drugs and other molecules (see "Using Bacteria to Manufacture Useful Complex Molecules," page 194).

One group of bacteria, the actinomycetes, is especially important in the production of antibiotics such as streptomycin, aureomycin, neomycin, and terramycin. Other actinomycetes fix nitrogen in nodules formed on the roots of certain plants and so play a major ecological role as well. Almost all cheeses are produced as a result of bacterial fermentation of lactose into lactic acid, which coagulates milk proteins. In a similar fashion, bacteria are used commercially in the production of acetic acid and vinegar, various amino acids, and enzymes.

Bacterial Form

Bacteria vary greatly in form and cellular organization (see Figure 11–3). Among the simpler bacteria, the straight, rod-shaped ones are known as *bacilli*, spherical ones are called *cocci*, and long coiled ones are called *spirilli* (Figure 11–4). Spherical bacteria may stick together in pairs after division (diplococci), they may occur in clusters (staphylococci), or they may form chains (streptococci). The organism that causes pneumonia is a diplococcus. The staphylococci are responsible for many serious infections characterized by boils or abscesses.

Rod-shaped bacilli are more commonly found alone than are the spherical cocci. When they do remain

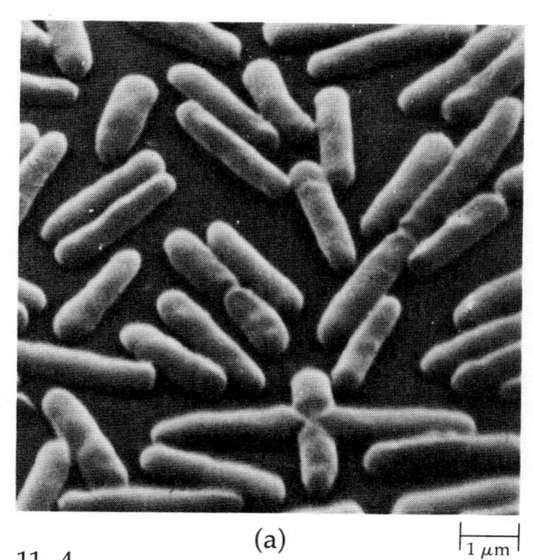

(a) 1 μm

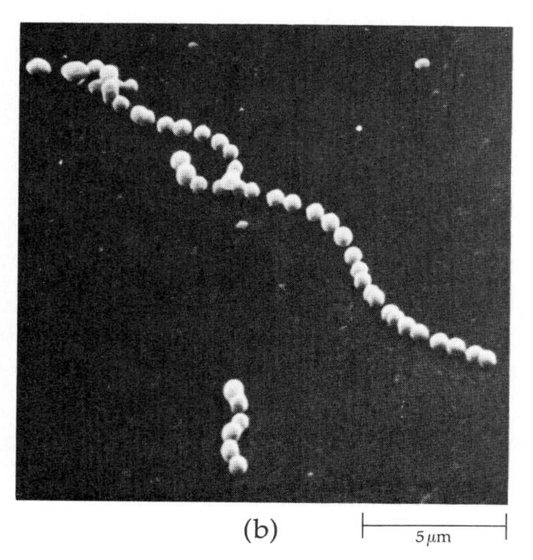

(b) 5 μm

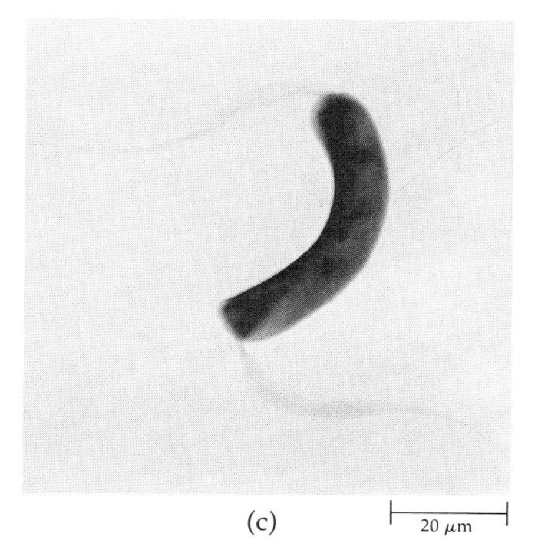

(c) 20 μm

11–4
The three major forms of bacteria: (a) *bacilli;* (b) *cocci; and* (c) *spirilli. Cell shape is a relatively constant feature in most species of bacteria. The rod-shaped bacteria include those microorganisms that cause lockjaw* (Clostridium tetani), *as well as the familiar* Escherichia coli. *The bacilli also cause many plant diseases, including fire blight of pears and apples* (Erwinia amylovora) *and bacterial wilt of tomatoes, potatoes, and bananas* (Pseudomonas solanacearum). *Among the cocci are* Diplococcus pneumoniae, *the cause of bacterial pneumonia;* Streptococcus lactis, *a common milk-souring agent; and* Nitrosococcus nitrosus, *soil bacteria that oxidize ammonia to nitrites. The spirilli, which are less common, are helically coiled bacteria.*

11–5
Four common genera of cyanobacteria: (a) Oscillatoria, *in which the only form of reproduction is by means of fragmentation of the filament.* (b) Gloeotrichia, *a filamentous form with a basal heterocyst (see page 204).* Gloeotrichia *is capable of forming akinetes—thick-walled cells resistant to environmental changes—just above the heterocysts.* (c) *Gelatinous "balls" of* Nostoc commune, *each containing hundreds of filaments, occur frequently in freshwater habitats.* (d) Thiothrix, *a genus that lacks chlorophyll, obtains energy by the oxidation of H_2S. The bacterial filaments, filled with sulfur droplets, are attached to the substrate at the base (center) and so form a characteristic rosette.*

together, they spread out end to end, in filaments, because they always divide transversely. Because these filaments are funguslike in appearance, the combining form *myco-* (from the Greek word for fungus) is often a part of the name of these organisms and is a clue to their appearance. The actinomycete *Mycobacterium tuberculosis,* for example, is a rod-shaped bacterium that forms a filamentous funguslike growth in culture, although not in the host.

Cyanobacteria often form filaments and may grow in large masses up to 1 meter or more in length. Some are unicellular, a few form branched filaments, and a very few form plates or irregular colonies (Figure 11–5). Any cell of a cyanobacterium may divide, and the resulting subunits separate to form new colonies. As in other filamentous or colonial bacteria, the cells are attached to one another only by their outer walls or by gelatinous sheaths, so that each cell actually leads an independent life.

Although their cells lack cilia, flagella, or any other type of locomotive organelle, some filamentous cyanobacteria are capable of motion. This motion may consist of gliding, as in some other groups of bacteria (Figure 11–3d), or may be combined with rotation around a longitudinal axis. Short segments that break off from the parent may glide away to a new site at rates as rapid as 10 micrometers per second. Movement may be connected with the extrusion of mucilage through small pores in the cell wall, together with the production of contractile waves in one of the surface layers of the wall. Some cyanobacteria show intermittent jerky movements.

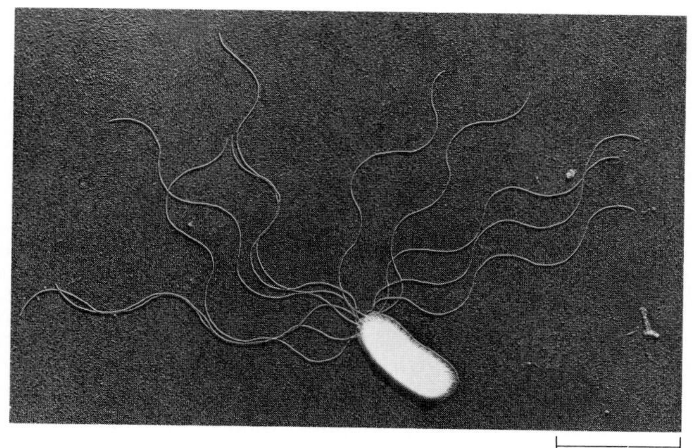

11–6
Flagella on a cell of Pseudomonas marginalis, *a common bacterium that is widespread in soils. It causes the soft-rot disease found in many fleshy and leafy vegetables.*

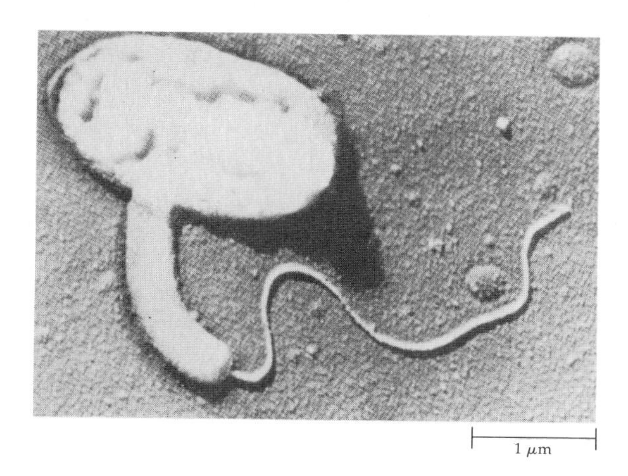

11–7
A cell of Bdellovibrio bacteriovorus, *which parasitizes other bacteria. These bacteria, which are abundant in soil and sewage, move by means of a single posterior flagellum. The parasitic cell shown here is attacking a single rod-shaped cell of* Erwinia amylovora *(top), the bacterium that causes fire blight of pears and apples.*

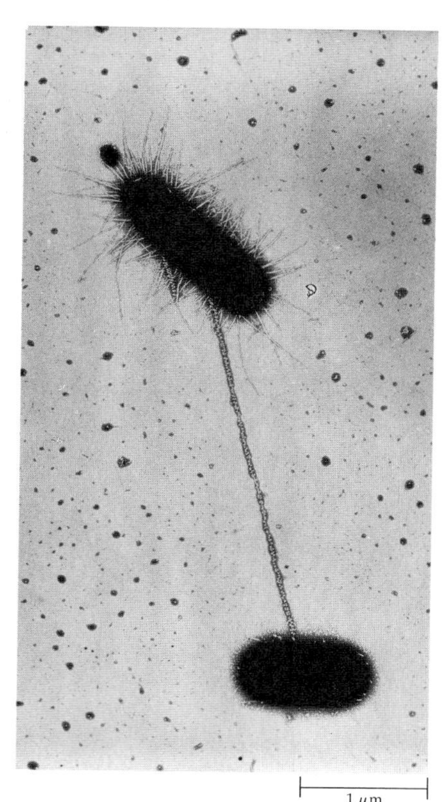

11–8
Conjugating cells of Escherichia coli. *The DNA is presumably passing from one cell to the other through a special hollow pilus, which acts as a conjugation bridge. Many ordinary pili can be seen on the upper bacterial cell.*

Flagella and Pili

Certain kinds of bacteria have very slender, rigid, helical flagella that they rotate in swimming from place to place. These flagella are relatively simple in structure and do not resemble, except in general shape, the 9-plus-2 flagella found in eukaryotes.

Bacterial flagella are long (3 to 12 micrometers), slender, and wavy. Because they are only 10 to 20 nanometers in diameter, they are usually too fine to be seen by ordinary microscopic techniques. In some bacteria, they are well distributed over the cell, and in others, they are restricted to one or both ends of the cell (Figures 11–6 and 11–7). Bacterial flagella seem to be composed entirely of a single special type of protein known as flagellin.

Pili are shorter (up to several micrometers) and straighter than flagella and are only about 7.5 to 10 nanometers in diameter (Figure 11–8). Like flagella, they are formed from bodies in the cytoplasm, and they consist entirely of protein, although their proteins are different from those of the flagella and their structure is distinct. Special hollow pili are formed on bacterial cells during conjugation; their exact function is not known (Figure 11–8). They may serve as a bridge for the transfer of DNA; they may pull the conjugating cells together; or they may have some other, unknown function. Ordinary pili evidently help bacteria to find and attach themselves to appropriate surface membranes. A better understanding of the function of bacterial pili may aid in combating bacterial diseases.

Cell Wall

The bacterial cell wall is chemically complex. Unlike the cell walls found in plants, bacterial walls lack cellulose. Instead, there is a network of molecules of another polysaccharide connected by polypeptide cross-links. In many bacteria, this network, called the peptidoglycan layer, makes up the basic structure of the cell wall. The walls in these bacteria are 15 to 80 nanometers thick. In other bacteria, large molecules of lipopolysaccharide—a polysaccharide chain with lipids attached to it—are deposited over the peptidoglycan layer. Such bacteria have cell walls that are only about 10 nanometers thick.

A Danish microbiologist, Hans Christian Gram, discovered that bacterial cell walls that lack the lipopolysaccharide layer retain a purple stain (crystal violet). This staining technique was first used to detect the presence of bacteria in animal tissues: those bacteria that pick up that stain are known as gram-positive, whereas the others are gram-negative. This fundamental difference between the two groups affects other characteristics of bacteria, such as their patterns of resistance to antibiotics. Actinomycin, for

---

SENSORY RESPONSES IN BACTERIA

Bacterial cells, despite their relative simplicity, are able to respond to stimuli, as evidenced by their capacity to move in the direction of increasing concentrations of oxygen or food molecules. Motile bacteria, including Escherichia coli, *are able to move toward or away from chemical stimuli by means of the rotation of their flagella. Bacteria have chemosensors for nutritious substances, such as sugars. They also have chemosensors for molecules that repel them; they do not simply sense the repellents by experiencing some harm that they cause. Julius Adler and his research group at the University of Wisconsin have calculated that* E. coli *has some 20 different chemosensors, about 12 for attractants and 8 for repellents. The proteins (chemoreceptors) responsible for binding the sugars galactose, maltose, and ribose are found in the periplasmic space between the cell wall and the plasma membrane. Other chemosensors—also proteins—have been shown to be closely associated with the plasma membrane. It is not known how the information about what part of a specific field of chemoreceptors is occupied by an attractant or a repellent is transmitted to the flagella, nor is it known how the bacterium integrates information from various kinds of chemoreceptors in producing a certain type of motion.*

Individuals of Escherichia coli *move rapidly in gently curved lines, or "runs," each lasting approximately a second. A series of runs is interrupted frequently by periods of tumbling lasting about a tenth of a second. After each tumble, the bacterium starts off in a new direction on its next run. The frequency of the tumbling controls the length of the runs and hence the overall direction of movement. If the bacterium is swimming in the "right" direction, tumbling is inhibited; if it is swimming in the "wrong" direction, the frequency of tumbling is increased. The end result is that the bacteria migrate toward the source of an attractant and are dispersed away from the source of a repellent. The presence of an attractant causes the flagella to rotate in a counterclockwise direction; the addition of a repellent causes a clockwise rotation. Thus sustained swimming is brought about by counterclockwise rotation of the flagella, and tumbling is brought about by clockwise rotation. In sustained swimming, the flagella work together in an organized bundle at the rear of the organism, even though they originate at sites all over the cell's surface. When the flagella rotate in a clockwise direction, the bundle falls apart and the cell tumbles. Under such circumstances, the flagella rotate independently rather than in coordination.*

Specific structural adaptations that aid in the sensory responses of bacteria have been discovered. For instance, Halobacterium halobium, *one of the nonsulfur photosynthetic bacteria, is often found in sunny places in concentrated brine. These bacteria have patches of a purple pigment in their plasma membrane that are analogous to the visual pigment of the human eye. These patches are in effect light-driven pumps which enable the bacterium to pump out protons. This process seems to lead directly to ATP production by means of ATPase located in the nonpurple areas of the plasma membrane. It is thus the only living unit other than chlorophyll-containing systems that is able to convert sunlight into chemical energy through the process of photosynthesis. The potential role of this organism in harnessing solar energy is obvious. It is mobile and responds phototactically to violet light, suggesting that its pigment detects light as does the analogous visual pigment.*

Recently, a species of spirillum from a swamp near Woods Hole, Massachusetts, has been shown to be able to orient and swim in relationship to a magnetic field. According to studies by Richard Blakemore of the University of New Hampshire, this bacterium accumulates about 20 opaque, roughly cubic granules of magnetite Fe_2O_3 *from its iron-rich environment, thus constructing what amounts to a magnet. It is apparently by virtue of this "magnet" that the bacterium is able to respond to the magnetic field of the earth. The bacterial flagella are opposite the north-seeking ends of their internal magnets. The ability to swim north (in the Northern Hemisphere) provides the additional ability to swim "down" toward the earth and richer sources of food rather than moving about randomly.*

These examples illustrate a series of remarkably complex responses for such small and simply constructed organisms. They foreshadow the kinds of increasingly complex responses found among eukaryotes.

example, is a relatively large antibiotic molecule that apparently cannot pass through the cell wall in gram-negative bacteria. In gram-positive bacteria, it passes through the cell wall and disrupts protein synthesis by binding to the DNA double helix. This chemical activity also makes it toxic to the host cells and thus limits its clinical use.

The adhesive properties of individual bacterial cells are determined by a layer of tangled fibers of polysaccharides called the *glycocalyx.* Although often absent in laboratory cultures of bacteria, in nature the glycocalyx plays a key role in the initiation and progression of bacterial infection by mediating the adhesion of the bacteria to different substrates. By means of enzymatic reactions that occur in the glycocalyx, certain bacteria are able to colonize the intact enamel surfaces of teeth, as well as many other seemingly resistant substrates.

Outside the bacterial cell wall, there is often found a gelatinous layer—the capsule—which is apparently secreted by the bacterial protoplast through the wall (Figure 11–9).

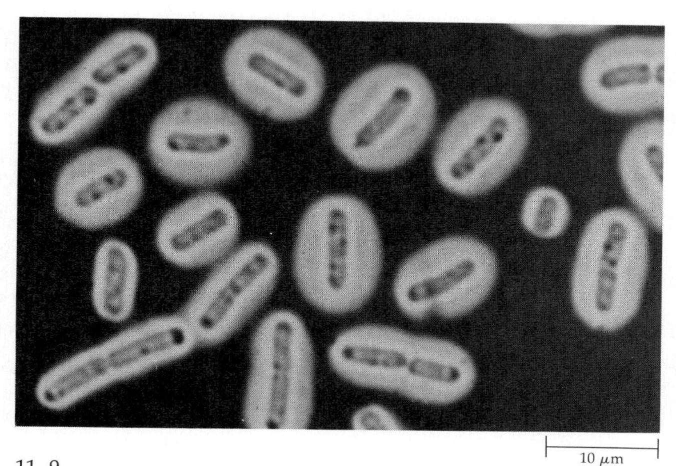

11–9
A photomicrograph of Bacillus megaterium *dispersed in India ink. Capsules stand out as translucent halos.*

Cytoplasm

As in all cells, the bacterial cytoplasm is bounded by a plasma membrane. Within this membrane, which has many enzymes localized on its inner surface, is cytoplasm with a number of ribosomes and granular inclusions, as well as one or more bodies of chromatin (Figure 11–10). Bacterial cells usually contain at least two chromatin bodies, because cell division lags behind the division of the genetic material. Each of the bodies consists of a closed loop of double-stranded DNA about 700 to 1000 times as long as the bacterial cell in which it occurs.

The cyanobacteria that occur in freshwater or marine habitats commonly contain bright, irregularly shaped structures known as gas vacuoles. These vacuoles provide and regulate the buoyancy of the organisms, thus allowing them to float at proper levels in the water. When numerous organisms become unable to regulate their gas vacuoles properly, they float to the surface of the water and form visible masses called "blooms."

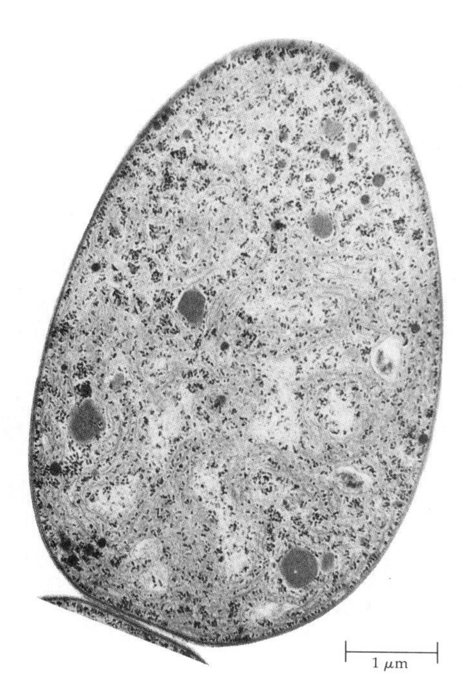

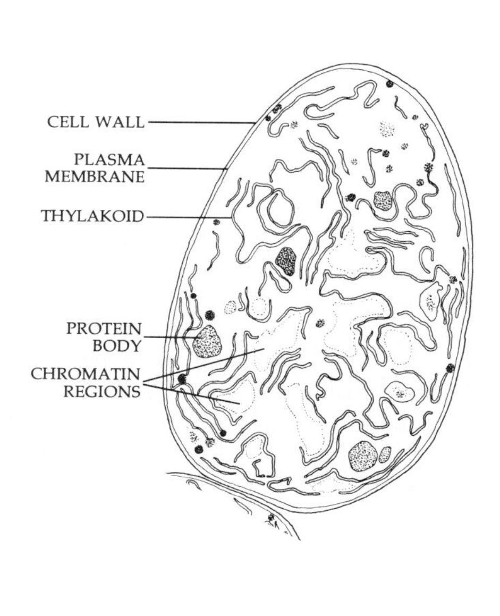

11–10
A cell of the cyanobacterium Anabaena azollae, *showing the major features visible with the electron microscope. The gelatinous sheath of this cell has been destroyed in preparing the specimen for electron microscopy. This organism is associated with the common floating water fern* Azolla *and is responsible for nitrogen fixation in it (see Chapter 27).*

11–11
Cell division in Anabaena. *This electron micrograph shows a chain of cells held together along incompletely separated walls, as well as a single cell undergoing cell division. The first cell on the left end of the chain is a heterocyst (see page 204). The sort of cell division shown here, in which the margins of the cell grow inward, is characteristic of all organisms, except plants and a few genera of algae, in which a cell plate is formed.*

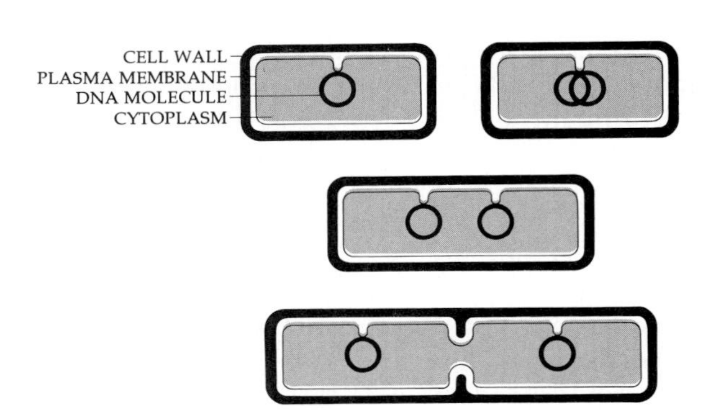

11–12
A schematic diagram of attachment of bacterial DNA molecules to the plasma membrane, which leads to the distribution of one DNA molecule to each daughter cell. Such attachment serves the same function as that of the spindle fibers during mitosis in eukaryotes.

Bacterial Genetics

Cell Division

The chief mode of reproduction in the bacteria is asexual; each cell simply increases in size and splits into two cells. In the process of fission, the plasma membrane and cell wall grow inward and eventually divide the cell in two. The new wall is thicker than ordinary cell walls, and it soon splits from the outside toward the center of the old cell, separating the two daughter cells (Figure 11–11). Chains of bacteria formed when this new cell wall splits incompletely may break into multicellular fragments, which, in cyanobacteria, are called *hormogonia.*

In many bacteria, the DNA molecules, following replication and separation, become attached to structures extending inward from the plasma membrane. These structures play a role in cell division comparable to that of the spindle fibers in mitosis (Figure 11–12).

Some bacteria have the ability to form thick-walled spores that are resistant to heat and dehydration. Bacterial spores may germinate to produce new individuals after decades or even centuries. In cyanobacteria, these spores are called *akinetes* (Figure 11–13).

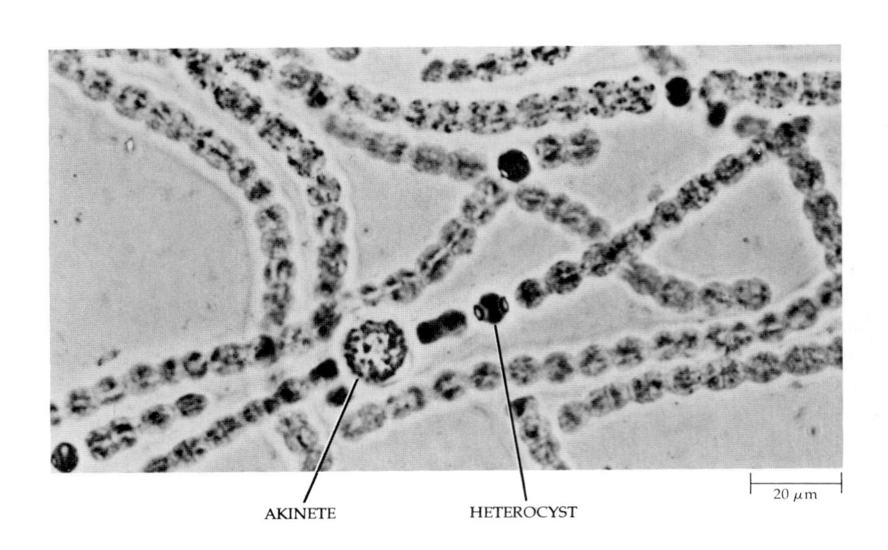

11–13
Filaments of Anabaena, *which is a nitrogen-fixing cyanobacterium, are composed of barrel-shaped cells held in a gelatinous matrix. Nitrogen fixation takes place within specialized cells called heterocysts. Like* Gloeotrichia *(see Figure 11–5b),* Anabaena *forms akinetes (spores). Electron micrographs of* Anabaena *also can be seen in Figures 11–10 and 11–11.*

Genetic Recombination

Although complex sexual reproduction that involves alternating meiosis and syngamy is generally limited to eukaryotes, genetic recombination does occur in many kinds of bacteria. This genetic recombination involves the transfer of a portion of a DNA molecule from one bacterial cell to another. This DNA fragment may simply act in concert with the DNA molecule of the cell it enters, or with both cells, to produce messenger RNA, or it may actually be incorporated into the DNA molecule of the recipient cell to be passed on to the daughter cells with the rest of the hereditary material. The process of recombination is the same whether the fragments are passed from cell to cell by direct contact, are carried into a cell by a virus, or enter cells in solution as "naked" DNA.

In the cells of many bacteria, in addition to the large circular molecule of DNA (sometimes called the bacterial "chromosome"), there are relatively small fragments of DNA that usually form closed circles. Some of these fragments, known as *plasmids*, can be integrated into and then replicated with the bacterial chromosome. In some bacterial strains, conjugation and the transmission of such fragments occurs frequently (Figure 11–8). Recombination is important in the spread of such characteristics as resistance to antibiotics from one kind of bacterium to another.

Genetic material can also be passed from one strain of bacteria to another strain by a process known as *transduction*. In this process, bacterial viruses, or bacteriophages (see Figure 11–1), may incorporate small portions of bacterial chromosomes into their own genetic material and then carry it to another bacterium. Both the viral DNA and the bacterial DNA may be incorporated into the chromosomes of the new host bacterial strain.

Another kind of genetic recombination in bacteria is known as *transformation*. In the causative agent of pneumonia, *Diplococcus pneumoniae*, there are two types of colonies: rough (R) and smooth (S). The strains that produce S colonies readily produce capsules and are virulent. As early as 1928, F. Griffith showed that it was possible to transform a harmless R strain into a virulent S strain by exposing the R strain to heat-killed S strain cells (see Figure 7–2, page 118). When bacterial cells are broken down by chemicals or by heat, fragments of DNA may pass into other cells. The demonstration that DNA was the genetically active material involved in such transformations was the first direct evidence for the genetic role of DNA. Transformation is now known to occur in many different groups of bacteria.

Mutation is a far more important source of variability in bacteria than genetic recombination (Figure 11–14). For a given gene, it has been calculated that there will be about 1 mutant cell per 10^7 (10 million) individuals. The amount of DNA in *Escherichia coli* is equivalent to approximately 5000 genes. Thus, in a culture of *E. coli*, there is about 1 mutant cell per 2000 individuals; 0.05 percent of the individuals in the culture will have a mutant phenotype in each cell division. In a culture that approaches 10^9 cells—one that has divided about 30 times—the frequency of mutants will be about 30 × 0.05 percent, or as high as 1.5 percent. Bacteria multiply very rapidly. For *E. coli*, the population may, under optimal conditions, double every 12.5 minutes. Thus the number of mutant individuals produced in such a population is very high. The rapid generation time of bacteria combined with these mutations is responsible for their extraordinary adaptability.

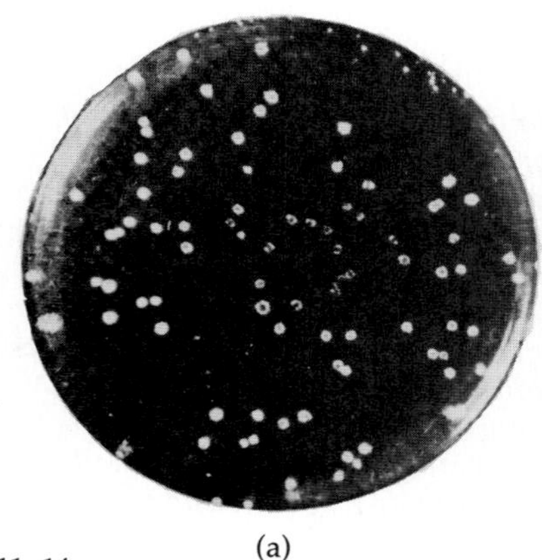

(a)

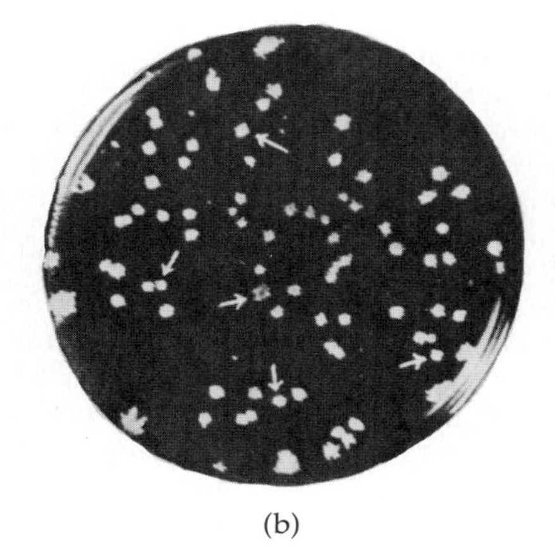

(b)

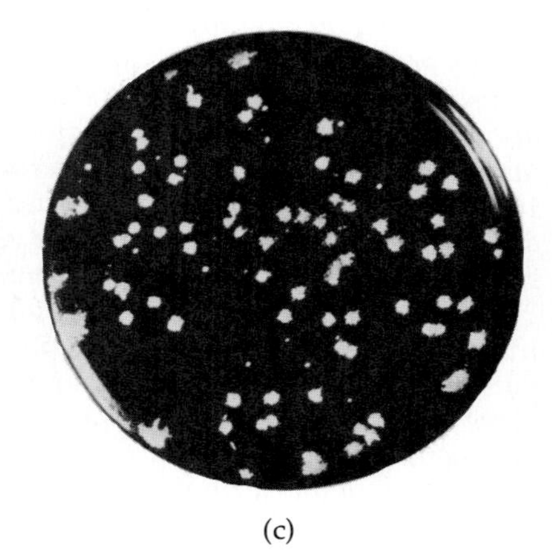

(c)

11–14
Replica plating as a means for detecting the presence of mutations in bacteria. The initial plate is shown in (a). *Colonies were transferred from* (a) *to* (b) *and* (c) *by means of a sterile velveteen disk that was pressed on plate* (a) *and then pressed on* (b) *and* (c), *thus placing colonies in the same relative positions.* (a) *and* (b) *show growth of the colonies on a complete medium, whereas a number of growth factors are lacking in* (c). *Under these conditions, the colonies indicated by arrows in* (b) *did not grow on plate* (c); *they are mutants that were already present but undetected in* (a).

USING BACTERIA TO MANUFACTURE USEFUL COMPLEX MOLECULES

During the past few years, the development of techniques for "genetic engineering" has made it possible to insert foreign genes from higher organisms into bacterial chromosomes. The resulting organisms—known as recombinant bacteria—can then synthesize nonbacterial proteins and other compounds. Since they can be grown easily and inexpensively, recombinant bacteria are an efficient source for producing proteins and other complex molecules.

In one technique, a small circular fragment of bacterial DNA called a plasmid is the vehicle for introducing the new gene into the bacterium. The eukaryotic genes can be localized by restriction enzymes, which break the DNA of the subject at points where particular base sequences occur. By using a combination of different specific restriction enzymes, any given DNA sample can be broken into a characteristic set of short pieces, and a given gene can be isolated that contains a particular set of DNA fragments. A recombinant bacterial plasmid is then constructed by breaking it open with restriction enzymes and connecting the "foreign" gene to the plasmid with the enzyme ligase, which attaches DNA and closes the new plasmid.

When this has been accomplished, the recombinant plasmids are purified by mixing them with bacterial cells that have been made permeable by placing them in a dilute solution of calcium chloride; the recombinant plasmids enter and become stabilized within some of the cells. The bacterial cells carrying the recombinant plasmid can be recognized by the properties of the genes present on the plasmid. Bacteria that have the modified plasmid can thus be selected and grown in the laboratory in whatever numbers are desired.

If the foreign genes are from eukaryotes, however, they are broken by long stretches of DNA that do not function as part of the gene. In order to put together a DNA sequence that will lead directly to the manufacture of the desired protein in a recombinant prokaryotic system, it is necessary to arrange into one continuous sequence the parts of the gene that are separated in the eukaryote.

This arrangement may be accomplished in one of two ways. If the protein is small, as is the case in many human hormones, the DNA sequence may be synthesized and inserted into the recombinant bacterial plasmid. If it is 100 bases or more in length, however, synthesis is more difficult. In this case, mRNA for the gene will be suitable, since it does not contain the intervening DNA sequences. The specific mRNA desired can be found in high concentrations in cells actively engaged in manufacturing the particular protein. With the aid of an enzyme called reverse transcriptase, a single strand of RNA can be copied to make a complementary strand of DNA. Reverse transcriptase is obtained from RNA viruses, which are described on page 208. The single-stranded DNA can be made into a double strand with the aid of DNA-copying enzymes, and the particular DNA segment desired can then be purified and introduced into a recombinant bacterial system.

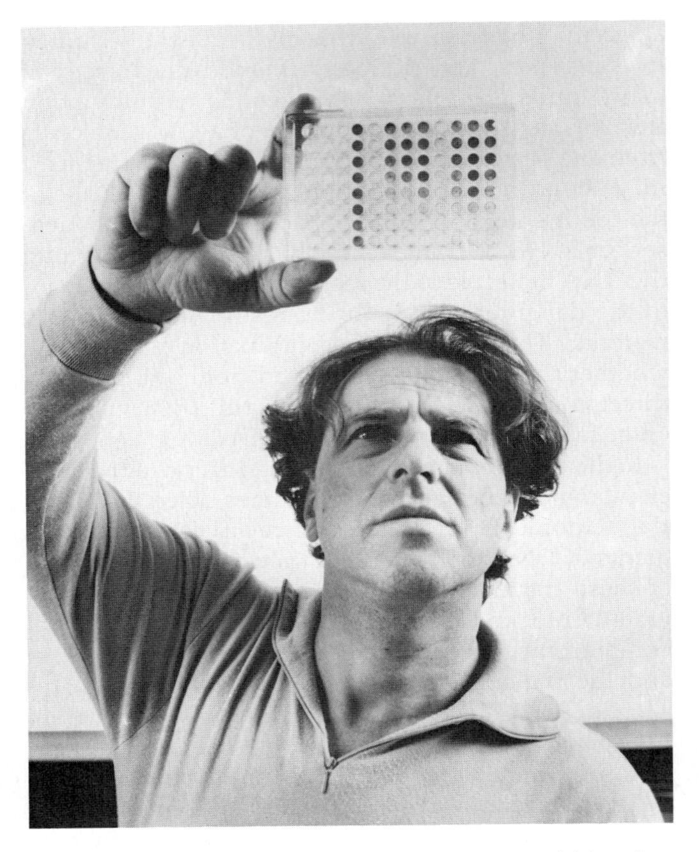

Dr. Charles Weissmann and his colleagues at the University of Zürich discovered the gene that controls the synthesis of interferon—a natural antiviral agent produced in minute quantities in the human body. The bacterium *Escherichia coli* and gene-splicing techniques were used to produce recombinant bacteria capable of synthesizing human interferon.

With the application of these methods of genetic engineering, rare and very expensive hormones and other proteins can be produced economically for use in the treatment of human disease or for other purposes. These techniques have already been used in the formation of bacterial strains that can synthesize hormones (such as insulin and human growth hormone), enzymes, vaccines, antibodies, and even interferon, an extremely potent antiviral substance in humans. Interferon is a complex protein found in the human body in extremely small amounts. Charles Weissmann of Zürich found the gene for interferon by randomly splicing large amounts of DNA into Escherichia coli *chromosomes and then screening the resulting strains—in the end, he had tested a total of 20,000 strains of* E. coli. *The implications of this work for controlling virus-caused diseases are enormous.*

A novel form of genetic exchange is responsible for a plant tumor called crown gall disease. A common soil bacterium, Agrobacterium tumefaciens, *can infect wound sites in plants and cause crown galls—swellings that contain cancerous, rapidly growing plant cells. The uncontrolled cell division of these tumorous plant cells continues even after the bacteria are eliminated. Thus, autonomous growth is a new genetic characteristic that has been acquired by the tumor cells, and when they divide, the trait is inherited by their daughter cells. Such tumor cells can be cultivated on agar as sterile tissue culture lines, free from bacteria, and they retain the tumorous growth habit of rapid uncontrolled cell division. Armin C. Braun of Rockefeller University first suggested that a tumor-inducing principle passes from the bacteria to the plant cells and is responsible for their permanently altered pattern of growth.*

Mary-Dell Chilton of Washington University and her co-workers, in studies of Agrobacterium tumefaciens *strains that cause plant tumors, have brought to light the existence of circular plasmids that are apparently the true infecting agents of crown gall disease. The plasmids are large, about 100 genes in size, and about a quarter of these genes appear related to tumor-inducing activities. About one-tenth of the plasmid plays a direct role in tumor induction; it is transferred to the plant cell and becomes incorporated into plant nuclear DNA, where it replicates in synchrony with plant genes. Indeed, this foreign DNA element, called T-DNA, is a very effective invader, for it behaves like plant genes and produces mRNA and protein products. At least some of these T-DNA elements are presumably responsible for the altered growth pattern of the tumor cells.*

Studies of tumor tissue by George Morel in Versailles, France, revealed that unique amino acid derivatives were present in the tumor cells in high concentrations. These compounds were never found in normal plant cells. Different strains of Agrobacterium *(containing different types of tumor-inducing plasmids) caused tumor lines to produce different types of amino acid derivatives. These unique compounds—called "opines"—play a central role in the ecology of crown gall disease, for* Agrobacterium *(but not other bacteria or fungi) can use them as nutrients, and the ability to use the opines in this manner is conferred on the tumor-inducing plasmid by genes. Thus the "purpose" of invading the plant cell with T-DNA is to trick the plant cell into producing specific nutrients for* Agrobacterium. *Another ancillary benefit to the bacterium is that the opines act as inducers of conjugation; in the crown gall tissue,* Agrobacterium *strains exchange their tumor-inducing plasmids freely, an activity that is very rare in other circumstances.*

Crown gall researchers are intrigued by the fact that DNA from a bacterium can somehow recombine with DNA from a higher plant cell. Prior to the discovery of this process, it was thought that such completely unrelated DNAs would never recombine in nature. A second aspect of considerable interest is the fact that "bacterial" genes can function in a plant cell. Presumably the T-DNA portion of the tumor-inducing plasmid produces chemical signals recognizable by plant enzymes; it is interesting to speculate how this molecular mimicry could have evolved.

Although the modus operandi of T-DNA is not fully understood, it is already clear that this DNA element can be exploited in the genetic engineering of plant cells. If new DNA is placed in the middle of T-DNA, the T-DNA is able to carry it along into the plant cell; the T-DNA acts as a "vector" at the molecular level. Genetic engineers hope to exploit this ability of T-DNA as a means of introducing desirable genes into crop plants. Thus, what began as a problem in plant pathology has become an exciting prospect for plant improvement involving the direct manipulation of plant genetic information.

An example of naturally occurring "genetic engineering." The tumorlike growth on this stem of the houseplant *Begonia semperflorens* is a crown gall. This disease is caused by an unusual form of parasitism that amounts to a kind of "genetic colonization" of the plant by the common soil bacterium *Agrobacterium tumefaciens.*

Bacterial Metabolism

Heterotrophs

Most bacteria are heterotrophs—organisms that cannot make organic compounds from simple inorganic substances but must obtain them from organic compounds found in other organisms. The largest group of heterotrophic bacteria are the *saprobes*. Saprobes are organisms that obtain their nourishment from dead organic matter. Saprobic bacteria and fungi are responsible for the decay and recycling of organic material in the soil; many of the characteristic odors associated with soil come from substances produced by heterotrophic bacteria.

Photosynthetic Bacteria

There are four groups of photosynthetic bacteria: (1) the cyanobacteria, (2) the green sulfur bacteria, (3) the purple sulfur bacteria (see Figure 6–2, page 97), and (4) the purple nonsulfur bacteria. (The colors of the third group may actually range from purple to red or brown.) Like the green plants, photosynthetic bacteria contain chlorophyll. The cyanobacteria contain chlorophyll *a*, as do the photosynthetic eukaryotes. The chlorophylls present in the other groups of photosynthetic bacteria differ in several ways from chlorophyll *a*, but they all have the same basic structure (see Figure 6–8, page 103).

The colors of these groups of bacteria are due to the presence of several different accessory pigments that function in photosynthesis. In the purple bacteria, these pigments are yellow and red carotenoids; in the cyanobacteria, in addition to carotenoids, there is always a blue pigment, phycocyanin, and generally a red one, phycoerythrin. The main carbohydrate storage product of the cyanobacteria is a polysaccharide known as cyanophycean starch, which is probably identical with glycogen.

Photosynthesis in the cyanobacteria is identical to that found in photosynthetic eukaryotes. In the green sulfur bacteria, the sulfur compounds play the same role in photosynthesis that water does in organisms that contain chlorophyll *a*. That is,

$$CO_2 + 2H_2S \xrightarrow{\text{light}} (CH_2O) + H_2O + 2S$$

As discussed in Chapter 6, an understanding of the course of photosynthesis in the purple sulfur bacteria was the key that led C. B. van Niel to propose the following generalized equation for photosynthesis,

$$CO_2 + 2H_2A \xrightarrow{\text{light}} (CH_2O) + H_2O + 2A$$

in which H_2A is a generalized hydrogen donor. In the photosynthetic purple nonsulfur bacteria, other compounds, including alcohols, fatty acids, and keto acids, serve as electron (hydrogen) donors for the photosynthetic reaction.

Because of their requirement for hydrogen sulfide or a similar substrate, the photosynthetic sulfur bacteria are able to grow only in habitats that contain large amounts of decaying organic material, which is distinguishable to us by its sulfurous odor. In this type of bacteria, elemental sulfur may accumulate in deposits within the cell (see Figure 11–5d).

In 1975, Ralph A. Lewin of the Scripps Institution of Oceanography announced an exciting discovery: a prokaryote containing chlorophylls *a* and *b* and carotenoids. This organism, named *Prochloron*, has the photosynthetic pigments of eukaryotic green algae and plants (Figure 11–15). As far as is known, it lives in association with colonial ascidians (sea squirts) along seashores in the tropics and subtropics. Biochemically, it has the characteristics that would be expected in the prokaryotic group that gave rise to the chloroplasts of the green algae by symbiosis, a relationship that will be discussed further in Chapter 14.

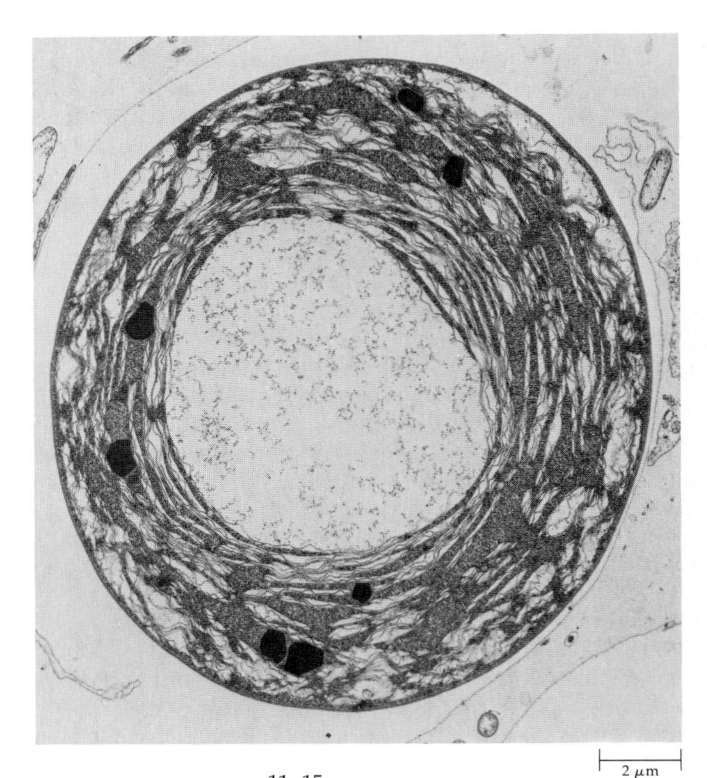

11–15
A single cell of Prochloron *from the cloacal cavity of* Diplosoma virens, *the sea squirt.* Prochloron *is a photosynthetic bacterium with chlorophylls* a *and* b *and carotenoids like those found in the green algae and plants.*

EVOLUTION OF PHOTOSYNTHESIS

The oldest known sedimentary rocks, from Greenland, are about 3.8 billion years old. The oldest fossil-bearing ones, however, are about 3.5 billion years old; they are found in Western Australia (see Figure 11–2). Fossil-bearing rocks from South Africa that range from 3 to 3.4 billion years old are also known. Organic spheroidal bodies similar to cyanobacteria have been found in South African strata about 3.2 billion years old. Fossil and chemical evidence from the same series of deposits indicates that photosynthesis occurred as early as 3.3 billion years ago.

There are two kinds of such chemical evidence. First, modern plants, in the process of photosynthesis, selectively accumulate ^{12}C in preference to its heavier isotope ^{13}C. Thus the ratio of ^{12}C to ^{13}C is higher in organic material that accumulated as a result of photosynthesis than in organic material formed in other ways. Such ^{12}C enrichment is not found in South African rocks dated at 3.34 billion years old but appears dramatically in those about 3.3 billion years old and is characteristic of those that are more recent in origin, thus indicating a possible date for the initiation of photosynthesis. The second kind of chemical evidence is that compounds have been found in the rock that are probably breakdown products of the chlorophyll molecule itself. Further evidence of the early occurrence of photosynthesis is provided by the similarity between the accumulations of calcium carbonate found in Rhodesian limestone that are about 2.7 billion years old and those produced by modern cyanobacteria. The cyanobacteria utilize chlorophyll a *in photosynthesis, and in the process, they evolve O_2.*

Chemoautotrophic Bacteria

Unlike many of the photosynthetic bacteria, chemoautotrophic bacteria require the presence of oxygen and do not utilize the energy of sunlight. The energy used to drive their synthetic reactions is obtained from the oxidation of inorganic molecules such as nitrogen, sulfur, and iron compounds, or from the oxidation of gaseous hydrogen.

One group of chemoautotrophic bacteria that has attracted a great deal of attention in recent years is the methanogenic (methane-producing) bacteria. These strictly anaerobic bacteria produce methane (CH_4) from CO_2 and H_2 and obtain their energy by means of this process. Methane-producing bacteria are rather diverse morphologically, but recently it has been shown—principally by Carl Woese of the University of Illinois and his colleagues—that the sequence of bases in a portion of their ribosomal RNA (rRNA) resemble one another closely. This is taken as evidence that these bacteria are closely related by descent. What is very surprising, however, is that these sequences differ rather sharply from those in the corresponding rRNA molecule in the other prokaryotes and also from that found in eukaryotes. In these sequences, for example, the cyanobacteria resemble *E. coli* or the actinomycetes much more closely than any of them resembles the methanogenic bacteria. The cell walls of methanogenic bacteria do not contain muramic acid, as do those of all other bacteria, and their metabolism appears to be fundamentally different. In view of this evidence, it has been suggested that this group might have originated more than 3 billion years ago, when there was an anaerobic atmosphere rich in CO_2 and H_2, and that they have persisted in certain favorable habitats. Their distinctiveness from other living prokaryotes is so great that it has been postulated that the methanogenic bacteria should be considered as a separate kingdom—the Archebacteria.

Bacterial Ecology

Soil Bacteria

Different groups of microorganisms are involved in specific stages of the processes of decomposition and recycling that occur in the soil, and such natural communities tend to be organized in a complex fashion (see Figure 11–16). Many bacteria and fungi break down carbon-containing compounds, releasing CO_2 into the atmosphere. The most important organic compounds originating from plants in nature are cellulose and lignin, and secondary ones are pectic substances, starch, and sugars. It has been estimated that more than 90 percent of the CO_2 production in the biosphere results from the activity of bacteria and fungi.

Some microorganisms break down proteins into peptides, which are subsequently broken into their constituent amino acids. Many microorganisms have the ability to break down amino acids, with the consequent release of ammonium ions (NH_4^+), by a process called ammonification (see Chapter 27). Ammonia can be oxidized to nitrite ions (NO_2^-) by the chemoautotrophic bacteria *Nitrosomonas*, and the nitrites are oxidized to nitrates (NO_3^-) by *Nitrobacter*. The conversion of ammonia to nitrites and nitrates constitutes the process of *nitrification*. This process releases energy, which is used to reduce carbon dioxide to carbohydrate. Several other kinds of bacteria are capable of reversing the process and changing the nitrates back into nitrites and, ultimately, ammonia.

Denitrification, the conversion of nitrates into nitrogen gas or nitrous oxide, results in the loss of nitrogen from the soil. The reverse of this process, which is extremely important biologically, is *nitrogen fixation.* Several genera of bacteria, as well as some fungi (yeasts), are capable of nitrogen fixation. Outstanding among them is the symbiotic bacterium *Rhizobium* (see Chapter 27), which forms nodules on the roots of legumes and a few other plants. The actinomycetes are involved in nodule-formation in many woody plants, such as alders *(Alnus)*, wax myrtles *(Myrica)*, and mountain lilacs *(Ceanothus)*. Actinomycetes are efficient nitrogen-fixers; they actually influence the accumulation of nitrogen in soils considerably more than the legumes, which are important chiefly in disturbed habitats. In addition, certain bacteria are regularly associated with the roots and even the leaves of plants, where they utilize the carbohydrate-rich exudates from the plants and in turn provide nitrogen in a usable form. Many free-living bacteria and symbiotic cyanobacteria play an important role in nitrogen fixation, and certain genera of these free-living bacteria are loosely associated with plants.

Sulfur is made available to plants (which cannot utilize elemental sulfur) by chemoautotrophic bacteria such as *Thiobacillus,* which oxidize elemental sulfur to sulfates:

$$2S + 2H_2O + 3O_2 \longrightarrow 4H^+ + 2\,SO_4^{2-}$$

Sulfates are accumulated by plants, and the sulfur contained in them is incorporated into the structure of proteins. The degradation of proteins, which is discussed as one aspect of the nitrogen cycle in Chapter 27, liberates amino acids, some of which contain sulfur (Figure 11–17). A number of bacteria are capable of breaking down these amino acids, thus releasing hydrogen sulfide (H_2S). Sulfates are also reduced to H_2S by certain soil microorganisms, such as *Desulfovibrio*.

11–16
The complexity of interactions that occur among soil organisms is suggested by this photograph, which shows bacteria growing on an agar plate to which penicillin has been added. Down the center of the plate is a dense colony of Staphylococcus epidermidis, *a strain of bacteria that is resistant to penicillin because it produces an enzyme called penicillinase, which breaks down the antibiotic. On either side of the* Staphylococcus *colony can be seen tiny colonies of* Neisseria gonorrhoeae, *the causative agent of gonorrhea.* Neisseria *is susceptible to penicillin but is able to grow in the part of the medium where the resistant* Staphylococcus *has broken down the antibiotic.*

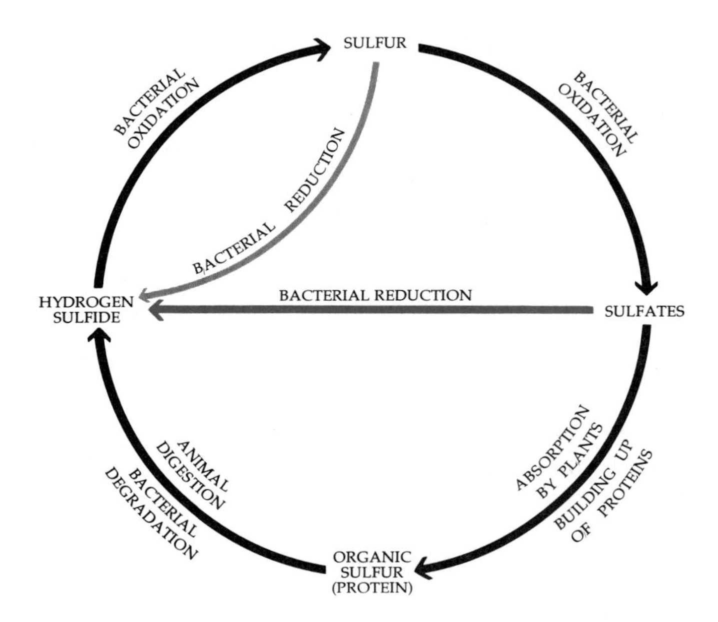

11–17
The sulfur cycle. Without the activities of bacteria and fungi, carbon, nitrogen, sulfur, and other elements would remain locked in the molecules into which they are incorporated. Soil-dwelling heterotrophs break down complex organic molecules into substances that can enter other biological cycles. Without the recycling of organic substances, all organisms would soon be overwhelmed by the products of their own metabolism. The release of large quantities of sulfur into the atmosphere by burning fossil fuels is placing a severe strain on the natural mechanisms of the sulfur cycle.

Parasitic and Symbiotic Bacteria

Some heterotrophic bacteria break down organic material while it is still incorporated in the bodies of living organisms. The disease-causing (pathogenic) bacteria belong to this group, as do a number of other nonpathogenic forms. Some of these bacteria have little effect on their hosts, and some can actually prove beneficial. For example, if all the bacterial inhabitants of the human intestinal tract are eliminated—as can happen, for example, following prolonged antibiotic therapy—the tissues are much more vulnerable to disease-causing bacteria and fungi. Furthermore, as will be discussed in Chapter 27, the relationship between the symbiotic bacteria of the genus *Rhizobium* and the legumes they inhabit is of mutual benefit.

In general, microbes cause diseases when they interfere with one or more essential physiological processes of the organisms they invade. When these effects are severe, the disease is considered serious. If death of the host organism results prematurely, the relationship will not be as favorable as possible for the parasite, which will then have to find another host individual.

Human Diseases

Some human diseases are caused by airborne bacteria. Among the better known are bacterial pneumonia, caused by *Diplococcus pneumoniae,* whooping cough, caused by *Bordetella pertussis,* and diphtheria, caused by *Corynebacterium diphtheriae.* The latter organism produces a powerful toxic substance that circulates rapidly throughout the body and causes serious damage to the heart muscle, nervous tissue, and kidneys. Diphtheria is now rare because most children are immunized against it in infancy. Other serious diseases are caused by airborne bacteria of the genus *Streptococcus,* which are associated with scarlet fever, rheumatic fever, and other infections. Tuberculosis, caused by *Mycobacterium tuberculosis,* is still a leading cause of death in humans, despite improved methods of detection, but the number of reported cases in the United States decreased to about 30,000 per year by the 1970s.

A number of other diseases of bacterial origin are spread in food or water. Examples are typhoid fever and paratyphoid (caused by bacteria of the genus *Salmonella),* and bacillary dysentery (caused by *Shigella dysenteriae).* Others, like typhus, which is caused by *Rickettsia prowazekii* are spread by insect vectors (Figure 11–18). Undulant fever, caused by bacteria of the genus *Brucella,* affects both cattle and humans and is usually contracted through eating or drinking milk or milk products from an infected cow. Pasteurization of milk destroys *Brucella,* and the disease has become rare in those portions of the world where this process is employed.

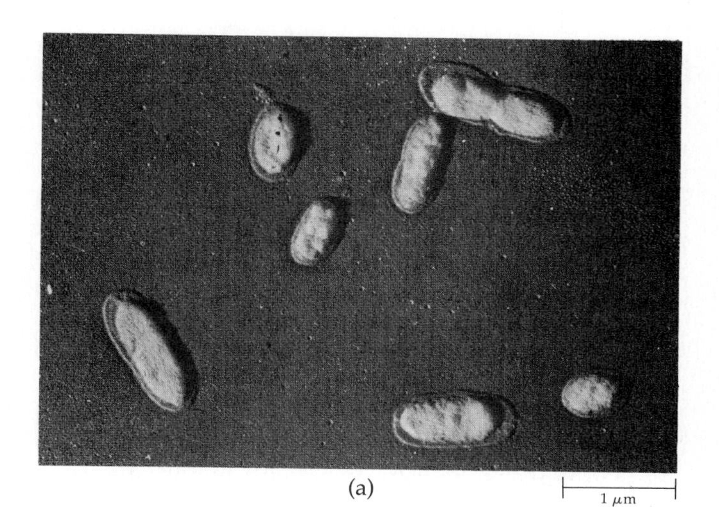

(a)

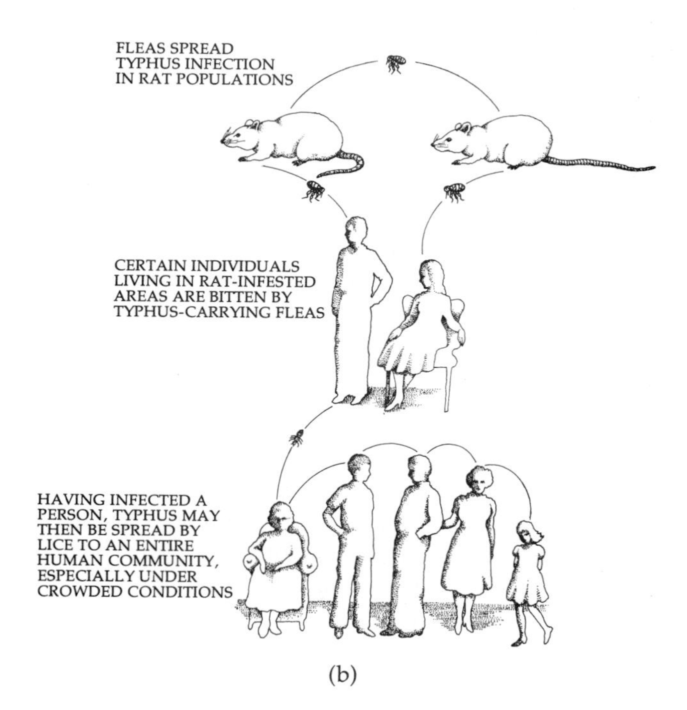

(b)

11–18
Typhus has often played a crucial role in human history. At the siege of Granada in 1489, over 17,000 Spanish soldiers died of typhus, as compared to 3000 killed in combat. (a) *Typhus is caused by* Rickettsia prowazekii, *one of the Rickettsiae, a group of small bacterialike organisms that are not able to grow on a cell-free medium.* (b) Rickettsia *is spread from rats to humans by fleas. It may then be transmitted under crowded conditions by body lice. More humans have died of rickettsial diseases than any other form of illness except malaria. Bacteria similar to* Rickettsia *have recently been associated with plant diseases, including almond leaf scorch and Pierce's disease of grapevines.*

Legionnaire's disease, spread in water, was first detected in 1976 when a mysterious lung ailment killed 34 members of the American Legion attending a conference in Philadelphia. It is now known to be caused by a rod-shaped bacterium that has been given the name *Legionella pneumophila.* These bacteria enter and multiply rapidly in the monocytes, a kind of white blood cell that normally plays a major defensive role against most microorganisms. It is now estimated that *Legionella pneumophila* causes about 25,000 cases of a severe and sometimes fatal form of pneumonia in the United States each year.

Bacteria play an important role in the spoilage of food and other stored organic products, and some of these organisms are very pathogenic to humans. Food poisoning by *Clostridium botulinum* is rare but extremely dangerous (Figure 11–19). *Staphylococcus* food poisoning is fairly common but, fortunately, much less serious. In recent years, a number of virulent strains of *Staphylococcus* have been associated with serious infections. Many of these are resistant to penicillin, and some produce the enzyme penicillinase, which breaks down the antibiotic. These resistant bacteria frequently come into contact with penicillin-producing fungi and often grow with them in nature, so this resistance confers a natural advantage (see Figure 11–16).

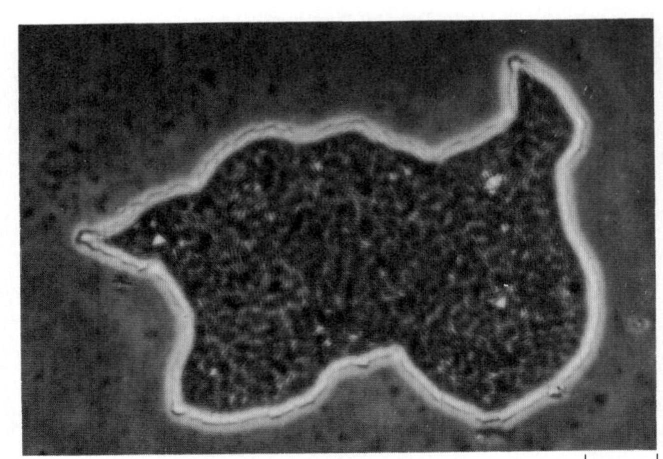

11–19
A colony of Clostridium botulinum. *The type of food poisoning (botulism) suffered by consumers of contaminated canned foods is associated with the survival of the spores of this bacterium. They are extremely resistant to heat and occasionally persist in food that has not been treated properly.* Clostridium botulinum, *which grows only in the absence of oxygen, produces its powerful toxin inside the can or jar. The toxin can be destroyed by boiling for about 15 minutes. The toxin is the most powerful known; 1 gram would be enough to kill 14 million people. Food botulism is rare in the United States; between 1977 and 1979, there were 146 reported cases and 8 reported deaths.*

Plant Diseases

Many economically important diseases of plants are also associated with bacteria. Bacterial diseases of plants can occur in any adequately moist or warm place. Almost all kinds of plants can be affected by bacterial diseases, and the plant-pathogenic bacteria can be extremely destructive (Figure 11–20).

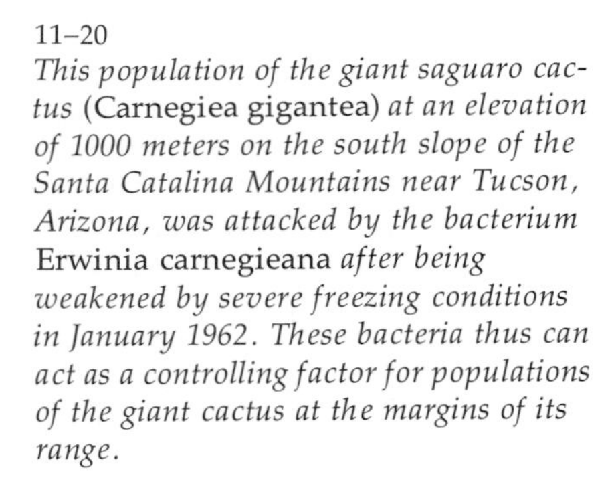

11–20
This population of the giant saguaro cactus (Carnegiea gigantea) *at an elevation of 1000 meters on the south slope of the Santa Catalina Mountains near Tucson, Arizona, was attacked by the bacterium* Erwinia carnegieana *after being weakened by severe freezing conditions in January 1962. These bacteria thus can act as a controlling factor for populations of the giant cactus at the margins of its range.*

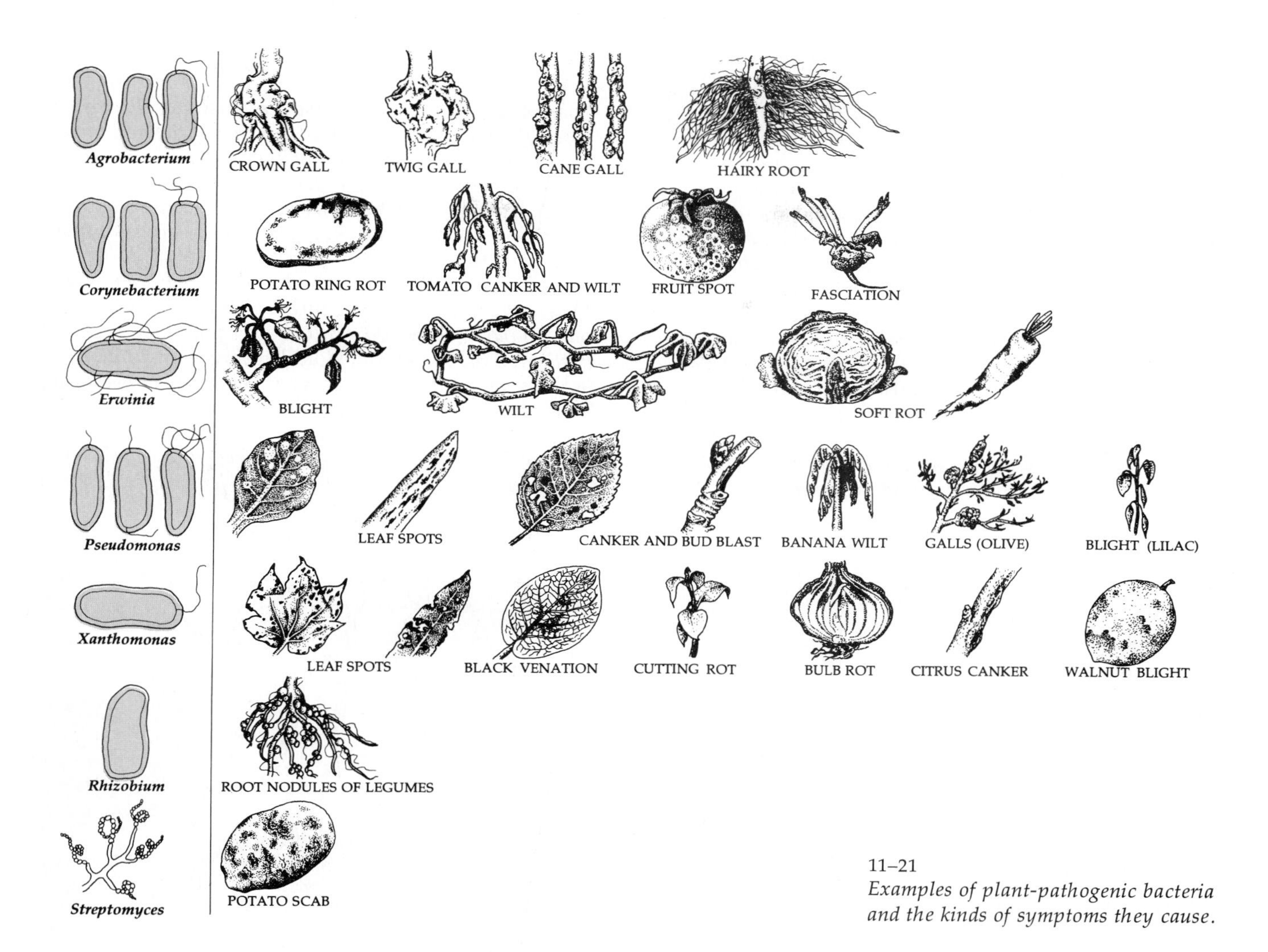

11–21
Examples of plant-pathogenic bacteria and the kinds of symptoms they cause.

Virtually all plant-pathogenic bacteria are rod-shaped, and almost all develop primarily in the host plant as parasites. The symptoms caused by plant-pathogenic bacteria are quite varied, with the most common appearing as spots of various sizes on stems, leaves, flowers, and fruits (Figure 11–21). Almost all such bacterial spots are caused by two closely related genera, *Pseudomonas* and *Xanthomonas.*

Some of the most destructive diseases of plants—such as blights, soft rots, and wilts—are also linked to bacteria. Blights are characterized by rapidly developing necroses (dead, discolored areas) on stems, leaves, and flowers. Fire blight in apples and pears may result in the death of young trees within a single season. The pathogen of fire blight is the rod-shaped bacterium *Erwinia amylovora.* Bacterial soft rots occur most commonly in the fleshy storage organs of vegetables, such as potatoes and onions, or in fleshy fruits, such as tomatoes and eggplants. The most destructive soft rots are caused by bacteria of the genus *Erwinia.* Bacterial vascular wilts affect only herbaceous plants. The bacteria invade the vessels of the xylem, where they move about and multiply. As a consequence, the bacteria interfere with the movement of water and inorganic nutrients, resulting in the wilting and death of the plant. The bacteria commonly destroy portions of the vessel walls and can even cause the vessels to rupture. They then spread to the adjacent parenchyma tissues, where they continue to multiply. In some bacterial wilts, the bacteria ooze to the surface of the stems or leaves through cracks formed over cavities filled with cellular debris, gums, and bacteria. More commonly, however, the bacteria do not reach the surface of the plant until the plant has been killed by the disease. Among the most important examples are the bacterial wilt of alfalfa and bean plants (each caused by a different species of *Corynebacterium*), the bacterial wilt of cucurbits (caused by *Erwinia tracheiphila*), and the black rot of crucifers (caused by *Xanthomonas campestris*).

MYCOPLASMAS

Mycoplasmas—the general name for a group of prokaryotic organisms that represent the smallest known cells—can be as small as 0.1 micrometer in diameter. Each mycoplasma cell consists of a plasma membrane, DNA, RNA, ribosomes, soluble proteins, sugars, and lipids; they lack nuclei and cellular organelles and do not have cell walls. An individual mycoplasma probably contains fewer than 650 genes, which is about one-fifth of the number found in a single common bacterium.

Mycoplasmas are known by various names, including pleuropneumonialike organisms (PPLOs), which refers to a type of disease that they can cause in animals. The mycoplasmas isolated thus far have been classified into six genera, with about 50 of the 60 recognized species assigned to the genus Mycoplasma. *By sequencing a portion of the ribosomal RNAs in these organisms, C. R. Woese and his colleagues at the University of Illinois have been able to determine that all but one of the genera are related to one another and that they are probably derived from a single bacterial line that includes* Bacillus *and* Lactobacillus. *The remaining genus,* Thermoplasma, *apparently acquired the characteristics of mycoplasmas independently.*

In addition to the mycoplasmas identified in animal tissue, mycoplasmalike organisms (MLOs) have also been found in more than 200 plant species and have been implicated in, or associated with, more than 50 plant diseases, many with symptoms of yellowing and stunting. Most attempts to culture MLOs obtained from plant tissue have been unsuccessful.

Some spiroplasmas—long, thin, helical mycoplasmas that may be up to 10 micrometers in length and less than

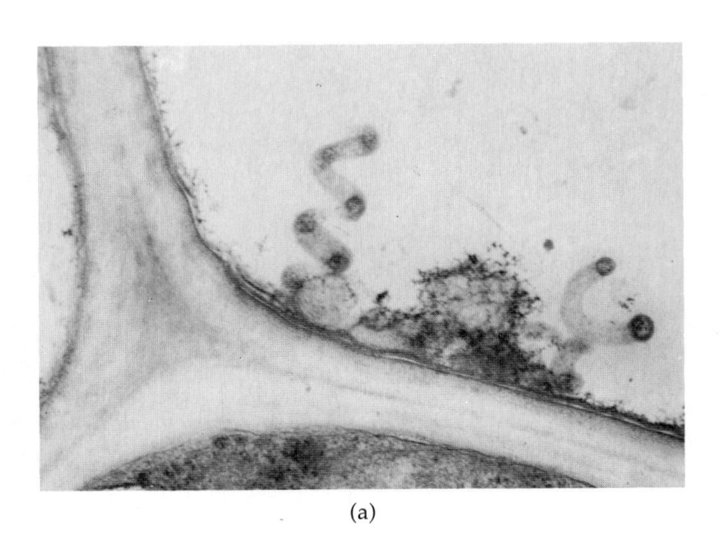

(a)

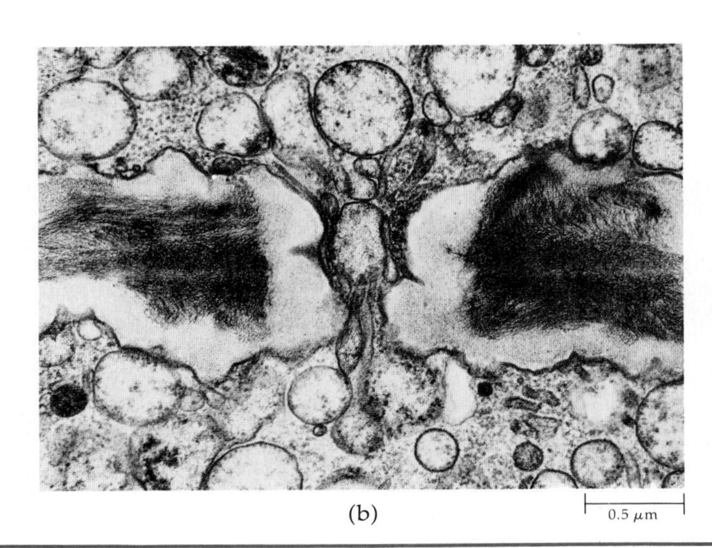

(b)

Ecology of Cyanobacteria

The cyanobacteria, or blue-green algae, deserve special discussion because of their varied ecological roles. Owing to their wide geographical ranges and extreme variability of habitats, their taxonomy is complicated. Although over 7500 names have been proposed for species of this group, experimental work carried out in the past decade suggests that there may actually be as few as 200 distinct nonsymbiotic species. As an example, *Microcoleus vaginatus* occurs in wet soil or fresh or brackish water from northern Greenland to Antarctica, and from the floor of Death Valley to the top of Pike's Peak. Studies of the organism have revealed that under changing environmental conditions, a single colony may undergo such extreme changes that its members come to resemble dozens of different species. It is clear that extensive studies will be necessary before we can begin to understand the extent of the variability in these seemingly simple organisms.

Like other bacteria, cyanobacteria sometimes grow in extremely inhospitable environments, from the near-boiling water of hot springs to the frigid lakes of Antarctica, where they have recently been reported to form luxuriant mats, 2 to 4 centimeters thick, in water beneath more than 5 meters of permanent ice. The greenish color of some polar bears in zoos has recently been shown to be due to colonies of cyanobacteria that develop within the hollow hairs of their fur! Cyanobacteria were the first colonists on the new island of Surtsey, near Iceland, following its eruption

0.2 micrometers in diameter—have been cultured on artificial media. These include Spiroplasma citri, *the agent of stubborn and little-leaf diseases of citrus, and spiroplasmas isolated from stunted corn plants* (a). *Spiroplasmas exhibit vigorous whirling and flexing movements in culture. Proof of pathogenicity—that the organisms isolated from diseased tissues actually caused the disease—has been obtained only with a few spiroplasma isolates.*

MLOs are generally confined to the sieve tubes of the phloem. The organisms are believed to move passively from one sieve-tube member to another through the sieve-plate pores along with the solution of sugar being transported by the phloem. The electron micrograph in (b) *shows MLOs apparently traversing a sieve-plate pore in a young inflorescence of a coconut palm* (Cocos nucifera) *affected by lethal yellowing disease. This disease has devastated many genera of palms in Florida. Motile spiroplasmas, however, may also be able to spread actively throughout the tissue.*

The transmission of MLOs from plant to plant is accomplished largely by insect vectors. One of the best-known MLO-caused plant diseases is aster yellows, which commonly is transmitted from plant to plant by the six-spotted aster leafhopper, Macrosteles fascinfrons. *In* (c), *which shows three Chinese asters* (Callistephus chinensis), *the plant on the left is healthy, that in the center has diseased leaves, and the one on the right is severely infected. In* (d), *MLOs can be seen in the aster stem cell on the left, while the cell on the right is free of MLO infection.*

(c)

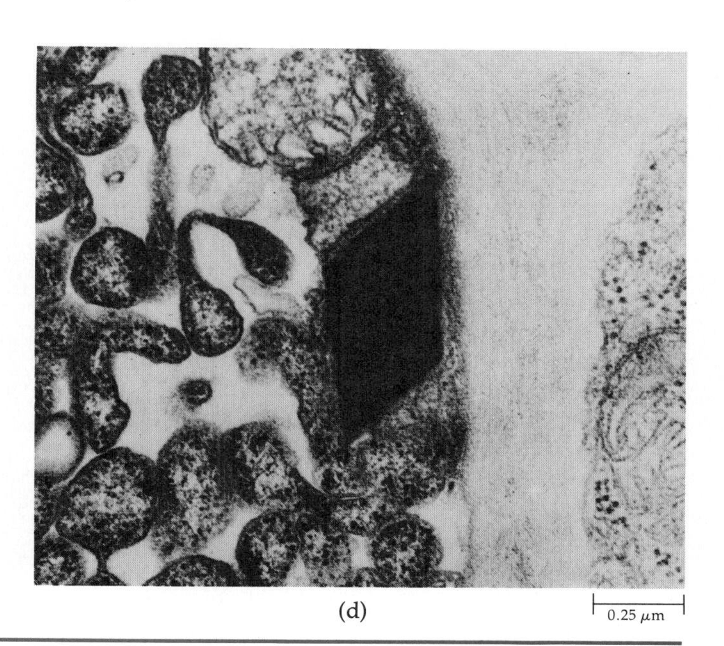

(d)

and appearance above the sea. On the other hand, they are absent in acidic waters, where eukaryotic algae are abundant.

Many cyanobacteria have a mucilaginous sheath or coating, which is often deeply pigmented, particularly in species that spread up onto the land. The pigments include a light golden yellow, brown, red, emerald green, blue, violet, and blue-black. In addition, the carotenoids and phycobilins modify the color of the cells in which they occur. Despite their name, only about half of the species of the "blue-green algae" are actually blue-green in color. Indeed, the Red Sea was named because of the dense concentrations, or "blooms," of the marine planktonic species of *Trichodesmium* that frequently occur in it. Blooms may occur in polluted fresh water as the result of a concentration of different species in this group, which can produce toxins harmful to fish and mammals.

Many marine species of this group are associated with limestone and lime-rich substances such as coralline algae (see Figure 14–29b, page 273) and the shells of molluscs. A number of freshwater forms, particularly those that occur around hot springs, often deposit thick layers of lime in their colonies. In Yellowstone National Park, the filamentous *Mastigocladus* occurs in hot water at temperatures up to 55°C, and the unicellular *Synechococcus* occurs in temperatures up to 73° to 75°C. In soil, including desert soils, they are abundant, with 20,000 to 50,000 individuals per gram being a representative figure.

Many cyanobacteria can fix nitrogen, and in Southeast Asia, rice is grown on the same land for years without the addition of fertilizers because of the rich growth of nitrogen-fixing bacteria, especially cyanobacteria, in the rice paddies. Here they often occur in association with the floating water fern *Azolla*, which has a nearly obligate relationship with them. Because of these nitrogen-fixing capacities, cyanobacteria—as well as lichens, liverworts, and other organisms with which they may be symbiotic—are able to colonize bare areas of rock and soil. In the sea, species of *Trichodesmium* are the major contributors to the fixation of atmospheric nitrogen, a process that annually amounts to about 10^8 metric tons worldwide.

Nitrogen fixation occurs only within *heterocysts*, enlarged cells formed in filamentous cyanobacteria (see Figures 11–11 and 11–13), in which the thylakoids are reorganized into a concentric or reticulate pattern. The formation of heterocysts in *Nostoc* and other genera is inhibited by the presence of ammonia or nitrates, but when these nitrogen-containing substances fall below a threshold, heterocysts begin to appear.

Cyanobacteria may occur as symbionts in some members of the following groups: sponges, amoebas, flagellate protozoa, certain diatoms, green algae that lack chlorophyll, other cyanobacteria, vascular plants, and oomycetes. When they occur as symbionts, they commonly lack a cell wall. Under such circumstances, they are functionally chloroplasts. In fact, in these forms, the prokaryotic symbiont divides at the same time as the host cell, by a process similar to chloroplast division. The relationship of cyanobacteria to the evolution of particular groups of eukaryotic algae is examined further in Chapter 14. In some of the associations mentioned, such as with sponges, the cyanobacteria play an important role in fixing atmospheric nitrogen. Cyanobacteria commonly form the photosynthetic component of lichens and can also occur in various bryophytes and vascular plants, where they may perform a nitrogen-fixing function. In turn, cyanobacteria may be parasitized by such organisms as chytrids (Chapter 13) and viruses.

THE VIRUSES

The viruses do not fit easily into any of the traditional categories with which living organisms are classified, and the problem of categorizing them is made even more difficult by the fact that there is considerable doubt about whether or not they should be considered as living. Viruses are discussed here because of their historical relationship to bacteria and because they represent one of the most valuable tools of modern genetics. They are of great importance as agents of disease, being the causative agents for smallpox, chicken pox, measles, German measles, mumps, influenza, colds (often complicated by secondary infections by bacteria), infectious hepatitis, yellow fever, polio, and rabies. Viruses are also responsible for many important diseases of domestic animals and plants. Although many viral diseases can be prevented by immunization, once established they are relatively difficult to control because they do not respond to antibiotics.

Many kinds of viruses are disease-causing agents in insects and thus are under active investigation as biological control agents. Insects destroy about a third of the world's food supply and an even larger proportion of the food supply of third-world countries. "Viral insecticides" affect a wide variety of insects but not other organisms and therefore do not have the undesirable biological side effects of many conventional insecticides. Already, there have been some notable successes in the application of viruses as biological control agents: for example, the corn earworm *(Heliothis zea)* has been effectively controlled by a virus in the United States.

The Nature of Viruses

The existence of viruses was first recognized when it was found that the causative agents of certain diseases could pass through the porcelain filters commonly used to trap bacteria. Viruses range in size from about 17 nanometers to more than 300 nanometers. Thus viruses are comparable to molecules in size, a hydrogen atom being about 0.1 nanometer in diameter and a large protein molecule being a few hundred nanometers in its greatest dimension.

Viruses are parasites that can multiply only within a living host cell and many are highly specific with regard to the type of cell in which they can multiply. They essentially "take over" the direction of the metabolism in the host cell, using their own nucleic acids to "command" the host to produce more virus particles. They compete with the genetic material of the host cell in regulating cell functions (see Figure 7–3, page 119). Cold viruses multiply in the mucous membranes of the respiratory tract, producing the all-too-familiar cold symptoms. Measles viruses and other rash-causing viruses multiply in the cells of the skin. The polio virus—only about 28 nanometers in diameter—multiplies in the intestinal tract and sometimes in the nerve cells. Even bacterial cells have their own set of viral parasites; indeed, one of the techniques for rapid identification of unknown bacteria is to expose them to a spectrum of known bacterial viruses—the bacteriophages—and observe which type destroys them (see Figure 11–1).

Plant cells that are completely free of viruses may

actually be exceptional, and the implications of this statement for agriculture may be profound. Many plants may have chronic infections of viruses, which lower their general vigor but only rarely become acute. For example, when virus-free strains of rhubarb were developed in Great Britain, their yield was 60 to 90 percent greater than plants with the "normal" virus infections. A number of cultivated plants with variegated foliage owe this characteristic to viral infections. It has been suggested that every species of organism, including the prokaryotes, may have at least one specific virus associated with it, in which case there may be literally millions of strains of viruses in existence.

Until the 1930s, viruses were considered to be extremely small bacteria. Evidence against this point of view began to accumulate in 1933, when Wendell Stanley prepared an extract of a common virus—the tobacco mosaic virus—from infected plants and purified it. The purified virus precipitated in the form of crystals. Crystallization is one of the chief tests for the presence of a single, uncontaminated chemical compound, and so, clearly, viruses were not composed of the complex variety of organic compounds that characterizes even so small a living thing as a bacterial cell. But when these needlelike crystals were put back into the solution and reapplied to a tobacco leaf, symptoms characteristic of tobacco mosaic virus reappeared.

The tobacco mosaic virus was subsequently identified as a large nucleoprotein (about 300 nanometers long), that is, a protein in combination with a nucleic acid (Figure 11–22). In the case of tobacco mosaic viruses, as well as most other plant viruses, the nucleic acid—which is the genetic material—is RNA instead of DNA. In other viruses, DNA serves as the genetic material, again combined with a protein. Viruses are the only "organisms" that do not contain both DNA and RNA; some DNA viruses, however, do have RNA sequences.

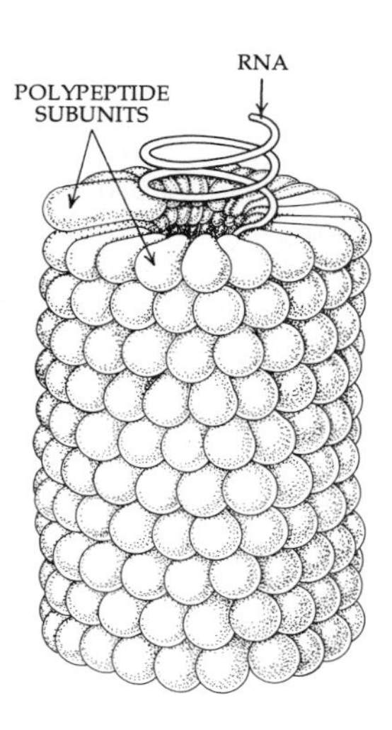

11–22
A diagram of a portion of a tobacco mosaic virus (TMV) particle. This virus has a central core of RNA and a protein coat composed of 2200 identical protein molecules, each containing 158 amino acid residues folded into spindle-shaped subunits. The RNA fits into a groove at the narrower end of the "spindle." Before the structure of any virus was known, Watson and Crick (see Chapter 7) predicted that the protein coats of viruses were composed of a large number of identical subunits. Can you explain the basis of their prediction?

The Structure of Viruses

By the use of electron microscopy, the structure of a large number of viruses has now been elucidated. The most common architectural arrangement is an icosahedron, a 20-sided figure, which is the most efficient symmetrical arrangement that subunits can take to form an outer shell with maximum internal capacity. (The geodesic domes of R. Buckminster Fuller are constructed along the same principles.) Among the icosahedral viruses are those that cause colds (Figure 11–23), polio, chicken pox, fever blisters, human warts, many types of cancer in mice and other animals, and many diseases of plants.

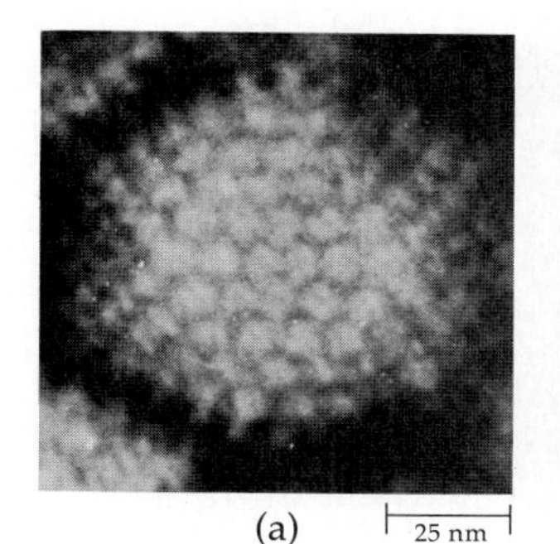

11–23
(a) *Adenovirus, one of the many viruses that cause colds in humans. This virus is an icosahedron. Each of its 20 sides is an equilateral triangle made up of protein subunits. There are 252 subunits in all.*
(b) *A model of the adenovirus, made up of 252 tennis balls.*

VIRAL INFECTIONS

The influenza virus (shown in the electron micrograph below) mutates rapidly. Changes in its genetic material (which is RNA) result in changes in its protein coat. Because immunity to a virus is, in effect, immunity to the specific proteins of its protective coat, new viral strains are able to infect previously immune populations. The virus is surrounded by a lipoprotein envelope through which stubby protein spikes protrude. The length of time between epidemics presumably reflects the time required for a new strain to become established. Influenza is the only infectious disease that appears periodically in life-threatening global epidemics; in the winter of 1968–69, more than 51 million cases of Hong Kong flu were reported in the United States alone, with at least 20,000 and perhaps as many as 80,000 deaths which could be attributed to this cause. The great influenza pandemic of 1918–19, which moved quickly around the world in three waves, caused more than 20 million deaths.

There is accumulating evidence that pandemic strains of influenza viruses in humans originate primarily after recombination between human influenza viruses and similar viruses from other animals, or because viruses from other animals are transmitted to humans and acquire the capacity to cause and spread disease. There is little evidence that mutation in existing strains plays any role in the origin of new virulent strains.

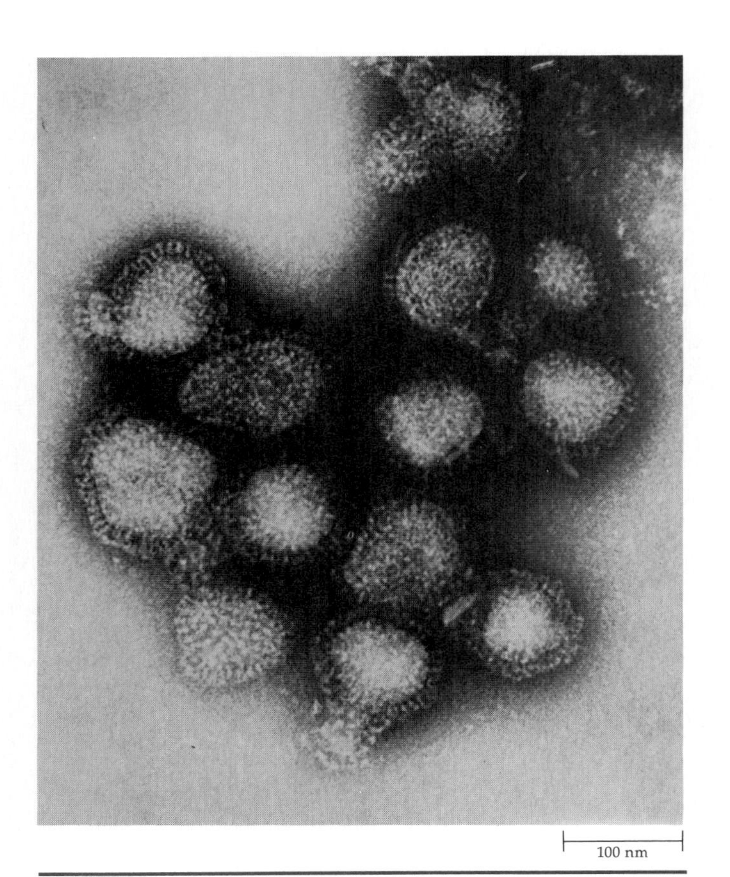

Bacteriophages, although only about 100 nanometers long, are more complex in structure (Figures 11–1 and 11–24). Each particle is fashioned of at least five separate proteins: the repeating protein subunits that make up the hexagonal head, the tail core, the submolecules of the contractile sheath, the base plate of the tail, and the tail fibers. The long DNA molecule is coiled within the hexagonal head of the bacteriophage. At least one additional protein is produced; this is the enzyme lysozyme which causes the breakdown (lysis) of the bacterial cell at the end of the infectious cycle. Clear spots called plaques appear in bacterial colonies in the areas where active lysis by phages is occurring.

Some of the larger viruses, such as those that cause smallpox and influenza, have outer envelopes that appear to be composed of portions of the host plasma membrane carried off by the virus particle at the completion of infection.

Plant viruses, although variable in shape and size, are usually described as elongated (rigid or flexible rods), as rhabdoviruses (bacilliform), or as spherical (Figure 11–25). Most of the elongated viruses appear as long, thin, flexible threads, and many of them seem to consist of particles of different lengths. The rhabdoviruses are about three to five times longer than they are wide, whereas most, if not all, of the spherical plant viruses are polyhedral (many-sided).

The Replication of Viruses

The mode of replication of particular viruses depends on their genetic constitution. The protein coat of a virus determines its attachment to the host plasma membrane and its entry into the host cell. But all viruses shed this coat before replication is initiated. In some, such as the bacteriophages, or bacterial viruses, it is left outside the host cell; in some, it is shed within; and in others, it is digested by the enzymes of the host cell. When this has been accomplished, one of two things happens:

1. A reaction may occur that prevents virus multiplication within the cell, and the viral DNA may be inserted in a linear fashion into the bacterial chromosome. In such a state, the virus is called a *prophage*. Phages that are capable of existing in prophage form are called *temperate* phages. The viruses involved in transduction are temperate phages. Temperate phages do not destroy their host cells unless they "escape" from the host chromosome, and they are in turn virtually unassailable by the host's immune defense system. This phenomenon has been demonstrated only in DNA bacteriophages.

11–24

T4 bacteriophage as seen with the aid of an electron microscope (a) *and as a model* (b). *The type T bacteriophages are a group of viruses that attack* E. coli. *The tail is a hollow core encased in a contractile sheath containing molecules of ATP that provide the energy for the contraction. At the tip of the tail is a base plate, from which radiate six long fibers. The fibers attach to the bacterial cell wall, drawing the base plate to it. The tail protein then contracts, forcing the hollow tail core into the cell like a syringe. The DNA molecule, which is some 650 times longer than the protein head, then passes into the bacterial cell, leaving the protein coat outside.*

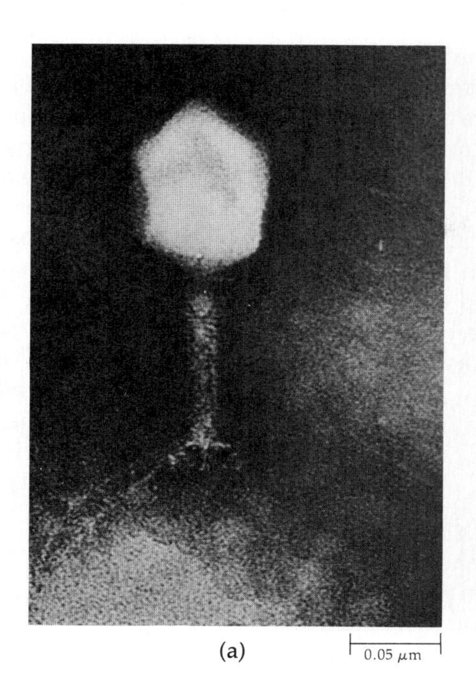

(a)

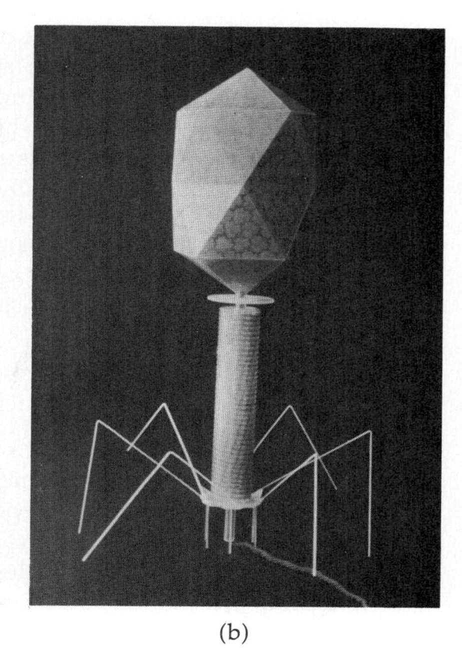

(b)

11–25

Examples of different shapes of plant viruses. (a) *A mixture of the rare Tulare apple mosaic virus and the tobacco mosaic virus. The Tulare apple mosaic virus has a shape typical of spherical viruses. The tobacco mosaic virus is a rigid, rod-shaped structure containing RNA and a protein coat.* (b) *Particles of a rhabdovirus in a cell from a pepper plant* (Capsicum frutescens). *This is the typical shape for those viruses that form bullet-shaped, or bacilliform, particles. The rhabdovirus particles are seen here in the perinuclear space, between the inner and outer membranes of the nuclear envelope. As the virus migrates to the perinuclear space and passes through the inner nuclear membrane, it acquires a portion of that membrane, which then surrounds the particle.* (c) *Particles of a flexible, rod-shaped virus in a cell from the Christmas cactus* (Zygocactus truncatus).

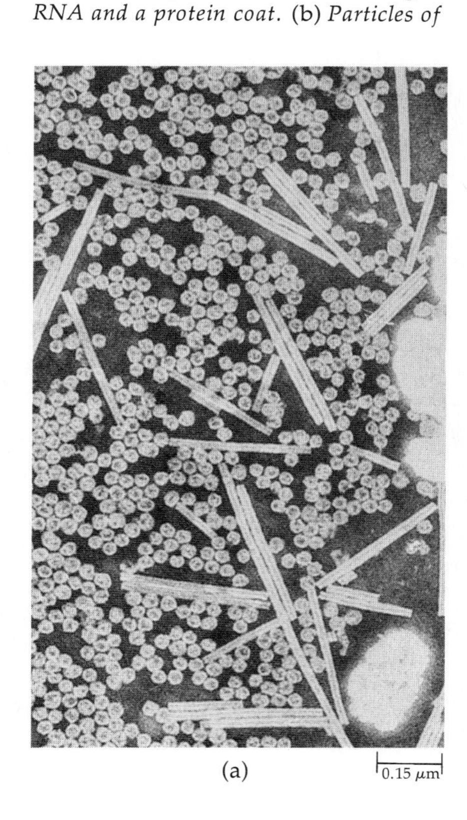

(a)

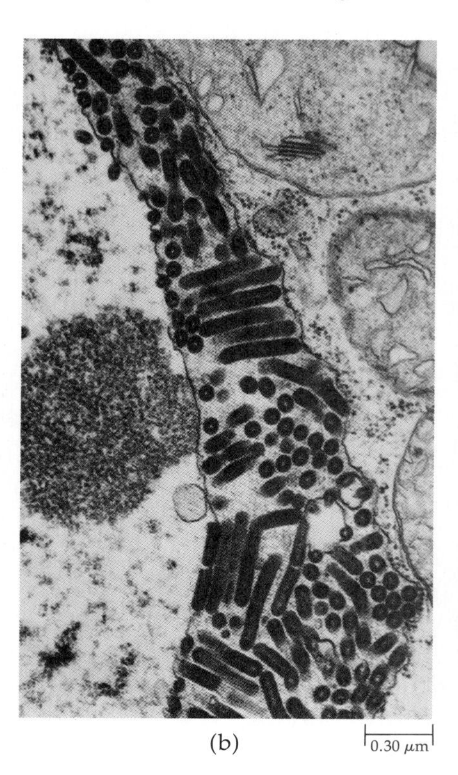

(b)

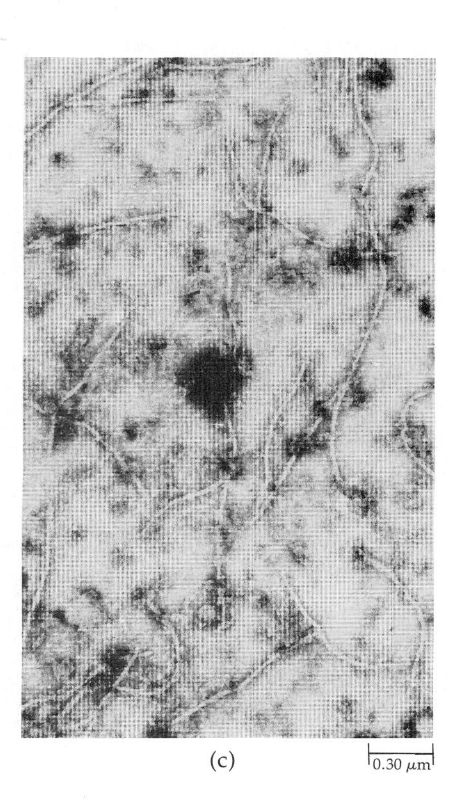

(c)

2. The virus may multiply. Virus multiplication occurs in three steps. First, the viral nucleic acids direct the host cells to produce new viral enzymes. Second, the required viral nucleic acids and structural proteins are synthesized, each in the appropriate amount. Finally, these materials are assembled into virus particles. These steps generally overlap in time, often involve extensive genetic regulation, and lead to the production of many—often thousands of—new virus particles per cell. When the process of virus multiplication is complete, the new virus particles escape from the host cell, which is by then generally dead.

In viruses that contain DNA—such as the vaccinia virus, an organism that causes a poxlike disease in cattle—the viral DNA simply directs the synthesis of a series of different mRNA molecules, which in turn direct the production of different proteins. The DNA is usually double-stranded, but single-stranded DNA occurs in some very small bacterial viruses.

In most RNA viruses, such as the tobacco mosaic virus, the RNA is single-stranded. This viral RNA replicates itself, presumably by directing the formation of a complementary strand which then serves as the template for new viral RNA molecules. The viral RNA also takes over the ribosomes of the host cell and acts as mRNA. In this role, it is responsible for the synthesis of enzymes and virus coat proteins.

Viral activity can profoundly affect the metabolism of the host cell. In the bacterium *Clostridium,* it has been shown for some strains that the production of the lethal toxins associated with botulism occurs only with the active and continued participation of specific bacteriophages. Noninfected bacterial cells do not produce the toxin. Even more surprisingly, infection by other specific bacteriophages causes the same bacterial strain to produce the toxins associated with gas gangrene and many other diseases in animals. The causative organisms associated with botulism and gas gangrene had hitherto been considered to be different species of *Clostridium;* they are now understood, at least in part, to be the same species infected by different bacteriophages.

Viruses and Cancer

It is well known that viruses cause cancer and cancerlike growths in many animals and plants (Figure 11–26), and the possibility that viruses are a cause of cancer in humans has been much discussed. Viruses can change from infectious to noninfectious forms, and they mutate readily to produce strains with new sets of characteristics. There is substantial evidence linking a series of viruses known as the Epstein-Barr viruses with two types of human cancer; the same viruses also cause infectious mononucleosis.

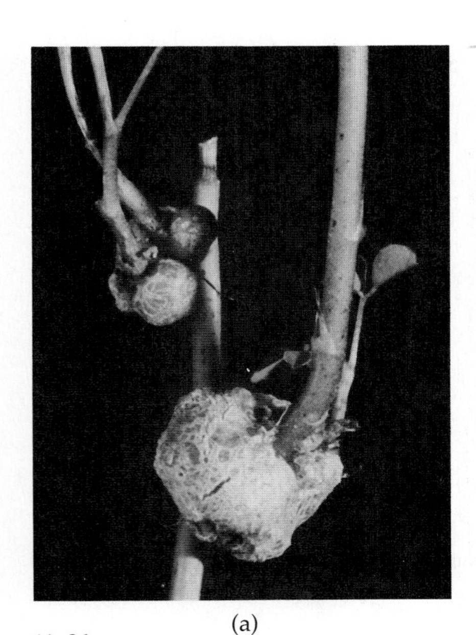
(a)

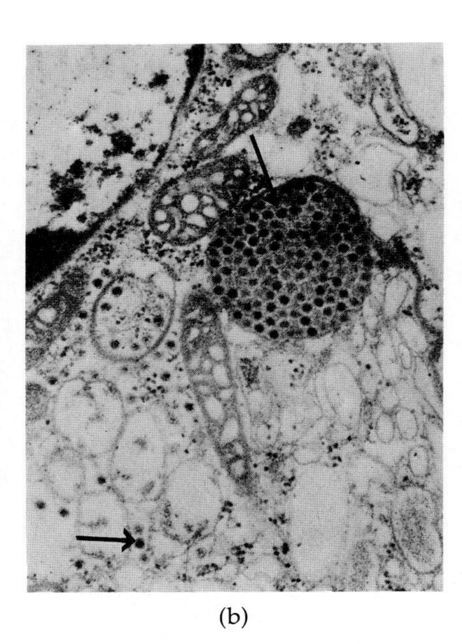
(b)

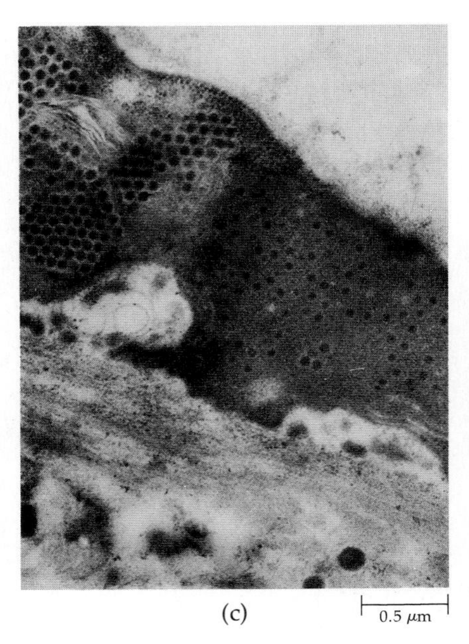

(c)

11–26
(a) *Tumors produced by the wound tumor virus in sweet clover* (Melilotus alba). (b) *Wound tumor virus particles (indicated by arrows) in an electron micrograph of a cell of the host plant.* (c) *The clover leafhopper* (Agallia constricta) *is the insect transmission vector of the wound tumor virus. Electron micrograph of an epidermal cell of the clover leafhopper. Viruses are being produced in the honeycomblike areas at the upper left. Individual viruses can be seen in the dark area below.*

Like other similar viruses, the Epstein-Barr viruses can persist in cells in a repressed state without inducing the manufacture of progeny viruses and the disease symptoms with which they are associated. Viruses affect their hosts differently depending on the genetics of the host and its physiological condition, and the presence of one kind of virus in a cell sometimes enhances the effect of another. Physical or chemical injury or exposure to X-rays can also play a role in activating certain viruses, which suggests that known cancer-causing substances, such as tobacco tar or radioactive materials, may exert their effects by activating a virus. The fact that there are many different forms of cancer, which may have different causes, makes it difficult to prove conclusively whether or not viruses are a cause, but this subject is under active study in research centers all over the world.

The Origin of Viruses

Because of the simplicity of viral structure, some workers have suggested that viruses represent the direct descendants of the first self-replicating units from which the earliest cells eventually evolved. This is clearly not so, for viruses exist today only by virtue of their ability to insert themselves into the machinery of their host cells. They compete with the nucleic acids of the host cells and take over the genetic and metabolic activities of the cells in directing the formation of new virus particles. They must have come into being after the evolution of cells in which the genetic code was already established.

Viruses are essentially similar to bacterial chromosomes (double-stranded DNA viruses) or molecules of messenger RNA (single-stranded and double-

---

VIROIDS

Viroids are the smallest known agents of infectious disease; they are much smaller than the smallest genomes of autonomously replicating viruses. Viroids are thus far known only from plants, and they consist of small molecules of RNA (molecular weight 1.1 to 1.3 $\times$ 10^5) that replicate autonomously in susceptible cells. The first viroid to be characterized, in 1971, was the causative agent of potato spindle tuber disease; it is called PSTV. Its RNA consists of rods about 50 nanometers long, or closed circles of similar dimension, that can be seen only with the aid of an electron microscope. The RNA is single-stranded, and there is no associated DNA or protein. Individual strains of PSTV differ in their virulence, thus indicating some genetic variation. PSTV consists of a single-stranded thread of RNA comprised of 359 bases. In infected cells, PSTV is associated almost exclusively with the nuclei, but on the basis of present evidence, it appears unlikely that viroid RNA can serve as mRNA. Whether the mechanism of PSTV replication involves transcription from RNA or DNA templates is not known, but the available evidence appears to favor RNA-directed viroid replication. The method might be similar to that involved in influenza virus RNA replication, which is now known to require host mRNA as a primer.

Because of the nuclear location and replication of viroids and their apparent inability to act as mRNAs, it has been suggested that they may cause their symptoms by interference with gene regulation in the infected host cells. In infected tissue, certain proteins occur in larger quantities than in healthy tissue. Although molecular sequences complementary to PSTV have not been found in uninfected host species, it has been postulated that PSTV may have originated from altered host genetic material. Specifically, it is postulated that PSTV originated from genes normally present in certain species of the potato family (Solanaceae), which are its primary hosts.

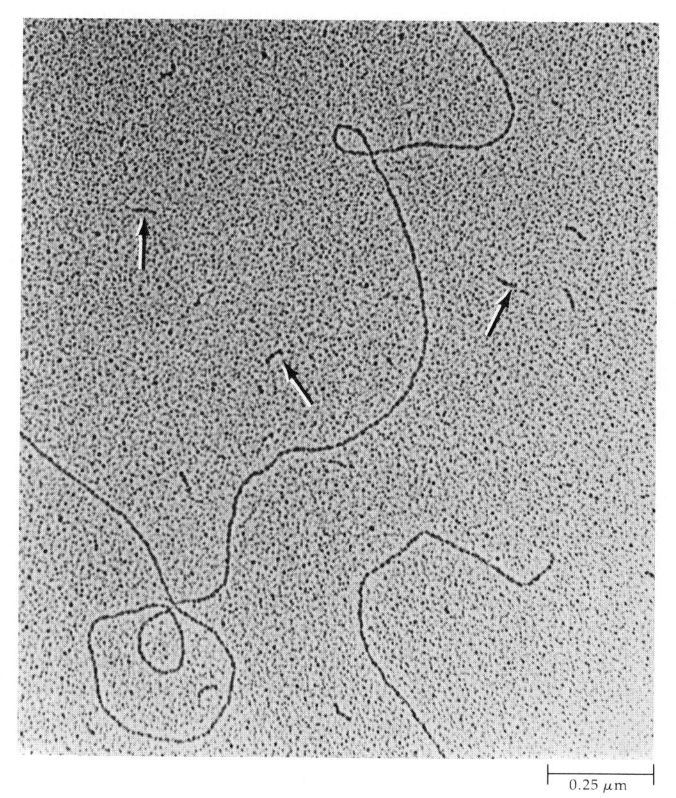

Electron micrograph of potato spindle tuber viroid (arrows) mixed with a double-stranded DNA, coliphage T_7 DNA.

---

stranded RNA viruses) packaged in a protein overcoat. The phenomenon of transformation, discussed earlier, suggests the likely mode for the origin of viruses. Fragments of DNA or RNA occasionally find their way into cells in one way or another. When they do, they may exert an effect upon the genetic processes of the host cell. If they can replicate themselves and spread from cell to cell, they may persist. If they can produce a protective protein wall, they become, in effect, viruses. It is also possible that the largest and most complex viruses, the pox viruses, evolved from degenerate bacteria.

Viruses have probably evolved on many different occasions over the past 3 billion years. In fact, current evidence seems to suggest that they are still evolving today.

SUMMARY

Bacteria comprise the living prokaryotic organisms. Prokaryotes lack an organized nucleus and cellular organelles and do not reproduce sexually. Their genetic material is incorporated into a single circular molecule of DNA. Prokaryotes, except for mycoplasmas and spiroplasmas, have rigid cell walls and are the only organisms in which polypeptides are incorporated into the basic structure of the cell wall. No prokaryote is truly multicellular; although the cells may not divide completely, thus forming filaments or masses, there are no protoplasmic connections between them.

Heterotrophic bacteria share with the fungi the role of decomposers in the world ecosystem. They are metabolically versatile: most bacteria are heterotrophic, but some are photosynthetic, and others are chemoautotrophic. Some are aerobic, others are obligate anaerobes, and still others are facultative anaerobes. A number of genera play important roles in the cycling of nitrogen, sulfur, and carbon. Many bacteria are important as pathogens to either animals or plants.

Photosynthetic bacteria comprise four distinct groups, one of which, the cyanobacteria, contains the same chlorophyll *a* that can be found in photosynthetic eukaryotes. The chemosynthetic autotrophs derive their energy from the oxidation of inorganic molecules. Those that oxidize proteins and amino acids convert ammonium into nitrates, thus playing an important role in the nitrogen cycle. In addition, many cyanobacteria are able to fix atmospheric nitrogen.

Bacteria have rigid cell walls made up of a polysaccharide connected by polypeptide cross-links. Bacterial cells may be rod-shaped (bacilli), spherical (cocci), or coiled (spirilli). If the cell wall does not divide completely, they may adhere in groups or in filamentous or solid masses; such forms are given special names. They may have flagella and be motile, or have shorter structures known as pili; the flagella may be located all over the cell or restricted to one or both ends.

Genetic recombination in bacteria involves the transfer of DNA from one cell to another. This may come about by conjugation, by transformation (the passive incorporation of fragments of DNA in the medium), or by transduction (injection by bacteriophage). However, mutation, combined with a high reproductive rate, is a much more important source of variability in bacteria than recombination.

Viruses are infectious agents composed of an inner core of nucleic acid—either RNA or DNA—and an outer protective coating of protein. They cannot reproduce themselves outside of living cells. In DNA viruses, the viral DNA competes with the DNA of the host cell in directing its activities. In RNA viruses, the viral RNA, which is usually single-stranded, acts as mRNA in the host cell, becoming associated with the ribosomes and serving as a template for the synthesis of proteins. A common architectural arrangement of viruses is the icosahedron, a 20-sided structure.

SUGGESTIONS FOR FURTHER READING

AGRIOS, GEORGE N.: *Plant Pathology,* 2nd ed., Academic Press, Inc., New York, 1978.

A modern discussion of the field, based on a biochemical approach to the problems of parasitism; includes many detailed drawings of the disease cycles.

ALEXANDER, MARTIN: *Microbial Ecology,* John Wiley & Sons, Inc., New York, 1971.

A basic text on microbial ecology, straightforward in presentation but not overly simplistic.

BARILE, M. F., et al.: *The Mycoplasmas, Vol. III: Plant and Insect Mycoplasmas,* Academic Press, Inc., New York, 1979.

A comprehensive account of a rapidly developing field.

BROCK, THOMAS D.: *Biology of Microorganisms,* 3rd ed., Prentice-Hall, Inc., Englewood Cliffs, N.J., 1979.

An interesting and often entertaining presentation of microbiology, including algae and protozoa, with emphasis on the whole cell and its ecology.

BUCHANAN, ROBERT E., and NORMAN E. GIBBONS (Eds.): *Bergey's Manual of Determinative Bacteriology,* 8th ed., The Williams and Wilkins Co., Baltimore, 1974.

A comprehensive treatise on the identification and classification of the prokaryotes.

CAMPBELL, R.: *Basic Microbiology*, Vol. 5, *Microbial Ecology*, Blackwell Scientific, London, 1977.*

A useful guide to the interactions between microorganisms and their environment, as well as their interactions with one another and with other organisms.

CARR, N. G., and B. A. WHITTON (Eds.): *The Biology of Blue-Green Algae*, University of California Press, Berkeley and Los Angeles, 1973.

A comprehensive text, with contributions by many experts in the field.

DAVIS, BERNARD D., et al.: *Microbiology*, 2nd ed., Harper & Row, Publishers, Inc., New York, 1973.

A comprehensive account of all aspects of the biology of bacteria and viruses, biochemically oriented and written from a medical point of view.

DICKINSON, C. H., and J. A. LUCAS: *Plant Pathology and Plant Pathogens*, Halsted Press, John Wiley & Sons, Inc., 1977.*

A well-written introduction to plant diseases, a subject of considerable general interest and economic importance.

FOGG, G. F., W. D. STEWART, P. FAY, and A. E. WALSBY: *The Blue-Green Algae*, Academic Press, Inc., New York, 1974.

A useful synthesis of current knowledge of the cyanobacteria, including an extensive bibliography.

GIBBS, A. J., and B. D. HARRISON: *Plant Virology: The Principles*, Halsted Press, John Wiley & Sons, Inc., 1979.

A very useful overall review of the basic nature of plant-virus interactions; well illustrated.

GOODHEART, CLYDE R.: *An Introduction to Virology*, W. B. Saunders Company, Philadelphia, 1969.

A general account of all aspects of virology, designed as a textbook for advanced undergraduates; particularly well illustrated.

LURIA, S. E., et al.: *General Virology*, 3rd ed., John Wiley & Sons, Inc., New York, 1978.

A comprehensive account of the viruses, perhaps the best general description of the group.

PELCZAR, MICHAEL J., ROGER D. REID, and E. C. S. CHAN: *Microbiology*, 4th ed., McGraw-Hill Book Company, New York, 1977.

A standard text that deals with the morphology, biochemistry, and ecological roles of prokaryotes, viruses, and selected eukaryotes.

STANIER, R. Y., E. A. ADELBERG, and J. L. INGRAHAM: *General Microbiology*, 4th ed., Macmillan Publishing Co., Inc., New York, 1977.

An introduction to the biology of microorganisms, with special emphasis on the properties of bacteria.

STEVENSON, L. HAROLD, and R. R. COLWELL (Eds.): *Estuarine Microbial Ecology*, University of South Carolina Press, Columbia, 1973.

A collection of papers dealing with the microbial ecology of estuaries and coastal waters.

*Available in paperback.

CHAPTER 12

The Fungi

12–1
A mycelium of a basidiomycete on a fallen tree trunk.

The fungi are as distinct from the algae, bryophytes, and vascular plants as they are from the animals. They are discussed in this text both because of their intrinsic interest and because they have been traditionally grouped with the plants. We treat them, however, as a distinct kingdom—the Fungi—which is one of the five main groups of living organisms discussed in Chapter 10.

The fungi, together with the heterotrophic bacteria and a few other groups of heterotrophic organisms, are the decomposers of the biosphere (Figure 12–1), and their activities are as necessary to the continued existence of the world as are those of the food producers. Decomposition releases carbon dioxide into the atmosphere and returns nitrogenous compounds and other materials to the soil where they can be used again by green plants and eventually by animals. It is estimated that, on the average, the top 20 centimeters of fertile soil may contain nearly 5 metric tons of fungi and bacteria per hectare (2.47 acres).

As decomposers, fungi often come into direct conflict with human interests. A fungus makes no distinction between a rotten tree that has fallen in the forest and a railroad tie; it is just as likely to attack one as the other. Equipped with a powerful arsenal of enzymes that break down organic products, fungi are often nuisances and are sometimes highly destructive. This is especially true in the tropics, because warmth and dampness promote fungal growth; it is estimated that during World War II less than 50 percent of the military supplies sent to tropical areas arrived in usable condition. Fungi attack cloth, paint, cartons, leather, waxes, jet fuel, insulation on cables and wires, photographic film, and even the coating of the lenses of optical equipment—in fact, almost any conceivable substance. Even in temperate regions, they are the scourge of food producers and sellers alike, for they grow on bread, fresh fruits, vegetables, meats, and other products. Fungi reduce the nutritional value, as well as the palatability, of such foodstuffs. They also produce extremely poisonous toxins, some of which—the aflotoxins—are extremely carcinogenic and show their effects at concentrations as low as a few parts per billion.

The economic importance of fungi as commercial pests is enhanced by their ability to grow under a wide range of conditions. Thus some strains of *Cladosporium herbarum,* which attack meat in cold storage, can grow at −6°C. In contrast, one species of *Chaetomium* grows optimally at 50°C and can even be grown at 60°C if transferred gradually to that temperature.

The abilities of fungi that make them such important commercial pests also make them commercially valuable. Many fungi, especially the yeasts, are useful because of their ability to produce substances such as ethanol and carbon dioxide (which plays a central role in baking). Others are of interest as sources of antibiotics, including penicillin—the first antibiotic to be used widely—and as potential sources of proteins.

Associations between fungi and other organisms are extremely diverse. For example, about four-fifths of all land plants form associations between their roots and fungi called mycorrhizae. These associations play a critical role in plant nutrition and distribution. Lichens are a large group of fungi that contain algal or cyanobacterial cells from which they obtain nutrients; consequently, the lichens are able to occupy the least hospitable habitats. Many fungi attack living organisms, rather than dead ones (Figure 12–2), and sometimes in highly unusual ways (see "Predaceous Fungi," page 224). They are the most important single cause of plant diseases; well over 5000 species of fungi attack economically valuable crop and garden plants, as well as many wild plants. In attacks on living trees, fungi cause the annual loss of over 20,000 cubic meters of timber in the state of California alone. Other fungi are the cause of serious diseases in humans and domestic animals.

Some 100,000 species of fungi have been described, and it is estimated that as many as 200,000 more may await discovery. There may actually be as many species of fungi as there are species of plants, although far fewer have been described thus far. The fungi have no direct evolutionary connection with the plants and apparently were derived independently from a different group of single-celled eukaryotes.

Traditionally, the fungi have been considered to include not only the related divisions included in this chapter but also the heterotrophic protista discussed in Chapter 13. Because there is little evidence for a direct relationship between the fungi and the several groups of protista that have traditionally been included with them, they are separated in this book for clarity of discussion.

BIOLOGY OF THE FUNGI

The fungi are primarily terrestrial. Although some fungi are unicellular, most are filamentous, and structures such as mushrooms consist of a great many such filaments, packed tightly together. Fungal filaments are known as *hyphae,* and a mass of hyphae is called a *mycelium* (see Figures 12–1 and 12–3). Growth of the hyphae occurs at their tips, but proteins are synthesized throughout the mycelium and are carried to the tips of the hyphae by cytoplasmic streaming, a phenomenon that is particularly well developed among the fungi. Within 24 hours, an individual fungus may produce more than a kilometer of new mycelium. The words "mycelium" and "mycologist" (a scientist who studies the fungi) are derived from the Greek word *myketos,* meaning "fungus."

12–2
A hornet killed by a parasitic fungus.

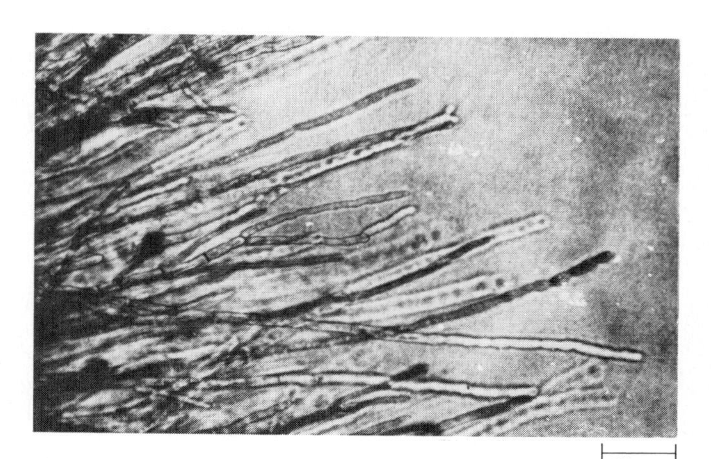

12–3
A mycelium of a species of Aspergillus. *The dark bodies within some of the hyphae are asexual spore-producing structures.*

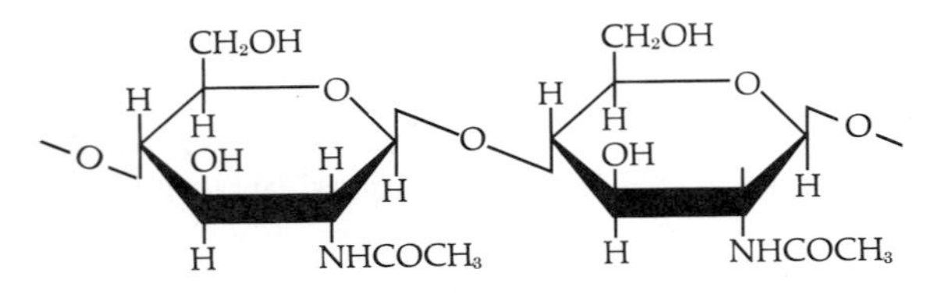

12–4
The structure of chitin, which consists of beta-1,4-linked N-acetylglucosamine *units. A similar linkage is found in cellulose and in the cell walls of bacteria, suggesting that such a linkage provides a particularly strong, compact polysaccharide. Chitin is characteristic of the cell walls of many fungi, as well as the exoskeletons of arthropods.*

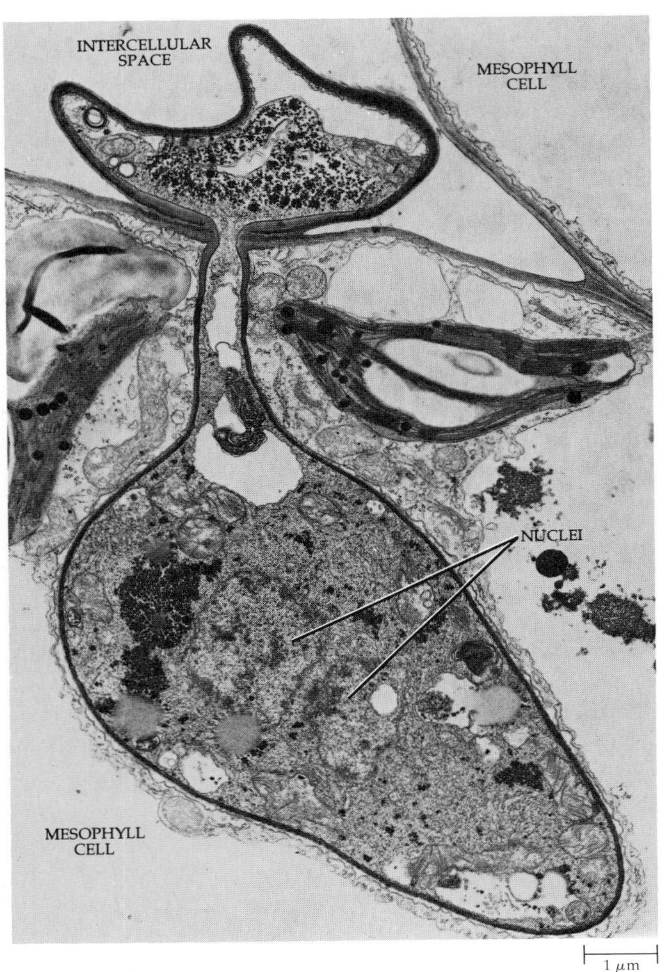

12–5
A haustorium of Melampsora lini, *a rust fungus, growing in a mesophyll cell in a leaf of common flax* (Linum usitatissimum).

In the plants and many groups of protista, the cell wall is built on a framework of cellulose microfibrils that is interpenetrated by a matrix of noncellulosic molecules, such as hemicelluloses and pectic substances. In the fungi, the cell wall is composed primarily of another polysaccharide—*chitin*—which is the same material found in the hard shells, or exoskeletons, of arthropods, such as insects, arachnids, and crustaceans (Figure 12–4). Chitin is far more resistant to microbial degradation than is the cellulose found in most eukaryotic cell walls.

With their rapid growth and filamentous form, fungi have a relationship to their environment that is very different from that found in any other group of organisms. The surface-to-volume ratio of fungi is very high, which means that they are in intimate contact with the environment. With a few exceptions, no part of a fungus is more than a few micrometers from the external environment, being separated from it only by a thin cell wall and the plasma membrane. As a consequence, a fungus with an extensive mycelium can have a profound effect on certain aspects of its surroundings, for example, binding soil particles together. Hyphae often fuse, even when they have grown from different spores, and thus increase the intricacy of the network.

The maintenance of this sort of intimate relationship between fungus and environment requires that all parts of the fungus are metabolically active, and the sorts of quiescent layers of tissue found, for example, in the higher plants are absent in the fungi. Enzymes and other substances secreted by fungi have an immediate effect on the surroundings and are of great importance for the maintenance of the fungus itself.

All fungi are heterotrophic. They obtain their food either as saprobes (organisms that live on dead organic material) or as parasites (organisms that feed on living matter). In either case, the food is ingested by absorption after it has been partially digested by enzymes that are secreted outside the fungal cell. Some fungi, including a number of yeasts, can release energy by anaerobic respiration, such as in the production of ethyl alcohol from glucose. Glycogen is the primary storage polysaccharide in some fungi, as it is in animals; lipids serve an important storage function in other fungi.

Saprobic fungi sometimes have a form of somewhat specialized hyphae, known as *rhizoids,* that anchor them to the substrate. Parasitic fungi often have specialized hyphae, called *haustoria* (singular: haustorium), that penetrate and absorb nourishment directly from the cells of other organisms (Figure 12–5).

All fungi have cell walls and most produce spores of some kind. They are nonmotile throughout their life cycle, lacking cells with cilia or flagella. Fungal spores are dispersed by the wind.

THE EVOLUTION OF FUNGI

The first fungi were probably unicellular eukaryotic organisms that apparently no longer have any living counterparts. From these organisms were derived the *coenocytic* fungi, in which many nuclei are found in a common cytoplasm (*coenocytic* means "contained in a common vessel," or multinucleate). The living coenocytic fungi are classified as members of the division Zygomycota and are commonly known as zygomycetes.

In members of the division Ascomycota, called ascomycetes, and the division Basidiomycota, or basidiomycetes, the mycelia are septate—divided by cell walls—but the septa, or cross walls, are perforated (Figure 12–6). In some fungi, the cytoplasm and its contents stream quite freely along the hyphae. Ascomycetes seem to have been derived from coenocytic forms similar to the zygomycetes, whereas the basidiomycetes were derived from the ascomycetes.

The oldest fossils that resemble fungi occur in strata about 900 million years old, but the oldest that have been identified with certainty as fungi are from the Ordovician period, 450 to 500 million years ago (see Appendix D). Zygomycetes were associated with the underground portions of the earliest vascular plants in the Silurian period, some 400 million years ago. Fungi may be among the oldest eukaryotes; the major groups were in existence by the close of the Carboniferous period, some 300 million years ago.

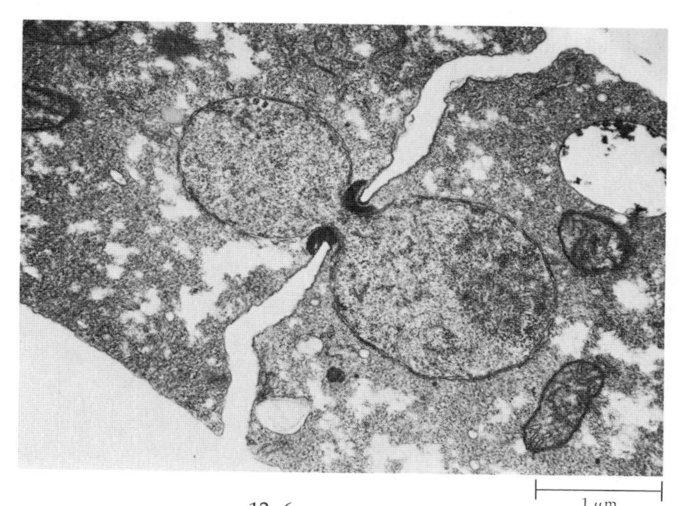

12–6
Neurospora crassa, *an ascomycete. Electron micrograph of a perforated cross wall with a nucleus passing through the septal pore.*

Mitosis and Meiosis

Fungi have many biological peculiarities that are just beginning to be understood. One of the most intriguing characteristics involves the process of nuclear division. In fungi, the processes of mitosis and meiosis are different from those that occur in plants, animals, and most protista. The nuclear envelope does not dissociate and re-form but is constricted near its midpoint between the two daughter nuclei, and the spindle apparatus is formed within the nuclear envelope. Centrioles are lacking in all fungi. Most interesting, however, is the recent discovery that the chromosomes of fungi—as far as is known—either lack or have only very small quantities of histone proteins. In this respect they resemble the otherwise dissimilar dinoflagellates. This unique combination of features tends to confirm the view that fungi are not directly related to other living eukaryotes and therefore deserve their present status as a separate kingdom.

FUNGAL REPRODUCTION

The reproductive structures of fungi are separated from the hyphae by complete septa. These reproductive structures are called *gametangia* if they are directly involved in the production of gametes, and *sporangia* if they are involved in the production of asexual spores. The gametes are equal in size, and the fungi are therefore *isogamous.* Meiosis immediately follows the formation of the zygote in all fungi; in other words, meiosis in fungi is zygotic (see Figures 10–9 and 10–10, page 180).

Nonmotile spores are the characteristic means of reproduction in fungi. Some of the spores are very small; they can remain suspended in the air for long periods, thus being carried to great heights and for great distances. This presumably is an important factor in the very wide distribution of many fungi. The spores of some fungi are spread by adhering to the bodies of insects and other animals. The bright colors and powdery textures often seen in association with many types of molds are produced by the spores (see Figure 12–14, page 223).

Heterokaryosis and Parasexuality

Heterokaryosis

Among the genetic characteristics that distinguish the fungi from other groups of organisms, none is more significant than the phenomenon of *heterokaryosis,* which was discovered by the mycologist H. Burgeff in 1912. A strain of fungus is heterokaryotic if the nuclei found in a common cytoplasm are genetically different, either because of mutation or the fusion of genetically distinct hyphae, which appears to be a common occurrence in nature. If the nuclei are genetically similar, the strain is *homokaryotic*.

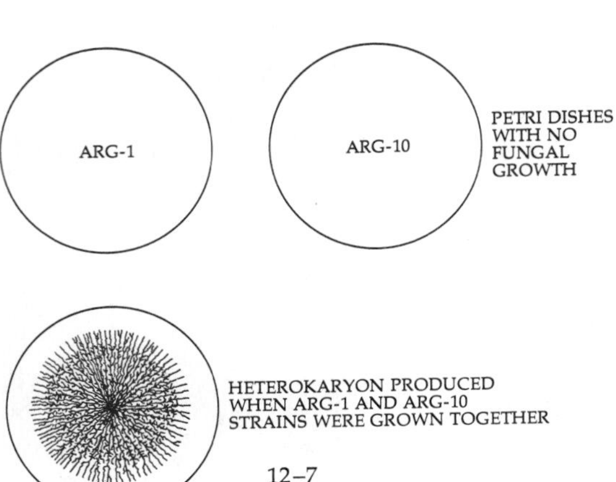

12–7
Arg-1 and Arg-10 are mutant strains of Neurospora, *each of which lacks a different enzyme involved in the production of arginine. Neither will grow on a minimally nutritious medium. However, a heterokaryon formed between the two strains grows readily on the minimal medium. By combining the genetic information contained in the original mutant nuclei, the heterokaryon can synthesize the missing amino acid.*

Heterokaryosis is extremely important in the genetics and evolution of the fungi. If the genetically different nuclei in heterokaryotic fungi are segregated, phenotypically distinct hyphae can be created. Thus, even if only two different sorts of nuclei are present, three distinct phenotypes (one the same as the original one, containing both types of nuclei) can be derived from that particular mycelium.

The results of heterokaryosis are somewhat similar to those of the diploid condition in other organisms; that is, the appearance and physiological characteristics of a heterokaryotic organism are determined by the interaction of the genetically different nuclei. As recessive mutations accumulate in some of the nuclear lines, they may be compensated for by the genes present in other nuclei. Finally, because many of the nuclear lines are unable to exist or, at least, to compete successfully, in a homokaryotic state, heterokaryotic strains are favored by selection (Figure 12–7).

Parasexuality

The *parasexual cycle* in fungi was discovered in 1952 by two investigators at the University of Glasgow. Working with *Aspergillus nidulans,* G. Pontecorvo and J. A. Roper found that haploid nuclei may fuse in a heterokaryotic mycelium to produce diploid nuclei, some of which are heterozygous (produced by the fusion of genetically different nuclei). It is estimated that *Aspergillus nidulans* contains only 1 diploid heterozygous nucleus for every 1000 haploid nuclei.

Within the diploid nucleus, the chromosomes may become associated. If this occurs, some crossing-over can take place, although such a phenomenon is infrequent. Haploid nuclei may or may not re-form. If they do re-form, they may be genetically different from either of the parent haploid nuclei that fused initially. These new haploid nuclei can then take part in other new heterokaryotic combinations.

It has recently been demonstrated that parasexual cycles exist in several other groups of fungi. The significance of such cycles in nature has yet to be assessed completely. However, the parasexual cycle appears to be a flexible and commonly occurring system, especially among those fungi that either do not reproduce sexually or in which sexual reproduction is an infrequent phenomenon.

THE MAJOR GROUPS OF FUNGI

Division Zygomycota

Most zygomycetes live on decaying plant or animal matter in the soil; some are parasites of plants, insects, or small soil animals. There are approximately 600 described species of zygomycetes. The terms "Zygomycota" and "zygomycetes" refer to the chief characteristic of the division—the production of sexual resting spores called *zygospores* (Figure 12–8). A zygospore is produced following the fusion of gametes of equal size which are formed within gametangia. Asexual reproduction by means of spores produced in more or less specialized sporangia borne on the hyphae is almost universal in zygomycetes. Most members of the division have coenocytic (multinucleate) hyphae.

One of the most common members of this division is *Rhizopus stolonifer,* a black bread mold that forms cottony masses on the surface of moist bread exposed to air. The life cycle of *R. stolonifer* is illustrated in Figure 12–9. The mycelium of *Rhizopus* is composed of three different types of haploid hyphae. The bulk of the mycelium consists of rapidly growing hyphae that are coenocytic (multinucleate) and aseptate (not divided by cross walls into cells or compartments). From these, arching hyphae called *stolons* are formed. The stolons form rhizoids wherever their tips come into contact with the substrate. Sporangia form on the tips of the sporangiophores ("sporangia bearers"), which are erect branches formed directly above the rhizoids. Each sporangium begins as a swelling into which a number of nuclei flow, and it is eventually cut off from the sporangiophores by the formation of a septum. The protoplasm within is cleaved, and a cell wall is formed around each spore. The sporangium becomes black as it matures, giving the mold its characteristic color. Each spore, when liberated, can germinate to produce a new mycelium.

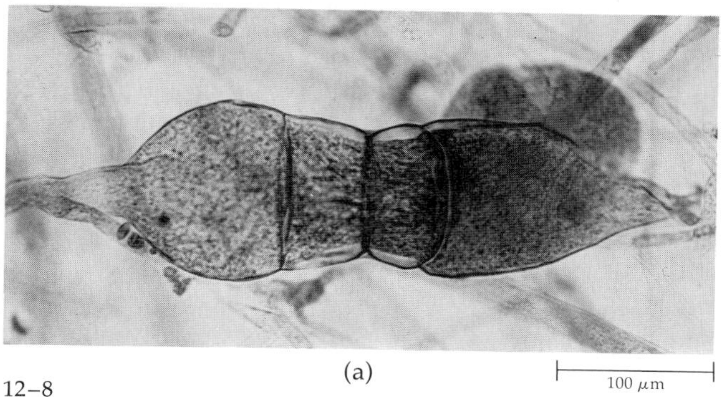

(a) 100 μm

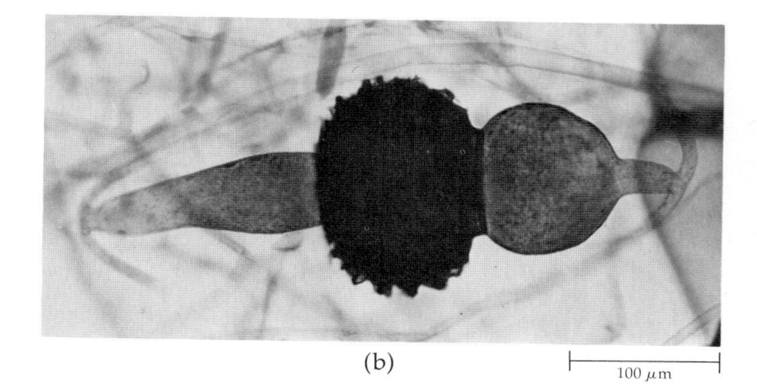

(b) 100 μm

12–8
Rhizopus stolonifer *is a black bread mold that colonizes the surface of moist substrates exposed to air.* (a) *Gametangia, the gamete-producing structures.* (b) *A zygospore, or sexual resting spore (the dark mass in the center).*

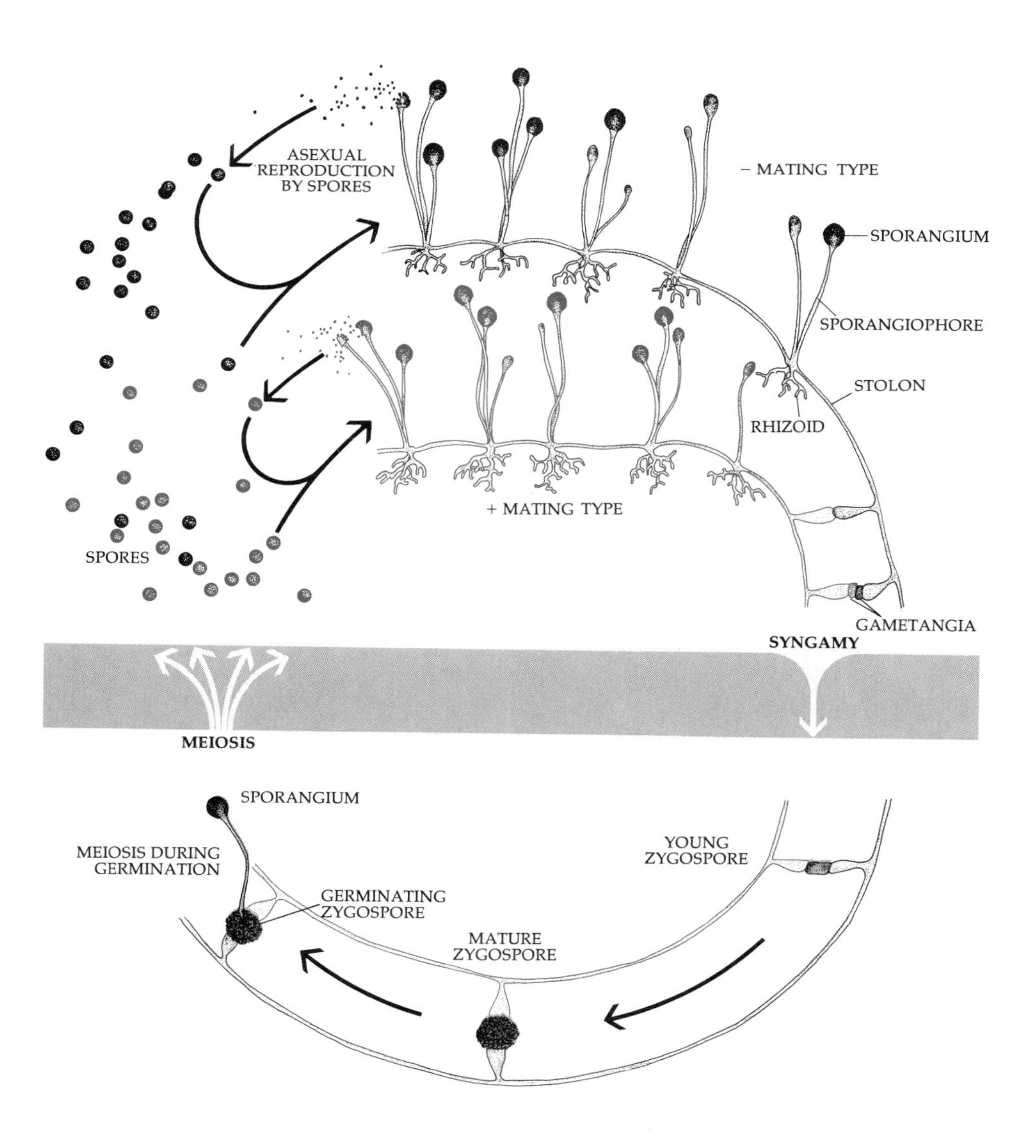

12–9
In Rhizopus stolonifer, *sexual reproduction occurs only between different mating strains, which have been traditionally labeled + and − types. (Although the mating strains are morphologically indistinguishable, they are shown here in two colors.) When the two strains are in close proximity, hormones are produced that cause their hyphal tips to come together and develop into gametangia, which become separated from the rest of the fungal body by the formation of septa (see Figure 12–8a). The walls between the two touching gametangia dissolve, and the two multinucleate protoplasts come together. The + and − nuclei fuse in pairs to form a young zygospore with several diploid nuclei. The zygospore then develops a thick, rough, black coat and becomes dormant, often for several months. Meiosis occurs at the time of germination. The zygospore cracks open and produces a sporangium that is similar to the asexually produced sporangium, and the cycle begins again.*

PHOTOTAXIS IN A FUNGUS

In Pilobolus, *a zygomycete that grows on dung, the sporangia are shot toward the light. The sporangium is oriented toward the light so that all light rays entering the subsporangial swelling converge on a basal photoreceptive area. Light focused elsewhere promotes maximum growth of the sporangiophore on the side away from the light. The high turgor pressure of the sap in the vacuole of the subsporangial swelling eventually causes it to split, which blasts the sporangium off to a distance of 2 meters or more. After the sporangium has been fired off, the sporangiophore collapses. The sporangium adheres where it lands, and if this happens to be on a blade of grass, it may be eaten by an herbivore. It then passes through the digestive tract of the herbivore unharmed and is deposited in the dung to begin the cycle anew.*

(a) *A dense colony of* Pilobolus *growing on horse manure, with the sporangia oriented toward the light.* (b) *The top of a single sporangiophore.* (c) *A sporangium being shot off into space as the subsporangial swelling collapses. The sporangium is trailing a stream of vacuolar sap from the collapsed subsporangial swelling.*

SPORANGIUM
SUBSPORANGIAL SWELLING
WATER DROPLET
SPORANGIOPHORE
SUN'S RAYS
SPORANGIUM TRAJECTORY

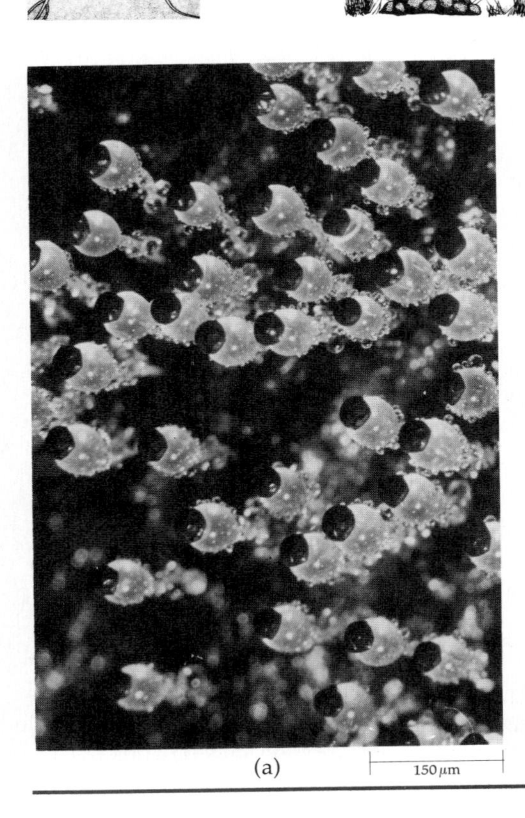

(a) 150 μm

(b) 75 μm

(c)

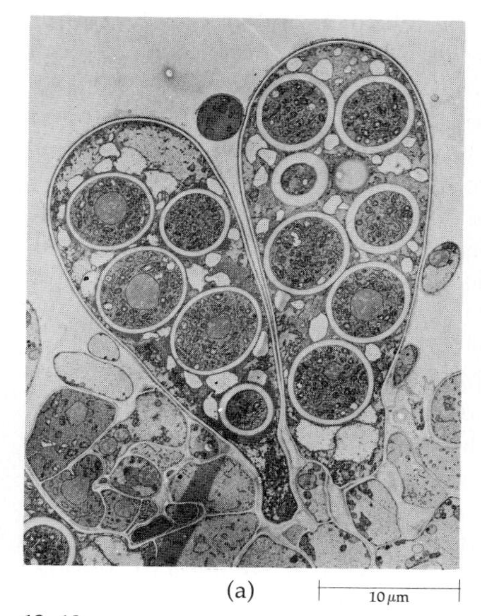

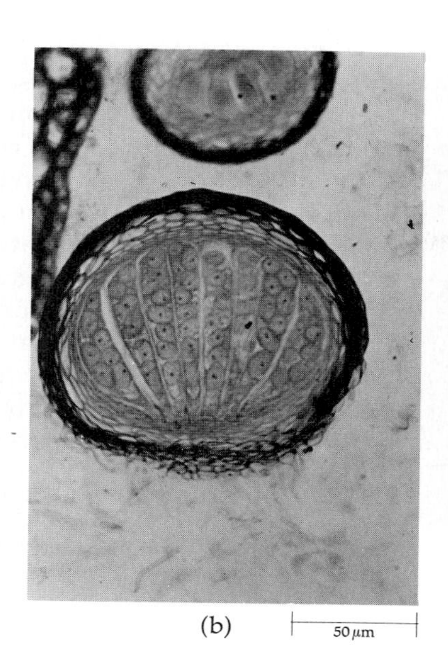

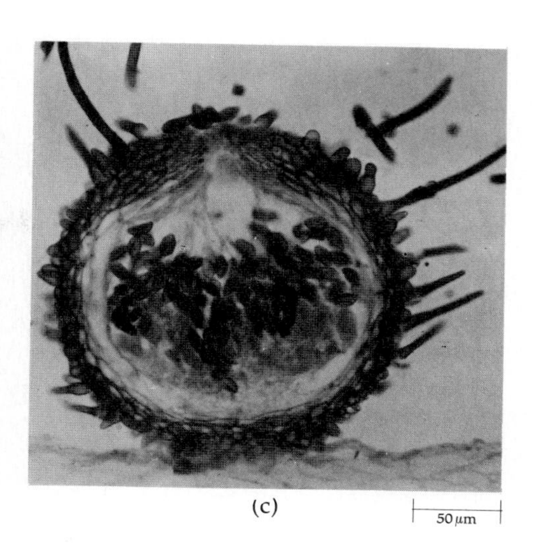

12–10
(a) *An electron micrograph showing two asci of* Ascodesmis nigricans *in which ascospores are maturing.* (b) *Ascocarp of* Erysiphe aggregata, *showing the enclosed asci and ascospores. This completely enclosed type of ascocarp is called a cleistothecium.* (c) *An ascocarp of* Chaetomium erraticum, *showing the enclosed asci and ascospores. Note the small pore at the top. This sort of ascocarp, with a small opening, is known as a perithecium.*

Division Ascomycota

The ascomycetes comprise about 30,000 described species, including a number of familiar and economically important fungi. Most of the blue-green, red, and brown molds that cause food spoilage are ascomycetes, including the salmon-colored bread mold *Neurospora,* which has played an important role in the development of modern genetics. Ascomycetes are the cause of a number of serious plant diseases, including powdery mildews that attack fruits, chestnut blight (caused by the fungus *Endothia parasitica,* accidentally introduced from northern China), and Dutch elm disease (caused by *Ceratocystis ulmi,* a fungus native to certain European countries). Yeasts are also ascomycetes, as are the edible morels and truffles (see Figure 12–20, page 228). As a whole, this group of fungi is relatively poorly known, and thousands of additional species—some undoubtedly of great economic importance—await scientific description.

Characteristics of the Ascomycetes

Ascomycetes, with the exception of the unicellular yeasts, are filamentous when they are growing. In general, their hyphae are septate, or divided by cross walls. The septa are perforated, however, and the cytoplasm and its nuclei can move freely through the septal pores (see Figure 12–6). The hyphal cells of the vegetative mycelium may be either uninucleate or multinucleate. Some species of ascomycetes are homothallic (self-fertile) and can produce sexual structures from a single genetic strain; others are heterothallic and require a combination of + and − strains.

In the majority of ascomycetes, asexual reproduction is by formation of specialized spores, known as *conidia* (a Greek word meaning "fine dust"), which are cut off from the tips of modified hyphae called *conidiophores* ("conidia bearers"). Conidia are usually multinucleate.

Sexual reproduction in ascomycetes always involves the formation of an *ascus* (plural: asci), a saclike structure that is characteristic of this division and distinguishes the ascomycetes from all other fungi (Figure 12–10). Ascus formation usually occurs within a complex structure composed of tightly interwoven hyphae—the *ascocarp.* Many ascocarps are macroscopic. An ascocarp may be open and more or less cup-shaped (an apothecium; see Figure 12–20b, page 228), closed and spherical in shape (a cleistothecium; see Figure 12–10b), or flask-shaped, with a small pore through which the ascospores escape (a perithecium; see Figure 12–10c). The asci are usually borne on the inner surface of the ascocarp. The layer of asci is called the *hymenium,* or hymenial layer (see Figure 12–11).

Figure 12–12 on page 221 illustrates the characteristic life cycle of an ascomycete. (Frequent reference to this illustration should prove helpful with the follow-

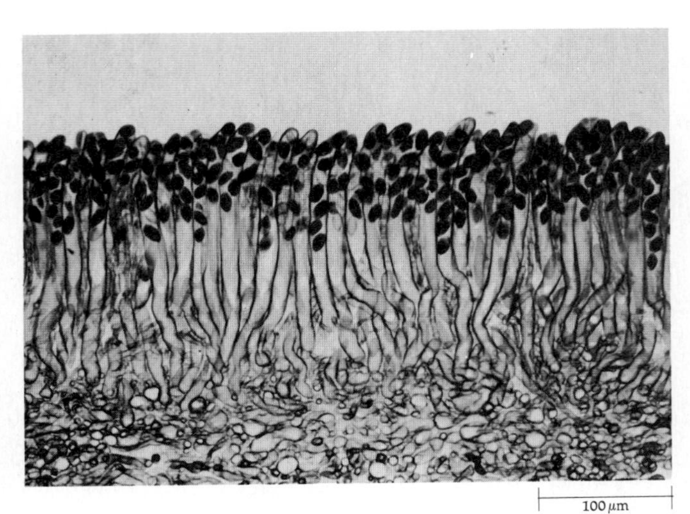

12–11
A section through the hymenial layer of Morchella *showing asci with ascospores.*

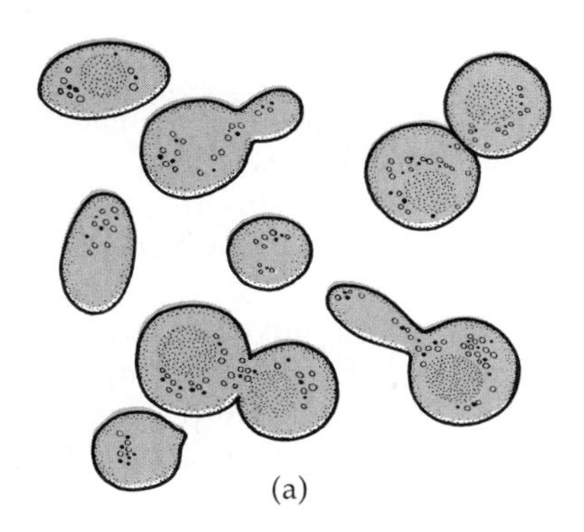

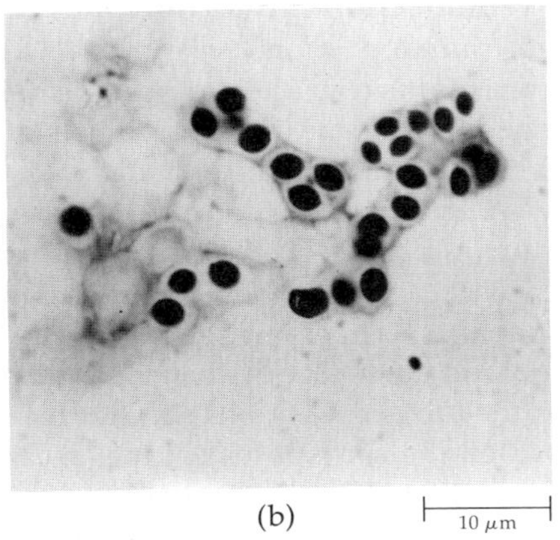

12–13
Yeasts. (a) *Budding cells of bread yeast,* Saccharomyces cerevisiae. (b) *Asci with ascospores of* Schizosaccharomyces octosporus.

ing material.) The mycelium is initiated with germination of an ascospore, and soon after, the mycelium begins to form conidiophores which bear conidia. Many crops of conidia are produced during the growing season, and it is the conidia that are primarily responsible for propagation of the fungus.

Ascus formation occurs on the same mycelium that produces conidia and is preceded by the formation of generally multinucleate gametangia called antheridia and ascogonia. The male nuclei of the antheridium pass into the *ascogonium* via the *trichogyne*, which is an outgrowth of the ascogonium. *Plasmogamy*, the fusion of the two protoplasts, has now taken place. The male nuclei may then pair with the genetically different female nuclei within the common cytoplasm but they do not fuse with them. Hyphal filaments now begin to grow out of the ascogonium and elongate into *ascogenous hyphae*. As the ascogenous hyphae develop, pairs of nuclei migrate into them and simultaneous mitotic divisions occur in the hyphae and ascogonium. Cell division in the developing ascogenous hyphae occurs in such a way that the resultant cells are *dikaryotic* (contain two haploid nuclei).

The ascus first forms at the tip of a developing dikaryotic, ascogenous hypha. In the formation of an ascus, one of the binucleate cells of the dikaryotic hypha grows over to form a hook, or crozier. In this hooked cell, the two nuclei divide in such a way that their spindle fibers are parallel and more or less vertical in orientation. Two of the daughter nuclei are close to one another at the top of the hook; one of the others is near the tip and the other is near the basal septum of the hook. Two septa are then formed; these divide the hook into three cells, of which the middle one becomes the ascus. It is in this middle cell that *karyogamy* occurs: the two nuclei fuse to form a diploid nucleus (zygote), the only diploid nucleus in the life cycle of the ascomycetes. Soon after karyogamy, the young ascus begins to elongate. The diploid nucleus then undergoes meiosis, which is generally followed by one mitotic division, giving a total of four or eight nuclei. These haploid nuclei are then cut off in segments of the cytoplasm to form *ascospores*. In most ascomycetes, the ascus becomes turgid at maturity and finally bursts, sending its ascospores explosively into the air. Although most shoot the ascospores only about 2 centimeters from the ascus, some species propel them as far as 30 centimeters.

Unicellular Ascomycetes: The Yeasts

Yeasts are predominantly unicellular and reproduce asexually by fission or by the pinching off of small buds (Figure 12–13a) rather than by spore formation. Sexual reproduction in yeasts occurs when either two cells or two ascospores unite and form a zygote. The

12–12
The typical life cycle of an ascomycete. Asexual reproduction occurs by way of specialized spores, known as conidia. Sexual reproduction involves the formation of asci and ascospores. Karyogamy is followed immediately by meiosis in the ascus.

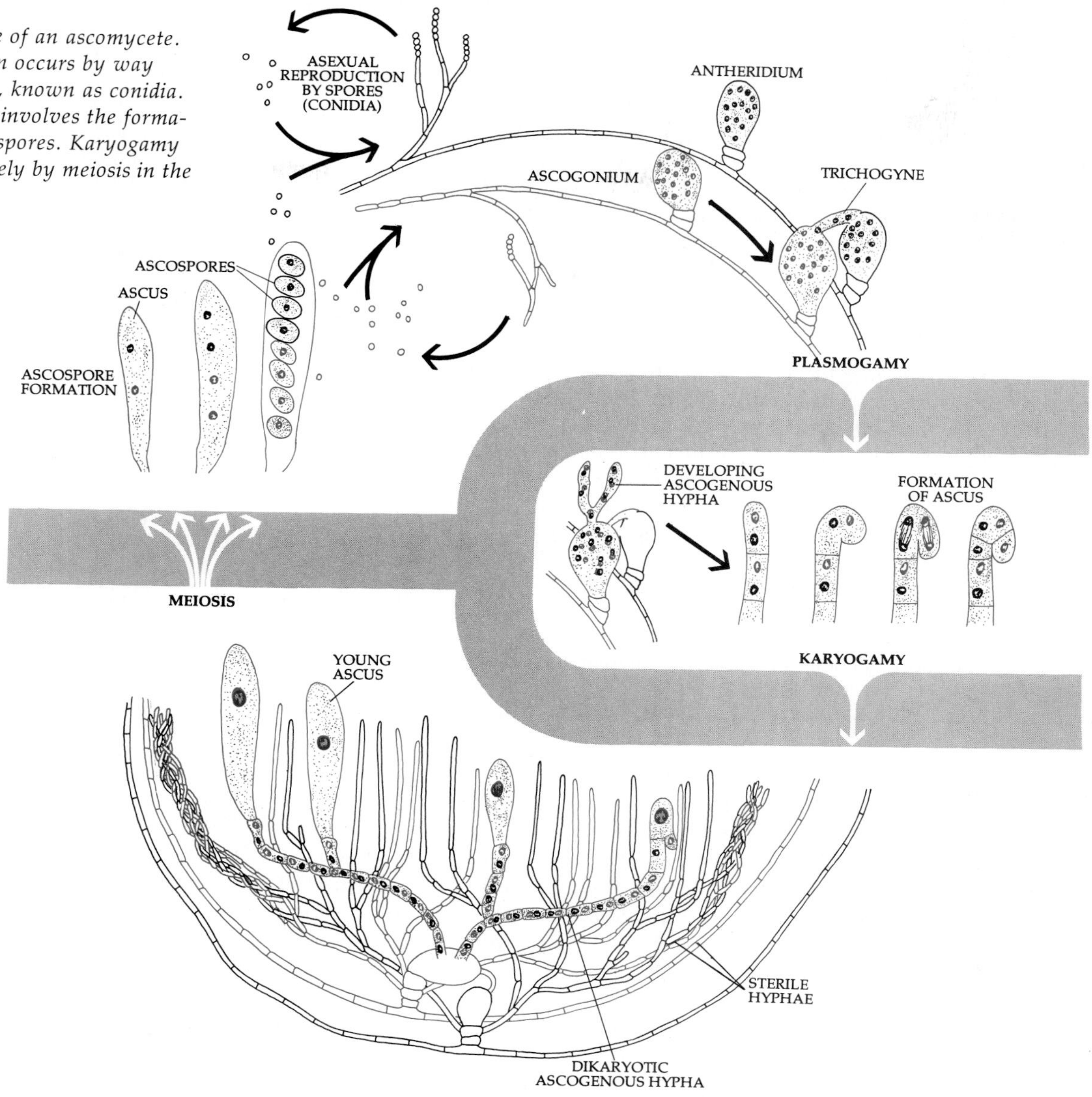

zygote may produce diploid buds or may undergo meiosis to produce four haploid nuclei. There may be a subsequent mitotic division. Within the zygote wall, which is now an ascus, walls are laid down around these nuclei so that eight ascospores are formed, which are liberated when the ascus wall breaks down (Figure 12–13b). The ascospores either bud asexually or fuse with another cell to repeat the sexual process.

The yeasts are important to humans because of their ability to ferment carbohydrates, breaking down glucose to produce ethyl alcohol and carbon dioxide in the process. Thus yeasts are utilized by brewers and vintners and also by bakers; alcohol is the important product for the former and carbon dioxide for the latter. Many domestically useful strains of yeast have been developed by selection and breeding. Wild species are also of great importance, as in the fermentation of wine, in which the yeast species that are naturally present are often supplemented by pure cultures of other specific strains (see Figure 5–12, page 94). On the other hand, only pure yeast cultures are used in brewing beer because the medium is sterilized by heating prior to fermentation. Some of the flavors of wine come directly from the grape, but most arise from the direct action of the yeast. Most of

ERGOTISM

A number of ascomycetes are parasitic on higher plants. The disease of plants called ergot is caused by Claviceps purpurea, *a parasite of rye* (Secale cereale) *and other grasses. Although ergot seldom causes serious damage to the crop of rye, it is dangerous because a small amount mixed with rye grains is enough to cause severe illness among domestic animals or among the people who eat bread containing flour from such grain. Ergotism, the toxic condition caused by eating grain infected with ergot, is often accompanied by gangrene, nervous spasms, psychotic delusions, and convulsions. It occurred frequently during the Middle Ages, when it was known as St. Anthony's fire. In one epidemic in 994, more than 40,000 people died. In 1722, ergotism struck down the cavalry of Czar Peter the Great on the eve of battle for the conquest of Turkey and thus changed the course of history. As recently as 1951, there was an outbreak in a small French village in which 30 people became temporarily insane, imagining that they were pursued by demons and snakes; 5 of the villagers died. Ergot, which causes muscles to contract and blood vessels to constrict, is used in medicine. Dark resting structures of* Claviceps *can be seen among the spikelets of rye in the accompanying photograph.*

the yeasts important in the production of wines, cider, sake, and beer are strains of a single species—*Saccharomyces cerevisiae*—although other species also play a role. This yeast is now virtually the only one used in baking bread (Figure 12–13a). Some species of yeast are important as human pathogens, causing such diseases as thrush and cryptococcosis. A number of yeasts, especially *S. cerevisiae,* are also important laboratory organisms for genetic research.

Superficially, most yeasts are simple ascomycetes. However, a few genera are apparently basidiomycetes—such as *Cryptococcus,* the causative agent of cryptococcosis, and some nonpathogenic species of *Candida.* The simplified and usually unicellular structure of yeasts evidently has been derived from that of the more complex mycelium-forming fungi. The resulting organisms are so reduced that it is difficult to determine their affinities to the different groups of fungi from which they ultimately have been derived. The yeasts include some 39 genera, with approximately 350 species. They are found in a wide range of terrestrial and aquatic habitats in which a suitable carbon source is available.

The Fungi Imperfecti

The class Fungi Imperfecti, or Deuteromycetes, is a large group comprising some 25,000 described species of fungi in which the sexual phase (the perfect phase) is unknown (Figure 12–14). The sexual phase may be unknown either because these fungi have not been sufficiently studied or because it has been lost in the course of evolution. Most fungi lacking a sexual phase (the "imperfect fungi") are ascomycetes that reproduce only by means of conidia. A few are basidiomycetes, as shown by the presence of the septa and clamp connections characteristic of that group (see Figure 12–27, page 231). Parasexual cycles are common and seem to compensate partially for the lack of sexual reproduction in maintaining variability. In general, the classification of the Fungi Imperfecti is based upon the mode of formation of their conidia (Figure 12–15).

There are many thousands of Fungi Imperfecti, including a number that are of great economic importance. For example, certain members of the genus *Penicillium* give some types of cheese the flavor, odor, and character so highly prized by gourmets. One such mold, *P. roquefortii,* was first found in caves near the French village of Roquefort. Legend has it that a peasant boy left his lunch, a fresh piece of mild cheese, in one of these caves and on returning several weeks later found it marbled, tart, and fragrant. Only cheeses from the area around these particular caves are permitted to bear the name of Roquefort. Another species of the same genus, *P. camembertii,* gives Camembert cheese its special qualities. In the Orient, soy paste (miso) is produced by fermenting soybeans with *Aspergillus oryzae;* soy sauce (shoyu) is produced by fermenting them with a mixture of *A. oryzae* and

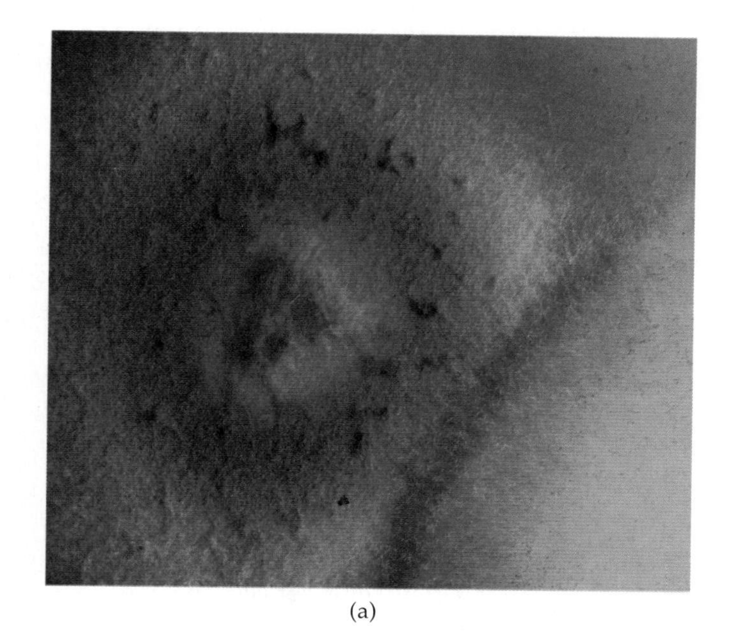

(a)

(b)

12–14
Penicillium *and* Aspergillus—*two of the common genera of Fungi Imperfecti.* (a) *A culture of* Penicillium dupontii, *showing the beautiful colors produced during growth and spore development.* (b) *A culture of* Aspergillus niger, *on which the conidia were placed in the form of a double cross. Notice the concentric growth pattern produced by successive "pulses" of spore production.*

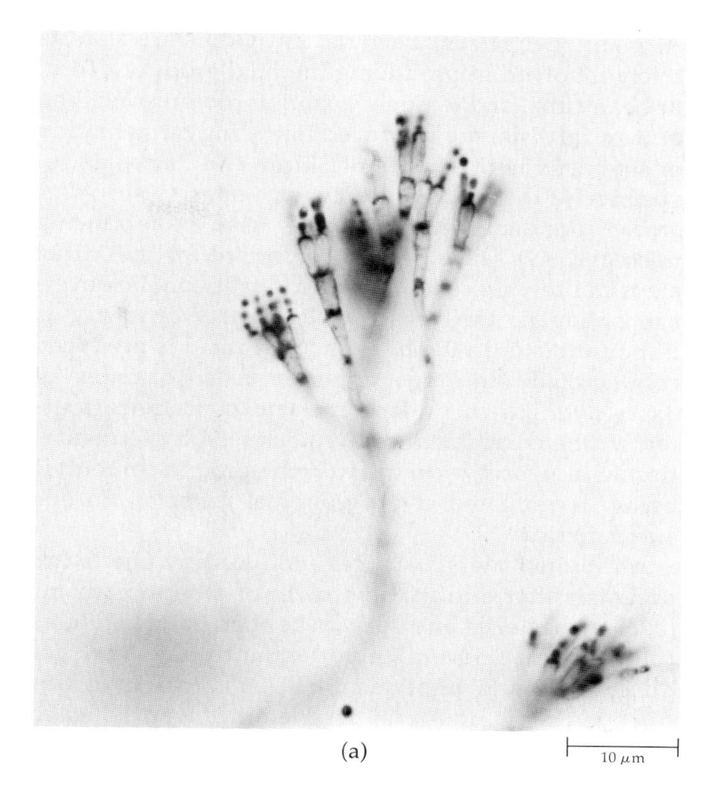

(a)

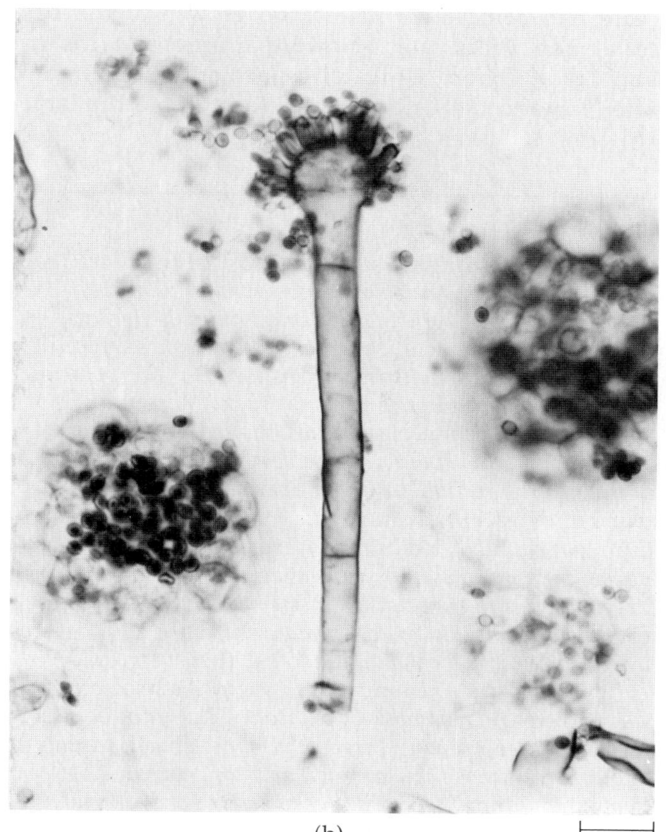

(b)

12–15
The conidiophores of imperfect fungi are used in their classification. (a) Penicillium *(branching) and* (b) Aspergillus *(tightly clumped).*

A. soyae. Lactic acid bacteria and yeasts are also actively involved in producing the final product. (Tofu, or bean curd, and tempeh, a similar protein-rich food of tropical Asia, are produced following fermentation of soybeans with species of *Mucor* and *Rhizopus*, respectively; these fungi are zygomycetes.) *Aspergillus oryzae*, the same fungus that is used in producing miso and soy sauce, is also important for the initial steps in brewing sake, the traditional alcoholic beverage of Japan; the yeast *Saccharomyces cerevisiae* is important later in the process. Citric acid is produced commercially in large amounts from colonies of *Aspergillus* grown under very acid conditions. Currently, the enrichment of livestock feed by fermentation with *A. oryzae* (to increase the protein content) is being investigated at experimental farms in Europe and America.

Antibiotics are substances produced by one living organism that inhibit the growth of other organisms (such as bacteria) and so may be therapeutically useful to humans. Several important antibiotics are produced by Fungi Imperfecti. The first antibiotic was discovered by Sir Alexander Fleming, who noted in 1928 that a strain of *Penicillium* that had contaminated a culture of *Staphylococcus* growing on a nutrient agar plate had completely halted the growth of the bacteria. Ten years later, Howard Florey and his associates at Oxford University purified penicillin and later came to the United States to promote the large-scale production of the drug. The demand during World War II was so great that the production of penicillin was increased from a few million units per month in 1942 to more than 700 billion units in 1945. Penicillin is effective in curing a wide variety of bacterial diseases, including pneumonia, scarlet fever, syphilis, gonorrhea, diphtheria, rheumatic fever, and many others. Many hundreds of antibiotics have now been discovered, and some of these undoubtedly also play a significant ecological role in nature.

Another group of Fungi Imperfecti, the dermatophytes, are the cause of ringworm and athlete's foot (or ringworm of the foot), diseases that are especially prevalent in the tropics. During World War II, more soldiers had to be sent back from the South Pacific because of skin infections than because of wounds received in battle. Another fungus, *Candida albicans*, which is yeastlike under certain conditions, causes thrush and other infections of the mucous membranes. Fungal spores are continually inhaled, and some cause internal diseases, which may be serious or fatal, especially if they attack the lungs.

In recent years, fungal diseases in humans have become more prevalent. Chemicals are now routinely administered to transplant patients to suppress their normal immune reactions so that they will be able to accept transplanted organs and tissues. These same chemicals, however, also reduce the patients' de-

PREDACEOUS FUNGI

Among the most highly specialized of the fungi are the predaceous fungi; they have developed a number of mechanisms for capturing the small animals they use as food. Some secrete a sticky substance on the surface of their hyphae in which passing protozoa, rotifers, small insects, or other animals become stuck. More than 50 species of Fungi Imperfecti capture small roundworms (nematodes) that abound in the soil. In the presence of a population of roundworms (or even water in which the worms have been growing), the hyphae of the fungi produce loops which swell rapidly, closing the opening like a noose, when a nematode rubs against its inner surface. Presumably the stimulation of the cell wall increases the amount of osmotically active material in the cell, causing water to enter the cells and increase their turgor.

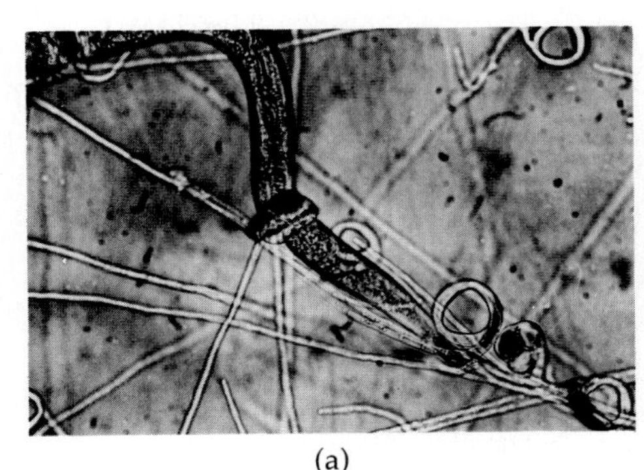

(a)

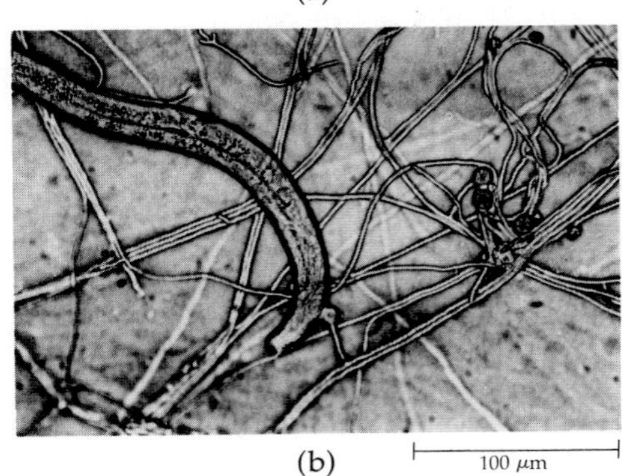

(b)

(a) *The predaceous imperfect fungus* Arthrobotrys dactyloides *has trapped a nematode. The traps consist of rings, each comprising three cells, which swell rapidly to about three times their original size and garrote the nematode. Once the worm has been trapped, fungal hyphae grow into its body and digest it. When triggered, the ring cells can expand completely in a period of less than a tenth of a second.* (b) *Another nematode-trapping fungus,* Dactylella drechsleri, *traps the worms by means of small adhesive knobs.*

fenses and make them much more susceptible to fungal diseases. Some chemical treatments, such as those for acute leukemia, also seem to lower the defenses against fungal infection, and therefore more medical attention is now being directed toward the prevention and treatment of fungal diseases.

The Lichens

The lichens are a large group of ascomycetes* that grow only in intimate association with green algae or cyanobacteria. Obtaining nutrition from these photosynthetic organisms, lichens have invaded the harshest habitats (Figure 12–16) and have diversified into about 25,000 species of varied appearance. Some 26 genera of photosynthetic organisms are found in combination with these ascomycetes. The most frequent are the green algae *Trebouxia* and *Trentepohlia* and the cyanobacterium *Nostoc;* one of these three organisms is found in about 90 percent of all lichens.

Lichens are extremely widespread in nature; they occur from arid desert regions to the arctic and grow on bare soil, tree trunks, sunbaked rocks, fence posts, and windswept alpine peaks all over the world (see Figures 12–16, 12–17, and 12–18). One species, *Verrucaria serpuloides,* is a permanently submerged marine lichen. They are often the first colonists of newly exposed rocky areas. In Antarctica, there are

*There are also about a dozen species of basidiomycetes that form associations with algae, but they are closely related to free-living groups of basidiomycetes and do not resemble the other lichen-forming fungi.

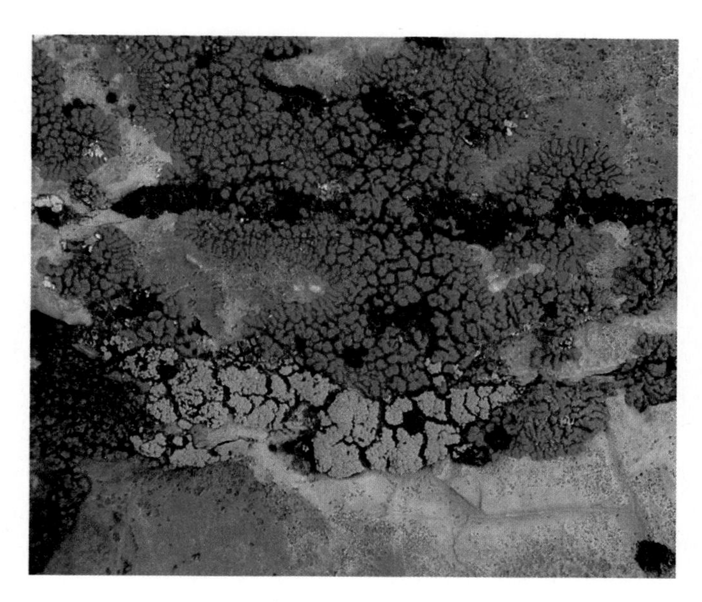

12–16
A mosaic of crustose ("encrusting") lichens growing on a bare rock surface.

(a)

(b)

12–17
(a) Parmelia perforata, *a foliose ("leafy") lichen growing on a dead tree branch in Mississippi.* (b) *Old man's beard* (Usnea), *a hanging lichen that often occurs in masses on the limbs of trees.*

more than 350 species of lichens but only two species of vascular plants. Seven of the species of lichens occur in the Queen Maud Mountains at 86°03′ south latitude.

The colors of lichens range from white to black, through shades of red, orange, brown, yellow, and green, and they contain many unusual chemical compounds. Some of them are so tiny that they are almost invisible to the unaided eye; others, like the reindeer "mosses," may cover kilometers of land with ankle-deep growth. Except for the reindeer mosses, the lichens are of little direct economic importance to humans. However, they have long been the subject of biological investigations because of the intriguing nature of the association between a fungus and its included algae or cyanobacteria. The fungal partner apparently plays the major role in determining the form of the lichen; recently, however, it has been demonstrated that a single fungus with different types of algae can produce morphologically very different individuals that have traditionally been placed in different genera. The algae or cyanobacteria found in lichens also occur as free-living species, whereas the lichen fungi are generally found in nature only in lichens.

Some lichens produce special fragments, known as *soredia,* which are composed of fungal hyphae and algae or cyanobacteria (Figure 12–19). Lichen fungi also often form ascocarps, which are similar in every way to those of nonlichen fungi except that in lichens they may endure and produce spores continuously over a number of years.

Biology of the Lichens

How can the lichens survive under environmental conditions so adverse to any other form of life? At one time, it was thought that the secret of the lichen's success was that the fungal tissue protected the alga or cyanobacterium from drying out. Actually, one of the chief factors in lichen survival seems to be the fact that they dry out very rapidly. Lichens are frequently very desiccated in nature, with a water content ranging from only 2 to 10 percent of their dry weight. When the lichen dries out, photosynthesis ceases; in this state of "suspended animation," even blazing

(a)

(b)

(c)

12–18
Several fruticose ("shrubby") lichens. (a) *Goldeye lichen,* Teloschistes chrysophthalmus. (b) *British soldiers,* Cladonia cristatella, *are 1 to 2 centimeters tall.* (c) Cladonia subtenuis, *commonly called reindeer "moss," is actually a lichen.*

12–19
(a) *A cross section of the lichen* Lobaria verrucosa. *The simplest lichens consist of a crust of fungal hyphae entwining colonies of algae. In the more complex lichens, however, the hyphae and the algal cells are organized in a thallus with a definite growth form and characteristic internal structure. The lichen shown here has four distinct layers: (1) the upper cortex, a protective surface of heavily gelatinized fungal hyphae; (2) the algal layer, which consists of algal cells and loosely interwoven, thin-walled hyphae; (3) the medulla, which is a thick layer of loosely packed, colorless, weakly gelatinized hyphae. This layer, which makes up about two-thirds of the thickness of the thallus, appears to serve as a storage area, with enlarged cells in the fungal hyphae; and (4) the lower cortex, which is thinner than the upper one and is covered with fine projections that attach it to the substrate. Soredia are fragments of algal cells and fungal hyphae by which lichens propagate.* (b) *Scanning electron micrograph of* Cladonia cristatella, *the lichen shown in Figure 12–18b.*

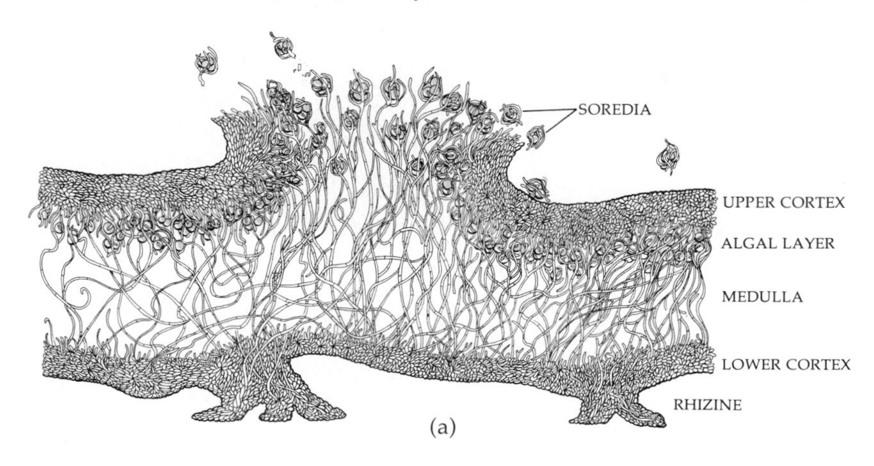

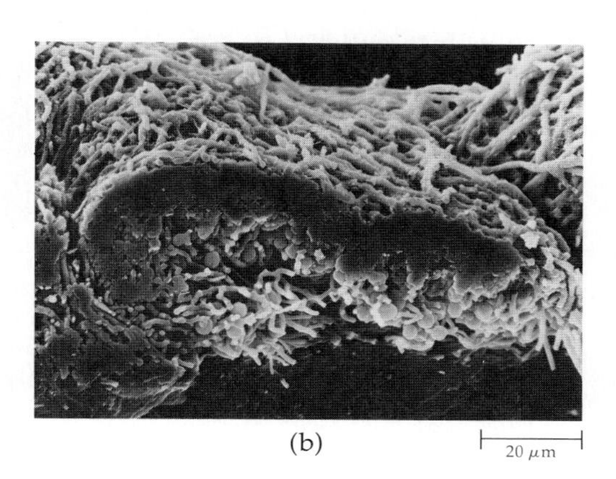

sunlight or great extremes of heat or cold can be endured by some species of lichens. Cessation of photosynthesis depends, in large part, on the fact that the upper cortex of the lichen becomes thicker and more opaque when dry, cutting off the passage of light energy. A wet lichen is destroyed by light intensities or temperatures that do not harm a dry lichen.

When a lichen is wetted by rain, it absorbs 3 to 35 times its own weight in water in a very short time. If a dry, brittle lichen is submerged in water, it becomes soft and pliable within a few minutes. This is the simple physical process of imbibition—the lichen takes up water just as blotting paper does.

The lichen reaches its maximum vitality, as judged by the rate of photosynthesis, after it has been soaked with water and has begun to dry. Its rate of photosynthesis reaches a peak when the water content is 65 to 90 percent of the maximum it can hold; below this level, if the lichen continues to lose water, the rate of photosynthesis decreases. In many environments, the water content of the lichen varies markedly in the course of a day, with most photosynthesis taking place only during a few hours, usually in the early morning after wetting by fog or dew. As a consequence, lichens are geared to an extremely slow rate of growth, their radius increasing at a rate of about 0.1 to 10 millimeters a year. Calculated on this basis, some mature lichens may be as much as several thousand years old. They achieve their most luxuriant growth development along seacoasts and on fog-shrouded mountains.

Lichens apparently absorb some mineral nutrients from their substrate (this is suggested by the fact that particular species are characteristically found on particular kinds of rocks or soil or tree trunks), but most of the elements absorbed by lichens enter them from the surrounding air and in rainfall. Lichens absorb elements from rainwater rapidly and concentrate them. Because they have no means of excreting these elements, lichens are particularly sensitive to toxic compounds. Absorption of toxic compounds by lichens causes degradation of their chlorophyll. Lichen growth is a very sensitive indication of the toxic components of polluted air, especially sulfur dioxide.

In at least one instance, the capacity of lichens to absorb substances from rainfall led to surprising consequences. In studies of fallout following atomic bomb testing, it was found that Alaskan Eskimos and Scandinavian Laplanders had unusually high levels of radioactivity in their bodies. These findings were totally unexpected because it had been calculated that the amount of fallout reaching the ground at the poles was much less than in the temperate regions. The significant factor was the reindeer mosses (actually lichens), which had absorbed the isotopes from the atmosphere and concentrated them. As their name implies, the reindeer mosses are the chief source of food for the reindeer and caribou, which in turn are the chief source of food for the Eskimos and Laplanders, who thus became the repositories of radioactive strontium and cesium.

(a)

(b)

(c)

12–20
Representative large ascomycetes. (a) *Common morel,* Morchella esculenta. *The true morels are among the choicest edible fungi. Mushroom gatherers look for them when the oak leaves are "the size of a mouse's ear."* (b) *Scarlet cup,* Sarcoscypha coccinea, *a beautiful fungus with an open ascocarp known as an apothecium.* (c) *The highly prized, edible ascocarp of a black truffle,* Tuber melanosporum. *In the truffles, this spore-bearing structure is produced below ground and remains closed, liberating its ascospores only when the ascocarp decays or is broken open by digging animals. Truffles are mycorrhizal, mainly on oaks, and are searched for by specially trained dogs and pigs. Neither morels nor truffles can be grown commercially.*

(a)

(b)

(c)

12–21
Representative large basidiomycetes. (a) *The fly agaric,* Amanita muscaria. *Mushrooms at various stages of growth; one has been picked to show the gills. This genus of mushrooms is often poisonous and can be recognized by the ring on the stalk and the cup around the base.* (b) *Puffballs,* Lycoperdon ericetorum. *The spores are being discharged from the pore at the top of each puffball and dispersed by the wind.* (c) *Corn smut is a familiar plant disease in which the fungus* Ustilago maydis *produces black, dusty-looking masses of spores in ears of corn.*

(a)

(b)

(c)

12–22
Common Homobasidiomycetes. (a) Suillus bovinus, *a bolete. In this group of mushrooms, the gills are replaced by pores.* (b) *Stinkhorn,* Phallus impudicus. *The basidiospores are released in a foul-smelling, sticky mass at the top of the fungus. Flies, such as the individuals of* Mydaea urbana *shown here, visit it for food and spread the spores, which adhere to their legs and bodies in great numbers.* (c) *A shelf fungus,* Ganoderma tsugae. *Members of this group are responsible for most wood rot.* (d) *An edible coral fungus,* Hericium coralloides. *The hymenium, an outer spore-bearing layer of basidia, is borne on all sides of the basidiocarp.*

(d)

(a)

(b)

12–23
(a) *Bird's-nest fungus,* Cyathus striatus. *The round structures in the "nests" of these puffballs contain the basidiospores, which are splashed out and dispersed by raindrops.* (b) *Earthstar,* Geastrum triplex. *The outer layer of a typical puffball is folded back in this genus.*

The Nature of the Relationship Between Fungi and Photosynthetic Organisms

What is the relationship between the two components of a lichen? It is clear that the fungus obtains nutrients from the alga or cyanobacterium, since the lichen behaves like a photosynthetic organism, dependent only on light, air, and mineral nutrients. In fact, the movement of nutrients from the included algae or cyanobacteria to the fungus has been established by tracing experiments using ^{14}C-labeled glucose. In those lichens that include cyanobacteria such as *Nostoc*, nitrogen fixation by the bacteria and transfer to the fungus are also important. In the lichen, the fungal hyphae form a close network around the included cells. Haustoria (the specialized hyphae of parasitic fungi), which are very common in lichens, penetrate the photosynthetic cells; other specialized organs, called appressoria, lie along the surface of the photosynthetic cells but do not penetrate the protoplast. The association with a fungus profoundly affects the nature of the metabolic output from the included photosynthetic cells. For example, green algae excrete large quantities of two particular alcohols, D-sorbitol and D-ribitol, only when they are associated with fungi to form lichens.

It is possible to separate the algal or cyanobacterial and fungal components of a particular lichen and grow them alone in pure culture. Under such conditions, the fungus grows in compact colonies, quite unlike the actual lichen. It requires a large number of complex carbohydrates for growth and does not ordinarily form fruiting bodies. In contrast, isolated lichen algae or cyanobacteria grow more rapidly when they are free-living than when they are in the lichen association. Thus it may be most appropriate to think of the lichen partnership as controlled parasitism of the alga or cyanobacterium by the fungus, rather than symbiosis. When grown together in sterile culture, the fungus seems first to bring its photosynthetic partner under control and then to establish the characteristic appearance of the mature lichen (Figure 12–24).

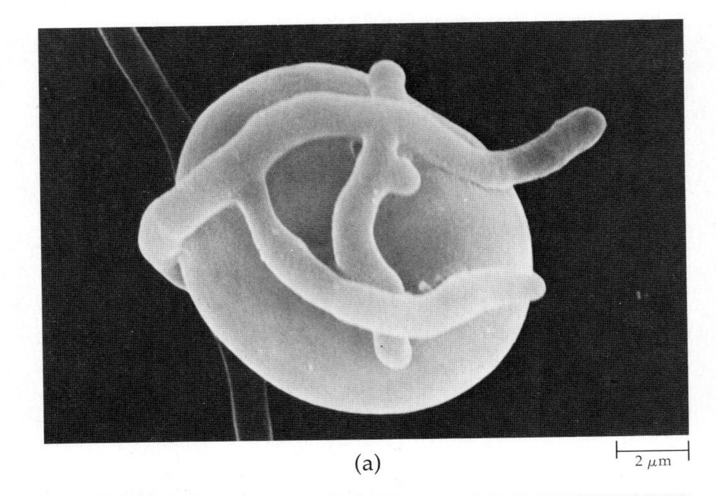

(a)

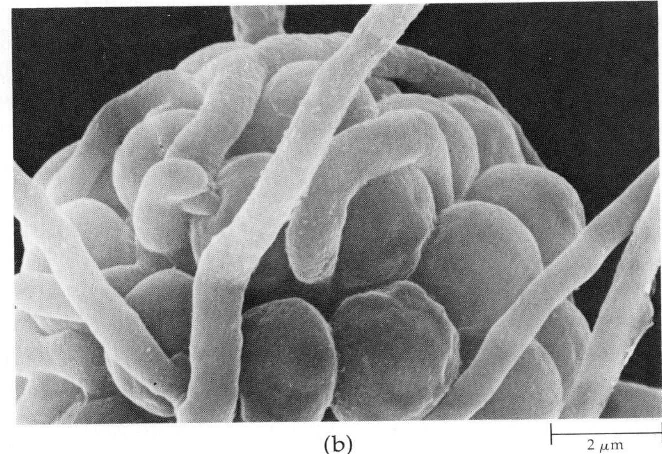

(b)

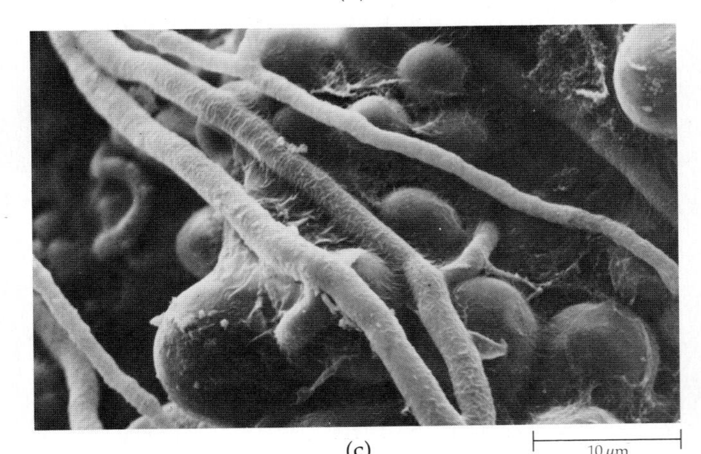

(c)

12–24
(a) *Scanning electron micrograph of early interactions between fungal and algal components of British soldiers* (Cladonia cristatella) *in sterile laboratory culture.* (b) *Algal cells in the same lichen, one of which is penetrated by a fungal haustorium.* (c) *Another lichen,* Lecidea albocaerulescens, *again showing fungal and algal components at an early stage. The associated green alga in both of these lichens is* Trebouxia.

Division Basidiomycota

The most familiar of all fungi are members of this large division which includes some 25,000 described species—not only mushrooms, toadstools, stinkhorns, puffballs, and shelf fungi but also two important plant pathogens: the rusts and smuts (see Figure 12–21, page 228). The basidiomycetes are distinguished from all other fungi by the production of *basidiospores,* which are borne outside a club-shaped, spore-producing structure called the *basidium* (Figure 12–25). The larger basidiomycetes are the best known of all fungi, although many more species undoubtedly await discovery.

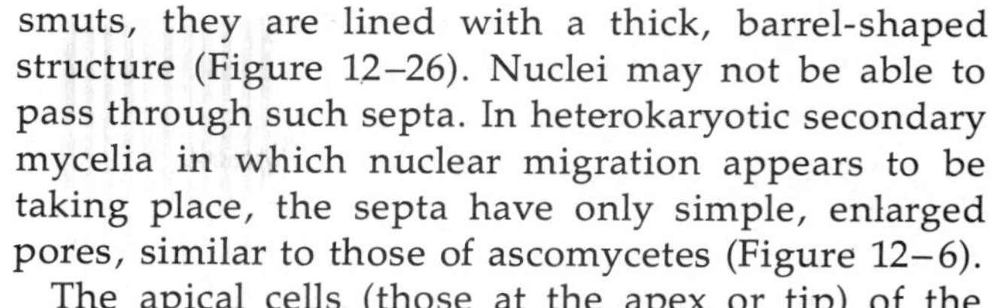

The mycelium of the basidiomycetes is always septate and in most species passes through three distinct phases—primary, secondary, and tertiary—during the life cycle of the fungus. When it germinates, a basidiospore produces the *primary mycelium.* Initially, this mycelium may be multinucleate, but septa are soon formed and the mycelium is divided into monokaryotic (uninucleate) cells. Commonly, the *secondary mycelium* is produced by fusion of primary hyphae from different mating types (in which case it is heterokaryotic) or it may arise when septa are not formed after nuclear formation (in which case it is homokaryotic). In either case, the result is formation of a dikaryotic (binucleate) mycelium, since karyogamy (fusion of gametic nuclei) does not immediately follow plasmogamy (fusion of protoplasts).

The septa in the secondary mycelia of basidiomycetes are perforated; however, except in the rusts and smuts, they are lined with a thick, barrel-shaped structure (Figure 12–26). Nuclei may not be able to pass through such septa. In heterokaryotic secondary mycelia in which nuclear migration appears to be taking place, the septa have only simple, enlarged pores, similar to those of ascomycetes (Figure 12–6).

The apical cells (those at the apex or tip) of the secondary mycelium usually divide by the formation of clamp connections (Figure 12–27). These clamp connections, which ensure the allocation of one nucleus of each type to the daughter cells, are highly characteristic of the basidiomycetes.

The *tertiary mycelium* arises directly from the secondary mycelium and forms *basidiocarps*—the fruiting bodies of the so-called higher fungi, such as mushrooms and puffballs. The formation of the basidiocarps may require light. Like the secondary mycelium, the tertiary mycelium is dikaryotic.

12–25
Scanning electron micrograph of a basidium and basidiospores, each attached to a sterigma in the Homobasidiomycete Aleurodiscus amorphus.

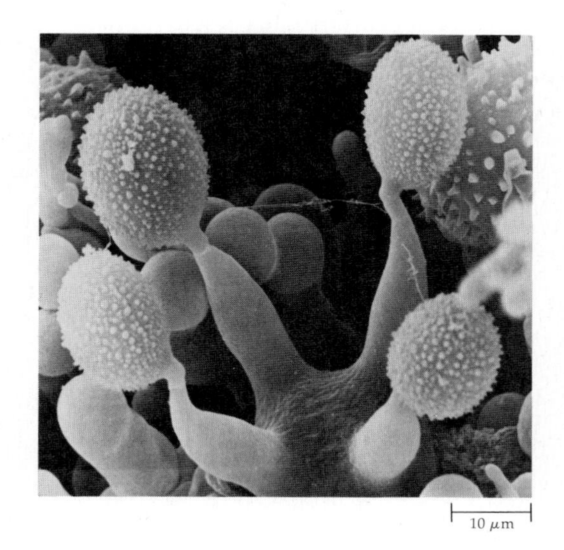

12–26
Characteristic septum of the secondary mycelium of a basidiomycete, shown in an electron micrograph of Laetisaria arvalis, *a common wood-rotting fungus.*

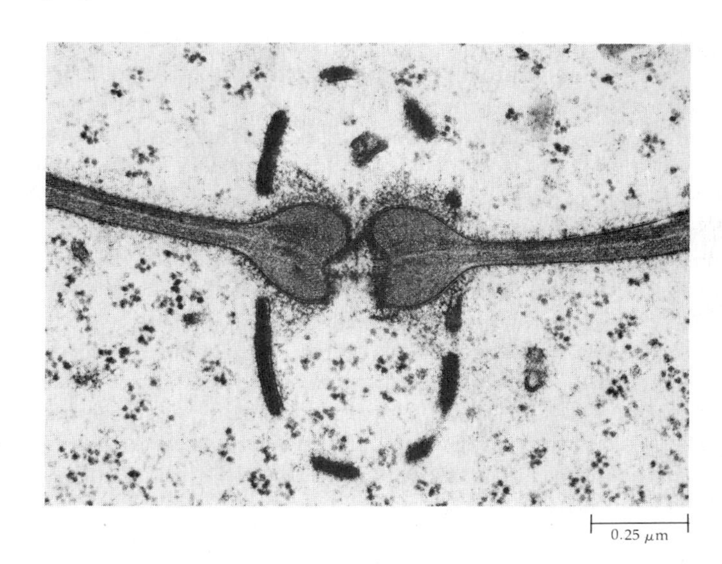

12–27
(a) *In the basidiomycetes, dikaryotic hyphae are distinguished by the presence of clamp connections over the septa. These clamp connections are temporary bridges formed during cell division; they presumably ensure the proper distribution of the two genetically distinct types of nuclei in the basidiocarp.* (b) *Clamp connections and characteristic septa in a hypha of* Coprinus lagopus.

(a)

(b) 0.4 μm

The basidiomycetes are divided into two classes: Homobasidiomycetes and Heterobasidiomycetes.

Class Homobasidiomycetes

The class Homobasidiomycetes contains the edible and poisonous mushrooms, the shelf fungi, stinkhorns, earthstars, false truffles, bird's-nest fungi, and puffballs (see Figures 12–22 and 12–23, page 229). All members of the Homobasidiomycetes produce basidiocarps, which are analogous to the ascocarps of the ascomycetes. The Homobasidiomycetes are also characterized by their club-shaped, aseptate basidia, each of which usually bears four basidiospores on minute projections called *sterigmata* (Figures 12–25 and 12–28).

What one recognizes as a mushroom or toadstool is a basidiocarp. ("Mushroom" is sometimes popularly used to designate the edible forms of basidiocarps and "toadstool" is used to designate the inedible ones, but mycologists do not recognize such a distinction and use only the term "mushroom." In this book, all such forms are referred to as mushrooms; this does not mean that they are all edible.) The mushroom consists of a *cap,* or *pileus,* that sits atop a *stalk,* or *stipe.* Early in its development—the "button" stage—the mushroom may be covered by a membranous tissue that ruptures as the mushroom enlarges. In some genera, remnants of this tissue are visible on the upper surface of the cap and, at the base of the stipe, as a cup, or *volva.* The lower surface of the cap consists of radiating strips of tissue called *gills* (Figure 12–28).

In relatively uniform habitats such as lawns and fields, the mycelium from which mushrooms are produced spreads underground, forming a ring which may grow as large as 30 meters in diameter. In an open area, the mycelium expands evenly in all directions, dying at the center and fruiting at the outer edges, where it grows most actively, because this is the area in which the nutritive material in the soil is most abundant. As a consequence, the mushrooms appear in rings, and as the mycelium grows, the rings become larger and larger in diameter. Such circles of mushrooms are known in European folk legend as "fairy rings" (Figure 12–29).

Among the Homobasidiomycetes, the best known are the gill fungi. *Agaricus campestris,* the common field mushroom, is particularly abundant. The closely related *A. bisporus* is one of the few mushrooms that can be cultivated commercially; some 60,000 metric tons are produced annually in the United States alone. The gill fungi also include the mushrooms of the genus *Amanita,* which includes the most highly poisonous of all mushrooms, as well as some that are edible. Even one bite of the "destroying angel," *Amanita phalloides,* can be fatal. Some basidiomycetes

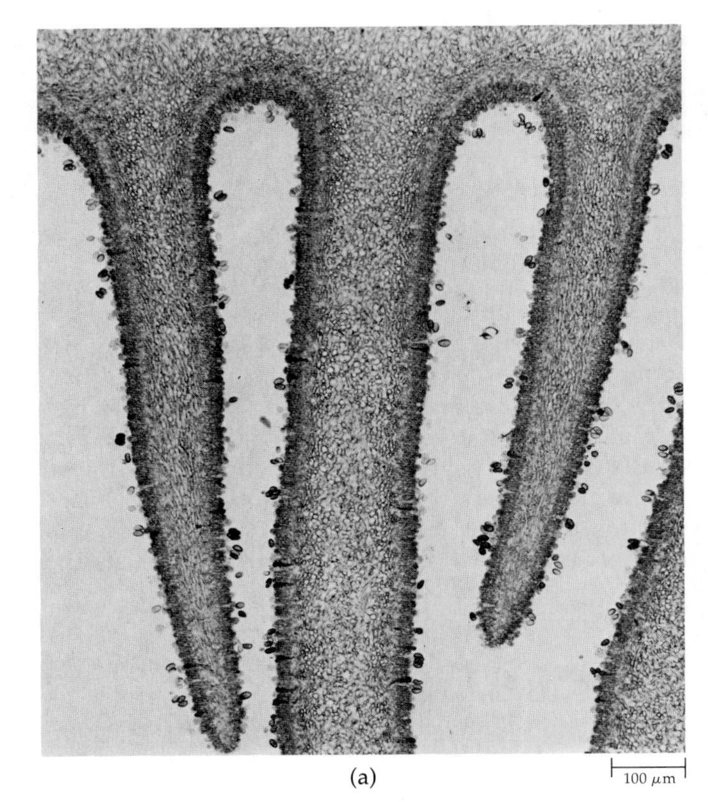

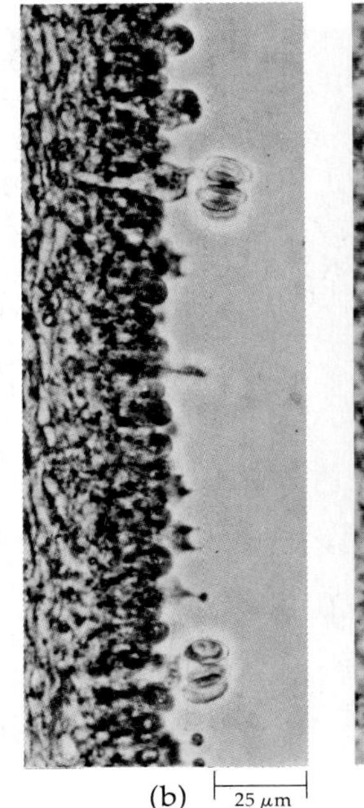

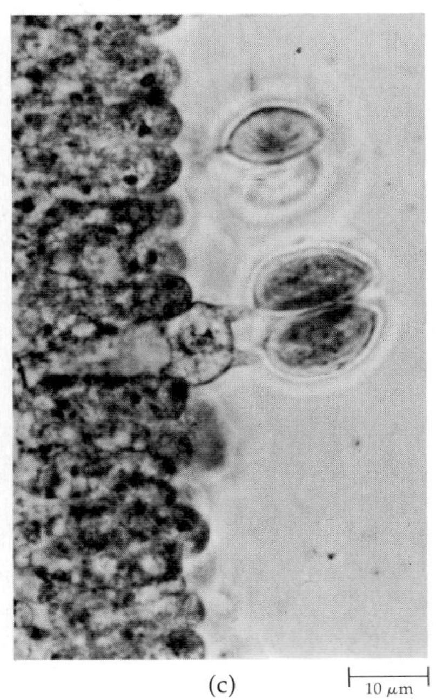

12–28
Coprinus, *a common mushroom.*
(a) *Section through the gills. The relatively dark margins of the gills constitute the hymenial layer.* (b) *Section through the hymenial layer, showing developing basidia and basidiospores.* (c) *Nearly mature basidiospores attached to the basidium by sterigmata.*

12–29
A "fairy ring" formed by the parasol mushroom, Lepiota procera.

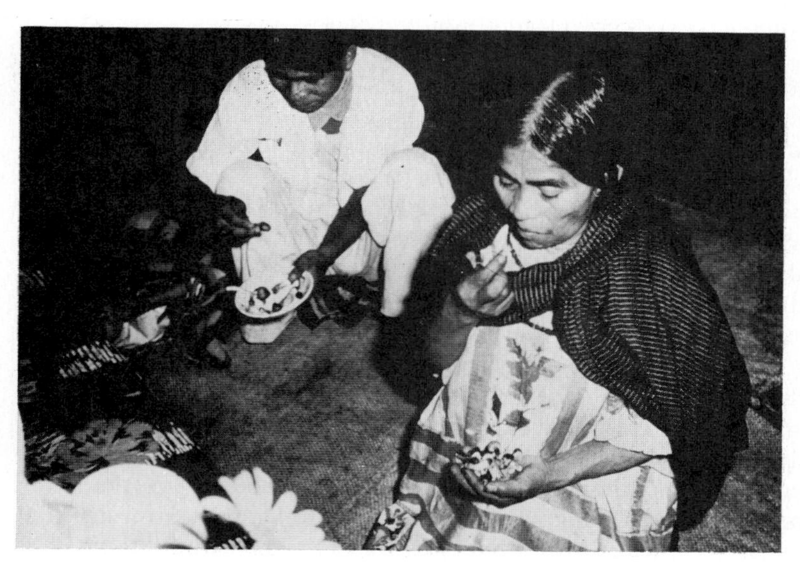

12–30
Mushrooms figure prominently in the religious ceremonies of several groups of Indians in southern Mexico and Central America; the Indians eat certain basidiomycetes for their hallucinogenic qualities. One of the most important of the mushrooms is Psilocybe mexicana, *shown here growing in a pasture near Huautla de Jiménez, Oaxaca, Mexico. The shaman María Sabına is eating* Psilocybe *in the course of a midnight religious ceremony. Psilocybin is the chemical responsible for the colorful visions experienced by those who eat these "sacred" mushrooms.*

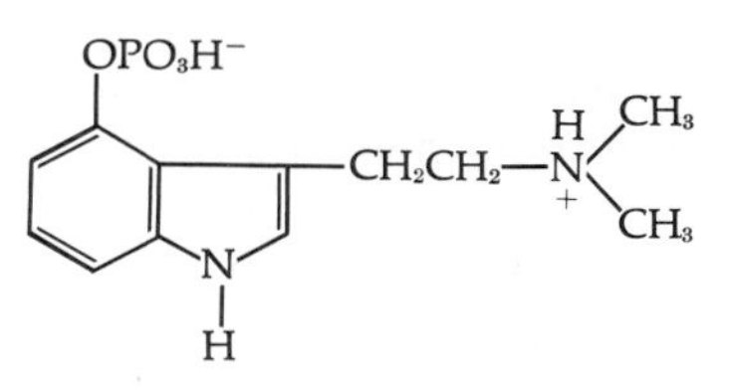

12–31
Life cycle of a Homobasidiomycete. The primary mycelia are produced from basidiospores. Secondary dikaryotic mycelia are formed from the primary mycelia. Some secondary mycelia are formed by the fusion of hyphae from different mating types, in which case the mycelia are heterokaryotic. The secondary mycelia divide and differentiate to form the tertiary mycelia that make up the basidiocarp.

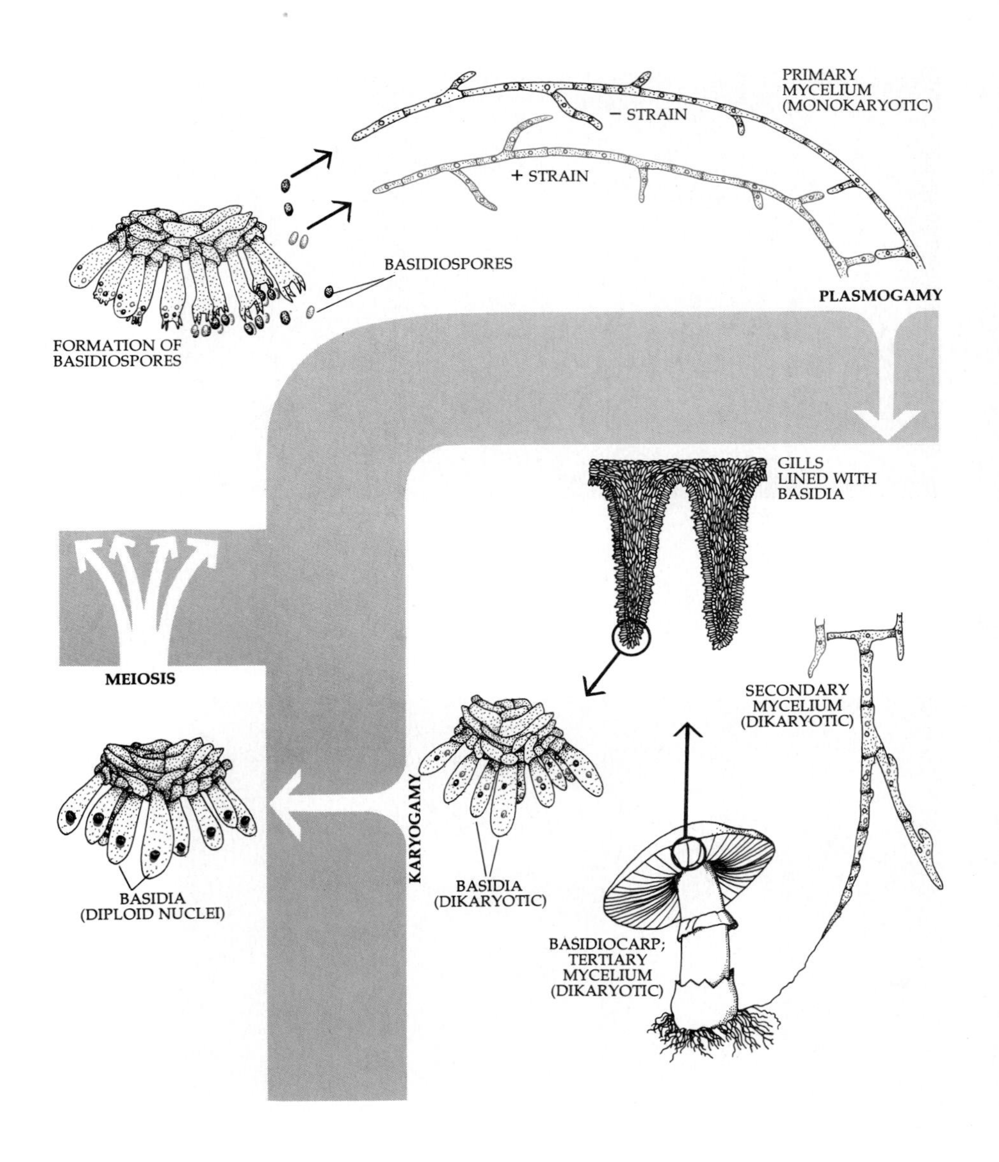

contain chemicals that impart hallucinogenic qualities to them (Figure 12–30).

In nature, most Homobasidiomycetes, including the mushrooms, reproduce primarily through the formation of basidiospores. A representative life cycle of a mushroom is shown in Figure 12–31. The gills of the mushroom are the sites of basidia and basidiospore formation. The basidia form on the surface layers, or hymenia, of the gills, each as a terminal cell of a dikaryotic hypha. Soon after the young basidium enlarges, karyogamy occurs. This is followed almost immediately by meiosis of each diploid nucleus, resulting in the formation of four haploid nuclei. Each of the four nuclei then migrates into a sterigma, which enlarges at its tip to form a uninucleate basidiospore. The reproductive capacity of a single mushroom is tremendous, with billions of spores being produced by a single basidiocarp. A single puffball may produce several trillion basidiospores.

Class Heterobasidiomycetes

The class Heterobasidiomycetes consists of the rusts, smuts, and jelly fungi (Figure 12–32). The jelly fungi are mostly saprobes and, like the Homobasidiomycetes, produce basidiocarps. The rusts and smuts are parasitic on vascular plants and do not form basidiocarps but produce their spores in clusters or masses called *sori* (see Figure 12–21c). As plant pathogens, the rusts and smuts are of tremendous economic importance, causing billions of dollars worth of damage to crops throughout the world each year. As a group, the Heterobasidiomycetes are characterized by septate (multicellular) basidia.

The life cycles of many rusts are complex, and these pathogens are a continual challenge to the plant pathologist whose task it is to keep them under control. Until recently, the rusts were thought to be obligate parasites on vascular plants, but now several

12–32
A jelly fungus, one representative of the Heterobasidiomycetes, growing on a dead limb in the Amazon forest of Brazil.

species have been maintained in artificial culture. Some smuts are also capable of completing their development under laboratory conditions.

Puccinia graminis, the cause of black stem rust of wheat, will serve to illustrate one of the life cycles of the parasitic Heterobasidiomycetes. It is one of some 7000 species of rusts. Numerous strains of *P. graminis* exist, and in addition to wheat, they parasitize other cereal grains such as barley, oats, and rye, and various species of wild grasses. *Puccinia graminis* is a continual source of economic loss for the wheat grower. In one year alone, the losses in Minnesota, North Dakota, South Dakota, and the prairie provinces of Canada amounted to nearly 8 million metric tons. As early as 100 A.D., Pliny described wheat rust as "the greatest pest of the crops." Today plant pathologists combat black stem rust largely by breeding resistant wheat varieties, but mutation and recombination in the rust make any advantage short-lived.

Puccinia graminis is heteroecious; that is, it requires two different hosts to complete its life cycle (Figure 12–33). Autoecious parasites require only one host. *Puccinia graminis* can grow indefinitely on its grass host, but then it reproduces only asexually. In order for sexual reproduction to take place, the rust must spend part of its life cycle on the common barberry,

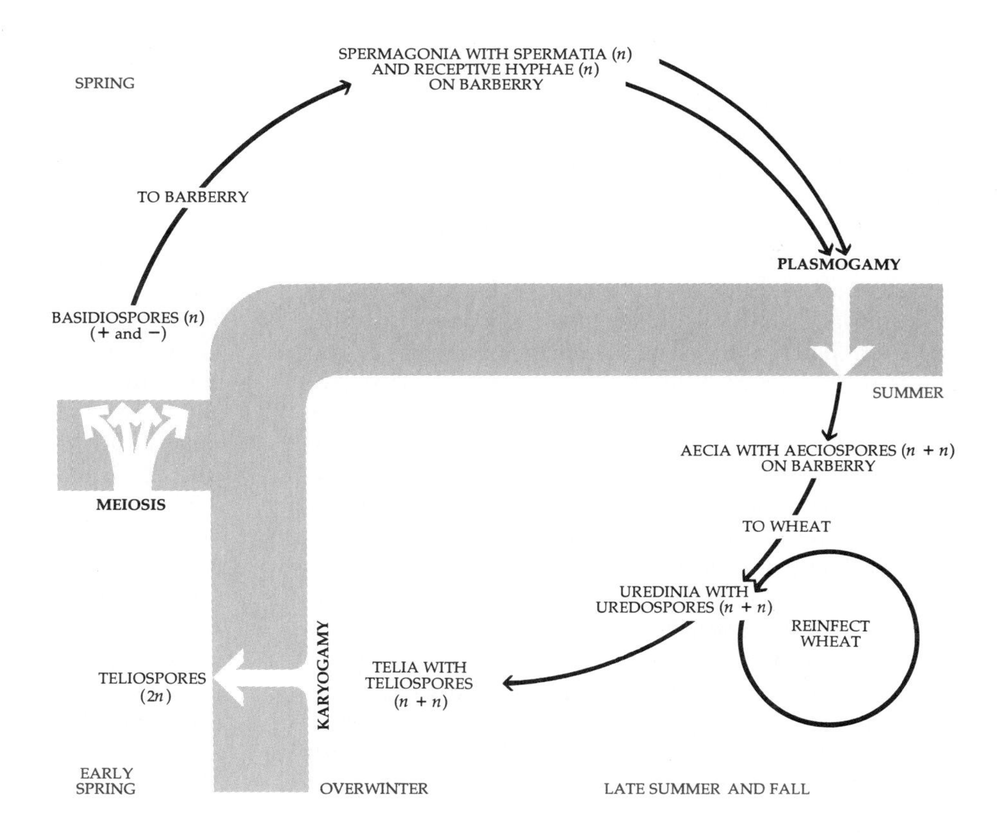

12–33
Life cycle of Puccinia graminis, *a rust.*

12–34
Some stages in the life cycle of Puccinia graminis, *the cause of black stem rust of wheat.* (a) *Transverse section of infected barberry leaf showing a spermagonium on the upper surface and two aecia on the lower surface. Each aecium contains many chains of aeciospores.* (b) *A single spermagonium on the upper surface of a barberry leaf. Note the hyphae projecting through the opening at the top and the spermatia-bearing hyphae lining its inner wall.* (c) *Section of a uredinium, with dikaryotic uredospores, on a wheat stem.* (d) *Section of a telium, with the two-celled, dikaryotic teliospores, on a stem of wheat.*

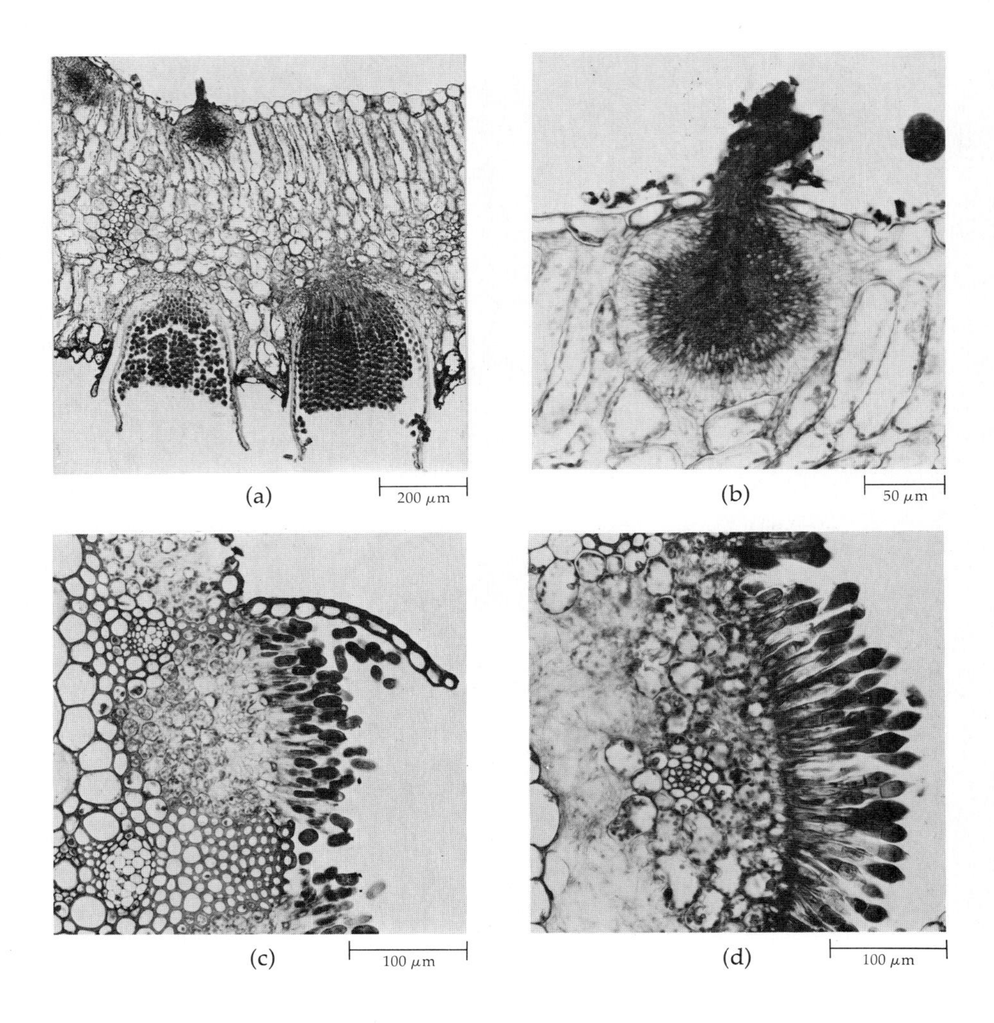

Berberis vulgaris, and part on a grass host. One method of attempting to control *P. graminis* has been to eradicate the barberry. The crown colony of Massachusetts passed a law ordering "whoever . . . hath any barberry bushes growing in his or their land . . . shall cause the same to be extirpated or destroyed on or before the thirteenth day of June, A.D. 1760."

Infection of the barberry occurs in the spring, when uninucleate basidiospores infect the plant by forming flask-shaped *spermagonia,* primarily on the upper surfaces of the leaves (Figures 12–34a and 12–34b). The form of *P. graminis* that grows on barberry consists of separate + and − strains, so that the basidiospores and the spermagonia derived from them are either + or −. Each spermagonium contains two types of hyphae: one that produces chains of small cells called *spermatia* and the other being the so-called *receptive hyphae.* The spermatia are discharged from an opening in the spermagonium. If the + spermatium of one spermagonium comes in contact with the − receptive hypha of another spermagonium, or vice versa, plasmogamy occurs and dikaryotic hyphae are produced. Aecial initials are produced from the dikaryotic hyphae that extend downward from the spermagonium. *Aecia* are then formed primarily on the lower surface of the leaf, where they produce chains of *aeciospores* (Figure 12–34a). The dikaryotic aeciospores then infect the wheat.

The first external manifestation of infection on the wheat is the appearance of rust-colored, linear streaks on the leaves and stems (the red stage). These streaks are *uredinia,* containing unicellular, dikaryotic *uredospores* (Figure 12–34c). Uredospores are produced throughout the summer and reinfect the wheat. In late summer and early fall, the red-colored sori gradually become darkened in color and become *telia* (Figure 12–34d) with two-celled, dikaryotic *teliospores* (the black stage). The teliospores are overwintering spores, which infect neither the wheat nor the barberry. In early spring, some time prior to germination, the two haploid nuclei fuse with one another to form a diploid nucleus. With the onset of germination, meiosis takes place. A short, cylindrical basidium, into which the four haploid nuclei migrate, emerges from each of the teliospore cells. Septations are laid down between the nuclei, which then migrate into the sterigmata and develop into basidiospores. Thus, the year-long cycle is completed.

MYCORRHIZAE

Certain fungi play a crucial role in the mineral nutrition of higher plants. If seedlings of many forest trees are grown in a sterile nutrient solution and then are transplanted to grassland soils, they fail to grow and eventually die from malnutrition sometimes, even if a soil analysis shows that there are abundant nutrients in the soil (Figure 12–35). If a small amount (0.1 percent by volume) of forest soil containing fungi is added to the soil around the roots of the seedlings, however, they will grow promptly and normally. The restoration of normal growth is caused by the functioning of *mycorrhizae* ("fungus-roots"), which are intimate and mutually beneficial symbiotic associations between roots and fungi.

Mycorrhizae occur in most groups of vascular plants. Only a few families of flowering plants characteristically lack mycorrhizae or form them very rarely; these include the mustard family (Brassicaceae) and the sedge family (Cyperaceae).

Many plants seem to grow normally when they are well supplied with essential elements, especially phosphorus, even if mycorrhizae are lacking; however, if the essential elements are limited, plants grow poorly or not at all when they lack mycorrhizae. The role of mycorrhizae in the direct transfer of phosphorus from the soil to the roots of plants has been verified experimentally. In turn, the plant supplies organic carbon to the symbiotic fungus. Mycorrhizae are of particular importance in the tropics, where the soil tends to be thin and most of the minerals are held in the standing crop of living plants. In such a habitat, recycling from the leaf layer is essentially direct (Figure 12–36). A better understanding of mycorrhizal relationships is certainly one key to lessening the amounts of fertilizer that must be applied to agricultural lands to insure satisfactory crops.

There are two major types of mycorrhizae found in nature: *endomycorrhizae* and *ectomycorrhizae*. Of these two, endomycorrhizae are by far the more common, occurring in about 80 percent of all vascular plants. The fungal component is a zygomycete, with only about 30 species of the fungi involved in such associations worldwide. Thus, the relationship is not highly specific. The fungal hyphae penetrate the cortical cells of the plant root, where they form coils, swellings, or minute branches (Figure 12–37). The hyphae also extend out into the surrounding soil. Because two of the characteristic intracellular swellings are called vesicles and arbuscles, endomycorrhizae are often called vesicular-arbuscular mycorrhizae (abbreviated to V/A mycorrhizae).

Ectomycorrhizae are characteristic of certain groups of trees and shrubs, primarily those found in temperate regions. These groups include the beech family

12–35
*Mycorrhizae and tree nutrition. Nine-month-old seedlings of white pine (*Pinus strobus*) were raised for two months in a sterile nutrient solution and then transplanted to prairie soil. The seedlings on the left were transplanted directly. The seedlings on the right were grown for two weeks in forest soil containing fungi before being transplanted to the grassland soil.*

12–36
Network of roots and fungi lifted off a decomposing leaf in the Amazon forest. Such a network allows less than a thousandth of the nutrients reaching the forest floor to penetrate more than 5 centimeters into the soil. Mycorrhizal fungi transfer most of the nutrients directly back to the roots of the plants from which the leaves fell. When a tropical forest is cleared, this relationship is destroyed, and the soil generally becomes infertile quite rapidly.

(Fagaceae), the willow family (Salicaceae), and the pine family (Pinaceae), as well as certain groups of tropical trees that form dense stands including only one or a few species (Figure 12–38). They are often associated with trees at or near timberline.

In ectomycorrhizae, the fungus surrounds but does not penetrate living cells in the roots (Figure 12–39). The extensive mycelium extends far out into the soil and plays an important role in transferring nutrients to the plant. Root hairs are often absent, their function evidently having been replaced by the fungal hyphae. Ectomycorrhizae are mostly formed with basidiomycetes, but some involve associations with ascomycetes, including truffles. At least 5000 species of fungi are involved in ectomycorrhizal associations, often with a high degree of specificity.

Two other specialized types of mycorrhizae are those characteristic of the heather family (Ericaceae) and a few closely related groups, and those associated with the orchids (Orchidaceae). In Ericaceae, the fungus may constitute up to 80 percent of the mycor-

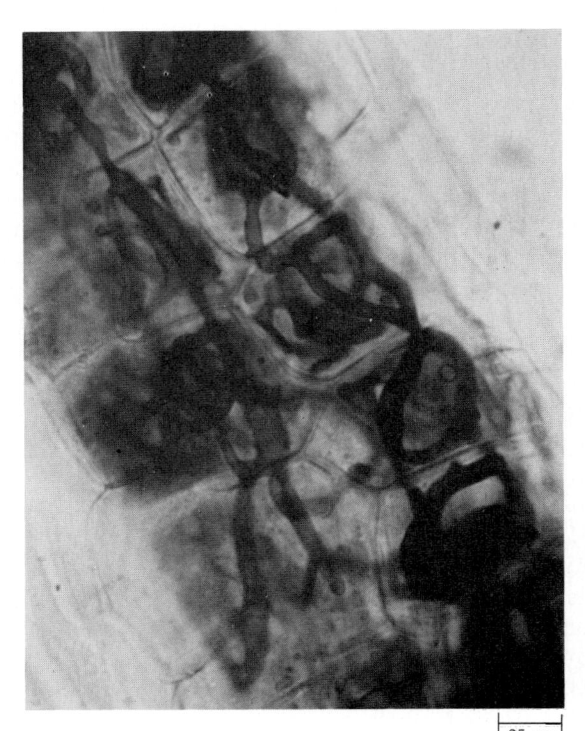

12–37
Surface view of an endomycorrhizal rootlet of fescue (Festuca), *a perennial grass, showing coiled hyphae in some of the cortical cells. Mycorrhizal associations are especially important for grasses growing on nutrient-poor soils or at high elevations.*

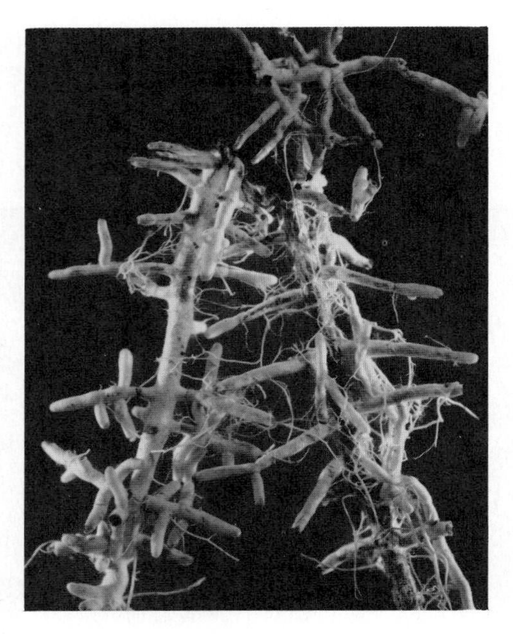

12–38
Ectomycorrhizal rootlets from a western hemlock (Tsuga heterophylla). *In such ectomycorrhizae, the fungus commonly forms a sheath of hyphae, called a fungal mantle, around the root. Hormones secreted by the fungus cause the root to branch. This growth pattern and the hyphal sheath impart a characteristic branched and swollen appearance to the ectomycorrhizae. The narrow strands extending from the mycorrhizae are rhizomorphs—bundles of hyphae that function as extensions of the root system.*

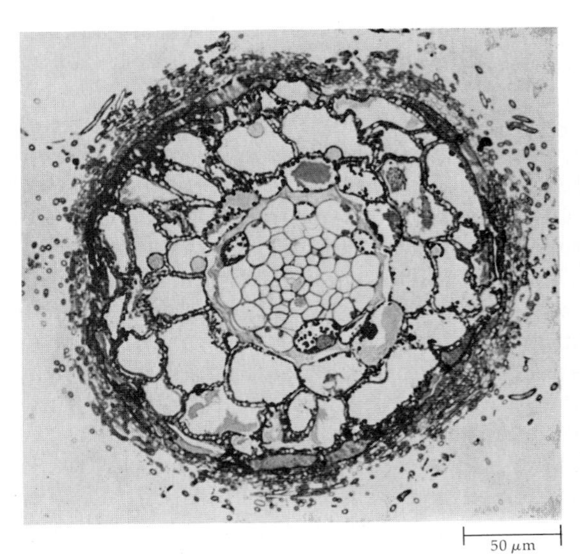

12–39
Transverse section of ectomycorrhizal rootlet of Pinus. *The hyphae of the fungus are confined mostly to the spaces between the cortical cells.*

rhizae by weight. It forms a prominent web of hyphae over the root surface, from which fine lateral hyphae penetrate the cortical cells of the plant. Although this relationship is undoubtedly an energetically expensive one for the vascular plant, it may play an important role in allowing the Ericaceae to colonize acidic soils that are poor in nutrients. Both ascomycetes and basidiomycetes are involved in these associations. In some subtypes that occur in Ericaceae, the fungi ensheathe the root tissue and also occur within it; in others, the fungus is entirely internal.

The seeds of orchids germinate in nature only in the presence of suitable fungi. This mycorrhizal association is unique in that the fungus, which is internal, also supplies the host plant with carbon, at least during its seedling stages. The fungi involved in such associations are often members of the Tulasnellales, an order of basidiomycetes that also includes many plant parasites and saprobes. Mature autotrophic orchid plants appear to be nonmycorrhizal.

A study of the fossils of early vascular plants (discussed in more detail in Chapter 16) has revealed that endomycorrhizal associations were as frequent then as they are in modern vascular plants. This finding has led K. A. Pirozynzki and D. W. Malloch to suggest that the evolution of mycorrhizal associations may have been a critical step allowing colonization of the land by plants. Given the relatively sterile soils available at the time of the first colonization of the land, the role of mycorrhizal fungi (undoubtedly zygomycetes) may have been of crucial significance, particularly in facilitating the uptake of phosphorus and other nutrients. A similar relationship has been demonstrated among contemporary plants colonizing extremely nutrient-poor soils, such as slag heaps: those individuals that have endomycorrhizae have a much better chance of surviving. Thus it may not have been a single organism but rather a symbiotic association of organisms, comparable to that of a lichen, that initially invaded the land. Just as lichens inhabit sites too extreme for either partner in isolation, so may have the ancestors of the first vascular plants.

SUMMARY

The fungi, together with the heterotrophic bacteria, are the principal decomposers of the biosphere, breaking down organic products and restoring carbon, nitrogen, and other components to the soil and air. Fungi are rapidly growing nonphotosynthetic organisms that characteristically form filaments called hyphae, which may be septate or aseptate. In most fungi, the hyphae are highly branched, forming a mycelium.

Fungi, which are almost all terrestrial, reproduce by means of spores that are dispersed by the wind. Among their genetic peculiarities are the phenomena of heterokaryosis and parasexuality. A heterokaryotic fungus is one in which the nuclei found in a common cytoplasm are genetically different. Parasexuality involves the fusion of haploid nuclei to form diploid nuclei, crossing-over between associated chromosomes, and their ultimate separation to re-form the haploid nuclei. It provides an alternative means for genetic recombination and generally is found among those fungi that do not reproduce sexually.

Glycogen is the primary storage polysaccharide of the fungi. The primary component of fungal cell walls is chitin. Most fungi are saprobes; that is, they live on organic materials from dead plants and animals. Rhizoids are specialized hyphae that serve to anchor some saprobic fungi to the substrate. Parasitic fungi often have specialized hyphae (haustoria) that penetrate the living cells of other organisms.

In addition to their role as decomposers, the fungi are economically important to humans as destroyers of foodstuffs and other organic materials. The group also includes the yeasts, *Penicillium* and other producers of antibiotics, cheese molds, and edible mushrooms.

The division Zygomycota contains the simplest fungi, which are commonly called zygomycetes. The members of this division have coenocytic mycelia. Their asexual spores are generally formed in sporangia—saclike structures in which all of the protoplasm becomes converted into spores.

Division Ascomycota, the members of which are commonly called ascomycetes, has some 30,000 described species, more than any other group of fungi. Their distinguishing characteristic is the ascus—a saclike structure in which the meiotic spores, known as ascospores, are formed. In the characteristic life cycle, male and female gametangia fuse to produce a filament that is dikaryotic (containing two paired haploid nuclei). The ascus forms at the tip of a dikaryotic hypha. Asexual reproduction may also take place by spore formation; typically the ascomycetes form a type of asexual spore known as a conidium. The yeasts are single-celled ascomycetes in which asexual reproduction occurs by fission or budding, and sexual reproduction occurs by the formation of asci not enclosed in fruiting bodies.

An artificial group of fungi—Fungi Imperfecti—includes many thousands of species with no known sexual cycle. Most Fungi Imperfecti have the hyphal characteristics of ascomycetes, and a few have those of basidiomycetes.

Lichens are ascomycetes that are obligate parasites on the green algae or cyanobacteria they include. The combination is morphologically and physiologically different from either organism as it exists separately. The ability of the lichen to survive under adverse environmental conditions is related to its ability to withstand desiccation and remain dormant when dry.

The division Basidiomycota, the members of which are called basidiomycetes, includes the most advanced of the fungi. They are also the most familiar and include mushrooms, toadstools, and several important plant pathogens. Their distinguishing characteristic is the production of basidia. Like the ascus, the basidium is produced at the tip of a dikaryotic hypha, and is the structure in which meiosis occurs. The resulting basidiospores are typically four in number. They differ from ascospores in that they are formed outside of the basidium. In the class Homobasidiomycetes the basidia are incorporated into complex fruiting bodies. Most of the Homobasidiomycetes do not form asexual spores. In the Heterobasidiomycetes, the other principal class of Basidiomycota, complex fruiting bodies are not present.

Mycorrhizae—symbiotic associations between the roots of plants and fungi—characterize all but a few families of vascular plants. Of the two major types of mycorrhizae, endomycorrhizae, in which the fungal partner is a zygomycete, occur in about 80 percent of all vascular plants. In such associations, the fungus actually penetrates the cortical cells of the host. In the second major type, ectomycorrhizae, the fungus does not penetrate the cells of the host but forms a feltlike layer over its roots. Both basidiomycetes and ascomycetes are involved in ectomycorrhizal associations. Mycorrhizal associations are important in obtaining phosphorus, and probably other nutrients, for the plant, and in providing organic carbon for the fungus.

SUGGESTIONS FOR FURTHER READING

AHMADJIAN, VERNON: *The Lichen Symbiosis*, Blaisdell Publishing Co., Waltham, Mass., 1967.

In this small but fascinating book, Ahmadjian presents work on the nature of the relationship between the fungal and algal components of a lichen.

ALEXOPOULOS, CONSTANTINE J., and CHARLES W. MIMS: *Introductory Mycology*, 3rd ed., John Wiley & Sons, Inc., New York, 1979.

A thorough account of the fungi and heterotrophic groups of protista; taxonomically oriented.

BATRA, LEKH R. (Ed.): *Insect-Fungus Symbiosis: Nutrition, Mutualism and Commensalism*, John Wiley & Sons, Inc., New York, 1979.

A fascinating collection of articles about the many and varied interactions between these abundant and successful organisms.

CHRISTENSEN, CLYDE M.: *The Molds and Man: An Introduction to the Fungi*, 3rd ed., University of Minnesota Press, Minneapolis, 1965.*

A highly informative account of the fungi and especially of their complex and important interrelationships with humans.

COOKE, RODERIC: *The Biology of Symbiotic Fungi*, John Wiley & Sons, Inc., New York, 1977.

A valuable discussion of interactions between fungi and other organisms, as well as between different fungi.

HALE, MASON E.: *The Biology of Lichens*, 2nd ed., University Park Press, Baltimore, Md., 1975.

An outstanding, concise summary of all aspects of the morphology, physiology, systematics, ecology, and economic uses and applications of the lichens.

HARLEY, J. L.: *The Biology of Mycorrhizae*, 3rd ed., Plant Science Monographs, Leonard Hill, London, 1980.

The standard work in the field; an invaluable reference.

LARGE, E. C.: *The Advance of the Fungi*, Dover Publications, Inc., New York, 1962.*

A fascinating popular account of the closely interwoven histories of fungi and humans, first published in 1940.

MARKS, G. C., and T. T. KOZLOWSKI (Eds.): *Ectomycorrhizae: Their Ecology and Physiology*, Academic Press, Inc., New York, 1973.

A detailed and valuable treatment of ectomycorrhizae, with each chapter written by a specialist in the subject.

MILLER, ORSON K., JR.: *Mushrooms of North America*, E. P. Dutton & Co., Inc., New York, 1977.

A beautifully illustrated guide to the major groups and species of North American fleshy fungi.

PHAFF, H. J., et al.: *The Life of Yeasts*, 2nd ed., Harvard University Press, Cambridge, Mass., 1978.

An authoritative account of the properties and uses of yeasts.

ROSS, IAN K.: *Biology of the Fungi: Their Development, Regulation, and Associations*, McGraw-Hill Book Company, New York, 1979.

An excellent biologically oriented overview of the fungi; highly recommended.

SHURTLEFF, WILLIAM, and AKIKO AOYAGI: *The Book of Miso*, Autumn Press, Brookline, Mass., 1976.*

A fascinating account of the production of soy paste in the Orient, with over 200 unusual recipes; The Book of Tofu *(1975), in the same series, is also recommended.*

* Available in paperback. (See also references on page 250.)

CHAPTER 13

Heterotrophic Protista: Water Molds and Slime Molds

13–1
Fruiting bodies, or sporangia, of the plasmodial slime mold Physarum viride.

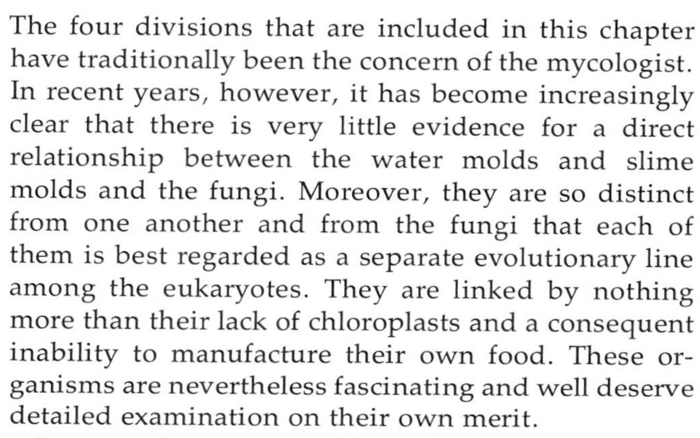

The four divisions that are included in this chapter have traditionally been the concern of the mycologist. In recent years, however, it has become increasingly clear that there is very little evidence for a direct relationship between the water molds and slime molds and the fungi. Moreover, they are so distinct from one another and from the fungi that each of them is best regarded as a separate evolutionary line among the eukaryotes. They are linked by nothing more than their lack of chloroplasts and a consequent inability to manufacture their own food. These organisms are nevertheless fascinating and well deserve detailed examination on their own merit.

Because the water molds and slime molds have nearly always been grouped with the fungi, there is a certain amount of overlap in the names and descriptions used; fungal terms are often applied to the heterotrophic protista. For instance, although they are no longer considered to be fungi, some of these "molds" are said to possess "mycelia" and "hyphae." Bear in mind that such similarities in terminology have a basis in tradition, rather than in a common biology or a shared evolutionary history.

Two of these divisions, Oomycota, the water molds and related organisms, and Chytridiomycota, the chytrids, consist primarily of aquatic organisms. The other two divisions, Acrasiomycota, the cellular slime molds, and Myxomycota, the plasmodial slime molds, are primarily terrestrial. The Chytridiomycota and Oomycota have coenocytic (multinucleate) hyphae, which can be extensive and highly branched in some members of the latter group. The Oomycota produce motile asexual spores—zoospores—with two flagella, one tinsel and one whiplash, whereas those of the Chytridiomycota have a single posterior flagellum of the whiplash type. The Myxomycota spend most of their nonreproductive phases as streaming masses of cytoplasm. The Acrasiomycota are amoebalike organisms that do not form motile spores at any stage of their life cycle.

DIVISION OOMYCOTA

The division Oomycota, with about 475 species, is a very distinctive group whose members are commonly called oomycetes. The cell walls of these organisms are composed largely of cellulose or celluloselike polymers, thus differing markedly from the cell walls of the fungi. The chromosomes of oomycetes contain histone proteins and resemble those of most eukaryotic organisms; they differ greatly from the chromosomes found in fungi. Compared to other eukaryotes, meiosis and mitosis, as far as is known, are not distinctive processes, and centrioles are present. The organisms in this division range from unicellular forms to highly branched, coenocytic filamentous ones.

Most species of oomycetes can reproduce both sexually and asexually. In oomycetes, sexual reproduction involves an *oogonium*, which contains many eggs, and an *antheridium*, containing numerous male nuclei. It results in the formation of a thick-walled zygote, the *oospore*, which serves as a resting spore (Figure 13–2). As mentioned previously, asexual reproduction is by means of motile zoospores that have two flagella—one tinsel and one whiplash.

One large group of the division Oomycota is aquatic. The members of this group—the so-called water molds—are abundant in fresh water and are easy to isolate from it. Most of them are saprobic, but a few are parasitic, including species that cause diseases of fish and fish eggs.

In some water molds, such as *Saprolegnia* (Figure 13–3), sexual reproduction can occur with male and female sex organs borne on the same individual; in other words, these water molds are *homothallic*. Other water molds, such as some species of *Achlya* (Figure 13–2), are *heterothallic*—male and female sex organs are borne by different individuals. Both *Saprolegnia* and *Achlya* can reproduce sexually and asexually.

Another group of oomycetes is primarily terrestrial, although the organisms still form motile zoospores when open water is available. Among this group—the Peronosporales—are several forms that are economically important. As C. J. Alexopoulos has said, "At least two of them have had a hand—or should we say hypha!—in shaping the economic history of an important portion of mankind."

One of these is *Plasmopara viticola*, the cause of downy mildew of grapes. Downy mildew was accidentally introduced into France in the late 1870s on American stock that had been imported because of its resistance to other diseases. The mildew soon threatened the entire French wine industry. It was eventually brought under control by a combination of good fortune and skillful observation. French vineyard owners in the vicinity of Medoc customarily put a disagreeable mixture of copper sulfate and lime on vines growing along the roadside to discourage passersby from picking them. A professor from the University of Bordeaux, who was studying the problem of downy mildew, noticed that these plants were free from symptoms of the disease. After conferring with the vineyard owners, the professor prepared his

(a) 100 μm

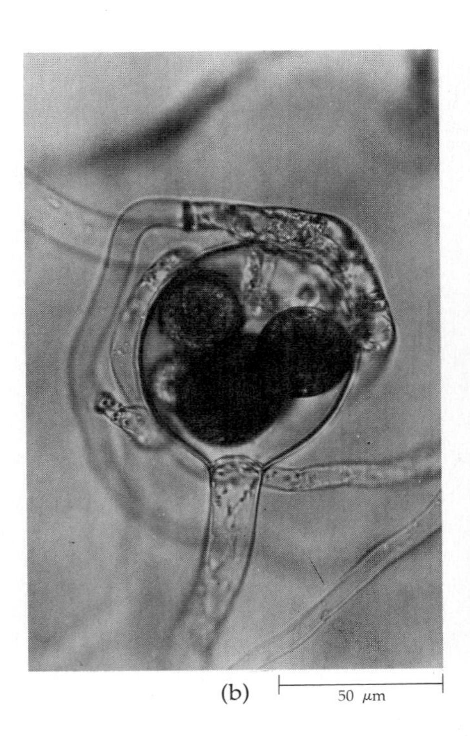

(b) 50 μm

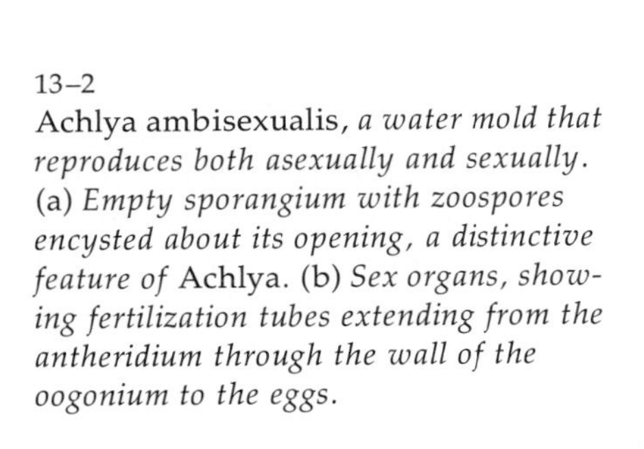

13–2
Achlya ambisexualis, *a water mold that reproduces both asexually and sexually.* (a) *Empty sporangium with zoospores encysted about its opening, a distinctive feature of* Achlya. (b) *Sex organs, showing fertilization tubes extending from the antheridium through the wall of the oogonium to the eggs.*

13–3
Life cycle of Saprolegnia, *an oomycete. The mycelium of this water mold is diploid. Reproduction is mainly asexual. Biflagellate zoospores released from a sporangium swim for a while and then encyst. Each eventually gives rise to a secondary zoospore, which also encysts and then germinates to produce a new mycelium.*

During sexual reproduction, oogonia and antheridia are formed on the same hyphae. Meiosis apparently occurs within these structures. The oogonia are enlarged cells in which a number of spherical eggs are produced. The antheridia develop from the tips of other filaments of the same individual and produce numerous male nuclei. In mating, the antheridia grow toward the oogonia and develop tubular processes called fertilization tubes, which penetrate the oogonia. Male nuclei travel down the fertilization tubes to the female nuclei and fuse with them. Following each nuclear fusion, a thick-walled zygote, called the oospore, is produced. On germination, the oospore develops into a hypha, which then produces a sporangium, beginning the cycle anew.

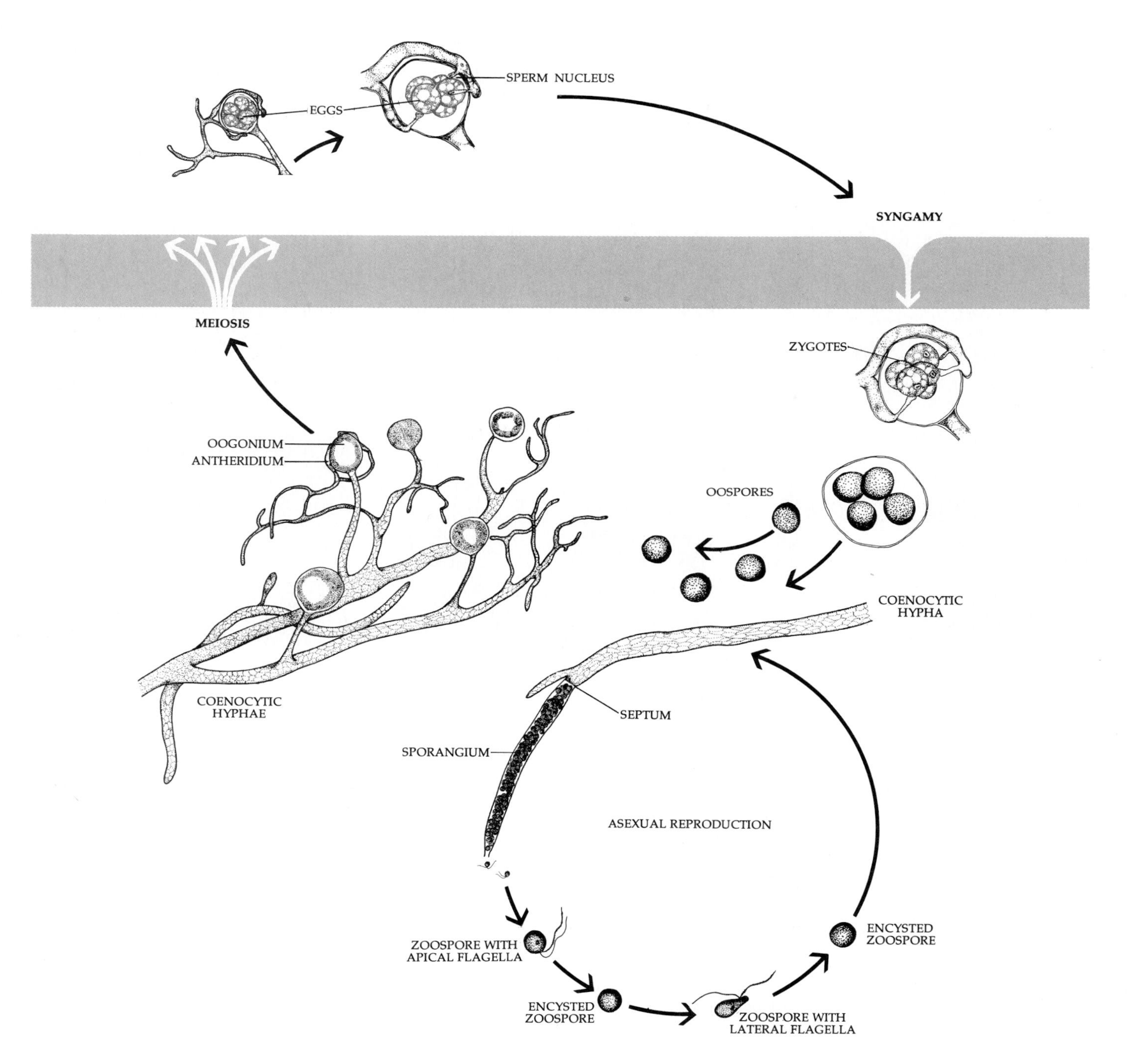

HORMONAL CONTROL OF SEXUALITY IN A WATER MOLD

Achlya ambisexualis *is a heterothallic water mold—it produces male and female sex organs on different individuals. The late John R. Raper of Harvard University studied the hormonal control of sexuality in these organisms. The female vegetative hyphae secrete a substance that induces the initial development of antheridia on the hyphae of the male organism.* (a) *Appearance of undifferentiated hyphae before the addition of this substance—characterized and named antheridiol by Alma Barksdale of the New York Botanical Garden.* (b) *Antheridial branches produced within two hours after the addition of crystalline antheridiol.* (c) *Antheridial branches growing toward a plastic particle containing antheridiol.*

After the developing antheridia appear, the male organism secretes its own hormone, called oogoniol, which induces the formation of oogonia on the hyphae of the female organism. With the appearance of young oogonia, the antheridial hyphae are attracted to them and distinct, mature antheridia develop. Raper attributed this latter response to a substance he called "hormone C," presumably secreted by the female, but subsequent research suggested that antheridiol elicits this developmental reaction. After the antheridia are formed, there is a differentiating response in the female oogonia, presumably hormonally induced, which leads to the maturation of the oogonium and its enclosed female gametes (eggs). Thus, even in the protista, sexual reproduction can involve a highly coordinated sequence of events that are hormonally controlled.

(a)

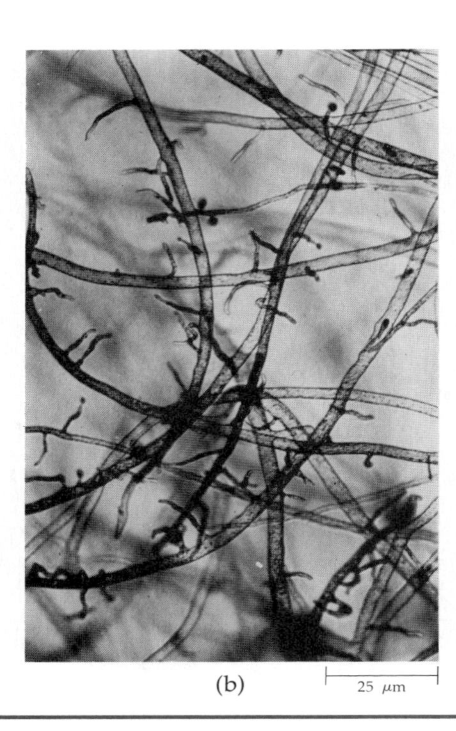

(b)

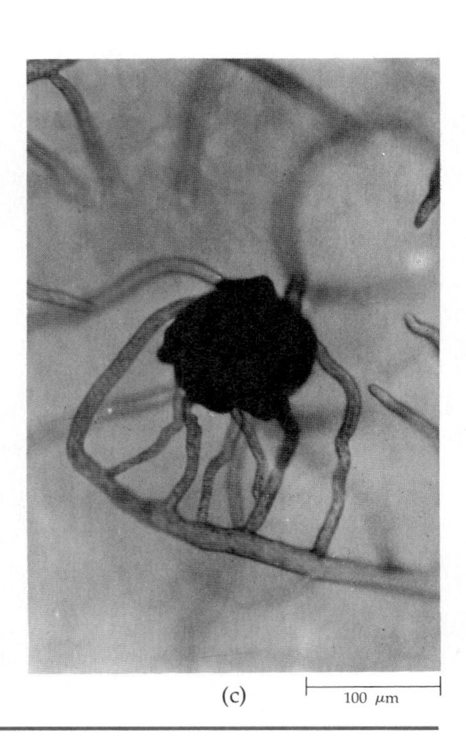

(c)

own mixture of chemicals—the Bordeaux mixture—which was made generally available in 1882. The Bordeaux mixture was the first chemical used in the control of a plant disease.

Another economically important member of this group is the genus *Phytophthora* (which means "plant destroyer"). *Phytophthora,* with about 35 species, is a particularly important plant pathogen that causes widespread destruction on many crops, including cacao, pineapples, tomatoes, rubber, papayas, onions, strawberries, apples, soybeans, tobacco, and citrus. A widespread member of this genus, *P. cinnamomi,* which occurs in soil, has killed or rendered unproductive millions of avocado trees in southern California and elsewhere in recent years. It has also destroyed tens of thousands of hectares of valuable eucalyptus timberland in Australia. The zoospores of *P. cinnamomi* are attracted to the plants they infect by chemicals exuded by the roots. This oomycete also produces resistant spores that can survive for up to six years in moist soil. Extensive breeding efforts are now under way with avocados and other susceptible crops to produce strains resistant to this oomycete.

The best-known member of *Phytophthora,* however, is *P. infestans* (Figure 13–4), the cause of the late blight of potatoes, which produced the great potato famines in Ireland. The famine of 1845–47, which was caused by this plant pathogenic organism, was responsible for over 1 million deaths from starvation and initiated large-scale emigration from Ireland to the United States; within a decade, the population of Ireland dropped from 8 million to 4 million. Virtually the entire Irish potato crop was wiped out in a single week in the summer of 1846.

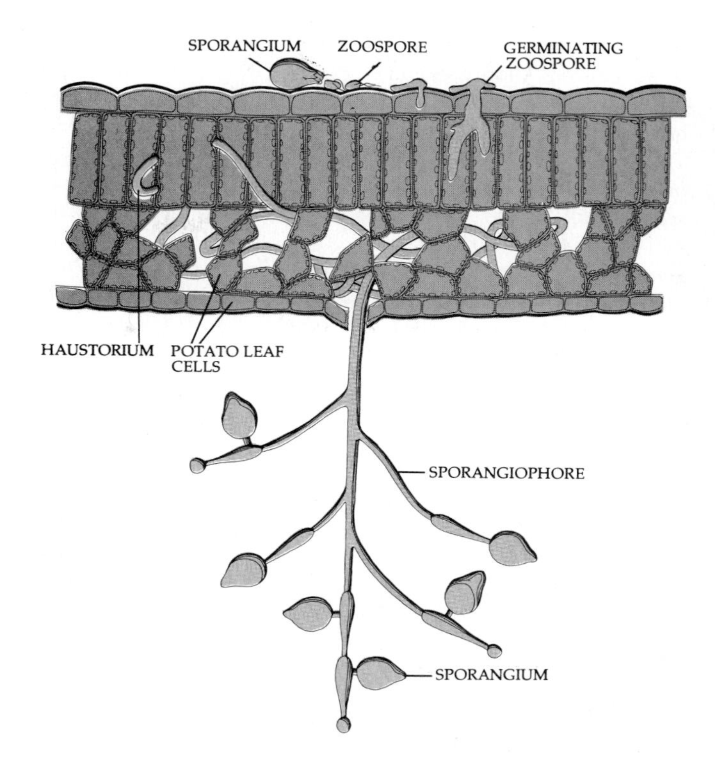

13–4
Phytophthora infestans, *the cause of the late blight of potatoes. The cells of the potato leaf are shown in gray. In the presence of water, zoospores are released from the sporangia and swim to the germination site (as shown here), or the sporangia germinate directly through a germ tube.*

DIVISION CHYTRIDIOMYCOTA

The Chytridiomycota, or chytrids, are a predominantly aquatic group of about 750 species. These organisms are extremely varied in form, in the nature of their sexual interactions, and in their life histories. They have cell walls that seem to consist mainly of chitin, although other polymers are also found. The mitotic and meiotic cycles, as far as is known, are like those in Oomycota, and all chytrids have coenocytic bodies with few septa at maturity.

The principal characteristic by which all chytrids are recognized as belonging to one division of the kingdom Protista is that their motile cells (zoospores and sperm) all possess a single, posterior, whiplash flagellum at maturity. This characteristic alone separates them from all other groups with which they have been compared, and they are probably more closely related to protozoa than to fungi or plants.

Some chytrids are simple, unicellular organisms that do not develop a mycelium. In these chytrids, the whole organism is transformed into a reproductive structure at the appropriate time. Some of them have slender rhizoids that extend into the substrate and serve as an anchor (Figure 13–5). Different species of chytrids are parasitic on or in algae, aquatic oomycetes, spores, pollen grains, or other parts of vascular plants; they are also saprobic on such substrates as dead insects. The genus *Coelomomyces,* for example, is an obligate parasite of the larvae of mosquitoes and other flies. It exhibits a life cycle comparable to that of the rust fungi, in that it has an alternation of hosts between the small aquatic crustaceans known as copepods and the mosquito larvae. *Coelomomyces* is being investigated as a possible biological control for mosquitoes.

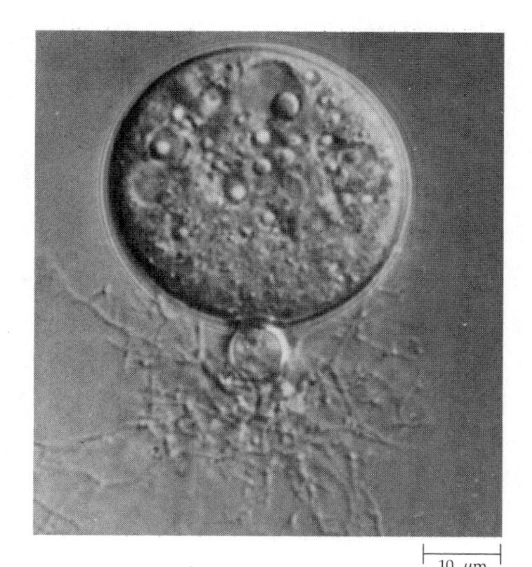

13–5
Chytridium confervae, *a common chytrid, as seen with the aid of Normarski system optics. Note the slender rhizoids extending downward from this single-celled organism.*

Other chytrids are much more complex in their structure and reproduction. Consider, for example, the life cycle of the genus *Allomyces.* Some species have an isomorphic alternation of generations comparable to that shown in Figure 13–6, whereas in other species the alternation is heteromorphic, in which the haploid and diploid individuals do not resemble one another closely. Alternation of generations is characteristic of the plants and of many algae but is otherwise found only in *Allomyces* and one other closely

13–6
In the chytrid Allomyces macrogynus, *whose life cycle is shown here, there is an isomorphic alternation of generations. The haploid and diploid individuals are indistinguishable until they begin to form reproductive organs. The haploid individuals produce colorless female gametangia (shown in black) and orange male gametangia (shown in color), in equal numbers. The male gametes, which are about half the size of the female gametes, are attracted by sirenin, a hormone produced by the female gametes. The zygote loses its flagella, becomes round, and soon germinates to produce a diploid individual. This diploid individual forms two kinds of sporangia: (1) asexual sporangia—colorless structures that release diploid zoospores—which in turn germinate and repeat the diploid generation; and (2) sexual sporangia—thick-walled, reddish-brown structures able to withstand severe environmental conditions. After a period of dormancy, meiosis occurs in these sexual sporangia, resulting in the formation of haploid zoospores. These zoospores develop into haploid individuals that produce gametangia at maturity.*

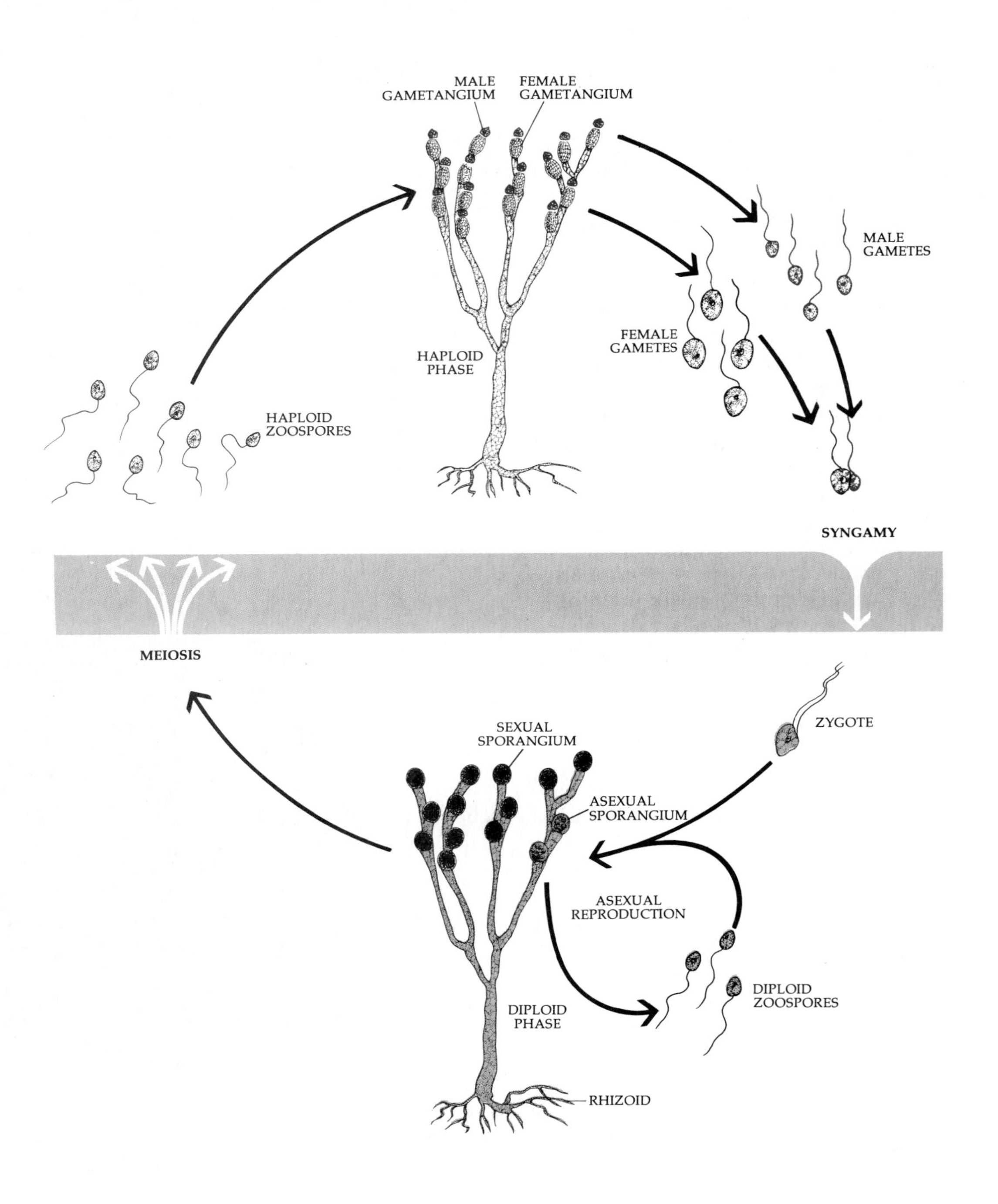

related genus of chytrids (see Figure 10–9, page 180). In terms of its life cycle, morphology, and physiology, *Allomyces* is one of the best known of the protista and has been used extensively in biological experiments.

DIVISION ACRASIOMYCOTA

The cellular slime molds—a group of about 26 species included in seven genera—are probably more closely related to the amoebas, which have been traditionally placed with the protozoa, than to any other group. One member of this division, *Dictyostelium discoideum,* was discussed as a model of multicellular differentiation in Chapter 7 (see pages 128–30). In this species and in other members of the division, the individual amoebas (myxamoebas) retain their individuality even in the swarming stage of their life cycle, when they come together to form compound sporangia (Figure 13–7). Whether sexual reproduction occurs in the cellular slime molds is not certain. Their cell walls contain cellulose but not chitin, and, in general, mitosis is normal.

The cellular slime molds were once considered rare curiosities. Then, in 1935, Kenneth B. Raper discovered *Dictyostelium discoideum,* and his subsequent research on this organism brought a great deal of attention to the cellular slime molds. It is now known that they are widely found in soils, from which they are easily isolated. One of the best sources is surface soil and well-decomposed leaf litter from old deciduous forests. The amoebas feed on bacteria in the soil and thus are ecologically important. They have proven to be useful experimental tools in studies of differentiation and molecular biology.

DIVISION MYXOMYCOTA

This division comprises the plasmodial slime molds, or myxomycetes, a group of about 450 species that seems to have no direct relationship to the cellular slime molds. During their nonreproductive stages, the plasmodial slime molds are thin streaming masses of protoplasm that creep along in amoeboid fashion. Lacking a cell wall, this "naked" mass of protoplasm is called a *plasmodium*. As these plasmodia travel, they engulf and digest bacteria, yeast cells, fungal spores, and small particles of decayed plant and animal matter. The successful culture of plasmodia in media that do not contain particulate matter suggests that they can also obtain food by direct absorption.

The plasmodium may grow to weigh as much as 20 to 30 grams, and because slime molds are spread out thinly, this amount can cover an area of up to several square meters (see Figure 10–6b, page 172). The plasmodium contains many nuclei but is not divided by cell walls. As it grows, the nuclei divide repeatedly and synchronously; that is, all the nuclei in a plasmodium divide at the same time. Centrioles are present and mitosis is normal, although the chromosomes are very small.

Typically, the moving plasmodium is fan-shaped, with flowing protoplasmic tubules that are thicker at the base of the fan and spread out, branch, and become thinner toward their outer ends. The tubules are composed of slightly solidified protoplasm through which the more liquified protoplasm flows rapidly. The foremost edge of the plasmodium consists entirely of a thin film of gel separated from the substrate only by a plasma membrane and a slime sheath of unknown chemical composition.

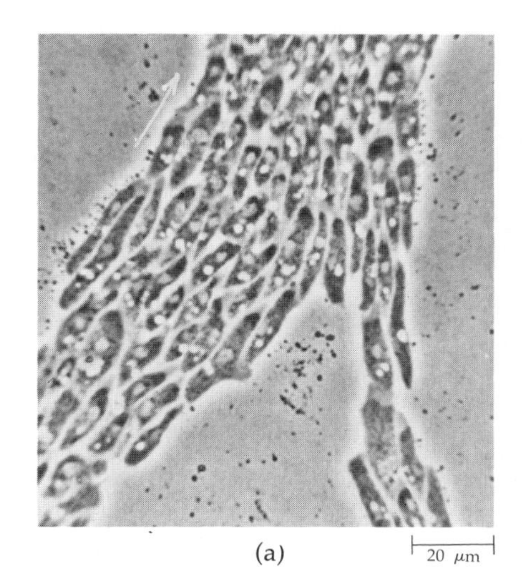

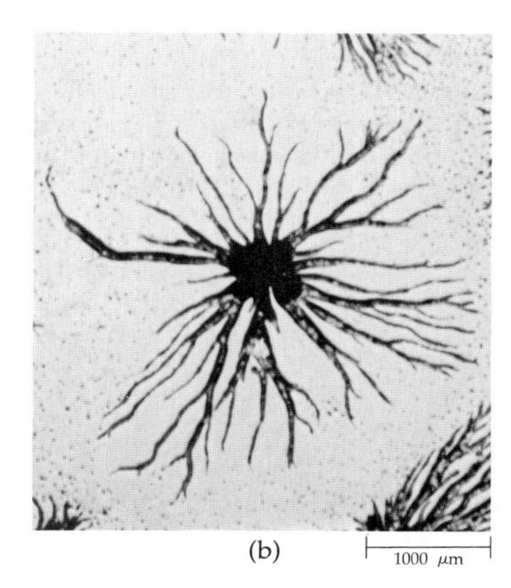

13–7
Dictyostelium discoideum. (a) *Aggregating myxamoebas. Note that each amoeba has retained its individuality. The arrow indicates the direction in which the stream of amoebas is moving.* (b) *Low-magnification view of a large number of myxamoebas, showing the overall pattern of aggregation.*

Plasmodial growth continues as long as an adequate food supply and moisture are available. Generally, when either of these is in short supply, the plasmodium migrates away from the feeding area. At such times, they may be found crossing roads or lawns, climbing trees, or in other unlikely places. In many species, when the plasmodium stops moving, it divides into a large number of small mounds. The mounds are similar in size and volume, so their formation is probably controlled by chemical effects within the plasmodium. Each mound produces a mature sporangium, usually borne at the tip of a stalk; this sporangium is often extremely ornate (Figures 13–8c and 13–8d). Meiosis occurs in the young diploid spores after cleavage and wall formation, and four nuclei are produced. After meiosis, three of the four nuclei disintegrate, leaving the spore with a single haploid nucleus. In some members of this group, discrete sporangia are not produced, and the entire plasmodium may develop either into a *plasmodiocarp* (Figure 13–8a), which retains the former shape of the plasmodium, or into an *aethalium* (Figure 13–8b), in which the plasmodium forms a large mound that differentiates almost like a single large sporangium.

The spores of myxomycetes are very resistant to environmental extremes; some have germinated after having been kept in the laboratory for more than 60 years. Thus spore formation in this group seems to make possible not only genetic recombination but also survival under adverse conditions.

Under favorable conditions, the spores split open and the protoplast slips out. The protoplast may remain amoeboid, or it may develop one or two whiplash flagella; the amoeboid and flagellate stages are readily interchangeable. The amoebae feed by the ingestion of bacteria and organic material and multiply by mitosis and cell cleavage. If the food supply is

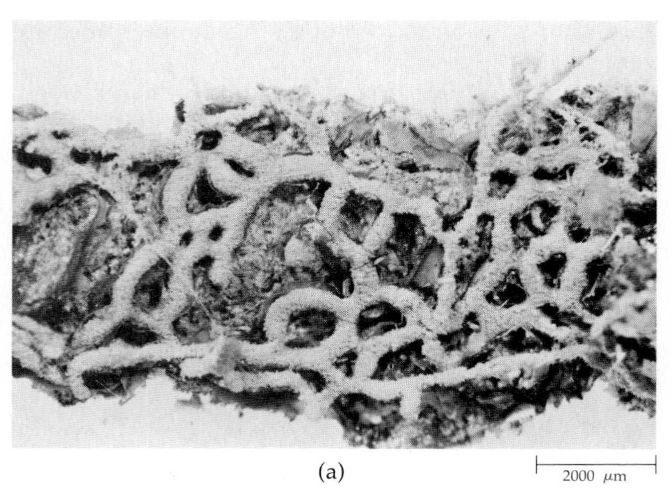

(a) 2000 μm

(b) 10,000 μm

13–8
Spore-producing structures in the division Myxomycota. (a) *Plasmodiocarp of* Hemitrichia serpula. (b) *Aethalia of* Lycogala *growing on bark.* (c) *Sporangia of* Arcyria. (d) *Sporangia of* Stemonitis splendens. *(See also Figure 13–1 on page 241.)*

(c)

(d)

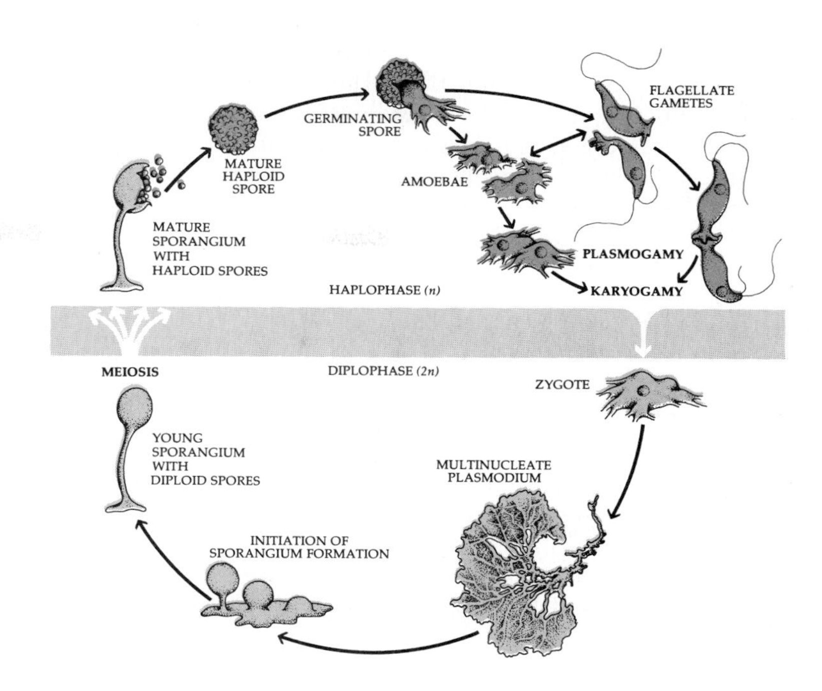

13–9
Life cycle of a typical myxomycete. Sexual reproduction in the plasmodial slime molds consists of three distinct phases: plasmogamy, karyogamy, and meiosis. ***Plasmogamy*** *consists of the union of two protoplasts, bringing two haploid nuclei together in the same cell.* ***Karyogamy*** *is the fusion of these two nuclei, resulting in the formation of a diploid zygote and the initiation of the so-called diplophase of the life cycle. The plasmodium is a multinucleate, free-flowing mass of protoplasm that can pass through a silk cloth or a piece of filter paper and remain unchanged.* ***Meiosis*** *restores the haploid condition and initiates the haplophase of the cycle.*

used up, or conditions are otherwise unfavorable, the amoebae may cease moving about, become round, and secrete a thin wall to form a microcyst. These microcysts can remain viable for a year or more, resuming activity when favorable conditions return.

After a period of growth, plasmodia appear in the amoeba population. Their appearance is governed by a number of factors, including cell age, environment, density of amoebae, and chemical inducers that play a role similar to those discussed for the cellular slime mold *Dictyostelium discoideum* (see pages 128–30). One method of plasmodium formation is by the fusion of gametes, which are usually genetically different from one another and are ultimately derived from different haploid spores. These gametes are simply some of the amoebae or flagellate cells which are now playing a new role. In some species and strains, however, the plasmodium is known to form directly from a single amoeba; such plasmodia are usually haploid, like the amoebae that originally gave rise to them.

The life cycle of a typical plasmodial slime mold is summarized in Figure 13–9.

SUMMARY

The four major groups of nonphotosynthetic protista that have often been treated as fungi are the divisions Oomycota, Chytridiomycota, Acrasiomycota, and Myxomycota. They have a more or less typical mitotic cycle, and centrioles are present. The first two groups are primarily aquatic; the latter two are terrestrial.

Members of the division Oomycota have coenocytic hyphae with septa formed only when sporangia or gametangia are mature. The sperm and zoospores of this group have two flagella—one tinsel and one whiplash. The cell walls are composed largely of cellulose or celluloselike polymers. Sexual reproduction is oogamous. Two of the partly terrestrial members of this division are *Phytophthora*—a very important causative agent of plant diseases, including the blight that caused the potato famine in Ireland—and *Plasmopara viticola,* which causes downy mildew of grapes.

The Chytridiomycota, or chytrids, are unicellular or coenocytic organisms that are aquatic; in most species that have been examined, the cell walls are rich in chitin. The motile spores and gametes have a single posterior whiplash flagellum. *Allomyces* and one closely related genus are the only nonphotosynthetic organisms known to have an alternation of generations similar to that of plants and many algae.

The cellular slime molds, division Acrasiomycota, are a small group of amoebalike organisms that swarm together at one stage of their life cycle to form compound sporangia. Flagellate cells are not known. It is not known if sexual reproduction occurs in this group.

The plasmodial slime molds, division Myxomycota, exist mainly as streaming, multinucleate masses of protoplasm called plasmodia, which are usually diploid. These plasmodia ultimately form sporangia in which diploid spores are formed. Meiosis occurs within each of these spores, and three of the resulting nuclei disintegrate, leaving one haploid nucleus in each spore. Under favorable conditions, these spores split open, producing amoebae which may become flagellated. These amoebae or flagellate cells may function as gametes. Plasmodium formation often, but not always, follows fusion of the gametes.

SUGGESTIONS FOR FURTHER READING

AINSWORTH, GEOFFREY C., and A. S. SUSSMAN (Eds.): *The Fungi: An Advanced Treatise,* 4 vols., Academic Press, Inc., New York, 1965–73.

This huge work contains summaries of all aspects of fungal biology contributed by experts in each field.

ALEXOPOULOS, CONSTANTINE J., and CHARLES W. MIMS: *Introductory Mycology,* 3rd ed., John Wiley & Sons, Inc., New York, 1979.

A thorough account of the fungi and heterotrophic groups of protista; taxonomically oriented.

BATRA, LEKH R. (Ed.): *Insect-Fungus Symbiosis: Nutrition, Mutualism and Commensalism,* John Wiley & Sons, Inc., New York, 1979.

A fascinating collection of articles about the many and varied interactions between these abundant and successful organisms.

BONNER, JOHN T.: *The Cellular Slime Molds,* 2nd ed., Princeton University Press, Princeton, N.J., 1968.

A record of experimental work with a small but fascinating group of organisms.

CHRISTENSEN, CLYDE M.: *The Molds and Man: An Introduction to the Fungi,* 3rd ed., University of Minnesota Press, Minneapolis, 1965.*

A highly informative account of the fungi and especially of their complex and important interrelationships with humans.

COOKE, RODERIC: *The Biology of Symbiotic Fungi,* John Wiley & Sons, Inc., New York, 1977.

A valuable discussion of interactions between fungi and other organisms, as well as between different fungi, covering a wide range of topics not previously presented together.

GRAY, W. D., and CONSTANTINE J. ALEXOPOULOS: *Biology of the Myxomycetes,* Ronald Press Co., John Wiley & Sons, Inc., New York, 1968.

A well-written book about the biology of the plasmodial slime molds, emphasizing ultrastructural, biochemical, and physiological aspects.

LARGE, E. C.: *The Advance of the Fungi,* Dover Publications, Inc., New York, 1962.*

A fascinating popular account of the closely interwoven histories of fungi and humans, first published in 1940.

ROSS, IAN K.: *Biology of the Fungi: Their Development, Regulation, and Associations,* McGraw-Hill Book Company, New York, 1979.

An excellent overall review of the fungi, biologically oriented, that will lead the reader to a deeper appreciation of the group; highly recommended.

*Available in paperback.

CHAPTER 14

Autotrophic Protista: The Algae

14–1
Algae on the rocks at low tide along the coast of North Carolina.

The open sea, the shore, and the land are the three life zones that make up our biosphere. The algae play a role in these first two ancient dwelling places, the sea and the shore, comparable to the role played by plants in the far younger but more familiar terrestrial world (Figure 14–1).

In the open sea, minute photosynthetic cells and tiny animals are found together as free-floating *plankton* (from the Greek word for "wanderer"). The planktonic algae, or *phytoplankton,* are the beginning of the food chain for all the animals that live in the deep waters. Phytoplankton are usually composed of single cells—some quite simple in appearance, others with intricate and delicately detailed forms sometimes joined into colonies or filaments (Figure 14–2). The smallest members of the phytoplankton, so small that they normally pass through the openings of a plankton net, are called *nannoplankton* (see page 278). Planktonic animals, or *zooplankton,* include an abundance of microscopic species, particularly Crustacea, such as tiny shrimp, copepods, and their relatives. In addition, large numbers of the larval, or immature, forms of nonplanktonic animal groups are represented.

The animal plankton feed on the phytoplankton (and on each other); small fish and some larger ones, as well as most of the great whales, feed on the phytoplankton and zooplankton; and still larger fish feed on the smaller ones. In this way, the "great meadow of the sea," as the phytoplankton are sometimes called, can be likened to the meadows of the land, serving as the source of animal nourishment.

Along the rocky shore can be found the larger, more complex algae, typically arranged in fairly distinct visible bands or layers in relation to intertidal levels. Their complexity reflects their ability to survive in this difficult life zone, where twice each day they are subject to great fluctuations of humidity, temperature, salinity, and light, in addition to the pounding action of the surf and the abrasive action of suspended sand particles churned up by the waves.

14–2
Phytoplankton. The organisms shown here are dinoflagellates and filamentous and unicellular diatoms.

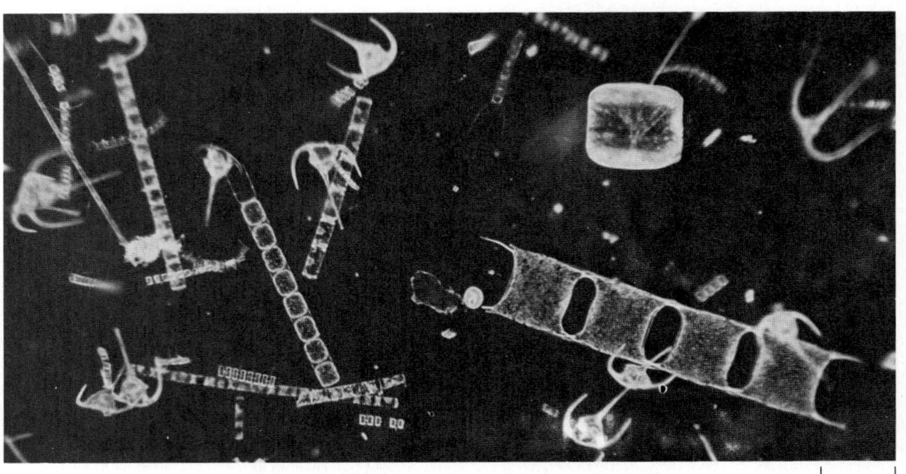

Anchored offshore beyond the zone of waves, algae provide shelter for a rich diversity of microscopic organisms, as well as for larger fish and invertebrate animals that feed on the microorganisms and on each other. These extensive algal growths may be so dense as to be called "jungles." For example, off the coast of California, there are kelp-bed jungles composed of giant brown kelps whose broad, 15-meter-long fronds are buoyed upward on sinuous stalks 30 meters or more from their holdfasts on the bottom. Many large carnivores, including sea otters and tuna, find food and refuge in these kelp beds, which are also harvested by humans for food and fertilizer.

From apparent oceanic origins, many species of algae became adapted to the "newer" land life zone with its various types of freshwater habitats. Freshwater algae include phytoplankton and macroscopic forms found in lakes, ponds, rivers, and streams, as well as those that inhabit swimming pools and domestic water supplies such as wells and reservoirs. In these aquatic habitats, the algae form the nutritional basis for all the animals, just as they do in the sea. Efforts are under way to develop commercial methods for culturing some of these algae to provide protein for humans; the yields of protein are about two to three times those of conventional crops, especially in sunny tropical regions such as Thailand and India. Many algae are found in soil, where they can be quite abundant; on tree trunks or wooden boards (such as the untreated planks of a deck or patio); on bare rocks; and even in the arid soils of the harshest deserts, where they may "come to life" only once or twice a year, after a rainfall. Certain species of algae occupy extremely rigorous environments, such as the near-boiling hot springs at Yellowstone National Park, the highly saline Great Salt Lake of Utah, and the constant cold of ice fields and glaciers. Even in suspended droplets of water in the atmosphere, certain species may thrive and multiply, falling to earth only during storms.

CHARACTERISTICS OF THE ALGAE

A number of distinct lines of algae had evolved by the early Paleozoic era, more than 450 million years ago. Most of the approximately 22,000 described species now in existence can be grouped into the six divisions discussed here. The term "algae" has been abandoned as a formal term in modern classification, because the various groups are not directly related to one another. Three divisions—Chrysophyta, Pyrrophyta, and Euglenophyta—consist almost entirely of unicellular organisms. The other divisions—Chlorophyta, Phaeophyta, and Rhodophyta—include groups that are multicellular.

Nearly all members of these divisions are photosynthetic. In general, they all have a relatively simple construction, which may be a single cell, a filament of cells, a plate of cells, or a solid body more or less comparable to that found in the vascular plants. The cell walls of algae, in general, have a polysaccharide matrix in which cellulose sometimes plays a major role. When algal cells divide, the plasma membranes generally pinch inward from the margin of the cell (furrowing), just as in animals, fungi, and protozoa. However, cell plates like those of higher plants have been found in one brown alga and a few genera of filamentous green algae. Most algal cells—except those of the red algae—have centrioles that may function in mitosis as they do in animal cells. These centrioles also become the basal bodies for flagella of any motile cells that are produced (see pages 30–31).

Multicellular algae do not have a complex array of tissues like that found in the vascular plants. In certain kelps (brown algae of the division Phaeophyta), however, a central conducting strand consisting of cells that resemble sieve elements is found in the stalk. The reproductive structures of the algae are generally single cells, not multicellular structures with sterile protective jackets such as are found in the bryophytes and vascular plants (see Chapter 15).

Table 14–1 *Comparative Summary of Characteristics in Six Divisions of Eukaryotic Algae*

DIVISION	NUMBER OF SPECIES	PHOTOSYNTHETIC PIGMENTS	CARBOHYDRATE FOOD RESERVE	FLAGELLA	CELL WALL COMPONENT	REMARKS
Chlorophyta (green algae)	7000	Chlorophyll *a*, chlorophyll *b*, carotenoids	Starch	None or 2 (or more); apical or lateral; equal; whiplash	Polysaccharides, sometimes cellulose	Mostly freshwater, but some marine
Phaeophyta (brown algae)	1500	Chlorophyll *a*, chlorophyll *c*, carotenoids, including fucoxanthin	Laminarin, mannitol	2, lateral, tinsel forward, whiplash behind; in reproductive cells only	Cellulose matrix with alginic acids (polysaccharides)	Almost all marine, flourish in cold ocean waters
Rhodophyta (red algae)	4000	Chlorophyll *a*, carotenoids, phycobilins	Floridean starch	None	Cellulose, pectic materials, calcium carbonate in many	Marine, some freshwater, many tropical species
Chrysophyta (diatoms and golden-brown algae)	11,500	Chlorophyll *a*, chlorophyll *c*, carotenoids, including fucoxanthin	Leucosin	None or 1 or 2, apical, whiplash or tinsel, equal or unequal	Pectic compounds with siliceous material	Marine and freshwater
Pyrrophyta (dinoflagellates)	1100	Chlorophyll *a*, chlorophyll *c*, carotenoids	Starch	None or 2, lateral, tinsel	Cellulose, other materials	Marine and freshwater; sexual reproduction rare
Euglenophyta (euglenoids)	800	Chlorophyll *a*, chlorophyll *b*, carotenoids	Paramylon	1 (to 3), apical, tinsel (one row of hairs)	No cell wall; have a proteinaceous pellicle	Mostly freshwater; sexual reproduction unknown

The divisions of algae differ widely from one another in the nature of their flagella (when present) and in their biochemical characteristics, especially with respect to differences in pigmentation, nature of reserve foods, and cell wall components (Table 14–1). The names of some divisions are derived from the colors of the predominant accessory pigments, which mask the grass-green of the chlorophylls. A wide variety of carotenoids is found in the algae. The xanthophyll fucoxanthin gives the brown algae their characteristic color and name; it is also found in the golden-brown algae and diatoms (division Chrysophyta). The red algae (Rhodophyta) owe their colors to several kinds of phycobilins—accessory pigments like those found in the cyanobacteria—which, unlike the carotenoids, are water soluble. In the green algae, the color of the chlorophylls is usually not masked by accessory pigments. However, some types—such as the species of *Chlamydomonas* found on the surface of snow, or *Trentepohlia*, a filamentous alga that grows on tree branches—produce large amounts of carotenoids as a shield against intense light; because of the presence of these accessory pigments, they often appear red or rust-colored.

A rich diversity of storage products is found in the different divisions of algae, with most having distinctive carbohydrate food reserves, often in addition to lipids. The green algae and dinoflagellates store their carbohydrates as starch, like the bryophytes and vascular plants. In the brown algae, another glucose polymer, laminarin, takes the place of starch. The carbohydrate reserves of the red algae are biochemically similar to starch as it occurs in green algae and plants, whereas that of the golden-brown algae is similar to laminarin. One of the most characteristic energy-rich molecules found in the cells of the brown algae is mannitol, a simple alcohol derived from the sugar mannose.

SYMBIOTIC ALGAE

Green algae, most resembling the genus Chlorella, *can be found living within many freshwater protozoa, sponges, hydra, and some flatworms. Most of these algae reproduce by simple cell division and are found in the vacuoles of the host cell. These vacuoles divide when the algae divide.*

Another green alga, Platymonas convolutae, *is found mostly in the subepidermal cells of the marine flatworm* Convoluta roscoffensis. *Within the cells of the flatworm,* Platymonas *has no cell wall and is irregularly shaped; its plasma membrane, greatly increased in surface area by fingerlike projections, is in more or less direct contact with the vacuole membrane of the host cell. When removed from the flatworm and cultured,* Platymonas *has a cell wall, four flagella, and an eyespot, all of which are missing when it is symbiotic.*

The most direct relationship between green algae and invertebrates involves certain nudibranchs, which are marine mollusks, and the chloroplasts of some siphonaceous green algae, such as Codium. *The chloroplasts are found in cells that line the entire respiratory chamber of the nudibranch. In the presence of light these chloroplasts carry on photosynthesis so efficiently that individuals of the nudibranch* Placobranchus ocellatus *are reported to evolve oxygen more rapidly than it is consumed by the mollusk.*

Certain dinoflagellates (division Pyrrophyta)—which resemble the nonflagellate cells of Chlorella *when they occur as symbionts—inhabit the cells of various marine sponges, coelenterates, mollusks, flatworms, and protozoa. In the giant clams of the family Tridachnidae, the dorsal surface of the inner lobes of the mantle may appear chocolate-brown as a result of the presence of symbiotic algae of this group, which are found in the blood sinuses and probably occur mainly within amoeboid blood cells. Dinoflagellates are also important symbionts in reef-building corals. Coral tissues may contain as many as 30,000 symbiotic dinoflagellates per cubic millimeter. These yellow-brown algae occur within cells in the lining of the gut of the coral polyps, and the entire biological productivity of the coral-reef ecosystem depends upon this symbiotic relationship. Since the algae require light for photosynthesis, the corals that contain them grow only in ocean waters less than 100 meters deep. Many of the variations in the shapes of coral are related to the light-gathering properties of different geometrical arrangements, somewhat similar to the ways that various branching patterns of trees act to expose their leaves to sunlight. It is because of the photosynthetic activities of these symbiotic algae that coral reefs can flourish in nutrient-poor tropical waters.*

Recently, several kinds of diatoms, growing without their characteristic shells, have been found as symbionts in large marine protozoans (Foraminifera). The diatoms apparently contribute significantly to the nutrition of these organisms in the nutrient-poor tropical seas, just as the dinoflagellates contribute to the nutrition of the reef-forming coral polyps.

50 μm

Each bell-shaped cell of the protozoan *Vorticella* contains numerous cells of the symbiotic green alga *Chlorella*.

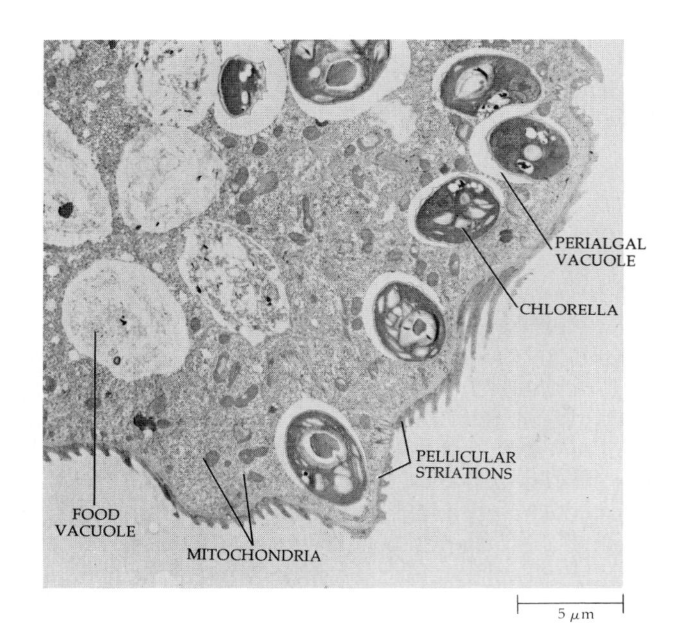

Electron micrograph of a *Vorticella* containing cells of *Chlorella*. Each algal cell is found in a separate vacuole bounded by a single membrane. The protozoan provides the algae with protection and shelter, while the algae probably produce a carbohydrate that serves as nourishment for the host.

Symbiosis and the Origin of the Chloroplast

In all organisms that produce oxygen during photosynthesis—the cyanobacteria, eukaryotic algae, and plants—chlorophyll *a* is involved in the process. The chlorophyll *a* is found in the chloroplasts of all these organisms except the cyanobacteria, which lack chloroplasts.

As discussed in Chapter 1, chloroplasts seem to have originated when symbiotic prokaryotes resembling modern cyanobacteria became permanent components of eukaryotic cells. The ease and frequency with which symbiotic relationships involving algae occur can be seen in the wide variety of such relationships that exist today. These modern examples include symbioses of algae with invertebrates, fungi (mostly in lichens), other algae, vertebrates, bryophytes, and vascular plants (in which the algae often grow in cavities in stems, petioles, and leaves). Among invertebrates, for example, some 150 genera belonging to eight different phyla are known to have algal symbionts growing within their cells.

THE GREEN ALGAE: DIVISION CHLOROPHYTA

The green algae are the most diverse of all the algae, both in form and in life history. The group comprises at least 7000 species. Although most green algae are aquatic, they are found in a wide variety of habitats, including the surface of snow, on tree trunks, in the soil, and in symbiotic relationship with lichens, protozoa, and hydra. Of the aquatic species, a few groups are entirely marine, but the great majority are found in fresh water. Many green algae are microscopic, but some of the marine forms are large; *Codium magnum* of Mexico, for example, sometimes attains a breadth of 25 centimeters and a length of more than 8 meters. The green algae have a long fossil record, with some simple forms reported from rocks nearly a billion years old.

The Chlorophyta resemble plants in several important characteristics. They have chlorophylls *a* and *b*, store their food as starch, and have firm cell walls composed, in most genera, of polysaccharides such as cellulose, with hemicelluloses and pectic substances incorporated into the cell wall structure. For these reasons, they are believed to be directly related to the evolutionary line from which plants evolved.

Most of the green algae have a unique mode of cell division involving a phycoplast (Figure 14–3). In these algae, the daughter nuclei move toward one another as the nonpersistent spindle collapses, and a new system of microtubules, the *phycoplast*, develops parallel to the plane of cell division. Presumably, the role of the phycoplast is to keep the two daughter nuclei apart while cell division (cytokinesis) takes place either by furrowing (cleavage) or cell-plate formation. The nuclear envelope persists throughout mitosis, and chromosome division occurs within it. If motile cells are present, the flagella are inserted at their anterior end. Internally, the motile cells have a system of microtubules that originates near the basal bodies of the flagella and radiates downward into the zoospore or gamete. This internal system of microtubules is a cross-shaped arrangement of four narrow bands of microtubules. Because of these cellular features, as well as several other cytological and biochemical features, not described here, Jeremy Pickett-Heaps and Harvey Marchant suggested in 1972 that this line, now designated as the class Chlorophyceae, be separated from a second group, the class Charophyceae, within the green algae.

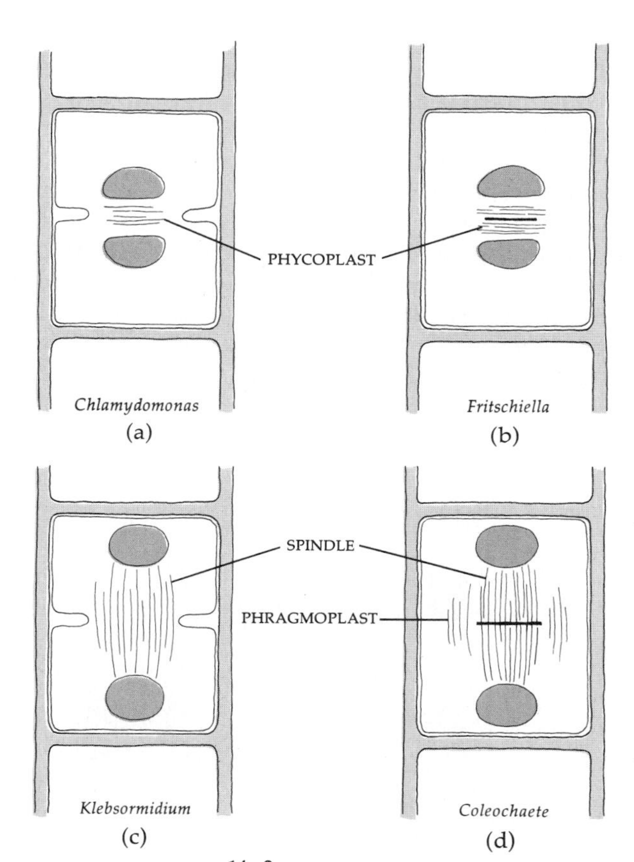

14–3
Cell division in two classes of the division Chlorophyta. In the class Chlorophyceae, (a) *and* (b), *the spindle is absent (nonpersistent) and the daughter nuclei, which are relatively near one another, are separated by phycoplasts.* (a) *Cell division by furrowing;* (b) *cell division by cell-plate formation. In the class Charophyceae,* (c) *and* (d), *the spindle is persistent and the daughter nuclei are relatively far apart.* (c) *Cell division by furrowing;* (d) *phragmoplast present and cell division by cell-plate formation.*

In the Charophyceae, the spindle apparatus persists during cell division, and in some members of the class, a *phragmoplast*—in which the microtubules are oriented perpendicular to the plane of cell division—is formed (Figure 14–3). The spindle remains until it is "broken," either by the growing cell plate—which originates in the central region of the cell and grows outward to the margins of the cell—or by furrowing. In the Charophyceae, the nuclear envelope disintegrates at the onset of mitosis. If motile cells are present, the flagella are often inserted laterally; the motile cells are asymmetric. Internally, the cells possess a system of microtubules in the form of a flat, broad band. This band originates in a multilayered structure (MLS) near the laterally located basal bodies (see Essay, page 282). The same kind of structure is also found in the motile sperm cells of plants. Clearly, then, the Charophyceae are more closely related to the plants than are the Chlorophyceae, so the ancestor of the plants should be sought among the Charophyceae.

The classification of green algae into the Chlorophyceae and Charophyceae is based upon the examination of a small sample of green algae by means of electron microscopy and other modern techniques. The green algae comprise a large group, and accumulating evidence suggests that other classes might eventually be recognized. Meanwhile, the structural and biochemical differences between Chlorophyceae and Charophyceae are impressive, so these two recognized classes will be used in this book as an aid in organizing information about the green algae.

Class Chlorophyceae

Most of the green algae belong to this diverse group, which employs a mode of cell division involving a phycoplast. This unique characteristic would indicate that no other groups of organisms have arisen from members of this class of green algae. The Chlorophyceae include unicellular, colonial, filamentous, and parenchymatous (three-dimensional) organisms that occur in all habitats where green algae can grow. We shall first examine some of the unicellular Chlorophyceae and then a few of the genera that have more complex forms.

Motile Unicellular Chlorophyceae

Among the least complex of the green algae is the unicellular biflagellate organism *Chlamydomonas*. *Chlamydomonas*, which is one of the most common freshwater green algae, is small (usually less than 25 micrometers long), green, and rounded or pear-shaped (Figure 14–4). It moves very rapidly, with a characteristic darting motion imparted by the beating of the two equal whiplash flagella that protrude from its smaller, anterior pole. The flagella propel the organism through the water by beating in opposite directions.

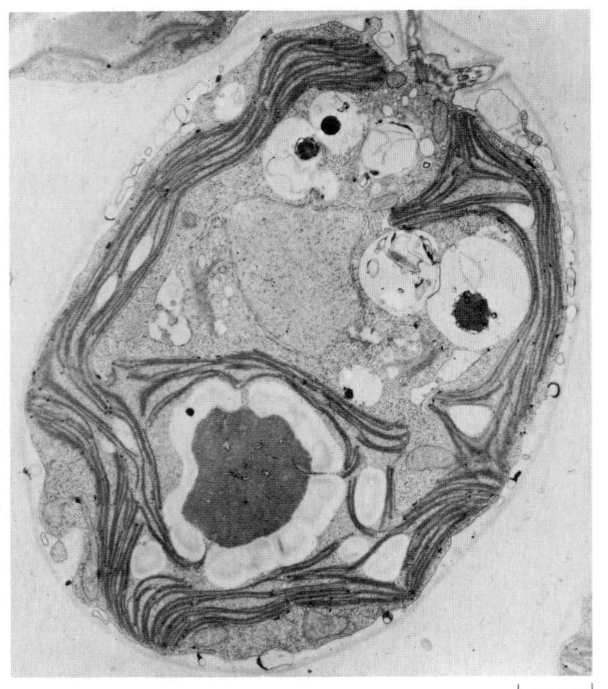

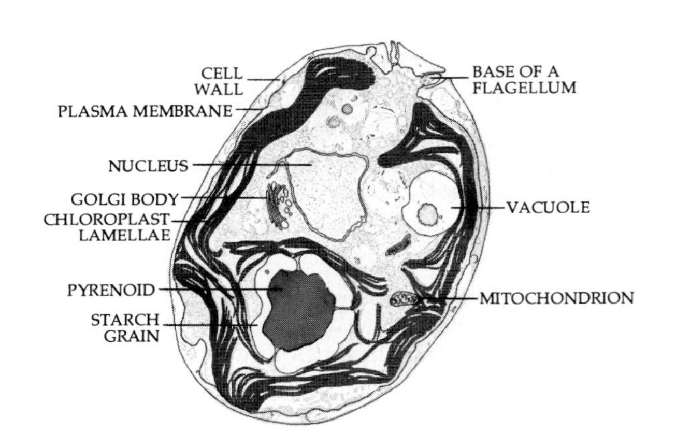

14–4
Chlamydomonas, *a unicellular green alga. Only the bases of the flagella can be seen in this electron micrograph.*

14–5
Life cycle of Chlamydomonas. *Sexual reproduction occurs when gametes of different mating types come together, cohering at first by their flagellar membranes and then by a slender protoplasmic thread—the conjugation tube. The protoplasts of the two cells fuse completely (plasmogamy), followed by the union of their nuclei (karyogamy). A thick wall is then formed around the diploid zygote. After a period of dormancy, meiosis occurs and four haploid cells emerge. Asexual reproduction of the haploid individuals by cell division is the most frequent mode of reproduction.*

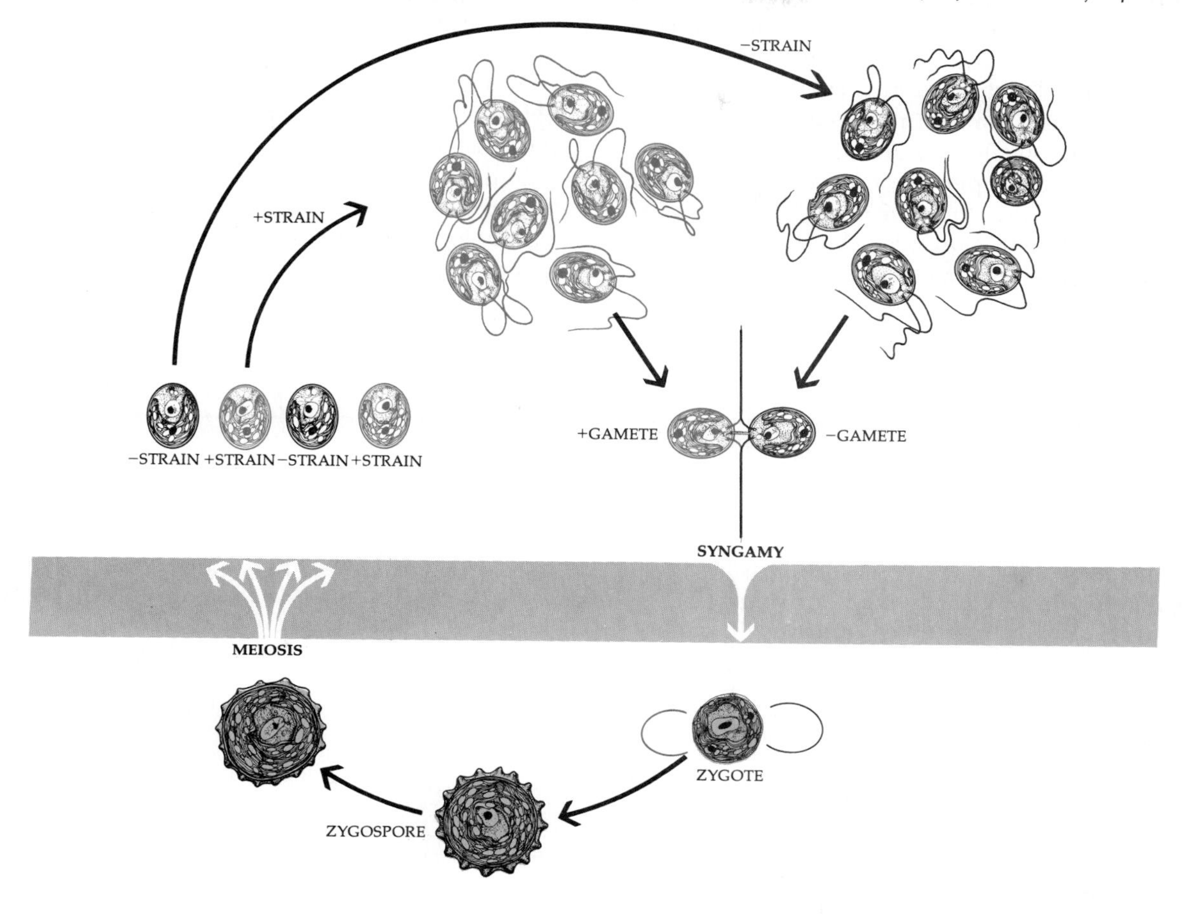

Each cell of *Chlamydomonas* has a single massive chloroplast, containing a red pigment body—the *stigma*, or eyespot—which may be a shading device associated with a site of light perception. The chloroplast also contains a roughly spherical body called the *pyrenoid.* Pyrenoids are commonly found among the algae but are lacking in nearly all plant chloroplasts. Recent research has shown that in two other green algae the pyrenoid is a region in the chloroplast where the enzyme RuBP carboxylase (see page 107) is concentrated. Pyrenoids may also be associated with the conversion of sugars to starch, for deposits of starch are usually found surrounding them.

The uninucleate protoplast of *Chlamydomonas* is surrounded by a thin glycoproteinaceous wall rich in hydroxyproline, inside of which is the plasma membrane. There is no cellulose in the wall of *Chlamydomonas.* At the anterior end of the cell, there are two contractile vacuoles.

Under some environmental conditions, cells of *Chlamydomonas* become nonmotile. Cells in this state are usually nonflagellated, and their walls become gelatinous. When conditions change, the flagella may reappear and the cells again become free-swimming.

Chlamydomonas reproduces both sexually and asexually. During asexual reproduction, the nucleus—which is haploid—usually divides twice mitotically, resulting in the production of four daughter cells within the parent cell wall. Each cell then secretes a wall around itself and develops flagella. The cells secrete an enzyme that breaks down the original mother wall, and the daughter cells can then escape, although fully formed daughter cells are often retained for some time within the parent cell wall. In ancient flagellates, such aggregations of daughter cells may have been the forerunners of colonial organisms.

Sexual reproduction, which occurs in some species of *Chlamydomonas,* involves the fusion of individuals belonging to different mating types (Figure 14–5). The vegetative cells are induced to form gametes by nitrogen starvation. The gametes, which resemble the vegetative cells, first become aggregated in clumps.

Within these clumps, pairs are formed, which stick together first by their flagellar membranes and later by a slender protoplasmic thread that connects them at the base of their flagella. As soon as this protoplasmic connection is formed, the flagella become free, and one or both pairs of flagella beat, thus propelling the zygote—the product of sexual union—through the water. The two gametes fuse completely. Soon the four flagella shorten and eventually disappear, and a thick cell wall forms around the diploid zygote. This thick-walled, resistant zygote (zygospore) then undergoes a period of dormancy. Meiosis occurs at the end of the dormant period, resulting in the production of four haploid cells, each of which develops two flagella and a cell wall. These cells can either divide asexually or mate with a cell of another mating strain to produce a new zygote. Thus, *Chlamydomonas* exhibits zygotic meiosis (see Figure 10–10a, page 180), and the haploid phase is the dominant phase in its life cycle.

In most species of *Chlamydomonas,* the cells of the two different mating types, conventionally designated + and −, are identical in size and structure (*isogamy*). In addition to these isogamous species, there are other species of *Chlamydomonas* in which *anisogamy* (larger female gametes) and *oogamy* (immobile female gametes) occur (Figure 14–6). Thus, the entire range of differences between gametes that occurs among algae is exhibited in the various species of the single genus *Chlamydomonas.*

RHIZOPLASTS: MUSCLELIKE STRUCTURES IN AN ALGA

The unicellular motile green alga Tetraselmis *possesses unique structures that contract and expand in response to changes in concentration of calcium ions in the cell. These striated structures, clearly visible at the right end of the alga in this electron micrograph, are called rhizoplasts; they resemble the muscles of animals both in their organization and in their response to calcium ion concentration. The rhizoplasts are connected to the basal bodies of the flagella and are thought to play a role in the control and function of the flagella. A rhizoplastlike structure is present in many protozoa. This surprising discovery makes the great diversity of green algae apparent once more. It also indicates the need for a thorough examination of the fine structure of the microscopic green algae before conclusions are drawn about their interrelationships.*

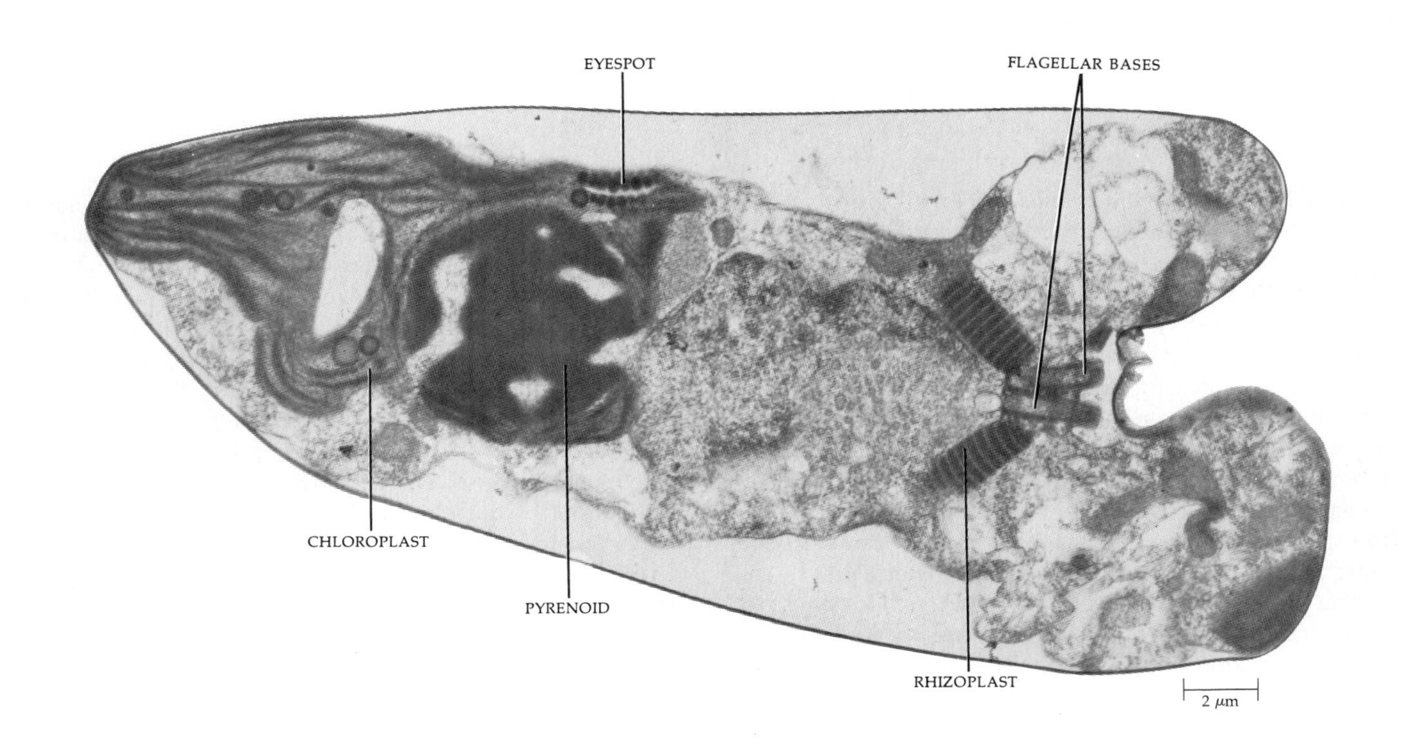

Historically, *Chlamydomonas* has been considered primitive among the green algae. It is unicellular and contains centrioles, basal bodies, and flagella. Moreover, its nuclear envelope does not break down during mitosis. By contrast, the presence of a phycoplast as a part of cell division in *Chlamydomonas* suggests that this genus is more advanced than had originally been believed, making it unlikely that *Chlamydomonas* represents the ultimate ancestral green flagellate.

Some of the other genera of the class Chlorophyceae will be discussed under the following four headings: (1) nonmotile unicellular genera, (2) motile colonial genera, (3) coenocytic (multinucleate) genera, and (4) filamentous and parenchymatous noncoenocytic genera. The genera included under each of these headings are grouped together for convenience and for clarity of presentation; they may or may not ultimately prove to be directly related to one another in an evolutionary sense.

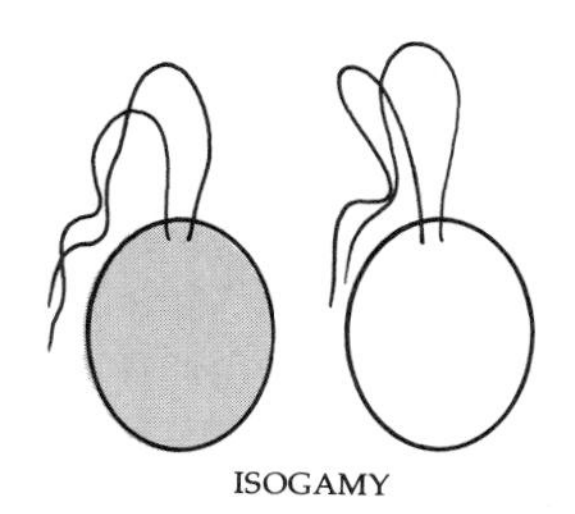

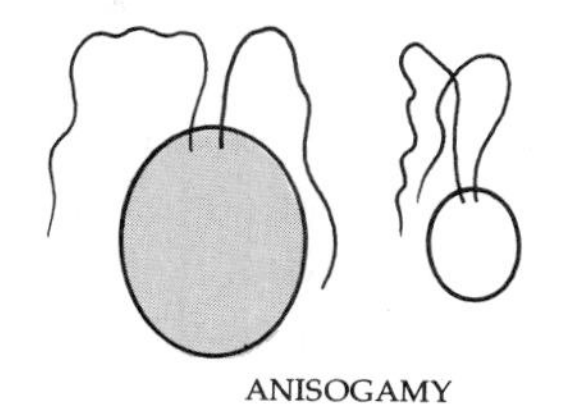

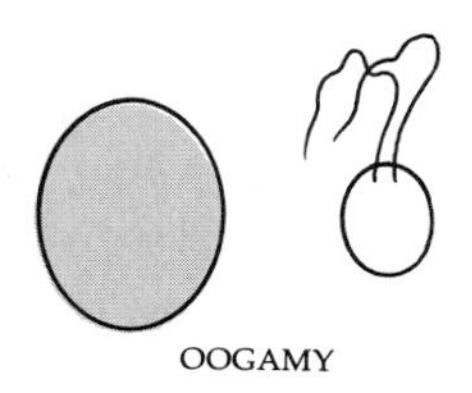

14–6
Types of sexual reproduction, based on gamete form; each is found in at least one species of Chlamydomonas. ***Isogamy****—the gametes are equal in size and shape.* ***Anisogamy****—one gamete, conventionally termed male, is smaller than the other.* ***Oogamy****—the female gamete is nonmotile.*

Nonmotile Unicellular Chlorophyceae

Chlorella is a highly plastic unicellular alga that closely resembles *Chlamydomonas* except for the absence of flagella. It was the first alga to be isolated in axenic (pure) culture and has been used widely for experimental studies of photosynthesis. In nature, *Chlorella* is widespread in both fresh and salt water and in soil. Each *Chlorella* cell contains a single cup-shaped chloroplast, with or without a pyrenoid, and a single minute nucleus. The only known method of reproduction in *Chlorella* is asexual, each haploid cell dividing mitotically two or three times to give rise to either four or eight nonmotile cells.

Chlorella is currently under investigation as a potential food source for humans. Pilot farms have been established in the United States, Germany, Japan, and Israel. The Japanese have been able to process *Chlorella* into a tasteless white powder, which is rich in vitamins and protein and can be mixed with flour for the preparation of baked goods. It is too costly to market at present, but it may represent an important supplemental food source in the future.

Eremosphaera, one of the largest unicellular green algae, can be seen with the naked eye (Figure 14–7). It is found in acid water on the bottom of swamps and quiet ponds. Each cell contains many pyrenoid-bearing chloroplasts and a single large nucleus suspended in the center of the cell by numerous radiating strands of cytoplasm. *Eremosphaera* reproduces by both asexual and sexual means. The asexual reproduction is similar to that of *Chlorella*. Sexual reproduction involves the union of biflagellate sperm with nonmotile cells that function as eggs. As in *Chlamydomonas*, meiosis is zygotic.

14–7
Eremosphaera, *one of the largest of the unicellular green algae. Numerous chloroplasts can be seen in strands of cytoplasm radiating from the center of the cell.*

Motile Colonial Chlorophyceae

In these genera, *Chlamydomonas*-like cells adhere in motile colonies that are propelled by the beating of the flagella of the individual cells (Figure 14–8). The members of this group have often been called the "volvocine line" after the genus *Volvox,* the largest, most complex organism exhibiting this colonial existence (Figure 14–9). In the colonies of some algae of this type, the cells are connected by cytoplasmic strands that provide for the integration of the whole organism. The simplest member of this group is *Gonium.* The *Gonium* colony consists of separate cells held together within a gelatinous matrix. Each colony is made up of 4, 8, 16, or 32 cells (depending on the species) arranged in a slightly curved, shield-shaped disk. The flagella of each cell beat separately, pulling the entire colony forward. Each cell in *Gonium* can divide to produce an entire new colony.

A closely related colonial organism is *Pandorina,* which forms a tightly packed ovoid or ellipsoid, usually consisting of 16 or 32 cells held together within a matrix. The colony is polar, the eyespots being larger in the cells at one end of the colony. Each cell has two flagella, and because all the flagella point outward, *Pandorina* rolls through the water like a ball. When the cells attain their maximum size, the colony sinks to the bottom and each of the cells divides to form a daughter colony. The latter remain together until all have developed flagella. The parent matrix then breaks open like Pandora's box (which suggested its name), releasing new daughter colonies.

Eudorina is also a spherical colony made up of green flagellates, the number of which is 32, 64, or 128 in the most common species. It differs from *Gonium* and *Pandorina* in that some of the cells in the colony, which are smaller than the others and are located in the anterior part of the colony (relative to the direction of movement), are incapable of reproducing to form new colonies. Here we see a beginning of specialization of function.

The most spectacular of these colonial green algae, however, is *Volvox. Volvox* consists of a hollow sphere made up of a single layer of 500 to 60,000 vegetative, biflagellate cells and a small number of reproductive cells. The number of cells in the sphere varies from species to species. When *Volvox* whirls through the water, it appears like a spinning universe of individual stars fixed in an invisible firmament.

Volvox exhibits polarity; that is, it has anterior and

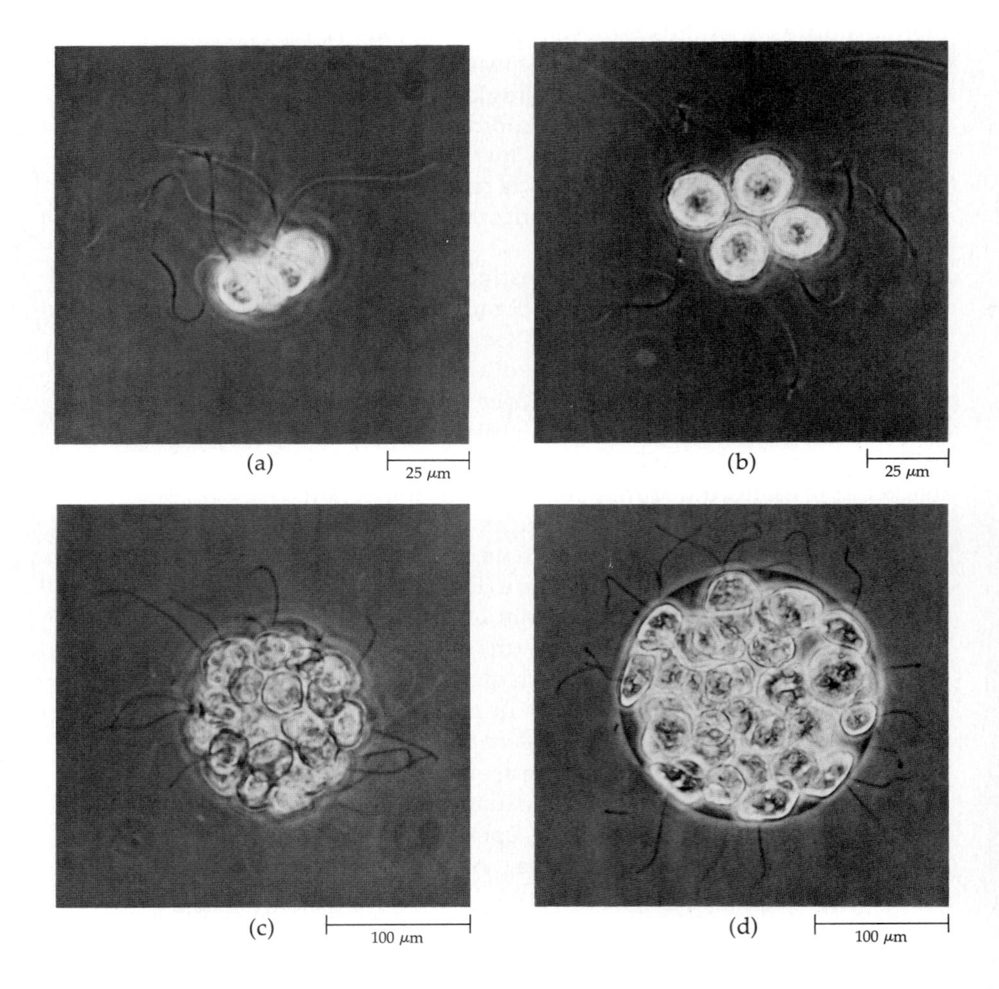

14–8
Several colonial Chlorophyceae. (a) *and* (b) *Two views of* Gonium, (c) Pandorina, *and* (d) Eudorina. *In these algae, cells similar to those of* Chlamydomonas *adhere in a gelatinous matrix to form multicellular colonies propelled by the beating of the flagella of the individual cells. Varying degrees of cellular specialization are found in different genera.*

posterior poles. The flagella of each cell beat in such a way as to spin the entire colony around its axis in a clockwise direction as it moves forward (usually toward the light at most stages of the life cycle).

Most of the cells of a *Volvox* spheroid are strictly vegetative, and the few that do engage in asexual reproduction are usually confined to the posterior hemisphere in characteristic pattern. In some members of the genus, the cells that will fulfill the reproductive function are not clearly distinguishable in juvenile spheroids, but they later become apparent as they undergo successive cycles of growth and division. Ultimately, these dividing cells give rise to new daughter spheroids, which "hatch" from the parental spheroid by releasing an enzyme that dissolves the transparent matrix of the spheroid. In more advanced members of the genus, however, asexual reproductive cells are set aside early in each reproductive cycle and are at all times structurally and functionally distinct from the vegetative, or somatic, cells. In these species, such as *V. carteri* (Figure 14–9), there is a true "division of labor" between two mutually dependent cell types. The nonmotile reproductive cells depend on the motile but nonreproductive somatic cells to carry them to the surface, where they can obtain adequate light and CO_2 for photosynthesis. Thus, the more advanced species of *Volvox* are not simply colonial like the simpler members of the genera; instead, they are multicellular and differentiated in the same sense as the higher plants and animals.

Sexual reproduction in *Volvox* is always oogamous, but substantial variation in details of reproductive pattern are seen among the various species. In some species, both eggs and sperm may be produced within a single spheroid. In other species, members of a single genetically uniform clone (a colony derived from a single colony) may become either sexual males or females, but mixed spheroids are not observed. In the more advanced species, however, sex is genetically determined, and each clone derived by asexual reproduction is either male or female. In all species that have been studied, sexual reproduction is synchronized within the population of colonies by a sexual inducer molecule, which is produced by a spheroid that has itself become sexual by some other, as yet poorly understood, mechanism. One male colony of *V. carteri* may produce enough inducer to "turn on" over half a billion other males and females.

Volvox is one of the simplest multicellular organisms to show clear division of labor, and the sexual inducers of *Volvox* are among the most potent biologically active substances known; thus, the genus (particularly *V. carteri*) has become the focus of a great deal of research in recent years. Like all of its relatives, *Volvox* is haploid; therefore, mutant genes are not masked by dominant alleles, and mutations affecting development can be readily detected. Hundreds of strains with specific, heritable, developmental defects have been isolated and are being studied with the hope of learning how specific genes regulate cellular differentiation.

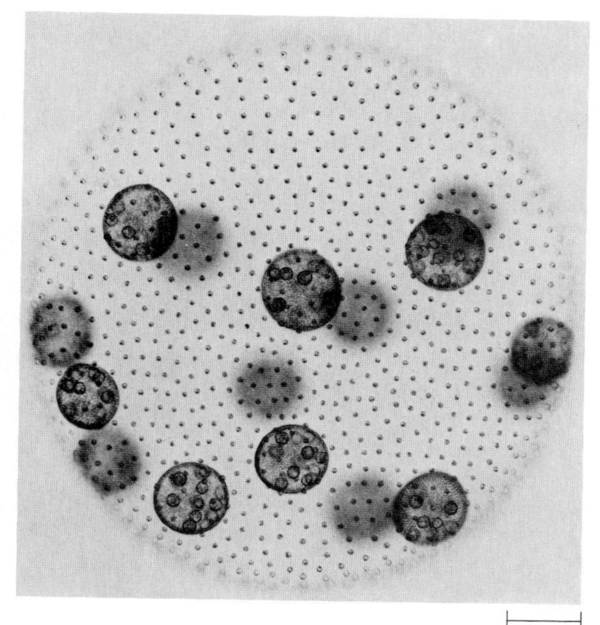

14–9
Asexual reproduction in Volvox carteri. *The approximately 2000 small cells at the periphery of the transparent sphere are the* Chlamydomonas-*like somatic cells; in this species, somatic cells are not interconnected in mature colonies as they are in some members of the genus. In the specimen shown here, the 16 reproductive cells in the interior of the spheroid have already undergone division to produce juvenile spheroids, each containing about 2000 tiny somatic cells and 16 reproductive cells. Eventually, each juvenile colony digests a passageway out of the parent colony, swims away, and the cycle then repeats.*

To summarize, increasing specialization in the members of the colonial Chlorophyceae is evident in several ways. First, there is an increase in the cell number and in the size of the colonies. Second, there is increasing specialization in cell morphology and function. Finally, there is increasing sexual specialization paralleling that within the genus *Chlamydomonas.* Thus, *Gonium* and *Pandorina* have isogamous reproduction, whereas species of *Eudorina* and *Volvox* are oogamous. Nevertheless, this line clearly represents an evolutionary "dead end," in that it has not given rise to a more complex group of organisms.

14–10
Scanning electron micrograph of a flattened young colony of Hydrodictyon reticulatum.

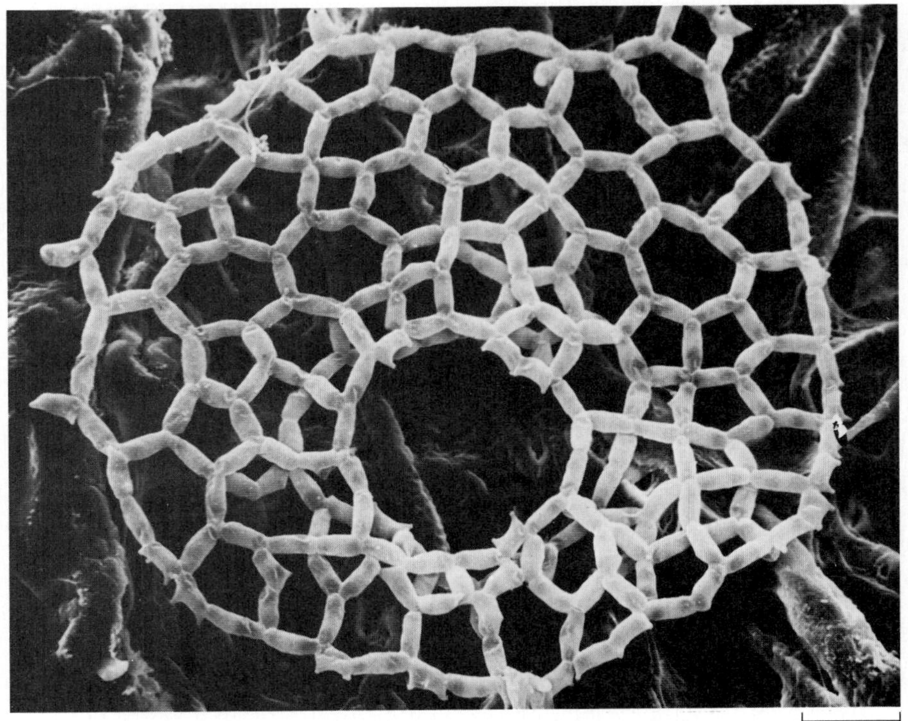

Coenocytic (Multinucleate) Chlorophyceae

A number of genera of this class have the tendency to form large, coenocytic cells; they are often termed the siphonous green algae. Many of them form motile zoospores or gametes during their life cycles. As a group, these genera are diverse, and their relationships with one another and with other green algae are not well understood.

Hydrodictyon, the "water net," is an example of a nonmotile colony (Figure 14–10). Under favorable conditions it accumulates in massive aggregates in ponds, lakes, and gentle streams. Each colony consists of many cylindrical cells arranged in the form of a large hollow cylinder. Initially uninucleate, each cell eventually becomes multinucleate. At maturity, each cell contains a large central vacuole and peripheral cytoplasm in which the nuclei and a large reticulate chloroplast with numerous pyrenoids are located. *Hydrodictyon* reproduces asexually through the formation of uninucleate, biflagellate zoospores. Eventually, the zoospores form into groups of four to nine (most typically six) within the cylindrical parent cell, lose their flagella, and form daughter colonies. Sexual reproduction in *Hydrodictyon* is isogamous, and its meiosis is zygotic.

The marine order Siphonales—the members of which have very large, branched, coenocytic cells that are rarely septate—develop as a result of repeated nuclear division without the formation of cell walls (Figure 14–11a). *Codium magnum,* previously noted to be among the largest of the green algae, is a member of this order; it may reach more than 8 meters in length. Only in the reproductive phase of this genus are cells produced that have definite walls. One siphonous green alga, *Valonia,* which is common in tropical waters, has been widely used in studies of cell walls and in physiological experiments requiring large amounts of cell sap. *Valonia,* which reaches the size of a hen's egg, appears to be unicellular but is actually a large multinucleate vesicle, with separate rhizoids and young branches (Figure 14–11b). One of the best known of the siphonous algae is *Acetabularia,* which has been widely used in experiments dealing with the genetic control of differentiation. The Siphonales are primarily diploid, with the gametes being the only haploid cells in the life cycle.

The final evolutionary line that we shall mention among the siphonous group consists of filamentous algae that have large, coenocytic cells separated by septa. Of these, *Cladophora* (Figure 14–12) is widespread both in fresh and salt water. Its filaments commonly grow in dense mats, either free-floating or attached to rocks or vegetation. They elongate and branch near the ends. Each cell contains many nuclei and a single peripheral, netlike chloroplast with many pyrenoids. Marine *Cladophora* have an isomorphic alternation of generations like that of *Allomyces* (see page 246). The zoospores are quadriflagellated and the gametes are biflagellated.

(a)

(b)

14–11
(a) *A species of* Codium, *abundant along the Atlantic coast.* (b) Valonia, *another siphonous green alga common in tropical waters.*

14–12
Cladophora *is a green alga that is widespread in fresh and salt water.*

A COMPLEX SEMITERRESTRIAL GREEN ALGA

Fritschiella *is one of the few green algae in which cell divisions occur in three planes, producing a solid vegetative body. These solid bodies, in which the cells are connected by plasmodesmata, resemble those of plants. The degree of differentiation in various tissues, however, is much less complex. Unlike the plants, but like the other members of the class Chlorophyceae,* Fritschiella *forms a phycoplast during cell division. It is semiterrestrial, occurring on damp surfaces such as tree trunks, moist walls, and leaf surfaces. Its branched, filamentous body consists of subterranean rhizoids, a prostrate system near the soil surface, and two kinds of erect branches. Although* Fritschiella *cannot be on the evolutionary line leading directly to the plants, it provides a very interesting example of analogous specialization in adapting to the conditions of life on land.*

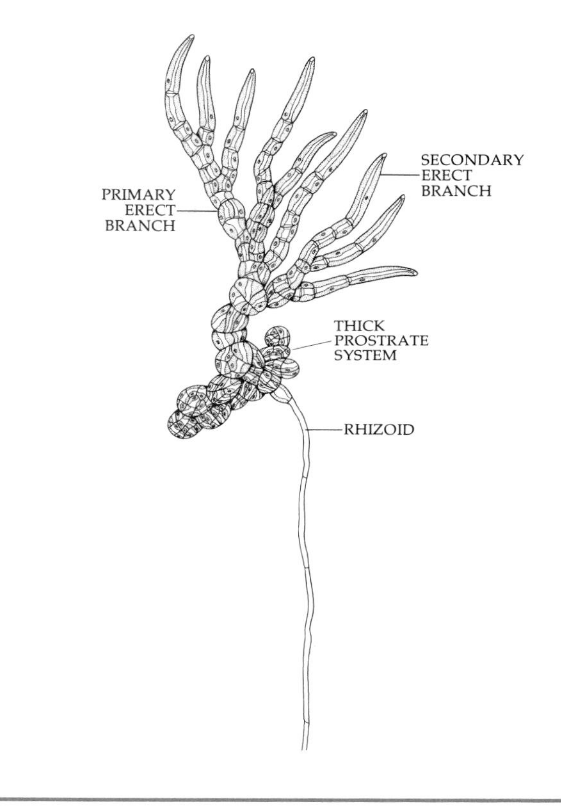

Filamentous and Parenchymatous Chlorophyceae

Several genera of Chlorophyceae consist of relatively small, mostly uninucleate cells. Among them are a number of filamentous genera; some are flat sheets of cells because the cells divide in two planes; several others are parenchymatous because the cells divide in three planes. These genera have revealed a great diversity of types of cell division when studied by means of transmission electron microscopy, and many are not directly related to one another.

The filamentous genera that we shall consider are *Ulothrix* and *Oedogonium*, two very different kinds of organisms. Both of these genera have a phycoplast like other members of the class Chlorophyceae, but they differ from the genera considered earlier in that their cells divide by the formation of a cell plate (Figure 14–13). In this respect, they resemble plants; however, because they have a phycoplast, it seems unlikely that they are on the evolutionary line leading to the plants. Mitosis in both genera occurs within the nuclear envelope, which persists throughout the mitotic cycle. In addition, the cells in both *Ulothrix* and *Oedogonium* are connected by plasmodesmata.

Ulothrix, an alga of cold-water streams, ponds, and lakes, has its submerged filaments attached to stones or other objects by means of a special basal cell called a holdfast (Figure 14–14). All cells of the filament are essentially similar in appearance and contain a single, ringlike chloroplast, pyrenoid, and nucleus. Asexual reproduction occurs through formation of zoospores with four flagella; sexual reproduction occurs by means of biflagellate isogametes. Meiosis occurs prior to the germination of the zygote; the *Ulothrix* filament is haploid.

Oedogonium, an unbranched filamentous alga, is a very unusual organism. The individuals are attached to the substrate by a holdfast. The cells themselves are uninucleate, with a netlike chloroplast that contains pyrenoids around its periphery.

Cell division in *Oedogonium* is most unusual. As a cell divides, a doughnut-shaped ring of wall material forms near its upper (apical) end. Following division of the nucleus, one of the daughter nuclei migrates into the upper end of the cell and the wall ruptures violently and precisely at this ring. The ring material is then drawn out into a cylinder forming the wall of the upper daughter cell, which is nearer the apex of the filament. A new cross wall forms in the plane of the phycoplast, just past the edge of the ruptured parental wall. As a consequence of ring expansion,

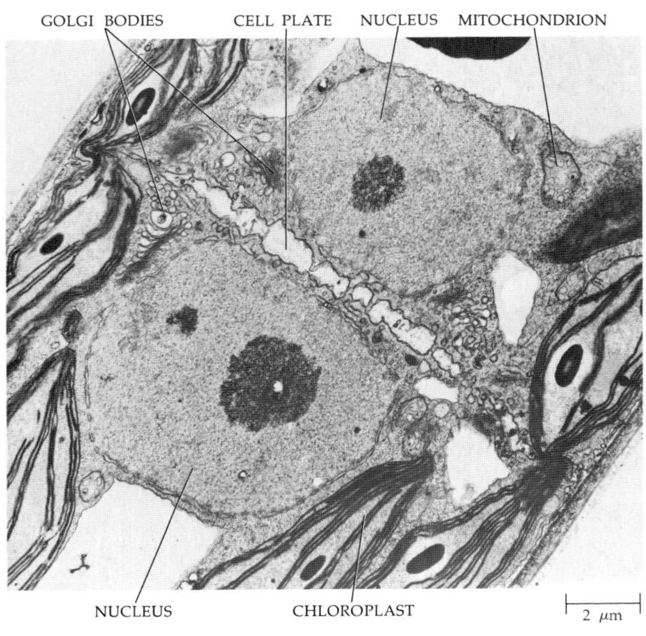

14–13
Ulothrix. *Cell division in this filamentous green alga occurs by cell-plate formation, which is nearly complete at the stage shown in this electron micrograph. The nuclei are close together at cytokinesis because the spindle collapses at telophase of the preceding mitosis. The phycoplast is not visible at this relatively low magnification.*

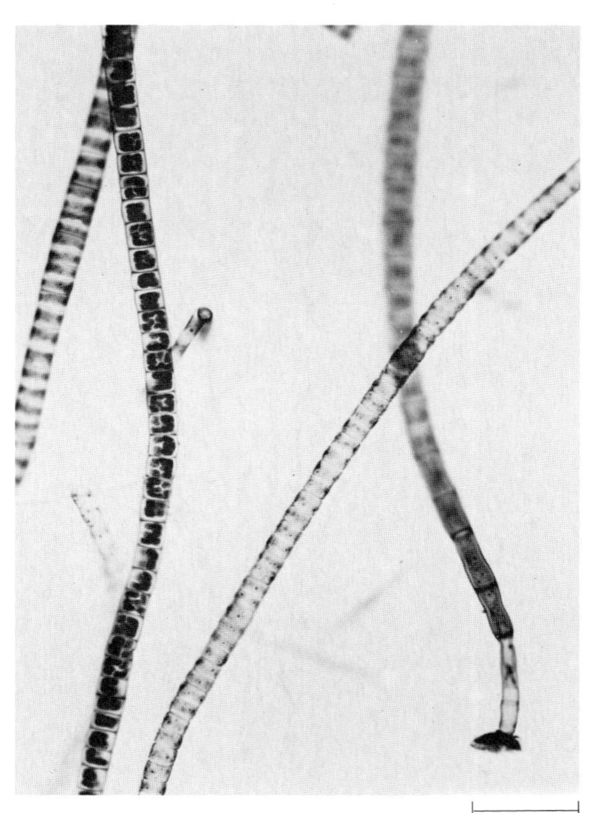

14–14
Ulothrix, *an unbranched, filamentous green alga. The filament on the left with dense contents consists of sporangia in which zoospores are forming. The other filaments are vegetative. A holdfast can be seen on the filament on the right.*

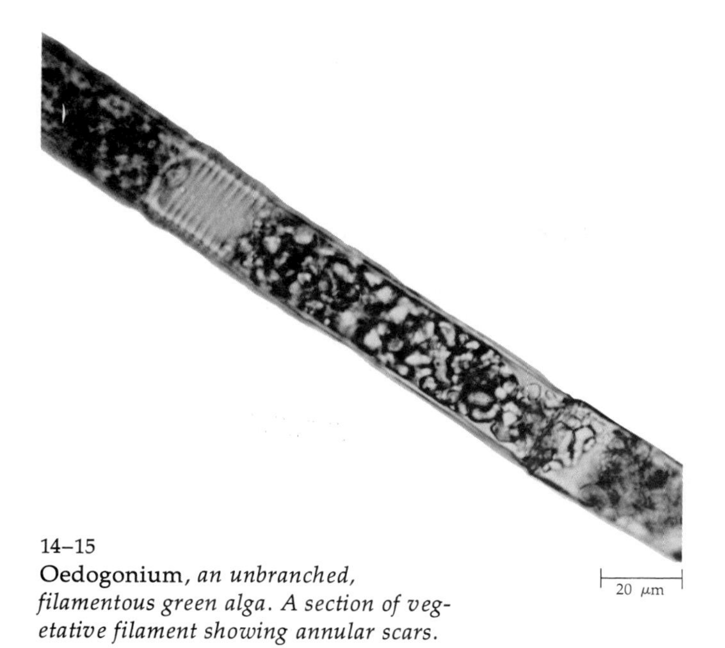

14–15
Oedogonium, *an unbranched, filamentous green alga. A section of vegetative filament showing annular scars.*

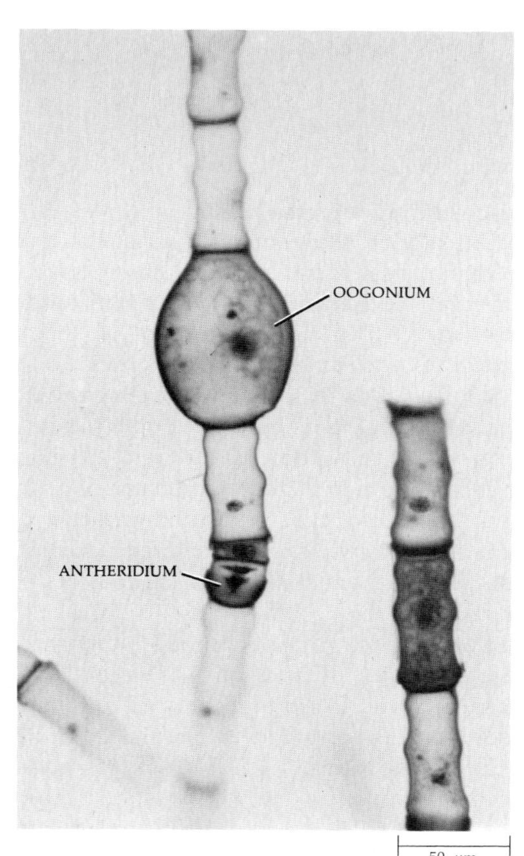

14–16
Sexual reproduction in Oedogonium *is oogamous. Each oogonium produces a single egg, whereas each antheridium produces two multiflagellate sperm.*

the rims of the original mother-cell wall are bent outward, giving rise to the characteristic "caps," or annular scars (Figure 14–15). These scars reflect the number of divisions that have occurred earlier.

Asexual reproduction in *Oedogonium* occurs by means of zoospore formation, with a single zoospore produced per cell. Each zoospore has a crown of about 120 flagella. Sexual reproduction is oogamous (Figure 14–16). Each male gametangium (antheridium) produces two multiflagellate sperm, and each female gametangium (oogonium) produces a single egg. Meiosis is zygotic, as in *Ulothrix.*

Ulva, commonly known as sea lettuce, occurs frequently along the seashore (Figure 14–17). Cell division in *Ulva* occurs in three planes, but only once in one of them, giving rise to a glistening, flat thallus (a simple, relatively undifferentiated vegetative body) which is two cells thick and up to a meter or more long in exceptionally large individuals. The thallus is anchored to the substrate by a holdfast produced by protuberances of the basal cells. Each cell of the thallus contains a single nucleus and chloroplast. *Ulva* is anisogamous and has an isomorphic alternation of generations (see Figure 14–18) similar to that in *Cladophora* and *Allomyces.* Cell division in *Ulva* is by furrowing (cleavage), and in this respect it differs from *Ulothrix* and *Oedogonium.* It lacks a true phycoplast and differs from other members of the class Chlorophyceae in this respect, but its nuclear envelope persists throughout mitosis.

14–17
Sea lettuce (Ulva lactuca) *growing in a tidal pool with mussels, limpets, and coralline red algae.*

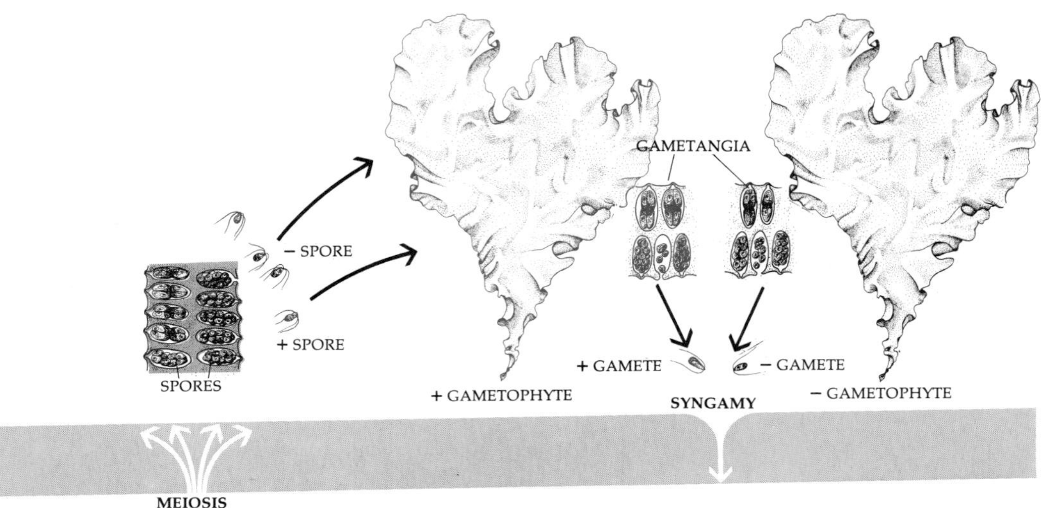

14–18
Alternation of generations in sea lettuce, Ulva. *The gametophyte and sporophyte are indistinguishable except for their reproductive structures.*

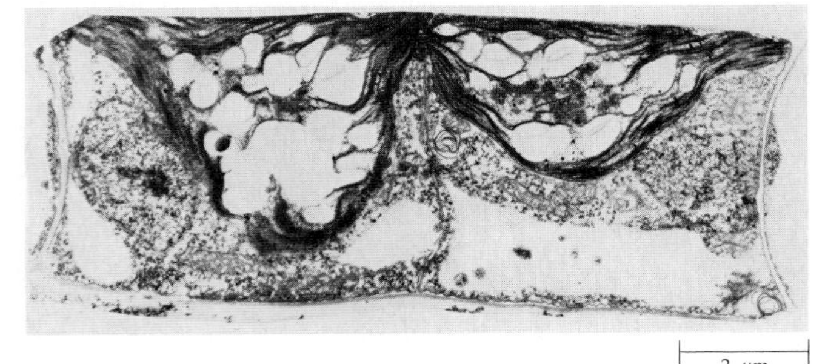

14–19
Klebsormidium subtillissimum. *Cell division in this filamentous alga is distinct (see Figure 14–3, page 255). At the late stage shown here, the nuclei are far apart and formation of the cross wall is nearly complete.*

Class Charophyceae

This exceedingly diverse group of green algae includes unicellular, filamentous, and parenchymatous genera that appear to have little in common, although fundamental microscopic and biochemical features indicate their affinities. In many of the Charophyceae, the nuclear envelope disintegrates at the onset of mitosis, as in the plants. All of them have a persistent spindle, and some develop a phragmoplast that aids in the formation of the new cell plate. The motile cells have two flagella which are often laterally inserted. Charophyceae manufacture the enzyme glycolate oxidase, which is associated with photorespiration; this enzyme is otherwise known to occur only in plants.

Klebsormidium is an unbranched filamentous green alga that occurs in soil and in fresh water. This genus was formerly considered to be related to *Ulothrix,* but because it differs from *Ulothrix* radically in its mode of cell division (Figure 14–19), lack of plasmodesmata, zoospore structure, lack of holdfasts, and many other features, this earlier notion of their close relationship has been discarded. What has been learned about these fundamental differences clearly illustrates the need for electron microscopic and biochemical studies in evaluating the relationships of such organisms. Sexual reproduction is unknown in *Kleb-*

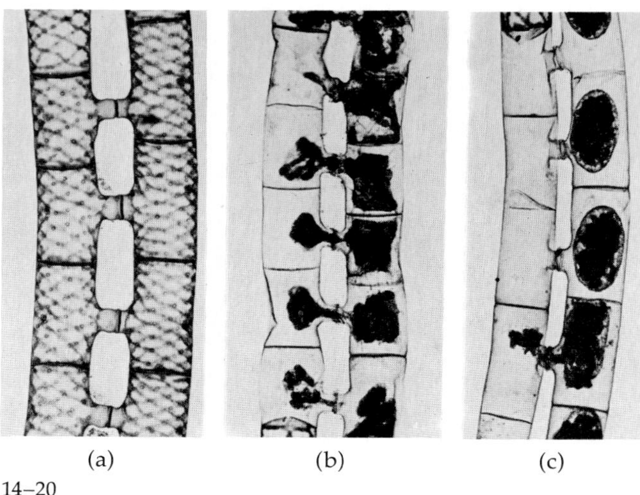

14–20
(a) *Sexual reproduction in* Spirogyra *follows the formation of conjugation tubes between the cells of adjacent filaments.* (b) *The contents of the cells of the − strain pass through these tubes into the cells of the + strain.* (c) *Syngamy occurs within these cells; the resulting zygote develops a thick, resistant cell wall and is termed a zygospore. The vegetative filaments of* Spirogyra *are haploid, and meiosis occurs during germination of the zygospores.*

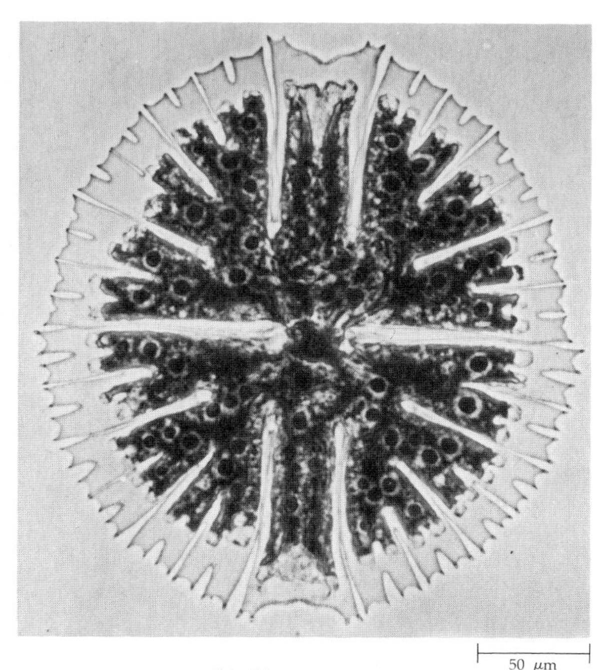

14–21
In this desmid, a species of Micrasterias, *each cell is deeply constricted, which is a characteristic of this group of freshwater green algae.*

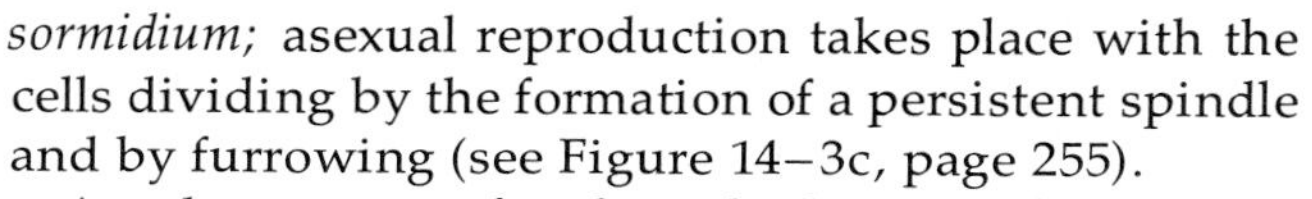

sormidium; asexual reproduction takes place with the cells dividing by the formation of a persistent spindle and by furrowing (see Figure 14–3c, page 255).

Another genus of unbranched green algae is the well-known *Spirogyra* (Figure 14–20), which is often found as frothy or slimy floating masses in small bodies of water. Each filament is surrounded by a watery sheath that is slimy to the touch. Like *Klebsormidium, Spirogyra* lacks plasmodesmata. The name *Spirogyra* refers to the helical arrangement of the one or more ribbonlike chloroplasts found within each uninucleate cell. The chloroplasts contain numerous pyrenoids. The only type of asexual reproduction that occurs in *Spirogyra* is fragmentation, and there are no motile cells at any stage of the life cycle. Morphologically, sexual reproduction in *Spirogyra* must be termed isogamous. However, during sexual reproduction one of the isogametes behaves like a male gamete by migrating across a conjugation tube to fuse with the other isogamete. Meiosis is zygotic.

The desmids are a large group of freshwater green algae related to *Spirogyra,* and like that genus, they lack flagellate cells. Some are multicellular, but most are unicellular, with a very peculiar cell construction. The cell wall is in two sections with a narrow constriction—the isthmus—between them (Figure 14–21). Some estimates place the number of species of desmids as high as 10,000. Cell division in the desmids, as in *Spirogyra,* is by a persistent spindle and furrowing.

The two groups of green algae that will now be considered are the only ones that resemble plants fully in the details of their cell division; that is, they have a phragmoplast and a nuclear envelope that disintegrates at the start of mitosis. In addition, they are oogamous, as are plants. Neither group, however, is the direct ancestor of the plants, because each is too specialized in certain respects. However, the existence of green algae that have features so similar to those of the plants does suggest that the ancestors of plants were part of a large and diverse group of advanced green algae, of which only a few members have survived.

Coleochaete is a complex freshwater alga that reproduces asexually by zoospores and has oogonia protected by a layer of sterile cells. Individuals, which grow on submerged plants, may consist of a mass of dichotomously branched filaments, or they may form discoid (disc-shaped) colonies. The vegetative cells are uninucleate and contain one large chloroplast and one or more pyrenoids. Cell division may occur at the apices of the filaments or the margins of the discs, depending upon the form of the organism (see "An Advanced Green Alga," page 282).

14–22
(a) Nitella hyalina, *showing clearly the determinate growth pattern.* (b) *A segment of* Chara *showing gametangia. The top structure is an oogonium, and that below is an antheridium.*

(a)

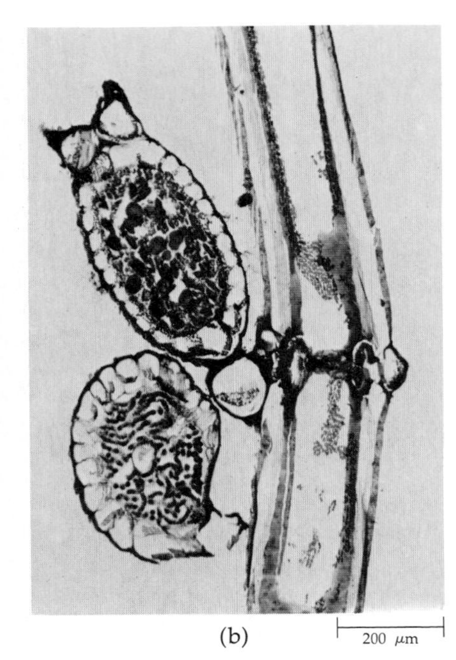

(b)

The stoneworts, order Charales, are very distinctive green algae that occur in fresh to brackish water and have heavily calcified cell walls (Figure 14–22). Consequently, although there are only about 250 living species, the stoneworts are well represented in the fossil record. In growth pattern, the stoneworts are complex, with nodal regions from which whorls of short branches arise, and coenocytic internodal regions. Their sperm are produced in multicellular antheridia that are more complex than those found in any other group of protista; their eggs are borne in oogonia of simpler construction than the antheridia. Sperm are the only flagellate cells in the stonewort life cycle.

THE BROWN ALGAE: DIVISION PHAEOPHYTA

The brown algae (see Table 14–1, page 253), an almost entirely marine group, comprise most of the conspicuous seaweeds of temperate regions. Although there are only about 1500 species, the brown algae are of considerable interest because they dominate rocky shores throughout the cooler regions of the world (Figure 14–23), and some, like the kelps, often form extensive beds offshore. In clear water, kelps flourish from low-tide level to a depth of 20 to 30 meters. On gently sloping shores, they may extend 5 to 10 kilometers from the coastline. Even in the tropics, where the brown algae are less common, there are the immense floating masses of *Sargassum* (Figure 14–24) in such areas as the Sargasso Sea in the Atlantic Ocean northeast of the Caribbean Islands.

The brown algae are also of interest because of their size and often elaborate internal tissue differentiation. The giant kelps *Macrocystis* and *Nereocystis* may have fronds more than 100 meters long under unusually favorable circumstances. Many of the brown algae, such as *Sargassum,* approach the vascular plants in complexity of organization of their vegetative parts. Many kelps exhibit conspicuous differentiation into holdfast, stipe (stalk), and blade regions (see Figure 14–23b) and have well-defined meristematic regions—areas of sustained cell division—within their bodies. In the kelps, which have regular cell division in three planes, the meristematic regions are generally located within the blade (Figure 14–25), although in some genera there is a single apical cell, or a group of apical cells, that constitutes the region of active cell division. All the cells have centrioles.

In the kelps, growth is rapid and productivity is high, particularly among those that are not exposed to the atmosphere; a recent study of three species from Nova Scotia showed that all renewed the tissue in their blades completely between one and five times each year. Experimental planting of kelps on a commercial scale have been initiated in recent years in an effort to produce fuel and food. The giant kelp *Macrocystis* has appeared especially promising in tests off the coast of southern California, for the growth of its blades is quite rapid.

The vegetative structure of the kelps is complex; they even have strands of elongated conducting cells in the center of the stipe that are similar to phloem (Figure 14–26). These cells have sieve plates similar to those of the sieve-tube elements of vascular plants,

(a)

(b)

(c)

14–23
Brown algae. (a) *Kelp exposed at low tide.* Laminaria saccharina *across the foreground and beneath the other species,* L. ochroleuca, *the uprising stalk in the center, and* L. digitata *in the background.* (b) *Detail of* L. digitata, *showing holdfasts, stipes, and the bases of several fronds.* (c) *Rockweed* (Fucus vesiculosus) *densely covers many rocky shores that are exposed at low tide. When submerged, the air-filled bladders on the blades carry them up toward the light. Photosynthetic rates of frequently exposed marine algae are one to seven times as great in air as in water, whereas they are higher in water for those rarely exposed, which accounts in part for the vertical distribution of seaweeds in intertidal areas.*

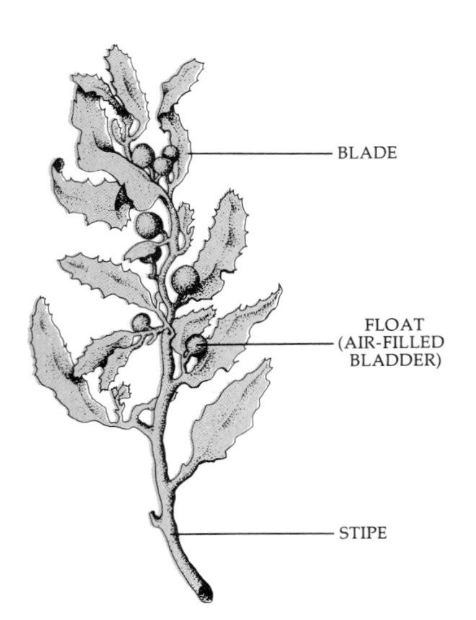

14–24
The brown alga Sargassum *has a complex pattern of organization of its vegetative parts.*

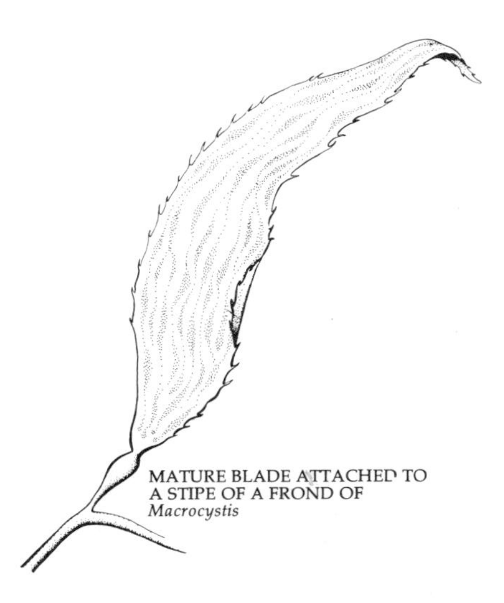

14–25
The leaflike blade of Macrocystis *is produced by the activity of a meristematic region at the junction of the stipe and the terminal blade. The latter splits as it expands, and segments form the mature blades, or "leaves."*

14–26
Some brown algae, such as the giant kelp Macrocystis pyrifera, *have evolved sieve tubes comparable to those found in vascular plants.* (a) *Cross section showing a sieve plate.* (b) *A longitudinal section of part of a stalk, with sieve tubes. In this view, the sieve plates appear thickened because they are covered with the wall substance callose.*

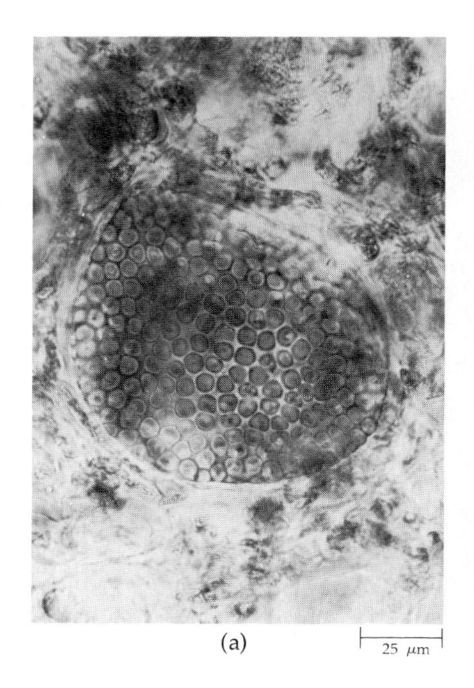

although they must have evolved independently. Marine algae do not need a mechanism for internal water transport, and it is thus not surprising that there are no cells analogous to tracheids or vessel elements. They must, however, conduct foods from the region of synthesis to other, poorly illuminated regions farther below the surface of the water. Recently, it has been shown experimentally that the sieve elements in the center of the stipes are the site of this translocation, which occurs in the giant kelp *Macrocystis* at a rate of about 60 centimeters per hour. In many of the kelps that are relatively thick, lateral translocation from the outer photosynthetic layers to the inner cells also takes place. The alcohol mannitol is the primary material that is translocated, along with amino acids.

Although it has been suggested that the Phaeophyta evolved from motile unicellular organisms, neither unicellular nor colonial organisms are found among the brown algae. Various brown algae are filamentous (branched or unbranched), and some form thin sheets; thus the diversification in this group parallels that in the multicellular green algae.

In addition to chlorophyll *a*, brown algae have chlorophyll *c* as an accessory pigment; chlorophyll *c* differs slightly but significantly from chlorophyll *b*. Brown algae also contain various carotenoids, including an abundance of the xanthophyll fucoxanthin, which gives the members of this group their characteristic dark brown or olive-green color. Instead of starch, laminarin is the characteristic carbohydrate storage product, but lipids are also accumulated. Another storage product is the alcohol mannitol. Motile cells in this group (zoospores and gametes) have two lateral flagella; the anterior one is of the tinsel variety, and the shorter posterior one is a whiplash flagellum. Cell walls in the brown algae are similar to those in plants and many green algae, being based on a matrix of cellulose microfibrils, but there is a well-developed outer layer of mucilaginous compounds. Alginic acids (algins)—gummy substances of considerable commercial importance as a stabilizer and emulsifier in paint and as a coating for paper—are abundant in the middle lamella and are also found elsewhere in the cell wall.

The life cycles of most brown algae are essentially similar to those of marine species of *Cladophora* and of *Ulva*; that is, they involve alternation of generations. Recently, the development of relatively simple culture methods for marine algae have allowed rapid advances in our knowledge. Meiosis is sporic, and both generations (gametophyte and sporophyte) are free-living (see Figure 10–10c, page 180). In some species of brown algae, the alternation is isomorphic, and in others it is heteromorphic. Heteromorphic alternation of generations is exhibited by the common kelp *Laminaria* (Figure 14–27), whereas the filamentous genus *Ectocarpus*, with similar sporophytic and gametophytic generations, illustrates isomorphic alternation. The molecules that are secreted by the female gametes and serve to attract the male gametes in several genera of brown algae are olefinic hydrocarbons. *Fucus* serves to illustrate yet another very unusual type of life cycle (Figure 14–28). Here meiosis is gametic, and the zygote grows directly into the new diploid organism (see Figure 10–10b, page 180). Zygotic meiosis, which is very common in the green algae, has not been observed in the brown algae.

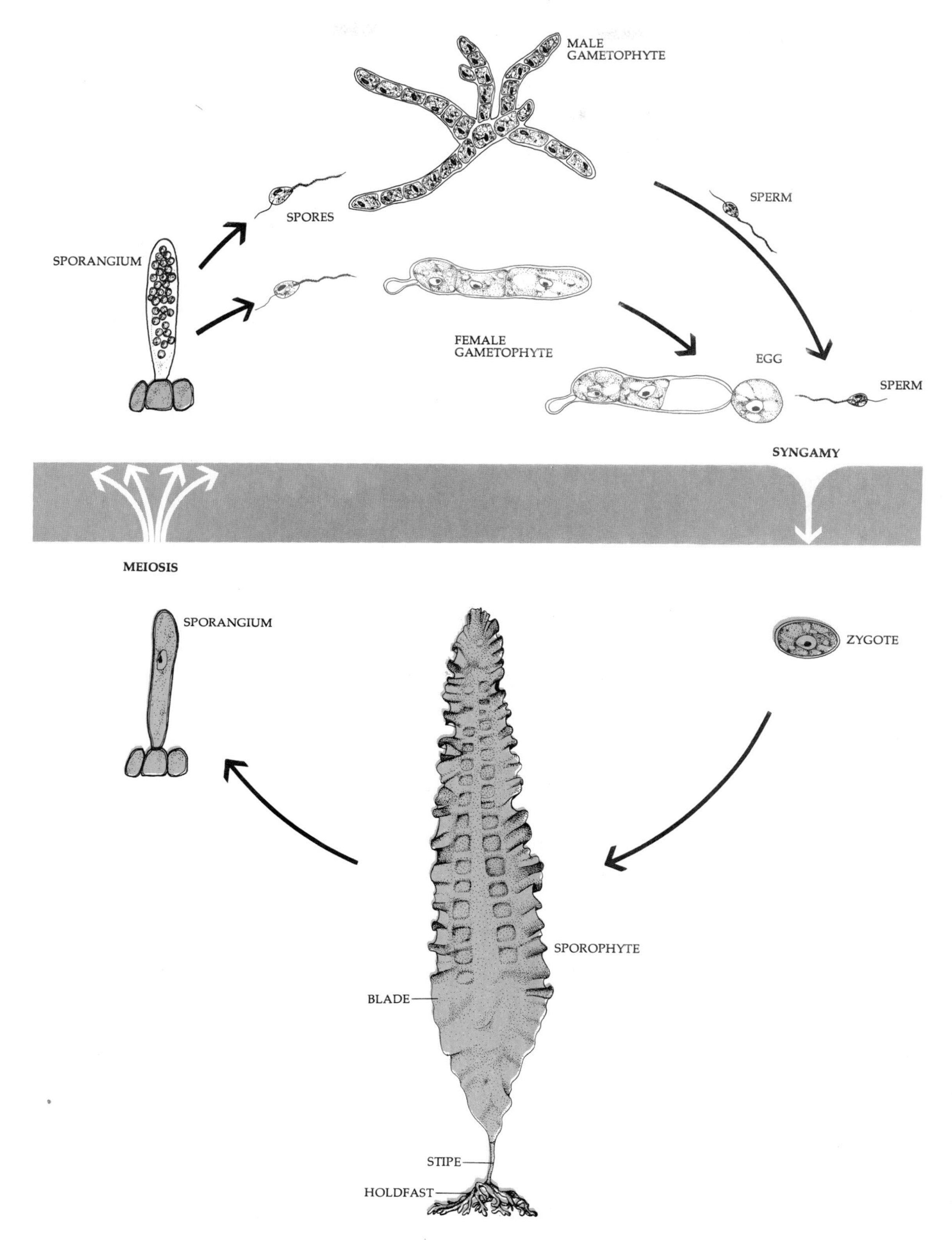

14–27
Life cycle of the kelp Laminaria. *Like most of the brown algae,* Laminaria *has an alternation of generations in which the sporophyte is conspicuous. Motile haploid zoospores are produced in the sporangia following meiosis. From these zoospores grow the small, filamentous gametophytes, which in turn produce the motile sperm and nonmotile eggs. In the simpler brown algae, the sporophyte and gametophyte are often similar.*

14–28
In Fucus, *gametangia are formed in specialized hollow chambers (conceptacles), which are found in fertile areas, called receptacles, at the tips of the branches of diploid individuals. There are two types of gametangia—oogonia and antheridia. Meiosis is followed immediately by mitosis to give rise to 8 eggs per oogonium and 64 sperm per antheridium. Eventually the eggs and sperm are set free in the water, where fertilization takes place. The life cycle of* Fucus *superficially resembles that found in some higher animals. Meiosis is gametic, and the zygote grows directly into the new diploid individual.*

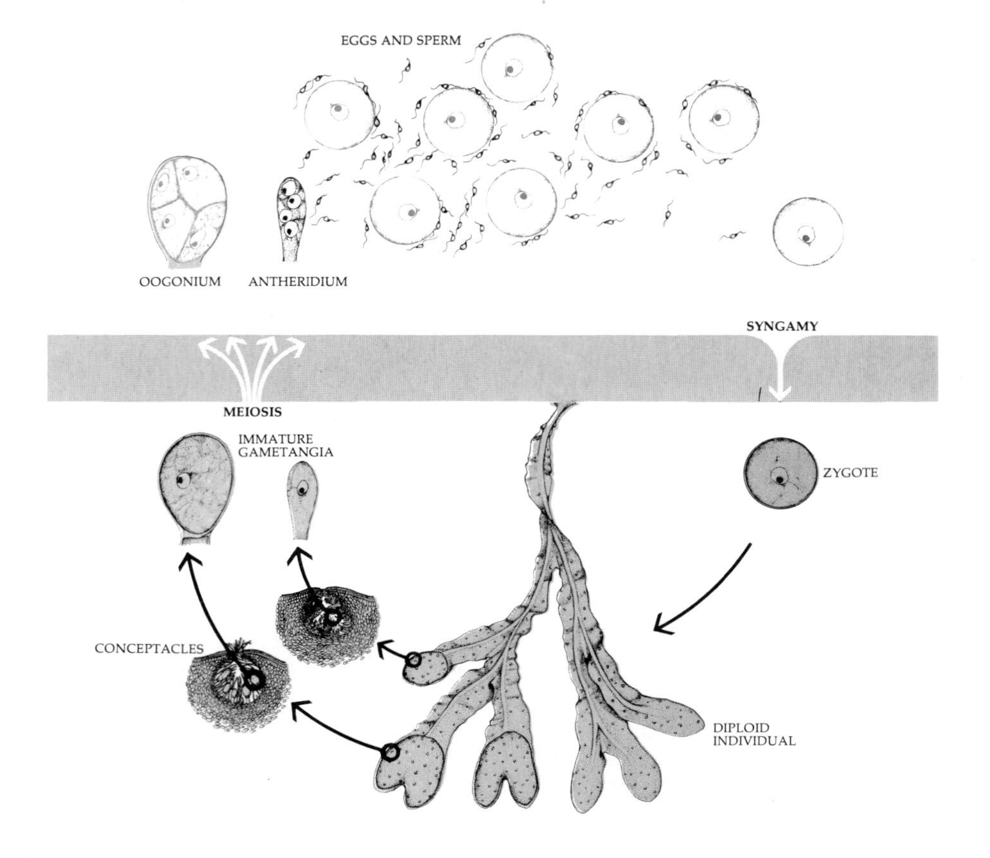

THE RED ALGAE: DIVISION RHODOPHYTA

The red algae have no flagellate cells, are structurally complex (Figure 14–29), and have complex life cycles. In contrast to the green and brown algae, which have centrioles in all cells, the red algae lack centrioles entirely. There are some 4000 species of red algae, and they are particularly abundant in tropical and warm waters, although many are found in the cooler regions of the world (see Table 14–1, page 253). Fewer than 100 species are found in fresh water, but in the sea the number of species is greater than that of all other groups of seaweeds combined. In Antarctic waters, the larger brown algae are absent and their place is taken by a kelplike red alga, *Himantothallus,* which can grow to more than 10 meters in length.

The water-soluble phycobilins, which mask the color of chlorophyll *a* and give the red algae their distinctive color, are accessory pigments and are particularly well suited to the absorption of the green, violet, and blue light that penetrates into deep water. The red algae are found at greater depths than any other groups of algae, some having been found at depths of up to 175 meters. Because phycobilins and chlorophyll *a* are also found in the cyanobacteria, it is considered likely that the chloroplasts of red algae are derived from symbiotic relationships between the algae and ancient organisms similar to modern cyanobacteria. The reserve carbohydrate of red algae is called floridean starch, which is similar to the amylopectin of plants. The grains of floridean starch are found in the cytoplasm rather than in the plastids.

The cell walls of most red algae include a rigid inner component composed of cellulose microfibrils and a mucilaginous outer component usually composed of sulfated polysaccharides, such as agar and carrageenan (see "The Economic Uses of Algae," page 275). It is these latter components that give the red algae their characteristic flexible, slippery texture. In addition, many red algae have the capacity to deposit calcium carbonate in their cell walls. Such algae are called coralline algae, and they play an important role in building coral reefs (see Figure 14–29b). In the fossil record, such calcareous algae are richly represented from Precambrian time to the present, with marine coralline red algae first appearing in the late Mesozoic era.

Red algae usually grow attached to rocks or other algae; there are a few floating forms and a few that are unicellular or colonial. Most red algae are composed of filaments, although this basic body plan is often difficult to distinguish because the filaments may be packed together very tightly. In many red algae, pit connections are a distinctive feature of the cell walls. These are actually lens-shaped plugs held in the walls by their equatorial grooves.

(a)

(b)

(c)

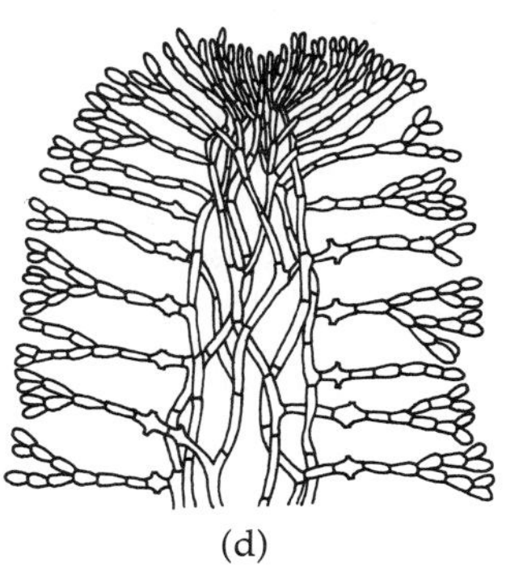
(d)

14–29
(a) *In* Pleonosporium dasyoides, *the basically filamentous structure of the red algae is clearly evident. It appears that multicellularity in the red algae evolved separately from that in the green and brown algae, where no similar patterns or organization are found.* (b) *The reef-building coralline alga,* Porolithon craspedium, *from the Marshall Islands in the South Pacific.* (c) *Irish moss* (Chondrus crispus) *an important source of carrageenan and other colloids.* (d) *Detail of structure of one of the simple red algae* Cumegloia, *showing its obviously filamentous form.*

The simpler red algae are filamentous, and cell division takes place mostly within the cells of the filament. In the more complex members of the division, the algal body consists of branched filaments that grow principally by the division of apical cells at the tips of the branches. In many red algae, these branched filaments are densely interwoven and are held together by copious mucilage. They give the impression of a complex thallus that is superficially comparable to that of the brown algae. The basic life cycle of the red algae involves an alternation of generations (Figure 14–30). In most red algae, the gametophyte and sporophyte are isomorphic, but an increasing number of heteromorphic cycles are being discovered.

14–30
Life cycle of Polysiphonia, *a red alga that is widely distributed in marine waters. The gametophytes are derived from haploid tetraspores and are unisexual. The sex organs arise near the tips of the branches. The male sex organs, or spermatangia, occur in dense clusters. Each spermatangium functions directly as a spermatium, or nonmotile sperm. The female sex organ, or carpogonium, develops a long, fingerlike outgrowth—the trichogyne. The enlarged basal portion of the carpogonium contains the nucleus and functions as an egg. Spermatia are carried passively to the trichogyne by water currents. When a spermatium becomes attached to a trichogyne, the walls between spermatium and trichogyne break down at their point of contact. The spermatial nucleus then enters the trichogyne and migrates to the egg nucleus, with which it fuses.*

The series of events following fertilization, which is extremely complicated, results in formation of carposporangia, which produce carpospores. When mature, the carpospores are liberated through an opening in the pericarp. Upon germination, each carpospore gives rise to a tetrasporophyte, which is similar in size and appearance to the gametophytes. The tetrasporophytes produce tetrasporangia, each of which, upon meiosis, gives rise to four haploid tetraspores, and the cycle begins anew.

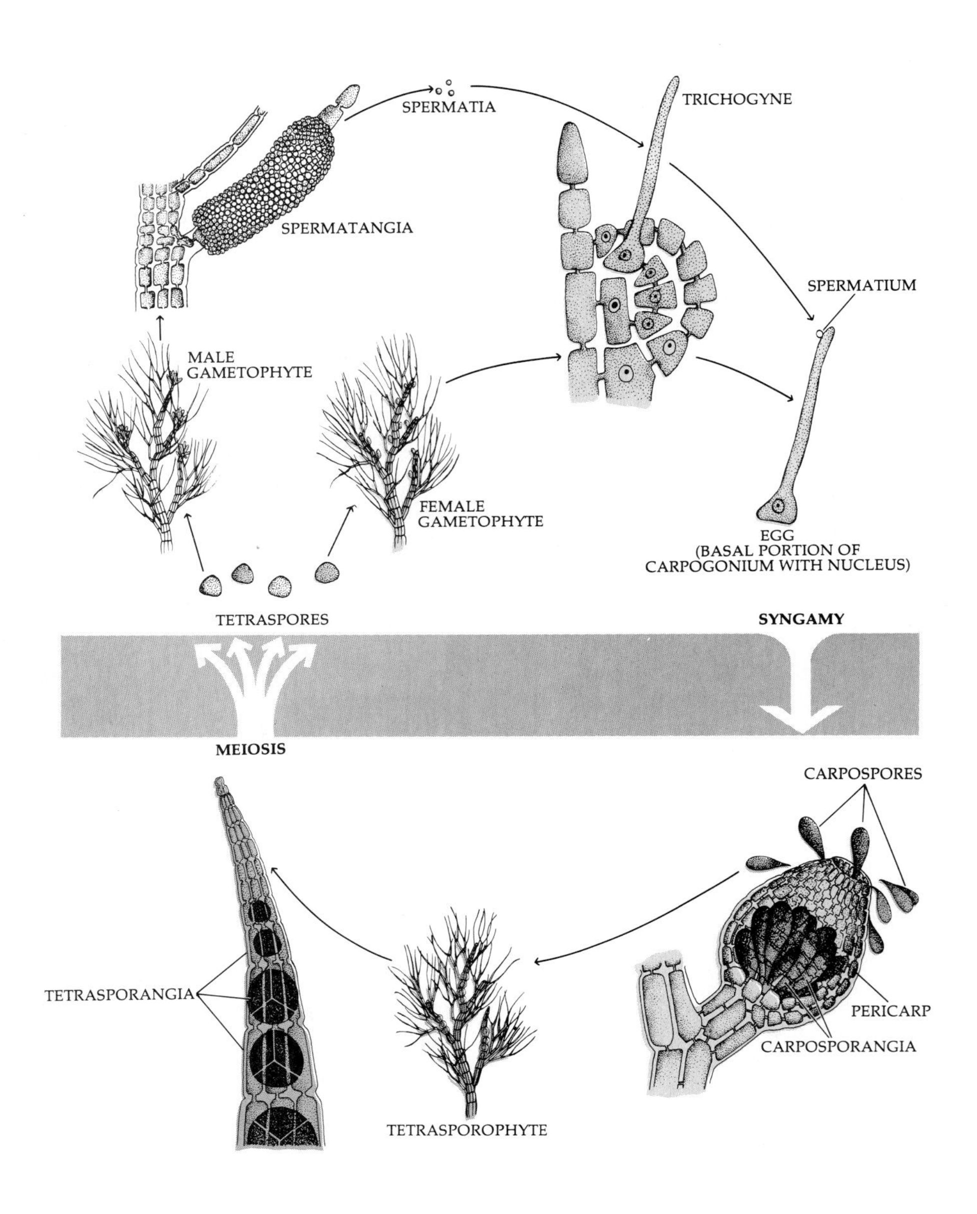

THE ECONOMIC USES OF ALGAE

People of various parts of the world, especially in the Far East, eat both red and brown algae. Porphyra *("nori"), one of the red algae, is eaten by many of the inhabitants of the north Pacific Basin and has been cultivated in Japan and China for centuries. The nori industry presently employs more than 30,000 persons in Japan alone. Various other red algae are eaten on the islands of the Pacific and also on the shores of the north Atlantic. Seaweeds are generally not of high nutritive value as a source of carbohydrates, because humans, like most other animals, lack the necessary enzymes to break down the cellulose in the cell walls. Seaweeds do, however, provide necessary salts, as well as a number of important vitamins and trace elements, and so are valuable supplementary foods. In many north temperate regions, kelp has been harvested for its ash, which is rich in sodium and potassium salts and is therefore valuable for industrial processes. Iodine is also produced commercially from kelp. Algae are often harvested and used directly for fertilizer.*

Alginates, which are a group of substances derived from such kelp as Macrocystis, *are widely used as thickening agents and colloid stabilizers in the food, textile, cosmetic, pharmaceutical, paper, and welding industries. Off the west coast of the United States,* Macrocystis *kelp beds can be harvested several times a year by cropping them just below the surface. Attempts are currently under way to cultivate this giant kelp on a commerical scale.*

One of the most useful direct commercial applications of any alga is the preparation of agar, which is made from a mucilaginous material extracted from the cell walls of a number of genera of red algae. Agar is used to make the capsules that contain vitamins and other drugs, as a dental-impression material, as a base for cosmetics, and as a culture medium for bacteria and other microorganisms. It is also employed as an antidrying agent in bakery goods, in the preparation of rapid-setting jellies and desserts, and as a temporary preservative for meat and fish in tropical regions. Agar is produced in many parts of the world, but Japan is the principal source. A similar algal colloid called carrageenan is used in preference to agar for the stabilization of emulsions such as paints, cosmetics, and dairy products. It has recently been proposed by phycologist Clinton Dawes of the University of South Florida that the common Florida red alga Eucheuma isiforme *could be cultivated to provide carrageenan on a commercial scale.*

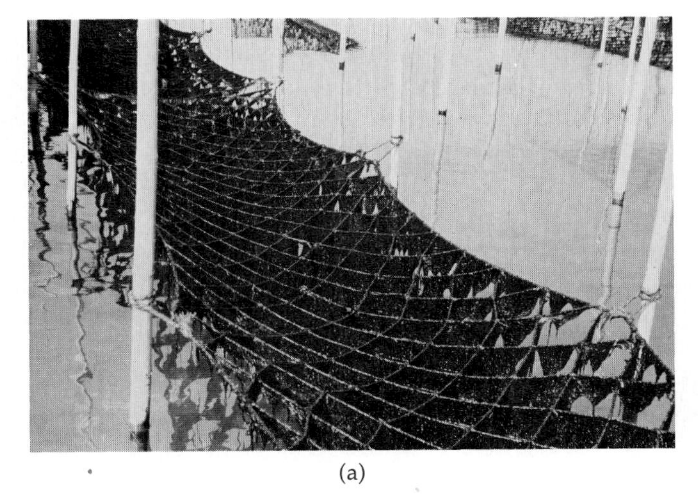

(a)

(b)

(c)

(a) A fixed system for the cultivation of *Porphyra*. Exposed at low tide, the nori net is heavily laden with the algae ready for harvest. (b) *Porphyra* being collected from a floating system with a harvesting machine that cuts the fronds from a net like a lawn mower. (c) Artificial "seeding" of conchospores—haploid spores of *Conchocelis*, a filamentous red alga—onto nets in indoor concrete tanks.

THE DIATOMS AND GOLDEN-BROWN ALGAE: DIVISION CHRYSOPHYTA

The diatoms and golden-brown algae (see Table 14–1, page 253) are unicellular organisms that are exceedingly important components of phytoplankton—the microscopic photosynthetic organisms that are suspended in the water. As such, they are a primary source of food for water-dwelling animals, both in marine and freshwater habitats. It is estimated that there may be as many as 11,500 living species in this division.

Like the brown algae, the Chrysophyta have chlorophylls *a* and *c*, the color of which is largely masked by the abundance of the accessory pigment fucoxanthin. The carbohydrate food reserve of the Chrysophyta is leucosin, which is similar in structure to the laminarin of the brown algae. In Chrysophyta, the products of photosynthesis are normally stored as large oil droplets. In one respect, the Chrysophyta differ greatly from the brown algae: their cell walls consist mainly of pectic compounds, which are often impregnated with siliceous materials and thus are very rigid. The walls contain no cellulose. On the basis of the existing evidence, it is difficult to decide whether the Chrysophyta are directly related to the brown algae. The two major classes of the Chrysophyta are Bacillariophyceae and Chrysophyceae, which are quite distinct from one another, plus a minor class—the yellow-green algae.

The Diatoms: Class Bacillariophyceae

Most species of Chrysophyta belong to this large group of unicellular organisms. Counting the great number of extinct species, as well as the nearly 11,500 living ones, the diatoms account for at least 40,000 valid species, and there may be as many as 100,000. There are often tremendous numbers of species in very small areas. In two small samples of mud from the ocean shoreline near Beaufort, North Carolina, for example, 369 species were identified. Most species of diatoms occur in plankton, but some are bottom dwellers or grow on other algae or plants. They occur both in fresh and salt water.

Diatoms have thin double shells, or frustules, made of polymerized, opaline silica ($SiO_2 \times nH_2O$), the two halves (valves) of which fit together, one on top of the other. The delicate markings of these shells, by which the species are identified, have been traditionally used by microscopists to test the quality of their lenses. Electron microscopy has shown that these fine tracings on the diatom shells are actually composed of a large number of minute, intricately shaped depressions, pores, or passageways that connect the living protoplasm within the shell with the outside environment (Figure 14–31). The most conspicuous features within the protoplast of diatoms are the brownish plastids that contain chlorophylls *a* and *c*, as well as fucoxanthin, like the brown algae. Diatoms reproduce primarily by cell division—an asexual process (Figure 14–32).

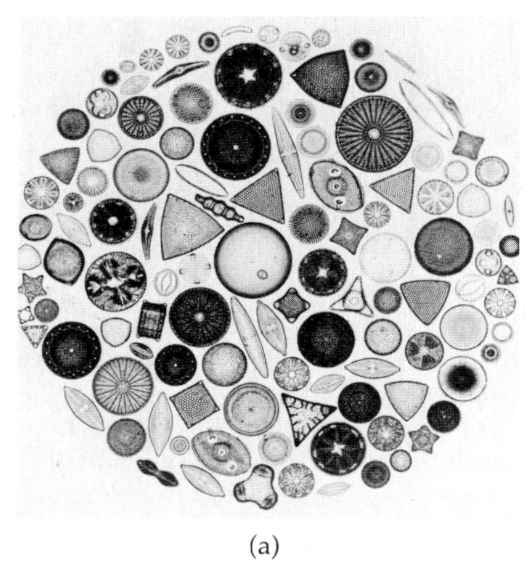

(a)

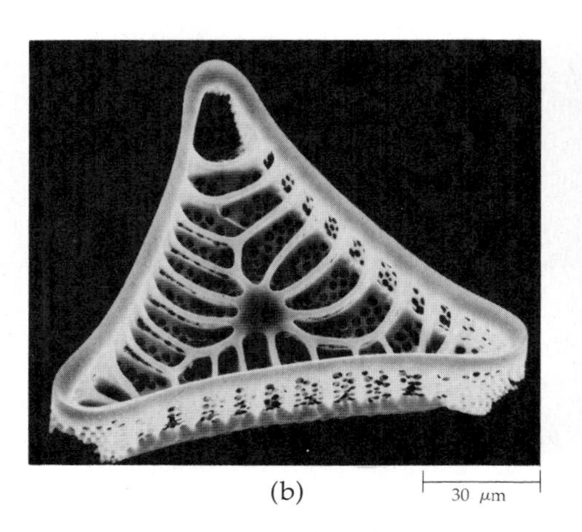

(b)

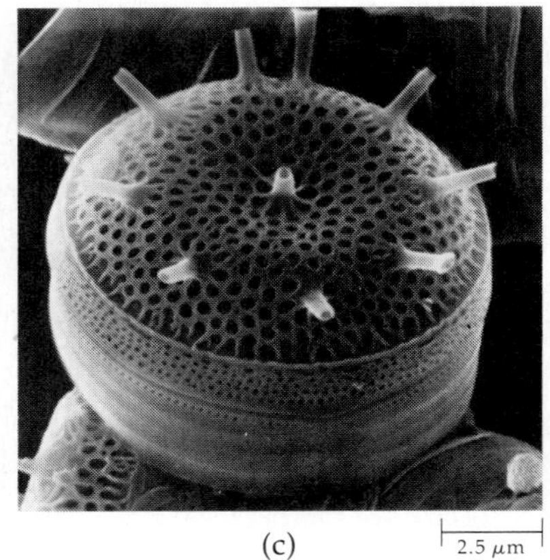

(c)

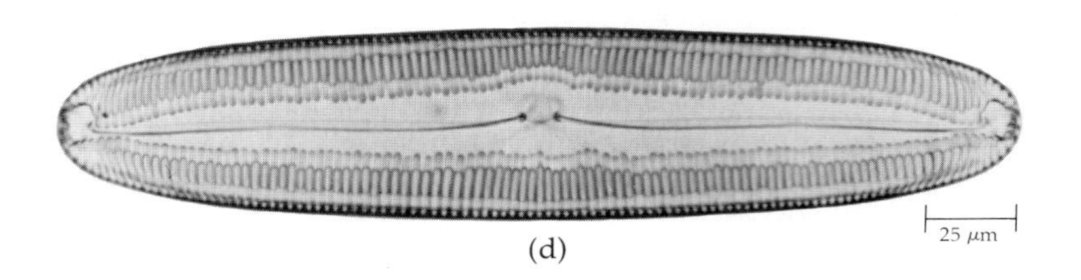

(d)

14–31
(a) *A selected array of marine diatoms, as seen with a light microscope. Scanning electron micrographs of* (b) *one valve of an* Entogonia *shell and* (c) Thalassioria nordenskioeldii, *a centric diatom.* (d) Pinnularia, *a pennate diatom, as seen with a light microscope.*

14–32

Reproduction in the diatoms is mainly asexual and occurs by cell division. Each daughter cell receives one of the valves of the previous shell and constructs a second valve. The old valve always forms the large part (the lid) of the silica "box," with the new valve fitting inside it. As a consequence, one cell of each new pair tends to be smaller than the one before it. In some species, the shells are expandable and are enlarged by the growing protoplasm within them. In other species, however, the shells are more rigid. When the individuals of these species have decreased in size to about 30 percent of the maximum diameter, sexual reproduction may be triggered. Certain cells function as male gametangia; they each produce four sperm through meiosis. Other cells function as female gametangia; in them, three of the four products of meiosis are nonfunctional, so that one egg is produced per cell. After fertilization, the resulting auxospore, or zygote, expands to the full size characteristic of the species. The walls formed by the auxospores are often different from those of the asexually reproducing cells of the same species. Once the auxospore is mature, it divides and produces new shells identical in all their intricate markings to the previous ones. This diagram illustrates sexual reproduction in a centric diatom.

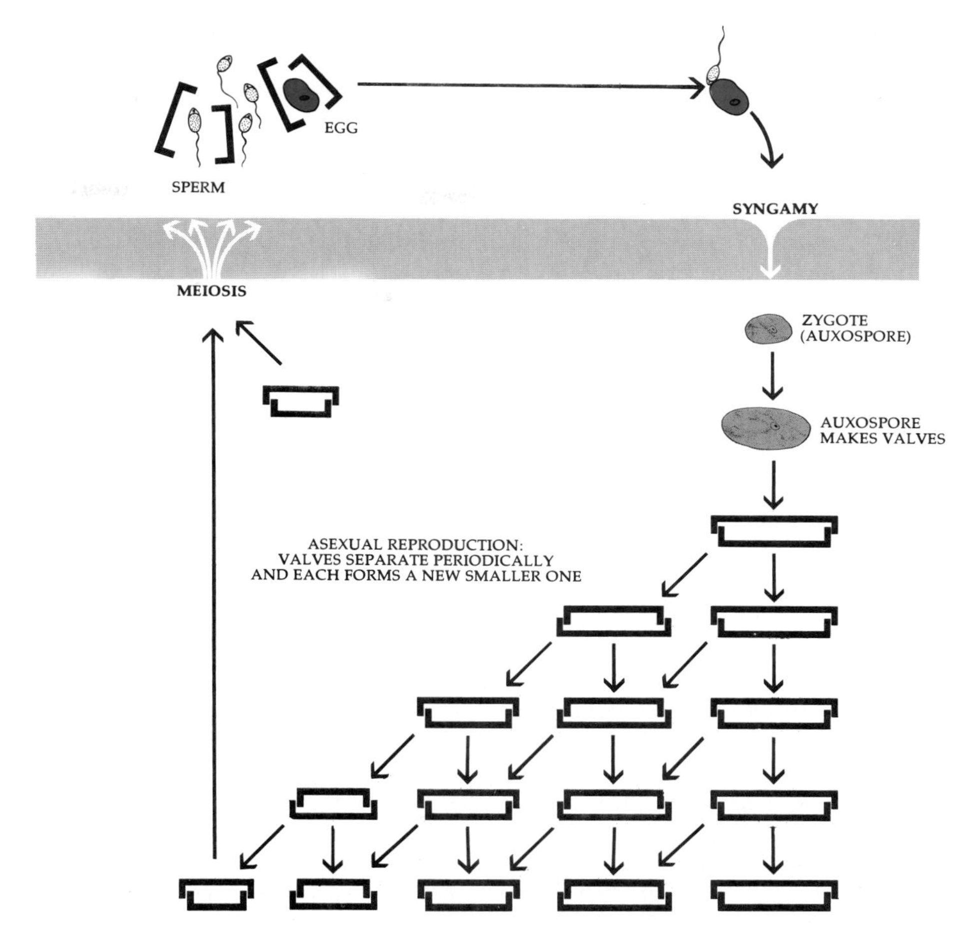

On the basis of symmetry, two types of diatoms are recognized: those that are bilaterally symmetrical—the pennate diatoms (Figure 14–31d)—and those that are radially symmetrical—the centric diatoms (Figure 14–31c). Centric diatoms are most abundant in marine waters. When sexual reproduction occurs in the centric diatoms, it is oogamous. The sperm have single anterior whiplash flagella and are the only flagellate cells found in the diatoms at any stage of the life cycle. In the pennate diatoms, sexual reproduction is isogamous, and both gametes are nonflagellated.

Although most species of diatoms are autotrophic, many species—primarily pennate ones that occur on the bottom of the sea in relatively shallow habitats—need not be photosynthetic and can exist by absorbing food. A few species are obligate heterotrophs, which means that they are no longer capable of providing their own food through the process of photosynthesis. On the other hand, some diatoms, lacking their characteristic shells, live symbiotically in large marine protozoa (Foraminifera) and provide food for these heterotrophic organisms.

Despite their lack of flagella or other locomotor organelles, many species of diatoms—but only pennate ones—are motile. Their locomotion results from a rigorously controlled secretion that occurs in response to a wide variety of physical and chemical stimuli. All motile diatoms seem to possess a fine groove along each shell, called the raphe, which is basically a pair of pores connected by a complex slit in the silicate wall of the diatom. Many nonmotile diatoms are attached to one another, with their shells arranged in long filaments; and the raphe apparently evolved as a locomotor device by modification of the apical pores that secrete the substances that unite these nonmotile diatoms into filaments.

A diatom moves in response to an external stimulus—such as mechanical disturbance, light, heat, or a toxic chemical—by initiating contractions in contractile bundles that lie adjacent to the raphe system. This contraction moves dehydrated crystalline bodies to reservoir areas adjacent to pores in the raphe, and the crystalline bodies are then discharged into the pores. Here they take up water and expand into twisting fibrils. The fibrils move along the raphe until they strike a surface. They immediately adhere to anything they touch and then contract. If the object to which they adhere is large enough, the diatom moves toward it, depositing a trail of secreted material analogous to the slime trail of a snail. If the object is small, it is transported along the raphe, and the diatom remains stationary. Even motile diatoms are usually at rest. Each can move only a limited distance, for they have available at any one time only a limited supply of the required crystalline bodies.

The silica shells of diatoms have collected over millions of years and form the fine, crumbly substance known as diatomaceous earth, which is used as an abrasive in silver polish and for filtering and insulating materials. The paint used on the center lines of roads is impregnated with diatomaceous earth because of its reflective characteristics. It is estimated that 1 cubic centimeter of diatomaceous earth contains some 4.6 million diatom shells. In the Santa Maria, California, oilfields there is a subterranean deposit of diatomaceous earth that is 900 meters thick, and near Lompoc, California, more than 270,000 metric tons of diatomaceous earth are quarried annually for industrial use.

Diatoms first became abundant in the fossil record some 100 million years ago, during the Cretaceous period; many of the fossil species are identical to those still living today, which indicates an unusual persistence through geological time.

The Golden-Brown Algae: Class Chrysophyceae

The second major class of Chrysophyta—the golden-brown algae—consists of about 1500 species (Figure 14–33). Until recently, they were thought to be primarily a freshwater group, but Chrysophyceae have been found to be of extraordinary importance in the marine plankton, particularly the nannoplankton. The nannoplankton are such small components of the plankton that they pass through an ordinary plankton net, which has mesh openings of 40 to 76 micrometers. Some of the nannoplankton consist of minute dinoflagellates and diatoms, but representatives of the Chrysophyceae are often abundant. In fact, it is now thought that the Chrysophyceae may be the major food-producing organisms present in the world's oceans.

Some of the golden-brown algae lack cell walls, whereas others have a well-defined wall rich in pectic substances. Many species have scales or skeletal structures, which may be superficial or internal and often are exceedingly elaborate. These structures may be siliceous or organic. Many golden-brown algae are motile and have two flagella (see Figure 1–24, page 30); others are amoeboid and lack flagella. Except for the presence of chloroplasts, the amoeboid cells are indistinguishable from amoeboid protozoa, and the two groups may be closely related. Reproduction in most golden-brown algae is asexual and involves zoospore formation.

The Yellow-Green Algae: Class Xanthophyceae

Yellow-green algae are related to the other classes in the division Chrysophyta by their lack of chlorophyll *b*; they differ from them by an additional lack of the accessory pigment fucoxanthin. Because of their greenish to yellow-green appearance, Xanthophyceae are often confused with chlorophycean algae.

One of the better-known members of the class is *Vaucheria*, the "water felt," a coenocytic, little-branched filamentous alga. *Vaucheria* reproduces both asexually, by the formation of large, compound zoospores that are multiflagellated, and sexually, by oogamy (Figure 14–34). *Vaucheria* is widespread in freshwater, brackish, and marine habitats, and often is found on mud that is alternately immersed in water and exposed to air.

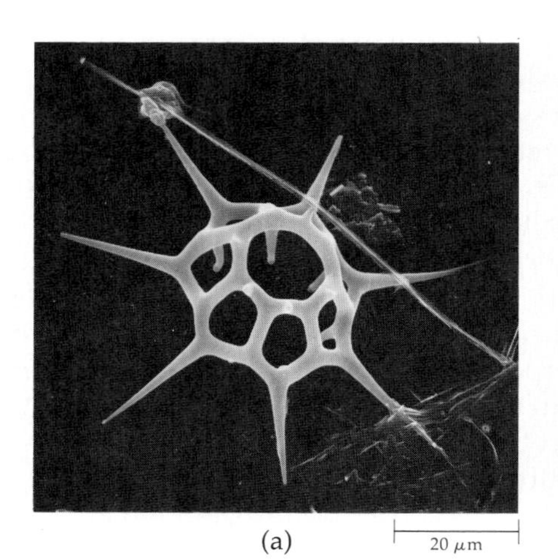

14–33
Scanning electron micrographs of two marine genera of golden-brown algae. (a) Distephanus speculum, *a cold-water species that has a siliceous skeleton, is encased in an amoeboid protoplast containing many small chloroplasts.* (b) *A species of the genus* Gephyrocapsa, *extremely minute members of the group known as coccolithophorids. Although these nannoplankton are quite abundant, they are so small that they are not caught by conventional plankton nets; they also dissolve in acid fixatives, so they are both difficult to find and to study.*

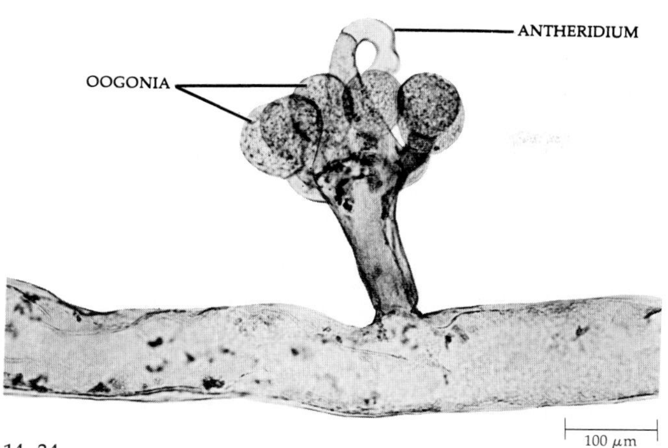

14–34
Vaucheria, *the "water felt," is a coenocytic, filamentous member of the division Chrysophyta.* Vaucheria *is oogamous, producing oogonia and antheridia. The antheridium shown here is empty.*

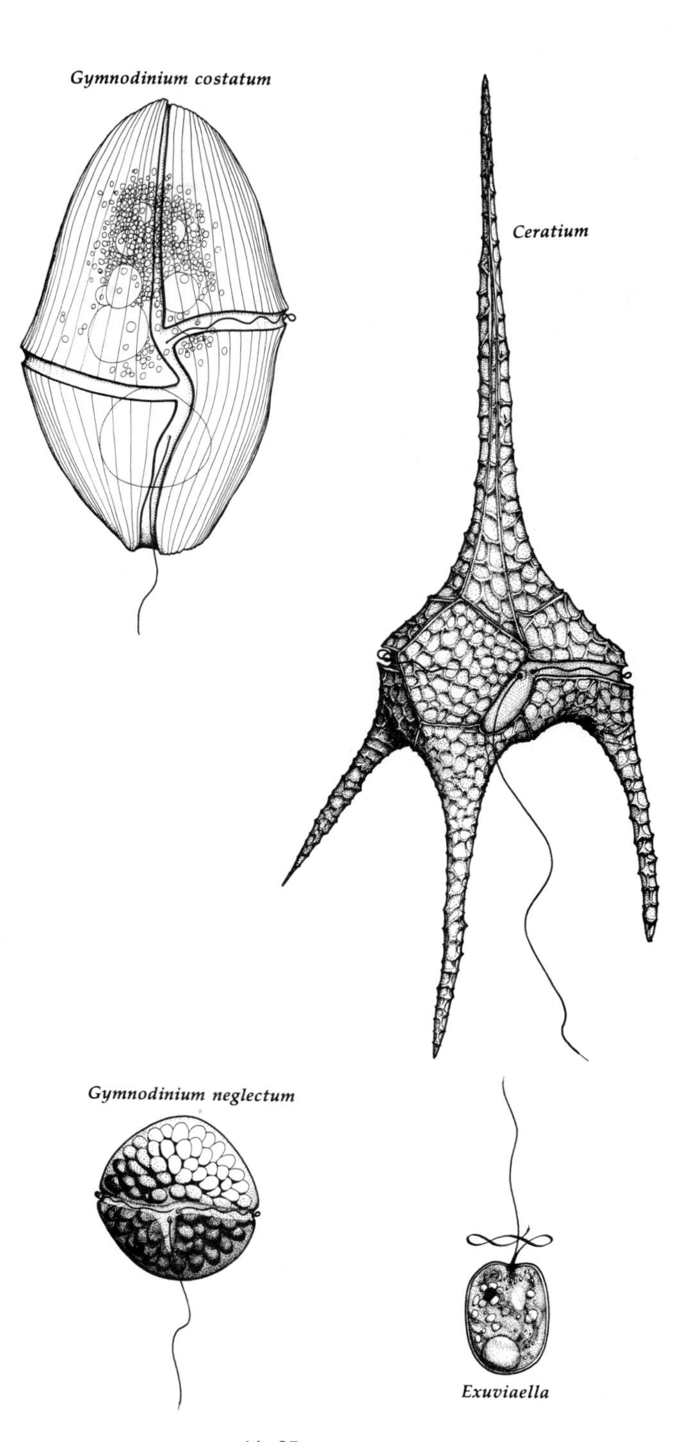

14–35
The "armor" of some dinoflagellates consists of cellulose plates completely enclosed by the plasma membrane. Those genera that appear to be naked actually have thinner cellulose plates that do not differ fundamentally from those of the armored ones.

THE DINOFLAGELLATES: DIVISION PYRROPHYTA

The dinoflagellates, or "whirling whips," are for the most part unicellular biflagellates (see Table 14–1, page 253). More than 1000 species are known, many of them of great importance in the marine plankton; others occur in fresh water. In the dinoflagellates, the flagella beat within two grooves; one groove circles the body like a belt, and the second groove is perpendicular to the first. The beating of the flagella in their respective grooves causes the organism to spin like a top as it moves. The encircling flagellum is ribbonlike. There are also numerous nonmotile dinoflagellates, some of which are nonflagellated.

Many of the dinoflagellates are bizarre in appearance, with stiff cellulose plates forming a wall (theca), which often looks like a strange helmet or part of an ancient coat of armor (Figures 14–35 and 14–36). The plates of the wall are in vesicles inside the plasma membrane, and not outside it, as is the cell wall of most algae. Most dinoflagellates contain chlorophylls *a* and *c*, which are generally masked by carotenoid pigments. The carbohydrate food reserve is starch. Some of the species do not contain chlorophyll and are heterotrophic, but their structure clearly allies them to the other members of the division. Even the autotrophs usually have strong requirements for vitamin B_{12}, like many diatoms, and cannot be regarded as primary producers without qualification. Some di-

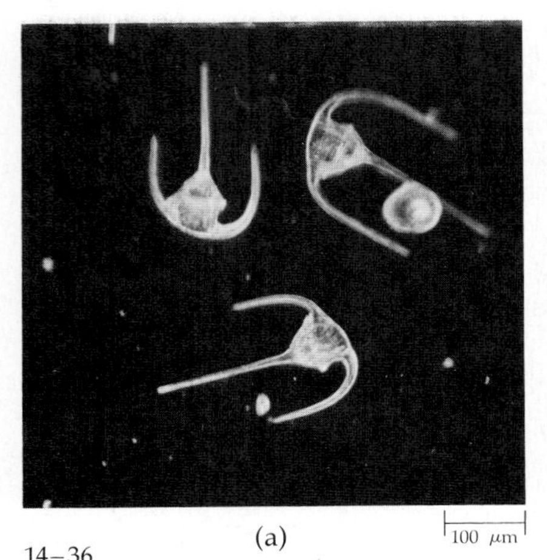

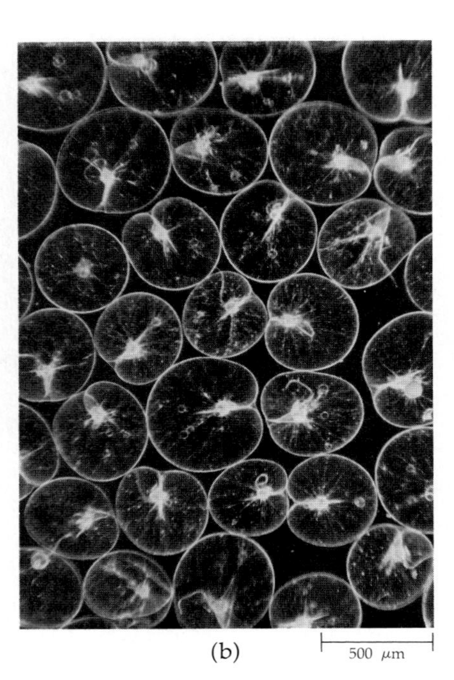

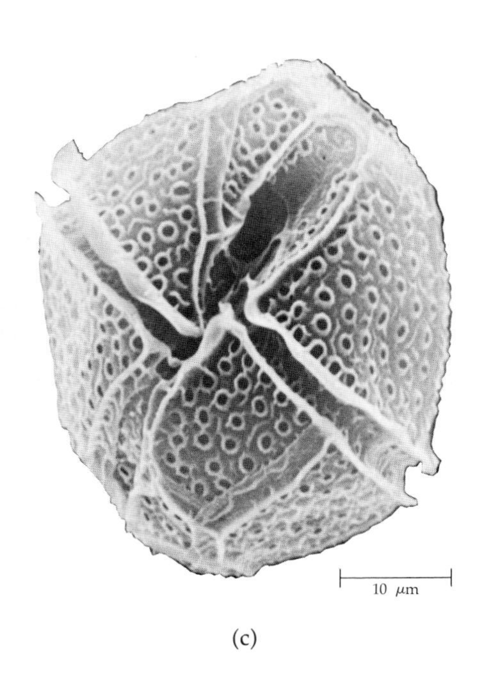

14–36
(a) Ceratium tripos, *an armored dinoflagellate.* (b) Noctiluca scintillans, *a bioluminescent marine dinoflagellate.* (c) Gonyaulax polyedra, *the dinoflagellate responsible for the spectacular red tides along the coast of southern California.*

noflagellates are capable of ingesting other cells. It is assumed that the colorless forms are grazers that obtain their nutrition by ingesting other cells or small particulate organic material.

Cyanobacteria are common symbionts in tropical dinoflagellates. Dinoflagellates in turn occur as symbionts in many other kinds of organisms, including sponges, jellyfish, sea anemones, corals, gastropods, turbellarians, and certain groups of protozoa. In their role as symbionts, the dinoflagellates lack thecae; they appear as golden spherical cells called zooxanthellae. Zooxanthellae are primarily responsible for the photosynthetic productivity that makes possible the growth of coral reefs in tropical waters, which are notoriously poor in nutrients.

Dinoflagellates also play an important role in human affairs. During the winter and spring of 1974, the west coast of Florida was ravaged by its 25th major red tide since 1844. Hundreds of thousands of dead fish littered the beaches, and millions of tourist dollars were lost. Such red tides are caused by unusual outbreaks, or "blooms," of red dinoflagellates, which color the sea red or brown. These dinoflagellates are ingested not only by fish, which may be poisoned directly, but also by shellfish, such as mussels and clams, which are generally not harmed. Shellfish accumulate and concentrate the toxins produced by these organisms and, depending on the species of dinoflagellate, may become dangerous for human consumption. In the fall of 1972, for example, the New England coast from Maine to Cape Cod experienced its first red tide. Twenty-six people were poisoned by shellfish contaminated by *Gonyaulax excavata* and public confidence was shaken to such an extent that the Massachusetts shellfish industry had returned to only about two-thirds of its former level four years later. Red tides are a recurrent problem. In the worst outbreak since 1972, the entire coast of Maine was legally closed to shellfishing from mid-August to mid-October 1980. Commercial fishermen were unable to harvest an estimated $7 million worth of clams, oysters, and mussels.

The poisons produced by some dinoflagellates, such as *Gonyaulax catanella,* are extraordinarily powerful nerve toxins. The chemical nature and biological activity of most of these toxins are relatively well known. On the other hand, the factors that cause the red tides themselves are poorly understood; levels of nutrients and certain trace metals, sewage runoff, ocean salinity and temperature, winds, light, and many other factors all seem to play some role. The periodic recurrence of red tides is compounded by the observation that once a "bloom" has occurred, if conditions shift for the worse, the dinoflagellates may lose their flagella and form resting cysts that sink to the bottom and lie dormant until conditions are more favorable for their renewed activity.

The chief method of reproduction in dinoflagellates is by longitudinal cell division, with each daughter cell receiving one of the flagella and a portion of the theca and then constructing the missing parts in a very intricate sequence (see "Mitosis in Dinoflagellates," page 283). Some nonmotile species form zoospores. In some of these species only the zoospore has the typical dinoflagellate structure, whereas the mature forms have cells with no flagella or may even be joined together in filaments. Sexual reproduction has been found in a number of dinoflagellates; it is generally isogamous, but anisogamy is also present.

THE EUGLENOIDS: DIVISION EUGLENOPHYTA

There are more than 800 species of euglenoids, most of which are found in fresh water, especially water rich in organic material (see Table 14–1, page 253). They range from less than 10 micrometers to more than 500 micrometers (0.5 millimeter) in length and are quite variable in form. All are unicellular except for the colonial genus *Colacium.* In the course of their evolution, some of the euglenoids have acquired chloroplasts that are biochemically similar to, but structurally different from, the chloroplasts of green algae. About a third of the approximately 40 genera of euglenoids have chloroplasts, and these have chlorophylls *a* and *b* together with several carotenoids. Euglenoids store their carbohydrate food reserves in the form of paramylon—a polysaccharide that is not found in any other group of organisms and that, unlike starch, is formed outside the chloroplast. Among the genera that lack chloroplasts, some absorb organic matter and others ingest it.

The euglenoids reproduce by cell division, with the individual cells remaining motile throughout the process. The nuclear envelope remains intact during mitosis, as it does in most of the green algae, the dinoflagellates, many fungi, and some ciliated protozoa. The centrioles function as basal bodies and organize a normal spindle apparatus within the nuclear envelope. The chromosomes, like those of dinoflagellates, remain condensed during interphase and throughout the mitotic cycle, but they do include histone proteins. No sexual reproduction is known to occur among the euglenoids.

The division takes its name from *Euglena,* a common genus. Many species of *Euglena* are elongated, as shown in Figure 14–37. The cell is complex and contains numerous small chloroplasts. A long, emergent flagellum with very fine hairs along one side is present, as well as a short, nonemergent flagellum. The emergent flagellum is usually held in front of the cell like a spinning lasso.

In *Euglena,* the flagella are attached at the base of the flask-shaped opening—the *reservoir*—at the anterior end of the cell. The contractile vacuole, which collects excess water from all parts of the cell, discharges it into the reservoir. The cell is delimited by a plasma membrane inside of which is a series of flexible, interlocking proteinaceous strips that are arranged helically. These strips form a structure called the *pellicle.* Unlike the stiff walls of plant cells, the flexible pellicle permits *Euglena* to change its shape, providing an alternative means of locomotion for mud-dwelling forms.

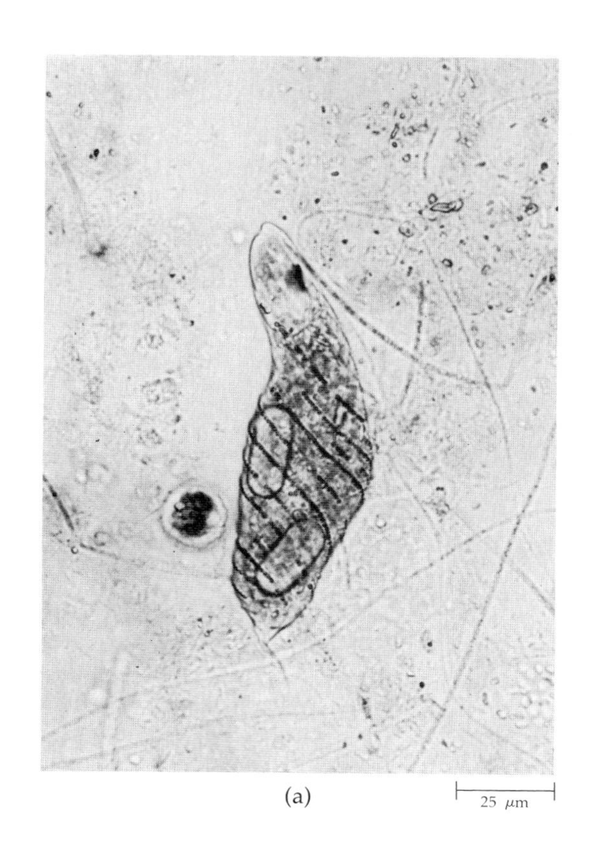

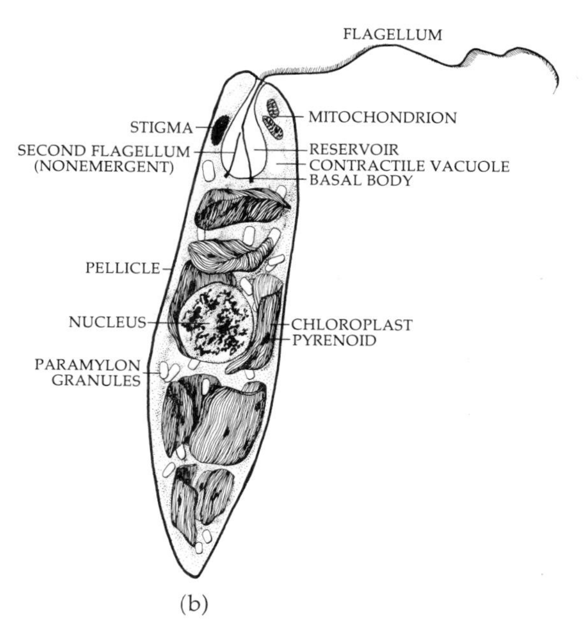

14–37
(a) Euglena, *showing two storage bodies of paramylon and the nature of the pellicle.* (b) *The structure of* Euglena, *as interpreted from electron micrographs.*

If one leaves a culture of *Euglena* near a sunny window, a clearly visible green cloud will form in the water, and this will move as the light changes, always toward a spot that is bright (but not too bright). If the light is too bright, the individuals of *Euglena* will swim away from it. *Euglena,* like *Chlamydomonas,* is probably able to orient with respect to light because of the presence of a pair of special structures—the stigma and an associated photoreceptor, which is a swelling at the base of the flagellum. The "shading" of this photoreceptor by the pigmented stigma when the individual euglenoid is in certain orientations with respect to light probably aids in its directional movement. Only a few nongreen euglenoids have a stigma.

Some species of *Euglena* can survive if they are kept in the dark, where they cannot photosynthesize, as long as they are provided with a carbon source in addition to their other vitamin and mineral requirements. If some strains of *Euglena* are kept in the light, at an appropriate temperature and in a rich medium, the cells may replicate faster than the chloroplasts, producing nonphotosynthetic cells which can nonetheless survive indefinitely in a suitable medium.

COLEOCHAETE: AN ADVANCED GREEN ALGA

Historically the green alga Coleochaete *has been singled out as a possible land plant relative, because some species are parenchymatous and grow by a peripheral meristem, reproduction is oogamous, and the zygotes are protected by a layer of sterile vegetative cells. More recent work has demonstrated that* Coleochaete *and a few other green algae resemble the land plants in the presence of a phragmoplast at cytokinesis, the photorespiratory enzyme glycolate oxidase localized in peroxisomes, and a multilayered structure (MLS) associated with the flagella of reproductive cells. Certain fossil evidence suggests that* Coleochaete *is an "archaic" green alga; that is, it is an extant form that exhibits many of the characteristics of its ancestors, which lived at the time when the land plants first appeared. However, the fossil record has not revealed the nature of the organisms that were transitional between green algae and plants as they exist today, probably because nonlignified tissues were not well preserved. Nevertheless,* Coleochaete *and its relatives have become valuable sources of clues as to how green plants became structurally and biochemically adapted to the land environment.*

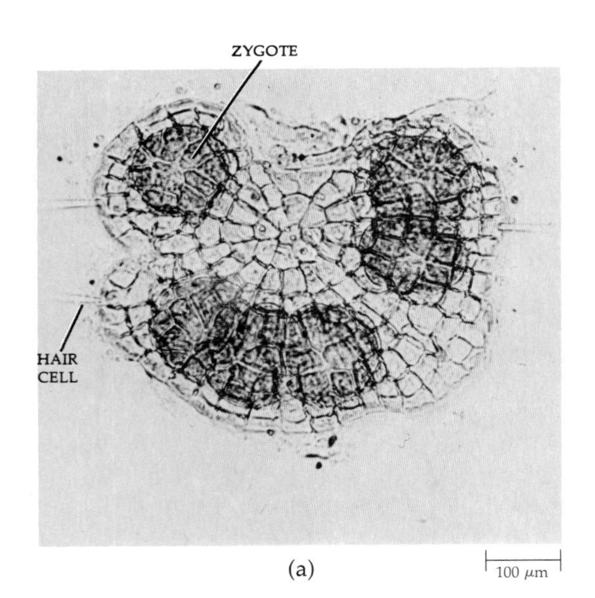

(a)

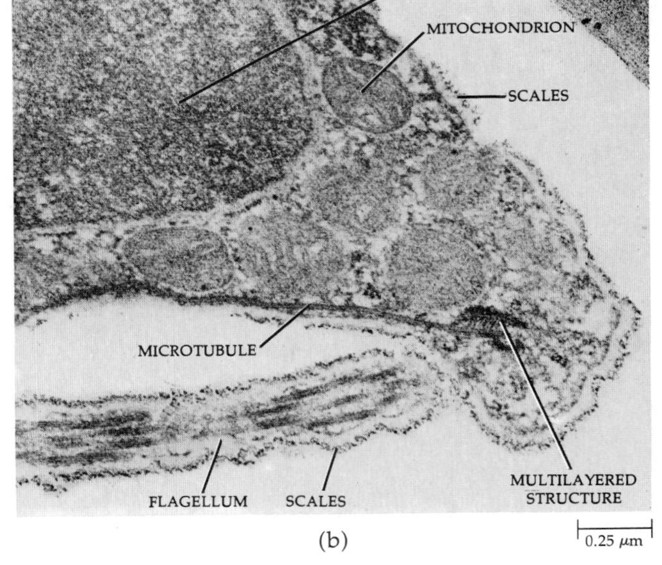

(b)

(a) *Coleochaete* collected from stems of aquatic flowering plants growing in the shallow littoral zone of a lake. This is a parenchymatous disk that is generally one cell thick. The large cells are zygotes that are protected by a cellular covering. The hair cells extending from the disk are ensheathed at the base. *Coleochaete* means "sheathed hair." These hairs are thought to discourage aquatic animals from feeding on the alga. (b) An electron micrograph of the anterior portion of a spermatozoid of *Coleochaete.* Note the multilayered structure (MLS) near the mitochondria. A layer of microtubules extends from the MLS down into the cell posterior and serves as a cytoskeleton for these wall-less cells. The flagellar and cellular membranes are covered by a layer of small diamond-shaped scales. Scales have been very useful in determining phylogenetic relationships in green algae.

MITOSIS IN DINOFLAGELLATES

Dinoflagellates have a unique type of mitosis which appears to combine features of both eukaryotes and prokaryotes. Indeed, their cellular organization is so distinctive that it is increasingly being called "dinokaryotic" in order to distinguish it from both the eukaryotic and prokaryotic types of organization. For example, as shown in (a), *the chromosomes are always visible and do not contract prior to mitosis. There are extremely large amounts of DNA per cell. The chromosomes are attached to the nuclear envelope, which persists during mitosis. The dinoflagellates have a much lower ratio of protein to DNA than other eukaryotes, and the typical eukaryotic histones are absent, although a histonelike protein is present.*

Cytoplasmic channels, as shown running down through the center of (b), *invade the dividing nucleus at the time of mitosis. The chromosomes remain attached to the nuclear envelope and are carried along on the sides of these channels, which contain bundles of microtubules similar to those of a eukaryotic spindle apparatus. The microtubules are all oriented in one direction and presumably regulate the separation of the portions of the nuclear envelope with its attached chromosomes. The organism illustrated in both* (a) *and* (b) *is* Cryptothecodinium cohnii.

At least two species of dinoflagellates have binucleate cells, in which one nucleus is dinokaryotic and the second is eukaryotic. In this second nucleus—which was presumably obtained by the host after the invasion of a chrysophytelike symbiotic organism—the chromosomes do not become condensed at any stage during the cell cycle, an observation that indicates they are not identical with "typical" eukaryotic chromosomes.

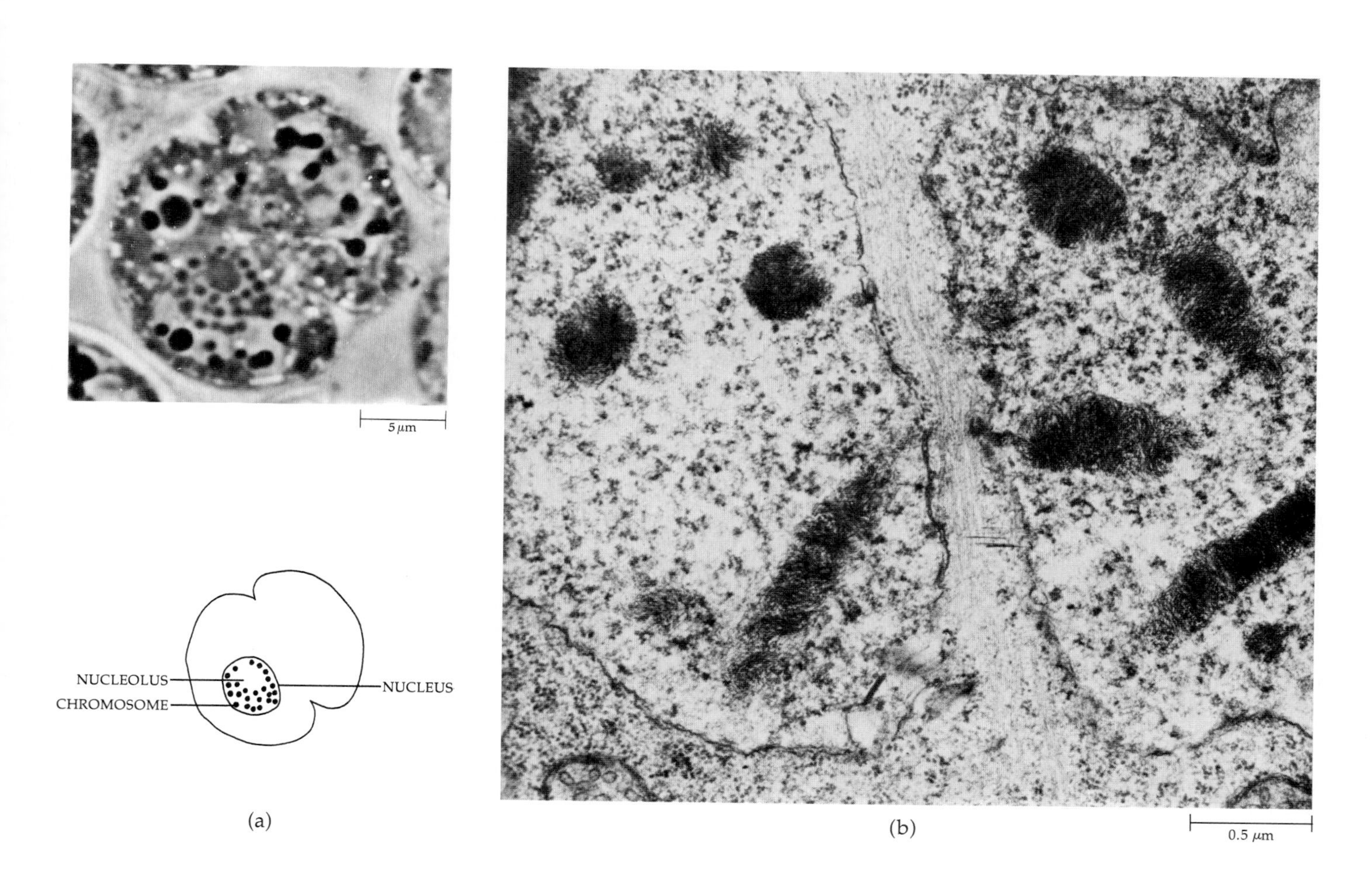

(a) (b)

DERIVATION OF THE DIVISIONS OF ALGAE

All eukaryotic cells—plant, animal, protist, or fungal—have endoplasmic reticulum, Golgi bodies, large ribosomes, mitochondria, microtubules, and a complex nucleus. The nuclear apparatus includes a double-membraned nuclear envelope that is often continuous with the endoplasmic reticulum; chromosomes that contain DNA in combination with proteins and that undergo mitosis and meiosis with the formation of a spindle apparatus; and one or more nucleoli. Each of the major lines of eukaryotes, with the exception of the Rhodophyta, includes forms that have flagella with the distinctive 9-plus-2 organization described in Chapter 1. It would thus seem logical to assume that the common ancestor of all living eukaryotes exhibited all these characteristics. If such an organism were living today, it would be classified as a protozoan.*

The red algae (division Rhodophyta)—like the cyanobacteria—have chlorophyll a, *carotenoids, and phycobilins. Phycobilins are found only in these two groups (and in the division Cryptophyta, a poorly understood and quite distinct group of algae not treated in this text). In cyanobacteria, these photosynthetic pigments are found on membranes throughout the cell, whereas in red algae they are confined to the chloroplasts. The arrangement of the thylakoids in cyanobacteria is very similar to that found in red algal chloroplasts. The simplest explanation of these relationships would seem to be that the chloroplasts of red algae are, in effect, cyanobacteria that became symbiotic in the cells of some primitive protozoan long ago.*

The green algae (division Chlorophyta) have chlorophylls a *and* b *and carotenoids, as do the plants. Until recently, no prokaryote that had this combination of photosynthetic pigments was known; now, however, the photosynthetic bacterium* Prochloron *has been found to have these characteristics (see page 196). Whether it is a member of the prokaryotic line that gave rise to the chloroplasts of green algae and thus to those of the plants is unknown, but it certainly illustrates the kind of organism that may have been involved.*

The euglenoids (division Euglenophyta) have the same photosynthetic pigments as the green algae and plants, and their chloroplasts are also somewhat similar in ultrastructure. In other respects, however, euglenoids are among the most distinctive of the algal divisions. They are essentially protozoa: food is stored outside the chloroplast, the eyespot is not located in the chloroplast, and there is a flexible proteinaceous pellicle in place of a cell wall. On the basis of these characteristics, it has been suggested that the prokaryote that gave rise to the chloroplasts of green algae also independently gave rise to those of the euglenoids. A more highly favored idea is that the euglenoids acquired the chloroplasts of green algae by ingesting the whole algal cell. Following the incorporation of such cells, there would have been a progressive loss of the other elements of the symbiotic green alga, ultimately leaving only the chloroplast and the plasma membrane. Such a relationship might explain why the chloroplasts of euglenoids are surrounded by three membranes, and those of green algae by two, when they are otherwise remarkably similar.

These few examples illustrate the likelihood that all chloroplasts did not originate in a single symbiotic event. A more reasonable explanation would seem to be that multiple symbiotic events involving diverse photosynthetic prokaryotes, as well as transfers of chloroplasts from one group of eukaryotes to another, have resulted in the evolution of chloroplasts in different algal groups. These algal groups literally have nothing in common except the fact that they are predominantly aquatic photosynthetic eukaryotes.

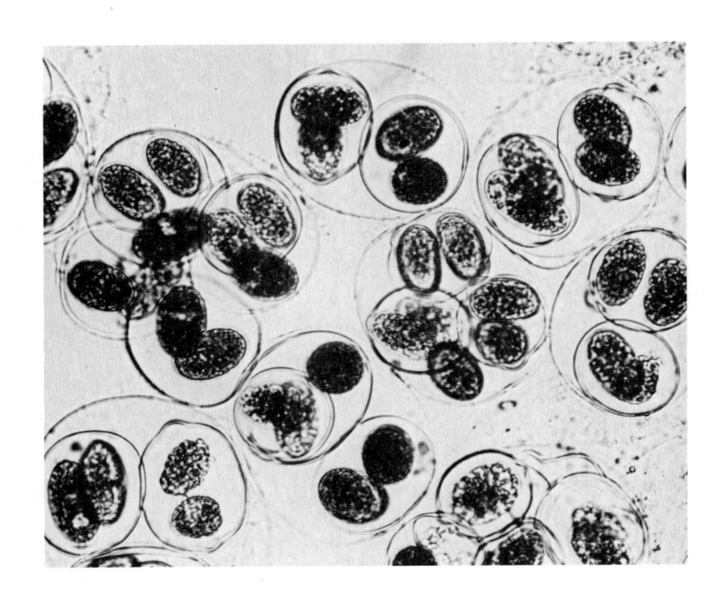

Glaucocystis consists of colorless eukaryotic host cells that contain blue-green "cyanelles," which function as chloroplasts and divide whenever the host cell does. Cyanelles are believed to be cyanobacteria that lost their cell walls and became incapable of independent existence as they adapted to an endosymbiotic life. The existence of *Glaucocystis* and similar forms supports the endosymbiotic theory of the origin of eukaryotic cells and organelles.

*In dinoflagellates and fungi, histone proteins are not combined with the chromosomal DNA, and the evidence for other kinds of proteins is limited.

SUMMARY

The algae are a large and diverse group of some 24,000 species of photosynthetic eukaryotic organisms—grouped here into six divisions—that are viewed as having developed from a series of parallel evolutionary lines. The algae are largely aquatic and are of great importance in marine habitats, playing an ecological role comparable to that of the green plants in land habitats. In the open sea, algae are generally found in the form of plankton. Larger, more complex algae are found along the shore. Many species are found in fresh water, and some occur on land.

The algae, in general, have a relatively simple body composed of one or several cells. Members of four of the six divisions may contain cellulose in their cell walls. Most algae have unicellular reproductive structures and lack vascular tissues. Despite these general similarities, certain biochemical and structural differences among the groups indicate that they are not closely related in an evolutionary sense. Among the principal criteria for grouping the algae are types of pigments, carbohydrate storage products, flagella (if present), and cell wall components.

The various divisions of algae appear to have had their origins in symbiotic relationships between nonphotosynthetic, protozoalike eukaryotic cells and one or more groups of photosynthetic prokaryotes. Members of three of the divisions are multicellular: the Chlorophyta, the Phaeophyta, and the Rhodophyta. The first and third also include unicellular forms. The other three divisions—Chrysophyta, Pyrrophyta, and Euglenophyta—are composed almost entirely of unicellular organisms.

The Chlorophyta (green algae) are the largest and most diverse group and the one from which the plants probably evolved. It is thought that *Chlamydomonas,* a common unicellular form, resembles a primitive, ancestral green alga, except in the presence of a phycoplast associated with cell division. Other green algae, the members of the class Charophyceae, have a form of cell division like that of the plants, involving a phragmoplast. Several evolutionary lines can be traced from a cell that resembles *Chlamydomonas.*

The Phaeophyta (brown algae) include the largest and most complex of the marine algae. In many types the vegetative body is well differentiated into holdfast, stipe (stalk), and blade. Some approach the vascular plants in the complexity of their food-conducting tissues. Members of this division contain chlorophylls *a* and *c* and have abundant quantities of the xanthophyll fucoxanthin, which gives them their olive-green to dark brown color. The sporophyte is usually larger than the gametophyte.

The Rhodophyta (red algae) are a large group particularly common in warmer marine waters. They nearly always grow attached to a substrate, and some grow at great depths (down to 175 meters). They contain phycobilins, which give them their characteristic colors.

The Chıysophyta are important components of freshwater and marine phytoplankton. Most of the known species are unicellular organisms known as diatoms. Diatoms are characterized by fine double silica shells; because of their shells, abundant fossil records of these organisms have been found. A second group is the golden-brown algae (class Chrysophyceae), a number of which are small cells (1 to 3 micrometers) that are found in large quantities in nannoplankton. The Chrysophyta are very important as symbiotic organisms in the sea. A third class—the Xanthophyceae—consists of a few organisms that lack chlorophyll *b*, as well as the accessory pigment fucoxanthin.

The Pyrrophyta (dinoflagellates) are unicellular biflagellates, many of which are marine organisms. The dinoflagellates are characterized by two flagella that beat in different planes, causing the organism to spin. Dinoflagellates often have stiff cellulose walls that can assume bizarre shapes.

Euglenophyta are a small group of organisms, most of which occur in fresh water and are unicellular. They contain chlorophylls *a* and *b* and store carbohydrates in an unusual polysaccharide substance called paramylon. Euglenoids lack a cell wall but have a flexible series of protein strips, which make up the pellicle, inside the plasma membrane. The cells are highly differentiated, containing chloroplasts, a contractile vacuole, and flagella. No sexual reproduction is known. The Euglenophyta show close affinities to nonphotosynthetic protozoan forms.

SUGGESTIONS FOR FURTHER READING

BOLD, HAROLD C., and MICHAEL J. WYNNE: *Introduction to the Algae: Structure and Reproduction,* Prentice-Hall, Inc., Englewood Cliffs, N.J., 1978.

A detailed reference work on the algae which contains a wealth of information on all groups; taxonomically oriented.

BONEY, A. D.: *A Biology of Marine Algae,* Hutchinson and Co., Ltd., London, 1966.*

An excellent treatise on marine algae which is ecologically and physiologically oriented.

CHAPMAN, A.R.O.: *Biology of Seaweeds: Levels of Organization,* University Park Press, Baltimore, Md., 1979.*

A biologically oriented view of the green, brown, and red marine algae; the discussion ranges from cellular ultrastructure and function to community ecology.

DAWSON, ELMER Y.: *Marine Botany*, Holt, Rinehart and Winston, Inc., New York, 1966.

A short, lively text that covers marine bacteria, fungi, and sea grasses, as well as algae.

PICKETT-HEAPS, JEREMY D.: *Green Algae: Structure, Reproduction and Evolution in Selected Genera*, Sinauer Associates, Inc., Sunderland, Mass., 1975.

A magnificently illustrated treatise on the green algae which contains a wealth of new information.

SLEIGH, M.: *The Biology of Protozoa*, University Park Press, Baltimore, Md., 1975.*

A book on the general biology of the protozoa, including chapters on structure, metabolism, reproduction, and ecology, with many excellent illustrations.

TRAINOR, F. R.: *Introductory Phycology*, John Wiley & Sons, Inc., New York, 1978.

An outline of the systematics of algae; clearly written and well illustrated.

*Available in paperback.

CHAPTER 15

The Bryophytes

15–1
Cloud forest near Vera Cruz, Mexico. The trunks and branches of the trees are covered with a thick layer of mosses and liverworts.

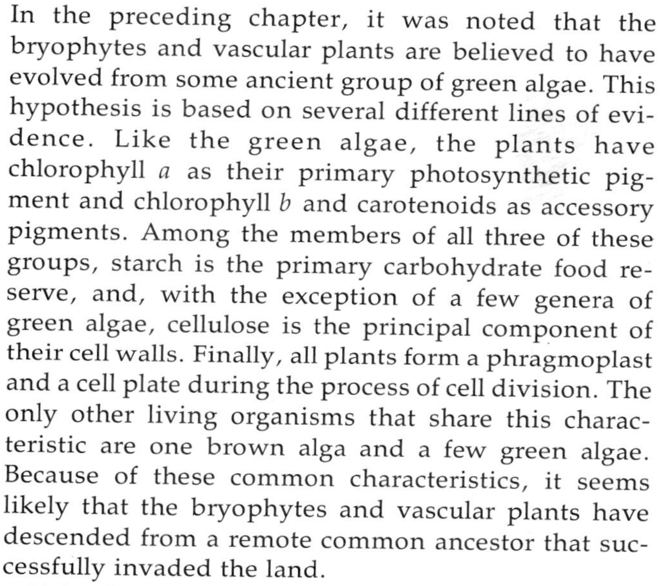

In the preceding chapter, it was noted that the bryophytes and vascular plants are believed to have evolved from some ancient group of green algae. This hypothesis is based on several different lines of evidence. Like the green algae, the plants have chlorophyll *a* as their primary photosynthetic pigment and chlorophyll *b* and carotenoids as accessory pigments. Among the members of all three of these groups, starch is the primary carbohydrate food reserve, and, with the exception of a few genera of green algae, cellulose is the principal component of their cell walls. Finally, all plants form a phragmoplast and a cell plate during the process of cell division. The only other living organisms that share this characteristic are one brown alga and a few green algae. Because of these common characteristics, it seems likely that the bryophytes and vascular plants have descended from a remote common ancestor that successfully invaded the land.

If this hypothesis is correct, the bryophytes and the vascular plants must have diverged long ago. The bryophytes first appear in the fossil record in the Devonian period, more than 350 million years ago; these ancient fossils are quite similar to species living today. Fossil records of vascular plants date from the Silurian period, some 400 million years ago, and so the hypothetical common ancestor of the bryophytes and the vascular plants—a relatively complex green alga—presumably invaded the land at a somewhat earlier time. Apparently, the common ancestor of the bryophytes and vascular plants had a well-developed alternation of generations, for this is the type of life cycle exhibited by all bryophytes and vascular plants. Its gametophytes probably bore multicellular gametangia because the gametangia of all bryophytes and vascular plants are multicellular structures. Being an aquatic organism, its sporophytes would have lacked vascular tissue, although conducting strands of elongate, thin-walled cells may have been present. It was probably endomycorrhizal, with a zygomycete partner, as discussed on page 239.

The organisms that were successful in making the transition from water to land had developed features that prevented them from drying out. One of these was the development of a sterile jacket layer around the sperm-producing and egg-producing cells of the male and female gametangia, called *antheridia* and *archegonia*, respectively. Similarly, a sterile jacket layer was formed around the spore-producing cells of the *sporangia*.

The invasion of the land by the ancestors of plants was also correlated with retention of the zygote within the female gametangium and its development there into an embryo. Thus, during its critical early stages of development, the young embryo, or sporophyte, is protected by the female gametophyte. In the algae, by contrast, the zygote is relatively independent of the female gametophyte.

The aerial parts of most vascular plants are covered with a fatty protective layer—the cuticle—which helps prevent drying out. This cuticle is closely correlated with the presence of stomata—specialized pores that function primarily in the regulation of gas exchange. Many bryophytes apparently lack a cuticle, but the sporophytes of the hornworts and mosses contain stomata. In the hornworts, however, the stomata apparently remain open until late in development, and those of the mosses close only after the moss sporophytes are completely dried up.

All plants are oogamous, and all possess a heteromorphic alternation of generations.

CHARACTERISTICS OF THE BRYOPHYTES

The bryophytes—the liverworts, hornworts, and mosses—are relatively small plants, many less than 2 centimeters long, with most less than 20 centimeters long. They are most common in warm and moist areas, especially in the tropics and subtropics, where a variety of species and a luxuriance of individuals often can be found. Bryophytes, however, are not confined to such tropical habitats. Many species of mosses are found in relatively dry deserts, and several can form extensive masses on dry exposed rocks where very high temperatures can be reached. Mosses sometimes dominate the terrain to the exclusion of other plants over large areas of the far north and far south, as well as on rocky slopes above timberline in the mountains. A significant number of mosses are able to withstand the long periods of severe temperatures on the Antarctic continent. Like the lichens, the bryophytes are remarkably sensitive to air pollution, especially sulfur dioxide, and they are often absent or represented by only a few species in highly polluted areas. A number of bryophytes (both mosses and liverworts) are aquatic, and some are even found on wave-splashed rocks, although none are truly marine. There are some 16,000 species, more than any other group of plants except the flowering plants; a representative moss is shown in Figure 15–2.

Two very important characteristics distinguish the bryophytes from vascular plants. One of these is the absence of vascular tissues—xylem and phloem—in the bryophytes. Accordingly, all bryophytes, strictly speaking, lack true leaves, true stems, and true roots. Nevertheless, the terms "leaf" and "stem" are commonly used when referring to the leaflike and stemlike structures of the gametophytes of leafy liverworts and mosses, and this practice will be followed in this book.

It is pertinent to point out that the stems of the gametophytes of many moss genera contain a central strand of water-conducting cells known as *hydroids*. Hydroids are elongated cells with oblique end walls that lack the specialized wall thickenings characteristic of the tracheids of vascular plants. They also lack a nucleus at maturity and so appear empty. In some genera, food-conducting cells known as *leptoids* surround the strand of hydroids (Figure 15–3), much as phloem surrounds a strand of xylem in some vascular plants. Leptoids are elongate cells that have nuclei and living protoplasts. They closely resemble the least specialized conducting cells found in the phloem of primitive vascular plants. Strands of hydroids may be present in the sporophyte stalks of mosses, but leptoids have been found in the sporophytes of only a few genera.

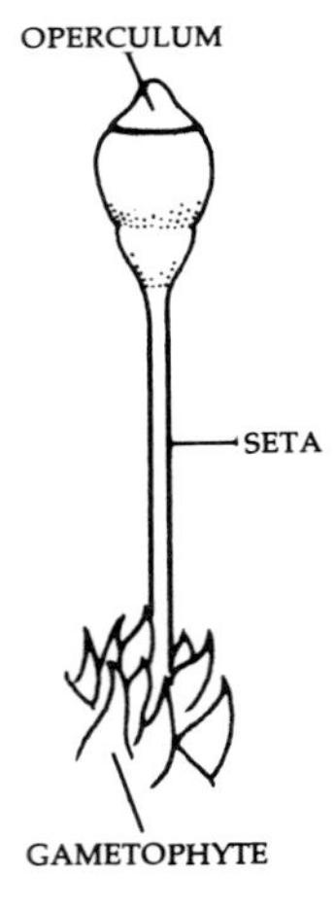

15–2
Urn moss, Physcomitrium turbinatum. *The "urn" at the end of the stalk is the spore capsule. The protective covering—the calyptra—has fallen away from several of the capsules to reveal the operculum, or lid.*

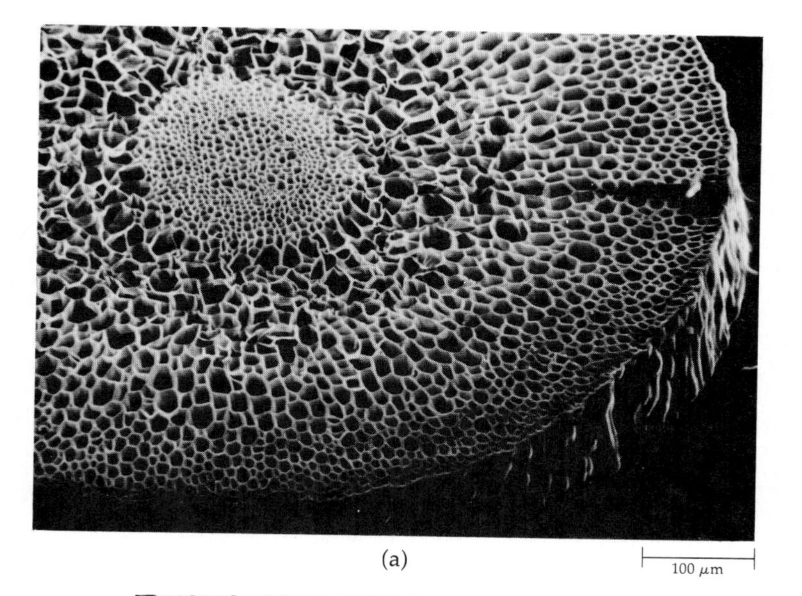

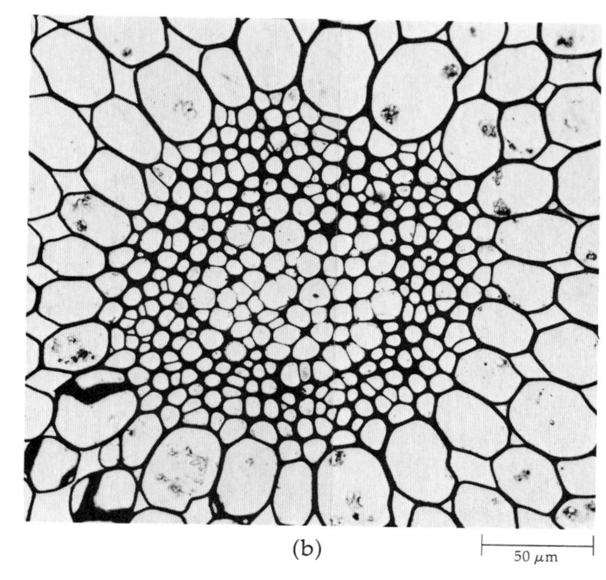

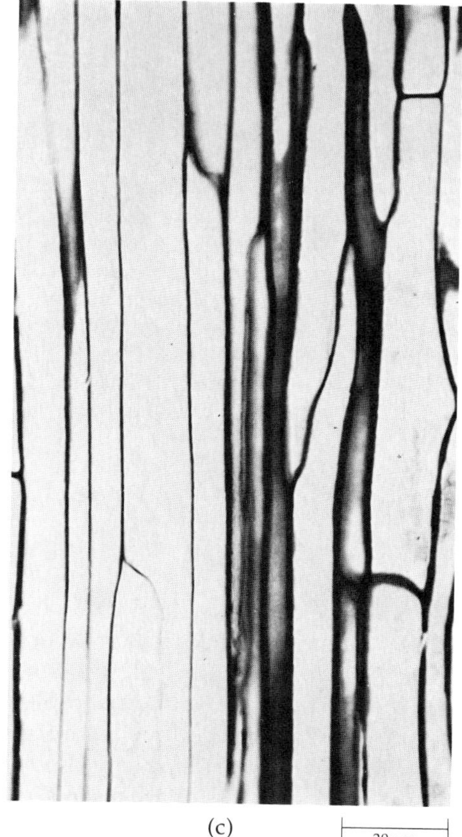

15–3
Conducting strands in the seta of the sporophyte of the moss Dawsonia superba. (a) *General organization of the seta as seen in transection with the scanning electron microscope.* (b) *Transection showing central column of hydroids surrounded by a sheath of leptoids, and parenchyma of the cortex.* (c) *Longitudinal section of a portion of central strand, showing (from left to right) hydroids, leptoids, and parenchyma.*

In most bryophytes, the gametophyte is attached to the substrate by means of elongate single cells or filaments of cells called *rhizoids.* The rhizoids generally serve only to anchor the plants, since absorption of water and inorganic ions commonly occurs directly and rapidly throughout the gametophyte. Rootlike organs are lacking, although the underground stems of a few mosses are structurally quite complex.

The second important distinguishing characteristic of the bryophytes is the nature of their alternation of generations: In bryophytes, the gametophytes are always nutritionally independent, whereas the sporophytes are permanently attached to the gametophytes and vary in their dependence upon them. In other words, the gametophyte is the conspicuous and dominant generation in the bryophytes. Among vascular plants, the sporophyte is the conspicuous and dominant generation (see Figure 10–10c, page 180).

Bryophytes occur in habitats that are moist, at least seasonally. (The bryophytes are sometimes referred to as the amphibians of the plant kingdom.) In order for fertilization to occur in bryophytes, the biflagellate sperm must swim through water to reach the egg inside an archegonium. The archegonium, which may be stalked, is flask-shaped and has a long neck and a basal swollen portion, the *venter,* which encloses a single egg (Figure 15–4a). The central cells of the neck, the *neck canal cells,* disintegrate when the archegonium is mature, leaving a column of fluid through which the sperm swim to the egg. The elongate or spherical antheridium is commonly stalked and consists of a one-cell-thick sterile jacket layer surrounding numerous *spermatogenous cells* (Figure 15–4b). Each spermatogenous cell forms a single biflagellate sperm.

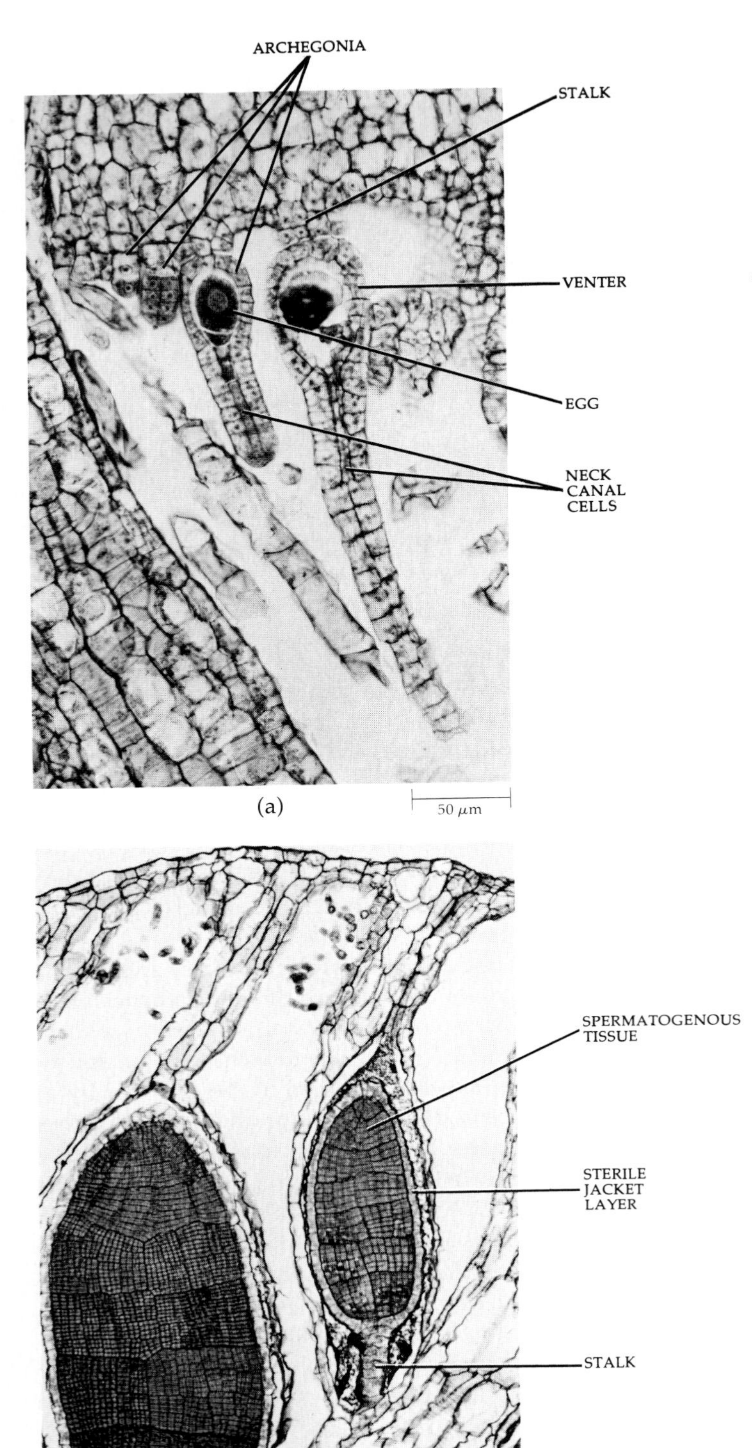

15–4
Gametangia of Marchantia. (a) *Portion of an archegoniophore (the stalk supporting the archegonia) showing several archegonia at different stages of development.* (b) *Portion of antheridiophore showing developing antheridia.*

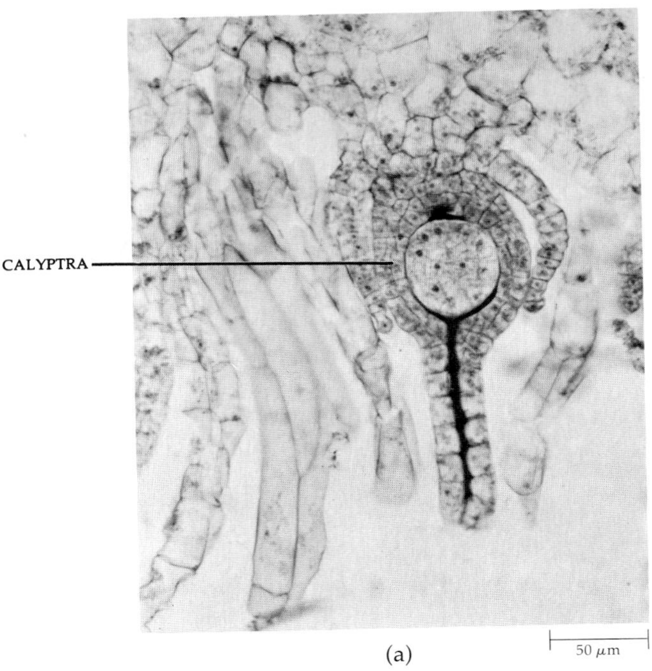

15–5
Marchantia. *Stages in the development of the sporophyte.* (a) *The embryo, or young sporophyte, is an undifferentiated spherical mass of cells within the enlarged venter, or calyptra.* (b) *The foot, seta, and capsule of the maturing sporophyte are now distinct.* (c) *Nearly mature sporophyte. Note in the micrograph the elaters (helical, filamentous structures that assist in the dispersal of spores) within the spore-filled capsule.*

The zygote is retained within the venter of the archegonium, where it develops into an embryo. For a period, the venter undergoes cell division, keeping pace with the growth of the young sporophyte within the archegonium. The enlarged archegonium is called a *calyptra* (Figure 15–5a). At maturity the sporophyte of many bryophytes consists of a *foot*—which remains embedded in the archegonium—a *seta*, or *stalk*, and a *capsule*, or *sporangium* (Figures 15–5b and 15–5c). The setae of moss sporophytes typically contain a central strand consisting entirely of water-conducting tissue, although food-conducting cells are also present in a few genera.

Generally, the cells of the young and maturing sporophyte contain chlorophyll and carry out photosynthesis, but by the time meiosis occurs in the capsule and the spores are produced, the chlorophyll has usually disappeared. In mosses, the calyptra is commonly lifted upward with the capsule as the seta

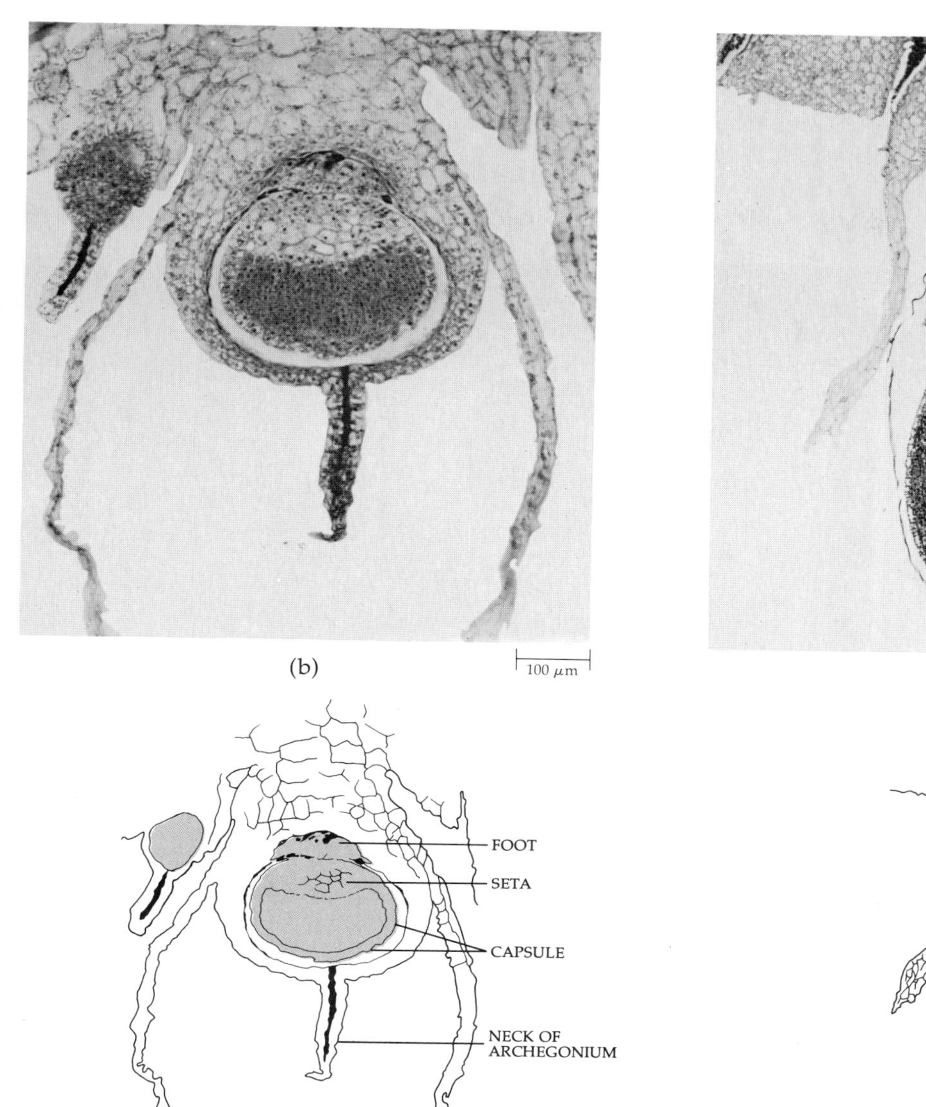

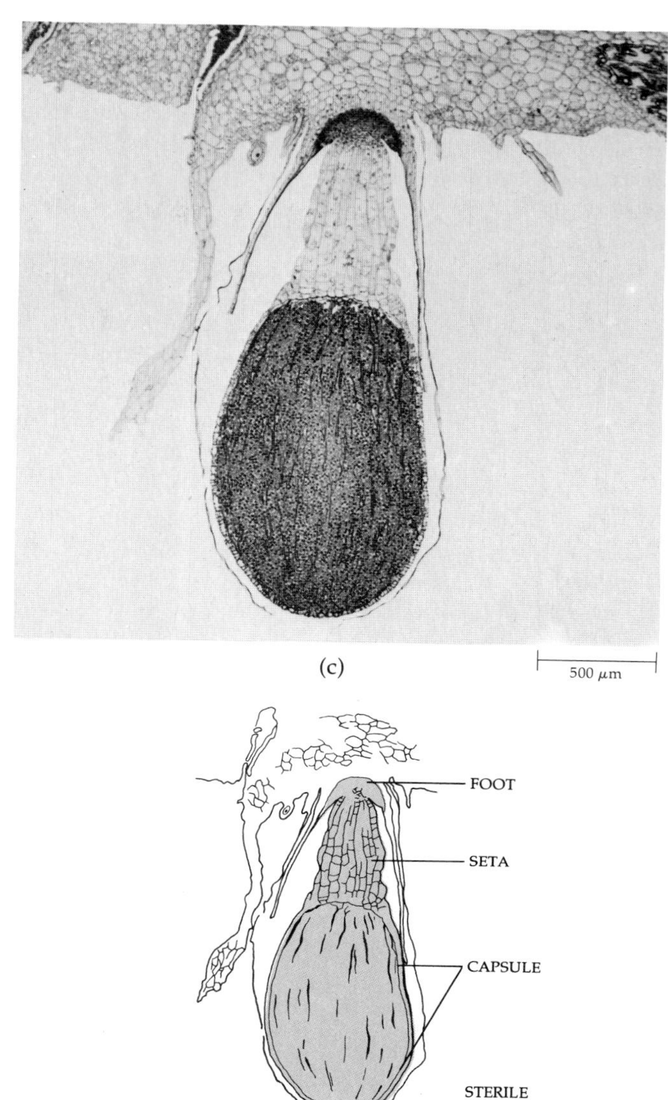

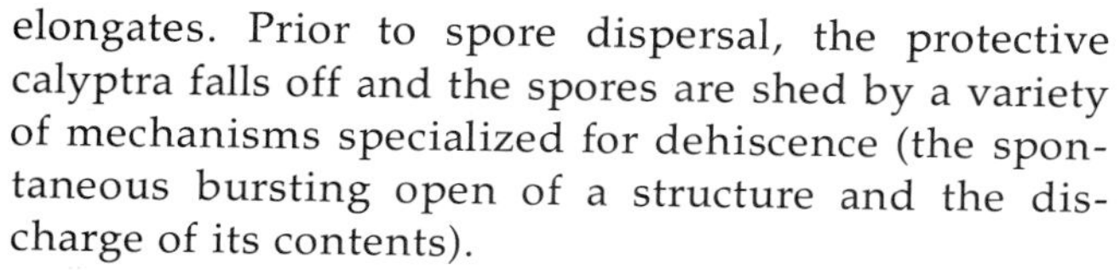

elongates. Prior to spore dispersal, the protective calyptra falls off and the spores are shed by a variety of mechanisms specialized for dehiscence (the spontaneous bursting open of a structure and the discharge of its contents).

The bryophytes are traditionally divided into three classes: Hepaticopsida (the liverworts, 6000 species), Anthocerotopsida (the hornworts, 100 species), and Muscopsida (the mosses, 9500 species).

THE LIVERWORTS: CLASS HEPATICOPSIDA

Liverworts, or hepatics, are small plants that are generally less conspicuous than mosses. Their name dates from the ninth century, when it was thought, because of the liver-shaped outline of the gametophyte in some genera, that these plants might be useful in treating diseases of the liver. According to the medieval "Doctrine of Signatures," the outward appearance of a body signaled the possession of special properties. The ending "-wort" simply means "herb" and so appears as part of many plant names.

The gametophytes of some liverworts are flattened, dorsiventral thalli (undifferentiated plant bodies having distinct upper and lower surfaces), which grow from an apical meristem. The gametophytes of most species, however, are leafy and grow from a single apical cell, which resembles an inverted pyramid with a base and three sides. Daughter cells are cut off from the sides of this single cell. The rhizoids of liverworts are single-celled, unlike those of mosses, which contain several cells each. The gametophytes develop directly from spores. The sporophytes of most liverworts are generally less complex than those of mosses, and their capsules have very different mechanisms for the release of spores, compared to those of the mosses.

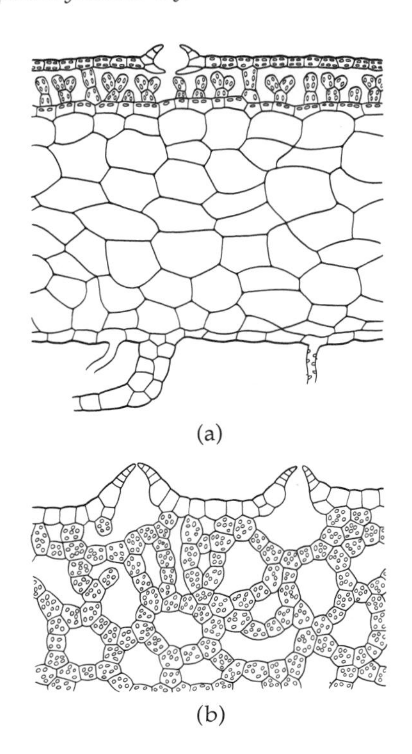

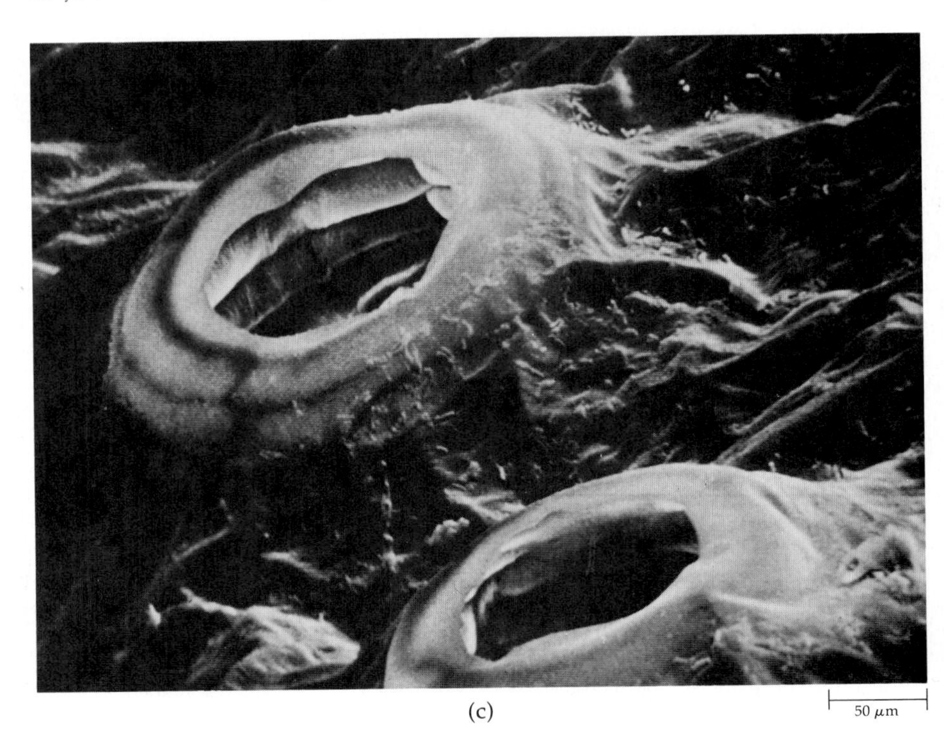

(a) (b) (c)

15–6
Transverse sections of two thallose liverworts, (a) Marchantia polymorpha *and* (b) Reboulia hemisphaerica *(upper portion only). Numerous chloroplast-bearing cells are evident. In* Reboulia, *the photosynthetic layer is thicker than in* Marchantia. *In both there are several layers of colorless cells below the photosynthetic ones, as well as rhizoids that anchor the plant body to the substrate. Pores permit the exchange of gases in the air-filled chambers that honeycomb the photosynthetic dorsal layer, functioning in the same way as the stomata of higher plants. Unlike stomata, however, they do not open and close. Pores in* Marchantia *are shown in the scanning electron micrograph in* (c).

Thallose Liverworts

Thallose liverworts are a diverse group of nonleafy hepatics. They can be found on moist, shaded banks and in other suitable habitats, such as flowerpots in a cool greenhouse. The thallus is many cells thick—about 30 at the midrib and approximately 10 in the thinner portions—and is sharply differentiated into a thin, chlorophyll-rich upper (dorsal) portion and a thicker, colorless lower (ventral) region (Figure 15–6). The lower surface bears two kinds of rhizoids, as well as rows of scales. The upper surface is divided into raised regions, each of which marks the limits of an underlying air chamber and has a large pore that leads to this chamber. In some genera, each chamber has a number of filamentous strands of tissue that are rich in chloroplasts, but the walls and floor of the chamber are colorless. There is also some differentiation among the layers of colorless cells, particular cells being specialized for storage of starch.

On the basis of sporophyte structure, *Riccia* and *Ricciocarpus* are among the simplest of liverworts (Figure 15–7). *Ricciocarpus,* which grows in water or on damp soil, is bisexual—that is, both sex organs arise on the same plant. Some species of *Riccia* are aquatic, although most are terrestrial. *Riccia* gametophytes may be either unisexual or bisexual. The sporophytes are deeply embedded within the dichotomously branched gametophytes of *Riccia* and *Ricciocarpus* and consist of little more than a sporangium. No special mechanism for spore dispersal occurs in these sporophytes. When the portion of the gametophyte containing mature sporophytes dies and decays, the spores are liberated.

One of the most familiar liverworts is *Marchantia,* a fairly widespread, terrestrial genus that grows on moist soil and rocks (Figure 15–8). Its dichotomously branched gametophytes are larger than those of *Riccia* and *Ricciocarpus.* Unlike the latter two genera, in which the sex organs are distributed along the dorsal surface of the thallus, *Marchantia* has its gametangia restricted to specialized erect structures called gametophores. The gametophytes of *Marchantia* are strictly unisexual, and the male and female gametophytes can readily be identified by their gametophores, which are structurally quite distinct from one another. The antheridia are borne on disk-headed stalks called *antheridiophores,* and the archegonia are borne on umbrella-headed stalks called *archegoniophores* (Figure 15–8).

15–7
Gametophytes of Ricciocarpus natans, *"an amphibious liverwort." The system of branching in this liverwort is dichotomous; that is, the main axis and subsequent axes fork repeatedly into two branches. The branching is less prominent in the aquatic form* (a) *than in the terrestrial form* (b).

(a)

(b)

15–8
The thallose liverwort Marchantia. *The antheridia and archegonia are elevated on specialized stalks above the thallus.*

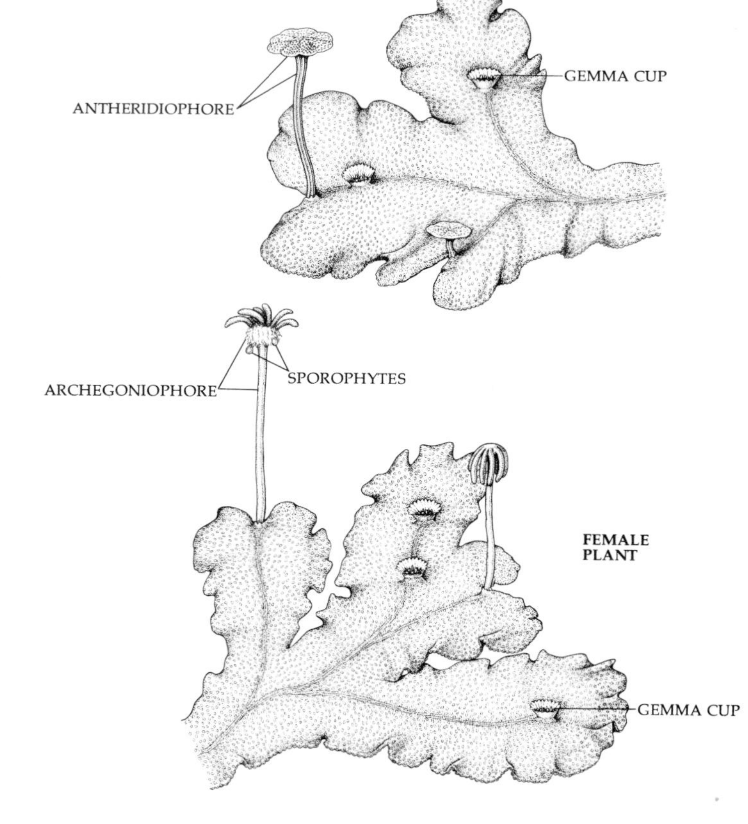

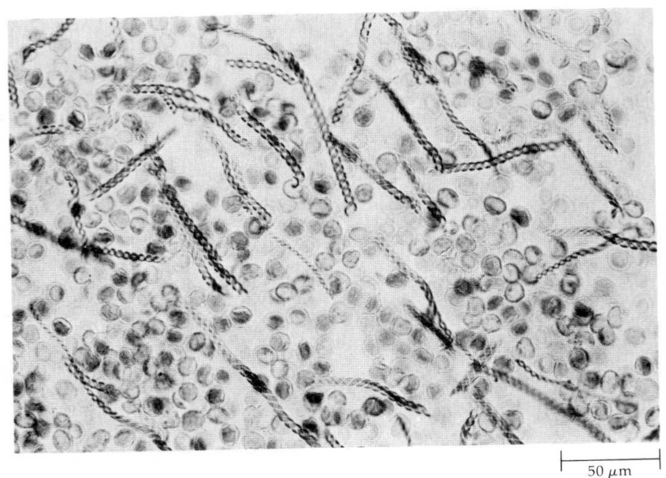

15–9
Mature spores and elaters from a capsule of Marchantia.

The sporophytes of *Marchantia* are more highly differentiated than those of *Riccia* and *Ricciocarpus*, consisting of a foot, a short stalk, or seta, and a capsule, or sporangium (Figure 15–5c). In addition to spores, the mature sporangium contains elongate cells, called *elaters*, which have helically arranged hygroscopic (moisture-absorbing) wall thickenings (Figure 15–9). The walls of these cells are sensitive to very slight changes in humidity, and they produce a twisting action that aids in spore dispersal after the capsule bursts open (dehisces) into a number of petallike segments.

Fragmentation constitutes the principal means of asexual reproduction in the liverworts. Another fairly widespread means of asexual reproduction in the liverworts and mosses is the production of *gemmae*—minute bodies that can give rise to new plants. In *Marchantia*, the gemmae are produced in special cup-

SPORE DISCHARGE IN LIVERWORTS

Three methods of spore discharge in liverworts. Most liverworts are similar to Cephalozia *in their method of spore discharge, whereas some resemble* Marchantia. *The method employed by* Frullania *is less common.*

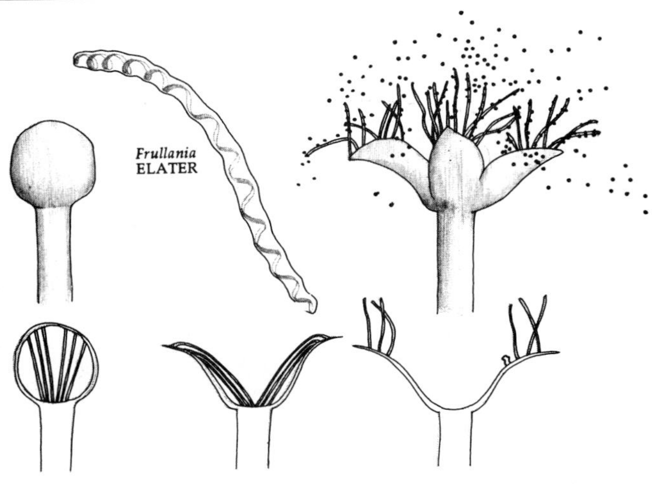

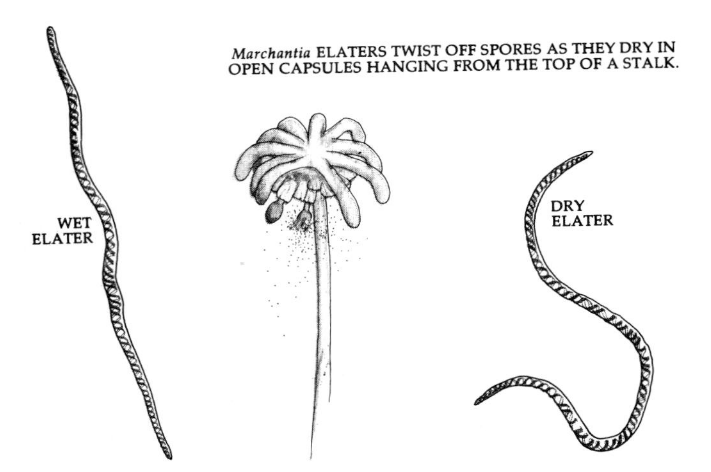

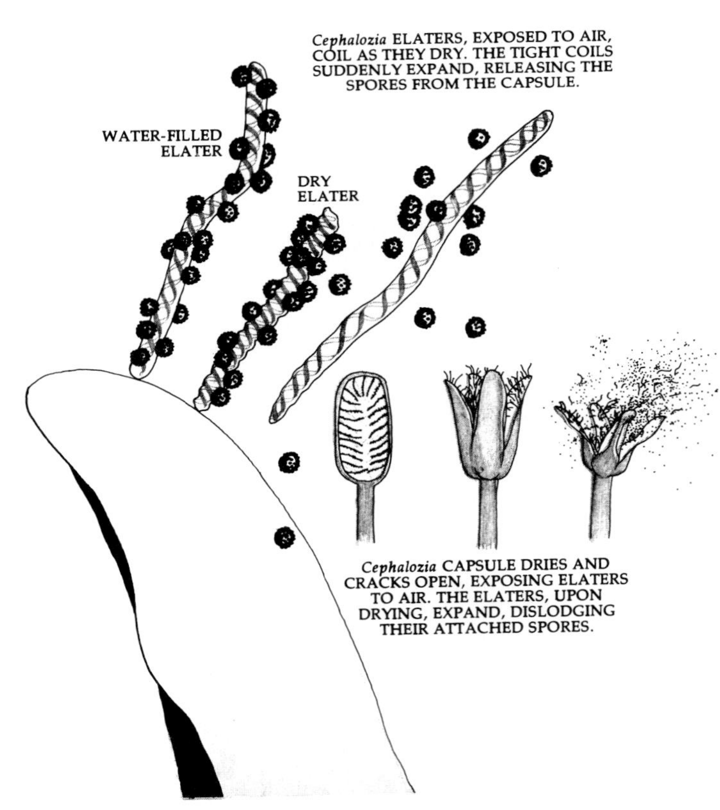

15–10
The liverwort Marchantia, *with cup-shaped gemma cups containing gemmae, which are splashed out by the rain and grow in the vicinity of the parent plant.*

like structures—called gemma cups—located on the dorsal surface of the gametophyte (Figures 15–8 and 15–10). Gemma cups are not produced by *Riccia* and *Ricciocarpus.*

Leafy Liverworts

The leafy liverworts are a diverse group that includes more than two-thirds of the 6000 known species of liverworts (Figure 15–11). The leafy liverworts are especially abundant in the tropics and subtropics, in regions of heavy rainfall or high humidity, but they are also present in large numbers in temperate regions. The plants are usually well branched and form small mats.

The leaves of liverworts, like those of mosses, generally consist of only a single layer of nondifferentiated cells. The leaves of many genera are arranged in two rows, with a third row of reduced leaves along the lower surface of the gametophyte. The leaves are often two-lobed, and each grows by means of two distinct apical growing points. In *Frullania,* a rather common liverwort that grows on bark, the leaves consist of a large, undivided dorsal lobe and a small, helmet-shaped ventral lobe (Figure 15–11c).

In the leafy liverworts, the antheridia generally occur in a packetlike swelling, the *androecium,* which develops on the lower portion of a modified leaf. The developing sporophyte, as well as the archegonia from which it develops, are characteristically surrounded by a tubular sheath, the *perianth* (Figure 15–11c).

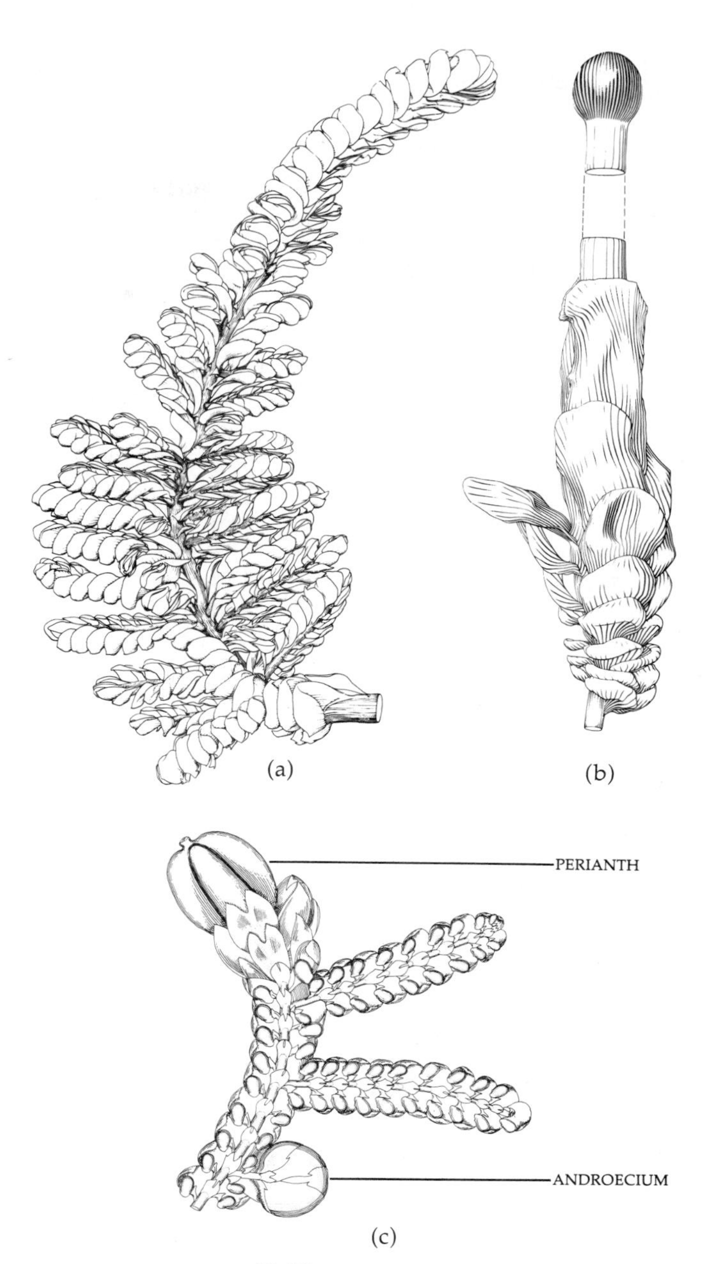

15–11
Leafy liverworts. (a) Clasmatocolea puccioana, *showing the characteristic arrangement of the leaves.* (b) *The end of a branch of* C. humilis. *The capsule and the long stalk of the sporophyte are visible.* (c) *A portion of a branch of* Frullania, *showing the characteristic arrangement of its leaves.*

THE HORNWORTS: CLASS ANTHOCEROTOPSIDA

The class Anthocerotopsida consists of only about 100 species in six genera, the most familiar being *Anthoceros,* which is distributed worldwide and generally occurs in moist, shaded habitats. The gametophyte of *Anthoceros* superficially resembles that of the thallose liverworts (Figure 15–12). However, each cell usually has a single large chloroplast, like those of many algae, rather than the many small, discoid ones found in the cells of other bryophytes and the vascular plants. Each chloroplast contains a pyrenoid, making the resemblance to algae even more striking. The gametophytes have a strong dorsiventral orientation; they are often rosettelike and are usually less than 2 centimeters in diameter. In addition, *Anthoceros* have extensive internal cavities filled with mucilage, rather than with air, as in the gametophytes of thallose liverworts. These mucilage-filled cavities are often inhabited by cyanobacteria of the genus *Nostoc,* which supply nitrogen through nitrogen fixation to their host plants.

15–12
(a) *Gametophyte of* Anthoceros, *a hornwort, showing attached sporophytes.* (b) *Stoma in one of the sporophytes, which are photosynthetic.* (c) *Developing spores and* (d) *mature spores.*

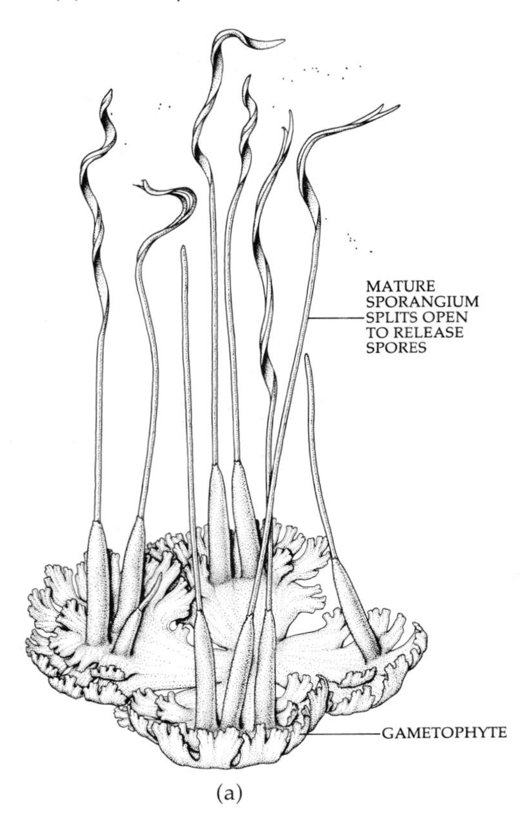

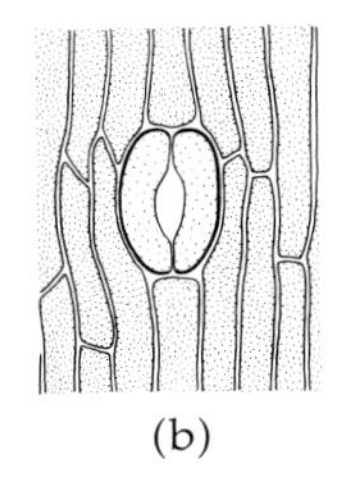

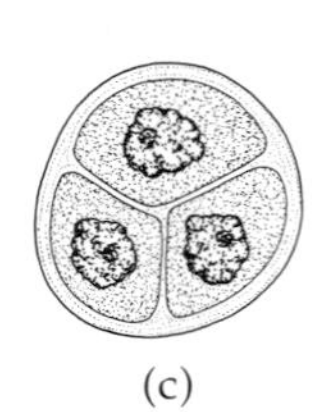

Some species of *Anthoceros* are unisexual and others are bisexual. The antheridia and archegonia are sunken on the dorsal surface of the gametophyte. Numerous sporophytes may develop on the same gametophyte (Figure 15–12).

The sporophyte of *Anthoceros* consists of a foot and a long cylindrical sporangium (Figure 15–13). Very early in its development, a meristem, or zone of actively dividing cells, develops between the foot and sporangium; this meristem is active as long as conditions are favorable for growth. As a result, the sporophyte continues to elongate for a prolonged period of time. Dehiscence of the sporangium begins at its tip and spreads toward its base as the spores mature. The dehiscing sporangium splits longitudinally into ribbonlike halves (see Figure 15–12). The sporophyte contains several layers of photosynthetic cells. Its surface is covered with a cuticle and contains stomata.

THE MOSSES: CLASS MUSCOPSIDA

Many groups of plants contain members that are commonly called "mosses" (reindeer "mosses" are lichens, club "mosses" and Spanish "moss" are vascular plants, and sea "moss" and Irish "moss" are algae), but the genuine mosses are members of the class Muscopsida, which consists of three subclasses: Bryidae (the "true" mosses), Sphagnidae (the peat mosses), and Andreaeidae (the granite mosses).

The True Mosses

The gametophytes of all mosses are represented by two distinct phases: the *protonema* (plural: protonemata, meaning "first threads"), which arises directly from a germinating spore, and the *leafy gametophyte.* In the true mosses the protonema is a uniseriate (the cells occurring in a single layer), branching filament, which superficially resembles a filamentous green alga (Figure 15–14). Some branches of the protonema penetrate the substratum and become colorless rhizoids, whereas others give rise to minute budlike structures that develop into the leafy gametophytes (Figure 15–15). Protonemata are likewise found in some liverworts.

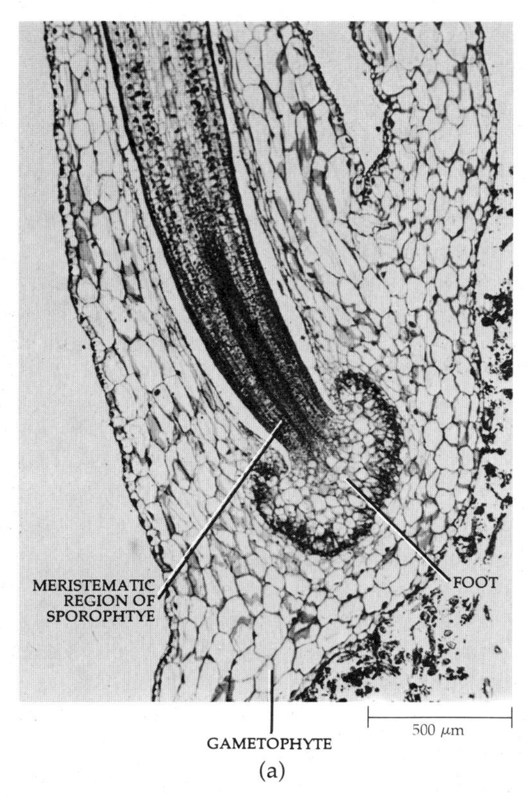

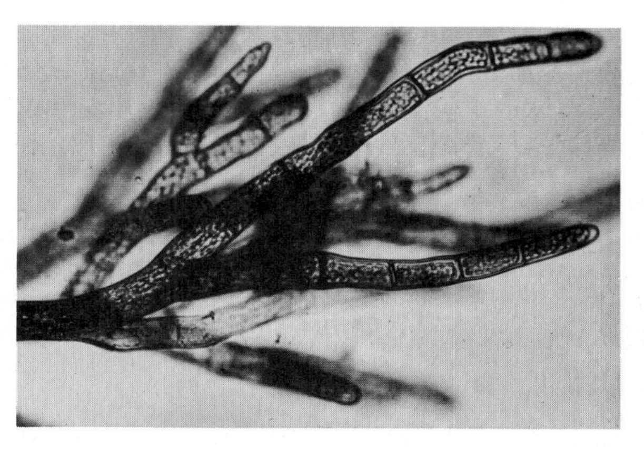

15–14
Protonema of a moss. Protonemata, or young gametophytes, are characteristic of mosses and some liverworts. They often resemble filamentous green algae.

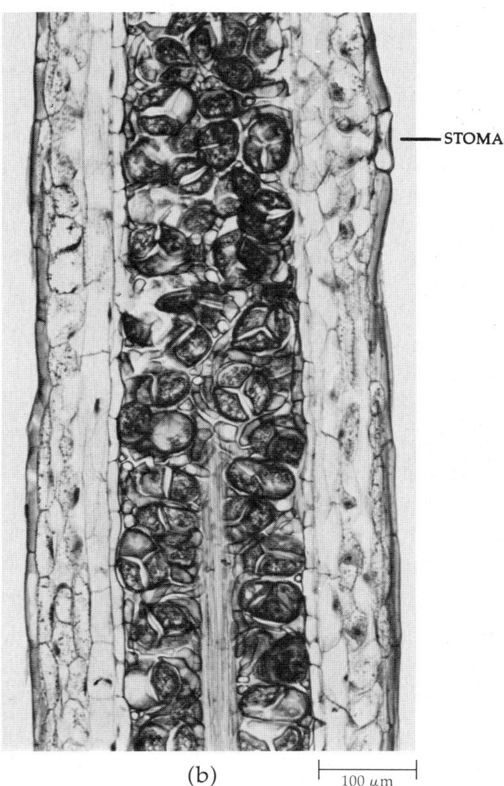

15–13
(a) *Longitudinal section of the sporophyte of* Anthoceros, *showing its foot embedded in tissue of the gametophyte.* (b) *Longitudinal section of the sporangium, showing tetrads of spores.*

15–15
Leafy male gametophytes of the moss Polytrichum piliferum, *showing the mature antheridial heads.*

In the true mosses, the gametophyte is leafy and usually upright, rather than being dorsiventrally flattened, as it is in the leafy liverworts. It grows from an apical initial cell that is similar to that of the leafy liverworts, that is, like an inverted pyramid with three sides. Although three ranks of leaves are produced initially, subsequent twisting of the axis results in displacement of these ranks to give the appearance of a spiral leaf arrangement. In some genera (*Fontinalis,* an aquatic moss, for example) the three-ranked condition of the leaves is obvious in the mature gametophytes.

The gametophytes of mosses range from a few millimeters to 50 centimeters or more in length, and they exhibit varying degrees of differentiation and complexity. All have multicellular rhizoids, and the leaves are normally only one cell layer thick except at the midrib (which is lacking in some genera). In some mosses, such as *Dawsonia superba* (see Figure 15–3), and the common *Polytrichum,* there is often a central strand of elongate cells in the stem that may function in conducting water, but many other genera lack such specialized tissues.

Two patterns of growth are common among the gametophytes of mosses (Figure 15–16). In the first, the "cushiony" mosses, the gametophytes are erect and little-branched, usually bearing terminal sporophytes. In the second, the gametophytes are much-branched and "feathery," and the plants are creeping; the sporophytes are borne laterally. This second type of growth pattern is found in those mosses that hang in lush masses from the branches of trees in moist regions. In some mosses the protonema is persistent and assumes the major photosynthetic role, while the leafy shoots of the gametophyte are minute.

At maturity, gametangia are produced by most leafy gametophytes, either at the tip of the main axis or on a lateral branch. In some genera the gametophytes are unisexual (Figure 15–17), whereas in others both archegonia and antheridia are produced by the same plant.

Sporophytes in the mosses are often small in relation to the gametophytes (Figure 15–18). The capsules are generally elevated on a seta, which may exceptionally reach 15 to 20 centimeters in length, although some lack a stalk entirely. A short foot at the base of the seta is embedded in the tissue of the gametophyte. In mosses, the seta usually elongates early in the development of the sporophyte (in contrast to the situation in liverworts), and the sporophyte is important in photosynthesis. The sporophyte of a moss is therefore much less nutritionally dependent upon the gametophyte than is the sporophyte of a liverwort. Stomata, which are lacking on the sporophytes of liverworts, are normally present on the sporophytes of mosses. Moss sporophytes have a high degree of internal organization, and a central strand of elongate cells is normally present in the seta of many genera.

When the sporophyte of a moss is mature, it gradually loses its ability to photosynthesize and turns yellow, then orange, and then brown. Eventually the lid, or *operculum,* of the capsule bursts off, revealing a ring of teeth—the *peristome*—surrounding the opening (Figure 15–19). The peristome plays a role in regulating spore discharge (Figure 15–20). The peristome is a characteristic of the subclass Bryidae that is lacking in the other two subclasses. A representative, general life cycle of a "true" moss is shown in Figure 15–21.

(a)

(b)

15–16
Two common growth forms found in moss gametophytes. (a) *"Cushiony" form, in which the gametophytes are erect and have few branches, as in* Polytrichum juniperinum. *Spore capsules atop long, slender seta—sporophytes—can be seen rising above the gametophytes.* (b) *"Feathery" form, with matted, creeping gametophytes, as in* Thuidium delicatulum.

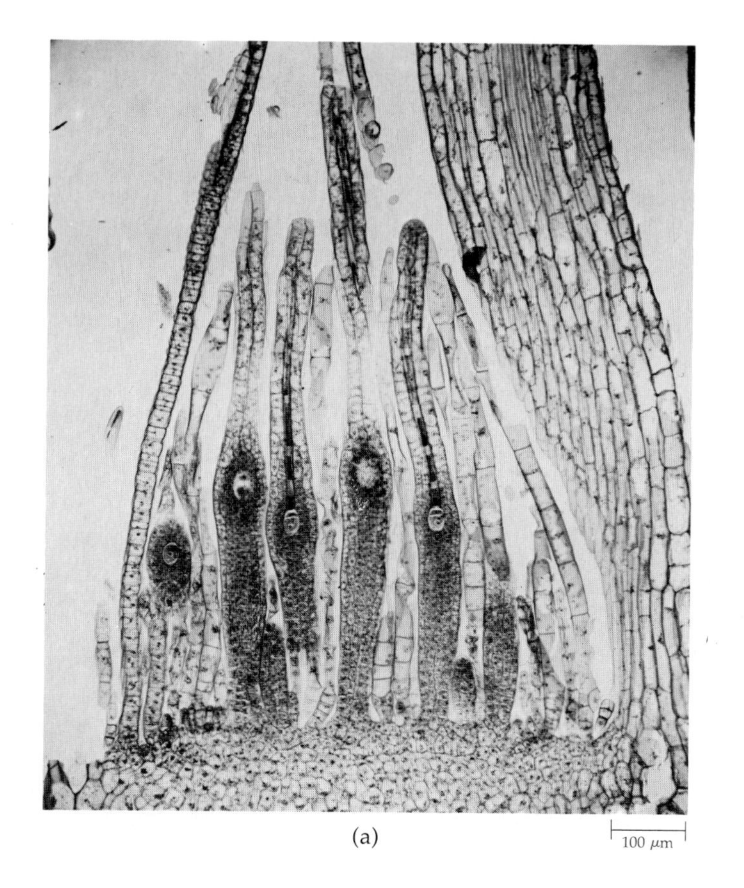

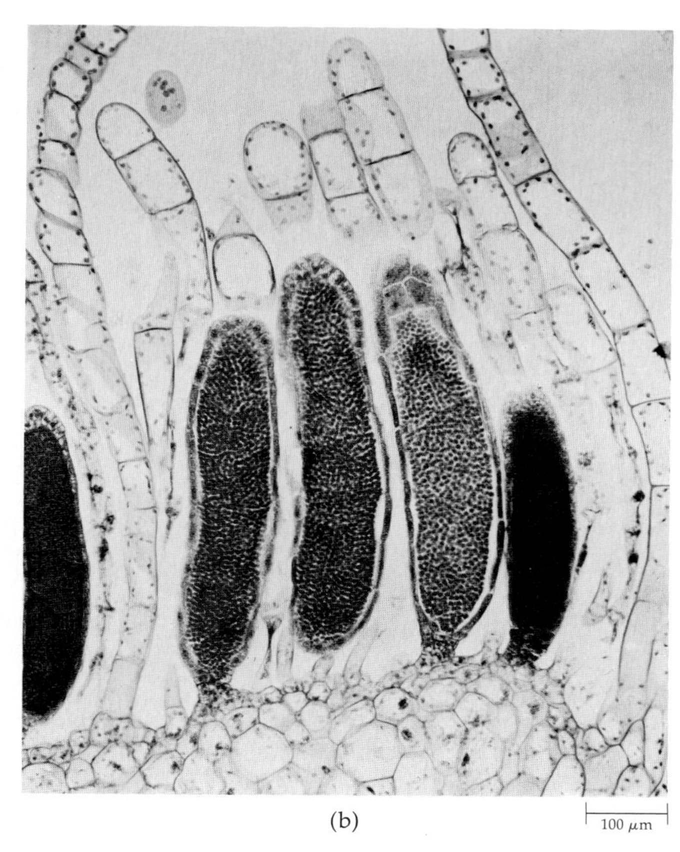

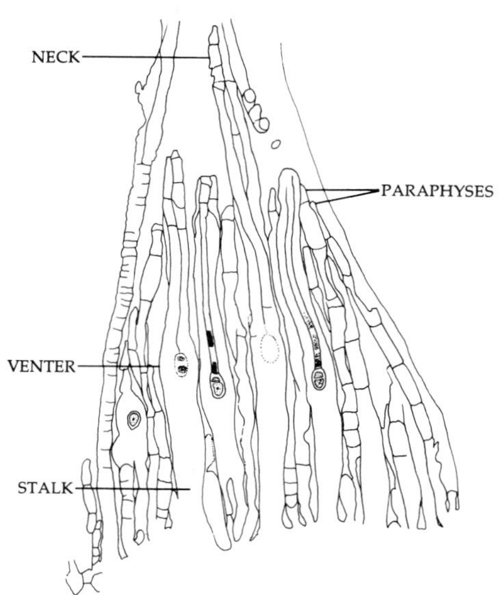

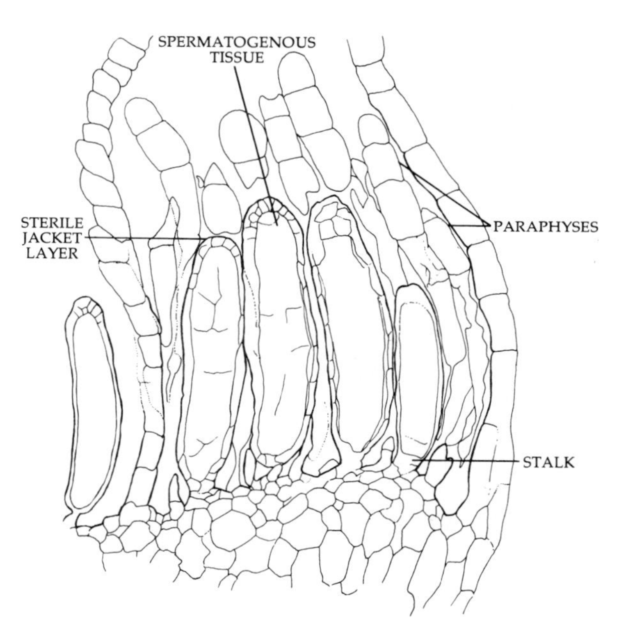

15–17
Gametangia of Mnium, *a moss.* (a) *Longitudinal section through archegonial head showing archegonia surrounded by sterile structures called paraphyses.* (b) *Longitudinal section through antheridial head showing antheridia surrounded by paraphyses.*

(a)

(b)

(c)

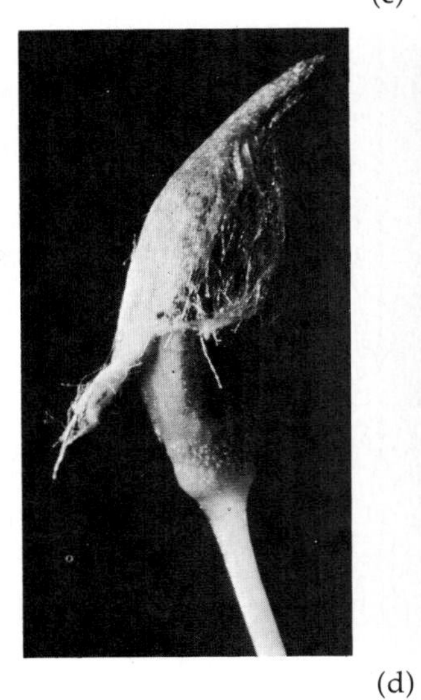

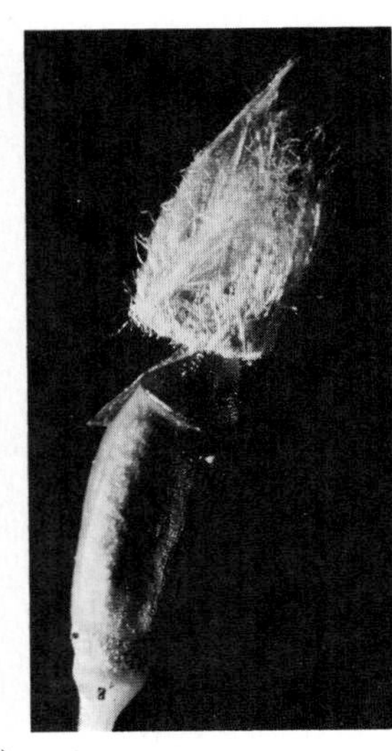

(d)

15–18
Sporophytes of mosses. (a) Ptilium crista-castrensis, *a "feathery" moss that often carpets the boreal coniferous forests. The female plant on the left bears two capsules elevated on long, twisted setae.* (b) Sphagnum, a *genus confined chiefly to waterlogged habitats. Many of the capsules, such as the two on the left, have burst.* (c) *Spore-bearing setae of the hairy moss,* Pogonatum brachyphyllum. (d) *Seta and capsule of the sporophyte of a moss, with the calyptra (the enlarged archegonium) partially removed on the left, and totally removed on the right, revealing the lid, or operculum, of the capsule.*

15–19
Scanning electron micrograph of part of the capsule opening of a moss, Bryum capillare, *showing 9 of the 16 outer peristome teeth.*

15–20
Methods of spore discharge in mosses. (a) Brachythecium *has a peristome consisting of two rings of teeth, which open to release the spores in response to changes in moisture. The outer set of peristomal teeth interlock with the inner set under damp conditions. As the capsule dries out, the outer teeth pull away, allowing the dispersal of spores by the wind.*
(b) *As the* Sphagnum *capsule dries, the internal tissues are replaced by gases; the capsule shrinks and eventually bursts, releasing a gaseous cloud of spores.*
(c) *The* Andreaea *sporangium contracts and splits as it dries out, allowing the spores to fall out. Most mosses are similar to* Brachythecium *in the discharge of spores.*

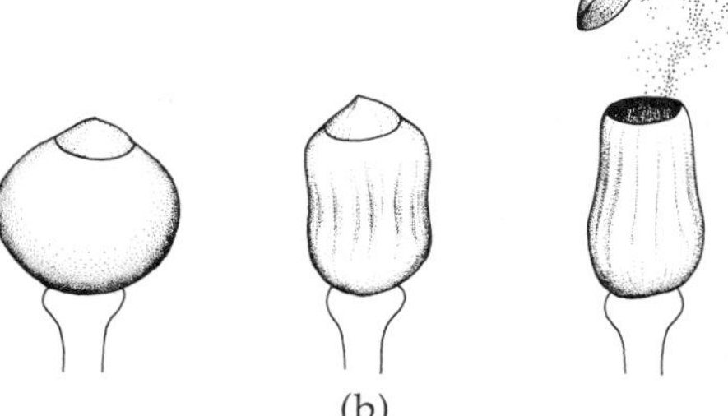

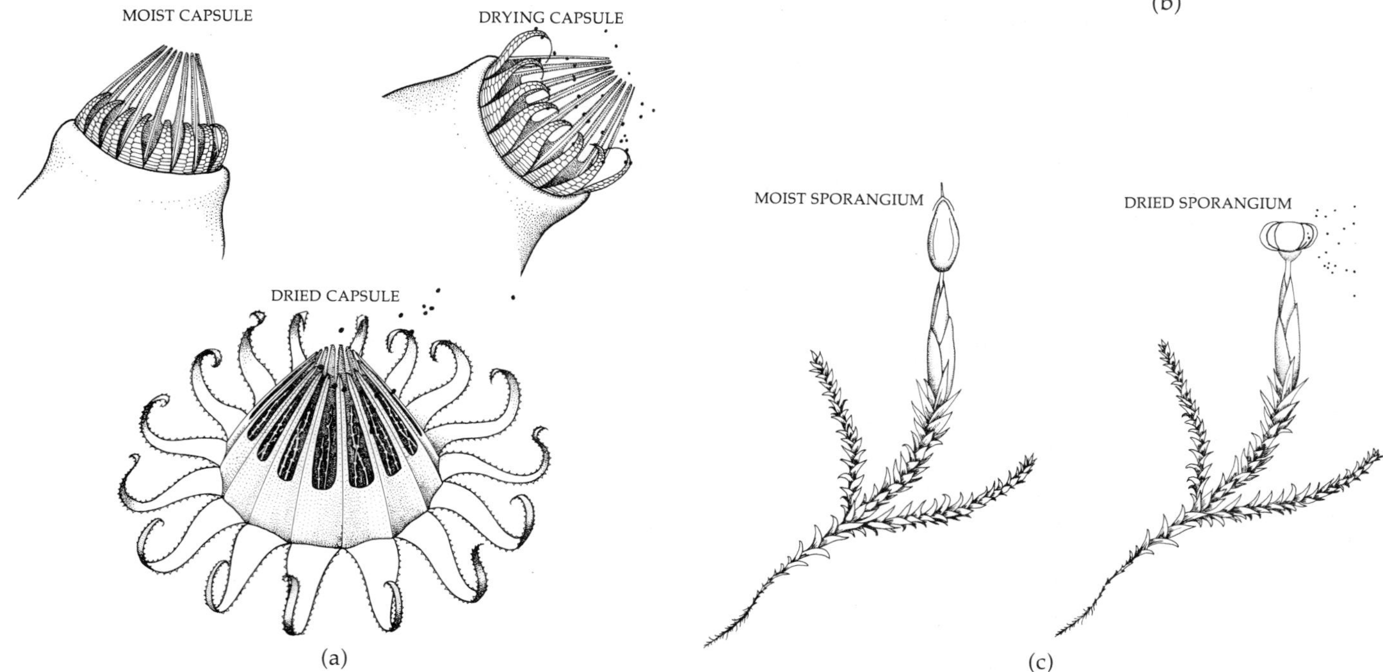

15–21
In a representative moss life cycle, spores are released from a capsule, which opens when a small lid (operculum) is lost. The haploid spore germinates to form a branched, filamentous protonema, from which a leafy gametophyte develops. Sperm, which are released by the mature antheridium, are attracted into the neck canal of an archegonium, where one fuses with the egg cell to produce the zygote. The zygote divides mitotically to form the sporophyte, and at the same time, the venter of the archegonium divides to form the protective calyptra. The sporophyte consists of a capsule, which may be raised on a seta, also part of the sporophyte, and a foot. Meiosis occurs within the capsule, resulting in the formation of haploid spores.

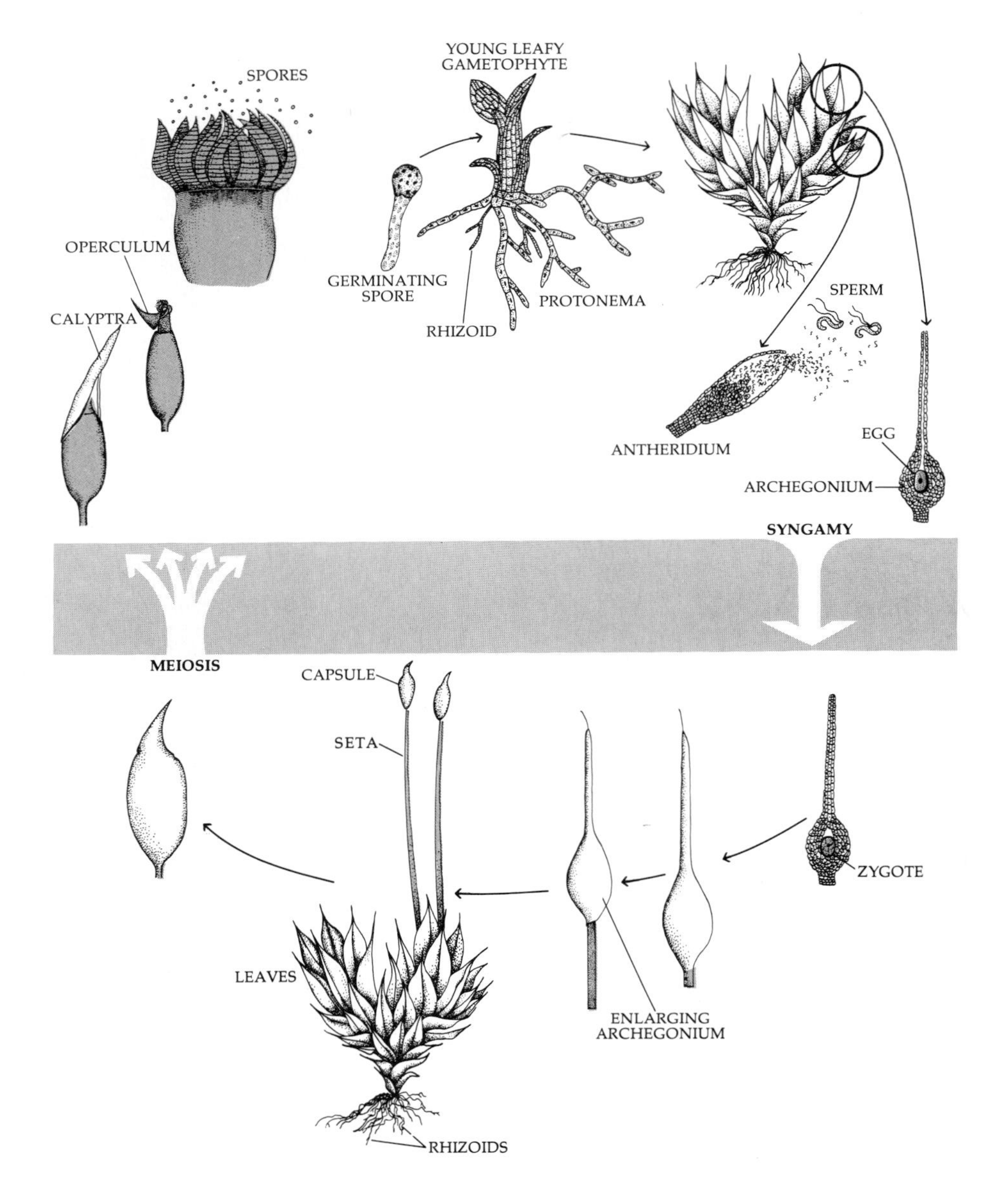

Asexual reproduction in the true mosses is accomplished largely by fragmentation. Virtually any part of the gametophyte is capable of regeneration, including the sterile parts of the sex organs. Many species produce gemmae, which give rise to new gametophytes.

The Peat Mosses

The more than 300 species of peat mosses in the genus *Sphagnum* form a very distinct group that doubtless has been separated from the main line of moss evolution for a long time. The stems of the gametophyte of *Sphagnum* bear clusters of branches, often five at a node, which are more densely tufted near the apex of the stem. The gametophytes of *Sphagnum* arise from a protonema that is platelike instead of filamentous. The leaves in *Sphagnum* lack midribs, and the mature plants lack rhizoids. In the boggy places where they grow, plants of *Sphagnum* are nearly always turgid and thus erect. In the leaves, there is considerable cellular differentiation: there are large dead cells with many pores, which allow for the direct entry of water, and between them there are rows of small, thick-walled living cells with many chloroplasts. Because of the presence of anthocyanins, plants of *Sphagnum* are often tinged with red or purple.

The sporophytes of *Sphagnum* (see Figure 15–18b) are also distinctive. The capsules are nearly spherical and are raised on a seta, the *pseudopodium,* that is part of the gametophyte; the sporophyte itself has a very short stalk. Spore discharge in *Sphagnum* is spectacular. As the capsule matures, its internal tissues shrink, leaving a gas-filled space. Contraction of the maturing capsule results in increased internal pressure, which eventually results in the operculum being blown off and the escaping gas carries a cloud of spores out of the capsule (see Figure 15–20). The most distinctive features of the subclass Sphagnidae are the lack of a peristome and the peculiar morphology of the gametophyte.

Ecology of Sphagnum

The peat mosses play an important ecological role in cold and temperate regions all over the world in that they form extensive peat bogs. The mosses contribute to the acidity of their own environment by selective ion absorption, and the pH (4.4) in the center of the tufts is often much lower than that of the surrounding soil and water (6.0). Because of its superior absorptive qualities and high acidity, peat moss is mixed with soil by gardeners to increase its water-holding capacity and to lower the soil pH. In Ireland and some other northern regions, dried peat is widely used as fuel, as well as for domestic heating.

The Granite Mosses

The genus *Andreaea* consists of approximately 50 species of small, blackish-green or olive-brown tufted rock mosses, which in their own way are as peculiar as *Sphagnum.* Although the gametophyte closely resembles that of the true mosses, it arises from a protonema that is platelike instead of filamentous. The sporophyte lacks a true seta, and it is elevated above the leaves on a stalk of gametophytic tissue. The minute capsules of *Andreaea* are marked by four vertical lines of weaker cells along which the capsule splits. The capsule remains intact above and below the dehiscence lines. The resulting four valves are very sensitive to the humidity of the surrounding air, opening widely when dry and closing when moist. This mechanism of spore discharge is different from that of any other moss (see Figure 15–20).

Andreaeobryum, a second genus of granite mosses, was discovered in Alaska in 1976. It differs from *Andreaea* in that its sporophyte has a true seta and its capsule splits to the apex.

SUMMARY

The Bryophyta are a very distinct division of plants divided into three classes: the Hepaticopsida (liverworts), Anthocerotopsida (hornworts), and Muscopsida (mosses). In their pigments and food reserves, they resemble the green algae and the vascular plants. Some of the hornworts and most of the mosses have stomata on their sporophytes like the vascular plants, and some mosses and liverworts have a cuticle on their leaves. Although most bryophytes lack specialized vascular strands, absorbing water directly through the leaves and stems, they are believed to share a remote common ancestor with the vascular plants.

All bryophytes have a well-defined heteromorphic alternation of generations in which the gametophyte is the dominant generation. The male sex organs, antheridia, and female sex organs, archegonia, are multicellular and have sterile jacket layers. Each archegonium contains a single egg. Numerous sperm are produced by each antheridium; the sperm are free-swimming and biflagellated. The sporophyte develops within the archegonium but usually grows out of it, although remaining permanently attached to the gametophyte. In most bryophytes, the sporophyte is differentiated into a foot, a seta (stalk), and a capsule (sporangium). The sporophyte of the hornworts is unique in the presence of a basal meristem, which adds tissue to the sporangium over a prolonged period. The sporophyte of the hornworts is nutritionally quite independent of the gametophyte; the same is also true in mosses. In liverworts, the sporophyte is nutritionally more dependent, sometimes remaining enclosed by gametophyte tissue at maturity. Upon germinating, the spores of mosses produce a filamentous or platelike gametophyte known as a protonema, from which the leafy gametophytes arise. Protonemata are found in some liverworts but are lacking in the hornworts.

SUGGESTIONS FOR FURTHER READING

CONARD, HENRY S., and PAUL L. REDFEARN, JR.: *How to Know the Mosses and Liverworts,* 2nd ed., William C. Brown Co., Dubuque, Iowa, 1979.*

A good beginner's guide for use in identifying many of the more common bryophytes. It includes a profusely illustrated key and a good glossary.

DOYLE, W. T.: *The Biology of Higher Cryptogams,* The Macmillan Company, New York, 1970.*

A concise account of the biology of the bryophytes and lower vascular plants with an emphasis on developmental and evolutionary relationships.

HÉBANT, CHARLES: *The Conducting Tissues of Bryophytes,* International Scholarly Book Services, Inc., Forest Grove, Oreg., 1978.

A comprehensive and well-illustrated review, summarizing the main results of research performed on the conducting tissues of bryophytes.

WATSON, ERIC V.: *The Structure and Life of Bryophytes,* 3rd ed., Humanities Press, Inc., Atlantic Highlands, N.J., 1971.*

A concise and well-written treatment of the entire group.

*Available in paperback.

CHAPTER 16

The Vascular Plants: An Introduction

16–1
Reconstruction of a Carboniferous swamp forest. Most of the tall treelike plants shown here belong to genera that are now extinct. A number of smaller herbaceous plants are related to modern species of Lycopodium *and* Selaginella.

The vascular plants, like all living things, had aquatic ancestors, and the story of their evolution is inseparably linked with their progressive occupation of the land. One of the key events in their early invasion of the land was the development of spores with durable protective walls that prevented them from drying out. This permitted the dispersal of the spores over the surface of the land by wind. Another important evolutionary advance was marked by the origin of cutin, a fatty substance that protects the plant body from excessive evaporation. The evolution of efficient conducting systems, consisting of xylem and phloem, solved the problem of water and food conduction on the land. Eventually, the below-ground parts of the sporophyte evolved into roots that function in absorption and anchorage, and portions of the above-ground parts evolved into leaves, which are the primary sites of photosynthesis. Meanwhile, the gametophytic generation underwent a progressive reduction in size and became increasingly more protected by and dependent upon the sporophyte. Finally, a number of lines developed seeds—structures that protect the embryonic sporophyte from the rigors of a life on land, nourish it, and enable it to withstand unfavorable conditions.

As discussed in Chapter 15, the common ancestor of the bryophytes and vascular plants was probably a relatively complex, multicellular green alga that invaded the land more than 400 million years ago, probably as part of an endomycorrhizal relationship (see page 239). This green alga presumably had a life cycle in which the sporophyte was the dominant generation.

Because of their adaptations for existence on land, the vascular plants are the dominant land plants of the biosphere. There are nine divisions with living representatives, including about 250,000 living species. In addition, there are several divisions that consist entirely of extinct vascular plants. In this chapter, we shall discuss three divisions of extinct forms and describe some of the evolutionary advances of the vascular plants as a whole. In Chapters 17 and 18, the divisions with living representatives will be presented.

EARLY VASCULAR PLANTS

Division Rhyniophyta

The earliest known vascular plants belong to the division Rhyniophyta, a group that dates back to some 400 million years ago, in the Silurian period. They were seedless plants, consisting of simple, dichotomously branching axes and terminal sporangia. Their plant bodies were not differentiated into roots, stems, or leaves, and their sporangia produced only one type of spore.

Cooksonia, a member of the Rhyniophyta believed to have inhabited mud flats, has the distinction of being the oldest known vascular plant (Figure 16–2). Its leafless aerial stems branched dichotomously and terminated in globose (spherical) sporangia that produced spores with protective walls permeated with cutin. Although nothing is known about its basal parts, it is likely that *Cooksonia* possessed a rhizome, or underground stem, that gave rise to the aerial branches. Tracheids, the type of water-conducting cells typical of primitive vascular plants, have been identified in macerated bits of the axis.

The best-known representative of the Rhyniophyta is *Rhynia* (Figure 16–3a). Probably a marsh plant, its leafless, dichotomously branching aerial stems were attached to a rhizome with tufts of water-absorbing rhizoids. The aerial stems of *Rhynia* were covered with a cuticle, contained stomata, and served as the principal photosynthetic organs.

The internal structure of *Rhynia* was similar to that of many of today's vascular plants. A single layer of superficial cells—the epidermis—surrounded the photosynthetic tissue of the cortex, and the center of the axis consisted of a solid strand of xylem surrounded by one or two layers of cells, which may or may not have been phloem cells. Apparently, the first xylem cells to mature occupied a central portion of the strand, while the last to mature were peripheral.

Division Zosterophyllophyta

During the early and middle Devonian period, from about 395 to about 350 million years ago, a second group of vascular plants, the Zosterophyllophyta, appeared. Like the Rhyniophyta, these plants were leafless and dichotomously branched. It is possible that the group was aquatic. The aerial stems were covered with a cuticle, but only the upper ones contained stomata, indicating that the lower branches may have been embedded in mud. In *Zosterophyllum*, the lower branches frequently produced lateral branches that forked into two axes, one that grew upward, the other downward (Figure 16–3b). The downward-growing branches may have permitted the plant to spread outwardly from the center by providing support.

Unlike the Rhyniophyta, the sporangia of the Zosterophyllophyta were borne laterally on short stalks. Only one type of spore was produced. Although the internal structure of the Zosterophyllophyta was essentially similar to that of the Rhyniophyta, the first xylem cells to mature in the Zosterophyllophyta were located around the periphery of the xylem strand, and the last to mature were located in the center.

The Zosterophyllophyta are considered by some paleobotanists as the likely progenitors of the lycopods (Chapter 17), whose sporangia were at first borne laterally and whose xylem also differentiated centripetally (from the outside of the strand to the center). Regardless of whether or not the lycopods evolved from the Zosterophyllophyta, the two groups have features in common that set them apart from the Rhyniophyta and Trimerophyta.

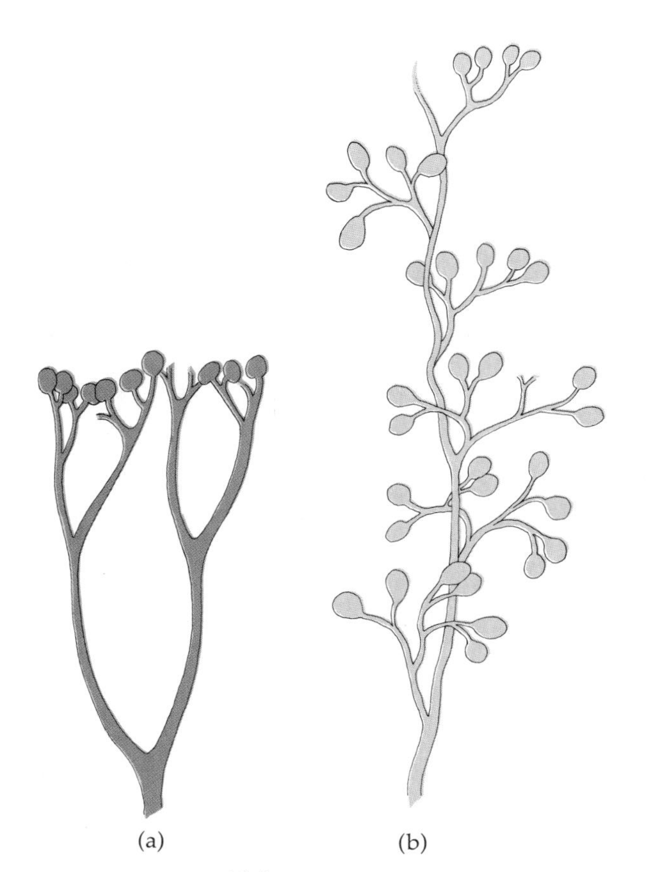

16–2
Reconstruction of the oldest known vascular plants. (a) Cooksonia caledonica. (b) Cooksonia hemisphaerica.

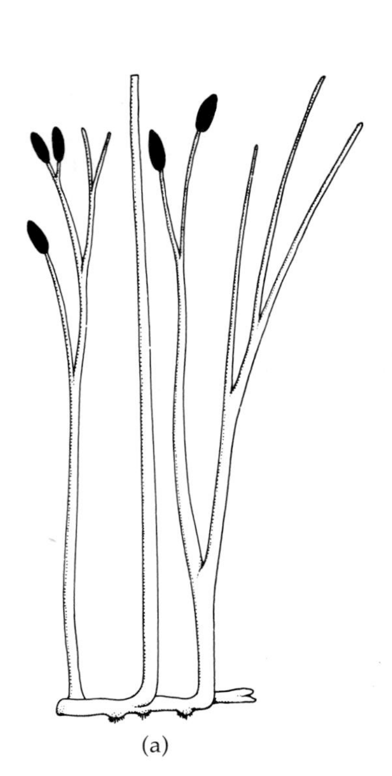
(a)

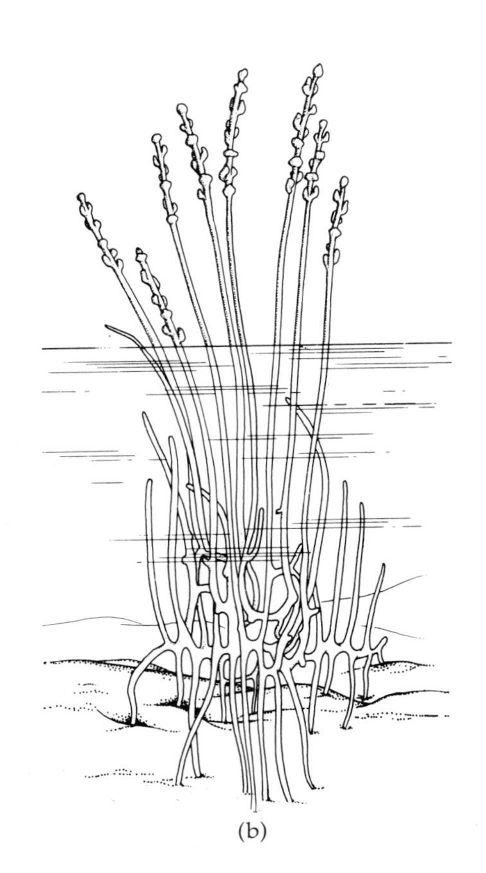
(b)

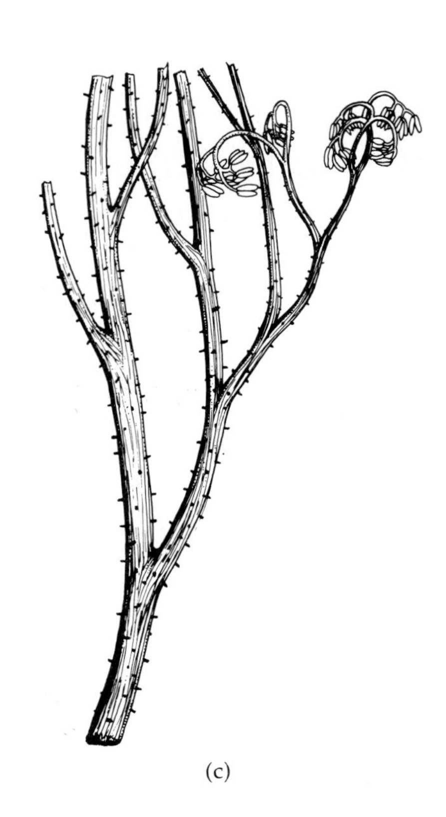
(c)

16–3
Early vascular plants. (a) Rhynia major *is a member of the Rhyniophyta, the simplest of the known vascular plants. The stem was leafless and dichotomously branched. The sporangia, which were terminal, apparently released their spores by splitting longitudinally.* (b) *In* Zosterophyllum, *the best-known genus of Zosterophyllophyta, the sporangia, which were aggregated into a terminal spike, split along definite slits that formed around the outer margin. The Zosterophyllophyta were larger than the Rhyniophyta, but like the latter, they were mostly dichotomously branched plants that were either naked, spiny, or toothed.* (c) *Trimerophyta was comprised of plants that had a strong central axis with smaller side branches. The side branches were dichotomously branched and often had terminal masses of paired sporangia that tapered at both ends. The best-known genera included in this group are* Psilophyton *and* Trimerophyton. *A reconstruction of* P. princeps *is shown here.*

Division Trimerophyta

The division Trimerophyta is considered by some paleobotanists to have evolved directly from the Rhyniophyta and to represent the ancestral stock of the ferns, the progymnosperms, and perhaps the horsetails as well. The Trimerophyta (Figure 16–3c) appear in mid-Devonian strata about 360 million years old.

Although the Trimerophyta were evolutionarily more advanced than the Rhyniophyta, they still lacked leaves. The main axis formed lateral branch systems that dichotomized several times. Some of the smaller branches terminated in elongate sporangia in which only one type of spore was produced, while others were entirely vegetative. Besides their more complex branching pattern, the Trimerophyta had a more massive vascular strand than the Rhyniophyta. Together with a wide cortex composed of thick-walled cells, the large vascular strand probably was capable of supporting a fairly large plant. As in the Rhyniophyta, the xylem of the Trimerophyta matured first in the center.

Figure 16–4 illustrates the possible course of evolution of the other divisions of vascular plants from the Zosterophyllophyta and Rhyniophyta.

ORGANIZATION OF THE VASCULAR PLANT BODY

From the previous discussion it is clear that the sporophytes, or plant bodies, of the early vascular plants were dichotomously branched axes that lacked leaves and roots. With evolutionary specialization, morphological and physiological differences arose between various parts of the plant body, bringing about the differentiation of roots, stems, and leaves—the organs of the plant (Figure 16–5). Collectively, the roots make up the *root system,* which serves to anchor the plant in the ground and to absorb water and minerals from the soil. The stems and leaves together make up the *shoot system,* with the stems functioning to raise the highly specialized photosynthetic organs—the leaves—toward the sun and to conduct water to the leaves and the end products of photosynthesis away from the leaves.

The different kinds of cells of the plant body are organized into tissues, and tissues are organized into still larger units called tissue systems. Three tissue systems—*dermal, vascular,* and *ground tissues*—occur in all organs of the plant; they are continuous from organ to organ and reveal the basic unity of the plant body. The dermal tissue system makes up the outer protective covering of the plant. The vascular tissue system comprises the conductive tissues—xylem and phloem—and is embedded in the ground tissue system (Figure 16–5). The principal differences in the structures of root, stem, and leaf lie primarily in the relative distribution of the vascular and ground tissue systems, as will be discussed in Section 5.

Primary and Secondary Growth

Primary growth may be defined as that growth which occurs relatively close to the tips of roots and stems. It is initiated by the apical meristems and is primarily involved with extension of the plant body. Tissues arising during primary growth are known as *primary tissues;* the part of the plant body composed of these tissues is called the *primary plant body* (Figure 16–5). The most primitive of vascular plants consisted en-

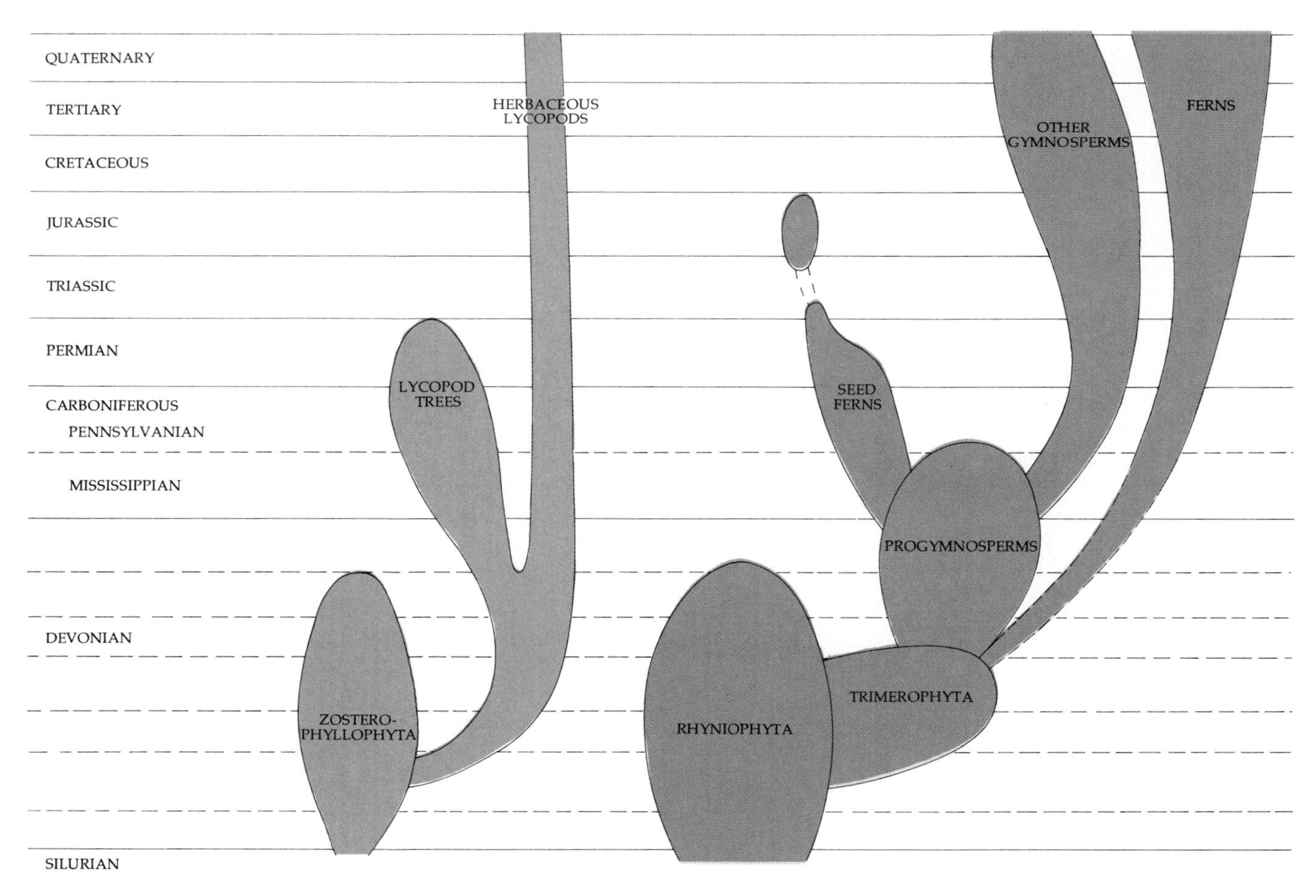

16–4
Possible course of evolution of vascular plants from the Zosterophyllophyta and Rhyniophyta. The length of Devonian time is exaggerated, and the possible relationships with the angiosperms, or flowering plants, are not indicated.

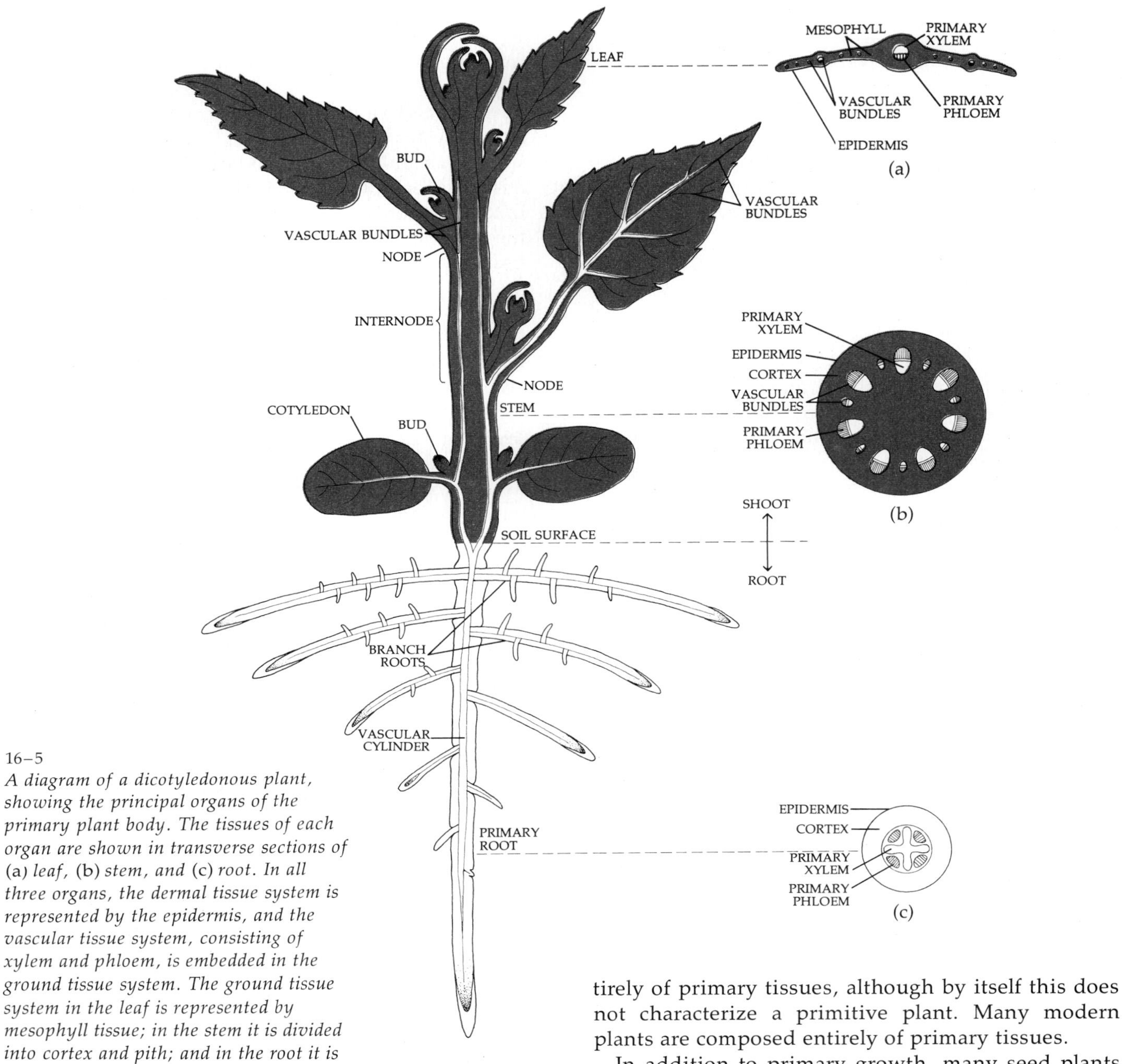

16–5
A diagram of a dicotyledonous plant, showing the principal organs of the primary plant body. The tissues of each organ are shown in transverse sections of (a) *leaf,* (b) *stem, and* (c) *root. In all three organs, the dermal tissue system is represented by the epidermis, and the vascular tissue system, consisting of xylem and phloem, is embedded in the ground tissue system. The ground tissue system in the leaf is represented by mesophyll tissue; in the stem it is divided into cortex and pith; and in the root it is represented by the cortex only. The leaf is specialized for photosynthesis; the stem for support of the leaves and for conduction; and the root for aborption and anchorage.*

tirely of primary tissues, although by itself this does not characterize a primitive plant. Many modern plants are composed entirely of primary tissues.

In addition to primary growth, many seed plants undergo additional growth, which brings about an increase in thickness of the stem and root. Such growth is termed *secondary growth* and involves the activity of a lateral meristem, the *vascular cambium*, which produces *secondary vascular tissues, secondary xylem*, and *secondary phloem* (see Figure 24–6, page 479). The production of secondary vascular tissues is commonly supplemented by the activity of the *cork cambium,* which forms a *periderm* composed mostly of *cork* tissue. The periderm replaces the epidermis as the dermal tissue system of the plant. The secondary vascular tissues and periderm make up what is known as the *secondary plant body*. Although only a few living seedless vascular plants exhibit such growth, secondary growth by a vascular cambium appeared in the mid-Devonian period about 360 million years ago among representatives of several unrelated groups.

Tracheary Elements

The presence of phloem has yet to be demonstrated unequivocally in any Devonian plants. This is because *sieve elements,* the conducting cells of the phloem, have soft walls and often collapse after they die, so they are generally poorly preserved in the fossil record. In contrast, *tracheary elements,* the conducting cells of the xylem, have rigid, persistent walls. Indeed, it is the presence of tracheary elements that is primarily used in the identification of a fossil as a vascular plant.

In the Silurian and Devonian fossils, the tracheary elements are elongated cells with long tapering ends. Such tracheary elements, called *tracheids,* are the first type of water-conducting cell to appear in the fossil record and the only type of water-conducting cell found in most living seedless vascular plants and gymnosperms. The tracheids not only provided channels for the passage of water and minerals through the plant but also provided support for the plant axis. Much of the rigidity of the cell walls of tracheids and other xylem cells is due to the presence of lignin in their walls. With the evolution of this cell wall component, the potential to increase greatly in size and to adapt an upright habit was acquired by the lignin-containing plants.

Extensive comparative studies of the tracheary elements in a wide range of vascular plants have clearly established that the tracheid is a more primitive (less specialized) type of cell than the *vessel member*—the principal water-conducting cell in the xylem of angiosperms. It is quite clear that vessel members evolved independently from tracheids in several groups of vascular plants, including the dicotyledons and monocotyledons (the two groups of angiosperms), the Gnetophyta (the vessel-containing gymnosperms), several unrelated species of ferns, and certain species of *Selaginella* (of the Lycophyta) and *Equisetum* (of the Sphenophyta). The evolution of the vessel member is an excellent example of convergent evolution, or the independent development of similar structures by unrelated or only distantly related organisms (see "Convergent Evolution," page 473).

Steles

The primary vascular tissues—primary xylem and primary phloem—and the pith, if present, collectively constitute what is known as the central cylinder, or *stele,* of the stem and root of the primary plant body. Several types of stele are recognized, three of which are the protostele, the siphonostele, and the eustele.

The *protostele* consists of a solid strand of vascular tissue, with either the phloem surrounding the xylem or interspersed within it (Figure 16–6). The protostele is regarded as the most primitive type. It is found in *Rhynia, Zosterophyllum,* and *Psilophyton.* Protosteles also occur in the Psilophyta, Lycophyta, and in the juvenile stems of some other groups. In addition, it is the type of stele found in most roots.

The *siphonostele,* or tubular stele, is characterized by the presence of a central column of ground tissue, called the *pith,* surrounded by the vascular tissue (Figure 16–6). The phloem may be either external to the xylem or both external and internal to it. The siphonostele occurs in the stems of many ferns and of certain gymnosperms and angiosperms.

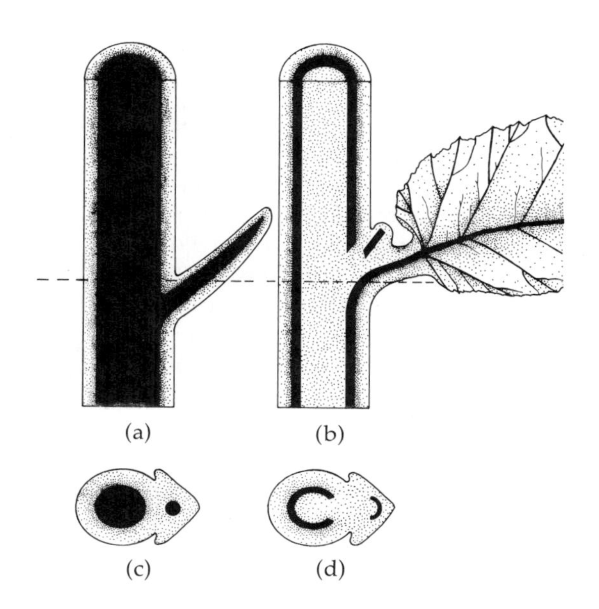

16–6
Longitudinal sections through nodal regions (parts of the stem to which leaves are attached) of stems possessing leaves that are (a) *microphyllous (relatively small with a single strand of vascular tissue) and* (b) *megaphyllous (relatively large, usually with a complex system of veins). With the exception of the single-veined leaves of* Equisetum, *microphylls are borne on stems with protosteles, and megaphylls are borne on stems with either siphonosteles or eusteles.* (c) *and* (d) *Transverse views through the planes represented by the dashed lines in* (a) *and* (b), *respectively. Note the presence of pith and leaf gap in* (b) *and* (d) *and their absence in* (a) *and* (c).

16–7
(a) *A primitive protostele, with diverging appendages, the evolutionary precursors of leaves.* (b) *A three-ribbed protostele, divided into a three-stranded vascular system with a pith.* (c) *The strands are more numerous and the pith is larger. The vascular strands of the appendages in* (a) *and* (b) *are now identified as leaf traces.* (d) *The stem bundles and their associated leaf traces—together called sympodia—have an undulating course, and the ground tissue regions above the diverging leaf traces resemble leaf gaps.*

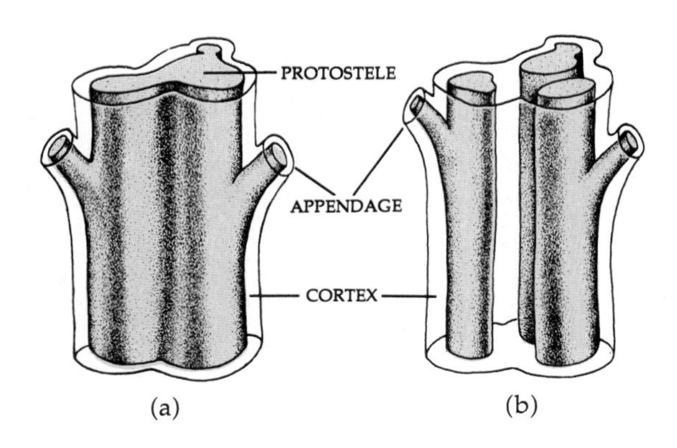

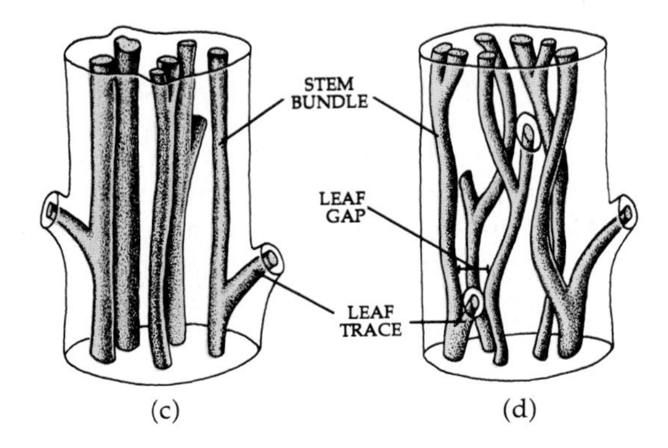

The *eustele,* which is found in *Equisetum* and in certain gymnosperms and angiosperms, consists of a system of strands surrounding a pith and separated from one another by ground tissue.

Until fairly recently, it was believed that the eustele of the seed plants evolved from the type of siphonostele found in ferns by its dissection (division) into strands. Evidence obtained from comparative studies of the vascular systems of conifers and dicotyledons and from surveys of the fossil record suggest, however, that the eustele of seed plants evolved from a more ancient protostele, such as the kind found in early fossil gymnosperms. The probable sequence in the evolution of a eustele from a protostele is illustrated in Figure 16–7.

Origins of Roots and Leaves

Although the fossil record reveals little information on the origins of roots as we know them today, it seems reasonable to assume that they evolved from the subterranean portions, or rhizomes, of the primitive axislike plant body. For the most part, roots are relatively simple structures that have retained many of the primitive structural characteristics no longer present in the stems of modern plants.

Leaves are the principal lateral appendages of the stem and, regardless of their ultimate size or structure, they arise as protuberances (leaf primordia) from the apical meristem of the shoot. Two morphologically distinct types of leaves—microphylls and megaphylls—are recognized, and they clearly seem to have evolved in different ways.

Microphylls are relatively small leaves that contain only a single strand of vascular tissue (Figure 16–6). With the exception of the single-veined leaves of *Equisetum*, microphylls are associated with stems possessing protosteles, as in the Psilophyta and Lycophyta. Examination of a microphyllous shoot reveals that the strand of vascular tissue that diverges from the protostele into the leaf, the *leaf trace*, does so without interrupting the pattern of the stele. Even though the name *microphyll* means "small leaf," some species of *Isoetes* have fairly long leaves that nonetheless are typical microphylls. In addition, certain Carboniferous and Permian members of the Lycophyta had microphylls that reached lengths of 1 meter or more.

Microphylls are believed to have evolved as superficial lateral outgrowths of the stem (Figure 16–8). At first these structures were small scalelike or spinelike outgrowths devoid of vascular tissue. Gradually, rudimentary leaf traces developed, which initially extended only to the base of the outgrowth. Finally, the leaf traces extended into the appendage, resulting in formation of the primitive microphyll.

Megaphylls, as the name implies, commonly are large leaves. With few exceptions, megaphylls are associated with stems possessing either siphonosteles or eusteles. In contrast to the single-veined microphyll, the blade, or *lamina*, of most megaphylls has a complex system of veins. In addition, the leaf traces of megaphylls are associated with *leaf gaps* in the stele of the stem. The leaf gap is a region of ground tissue in the stele resulting from the divergence of the leaf trace away from the stele toward the leaf.

Megaphylls are believed to have evolved from entire branch systems by the series of steps shown in Figure 16–8. The earliest plants had a leafless, dichotomously branching axis, and unequal branching resulted in more aggressive branches "overtopping" the weaker ones. This was followed with a flattening-out, or "planation," of the subordinated lateral branches. The final step was fusion, or "webbing," of the separate lateral branches to form a primitive lamina.

Although the leaves of *Equisetum* are microphylls by definition, some fossil members of the Sphenophyta had relatively large leaves with dichotomous venation. Possibly the small leaves of *Equisetum* were derived from larger, more complex ones.

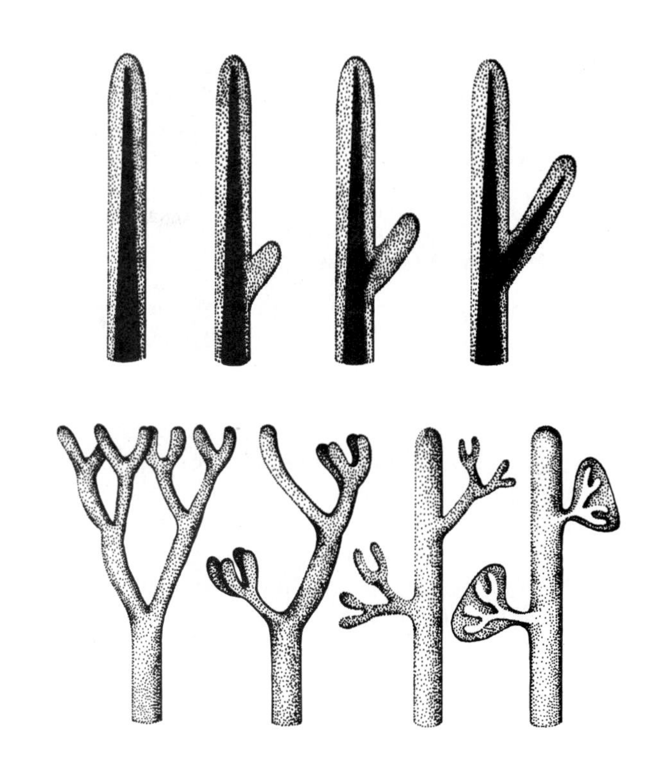

16–8
According to one widely accepted theory, microphylls (above) evolved as outgrowths of the main axis of the plant. Megaphylls (below) evolved by fusion of branch systems.

REPRODUCTIVE SYSTEMS

All vascular plants are oogamous and all have an alternation of generations (Figure 16–9). In the vascular plants, the sporophyte is the dominant phase in the life cycle; it is larger and structurally much more complex than the gametophyte. Almost all of the preceding remarks in this chapter have been devoted to the vegetative structure of the sporophyte. We shall now consider some general features of the reproductive structures of the vascular plants.

Homospory and Heterospory

As previously noted, the early vascular plants produced only one kind of spore. Such vascular plants are said to be *homosporous*. Among living vascular plants, homospory is found in the Psilophyta, Sphenophyta, some of the Lycophyta, and almost all the ferns. Following meiosis, homosporous plants produce only one kind of spore within the sporangium. Upon germination such spores produce bisexual gametophytes, that is, gametophytes that bear both antheridia and archegonia.

Heterospory—the production of two types of spores in two different kinds of sporangia—is found in some of the Lycophyta, in a few ferns, and in all seed plants. Heterospory has arisen many times in unrelated groups during the evolution of vascular plants. It was common as early as the Devonian period, more than 350 million years ago. The two types of spores are called *microspores* and *megaspores*, which are produced in *microsporangia* and *megasporangia*, respectively. Although "micro-" implies smallness and "mega-" implies largeness, megaspores are not always larger than microspores, especially in the seed plants. The two types of spores are actually defined on the basis of function, not relative size. Microspores give rise to male gametophytes (microgametophytes),

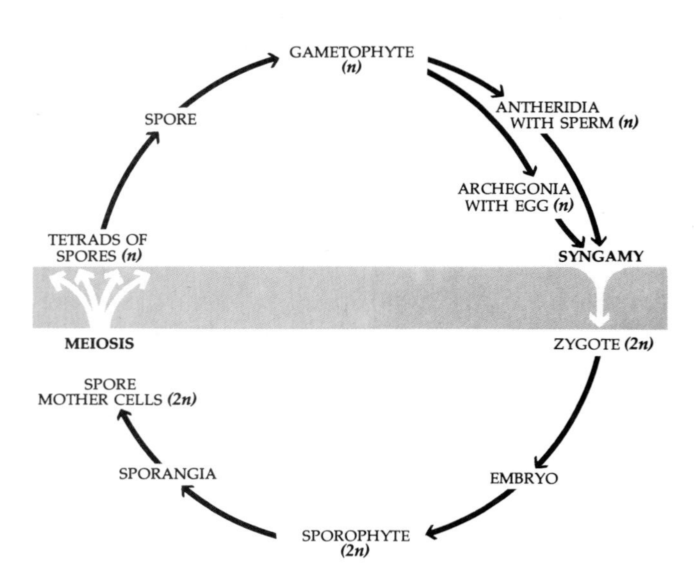

16–9
Generalized life cycle of a vascular plant.

and megaspore give rise to female gametophytes (megagametophytes). Both of these types of unisexual gametophytes are much reduced in size as compared with the gametophytes of homosporous vascular plants.

Gametophytes and Gametes

The relatively large gametophytes of homosporous plants are independent of the sporophyte for their nutrition, although the subterranean gametophytes of some species—such as those of *Psilotum* and a number of species of club mosses *(Lycopodium)*—are clearly heterotrophic and depend upon endomycorrhizal fungi for their nutrients. Other species of *Lycopodium*, like most ferns and the horsetails, have free-living, photosynthetic gametophytes. In contrast, the gametophytes of heterosporous vascular plants are dependent upon the sporophyte for their nutrition.

The evolution of the gametophyte of vascular plants is characterized by a progressive reduction in size and complexity, and the gametophytes of angiosperms are the most greatly reduced of all. The mature megagametophyte of angiosperms commonly consists of only seven cells, one of them an egg cell. When mature, the microgametophyte contains only three cells, and two of them are sperm. Archegonia and antheridia, which are found in all seedless vascular plants (Figure 16–10), are absent in all angiosperms and in a few gymnosperms. In the seedless vascular plants, including the ferns, the motile sperm swim through water to the archegonium. These plants must therefore grow in habitats where water is at least occasionally plentiful. In the gymnosperms and angiosperms, entire microgametophytes *(pollen grains)* are carried to the vicinity of the megagametophyte, where they produce special structures called *pollen tubes,* which bring the sperm near the egg. This transfer of the pollen grains is called *pollination.*

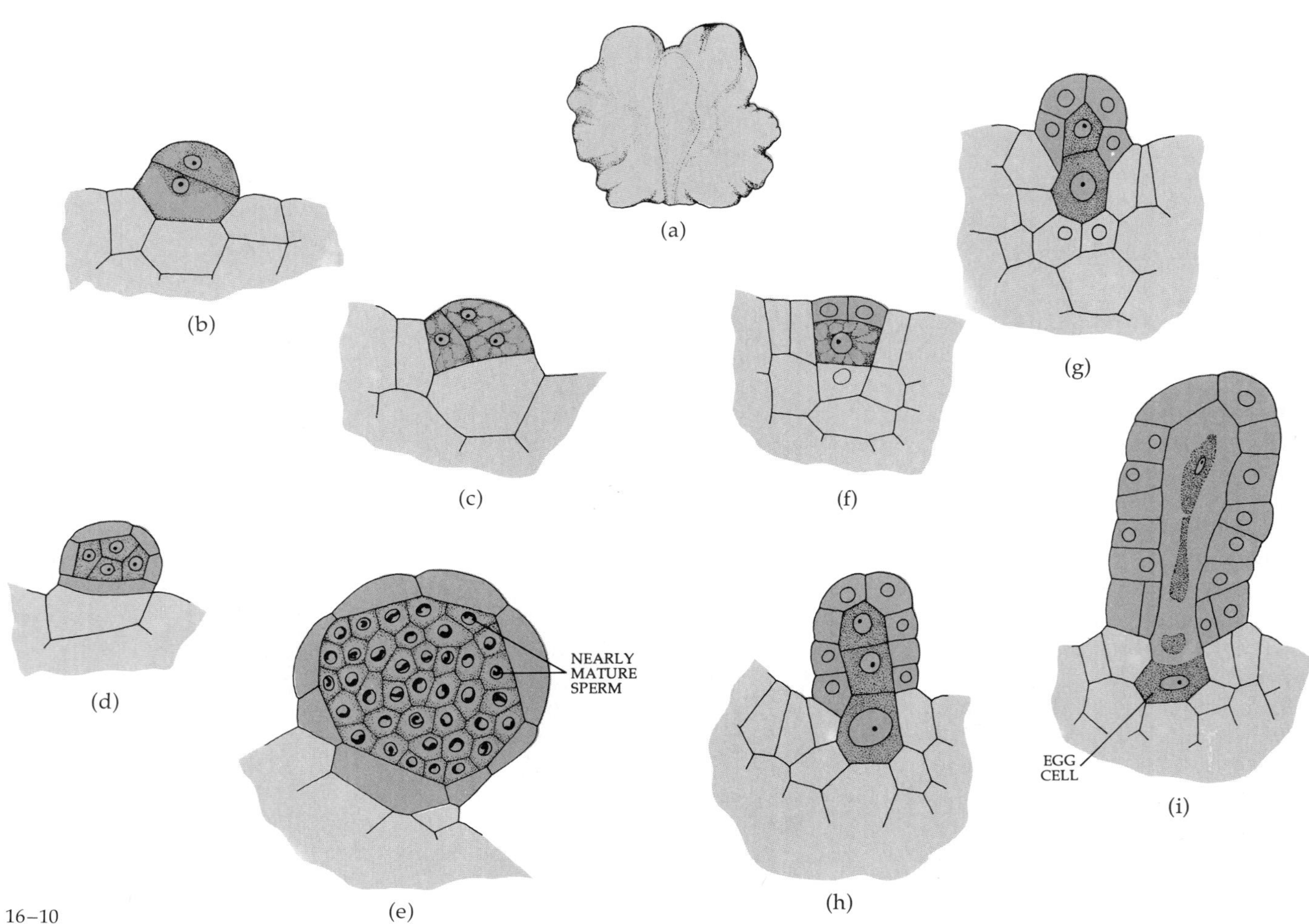

16–10
(a) *Gametophyte of the fern* Gleichenia dichotoma. *Successive stages in the development of the gametangia:* (b) *through* (e) *represent antheridium development;* (f) *through* (i) *represent archegonium development. Multicellular gametangia are characteristic of many of the seedless vascular plants.*

16–11
Seedlike structures in a number of Paleozoic plants, showing some stages in the evolution of seeds. In Genomosperma kidstonii *(the Greek word* genomein *means "to become," and* sperma *means "seed"), eight fingerlike projections arise at the base of the megasporangium and are separated for their entire length. In* G. latens *the integumentary lobes are fused from the base of the megasporangium for about a third of their length. In* Eurystoma angulare *fusion is almost complete, while in* Stamnostoma huttonense *it is complete.*

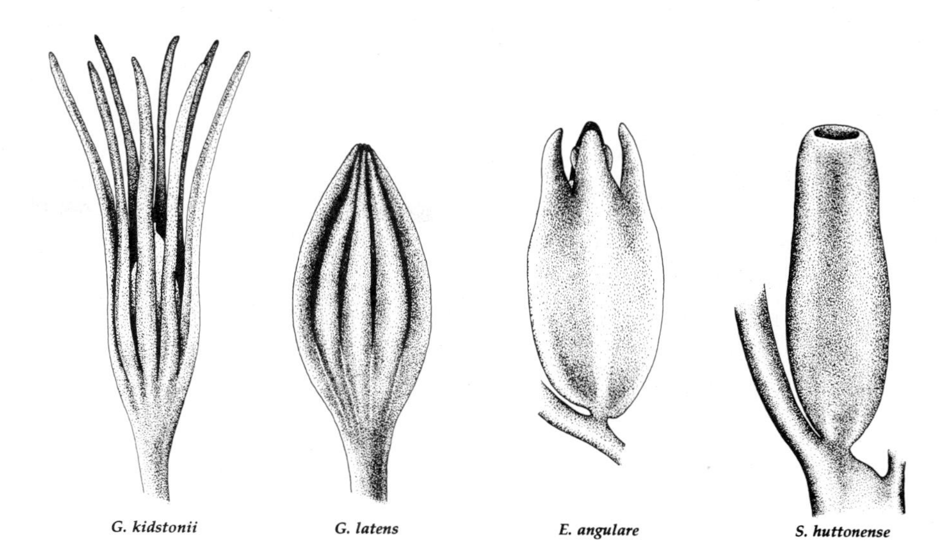

Evolution of the Seed

One of the most dramatic innovations exhibited by the vascular plants is the evolution of the seed (Figure 16–11). In plants that have seeds, the highly reduced megagametophyte is retained within the megaspore. The megaspore, in turn, is retained within the megasporangium, which is called the *nucellus* in seed plants. The megasporangia of seed plants, unlike those of other heterosporous plants, are enveloped by one or two layers of tissue, the *integuments*. This entire structure—the nucellus plus its integument or integuments—is known as the *ovule* (Figure 16–12).

Following fertilization, the integuments develop into a seed coat, and a seed is formed. In other words, it is the ovule that develops into a seed. In most modern seed plants, an embryo, or young sporophyte, develops within the seed before dispersal. In addition, all seeds contain stored food of some kind.

The evolution of the seed is one of the principal factors responsible for the dominance of seed plants in today's flora, because it permits the young sporophyte, protected by the seed coat and provided with a supply of stored food, to remain dormant until conditions are favorable for it to resume growth.

The oldest known seeds are found in late Devonian strata that are about 350 million years old (Figure 16–13). During the remainder of the Paleozoic era, and in the Mesozoic era that followed, a wide variety of seed types evolved.

The angiosperms, which are overwhelmingly the most successful vascular plants at the present time, have not been identified with certainty in the fossil record prior to about 127 million years ago, in the early part of the Cretaceous period. Although they actually must be somewhat older, they are still relative newcomers in the broad picture of vascular plant evolution.

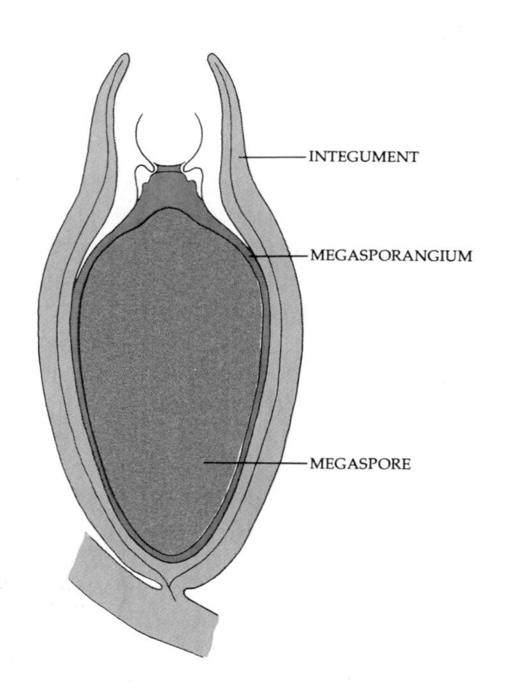

16–12
Sectional view of the ovule or seed of Eurystoma angulare, *showing the spatial relationship of the integument, the megasporangium (nucellus), and the megaspore.*

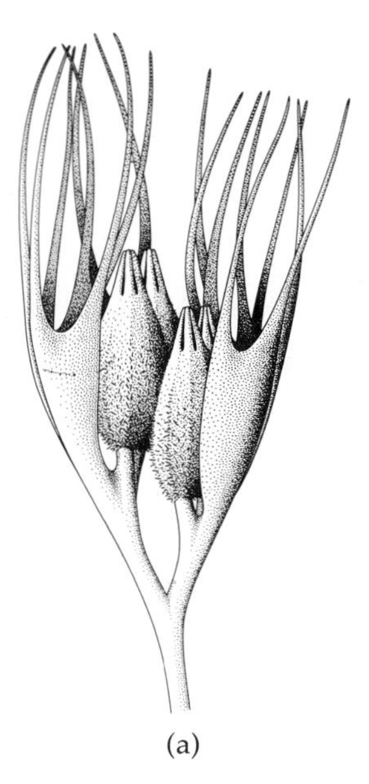

(a)

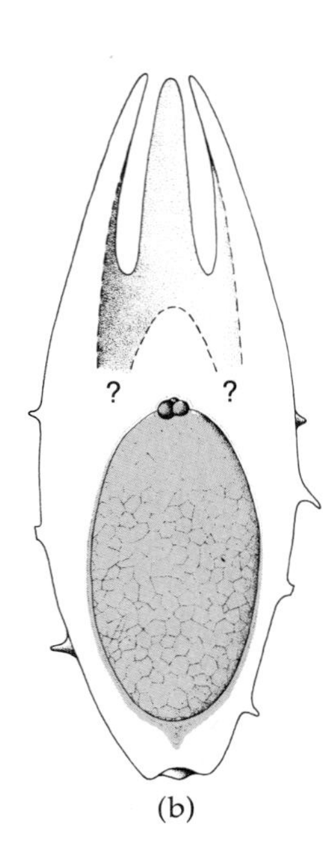

(b)

(c)

16–13
(a) *Reconstruction of a cupule complex of the late Devonian plant* Archeosperma arnoldii *with seedlike structures. The cupules are arranged in pairs, and each cupule contains two seeds.* (b) *Diagram showing the position of the megaspore within the seed.* (c) *A megaspore released from a seed by maceration. These fossils demonstrate that seed plants existed some 350 million years ago.*

SUMMARY

Vascular plants are those plants containing xylem and phloem and exhibiting an alternation of generations in which the sporophyte is the dominant and nutritionally independent phase.

The earliest known vascular plants belong to the division Rhyniophyta and are found in strata of the Silurian period some 400 million years old. The plant bodies of the Rhyniophyta and their contemporaries were simple, dichotomously branching axes lacking roots and leaves. With evolutionary specialization, morphological and physiological differences arose between various parts of the plant body, bringing about the differentiation of root, stem, and leaf.

The plant bodies of many vascular plants consist entirely of primary tissues. Today, secondary growth is confined largely to the seed plants, although it occurred in several unrelated fossil groups. The primary vascular tissues and associated ground tissues exhibit three basic arrangements: (1) the protostele, which consists of a solid core of vascular tissue; (2) the siphonostele, which contains a pith, surrounded by vascular tissue; and (3) the eustele, which consists of a system of strands surrounding a pith and separated from one another by ground tissue.

Roots evolved from the underground portions of the primitive plant body. Leaves originated in more than one way. Microphylls, single-veined leaves whose leaf traces are not associated with leaf gaps, evolved as superficial lateral outgrowths of the stem. Megaphylls, leaves with complex venation and leaf traces associated with leaf gaps, evolved from branch systems. Microphyllous shoots are associated with protosteles, and megaphyllous shoots are associated with siphonosteles and eusteles.

Vascular plants are either homosporous or heterosporous. Homosporous vascular plants produce only one type of spore, which gives rise to a bisexual gametophyte. Heterosporous plants produce microspores and megaspores, which germinate and give rise to male gametophytes and female gametophytes, respectively. The gametophytes of heterosporous plants are much reduced in size, compared with those of homosporous plants. In the history of the vascular plants, heterospory has evolved a number of times. There has been a long, continuous evolutionary trend toward a reduction in the size and the complexity of the gametophyte, which culminated in the angiosperms. Primitive vascular plants have archegonia and antheridia, but these have been lost in a few gymnosperms and in all angiosperms.

A seed is a structure formed by maturation of an ovule. Following fertilization, an embryo, or young sporophyte, is formed and the integument or integuments of the ovule develop into a seed coat. The oldest known seed plants are from the late Devonian period, about 350 million years ago.

SUGGESTIONS FOR FURTHER READING

BANKS, HARLAN P.: *Evolution and Plants of the Past,* Wadsworth Publishing Co., Inc., Belmont, Calif., 1970.*

An introduction to plant evolution as it is revealed by the fossil record.

BIERHORST, DAVID W.: *Morphology of Vascular Plants,* The Macmillan Company, New York, 1971.

A detailed, profusely illustrated study of the morphology of the vascular plants.

DOYLE, W. T.: *The Biology of Higher Cryptogams,* The Macmillan Company, New York, 1970.*

A concise account of the biology of the bryophytes and seedless vascular plants, with an emphasis on development and evolutionary relationships.

ESAU, KATHERINE: *Plant Anatomy,* 2nd ed., John Wiley & Sons, Inc., New York, 1965.

The standard work in the field; a well-illustrated book that considers all aspects of plant anatomy.

———: *Anatomy of Seed Plants,* 2nd ed., John Wiley & Sons, Inc., New York, 1977.

A shorter book than the preceding entry; an excellent textbook and reference.

FAHN, ABRAHAM: *Plant Anatomy,* 2nd ed., Pergamon Press, Inc., Elmsford, N.Y., 1974.*

A well-illustrated, up-to-date textbook considering all aspects of plant anatomy.

FOSTER, ADRIANCE S., and ERNEST M. GIFFORD: *Comparative Morphology of Vascular Plants,* 2nd ed., W. H. Freeman and Company, San Francisco, 1974.

A well-organized, general account of the vascular plants, containing much interpretative material.

RADFORD, ALBERT E., *et al.: Vascular Plant Systematics,* Harper & Row, Publishers, Inc., New York, 1974.*

An invaluable source book, including glossaries, techniques, bibliographies, indices, and useful discussions of all the various aspects of plant systematics—the scientific study of the kinds and diversity of plants, and of the relationships among them.

TAYLOR, THOMAS N.: *Paleobotany: An Introduction to Fossil Plant Biology,* McGraw-Hill Book Company, New York, 1981.

A readable and up-to-date survey that conveys both traditional knowledge and the excitement of new discoveries in paleobotany.

*Available in paperback.

17–1
A reconstruction of a Carboniferous swamp forest, with Lepidodendron *(the tall trees; left to middle),* Sigillaria *(the sparsely branched trees with terminal tufts of leaves; middle),* Calamites *(the giant horsetails) and* Cordaites *(right), and tree ferns (left to middle foreground). Both* Lepidodendron *and* Sigillaria *were lycopod trees.* Cordaites *was a much-branched tree that produced cone-like structures with seeds.* *(See "Coal Age Plants," page 330.)*

CHAPTER 17

The Seedless Vascular Plants

The vascular plants can be divided artificially into two major groups, the seedless vascular plants and the seed plants. In the previous chapter, three entirely fossil divisions of seedless vascular plants were discussed. In this chapter, we shall consider the four major divisions of seedless vascular plants with living representatives: the Psilophyta, the Lycophyta, the Sphenophyta, and the Pterophyta. The Pterophyta, the ferns, is the largest group, with some 12,000 species throughout the world.

DIVISION PSILOPHYTA

The Psilophyta are represented by two living genera, *Psilotum* and *Tmesipteris*. *Psilotum* is tropical and subtropical in distribution and occurs in the United States in Florida, Louisiana, Arizona, Texas, Hawaii, and Puerto Rico. *Tmesipteris* is restricted in distribution to Australia, New Caledonia, New Zealand, and other islands of the South Pacific. At one time, these genera were included in the same group as *Rhynia* and *Psilophyton*. However, new paleobotanical information has brought about the reclassification of these two fossil genera. In addition, it has recently been suggested that *Psilotum* and *Tmesipteris* are, in fact, primitive ferns. Because of the simplicity of their sporophytes, they are placed in a separate division in this book.

Psilotum is unique among living vascular plants in that it lacks both roots and leaves. The sporophyte consists of a dichotomously branching aerial portion with small scalelike outgrowths and a branching underground portion, or system of rhizomes with many rhizoids (Figure 17–2). A mycorrhizal fungus is present in the outer cortical cells of the rhizomes. The stele of *Psilotum* (Figure 17–3) has been interpreted as a protostele by some morphologists and as a siphonostele by others. The problem lies in the nature of the tissue in the center of the stele—that is, whether it is xylem or sclerified ground tissue of a pith.

17–2
Whisk fern, Psilotum nudum. (a) *View of sporophyte showing rhizomes and dichotomously branching aerial branches.* (b) *Detail of aerial branches, showing sporangia (in united groups of three) and scalelike outgrowths.*

(a)

(b)

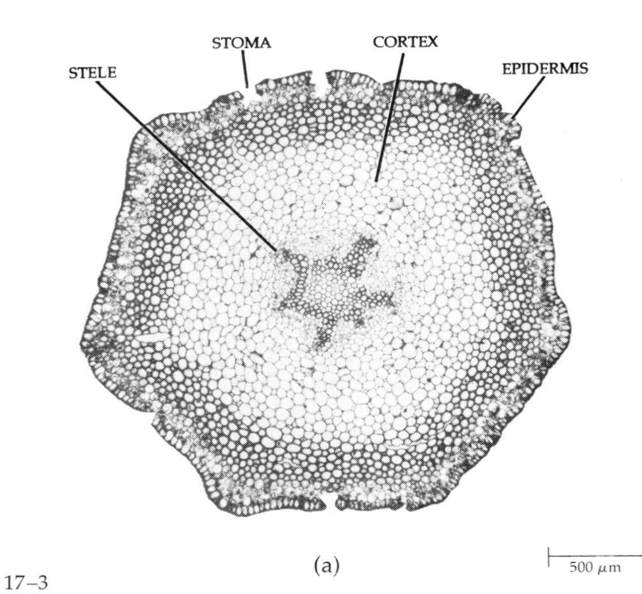

(a)

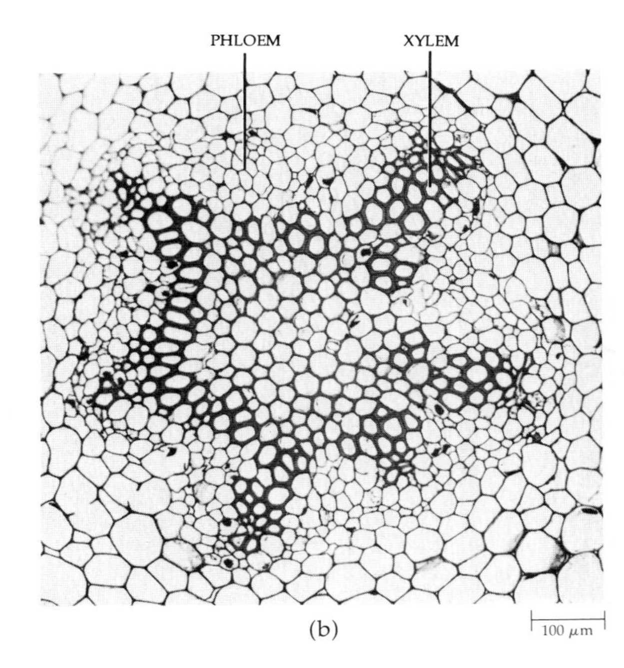

(b)

17–3
Psilotum nudum. (a) *Transverse section of stem, showing mature tissues.* (b) *Detail of stele, showing xylem and phloem.*

Psilotum is homosporous, the spores being produced in sporangia borne on the ends of short, lateral branches (see Figure 17–4). Upon germination, the spores give rise to bisexual gametophytes, which resemble portions of the rhizome (Figure 17–5). Like the rhizome, the subterranean gametophyte contains a symbiotic fungus. In addition, some gametophytes contain vascular tissue. The sperm of *Psilotum* are multiflagellated and require water to swim to the egg. Initially, the sporophyte is attached to the gametophyte by a foot, a structure that absorbs nutrients from the gametophyte. Eventually it becomes detached from the foot, which remains embedded in the gametophyte.

17–4
Psilotum nudum. *Longitudinal section through portions of two sporangia, showing mature spores. A scalelike appendage subtends the sporangia.*

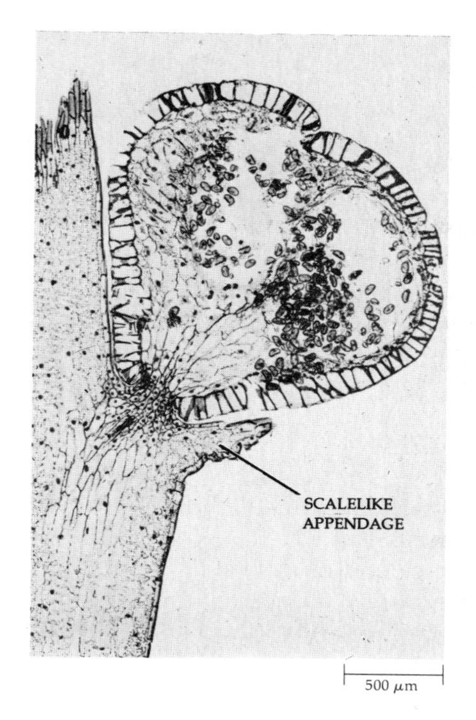

ANTHERIDIA
ARCHEGONIA
RHIZOIDS
(a)
ANTHERIDIA
RHIZOIDS
ARCHEGONIA
(b)

17–5
Psilotum nudum. (a) *Gametophyte, which is bisexual—that is, it bears both antheridia and archegonia.* (b) *Detail of gametophyte.*

17–6
(a) Tmesipteris parva, *growing on the trunk of the tree fern* Cyathea australis *in New South Wales, Australia.* (b) T. lanceolata, *from New Caledonia.*

(a)

(b)

Tmesipteris grows as an epiphyte on tree ferns and other plants (Figure 17–6). The leaflike appendages of *Tmesipteris* are larger than the scalelike outgrowths of *Psilotum,* and some morphologists interpret them to be flattened branchlets rather than as microphylls. In other respects, *Tmesipteris* is essentially similar to *Psilotum.*

DIVISION LYCOPHYTA

The five living genera (only three of which are considered here; *Phylloglossum* and *Stylites* are not discussed) and approximately 1000 living species of Lycophyta are the representatives of an evolutionary line that extends back to the Devonian period. It seems likely that the progenitors of the Lycophyta were *Zosterophyllum*-type plants (see Figure 16–3, page 306). The lycopods early split into two major groups. One group remained herbaceous and is still represented in today's flora. The second group, the lepidodendrids, became woody and treelike and were among the dominant plants of the coal-forming forests of the Carboniferous period (see Figure 17–1). Some of them bore seedlike structures analogous to those of modern seed plants. The lepidodendrids became extinct in the Permian period, about 280 million years ago.

Lycopodium

Perhaps the most familiar living representatives of the Lycophyta are the club mosses, *Lycopodium*. The approximately 200 species of this genus extend from arctic regions into the tropics, but they rarely form conspicuous elements in any plant community. Most tropical species are epiphytes and are thus rarely seen, but several of the temperate species form mats which may be evident on forest floors. Because they are evergreen, they are most noticeable in winter.

The sporophyte of *Lycopodium* consists of a branching rhizome from which aerial branches and adventitious roots arise (see Figure 17–7). The leaves of *Lycopodium*, which are usually spirally arranged, are microphylls, and the stem and root are protostelic (Figure 17–8).

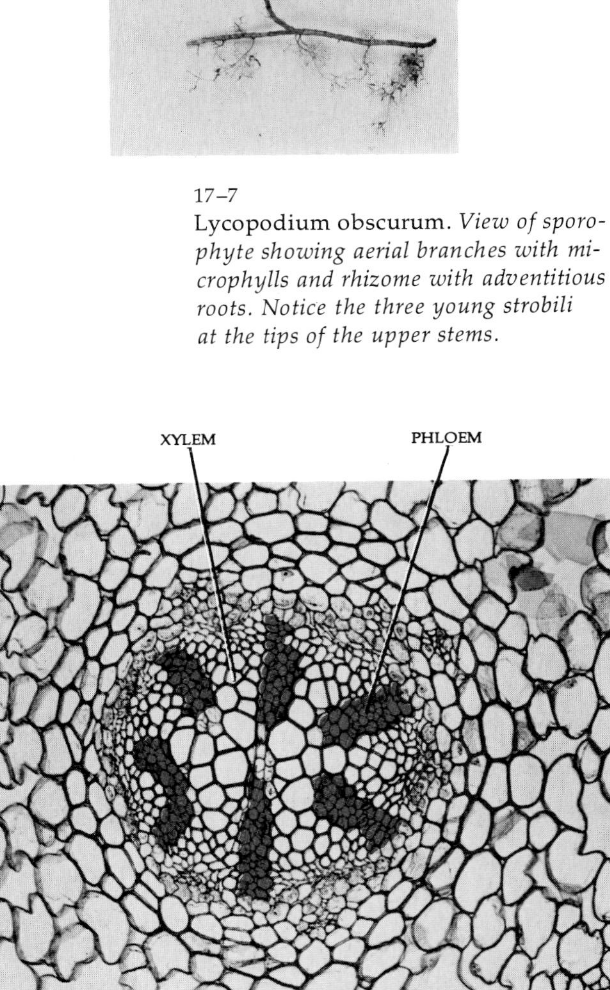

17–7
Lycopodium obscurum. *View of sporophyte showing aerial branches with microphylls and rhizome with adventitious roots. Notice the three young strobili at the tips of the upper stems.*

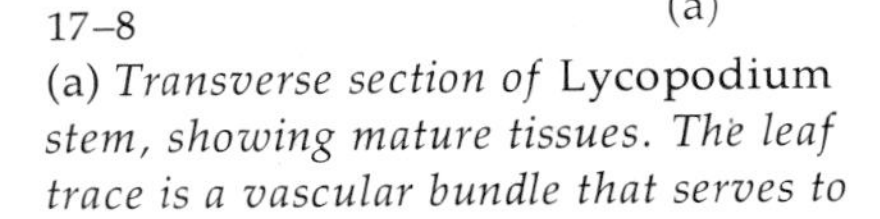

17–8
(a) *Transverse section of* Lycopodium *stem, showing mature tissues. The leaf trace is a vascular bundle that serves to connect the vascular tissues of the stem with the vascular tissues at the base of a leaf.* (b) *Detail of protostele, showing xylem and phloem. Both the stem and root of* Lycopodium *show this protostelic organization.*

Lycopodium is homosporous. The sporangia occur singly upon the upper surface of fertile microphylls called *sporophylls*—modified leaf or leaflike organs that bear sporangia (Figure 17–9). In some species, the sporophylls are similar to ordinary microphylls and are interspersed among the sterile microphylls (Figure 17–10a). In others, the nonphotosynthetic sporophylls are grouped into *strobili*, or cones, at the ends of the aerial branches (Figure 17–10b).

Upon germination, the spores of *Lycopodium* give rise to bisexual gametophytes which, depending on the species, are either green, irregularly lobed masses or branching, subterranean, and nonphotosynthetic structures. Like the gametophytes of *Psilotum* and *Tmesipteris*, those of the species of *Lycopodium* with subterranean gametophytes contain a symbiotic fungus. The development and maturation of archegonia and antheridia in a *Lycopodium* gametophyte may require from 6 to 15 years. Some subterranean forms reportedly live for as long as 25 years, and they may even produce a series of sporophytes in successive archegonia as they continue to grow.

Water is required for fertilization, the biflagellate sperm swimming through water to the archegonium and then down its neck. Following fertilization, the zygote develops into an embryo, which grows within the venter of the archegonium. The young sporophyte may remain attached to the gametophyte for a long time (see Figure 17–12, page 323), but eventually it becomes an independent plant.

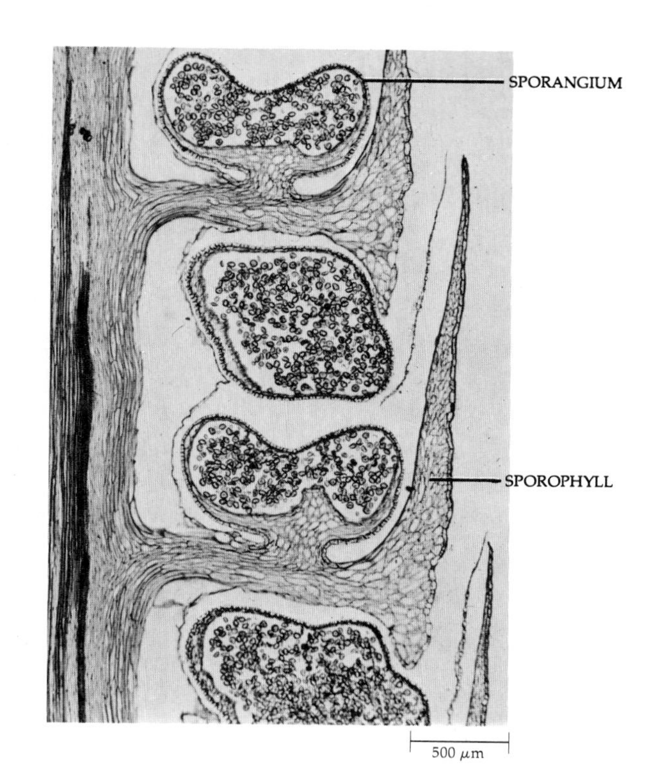

17–9
Lycopodium. *Longitudinal section showing portion of mature strobilus.* Lycopodium *is homosporous, producing only one type of spore. Each sporophyll bears a single sporangium. Notice that all of the spores are of similar size.*

17–10
Some species of Lycopodium *have their sporophylls grouped in strobili, whereas others have them interspersed among sterile microphylls.* (a) Lycopodium lucidulum *lacks strobili.* (b) *Strobili of* Lycopodium clavatum.

(a)

(b)

17–11

Life cycle of Lycopodium. Lycopodium *is homosporous—that is, the sporophyte produces only one sort of spore as a result of meiosis. The gametophyte developing from this spore then produces both archegonia, each of which contains an egg, and antheridia, each of which produces many biflagellate motile sperm. The gametophytes of some species of* Lycopodium *are subterranean and require the presence of a mycorrhizal fungus for normal growth. Water is necessary for fertilization to occur, the sperm swimming through it to the archegonium. The zygote is formed and begins its development within the archegonium, as in some seed plants.*

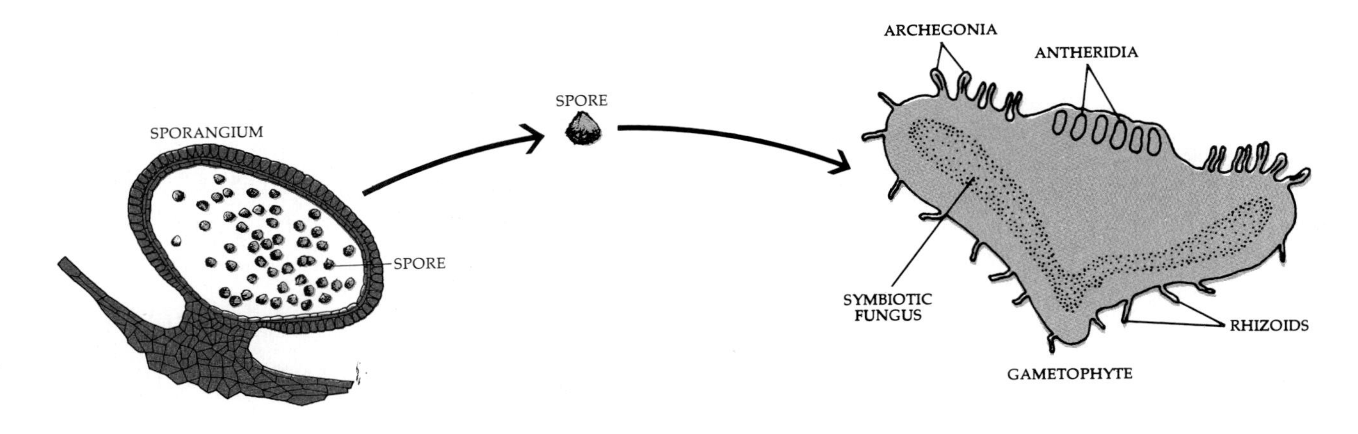

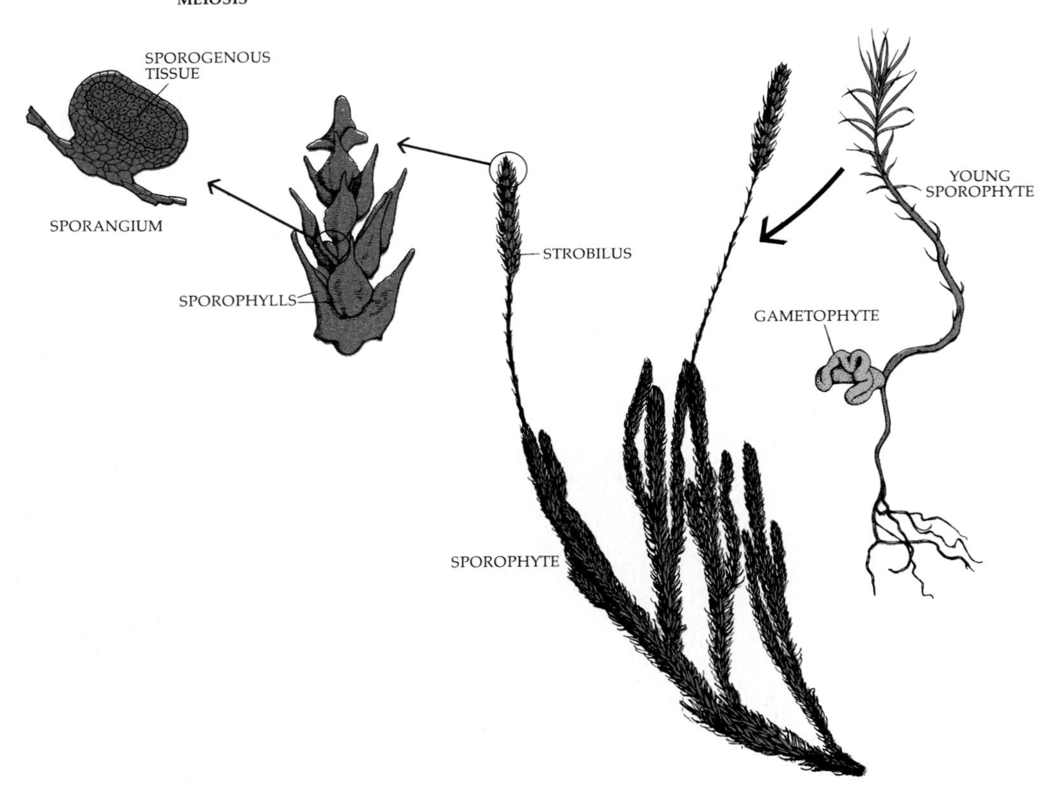

17–12
Lycopodium. *Gametophyte with young sporophyte attached to it.*

(a)

(b)

17–13
(a) Selaginella kraussiana, *a prostrate, creeping plant. Adventitious roots can be seen arising from stems;* (b) S. rupestris *with strobili.*

The life cycle of *Lycopodium*, the evergreen club mosses that comprise one of the five living genera of the division Lycophyta considered here, is summarized in Figure 17–11.

Selaginella

Among the living genera of lycopods, *Selaginella* has the most species, with about 700. These are mainly tropical in distribution. Many grow in moist situations; a few inhabit desert regions, becoming dormant during the driest part of the year. Among the latter is the so-called resurrection plant, *Selaginella lepidophylla*, which is native to Mexico and ranges north to Texas and New Mexico.

Basically, the herbaceous sporophyte of *Selaginella* is similar to that of *Lycopodium*, bearing microphylls and having sporophylls arranged in strobili (Figure 17–13). Unlike *Lycopodium*, in *Selaginella* a small scalelike outgrowth, called a ligule, develops near the base of the upper surface of each microphyll and sporophyll (Figure 17–14). The stem and root are protostelic (Figure 17–15).

Whereas *Lycopodium* is homosporous, *Selaginella* is heterosporous. This constitutes the most significant difference between the two genera. Each sporophyll bears a single sporangium on its upper surface. Megasporangia are borne by *megasporophylls*, and microsporangia are borne by *microsporophylls*. Both kinds of sporangia occur in the same strobilus (Figure 17–14).

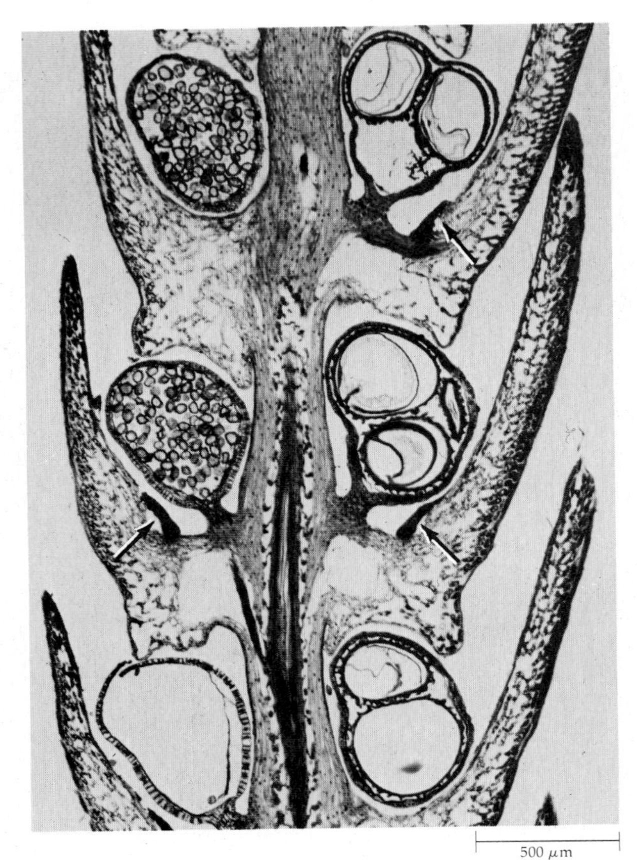

17–14
Selaginella. *Portion of strobilus, showing microsporangia with microspores on left and megasporangia with megaspores on right. Ligules (indicated by arrows) can be seen at the base of some sporophylls.*

Being heterosporous, *Selaginella* produces unisexual gametophytes (see Figure 17–17a, page 326). The male gametophytes (microgametophytes) develop from microspores. Four microspores are produced by meiosis from each microspore mother cell. The male gametophyte develops within the microspore and lacks chlorophyll. At maturity the male gametophyte consists of a single prothallial, or vegetative, cell and an antheridium, which gives rise to many biflagellate sperm. The microspore wall must rupture for the sperm to be liberated.

During development of the female gametophyte (megagametophyte), the megaspore wall ruptures, and the gametophyte protrudes through the rupture to the outside. This is the portion of the female gametophyte in which the archegonia develop. It has been reported that the female gametophytes sometimes develop chloroplasts, although it is more likely that the gametophytes derive their nutrition largely from food stored within the megaspores.

Water is required for the sperm to swim to the archegonia and fertilize the eggs. Commonly this occurs after the gametophytes have been shed from the strobilus. During development of the embryos of both *Lycopodium* and *Selaginella*, a structure called a suspensor is formed. Although inactive in *Lycopodium* and some species of *Selaginella*, in other *Selaginella* species the suspensor serves to thrust the developing embryo deep within the nutrient-rich tissue of the female gametophyte. Gradually, the developing sporophyte emerges from the gametophyte (see Figure 17–17b, page 326) and becomes an independent plant.

The life cycle of *Selaginella*, a second genus of the division Lycophyta, is summarized in Figure 17–16.

17–15
Selaginella. (a) *Transverse section of stem, showing mature tissues. The protostele is suspended in the middle of the hollow stem by elongate cortical cells (endodermal cells), called trabeculae. Only a portion of each trabecula can be seen here.* (b) *Detail of protostele.*

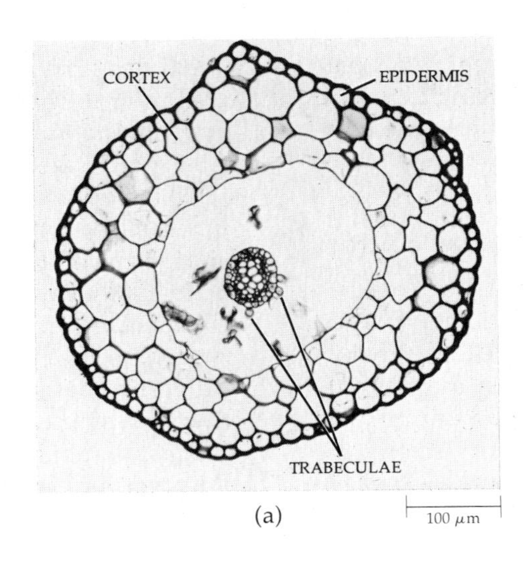

(a)

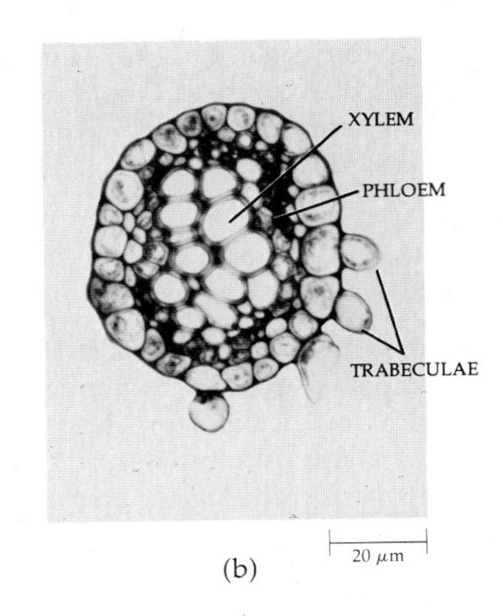

(b)

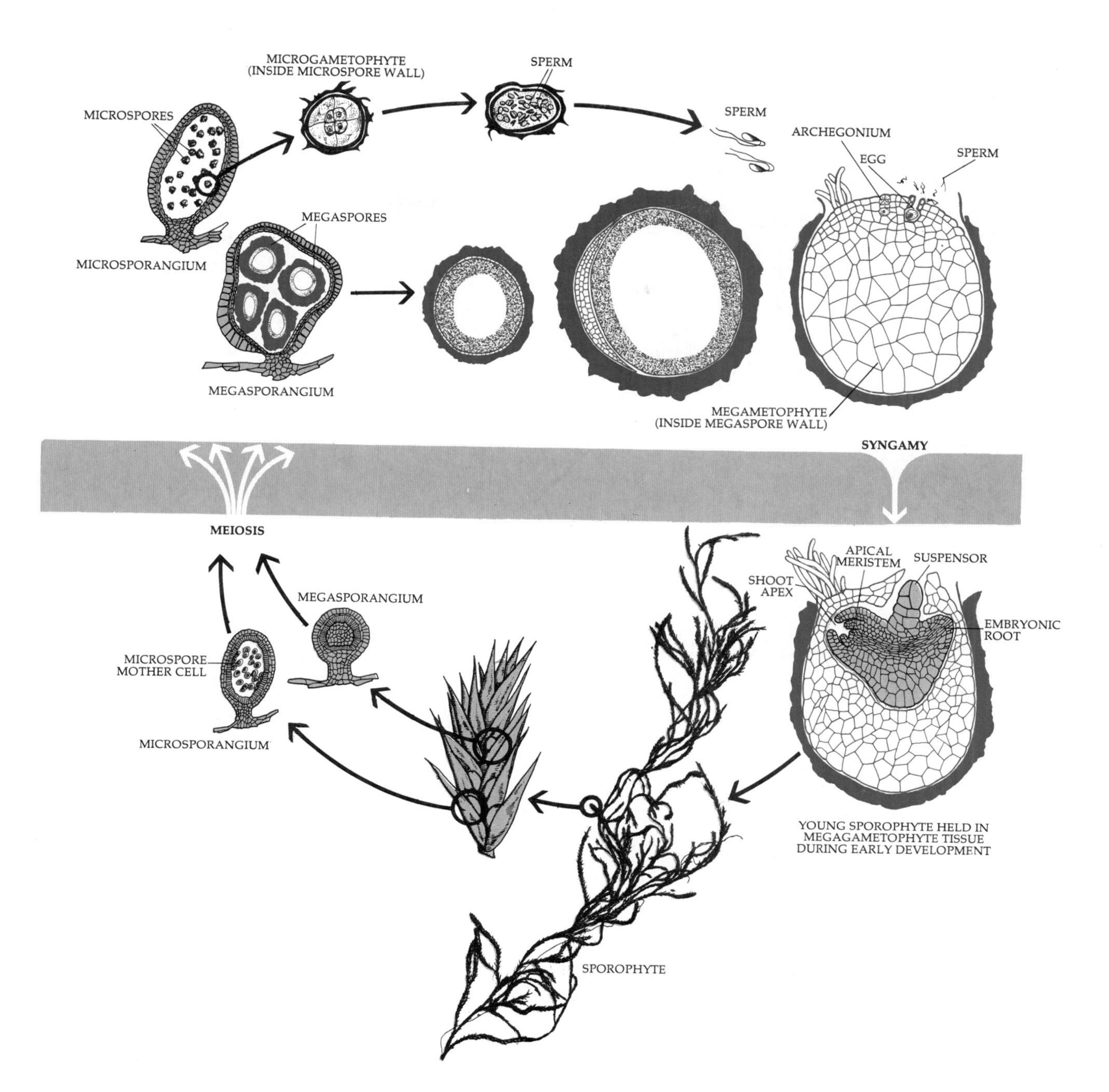

17–16
The life cycle of Selaginella, *which is heterosporous. Two kinds of sporangia are borne on the sporophyte and give rise to two morphologically distinct kinds of gametophytes, both much smaller than the sporophyte. As in the seed plants, the young sporophyte is enclosed by the tissues of the megagametophyte, and the major source of food for the developing embryo is material stored in the megaspore. However, there is no dormant period in the development of the embryo of* Selaginella, *as there is in many seed plants.*

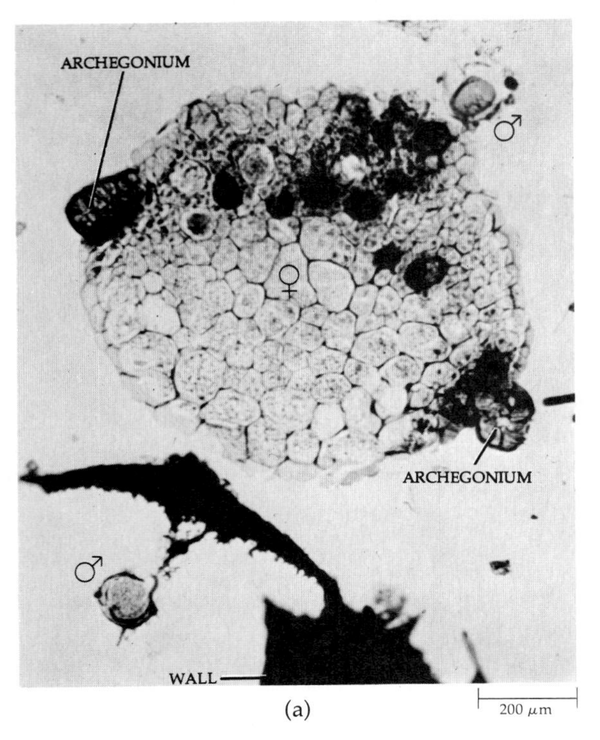

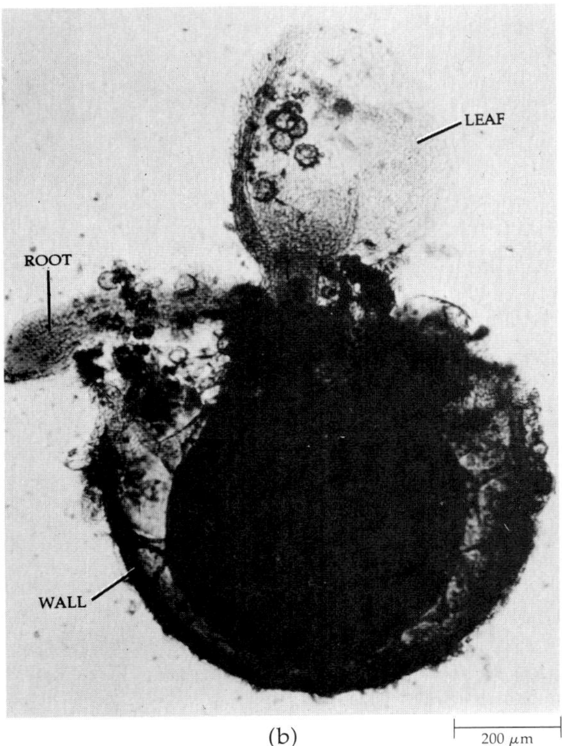

17–17
Selaginella. (a) *Female and male gametophytes. The much larger female gametophyte has been removed from the megaspore; a portion of megaspore wall can be seen below.* (b) *Young sporophyte emerging from female gametophyte. Notice the many microspores on the leaf and root of the sporophyte.*

Isoetes

A very distinctive member of the Lycophyta is *Isoetes*, the quillwort. The sporophyte of this aquatic plant consists of a short, fleshy underground stem (corm) bearing quill-like leaves on its upper surface and roots on its lower portion (Figure 17–18). Although the leaves of *Isoetes* are relatively long, they are considered to be microphylls—they are univeined, and their leaf traces are not associated with leaf gaps. Each leaf is a potential sporophyll.

Like *Selaginella, Isoetes* is heterosporous. The megasporangia are borne at the base of certain leaves (megasporophylls), and the microsporangia are borne at the base of other leaves (microsporophylls) nearer the center of the plant (Figure 17–19). A ligule is found just above the sporangium of each sporophyll.

One of the distinctive features of *Isoetes* is the presence of a specialized cambium that adds secondary tissues to the corm. Externally the cambium produces only parenchyma tissue, while internally it produces a peculiar vascular tissue consisting of sieve elements, parenchyma cells, and tracheids in varying proportions.

17–18
Isoetes muricata. *View of sporophyte showing quill-like leaves, stem, and roots.*

17–19
A diagram of a vertical section of an Isoetes *plant. Leaves are borne on the upper surface, and roots on the lower surface, of a short, fleshy underground stem. Some leaves (megasporophylls) bear megasporangia and others (microsporophylls) bear microsporangia. The microsporophylls are located nearer the center of the plant.*

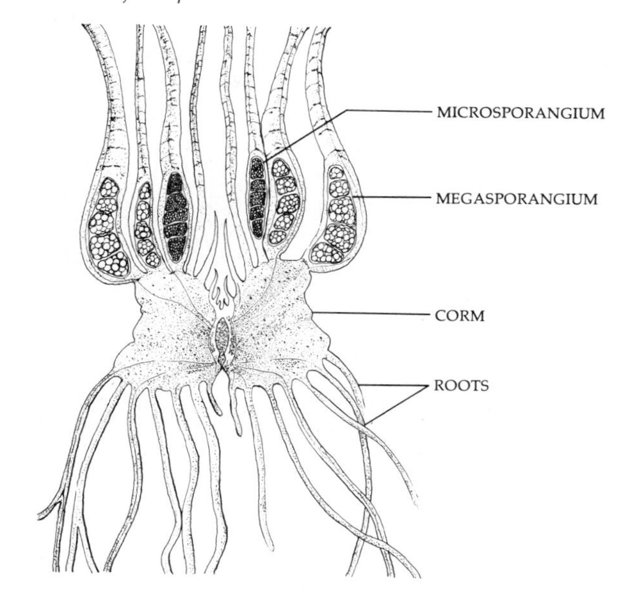

DIVISION SPHENOPHYTA

Like the Lycophyta, the Sphenophyta, or horsetails, extend back to the Devonian period. The sphenophytes reached their maximum abundance and diversity late in the Paleozoic era, about 300 million years ago. During the late Devonian and Carboniferous periods, they were represented by *Calamites*—trees that reached over 20 centimeters in diameter and 15 meters or more in height. Today the Sphenophyta are represented by a single herbaceous genus, *Equisetum*, which contains 15 species.

Equisetum is widespread in distribution and is often found in moist or damp places, by streams, or along the edge of woods (Figure 17–20). The horsetails are easily recognized because of their conspicuously jointed stems and rough texture. The small, scalelike leaves, which by definition are microphylls, are whorled at the nodes. When present, the branches arise laterally at the nodes and alternate with the leaves. The internodes (the portions of the stems between successive nodes) are ribbed, and the ribs are tough and strengthened with siliceous deposits in the epidermal cells. The unbranched species have been used in cleaning pots and pans, particularly in colonial and frontier times, and have thus earned the name "scouring rushes." The roots are adventitious, arising at the nodes of the rhizomes.

17–20
Equisetum. (a) *Fertile shoots of* E. arvense, *each with a terminal strobilus. Notice the whorls of scalelike leaves at each node.* (b) *Much-branched vegetative shoots of* E. hyemale.

(a)

(b)

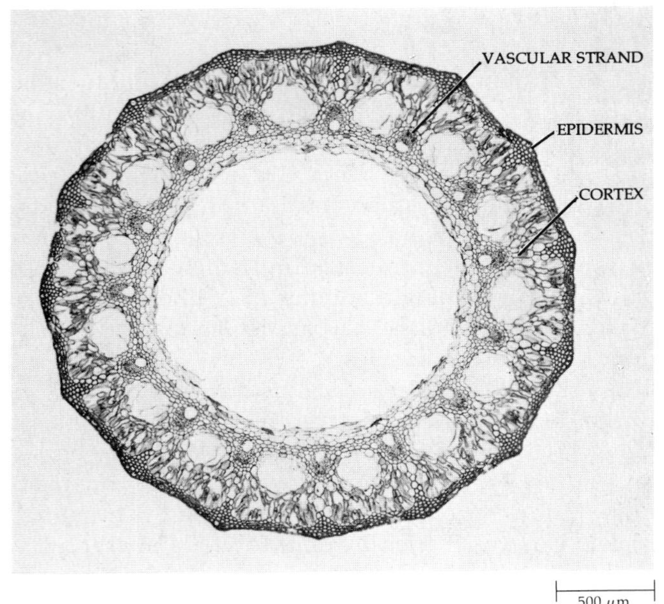

17–21
Equisetum. *Transverse section of stem, showing mature tissues.*

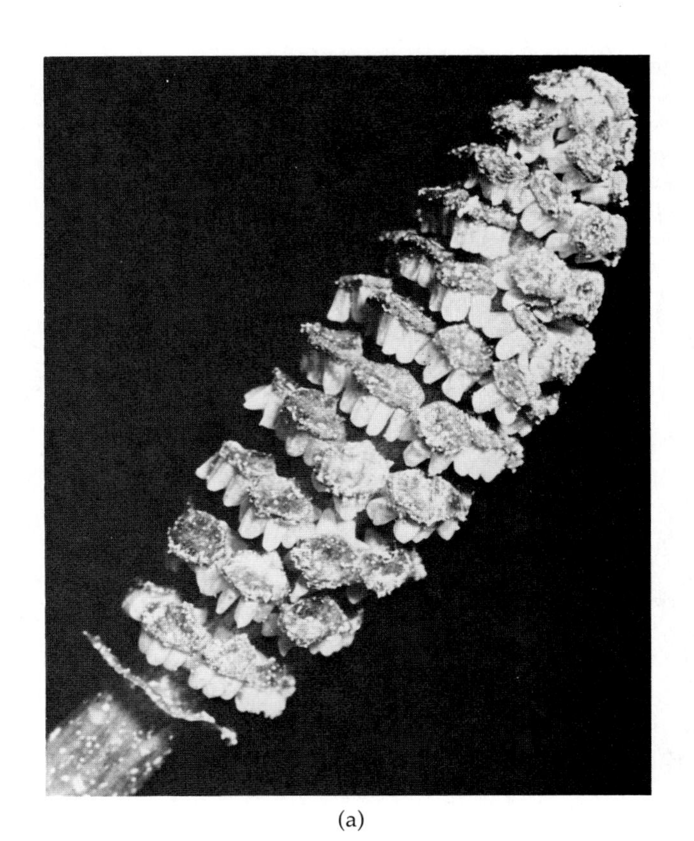

(a)

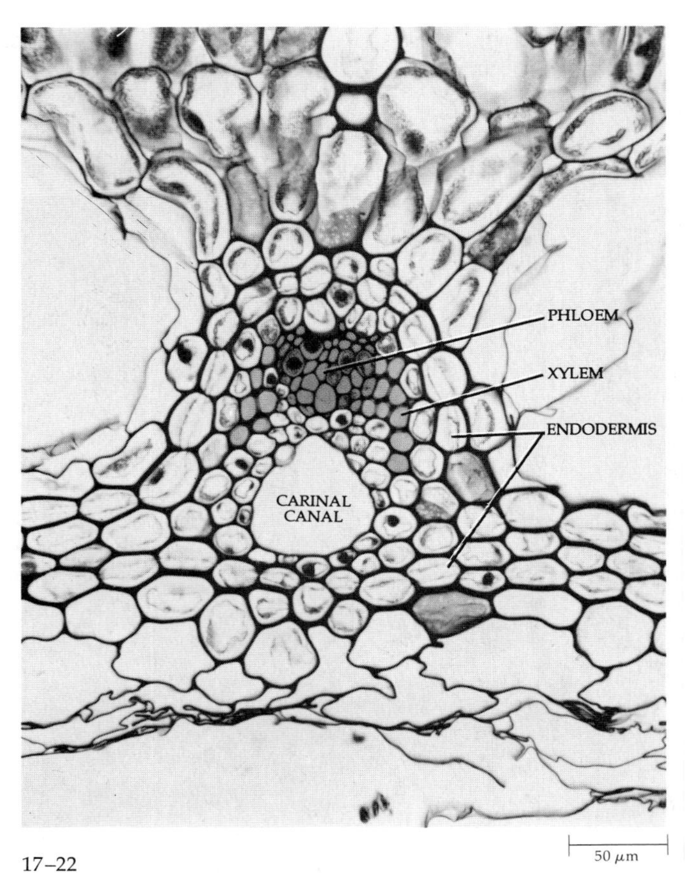

17–22
Equisetum. *Detail of vascular strand, or bundle, showing xylem and phloem.*

(b)

500 μm

17–23
Equisetum arvense. (a) *Mature, elongated strobilus, in which the sporangiophores have separated, revealing the sporangia. The flattened tops of the sporangiophores appear dark and the sporangia attached to their undersides are light in appearance.* (b) *Longitudinal section of strobilus, showing sporangiophores and immature sporangia.*

17–24
Spores of Equisetum telmateia *in dry (left) and moist (right) conditions. If a spore falls on soil that is too dry to permit germination, its elaters remain outspread and it is carried off again by the wind.*

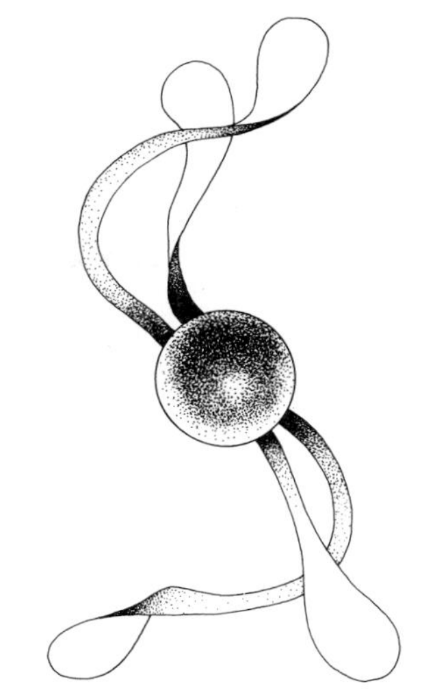

The aerial stems of *Equisetum* arise from branching underground rhizomes and, although they may die back during unfavorable seasons, the rhizomes are perennial. Anatomically, the structure of the aerial stem is quite complex (Figure 17–21). At maturity, the internodes contain a hollow pith, surrounded by a ring of smaller canals, called carinal canals. Each of these smaller canals is associated with a strand of primary xylem and primary phloem (Figure 17–22).

Equisetum is homosporous. Sporangia are borne in groups of five to ten along the margins of small, umbrellalike structures, known as *sporangiophores* (sporangia-bearing branches), which are clustered into strobili at the apex of the stem (Figure 17–23). The fertile stems of some species do not contain chlorophyll, and these stems are sharply distinct from the vegetative ones, often appearing before them early in spring (see Figure 17–20a). In other species, the strobili are borne at the tips of normal vegetative stems. When the spores are mature, the sporangia contract and split along their inner surface, releasing numerous spores. Elaters, which arise from the outer layer of the spore wall, coil when moist and uncoil when dry, thus presumably playing a role in spore dispersal (Figure 17–24).

The gametophytes of *Equisetum* are green and free-living, most being about the size of a pinhead. The gametophytes are of two types: either bisexual or strictly male. In the bisexual gametophytes, the archegonia develop before the antheridia (Figure 17–25). This developmental pattern increases the probability of cross-fertilization. The sperm are multiflagellated and require water to swim to the eggs. The eggs of several archegonia on a single gametophyte may be fertilized and develop into embryos, or young sporophytes.

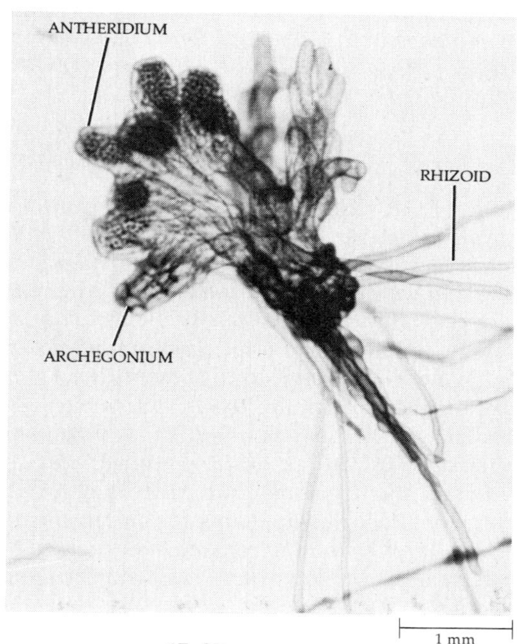

17–25
Equisetum. *Bisexual gametophyte, showing male and female gametangia.*

COAL AGE PLANTS

The amount of carbon dioxide involved in photosynthesis on an annual basis is about 100 billion metric tons. The amount returned as carbon dioxide as a result of oxidation of these living materials is about the same, differing only by 1 part in 10,000. This very slight imbalance is caused by the burying of organisms in sediment or mud under conditions in which oxygen is excluded and decay is only partial. This accumulation of partially decayed material is known as peat. The peat may eventually become covered with sedimentary rock and so placed under pressure. Depending on time, temperature, and other factors, peat may become compressed into soft or hard coal, petroleum, or natural gas—the so-called fossil fuels.

During certain periods in the earth's history, the rate of fossil fuel formation was greater than at other times. One such period was the Carboniferous, some 300 million years ago (see Figure 17–1). The lands were low, covered by shallow seas or swamps, and, in what are now temperate regions of Europe and North America, conditions were favorable for year-round growth. Euramerica was tropical to subtropical with the equator then arcing across the Appalachians, over northern Europe, and through the Ukraine. Five groups of plants dominated the swamplands, and three of them were seedless vascular plants: Lycophyta, Sphenophyta, and Pterophyta. The other two were gymnospermous types of seed plants—the seed ferns (Pteridospermales) and cordaites (Cordaitales). Because of their prominence in the Coal Age flora, the seed ferns and cordaites will be considered in this discussion, rather than in Chapter 18, "The Seed Plants."

LYCOPOD TREES

Two-thirds of the "Age of Coal" in the late Carboniferous period (Pennsylvanian) was dominated by lycopod trees, or lepidodendrids, most of which grew to heights of 10 to 25 meters and were sparsely branched (a). *After the plant attained over half of its total height, the trunk branched dichotomously. Successive branching produced progressively smaller branches until, finally, the tissues of the branch tips lost their ability to grow further. The branches bore long leaves that left leaf scars on cushions when they abscised, or fell off the branches. Although treelike in size, the lepidodendrids were essentially giant herbs, largely supported by a massive cortex surrounding a relatively small amount of xylem. Inasmuch as their root systems were shallow, these tall swamp plants were easily blown over by winds.*

Like Selaginella, *the lycopod trees were heterosporous, and their sporophylls were aggregated into cones. Some of the lepidodendrids produced seedlike structures.*

As the swamplands began to dry up and the climate began to change toward the end of the Carboniferous period, the lycopod trees vanished almost overnight, geologically speaking. Herbaceous lycopods essentially similar to Lycopodium *and* Selaginella *existed in the Carboniferous period and, as mentioned previously, are the only types still represented in today's flora.*

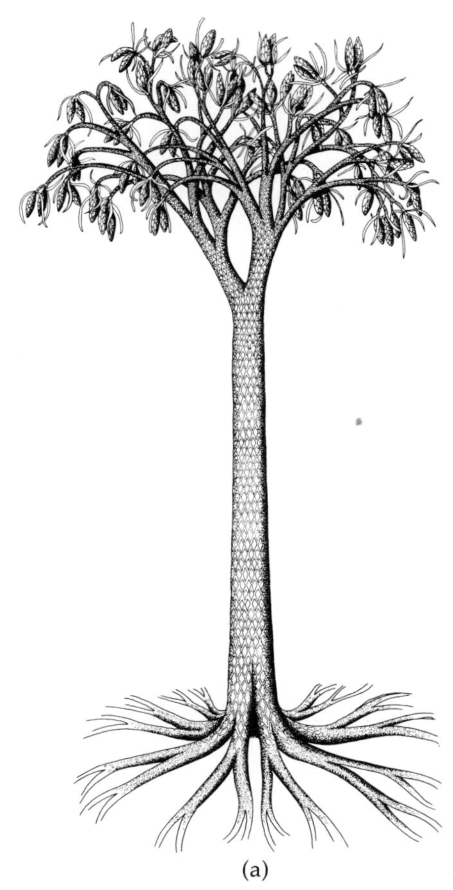

(a)

One of the dominant trees of the late Carboniferous period was *Lepidodendron,* which grew to heights of 10 to 25 meters. Preservation of the lepidodendrids has been so good that paleobotanists have learned a great deal about the plant's development.

CALAMITES

Calamites, *or giant horsetails, were plants of treelike proportions, reaching heights of 18 meters or more (see Figure 17–1). Like the plant body of* Equisetum, *that of* Calamites *consisted of a branched aerial portion and an underground rhizome system. In addition, the leaves and branches were whorled at the nodes. Even the stems were remarkably similar to those of* Equisetum, *except for the presence of secondary xylem, which accounted for most of the great diameter of the stems (trunks up to one-third meter in diameter).*

The fertile appendages, or sporangiophores, of Calamites *were aggregated into cones. Although most were homosporous, a few giant horsetails were heterosporous. Like the lycopod trees, the giant horsetails scarcely survived the Carboniferous period.*

FERNS

The ferns represented in the fossil record are recognizable as members of today's primitive fern families. The "Age of Ferns" in the late Carboniferous period was dominated by tree ferns such as Psaronius. *Up to 7.5 meters tall and very wide at the base of the trunk,* Psaronius *tapered toward the top and ended in an aggregate of large, pinnately compound fronds. Adventitious roots arose high on the stem and then grew downward either through the outer part of the trunk or along its surface. Some of the roots eventually reached the ground and acted as props for the main plant body.*

SEED PLANTS

The two remaining plant groups that dominated the swamplands of Euramerica were the seed ferns and the cordaites. Remnants of seed ferns are common in rocks of Carboniferous age. Their large, pinnately compound fronds are so fernlike that they were long regarded as ferns (b). *Then in 1905, F. W. Oliver and D. H. Scott demonstrated that these plants bore seeds and so were gymnosperms. The plants grew to about 5 meters in height, with the fronds borne at the top of a slender stem, or trunk. Microsporangia and seeds were borne on the fronds. The seed ferns survived into the Mesozoic era. It has been suggested that the seed ferns and the ferns evolved from a common ancestor, but Figure 16–4 (page 307) shows a more probable hypothesis.*

The cordaites were widely distributed during the Carboniferous period both in swamps and in drier environments (see Figure 17–1). They were tall (15 to 30 meters), much-branched trees that formed extensive forests. Their long (up to 1 meter), straplike leaves were spirally arranged at the tips of the youngest branches (c). *The center of the stem was occupied by a large pith, and a vascular cambium gave rise to a complete cylinder of secondary xylem. The root system, located at the base of the plant, also contained secondary xylem. Pollen-bearing cones and seed-bearing conelike structures were borne on separate branches. The cordaites were virtually eliminated during the Permian period.*

IN CONCLUSION

The dominant plants of the Carboniferous period—the lycopod trees and giant horsetails—disappeared from the face of the earth shortly after this period, in the Permian era. The Permian era was a time of worldwide drought and extensive glaciation. Only the herbaceous relatives of the woody groups of plants that were so abundant in the Carboniferous period continued to flourish and exist today, as do several families of ferns that appeared in the Carboniferous period. Both the seed ferns and cordaites eventually disappeared. Only one group of Carboniferous gymnosperms, the conifers (not a dominant group at the time), survived and went on to produce new types during the Permian era (see Figure 16–4, page 307). The living conifers are discussed in detail in Chapter 18.

(b)
One of the most interesting of all gymnosperm groups is the order Pteridospermales, the seed ferns. The remnants of these bizarre plants are common in rocks of Carboniferous age and have been well known to paleobotanists for a century or more. Their vegetative parts are so fernlike that for many years they were grouped with the ferns. This drawing is a reconstruction of the Carboniferous seed fern *Medullosa noei.* The plant is about 5 meters tall.

(c)
Tip of a young branch of the primitive conifer *Cordaites*, with long, straplike leaves and cones.

(a)

(c)

(e)

(b)

(d)

17–26
Ferns. (a) *Cinnamon fern,* Osmunda cinnamomea. (b) *Tree ferns,* Sphaeropteris glauca, *growing in Java.* (c) *Hart's tongue fern,* Phyllitis scolopendrium, *growing on mossy limestone at Owen Sound, Ontario.* (d) Salvinia natans, *and* (e) Marsilea, *both of which are heterosporous water ferns.*

17–27
Adiantum, a homosporous fern. Transverse section of rhizome with siphonostele. Note the wide leaf gap.

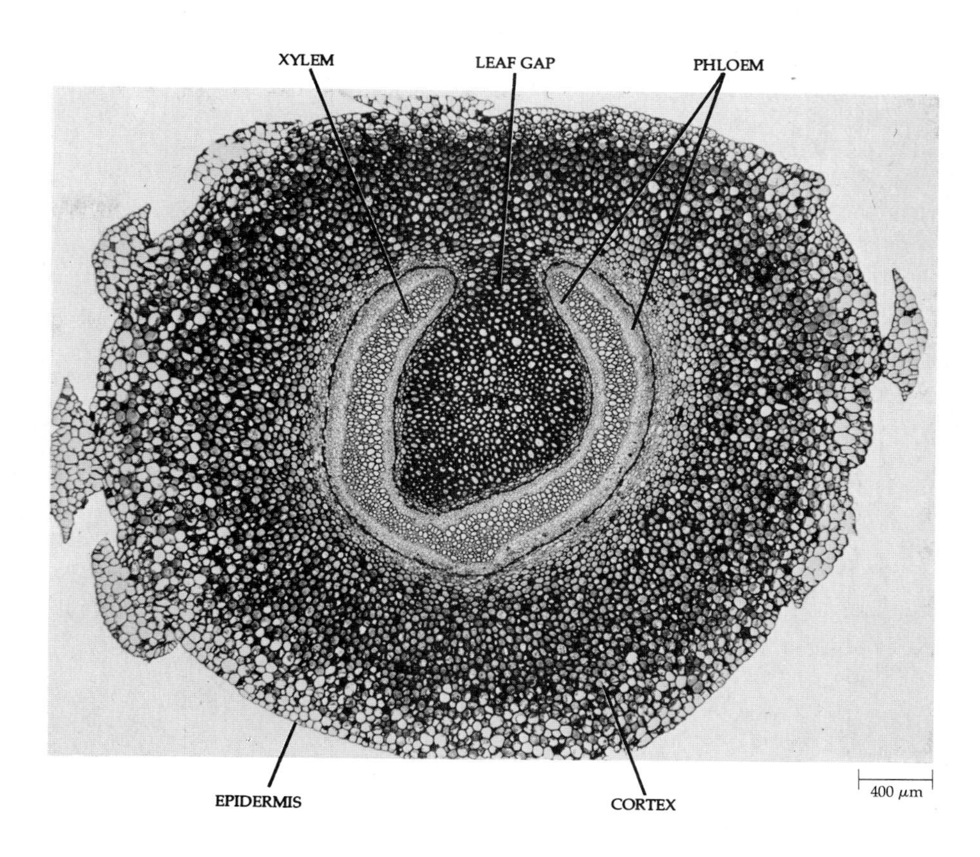

DIVISION PTEROPHYTA

Ferns have been relatively abundant in the fossil record from the Carboniferous period to the present but are not known from the Devonian period. About two-thirds of the approximately 12,000 living species are found in tropical regions, with the other third inhabiting temperate regions of the globe, including desert areas.

In both form and habitat, ferns exhibit great diversity (Figure 17–26). Some ferns are quite "unfernlike" in appearance; for example, *Salvinia,* an aquatic fern, has leaves about 2 centimeters long (Figure 17–26d). At the other extreme in size are tree ferns (Figure 17–26b), such as those of the genus *Cyathea,* some of which have been recorded to reach heights of more than 24 meters and to have leaves of 5 meters or more in length. Although the trunks of such tree ferns may be 30 centimeters or more in diameter, their tissues are entirely primary in origin. Only *Botrychium,* the relatively small "grape fern," is known to have a vascular cambium.

Most garden and woodland ferns of temperate regions consist of fleshy, underground siphonostelic rhizomes (Figures 17–27 and 17–28), which produce new sets of leaves each year. The roots are adventitious, arising from the rhizomes, near the bases of the leaves. The leaves, or *fronds*, are megaphylls, and represent the most conspicuous part of the sporophyte.

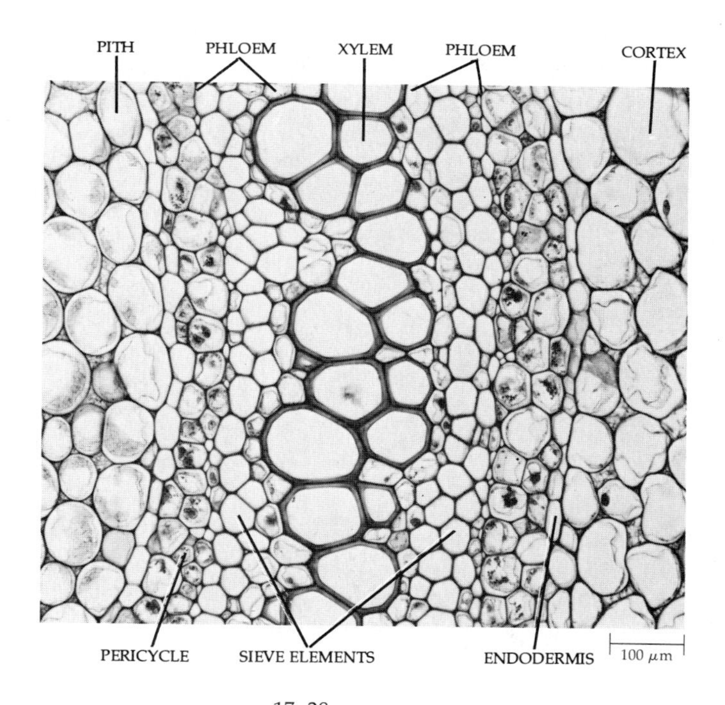

17–28
Transverse section of part of the vascular region of a rhizome of Dicksonia, a homosporous fern. The phloem is composed mainly of sieve elements; the xylem is composed entirely of tracheids.

The ferns are unique among seedless vascular plants in their possession of megaphylls. Commonly, the fronds are compound; that is, the lamina is divided into leaflets, or *pinnae*, which are attached to the *rachis*, an extension of the leaf stalk, or petiole. In nearly all ferns, the young leaves are coiled in the bud and are commonly referred to as "fiddleheads" (Figure 17–29). This type of leaf development is known as *circinate vernation*. It results from more rapid growth on the lower than the upper surface of the leaf early in development, and is mediated by the hormone auxin produced by the young pinnae on the inner side of the fiddlehead.

All but a few genera of ferns are homosporous. The sporangia are variously placed—on the lower surface of the leaves, on specially modified leaves, or on separate stalks. The sporangia commonly occur in clusters called *sori*. In many genera, the sori are covered by specialized outgrowths of the leaf, the *indusia*, which may shrivel when the sporangia are ripe (Figures 17–30 and 17–31). At this time, the mature spores—the result of meiosis in the spore mother cells—are ejected through a crack in the so-called *lip cells* (collectively called the stomium) of the sporangium. The

17–29
"Fiddleheads" of the cinnamon fern (Osmunda cinnamomea). *Years ago, the young coiled leaves of such ferns were served as a delicacy in the best hotels in New England and New York, and they are still available locally.*

(a)

(b)

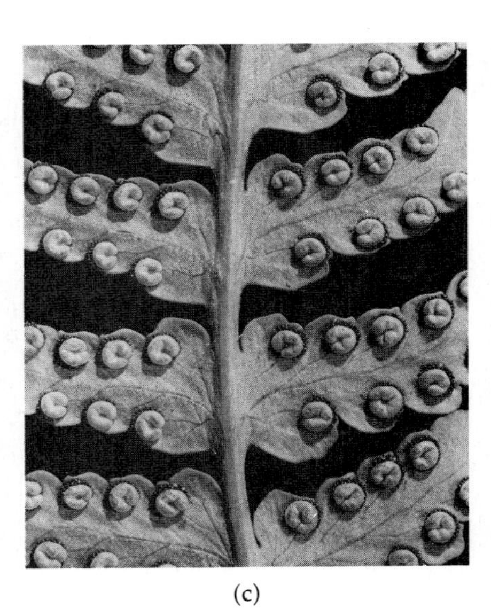

(c)

(d)

17–30
Sori are clusters of sporangia found on the undersides of the leaves of ferns. (a) *In* Polypodium virginianum *and other ferns of this genus, the sori are bare.* (b) *In* Pellaea glabella, *shown here, as well as in the bracken fern* (Pteridium aquilinum) *and the maidenhair ferns* (Adiantum), *the sori are located along the margins of the leaf blades, which are rolled back over them.* (c) *In the evergreen wood fern* (Dryopteris marginalis), *the sori, which are also located near the margins of the leaf blades, are completely covered by kidney-shaped indusia.* (d) *In the rattlesnake fern* (Botrychium virginianum), *the globular sporangia are borne on stalks that differ greatly from the leaves. Many sporangia have specialized dehiscence mechanisms and walls that consist of only a single layer of cells. However, in some more primitive groups of ferns, such as that to which* Botrychium *belongs, the sporangia have walls made up of several layers of cells and split along a simple line of thin-walled cells.*

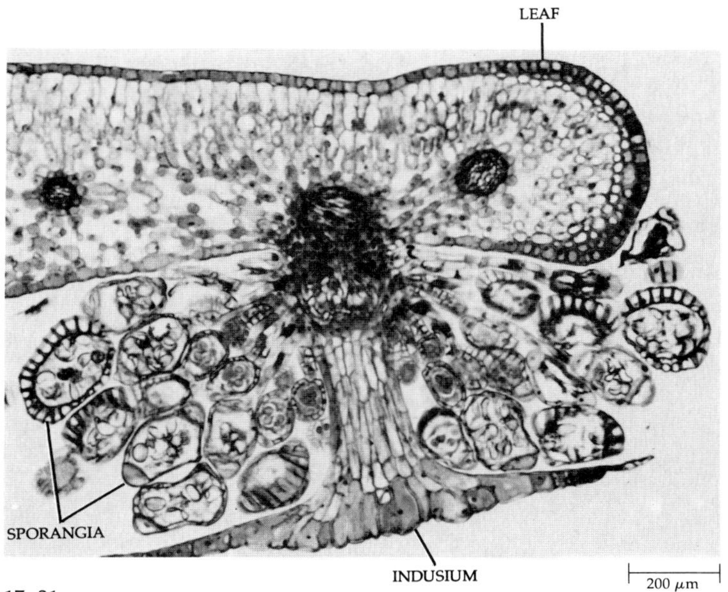

17–31
Cyrtomium falcatum, *a homosporous fern. Transverse section of a leaf, showing a sorus on the lower surface. The sporangia are in different stages of development and are protected by an umbrellalike indusium.*

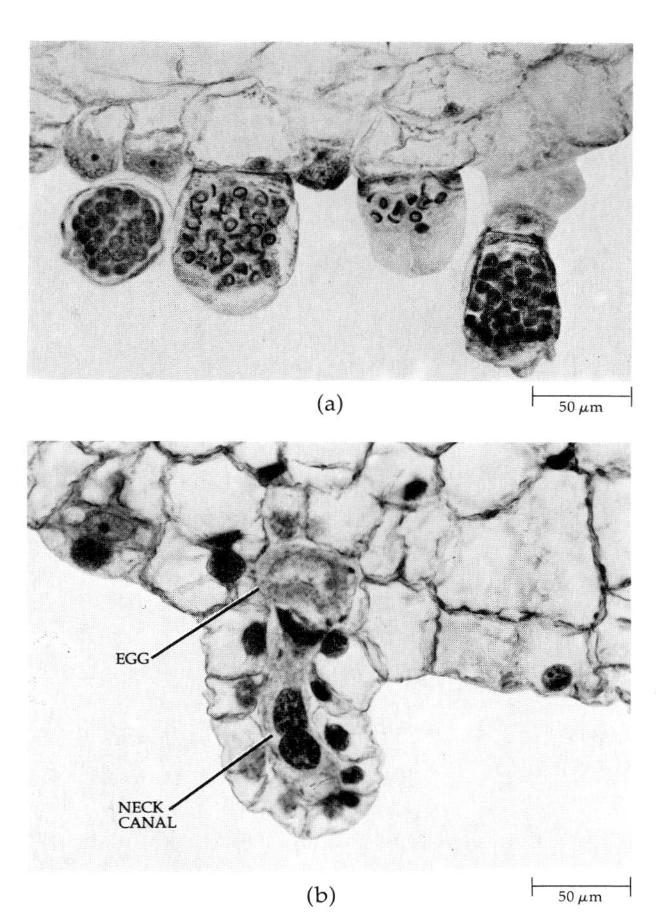

17–32
Gametangia of Osmunda. (a) *The sperm are nearly mature in the two middle antheridia.* (b) *Archegonium.*

sporangia are stalked, and each contains a special layer of unevenly thick-walled cells called an *annulus*. Contraction of the annulus causes tearing of the lip cells. Sudden expansion of the annulus then results in a catapultlike discharge of the spores.

Heterospory in ferns is restricted to two specialized groups of water ferns, although a number of extinct ferns were heterosporous.

The spores of most homosporous ferns give rise to free-living bisexual gametophytes. The gametophyte begins development as a small, pale green, algalike chain of cells called the germ filament, or protonema. It then develops into a flat, heart-shaped, membranous structure, the *prothallus*, with numerous rhizoids on its ventral (lower) surface. Both antheridia and archegonia develop on the ventral surface of the prothallus (Figure 17–32). The antheridia generally appear earlier than archegonia, primarily among the rhizoids. The archegonia then are formed near the notch. Water is required for the multiflagellate sperm to swim to the eggs in both homosporous and heterosporous ferns (Figure 17–33).

Early in its development, the embryo, or young sporophyte, receives nutrients from the gametophyte via a foot (Figure 17–34). However, development is rapid, and the sporophyte soon becomes an independent plant, at which time the gametophyte disintegrates.

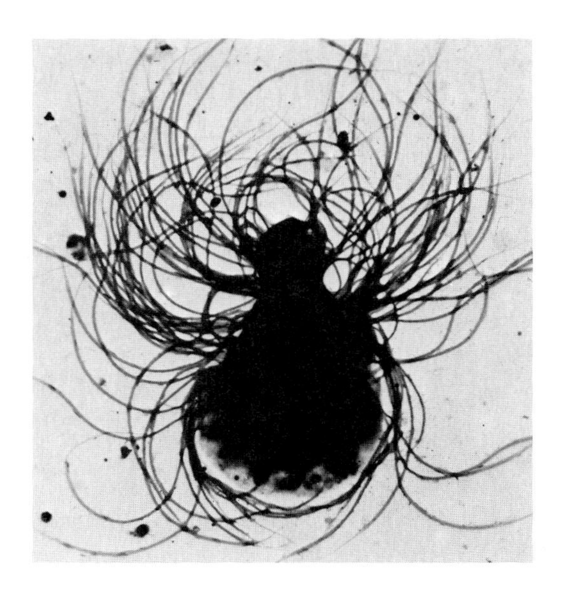

17–33
Mature sperm of the heterosporous fern Marsilea. *Each sperm possesses more than 100 flagella.*

17–34
Young fern sporophyte attached to a gametophyte.

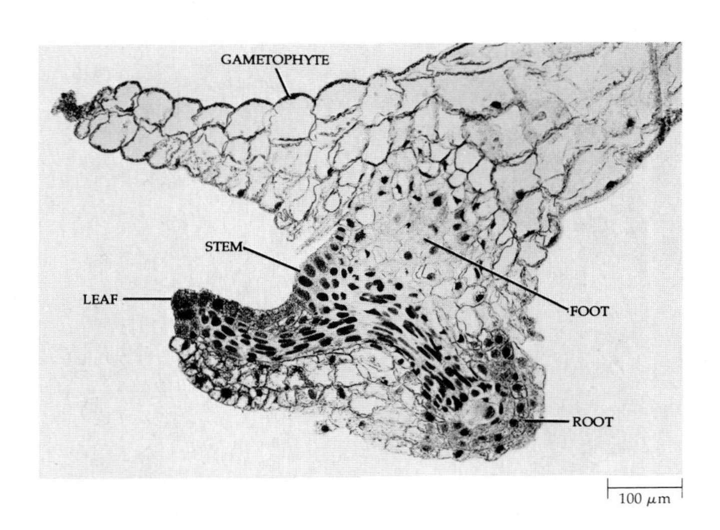

17–35
Life cycle of a homosporous fern. Following meiosis, spores are produced in the sporangia and are eventually dispersed. The gametophytes are green and nutritionally independent in most species. Many are only one-cell-layer thick and somewhat heart-shaped with an apical notch; others are thicker and may be more irregular in form. From the lower surface of the gametophyte, specialized cellular filaments known as rhizoids extend downward into the substrate.

The lower surface of the gametophyte bears flask-shaped archegonia, whose swollen lower portions are sunken into the gametophyte tissue. The necks of the archegonia are composed of several tiers of cells. Antheridia are also borne on the lower surface of the gametophyte and have a sterile jacket layer. Numerous spirally coiled, multiflagellate sperm are produced within the antheridia. When the sperm are mature and there is an adequate supply of water, the antheridia burst, releasing the sperm, which then swim into the neck of the archegonium.

In the lower portion of the archegonium, the egg is fertilized and the resulting zygote begins to divide immediately. The young embryo grows and differentiates directly into the adult sporophyte, obtaining its nutrition from the gametophyte for a time but soon achieving a level of photosynthesis sufficient to maintain itself. After the young sporophyte becomes rooted in the soil, the gametophyte disintegrates.

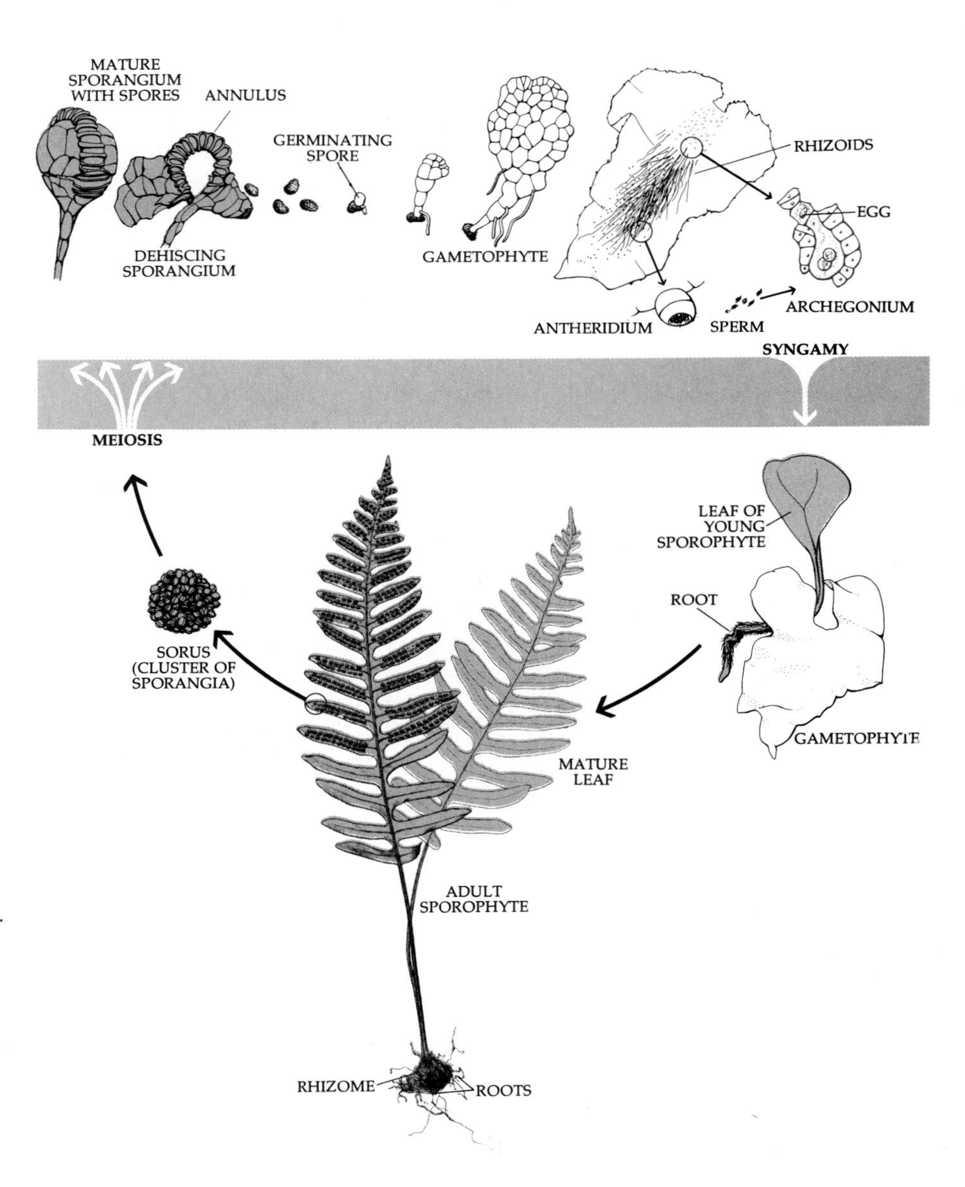

The life cycle of a homosporous fern is summarized in Figure 17–35.

SUMMARY

The living seedless vascular plants are classified in four divisions: the Psilophyta (*Psilotum* and *Tmesipteris*), the Lycophyta (which includes *Lycopodium, Selaginella,* and *Isoetes*), the Sphenophyta (*Equisetum*), and the Pterophyta (ferns). Most of the seedless vascular plants are homosporous. Heterospory is exhibited by *Selaginella, Isoetes,* and some water ferns.

The life cycles of the seedless vascular plants are essentially similar to one another: a heteromorphic alternation of generations in which the sporophyte is dominant and free-living. The gametophytes of the homosporous species are bisexual, producing both antheridia and archegonia, and are independent of the sporophyte for their nutrition. Those of heterosporous species are much reduced in size, unisexual, and dependent upon stored food derived from the sporophyte for their nutrition. All of the seedless vascular plants have motile sperm, and the presence of water is necessary for them to swim to the eggs.

The Psilophyta differ from other living seedless vascular plants in their lack of leaves (with the possible exception of *Tmesipteris*) and roots. The Lycophyta are characterized by the presence of microphylls. In addition, the small, scalelike leaves of *Equisetum* are, by definition, microphylls. Probably the most distinctive characteristic of the ferns is the presence of megaphylls.

Two of the four groups of seedless vascular plants (Lycophyta and Sphenophyta) extend back to the Devonian period. Among the seedless vascular plants, only the ferns, which first appear in the fossil record in the Carboniferous period, are represented by a large number of living species.

Five groups of vascular plants dominated the swamplands of the Carboniferous period ("Age of Coal"), and three of them were seedless vascular plants: Lycophyta, Sphenophyta, and Pterophyta. The other two were gymnospermous types—the seed ferns and the cordaites.

SUGGESTIONS FOR FURTHER READING

BANKS, HARLAN P.: *Evolution and Plants of the Past,* Wadsworth Publishing Co., Inc., Belmont, Calif., 1970.*

An introduction to plant evolution as it is revealed by the fossil record.

BIERHORST, DAVID W.: *Morphology of Vascular Plants,* The Macmillan Company, New York, 1971.

An encyclopedic and profusely illustrated presentation of the morphology of the vascular plants by one of their leading contemporary students.

FOSTER, ADRIANCE S., and ERNEST M. GIFFORD: *Comparative Morphology of Vascular Plants,* 2nd ed., W. H. Freeman and Company, San Francisco, 1974.

A well-organized, general account of the vascular plants, containing much interpretative material.

*Available in paperback.

CHAPTER 18

The Seed Plants

18–1
Reconstruction of the progymnosperm Archaeopteris, *which is common in the fossil record from eastern North America. Specimens of* Archaeopteris *attained a height of 20 meters or more.*

As mentioned in Chapter 16, the seed constitutes one of the most dramatic innovations to arise during the evolution of vascular plants, and it seems to be one of the factors responsible for the dominance of seed plants in today's flora. The reason is simple: the seed has survival value.

The oldest known seeds are from the late Devonian period, some 350 million years ago. During the next 50 million years, a wide array of seed-bearing plants evolved, including the seed ferns, cordaites, and conifers, all of which are gymnosperms.

Within the seeds of modern gymnosperms and angiosperms, the megagametophyte is held within a fleshy covering known as the *nucellus*, which is the morphological equivalent of the megasporangium. This, in turn, is enveloped by one or more additional layers that make up the *integument*, which completely encloses the megasporangium except for an opening at the apex called the *micropyle*. It is the integument that develops into the *seed coat*. In most modern seed plants, the young sporophyte, or embryo, develops within the seed before dispersal; in ancient seed plants, the seeds were generally shed before the embryo developed. To date, the oldest known embryo-containing seeds, from the early Permian period, some 270 million years ago, have been found in western Texas. Perhaps the development of the embryo before dispersal gives the seed a better chance of survival in cold and harsh conditions, and certainly the Permian was a period of climatic extremes.

In addition to a megaspore or an embryo and a seed coat, all seeds contain a form of stored food. The megasporangium plus its integument or integuments is called an *ovule*, and a seed may be defined as a mature ovule.

The evolution of the seed provided valuable means of adaptation to life on land; within the seed coat, the embryo can often remain dormant until conditions are favorable for germination and thus may be able to survive drought, frost, or other unfavorable environ-

18–2
Radial view of the secondary xylem, or wood, of the progymnosperm Callixylon newberryi. *This fossil wood is remarkably similar to that of gymnosperms.*

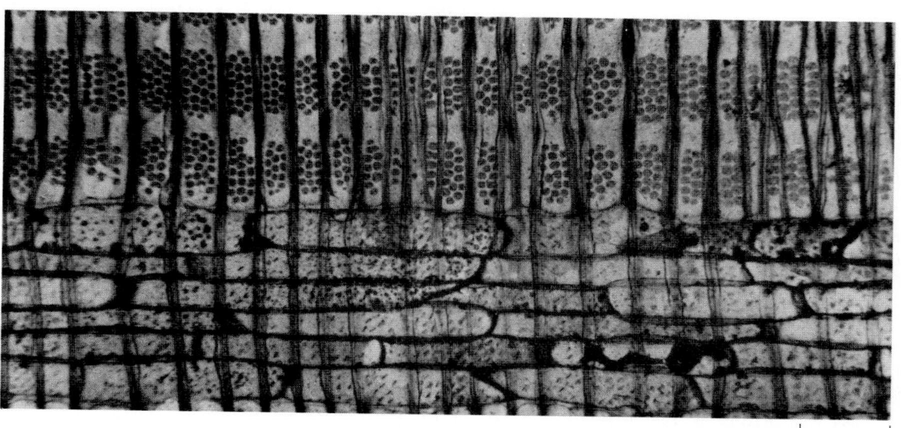

mental circumstances. The supply of nutrients in the stored food is generally more than adequate to nourish the young plant until it becomes an independent entity.

The seed plants, all of which possess megaphylls, include five divisions with living representatives: the Cycadophyta, the Ginkgophyta, the Coniferophyta, the Gnetophyta, and the Anthophyta. Before beginning our discussion of seed plants, we shall briefly examine one more group of seedless vascular plants, the entirely fossil progymnosperms. They are discussed at this time, rather than in Chapter 17, because they are the likely progenitors of the gymnosperms.

THE PROGYMNOSPERMS

In the Paleozoic era, there existed a group of plants called the progymnosperms, which had characteristics intermediate between those of the Trimerophyta and the gymnosperms. Although the progymnosperms reproduced by freely dispersed spores, they produced secondary xylem remarkably similar to that of gymnosperms (Figure 18–2). The progymnosperms and Paleozoic ferns appear to have evolved from the more ancient trimerophytes (see Figure 16–4, page 307), from which they differed primarily by having more elaborate branch systems and correspondingly more complex vascular systems. One kind of progymnosperm, the *Aneurophyton*-type, which occurred in the Devonian period some 350 to 370 million years ago was characterized predominantly by three-dimensional branching (Figure 18–3). The primary xylem existed in the form of a protostele. The organization of these plants is similar to that of some early seed ferns, leading some paleobotanists to suggest that the branch systems of the *Aneurophyton*-type progymnosperm may have been the precursors of the fernlike leaves of early seed ferns.

18–3
Reconstruction of a portion of the branch system of Triloboxylon ashlandicum, *an* Aneurophyton-*type progymnosperm. The main axis bears vegetative branches at the top and bottom and fertile organs with sporangia in between.*

The second major kind of progymnosperm, the *Archaeopteris*-type, also appeared in the Devonian period, about 360 million years ago, and extended into the early Mississippian period, about 340 million years ago (see Figure 18–1). This group is considered advanced because their lateral branch systems were flattened in one plane and bore laminar structures considered to be leaves (Figure 18–4). These leafy branch systems resemble those of early conifers. The larger branches of *Archaeopteris*-type progymnosperms had a pith. Although the progymnosperms typically were homosporous, some species of *Archaeopteris* were heterosporous.

The progymnosperms clearly represent a group of plants that reached an evolutionary level close to that of the gymnosperms. Moreover, morphological evidence amassed during the past several years strongly supports the contention that the gymnosperms evolved from the progymnosperms by the evolution of seeds in a number of different lines.

18–4
Reconstruction of a vegetative lateral branch system of the progymnosperm Archaeopteris macilenta. *The branch system is attached to a larger axis.*

THE GYMNOSPERMS

The gymnosperms include four divisions with living representatives: Cycadophyta (cycads), Ginkgophyta (maidenhair tree), Coniferophyta (conifers), and Gnetophyta (vessel-containing gymnosperms). The name *gymnosperm*, which means literally "naked seed," points to one of the principal characteristics of all members of this group of vascular plants: their ovules and seeds are borne exposed on the surface of sporophylls or analogous structures. The four divisions of gymnosperms probably represent the achievement of a particular stage in evolutionary progression by various descendants of the progymnosperms.

With few exceptions, the female gametophytes of gymnosperms produce several archegonia each. As a result, more than one egg may be fertilized, so several embryos may begin to develop within a single ovule—a phenomenon known as polyembryony. In most cases, only one embryo survives, so that relatively few fully developed seeds contain more than one embryo.

In the seedless vascular plants, water is required for the motile, flagellate sperm to reach and fertilize the eggs. In the gymnosperms, water is not required as a medium of transport of the sperm to the eggs. Instead, the partly developed male gametophyte, the *pollen grain*, is transferred bodily (in general, passively, by the wind) to the vicinity of a female gametophyte within an ovule. In gymnosperms, this process is called *pollination*. After pollination occurs, the male gametophyte produces a tubular outgrowth, the *pollen tube*. In the conifers and Gnetophyta, the sperm are nonmotile, and the pollen tubes convey them directly to the archegonia. In the cycads and *Ginkgo*, the sperm are multiflagellated, and the pollen tube apparently functions largely as a haustorial structure (analogous to the haustoria of parasitic fungi), which penetrates the ovule and absorbs nourishment directly from it. The pollen tube may grow for several months in the tissue of the nucellus, or megasporangium, before reaching a cavity above the female gametophyte. At that time, the pollen tube bursts and liberates its two sperm into the cavity. The sperm then swim to an archegonium, and one fertilizes the egg. With the development of sperm-conveying pollen tubes, the evolving vascular plants no longer were dependent upon water to assure fertilization.

The Conifers

By far the most numerous and widespread of the gymnosperm divisions living today, the Coniferophyta comprise some 50 genera with about 550 species. The tallest vascular plant, which is the redwood *(Sequoia sempervirens)* of coastal California and southwestern Oregon, is a member of this group (see Figure 29–1, page 578). These trees attain heights of up to 117 meters and diameters in excess of 11 meters. The conifers, which also include pines, firs, and spruces, are of great commercial value; their stately forests provide the wealth of vast regions of the North Temperate Zone. During the early Tertiary period, some genera were more widespread than they are now and they dominated huge expanses on all the northern continents.

The history of the conifers extends back at least to late Carboniferous times, some 290 million years ago. Their leaves have many drought-resistant features, and perhaps the origin of conifers ought to be sought during the relatively dry Permian period, which immediately followed the Carboniferous. At that time, increasing worldwide aridity must have provided a powerful evolutionary stimulus.

18–5
Longleaf pines, Pinus palustris, *growing in North Carolina.*

The Pines

The pines (genus *Pinus*) include among them perhaps the most familiar of all gymnosperms (Figure 18–5), because they dominate broad stretches of North America and Eurasia and are widely cultivated even in the Southern Hemisphere. There are some 90 species of pines, all of which are characterized by an arrangement of the leaves that is unique among living conifers. The conspicuous leaves of pines are needle-like. In the seedlings, they are borne singly and are

(a)

(b)

18–6
(a) *A seedling of longleaf pine,* Pinus palustris, *from Georgia, showing the juvenile leaves (long needles, borne singly) and the first mature leaves borne*

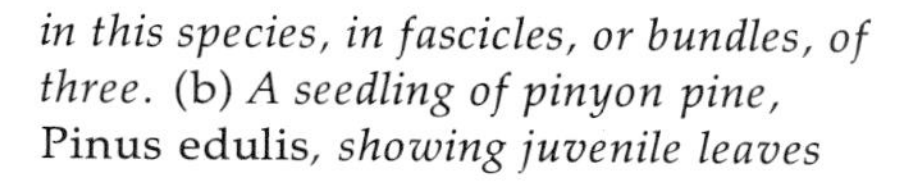

in this species, in fascicles, or bundles, of three. (b) *A seedling of pinyon pine,* Pinus edulis, *showing juvenile leaves and young tap root system. The mature leaves of this species are borne in fascicles of two needles.*

spirally arranged on the stems (Figure 18–6). After a year or two of growth, a pine begins to produce its leaves in bundles, or fascicles, each of which contains a specific number of long, needlelike leaves—ranging from one to eight, depending on the species. These fascicles, wrapped at the base by a series of short, scalelike leaves, are actually short shoots in which the activity of the apical meristem becomes suspended (Figure 18–7). Thus, a fascicle of needles in a pine is morphologically a determinate (restricted in growth) branch system. Under unusual circumstances, the apical meristem within a fascicle of needles in a pine may be reactivated and grow into a new shoot with indeterminate growth, or sometimes may even produce roots and grow into an entire pine tree (Figure 18–8).

The leaves of pines, like those of many other conifers, are impressively suited for growth under arid conditions (Figure 18–9). The epidermis is covered with a thick cuticle, beneath which are one or more layers of compactly arranged, thick-walled cells—the hypodermis. The stomata are sunken below the surface of the leaf. The mesophyll, or ground tissue of the leaf, consists of parenchyma cells with conspicuous wall ridges that project into the cells, like the pieces of a puzzle. Commonly, the mesophyll is penetrated by

18–8
One-year-old pines (Pinus radiata) *grown from rooted fascicles of needles. This experiment, carried out by Jochen Kummerow of San Diego State University, emphasizes that a fascicle of pine needles is actually a short shoot in which the activity of the apical meristem has been suspended but can be regenerated.*

(a)

(b)

18–7
(a) *Bristlecone pine,* Pinus longaeva, *at Bryce Canyon, Utah. Branch showing clusters of five mature needles and a mature ovulate cone.* (b) *Branch of red pine,* Pinus resinosa, *with young ovulate cones. Note the fascicles of needlelike leaves characteristic of the adult plant.*

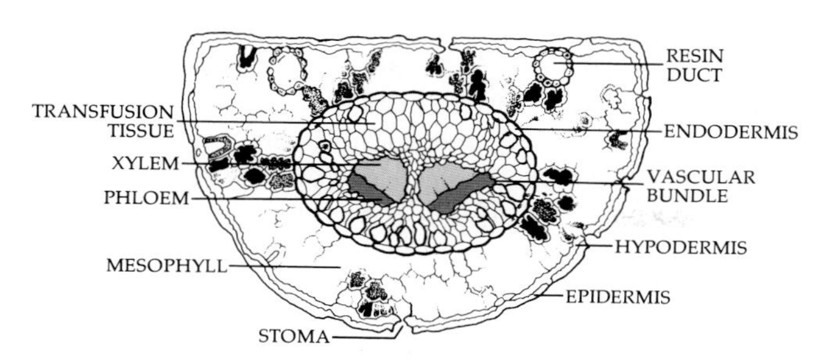

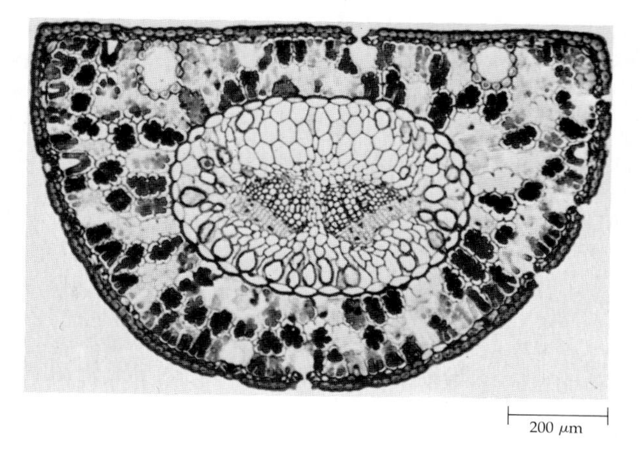

18–9
Pinus. *Transverse section of needle, showing mature tissues.*

two or more resin ducts. One vascular bundle, or two bundles side-by-side, are found in the center of the leaf. The veins are surrounded by transfusion tissue, composed of living parenchyma cells and short, nonliving tracheids. The transfusion tissue is believed to conduct materials between the mesophyll and vascular bundles. An endodermis surrounds the transfusion tissue, so that the transfusion tissue and mesophyll are not in direct contact with one another.

In the pines and other conifers, secondary growth begins early and leads to the internal formation of substantial amounts of secondary xylem (Figure 18–10). Secondary phloem is produced externally. The xylem consists primarily of tracheids and the phloem of sieve cells (typical food-conducting cells of gymnosperms and seedless vascular plants). Both tissues are traversed radially by narrow rays. With the initiation of secondary growth, the epidermis is eventually replaced with a periderm, which has its origin in the outer layer of cortical cells. As secondary growth continues, subsequent periderms are produced deeper in the bark.

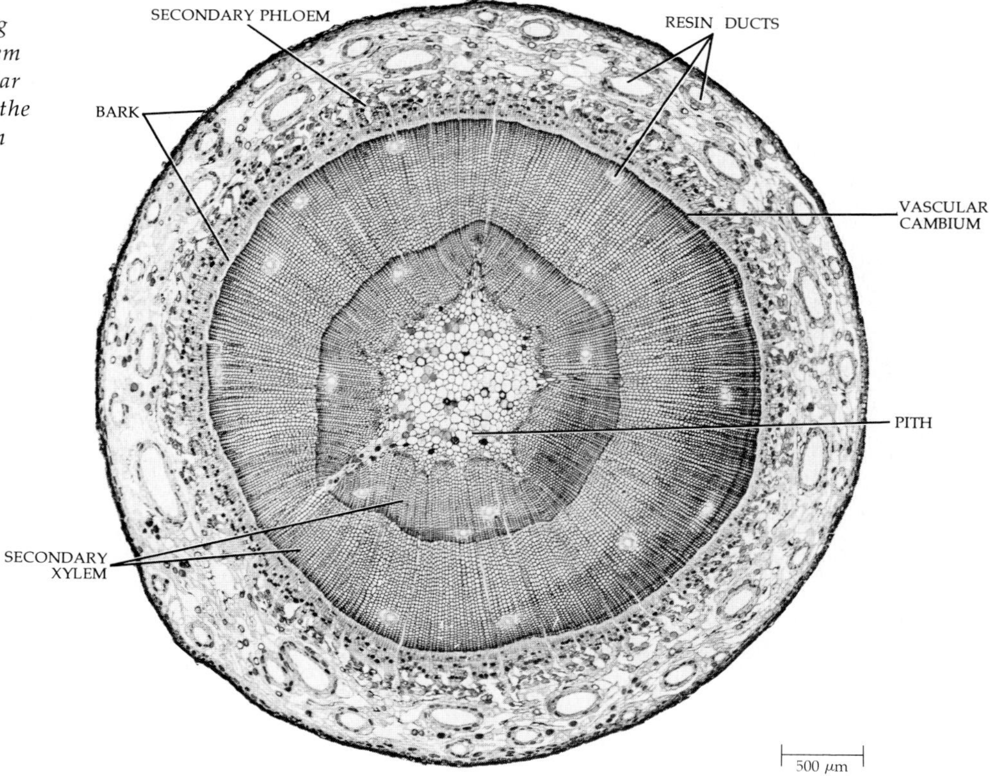

18–10
Pinus. *Cross-section of stem, showing secondary xylem and secondary phloem separated from one another by vascular cambium. All of the tissues, including the phloem, outside the vascular cambium comprise the bark.*

Reproduction in Pines. The microsporangia and megasporangia in pines and other conifers are borne in separate cones, usually on the same tree. Ordinarily the microsporangiate cones are borne on the lower branches, and the megasporangiate, or ovulate, cones are borne on the higher branches of the tree.

Microsporangiate cones in the pines are relatively small, usually 1 to 2 centimeters in length (Figure 18–11). The microsporophylls (Figure 18–12) are spirally arranged and more or less membranous in texture. Each bears two microsporangia. Each young microsporangium contains many microsporocytes, or microspore mother cells, which in early spring undergo meiosis to give rise to four haploid microspores each. Each microspore develops into a winged pollen grain, consisting of two *prothallial cells*, a *generative cell*, and a *tube cell* (Figure 18–13). This four-celled pollen grain comprises the immature male gametophyte. It is at this stage that the pollen grains are shed in enormous quantities, and some are carried by the wind to the ovulate cones.

Ovulate cones of pines are much larger and considerably more complex in structure than the pollen-bearing cones (Figure 18–14). The ovuliferous scales (cone scales), which bear the ovules, are not megasporophylls but are entire modified branch systems of determinate growth, properly known as *seed-scale complexes*. Each seed-scale complex consists of the ovuliferous scale—which bears two ovules on its upper surface—and a subtending sterile bract (Figure 18–15). The scales are arranged spirally around the axis of the cone. (The ovulate cone is, therefore, a compound structure, whereas the microsporangiate cone is a simple one, with the microsporangia directly attached to the microsporophylls.) Each ovule contains a multicellular nucellus (the megasporangium) surrounded by a massive integument with an opening, the micropyle, facing the cone axis. Each megasporangium contains a single megasporocyte, or megaspore mother cell, which ultimately undergoes meiosis to give rise to a linear series of four megaspores. However, only one of these megaspores is functional; the three nearest the micropyle soon degenerate.

Pollination in pines occurs in spring, the pollen adhering to a drop of sticky fluid about the micropyle.

18–11
Jack pine, Pinus banksiana. *Microsporangiate cones shedding pollen.*

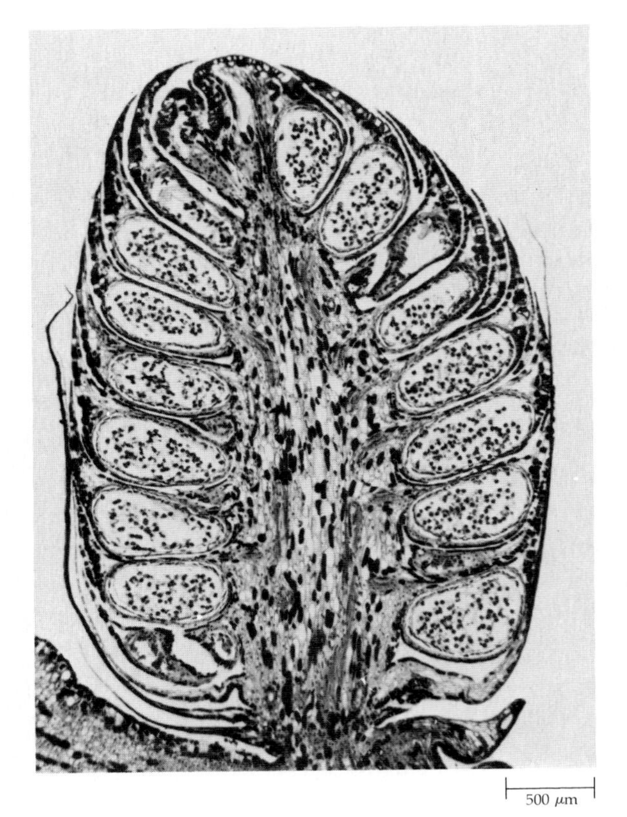

18–12
Pinus. *Longitudinal view of a pollen-producing cone, showing microsporophylls and microsporangia containing mature pollen grains.*

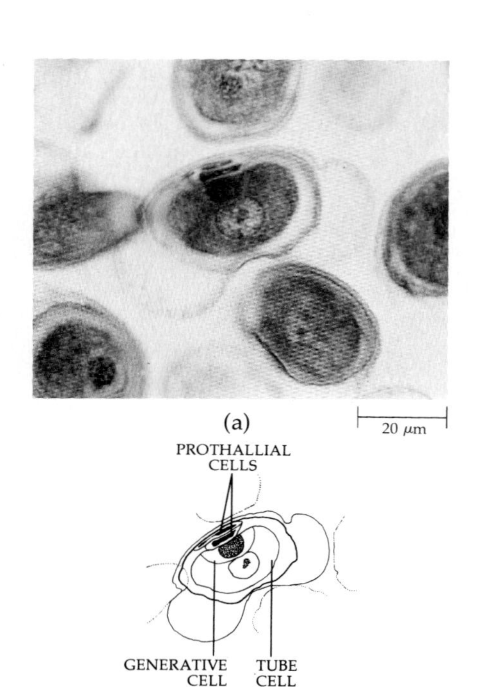

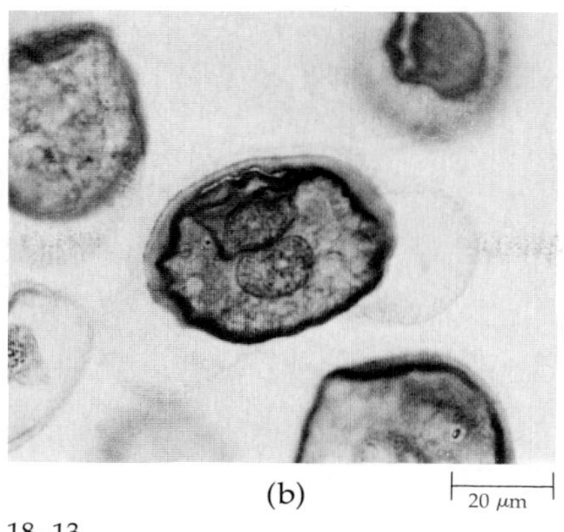

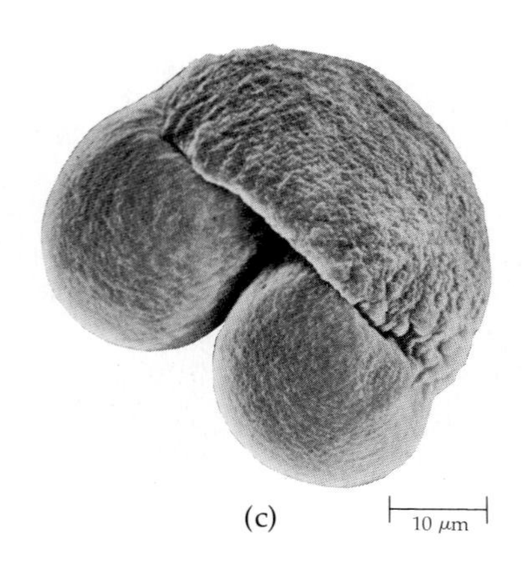

18–13
Pinus. (a) *Pollen grains with enclosed immature male gametophytes. Each gametophyte consists of two prothallial cells, a relatively small generative cell, and a relatively large tube cell.* (b) *A somewhat older pollen grain than that shown in* (a). *Here the prothallial cells, which have no apparent function, have degenerated.* (c) *Scanning electron micrograph of a pine pollen grain, with its two bladder-shaped wings. When the pollen grain germinates, the pollen tube emerges from the lower end of the grain between the wings.*

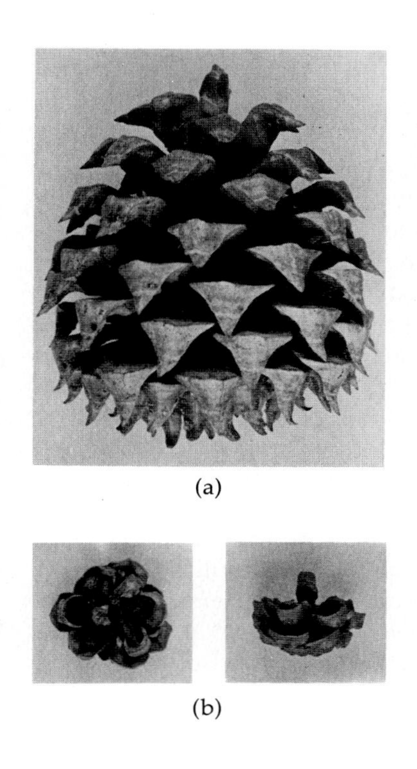

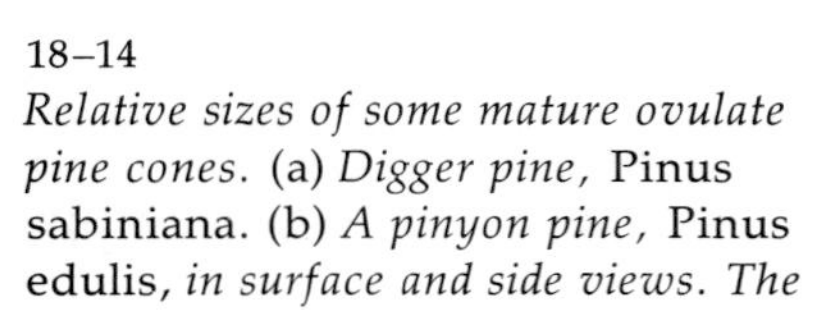

18–14
Relative sizes of some mature ovulate pine cones. (a) *Digger pine,* Pinus sabiniana. (b) *A pinyon pine,* Pinus edulis, *in surface and side views. The edible seeds of this and certain other pines are called "pine nuts."* (c) *Sugar pine,* Pinus lambertiana. (d) *Western yellow pine,* Pinus ponderosa. (e) *Eastern white pine,* Pinus strobus. (f) *Red pine,* Pinus resinosa.

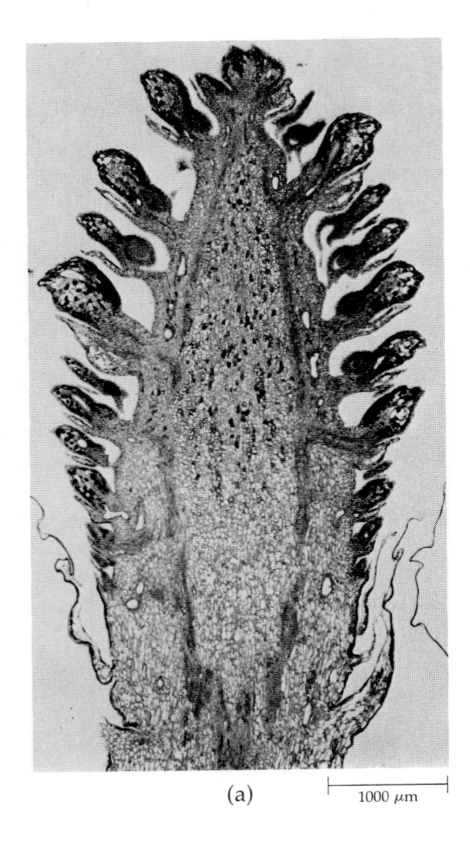

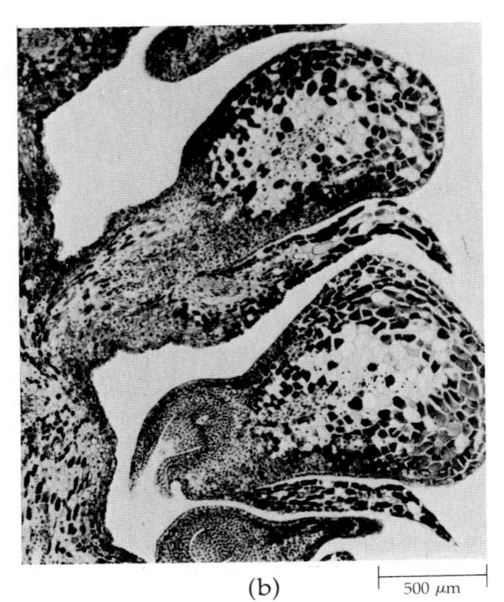

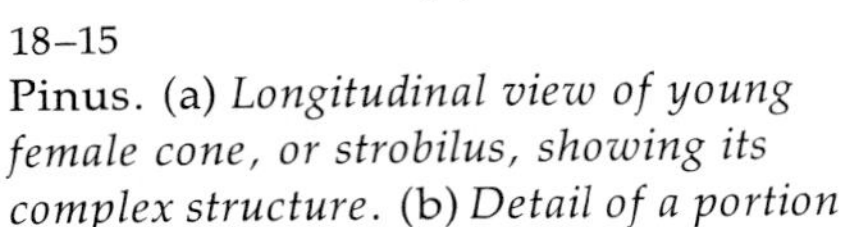

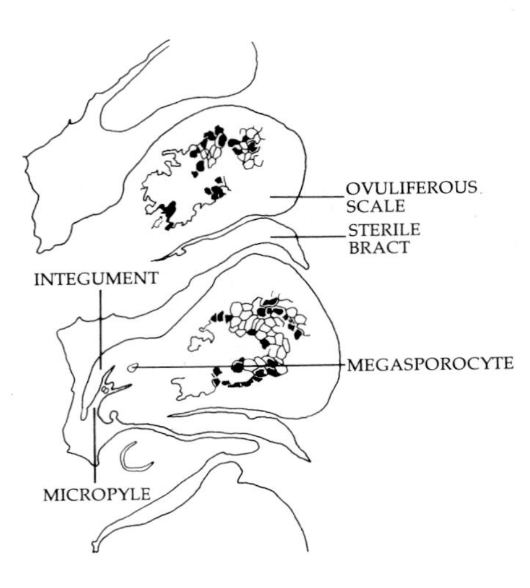

(a) 1000 μm

(b) 500 μm

18–15
Pinus. (a) *Longitudinal view of young female cone, or strobilus, showing its complex structure.* (b) *Detail of a portion of the strobilus. Note the megasporocyte (megaspore mother cell) surrounded by the nucellus.*

At this stage, the scales of the ovulate cone are widely separated. As the micropylar fluid evaporates, the pollen grain is drawn into the space between the micropyle and nucellus, or megasporangium. Following pollination, the scales grow together and afford a higher degree of protection for the developing ovules. Shortly after the pollen grain comes in contact with the nucellus, it germinates, forming a pollen tube. At this time, meiosis has not yet occurred in the megasporangium. About a month after pollination, the four megaspores are produced, only one of which develops into a megagametophyte. Development of the female gametophyte is very sluggish; it often does not begin developing until some 6 months after pollination, and it may then require another 6 months for completion. In the early stages of megagametophyte development, mitosis proceeds without immediate cell-wall formation. About 13 months after pollination, when the female gametophyte contains some 2000 free nuclei, cell-wall formation begins. Then, approximately 15 months after pollination occurs, archegonia, usually two or three in number, differentiate at the micropylar end of the megagametophyte (Figure 18–16). The stage is now set for fertilization to take place.

A year earlier, the pollen grain had begun to germinate, slowly digesting its way through the tissues of the nucellus on its way toward the developing female gametophyte. About 12 months after pollination, the generative cell of the four-celled male gametophyte undergoes division to give rise to two kinds of cell—a *stalk cell* and a *body cell*. Subsequently, before the pollen tube reaches the female gametophyte, the body cell divides to produce two sperm. The male gametophyte, or germinating pollen grain, is now mature. Bear in mind that an antheridium is not formed by the male gametophyte.

Some 15 months after pollination, the pollen tube reaches the egg cell of an archegonium, where it discharges much of its cytoplasm and both of its sperm into the cytoplasm of the egg (Figure 18–17). One sperm nucleus unites with the egg nucleus, and the other degenerates. Commonly, the eggs of all archegonia are fertilized and begin to develop into embryos (the phenomenon of polyembryony). However, only one embryo generally develops fully.

During early embryogeny, four tiers of cells are produced near the lower end of the archegonium. Each of the four cells of the lowermost tier (that is, the tier farthest from the micropylar end of the ovule) begins to form an embryo, while the four cells of the tier next to the bottom, the suspensor cells, elongate greatly and force the developing embryos into the female gametophyte. Thus, for a second time in the pine life cycle, polyembryony is exhibited. Once again, however, only one of the embryos develops fully. During embryogeny the integument develops into a seed coat.

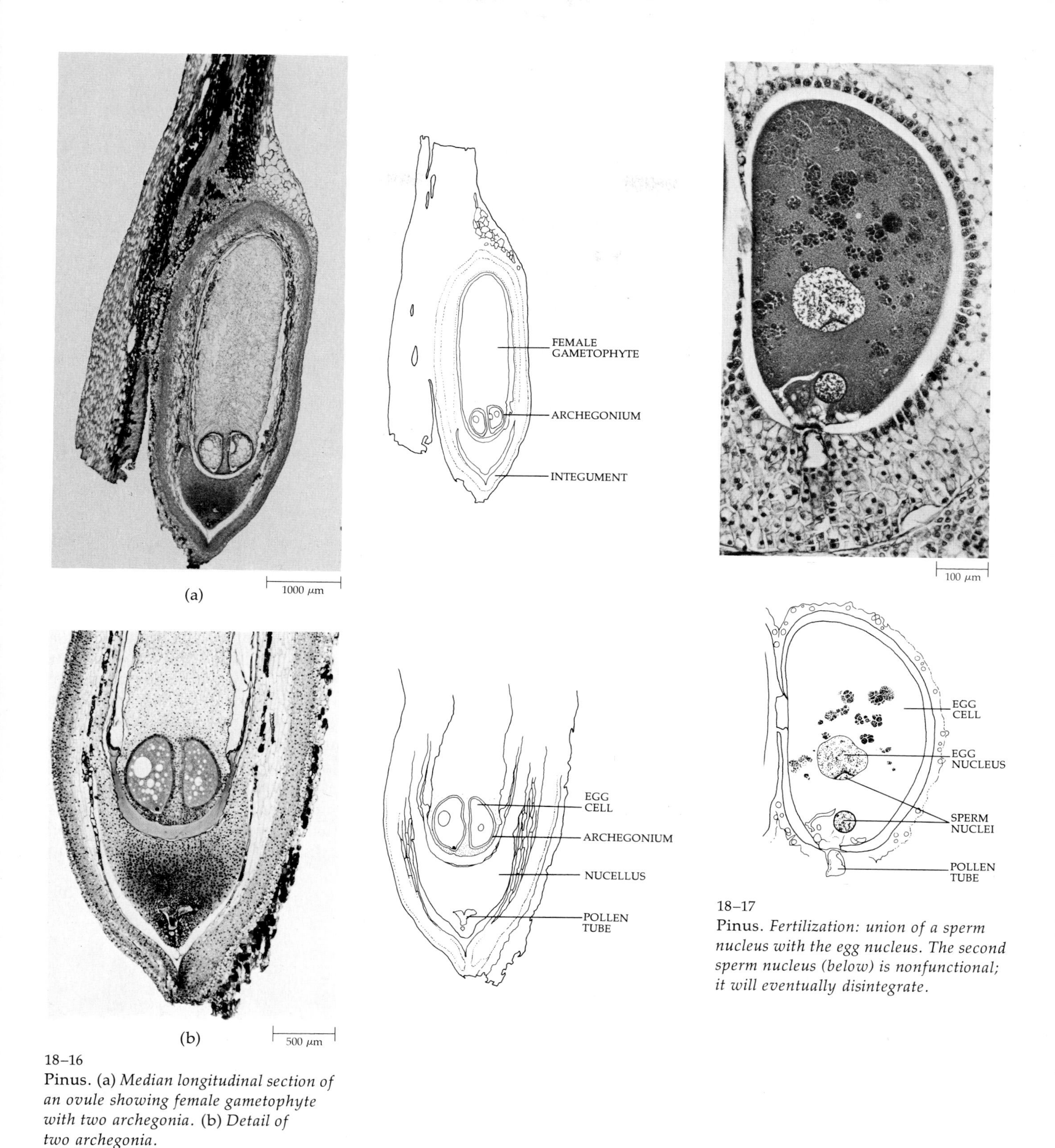

18–16
Pinus. (a) *Median longitudinal section of an ovule showing female gametophyte with two archegonia.* (b) *Detail of two archegonia.*

18–17
Pinus. *Fertilization: union of a sperm nucleus with the egg nucleus. The second sperm nucleus (below) is nonfunctional; it will eventually disintegrate.*

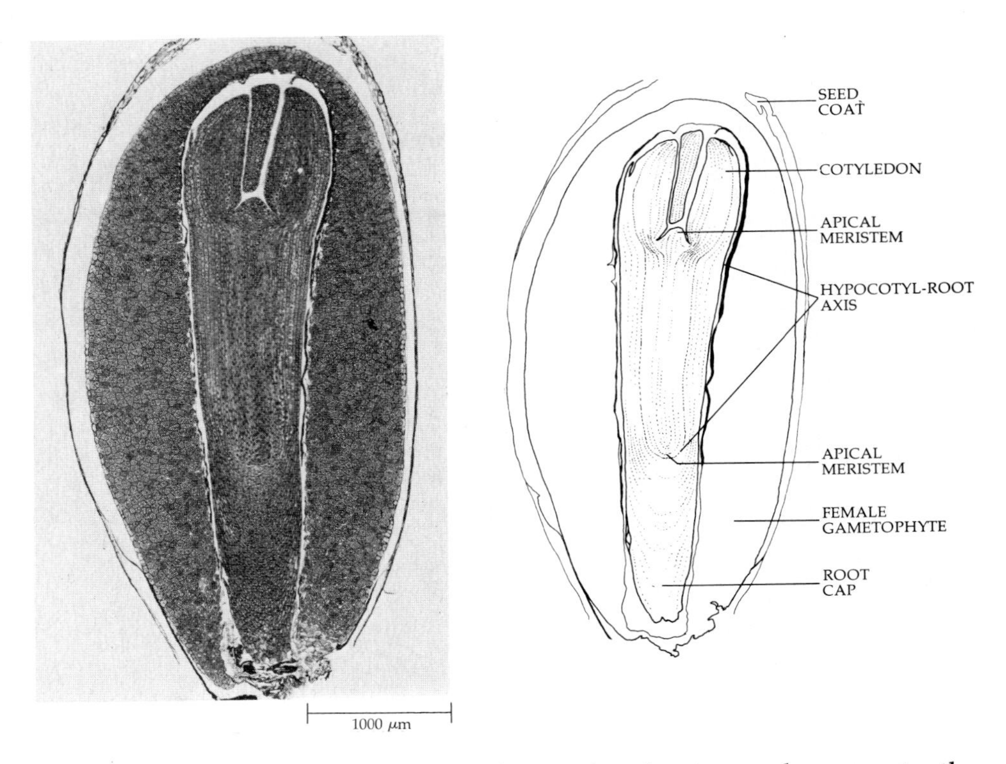

18–18
Pinus. *Longitudinal section of seed.*

The seed of a pine and other conifers is a remarkable structure, for it consists of a combination of two sporophytic generations—the seed coat and the embryo—and one gametophytic generation, which serves as reserve food or nutritive tissue (Figure 18–18). The seed coat and embryo are diploid, and the female gametophyte is haploid. The embryo consists of a hypocotyl-root axis, with a root cap and apical meristem at one end and an apical meristem and several (generally eight) cotyledons, or seed leaves, at the other. The integument consists of three layers, of which the middle layer becomes hard and serves as the seed coat.

The seeds of pines are often shed from the cones during the autumn of the second year following the initial appearance of the cones and the occurrence of pollination. At maturity, the cone scales separate; the winged seeds flutter through the air and are sometimes carried considerable distances by the wind. In some species of pines, such as jack pines *(Pinus banksiana),* the scales do not separate until the cones are subjected to extreme heat. When a forest fire sweeps rapidly through the pine grove and burns the parent trees, many of the fire-resistant cones are only scorched. These cones open and release the seed crop accumulated over many years, thus reestablishing the species in the recently burned soil. Sometimes the closed cones remain on the branches of the trees for many years and actually become buried in the wood as the growing trunks and limbs swell out around them (Figure 18–19).

The pine life cycle is summarized in Figure 18–20.

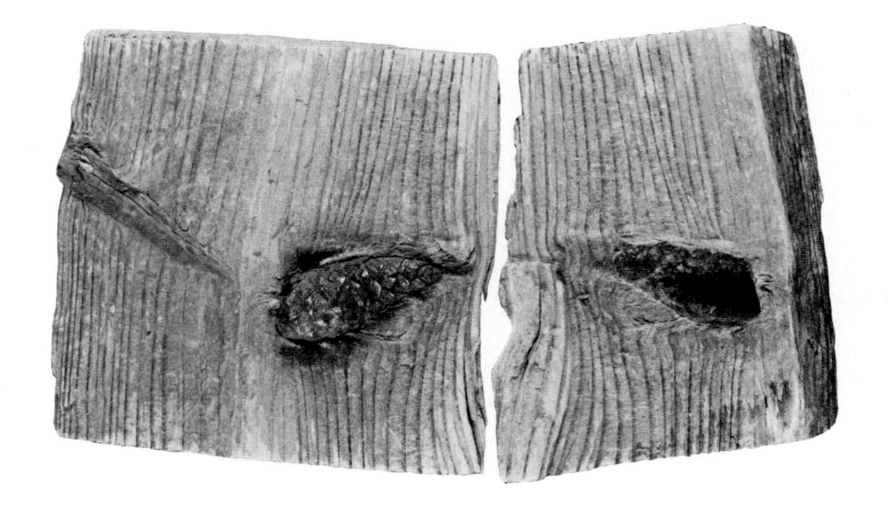

18–19
*Portion of a stem of jack pine (*Pinus banksiana*), with an ovulate cone embedded in wood.*

18–20
Life cycle of a pine. The gametophytes are much reduced and nutritionally dependent upon the sporophyte. They are enclosed within the tissue of the sporophyte. The ovule that encloses the megagametophyte matures after fertilization and becomes the seed. The elaborate suspensor, which is characteristic of the pines, disintegrates by the time the embryo is fully developed.

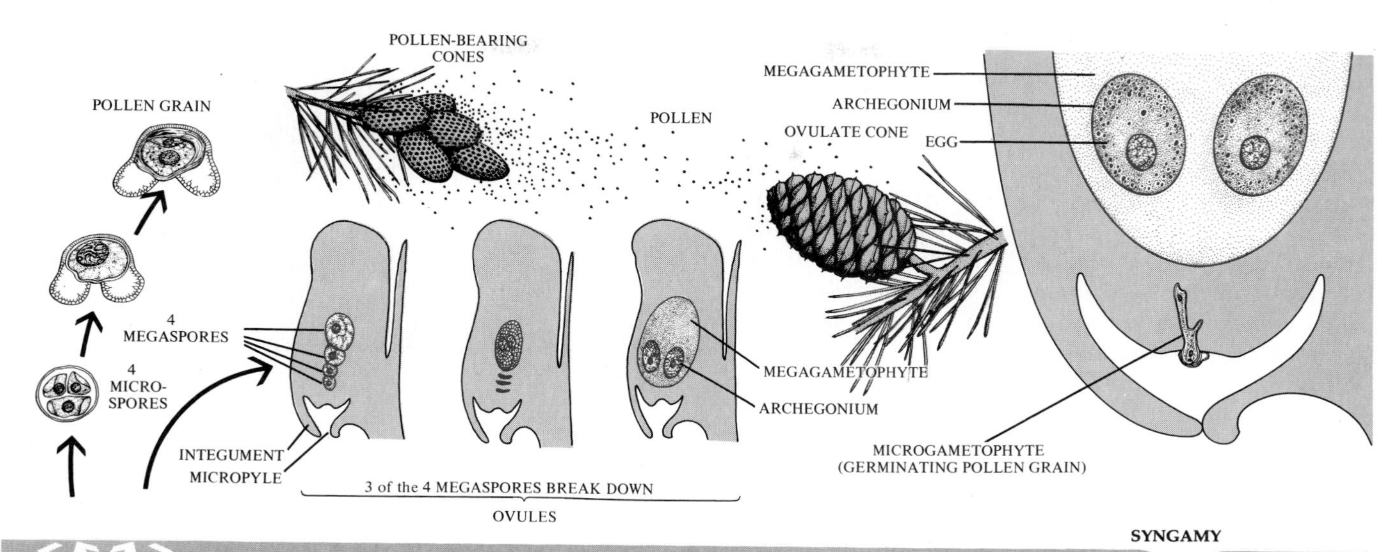

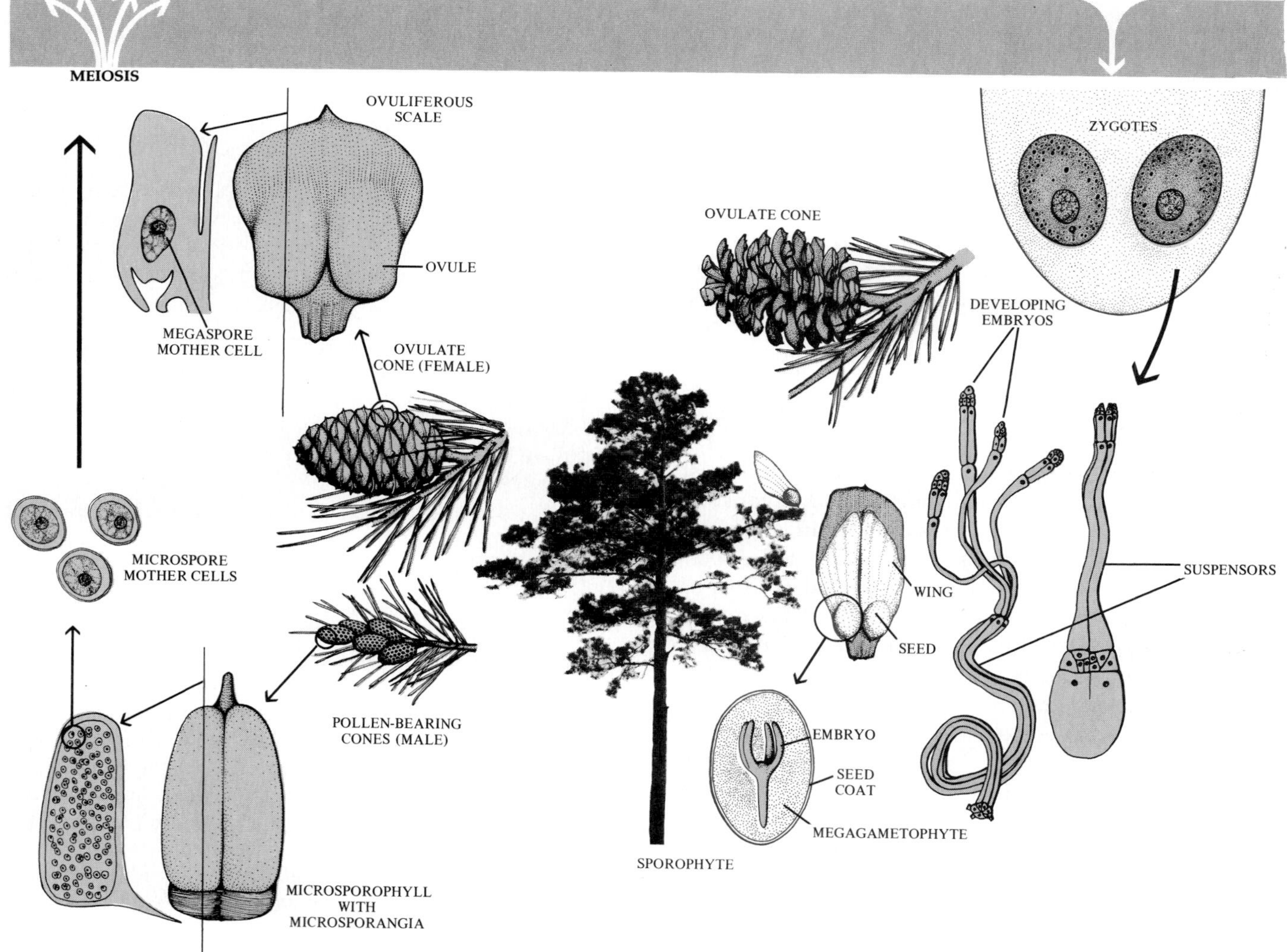

Other Conifers

Although other conifers (see Figures 18–21 through 18–28) lack the needle clusters of the pines and may also differ in a number of relatively minor details of their reproductive systems, the group is a fairly homogeneous one. Among its important representatives are the firs *(Abies)*, spruces *(Picea)*, hemlocks *(Tsuga)*, Douglas fir *(Pseudotsuga)*, cypresses *(Cupressus)*, and junipers *(Juniperus)*. In the yews (family Taxaceae), the ovules are not borne in cones but are solitary and surrounded by a fleshy, cuplike structure—the *aril* (see Figure 18–25a). A number of other genera of conifers are typically found only in the Southern Hemisphere. Some of them, like the Norfolk Island pine *(Araucaria excelsa)* and the monkey-puzzle tree *(A. araucana)*, are frequently cultivated.

One of the most interesting groups of conifers is the family Taxodiaceae, which includes the tallest vascular plant, *Sequoia sempervirens* (see Figure 29–1). The famous "big tree," *Sequoiadendron giganteum*, which forms spectacular, widely scattered groves along the west slope of the Sierra Nevada of California, also belongs to this group, as do the bald cypresses *(Taxodium)* found in the southeastern United States and in Mexico (Figure 18–26). All these genera were more widespread in the Tertiary period than they are at present (see Figure 18–28).

Another genus that was abundant in the Tertiary in both Eurasia and North America was *Metasequoia*, the dawn redwood (Figure 18–27). Indeed, *Metasequoia* was the most abundant conifer in western and arctic North America from the late Cretaceous period to the Miocene (in other words, through most of the Tertiary period). The genus *Metasequoia* was first described from fossil material by the Japanese paleobotanist Shigeru Miki in 1941. Three years later the Chinese forester Tsang Wang, from the Central Bureau of Forest Research of China, visited the village of Motao-chi in remote Sichuan Province in the southwest and discovered a huge tree of a sort he had never seen before. The natives of the area had built a temple around the base of the tree. Tsang collected specimens of the tree's needles and cones, and studies of these samples revealed that the fossil, *Metasequoia*, had "come to life." In 1948, paleobotanist Ralph Chaney of the University of California led an expedition down the Yangtze River and across three mountain ranges to valleys where a thousand dawn redwoods were growing, the last remnant of the once great *Metasequoia* forest. Thousands of seeds were subsequently distributed, largely through the efforts of the late Elmer Drew Merrill of the Arnold Arboretum in Boston, and this "living fossil" can be seen growing in parks and gardens all over the world.

18–21
Ovulate cone of balsam fir (Abies balsamea). *The upright cones, which are 5 to 10 centimeters long, do not fall to the ground whole, as they do in the pines; rather, they shatter and fall apart, scattering the winged seeds.*

18–22
European larch (Larix decidua). *The larch genus occurs all around the Northern Hemisphere; this species, which is a common European tree, has been used in the eastern United States for reforestation. Leaves of larch are needlelike, as in the pines. They are borne singly on short branch shoots and are spirally arranged. Unlike most conifers, the larches are deciduous; that is, they shed their leaves at the end of each growing season. One young ovulate cone and three young microsporangiate cones are shown here.*

18–23
Gowen cypress (Cupressus goveniana). *The small trees of this species—only about 6 meters in height at maturity—are extremely local, being found only near Monterey, California.*

(a)

(b)

18–25
In conifers of the yew family (Taxaceae), the seeds are surrounded by a fleshy cup, the aril, which attracts birds and other animals that eat them and thus spread the seeds. (a) *Species of the genus* Taxus, *the yews, which occur around the Northern Hemisphere, produce red, fleshy ovulate structures.* (b) *Sporophylls and sporangia of the pollen-bearing cones of the same species. Ovulate cones and pollen-bearing cones are found on separate individuals.*

18–24
The common juniper (Juniperus communis) *has spherical ovulate cones like those of the cypresses, but the scales are fleshy and fused together. Juniper "berries" give gin its distinctive taste and aroma.*

18–26
The bald cypress (Taxodium distichum) *is a deciduous member of the redwood family that grows in the swamps of the southeastern United States.*

18–27
The dawn redwood (Metasequoia glyptostroboides). *This tree, growing in Hubei Province in central China, is more than 400 years old.*

18–28
Fossil branchlet of Metasequoia, *about 50 million years old. The accompanying map shows the geographic distribution of some living and fossil members of the redwood family (Taxodiaceae).*

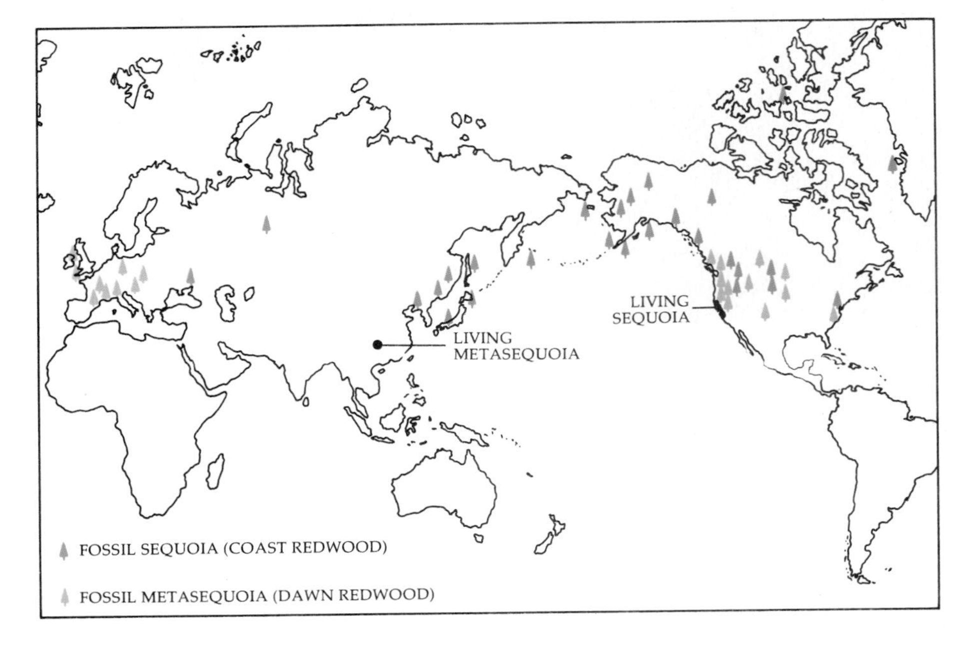

18–29
Ovulate and microsporangiate plants of Zamia pumila, *the only species of cycad native to the United States. The stems are mostly or entirely underground and, along with the rootstalks, were used by the Seminole Indians as food. The two large grey cones in the foreground are ovulate cones; the smaller brown cones are microsporangiate cones.*

(a)

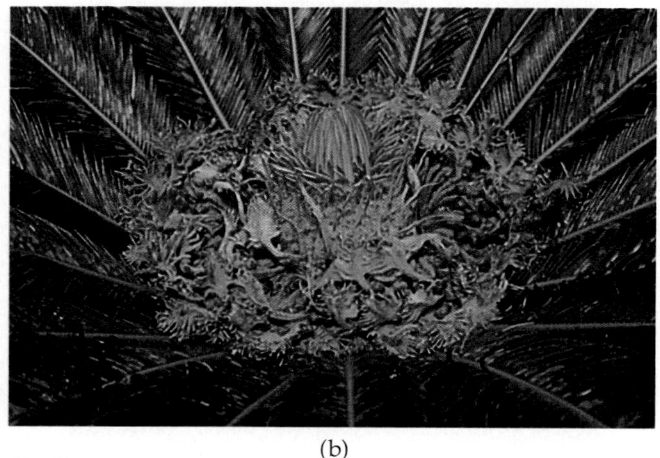

(b)

18–30
(a) Encephalartos altensteinii, *a cycad native to South Africa. This is a microsporangiate plant with male cones.* (b) *An ovulate plant of* Cycas revoluta, *with megasporophylls loosely arranged around a crown of young leaves. Seeds can be seen attached to the margins of the megasporophylls.*

Other Living Gymnosperms

Cycads

The other groups of living gymnosperms are remarkably diverse and scarcely resemble one another at all. Among them are the cycads, division Cycadophyta, which are a series of palmlike plants found mainly in the tropical and subtropical portions of the world. These bizarre plants were so numerous in Mesozoic times that this era is often called the "Age of Cycads and Dinosaurs." There are 10 genera with about 100 species of living cycads. *Zamia pumila,* which occurs commonly in the sandy woods of Florida, is the only cycad native to the United States (Figure 18–29).

Most cycads are fairly large plants; some reach 18 meters or more in height. Many have a distinct trunk that is densely covered with the bases of shed leaves. The functional leaves characteristically occur in a cluster at the top of the stem, giving the cycads an aspect similar to that of the palms (indeed, a common name for the cycads is "sago palms"). Unlike the palms, however, the cycads exhibit true, if sluggish, secondary growth from a vascular cambium; the central portion of their trunks is occupied by a great mass of pith. The reproductive units of cycads are more or less reduced leaves with attached sporangia that are loosely or tightly clustered into conelike structures near the apex of the plant. In cycads, pollen and seed cones are borne on different plants (Figure 18–30).

Ginkgo

The maidenhair tree, *Ginkgo biloba*, is easily recognized by its fan-shaped leaves with their openly branched, dichotomous (forking) pattern of veins (Figure 18–31). The leaves on the numerous spur shoots are more or less entire, whereas on the long shoots and in seedlings they are deeply lobed. Unlike most gymnosperms, *Ginkgo* is deciduous; its leaves turn a beautiful golden color before falling in autumn.

Ginkgo is the sole living survivor of an evolutionary line that probably extends back to the late Paleozoic era and was common during much of the Mesozoic era. The division Ginkgophyta was once widespread but seems to have been represented by relatively few species. There may indeed no longer be any truly native trees of the living species, but *Ginkgo biloba* has long been cultivated in the temple grounds of China and Japan and has been an important feature of the gardens of the temperate regions of the world for more than 150 years. Ginkgo is especially resistant to air pollution and so is commonly cultivated in urban parks.

As in the cycads, *Ginkgo* bears the ovules and microsporangia on different individuals. The ovules of *Ginkgo* are borne in pairs on the end of short stalks and ripen to produce seeds in autumn. In *Ginkgo*, fertilization within the ovules may not occur until after the ovules have been shed from the tree. Embryos are formed during the later stages of maturation of the seeds, which occur on the ground. The seeds have a rancid odor as a result of the butyric acid in their fleshy coats, and for this reason only male trees are usually cultivated in parks or in gardens. The trees are propagated vegetatively, and male trees can be obtained in this way. The microsporophylls are grouped in conelike structures, each microsporophyll bearing two microsporangia.

Gnetophyta

The division Gnetophyta is a small group (about 70 species) of gymnosperms consisting of three genera: *Gnetum*, *Ephedra*, and *Welwitschia*. Although the three genera have long been placed in a single order as a matter of convenience, each genus is now placed in a separate order, in recognition of the fact that each differs greatly from the others both structurally and reproductively.

Gnetum, a genus of about 30 species, consists of trees and climbing vines with large leathery leaves that closely resemble those of dicotyledons (Figure 18–32). It is found throughout the moist tropics.

Most of the approximately 35 species of *Ephedra* are profusely branched shrubs with inconspicuous small, scalelike leaves (Figure 18–33). With its small leaves and apparently jointed stems, *Ephedra* superficially resembles *Equisetum*. Most species of *Ephedra* inhabit arid or desert regions of the world.

Welwitschia is probably the most bizarre of vascular plants (Figure 18–34). Most of the plant is buried in sandy soil. The exposed part consists of a massive woody, concave disk that produces only two strap-shaped leaves. *Welwitschia* grows in desert areas of southwestern Africa.

(a)

(b)

18–31
(a) Ginkgo biloba, *the maidenhair tree.*
(b) Ginkgo *leaves and fleshy seeds attached to spur shoot.*

18–32
The large leathery leaves of Gnetum *resemble those of certain dicots. The species of* Gnetum *grow as shrubs or woody vines in tropical or subtropical forests.*

(a) (b)

18–33
Ephedra *is the only genus of the three living genera of Gnetophyta found in the United States.* (a) Ephedra californica, *a densely branched shrub that, like other members of the genus, has scalelike leaves.* (b) *Branches of* E. californica, *bearing microsporangiate strobili.*

18–34
A large, seed-producing plant of Welwitschia mirabilis, *growing in the Namib Desert of southern Africa.* Welwitschia *produces only two leaves, which continue to grow for the life of the plant. As growth continues, the leaves break off at the tips and split lengthwise; hence, older plants appear to have numerous leaves.*

The Gnetophyta have many angiospermlike features, such as the similarity of their strobili to the inflorescences of angiosperms, the presence of vessels in the xylem, and the lack of archegonia in *Gnetum* and *Welwitschia*. As a consequence, some botanists have considered them as possible connecting links between the gymnosperms and angiosperms. However, they are currently regarded as specialized end points of one line of gymnosperm evolution.

THE ANGIOSPERMS

The flowering plants, division Anthophyta, comprise about 235,000 species, by far the largest number of species of any plant group. In their vegetative structures, they are enormously diverse. In size, they range from species of *Eucalyptus*, trees well over 100 meters in height and nearly 20 meters in girth (Figure 18–35), to some duckweeds of the genus *Wolffia*, which are simple floating plants scarcely 1 millimeter in length (Figure 18–36). Some are vines that climb high into the canopy of the tropical rain forest, and others are epiphytes that grow in the canopy. Many flowering plants, or angiosperms, such as cacti, are adapted to grow in extremely arid regions of the world. Although the vast majority of flowering plants are free-living, both parasitic and saprophytic forms exist (Figure 18–37).

18–35
The giant eucalyptus, or red tingle (Eucalyptus jacksonii), *growing in the Valley of the Giants in Southwest Australia. The enormous stature of this angiosperm is evident by comparison to the man standing in the burnt-out base of the tree.*

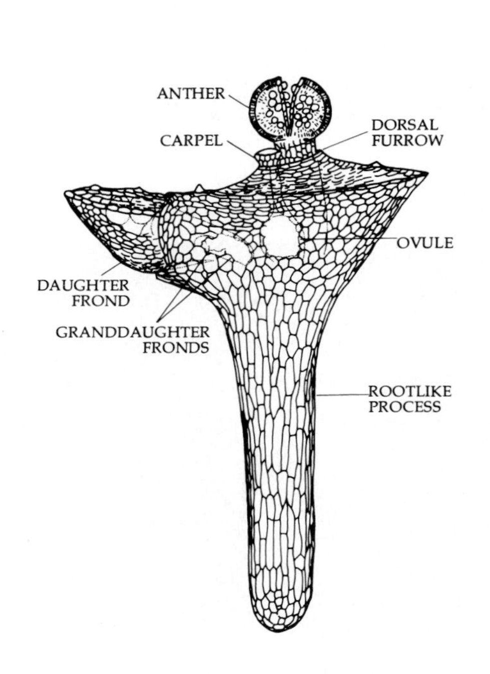

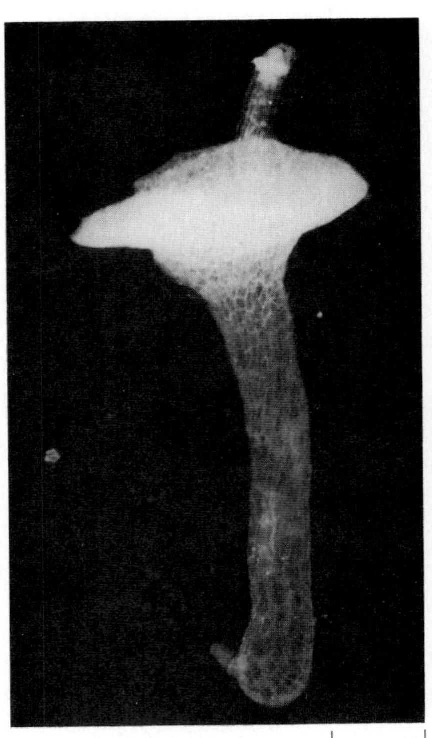

18–36
A duckweed, Wolffia microscopica, *at about 1 millimeter in length, is the smallest of the angiosperms. These tiny plants float submerged in water. The accompanying drawing diagrams the plant body and the extremely simple flower. The duckweeds are monocots.*

(a)

(b)

(c)

18–37
Nongreen angiosperms. (a) *Dodder* (Cuscuta cuspidata), *which has only small amounts of chlorophyll and is a parasite on other angiosperms such as the sweet bay* (Magnolia virginiana) *shown here. Plants of* Cuscuta, *which are bright orange or yellow, obtain most of their carbohydrate supply from their host plant.* Cuscuta *is a member of the morning glory family (Convolvulaceae).* (b) *Squaw-root* (Conopholis americana), *another parasitic angiosperm, forms large rounded knobs on the roots of trees, mostly oaks* (Quercus). (c) *Indian pipe* (Monotropa uniflora) *was long considered to be saprophytic (depending on decaying organic matter for its nutrition). It has recently been shown that these "saprophytes" have obligate relationships with mycorrhizal fungi that are associated with a second plant, in this case, a green, actively photosynthetic angiosperm. The fungus forms a bridge that transfers carbohydrates actively from the photosynthetic plant to the Indian pipe.* Monotropa *is entirely lacking in chlorophyll. Both dodder and the Indian pipe are dicots.*

(a)

(b)

(c)

18–38
Monocots. (a) *A member of the palm family, the coconut palm* (Cocos nucifera), *growing in Tehuantetec, Oaxaca, Mexico. The fruit of coconut is not a nut but a drupe (see Chapter 19).* (b) *Flowers and fruits of the banana plant* (Musa × paradisiaca). *The banana flower has an inferior ovary, and the tip of the fruit bears a large scar left by the fallen flower parts.* (c) *Rice* (Oryza sativa) *is a member of the grass family.*

(a)

(b)

(c)

18–39
Dicots. (a) *The water lily* (Nymphaea odorata). *The very fragrant* N. odorata *flower contains numerous petals and stamens and is regular, or radially symmetrical. The genus* Nymphaea *is widely distributed in tropical and temperate regions throughout the world.* (b) *Giant saguaro cactus* (Carnegiea gigantea). *The cacti, of which there are about 2000 species, are almost exclusively a New World family. The thick fleshy stems, which can serve to store water, have taken over the function of the leaves in photosynthesis.* (c) *Round-lobed hepatica* (Hepatica americana), *which flowers in deciduous woodlands in the early spring. The flowers have no petals but have six to ten sepals and numerous spirally arranged stamens and carpels.*

The Anthophyta includes two classes: the Monocotyledones (monocots), with about 65,000 species, and the Dicotyledones (dicots), with about 170,000 species. The similarities between these two groups are far greater than the differences; nonetheless the two classes are clearly recognizable natural units. Among the monocots are such familiar plants as the grasses, lilies, irises, orchids, cattails, and palms (Figure 18–38). The dicots include almost all the familiar trees and shrubs (other than conifers), many of the herbs, plus many other plants (Figure 18–39). The major differences between the monocots and dicots are summarized in Table 18–1.

In Section 5, we shall consider in some detail the structure and development of the plant body, that is, the sporophyte, of flowering plants. The remainder of this chapter will be concerned with the most distinctive characteristic of angiosperms—the flower—and with reproduction in flowering plants.

Table 18–1 *Main Differences Between Monocots and Dicots*

CHARACTERISTIC	DICOTS	MONOCOTS
Flower parts	In fours or fives (usually)	In threes (usually)
Pollen	Basically tricolpate (having three furrows or pores)	Basically monocolpate (having one furrow or pore)
Cotyledons	Two	One
Leaf venation	Usually netlike	Usually parallel
Primary vascular bundles in stem	In a ring	Scattered
True secondary growth, with vascular cambium	Commonly present	Absent

The Flower

The flower is a determinate shoot that bears sporophylls (Figure 18–40). The name "angiosperm" is derived from the Greek words *angeion,* meaning "vessel," and *sperma,* meaning "seed." Probably the most distinctive structure of the flower is the carpel—the "vessel"—containing the ovules which develop, after fertilization, into seeds.

Flowers may be clustered in various ways into aggregations called *inflorescences* (Figures 18–41 and 18–42). The stalk of an inflorescence or of a solitary flower is known as a *peduncle*. The stalk of an individual flower in an inflorescence is called a *pedicel*. The part of the flower stalk to which the floral parts are attached is termed the *receptacle*. Like any other shoot tip, the receptacle consists of nodes and internodes. In the flower, the internodes are very short, and consequently, the nodes are very close together.

Most flowers contain two sets of sterile appendages, the *sepals* and *petals*, which are attached to the receptacle below the fertile parts of the flower, the *stamens* and *carpels*. The sepals arise below the petals, and the stamens arise below the carpels. Collectively, the sepals form the *calyx*, and the petals form the *corolla*. Together, the calyx and corolla constitute the *perianth*. The sepals and petals are essentially leaflike in structure. Commonly, the sepals are green and the petals are brightly colored, although in many flowers both parts are similar in color (Figure 18–40).

The stamens—collectively the *androecium* ("house of man")—are microsporophylls. In the vast majority of living angiosperms, the stamen consists of a slender stalk, or *filament*, upon which is borne a two-lobed *anther* containing four microsporangia, or pollen sacs.

The carpels—collectively the *gynoecium* ("house of woman")—are megasporophylls that are folded lengthwise, enclosing one or more ovules. A given flower may contain one or more carpels. If more than one carpel is present, they may be separate or fused together, in part or entirely. Sometimes the individual carpel or the group of fused carpels is called a pistil. The word "pistil" comes from the same root as "pestle," the instrument with a similar shape that pharmacists use for grinding substances into a powder in a mortar.

In most flowers, the individual carpels or group of fused carpels are differentiated into a lower part, the *ovary*, which encloses the ovules, and an upper part, the *stigma*, which receives the pollen. In many flowers a more or less elongated structure, the *style*, connects the stigma with the ovary. If the carpels are fused, there may be a common style or stigma, or each carpel may retain a separate style and stigma. The common ovary of such fused carpels is generally (but not always) partitioned into two or more locules—chambers of the ovary in which the ovules occur.

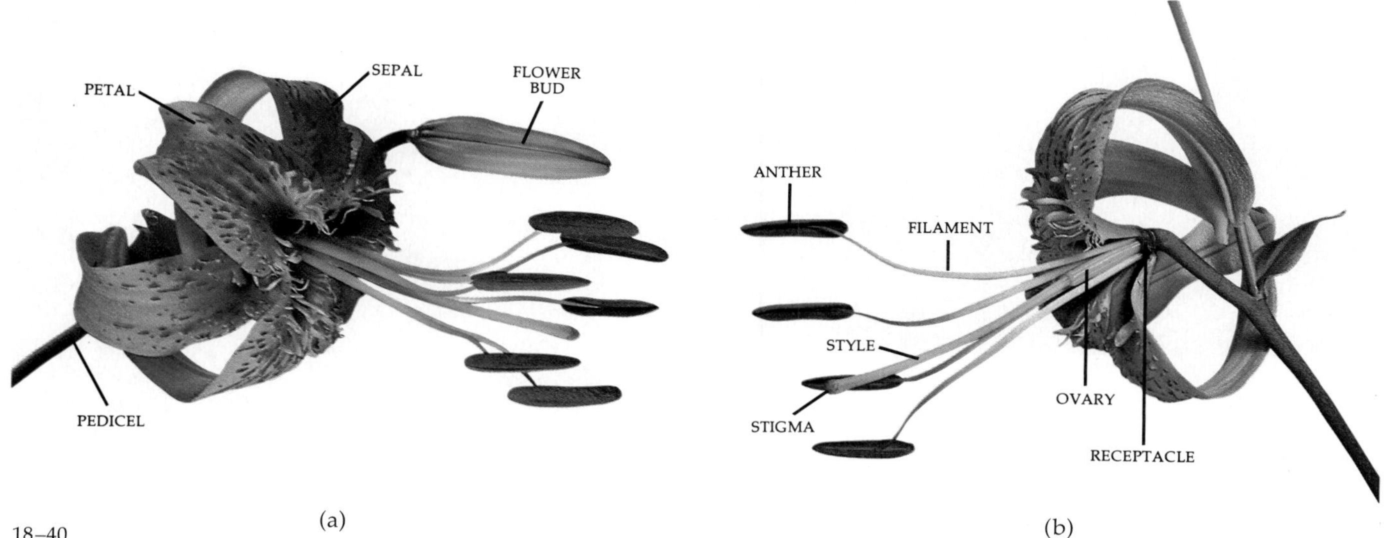

18–40
Parts of a lily (Lilium henryi) *flower.* (a) *An intact flower. In some flowers, such as lilies, the sepals and petals are similar to one another, and the perianth parts may then be referred to as tepals. Note that the sepals are attached to the receptacle below the petals.* (b) *A partly dissected flower with two tepals and two stamens removed to reveal the ovary. The gynoecium consists of the ovary, style, and stigma. The stamen consists of the filament and anther.*

18–41
Inflorescences of (a) *shooting star* (Dodecatheon pauciflorum), (b) *butter-and-eggs* (Linaria vulgaris), (c) *lupine* (Lupinus diffusus), (d) *bluebells* (Mertensia virginica), *and* (e) *water hemlock* (Cicuta maculata). *Using Figure 18–42 as a guide, can you identify the types of inflorescences shown here?*

18–42
Illustrations of some of the common types of inflorescences found in the angiosperms, accompanied by simplified diagrams (in color).

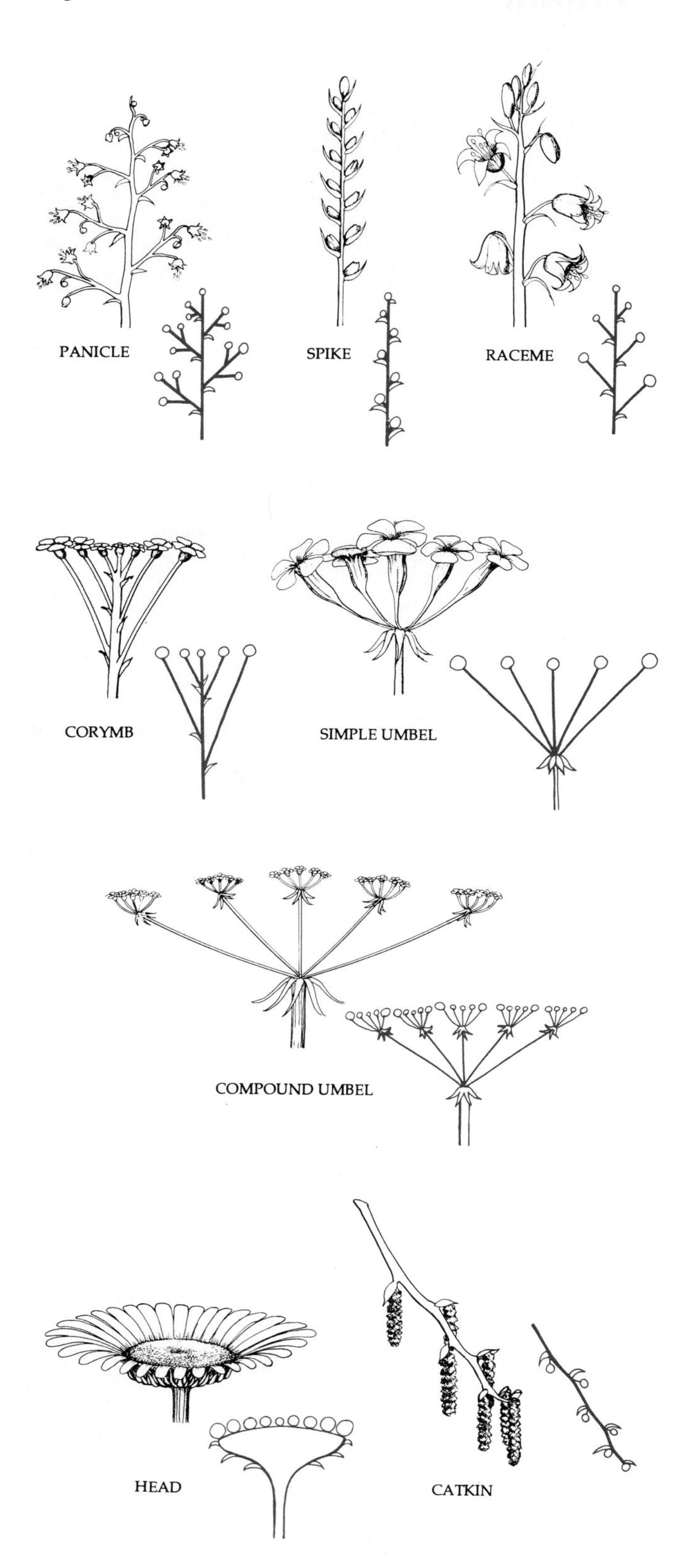

18–43
Types of placentation. (a) *Parietal.* (b) *Axile.* (c) *Free central. Basal placentation is not shown here.*

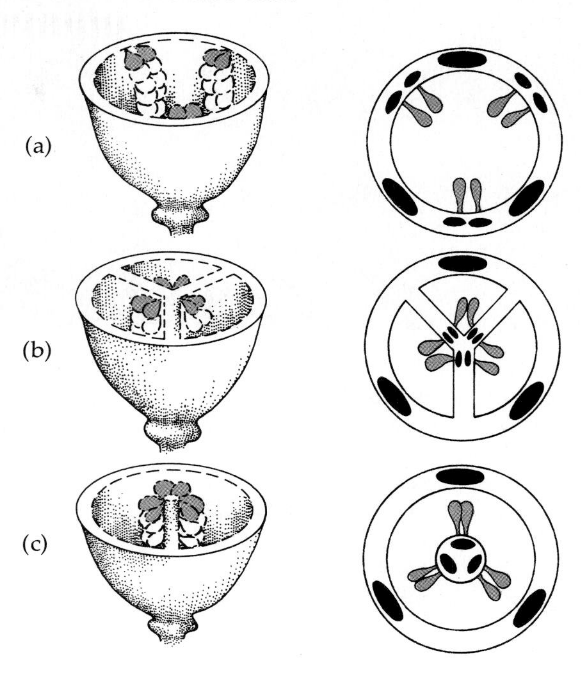

The portion of the ovary where the ovules originate and remain attached until maturity is called the *placenta*. The arrangement of the placentae (the placentation) and, consequently, the ovules, varies among different flowers (Figure 18–43). In some flowers, the placentation is *parietal*; that is, the ovules are borne on the ovary wall or on extensions of it. In other flowers, the ovules are borne on a central column of tissue in a partitioned ovary with as many locules as there are carpels. This is *axile* placentation. In still others, the placentation is *free central*, the ovules being borne upon a central column of tissue not connected by partitions with the ovary wall. And finally, in some flowers a single ovule occurs at the very base of a unilocular ovary. This is *basal* placentation.

Given the basic structure of the flower, many variations may exist. The majority of flowers contain both stamens and carpels, and such flowers are said to be *perfect*. If either stamens or carpels are missing, the flower is *imperfect*, and depending on the part that is present, the flower is said to be either *staminate* or *carpellate* (Figure 18–44). If both staminate and carpellate flowers occur on the same plant, as in corn and oaks, the species is said to be *monoecious*. If they are found on separate plants, the species is said to be *dioecious*, as in the willows and the American holly.

18–44
Staminate and carpellate flowers of a tan-bark oak (Lithocarpus densiflora). *Most members of the family Fagaceae, including the true oaks* (Quercus), *are monoecious, meaning that the staminate and carpellate flowers are separate but are borne on the same tree.*

(a)

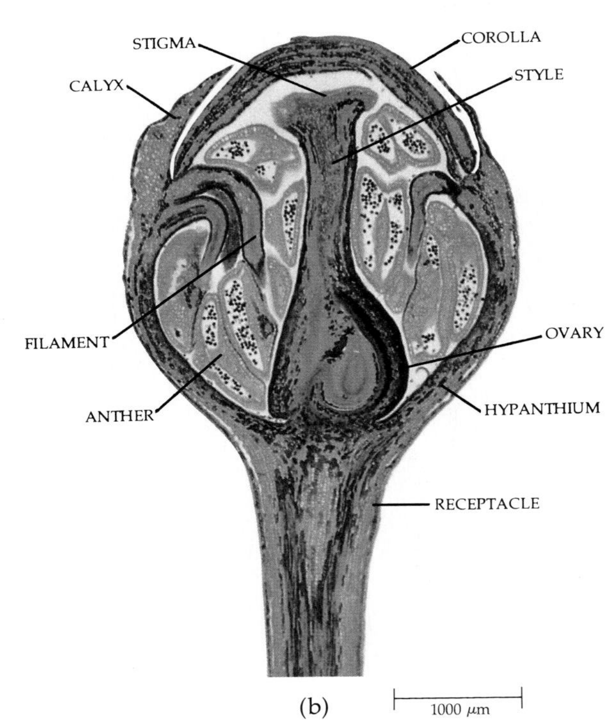

(b)

18–45
(a) *and* (b) *Cherry* (Prunus) *flowers exhibit perigyny—their sepals, petals, and stamens are attached to a hypanthium (a cup-shaped extension of the receptacle). In* (b) *the stamens are crowded in the hypanthium because the flower has not yet opened.* (c) *and* (d) *Apple* (Malus) *flowers exhibit epigyny—their sepals, petals, and stamens apparently arise from the top of the ovary. In* (d) *the flower is very nearly open, but the stamens are not yet erect.*

Any one of the floral parts—sepals, petals, stamens, or carpels—may normally be lacking from a flower. Flowers with all four floral parts are called *complete* flowers. If any part is lacking, the flower is said to be *incomplete*. Thus, an imperfect flower is also incomplete, but not all incomplete flowers are imperfect.

The arrangement of the floral parts may be either spiral on a more or less elongated receptacle, or similar parts—such as the petals—may be located at one level in whorls. The parts may be united with other members of the same whorl *(coalescence)* or with members of other whorls *(adnation)*. An example of adnation is the union of stamens with the corolla, which is fairly common. When the floral parts of the same whorl are not joined, the prefixes *apo-* (meaning "separate") or *poly-* may be used to describe the condition. For example, aposepalous or polysepalous indicates sepals that are not joined, or separate sepals. When the parts are coalesced, either *syn-* or *sym-* is used (for example, synsepaly and sympetaly indicate united sepals and united petals, respectively).

In addition to the floral parts being either spiral or whorled in arrangement, the level of insertion of the sepals, petals, and stamens on the floral axis varies in relation to the level of the ovary or ovaries. If the sepals, petals, and stamens are attached to the receptacle below the ovary, the ovary is said to be *superior*, and the flower is said to be *hypogynous*. In some flowers with superior ovaries, the petals and stamens are attached to the margin of a cup-shaped extension of the receptacle (the hypanthium). Such flowers are said to be *perigynous* (Figure 18–45a and b). In other

(c)

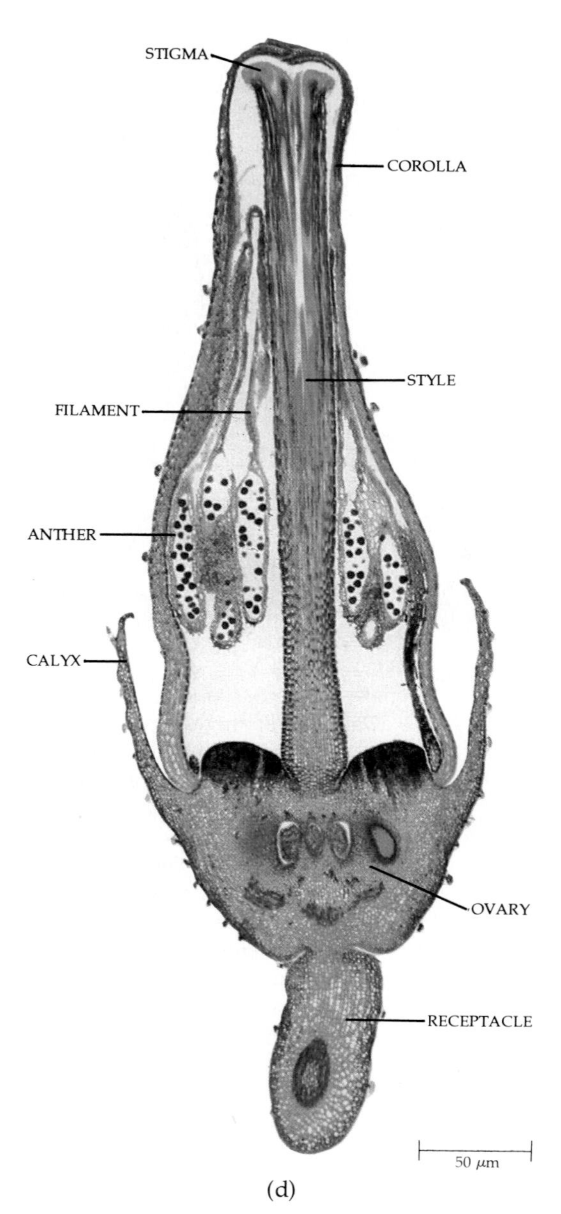

(d)

flowers the sepals, petals, and stamens apparently grow from the top of the ovary, which is *inferior*. Such flowers are said to be *epigynous* (Figure 18–45c and d).

Finally, with regard to variation in floral structure, mention should be made of symmetry. In some flowers, the corolla is made up of petals of similar shape that radiate from the center of the flower and are equidistant from each other; that is, they are radially symmetrical. Such flowers are said to be *regular*, or actinomorphic. In other flowers, one or more members of at least one whorl are of a different form than other members of the same whorl; that is, they are bilaterally symmetrical. These flowers are said to be *irregular*, or zygomorphic.

The Angiosperm Life Cycle

The gametophytes of angiosperms are very much reduced in size—more so than those of any other heterosporous plants, including the gymnosperms. The mature microgametophyte consists of only three cells, and the mature megagametophyte, held for its entire existence within the tissues of the sporophyte, in most instances consists of only seven cells. Both antheridia and archegonia are lacking. Pollination is indirect: pollen is deposited on the stigma, after which the pollen tube conveys two nonmotile sperm to the female gametophyte. After fertilization, the ovule develops into a seed, which is enclosed in the ovary. At the same time, the ovary develops into a fruit.

Microsporogenesis and Microgametogenesis

Microsporogenesis is the formation of microspores within the microsporangia, or pollen sacs, of the anther. Microgametogenesis is the development of the microspore into the microgametophyte, or pollen grain.

Initially, the anther consists of a mass of cells without visible differentiation, except for a partly differentiated epidermis. Eventually, four groups of fertile, or *sporogenous,* cells become discernible within the anther; each group is surrounded by several layers of sterile cells. The sterile cells develop into the wall of the pollen sac, including nutritive cells, which supply food to the developing microspores. The nutritive cells constitute the *tapetum*, the innermost layer of the pollen sac wall (Figure 18–46). The sporogenous cells become microsporocytes, or microspore mother cells, which divide meiotically, each diploid microsporocyte giving rise to a tetrad of haploid microspores. Microsporogenesis is completed with formation of the single-celled microspores.

During meiosis, each nuclear division may be followed immediately by cell-wall formation, or the four microspore protoplasts may be walled off simulta-

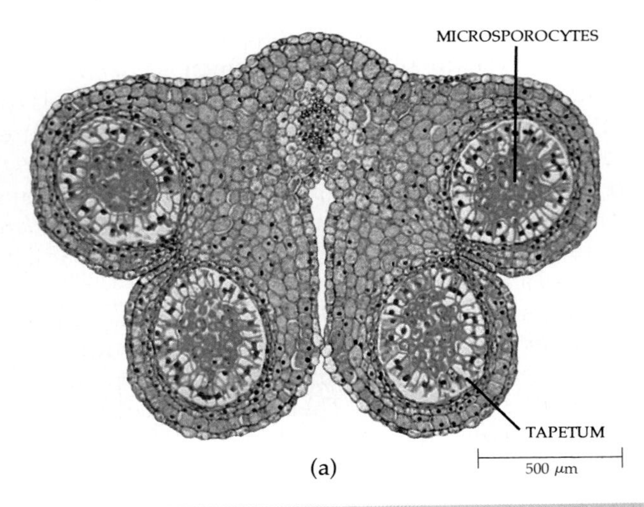

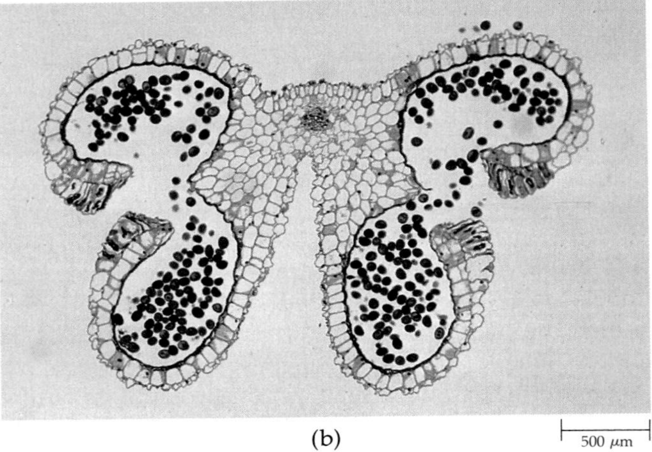

18–46
Two transverse sections of lily (Lilium) *anthers.* (a) *Immature anther, showing the four pollen sacs containing microsporocytes, or microspore mother cells, surrounded by the tapetum.* (b) *Mature anther containing pollen grains. The partitions between adjacent pollen sacs break down prior to dehiscence, as shown here.*

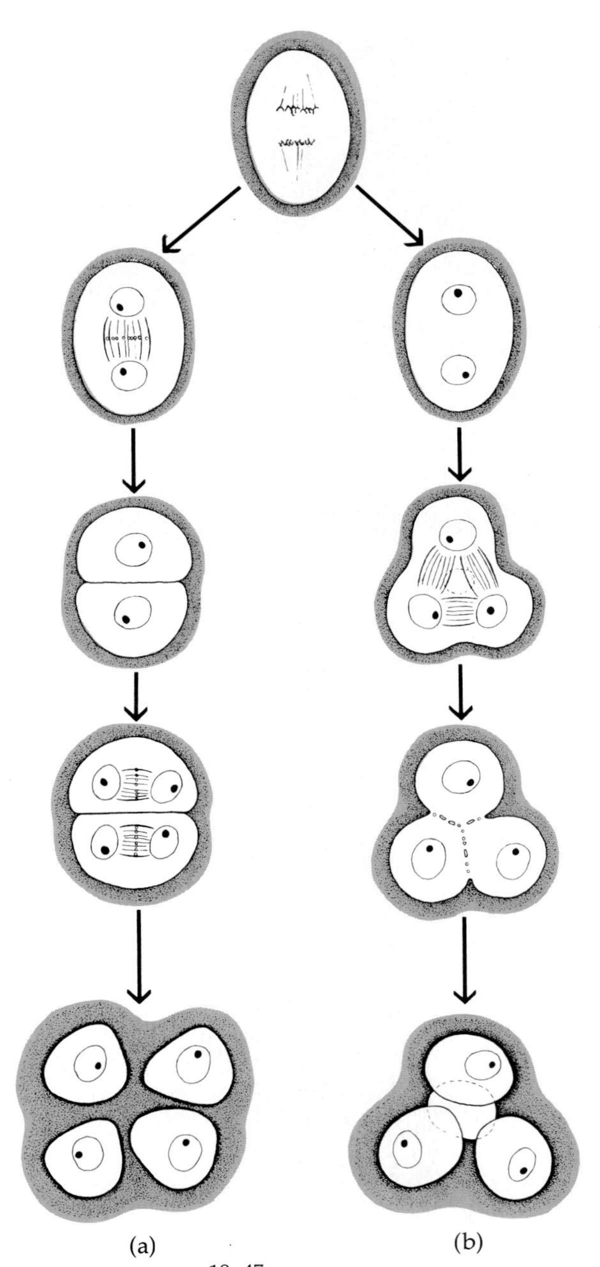

18–47
Diagrams of microsporogenesis leading to the formation of microspores. (a) *In the successive type of cytokinesis, commonly found in monocots, walls appear after both the first and second meiotic divisions.* (b) *In the simultaneous type of cytokinesis, common in dicots, no wall is formed after the first meiotic division. The four microspores are walled off simultaneously after the second meiotic division. Both types of cytokinesis lead to tetrads of microspores.*

neously after the second meiotic division (Figure 18–47). The first condition is common in monocots, the second in dicots. The entire tetrad and the individual grains within it are at first enclosed by callose walls which are not penetrated by plasmodesmata. Development of the pollen grain wall begins while the tetrads are still enclosed in callose. At the same time or after the developing pollen grains are released from the tetrad by dissolution of the callose walls (through activity of the enzyme callase), the grains increase in size. Following this period, the major features of the pollen grains are established (Figure 18–48). The pollen grains develop a resistant outer wall,

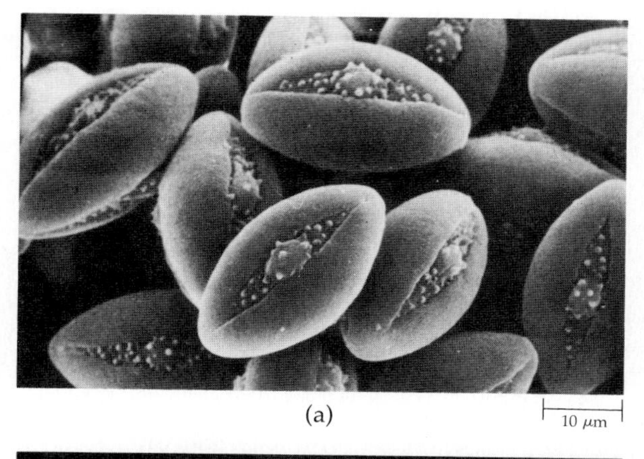

(a) 10 μm

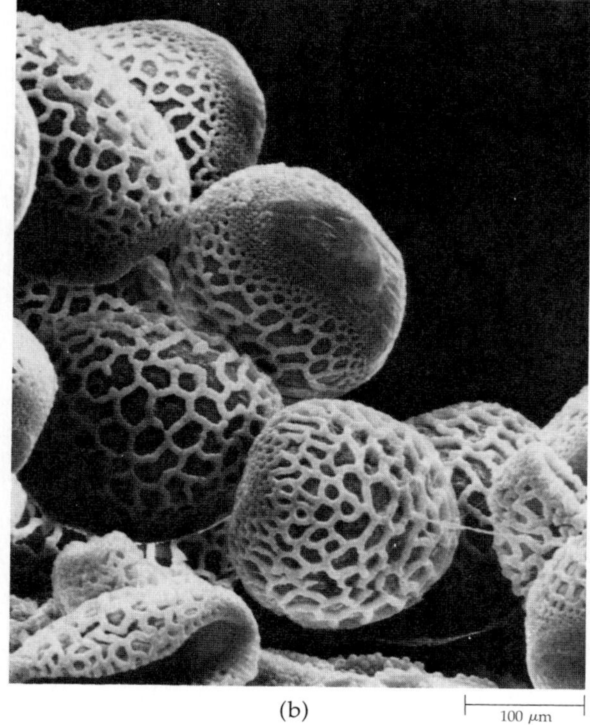

(b) 100 μm

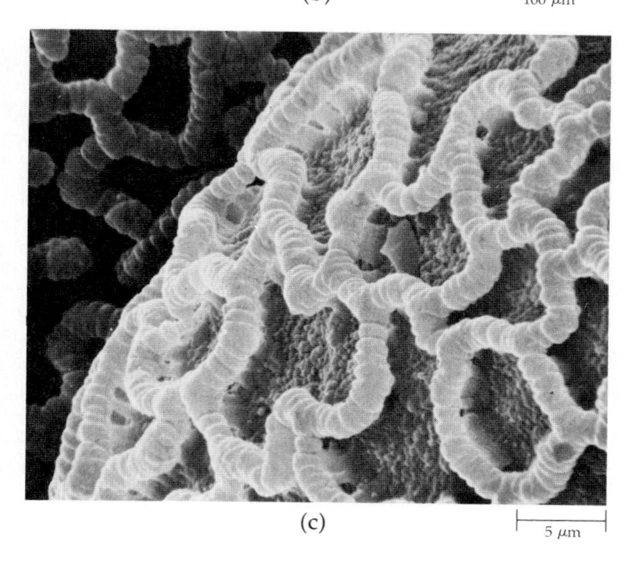

(c) 5 μm

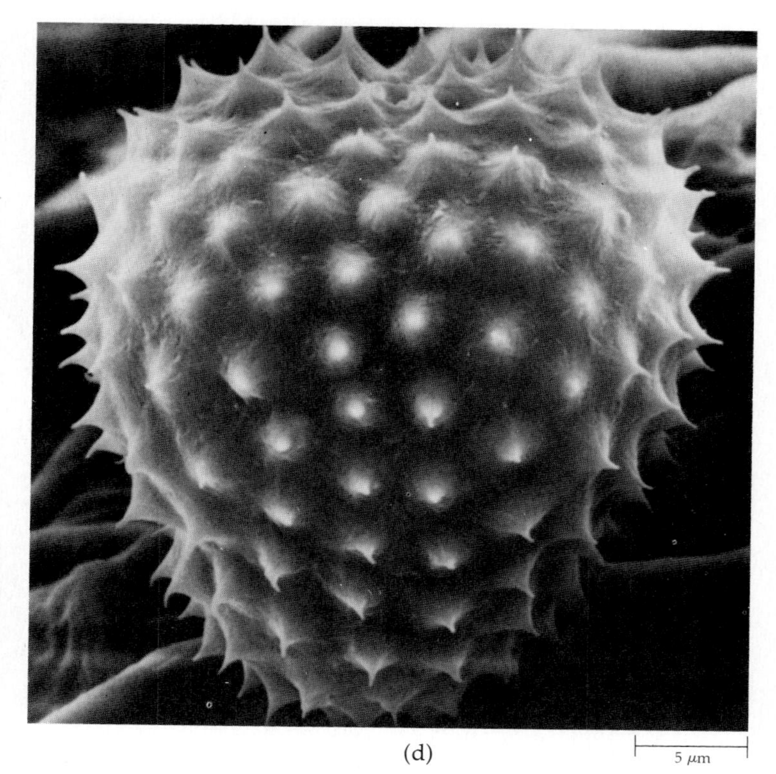

(d) 5 μm

18–48
The wall of the pollen grain serves to protect the male gametophyte on its often hazardous journey between the anther and the stigma. The outer layer, or exine, is composed chiefly of a substance known as sporopollenin, which appears to be a polymer composed chiefly of carotenoids. The exine, which is remarkably tough and resistant, is often elaborately sculptured.

The sculpturing of pollen grain walls is very precise and distinctly different from one species to another, as revealed in the following scanning electron micrographs of selected pollen grains. (a) *Pollen grains of the horse chestnut* (Aesculus hippocastanum). (b) *Pollen grains of a lily* (Lilium longiflorum). (c) *Detail of the surface of a lily* (L. longiflorum) *pollen grain.* (d) *Pollen grain of the Western ragweed* (Ambrosia psilostachya). *The pollen of ragweed is a primary cause of hay fever. Spiny pollen grains such as those of ragweed are common among Asteraceae.*

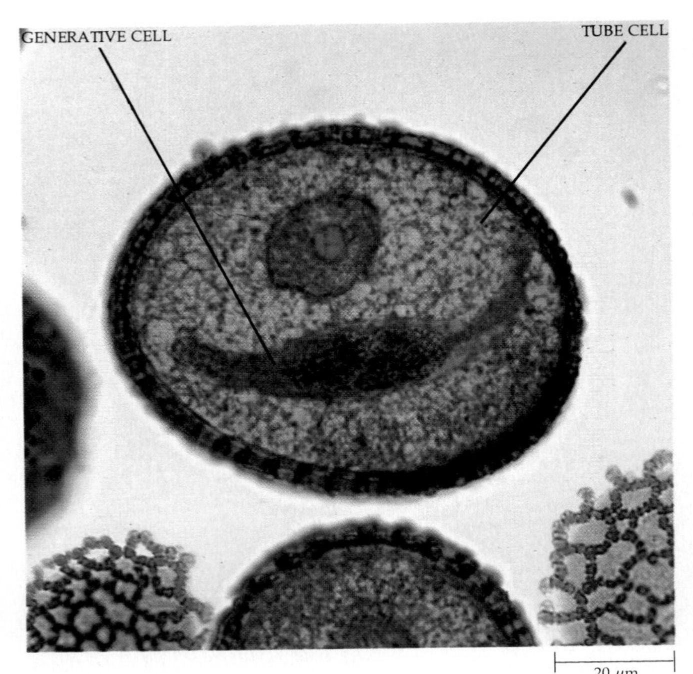

18–49
Mature pollen grain of Lilium, *containing a two-celled male gametophyte. The spindle-shaped generative cell will divide mitotically to give rise to two sperm; the larger tube cell will form the pollen tube.*

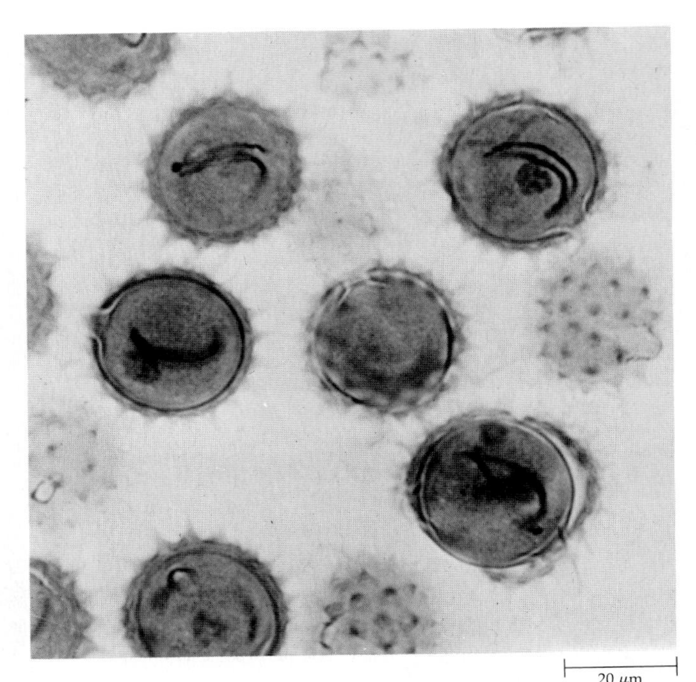

18–50
Mature pollen grains with three-celled male gametophytes of Silphium. *Prior to pollination, each pollen grain contains two filamentous sperm cells, which are suspended in the cytoplasm of the larger tube cell.*

called the *exine*, and a cellulosic inner wall, called the *intine*. The exine is composed of a very resistant substance known as *sporopollenin*, which apparently is derived partly from the tapetum and partly from the spores following the breakdown of the callose walls. The intine, which is composed of cellulose and pectin, is laid down by the protoplast of the spores.

Pollen grains, like spores, vary considerably in size and shape, ranging from less than 20 nanometers to over 250 nanometers in diameter. They also differ in the number and arrangement of the pores through which the pollen tubes ultimately grow. Nearly all families, many genera, and a fair number of species of flowering plants can be identified solely on the basis of specific characteristics of their pollen grains. In contrast with the larger portions of plants—such as leaves, flowers, and fruits—the pollen grains, because of the chemical nature of the exine, are extremely well represented in the fossil record. Thus pollen provides a valuable index to the kinds of plants and the nature of the climate that prevailed in the past.

Microgametogenesis in angiosperms is uniform and begins when the uninucleate microspore divides mitotically to form two cells within the original spore wall. One cell is called the *tube cell*, and the other is called the *generative cell* (Figure 18–49). In many species, the microgametophyte is in this two-celled stage at the time the pollen grains are liberated by dehiscence of the anther. In others, the generative nucleus divides prior to release of the pollen grains, giving rise to two male gametes, or sperm (Figure 18–50).

Megasporogenesis and Megagametogenesis

Megasporogenesis is the process of megaspore formation within the nucellus (megasporangium). Megagametogenesis is the development of the megaspore into the megagametophyte.

The ovule is a relatively complex structure, consisting of a stalk—the *funiculus*—bearing a nucellus enclosed by one or two integuments. Depending upon the species, one to many ovules may arise from the placentae, or ovule-bearing regions of the ovary wall. Initially, the developing ovule is entirely nucellus, but soon it develops one or two enveloping layers, the integuments, with a small opening, the micropyle, at one end (Figure 18–51).

Early in the development of the ovule, a single megasporocyte, or megaspore mother cell, arises in the nucellus (Figure 18–51a). The diploid megaspore mother cell divides meiotically to form four haploid megaspores, which are generally arranged in a linear tetrad. With this, megasporogenesis is completed. Of

18–51
Lilium. *Some stages in development of an ovule and embryo sac.* (a) *Two young ovules, each with a single, large megasporocyte, or megaspore mother cell. Integuments have not begun to develop.* (b) *Ovule now has integuments. The megasporocyte is in first prophase of meiosis.* (c) *Ovule with eight-nucleate embryo sac (only six of the nuclei can be seen here). The polar nuclei have not yet migrated to the center of the sac.*

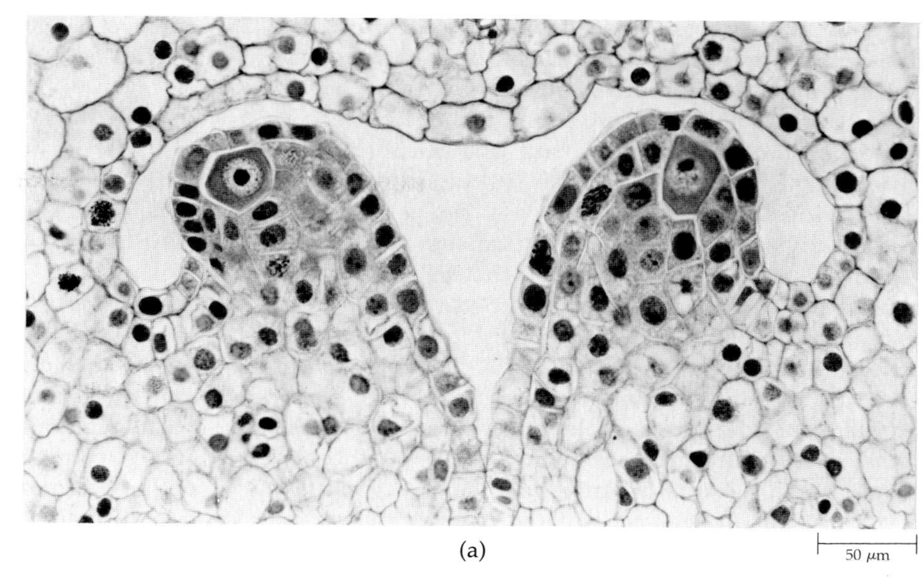

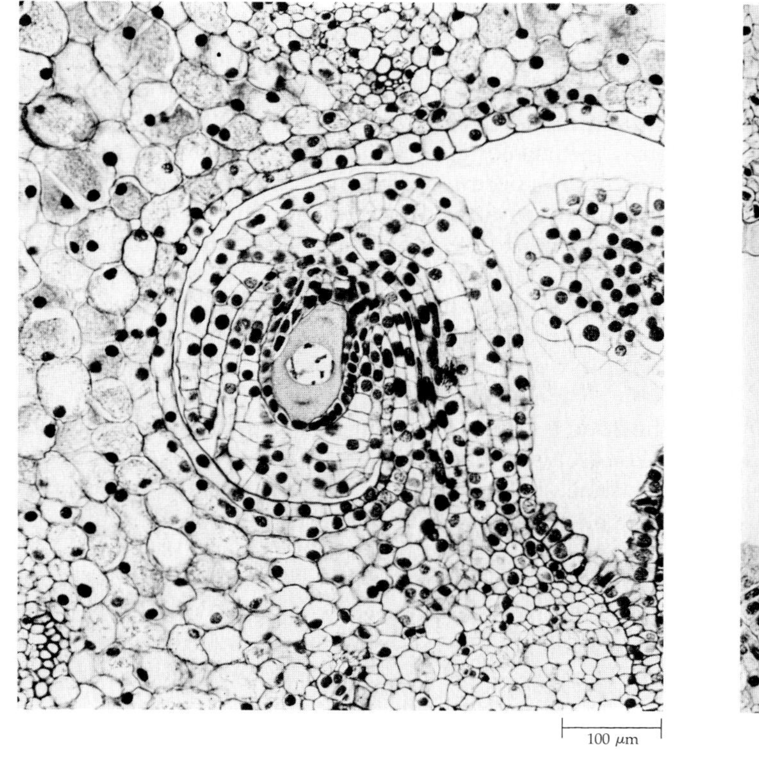

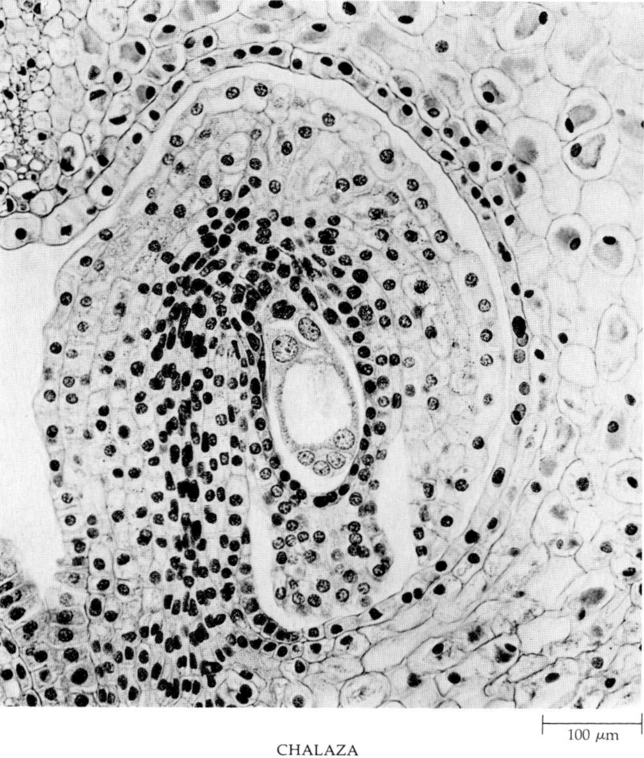

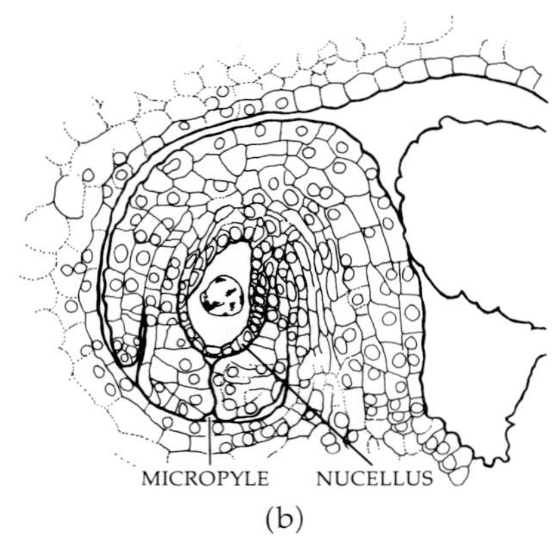

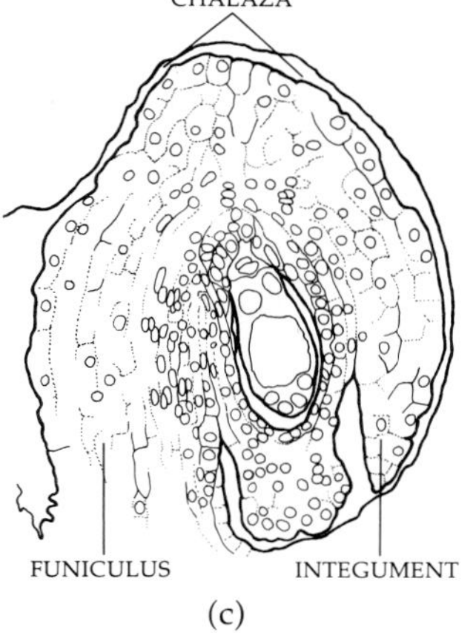

the four megaspores, three usually disintegrate, the one farthest from the micropyle surviving and developing into the megagametophyte.

The functional megaspore soon begins to enlarge at the expense of the nucellus, and its nucleus divides mitotically (Figure 18–51b). Each of the resulting nuclei divides mitotically, followed by yet another mitotic division of the four resultant nuclei. At the end of the third mitotic division, the eight nuclei are arranged in two groups of four, one group near the micropylar end of the megagametophyte and the other at the opposite (*chalazal*) end. One nucleus from each group migrates into the center of the eight-nucleate cell; these two nuclei are then known as the *polar nuclei*. The three remaining nuclei at the micropylar end become organized as the *egg apparatus*, consisting of an *egg cell* and two cellular *synergids*. Cell-wall formation also occurs around the other three nuclei left at the chalazal end, forming the so-called *antipodals*. The *central cell*, containing the polar nuclei, remains binucleate.

The eight-nucleate, seven-celled structure is the mature female gametophyte, or *embryo sac* (Figure 18–51c). The pattern of embryo sac development just described is the most common one. Other patterns of embryo sac development occur in about a third of the species of angiosperms that have been investigated.

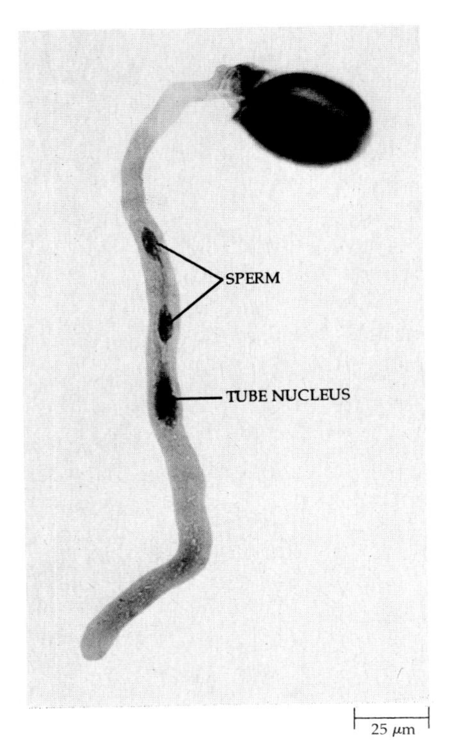

18–52
Mature male gametophyte of Solomon's seal (Polygonatum). *Both sperm and the tube nucleus can be seen in the pollen tube.*

Pollination and Fertilization

With dehiscence of the anthers, the pollen grains are transferred to the stigmas by a variety of vectors (see Chapter 19). Once in contact with the stigma, the pollen grains germinate and form a pollen tube. If the generative cell has not already divided, it soon does so to form the two sperm. The germinated pollen grain, with its tube nucleus and two sperm, constitutes the mature microgametophyte (Figure 18–52).

The stigma and style are modified both structurally and physiologically to facilitate germination of the pollen grain and growth of the pollen tube. The surface of many stigmas is essentially glandular tissue (*stigmatic tissue*) and excretes a solution that becomes sugary. The stigmatic tissue is connected with the ovule by *transmitting tissue*, which serves as a path through the style for the growing pollen tubes. Some styles contain open canals, in which the transmitting tissue lines the canal. In such styles, the pollen tubes grow either along or among the cells of the lining. In most angiosperms, however, the styles are solid, and one or more strands of transmitting tissue extend from stigma to ovules. The pollen tubes grow either between the cells of transmitting tissue or within their thick walls, depending upon the species.

Commonly, the pollen tube enters the ovule through the micropyle and penetrates one of the synergids, which is destroyed in the process. The two sperm and the tube nucleus are then released into the synergid through a subterminal pore that develops in the pollen tube. Ultimately, one sperm nucleus enters the egg cell and the other enters the central cell, where it unites with the two polar nuclei (Figure 18–53). Recall that in gymnosperms, only one of the two sperm is functional; one unites with the egg and the other degenerates. The involvement of both sperm in angiosperms—the union of one sperm with the egg and the other with the polar nuclei—is called *double fertilization* and represents one of the principal characteristics of this group. (As previously noted, "true" fertilization, or syngamy, involves only the union of gametes, in this case, sperm and egg.) With union of sperm and egg, a diploid zygote is formed. Union of the other sperm with the two polar nuclei, called *triple fusion*, results in formation of a triploid *primary endosperm nucleus*.

Development of Seed and Fruit

With double fertilization, a number of processes are initiated: the primary endosperm nucleus divides to form *endosperm*; the zygote develops into an embryo;

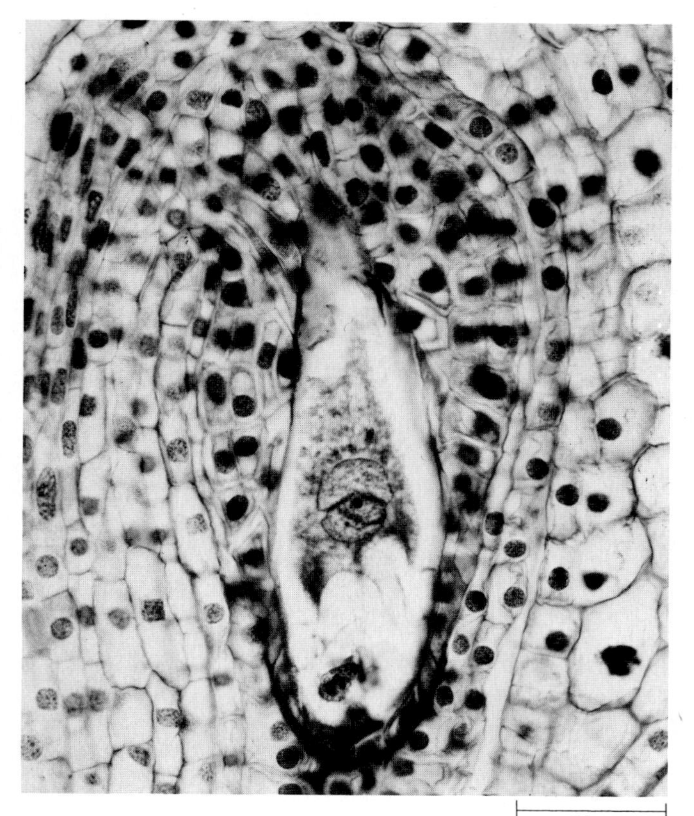

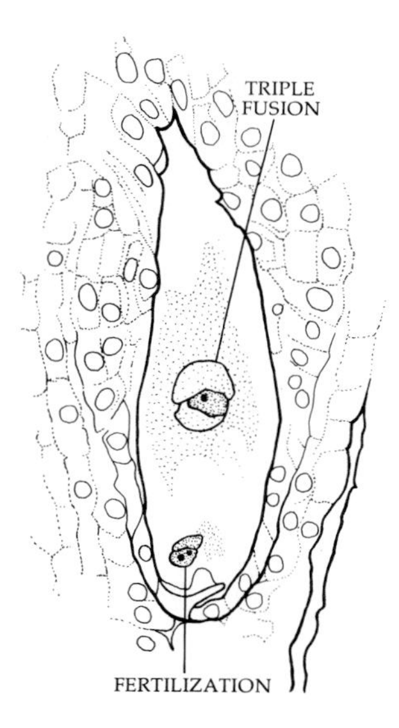

18–53
Lilium. *Double fertilization. Union of sperm and egg nuclei—"true" fertilization—can be seen in lower half of the micrograph. Triple fusion of the other sperm nucleus and the two polar nuclei has taken place above.*

the integuments develop into a seed coat; and the ovary wall and related structures develop into a fruit.

In contrast to the embryogeny of the majority of gymnosperms, which begins with a free nuclear stage, embryogeny in angiosperms is similar to that in lower vascular plants in that the first nuclear division of the zygote is accompanied by cell-wall formation. In the early stages of development, the embryos of dicotyledons and monocotyledons undergo similar sequences of cell division, both becoming spherical bodies. It is with the formation of the cotyledons that a distinction first appears between dicot and monocot embryos, for whereas the dicot embryo begins to develop two cotyledons, the monocot embryo forms only one. Embryogeny in angiosperms is presented in detail in Section 5.

Endosperm formation begins with the mitotic division of the primary endosperm nucleus and usually is initiated prior to division of the zygote. In some species a variable number of free nuclear divisions precede cell-wall formation (nuclear type endosperm formation), whereas in other species, initial and subsequent mitoses are followed by cytokinesis (cellular type endosperm formation). Although endosperm development may occur in a variety of ways, the function of the resulting tissue remains the same: to provide essential food materials for the developing embryo and, in many cases, the young seedling. In some seeds the nucellus proliferates and develops into a food-storing tissue known as perisperm. Some seeds may contain both endosperm and perisperm, as in the beet plant *(Beta)*. In many dicots and some monocots, however, most or all of these storage tissues are absorbed by the developing embryo before the seed becomes dormant. The embryos of such seeds commonly develop fleshy food-storing cotyledons. The principal food materials stored in seeds are carbohydrates, proteins, and lipids.

Angiosperm seeds differ from those of gymnosperms in the origin of their stored food. In gymnosperms the stored food is provided by the female gametophyte, whereas in angiosperms, it is provided by endosperm, which is neither gametophytic nor sporophytic tissue.

Concomitantly, with development of the ovule into a seed, the ovary (and sometimes other portions of the flower or inflorescence) develops into a fruit. As the ovary develops into a fruit, its wall, the pericarp, often thickens and becomes differentiated into distinct layers—the exocarp (outer layer), the mesocarp (middle), and the endocarp (inner), or into exocarp and endocarp only. These layers are generally more conspicuous in fleshy fruits than in dry ones. Fruits are discussed in greater detail in Chapter 19.

The angiosperm life cycle is summarized in Figure 18–54.

18–54
The life cycle of an angiosperm begins with the seed. The sporophyte eventually produces flowers. Within the anthers of the flower, microspore mother cells develop; these divide meiotically, each giving rise to four haploid microspores. Each microspore divides once to form a tube cell and a generative cell. This two-celled structure is the immature microgametophyte, or pollen grain. Either before or during germination, the generative cell divides to form two sperm which are conveyed to the egg apparatus by the pollen tube. The germinated pollen grain, with its tube nucleus and two sperm, constitutes the mature male gametophyte.

Within the ovule, a single megaspore mother cell develops and eventually divides meiotically, giving rise to four megaspores, three of which disintegrate. The fourth develops into the female gametophyte, which at maturity is a seven-celled, eight-nucleate structure known as the embryo sac.

The pollen grain germinates on the stigma, producing a pollen tube, which grows down the style into the ovary and enters the ovule through the micropyle. One sperm nucleus from the pollen tube fuses with the egg to produce the zygote. The second sperm nucleus fuses with the two polar nuclei of the embryo sac to produce the triploid primary endosperm nucleus. This phenomenon of double fertilization is found only among the angiosperms. The embryo develops within the embryo sac and the integuments of the ovule develop into a seed coat. Eventually, the seed is shed from the ovary.

The flower shown here is the Bermuda buttercup (Oxalis pes-caprae), *which is native to South Africa.*

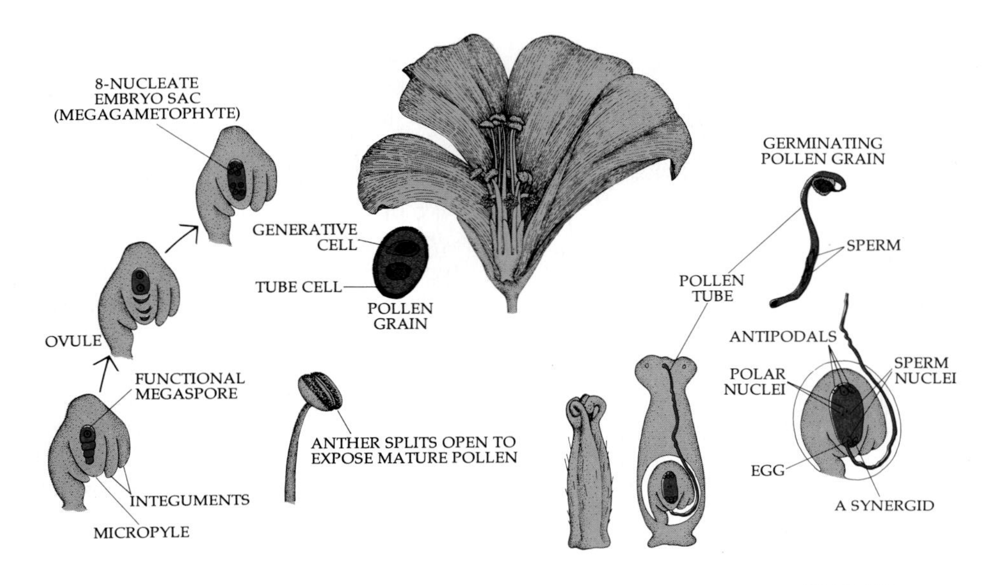

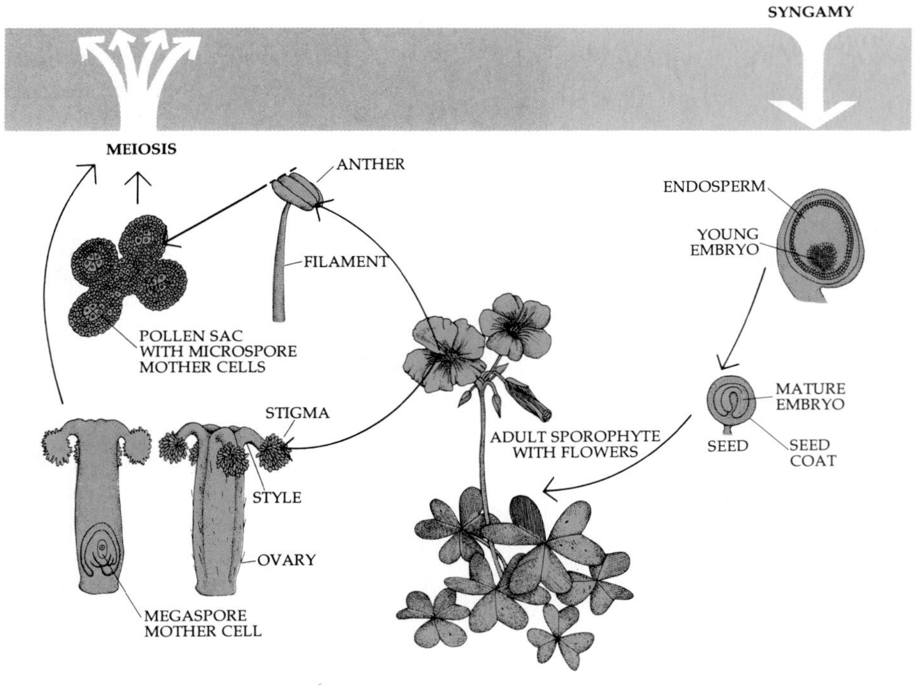

SUMMARY

The seed plants consist of the gymnosperms and the angiosperms. In addition to seeds, all seed plants bear megaphylls. The prerequisites of the seed habit include: heterospory; retention of a single megaspore within the megasporangium; development of the embryo, or young sporophyte, within the megagametophyte; and integuments. All seeds consist of a seed coat derived from the integument, an embryo, and stored food. In gymnosperms the stored food is provided by the haploid female gametophyte; in angiosperms it is provided by generally triploid endosperm. The oldest known seeds occur in strata of the late Devonian period. The likely progenitors of the gymnosperms are the progymnosperms, a Paleozoic group of seedless vascular plants.

The living gymnosperms consist of four divisions: Cycadophyta, Ginkgophyta, Coniferophyta, and Gnetophyta. The angiosperms, or flowering plants, belong to the division Anthophyta, which is divided into two large classes—the Monocotyledones and the Dicotyledones.

The life cycles of gymnosperms and angiosperms are essentially similar: a heteromorphic alternation of generations with large, independent sporophytes and greatly reduced gametophytes. In gymnosperms the ovules (megasporangia plus integuments) are borne exposed on the surfaces of the megasporophylls or analogous structures, whereas in the angiosperms the ovules are borne within the megasporophylls (the carpels). The distinctive reproductive structure of angiosperms—the flower—is characterized by the presence of carpels.

The gametophytes of angiosperms are more greatly reduced than those of gymnosperms. At maturity the female gametophyte of most gymnosperms is a multicellular structure with several archegonia. In angiosperms the female gametophyte, commonly called an embryo sac, is most often a seven-celled, eight-nucleate structure. Archegonia are lacking, and the egg cell is associated with two synergids; these three cells are called the egg apparatus.

The male gametophytes of both gymnosperms and angiosperms develop as pollen grains. Antheridia are lacking in both groups of seed plants. In gymnosperms the male gametes, or sperm, arise directly from the body cell, whereas in angiosperms, they arise directly from the generative cell. The germinated pollen grain, with its tube nucleus and two sperm, is the mature male gametophyte. Except for the Cycadophyta and Ginkgophyta, which have flagellate sperm, the sperm of seed plants are nonmotile and are conveyed to the egg by the pollen tube.

In seed plants, water is not necessary for the sperm to reach the eggs. Rather, the sperm are conveyed to the eggs by a combination of pollination and pollen tube formation. Pollination in gymnosperms is the transfer of pollen from microsporangium to megasporangium. In angiosperms, pollination is the transfer of pollen from anther to stigma.

In gymnosperms, one sperm of the male gametophyte (germinated pollen grain) unites with the egg of an archegonium. The second sperm has no apparent function, and it degenerates. In angiosperms, both sperm are functional: one unites with the egg (true fertilization, or syngamy), and the other unites with the two polar nuclei, resulting in the formation of a diploid ($2n$) zygote and a triploid ($3n$) primary endosperm nucleus, respectively. This phenomenon, which represents one of the principal characteristics of angiosperms, is called double fertilization.

With fertilization in gymnosperms, the ovules develop into seeds. In angiosperms, the ovules develop into seeds and the ovaries (and sometimes associated floral parts) develop into fruits, which enclose the seeds.

SUGGESTIONS FOR FURTHER READING

BIERHORST, DAVID W.: *Morphology of Vascular Plants,* The Macmillan Company, New York, 1971.

An encyclopedic and profusely illustrated treatise on the morphology of the vascular plants by one of their leading contemporary students.

FOSTER, ADRIANCE S., and ERNEST M. GIFFORD.: *Comparative Morphology of Vascular Plants,* 2nd ed., W. H. Freeman and Company, San Francisco, 1974.

A well-organized and general account of the vascular plants, containing much summary and interpretative material.

(a)

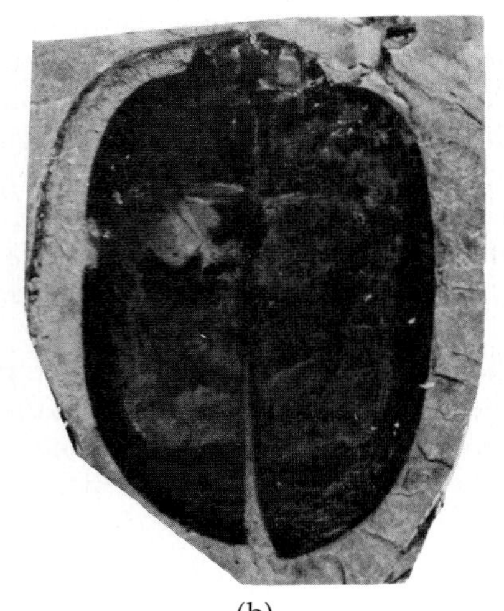

(b)

19–1
(a) *A longhorn beetle of the family Cerambycidae visiting the flower of a composite (Asteraceae) in California.* (b) *Carapace of a beetle from the same strata as* Liriophyllum, *an extinct angiosperm from about 100 million years ago (see "Early Angiosperms and Their Flowers," page 376). The evolution of the flowering plants is, to a large extent, the story of increasingly specialized relationships between flowers and their insect pollinators, in which the beetles played an important early role.*

CHAPTER 19

Evolution of the Flowering Plants

The angiosperms make up much of the visible world of modern plants. Trees, shrubs, lawns, gardens, fields of wheat and corn, wildflowers, fruits and vegetables on the grocery shelves, the bright splashes of color in a florist's window, the geranium on a fire escape, duckweed and pond lilies, eel and turtle grass, the saguaro cacti and the prickly pears—wherever you are, flowering plants are also there.

In a letter to a friend, Charles Darwin referred to the apparent sudden appearance of this great group of plants as "an abominable mystery." In the fossil strata, one first finds layer after layer of lycopod and sphenophyte trees, which, together with primitive gymnosperms, made up the lush vegetation of the Carboniferous forests (see Figure 17–1, page 316); next one finds ferns, cycadophytes, seed ferns, ginkgos, and conifers; and finally, during the first half of the Cretaceous period, the angiosperms. During this period, the flowering plants appeared, gradually assumed worldwide dominance in the vegetation, and by some 75 million years ago, many modern families and some modern genera already existed (Figure 19–2).

Why did the angiosperms rise to world dominance and then continue to diversify to such a spectacular extent? In this chapter, we shall attempt to answer this question, centering our discussion around three topics—the origin of the angiosperms, the evolution of the flower, and the role of certain chemical substances in angiosperm evolution.

ORIGIN OF THE ANGIOSPERMS

It is now generally agreed that the angiosperms evolved from some primitive gymnosperm, probably a shrub. No likely candidates are known from the Cretaceous period, but there are a number of gymnosperms represented earlier in the Mesozoic and Paleozoic periods that display certain combinations of angiospermlike traits. This in itself suggests an

(a)

(b)

(c)

19–2
Fossils of a gymnosperm (a) *and two angiosperms* (b) *and* (c) *from late Cretaceous deposits in Wyoming.* (a) *Twig and individual cone scales of* Araucarites longifolia, *an extinct species of conifer belonging to a family now restricted to the Southern Hemisphere.* (b) *Leaf of a fan palm,* Sabalites montana, *an extinct species distantly related to the palmettos of the southeastern United States.* (c) *Leaf of* Viburnum marginatum, *an extinct species belonging to the widespread genus of shrubs generally known as arrow-woods.*

earlier origin for the angiosperms than can be documented from the fossil record.

The first fossil remains that can definitely be assigned to the angiosperms date from the early part of the Cretaceous period, some 127 million years ago. These remains consist of pollen with a single pore similar to (but distinguishable from) the spores of ferns and the pollen of gymnosperms. It is likely that most angiosperms that existed more than 127 million years ago had pollen that could not be distinguished from the pollen of this group of gymnosperms. All monocots, as well as the most primitive dicots, have pollen similar to that which appears first in the known fossil record. By 120 million years ago, the three-pored angiosperm pollen characteristic of all but the most primitive dicots had appeared in the fossil record. By about 80 to 90 million years ago, angiosperms were more abundant than any other plants all over the world.

Paleobotanist Daniel Axelrod, of the University of California, has suggested that the early evolution of the angiosperms may have taken place away from the lowland basins where fossil deposits are normally found. In the hills and uplands of the tropics, the angiosperms could have evolved into a variety of forms without much chance of their remains being preserved in the fossil record. It was when they invaded the lowlands, about 120 million years ago, that they assumed the dominant role in the world's vegetation.

Very early in the history of the flowering plants, different lines evolved a number of adaptations that made them particularly resistant to drought and cold.

Among these were tough leaves, often reduced in size; vessel elements (water-conducting cells); and a tough, resistant seed that protects the young embryo from drying out. Many groups of angiosperms became deciduous; that is, they shed their leaves at particular times of the year when vegetative growth is not possible. This happened first in tropical areas that experienced periodic drought. The deciduous plants then spread to the north, where parts of the year were so cold that no water was available for growth. The sort of pollination system found in angiosperms, in which free water is not required, is also highly favored in dry habitats. The features that make many angiosperms particularly resistant to drought and cold are not found in all of the flowering plants, nor are they restricted to the angiosperms, but they have certainly played a major role in the diversification of the group.

About 127 million years ago—when angiosperm pollen first appears in the fossil record—Africa and South America were directly linked with one another and with Antarctica, India, and Australia in a great southern supercontinent called Gondwanaland (see Figure 19–3). Africa and South America began to separate at about this time, forming the southern Atlantic Ocean, but they did not move completely apart in the tropical regions until about 90 million years ago. India began to move northward at about the same time, colliding with Asia about 45 million years ago and thrusting up the Himalayas in the process. Australia began to separate from Antarctica about 55 million years ago, but their separation did not become complete until about 40 million years ago.

Within the central regions of West Gondwanaland, formed by what are now the continents of South America and Africa, the sort of arid-to-subhumid habitats envisioned by Axelrod and others as an important early site of angiosperm evolution would have been common. With the final separation of these two continents, at about the time the angiosperms became abundant in the fossil record worldwide, the world climate changed greatly. This was especially the case in these equatorial regions, which became milder, with fewer extremes of temperature and humidity. These changes are thought to be related to the evolutionary success of the early angiosperms, which increasingly dominated floras worldwide.

When India and Australia were in the far south, they were covered with vegetation typical of a cool temperate climate. Certain groups of plants and animals are common to southern South America and southeastern Australia-Tasmania at the present time, having achieved their ranges by migration across Antarctica long before the onset of full glaciation there, which occurred about 20 million years ago (Figure 19–4). As Australia moved northward during the past 55 million years, it reached the great zone of aridity flanking the tropics, and the sorts of arid plant communities that are now so widespread there expanded greatly. As it neared Asia, the northern edge of Australia reached truly tropical climates and was invaded by the plants and animals of tropical Asia. To a large extent, however, the plants and animals characteristic of Asia and Australia have remained geographically distinct, and the line separating the areas where each is predominant is known as "Wal-

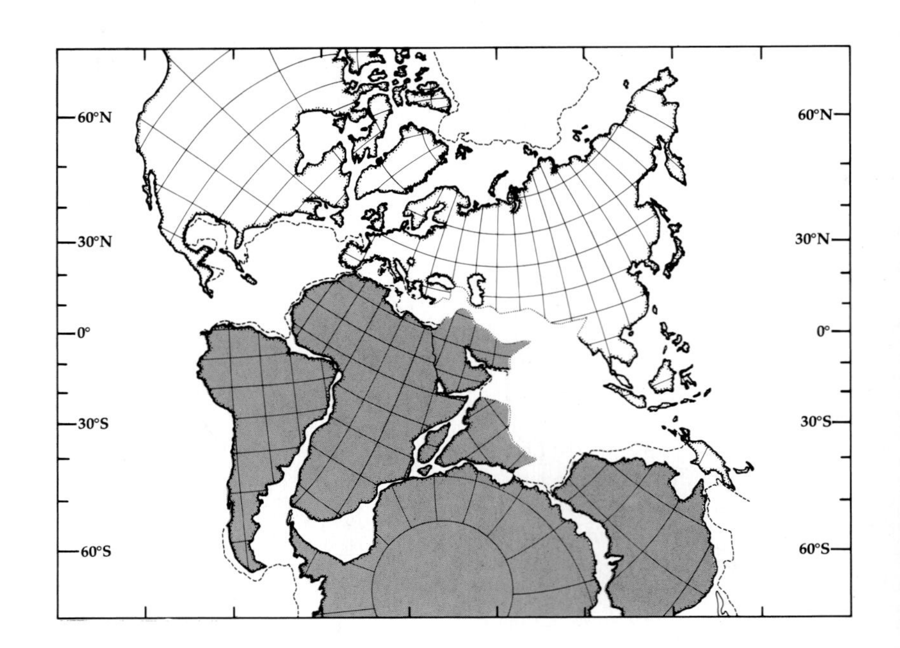

19–3
Relationship between the earth's southern land areas at the time of first appearance of the angiosperms in the fossil record. In the middle Cretaceous period, 110 ± 10 million years ago, South America was directly connected with Africa, Madagascar, and India, and via Antarctica, with Australia. These combined land masses, indicated here in color, formed the supercontinent of Gondwanaland.

(a)

(b)

19–4

(a) *This silver beech* (Nothofagus menziesii), *growing in Upper Caples Valley, Southland, New Zealand, is a relic of the Antarcto-Tertiary Geoflora forest that extended from what is now southern South America across Antarctica to Australia and New Zealand. These areas remained in proximity into middle Eocene time, about 45 million years ago.* (b) *The palm* Nypa fruticans, *shown growing along the edge of a tidal swamp. In Eocene time, when Africa and South America were in closer contact,* Nypa *was distributed widely throughout the warmer parts of the world. It is now confined to Southeast Asia, northern and northeastern Australia, and some Pacific Islands.*

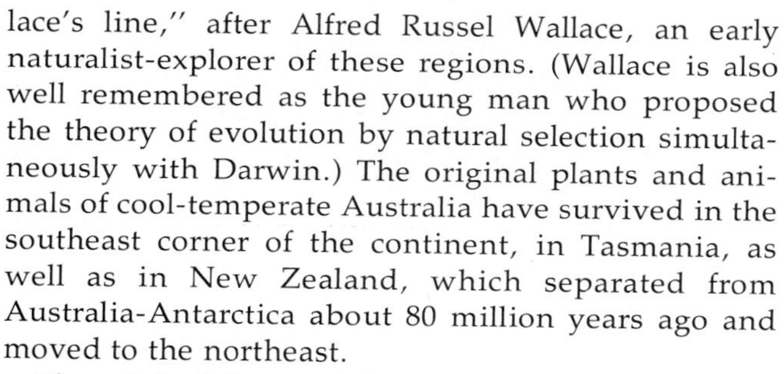

lace's line," after Alfred Russel Wallace, an early naturalist-explorer of these regions. (Wallace is also well remembered as the young man who proposed the theory of evolution by natural selection simultaneously with Darwin.) The original plants and animals of cool-temperate Australia have survived in the southeast corner of the continent, in Tasmania, as well as in New Zealand, which separated from Australia-Antarctica about 80 million years ago and moved to the northeast.

The original flora and fauna of India did not fare so well, and there are few remnants surviving. India moved much farther than Australia, crossing the south arid zone, the torrid tropics, and the north arid zone in the process. As a result, nearly all of India's original plants and animals became extinct and were eventually replaced by desert, tropical, and mountain flora and fauna from Eurasia.

Throughout the history of the angiosperms, new evolutionary lines have been produced as the world climate changed. Both in the origin of the angiosperms and in their subsequent diversification and reassortment into modern communities, the challenge of aridity has been met repeatedly. Drought has been a powerful selective force, with relatively few organisms able to live in the more arid communities. The structure of those angiosperms that are able to grow in arid lands is often profoundly different from that of their ancestors.

EARLY ANGIOSPERMS AND THEIR FLOWERS

Liriophyllum, *representing an extinct family of angiosperms, from the middle Cretaceous period (about 100 million years ago) in central Kansas. The reproductive axis of this plant* (a) *is elongate, up to 12 centimeters in length, with more than 50 spirally arranged, many-seeded carpels bearing a suture on their upper surface. Neither these reproductive axes nor the large, entire-margined, bilobed leaves* (b) *of* Liriophyllum *have been related to any present-day angiosperm family. Careful studies by David L. Dilcher of Indiana University and his students are revealing a great deal about the nature of early angiosperms and their flowers.*

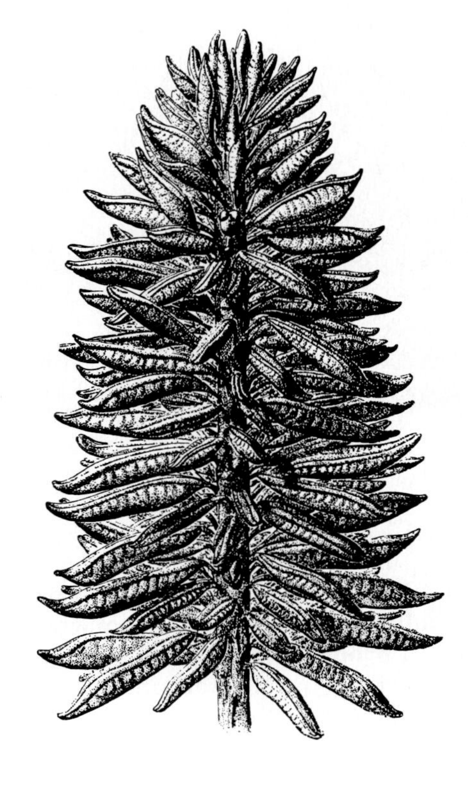

(a)

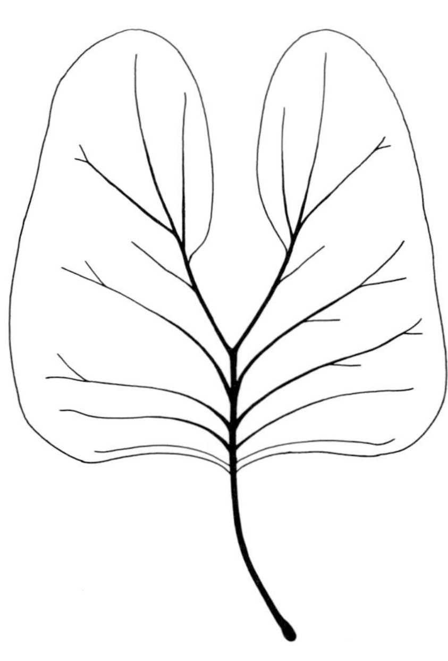

(b)

EVOLUTION OF THE FLOWER

The Parts of the Flower

The single characteristic that sets the angiosperms apart from all other groups is the flower. Most of what we surmise about the evolution of the flower is based on a comparative study of modern forms, because flowers, being delicate, are rarely found in the fossil record, and only recently has the critical study of fossil flowers been used as a guide to the evolution of angiosperms. As noted previously, the flower is a determinate shoot bearing various leaflike appendages; how the various parts of the flower are related to leaves will now be discussed in more detail.

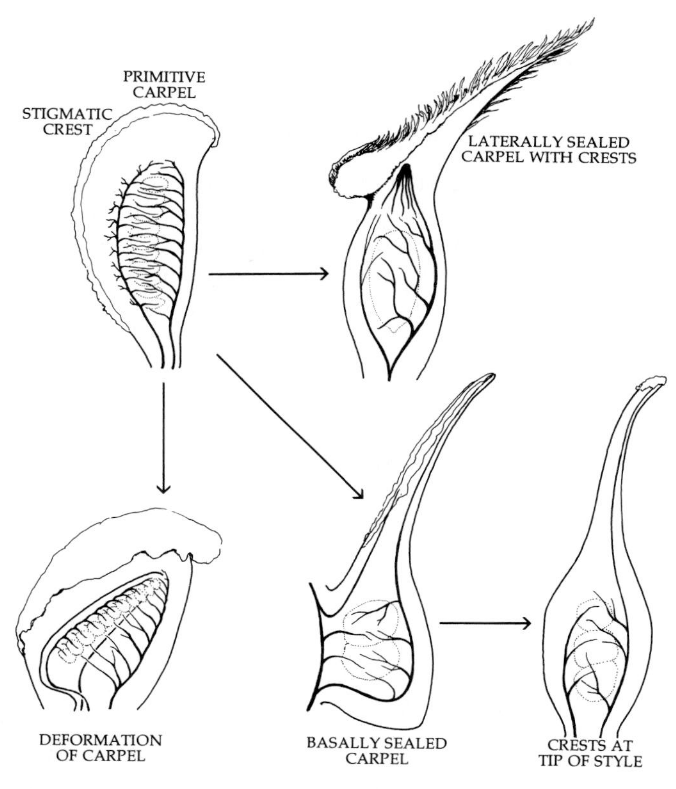

19–5
Some changes that have occurred during the evolution of the angiosperm carpel. The primitive carpel is conduplicate (folded together lengthwise). The folded blade encloses a number of ovules attached to the inner surface. The margins are not sealed and there is no localized stigmatic surface, although stigmatic hairs are extensively developed on the inner surfaces. In the carpels derived from this primitive example, the margins of the blade are more completely sealed, and the pollen-trapping stigmatic surface is localized in a crest.

The Carpel

The carpel in its most primitive form is a folded blade. As shown in Figure 19–5, in some generalized carpels there is no localized area for the entrapment of pollen grains. Both margins of the folded blade are covered with stigmatic hairs. The folded carpel surrounds the ovules, which are located on its inner surface. The carpels are closed in all the living angiosperms, although the different taxa show different stages in the closure process. Primitive angiosperms have large stigmatic surfaces located directly on the "unsealed" carpels. More specialized forms, including nearly all of those that exist today, have a much smaller stigma held well above the carpel on a style.

The ovules, which were probably in rows along and near the edges of the inner surface of the carpel in the original angiosperms, became arranged in various ways in more specialized carpels. There are many ovules in primitive angiosperms and few in advanced species. The gynoecium of the primitive angiosperm flower contained a number of separate carpels, which have been reduced in number or fused together or both during the course of evolutionary specialization (Figure 19–6).

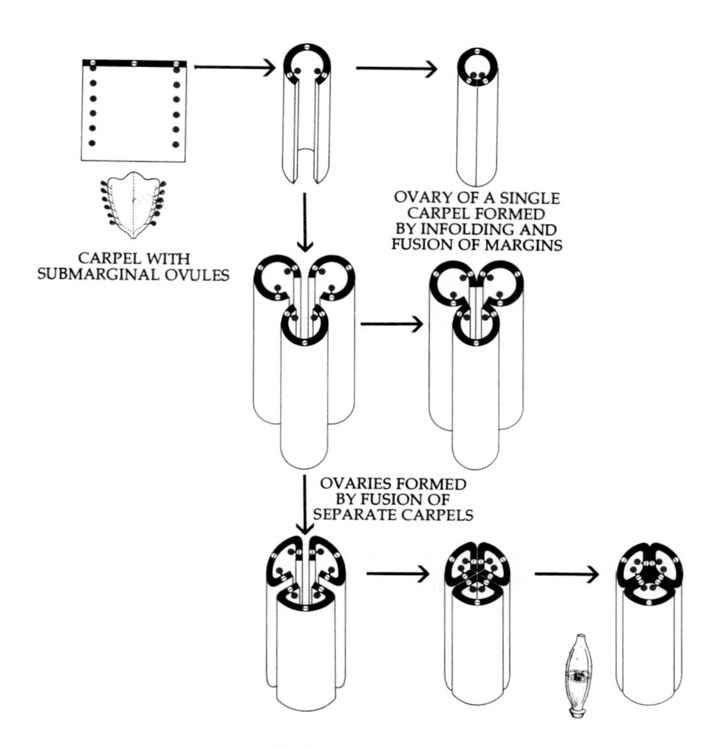

19–6
Presumed evolutionary development of the gynoecium.

The Androecium

Although the stamens of modern plants rarely resemble leaves, it is possible to find leaflike stamens among the magnolias and their relatives. Like these, the primitive stamen may have been blade-shaped, with sporangia near the center of the blade (Figure 19–7). According to one theory, the blade became differentiated into a slender stalk (the filament), with the sporangia near its apex. An alternative theory, which is also attractive, holds that the stamens were derived from slender branch systems bearing terminal sporangia. The branches gradually became fused and, in some instances, leaflike.

In some specialized flowers, the stamens, like the carpels, became fused. Stamens may be fused with one another into columnar structures, as in the pea, melon, and sunflower families; or they may be fused with the corolla, as in the phlox, snapdragon, and mint families.

In some evolutionarily advanced flowers, the stamens have become secondarily sterile; that is, they have lost their sporangia and have become transformed into specialized structures, such as nectaries, which are glands that secrete nectar, a sugary fluid that attracts pollinators and provides food for them. (It is important to note, however, that most nectaries are not modified stamens.) Stamens have also played a part in the evolution of petals.

The Perianth

The perianth consists of the sepals and the petals. The sepals of most flowers are green and photosynthetic. Supplied by more than one vascular strand, sepals resemble the leaves of the plant on which they occur. They are considered to have been derived directly from foliage leaves.

In a few families, such as the water lilies, the petals seem to have been derived from sepals. In most angiosperms, however, the petals appear to be stamens that have lost their sporangia and have become specially modified for their new role—that of drawing attention to the flower. Most petals have just one vascular strand, as do stamens, but sepals, like leaves, normally have three or more.

Petal fusion has occurred in the evolution of many groups of angiosperms, resulting in the familiar tubular corolla that is characteristic of many families. When a tubular corolla is present, the stamens often fuse with it and appear to arise from the corolla. In a number of the evolutionarily advanced families, the sepals are similarly fused into a tube.

Evolutionary Trends Among Flowers

A modern flower that may resemble the primitive flower in general structure is the magnolia (Figure 19–8). The magnolia flower has numerous carpels and stamens, and many perianth parts, all of which are well separated from one another; in addition, the spiral arrangement of these parts on the tip of the flower stalk, or receptacle, is still clearly evident. The very elongate receptacle and its conelike nature are, however, clearly secondary specializations within the magnolia family and are not shared by other primitive angiosperms.

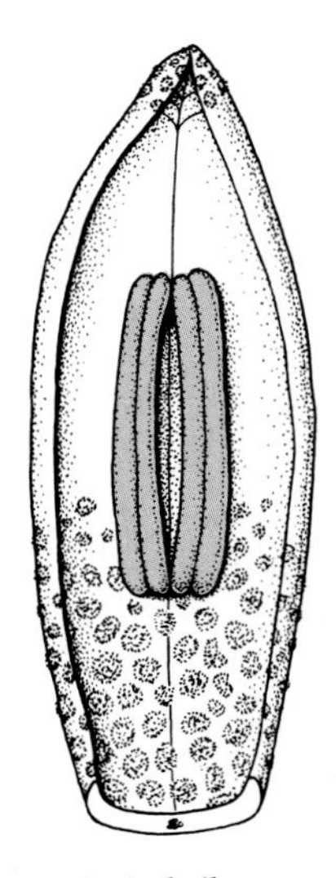

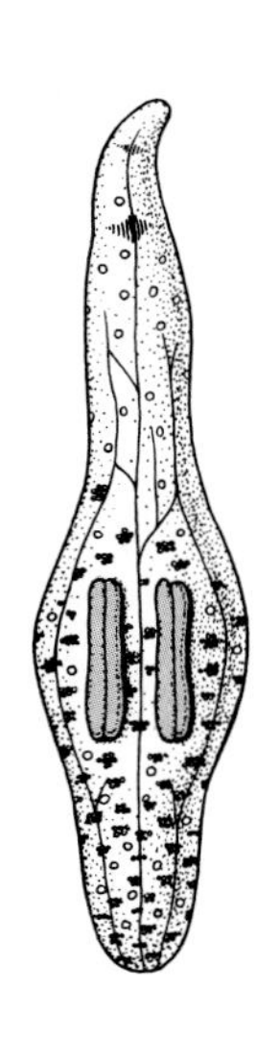

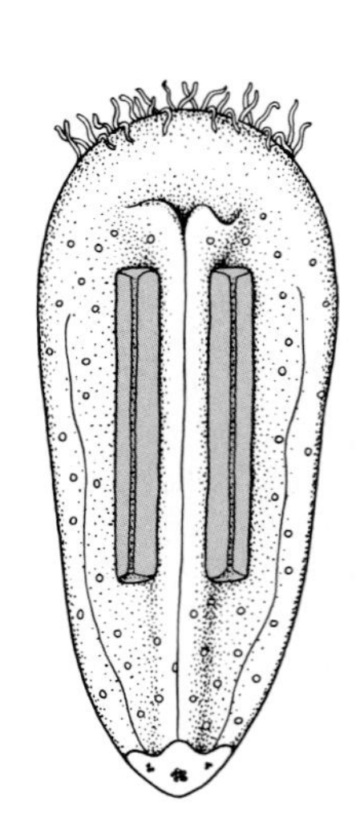

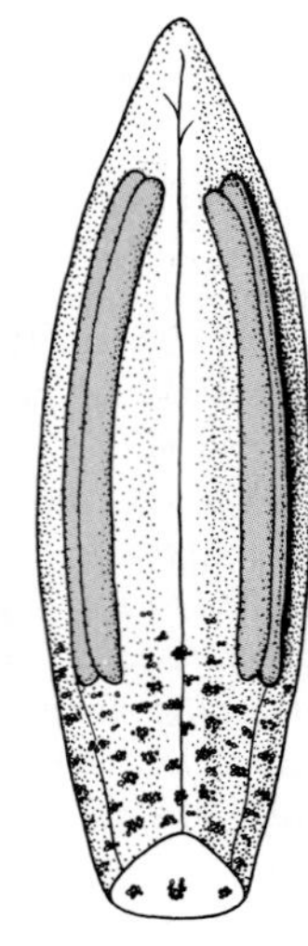

19–7
Stamens of primitive angiosperms. In these woody plants, the anthers are borne on the lower or upper surface of a leaflike microsporophyll. Examples of the former, Himantandra *and* Degeneria, *are here viewed from below;* Austrobaileya *and* Magnolia *are viewed from above. In most modern angiosperms, the stamens contain much less sterile tissue than those of primitive angiosperms, and the anther is borne at the end of a slender filament. Such stamens are not at all leaflike. These differences are difficult to explain and have led to the hypothesis that leaflike stamens such as these may actually have been derived through the fusion of branch systems bearing terminal sporangia.*

(a)

(b)

(c)

19–8
The flower structure of the southern magnolia (Magnolia grandiflora). *The cone-shaped receptacle is made up of spirally arranged carpels from which curved styles emerge. Below the styles in* (a) *and* (b) *are the cream-white stamens.* (a) *An unopened bud cut open to reveal the pollen-receptive stigmas; the anthers have not yet shed their pollen.* (b) *The floral axis of a second-day flower, showing stigmas that are no longer receptive and stamens that are shedding pollen.* (c) *Fruit, showing carpels and bright red seeds, each protruding on a slender stalk.*

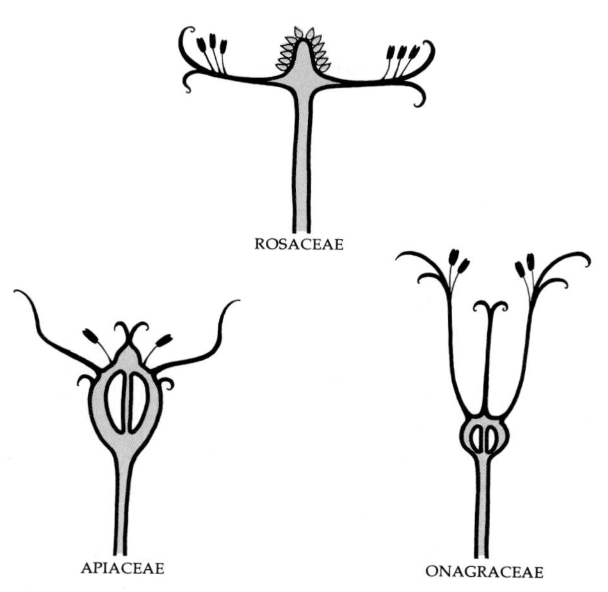

19–9
Types of flowers in three common families of dicots, showing changes in the position of the ovary. Many Rosaceae have superior ovaries with the flower parts attached below the ovaries; their flowers are perigynous. The Apiaceae and Onagraceae have inferior ovaries; that is, the flower parts are attached above them. The flowers of these two families are epigynous.

By comparing such a primitive flower with specialized ones, four main trends in flower evolution can be traced (see Figures 19–9 and 19–10).

1. From flowers with many parts, indefinite in number, flowers have evolved that have few parts, definite in number.
2. The number of kinds of parts has been reduced from four in the primitive flower to three, two, or sometimes one in the more advanced flowers. The shoot has become shortened so that the original spiral arrangement of parts is no longer evident. The floral parts have become fused.
3. The ovary has become inferior rather than superior in position.
4. The regularity (radial symmetry) of the primitive flower has given way to irregularity (bilateral symmetry) in the more advanced flowers.

Examples of Specialized Families

Among the most evolutionarily specialized of the flowers are the Asteraceae (Compositae), which are dicots, and the Orchidaceae, which are monocots. These are the two largest families of angiosperms in terms of the number of species.

19–10
Examples of specialized trends in flowers. (a) *Wintergreen,* Chimaphila umbellata. *The sepals (not visible) and petals are reduced to five each, the stamens to ten, and the five carpels are fused into a compound gynoecium with a single stigma.* (b) *Lotus,* Nelumbo lutea. *The undifferentiated sepals and the numerous petals and stamens are spirally arranged; the carpels are fused into a compound gynoecium.* (c) *Baby blue eyes,* Nemophila menziesii. *The sepals (not visible), petals, and carpels are fused, and the stamens are reduced to five and arise on the corolla.* (d) *A diagram of a cotton* (Gossypium) *flower, showing the column of stamens fused around the style.*

(a)

(b)

(c)

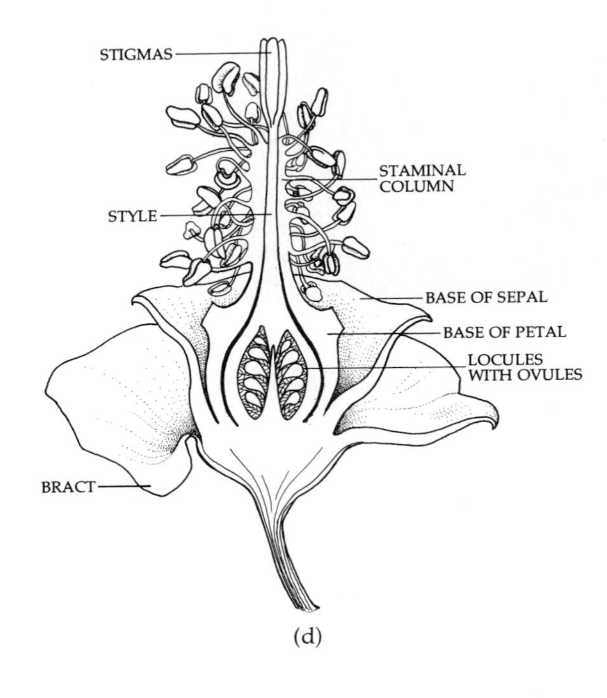

(d)

Asteraceae

In the Asteraceae (the composites), the epigynous flowers are relatively small and closely bunched together into a head. In each of the tiny flowers, there is an inferior ovary composed of two fused carpels with a single ovule in one locule (Figure 19–11).

In flowers of Asteraceae, the stamens are reduced to five in number and are usually fused to one another (coalesced) and to the corolla (adnate). The petals, also five in number, are fused to one another and also to the ovary, and the sepals are absent or reduced to a series of bristles or scales known as the pappus. As in the familiar dandelion, the pappus often serves as an agent of dispersal by wind (see Figure 19–33). In other Asteraceae, such as beggar-ticks *(Bidens),* the pappus may be barbed and serve for attachment to animals. In many Asteraceae, each head includes two types of flowers—disk flowers, which make up the central portion of the aggregate, and ray flowers, which are arranged on the outer periphery. In the ray flowers, which are sometimes completely sterile and often carpellate, the fused corolla forms a long strap-shaped "petal" in species such as the sunflower, daisy, and black-eyed Susan.

In general, the head in the composites has the ap-

19–11
(a) *A diagram of the organization of the head of a composite (family Asteraceae). The individual flowers are subordinated to the overall display of the head, which acts as a single large flower.* (b) *Sunflower,* Helianthus annuus. (c) *Thistle,* Cirsium pastoris. (d) *Gaillardia,* Gaillardia pulchella.

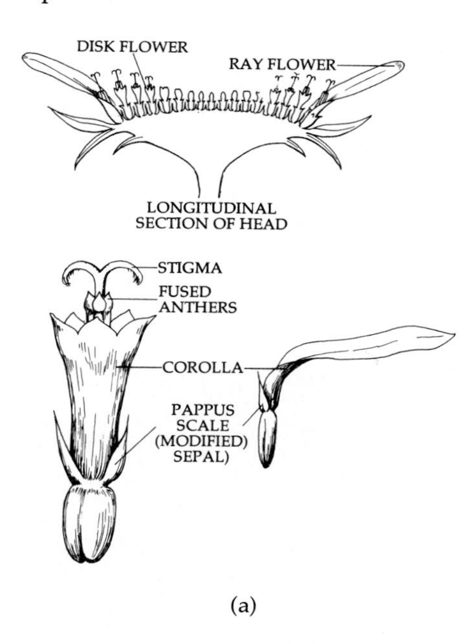

(a)

(b)

(c)

(d)

pearance of a single large flower. Unlike many single flowers, however, the head matures over a period of days, with the individual flowers opening serially in a centripetal spiral. As a consequence, the ovules in a given head may be fertilized by a number of different pollen donors. The success of this plan as an evolutionary strategy is attested to by the great abundance of the members of Asteraceae, which is the largest family of flowering plants, with 22,000 species.

Orchidaceae

Another successful flower plan is that of the orchids (Orchidaceae, the second largest plant family), which, unlike the composites, are monocots. There are at least 17,000 species of orchids, but the classification of many of the genera is poorly established. Most of the species are tropical; only about 140 occur in North America north of Mexico, for example. In the orchids, the three carpels are fused, and, as in the composites, the ovary is inferior. Unlike the composites, however, each ovary contains many thousands of minute ovules; consequently, each pollination event may result in the production of a large number of seeds. Usually only one stamen is present (in the lady-slipper orchids, there are two), and this is characteristically fused with the style and stigma into a single complex structure—the column. The entire contents of the anther are held and distributed as a unit—the pollinium. The three petals are modified so that the two lateral ones form wings and the third forms a cuplike lip that is often very large and showy. The sepals, also three in number, are often colored and similar to petals in appearance. The flower is always irregular (Figure 19–12).

(a)

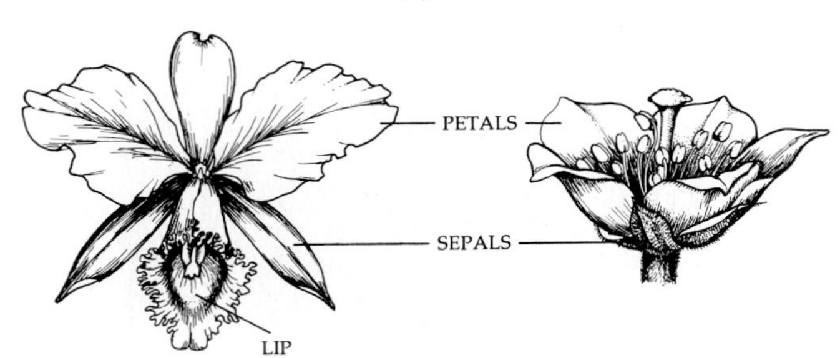

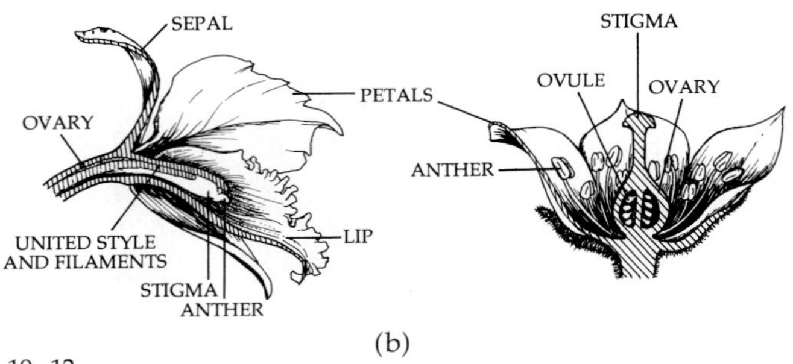

(b)

19–12
(a) Encyclia *orchids. Orchids are the most specialized family of monocots.*
(b) *A comparison of the parts of an orchid flower, shown on the left, with those of a radially symmetrical flower, shown on the right. The "lip" is a modified petal that serves as a landing platform for insects.*

GENETIC SELF-INCOMPATIBILITY

By recombining genes from different individuals, sexual reproduction produces variability in natural populations, giving organisms the potential for adaptation to their environment. Plants that regularly self-pollinate lose some of this capacity. Early in the history of the flowering plants, mechanisms evolved in many families that make cross-pollination almost mandatory, even when the flowers are bisexual or when both staminate and ovulate flowers are present on the same plant.

Two basic mechanisms are found among modern families of flowering plants. In the more common one, present in such economically important groups as grasses and legumes, the behavior of the pollen grain is determined by its own (haploid) genotype. If it carries a gene at the incompatibility locus that is identical to one at either of the two corresponding loci in the diploid stigma and style, then its entry is barred. If it carries a gene at this locus that is different from either of those

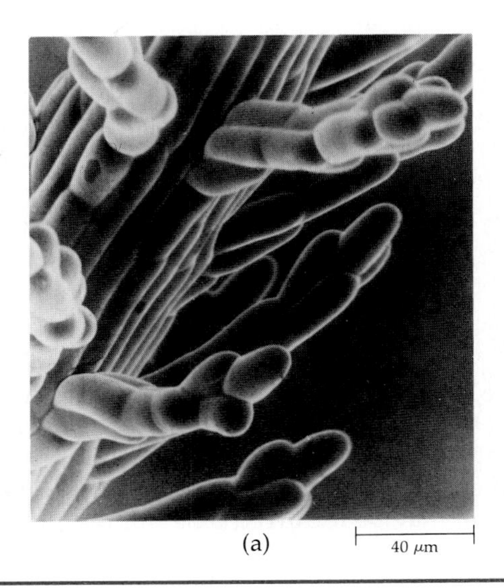

(a)

The Agents of Evolution

Land plants, unlike most animals, cannot go from place to place to find food or shelter and cannot move about to seek their mates. In general, they must satisfy these needs passively. The flowering plants, however, have evolved a set of features that allows them to "move" in seeking a mate; this set of features is embodied in the flower. By attracting insects and other animals and by directing their activities so that a high frequency of cross-pollination and cross-fertilization of the plants will result, the angiosperms have, in a sense, transcended their rooted condition and become just as motile as the higher animals in this one respect. How did this all come about?

The earliest seed-bearing plants—various groups of

present in the stigma tissue, then it is accepted.

In the second incompatibility system, found in the crucifer and composite families, the behavior of the pollen is determined by the pollen parent; again a "match" at the incompatibility locus determines acceptance or rejection. A moment's thought will show how effective such systems are likely to be in providing a barrier to self-pollination.

Although much remains to be learned about the physiology of these mechanisms, it is known that they depend upon "recognition" reactions between specific incompatibility proteins carried by the pollen grains and matching proteins produced in the stigmas or styles. In grasses, the reaction often occurs on the stigma surface. The scanning electron micrograph (a) *shows part of the stigma of the orchard grass* (Dactylis glomerata), *with numerous papillae, each capable of capturing several pollen grains. The transmission electron micrograph in* (b) *shows the wall of a papilla of another grass species, rye* (Secale cereale), *in section. Outside of the cuticle lie two other layers, an inner one composed of mucilaginous pectic materials, and an outer proteinaceous one. The incompatibility response is shown when the tip of the pollen tube comes in contact with the outer layer, or very shortly after. The micrographs in* (c) *and* (d), *taken with the fluorescence microscope, show stigmas of foxtail grass* (Alopecurus pratensis) *stained with a fluorescent stain for the wall polysaccharide, callose. In* (c) *the pollination has been a compatible one, and the tube is seen growing toward the ovary after penetrating the stigma. In* (d) *the pollination has been incompatible; after the tube tip has touched the surface protein layer, its growth has been halted and the tube has filled with callose, which shows that it has been rejected.*

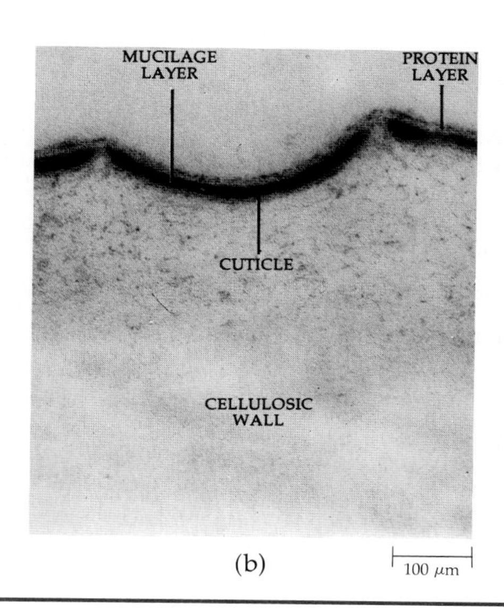

(b)

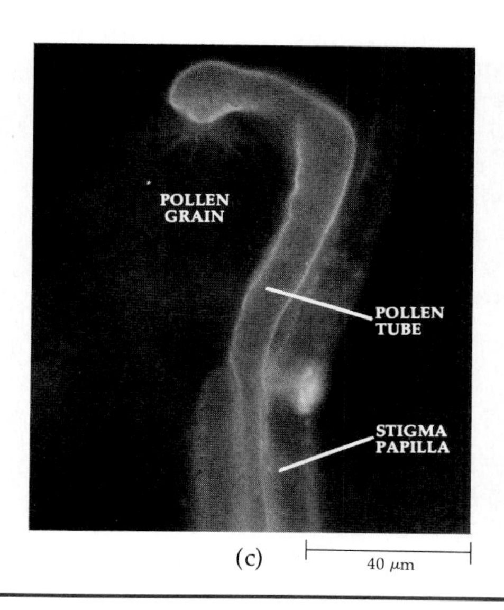

(c)

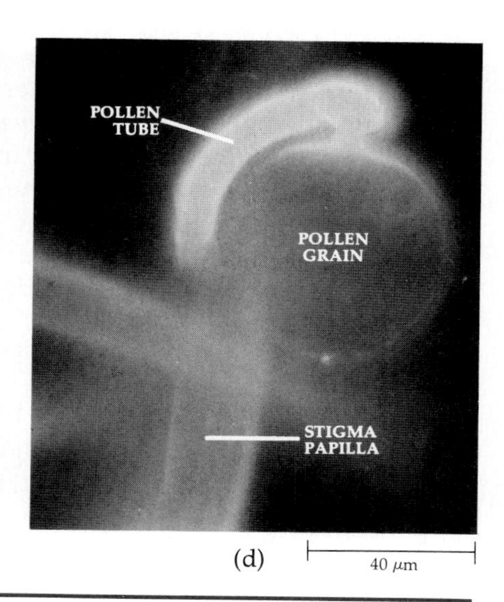

(d)

gymnosperms—were pollinated passively, by the action of the wind. The ovules, which were borne on the leaves or within the cones, exuded drops of sticky sap from their micropyles. These drops caught the pollen grains and drew them to the micropyle, just as in the modern conifers. Insects, probably beetles, feeding on the sap and resin of stems and on leaves, must have come across the protein-rich pollen grains and the sticky droplets from the ovules. Once they began returning regularly to these new-found sources of food, they began inadvertently to carry pollen from other plants. For some plants, even this desultory, aimless pollination by wandering insects must have been more effective than wind pollination alone.

The more attractive the plants were to the beetles, the more frequently they would be visited, and the more seeds they would produce. Any changes in the phenotype that made such visits more frequent or more efficient offered an immediate selective advantage. Several evolutionary developments are seen as a direct consequence of pollination by insects. For example, plants that had flowers that provided special sources of food for their pollinators had a selective advantage. In addition to edible flower parts, pollen, and sticky fluid around the ovules, plants evolved specialized glands, or nectaries, in their flowers. As mentioned previously, these glands secrete nectar, a nutritious sugary fluid that is attractive to insects.

Attraction of the insects to the flowers raised a new problem: protection of the ovule from predatory insects. The evolution by natural selection of the closed carpel may have been a consequence of this need.

Further changes in the shape of the flower, such as the evolution of the inferior ovary, may also have served to protect the ovules from being eaten.

Another important development was the bisexual flower. The presence of both carpels and stamens in a single flower (in contrast, for instance, to the microsporangiate and ovulate cones of conifers) serves to make each visit by the pollinator more effective, because the pollinator can both pick up and deliver pollen at each stop.

In the early part of the Tertiary period, 40 to 60 million years ago, such specialized groups of flower-visiting insects as bees and butterflies became abundant and diverse. The rise and diversification of these groups of insects was a direct result of the increasing diversity of angiosperms. In turn, the insects profoundly influenced the evolutionary course of the angiosperms and contributed greatly to their diversification.

If a given plant species is pollinated by only one or a few kinds of visitors, it tends to become specialized relative to the characteristics of these visitors. Many of the modifications that have evolved in angiosperm flowers were special adaptations in relation to the promotion of floral constancy by the visitors.

Some of the special modifications of the flower that came about during the course of evolution in response to specific pollinators will be described in the following pages.

Beetle-Pollinated Flowers

A number of modern species of angiosperms are pollinated solely or chiefly by beetles (Figure 19–13). The flowers of one distinct type are large and borne singly, like those of the magnolia, the lily, the California poppy, and the wild rose; the flowers of another type are small and aggregated in an inflorescence, such as those of the dogwood, elder, and spiraea. Members of some 16 families of beetles are frequent visitors to flowers, although, as a rule, they derive most of their nourishment from other sources, such as sap, fruit, dung, and carrion. In beetles, the sense of smell is much more highly developed than the visual sense, and beetle flowers are often white or dull in color and frequently have strong odors (Figure 19–14a). These odors are usually fruity, spicy, or similar to the foul odors of fermentation and are thus distinct from the sweeter odors of flowers pollinated by bees, moths, and butterflies. Some beetle flowers secrete nectar; in others, the beetles chew directly on the petals or on specialized food bodies (pads or clusters of cells on the surfaces of various floral parts), as well as eating the pollen. Most beetle flowers have the ovules well buried beneath the floral chamber, out of reach of the chewing jaws of the beetles.

19–13
A pollen-eating beetle (Asclera ruficollis) *at the open, bowl-shaped flowers of round-leaved hepatica* (Hepatica americana). *All species of this family of beetles (Oedemeridae) are obligate pollen-feeders as adults.*

19–14
(a) *Western skunk cabbage* (Lysichiton americanum) *is pollinated by small, actively flying beetles (Staphylinidae). The beetles are attracted by its very strong odor. Other species of this plant family (Araceae) have inflorescences with odors resembling dead fish or carrion, and some of these are pollinated by carrion flies. One such plant is* (b) *the skunk cabbage* Symplocarpus foetidus, *and another is shown in Figure 8–13 on page 143.* (c) *A number of plants in other families have similar odors and are pollinated by carrion flies; the foul-smelling milkweed* Stapelia schinzii *and its relatives, most of which are native to Africa, is one example.*

(a)

19–15
Bees have become as highly specialized as the flowers with which they have coevolved. Their mouth parts have become fused into a sucking tube containing a tongue. The first segment of each of the three pairs of legs has a patch of bristles on its inner surface. Those of the first and second pairs are pollen brushes that gather the pollen that sticks to the bee's hairy body. On the third pair of legs, the bristles form a pollen comb that collects pollen from these brushes and from the abdomen. From the comb, the pollen is forced up into pollen baskets, concave surfaces fringed with hairs on the upper segment of the third pair of legs. Shown here is a honeybee (Apis mellifera) *foraging in a flower of* Salvia.

Bee-Pollinated Flowers

Bees are the most important group of flower-visiting animals. Both male and female bees live on nectar, and the females also collect pollen to feed the larvae. Bees have mouth parts, body hairs, and other appendages specially fitted for collecting and carrying these food materials (Figure 19–15). As Karl von Frisch and other investigators of insect behavior have shown, bees can quickly learn to recognize colors, odors, and outlines. The color spectrum that a bee can see is somewhat different than ours; it can see ultraviolet, which is invisible to us, but it cannot distinguish shades of red, all of which appear black to a bee.

Bees have developed a high degree of constancy to particular kinds of flowers. Such constancy increases the efficiency of the bee, and in bees with narrowly restricted foraging habits, there are often conspicuous morphological and physiological adaptations related to the characteristics of the plant of choice. When constant to this degree, bees exert a powerful evolutionary force for specialization in the plants they visit. There are some 20,000 known species of bees, nearly all of which visit flowers for food.

Bee flowers—that is, flowers that coevolved with bees—have showy, brightly colored petals, which are usually blue or yellow. They often have a distinctive pattern by which the bee can quickly recognize them.

(b)

(c)

This pattern may include a "honey guide," special markings that indicate the position of the nectar (Figure 19–16). Bee flowers are never pure red, and, as special photographic techniques have shown, they often have distinctive markings normally invisible to humans (Figure 19–17).

In bee flowers, the nectary is characteristically situated at the base of the corolla tube and is usually set below the surface, accessible only to a specialized sucking organ. Bee flowers characteristically have a "landing platform" of some sort.

Some of the evolutionarily more advanced flowers, in particular the orchids, have developed complex passageways and traps that force the bee to follow a particular route into and out of the flower. This ensures that both anther and stigma come into contact with the bee's body at a particular point and in the proper sequence (see Figure 19–15).

An even more bizarre pollination strategy has been adopted by orchids of the genus *Ophrys*. The flower resembles a female bee, wasp, or fly rather closely. The males of these insect species emerge early in the spring, before their females. The orchids bloom early in the spring as well, and the male insects attempt to copulate with the orchid flower. During the course of their "sexual" visit, a pollinium may be deposited on the insect's body and thus transferred to the next flower that it visits.

A whole additional spectrum of flowers with different characteristics is pollinated by flies of various kinds, including mosquitoes. These insects feed on nectar from the flower but do not gather pollen or provision larval cells. Examples of flowers pollinated by mosquitoes and flies are shown in Figures 19–14b, 19–14c, and 19–19.

19–16
"Honey guides" on the flowers of the foxglove (Digitalis purpurea) *serve as distinctive signals to insect visitors.*

(a)

(b)

19–17
The spectrum of visible light is slightly different for the bee than for the human. The bee's eyes are sensitive to ultraviolet light. Therefore, a flower such as that of the marsh marigold (Caltha palustris), *which appears solid yellow to the human eye* (a), *is seen under ultraviolet light to have distinctive markings visible to the bee* (b).

19–18
A sweat bee (family Halictidae) gathering pollen from the stamens of a flower of a cactus (Echinocereus) *in Baja California, Mexico. The tall structures in the center of the flower are stigmas.*

(a)

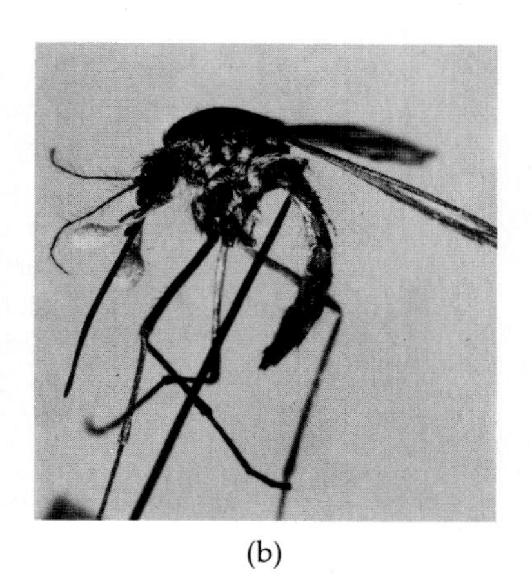

(b)

(c)

19–19
Pollination by mosquitoes and flies. (a) *Some small-flowered orchids, such as* Habenaria elegans, *in which the flowers are white or green and relatively inconspicuous, are visited and pollinated by mosquitoes in North Temperate and Arctic regions. The mosquitoes obtain nectar from the flowers.* (b) *A female mosquito of the genus* Aedes, *with an orchid pollinium attached to its head.* (c) *A fly on a flower of a lily* (Zigadenus fremontii). *Notice the conspicuous yellow nectaries.*

(a)

(b)

19–20
(a) *A skipper* (Poanes hobomok) *visiting the flower of a wild geranium* (Geranium maculatum). (b) *Painted lady* (Cynthia annabella), *a butterfly, on a composite. The long sucking tongues of moths and butterflies are coiled up at rest and extended when feeding. They vary in length from a few millimeters in some of the smaller moths, 1 to 2 centimeters in many butterflies, 2 to 8 centimeters in some hawkmoths of the North Temperate Zone, and up to 25 centimeters in tropical hawkmoths.*

Flowers Pollinated by Moths and Butterflies

Flowers that have coevolved with diurnal (active during the day rather than at night) moths and butterflies are similar in many respects to bee flowers because these insects are guided to flowers by a combination of sight and smell similar to that of the bees (Figure 19–20). At least some species of butterfly, however, may be able to see red as well as blue and yellow, and some butterfly flowers are red and orange.

Most moths are night fliers, and the typical moth flower—as seen, for example, in several species of tobacco *(Nicotiana)*—is white and has a heavy fragrance, a sweet penetrating odor that is emitted only after sunset. Other moth flowers, although not white, display colors that stand out against a dark background in the evening (for example, the yellow-flowered species of evening primrose, *Oenothera,* or the pink-flowered amaryllis, *Amaryllis belladonna*).

The nectary of a moth or butterfly flower is found at the base of a long slender corolla tube or a spur, where it is usually accessible only to the long tongues of moths and butterflies. Hawkmoths, for instance, do not usually enter flowers, as bees do, but hover above them, inserting their long tongues into the floral tube. Consequently, hawkmoth flowers do not have the landing platforms, traps, and elaborate internal machinery seen in some of the bee flowers. The more generalized moth-flower relationships, which involve smaller moths that do not consume nearly as much energy as the hawkmoths, also involve shorter corolla tubes and often smaller flowers, over which the moths scramble. One of the most specialized moth-flower relationships is shown in Figure 19–21.

Bird-Pollinated Flowers

Some birds regularly visit flowers to feed on nectar, floral parts, and flower-inhabiting insects; many of these birds also serve as pollinators. In North and South America, the chief bird pollinators are hummingbirds (Figure 19–22); in other parts of the world, flowers are visited regularly by representatives of other specialized bird families (Figure 19–23).

Although bird flowers have a copious thin nectar (some even drip with nectar when the pollen is ripe), they usually have little odor, which is a corollary of the fact that the sense of smell is poorly developed in birds. However, birds do have a keen color sense (much like our own), so most bird flowers are colorful, with red and yellow being the most common.

Bird-pollinated flowers include the red columbine, fuchsia, passion flower, eucalyptus, hibiscus, and many members of the cactus, banana, and orchid families. They are large or are parts of large inflores-

19–21
Yucca moth (Tegeticula yucasella) *scraping pollen from a yucca flower. The female moth visits the creamy white flowers by night and gathers pollen, which it rolls into a tight little ball and carries in its specialized mouth parts to another flower. In the second flower, it pierces the ovary wall with its long ovipositor and lays a batch of eggs among the ovules. It then packs the sticky mass of pollen through the openings of the stigma. Moth larvae and seeds develop simultaneously, with the larvae feeding on the developing yucca seeds. When the larvae are fully developed, they gnaw their way through the ovary wall and lower themselves to the ground, where they pupate until the yuccas bloom again. It is estimated that only about 20 percent of the seeds are usually eaten.*

19–22
A male Anna hummingbird (Calypte anna) *at a flower of the scarlet monkeyflower* (Mimulus cardinalis) *in southern California. Note the pollen on the bird's forehead, which is in contact with the stigma of the flower.*

19–23
A collared sunbird (Anthreptes collarii) *perching and feeding on a flower of a bird of paradise* (Strelitzia reginae) *in South Africa.*

cences, features that can be correlated with their importance as visual stimuli and their ability to hold large amounts of nectar.

Bird and other animal pollinators usually restrict their visits to the flowers of a particular plant species, but this is only one factor promoting outcrossing (cross-pollination between individuals of the same species). For outcrossing to result, it is also necessary that the pollinator not confine its visits to a single flower or to the flowers of a single plant. When flowers are visited regularly by large animals with a high rate of energy expenditure, such as birds, hawkmoths, or bats, they must produce large amounts of nectar to support the metabolic requirements of these animals and keep them coming back. On the other hand, if an abundant supply of this nectar is available to animals with a lower rate of energy expenditure, such as small bees or beetles, these will tend to remain at a single flower and, being satisfied there, will not move on to other plants and thus bring about outcrossing. Consequently, in the course of evolution, flowers that are regularly pollinated by animals operating at a high rate of energy consumption, such as hummingbirds, have tended to evolve flowers in which the nectar is held in tubes or is otherwise unavailable to smaller animals of lower energy consumption. Similarly, red color is a signal to birds but not to insects; red is outside the insects' visual spectrum. Birds, like ourselves, do not respond very strongly to odor clues. Thus, odorless, red flowers are inconspicuous to insects and do not tend to attract visits by them, an adaptation that is advantageous in view of their copious production of nectar.

Bat-Pollinated Flowers

Flower-visiting bats are found in tropical areas of both the Old World and the New World. Bats that derive all or most of their nourishment from flowers have slender and elongated muzzles and long extensible tongues, sometimes with a brushlike tip, and their front teeth are often reduced in size or are missing altogether.

Bat flowers are similar in many respects to bird flowers, being large, strong flowers with copious nectar (Figure 19–24). Because bats feed at night, bat flowers are usually dull in color and open only at night. They are often tubular or protect their nectar in other ways. Bats are attracted to the flowers largely through their sense of smell, and bat-pollinated flowers characteristically have very strong fermenting or fruitlike odors. Bats fly from tree to tree, lapping up nectar, eating pollen and other flower parts, and carrying pollen from flower to flower on their fur.

It has recently been discovered that some bats derive a significant portion of their dietary protein from the pollen they consume. As yet another example of coevolution, the pollen of the flowers they visit has been found to contain significantly higher levels of protein than is found in insect-pollinated flowers.

Wind-Pollinated Flowers

At about the turn of the century, it was thought by botanists that wind-pollinated flowers were the most primitive of angiosperm flowers and that those of all other angiosperms had been derived from them. The conifers—thought by some scientists at the time to be direct ancestors of the angiosperms—have small, colorless, odorless, unisexual cones and are pollinated by the wind. Similarly, the many examples of wind-pollinated flowers have dull colors, are relatively odorless, and do not have nectar; the petals are small or absent; and the sexes are often separated. However, studies of other characteristics of these wind-pollinated angiosperms, in particular, their wood (in which the vessel elements are often specialized), have convinced plant anatomists that wind-pollinated angiosperms evolved not from the

19–24
By thrusting its face into the tubular corolla of an organ-pipe cactus (Lemaireocereus) *flower, this bat of the genus* Leptonycteris *is able to lap up nectar with its long, bristly tongue. Pollen grains clinging to the bat's face and neck are transferred to the next flower it visits. Bat-pollinated flowers have dingy colors and a musty scent (similar to that produced by bats to attract one another), and they open at night.*

(a)

(b)

(c)

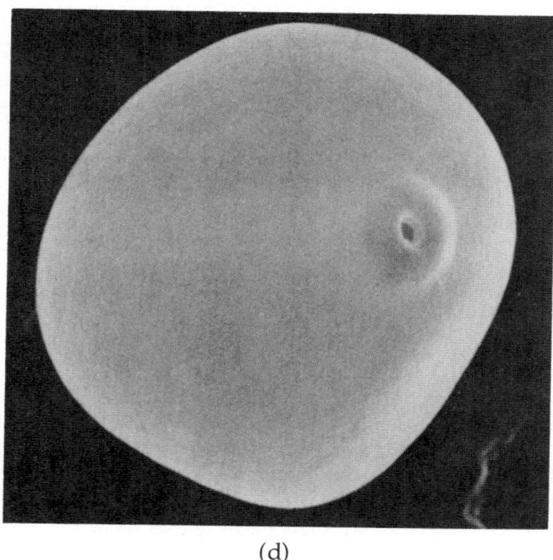

(d)

19–25
Unlike most angiosperms, the grasses are wind-pollinated. Corn (Zea mays) *has* (a) *staminate inflorescences (tassels) at the top of the stem and* (b) *ovulate inflorescences with long protruding stigmas (the "silk" on the ears of corn) lower on the stem.* (c) *Grasses characteristically have enlarged, feathery stigmas to catch the wind-blown pollen, which is shed by the hanging anthers, as seen here in a grass of the genus* Agropyron. (d) *Scanning electron micrograph of a pollen grain of corn* (Zea mays), *showing the smooth pollen wall found in most wind-pollinated plants and the single aperture characteristic of monocots.*

conifers but from insect-pollinated angiosperms. According to present evidence, the wind-pollinated angiosperms originated independently from different ancestral stocks. They are best represented in temperate regions and are relatively rare in the tropics. In temperate climates, many trees of the same species, for example, are often found close together, and the dispersal of pollen by wind can occur readily in early spring, when the trees are leafless. In the tropics, on the other hand, many more kinds of trees are found in a given area, and the distance to the next nearest individual of a given species may be quite great. Under these circumstances, pollination by insects and other animals that have the ability to seek out other individuals of the same plant species, sometimes over relatively great distances (up to more than 20 kilometers), is much more efficient than wind pollination.

Wind-pollinated angiosperms do not depend on insects to transport their pollen from place to place and therefore do not offer nutritious rewards to visitors. Wind pollination is very inefficient, however, and is successful only in situations where a large number of individuals grow close together. Nearly all of the pollen falls to the ground within a hundred meters of the parent plant, and if the individuals are widely scattered, the chance of a pollen grain finding a receptive stigma is very slim. Many wind-pollinated plants are dioecious (separate male and female flowers), like the cottonwoods, or are genetically self-incompatible, like the oaks; for such plants, outcrossing is obligatory. In the tropical forest, where the individual plants within a population are more widely dispersed, wind pollination characterizes very few plants.

19–26
Grass flowers (florets) usually develop in clusters. (a) *As a cluster matures, a single pair of dry, chaffy bracts—the glumes—separate a little, exposing the elongating spikelet, with from one or many florets (depending on the type of grass) attached to a central axis, or rachilla.* (b) *Each floret is surrounded by two distinctive bracts of its own, the palea and the lemma. These are forced apart, exposing the inner parts of the flower* (c), *by the swelling of the lodicules—small, rounded bodies at the base of the carpel—and are spread wide when the grass is in flower. The stamens, usually three in number, have slender filaments and long anthers; and the stigmas are typically long and feathery and so are efficient at intercepting the wind-borne pollen.*

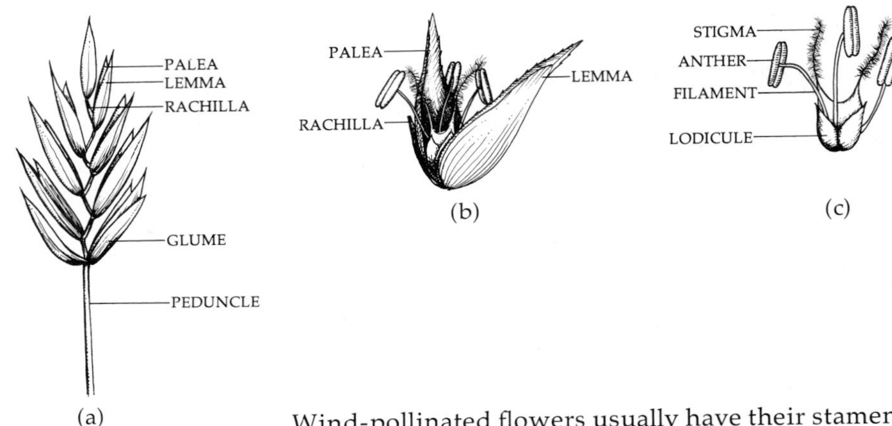

Wind-pollinated flowers usually have their stamens well exposed, where the pollen can be caught by the wind. In some, the anthers are suspended from long filaments hanging free from the flower (Figures 19–25 and 19–26). The abundant pollen grains, which are generally smooth and small, do not adhere to one another like the pollen grains of insect-pollinated species. The large stigmas are characteristically exposed, and they often have branches or feathery outgrowths adapted for intercepting wind-borne pollen grains. Most wind-pollinated plants have ovaries with single ovules (and hence single-seeded fruit), because each pollination consists of the meeting of one pollen grain with one stigma and leads to the fertilization of one ovule for each flower. Thus, the oak flower produces a single acorn and the grass flower produces one grain, but plants with very small flowers tend to have, in compensation, multiple inflorescences (Figure 19–27).

19–27
Most common species of trees in temperate regions are wind-pollinated. The staminate flowers of the paper birch (Betula papyrifera) *hang down in catkins—flexible, thin tassels several centimeters long. These catkins are whipped by passing breezes, and the pollen, when mature, is scattered about by the wind.*

Flower Colors

Surprisingly, the wide array of colors in the flowers of the angiosperms is produced by a very small number of pigments. Many red, orange, or yellow flowers owe their color to the presence of carotenoids similar to those that occur in the leaves. The most important pigments in floral coloration, however, are the *flavonoids*, which are compounds in which two six-carbon rings are linked by a three-carbon unit. Flavonoids probably occur in all angiosperms and are more sporadically distributed among the members of other groups of vascular plants, with a few reports even from other photosynthesizing organisms and animals. In the leaves, they block far ultraviolet radiation, which is highly destructive to nucleic acids and proteins, and usually selectively admit light of blue-green and red wavelengths, which is important in photosynthesis.

One major class of flavonoids that is important in flowers are the anthocyanins (Figure 19–28). Most red and blue plant pigments are anthocyanins, which are water-soluble and are found in vacuoles, unlike the carotenoids, which are fat-soluble and are found in plastids. The color of the anthocyanins depends on the acidity of the cell sap of the vacuole; for example, cyanidin is red in acid solution, violet in neutral solution, and blue in alkaline solution. Another group of flavonoids, very common in leaves and also in many flowers, are the flavonols. These are often colorless but may contribute to the ivory or white hues of certain flowers.

For all flowering plants, different mixtures of flavonoids and carotenoids (as well as changes in cellular pH), and differences in the structural and thus the reflective properties of the flower parts, produce the characteristic pigmentation. The bright fall colors of leaves comes about when large quantities of colorless flavonols are converted into anthocyanins as chlorophyll breaks down. In the all-yellow flower of the marsh marigold *(Caltha palustris)*, the ultraviolet-reflective outer portion is colored by carotenoids, whereas the ultraviolet-nonreflective inner portion is yellow to our eye because of the presence of a yellow chalcone, one of the flavonoids. To a bee or other insect, the outer portion of the flower would appear to be a mixture of yellow and ultraviolet, a color called "bee's purple," whereas the nonreflective inner portion would appear pure yellow (see Figure 19–17). Most, but not all, ultraviolet reflectivity in flowers is related to the presence of carotenoids, and thus ultraviolet patterns are more common in yellow flowers than in any other group.

In the goosefoot, cactus, and portulaca families and their relatives, the reddish pigments are not anthocyanins nor even flavonoids but a group of more complex aromatic compounds known as betacyanins. The red flowers of *Bougainvillea* and the red color of beets are due to the presence of betacyanins. No anthocyanins occur in these plants, and these three families have been shown by their biochemistry to be close relatives.

EVOLUTION OF FRUITS

The fruit is a mature ovary, which may or may not retain some additional floral parts. A fruit in which such additional parts are retained is known as an accessory fruit.

Fruits are generally classified as simple, multiple, or aggregate, depending on the arrangement of the carpels from which the fruit developed. *Simple* fruits develop from one carpel or several united carpels. *Aggregate* fruits, such as magnolia, raspberry, and strawberry, consist of a number of separate carpels of one gynoecium. *Multiple* fruits consist of the gynoecia of more than one flower. The pineapple, for example, is a multiple fruit consisting of an inflorescence with many previously separate ovaries fused on the axis on which pineapple flowers were borne (the other flower parts are squeezed between the expanding ovaries.)

Simple fruits are by far the most diverse of the three groups. When ripe, they may be soft and fleshy, dry and woody, or papery. There are three main types of fleshy fruit—the *berry*, the *drupe*, and the *pome*. In the berry—examples of which are tomatoes, dates, and grapes—there are one to several carpels, each of which is usually many-seeded. The inner layer of the fruit wall is fleshy. In the drupe, there are also one to several carpels, but each usually contains only a single seed. The inner layer of the fruit is stony and usually tightly adherent to the seed. Familiar drupes are the peach, cherry, olive, and plum. The coconut is a drupe whose outer layer is fibrous rather than fleshy, but in temperate regions we usually see only the seed with the adherent stony inner layer of the fruit (Figure 19–29). A highly specialized sort of fleshy fruit is the pome, which is characteristic of one subfamily of the rose family. The pome is derived from a compound inferior ovary in which the fleshy portion comes largely from the enlarged base of the perianth. Apples and pears are pomes.

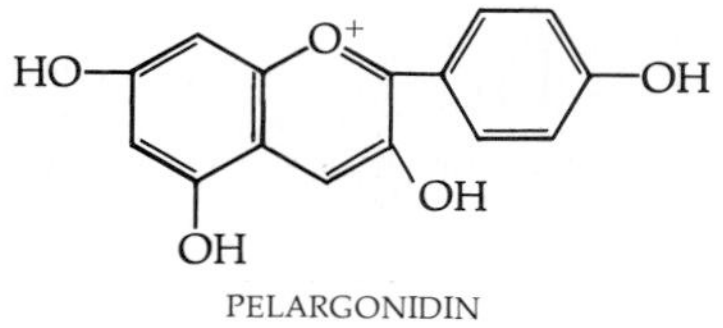

HO OH OH OH O+
PELARGONIDIN

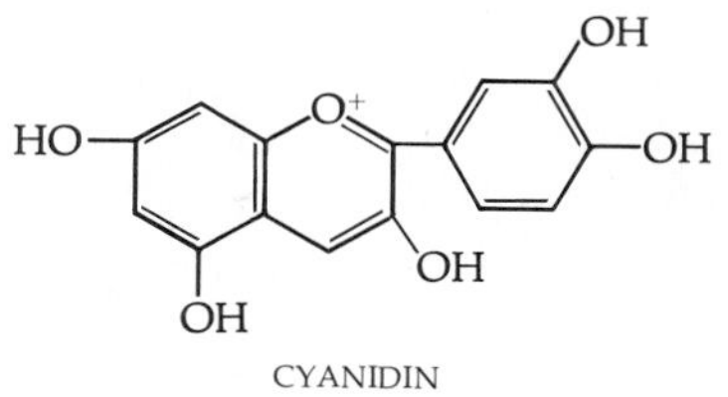

HO OH OH OH OH O+
CYANIDIN

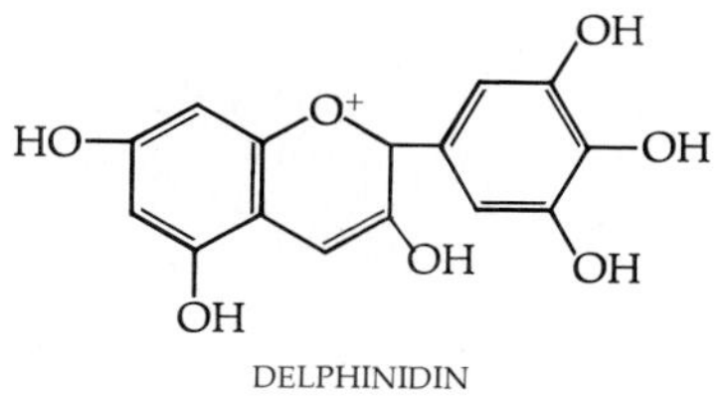

HO OH OH OH OH OH O+
DELPHINIDIN

19–28
Three anthocyanin pigments, the basic pigments on which flower colors in many angiosperms depend: pelargonidin (red), cyanidin (violet), and delphinidin (blue). Related compounds known as flavonols are yellow or ivory, and the carotenoids are red, orange, or yellow. Betalins are red pigments that occur in one group of dicots. Mixtures of these different pigments, together with changes in cellular pH, *produce the entire range of flower color in the angiosperms. Changes in flower color provide "signals" to pollinators, telling them which flowers have opened recently and are more apt to provide food.*

19–29
Fruit of the coconut (Cocos nucifera), *a drupe. Only the hard inner shell and its contents are present at maturity, the husk having rotted away. The coconut milk is nuclear endosperm. It acquires cell walls by the time of germination. The entire fruit floats easily in the sea, and coconuts have been widely dispersed by this means.*

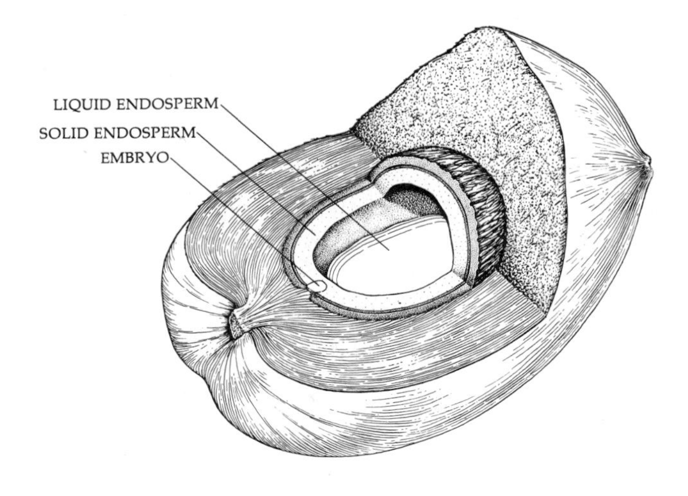

Dry simple fruits are classified as either *dehiscent* or *indehiscent* (Figures 19–30, 19–31, and 19–32). In the dehiscent fruits, the tissues of the mature ovary wall (the pericarp) break open, freeing the seeds. In the indehiscent fruits, the seeds remain in the fruit after the fruit is shed from the parent plant.

There are several kinds of dehiscent simple dry fruits. The *follicle* is derived from a single carpel that splits down one side at maturity, as in columbines and milkweeds (Figure 19–30a). In the pea family, the characteristic fruit is a *legume*. Legumes resemble follicles but are split along both sides. In the mustard family (Brassicaceae), the fruit is called a *silique* and is formed of two fused carpels. At maturity, two halves split off, leaving the seeds attached to a persistent central portion. The most common sort of dehiscent simple dry fruit is the *capsule*, which is formed from a compound ovary in plants with either a superior or an inferior ovary. Capsules shed their seeds in a variety of ways. In the poppy family, Papaveraceae, the seeds are often shed when the capsule splits longitudinally, but in some members of this family they are shed through holes near the top of the capsule (Figure 19–32).

Indehiscent simple dry fruits are found in a great variety of plant families. Most common is the *achene*, a small single-seeded fruit in which the seed lies free in the cavity except for its attachment by the funiculus. Achenes are characteristic of the buttercup and buckwheat families. Winged achenes, such as those found in the elm and ash, are commonly known as *samaras*. The achenelike fruit in the grass family (Poaceae), the *caryopsis*, is derived from a compound ovary, and the seed coat is firmly united to the fruit wall. In the Asteraceae, the achenelike fruit is derived from a compound, inferior ovary and is called a

19–30
Dehiscent fruits. (a) *Bursting follicles of milkweed* (Asclepias syriaca) *and* (b) *capsules of cottonwood* (Populus). *In both, the seeds have tufts of light silky hair which aid in their dispersal by the wind.*

(a)

(b)

19–31
Seed discharge in the dehiscent fruit of the dwarf mistletoe (Arceuthobium), *a parasitic angiosperm that is the most serious cause of loss to forest productivity in the western United States. Very high hydrostatic pressure that builds up in the fruit shoots the seeds up to 15 meters laterally. They have an initial velocity of about 100 kilometers per hour, and presumably this is one of the ways they are spread from tree to tree, although they are also sticky and are carried by birds.*

cypsela. Acorns and hazelnuts are examples of *nuts*, which resemble achenes but have a stony fruit wall and are derived from a compound ovary. Finally, in the parsley family (Apiaceae) and the maples, as well as a number of other unrelated groups, the fruit is a *schizocarp*, which is derived from a compound ovary but splits at maturity into two or more one-seeded portions.

Dispersal of Fruits

Just as flowers may be classified according to their pollinators, fruits may be grouped according to their dispersal agents.

Wind-Borne Fruits

Some plants have extremely light fruits or seeds and can be dispersed by wind because of this lightness. The dustlike seeds of all members of the orchid family, for example, are wind-borne. Other fruits have wings, sometimes formed from perianth parts. In the maple, for example, which has a gynoecium composed of two fused carpels, each carpel develops a long wing (Figure 19–33). The two carpels separate and fall when mature. Many of the Asteraceae—the dandelion, for example—develop a plumelike pappus, which aids in keeping the light fruits aloft (Figure 19–33). In some plants, it is the seed itself, rather

19–32
The samara, a winged fruit characteristic of the ash (Fraxinus) *and the elm* (Ulmus), *is indehiscent, retaining its single seed at maturity. Capsules and legumes are common types of dehiscent fruits—fruits that shed their seeds at maturity. In the poppy* (Papaver) *capsule, the ripe seeds are shaken out through pores in the dry fruit, like salt from a saltshaker.*

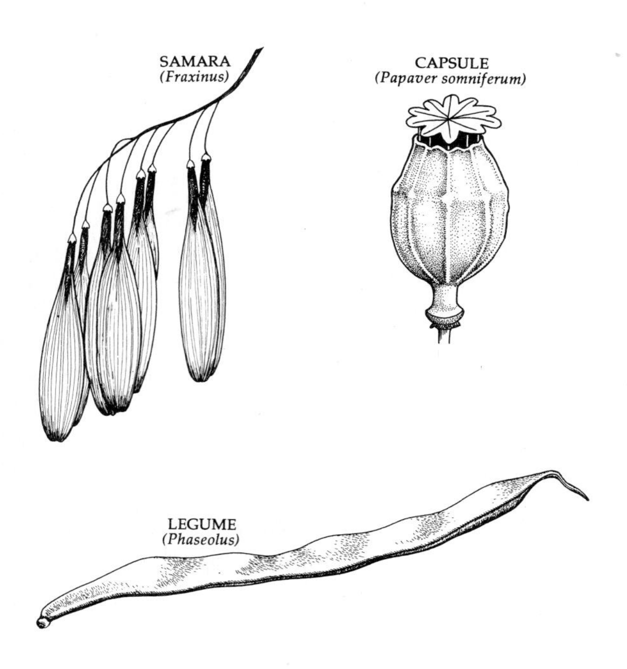

19–33
Wind-dispersed fruits. In maples (Acer), *each half of the schizocarp (a two-part indehiscent fruit that splits at maturity) has a long wing. The fruits of the dandelion* (Taraxacum) *and other composites have a modified calyx, called the pappus, which is adherent to the mature achene and may form a plumelike structure that aids in wind dispersal.*

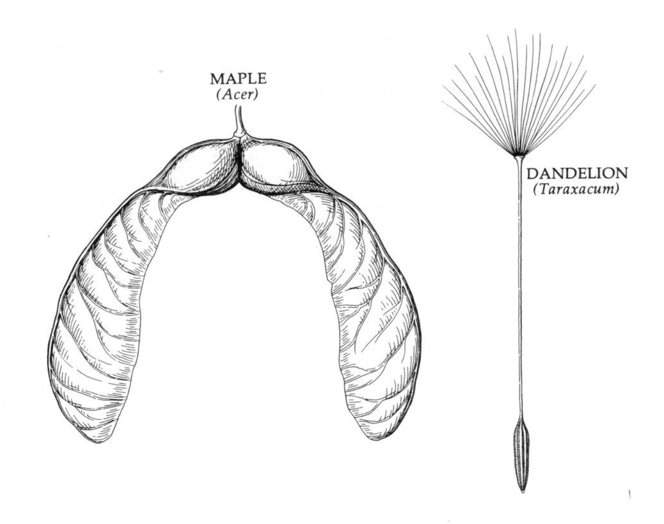

19–34
In the tumbleweed (Salsola kali), *the plant breaks off and is blown across the open country, scattering its seeds as it tumbles along.*

than the fruit, that carries the wing or plume; the familiar butter-and-eggs *(Linaria)* has a winged seed, and fireweed *(Epilobium)* and milkweed *(Asclepias)* have plumed seeds. In the willows and the poplars (Salicaceae), the seed coat is covered with woolly hairs. In the tumbleweeds *(Salsola kali)*, the whole plant or the fruiting portion of it is blown by the wind and scatters seeds as it moves (Figure 19–34).

Other plants shoot their seeds aloft. In touch-me-nots *(Impatiens)*, the valves of the capsules separate suddenly, throwing seeds for some distance. In the witch hazel *(Hamamelis)*, the endocarp contracts as the fruit dries, discharging the seeds sometimes as far as 15 meters.

Water-Borne Fruits

The fruits and seeds of many plants, especially those growing in or near water sources, are adapted for floating, either because air is trapped in some part of the fruit or because the fruit contains buoyant tissue. Some fruits are especially adapted for dispersal by ocean currents; notable among these is the coconut, which is why almost every newly formed Pacific atoll quickly acquires its own coconut tree. Rain is also a common means of fruit and seed dispersal and is particularly important for plants that live on hillsides or mountain slopes.

19–35
The seeds of fleshy fruits are usually dispersed by vertebrates that eat the fruits and drop the seeds as part of their feces. The tree sparrow (Spizella arborea) *shown here is eating the bright red fruits of winterberry* (Ilex montana), *a member of the holly family.*

Animal-Borne Fruits

The evolution of sweet and often highly colored fleshy fruits was clearly involved in the coevolution of animals and flowering plants. The majority of fruits in which much of the pericarp is fleshy—cherries, raspberries, grapes—are eaten by vertebrates. When such fruits are eaten by birds or mammals, the seeds that lie within them are spread by being passed unharmed through the digestive tract (Figure 19–35).

When fleshy fruits ripen, they undergo a series of characteristic changes, mediated by the hormone ethylene, which will be discussed in Chapter 25. Among these changes are a rise in sugar content, a softening of the fruit through the breakdown of pectic substances, and often a change in color from inconspicuous, leaflike green to bright red, yellow, blue, or black. The seeds of some plants, especially tropical ones, often have fleshy appendages, or arils, that have the bright colors characteristic of fleshy fruits and, like them, are aided in dispersal by vertebrates. In this way, fruits in which the fruit wall is dry, such as a capsule, can achieve the dispersal advantages exhibited by fleshy ones. Common examples are bittersweet *(Celastrus scandens)* and the strawberry bush *(Euonymus americanus)*.

Unripe fruits are often green or colored in such a way that they are hidden among the green leaves and thus are somewhat concealed from birds, mammals, and insects. They may actually be disagreeable to the taste—such as the acidic taste of unripe fruits of cherries *(Prunus)*—thereby discouraging animals from eating them before the seeds within are ripe. The changes in color that accompany ripening are the plant's "signal" that the fruit is ready to be eaten (the seeds are ripe and ready for dispersal).

It is no coincidence that red is such a prominent color among ripe fruits. The fruit is thereby still concealed from insects, which are too small to disperse the large seeds of fleshy fruits effectively. As discussed previously, pure red is invisible to insects and other invertebrates but is very conspicuous to birds and mammals (including ourselves).

A number of other angiosperms have fruits or seeds that are dispersed by adhering to fur or feathers (Figure 19–36). These have hooks, barbs, spines, hairs, or sticky coverings and so are transported, often for great distances, by animals.

BURDOCK
(Arctium minus)

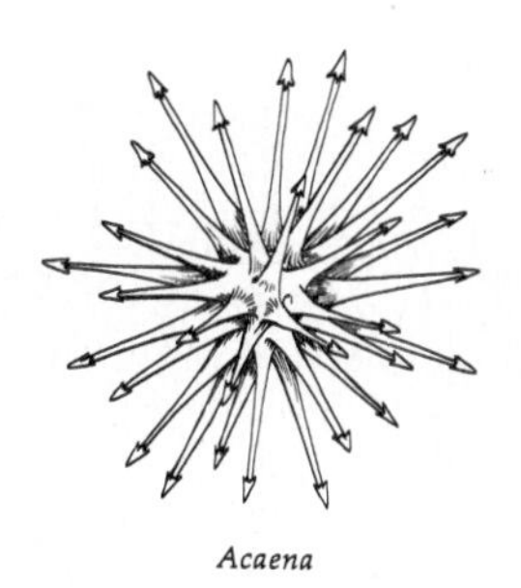

19–36
Two fruits with bristly spines that are dispersed by adhering to fur or feathers. In burdock (Arctium minus), *the entire head is dispersed. The spines of* Acaena *are on the outside of a floral cup that grows up around the carpels at maturity.*

BIOCHEMICAL COEVOLUTION

Another important factor in the evolution of angiosperms has to do with the so-called secondary plant substances. Once thought of as waste products, these include an array of chemically unrelated compounds such as alkaloids, quinones, essential oils (including terpenoids), glycosides (including cyanogenic substances and saponins), flavonoids, and even raphides

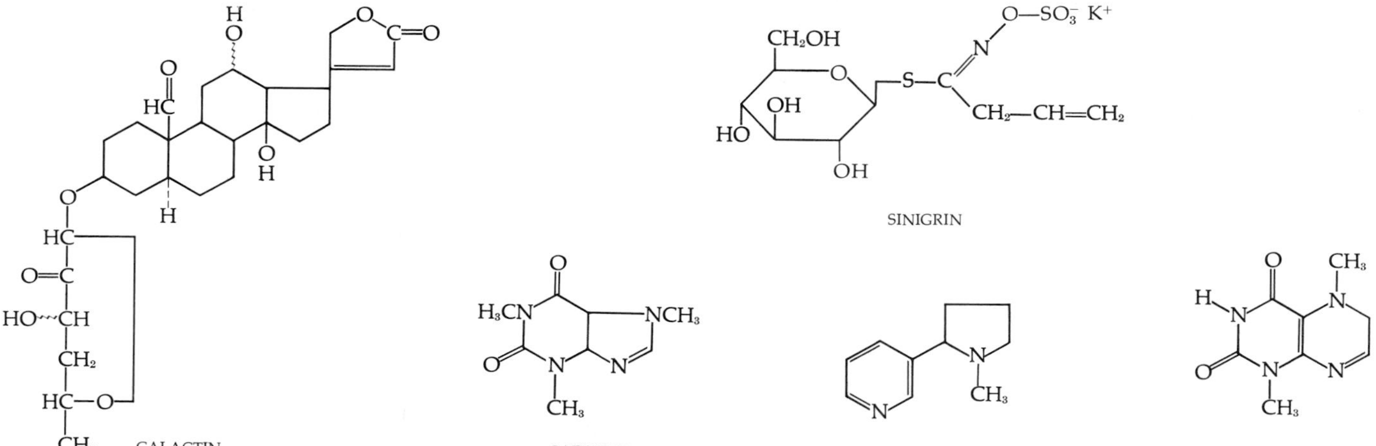

CALACTIN

SINIGRIN

CAFFEINE

NICOTINE

THEOBROMINE

19–37
Secondary plant substances: sinigrin, from black mustard, Brassica nigra; *calactin, a cardiac glycoside, from the milkweed* Asclepias curassavica; *nicotine, from tobacco,* Nicotiana tabacum, *a member of the nightshade family; caffeine, from* Coffea arabica *of the madder family; and theobromine, a prominent alkaloid in coffee, tea* (Thea sinensis), *and cocoa* (Theobroma cacao). *Nicotine, caffeine, and theobromine are alkaloids, members of a diverse class of nitrogen-containing ring compounds that are physiologically active in vertebrates. Various other, often related, compounds are found in other members of the same plant families. Secondary plant substances are often useful in plant classification.*

19–38
Poison ivy (Toxicodendron radicans) *produces a secondary plant substance, 3-pentadecanedienyl catechol, which causes an irritating and persistent rash on the skin of humans. The ability to produce this alcohol presumably evolved under the selective pressure exerted by herbivores. Fortunately, the plant is easily identifiable by its characteristic and unusual compound leaves with their three leaflets.*

(needlelike crystals of calcium oxalate). These compounds generally characterize whole families or groups of families of flowering plants, although they sometimes occur in families that are more or less unrelated in other respects (Figure 19–37).

These chemicals appear to play a major role in nature in restricting the palatability of the plants in which they occur (Figure 19–38). When a given family of plants is characterized by a distinctive group of secondary plant substances, it is apt to be eaten only by insects belonging to certain families. The mustard family (Brassicaceae), for example, is characterized by the presence of mustard oil glycosides and associated enzymes that break down these glycosides to release the pungent odors associated with cabbage, horseradish, and mustard. Certain groups of true bugs, beetles, and the larvae of some groups of moths feed only on the leaves of plants of this family. The larvae of most of the members of the butterfly subfamily Pierinae (which includes the cabbage butterflies and orange-tips) also feed only on these plants. On the other hand, other families of insects ignore plants of the mustard family and will not feed on them even if starved. The same chemicals that act as deterrents to most groups of insect herbivores often act as feeding stimuli for these narrowly restricted feeders. For example, certain moth larvae that feed on cabbage will extrude their mouth parts and go through their characteristic feeding behavior when presented with agar or filter paper containing juices pressed from these plants.

Thus, it is clear that the ability to manufacture these chemicals and retain them in their tissues is an important evolutionary step for the plants concerned and gives them biochemical protection from the predations of most herbivores. The evolution of the Brassicaceae doubtless took place in part behind just such a biochemical shield. From the standpoint of the insects, such protected plants, because they are not already heavily utilized by other plant feeders, represent an unexploited food source for any group of insects that can tolerate or break down the poisons manufactured by the plant. The main evolution of the butterfly group Pierinae probably occurred after their ancestors had acquired the ability to feed on plants of the mustard family by breaking down these toxic molecules.

The insects that are narrowly restricted in their plant feeding habits to groups of plants with secondary plant substances are often brightly colored, which serves as a signal to their predators that they carry the noxious chemicals in their bodies and hence are protected. For example, the assemblage of insects found feeding on a milkweed plant on a summer day include bright green chrysomelid beetles, bright red cerambycid beetles and true bugs, and orange and black monarch butterflies, among others. Milkweeds (family Asclepiadaceae) are richly endowed with alkaloids and cardiac glycosides, heart poisons that have potent effects in vertebrates, the main potential predators of these insects. If a bird ingests a monarch butterfly, severe vomiting and gastric distress will follow, and the orange-and-black pattern will be avoided by the predator in the future. Other insects, such as the viceroy butterfly (which also has an orange-and-black pattern), have evolved similar coloration and markings. They thus escape predation by capitalizing on their resemblance to the poisonous monarch. This phenomenon, known as *mimicry*, is ultimately dependent upon the plant's chemical defenses. Various drugs and psychedelic chemicals, such as the active ingredients in marijuana *(Cannabis sativa)* and the opium poppy, among others, are also secondary plant compounds that in nature play a role in discouraging the attacks of herbivores (Figure 19–39).

Still more sophisticated systems are known. When the leaves of potato or tomato plants are wounded, as by the Colorado potato beetle, the concentration of proteinase inhibitors, which interfere with the digestive enzymes in the gut of the beetles, rapidly increase in the tissues of the plant that have been ex-

19–39
Plants that produce hallucinogenic and medicinal compounds. (a) *Mescaline, from the peyote cactus* (Lophophora williamsii), *is used ceremonially by many Indian groups of northern Mexico and the southwestern United States.* (b) *Tetrahydrocannabinol (THC) is the most important active molecule in marijuana* (Cannabis sativa). (c) *Quinine, a valuable drug used in the treatment and prevention of malaria, is derived from tropical trees and shrubs of the genus* Cinchona. *All of these substances presumably serve to protect the plants in which they occur from the depredations of insects but are physiologically active in vertebrates, including humans.*

posed to air. Other plants manufacture molecules that resemble the hormones of insects or other predators and thus interfere with the predators' normal growth and development. One of the most useful of these is a complex molecule called diosgenin, obtained mainly from wild yams in Mexico, but with smaller amounts obtained in plants from India and China. Diosgenin is only two simple chemical steps away from the chemical 16-dehydropregnenolone (16D), the main active ingredient in oral contraceptives, and wild yams have been a major source of 16D. Unfortunately, they grow very slowly, and their source from the wild shows signs of exhaustion. The Soviets and Ecuadorians, among others, have recently been experimenting with species of *Solanum*, the potato and nightshade genus, that contain solasodine, a molecule that can be converted to 16D in two steps. Some of these plants are being developed into commercial crops.

As mentioned previously, pollination systems have developed a particular coevolutionary pattern in which all the possible pollination types have evolved not once but usually several times in each group. The resulting array of forms gives the angiosperms an extremely wide variety. In the case of biochemical relationships, however, the evolutionary steps appear to have been large and definitive, and whole families of plants can be characterized biochemically and associated with major groups of plant-eating insects. These biochemical relationships may have played a key role in the early success of the angiosperms.

SUMMARY

The angiosperms, presently the dominant group of vascular plants, first appear in the fossil record about 127 million years ago, in the early part of the Cretaceous period. The group became dominant worldwide some 80 to 90 million years ago, in the late Cretaceous period. Then the fossils of many modern families and some genera can be recognized. The pollen of primitive angiosperms may be indistinguishable from gymnosperm pollen or fern spores, so that it is difficult to be certain of the presence of angiosperms more than 127 million years ago, but the group is doubtless somewhat older.

Angiosperms may have evolved in the semiarid uplands and dry interior basins of West Gondwanaland, a supercontinent that included what are now South America and Africa. By the time these continents finally separated about 90 million years ago, their climates had changed considerably, and the angiosperms were nearing world dominance. Possible reasons for their success include various adaptations for drought resistance, as well as the evolution of efficient and often highly specialized pollination systems.

The flower is the most conspicuous feature of the angiosperms and has played a significant role in their evolution. The carpel is a leaflike structure that has undergone enfolding to enclose the ovules (which contain the megasporangia) and subsequent differentiation into a basal, swollen ovary, a stalklike style, and a stigma that is receptive to pollen. Similarly, the stamen has evolved from leaflike precursors or slender, branching systems with terminal sporangia to become specialized into the slender structure characteristic of most living angiosperms. Sepals are specialized leaves that protect the flower in bud, and petals in most angiosperms are sterilized stamens that have assumed a function in attracting insects. Some petals, however, are derived from sepals. The spiral arrangement of the flower parts of primitive angiosperms, coupled with their possession of numerous free parts, has given way to a whorled arrangement in most modern forms, with a definite number of parts which are often fused with one another or with the other whorls of the flower.

Examples of specialized families are Asteraceae (the composites), in which numerous highly specialized flowers are aggregated into a head, which functions as the attractive unit for insects; and Orchidaceae (the orchids), in which bizarre elaboration of the flower parts has resulted in a highly irregular flower with the most specialized pollination systems known.

Pollination by insects is basic in the angiosperms, and the first pollinating agents were probably beetles or similar insects. The closing of the carpel may have been a device to protect the ovules from being eaten in the course of such activities. More specialized groups of insects evolved later in the history of the angiosperms, and wasps, flies, butterflies, and moths have each left their mark on the morphology of certain angiosperm flowers. The bees, however, are the most specialized and constant of flower-visiting insects and have probably had the greatest effect on the evolution of angiosperm flowers. Each group of flower-visiting animals is associated with a particular group of floral characteristics related to the visual and olfactory senses of the animals they attract. Some angiosperms have become wind-pollinated, shedding copious quantities of small, nonsticky pollen and having well-developed, often feathery stigmas that are efficient in collecting such pollen from the air.

Flowers that are regularly visited and pollinated by animals with high energy requirements, such as hummingbirds, must produce large amounts of nectar. They must then protect and conceal these sources of nectar from other potential visitors with lower energy requirements, which might satiate themselves at a single flower or the flowers of a single plant and fail to move on to another plant of the same species to effect cross-pollination. Wind pollination is inefficient, and the individual plants must grow close together in large groups for the system to work, whereas insects, birds, or bats can carry pollen great distances from flower to flower.

Fruits are as diverse as the flowers from which they are derived and can be classified either morphologically, anatomically, or in terms of their methods of dispersal. They are basically mature ovaries, but, if additional floral parts are retained, the fruits are said to be accessory. Simple fruits are derived from one carpel or a group of united carpels, aggregate fruits from the free carpels of one flower, and multiple fruits from the fused carpels of several or many flowers. Dehiscent fruits split open to release the seeds, and indehiscent ones do not.

Wind-borne fruits or seeds are light and often have wings or tufts of trichomes that aid in their dispersal. Some plants shoot their seeds off explosively. Some seeds or fruits are borne away by water, in which case they must be buoyant and have resistant coats. Others are disseminated by animals, particularly vertebrates, and have evolved fleshy coverings that are tasty and often conspicuous to attract feeding animals. Others adhere to the coats of mammals or the feathers of birds and are distributed in this manner.

A third aspect of the evolutionary success and diversification of the angiosperms has been biochemical coevolution. Certain groups of angiosperms have evolved various secondary plant substances, such as alkaloids, which protect them from most foraging herbivores. Certain herbivores, however, normally those with narrow feeding habits, are found associated with these plants, from which their competitors are excluded. This pattern is a sure indication of a stepwise pattern of coevolutionary interaction, and it appears likely that the early angiosperms also may have been protected by their ability to produce some chemicals that functioned as poisons for herbivores.

SUGGESTIONS FOR FURTHER READING

ALSTON, R. E., and B. L. TURNER: *Biochemical Systematics,* Prentice-Hall, Inc., Englewood Cliffs, N.J., 1963.

Although increasingly out of date, this book still provides a useful outline of the classes of secondary plant substances.

FAEGRI, K., and L. VAN DER PIJL: *The Principles of Pollination Ecology,* 3rd ed., Pergamon Press, Inc., Elmsford, N.Y., 1979.*

A rigorous and scholarly examination of the worldwide variety of pollination systems, and their role in plant and animal ecology.

HEISER, CHARLES B., JR.: *Seed to Civilization: The Story of Food,* 2nd ed., W. H. Freeman and Company, San Francisco, 1981.

A broadly based and useful review of the evolution of domesticated plants and animals.

HEYWOOD, VERNON H. (Ed.): *Flowering Plants of the World,* Mayflower Books, Inc., New York, 1978.

The best available guide to the families of flowering plants for the student, a well-illustrated and accurate book that is a pleasure to use.

LEWIS, WALTER H., and P. F. ELVIN-LEWIS: *Medical Botany: Plants Affecting Man's Health,* John Wiley & Sons, Inc., New York, 1977.

A well-written review of plants that injure, heal, and nourish, or alter the conscious mind.

PIJL, L. VAN DER: *Principles of Dispersal in Higher Plants,* 2nd ed., Springer-Verlag, New York, 1972.

A brief and rather technical but informative book on all matters of seed and fruit dispersal.

PROCTOR, MICHAEL, and PETER YEO: *The Pollination of Flowers,* Taplinger Publishing Co., Inc., New York, 1973.

An outstanding and beautifully illustrated introduction to all aspects of pollination biology.

RADFORD, ALBERT E., *et al.*: *Vascular Plant Systematics,* Harper & Row, Publishers, Inc., New York, 1974.*

An invaluable source book, including glossaries, techniques, bibliographies, indices, and useful discussions of all the various aspects of plant systematics—the scientific study of the kinds and diversity of plants, and of the relationships among them.

STEBBINS, G. LEDYARD: *Flowering Plants: Evolution Above the Species Level,* Harvard University Press, Cambridge, Mass., 1974.

A scholarly and fascinating treatise on the origin and subsequent diversification of the angiosperms by one of the masters of evolutionary theory.

*Available in paperback.

SECTION 5 The Angiosperm Plant Body: Structure and Development

CHAPTER 20

Early Development of the Plant Body

In the previous section, we traced the long evolutionary development of the angiosperms from their presumed ancestor—a relatively complex, multicellular green alga—through a series of early vascular plants whose forked axes were the forerunners of the leaves and roots of the higher vascular plants.

In this section, we shall be concerned primarily with the end result of this evolutionary history—the flowering plant. This chapter thus begins where the story of the angiosperm life cycle ended in Section 4—with the seed, consisting of a seed coat, stored food, and an embryo. We shall follow the formation of the embryo because it is through this process, known as embryogeny, that the vegetative parts of the plant—root, stem, and leaf—have their origin and the organization of tissues is initiated.

THE MATURE EMBRYO AND SEED

The mature embryo of flowering plants consists of a stemlike axis bearing either one or two cotyledons (Figures 20–2 and 20–3). The cotyledons—sometimes referred to as the seed leaves—are the first leaves of the young sporophyte. As the names *dicotyledon* and *monocotyledon* imply, the embryos of dicotyledons have two cotyledons and those of monocotyledons have only one.

At opposite ends of the embryo axis are found the apical meristems of the shoot and root. As discussed previously, the apical meristems are found at the tips of all shoots and roots, and meristems are composed of meristematic cells—cells that are physiologically young and capable of repeated division. In the embryo, the apical meristem of the shoot terminates the *epicotyl*—the stemlike axis above *(epi-)* the cotyledons. In some embryos, the epicotyl consists of little more than the apical meristem (Figures 20–2b and 20–3b), whereas in others it bears one or more young leaves (Figures 20–2a and 20–3a). The epicotyl, together with its young leaves, is called a *plumule.*

20–1
Acorns germinating on the forest floor. The first structure to emerge from the germinating seed is the root, which will anchor the young plant to the soil and absorb water essential for growth of the developing seedling.

20–2
Seeds and stages in the germination of some common dicotyledons. (a) *Garden bean* (Phaseolus vulgaris)*; seed shown open and from external edge view.* (b) *Castor bean* (Ricinus communis)*; seed open, showing both flat and edge views of embryo.* (c) *Pea* (Pisum sativum)*; external view of seed only.*

Seed germination in both the garden bean and the castor bean is epigeous; that is, during germination the cotyledons are carried above ground by the elongating hypocotyl. Note that in both of these seedlings, the elongating hypocotyl forms a hook, which then straightens out, pulling the cotyledons and plumule above ground. By contrast, seed germination in the pea is hypogeous—the cotyledons remain underground. In the pea seedling, it is the epicotyl that elongates and forms a hook, which then straightens out and pulls the plumule above ground.

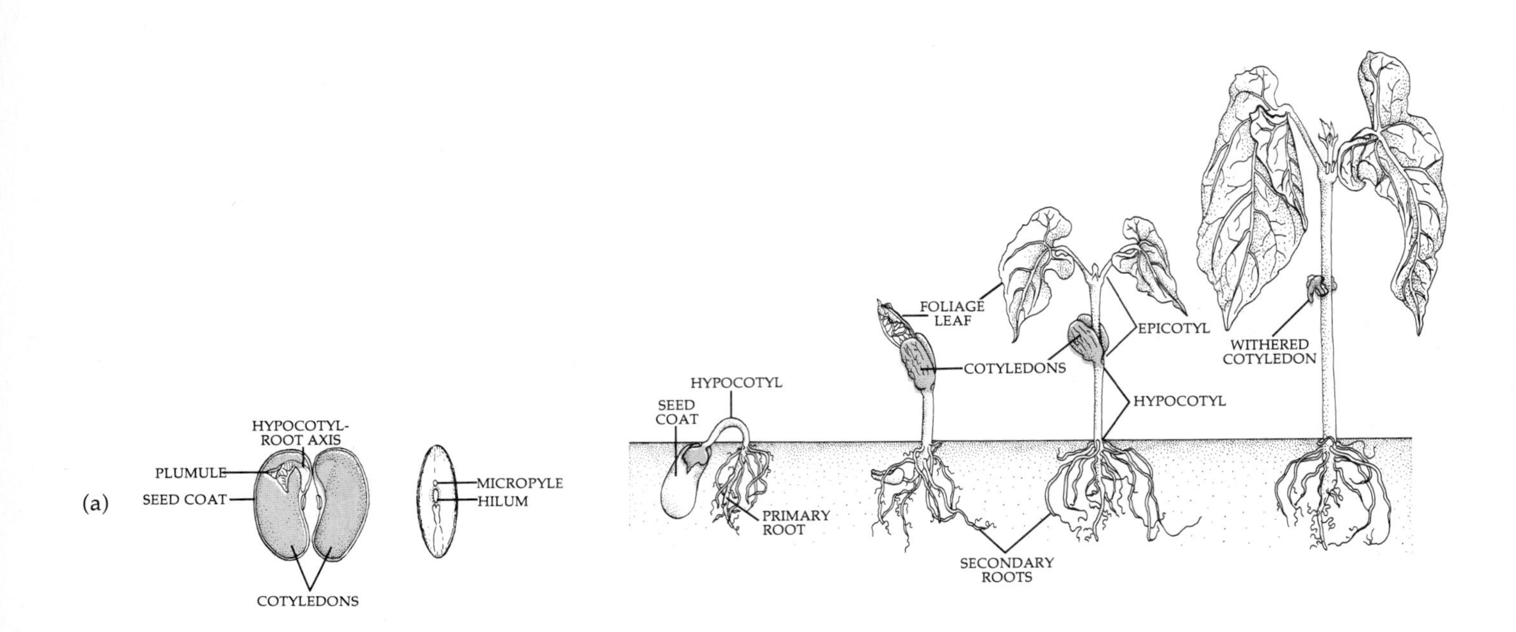

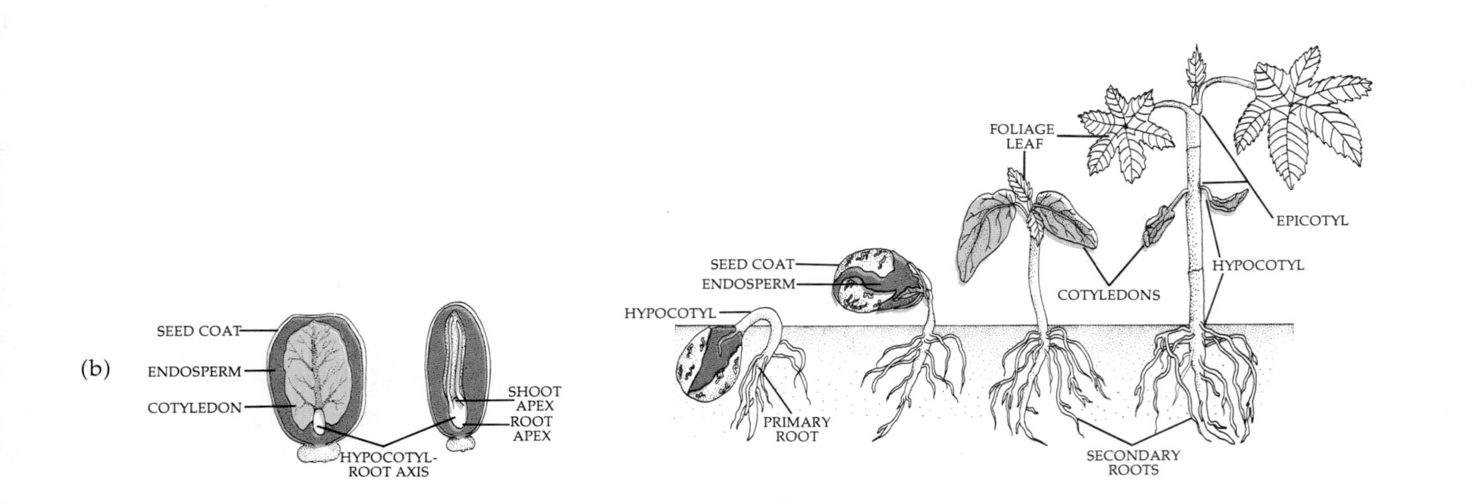

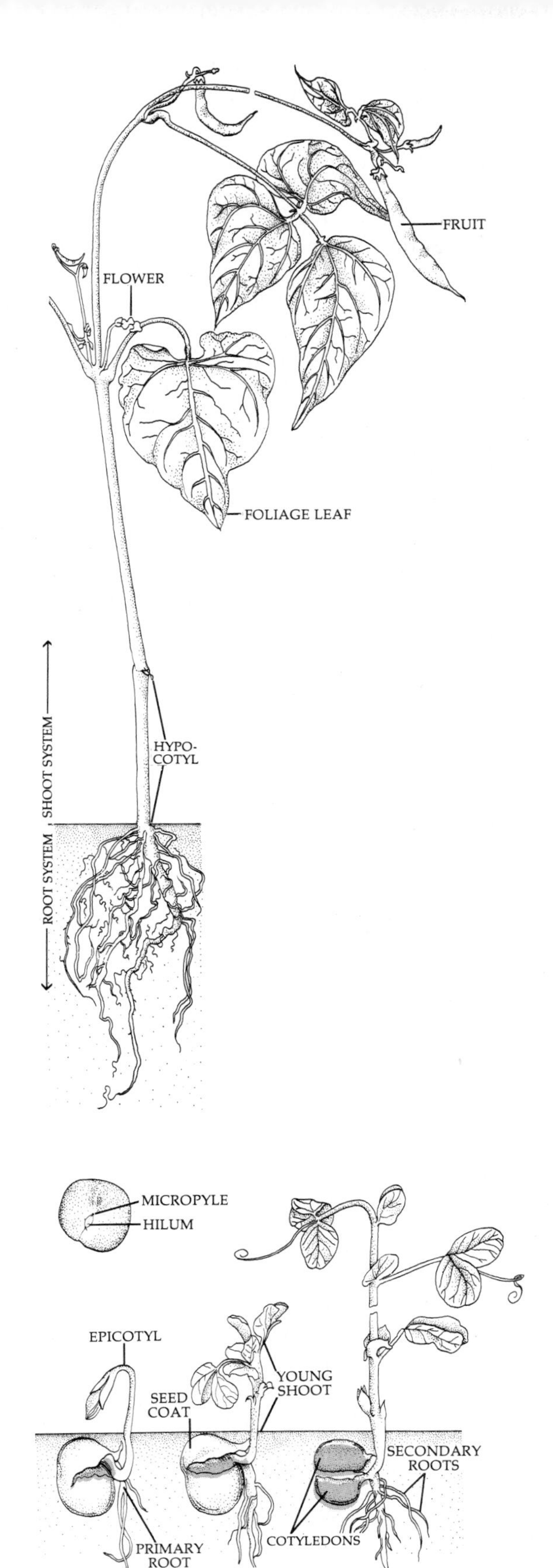

20–3

Seeds and stages in the germination of some common monocotyledons: (a) *corn* (Zea mays) *and* (b) *onion* (Allium cepa). *Both seeds shown in longitudinal section.*

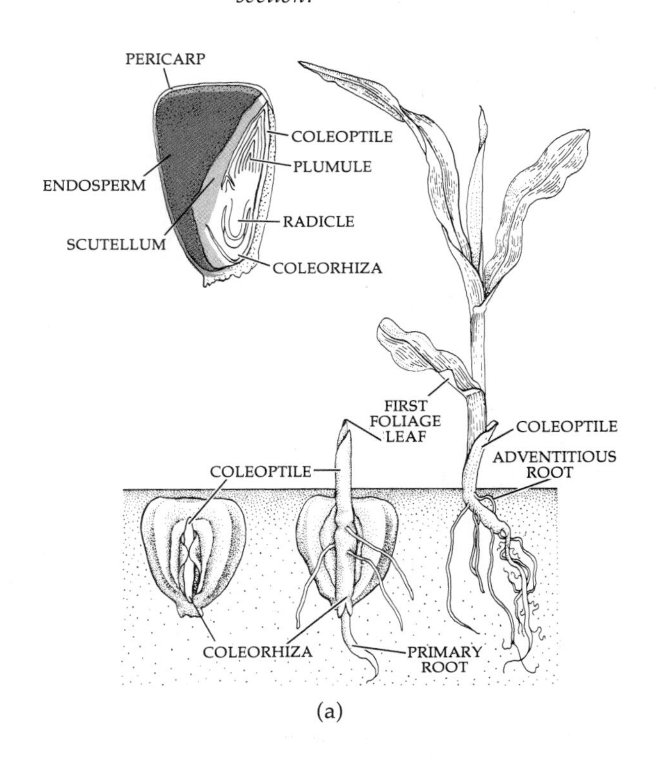

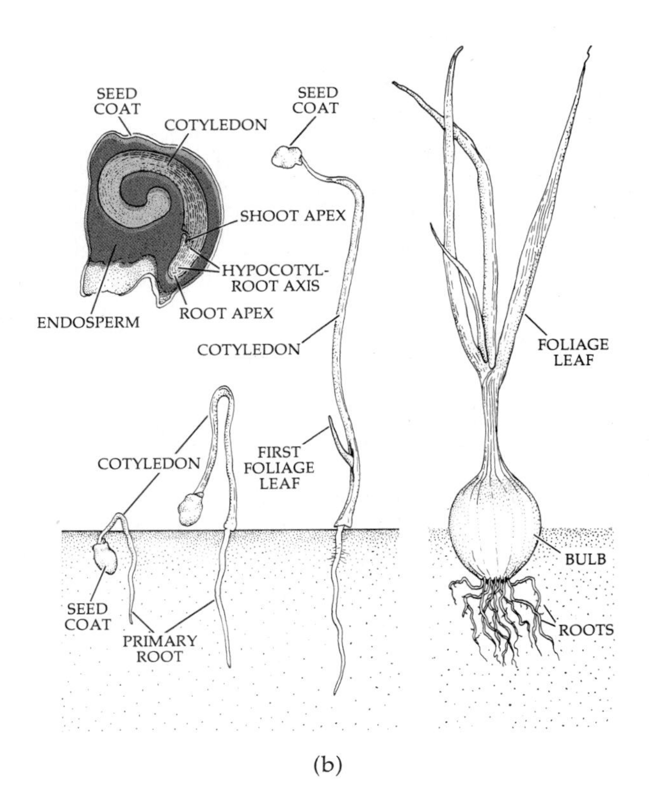

20–4
(a) *Longitudinal section of mature grain, or kernel, of wheat* (Triticum aestivum). *The covering layers of the wheat kernel consist largely of the pericarp. The seed coat, which becomes fused with the pericarp, disintegrates during development of the kernel.* (b) *Detail of the mature wheat embryo.*

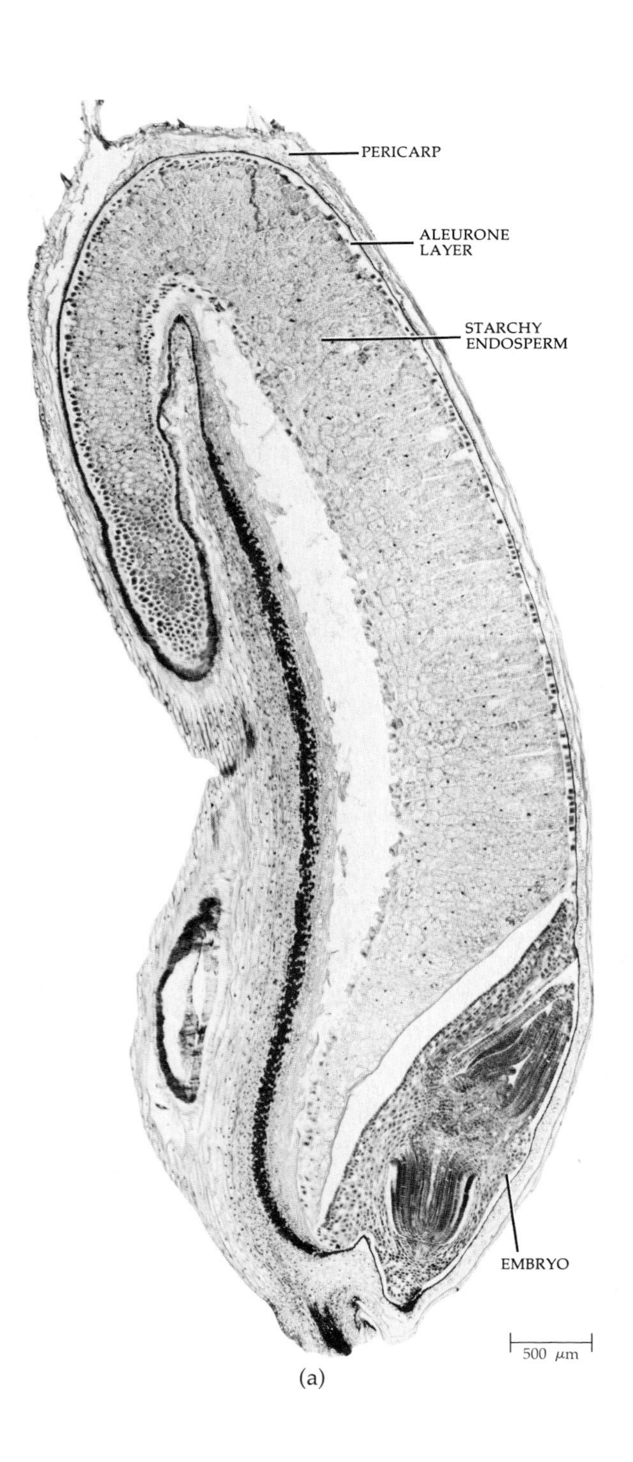

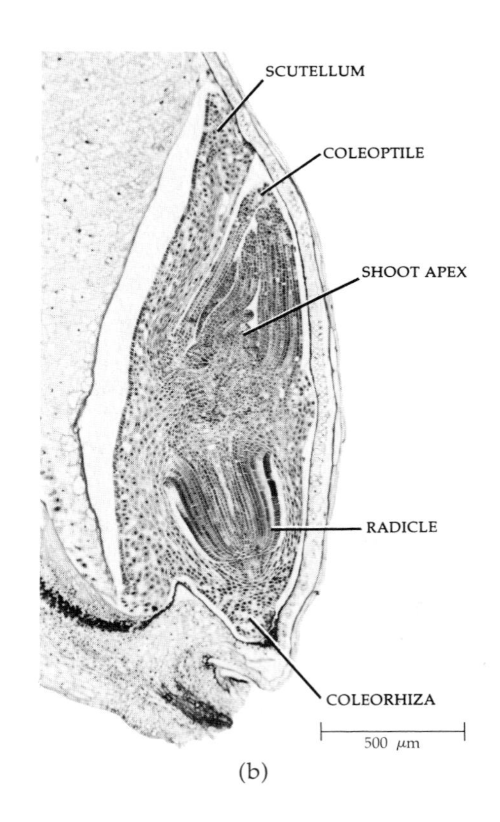

The stemlike axis below *(hypo-)* the cotyledons is referred to as the *hypocotyl*. At the lower end of the hypocotyl there may be an embryonic root, or *radicle*, with distinct root characteristics (Figure 20–4). In many plants, however, the lower end of the axis consists of little more than an apical meristem covered by a root cap. If a radicle cannot be distinguished in the embryo, the embryo axis below the cotyledons may be called the hypocotyl-root axis.

In the discussion of the development of the angiosperm seed in Chapter 18, it was noted that in many dicotyledons most or all of the endosperm and the perisperm (if present) is absorbed by the developing embryo and that the embryos of such seeds develop fleshy, food-storing cotyledons. The cotyledons of most dicotyledonous embryos are fleshy and occupy the largest volume of the seed. Familiar examples of seeds lacking endosperm are sunflower, walnut, pea, and bean (Figures 20–2a and 20–2c). In dicots with large amounts of endosperm, the cotyledons are thin and membranous (Figure 20–2b) and serve to absorb stored food from the endosperm.

In the monocotyledons, the single cotyledon, although somewhat fleshy, usually performs an absorbing rather than a food-storing function (Figure 20–3). Embedded in endosperm, the cotyledon absorbs food digested by enzymatic activity. The digested food is then moved by way of the cotyledon to the growing regions of the embryo. Among the most highly differentiated of monocot embryos are those of grasses (Figures 20–3a and 20–4). When fully formed, the grass embryo possesses a massive cotyledon, the *scutellum*, which is pressed close to the endosperm. The scutellum, like the cotyledons of most monocots, functions in the absorption of food stored in the endosperm. The scutellum is attached to one side of the axis of the embryo, which has a radicle at its lower end and a plumule at its upper end. Both the radicle and the plumule are enclosed by sheathlike, protective structures, called the *coleorhiza* and the *coleoptile*, respectively (Figures 20–3a and 20–4b).

All seeds are enclosed by a *seed coat*, which develops from the integuments of the ovule and provides protection for the enclosed embryo. The seed coat is usually much thinner than the integuments were originally. The thin, dry seed coat may have a papery texture, but in many seeds it is very hard and highly impermeable to water. The micropyle is often visible on the seed coat as a small pore. Commonly, the micropyle is associated with a scar, called the *hilum*, which is left on the seed coat after the seed has separated from its stalk, or *funiculus* (see the seeds on the left in Figures 20–2a and 20–2c).

FORMATION OF THE EMBRYO

The early stages of embryo development are essentially the same in dicotyledons and monocotyledons (Figures 20–5 and 20–6). Formation of the embryo begins with the division of the fertilized egg, or zygote, within the embryo sac of the ovule. In most flowering plants, the first division of the zygote is transverse, or nearly so, with regard to the long axis of the zygote (Figures 20–5a and 20–6a). With this division, the polarity of the embryo is established:

WHEAT

Like all grasses, bread wheat (Triticum aestivum) *is a monocotyledon, and its fruit—the grain or kernel—is one-seeded. The covering layers of the wheat kernel are composed of the pericarp and the remains of the seed coat. The endosperm and the embryo are located within the covering layers. The endosperm represents more than 80 percent of the bulk of the wheat kernel. The outermost layer of the endosperm is called the aleurone layer, which contains lipid and protein reserves. The aleurone layer surrounds the starchy endosperm, as well as the embryo.*

White flour is made from the starchy endosperm. In the milling of wheat, the bran—consisting of the covering layers and the aleurone layer—is removed. The bran constitutes about 14 percent of the kernel. Actually, the bran somewhat decreases the nutritional value of the wheat kernel; because it is mostly cellulosic, it cannot be digested by humans and tends to speed the passage of food through the intestinal tract, resulting in lower absorption. The embryo ("wheat germ"), which represents about 3 percent of the kernel, is also removed, because its high oil content reduces the length of time that the flour can be stored. The bran and wheat germ, which contain most of the vitamins found in wheat, are being used more and more for human consumption, as well as for livestock feed.

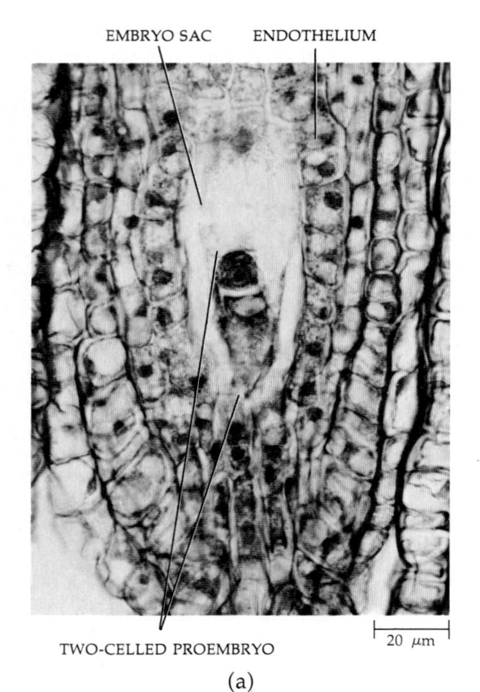

(a)

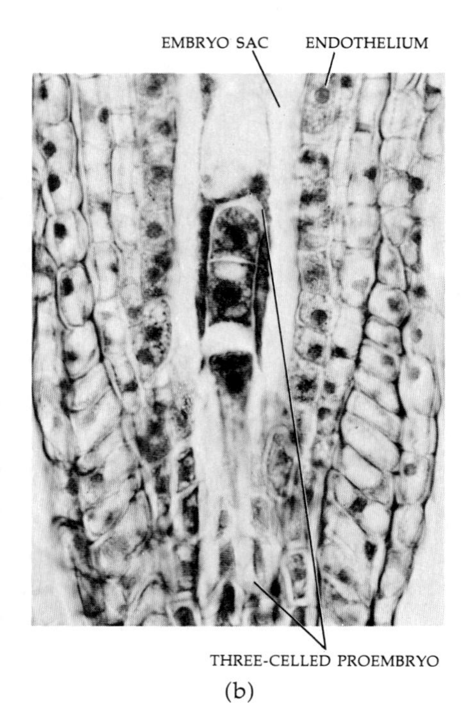

(b)

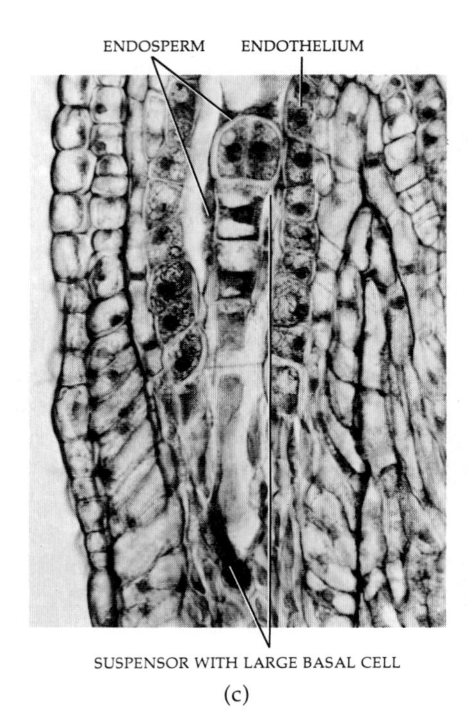

(c)

20–5
Stages in the development of the embryo of shepherd's purse (Capsella bursa-pastoris), *a dicot.* (a) *Two-celled stage, resulting from transverse division of the zygote.* (b) *Three-celled proembryo.* (c) *Six-celled proembryo. The suspensor is now distinct from the two terminal cells, which develop into the embryo proper.* (d) *Embryo proper is globular and has a protoderm. The large cell near the bottom is the basal cell of the suspensor.* (e) *Embryo at heart-shaped (emergence of cotyledons) stage.* (f) *Embryo at torpedo stage. In* Capsella *the embryos curve.* (g) *Mature embryo. The dark layer of cells lining the embryo sac in* (a) *through* (f) *is the endothelium, or innermost layer of the integument.*

the upper (chalazal) pole is the main seat for growth of the embryo; the lower (micropylar) pole produces a stalklike *suspensor*, which anchors the embryo at the micropyle.

Through an orderly progression of divisions, the embryo eventually differentiates into a nearly spherical structure, the embryo proper, and the suspensor (Figures 20–5b through d and 20–6b through d). Before this stage is reached, the developing embryo is often referred to as the *proembryo*.

Suspensors were mentioned in Chapters 17 and 18, in relation to the embryos of *Lycopodium, Selaginella,* and *Pinus,* as structures that merely push the developing embryos into nutritive tissues. Until recently, it was believed that the suspensors of angiosperm embryos played a similarly limited role. It now appears that the suspensors of angiosperms are also actively involved in absorption of nutrients from the endosperm. In addition, in some embryos, proteinaceous substances manufactured in the suspensor apparently are utilized by the embryo proper during periods of rapid growth.

When first formed, the embryo proper consists of a mass of relatively undifferentiated cells. Soon, however, changes in the internal structure of the embryo result in the initial development of the tissue systems of the plant. The future epidermis, the *protoderm*, is formed by periclinal divisions in the outermost cells of the embryo proper (Figures 20–5d and 20–6d). (Periclinal divisions are those in which the cell plates that form between the two new cells are parallel with the surface of the plant part in which they occur.) In addition, differences in the degree of vacuolation and density of the cells within the embryo result in the initiation of the *procambium* and *ground meristem*. The highly vacuolated, less dense ground meristem gives rise to the *ground tissue*, which surrounds the less vacuolated and denser procambium, the precursor of the vascular tissues, xylem and phloem. The protoderm, ground meristem, and procambium—the so-called *primary meristems*—are continuous between the cotyledons and the axis of the embryo (Figure 20–5f and g and Figure 20–6e and f).

Development of the cotyledons may begin either

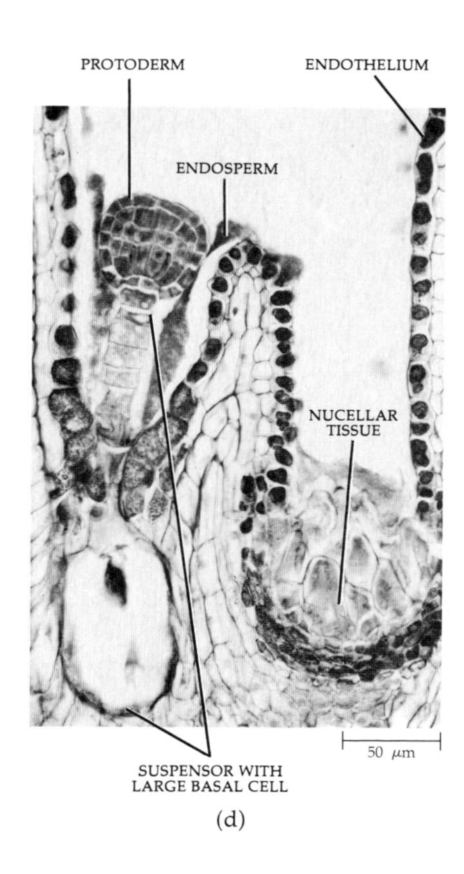

(d)

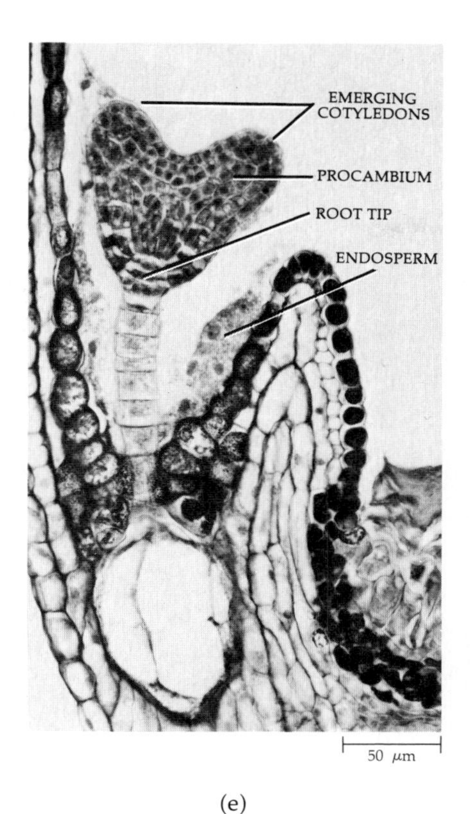

(e)

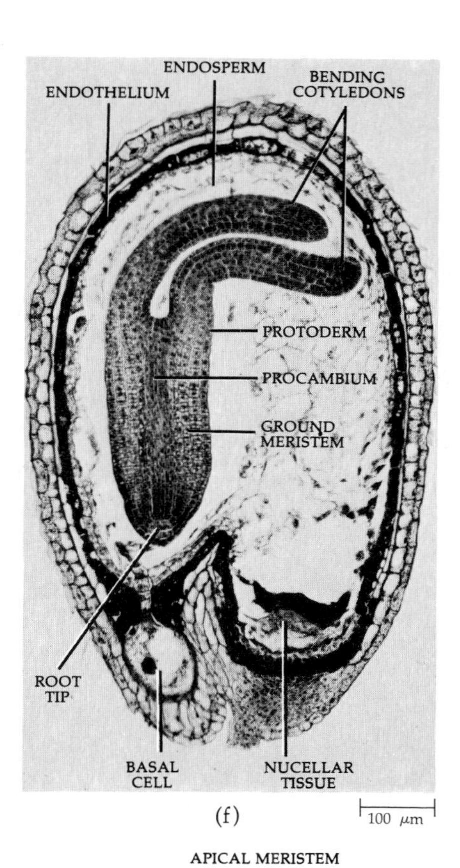

(f)

during or after initiation of the primary meristems. (The stage of embryo development preceding cotyledon development is often referred to as the globular stage.) With the initiation of the cotyledons, the globular embryo in dicots gradually assumes a two-lobed form. This stage of the development of the embryo in dicots is often called the heart-shaped stage (Figure 20–5e).

Inasmuch as the embryos of monocots form only one cotyledon, they do not have a heart-shaped stage. Instead, their embryos become cylindrical in shape (Figure 20–6e). As embryo development continues, the cotyledons and axis elongate (the so-called torpedo stage of embryo development), and the primary meristems extend along with them (Figures 20–5f and 20–6f). During elongation, the embryo either remains straight or becomes curved. The single cotyledon of the monocot often becomes so large in comparison with the rest of the embryo that it is the dominating structure (see Figure 20–3b). With continued enlargement of the embryo, the cells of the suspensor are gradually crushed.

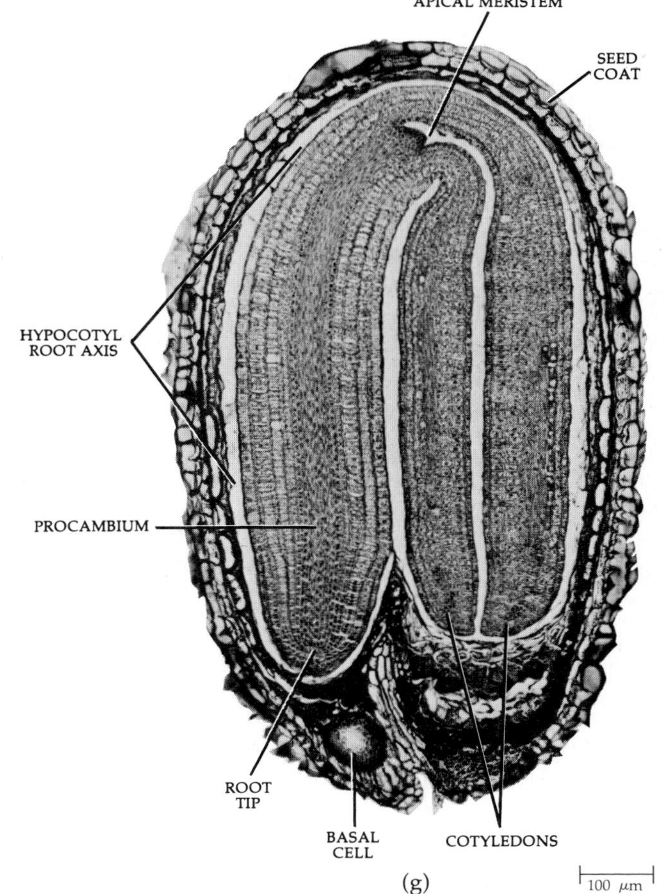

(g)

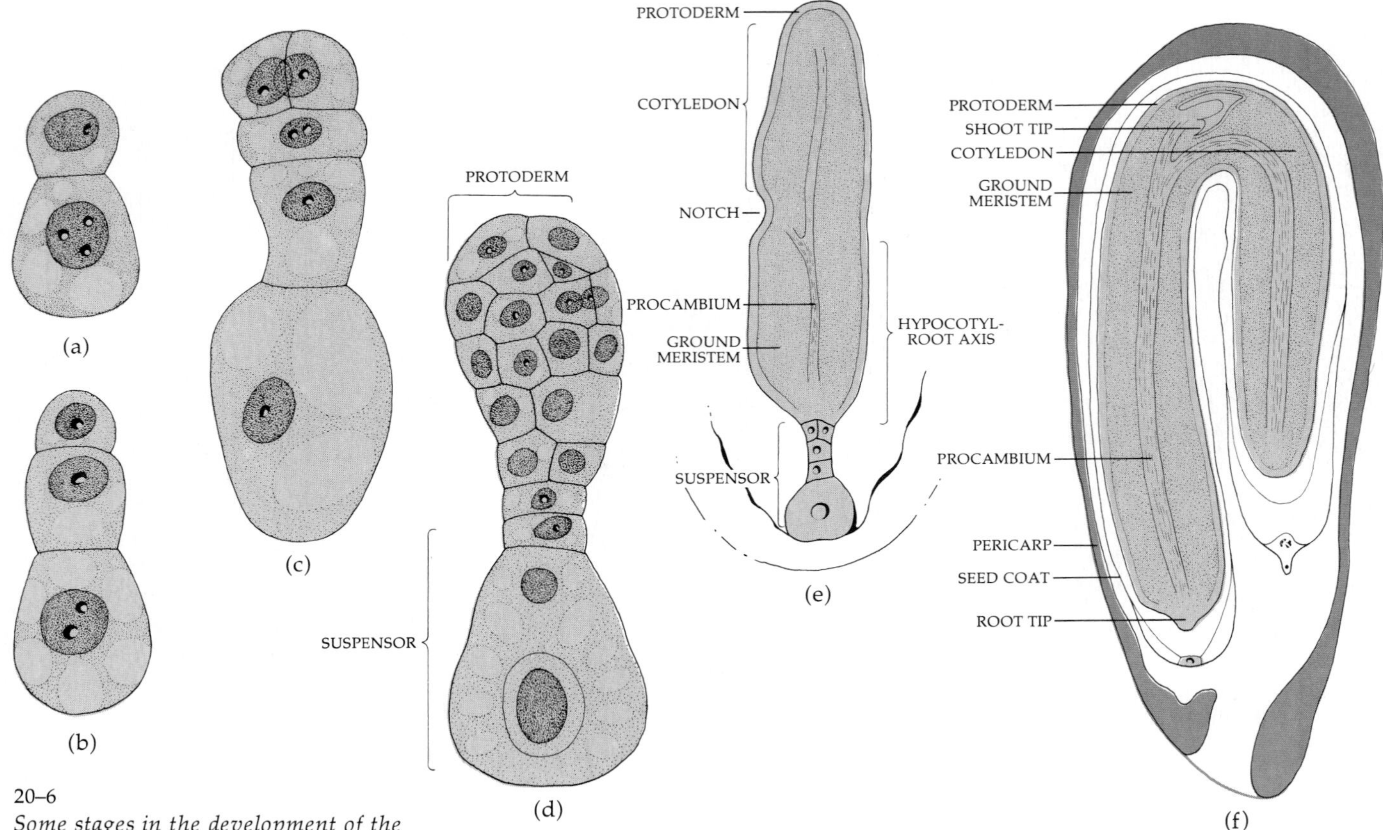

20–6
Some stages in the development of the embryo of arrowroot (Sagittaria), *a monocot. Early stages are shown in* (a) *through* (d). (a) *The two-celled stage, resulting from transverse division of the zygote.* (b) *The three-celled proembryo.* (c) *Disregarding the large basal cell, the proembryo is now at the four-celled stage. All four of these cells, through a series of divisions, contribute to formation of the embryo proper.* (d) *The protoderm has been initiated at the terminal end of the embryo proper. At this stage, the suspensor consists of only two cells, one of which is the large basal cell. In* (e) *and* (f) *late stages in embryo development are shown.* (e) *A depression, or notch (the site of the future shoot apex), has formed at the base of the emerging cotyledon.* (f) *Curving cotyledon; embryo approaching maturity.*

During the early stages of embryogeny, cell division takes place throughout the young sporophyte. However, as the embryo develops, the addition of new cells gradually becomes restricted to the *apical meristems* of the shoot and root. In dicotyledons, the apical meristem of the shoot arises between the two cotyledons (Figure 20–5g). In monocotyledons, it arises on one side of the cotyledon and is completely surrounded by a sheathlike extension from the base of the cotyledon (Figure 20–6f). The apical meristems of shoot and root are of great importance, since these tissues are the source of virtually all of the new cells responsible for the development of the seedling and the adult plant from the embryo.

Throughout the period of embryo formation, there is a continual flow of nutrients between the parent plant and tissues of the ovule, resulting in a massive buildup of food reserves within the endosperm, perisperm, or cotyledons of the developing seed. Eventually, the stalk, or funiculus, connecting the ovule to the ovary wall separates from the ovule, and the ovule becomes a nutritionally closed system. Finally, the seed becomes desiccated as it loses water to the surrounding environment, and the seed coat hardens, encasing the embryo and stored food in "protective armor."

REQUIREMENTS FOR SEED GERMINATION

The growth of the embryo is usually delayed while the seed matures and is dispersed. Resumption of growth of the embryo, or *germination* of the seed, is dependent upon many factors, both external and internal. Among the external, or environmental, factors, three are especially important: water, oxygen, and temperature.

Most mature seeds are extremely dry, normally containing only 5 to 20 percent of their total weight as water. Thus, germination is not possible until the seed imbibes the water required for metabolic activities. Enzymes already present in the seed are activated, and new ones are synthesized for the digestion and utilization of the stored foods accumulated in the cells of the seed during the period of embryo formation. The same cells that earlier had been synthesizing enormous amounts of reserve materials now completely reverse their metabolic processes. Cell enlargement and cell division are initiated in the embryo and follow patterns characteristic of the species. Further growth requires a continuous supply of water and nutrients. As the seed imbibes water, it swells and considerable pressure may develop within it (see boxed essay on "Imbibition," page 65.)

During the early stages of germination, respiration may be entirely anaerobic, but as soon as the seed coat is ruptured, the seed switches to aerobic respiration, which requires oxygen. If the soil is waterlogged, the amount of oxygen available to the seed may be inadequate for aerobic respiration to occur, and the seed will fail to germinate.

Although many seeds germinate over a fairly wide range of temperatures, they usually will not germinate below or above a certain temperature range specific for the species. The minimum temperature for many species is 0°–5°C; the maximum is 45°–48°C; and the optimum range is 25°–30°C.

Even when external conditions are favorable, some seeds will fail to germinate. Such seeds are said to be *dormant*. The most common causes of dormancy in seeds are the physiological immaturity of the embryo and the impermeability of the seed coat to water and, sometimes, to oxygen. Some physiologically immature seeds must undergo a complex series of enzymatic and biochemical changes, collectively called *after-ripening*, before they will germinate. In temperate regions, after-ripening is triggered by the low temperatures of winter. Thus, the necessity for a period of after-ripening helps prevent germination of the seed during the inclement period of winter when it would be unlikely to survive.

Dormancy is of great survival value to the plant. As in the example of after-ripening, it is a method of ensuring that conditions will be favorable for growth of the seedling when germination does occur. Some seeds must pass through the intestines of birds or mammals before they will germinate, resulting in wider dispersal of the species. Some seeds of desert species will germinate only when inhibitors in their coats are leached away by rainfall; this adaptation ensures that the seed will germinate only during those rare intervals when desert rainfall provides sufficient water for the seedling to mature. Similarly, others must be cracked mechanically, as by tumbling along in the rushing water of a gravelly streambed. Still other seeds lie dormant until the intense heat of a fire cracks the seed coat, thus promoting persistence of the species in areas frequently swept by fires.

FROM EMBRYO TO ADULT PLANT

When germination occurs, the first structure to emerge from most seeds is the radicle, or embryonic root; this enables the developing seedling to become anchored in the soil and to absorb water. The continuation of this first root, called the *primary root*, develops branch roots, or *lateral roots*, and these roots in turn may give rise to additional lateral roots. In this manner, a much-branched root system develops. Commonly, the primary root in monocots is short-lived, and the root system of the adult plant develops from *adventitious roots*, which arise at the nodes (the parts of the stem at which the leaves are attached) and then produce lateral roots.

The way in which the shoot emerges from the seed during germination varies from species to species. For example, after the root emerges from the seed of the garden bean *(Phaseolus vulgaris)*, the hypocotyl elongates and becomes bent in the process (Figure 20–2a). Thus, the delicate shoot tip is protected from injury by being pulled rather than pushed through the soil. When the bend, or hook, as it is called, reaches the soil surface, it straightens out and pulls the cotyledons and plumule up into the air. This type of seed germination, in which the cotyledons are carried above ground level, is called *epigeous*.

During germination and subsequent development of the seedling, the food stored in the cotyledons is digested and transported to the growing parts of the young plant. The cotyledons gradually decrease in size, wither, and eventually drop off. By this time, the seedling has become established; that is, it is no longer dependent upon the stored food of the seed for its nourishment. The seedling is now a photosynthesizing, autotrophic organism.

Germination of the castor bean *(Ricinus communis)* (Figure 20–2b) is essentially similar to that of the garden bean, except that in the castor bean the stored food is found in the endosperm. As the hook straightens out, the endosperm and often the seed coat are carried upward, along with the cotyledons and plumule. During this period, the digested foods of the endosperm are absorbed by the cotyledons and transported to the growing parts of the seedling. In both the garden bean and castor bean, the cotyledons become green upon exposure to light, but they do not play an important photosynthetic function.

In the pea *(Pisum sativum)*, the epicotyl is the structure that elongates and forms the hook. As the epicotyl straightens out, the plumule is raised above the soil surface. The cotyledons remain in the soil (Figure 20–2c), where they eventually decompose. This type of seed germination, in which the cotyledons remain underground, is called *hypogeous*.

In the large majority of monocot seeds, the stored food is found in the endosperm. In relatively simple monocot seeds, such as that of the onion *(Allium cepa)*, it is the single tubular cotyledon that emerges from the seed and forms the hook (Figure 20–3b). When the cotyledon straightens, it carries the seed coat and enclosed endosperm upward. Throughout this period, and for some time afterward, the embryo obtains much of its nourishment from the endosperm by way of the cotyledon. Furthermore, the green cotyledon in onion functions as a photosynthetic leaf, contributing significantly to the food supply of the developing seedling. Soon the plumule, enclosed within the protective, sheathlike base of the cotyledon, elongates and emerges from the cotyledon.

Our last example of seedling development is provided by corn *(Zea mays)*, a monocot that has a highly differentiated embryo (see Figure 20–7). Both radicle and plumule are enclosed in sheathlike structures, the coleorhiza and the coleoptile, respectively. The coleorhiza is the first structure to grow through the pericarp (mature ovary wall) of the corn grain. (In corn, the integuments disintegrate during development of the seed and fruit, and hence the pericarp functions as a "seed coat.") The coleorhiza is then followed by the radicle, or primary root, which elongates very rapidly and quickly penetrates the coleorhiza. After the primary root emerges, the coleoptile is pushed upward by elongation of the first *internode*. (An internode is the part of the stem between two successive nodes.) When the base of the coleoptile reaches the soil surface, its edges spread apart at the tip, and the first leaves of the plumule begin to emerge. In addition to the primary root, two or more adventitious roots, which arise from the cotyledonary node, grow through the pericarp and then bend downward.

Regardless of the manner in which the shoot emerges from the seed, the activity of the apical meristem of the shoot results in the formation of an orderly sequence of leaves, nodes, and internodes. Apical meristems, which develop in the axils (the upper angles between leaves and stems) of the leaves, produce axillary shoots and these, in turn, may form additional axillary shoots.

The period from germination to the time the seedling becomes established as an independent organism constitutes the most crucial phase in the life history of the plant. During this period, the plant is most susceptible to injury by a wide range of insect pests and parasitic fungi, and water stress can very rapidly prove fatal.

This type of growth of the root system and the shoot system is called vegetative growth. Eventually,

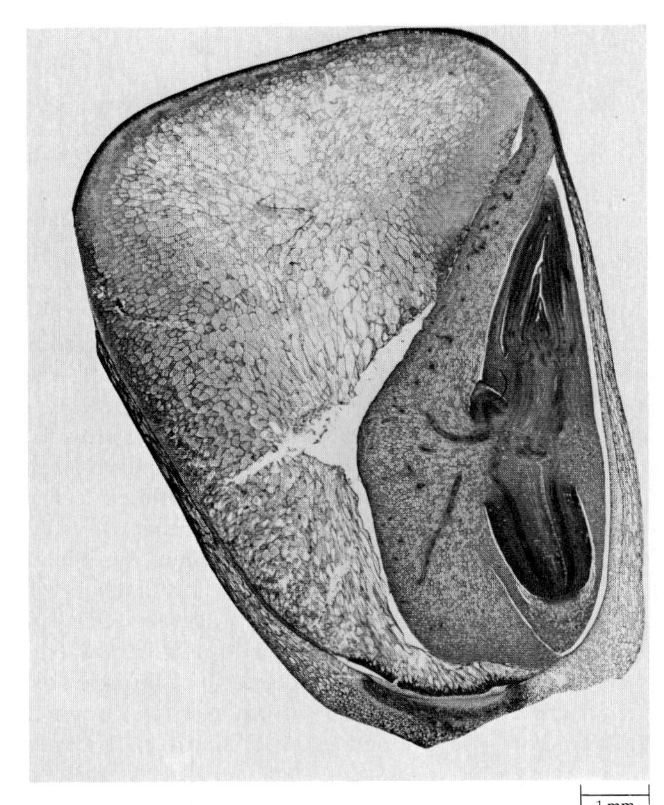

20–7
Longitudinal section of mature grain, or kernel, of corn (Zea mays). *By comparison with the corn grain diagram in Figure 20–3a, and with the photomicrographs of the wheat* (Triticum aestivum) *grain and embryo in Figure 20–4, you should be able to identify the pericarp, endosperm, and various parts of the embryo. Grain embryos commonly contain two or more seminal (seed) adventitious roots. Although at first pointed upward, these roots become inclined downward with further growth.*

one or more of the vegetative apical meristems of the shoot changes into a reproductive apical meristem, that is, a meristem that develops into a flower or an inflorescence (cluster of flowers). When the flowers are formed, the plant is prepared to repeat the life cycle.

SUMMARY

The seeds of flowering plants consist of an embryo, a seed coat, and stored food. When fully formed, the embryo consists basically of a hypocotyl-root axis bearing either one or two cotyledons and an apical meristem at the shoot apex and the root apex. The cotyledons of most dicots are fleshy and contain the stored food of the seed. In other dicots and in most monocots, the stored food resides in the endosperm, and the cotyledons function to absorb digested food from the endosperm. The digested food is then transported to growing regions of the embryo.

During embryo development, the shoot and root of the young plant are initiated as one continuous structure, beginning with the zygote. Through an orderly progression of divisions, the embryo differentiates into a suspensor and an embryo proper, within which the so-called primary meristems—the precursors of the epidermis, ground tissue, and vascular tissues—are formed. Development of the cotyledons may begin either during or after initiation of the primary meristems.

As the embryo develops, the addition of new cells is gradually restricted to the apical meristems. Germination of the seed—resumption of growth of the embryo—is dependent upon environmental factors, including water, oxygen, and temperature. Many seeds must pass through a period of dormancy before they are able to germinate.

Following a period of vegetative growth, one or more apical meristems of the shoot are changed into reproductive apical meristems, which develop into flowers or inflorescences.

CHAPTER 21

Cells and Tissues of the Plant Body

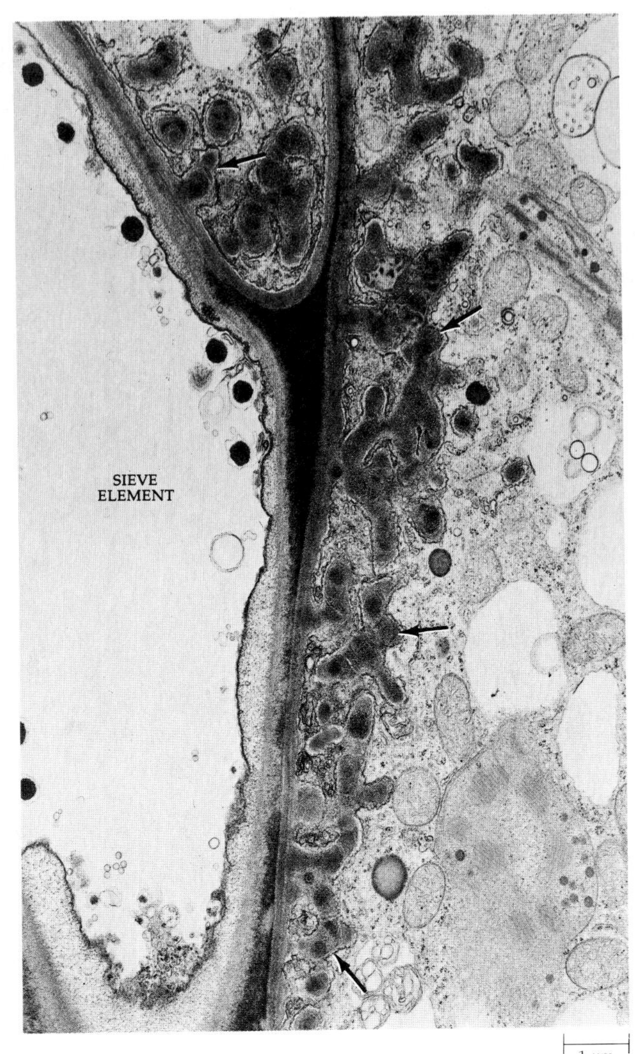

21–1
Transverse section from the phloem of the fern Platycerium bifurcatum, *showing portions of two transfer cells, with their numerous wall ingrowths (see arrows), bordering a relatively clear sieve element, or food-conducting cell.*

Following the development of the embryo, the formation of new cells, tissues, and organs becomes restricted almost entirely to the meristems—the perpetually young tissues that play such a central role in growth. As discussed in Chapter 16, there are two main types of meristems—apical meristems and lateral meristems. The apical meristems are involved primarily with extension of the plant body; they are found at the tips of shoots and roots. This type of growth is called *primary growth*, during which primary tissues are formed. The part of the plant body composed of these tissues is called the primary plant body.

The lateral meristems, the *vascular cambium* and the *cork cambium*, produce the secondary tissues, which make up the secondary plant body. The vascular cambium produces secondary xylem and secondary phloem; cork cambium produces mostly cork.

In both the apical and the lateral meristems, certain cells are able to divide repeatedly; after each division one of the sister cells remains in the meristem while the other moves into the plant body. The self-perpetuating cells that remain in the meristem are called *initials*; their sister cells are called *derivatives*. It is important to note that the derivatives commonly divide one or more times before they begin to differentiate into specific types of cells. Consequently, the meristem includes the initials and their immediate derivatives.

Cell division is not limited to the apical and lateral meristems. For example, the protoderm, procambium, and ground meristem, which are partly differentiated tissues, are called *primary meristems* because (1) they give rise to the primary tissues and (2) many of their cells remain meristematic for some time before they begin to differentiate into specific cell types. With the exception of their initiation in the embryo proper during embryogeny, the primary meristems are derived from the apical meristems.

Growth of the plant body involves both cell division and cell enlargement. As one progresses from younger to older meristematic tissues, the overall size of the cells increases; eventually, the main factor involved in the increase in size of the particular region of the root, stem, or leaf is cell enlargement.

Differentiation—the process by which cells become different from one another and from the meristematic cells from which they originated—often begins while the cell is still enlarging. At maturity, when differentiation is complete, some cells are living and others are dead. Among these living and dead cells are many different cell types (Figure 21–2). How cells that have a common origin come to be so different is generally considered one of the crucial questions of modern biology. Some factors involved in the control of cellular differentiation are discussed in Chapter 7 (pages 128–31) and in Chapter 25.

In the plant body, basic tissue patterns are established by early meristematic activity. The shape of the plant and organization of its tissues are influenced greatly by both cell division and cell enlargement. The acquisition of a particular shape or form is known as *morphogenesis* (from *morphe*, meaning "form," and *genere*, meaning "to create").

THE TISSUE SYSTEMS

Botanists have long recognized that the principal tissues of vascular plants are organized into larger units in all parts of the plant. These groups of tissues are known as *tissue systems*, and their presence in the root, stem, and leaf reveals both the basic similarity of the plant organs and the continuity of the plant body. The three tissue systems are: (1) the fundamental, or *ground tissue system*, (2) the *vascular tissue system*, and (3) the *dermal tissue system*.

The ground tissue system consists of the three so-called ground tissues—*parenchyma*, *collenchyma*, and *sclerenchyma*. Parenchyma is by far the most common of the ground tissues. The vascular tissue system consists of the two conducting tissues—*xylem* and *phloem*. The dermal tissue system is represented by the *epidermis*, the outer protective covering of the primary plant body, and later by the *periderm*, in the secondary plant body.

TISSUES AND THEIR COMPONENT CELLS

Tissues may be defined as groups of cells that are structurally and/or functionally distinct. Tissues composed of only one type of cell are called *simple tissues*, whereas those composed of two or more types of cells are called *complex tissues*. Parenchyma, collenchyma, and sclerenchyma are simple tissues; xylem, phloem, and epidermis are complex tissues.

Parenchyma

Parenchyma tissue, the progenitor of all other tissues, is composed of parenchyma cells. In the primary plant body, parenchyma cells commonly occur as continu-

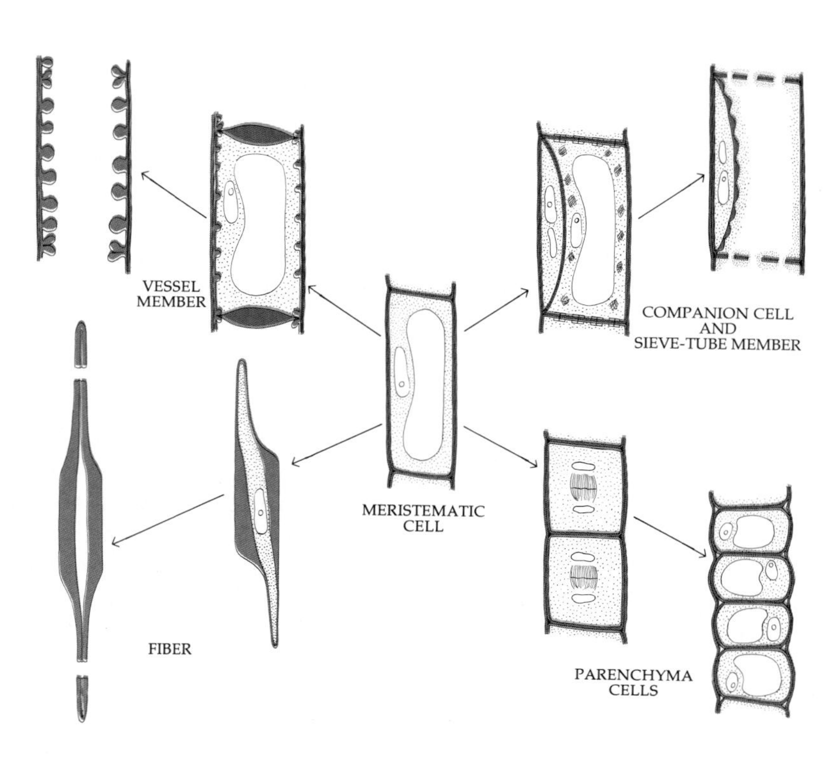

21–2
Diagram illustrating some of the cell types that may originate from a meristematic cell of the procambium or the vascular cambium.

ous masses in the cortex of stems (Figure 21–3) and roots, in the pith of stems, in leaf mesophyll (see Figure 21–22), and in the flesh of fruits. In addition, parenchyma cells occur as vertical strands of cells in the primary and secondary vascular tissues and as horizontal strands called rays in the secondary vascular tissues (see Chapter 24).

Characteristically found living at maturity, parenchyma cells are capable of cell division, and although their walls are commonly primary, some parenchyma cells also have secondary walls. Because of their ability to divide, parenchyma cells play an important role in regeneration and wound healing. It is these cells that initiate adventitious structures, such as adventitious roots on stem cuttings. Parenchyma cells are involved in such activities as photosynthesis, storage, and secretion—activities dependent upon living protoplasm. In addition, parenchyma cells may play a role in the movement of water and the transport of food substances in plants.

Transfer Cells

During the past few years, considerable attention has been given to a special kind of parenchyma cell containing ingrowths of the cell wall which often greatly increase the surface area of the plasma membrane (see Figure 21–1). These parenchyma cells, called *transfer cells*, are believed to play an important role in the transfer of solutes over short distances. Although such cells have long been known to exist in various plant parts, botanists have only recently realized that they are exceedingly common and probably serve a similar function throughout the plant body. Transfer cells occur in association with the xylem and phloem of small, or minor, veins in cotyledons and in the leaves of many herbaceous dicotyledons, and with the xylem and phloem of leaf traces at the nodes in both dicotyledons and monocotyledons. In addition, they are found in various tissues of reproductive structures (placentae, embryo sacs, endosperm) and in various glandular structures (nectaries, salt glands, the glands of carnivorous plants). Each of these locations is a potential site of intensive short-distance solute transfer.

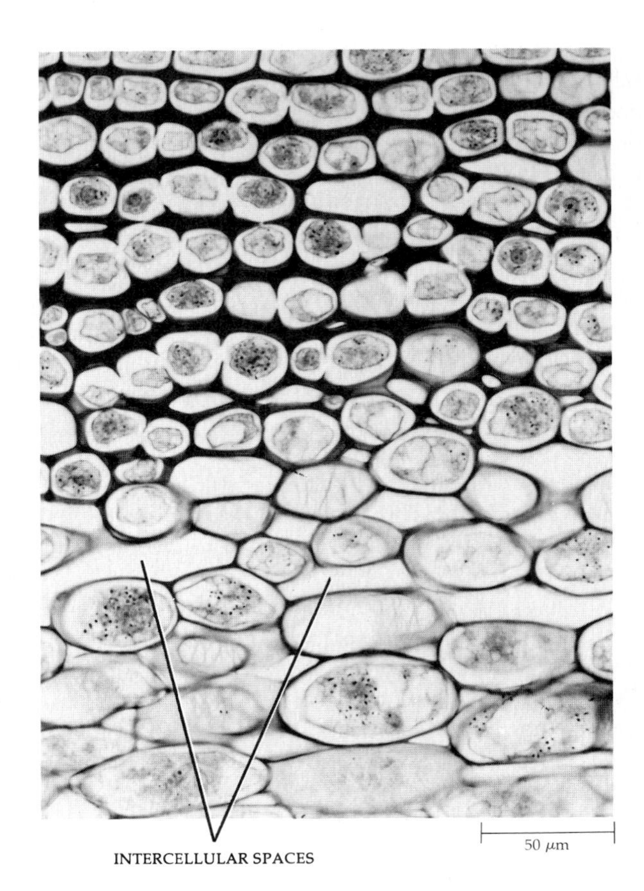

21–3
Transverse section of the cortex of an elderberry (Sambucus canadensis) *stem, showing collenchyma cells with unevenly thickened walls (above) and parenchyma cells (below). Clear areas between the cells are intercellular spaces. In some of the parenchyma cells, a meshwork of lines is visible in the facing walls. The light areas within the meshwork are primary pit-fields.*

Collenchyma

Collenchyma tissue is composed of collenchyma cells which, like parenchyma cells, are found living at maturity (Figures 21–3 and 21–5). Collenchyma tissue commonly occurs in discrete strands or as continuous cylinders beneath the epidermis in stems and petioles (leaf stalks). It also is found bordering the veins in dicot leaves. (The "strings" on the outer surface of celery stalks, or petioles, consist almost entirely of collenchyma.) The typically elongated collenchyma cells (Figure 21–4) contain unevenly thickened, nonlignified primary walls, making them especially well adapted for the support of young, growing organs (see the description of the primary wall on page 33). The name *collenchyma* derives from the Greek word *colla*, meaning "glue," which refers to their characteristic thick, glistening walls in fresh tissue (Figure 21–6). Being primary in nature, the walls of collenchyma cells are readily stretched and offer relatively little resistance to elongation of the plant part in which they are found. In addition, because collenchyma cells are living at maturity, they can continue to develop thick, flexible walls while the organ in which they are found is still elongating.

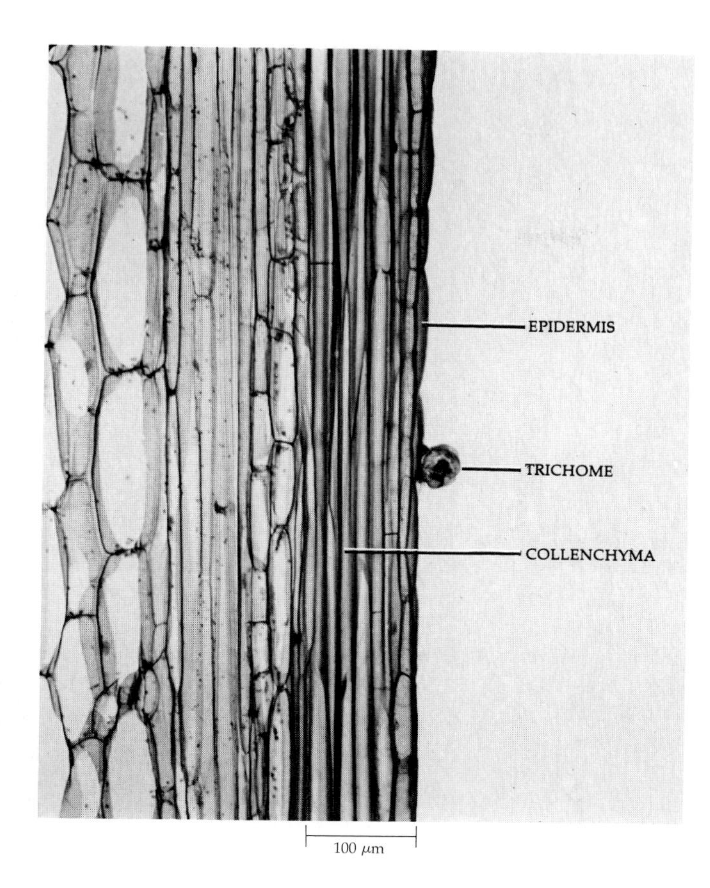

21–4
Longitudinal section showing the elongated collenchyma cells in the stem of a squash (Cucurbita maxima). *Trichomes are shown in Figure 21–21 on page 429.*

Sclerenchyma

Sclerenchyma tissue is composed of sclerenchyma cells, which may develop in any or all parts of the primary and secondary plant body; they often lack protoplasts at maturity. The term *sclerenchyma* is derived from the Greek *skleros,* meaning "hard," and the principal characteristic of sclerenchyma cells is their thick, often lignified secondary walls. Because of these walls, the sclerenchyma cells are important strengthening and supporting elements in plant parts that have ceased elongating (see the description of the secondary wall on page 34).

Two types of sclerenchyma cells are recognized: *fibers* and *sclereids*. Fibers are generally long, slender cells, which commonly occur in strands or bundles (Figure 21–7). The so-called bast fibers—such as hemp, jute, and flax—are derived from the stems of dicotyledons. Other economically important fibers—such as Manila hemp—are present in the leaves of monocots. Sclereids are variable in shape and are often branched (Figure 21–8), but compared with most fibers, sclereids are relatively short cells. Sclereids may occur singly or in aggregates throughout the ground tissue. They make up the seed coats of seeds, the shells of nuts, the stone, or endocarp, of stone fruits, and they give pears their characteristic gritty texture (Figure 21–9).

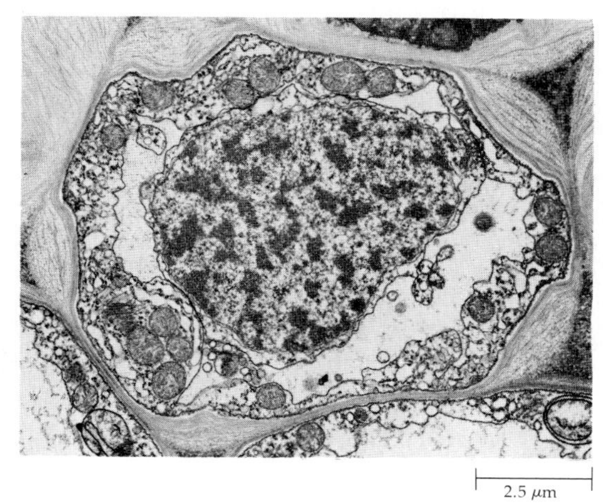

21–5
Electron micrograph of a mature collenchyma cell from the filament of a wheat (Triticum aestivum) *stamen. Notice the protoplasmic contents and unevenly thickened wall of this cell.*

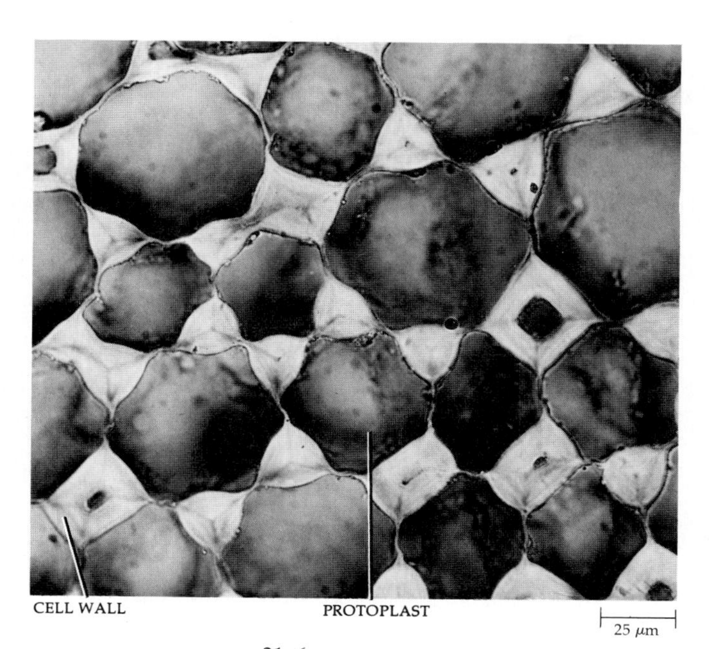

21–6
Transverse section of collenchyma tissue from a petiole in rhubarb (Rheum rhaponticum). *In fresh tissue like this, the unevenly thickened collenchyma cell walls have a glistening appearance.*

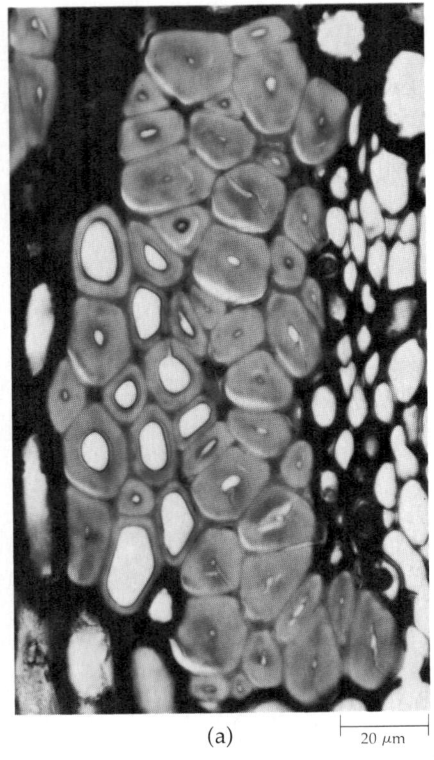
(a) 20 μm

(b) 50 μm

21–7
Primary phloem fibers from the stem of a linden, or basswood (Tilia americana), *seen here in both* (a) *cross-sectional and* (b) *longitudinal views. The secondary walls of these long, thick-walled fibers contain relatively inconspicuous pits. Only a portion of the fibers can be seen in* (b).

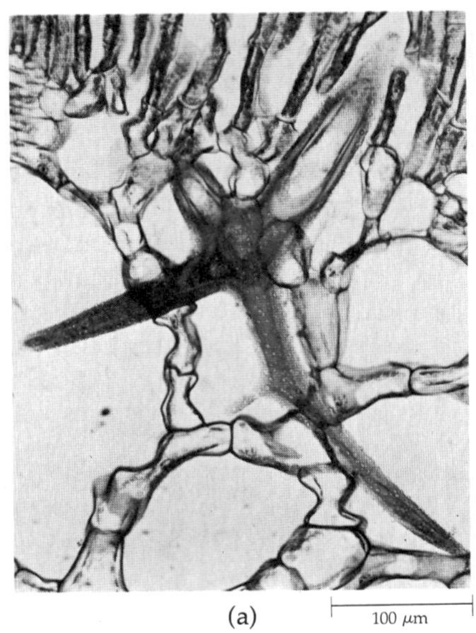
(a) 100 μm

(b) 100 μm

21–8
Branched sclereid from a leaf of the water lily (Nymphaea odorata), *as seen in* (a) *ordinary light and* (b) *polarized light. Numerous small angular crystals are embedded in the wall of this sclereid.*

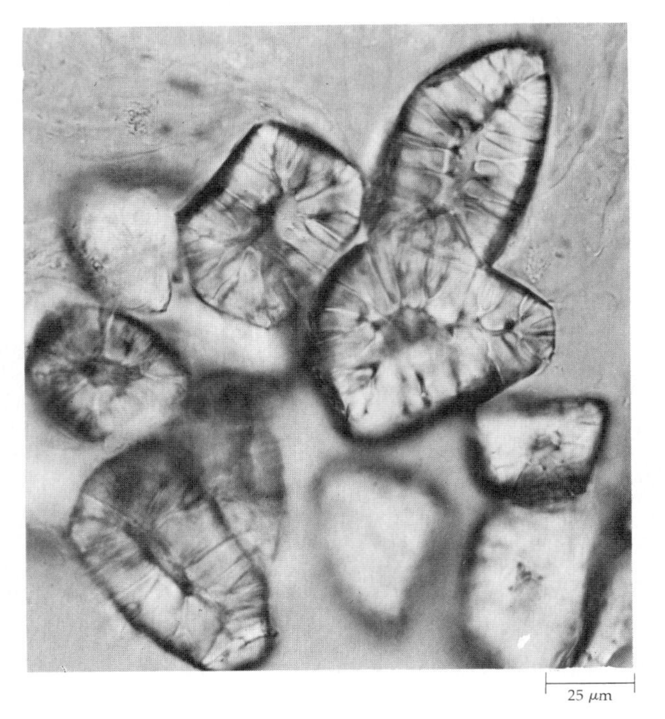
25 μm

21–9
Sclereids (stone cells) from the flesh of the pear (Pyrus communis) *fruit. The secondary walls contain conspicuous simple pits, with many branches. Branched pits such as these are called ramiform pits. During formation of the clusters of stone cells in the flesh of the pear fruit, cell divisions occur concentrically around some of the sclereids formed earlier. The newly formed cells differentiate as stone cells, adding to the cluster.*

Xylem

Xylem is the principal water-conducting tissue found in vascular plants. It is also involved in the conduction of minerals, in food storage, and in support. Together with the phloem, the xylem forms a continuous system of vascular tissue extending throughout the plant body (Figure 21–10). Xylem may be primary or secondary in origin. The primary xylem is derived from the procambium, and the secondary xylem is derived from the vascular cambium (see Chapter 24).

The principal conducting cells of the xylem are the *tracheary elements*, of which there are two types, the *tracheids* and the *vessel members*, or vessel elements. Both are elongated cells that have secondary walls and lack protoplasts at maturity (Figure 21–11a through d), and each type may have *pits* in their walls. In addition to the presence of pits, vessel members contain *perforations*, which are areas lacking both pri-

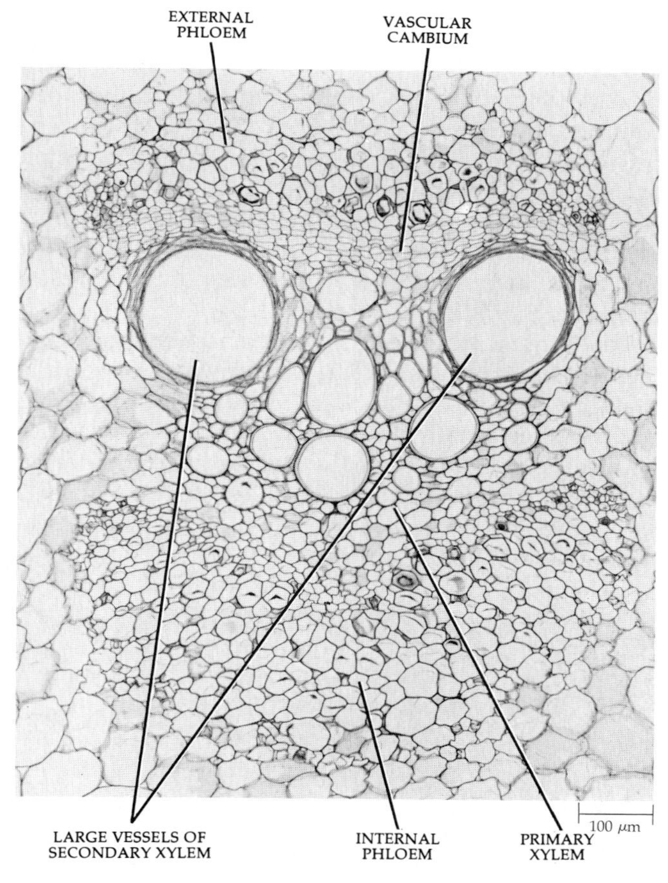

21–10
Transverse section of a vascular bundle from the stem of a squash (Cucurbita maxima). *Phloem occurs on both sides of the xylem, and a vascular cambium has developed between the external phloem and the xylem.*

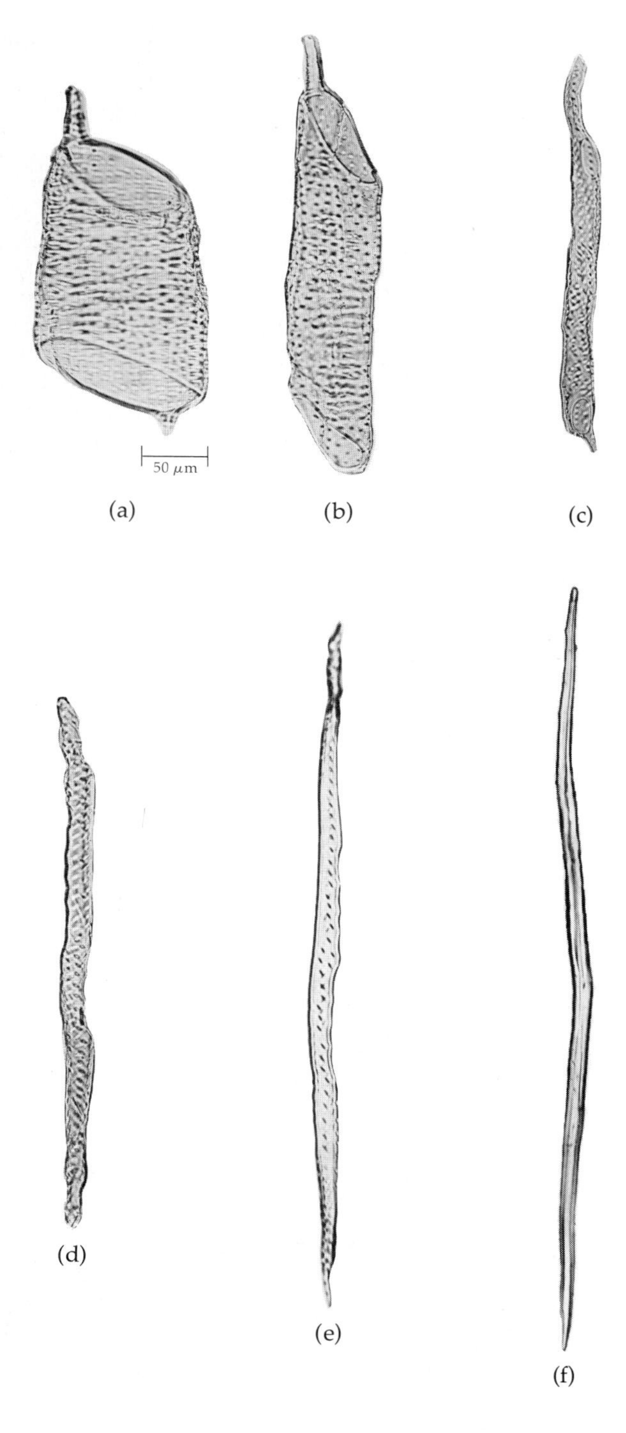

21–11
Cell types in the secondary xylem, or wood, of an oak tree (Quercus), *as illustrated from macerated tissue (tissue that has been broken down into its constituent cells).* (a) *and* (b) *Wide vessel members.* (c) *Narrow vessel member.* (d) *Tracheid.* (e) *Fiber-tracheid.* (f) *Libriform fiber (the longest fiber in oak wood). The spotted appearance of these cells is due to pits in the walls; pits are absent in* (f).

21–12
Scanning electron micrographs of the perforated end walls of vessel members from secondary xylem. (a) *A simple perforation plate, with its single large opening, seen here between two vessel members in linden, or basswood* (Tilia americana). (b) *The ladderlike bars of a scalariform perforation plate between vessel members of red alder* (Alnus rubra). *Pits can be seen in the wall below the perforation plate in* (a) *and in portions of the wall in* (b).

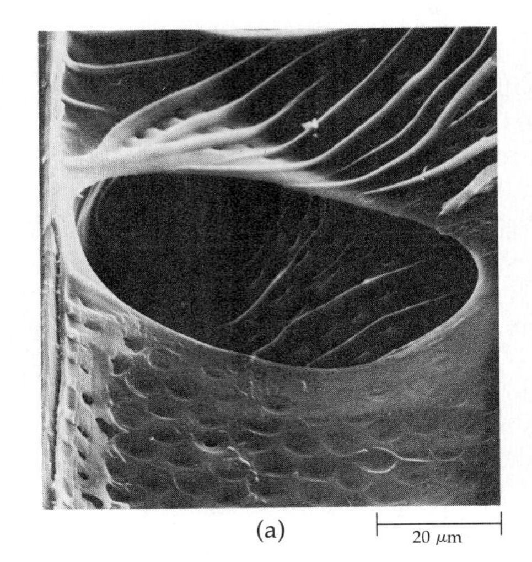

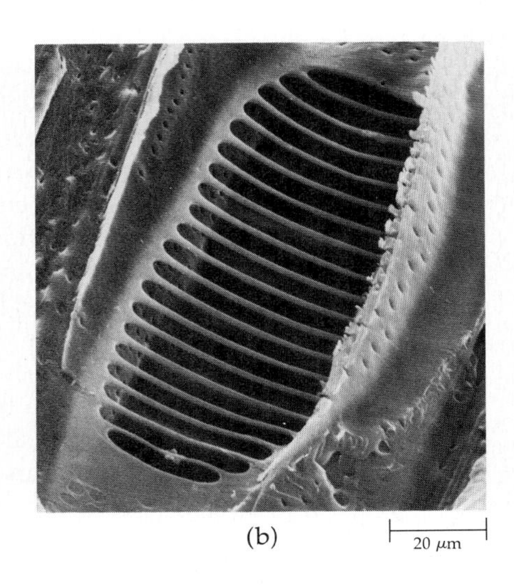

mary and secondary walls; perforations are literally holes in the cell wall. Perforations generally occur on the end walls, but they may also be found on the lateral walls. The part of the wall bearing the perforation or perforations is called the *perforation plate* (Figure 21–12).

Vessel members are joined into long, continuous columns, or tubes, called *vessels* (Figure 21–13). The pits of the long, tapering tracheids are concentrated on the overlapping ends of the cells. The tracheid is the only type of water-conducting cell found in most lower vascular plants and gymnosperms; the xylem of the great majority of angiosperms contains vessel members in addition to tracheids.

The term *tracheary element* dates back to the seventeenth century and the Italian physician Marcello Malpighi, one of the founders of plant anatomy. Malpighi was especially interested in finding similarities between plants and animals. During his examination of xylem, he observed air bubbles escaping from a

21–13
Scanning electron micrographs of vessel members from secondary xylem. (a) *External view of parts of three vessel members of red oak* (Quercus rubra). *Notice the rims between vessel members.* (b) *Parts of two vessel members of squash* (Cucurbita maxima), *cut lengthwise so that half the cylinder is seen from the inside. The arrow points to the boundary between the two elements. Notice the numerous pits in the walls of the vessel members.*

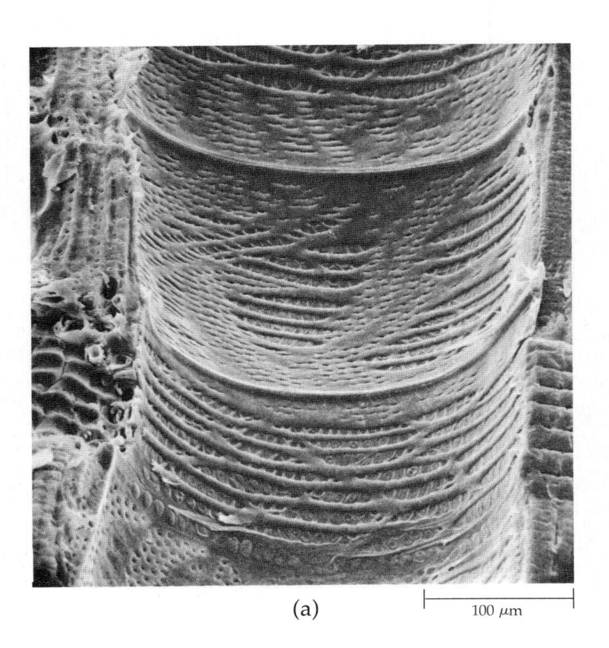

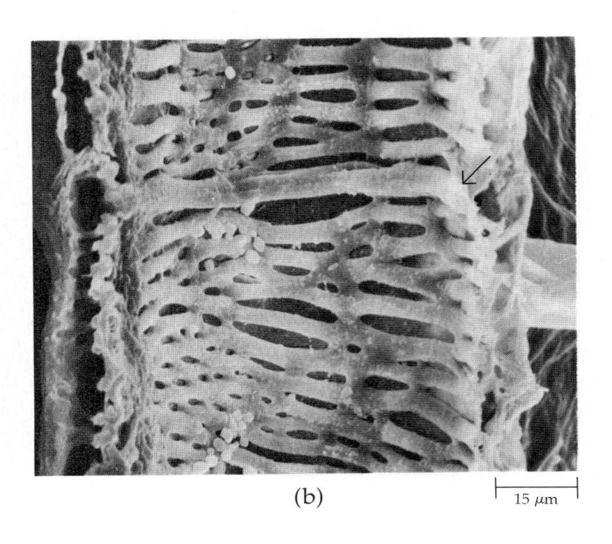

vessel with spiral wall thickenings. He immediately compared the vessel with the tracheae, or air ducts, of insects and later applied the same term to the vessels of the xylem. The term has been used ever since for the water-conducting cells of the xylem.

Vessel members are generally thought to be more efficient conductors of water than the tracheids, because water can flow relatively unimpeded from vessel member to vessel member through the perforations. When flowing through tracheids, water must pass through the membranes of the pit-pairs (see pages 34–35). However, the pit membranes probably offer relatively little resistance to the flow of water, because during the final stages of differentiation they are partially hydrolyzed, leaving only a highly permeable meshwork of cellulose microfibrils.

In the secondary xylem and late-formed primary xylem (metaxylem), the pitted secondary walls of the tracheids and vessel members cover the entire primary wall, except at the pits and the perforations of the vessel members (Figure 21–11a through d). Consequently, these walls are rigid and cannot be stretched. During the period of elongation or expansion of the roots, stems, and leaves, many of the first-formed tracheary elements of the early-formed primary xylem (protoxylem) have their secondary walls deposited in the form of rings or spirals (Figure 21–14). These thickenings, which may be annular (ringlike) or helical (spiral), make it possible for such tracheary elements to be stretched or extended, although they are frequently destroyed during the overall elongation of the organ. In the primary xylem, the nature of the wall thickening is greatly influenced by the amount of elongation. If little elongation occurs, pitted elements rather than extensible types appear. On the other hand, if much elongation takes place, many elements with annular and helical thickenings will appear.

In addition to tracheids and vessel members, xylem contains parenchyma cells that store various substances. The xylem parenchyma cells commonly occur in vertical strands. In the secondary xylem, they also are found in rays. Fibers (Figure 21–11e and f) and sclereids also occur in xylem. Many xylem fibers are living at maturity and serve a dual function of storage and support.

Phloem

Phloem is the principal food-conducting tissue found in vascular plants (Figure 21–10). The phloem may be primary or secondary in origin, and as with primary xylem, the first-formed primary phloem (protophloem) is frequently stretched and destroyed during elongation of the organ.

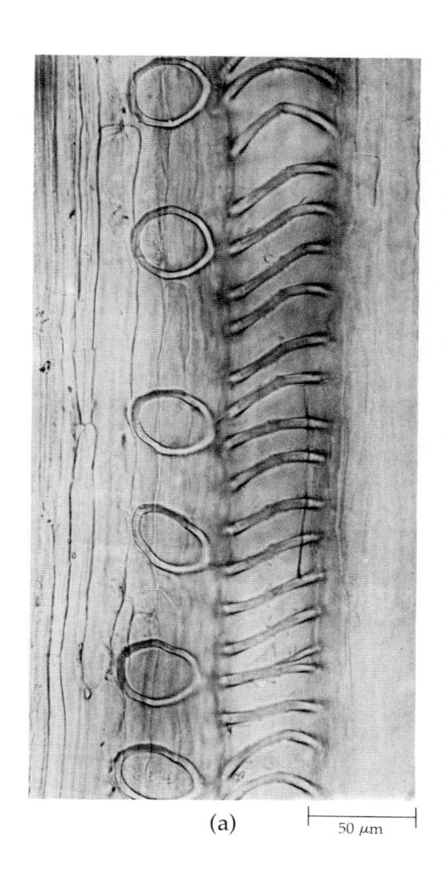

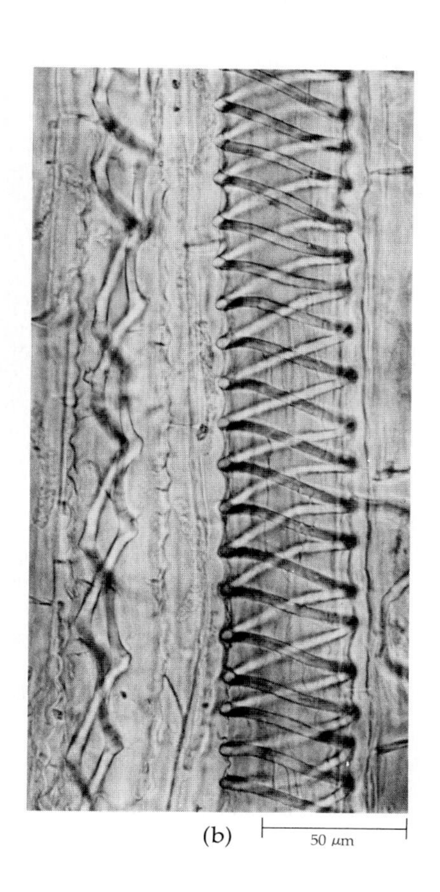

21–14
Parts of tracheary elements from the first-formed primary xylem (protoxylem) of the castor bean (Ricinus communis). (a) *Annular (the ringlike shapes at left) and helical wall thickenings in partly extended elements.* (b) *Double helical thickenings in elements that have been extended. The element on the left has been greatly extended, and the coils of the helices have been pulled far apart.*

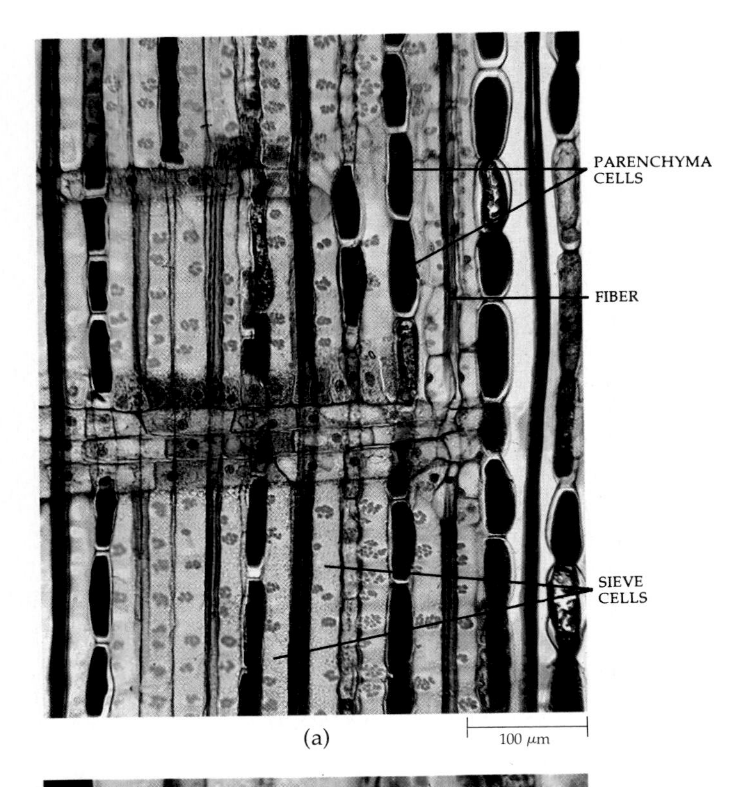

(a)

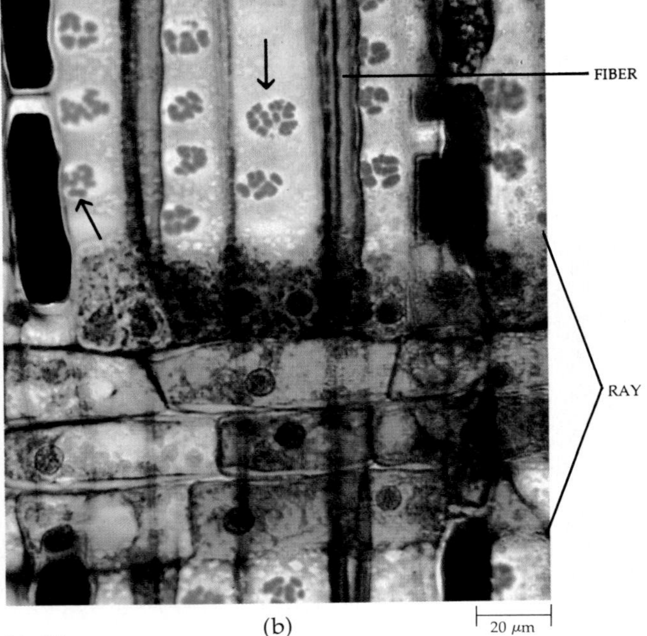

(b)

21–15
(a) *Longitudinal (radial) view of secondary phloem of yew* (Taxus canadensis), *showing vertically oriented sieve cells, strands of parenchyma cells, and fibers. Parts of two horizontally oriented rays can be seen traversing the vertical cells.* (b) *Detail of portion of the secondary phloem of yew, showing sieve areas, with callose (stained blue) on the walls of the sieve cells, and albuminous cells (see page 426), which here constitute the top row of cells in the ray.*

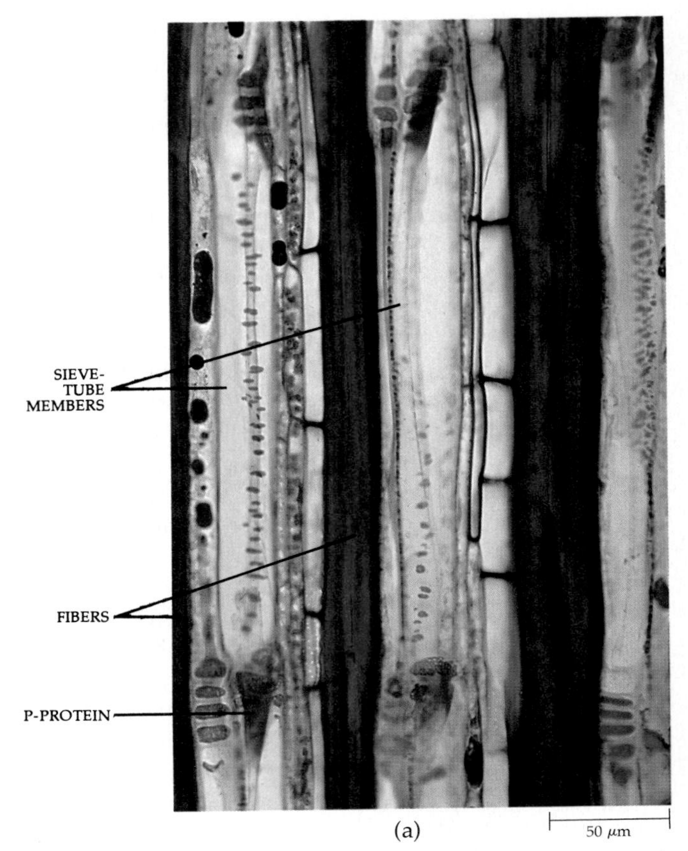

(a)

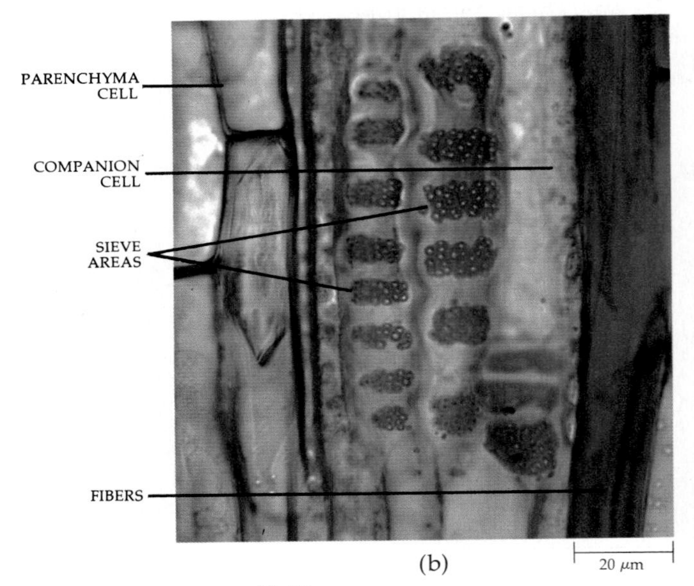

(b)

21–16
(a) *Longitudinal (radial) view of secondary phloem of linden* (Tilia americana), *showing sieve-tube members and conspicuous groups of thick-walled fibers.* (b) *Compound sieve plates of linden sieve-tube members. (Compound sieve plates consist of two or more sieve areas.) Each sieve area is composed of pores bordered by cylinders of callose, which are stained blue in this section.*

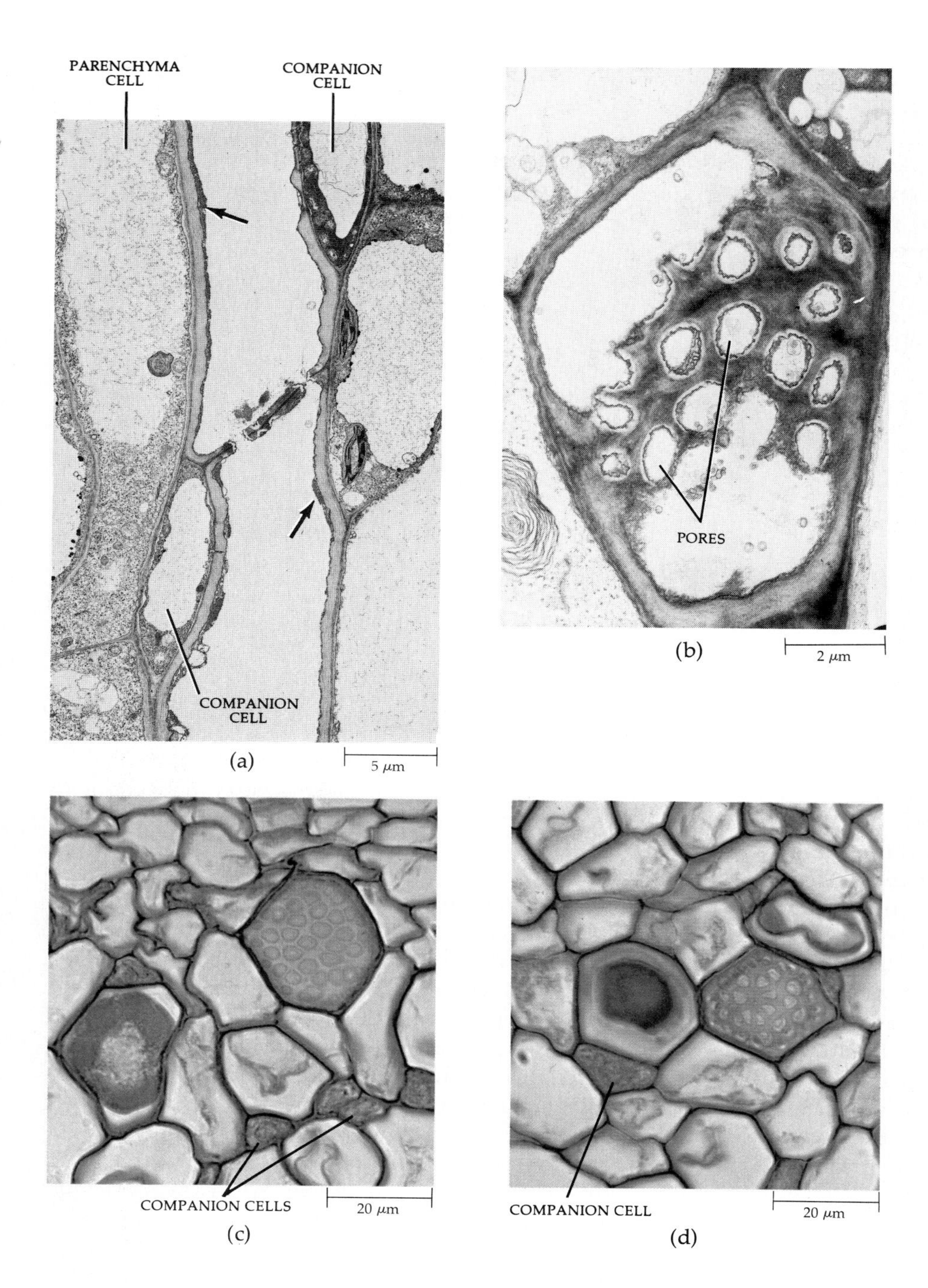

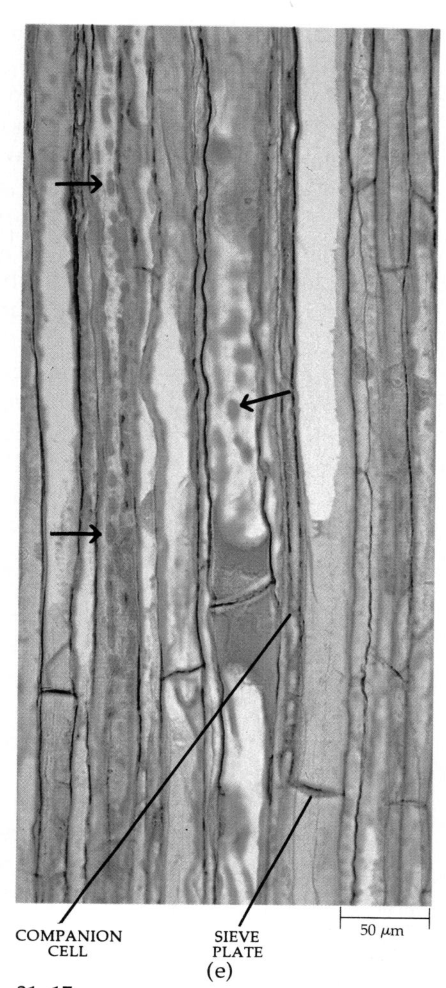

21–17
The phloem in the stem of the squash (Cucurbita maxima), *as seen in electron micrographs,* (a) *and* (b), *as well as in photomicrographs,* (c) *through* (e). (a) *Longitudinal view of parts of two mature sieve-tube members and a sieve plate, showing distribution of P-protein (see arrows) along the wall.* (b) *Face view of simple sieve plate (one sieve area per plate) between two mature sieve-tube members. The sieve-plate pores are open.* (c) *Transverse section, showing two immature sieve-tube members. Slime bodies, or P-protein bodies, can be seen in the sieve-tube member on the left, an immature sieve plate in the one to the right, above. The small, dense cells are companion cells.* (d) *Transverse section, showing some mature sieve-tube members. A slime plug can be seen in the sieve-tube member on the left; a mature sieve plate can be seen in the one on the right. The small, dense cells are companion cells.* (e) *Longitudinal section, showing mature and immature sieve-tube members. The arrows point to P-protein bodies in immature cells.*

The principal conducting cells of the phloem are the *sieve elements,* of which there are two types, the *sieve cells* (Figure 21–15) and the *sieve-tube members* (Figures 21–16 and 21–17). The term "sieve" refers to the clusters of pores, the *sieve areas,* through which the protoplasts of adjacent sieve elements are interconnected. In sieve cells, the pores are narrow and the sieve areas are rather uniform in structure on all walls. Most of the sieve areas are concentrated on the overlapping ends of the long, slender cells (Figure 21–15a). In sieve-tube members, the sieve areas on some walls have larger pores than those on other walls of the same cell. The part of the wall bearing the sieve areas with larger pores is called a *sieve plate*

(Figures 21–16b and 21–17a and b). Although sieve plates may occur on any wall, they generally are located on the end walls. Sieve-tube members occur end-on-end in longitudinal series called *sieve tubes*. Thus, the principal distinction between the two types of sieve elements is the presence of sieve plates in sieve-tube members and their absence in sieve cells. Sieve cells are the only type of food-conducting cell in most lower vascular plants and gymnosperms, whereas sieve-tube members occur in angiosperms, which lack sieve cells.

The walls of sieve elements are generally described as primary. In cut sections of phloem tissue, the pores of the sieve areas and sieve plates are generally blocked by a wall substance called *callose*, which is a polysaccharide composed of spirally wound chains of glucose residues. The presence of callose in the pores of conducting sieve elements long puzzled botanists; it seemed illogical for the pores to contain any substance that would impede the movement of other substances from cell to cell. It is now known that most, if not all, of the callose found in the pores of conducting sieve elements is deposited there in response to injury during preparation of the tissue for microscopy.

Unlike tracheary elements, sieve elements have living protoplasts at maturity (Figure 21–17). The protoplasts of mature sieve elements are unique among the living cells of the plant in that they either lack nuclei entirely or contain only remnants of them. In addition, most mature sieve elements lack a clear boundary between their cytoplasm and vacuoles. When young, the sieve element contains several vacuoles, each separated from the cytoplasm by a tonoplast, or vacuolar membrane. In the final stages of differentation, the tonoplasts disappear and the distinction between cytoplasmic and vacuolar contents no longer exists. At maturity, all of the remaining components of the sieve-element protoplast are distributed along the wall; they consist of the plasma membrane, a network of smooth endoplasmic reticulum next to the plasma membrane, and some plastids and mitochondria. Ribosomes, Golgi bodies, microtubules, and nuclei are lacking.

The protoplasts of the sieve-tube members of dicotyledons (and some monocotyledons) are characterized by the presence of a proteinaceous substance known as *slime*, or *P-protein*. (The "P" in P-protein stands for phloem.) P-protein has its origin in the young sieve-tube member in the form of discrete bodies, called slime bodies or P-protein bodies (Figure 21–17c and e). During the late stages of differentiation, the P-protein bodies elongate and disperse. In cut sections of phloem tissues, "slime plugs" of P-protein are usually found near the sieve plates (Figure 21–17d). Like callose, slime plugs are not found in undisturbed cells and so are considered to be the result of the surging of the contents of the sieve-tube members that occurs when the tissues and sieve tubes are severed from the plant. In normal, mature sieve-tube members, P-protein is apparently distributed along the walls and is continuous from one sieve-tube member to the next through the pores of the sieve areas and sieve plates. The sieve-plate pores are lined with P-protein, but not plugged by it (Figure 21–17a and b). The function of P-protein has not been determined. However, some botanists believe that, together with wound callose, P-protein serves to seal the sieve-plate pores at the time of wounding, thus preventing the loss of contents from sieve tubes.

Sieve-tube members are characteristically associated with specialized parenchyma cells called *companion cells* (Figure 21–17), which contain all of the components commonly found in living plant cells, including a nucleus. The sieve-tube member and its associated companion cells are closely related developmentally (they are derived from the same mother cell), and they have numerous connections with one another. Functionally, the companion cells are very important, for they are largely responsible for the active secretion of substances into (and their removal from) the sieve-tube members. This subject will be discussed further in Chapter 28 when the mechanism of phloem transport in angiosperms is considered in detail.

The sieve cells of gymnosperms are characteristically associated with specialized parenchyma cells called *albuminous cells* (Figure 21–15b). Although generally not derived from the same mother cell as their associated sieve cells, the albuminous cells are believed to perform the same roles as companion cells. Like the companion cell, the albuminous cell contains a nucleus, in addition to other cytoplasmic components characteristic of living cells.

The sieve elements in most species apparently are short-lived and die in less than a year after their origin. This is not true, however, of all sieve elements. In the secondary phloem of the basswood, or linden tree (*Tilia americana*), some sieve elements remain alive and presumably function as conducting elements for five to ten years. Sieve elements are known to remain alive for many years in perennial monocots. In certain palms, some sieve elements at the base of the main stem may be more than a century old. When the sieve elements die, their associated companion cells or albuminous cells also die, which is one more indication of the intimate relationship that exists between sieve elements and their companion cells or albuminous cells.

Other parenchyma cells occur in the primary and secondary phloem (Figures 21–15 through 21–17). They are largely concerned with the storage of various substances. Fibers (Figures 21–15 and 21–16) and sclereids may also be present.

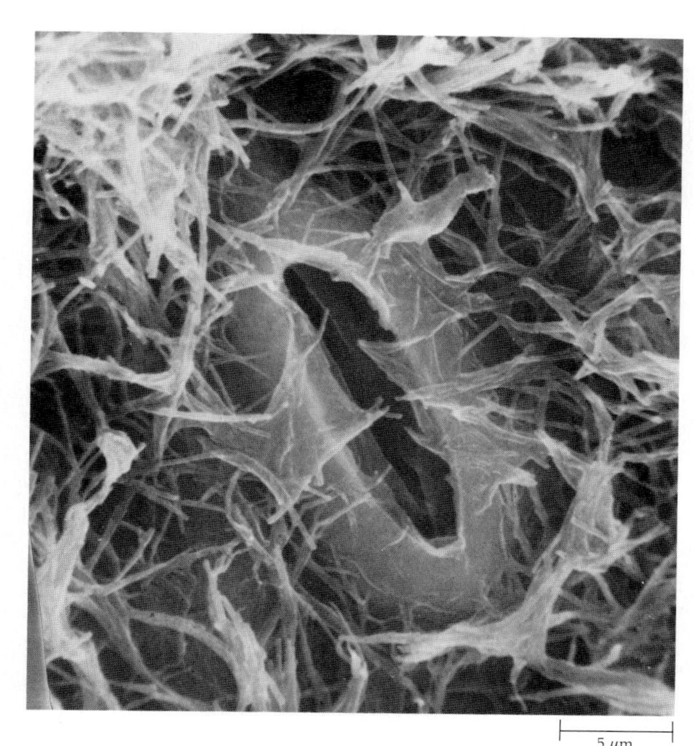

21–18
Surface view of lower epidermis of a Eucalyptus globulus *leaf, taken with the scanning electron microscope. A single stoma and numerous filamentous wax deposits can be seen here.*

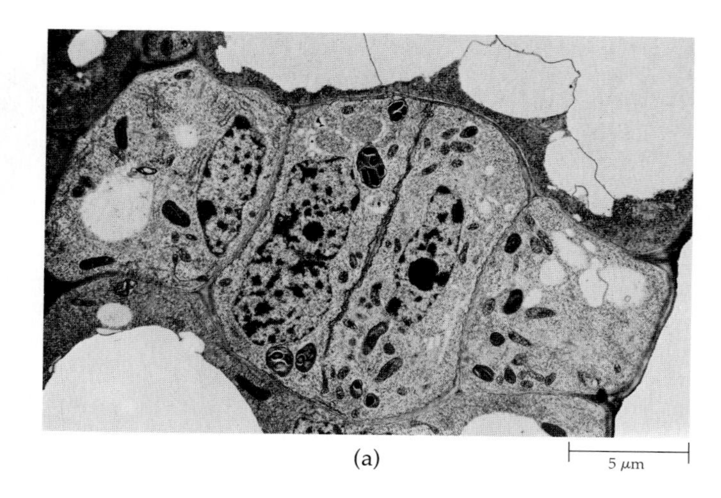

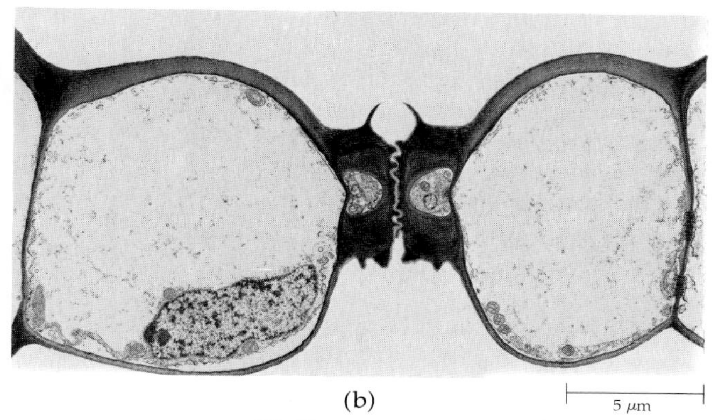

21–19
Electron micrographs of corn (Zea mays) *stomata.* (a) *Surface view of immature guard cells (center) and associated accessory, or subsidiary, cells (one on either side of the guard cells).* (b) *Transverse section through mature, thick-walled guard cells, each of which is attached to an accessory cell.*

Epidermis

The epidermis is the outermost layer of cells of the primary plant body, and it constitutes the dermal tissue system of leaves, floral parts, fruits, and seeds, and of stems and roots until they undergo considerable secondary growth. The epidermal cells are quite variable both functionally and structurally. In addition to the ordinary epidermal cells, which form the bulk of the epidermis, the epidermis may contain *stomata* (Figures 21–18 through 21–20), many types of appendages, or *trichomes* (Figure 21–21), and other kinds of cells specialized for specific functions.

In most plants, the epidermis is only one layer of cells in thickness. However, in some plants, divisions in the protoderm of the leaf are parallel with the surface (periclinal divisions) and an epidermis with several layers (a multiple epidermis) is formed (Figure 21–22). A multiple epidermis is found in the leaves of such familiar house plants as the rubber plant (*Ficus elastica*) and *Peperomia*. The multiple epidermis is believed to serve as a water-storage tissue.

The main mass of epidermal cells is closely knit and affords considerable mechanical protection to the plant part. The walls of the epidermal cells of the

21–20
(a) *Surface and* (b) *sectional views of developing stomata.* (c) *Diagram of a mature stoma, showing its relationship to the epidermis and underlying cells. The guard cells originate from unequal division of a protodermal cell. The smaller of the two cells is called a stoma mother cell, or guard-cell mother cell, and it is this cell that divides to give rise directly to the two guard cells* (a *and* b). *After the guard cells are formed, the intercellular substance in the median part of their common wall swells and then dissolves to form the pore. During this process the guard cells develop unevenly thickened walls. The walls next to the pores are generally thicker than those adjacent to other epidermal cells. While the stomata are being formed, they may be raised or lowered below the surface of the epidermis. Often there is a large air space, or substomatal chamber, just behind the stoma. Unlike ordinary epidermal cells, guard cells contain chloroplasts.*

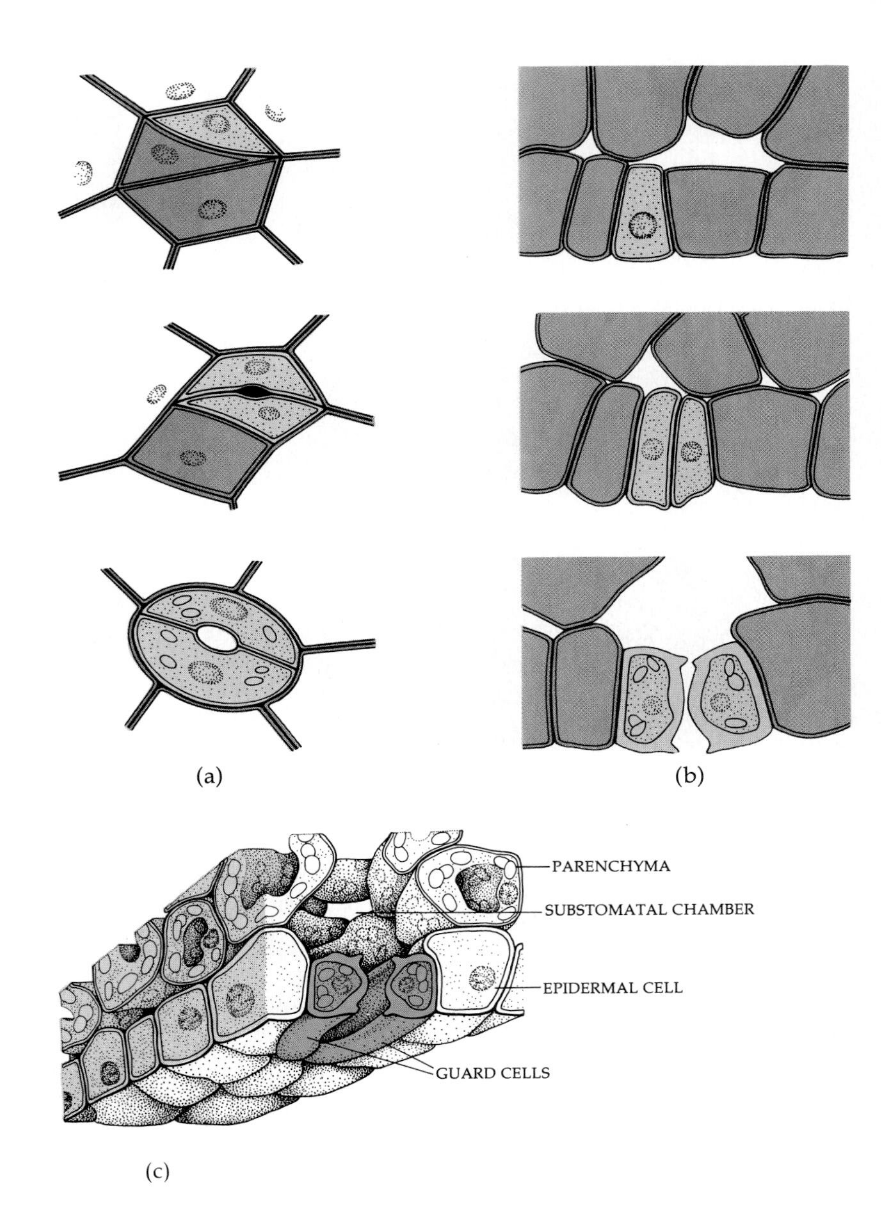

aerial parts contain cutin and are covered with a cuticle, which minimizes water loss. The cuticle may also be covered with wax, either in smooth sheets or as rods or filaments extending upward from the surface (Figure 21–18; see also Figure 2–9). It is this wax that is responsible for the whitish or bluish "bloom" on leaves.

Interspersed among the flat, tightly packed epidermal cells are specialized cells that contain chloroplasts. These are the *guard cells* (Figures 21–19 and 21–20), which regulate the small openings, or stomata (singular: stoma) in the aerial parts of the plant and, hence, control the movement of gases, including water vapor, into and out of those parts. (The mechanism of stomatal opening and closing is discussed in Chapter 28.) Although stomata occur on all aerial parts, they are most abundant on leaves. Stomata are often associated with epidermal cells that differ in shape from the ordinary epidermal cells. Such cells are termed accessory, or subsidiary, cells (Figure 21–19).

Periderm

In stems and roots having secondary growth, the epidermis is commonly replaced by a periderm. The periderm consists largely of protective *cork tissue*, which is nonliving and has walls that are heavily suberized at maturity; of the *cork cambium;* and of *phelloderm*, a living parenchyma tissue. The cork cambium forms cork tissue on its outer surface and phelloderm on its inner surface. The origin of the cork cambium is variable, depending on the species and plant part. The periderm will be considered in detail in Chapter 24.

21–21
Trichomes. (a) *Surface (above) and sectional (below) views of the peltate hair, or scale, of the olive* (Olea europaea) *leaf.* (b) *Dendroid (treelike) hair from the leaf of the sycamore, or plane tree* (Platanus orientalis). (c) *Water vesicle of the "ice plant"* (Mesembryanthemum crystallinum). (d) *Short, unbranched hair from tomato* (Lycopersicon esculentum) *stem.* (e) *Glandular hair from tomato stem.* (f) *Stinging hair of* Urtica. *The stinging hair consists of a long needlelike part and a broad base surrounded by other epidermal cells. When the hair is touched, the tip breaks off and poisonous cell contents (histamine and acetylcholine) are injected into the skin.*

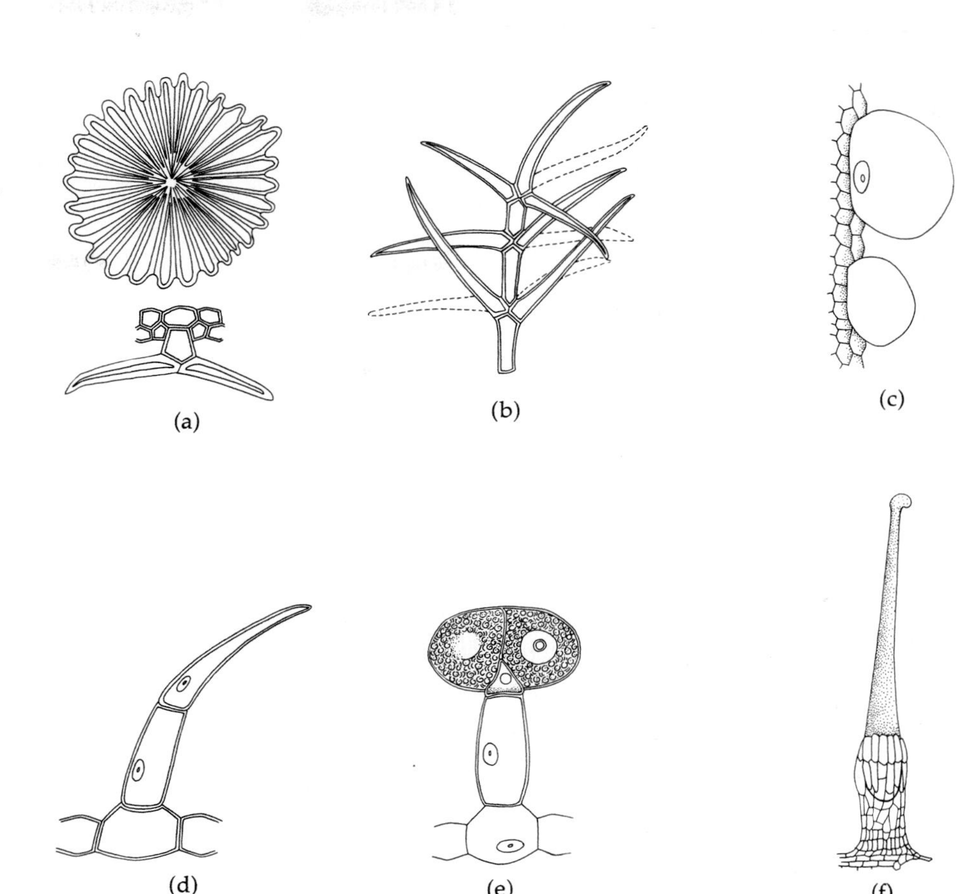

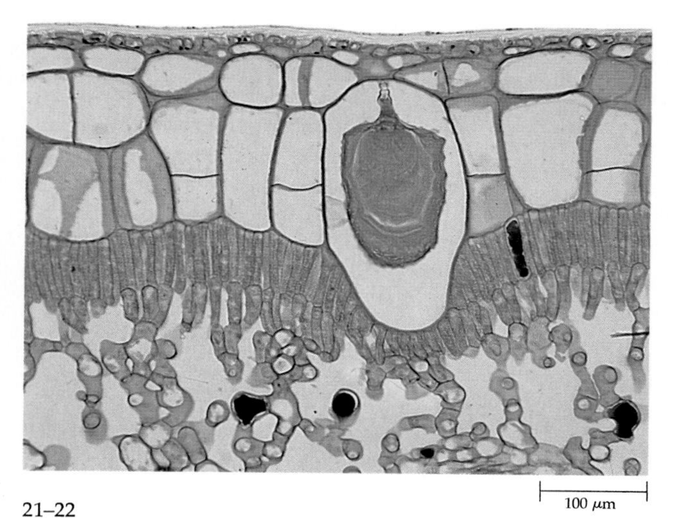

21–22
Transverse section showing upper portion of leaf blade of the rubber plant (Ficus elastica). *Notice the thick cuticle covering the multiple epidermis of mostly large cells. The club-shaped structure in the largest epidermal cell consists mostly of calcium carbonate deposited on a cellulose stalk. The cells below the large, clear epidermal cells are mesophyll cells.*

SUMMARY

A summary of plant tissues and their cell types is found in Table 21–1, and the characteristics, location, and function of these types of plant cells are given in Table 21–2.

Table 21–1 *Summary of Plant Tissues and Their Cell Types*

TISSUE	CELL TYPES
Epidermis	Generally parenchyma cells; guard cells and trichomes; sclerenchyma cells
Periderm	Generally parenchyma cells; sclerenchyma cells
Xylem	Tracheids; vessel members; sclerenchyma cells; parenchyma cells
Phloem	Sieve cells or sieve-tube members; albuminous cells or companion cells; parenchyma cells; sclerenchyma cells
Parenchyma	Parenchyma cells
Collenchyma	Collenchyma cells
Sclerenchyma	Fibers or sclereids

Table 21–2 *Summary of Cell Types*

CELL TYPE	CHARACTERISTICS	LOCATION	FUNCTION
Parenchyma	Shape: commonly polyhedral (many-sided); variable Cell wall: primary, or primary and secondary; may be lignified, suberized, or cutinized Living at maturity	Throughout the plant, as parenchyma tissue in cortex, as pith and pith rays, or in xylem and phloem	Such metabolic processes as respiration, digestion, and photosynthesis; storage and conduction; wound healing and regeneration
Collenchyma	Shape: elongated Cell wall: unevenly thickened; primary only—nonlignified Living at maturity	On the periphery (beneath the epidermis) in young elongating stems; often as a cylinder of tissue or only in patches; in ribs along veins in some leaves	Support in primary plant body
Fibers	Shape: generally very long Cell wall: primary and thick secondary—often lignified Often (not always) dead at maturity	Sometimes in cortex of stems, most often associated with xylem and phloem; in leaves of monocotyledons	Support
Sclereids	Shape: variable; generally shorter than fibers Cell wall: primary and thick secondary—generally lignified May be living or dead at maturity	Throughout the plant	Mechanical, protective
Tracheid	Shape: elongated and tapering Cell wall: primary and secondary; lignified; contains pits, but not perforations Dead at maturity	Xylem	Chief water-conducting element in gymnosperms and lower vascular plants; also found in angiosperms
Vessel member	Shape: elongated, generally not as long as tracheids Cell wall: primary and secondary; lignified; contains pits and perforations; several vessel members end-to-end constitute a vessel Dead at maturity	Xylem	Chief water-conducting element in angiosperms

CELL TYPE	CHARACTERISTICS	LOCATION	FUNCTION
Sieve cell	Shape: elongated and tapering Cell wall: primary in most species; with sieve areas; callose often associated with wall and pores Living at maturity; either lacks or contains remnants of a nucleus at maturity; lacks distinction between vacuole and cytoplasm	Phloem	Chief food-conducting element in gymnosperms and lower vascular plants
Albuminous cell	Shape: generally elongated Cell wall: primary Living at maturity; associated with sieve cell, but generally not derived from same mother cell as sieve cell; has numerous connections with sieve cell	Phloem	Believed to play a role in the movement of food into and out of the sieve cell
Sieve-tube member	Shape: elongated Cell wall: primary, with sieve areas; sieve areas on end wall with much larger pores than those on side walls; this wall part is termed a sieve plate; callose often associated with wall and pores Living at maturity; either lacks a nucleus at maturity, or contains only remnants of nucleus; contains a proteinaceous substance known as slime, or P-protein, in dicots and some monocots; several sieve-tube members in a vertical series constitute a sieve tube	Phloem	Chief food-conducting element in angiosperms
Companion cell	Shape: variable, generally elongated Cell wall: primary Living at maturity; closely associated with sieve-tube members; derived from same mother cell as sieve-tube member; has numerous connections with sieve-tube member	Phloem	Believed to play a role in the movement of food into and out of the sieve-tube member

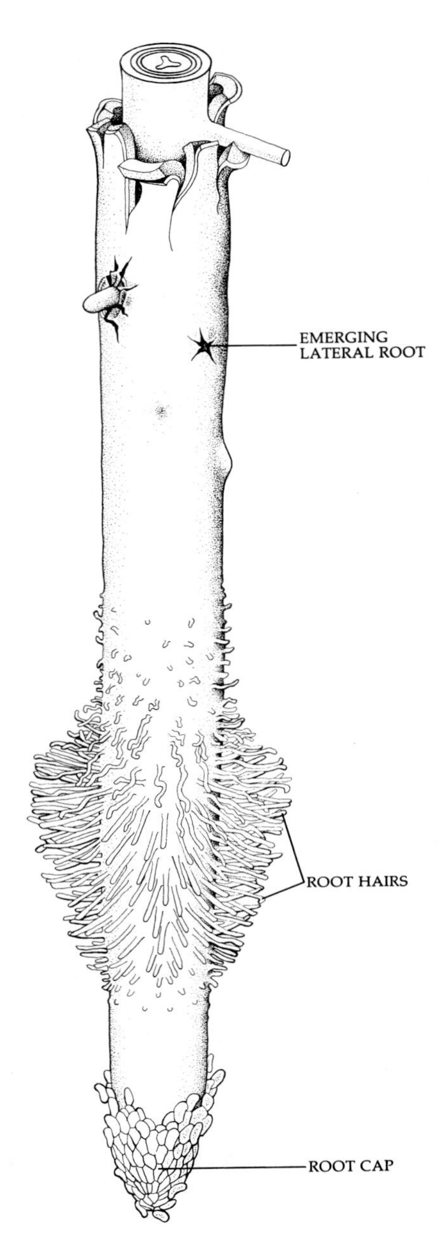

22–1
A portion of a dicot root, showing the spatial relationship between the root cap and region of root hairs, and (near the top) the sites of emergence of lateral roots, which arise from deep within the parent root. New root hairs arise just beyond the region of elongation at about the same rate as that at which the older hairs die off.

CHAPTER 22

The Root: Primary Structure and Development

In most vascular plants, the roots constitute the underground portion of the sporophyte and are involved primarily in anchorage and absorption (Figure 22–1). Two other functions associated with roots are storage and conduction. Most roots are important storage organs and some, such as those of the carrot, sugar beet, and sweet potato, are specifically adapted for the storage of food. Foods manufactured above ground, in photosynthesizing portions of the plant body, move through the phloem to the storage tissues of the root. This food may eventually be used by the root itself, but more often the stored food is digested and transported back through the phloem to the above-ground parts. In biennials (plants that complete their life cycle over a two-year period) such as the sugar beet and carrot, large food reserves accumulate in the storage regions of the root during the first year and then are used during the second year to produce flowers, fruits, and seeds. Water and minerals, or inorganic ions, absorbed by the roots move through the xylem to the aerial parts of the plant. In addition, hormones (particularly cytokinins and gibberellins) synthesized in meristematic regions of the roots are transported upward in the xylem to the aerial parts, which depend on the presence of such hormones to stimulate growth and development (Chapter 25).

ROOT SYSTEMS

The first root of the plant originates in the embryo and is usually called the *primary root.* In gymnosperms and dicotyledons, this root becomes a *taproot;* it grows directly downward, giving rise to branch roots, or *lateral roots,* along the way. The older lateral roots are found nearer the neck of the root (where the root and stem meet), and the younger ones are found nearer the root tip. This type of root system—that is, one that develops from a taproot and its branches—is called a *taproot system* (Figure 22–2a).

22–2
Two types of root systems. (a) *Taproot system of dandelion* (Taraxacum officinale). (b) *Fibrous root system of a grass.*

(a)

(b)

In monocotyledons, the primary root is usually short-lived, and the root system develops from adventitious roots that arise from the stem. These adventitious roots and their lateral roots give rise to a *fibrous root system*, in which no one root is more prominent than the others (Figure 22–2b). Taproot systems generally penetrate deeper into the soil than fibrous root systems. The shallowness of fibrous root systems and the tenacity with which they cling to soil particles make them especially well suited as ground cover for the prevention of soil erosion.

The extent of a root system—that is, the depth to which it penetrates the soil and spreads laterally—is dependent upon several factors, including the moisture, temperature, and composition of the soil. The bulk of most so-called "feeder roots" (roots actively engaged in the uptake of water and minerals) occurs in the upper meter of soil, and the majority of the feeder roots of most trees occur in the upper 15 centimeters of soil, the part of the soil normally richest in organic matter. Some trees, such as spruces, beeches, and poplars, rarely produce deep taproots, whereas others, such as oaks and many pines, commonly produce relatively deep taproots, making such trees rather difficult to transplant. The deepest known tree root was that of a pine growing on a highly porous sandy soil; it penetrated the soil about 6.5 meters. The lateral spread of tree roots is usually greater than the spread of the crown of the tree. The root systems of corn plants *(Zea mays)* often reach a depth of about 1.5 meters, with a lateral spread of about a meter on all sides of the plant. The roots of alfalfa *(Medicago sativa)* may extend to depths of up to 6 meters or more.

As a plant grows, it needs to maintain a balance between the total surface area available for the manufacture of food (the photosynthesizing surface) and the surface area available for the absorption of water and minerals. In a young plant, that is, one just becoming established, the total water- and mineral-absorbing surface usually far exceeds the photosynthesizing surface. However, this relationship tends to change in favor of the photosynthesizing surface as the plant ages, a fact that gardeners must consider. Even when plants are carefully transplanted, the balance between shoot and root is invariably disturbed. Most of the fine feeder roots are left behind when the plant is removed from the soil; cutting back the shoot helps to reestablish a balance between the root system and the shoot system.

One of the most detailed studies conducted on the extent of the root and shoot systems of any one plant was made on a 4-month-old rye plant *(Secale cereale)*. The total surface area of the root system, including root hairs, measured 639 square meters, or 130 times the surface area of the shoot system. What is even more amazing is the fact that the roots occupied only about 6 liters of soil.

ORIGIN AND GROWTH OF PRIMARY TISSUES

The growth of many roots is apparently a continuous process that stops only under such adverse conditions as drought and low temperatures. During their growth through the soil, roots follow the path of least resistance and frequently follow spaces left by earlier roots that have died and rotted.

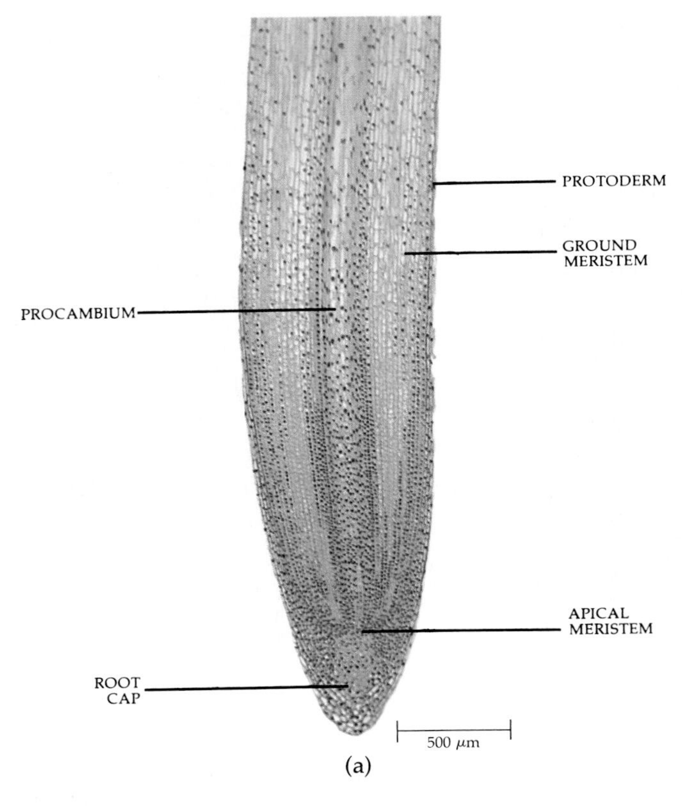

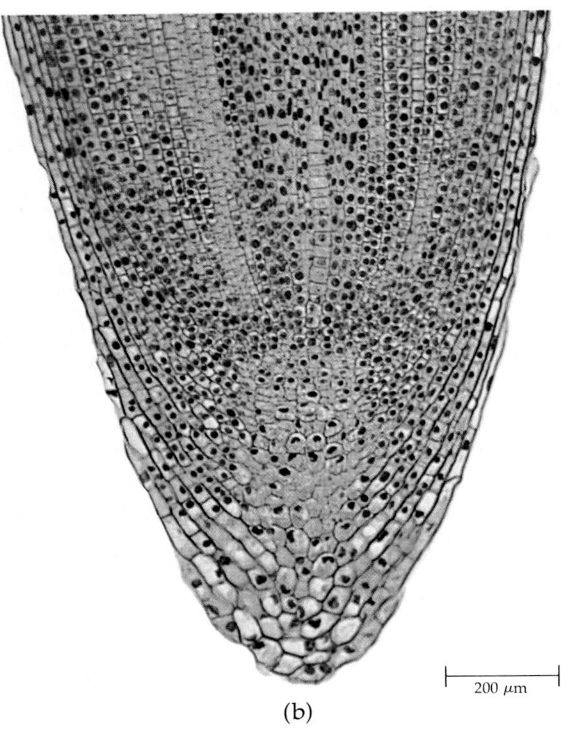

22–3
Longitudinal sections of onion (Allium cepa) *root tip.* (a) *Primary meristems can be distinguished close to the apical meristem.* (b) *Detail of the apical meristem. Compare the organization of this apical meristem with that of the corn root tip shown in Figure 22–4.*

The tip of the root is covered by a *root cap* (Figures 22–1, 22–3, and 22–4), a thimblelike mass of cells that protects the apical meristem behind it and aids the root in its penetration of the soil. As the root grows longer and the root cap is pushed forward, the cells on the periphery of the root cap are sloughed off. These sloughed-off cells form a slimy covering around the root and lubricate its passage through the soil. As quickly as root cap cells are sloughed off, new ones are added by the apical meristem. The longevity (from time of initiation to sloughing off) of root cap cells ranges from four to nine days, depending upon the length of the cap and the species.

The slimy substance of the root cap is a highly hydrated polysaccharide, probably a pectic substance, which is secreted by the outer root cap cells. The slimy substance accumulates in Golgi vesicles, which fuse with the plasma membrane and then release the slime into the wall. Eventually the slime passes to the outside, where it forms droplets.

In addition to protecting the apical meristem and aiding the root in its penetration of the soil, the root cap plays an important role in controlling the response of the root to gravity (geotropism, or gravitropism; see Chapter 26).

Growth Regions of the Root

Apart from the root cap, the most striking structural feature of the root apex is the arrangement of longitudinal files of cells that emanate from the apical meristem. The apical meristem is composed of relatively small (10 to 20 micrometers in diameter), many-sided cells—the initials and their immediate derivatives (Chapter 21)—with dense cytoplasm and large nuclei (Figures 22–3 and 22–4). The organization and number of initials in the apical meristems of roots are both quite variable.

Two main types of apical organization are found in the roots of seed plants. In one, the root cap, the vascular cylinder of xylem and phloem, and the cortex emerge from a common group of cells in the apical meristem (Figure 22–3). In the other, each region can be traced to an independent layer of cells (Figure 22–4). In the second type of organization, the epidermis has a common origin with either the root cap or the cortex.

Although the region of initials in the apical meristem of the root was once considered to be a region of active cell division, studies on the apical meristems of many roots indicate that this region is relatively inactive and that most cell division occurs a short distance beyond the relatively quiescent initials. The relatively inactive region of the apical meristem is known as the *quiescent center* (Figure 22–5).

As indicated by the term "relatively," this quiescent center is not totally devoid of divisions under

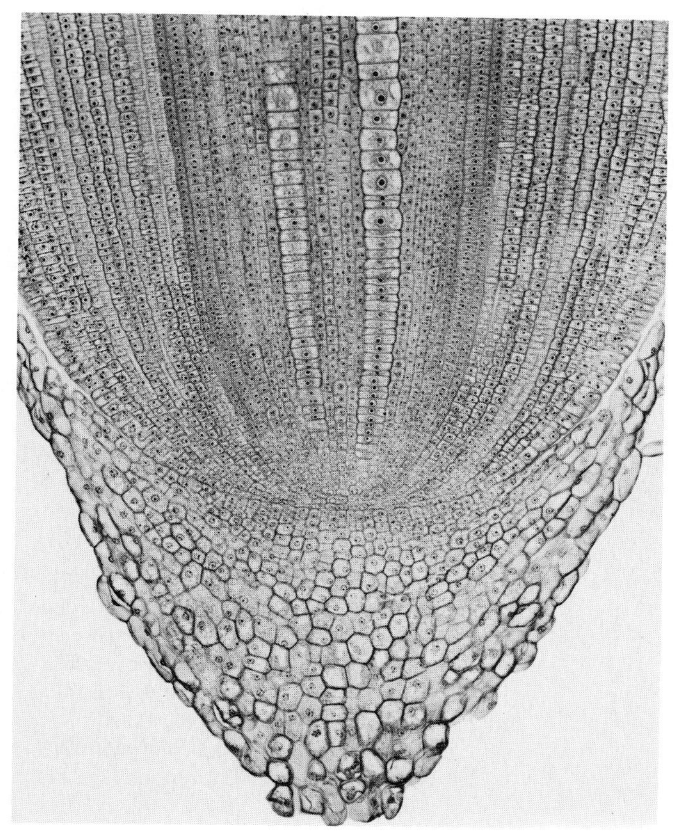

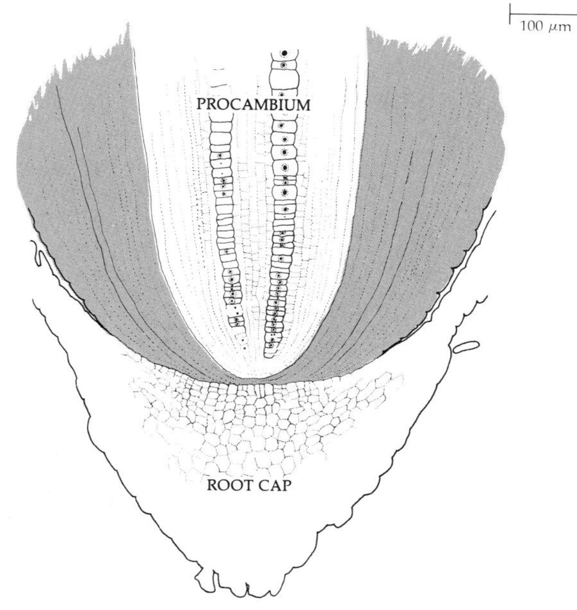

22–4
Apical meristem of a corn (Zea mays) *root tip. Notice the three distinct layers of initials. The lower layer gives rise to the root cap, below; the middle layer to the ground meristem or cortex; and the upper to the procambium or vascular cylinder.*

22–5
Apical meristem of a corn (Zea mays) *root tip, showing the quiescent center. To prepare this autoradiograph, the root tip was supplied for one day with thymine labeled with radioactive tritium* (^3H). *In the rapidly dividing cells around the quiescent center (within the dashed oval), the radioactive material was quickly incorporated into the nuclear DNA and left its marks (the dark grains) on this autoradiograph.*

normal conditions. Moreover, the quiescent center is able to repopulate the bordering meristematic regions when they are injured. In a recent study, for example, the isolated quiescent centers of corn *(Zea mays)* grown in sterile culture were found to be able to form whole roots without first forming callus, or wound, tissue. In another study of corn roots, a striking correlation was found between the size of the quiescent center and the complexity of the primary vascular pattern of the root. These and other studies indicate that the quiescent center plays an essential role in root development.

The distance beyond the apical meristem at which most cell division takes place varies from species to species and also within the same species, depending on the age of the root. The combination of the apical meristem and the nearby portion of root in which cell division does occur is called the *region of cell division* (Figure 22–6).

Behind the region of cell division, but not sharply delimited from it, is the *region of elongation*, which usually measures only a few millimeters in length (Figure 22–6). The elongation of cells in this region results in most of the increase in length of the root.

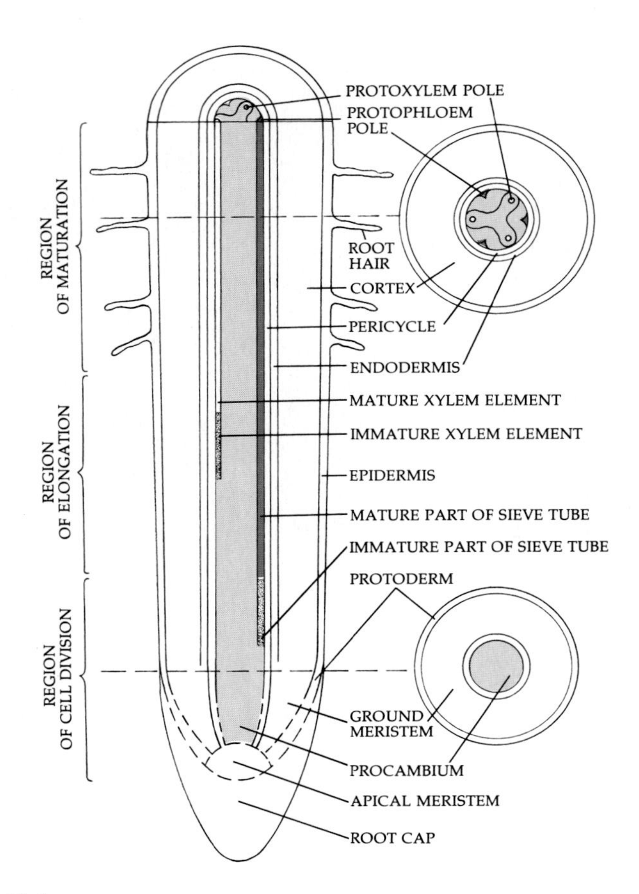

22–6
A diagram illustrating early stages in the primary development of a root tip. (Compare this figure with Figure 22–1).

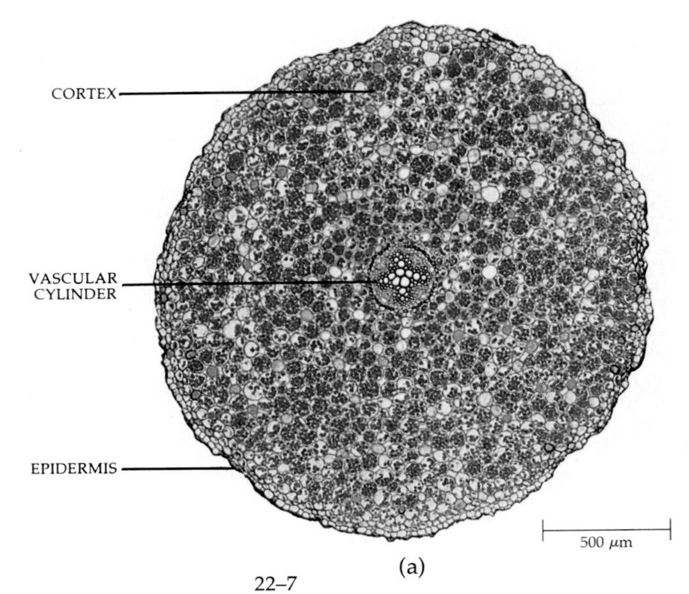

22–7
Transverse sections of the root of a buttercup (Ranunculus). (a) *Overall view of mature root.* (b) *Detail of immature vascular cylinder. Notice the intercellular spaces among the cortical cells.* (c) *Detail of mature vascular cylinder. Note the numerous starch grains in the cortical cells.*

Beyond this region the root does not increase in length. Thus, growth in length of the root occurs near the root tip and results in a very limited portion of the root constantly being pushed through the soil.

The region of elongation is followed by the *region of maturation*, in which most of the cells of the primary tissues mature (Figure 22–6). Root hairs are also produced in this region, and sometimes this part of the root is called the root-hair zone (see Figure 22–1).

It is important to note that there is a gradual transition from one region of the root to another. The regions are not sharply delimited from one another. Some cells begin to elongate and differentiate in the region of cell division, whereas others reach maturity in the region of elongation. For example, the first-formed elements of the phloem and xylem mature in the region of elongation and are often stretched and destroyed during elongation of the root.

The protoderm, procambium, and ground meristem can be distinguished in very close proximity to the apical meristem (Figures 22–3 and 22–6). These are the primary meristems that differentiate into the epidermis, the primary vascular tissues, and the cortex, respectively (Chapter 20).

22–8
Transverse sections of a corn (Zea mays) *root.* (a) *Overall view of mature root. Part of a lateral root can be seen at the lower right. The vascular cylinder, with its pith, is quite distinct.* (b) *Detail of immature vascular cylinder.* (c) *Detail of mature vascular cylinder.*

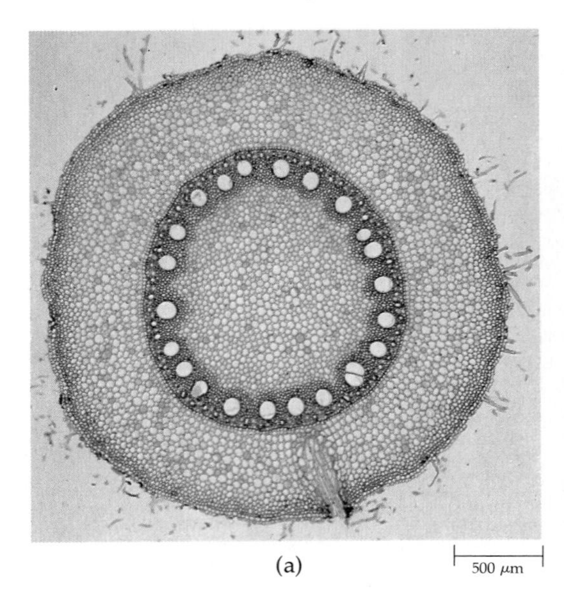

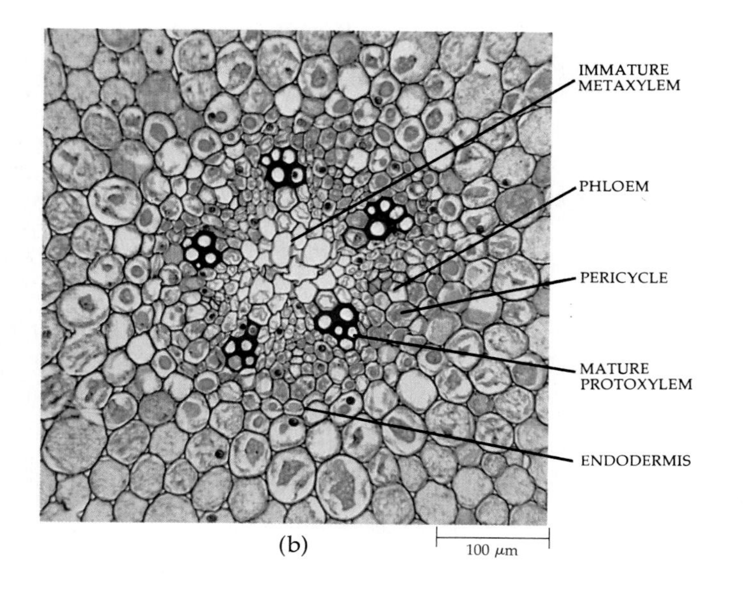

(b)

ENDODERMIS
PERICYCLE
PRIMARY PHLOEM
PRIMARY XYLEM
100 μm

(c)

PRIMARY STRUCTURE

Compared with that of the stem, the internal structure of the root is usually relatively simple. This is due in large part to the absence of leaves in the root and the corresponding absence of nodes and internodes. Thus, the arrangement of tissues shows very little difference from one level of the root to another.

The three tissue systems of the root in the primary stage of growth can be readily distinguished in both transverse (Figures 22–7 and 22–8) and longitudinal sections (Figure 22–3). The epidermis (dermal tissue system), the cortex (ground tissue system), and the vascular tissues (vascular tissue system) can be seen to be clearly distinguished from one another. In most roots, the vascular tissues form a solid cylinder (Figure 22–7), but in some they form a hollow cylinder around a pith (Figure 22–8).

Epidermis

The epidermis in young roots absorbs water and minerals, and this function is facilitated by *root hairs*—tubular extensions of the epidermal cells—which greatly increase the absorbing surface of the root (Figure 22–9). (See pages 237–39 in Chapter 12 for a discussion of the role of mycorrhizae in absorption.) As part of the previously mentioned study of a four-month-old rye plant, it was estimated that the plant contained approximately 14 billion root hairs, with an absorbing surface of 401 square meters. Placed end to end, these root hairs would extend well over 10,000 kilometers.

Root hairs are relatively short-lived and are confined largely to the region of maturation of the root. The production of new root hairs occurs just beyond the region of elongation (Figures 22–1 and 22–6) and

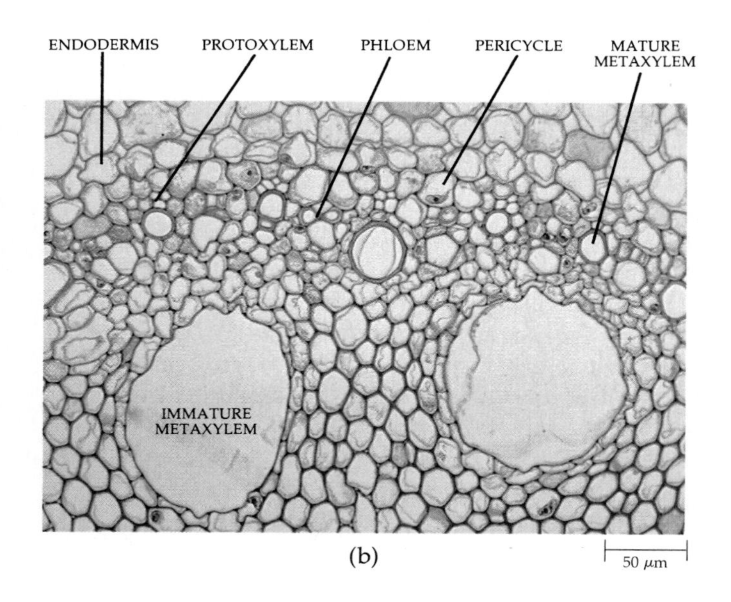

(b)

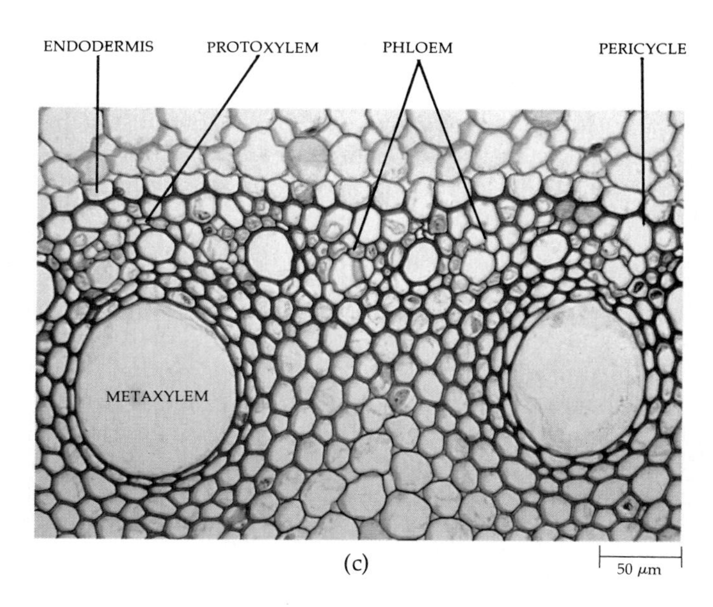

(c)

at about the same rate as that at which the older root hairs are dying off at the upper end of the root hair zone. As the tip of the root penetrates the soil, new root hairs develop immediately behind it, providing the root with surfaces capable of absorbing new supplies of water and minerals, or inorganic ions. (See Chapter 28 for a discussion of absorption of water and inorganic ions by roots.) Obviously, it is the new and growing roots—the feeder roots—that are primarily involved in the absorption of water and minerals. For this reason, great care must be taken by gardeners to remove as much soil as possible along with the root system when transplanting a plant or shrub. If the plant is simply "torn" from the soil, most of the feeder roots will be left behind and the plant will probably not survive.

The epidermal cells of the root, including those with root hairs, are parenchyma cells, which are closely packed together. In young portions of the root, the epidermis has at most a thin cuticle; hence, the cell wall offers little resistance to the passage of water and minerals into the root. In addition, the surface of many roots is covered by a slime-sheath, or *mucigel*, which enables the roots to establish a much more intimate contact with the particles of soil. The exact origin of mucigel has not yet been determined, but it has been suggested that it is formed, at least in part, by the root cap. The mucigel has been shown to provide an environment favorable to beneficial bacteria. It may also influence the availability of ions to the root, as well as providing the root with short-term protection from drying out (desiccation).

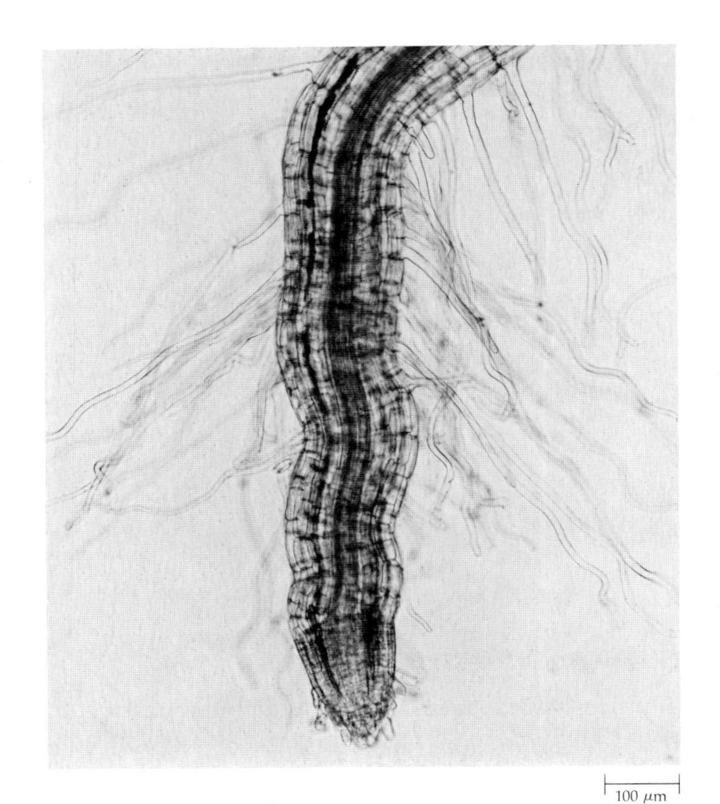

22–9
Root of a bent grass (Agrostis tenuis) *seedling, showing root hairs. Root hairs may be as much as 1.3 centimeters long and may attain their full size within hours. Each hair is comparatively short-lived, but the formation of new root hairs and the death of old ones continues as long as the root is growing.*

Cortex

As seen in transverse section (Figure 22–7a), the cortex occupies by far the greatest area of the primary body of most roots. The cells of the cortex store starch and other substances but usually lack chloroplasts. Roots that undergo considerable amounts of secondary growth—those of gymnosperms and most dicots—shed their cortex early. In such roots, the cortical cells remain parenchymatous. In monocots, by contrast, the cortex is retained for the life of the root, and many of the cortical cells develop secondary walls that become lignified.

Regardless of the degree of differentiation, the cortical tissue contains numerous intercellular spaces—air spaces essential for aeration of the cells of the root (see Figures 22–7 and 22–8). The cortical cells have numerous contacts with one another, and their protoplasts are connected by plasmodesmata. Thus substances moving across the cortex may move from cell to cell by way of the protoplasts and plasmodesmata or by way of the cell walls.

Unlike the rest of the cortex, the innermost layer is compactly arranged and lacks air spaces. This layer, the *endodermis* (Figures 22–7 and 22–8), is characterized by the presence of *Casparian strips* in its anticlinal walls (the walls perpendicular to the surface of the root). The Casparian strip is a bandlike portion of the primary wall that is impregnated with a fatty substance called suberin and is sometimes lignified. The protoplasts of endodermal cells are quite firmly attached to the Casparian strips (Figures 22–10 and 22–11). Inasmuch as the endodermis is compact and the Casparian strips are impermeable to water, all substances entering and leaving the vascular cylinder must enter the protoplasts of the endodermal cells,

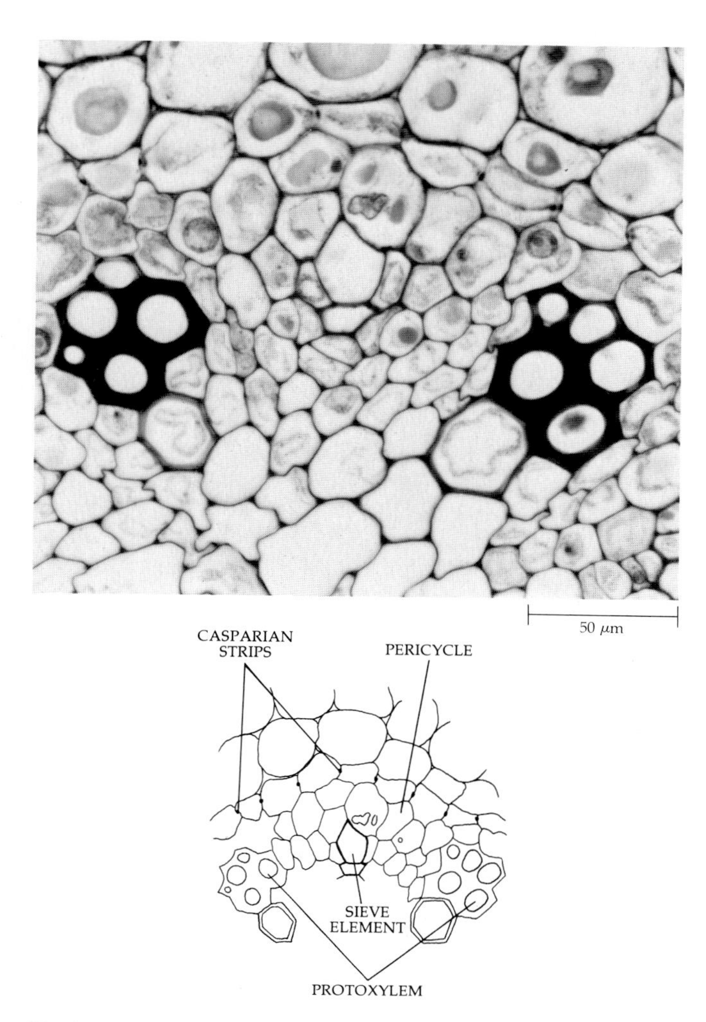

22–10
High-magnification view of a portion of an immature buttercup (Ranunculus) *root, showing Casparian strips in the endodermal cells. Notice that the plasmolyzed protoplasts of the endodermal cells cling to the strips.*

either by crossing their plasma membranes or by way of the numerous plasmodesmata connecting them with the protoplasts of neighboring cells of the cortex and vascular cylinder.

The effectiveness of the Casparian strip as a barrier to the movement of substances across the walls of the endodermis has been demonstrated in corn *(Zea mays)* roots that had absorbed the element lanthanum, a positively charged ion that cannot penetrate cell membranes. When the roots were examined with the electron microscope, the lanthanum was found only in the cell walls of the cortex; its progress across the root was abruptly and completely stopped by the Casparian strip.

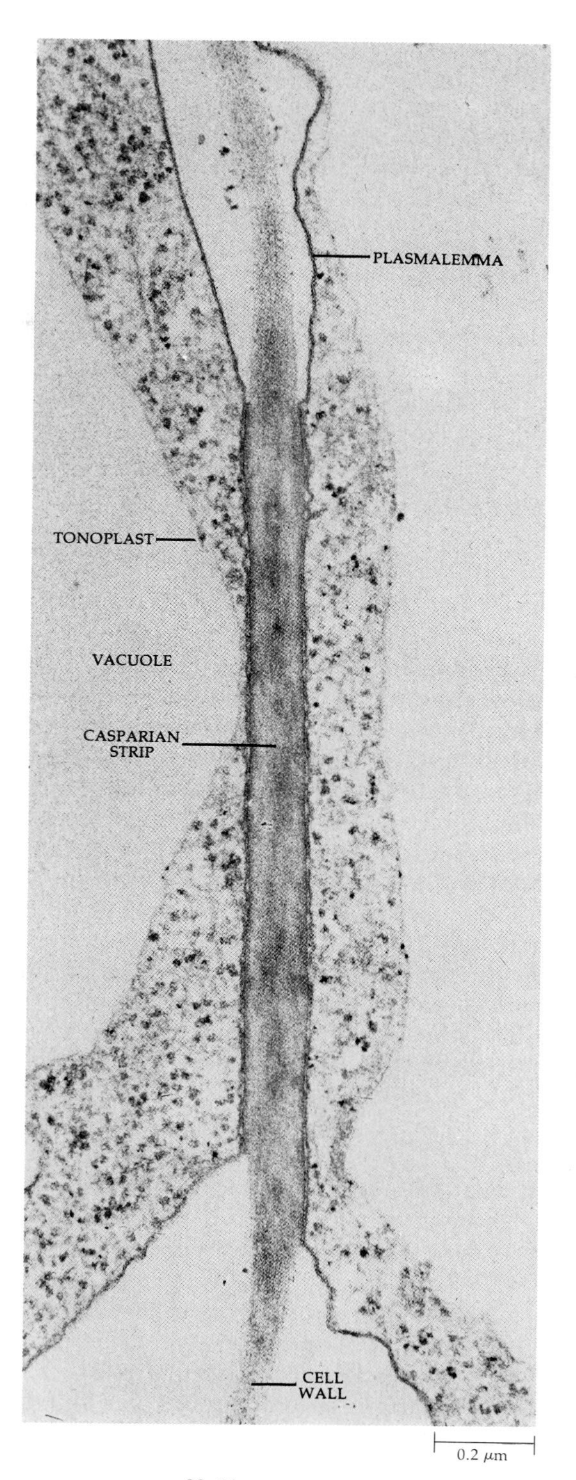

22–11
Transverse section of a radial wall between two endodermal cells. The region of the Casparian strip is more intensely stained than the rest of the wall. Notice how firmly the plasma membrane adheres to the wall in the region of the Casparian strip in each of these plasmolyzed cells.

22–12
Three-dimensional diagrams showing three developmental stages of an endodermal cell from a root that remains in a primary state. (a) *Initially, the endodermal cell is characterized by the presence of a Casparian strip in its anticlinal walls.* (b) *Then a layer of suberin is deposited internally over all wall surfaces.* (c) *Finally, the layer of suberin is covered by a thick, often lignified, secondary wall.*

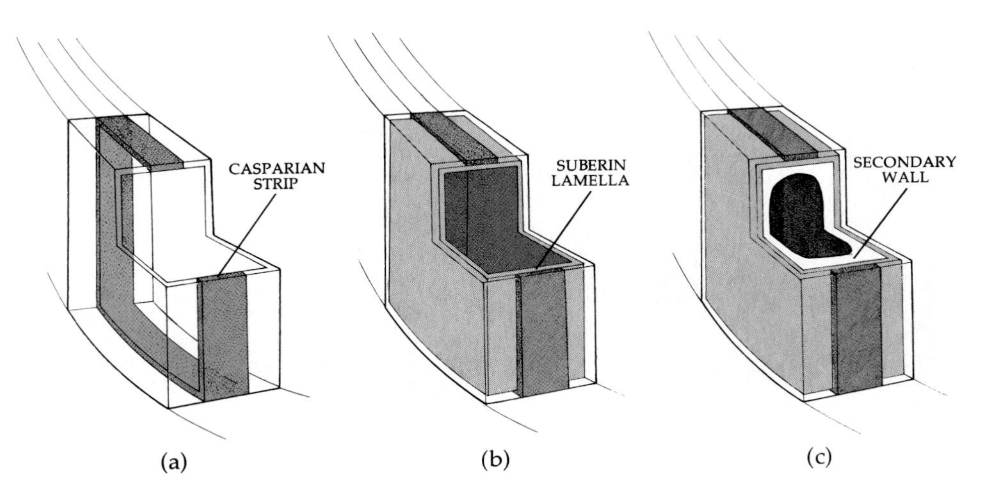

As mentioned previously, in roots that undergo secondary growth, the cortex and its endodermis are shed early. In roots in which the cortex is retained, a layer of suberin is eventually deposited internally over all wall surfaces of the endodermis (Figure 22–12). This is followed by the deposition of secondary walls, which may become lignified. These changes in the endodermis begin opposite the phloem strands and spread toward the protoxylem (Figure 22–7a). Opposite the protoxylem, some of the endodermal cells may never undergo such modifications but may remain thin-walled and retain their Casparian strips. Such cells are called *passage cells*, and in most species, they eventually become suberized.

Vascular Cylinder

The vascular cylinder of the root consists of vascular tissues and one or more layers of cells, the *pericycle*, which completely surrounds the vascular tissues (Figures 22–7 and 22–8). In the young root, the pericycle is composed of parenchyma cells with primary walls, but as the root ages, the cells of the pericycle may develop secondary walls (Figure 22–8).

The pericycle plays several important roles. In most seed plants, lateral roots arise in the pericycle. In plants undergoing secondary growth, the pericycle contributes to the vascular cambium and generally gives rise to the first cork cambium. Pericycle often proliferates, that is, gives rise to more pericycle.

The center of the vascular cylinder of most roots is occupied by a solid core of primary xylem from which ridgelike projections extend toward the pericycle (Figure 22–7). Nestled between the ridges of xylem are strands of primary phloem. Obviously, the vascular cylinder of the root is a protostele.

The number of ridges of primary xylem varies from species to species, sometimes even varying along the axis of a given root. If two ridges are present, the root is said to be *diarch*; if three are present, *triarch*; four, *tetrarch* (Figure 22–7); and if many are present, *polyarch* (Figure 22–8). The first *(proto-)* xylem elements to mature in roots are located next to the pericycle, and the tips of the ridges are commonly referred to as *protoxylem poles* (Figures 22–7 and 22–8). The metaxylem *(meta-,* meaning "after") occupies the inner portions of the ridges and the center of the vascular cylinder and matures after the protoxylem. The roots of some monocotyledons (for example, corn) have a pith (Figure 22–8), which some botanists interpret as potential vascular tissue.

ORIGIN OF LATERAL ROOTS

In most seed plants, lateral roots (branch roots) arise in the pericycle. Because lateral roots originate deep from within the parent root, they are said to be endogenous, meaning "originating within an organ" (Figures 22–1 and 22–13).

Divisions in the pericycle that initiate lateral roots occur some distance beyond the region of elongation in partially or fully differentiated root tissues. In angiosperm roots, derivatives of both the pericycle and endodermis commonly contribute to the new root primordium, although in many cases the derivatives of the endodermis are short-lived. As the young lateral root, or *root primordium*, increases in size, it pushes its way through the cortex (Figure 22–13c), possibly secreting enzymes that digest some of the cortical cells lying in its path. While still very young, the root primordium develops a root cap and apical

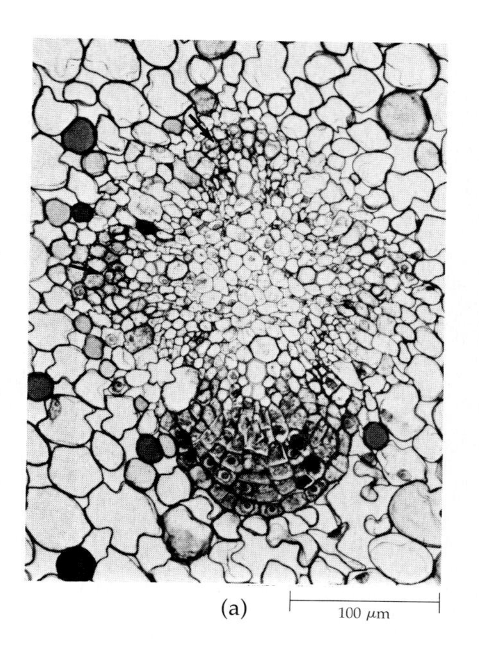

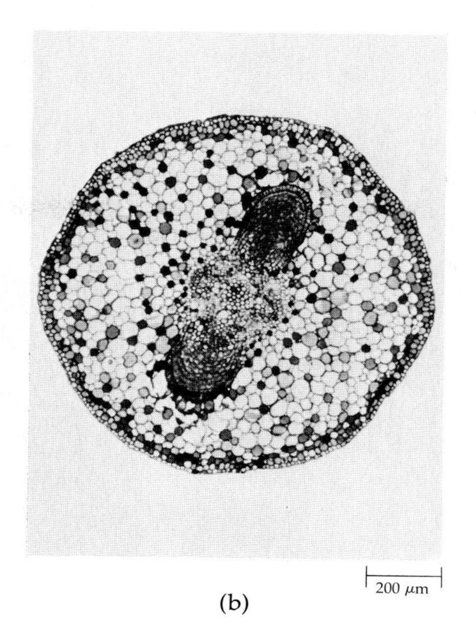

(c) 500 μm

22–13
Three stages in the origin of branch roots in a willow (Salix). (a) *One root primordium is present (below) and two others are being initiated in the region of the pericycle (see arrows). The vascular cylinder is still very young.* (b) *Two root primordia penetrating the cortex.* (c) *One lateral root has reached the outside, and the other nearly has.*

meristem, and the primary meristems appear. Initially, the vascular cylinders of lateral root and parent root are not connected to one another. The two vascular cylinders are joined later, when derivatives of intervening parenchyma cells differentiate into xylem and phloem.

AERIAL ROOTS

Aerial roots are adventitious roots produced from above-ground structures. The aerial roots of some plants serve as prop roots for support, as in corn (Figure 22–14). When they come in contact with the soil, they branch and function also in the absorption of water and minerals. Prop roots are produced from the stems and branches of many tropical trees, such as the red mangrove *(Rhizophora mangle)*, the banyan tree *(Ficus benghalensis)*, and some of the palms. Other aerial roots, as in the English ivy *(Hedera helix)*, cling to the surface of objects such as walls and provide support for the climbing stem.

Roots require oxygen for respiration, which is why most plants cannot live in soil in which there is not adequate drainage and which consequently lacks air spaces. Some trees that grow in swampy habitats develop roots that grow out of the water and serve not only to anchor but to aerate the plant. For example, the root system of the black mangrove *(Avicennia germinans)* develops negatively geotropic extensions called pneumatophores (air roots), which grow upward out of the mud and so provide adequate aeration (Figure 22–15). The "knees" of the bald cypress *(Taxodium distichum*; Figure 18–26, page 352) have had a similar function attributed to them, but this idea is now in doubt.

22–14
Prop roots of corn (Zea mays), *a type of adventitious root.*

22–15
Pneumatophores (air roots) of the black mangrove (Avicennia germinans) *protruding from the mud near the base of a tree (not shown).*

(a)

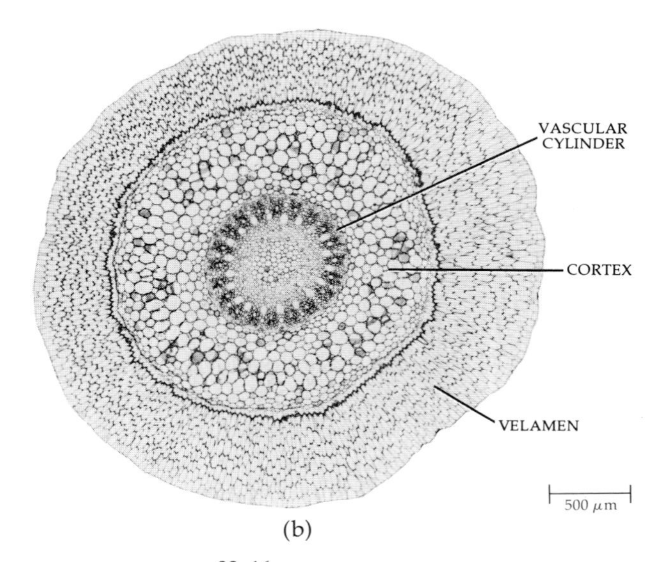

(b)

22–16
(a) *Aerial roots of an orchid* (Cattleya).
(b) *Transverse section of an orchid root, showing the multiple epidermis, or velamen.*

Special Adaptations

Many special adaptations of roots are found among epiphytes—plants that grow on other plants but are not parasitic on them. The epidermis of the orchid root, for example, is several layers thick (Figure 22–16) and, in some species, is the only photosynthetic organ of the plant. This multiple epidermis, called velamen, provides mechanical protection for the cortex, as well as reducing water loss. The velamen also may function in the absorption of water.

Among epiphytes, *Dischidia rafflesiana,* the "flower pot plant," has a very unusual modification. Some of its leaves are flattened, succulent structures, but others form hollow containers—the "flower pots"—that collect debris and rainwater (Figure 22–17). Ant colonies live in the pots and add to the nitrogen supply of the plant. Roots, formed at the node above the modified leaf, grow downward and into the pot, from which they absorb water and minerals.

ADAPTATIONS FOR FOOD STORAGE

Most roots are storage organs, and in some plants the roots are specialized for this function. Such roots are fleshy because of an abundance of storage parenchyma, which is permeated by vascular tissue. The development of some storage roots, such as that of the carrot *(Daucus carota),* is essentially similar to that of nonfleshy roots, except for a predominance of parenchyma cells in their secondary xylem and phloem.

The root of the sweet potato *(Ipomoea batatas)* develops in a manner similar to the carrot; however, in the sweet potato, additional vascular cambium cells develop within the secondary xylem around individual vessels or groups of vessels (Figure 22–18). These additional cambia (plural of cambium), while producing a few tracheary elements toward the vessels and a few sieve tubes away from them, mainly produce storage parenchyma cells in both directions. In the sugar beet *(Beta vulgaris)*, for example, most of the increase in thickness of the root results from the development of extra cambia (supernumerary cambia) around the original vascular cambium (Figure 22–19). These concentric layers of cambia, which superficially resemble growth rings in woody roots and stems, produce parenchyma-dominated xylem and phloem toward the inside and outside, respectively. The upper portion of most fleshy roots actually develops from the hypocotyl.

(a)

(b)

22–17
The epiphyte Dischidia rafflesiana, *or "flower pot plant."* (a) *A modified leaf, or "pot," which collects debris and rainwater.* (b) *Modified leaf cut open to show roots that have grown down into it.*

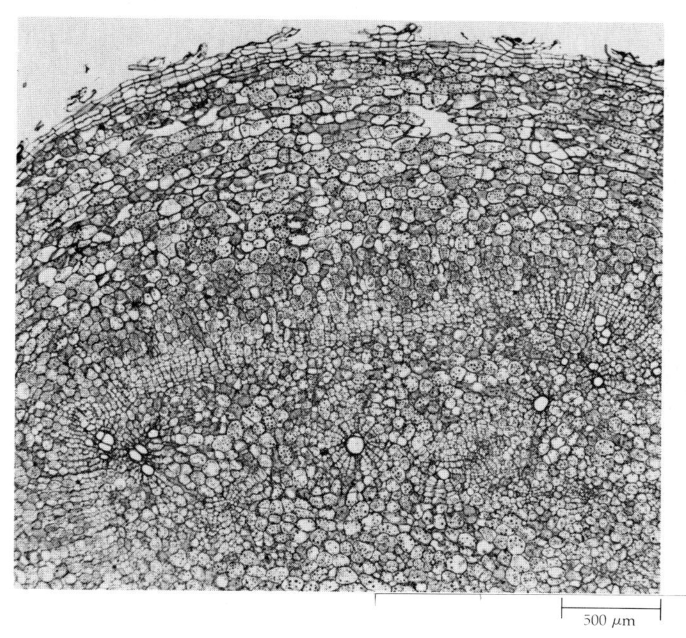

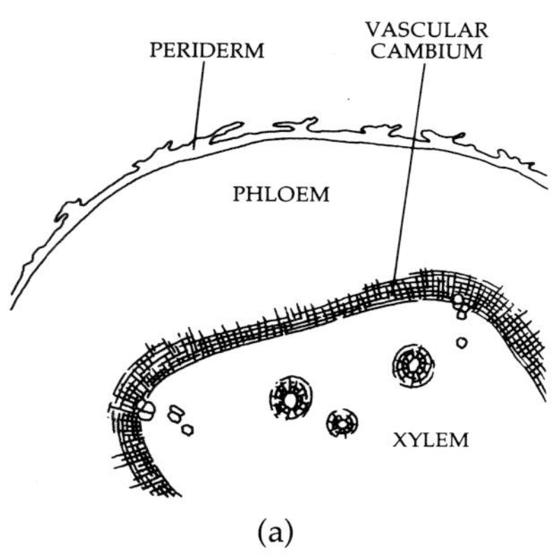

(a)

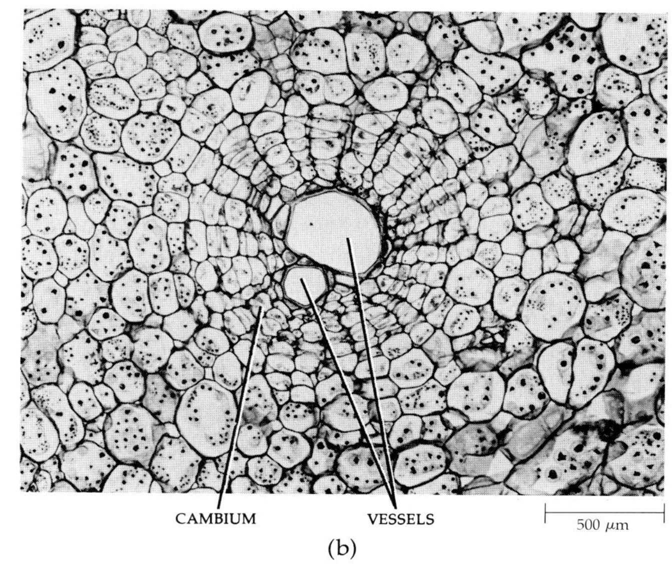

(b)

22–18
Transverse sections of the root of a sweet potato (Ipomoea batatas). (a) *Overall view.* (b) *Detail of xylem, showing cambium around vessels.*

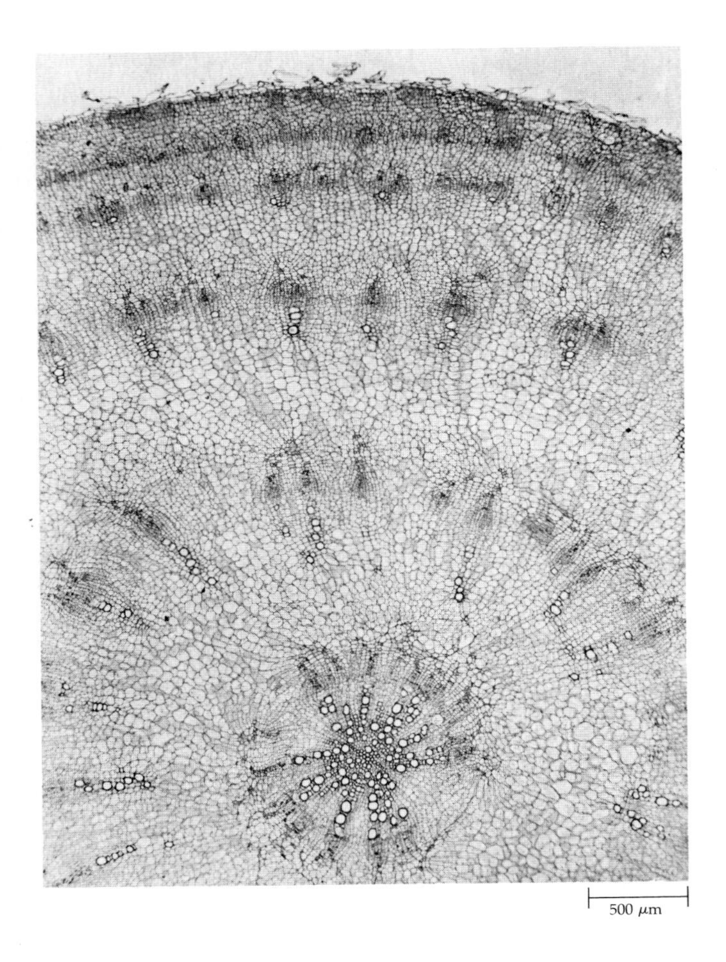

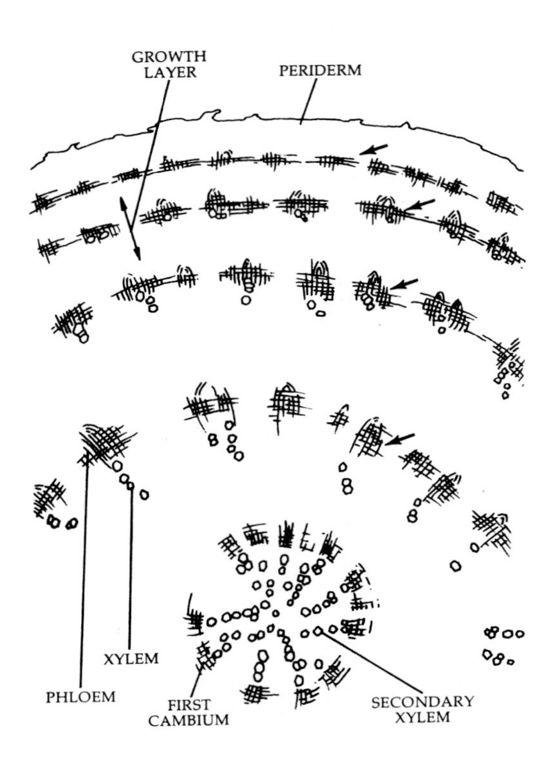

22–19
Transverse section of a sugar beet (Beta vulgaris) *root, with supernumerary cambia indicated by arrows.*

SUMMARY

Roots are organs specialized for anchorage, absorption, storage, and conduction. Gymnosperms and dicots commonly produce taproot systems, whereas monocots usually produce fibrous root systems. The extent of the root system is dependent upon several factors, but the bulk of most feeder roots is found in the upper meter of the soil.

The apical meristems of most roots contain a quiescent center; most meristematic activity, or cell division, occurs a short distance from the apical initials. During primary growth, the apical meristem gives rise to the three primary meristems—protoderm, ground meristem, and procambium—which differentiate into epidermis, cortex, and vascular cylinder, respectively. In addition, the apical meristem produces the root cap, which serves to protect the meristem and aid the root in its penetration of the soil.

Many epidermal cells of the root develop root hairs, which greatly increase the absorbing surface of the root. With the exception of the endodermis, the cortex contains numerous intercellular spaces. The compactly arranged endodermal cells contain Casparian strips on their anticlinal walls. Consequently, all substances moving between the cortex and vascular cylinder must pass through the protoplasts of the endodermal cells.

The vascular cylinder consists of pericycle and the primary vascular tissues, which are completely surrounded by the pericycle. The primary xylem usually occupies the center of the vascular cylinder and has radiating ridges that alternate with strands of primary phloem. Branch roots originate in the pericycle and push their way to the outside through the cortex and epidermis.

CHAPTER 23

The Shoot: Primary Structure and Development

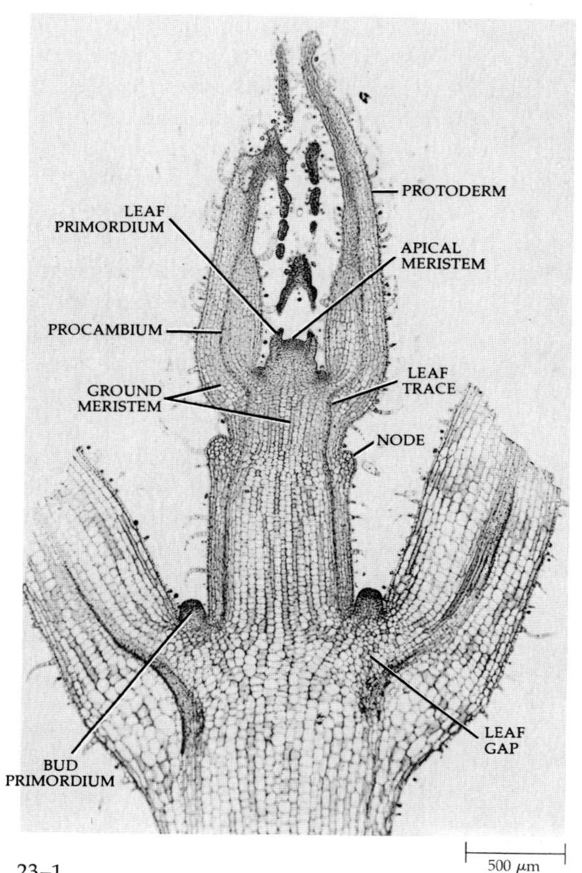

23–1
Longitudinal section of shoot tip of the common houseplant Coleus blumei, *a dicot. The leaves in* Coleus *are arranged opposite one another at the nodes, each successive pair at right angles to the previous pair; thus the leaves of the labeled node are at right angles to the plane of section.*

The shoot, which consists of the stem and its leaves, is initiated during development of the embryo, where it may be represented by a plumule, consisting of a stem (the epicotyl), one or more *leaf primordia* (rudimentary leaves), and an apical meristem. (The plumule can be thought of as the first bud of the plant.) With resumption of growth of the embryo during germination of the seed, new leaves develop from the apical meristem, and the stem elongates and differentiates into nodes and internodes. Gradually *bud primordia* form in the axils of the leaves (Figure 23–1), and eventually they follow a sequence of growth and differentiation more or less similar to that of the first bud. This pattern is repeated many times as the shoot system of the plant is produced.

Commonly, the growing terminal bud of a shoot inhibits development of lateral buds, a phenomenon known as *apical dominance* (see Chapter 25). As the distance between shoot tip and lateral buds increases, the retarding influence of the terminal bud is lessened and the lateral buds can proceed with their development. (This is why pinching off the shoot tips, a common practice of home gardeners, results in fuller and bushier plants.)

The two principal functions associated with stems are conduction and support. Substances manufactured in the leaves are transported through the stems by way of the phloem to sites of utilization of those substances, including growing leaves, stems, and roots, as well as developing flowers, seeds, and fruits. Much of the food material is stored in parenchyma cells of roots, seeds, and fruits, but stems are also important storage organs, and some, such as the underground stems of the white potato, are specifically adapted for storage. The principal photosynthetic organs of the plant—the leaves—are supported by stems, which act to place the leaves in favorable positions for exposure to light. In addition, most of the plant's loss of water vapor occurs through the leaves (Chapter 28). Water is conducted upward in the xylem from the roots and into the leaves via the stem.

ORIGIN AND GROWTH OF THE PRIMARY TISSUES OF THE STEM

The organization of the apical meristem of the shoot is more complex than that of the root. In addition to adding cells to the primary plant body, the apical meristem of the shoot is involved in the formation of leaf primordia, as well as the bud primordia (Figure 23–1) that develop into lateral branches. The apical meristem of the shoot also differs in its lack of a protective covering comparable to the root cap.

The vegetative shoot apex of most flowering plants has what is termed a *tunica-corpus* type of organization (Figure 23–2). The two regions—tunica and corpus—are usually distinguished by the planes of cell division present in them. The tunica consists of the outermost layer or layers of cells, which divide in planes perpendicular to the surface of the meristem (anticlinal divisions) and contribute primarily to surface growth. The corpus consists of a body of cells enclosed by the tunica layers. In the corpus, the cells divide in various planes and add bulk to the developing shoot. The corpus and each layer of tunica have their own initials.

In the shoot apices of many angiosperms, the bulk of the corpus corresponds to an area of conspicuously vacuolated cells called the zone of *central mother cells* (Figure 23–2). This zone of vacuolated cells is surrounded by the *peripheral meristem,* which originates partly from the tunica and partly from the corpus, or central mother cell zone. Beneath the central mother cells is the *pith meristem.* Cell divisions are relatively infrequent in the central mother cell zone; by contrast, the peripheral zone is very active mitotically. The protoderm always originates from the outermost tunica layer, whereas the procambium and part of the ground meristem (the cortex and sometimes part of the pith) are derived from the peripheral meristem. The rest of the ground meristem (all or most of the pith) is formed by the pith meristem.

Although the primary tissues of the stem pass through periods of growth similar to those of the root, the stem cannot be divided along its axis into regions of cell division, elongation, and maturation as in the case of roots. When actively growing, the apical meristem of the shoot gives rise to leaf primordia in such rapid succession that nodes and internodes cannot at first be distinguished. Gradually, growth begins to occur between the levels of leaf attachment; the elongated parts of the stem take on the appearance of internodes; and the portions of the stem at which the leaves are attached become recognizable as nodes (Figure 23–3). Thus, increase in length of the stem occurs largely by internodal elongation.

Commonly, the meristematic activity causing the elongation of the internode is more intense at the base of the developing internodes than elsewhere. If elongation of the internode takes place over a prolonged period, the meristematic region at the base of the internode may be called an *intercalary meristem* (a meristem region between two more highly differentiated regions). Certain elements of the primary xylem and primary phloem differentiate within the intercalary meristem and connect the more highly

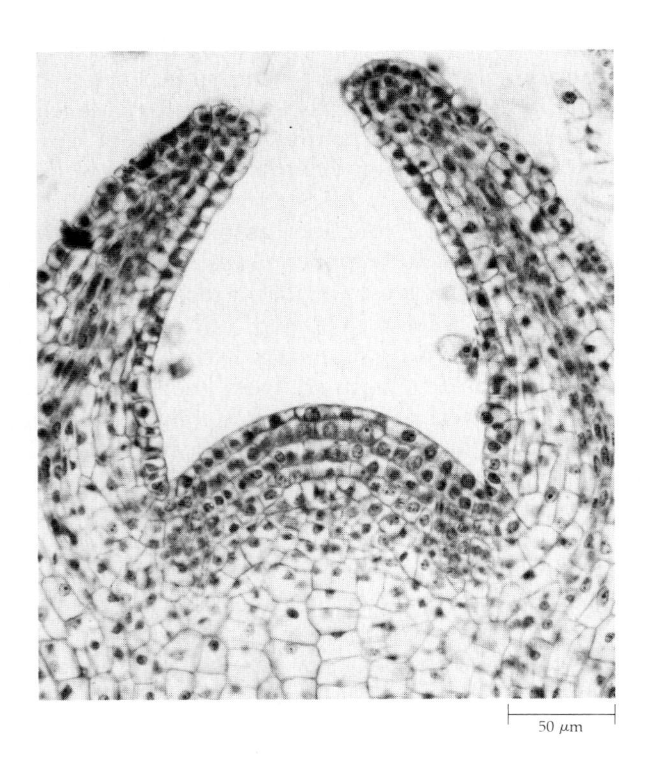

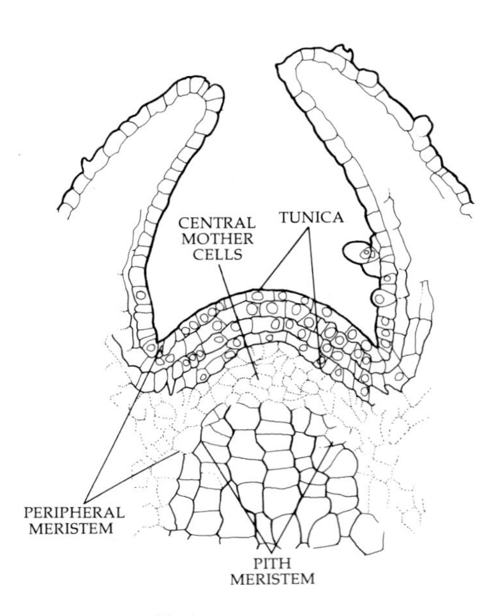

23–2
Detail of Coleus blumei *shoot apex, showing tunica-corpus organization. The zone of central mother cells roughly corresponds to the corpus.*

23–3
Stages in growth of the terminal bud and two lateral buds of the horse chestnut (Aesculus hippocastanum). (a) *The young shoots are tightly packed in the buds and are protected by bud scales.* (b) *The buds open to reveal the oldest rudimentary leaves.* (c) *Internodal elongation has separated the nodes from one another. The terminal bud of the horse chestnut is a mixed bud, containing both leaves and flowers, although the flowers are not visible here. The lateral buds produce only leaves.* (d) *Detail of the lower portions of the young shoots; the bud scales are separated and folded back.*

differentiated regions of the stem above and below the meristem.

As in the root, the apical meristem of the shoot gives rise to the primary meristems—protoderm, ground meristem, and procambium (Figure 23–1). These primary meristems in turn develop into the epidermis, ground tissue, and primary vascular tissues, respectively.

PRIMARY STRUCTURE OF THE STEM

Considerable variation exists in the primary structure of stems of seed plants, but three basic types of organization can be recognized: (1) In some conifers and dicots, the narrow, elongated procambial cells—and consequently the primary vascular tissues that develop from them—appear as a more or less continuous hollow cylinder within the ground tissue (Figure 23–4). The outer region of ground tissue is called the *cortex,* and the inner region is called the *pith.* (2) In other conifers and dicots, the primary vascular tissues develop as a cylinder of discrete strands separated from one another by ground tissue (Figure 23–5). The ground tissue separating the procambial strands (and later the mature vascular bundles) is continuous with the cortex and pith, and is called the *interfascicular parenchyma.* (Interfascicular means "between the bundles.") The interfascicular regions are often called *pith rays.* Narrow interfascicular regions also interconnect the cortex and pith in the first type of organization, but there they are inconspicuous. (3) In the stems of most monocotyledons and of some herbaceous dicotyledons, the arrangement of the procambial strands and vascular bundles is more complex. The vascular tissues do not appear as a single ring of bundles between a cortex and a pith. Instead, the vascular tissues commonly develop in more than one ring of bundles or as a system of strands scattered throughout the ground tissue. In the latter instance, the ground tissue often cannot be distinguished as cortex and pith (Figure 23–6).

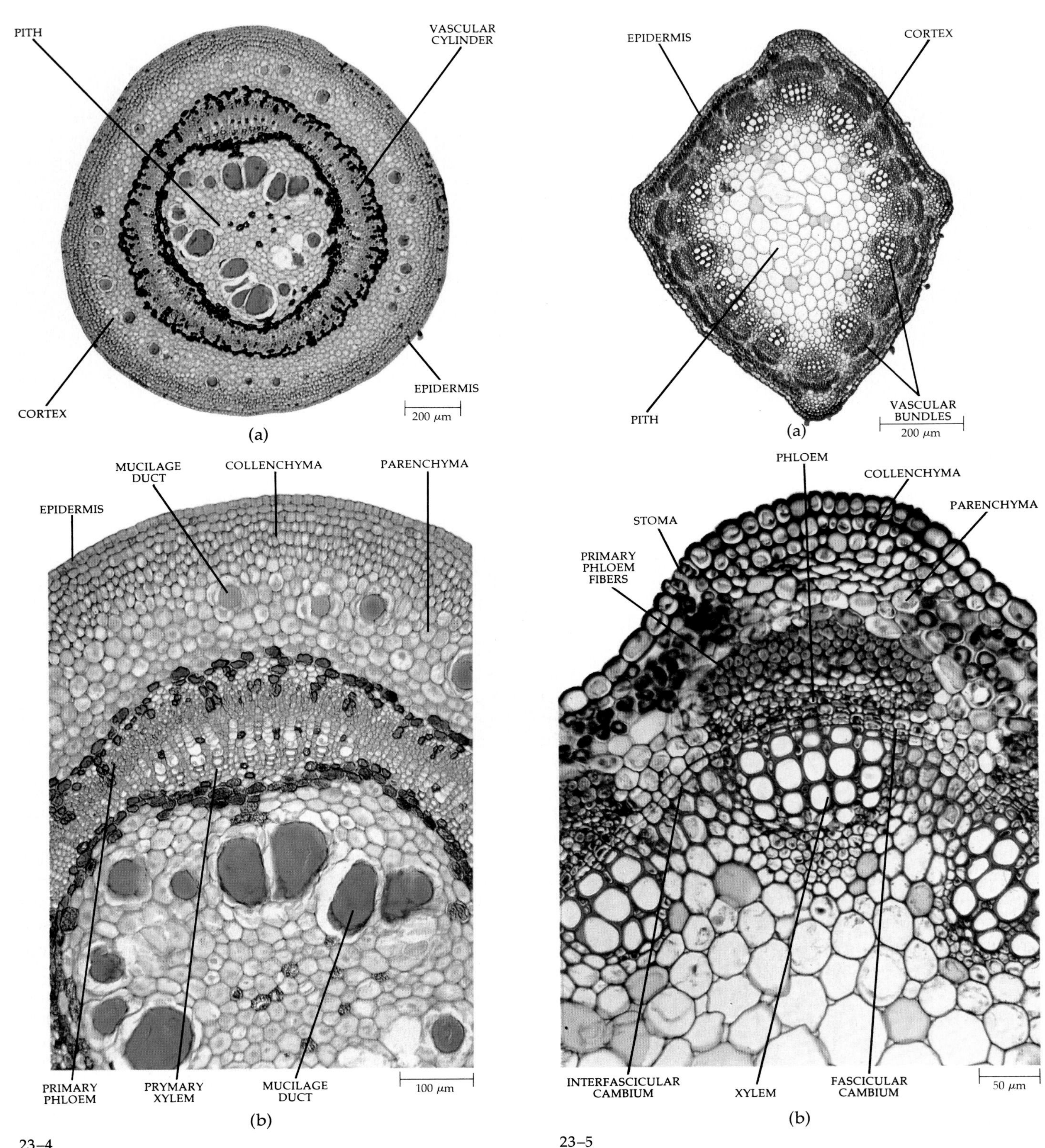

23–4
(a) *Transverse section of linden* (Tilia americana) *stem in a primary stage of growth.* (b) *Detail of a portion of the same linden stem.*

23–5
(a) *Transverse section of stem of alfalfa* (Medicago sativa), *a dicot with discrete vascular bundles.* (b) *Detail of a portion of the same alfalfa stem.*

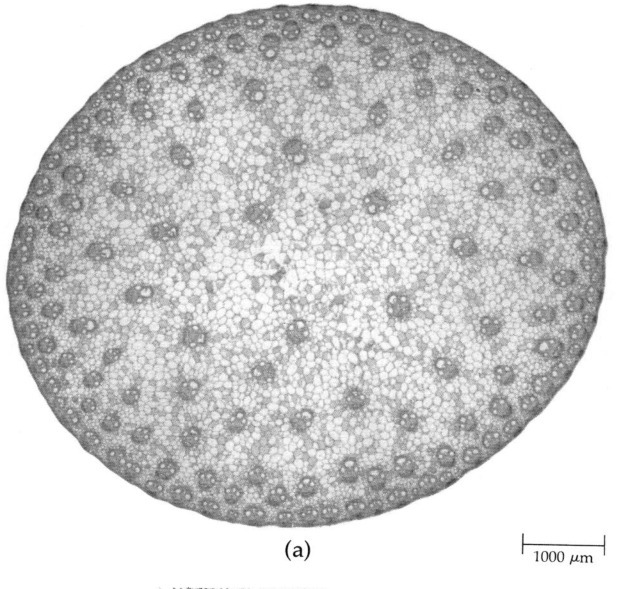

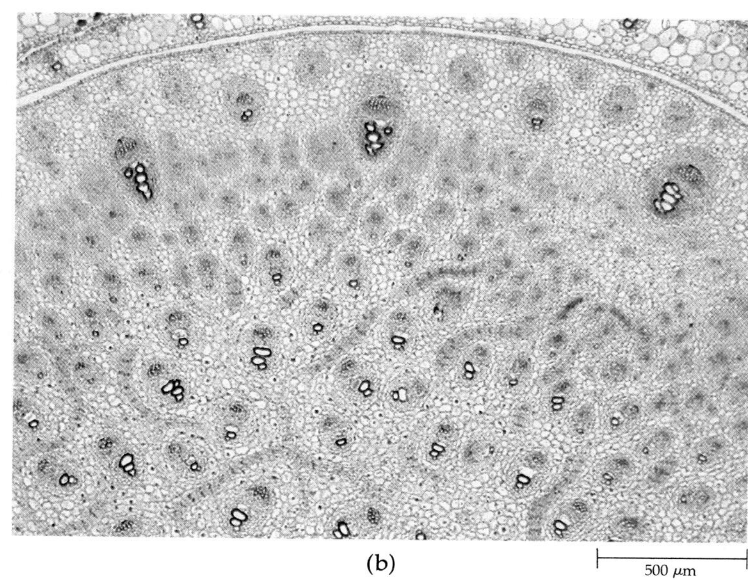

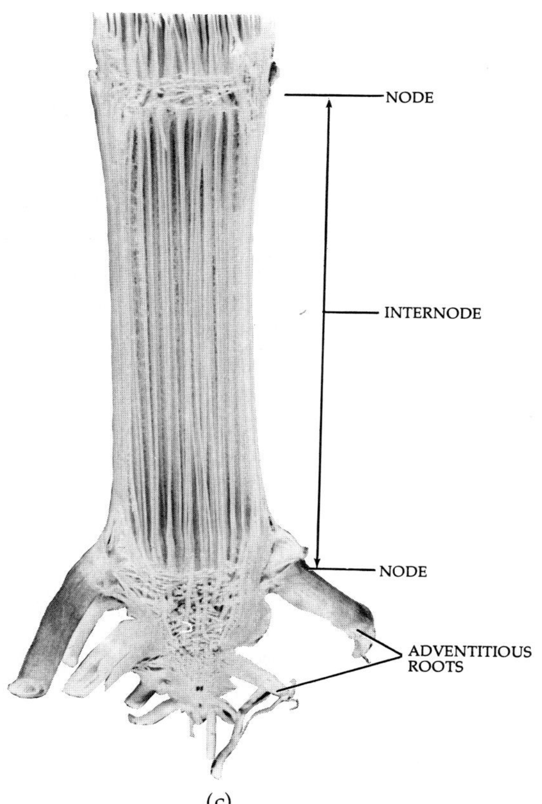

23–6
Stem of corn (Zea mays). (a) *Transverse section of the internodal region, showing numerous vascular bundles scattered throughout the ground tissue.* (b) *Transverse section of the nodal region of a young corn stem, showing horizontal procambial strands that interconnect with vertical bundles.* (c) *A mature stem split longitudinally; the ground tissue has been removed to expose the vascular system.*

The *Tilia* Stem

The stem of the linden *(Tilia americana)* exemplifies the first type of organization (Figure 23–4). As in most stems, the epidermis is a single layer of cells covered by a cuticle. The stem epidermis generally contains far fewer stomata than the leaf epidermis.

The cortex of the *Tilia* stem consists of collenchyma and parenchyma cells. The several layers of collenchyma cells, which provide support to the young stem, form a continuous cylinder beneath the epidermis. The rest of the cortex consists of parenchyma cells that will contain chloroplasts when mature. The innermost layer of cortical cells, which are dark in color, sharply delimits the cortex from the cylinder of primary vascular tissues.

In the great majority of stems, including those of *Tilia*, the primary phloem develops from the outer cells of the procambium, and the primary xylem develops from the inner ones. However, not all of the procambial cells differentiate into primary xylem and primary phloem. A single layer of cells between the primary xylem and the primary phloem remains meristematic and becomes the vascular cambium. *Tilia* is also an example of a woody stem—a stem that produces much secondary xylem. (Secondary growth in stems is discussed in Chapter 24.) After internodal elongation is completed in the *Tilia* stem, fibers develop in the primary phloem. These fibers are called *primary phloem fibers* (see Figure 24–10, page 482).

The inner boundary of the primary xylem in *Tilia* is sharply delimited by one or two layers of pith cells that are dark in appearance. The pith is composed primarily of parenchyma cells and contains numerous large ducts, or canals, containing mucilage (a slimy carbohydrate). Similar ducts are formed in the cortex.

As the cortical and pith cells increase in size, numerous intercellular spaces develop among them; these air spaces are essential for interchange of gases with the atmosphere. The cortical and pith parenchyma cells store various substances.

The *Sambucus* Stem

In the stem of the elderberry *(Sambucus canadensis)*, the procambial strands and primary vascular bundles form a system of discrete strands around the pith. The epidermis, cortex, and pith are essentially similar in organization to those of *Tilia*, so the following discussion of the elderberry stem will be used to explain in more detail the development of the primary vascular tissues of stems.

Figure 23–7a shows three procambial strands in which the primary vascular tissues have just begun to differentiate. The strand on the left is somewhat older than the two on the right and contains at least one mature sieve element and one mature tracheary element. Notice that the first mature sieve element appears in the outer part of the procambial strand (next to the cortex), and that the first mature tracheary element appears in the inner part (next to the pith). Comparing Figures 23–7a and 23–7c, we see that the more recently formed sieve elements appear closer to the center of the stem and that the xylem differentiates in the opposite direction.

The first-formed primary xylem and primary phloem elements (protoxylem and protophloem, respectively) are stretched during elongation of the internode and are frequently destroyed. As in the *Tilia* stem, fibers develop in the primary phloem after internodal elongation is completed (see Figure 24–9).

Like the stems of *Tilia*, those of *Sambucus* become woody. In *Tilia*, almost all the vascular cambium originates from procambial cells between the primary xylem and primary phloem, because the interfascicular regions are very narrow. In *Sambucus*, the interfascicular regions are relatively wide. Consequently, a substantial portion of the vascular cambium in *Sambucus* develops from the interfascicular parenchyma between the bundles.

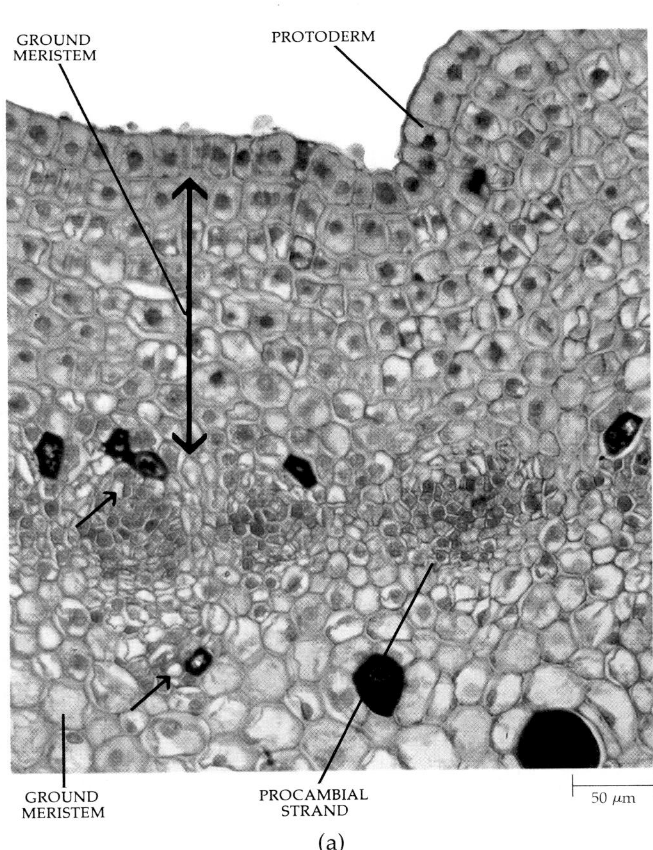

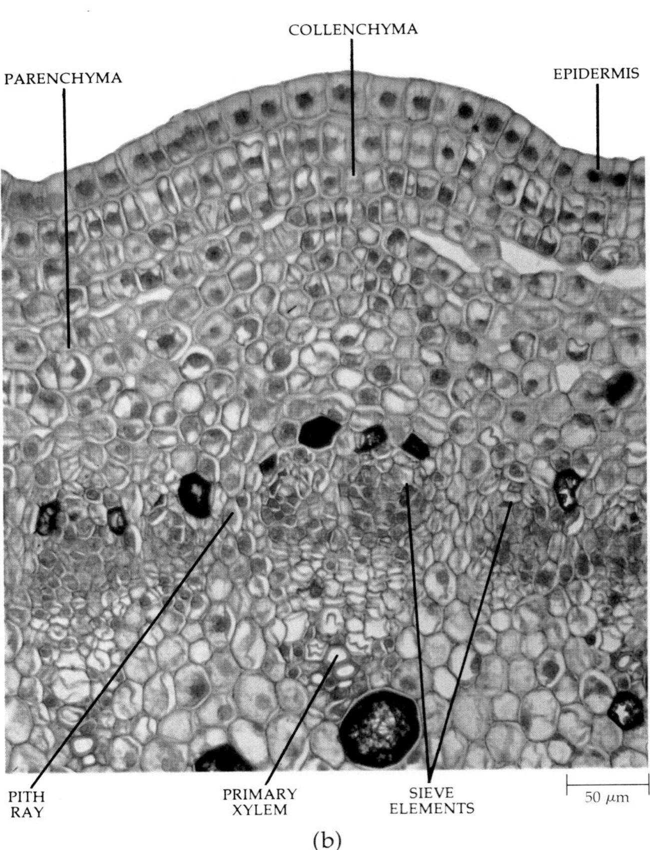

23–7
Transverse sections of the stem of the elderberry (Sambucus canadensis) *in a primary stage of growth.* (a) *A very young stem, showing protoderm, ground meristem, and discrete procambial strands. The procambial strand on the left contains one mature sieve element (arrow, above) and one mature tracheary element (arrow, below).* (b) *Primary tissues further along in development.* (c) *Stem near completion of primary growth. Fascicular and interfascicular cambia are not yet formed. (For further stages in the growth of the elderberry stem, see Figures 24–8 and 24–9.)*

The *Medicago* and *Ranunculus* Stems

The stems of many dicotyledons undergo little or no secondary growth and therefore are *herbaceous*, or nonwoody (see Chapter 24). Examples of herbaceous dicot stems can be found in alfalfa *(Medicago sativa)*, as well as in the buttercups *(Ranunculus)*.

Medicago is an example of an herbaceous dicotyledon that exhibits some secondary growth (Figure 23–5). The structure and development of the primary tissues of the *Medicago* stem are essentially similar in organization to those of *Sambucus* and other woody dicotyledons. The vascular bundles are separated by wide interfascicular regions and surround a large pith. The vascular cambium is partly procambial and partly interfascicular in origin, but secondary vascular tissues are formed mainly in the vascular bundles. The interfascicular cambium generally produces only sclerenchyma cells on the xylem side.

The herbaceous stem of *Ranunculus* is an extreme example, and its vascular bundles resemble those of many monocotyledons. The vascular bundles retain no procambium after the primary vascular tissues mature; consequently, the bundles never develop a vascular cambium and lose their potential for further growth. Vascular bundles such as those of *Ranunculus* (Figure 23–8) and the monocotyledons, in which all the procambial cells mature and the potential for further growth within the bundle is lost, are said to be "closed." Vascular bundles that do give rise to a cambium are said to be "open." In most dicotyledons, the vascular bundles are of the open type; they produce some secondary vascular tissues.

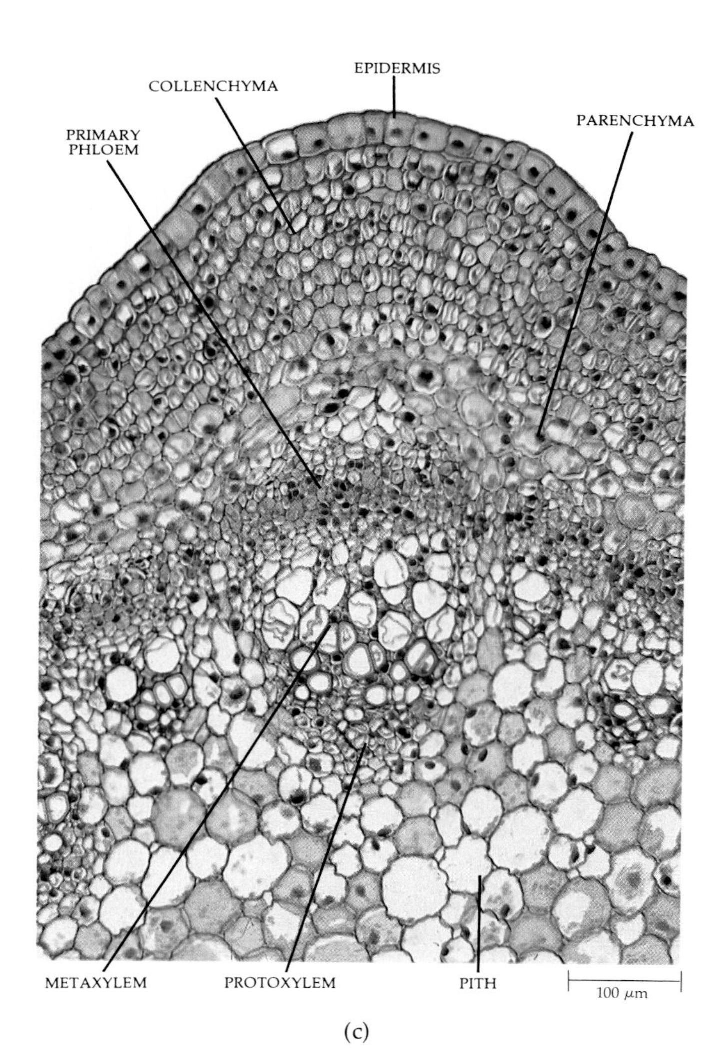

(c)

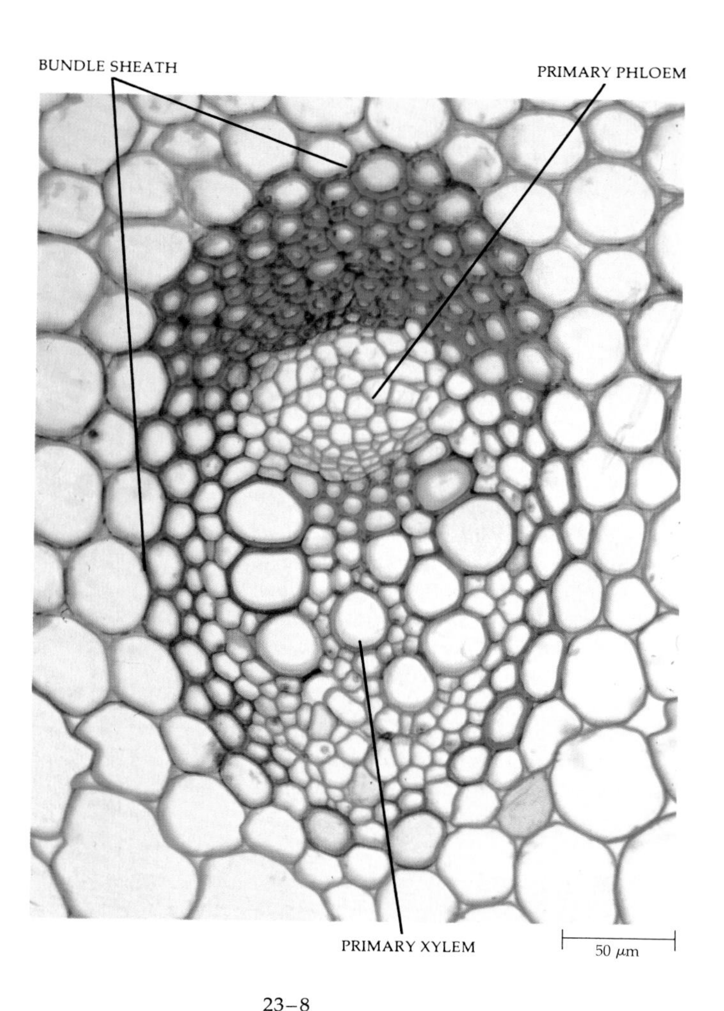

23–8
Transverse section of vascular bundle of the buttercup (Ranunculus), *an herbaceous dicot. The vascular bundles of the buttercup are closed, in that all of the procambial cells mature; consequently, no secondary tissues develop. The primary phloem and primary xylem are surrounded by a bundle sheath of thick-walled sclerenchyma cells. Compare the vascular bundle shown here with the mature vascular bundle of corn shown in Figure 23–9c.*

The *Zea* Stem

The herbaceous stem of corn (*Zea mays*) exemplifies the stems of monocots in which the vascular bundles form a system of strands scattered throughout the ground tissue (Figure 23–6). As in other monocots, the vascular bundles of corn are closed.

Figure 23–9 shows three stages in the development of a corn vascular bundle. As in the bundles of dicot stems, the phloem develops from the outer cells of the procambial strand, and the xylem develops from the inner cells. Also, differentiation of the phloem and the xylem are in opposite directions; as seen in transverse sections, the phloem differentiates from outside to inside, and the xylem differentiates from inside to outside. The first-formed phloem and xylem elements (protophloem and protoxylem) are stretched and destroyed during elongation of the internode. This results in the formation of a very large air space on the xylem side of the bundle (Figure 23–9c). The mature vascular bundle contains two large vessel members (the metaxylem vessels), and the phloem (metaphloem) tissue is composed of sieve-tube members

23–9
Three stages in the differentiation of the vascular bundles of corn (Zea mays), *as seen in transverse sections of the stem.* (a) *The protophloem elements and two protoxylem elements are mature.* (b) *The protophloem sieve elements are now crushed, and much of the metaphloem is mature. Three protoxylem members are now mature, and the two metaxylem vessel members are almost fully expanded.* (c) *Mature vascular bundle surrounded by a sheath of thick-walled sclerenchyma cells. The metaphloem is composed entirely of sieve-tube members and companion cells. The portion of the vascular bundle once occupied by the protoxylem elements is now a large air space. Note the wall thickenings of destroyed protoxylem elements bordering the air space.*

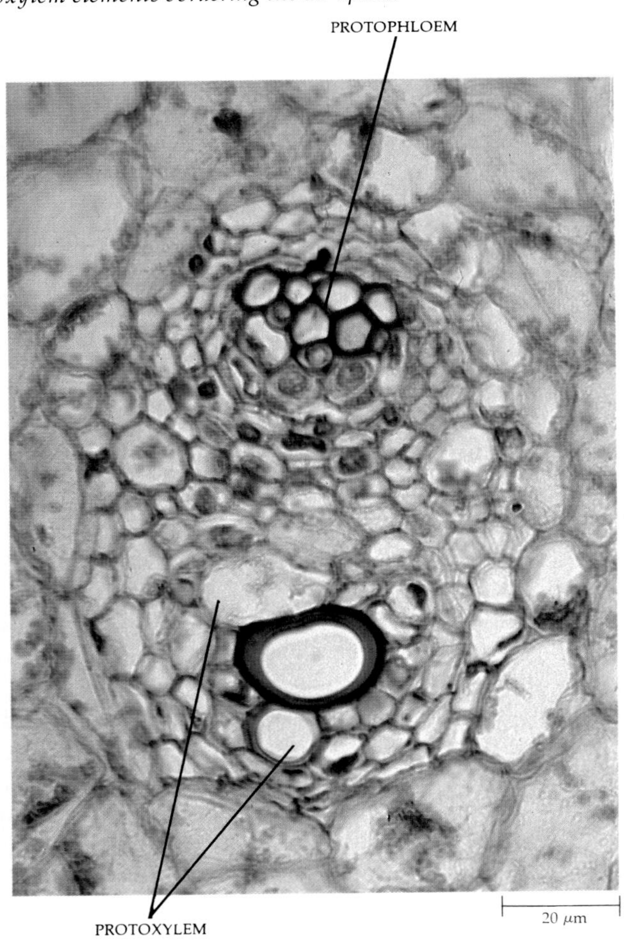

(a)

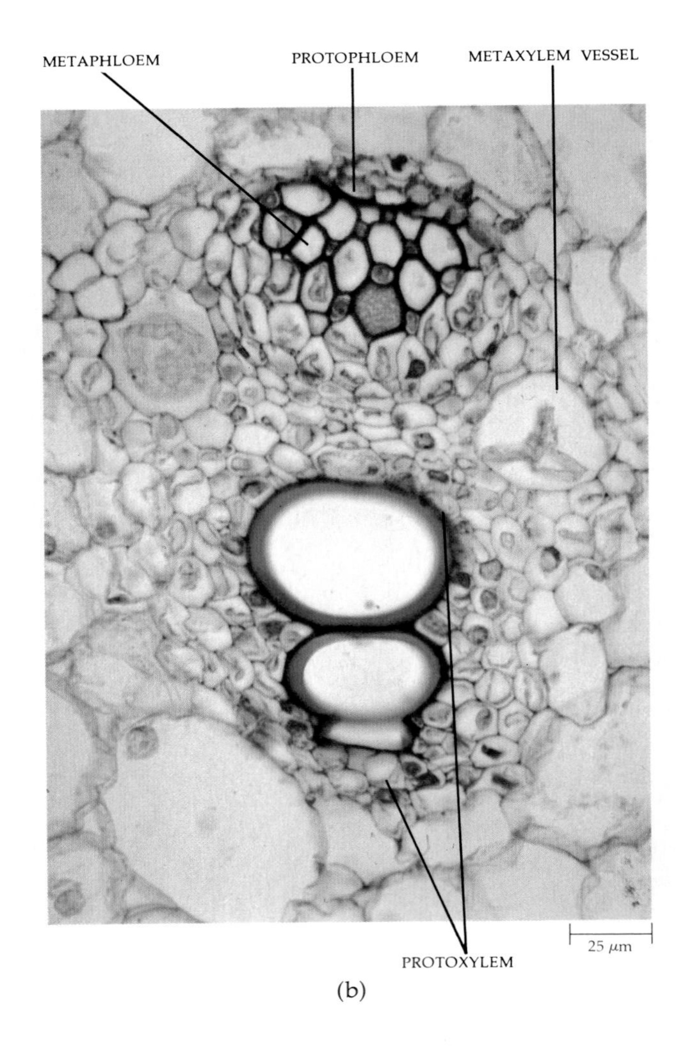

(b)

and companion cells. The entire bundle is enclosed in a sheath of sclerenchyma cells.

RELATION BETWEEN THE VASCULAR TISSUES OF THE STEM AND THE LEAF

The pattern formed by the vascular bundles in the stem reflects the close structural and developmental relationship that exists between the stem and its lateral appendages, the leaves. The term "shoot" serves not only as a collective term for these two vegetative organs but also as an expression of their intimate association.

The procambial strands of the stem arise behind the apical meristem just below the developing leaf primordia and sometimes are present below the sites of future leaf primordia even before they have begun to develop. As the leaf primordia increase upward in length, the procambial strands also differentiate upward within them. From its inception, the procambial system of the leaf is continuous with that of the stem.

At each node, one or more vascular bundles diverge from the cylinder of strands in the stem, cross the cortex, and enter the leaf or leaves attached at that node (Figures 23–10 and 23–11). A vascular bundle in the stem that extends from a longitudinal *stem bundle* to the base of a leaf, where it connects with the vascular system of the leaf, is called a *leaf trace*, and the wide gap or region of ground tissue found in the vascular cylinder, where the leaf trace diverges toward the leaf, is called a *leaf gap*. A single leaf may have one or more leaf traces connecting its vascular system with that in the stem. The number of internodes that leaf traces traverse before they join with stem bundles varies, so the traces vary in length.

If the stem bundles are followed either upward or downward in the stem, they will be found to be associated with several leaf traces. A stem bundle and its associated leaf traces are called a *sympodium* (plural: sympodia). In some stems, some or all of the sympodia are interconnected, whereas in others, all of the sympodia are independent units of the vascular system. Regardless, the pattern of the vascular system in the stem is a reflection of the arrangement of the leaves on the stem.

SIEVE ELEMENT
COMPANION CELL
AIR SPACE
METAXYLEM VESSEL
SHEATH
50 μm
(c)

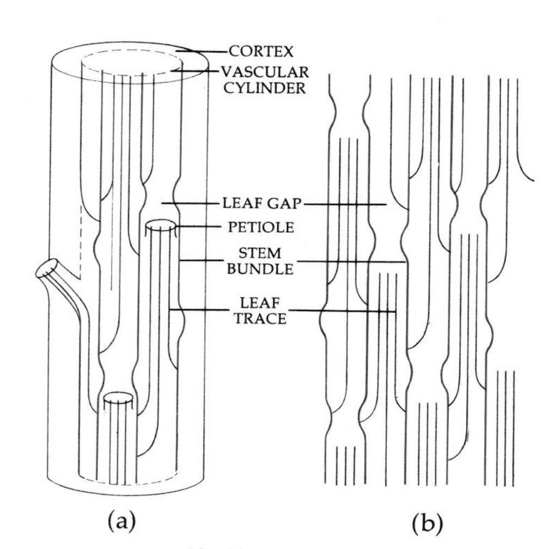

23–10
Diagrams of the primary vascular system of a stem of the oak-leaved goosefoot (Chenopodium glaucum), *in which discrete bundles form a cylinder of interconnected strands around a pith.* (a) *A three-dimensional diagram showing the arrangement of the bundles within the stem.* (b) *Vascular system spread out in one plane. Notice the leaf traces diverging outward from the vascular cylinder and the relation of the traces to the vascular bundles of the stem. In* C. glaucum *each leaf has three leaf traces.*

23–11
Diagrams of longitudinal (a) *and transverse* (b) *sections of part of the stem of tobacco* (Nicotiana tabacum), *illustrating the relationship of the vascular systems of the leaf and the stem. The stem of tobacco has a "continuous" vascular cylinder, portions of which are wider than others. At each node a single leaf trace diverges toward a leaf. In addition to leaf traces, branch traces occur at the nodes and are often closely associated with the leaf traces. In tobacco, two branch traces extend from the vascular system of the stem to the bud, and the branch gap is continuous with the leaf gap.*

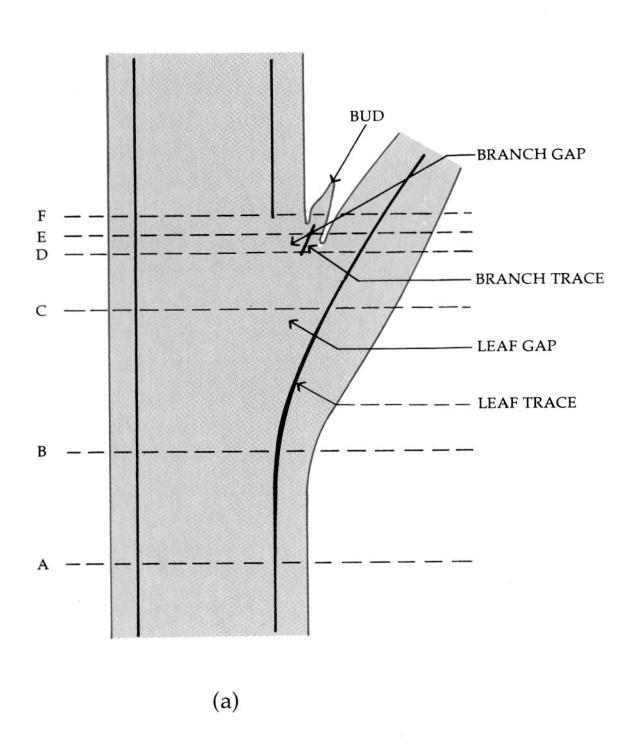

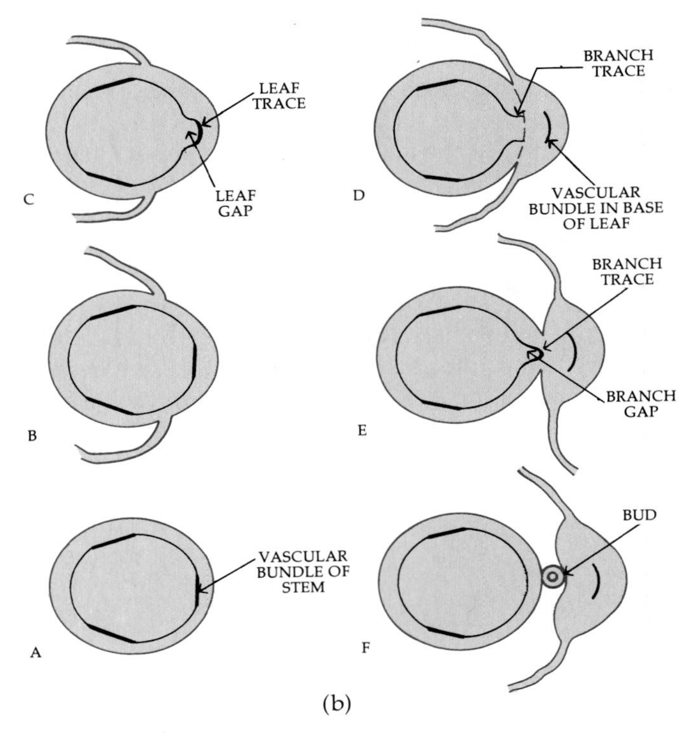

DEVELOPMENT OF THE LEAF

The first structural evidence of leaf initiation in most angiosperms is the appearance of periclinal divisions beneath the protoderm in the peripheral region of the shoot apex. A combination of cell enlargement and further divisions soon results in formation of a bulge, or *leaf buttress* (Figure 23–12a), as a center of mitotic activity is established at an exact time and location on the apex relative to previously initiated leaves. Either before or during buttress formation, a procambial strand appears beneath the young leaf primordium.

With continued upward growth, the leaf buttress develops into an erect, peglike structure—the leaf primordium (Figure 23–12b). In dicotyledons, this structure soon develops localized regions of meristematic activity on approximately opposite sides of its axis. These regions, which will initiate formation of the blade of the leaf, are called *marginal meristems* (Figures 23–13 and 23–14).

Expansion and increase in length of the leaf occur largely by *intercalary growth,* that is, by cell division and cell enlargement throughout the blade, with cell enlargement contributing the most. As a result of the activity of the marginal meristem, a certain number of layers of mesophyll cells become established early in blade development, although the number of layers may be increased during later development. Differences in rates of cell division and cell enlargement in the various layers of the blade result in the formation of numerous intercellular spaces and produce the mesophyll form characteristic of the leaf. Typically, the leaf stops growing first at the tip and last at the base. Compared with growth of the stem, the growth of most leaves is of short duration. The unlimited or prolonged growth of the vegetative apical meristems is described as being *indeterminate,* and the restricted type of growth exhibited by the leaf and by floral apices is said to be *determinate.*

As the peglike leaf primordium elongates, the procambial strand beneath develops upward into it (Figure 23–12c) and is continuous with the coarse veins emanating from the main vein, or midvein, of the blade (Figure 23–13). The smaller veins of the leaf are initiated at the tip of the leaf. They develop from the top to the bottom of the leaf in continuity with the coarser veins. Thus, the tip of the leaf is the first part to have a complete system of veins. This course of development reflects the overall maturation of the leaf, which is from the tip to the base of the leaf.

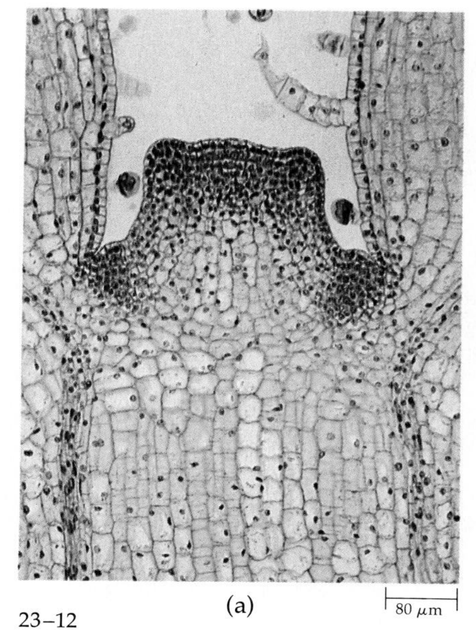

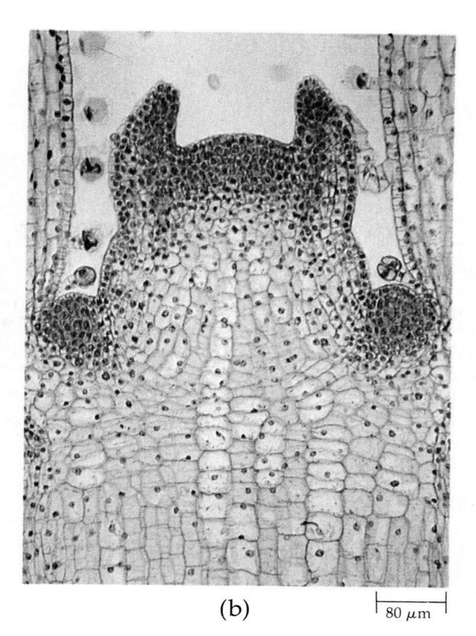

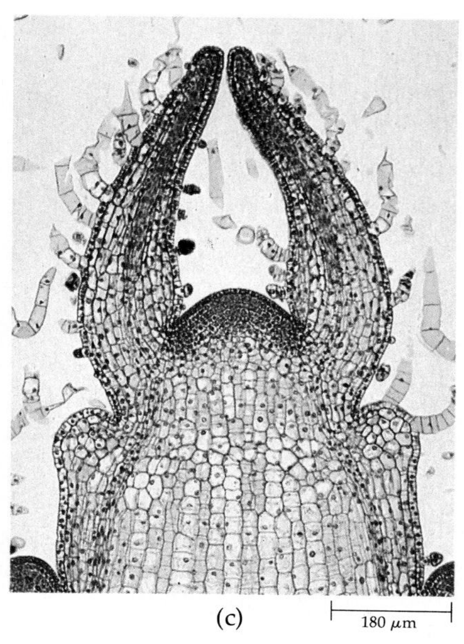

23–12
Some early stages of leaf development in Coleus blumei, *as seen in longitudinal sections of the shoot tip. The leaves in* Coleus *occur in pairs, opposite one another at the nodes (see Figure 23–1).* (a) *Two small bulges, or leaf buttresses, can be seen opposite one another on the flanks of the apical meristem. In addition, a bud primordium can be seen arising in the axil of each of the two young leaves, below.* (b) *Two erect, peglike leaf primordia have developed from the leaf buttresses. Notice the procambial strands extending upward into the leaf primordia. The bud primordia, below, are further along in development than those in* (a). *As the leaf primordia elongate* (c), *the procambial strands, which are continuous with the vascular bundles in the stem, continue to develop into the leaves. Trichomes, or epidermal hairs, develop from certain protodermal cells very early, long before the protoderm matures to become the epidermis.*

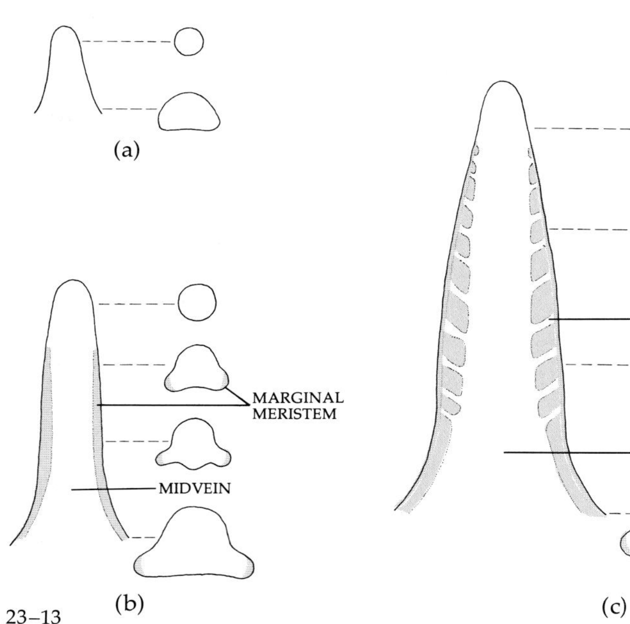

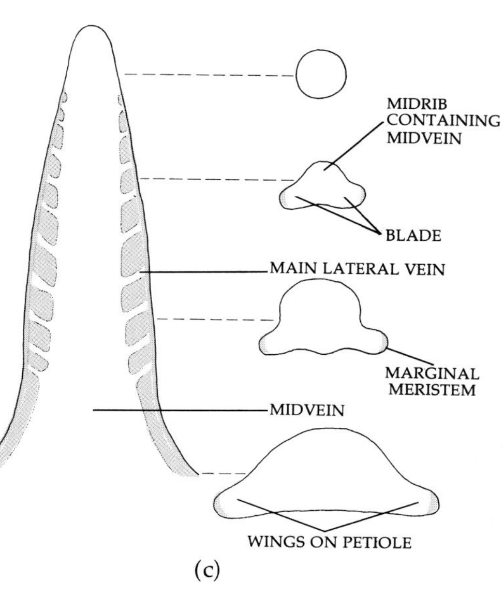

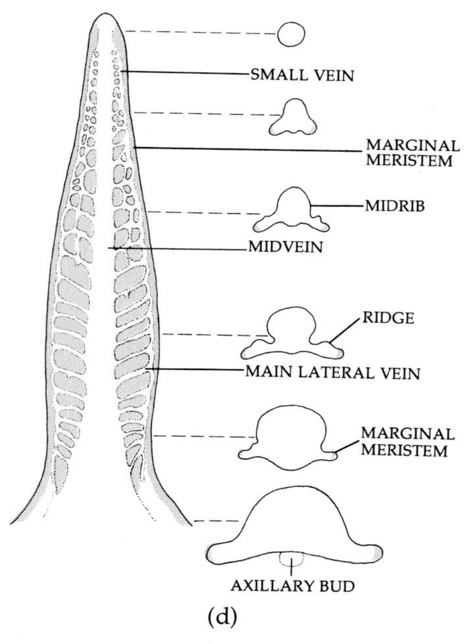

23–13
Diagrams of longitudinal and transverse sections illustrating some early stages of leaf development in tobacco (Nicotiana tabacum). (a) *A young, peglike leaf primordium without a blade, or lamina.* (b) *Marginal meristem activity has begun on opposite sides of the primordium to form the blade.* (c) *The main lateral veins (narrow, clear areas) can be seen extending from the midvein as the blade grows.* (d) *The primordium has increased in height. With further growth of the blade, some ridges appear in association with some of the larger veins. Small veins have begun to develop at the tip of the leaf. The bottom diagram in* (a) *through* (d) *is a transverse section through the "winged" petiole.*

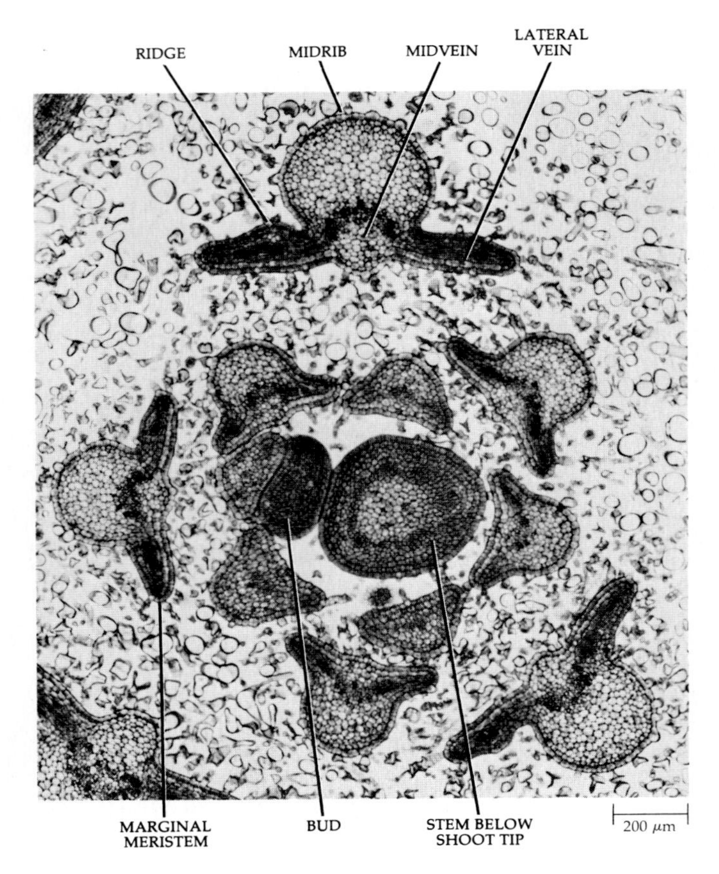

23–14
Transverse section of developing leaves of tobacco (Nicotiana tabacum) *grouped around the shoot tip, sectioned below the apical meristem. The younger leaves are nearer the axis. A leaf primordium at first lacks differentiation into midrib and blade. Some early stages in development of blade and midrib can be seen here. Compare these sections with the diagrams of transverse sections in Figure 23–13. (Portions of numerous trichomes surround the developing leaves seen here.)*

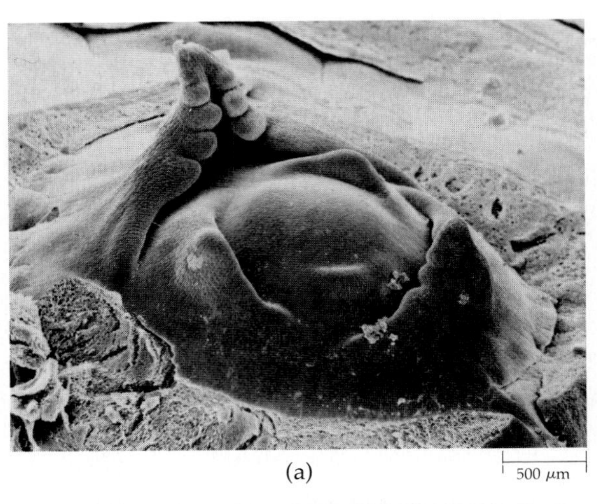

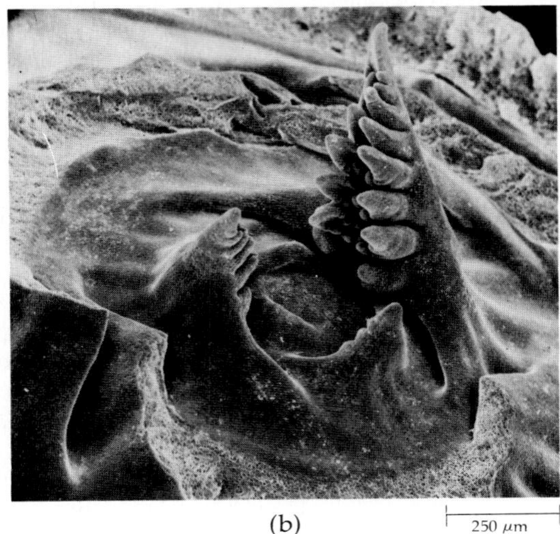

23–15
Scanning electron micrographs of celery (Apium graveolens), *showing early stages of leaf development in the shoot apex.* (a) *The nearer a developing leaf is to the center of the apex, the younger it is.* (b) *Two of the leaves shown here (right and left) have developed further than any shown in* (a).

Figure 23–15 provides three-dimensional views of the shoot apex and some early stages of leaf development in celery *(Apium graveolens).* The leaf primordia and young leaves are helically arranged on the shoot apex, with the younger leaves nearer the center of the apex. Two leaf buttresses can be seen, top right and left, near the center of the apex in Figure 23–15a. As a leaf primordium arises from a buttress, it broadens into a shell-like form while growing vigorously at its tip for a relatively short time to form the petiole-rachis part of what is to become a bipinnately compound leaf. Beneath the tip, protuberances arise on both sides of the rachis—first at the bottom and then in succession toward the tip—to form leaflets. Further protuberances develop from the earlier ones, becoming either lobes of the primary leaflets or distinctly separated secondary leaflets. These can be seen arising from the primary leaflets of the relatively large, rudimentary leaf in Figure 23–15b. Each leaflet resembles a simple leaf in its development.

MORPHOLOGY OF THE LEAF

Leaves vary greatly in form and in internal structure. In dicots, the leaf commonly consists of an expanded portion, the blade, or *lamina*, and a stalklike portion, the *petiole* (Figure 23–16). Small scalelike or leaflike structures called *stipules* develop at the base of some leaves (Figure 23–17). Many leaves lack petioles and are said to be *sessile* (Figure 23–18). In most monocots and certain dicots, the base of the leaf is expanded into a *sheath*, which encircles the stem (Figure 23–18b). In some grasses, the sheath extends the length of an internode. The arrangement of leaves on the stem may be *spiral* (alternate), *opposite* (in pairs), or *whorled* (three or more at a node). For example, mulberry *(Morus alba)* and oak *(Quercus)* leaves are spiral, maple *(Acer)* leaves are opposite, and Culver's-root *(Veronicastrum virginicum)* leaves are whorled (Figure 23–16).

The leaves of dicotyledons are either simple or compound. In simple leaves, the blades are not divided into distinct parts, although they may be deeply lobed (Figure 23–16). The blades of compound leaves are divided into leaflets, each usually with its own small petiole (which is called a petiolule). Two types of compound leaves can be distinguished: pinnately compound leaves and palmately compound leaves (Figure 23–19). In pinnately compound leaves, the leaflets arise from either side of an axis, the *rachis*, like the pinnae of a feather. (The rachis is an extension of the petiole.) The leaflets of a palmately compound leaf diverge from the tip of the petiole, and a rachis is lacking.

(a)

(b)

(c)

(d)

(e)

23–16
Several examples of simple leaves. (a) *Mulberry* (Morus alba). (b) *Culver's-root* (Veronicastrum virginicum). (c) *Sugar maple* (Acer saccharum). (d) *Silver maple* (Acer saccharinum). (e) *Red oak* (Quercus rubra). *Note the spiral, or alternate, arrangement of the leaves in mulberry, and the whorled arrangement of those in Culver's-root. Leaf arrangement in the maples is opposite, and in oak it is spiral, although only single leaves of these trees are shown here.*

23–17
The pinnately compound leaf of the pea (Pisum sativum). *Notice the stipules at the base of the leaf and the slender tendrils at the tip of the leaf. In the pea leaf, the stipules are often larger than the leaflets.*

23–18
Sessile leaves (leaves without a petiole) are often found among dicots, such as Moricandia, *a member of the mustard family* (a), *but are particularly characteristic of grasses and other monocots.* (b) *In corn* (Zea mays), *a monocot, the base of the leaf forms a sheath around the stem. The upper portion of the sheath has been pulled away from the stem to reveal the ligule, a small flap of tissue extending upward from the sheath.*

(a)

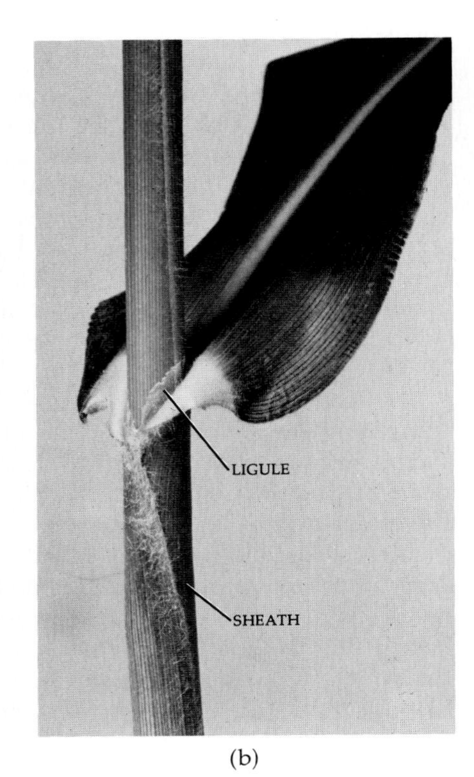

(b)

23–19
Some examples of compound leaves. A palmately compound leaf is shown in (a); *all the others are pinnately compound.* (a) *Red buckeye* (Aesculus pavia). (b) *Shagbark hickory* (Carya ovata). (c) *Green ash* (Fraxinus pennsylvanica *var.* subintegerrima). (d) *Black locust* (Robinia pseudo-acacia). (e) *Honey locust* (Gleditsia triacanthos). *In the honey locust, each leaflet is subdivided into smaller leaflets.*

(a)

(b)

Since leaflets are similar in appearance to simple leaves, it is sometimes difficult to determine whether the structure is a leaflet or a leaf. Two criteria may be used to distinguish leaflets from leaves: (1) buds are found in the axils of leaves—both simple and compound—but not in the axils of leaflets; and (2) leaves extend from the stem in various planes, whereas the leaflets of a given leaf all lie in the same plane.

STRUCTURE OF THE LEAF

Variations in the structure of angiosperm leaves are to a great extent related to the habitat, and the availability of water is an especially important factor affecting their form and structure. On the basis of their water requirements or adaptations, plants are commonly characterized as: *mesophytes* (plants that require abundant soil water and a relatively humid atmosphere), *hydrophytes* (plants that depend on an abundant supply of moisture or grow wholly or partly submerged in water), and *xerophytes* (plants that are adapted to arid habitats). Such distinctions are not sharp, however, and leaves often exhibit a combination of features that are characteristic of different ecological types. Regardless of their varying forms, the foliage leaves of angiosperms are specialized as photosynthetic organs and, like roots and stems, consist of dermal, ground, and vascular tissue systems.

Epidermis

The ordinary epidermal cells of the leaf, like those of the stem, are compactly arranged and covered with a cuticle that reduces water loss (see Chapter 21, page 427). Stomata may occur on both sides of the leaf but are usually more numerous on the lower surface (Figure 23–20). In leaves of hydrophytes that float on the surface of the water, stomata may occur in the upper epidermis only (Figure 23–21); the immersed leaves usually lack stomata entirely. The leaves of xerophytes generally contain greater numbers of stomata than those of other plants. Presumably these numerous stomata permit a higher rate of gas exchange during the relatively rare periods of favorable water supply. In many xerophytes, the stomata are sunken in depressions on the lower surface of the leaf (Figure 23–22). The depressions may also contain many epidermal hairs. Together these two features may serve to reduce water loss from the leaf. Epidermal hairs, or trichomes, may occur on either or both surfaces of a leaf. Thick coats of epidermal hairs may also retard water loss from leaves.

In the leaves of dicotyledons, the stomata are scattered and appear to be randomly arranged; their development is mixed—that is, mature and immature stomata occur side by side in a partially developed leaf. In monocotyledons, the stomata are arranged in rows parallel with the long axis of the leaf. Their development begins at the tips of the leaves and progresses downward.

(c)

(d)

(e)

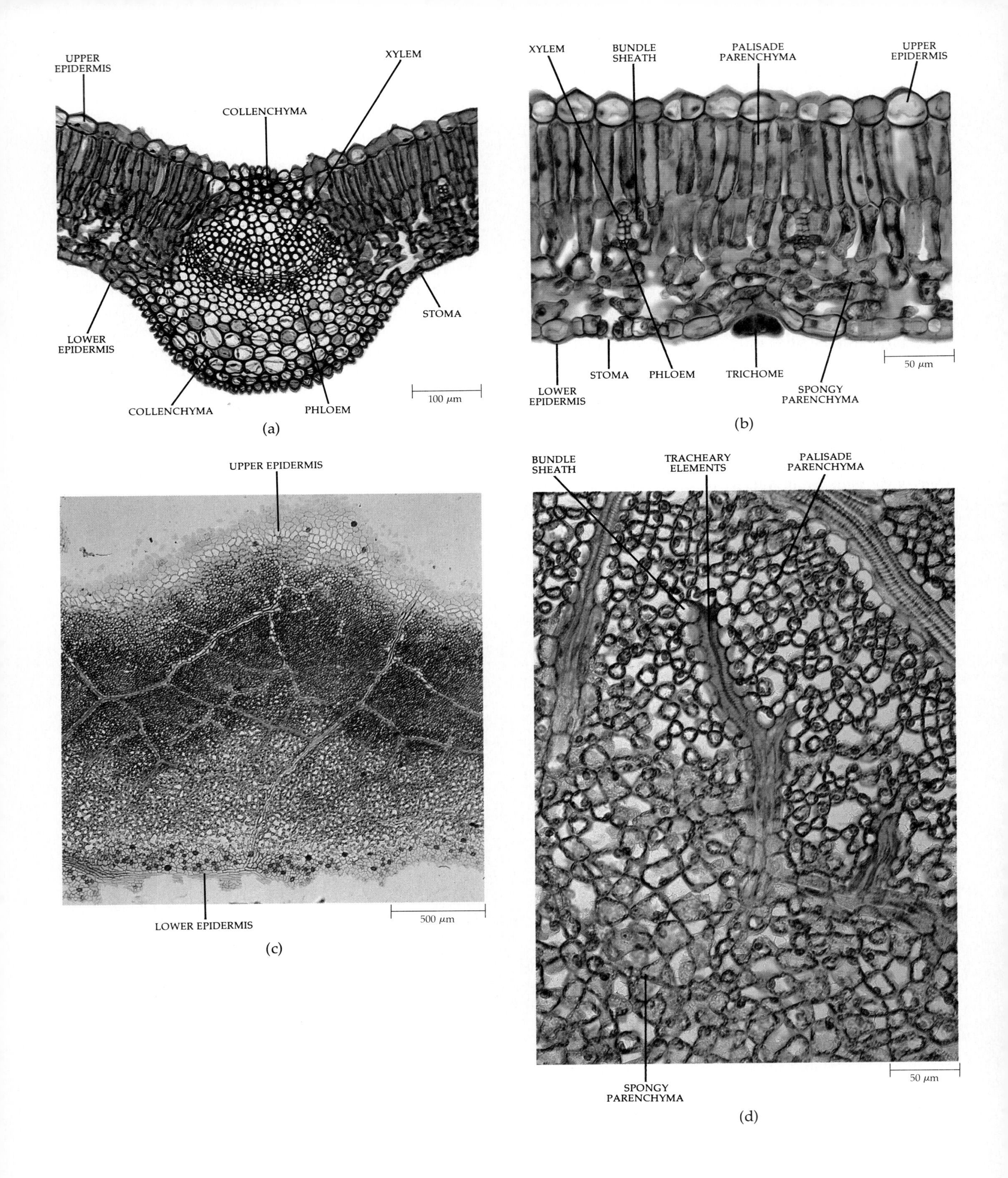
UPPER
EPIDERMIS
COLLENCHYMA
XYLEM
STOMA
LOWER
EPIDERMIS
COLLENCHYMA
PHLOEM
100 μm
(a)
XYLEM
BUNDLE
SHEATH
PALISADE
PARENCHYMA
UPPER
EPIDERMIS
LOWER
EPIDERMIS
STOMA
PHLOEM
TRICHOME
SPONGY
PARENCHYMA
50 μm
(b)
UPPER EPIDERMIS
LOWER EPIDERMIS
500 μm
(c)
BUNDLE
SHEATH
TRACHEARY
ELEMENTS
PALISADE
PARENCHYMA
SPONGY
PARENCHYMA
50 μm
(d)

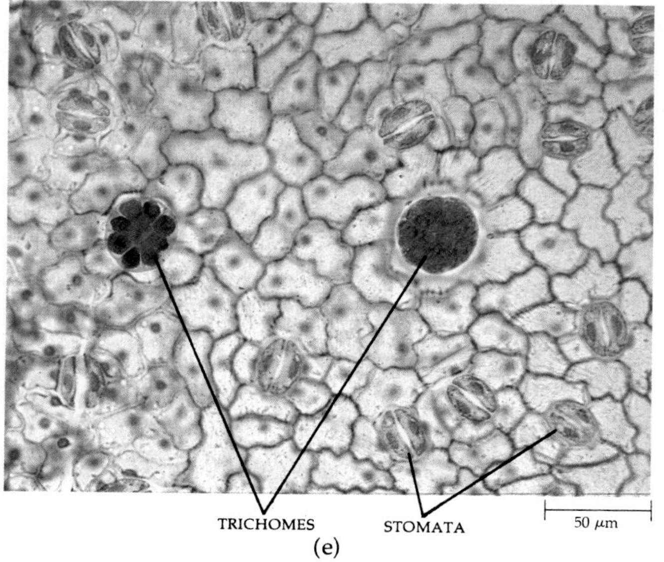

23–20
Sections of lilac (Syringa vulgaris) *leaf.* (a) *A transverse section through a midrib showing the midvein.* (b) *A transverse section through a portion of the blade. Two small veins (minor veins) can be seen in this view.* (c) *This is a paradermal section of the leaf—a section cut approximately parallel with the leaf surface. Progressing from the top to the bottom of this micrograph, the section cuts deeper into the leaf. Thus, part of the upper epidermis can be seen in the light area at the top of the micrograph, and part of the lower epidermis can be seen at the bottom. Notice the greater number of stomata in the lower epidermis. The venation in lilac is netted.* (d) *and* (e) *are enlargements of portions of* (c). *Portions of palisade parenchyma (above) and spongy parenchyma (below) are shown in* (d). *A vein ending, sectioned through some tracheary elements and surrounded by a bundle sheath, can be seen in the middle of this micrograph. A portion of the lower epidermis, with two trichomes and several stomata, in addition to many ordinary epidermal cells, is shown in* (e).

PALISADE PARENCHYMA
SCLEREID
UPPER EPIDERMIS
STOMA
SPONGY PARENCHYMA
TRICHOME
LOWER EPIDERMIS
VEIN
200 μm

23–21
Transverse section of the water lily (Nymphaea odorata) *leaf, which floats on the surface of the water and has stomata in the upper epidermis only. As is typical of hydrophytes, the vascular tissue in the* Nymphaea *leaf is much reduced, especially the xylem. The palisade parenchyma consists of several layers of cells above the spongy parenchyma.*

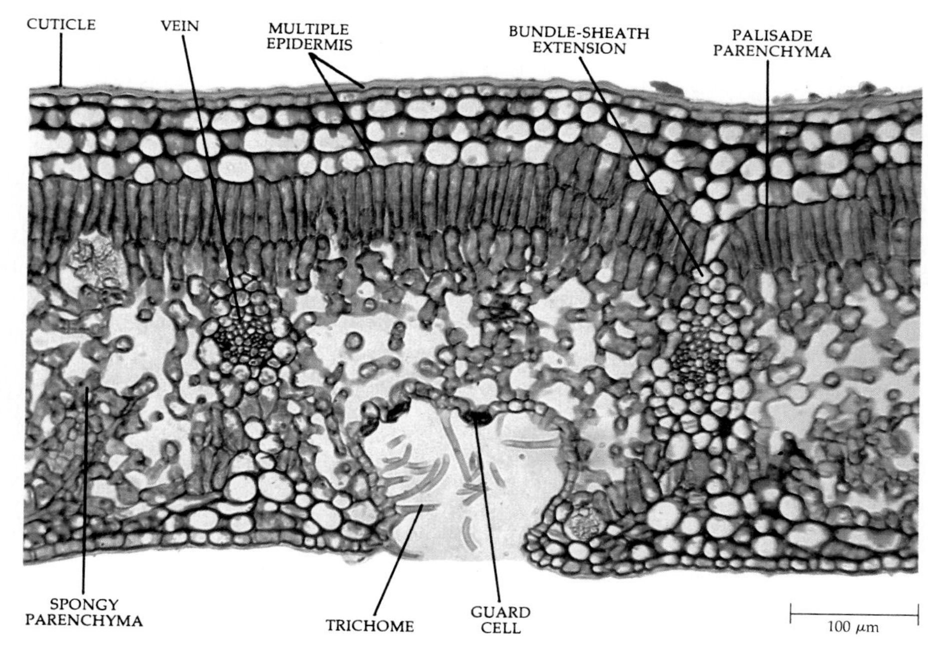

23–22
Transverse section of oleander (Nerium oleander) *leaf. Oleander is a xerophyte, and this is reflected in the structure of the leaf. Note the very thick cuticle covering the multiple (several-layered) epidermis on the upper and lower surfaces of the leaf. The stomata and epidermal hairs (trichomes) are restricted to invaginated portions of the lower epidermis called stomatal crypts.*

Mesophyll

It is the *mesophyll*—the ground tissue of the leaf—with its large volume of intercellular spaces and numerous chloroplasts, that is particularly specialized for photosynthesis. The intercellular spaces are connected with the outer atmosphere through the stomata, which facilitates rapid gas exchange, an important factor in photosynthetic efficiency. In mesophytes, the mesophyll is differentiated into *palisade parenchyma* and *spongy parenchyma*. The cells of the palisade tissue are columnar in shape, with their long axes oriented at right angles to the epidermis, and the spongy parenchyma cells are irregular in shape (Figures 23–20b and 23–20d). Although the palisade parenchyma appears more compact than the spongy parenchyma, most of the vertical walls of the palisade cells are exposed to intercellular space, and in some leaves the palisade surface may be two to four times greater than the spongy surface. Chloroplasts are also more numerous in palisade cells than in spongy cells. Therefore, most of the photosynthesis in the leaf apparently takes place within the palisade parenchyma.

The palisade parenchyma is usually located on the upper side of the leaf, and the spongy parenchyma is generally on the lower side (Figure 23–20). In leaves of xerophytes, palisade parenchyma often occurs on both sides of the leaf. In some plants, for instance, corn (see Figure 6–18, page 108) and other grasses (Figure 23–23), the mesophyll cells are of more or less similar shape, and a distinction between spongy and palisade parenchyma does not exist.

Vascular Bundles

The mesophyll of the leaf is thoroughly permeated by numerous vascular bundles, or *veins*, which are connected with the vascular system of the stem. In most dicots, the veins are arranged in a branching pattern, with successively smaller veins branching from somewhat larger ones. This type of vein arrangement is known as *netted venation* (Figure 23–24). Often the largest vein extends along the long axis of the leaf as a midvein which, with its associated ground tissue, comprises the so-called midrib of such leaves (Figure 23–20a). By contrast, most monocot leaves have many

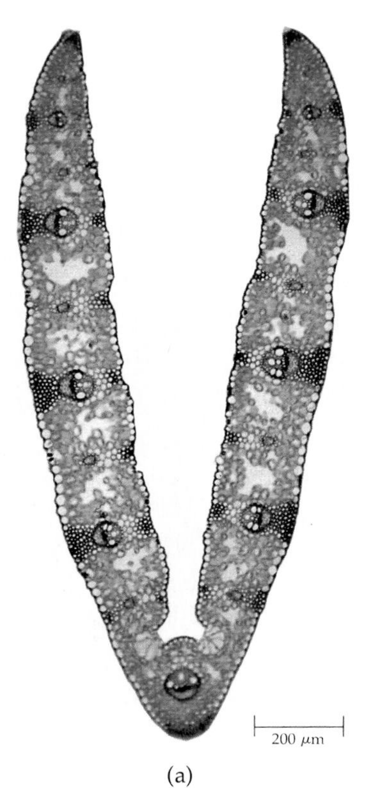

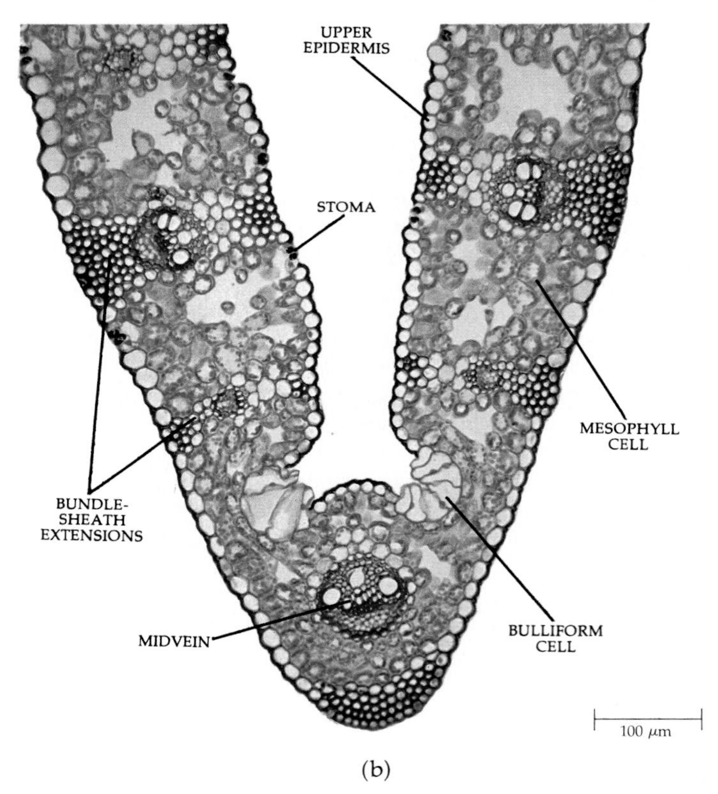

23–23
Transverse sections of annual bluegrass (Poa annua) *leaf.* (a) *Overall view of leaf.* (b) *Enlargement of portion of leaf including midvein. In the grass leaf, the mesophyll is not differentiated as palisade and spongy parenchyma. Bundle-sheath cells composed of sclerenchyma cells extend from the veins to both upper and lower epidermis. The epidermis of the grass leaf contains bulliform cells—large epidermal cells thought to play a part in the rolling and unrolling (folding and unfolding) of grass leaves. In the* Poa *leaf shown here, the bulliform cells are partly collapsed and the leaf is folded. An increase in turgor in the bulliform cells would presumably cause the leaf to unfold.*

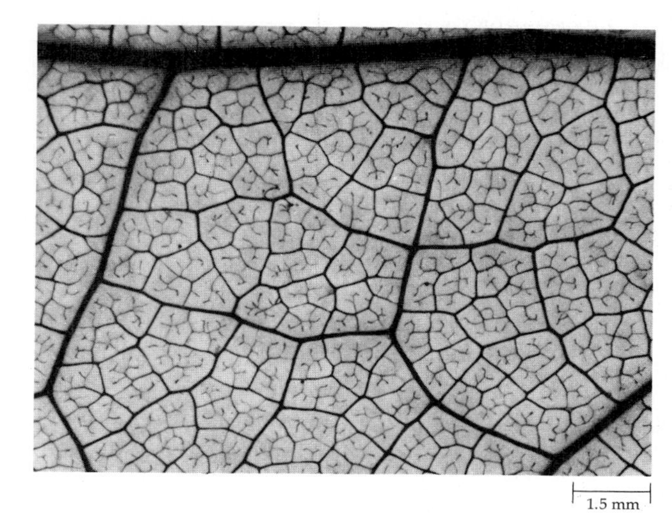

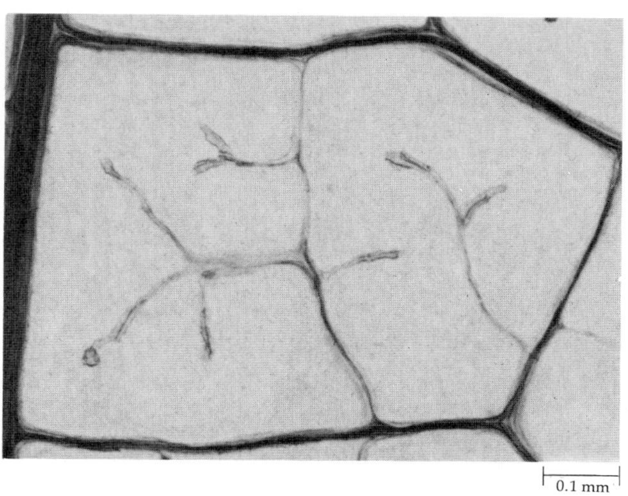

23–24
A leaf of the tulip tree (Liriodendron tulipifera), *cleared (with the chlorophyll removed) to show the veins, at three successive magnifications. No mesophyll cell of the leaf is far from a vein. Water and dissolved minerals are carried to the leaf through the xylem; organic molecules produced in the leaf are carried out of the leaf through the phloem.*

veins of fairly similar size that are oriented parallel to one another along the leaf. This vein arrangement is called *parallel venation*. In parallel-veined leaves, the longitudinal veins are interconnected by smaller veins, forming a complex network (Figure 23–25).

The veins contain xylem and phloem, which generally are entirely primary in origin. (The midvein and sometimes the coarser veins undergo secondary growth in some dicot leaves.) The vein-endings in dicot leaves often contain only tracheary elements, although both xylem and phloem elements may extend to the ends of the vein. Commonly, the xylem occurs on the upper side of the leaf, and the phloem occurs on the lower side (Figures 23–20a and 23–20b).

The vascular tissues of the veins are rarely exposed to the intercellular spaces of the mesophyll. The large veins are surrounded by parenchyma cells that contain few chloroplasts, whereas the small veins are enclosed by one or more layers of compactly arranged cells that form a *bundle sheath* (see Figures 23–20 through 23–23). The cells of the bundle sheath often resemble the mesophyll cells in which the small veins are located. The bundle sheaths extend to the ends of the veins, assuring that no part of the vascular tissue is exposed to air in the intercellular spaces and that all substances entering and leaving the vascular tissues must pass through the sheath (Figure 23–20d). Thus the bundle sheath performs a function analogous to that of the endodermis of the root.

In many leaves, the bundle sheaths are connected with either or both upper and lower epidermis by cells resembling the sheath cells (Figures 23–22 and 23–23). Such connections (actually, extensions of the sheaths) are called *bundle-sheath extensions*. Besides offering mechanical support to the leaf, in dicotyledons they apparently conduct water to the epidermis.

The epidermis itself provides considerable strength to the leaf because of its compact structure and its cuticle. In addition, the larger veins of dicot leaves are often bordered by collenchyma cells, which provide support to the leaf. In monocot leaves, the veins may be bordered by fibers. Collenchyma cells and fibers may also be found along the leaf margins of dicot and monocot leaves, respectively.

(a)

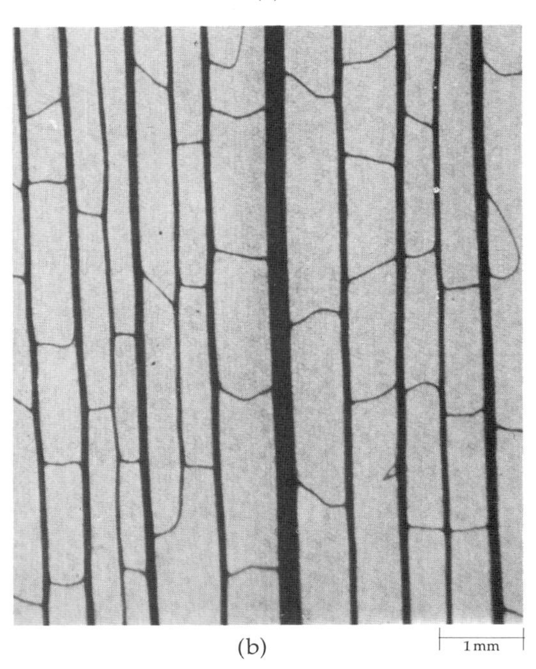

(b)

23–25
(a) *The basal leaves of this lady's slipper orchid* (Cypripedium acaule) *show the characteristic parallel venation of a monocot. This orchid is widespread in the eastern United States.* (b) *A portion of a cleared (chlorophyll removed) leaf of the orchid* Cochleanthus.

PLANTS AND AIR POLLUTION

The plant leaf, like the human lung, can function only when it is able to exchange gases with the surrounding air. As a consequence, the leaf, like the lung, is an organ that is exceedingly susceptible to air pollution.

Air pollution has many forms. Some pollutants are particulate. The particles may be organic, such as those present in the smoke produced by burning fossil fuels and garbage, or inorganic, such as cement kiln dusts, foundry dusts, and the lead compounds released in the combustion of leaded gasoline. As a major component of smog, these particles reduce the amount of sunlight reaching the earth's surface. Such particles also have direct ill effects on plants. They may clog stomata and prevent them from functioning, or they—particularly metallic particles—may act as plant poisons.

Fluorides, which enter the air as waste products from the manufacture of phosphates, steel, aluminum, and other industrial products, act as cumulative poisons, entering the leaf through the stomata and causing collapse of leaf tissue, apparently by inhibiting enzymes concerned with cellulose synthesis. Thousands of acres of Florida citrus groves have been damaged by fluorides discharged from phosphate fertilizer plants.

When ores containing sulfur are processed, sulfur dioxide is produced:

$$2CuS + 3O_2 \longrightarrow 2CuO + 2SO_2$$

Sulfur dioxide has a disagreeable pungent-sweet taste; it is unusual among air pollutants in that it can be tasted at lower concentrations than it can be smelled. Sulfur oxides are also produced by the burning of fossil fuels containing sulfur. In moist air, sulfur oxides react with water to form droplets of sulfuric acid, a strong corrosive acid and a component of acid rain. In several parts of the United States, virtual deserts have been created by the emission of sulfur dioxide combined with metals. As long ago as 1905, air pollution controls were instituted in areas surrounding copper smelters in Tennessee, but the surrounding area, once covered by luxuriant forest, remains barren to this day. Not only was all the vegetation killed but the acid leached the soil of nutrients. Similarly, in a copper-smelting area of the Sacramento Valley, California, all vegetation was killed over an area of 260 square kilometers, and growth was severely affected over an additional 320 square kilometers.

The most familiar type of air pollution to Californians is photochemical smog, which is produced by sunlight acting on automobile exhausts. The Los Angeles area offers an ideal setting for the formation of photochemical smog because the life of that area is heavily geared to the use of the automobile, and the mountains to the north and east form a "basin" that prevents the dispersal of reactants. Not only are many species of plants unable to survive in the city itself (as is true in many other major cities of the world) but smog moving out of the Los Angeles basin is damaging agricultural crops and killing pine forests in mountains as far away as 160 kilometers.

Sulfur dioxide injury on a blackberry *(Rubus)* leaf is characterized by areas of injured leaf tissue surrounded by healthy tissue.

One of the principal ingredients of photochemical smog is nitrogen dioxide, NO_2*, which is produced by any combustion process that occurs in air (dry air is 77 percent nitrogen) and so is present in automobile exhausts. Under the influence of light,* NO_2 *is split into* NO *and atomic oxygen. The latter is extremely reactive and forms ozone,* O_3*, by reaction with normal (molecular) oxygen.*

Similar reactions, powered by ultraviolet light, occur at the outer layers of the atmosphere, producing the ozone shield described on page 2. Ozone can also be produced by an electric spark and is the source of the "clean" smell after a lightning storm. Ozone is a highly toxic substance. In plants, it damages the thin-walled palisade cells, apparently affecting the permeability of the membranes of both the cells and their chloroplasts. Another component of photochemical smog is PAN (peroxyacetyl nitrate, $C_2H_3O_5N$*). It is several times more toxic than ozone but is normally present in much lower concentrations. Photosynthesis is reduced 66 percent by photochemical smog concentration of 0.25 part per million.*

At the present time, there is much concern over the adverse effects of acid rain on the environment. Under normal conditions, rain arising in a nonpolluted area should have a pH *of 5.6. With the burning of fossil fuels and the smelting of sulfide ores, the large quantities of sulfur oxides and nitrogen oxides emitted into the atmosphere react with water to form strong acids (sulfuric acid and nitric acid). Rain and snow formed in such areas have a* pH *of less than 5.6 and, by definition, are acidic. The phenomenon of acid precipitation is widespread today and occurs over large areas of western Europe, eastern United States, and southeastern Canada, where the yearly average* pH *of precipitation ranges from 4 to 4.5. Moreover, the rain of individual storms is often much more acid. Acid rains have been recorded in Scotland, Norway, and Iceland with* pH *values of 2.4, 2.7, and 3.5, respectively. The move to reduce local pollution problems through the building of tall stacks on smelters and industrial plants has created regional problems. Pollutants emanating from tall stacks are transported over long distances through the atmosphere. For example, more than 75 percent of the sulfur in rain that falls in the Scandinavian countries is believed to originate in the British Isles and central Europe. The effect of acid rain on vegetation is not well understood but is currently being investigated throughout the Northern Hemisphere. Acid rains have been shown to decrease growth of forest trees in Sweden. In addition, simulated acid rains have been shown to injure leaves and inhibit seed germination. The clearest effect of acid rain has been on fish populations, which have been virtually eliminated in acidified lakes in some parts of the world.*

Ozone injury on a tobacco *(Nicotiana)* leaf appears as stipples or flecks of dead tissues on the upper surface of the leaf. In cases of severe ozone injury, the flecks coalesce into larger lesions that are visible on both leaf surfaces.

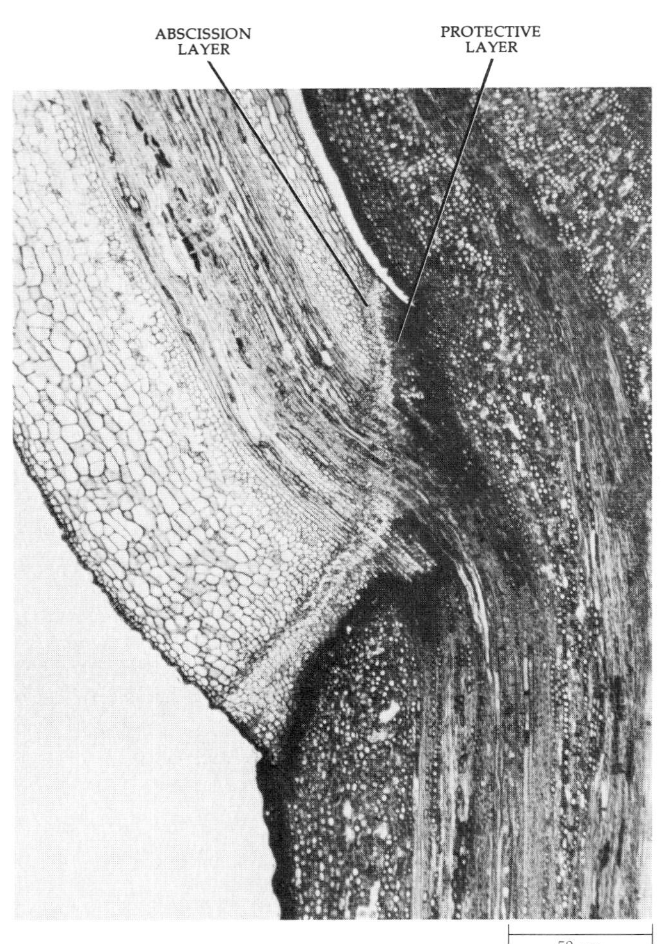

23–26
Abscission zone in maple (Acer) *leaf, as seen in longitudinal section through the base of the petiole.*

LEAF ABSCISSION

In many plants, the normal separation of the leaf from the stem—the process of *abscission*—is preceded by certain structural and chemical changes near the base of the petiole. These changes result in the formation of an *abscission zone* (Figure 23–26). In woody dicotyledons, two layers can be distinguished in the abscission zone: a separation (abscission) layer, and a protective layer. The separation layer consists of relatively short cells with poorly developed wall thickenings that make it structurally weak. A protective layer is formed by the deposition of suberin in the cell walls and intercellular spaces beneath the separation layer. After the leaf falls, the protective layer is recognized as a *leaf scar* on the stem (see Figure 24–18, page 487). Hormonal factors associated with abscission are discussed in Chapter 25.

TRANSITION BETWEEN VASCULAR SYSTEMS OF THE ROOT AND THE SHOOT

As discussed in previous chapters, the distinction between plant organs is based primarily on the relative distribution of the vascular and ground tissues. For example, in dicot roots the vascular tissues generally form a solid cylinder surrounded by the cortex. In addition, the strands of primary phloem alternate with the radiating ridges of primary xylem. In contrast, in the stem the vascular tissues often form a cylinder of discrete strands around a pith, with the phloem on the outside of the vascular bundles and the xylem on the inside. Obviously, somewhere in the primary plant body a change from the type of structure found in the root to that found in the shoot must take place. This change is a gradual one, and the region of the plant axis through which it occurs is called the *transition region.*

As described in Chapter 20, the shoot and root are initiated as a single continuous structure during the development of the embryo. Consequently, vascular transition occurs in the axis of the embryo or young seedling. This transition is initiated during the appearance of the procambial system in the embryo and is completed with the differentiation of the variously distributed procambial tissues in the seedling. Vascular continuity between the root and shoot systems is maintained throughout the life of the plant.

The structure of the transition region can be very complex, and much variation exists in the transition regions of different kinds of plants. In most gymnosperms and dicotyledons, vascular transition occurs between the root and cotyledons. Figure 23–27 depicts a type of transition region commonly found in dicotyledons. Notice the diarch (having two pro-

23–27
The transition region—the connection between root and cotyledons—in the seedling of a dicotyledon with a diarch root. In the root, the primary vascular system is represented by a single cylinder of vascular tissue. In the hypocotyl-root axis, the vascular system branches and diverges into the cotyledons, and the xylem and phloem become reoriented along the hypocotyl-root axis.

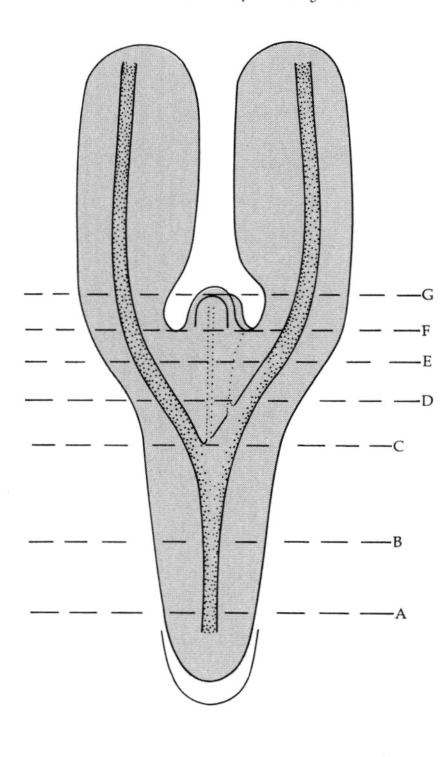

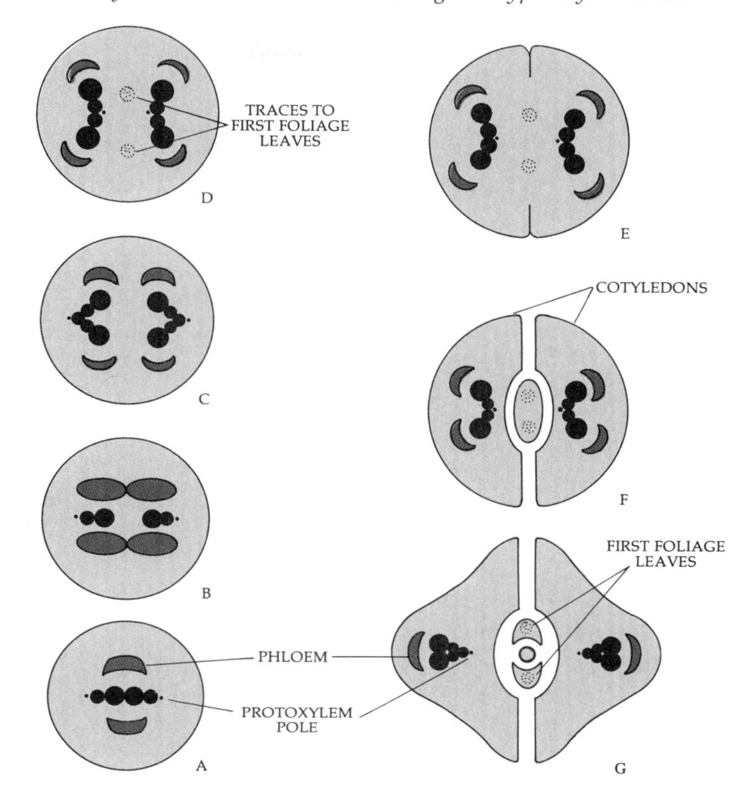

toxylem poles) structure of the root; the branching and reorientation of the primary xylem and the primary phloem, which in the upper part of the axis results in the formation of a pith; and the traces of the first leaves of the epicotyl.

DEVELOPMENT OF THE FLOWER

The development of the flower or inflorescence terminates the meristematic activity of the vegetative shoot apex.

During the transition to flowering, the vegetative shoot apex undergoes a sequence of physiological and structural changes and is transformed into a reproductive apex. Consequently, flowering may be considered as a stage in the development of the shoot apex and of the plant as a whole. Inasmuch as the reproductive apex exhibits a determinate growth pattern, flowering in annuals indicates that the plant is approaching completion of its life cycle. By contrast, flowering in perennials may be repeated. Various environmental factors, including the length of day and the temperature, are known to be involved in the induction to flowering (see Chapter 26).

The transition from a vegetative to a floral apex is often preceded by an elongation of the internodes and the early development of lateral buds below the shoot apex. The apex itself undergoes marked increase in mitotic activity, accompanied by changes in dimensions and organization: from a relatively small apex with a tunica-corpus type of organization, the apex becomes broad and domelike.

The initiation and early stages of development of the sepals, petals, stamens, and carpels are quite similar to those of leaves. Commonly, the initiation of the floral parts begins with the sepals, followed by the petals, then the stamens, and finally the carpels (Figure 23–28). This usual order of appearance of the floral parts may be modified in certain flowers, but the floral parts always have the same relative spatial relation to one another (Figures 23–29 and 23–30). The floral parts may remain separate during their development, or they may become united within different whorls (coalescence) and between whorls (adnation).

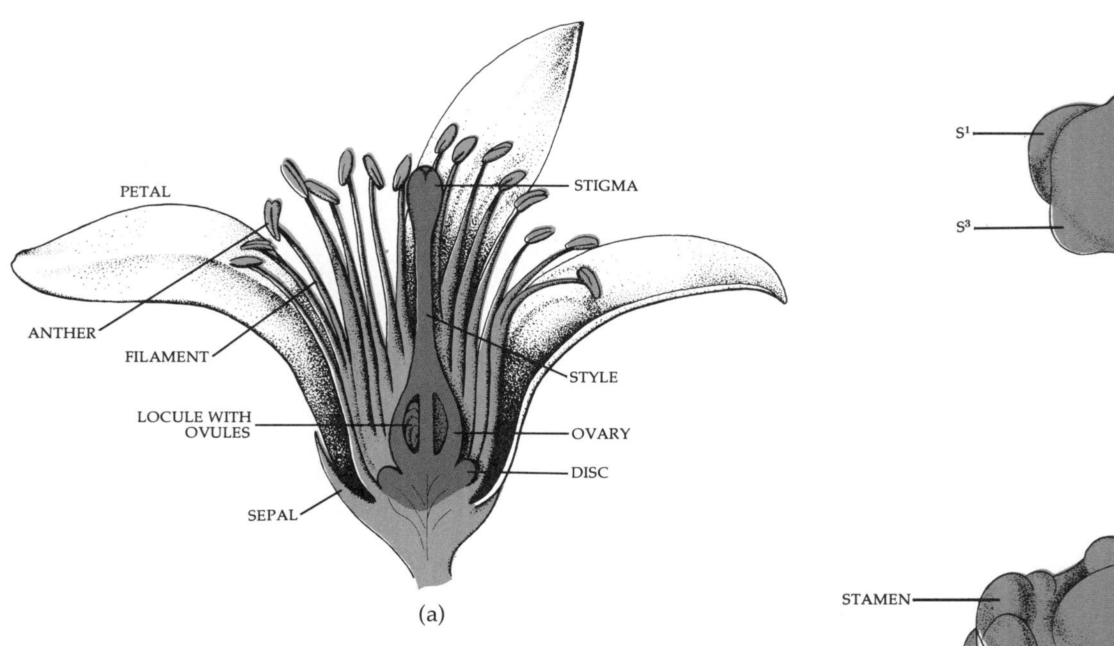

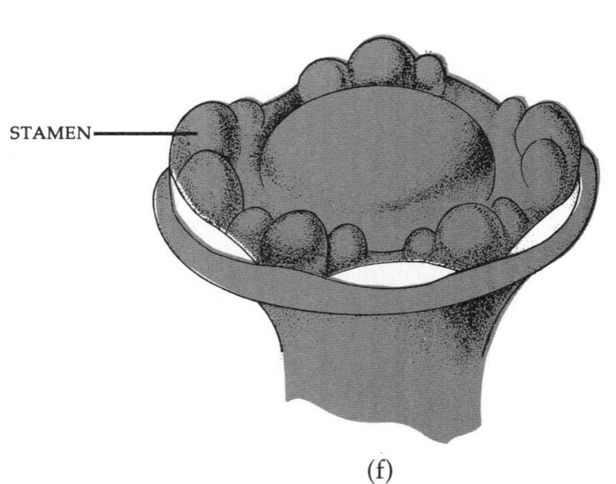

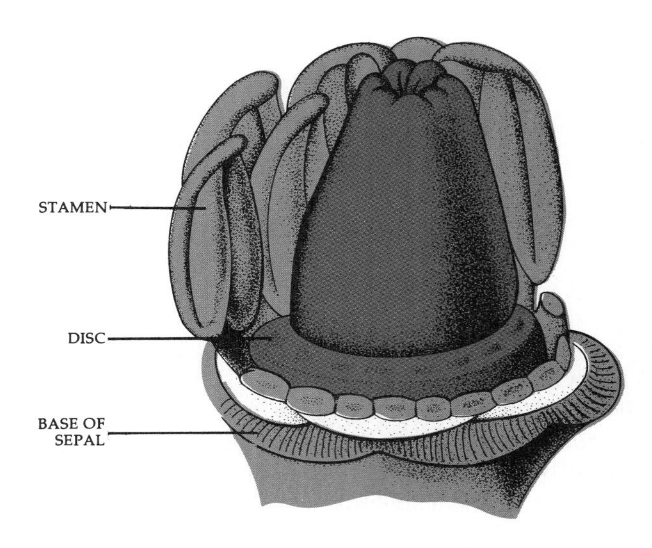

23–28
Development of an orange (Citrus aurantium) *flower.* (a) *Longitudinal section of the flower.* (b) *Successive appearance of the five sepals* (S1 *through* S5). (c) *Simultaneous appearance of the five petals, which alternate with the sepals.* (d) *Appearance of the first five stamens, which alternate with the petals and arise directly above the sepals.* (e) *Each of the original stamens is now accompanied by two others, one on its right, the other on its left. (The calyx has been removed.)* (f) *Flower a little older than in* (e). *Again, each group of stamens consists of three stamens, but they are more developed than in* (e). *In the center, the receptacle is elevated. (Both calyx and corolla have been removed.)* (g) *Between each group of three stamens, others are now borne, so that the androecium forms a whorl in which all the stamens touch each other. The stamens, being borne successively, are of unequal size; the five which are the first to appear are more elevated than the others.* (h) *Gynoecium at the moment of its appearance from the disc. Some stamens, the bases of removed stamens, carpel primordia, developing locules, and placental walls are shown here.* (i) *Gynoecium further along in development than in* (h). (j) *Gynoecium shortly after opening of the flower.* (k) *Gynoecium enclosed at its base by the disc.* (l) *Young stamen, with a short filament and a long anther.*

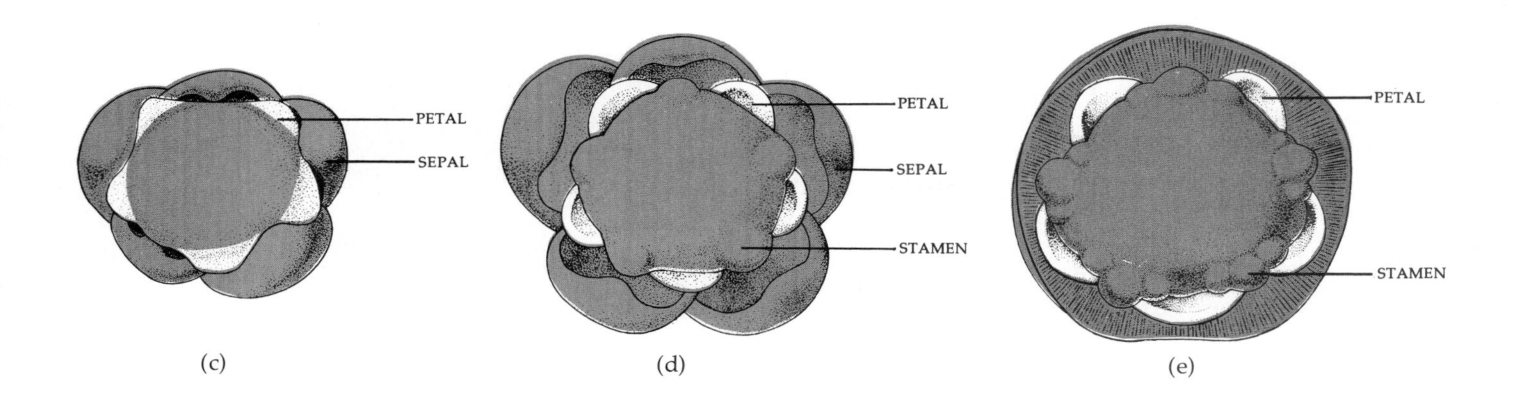
PETAL
SEPAL
(c)
PETAL
SEPAL
STAMEN
(d)
PETAL
STAMEN
(e)

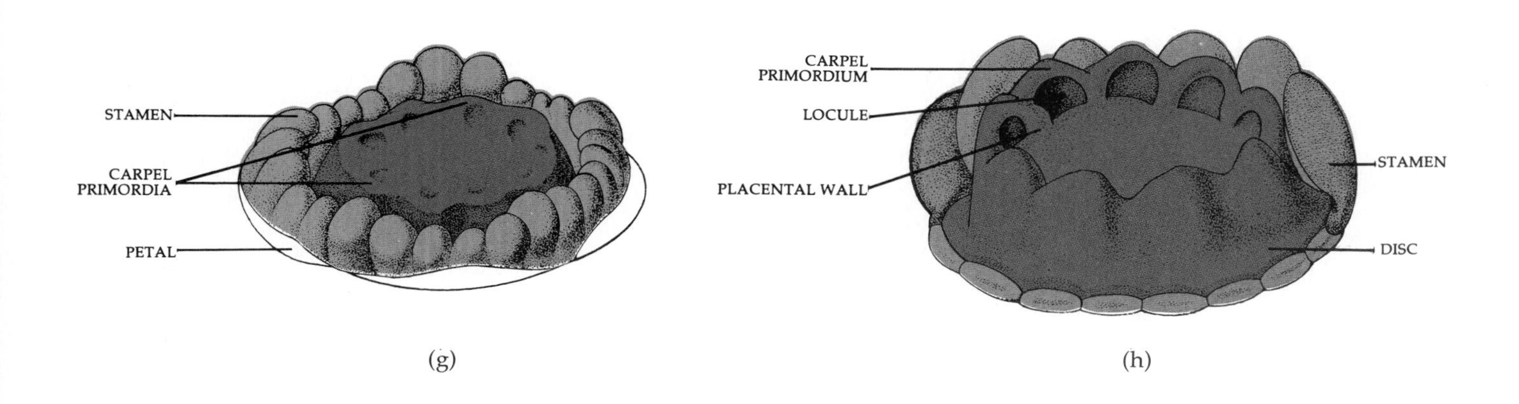
STAMEN
CARPEL
PRIMORDIA
PETAL
(g)
CARPEL
PRIMORDIUM
LOCULE
PLACENTAL WALL
STAMEN
DISC
(h)

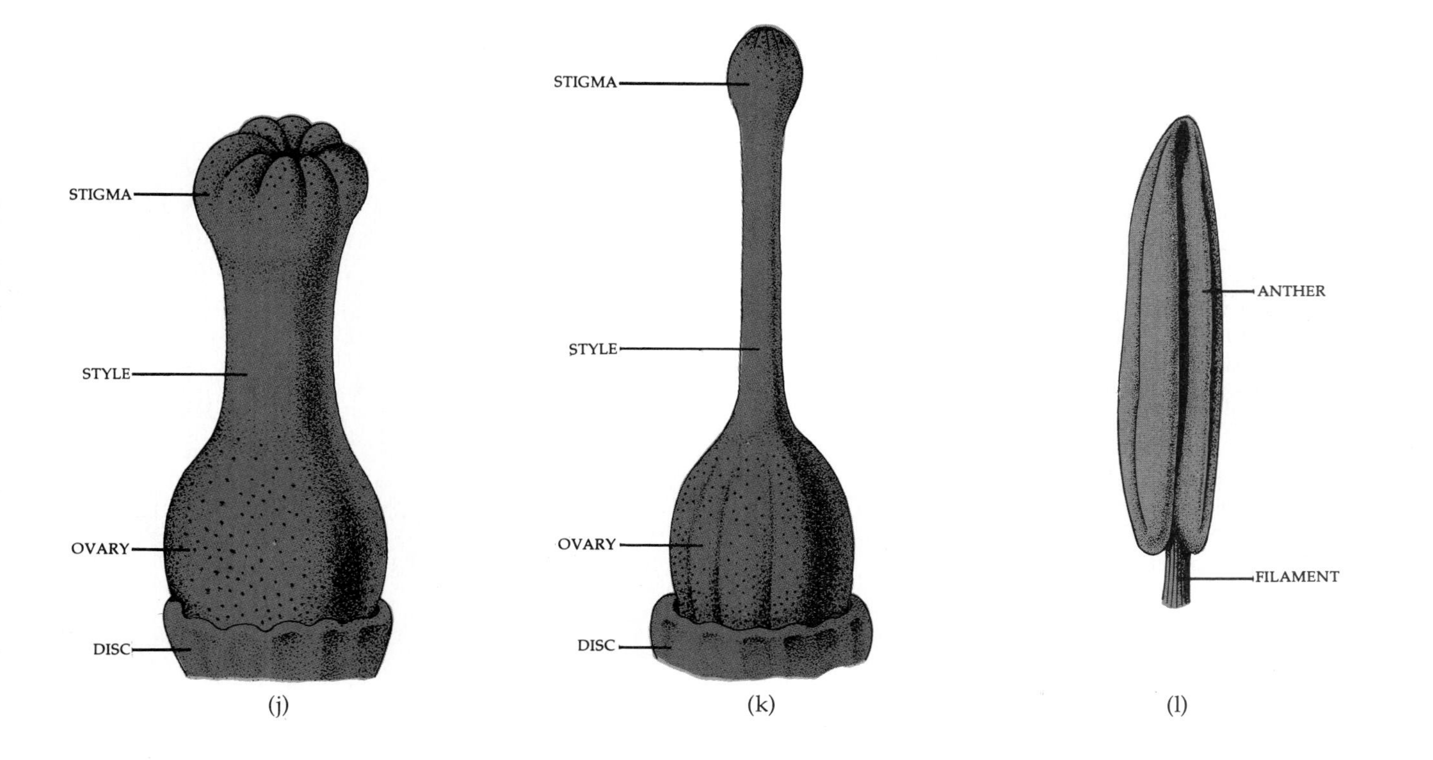
STIGMA
STYLE
OVARY
DISC
(j)
STIGMA
STYLE
OVARY
DISC
(k)
ANTHER
FILAMENT
(l)

23–29
Longitudinal section of young inflorescence of fleabane (Erigeron), *a member of the Asteraceae. The inflorescence contains numerous flower primordia. The older flower primordia are on the periphery of the inflorescence, the younger in the center.*

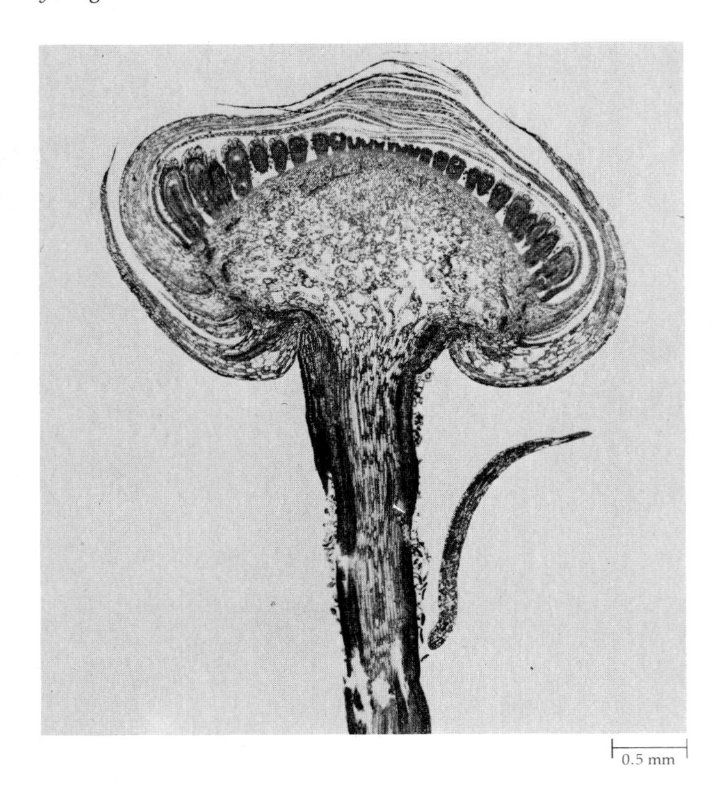

COROLLA

(a) 50 μm

COROLLA

STAMENS

(b) 50 μm

CARPELS COROLLA STAMEN

(c) 50 μm

23–30
Stages in development of the fleabane (Erigeron) *flower, as seen in longitudinal sections.* (a) *Flower primordium with developing corolla.* (b) *Initiation of stamens, which are adnate to the corolla.* (c) *Flower with developing corolla, stamens, and carpels.* (d) *and* (e) *Flowers with developing ovules and styles. The two carpels overarch the cavity containing the ovule and then become prolonged into a solid style with a two-part stigma. The calyx, or pappus, is initiated at about the same time as the anthers but develops slowly.* (f) *Flowers approaching maturity. Pollen grains can now be seen in the anthers. Notice that the fleabane flower has an inferior ovary.*

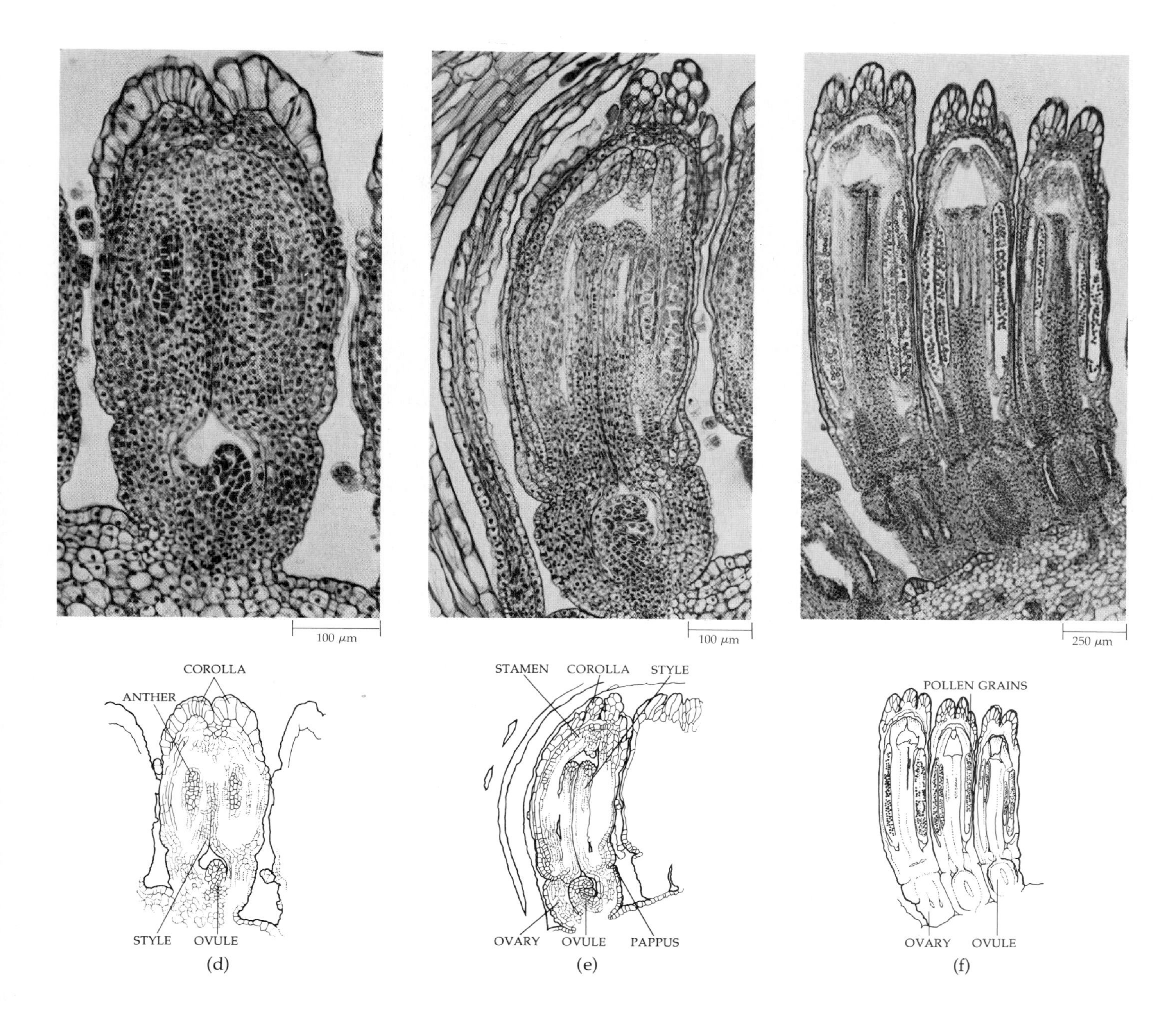

The basic structure of the flower and some of its variations were discussed in Chapter 18.

STEM AND LEAF MODIFICATIONS

Stems and leaves may undergo modifications and perform functions quite different from those commonly associated with these two components of the shoot. One of the most common modifications is the formation of <u>tendrils</u>, which aid in support. Some tendrils are modified stems. In ivy *(Hedera)*, for example, the tendrils produce enlarged, cuplike structures—holdfasts—at their tips. The tendrils of the grape *(Vitis)* (Figure 23–31), as well as those of Virginia creeper *(Parthenocissus quinquefolia)* are also modified stems that coil around the supporting structure. In the grape, the tendrils sometimes produce small leaves or flowers.

Most tendrils are leaf modifications. In legumes, such as the garden pea *(Pisum sativum)*, the tendrils constitute the terminal part of the pinnately compound leaf (Figure 23–17). Not all legumes form tendrils. One legume, the peanut *(Arachis hypogaea)*, has another interesting adaptation. After fertilization takes place, the stamens and corolla of the flower fall off, and the internode between ovary and receptacle begins to elongate. With continued elongation, the stalk bends downward and buries the developing fruit several centimeters into the ground, where it ripens. If the ovary is not buried, it withers and fails to mature.

23–31
The tendrils of grape (Vitis) *are modified stems.*

23–33
The branches of the spineless cactus Epiphyllum *resemble leaves but are actually modified stems called cladophylls.*

23–32
The filmy branches of the common edible asparagus (Asparagus officinalis) *resemble leaves. Such modified stems are called cladophylls.*

Branches that assume the form of and closely resemble foliage leaves are called *cladophylls.* The filmy, leaflike branches of asparagus *(Asparagus officinalis)* are familiar examples of cladophylls (Figure 23–32). The thick and fleshy aerial shoots ("spears") of asparagus are the edible portion of the plant. The scales found on the spears are true leaves. If the asparagus plants are allowed to continue growing, cladophylls develop in the axils of the minute, inconspicuous scales and then act as photosynthetic organs. In some cacti, the branches resemble leaves (Figure 23–33).

In some plants, the leaves are modified as spines, which are hard and dry and nonphotosynthetic. The terms "spine" and "thorn" are frequently used interchangeably, but technically thorns are modified branches that arise in the axils of leaves (Figure 23–34). Another term commonly used interchangeably with thorn and spine is "prickle." A prickle, however, is neither a stem nor a leaf but a small, more or less slender, sharp outgrowth from the cortex and epidermis. The so-called thorns on rose stems are prickles.

Among the most spectacular of modified or specialized leaves are those of the carnivorous plants, such as the pitcher plant, the sundew, and the Venus flytrap, which capture insects and digest them with enzymes secreted by the plant. The available nutrients are absorbed by the plant (see Chapter 27).

(a)

(b)

23–34

(a) *Spines, as in this* Cereus *cactus, are modified leaves.* (b) *Thorns are modified branches, which, as this photograph of a hawthorn* (Crataegus flava) *shows, arise in the axils of the leaves.*

CONVERGENT EVOLUTION

Comparable selective forces, acting on plants growing in similar habitats but different parts of the world, often cause totally unrelated species to assume a similar appearance. The process by which this happens is known as convergent evolution.

Let us consider some of the adaptive characteristics of plants growing in desert environments—fleshy, columnar stems (that provide the capacity for water storage), protective spines, and reduced leaves. Three fundamentally different families of flowering plants—the spurge family (Euphorbiaceae), the cactus family (Cactaceae), and the milkweed family (Asclepiadaceae)—have members that have evolved in this direction. The cactuslike representatives of the spurge and milkweed families shown here evolved from leafy plants that look quite different from one another.

SPURGE CACTUS

MILKWEED

The cacti occur (with one exception) exclusively in the New World. The comparably fleshy members of the spurge and milkweed families occur mainly in desert regions in Asia and especially Africa, where they play an ecological role similar to that of the New World cacti.

Although all the plants shown in this photograph have CAM *photosynthesis, all three are related to and derived from plants that have* C_3 *photosynthesis, which indicates that the physiological adaptations involved in* CAM *photosynthesis also arose as a result of convergent evolution* *(see pages 111–12)**.*

23–35
White potato (Solanum tuberosum).
(a) *Plant with tubers attached to a rhizome, or underground stem.* (b) *Detail of a rhizome. Notice the two very young tubers arising from the rhizome above the larger potatoes.*

(a)

(b)

Food Storage

Stems, like roots, serve food storage functions. Probably the most familiar type of specialized storage stem is the *tuber,* as exemplified by the Irish, or white, potato *(Solanum tuberosum).* In the white potato, tubers arise at the tips of *stolons* (slender stems growing along the surface of the ground) of plants grown from seed. However, when cuttings of tubers are used for propagation, the tubers arise at the ends of long thin *rhizomes,* or underground stems (Figure 23–35). Except for vascular tissue, almost the entire mass of the tuber inside the periderm ("skin") is storage parenchyma. The so-called "eyes" of the white potato are depressions containing groups of buds. The depression is the axil of a scalelike leaf.

A *bulb* is a large bud consisting of a small, conical stem with numerous modified leaves attached to it. The leaves are scalelike and contain thickened bases where food is stored. Adventitious roots arise from the bottom of the stem. Familiar examples of plants with bulbs are the onion (Figure 23–36b) and the lily.

Although superficially similar to bulbs, *corms* consist primarily of stem tissue. Their leaves commonly are thin and much smaller than those of bulbs; consequently, the stored food of the corm is found within the fleshy stem. Several well known plants, such as gladiolus (Figure 23–36c), crocus, and cyclamen, produce corms.

Kohlrabi *(Brassica oleracea* var. *caulorapa)* is one example of an edible plant with a fleshy storage stem. The short, thick stem stands above the ground and bears several leaves with very broad bases (Figure 23–36a). The common cabbage *(Brassica oleracea* var. *capitata)* is closely related to kohlrabi. The so-called "head" of cabbage consists of a short stem bearing numerous thick, overlapping leaves. In addition to a terminal bud, several well-developed axillary buds may be found within the head.

The leaf stalks, or petioles, of some plants become quite thick and fleshy. Celery *(Apium graveolens)* and rhubarb *(Rheum rhaponticum)* are familiar examples.

23–36
Examples of modified leaves or stems.
(a) *The fleshy storage stem of kohlrabi* (Brassica oleracea *var.* caulorapa).
(b) *An onion* (Allium cepa) *bulb, which consists of a conical stem with scalelike, food-containing leaves attached. The leaves are the part of the onion we eat.*
(c) *A gladiolus* (Gladiolus grandiflora) *corm, which is a fleshy stem with small, thin leaves.*

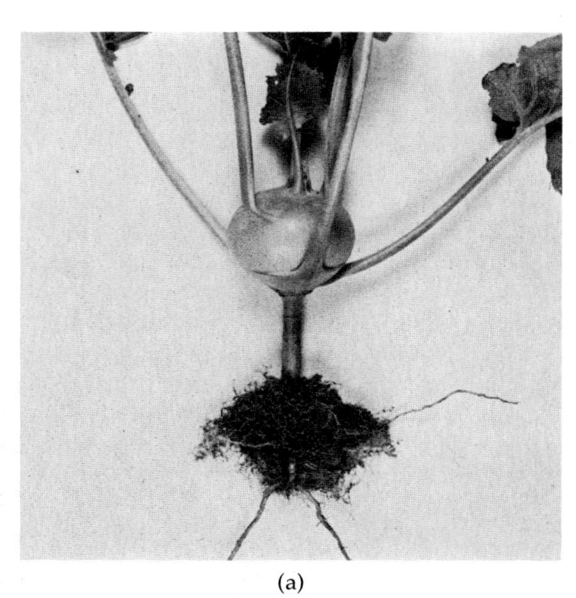

(a)

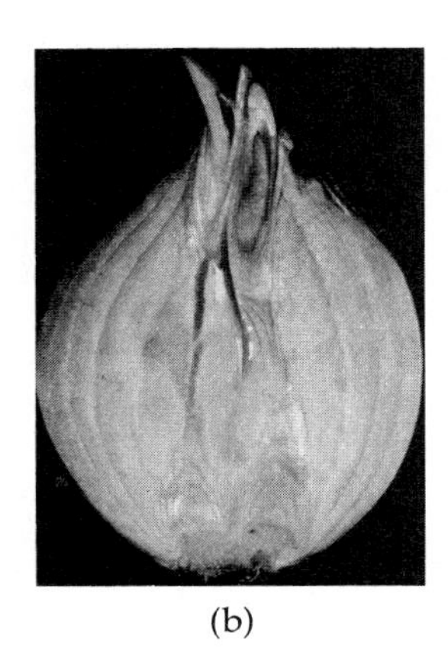

(b)

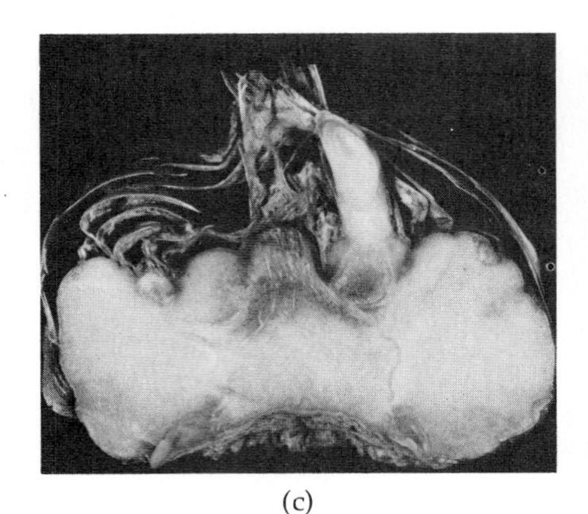

(c)

Water Storage: Succulency

Succulent plants are plants that have juicy tissues, that is, tissues specialized for the storage of water. Most of these plants, such as the cacti of the American deserts, the *Euphorbia* of similar appearance of the African deserts (see "Convergent Evolution," page 473), and the century plant *(Agave)*, normally grow in arid regions, where the ability to store water is necessary for their survival. The green, fleshy stems of the cacti serve both as photosynthetic and storage organs. The water-storing tissue consists of large, thin-walled parenchyma cells that lack chloroplasts.

In the century plant, the leaves are succulent. As in succulent stems, nonphotosynthetic parenchyma cells of the ground tissue constitute the water-storing tissue. Other examples of plants with succulent leaves are the ice plant *(Mesembryanthemum crystallinum)*, the stonecrops *(Sedum)*, and certain species of *Peperomia*. In the ice plant, large epidermal cells with appendages (trichomes), called water vesicles, which superficially resemble beads of ice, serve a water storage function (see Figure 21–21c, page 429). The water-storing cells of the *Peperomia* leaf are part of a multiple (several-layered) epidermis derived by anticlinal divisions of the protoderm (Figure 23–37).

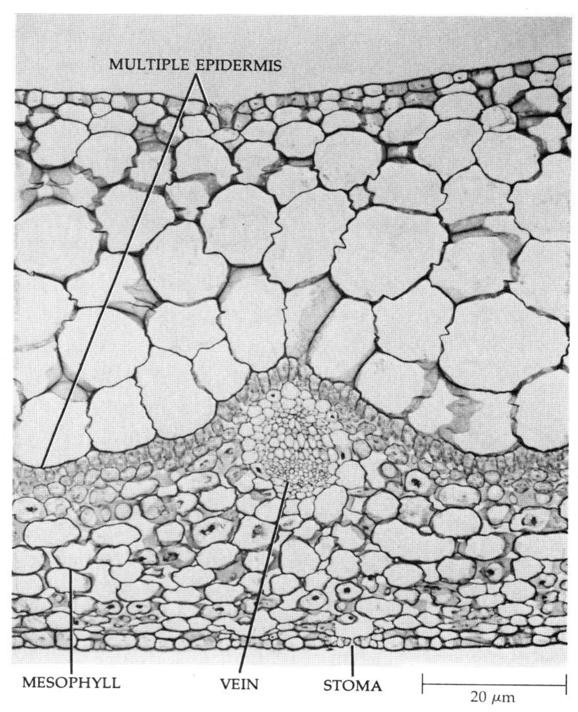

23–37
Transverse section of leaf blade of Peperomia. *The very thick multiple epidermis, visible on its upper surface, presumably functions as water-storage tissue.*

SUMMARY

The vegetative shoot apices of most flowering plants have a tunica-corpus type of organization consisting of one or more peripheral layers of cells (the tunica) and an interior (the corpus). Although the primary tissues of the stem pass through periods of growth similar to those of the root, the stem cannot be divided into regions of cell division, elongation, and maturation in the same manner as roots. The stem increases in length largely by internodal elongation.

As in the root, the apical meristem of the shoot gives rise to protoderm, ground meristem, and procambium, which develop into the primary tissues. Three basic variations exist in stems with regard to the relative distribution of ground and primary vascular tissues: the primary tissues may develop (1) as a more or less continuous hollow cylinder, (2) as a cylinder of discrete strands, or (3) as a system of strands scattered throughout the ground tissue. Regardless of the type of organization, the phloem is commonly located outside the xylem.

Leaves have their origin in the peripheral region of the shoot apex, and their position on the stem is reflected in the pattern of the vascular system in the stem. Leaves are determinate in growth; that is, their development is of relatively short duration. However, vegetative shoot apices may exhibit unlimited or indeterminate growth.

In dicotyledons, most leaves consist of a blade and petiole. The blades of some leaves are divided into leaflets. Stomata are commonly more numerous on the lower than the upper surface of the leaf. The ground tissue, or mesophyll, of the leaf is specialized as a photosynthetic tissue and, in mesophytes, is differentiated into palisade parenchyma and spongy parenchyma. The mesophyll is thoroughly permeated by air spaces and by veins, which are composed of xylem and phloem surrounded by a parenchymatous bundle sheath. The xylem commonly occurs on the upper side of the vein, whereas the phloem occurs on the lower side. In many plants, leaf abscission is preceded by formation of an abscission zone at the base of the petiole.

The change in the type of structure found in the root to that in the shoot occurs in a region of the plant axis of the embryo and young seedling called the transition region.

At flowering, the vegetative shoot apex is directly transformed into a reproductive apex.

Stems, like roots, may serve food-storage functions. Examples of fleshy stems are tubers, bulbs, and corms. Water-storing plants are known as succulents. The water-storage tissue of succulent plants is made up of large parenchyma cells. Stems or leaves or both may be succulent.

CHAPTER 24

Secondary Growth

24–1
A solitary shagbark hickory (Carya ovata) *in the winter condition. Plants have been able to achieve such great stature because of the ability of their roots and stems to increase in girth, that is, to undergo secondary growth. Most of the tissue produced in this manner is secondary xylem, or wood, which not only conducts water and minerals to the far reaches of the shoot but also provides great strength to the roots and stems.*

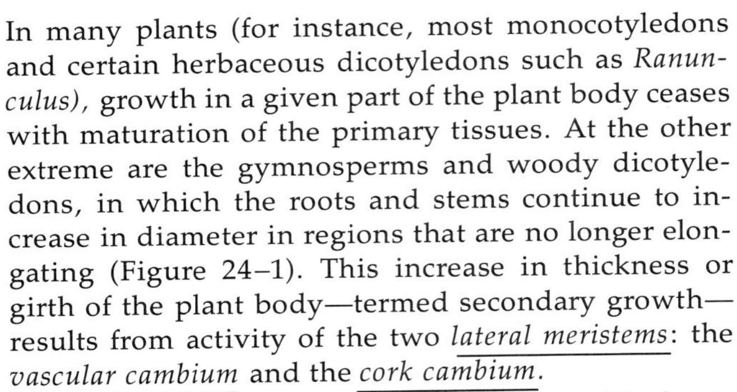

In many plants (for instance, most monocotyledons and certain herbaceous dicotyledons such as *Ranunculus*), growth in a given part of the plant body ceases with maturation of the primary tissues. At the other extreme are the gymnosperms and woody dicotyledons, in which the roots and stems continue to increase in diameter in regions that are no longer elongating (Figure 24–1). This increase in thickness or girth of the plant body—termed secondary growth—results from activity of the two *lateral meristems*: the *vascular cambium* and the *cork cambium.*

Herbs, or herbaceous plants, are plants with shoots that undergo little or no secondary growth. In temperate regions either the shoot or the entire plant lives for only one season, depending on the species. Woody plants—trees and shrubs—can live for many years. At the start of each growing season, primary growth is resumed, and additional secondary tissues are added to the older plant parts through reactivation of the lateral meristems. Although most monocots lack secondary growth, some (such as the palms) may develop thick stems by primary growth alone.

Plants are often classified according to their seasonal growth cycles as annuals, biennials, or perennials. In the *annuals*—which include many weeds, wild flowers, garden flowers, and vegetables—the entire cycle from seed to vegetative plant to flowering plant to seed again occurs within a single growing season, which may be only a few weeks in length. Only the dormant seed bridges the gap between one season of growth and the next.

In the *biennials*, two seasons are needed for the period from seed germination to seed formation. The first season of growth ends with the formation of a root, a short stem, and a rosette of leaves near the soil surface. In the second growing season, flowering, fruiting, seed formation, and death occur, completing the life cycle. In temperate regions, annuals and biennials seldom become woody, although both their stems and roots may undergo a limited amount of secondary growth.

Perennials are plants in which the vegetative structures live year after year. The herbaceous perennials pass unfavorable seasons as dormant underground roots, rhizomes, bulbs, or tubers. The woody perennials, which include vines, shrubs, and trees, survive above ground but usually stop growing during the unfavorable seasons. Woody perennials flower only when they become adult plants, which may take many years. For example, the horse chestnut, *Aesculus hippocastanum,* does not flower until it is about 25 years old. *Puya raimondii,* a large (up to 10 meters high) relative of the pineapple that is found in the Andes, takes about 150 years to flower. Many woody plants are deciduous, losing all their leaves at the same time and developing new leaves from buds when the season again becomes favorable for growth. In evergreen trees and shrubs, leaves are also lost and replaced but not simultaneously.

THE VASCULAR CAMBIUM

Unlike the many-sided initials of the apical meristems, which contain dense cytoplasm and large nuclei, the meristematic cells of the vascular cambium are highly vacuolated. They exist in two forms: as vertically elongated *fusiform initials,* and as horizontally elongated or squarish *ray initials*. The fusiform initials are much longer than they are wide and appear flattened or brick-shaped in transverse section. In the white pine, *Pinus strobus,* the fusiform initials average 3.2 millimeters in length; in the apple, *Malus sylvestris,* 0.53 millimeters (Figure 24–2); and in the black locust, *Robinia pseudo-acacia,* 0.17 millimeters (Figure 24–3).

Secondary xylem and secondary phloem are produced through periclinal divisions of the cambial initials and their derivatives. In other words, the cell plate that forms between the dividing cambial initials is parallel to the surface of the root or stem (Figure 24–4a). If the derivative of a cambial initial is divided off toward the outside of the root or stem, it eventually becomes a phloem cell; if it divides off toward the inside, it becomes a xylem cell. In this manner a long, continuous radial file, or row, of cells is formed, extending from the cambial initial outward to the phloem and inward to the xylem (Figure 24–5).

The xylem and phloem cells produced by the fusiform initials have their long axes oriented vertically and make up what is known as the *axial system* of the secondary vascular tissues. The ray initials pro-

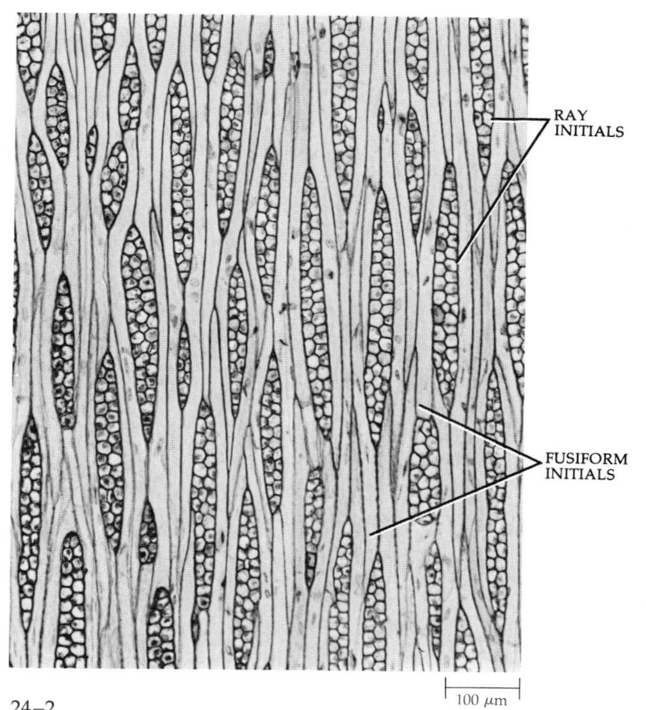

24–2
Tangential view of a portion of vascular cambium of the apple (Malus sylvestris) *tree. Tangential sections are cut at right angles to the rays, so we see the rays here in transverse section.*

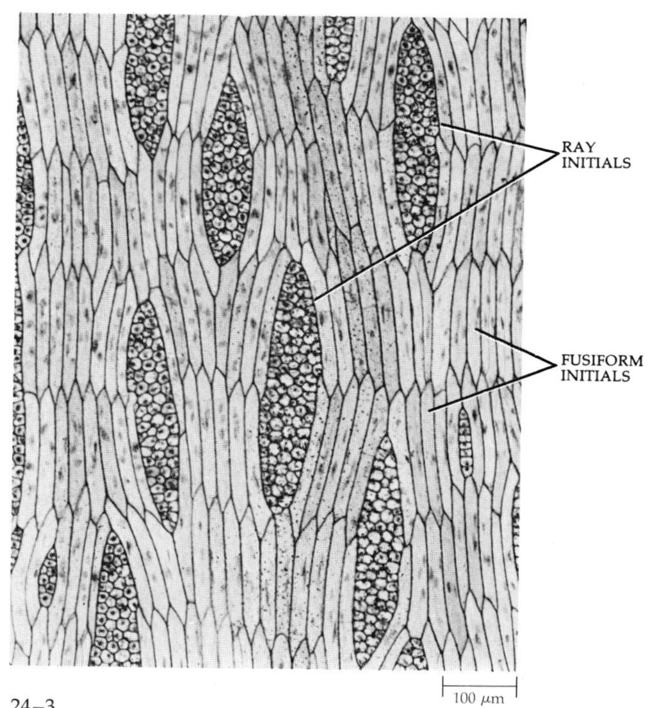

24–3
Vascular cambium of the black locust (Robinia pseudo-acacia) *tree, seen here in tangential view.*

24–4
Periclinal and anticlinal divisions of fusiform initials. (a) *Periclinal divisions are involved in the formation of secondary xylem and secondary phloem cells, and result in the formation of radial rows of cells (see Figure 24–5). When an initial divides periclinally, two cells appear, one behind (or in front of) the other.* (b) *Anticlinal divisions are involved in the multiplication of fusiform initials. When an initial divides anticlinally, two cells appear side by side where one was present previously.*

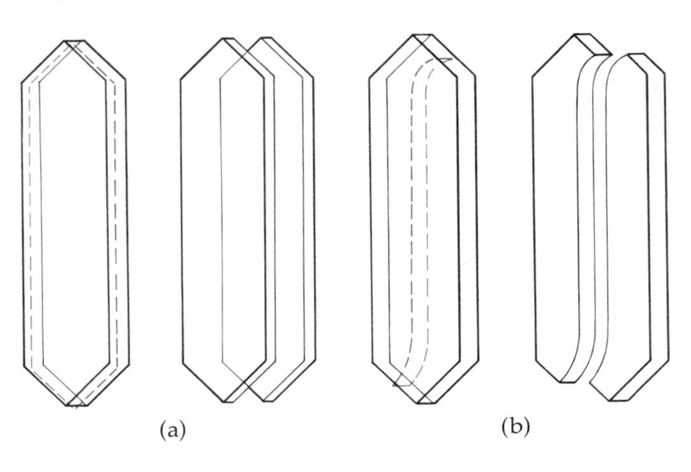

24–5
Diagram showing the relationship of the vascular cambium to its derivative tissues—secondary xylem and secondary phloem. The vascular cambium is made up of two types of cells—fusiform initials and ray initials—which give rise to the axial and radial systems, respectively, of the secondary vascular tissues. When the cambial initials produce secondary xylem and secondary phloem, they divide periclinally. Following division of an initial, one daughter cell (the initial) remains meristematic, and the other (the derivative of the initial) eventually develops into one or more cells of the vascular tissue. Cells produced toward the inner surface of the vascular cambium become xylem elements, and those produced toward the outer surface become phloem elements. The ray initials divide to form vascular rays, which lie at right angles to the derivatives of the fusiform initials. With the production of additional secondary xylem, the vascular cambium and secondary phloem are displaced in an outward direction. The diagrams (left to right) represent successively more mature stages.

duce horizontally oriented *ray cells,* which form the *vascular rays* or *radial system* (Figure 24–5). The rays are composed largely of parenchyma cells and are variable in length. Nutrients move from protoplast to protoplast via plasmodesmata *(symplastic movement),* passing from the secondary phloem through the vascular cambium and to the living cells of the secondary xylem by way of the vascular rays. By contrast, water passes from the secondary xylem to the cambium and the secondary phloem largely by way of the walls *(apoplastic movement)* of both the rays and cells of the axial system. The rays also serve as storage centers for such substances as starch and lipids.

In a restricted sense, the term "vascular cambium" is used to refer only to the cambial initials, of which there is one per radial file (Figure 24–5). However, it is often difficult, if not impossible, to distinguish between the initials and their immediate derivatives, which may remain meristematic for a considerable period of time. Even in the winter condition, when the cambium is dormant, or inactive, several layers of undifferentiated cells that are similar in appearance can be seen between the xylem and phloem. Consequently, some botanists use the term "vascular cambium" in a broader sense to refer to the initials and their immediate derivatives, which are indistinguishable from the initials. Others refer to this region of initials and derivatives as the *cambial zone*.

As the vascular cambium adds cells to the secondary xylem and the core of xylem increases in width, the cambium is displaced outward. In order to accommodate to this change, the vascular cambium undergoes an increase in circumference, which is accomplished by anticlinal divisions of the initials (Figure 24–4b). Along with an increase in the number of fusiform initials, new ray initials and rays are added, so that a fairly constant ratio of rays to fusiform cells is

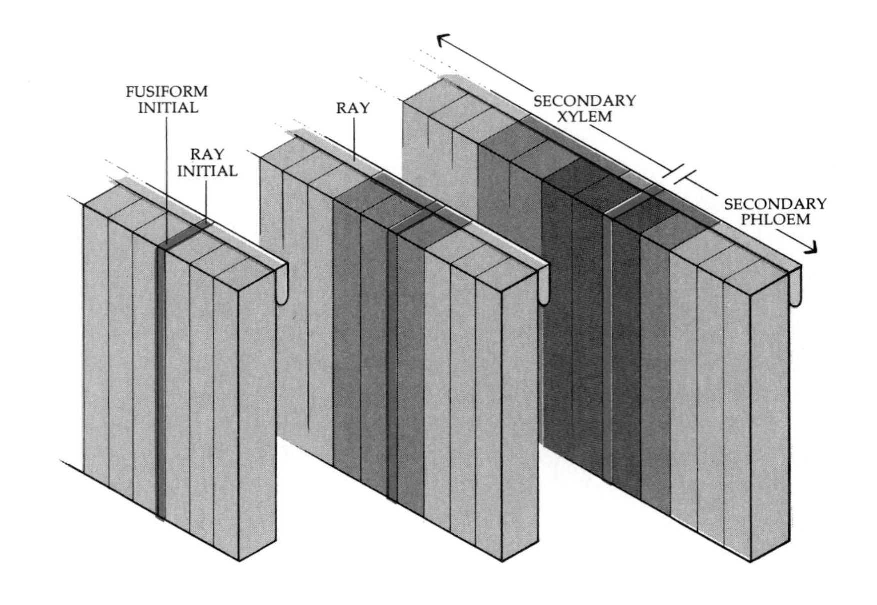

maintained in the secondary vascular tissues. Obviously, the developmental changes that occur in the cambium are exceedingly complex.

In temperate regions, the vascular cambium is dormant during winter and becomes reactivated in the spring. New growth layers, or increments, of secondary xylem and secondary phloem are laid down during the growing season. Reactivation of the vascular cambium is triggered by the expansion of the buds and resumption of their growth. Apparently, the hormone auxin, produced by the developing shoots, moves downward in the stems and stimulates resumption of cambial activity. Other factors are also involved in cambial reactivation and in continued normal growth of the cambium (Chapter 25).

EFFECT OF SECONDARY GROWTH ON THE PRIMARY PLANT BODY

Root

In roots, the vascular cambium is initiated by the meristematic procambial cells that remain between the primary xylem and primary phloem. Thus, depending on the number of phloem strands present in the root, two or more independent regions of cambial activity are initiated more or less simultaneously (Figure 24–6). Soon afterward, the pericyclic cells opposite the protoxylem poles divide periclinally, and the inner sister cells contribute to the vascular cambium. Now the cambium completely surrounds the core of xylem.

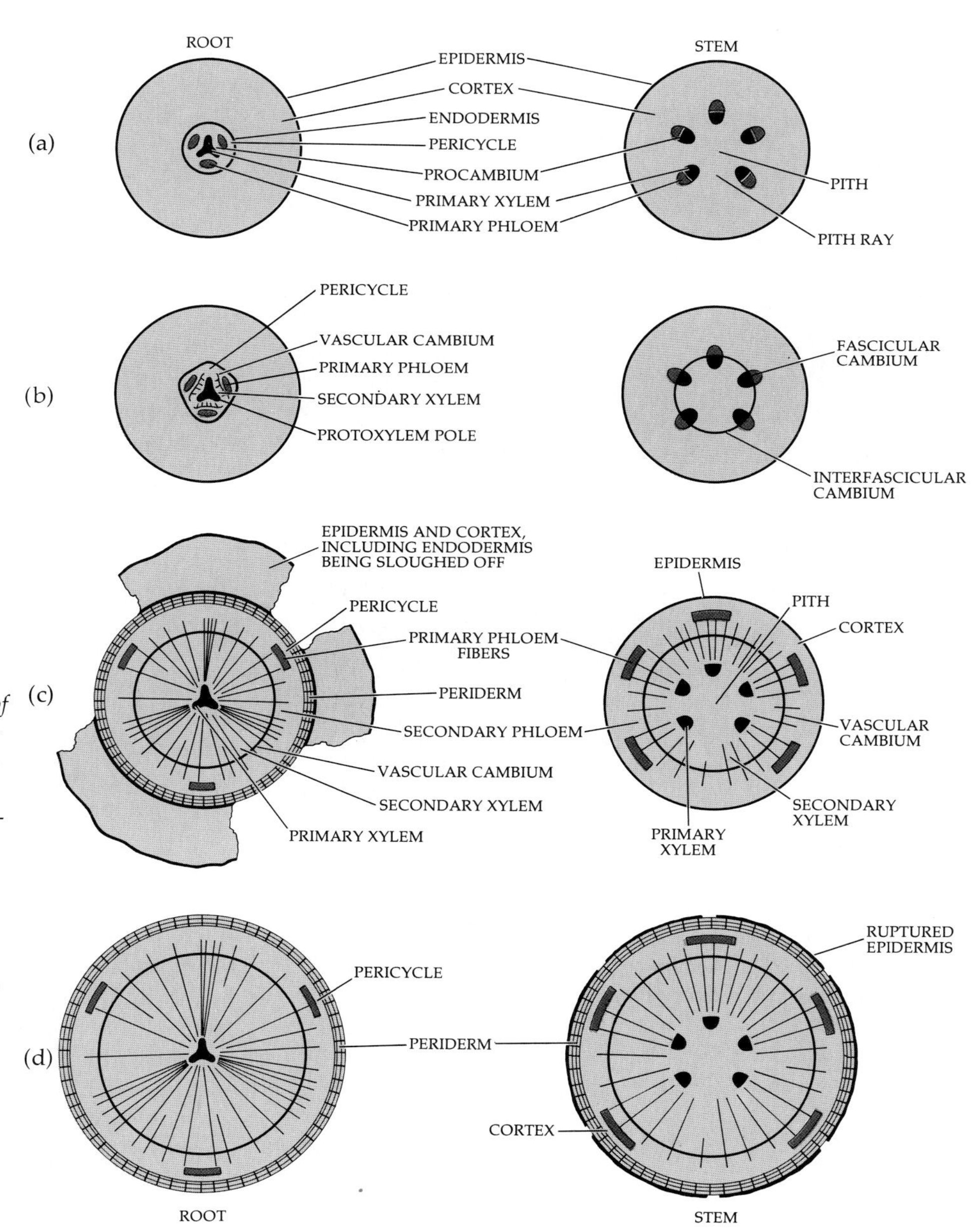

24–6
Comparison of primary and secondary structure in root and stem of a woody dicot. (a) *Root and stem at completion of primary growth. In the triarch root represented here, cambial activity has been initiated in three independent regions from procambium between the three primary phloem strands and the primary xylem.* (b) *Origin of vascular cambium. The pericyclic cells opposite the three protoxylem poles will also contribute to the vascular cambium. Some secondary xylem has already been produced by the newly formed vascular cambium of procambial origin.* (c) *After formation of some secondary xylem and secondary phloem in root and stem, and periderm formation in root.* (d) *At end of first year's growth, showing the effect of secondary growth—including periderm formation—on the primary plant body.*

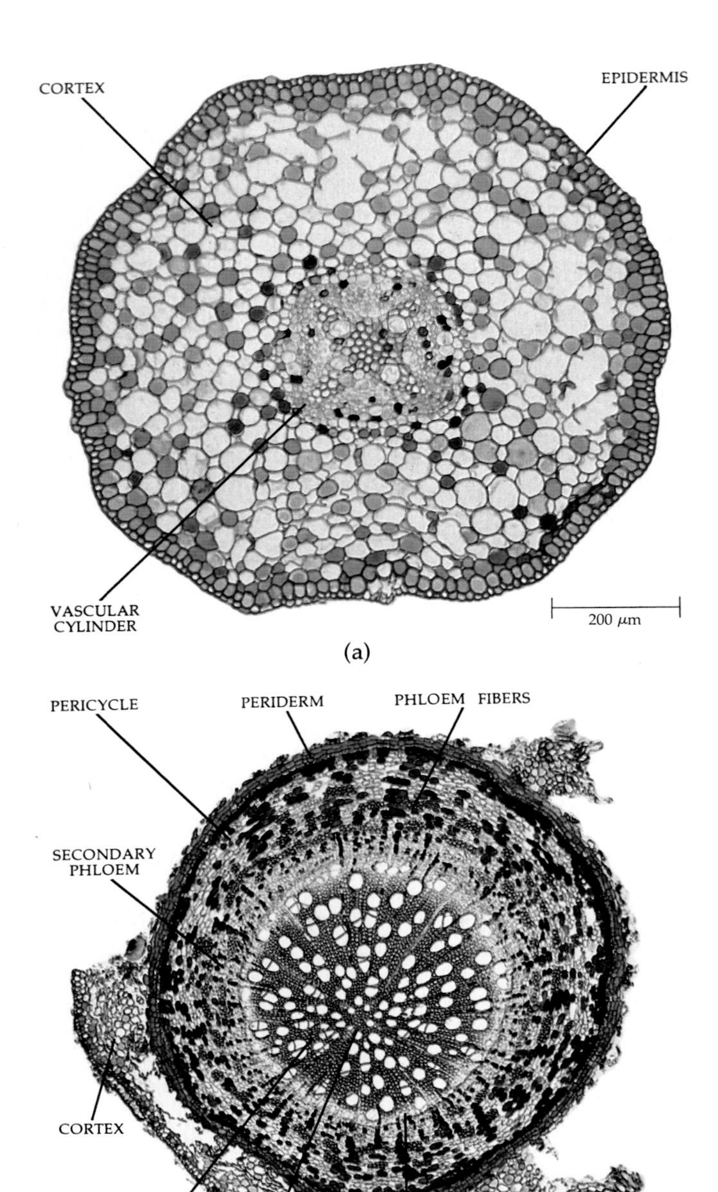

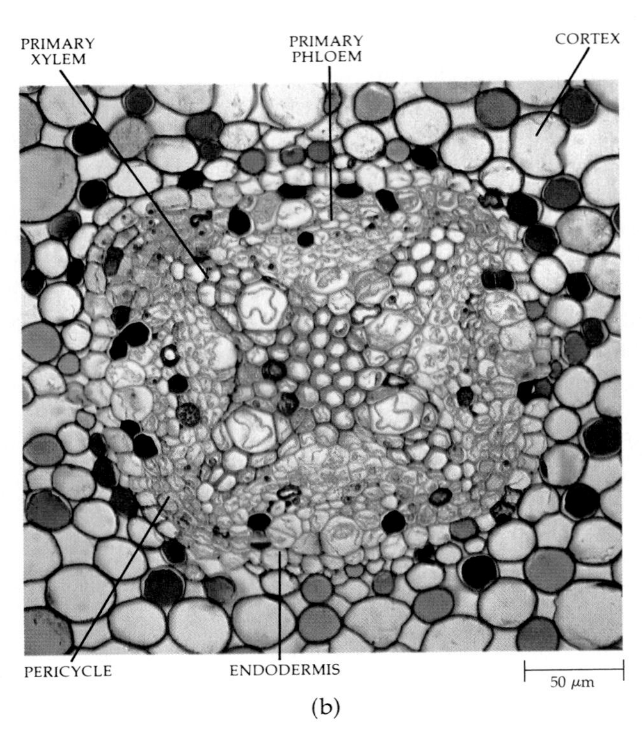

24–7
Transverse sections of the willow (Salix) *root, which becomes woody.* (a) *Overall view of root near completion of primary growth.* (b) *Detail of primary vascular cylinder.* (c) *Overall view of root at end of first year's growth, showing effect of secondary growth on primary plant body.*

As soon as it is formed, the vascular cambium opposite the phloem strands begins to produce secondary xylem, and in the process the strands of primary phloem are displaced outwardly from their positions between the ridges of primary xylem. By the time the cambium opposite the protoxylem poles is actively dividing, the cambium is circular in outline and the primary phloem has been separated from the primary xylem (Figure 24–6).

By repeated divisions toward the inside and outside, secondary xylem and secondary phloem are added to the root (Figure 24–6 and 24–7). In some roots, the vascular cambium derived from the pericycle forms wide rays, whereas narrower rays are produced in other parts of the secondary vascular tissues.

With increase in width of the secondary xylem and phloem, most of the primary phloem is crushed or obliterated. Primary phloem fibers may be the only remaining distinguishable components of the primary phloem.

Stem

As mentioned previously, the vascular cambium of the stem arises from the procambium that remains undifferentiated between the primary xylem and primary phloem, as well as from parenchyma of the interfascicular regions. That portion of the cambium arising within the vascular bundles is known as *fascicular cambium*, and that arising in the interfascicular regions, or pith rays, is called *interfascicular cambium*. The vascular cambium of the stem, unlike that of the root, is essentially circular in outline from its inception (Figure 24–6).

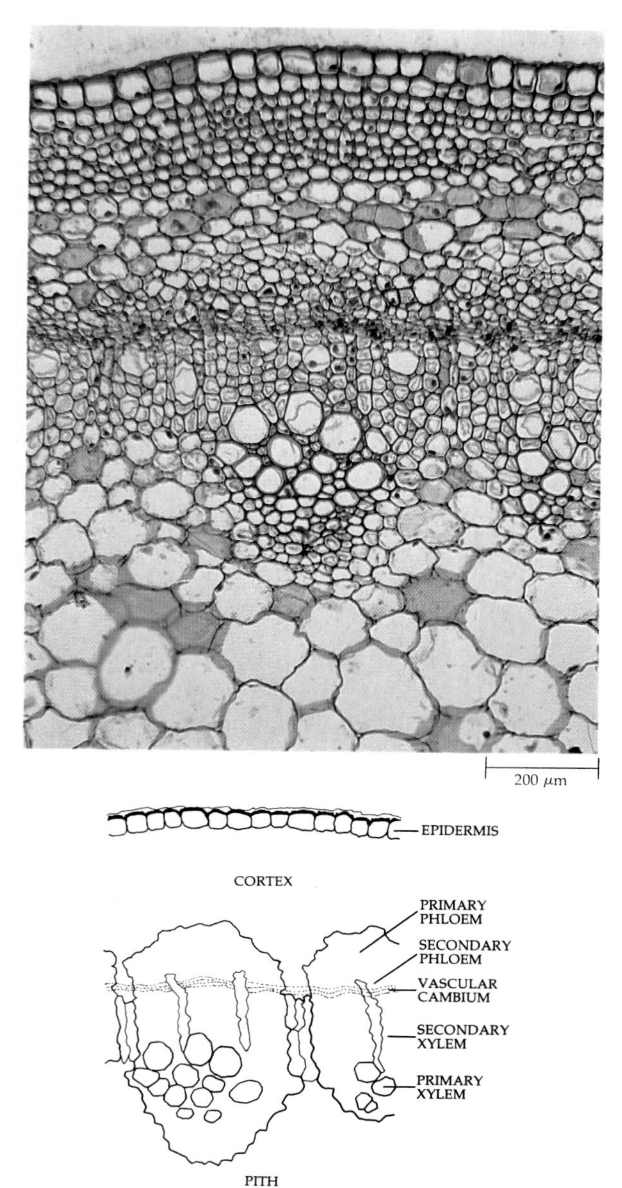

24–8
Transverse section of an elderberry (Sambucus canadensis) *stem in which a small amount of secondary growth has taken place. A cork cambium has not yet been formed.*

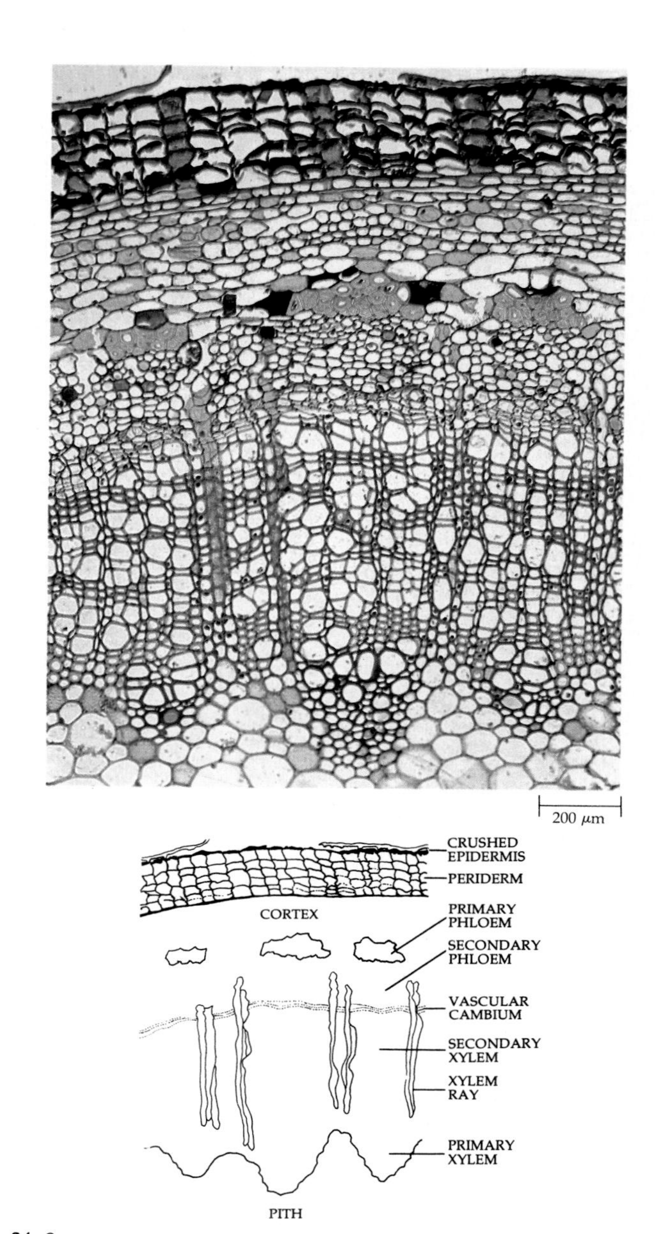

24–9
Transverse section of an elderberry (Sambucus canadensis) *stem at the end of first year's growth.*

In woody stems, the production of secondary xylem and secondary phloem results in the formation of a cylinder of secondary vascular tissues, with the rays extending radially through the cylinder (Figure 24–6). Commonly, much more secondary xylem than secondary phloem is produced in the stem in any given year; this situation is also true in the root. As in the root, with secondary growth the primary phloem is pushed outward and its thin-walled cells are destroyed. Only the thick-walled primary phloem fibers remain intact (Figure 24–9).

Figures 24–8 and 24–9 show an elderberry *(Sambucus canadensis)* stem in two stages of secondary growth. (See Chapter 23 for a description of primary growth in the *Sambucus* stem.) Only a small amount of secondary xylem and secondary phloem has been produced in the stem of Figure 24–8. Figure 24–9 shows a stem at the end of the first year's growth. Note that considerably more secondary xylem than secondary phloem has been formed. The thick-walled cells outside the secondary phloem are primary phloem fibers.

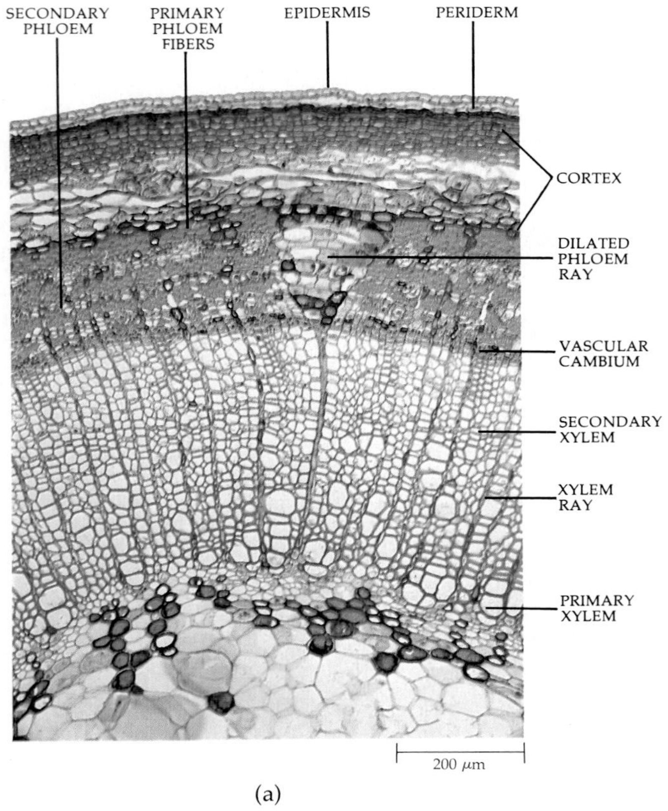

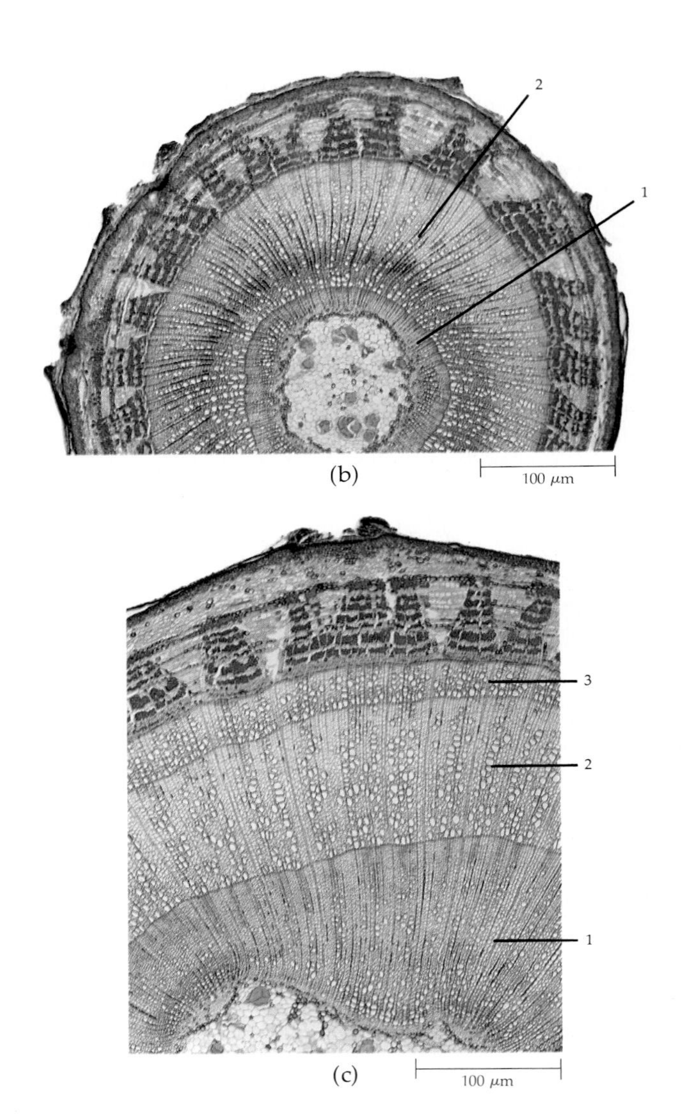

24–10
Transverse sections of linden (Tilia americana) *tree stems.* (a) *One-year-old stem.* (b) *Two-year-old stem.* (c) *Three-year-old stem. Numbers indicate growth rings in the secondary xylem.*

Figure 24–10 shows one-year-old, two-year-old, and three-year-old stems of linden, or basswood *(Tilia americana)*. In Chapter 23, the stem of *Tilia* was given as an example of one in which the primary tissues arise as an almost continuous hollow cylinder; thus most of the vascular cambium in the *Tilia* stem is fascicular in origin. Some of the rays in the secondary phloem of *Tilia* become very wide as the stem increases in girth. This is one way in which the tissues outside the vascular cambium keep up with the increase in girth of the core of xylem.

The vascular cambia and secondary tissues of root and stem are continuous with one another. There is no transition region in the secondary plant body as there is in the primary plant body (see page 466).

The Periderm

In most woody roots and stems, cork formation usually follows the initiation of secondary xylem and secondary phloem production, and the cork tissue replaces the epidermis as the protective covering on those portions of the plant. *Cork,* or phellem, is formed by a *cork cambium,* or phellogen, which may also form *phelloderm* ("cork skin"). The cork is formed toward the outer surface of the cork cambium, and the phelloderm is formed toward the inner surface (see Figures 24–11 and 24–12). Together, these three tissues—cork, cork cambium, and phelloderm—make up the *periderm*.

In most dicots and gymnosperms, the first periderm commonly appears during the first year of growth in those portions of the root or stem that are no longer elongating. In stems, the first cork cambium most commonly originates in a layer of cortical cells immediately below the epidermis (Figures 24–6 and 24–11), although in many species it originates in the epidermis. In roots, the first cork cambium arises through periclinal division of pericycle cells, the outer sister cells combining to form a complete cylinder of cork cambium. Afterwards, the remaining cells of the pericycle may proliferate below the periderm and

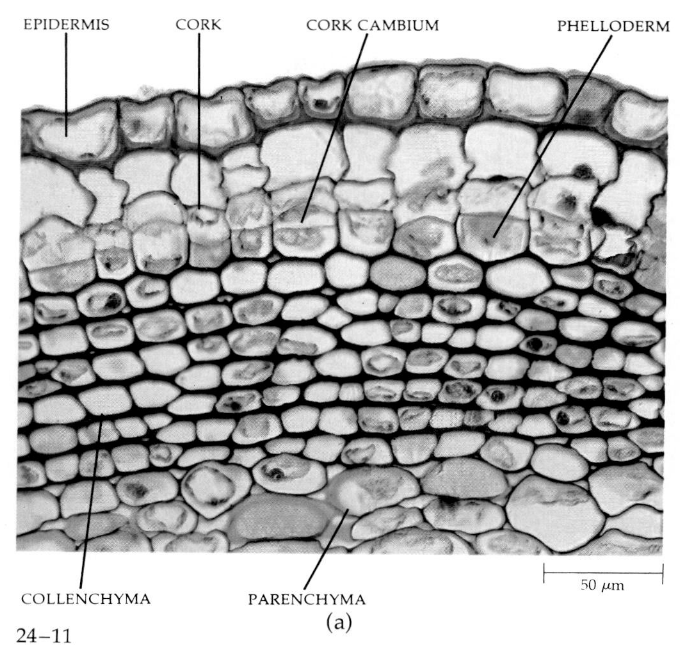

(a)

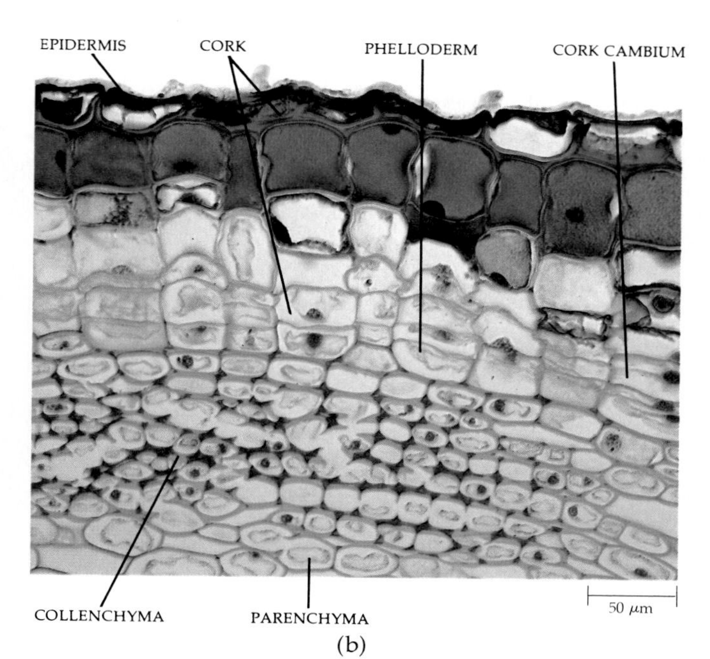

(b)

24–11
Some stages of periderm and lenticel development in elderberry (Sambucus canadensis), *as seen in transverse sections.* (a) *Newly formed periderm beneath the epidermis; collenchyma and parenchyma of cortex.* (b) *Periderm in more advanced stage of development.* (c) *Initiation of lenticel; collenchyma of cortex beneath the developing lenticel.* (d) *Well-developed lenticel. The phelloderm in* Sambucus *generally consists of a single layer of cells.*

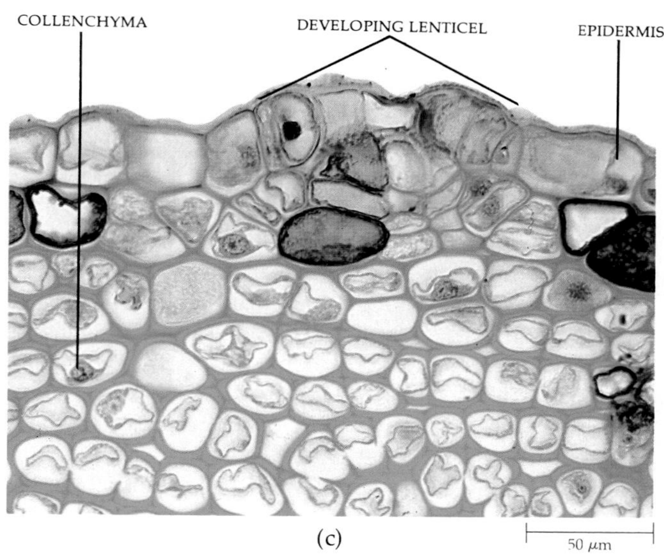

(c)

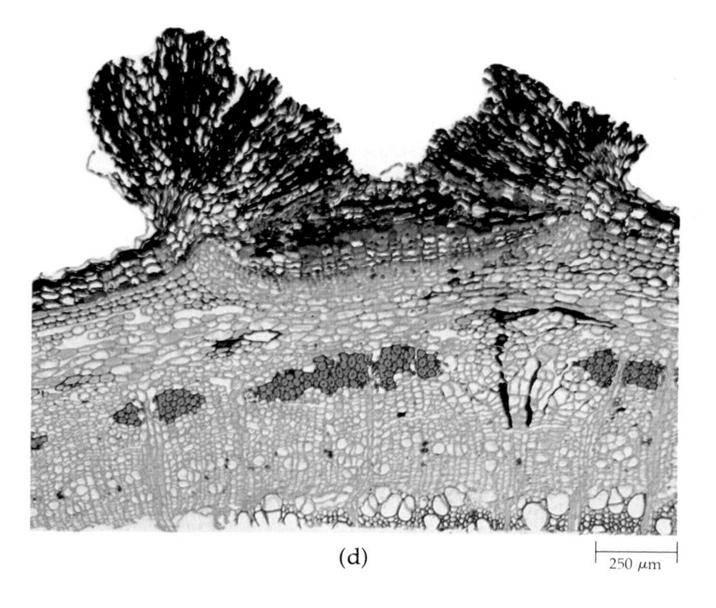

(d)

give rise to a tissue that resembles a cortex (Figures 24–6 and 24–7).

Repeated divisions of the cork cambium result in the formation of radial rows of compactly arranged cells, most of which are cork cells (Figures 24–11 and 24–12). During differentiation of the cork cells, their inner walls are lined by a relatively thick layer of a fatty substance—*suberin*—which makes the tissue highly impermeable to water and gases. The walls of the cork cells may also become lignified. At maturity, the cork cells are dead.

The cells of the phelloderm are living at maturity, lack suberin, and resemble cortical parenchyma cells. The phelloderm cells may be distinguished from cortical cells by their inner position in the radial rows of other periderm cells (Figure 24–12).

With the formation of the first periderm in the root, the cortex (including the endodermis) and the epidermis are isolated from the rest of the root. Being separated from the supply of water and minerals by the impermeable barrier of cork, the cortex and

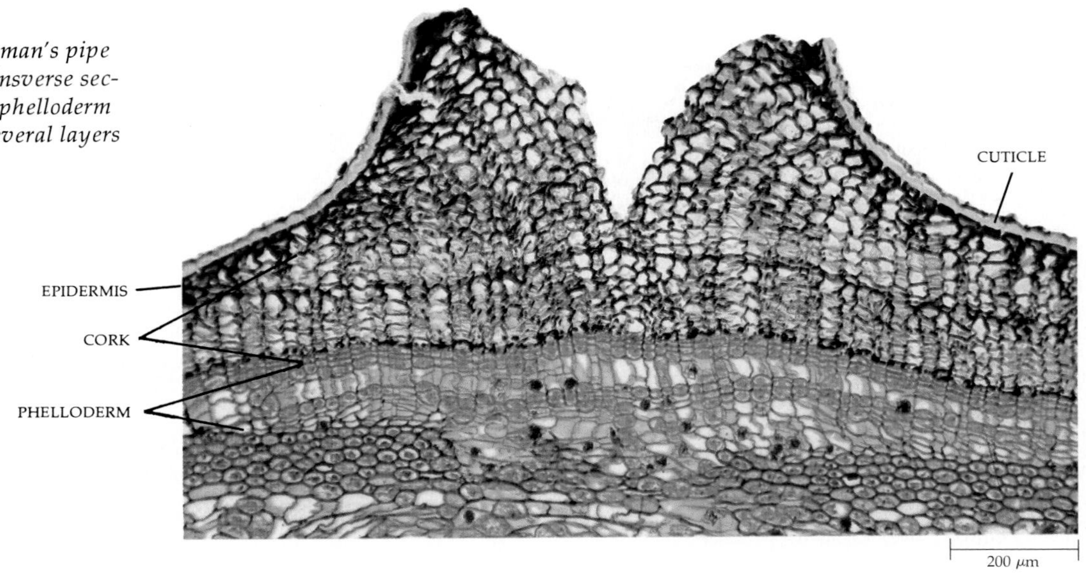

24–12
Lenticel of the stem of Dutchman's pipe (Aristolochia), *as seen in transverse section. Unlike* Sambucus, *the phelloderm of* Aristolochia *consists of several layers of cells.*

epidermis eventually die and are sloughed off. Because the first periderm of the stem usually arises immediately below the epidermis, the cortex of the stem is not sloughed off during the first year (see Figures 24–6 and 24–9), although the epidermis does dry up and peel off.

At the end of the first year's growth, the following tissues are present in a woody root (from outside to inside): remnants of the epidermis and cortex, periderm, pericycle, primary phloem (fibers and crushed soft-walled cells), secondary phloem, vascular cambium, secondary xylem, and primary xylem. The following tissues are present in the stem: remnants of the epidermis, periderm, cortex, primary phloem (fibers and crushed soft-walled cells), secondary phloem, vascular cambium, secondary xylem, primary xylem, and pith (Figure 24–6).

The Lenticels

In the preceding discussion, it was noted that the suberin-containing cork cells are compactly arranged and, as a tissue, present an impermeable barrier to water and gases. However, the inner tissues of the stem, like all metabolically active tissues, need to exchange gases with the surrounding air; similarly, the inner tissues of the root need to exchange gases with the surrounding air spaces between soil particles. In stems and roots containing periderms, this necessary gas exchange is accomplished by means of lenticels (Figures 24–11 and 24–12)—portions of the periderm in which the phellogen (cork cambium) is more active than elsewhere, resulting in the formation of a tissue with numerous intercellular spaces. In addition, the phellogen itself contains intercellular spaces in the region of the lenticels.

Lenticels begin to form during development of the first periderm (Figure 24–11) and, in the stem, generally appear below a stoma or group of stomata. On the surface of the stem or root, the lenticels appear as raised circular, oval, or elongated areas (see Figure 24–18, page 487). Lenticels are also formed on some fruits—for instance, the small dots on the surface of apples and pears are lenticels. As the roots and stems grow older, lenticels continue to develop at the bottom of cracks in the bark in newly formed periderms.

The Bark

The terms "periderm," "cork," and "bark" are often unnecessarily confused with one another. As previously discussed, cork is one of three parts of the periderm; it is a secondary tissue that replaces the epidermis in most woody roots and stems. The term bark refers to all the tissues outside the vascular cambium, including the periderm when present (Figures 24–13 and 24–14). When the vascular cambium first appears, and secondary phloem has not yet been formed, the bark consists entirely of primary tissues. At the end of the first year's growth, the bark includes any primary tissues still present, the secondary phloem, the periderm, and any dead tissues remaining outside the periderm.

Each growing season, the vascular cambium adds secondary phloem to the bark, as well as secondary xylem, or wood, to the core of the stem or root. Usually less secondary phloem is produced by the vascular cambium than secondary xylem. In addition, the soft-walled cells (sieve elements and various kinds of parenchymatic elements) of the old secondary phloem are commonly crushed (see Figures 24–15 through 24–17, on page 486). Eventually, the old secondary

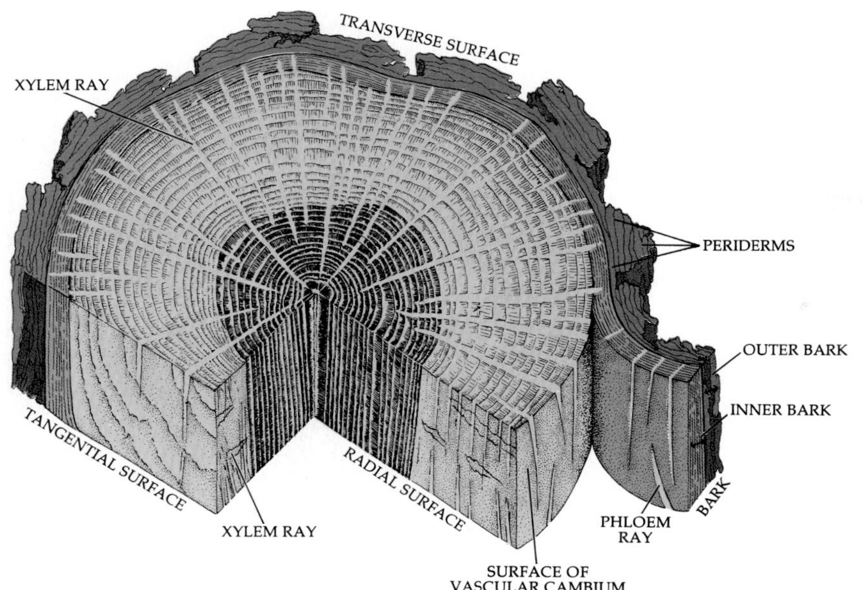

24–13
Diagram of part of a red oak (Quercus rubra) *stem, showing the transverse, tangential, and radial surfaces. The dark area in the center is heartwood. The lighter part of the wood is sapwood.*

phloem is separated from the rest of the phloem by newly formed periderms. As a result, considerably less secondary phloem accumulates in the stem or root than secondary xylem, which continues to accumulate year after year.

As the stem or root increases in girth, considerable stress is placed on the older tissues of the bark. In some plants, tearing of these tissues results in the formation of large air spaces. In many plants, the parenchyma cells of the axial system and rays divide and enlarge; in this manner, the old secondary phloem keeps up for a while with the increase in circumference of the plant part. It was noted earlier that certain rays in the *Tilia* stem become very wide as the stem increases in girth; these rays are called dilated rays.

The first-formed periderm may keep up with the increase in girth of the root or stem for several years, with the cork cambium exhibiting periods of activity and inactivity that may or may not correspond to the periods of activity of the vascular cambium. In the stems of apple *(Malus sylvestris)* and pear *(Pyrus communis)* trees, the first cork cambium may remain active for up to 20 years. In most woody roots and stems, additional periderms are formed as the axis increases in circumference. After the first periderm, subsequently formed periderms originate deeper and deeper in the bark (Figures 24–13 and 24–14) from parenchyma cells of the phloem no longer actively engaged in the transport, or translocation, of food substances. These parenchyma cells become meristematic and form new cork cambia.

All of the tissues outside the innermost cork cambium—all of the periderms, together with any cortical and phloem tissues included among them—make up the *outer bark* (Figures 24–13 and 24–14).

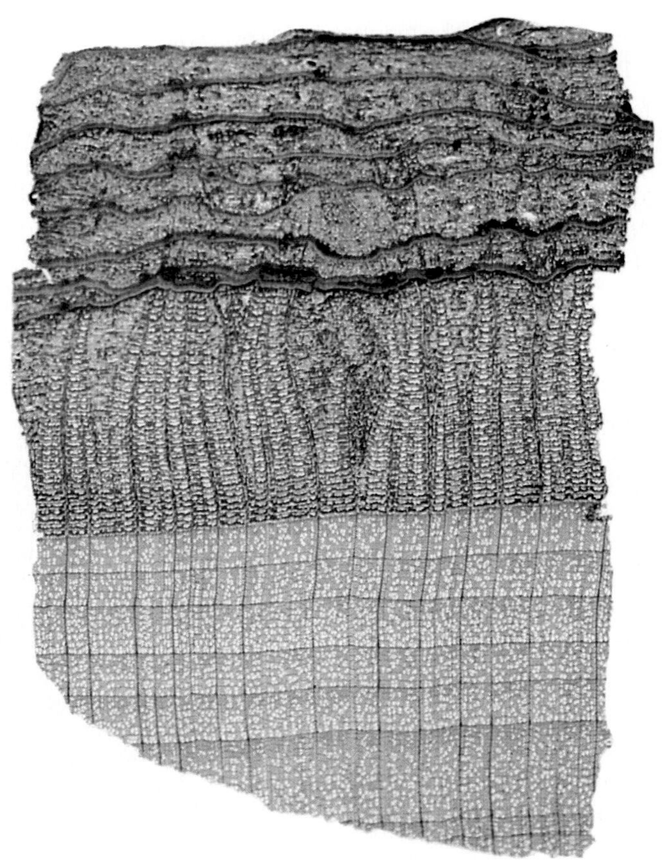

24–14
Transverse section of the bark and some secondary xylem from an old stem of linden (Tilia americana). *Several periderms can be seen traversing the mostly brownish outer bark in the upper third of the section. Below the outer bark is the inner bark, which is quite distinct in appearance from the more lightly stained xylem in the lower third of the section.*

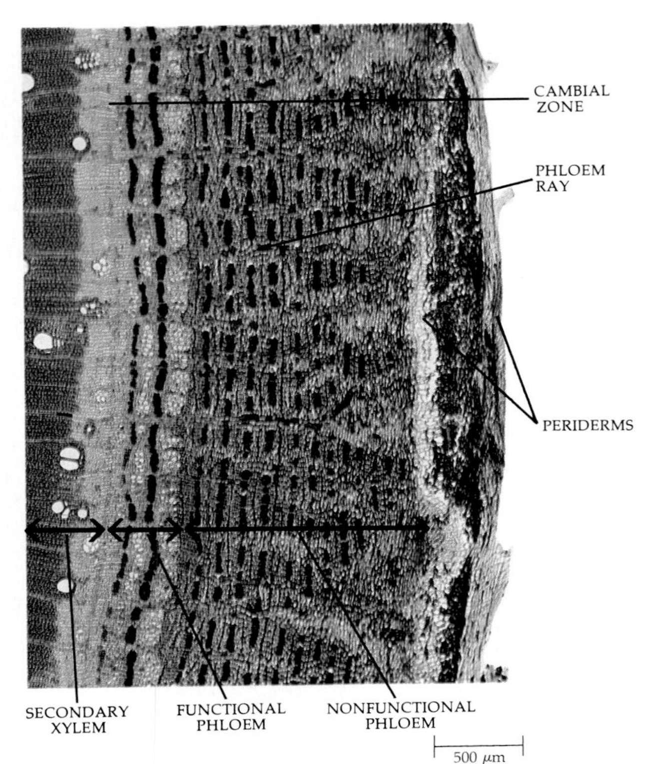

24–15
Transverse section of the bark of the black locust (Robinia pseudo-acacia) *stem, consisting mostly of nonfunctional phloem.*

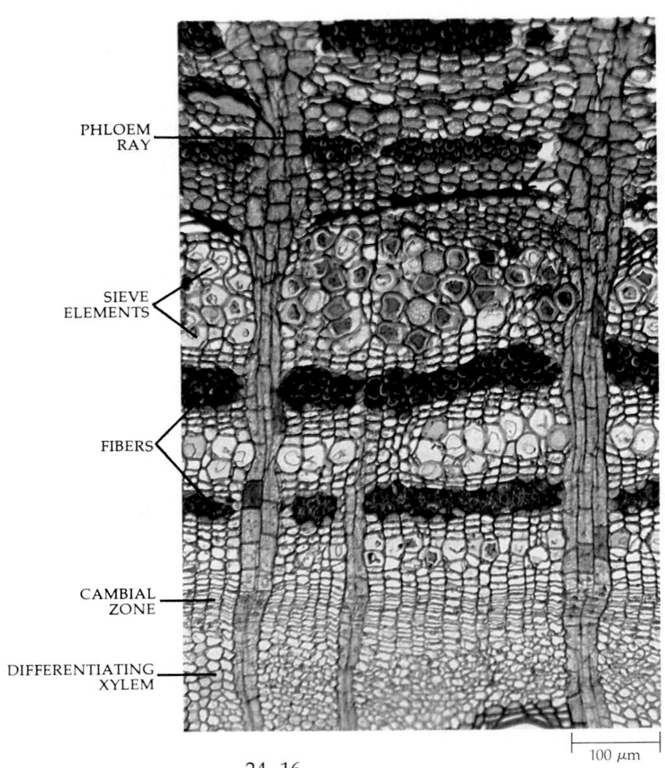

24–16
Transverse section of secondary phloem of the black locust, showing mostly functional phloem. Sieve elements (indicated by arrows) of the nonfunctional phloem have collapsed.

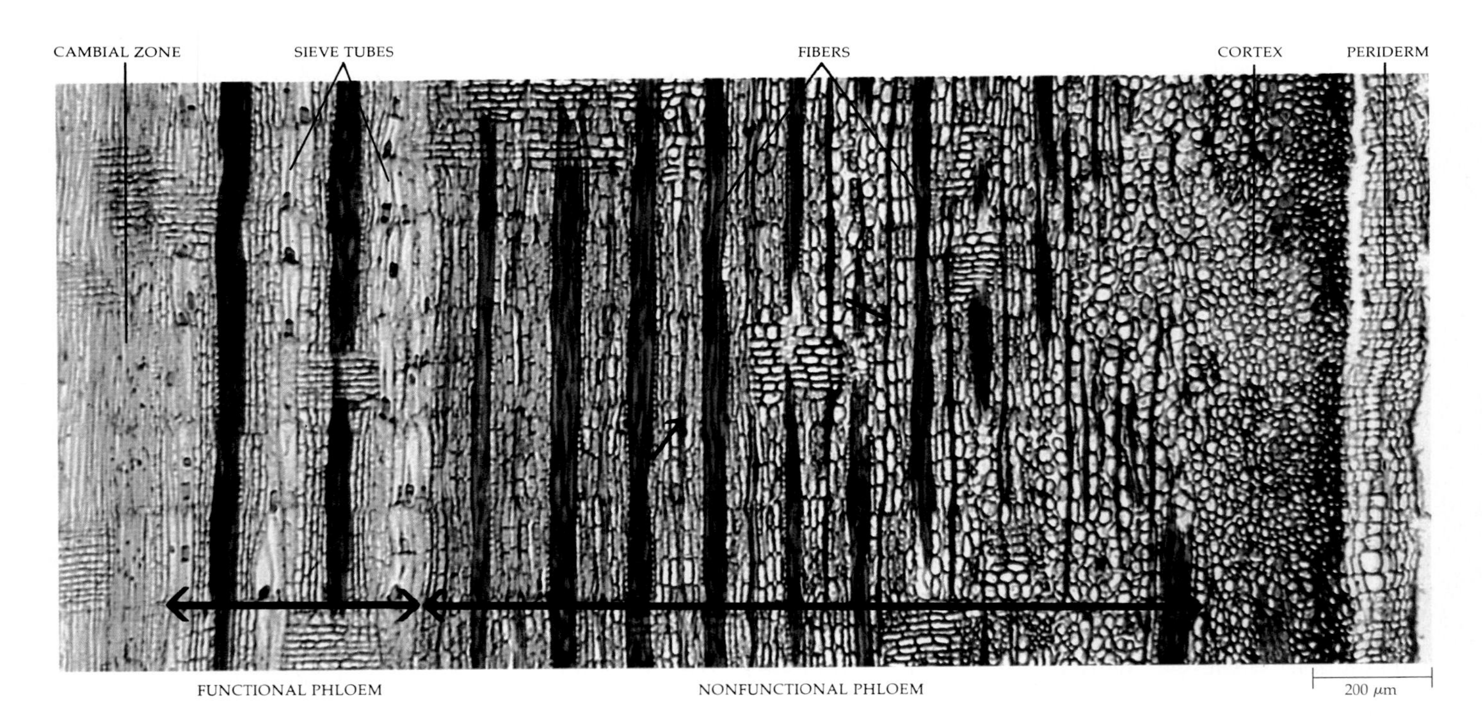

24–17
Radial section of the bark of the black locust. Most of the section consists of nonfunctional phloem, in which the sieve elements are collapsed (see arrows). In black locust, the functional phloem consists only of the current season's growth increment, which becomes nonfunctional in late autumn when its sieve elements die and collapse.

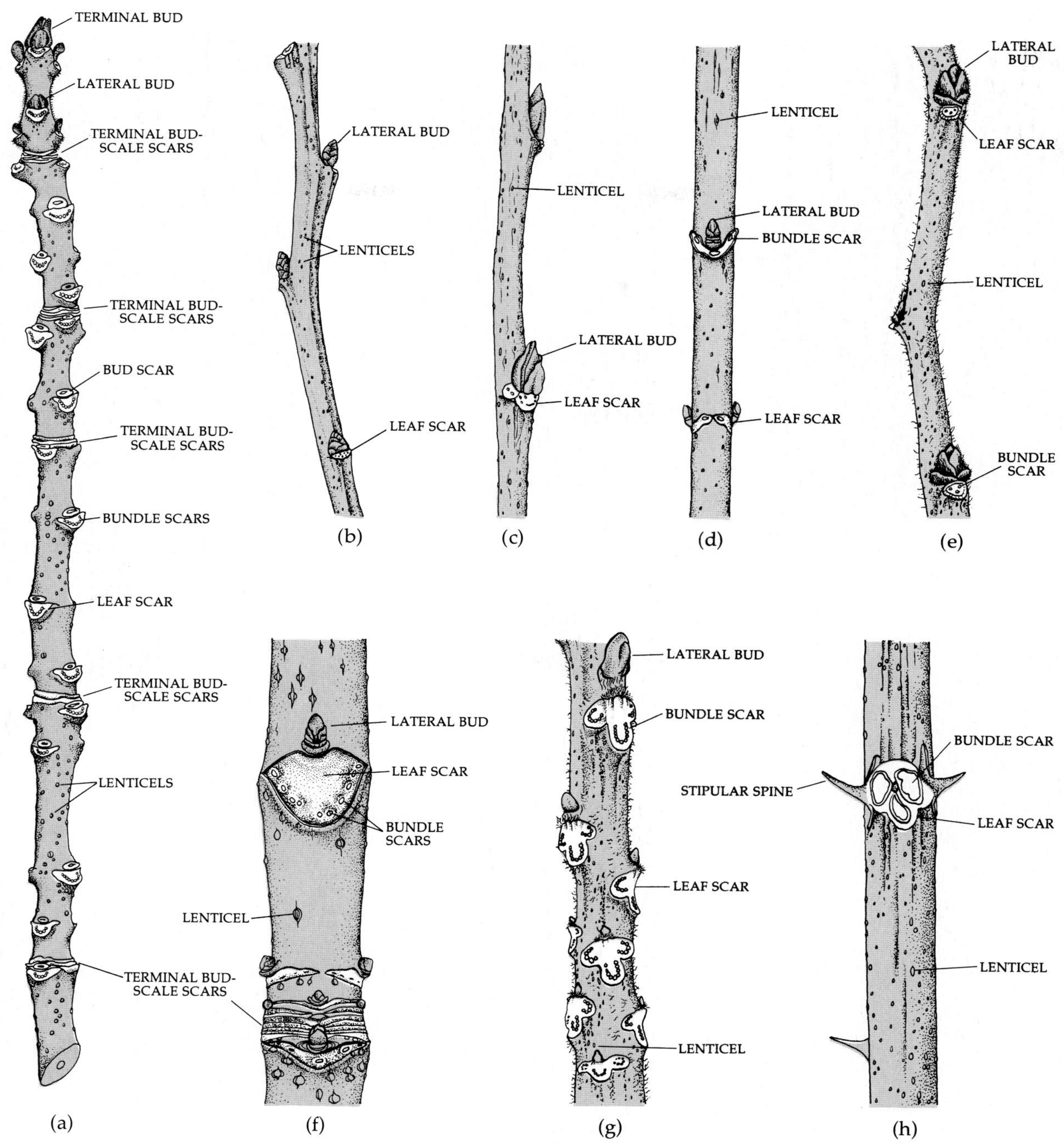

24–18

External features of woody stems. Examination of the twigs of deciduous woody plants reveals many important developmental and structural features of the stem. The most conspicuous structures on the twigs are the buds. Buds occur at the tips—the terminal buds—and in the axils of the leaves—the lateral, or axillary, buds of the twigs. In addition, accessory buds occur in some species. Commonly occurring in pairs, the accessory buds are located one each on either side of an axillary bud. In some species, the accessory buds do not develop if their associated axillary bud undergoes normal development. In others, the accessory buds give rise to flowers and the axillary bud gives rise to a leafy shoot.

After the leaves fall, leaf scars, with their bundle scars, can be seen beneath the axillary buds. The protective layer of the abscission zone produces the leaf scar. The bundle scars are the severed ends of vascular bundles that extended from the leaf traces into the petiole of the leaf, prior to abscission.

Groups of terminal-bud-scale scars reveal the locations of previous terminal buds, and until they are obscured by secondary growth, these groups of scars may be used to determine the age of portions of the stem. The portion of stem between two groups of such scars represents one year's growth. The lenticels appear as slightly raised areas on the stem.

(a) *Green ash* (Fraxinus pennsylvanica *var.* subintegerrima). (b) *White oak* (Quercus alba). (c) *Linden* (Tilia americana). (d) *Boxelder* (Acer negundo). (e) *American elm* (Ulmus americana). (f) *Horse-chestnut* (Aesculus hippocastanum). (g) *Butternut* (Juglans cinerea). (h) *Black locust* (Robinia pseudo-acacia).

(a)

(b)

(c)

(d)

24–19
Bark of four species of trees. (a) *Thin, peeling bark of the paper birch* (Betula papyrifera). *Elongated areas on surface of bark are lenticels.* (b) *Shaggy bark of the shagbark hickory* (Carya ovata). (c) *Scaly bark of the sycamore, or plane tree* (Platanus occidentalis). (d) *Deeply furrowed bark of the black oak* (Quercus velutina).

With maturation of the suberin-containing cork cells, the tissues outside them are separated from the supply of water and nutrients. Hence the outer bark consists entirely of dead tissues. The living part of the bark, which is inside the innermost cork cambium and extends inward to the vascular cambium, is called the *inner bark* (see Figures 24–13 and 24–14).

The manner in which new periderms are formed and the kinds of tissues isolated by them have a marked influence on the appearance of the outer surface of the bark (Figure 24–19). In some barks the newly formed periderms develop as discontinuous overlapping layers, resulting in formation of a scaly type of bark, called scale bark (see Figures 24–13 and 24–14). Scale barks, for example, are found on relatively young stems of pine *(Pinus)* and pear *(Pyrus communis)* trees. In other barks, the newly formed periderms arise as more or less continuous, concentric rings around the axis, resulting in formation of a ring bark. Grape *(Vitis)* and honeysuckle *(Lonicera)* are examples of plants with ring barks, which are less common than scale barks. The barks of many plants are intermediate between ring and scale barks.

Commercial cork is obtained from the bark of the cork oak, *Quercus suber,* which is native to the Mediterranean region. The first cork cambium of this tree has its origin in the epidermis, and the cork produced by it is of little commercial value. When the tree is about 20 years old, the original periderm is removed, and a new cork cambium is formed in the cortex, just a few millimeters below the site of the first one. The cork produced by the new cork cambium accumulates very rapidly and after about 10 years is thick enough to be stripped off the tree. Once again a new cork cambium arises beneath the previous one, and after about another 10 years the cork can be stripped again. This procedure may be repeated at about 10-year intervals until the tree is 150 or more years old. The spots and long dark streaks seen on the surfaces of commercial cork are lenticels.

In most woody roots and stems, very little secondary phloem is actually involved in the conduction of food. In most species, only the current year's growth increment, or growth ring, of secondary phloem is active in the long-distance transport of food through the stem. This is because the sieve elements are short-lived (Chapter 21); most of them die by the end of the same year in which they are derived from the vascular cambium. In some plants, such as black locust *(Robinia pseudo-acacia),* the sieve elements collapse and are crushed relatively soon after they die (see Figures 24–15 through 24–17).

The part of the inner bark actively engaged in the transport of food substances is called *functional phloem.* Although the sieve elements outside the functional phloem are dead, the phloem parenchyma cells (axial parenchyma) and parenchyma cells of the rays may remain alive and continue to function as storage cells for many years. That part of the inner bark is known as *nonfunctional phloem*. Only the outer bark is composed entirely of dead tissue.

THE WOOD: SECONDARY XYLEM

Apart from use of various plant tissues as food for humans, no single plant tissue has played a more indispensable role to human survival throughout recorded history than wood, or secondary xylem. Commonly, woods are classified as either *hardwoods* or *softwoods*. The so-called hardwoods are dicot woods, and the softwoods are conifer woods. The two kinds of woods have basic structural differences, but the terms "hardwood" and "softwood" do not accurately express the degree of density or hardness of the wood. For example, one of the lightest and softest of woods is balsa *(Ochroma lagopus)*, a tropical dicot. By contrast, the woods of some conifers, such as hemlock *(Tsuga)*, are harder than some hardwoods.

Conifer Wood

The structure of conifer wood is relatively simple compared with that of most dicots. The principal features of conifer wood are its lack of vessels (Chapter 21) and its relatively small amount of axial or wood parenchyma. Long, tapering tracheids constitute the dominant cell type in the axial system. In certain genera, such as *Pinus,* the only parenchyma cells of the axial system are those associated with *resin ducts.* Resin ducts are relatively large intercellular spaces lined with thin-walled parenchyma cells, which secrete resin into the duct. In *Pinus,* resin ducts occur in both the axial system and the rays (Figures 24–20 and 24–21). Wounding, pressure, and injuries by frost and wind can stimulate the formation of resin ducts in conifer wood, leading some investigators to suggest that all resin ducts are traumatic in origin. Resin apparently protects the plant from attack by decay-producing fungi and bark beetles.

The tracheids of conifers are characterized by large, circular, bordered pits that are most abundant on the ends of the cells, where they overlap with other tracheids (Figures 24–20 through 24–22). The pairs of pits (pit-pairs, see Chapter 1, page 35) between conifer tracheids are each characterized by the presence of a *torus* (plural: tori). The torus is a thickened central

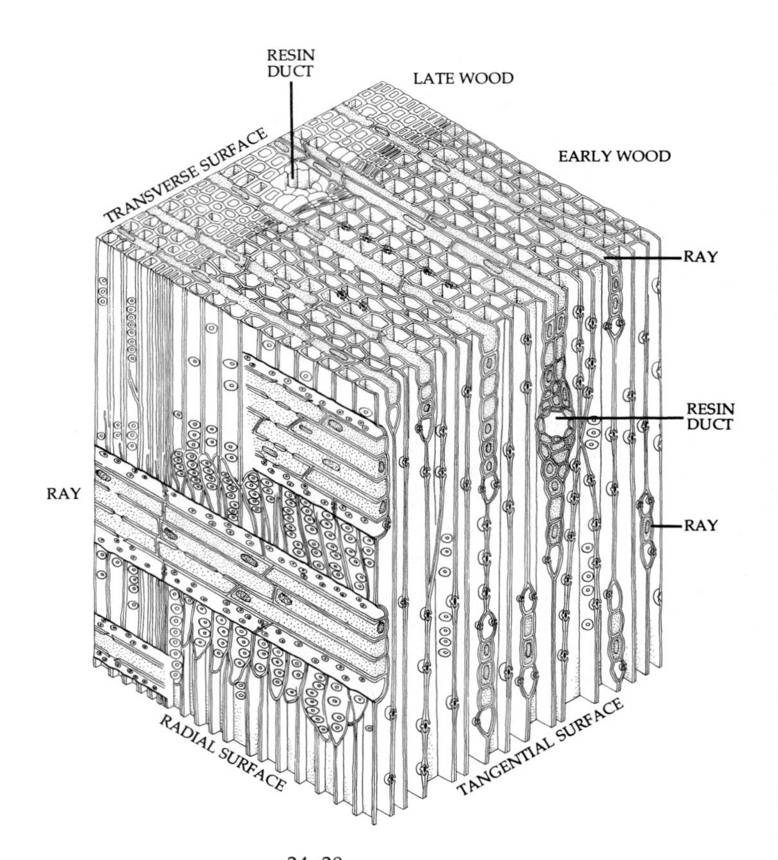

24–20
Block diagram of the secondary xylem of white pine (Pinus strobus). *With the exception of the parenchyma cells associated with the resin ducts, the axial system consists entirely of tracheids. The rays are only one cell wide, except for those containing resin ducts. Early and late wood are described on page 493.*

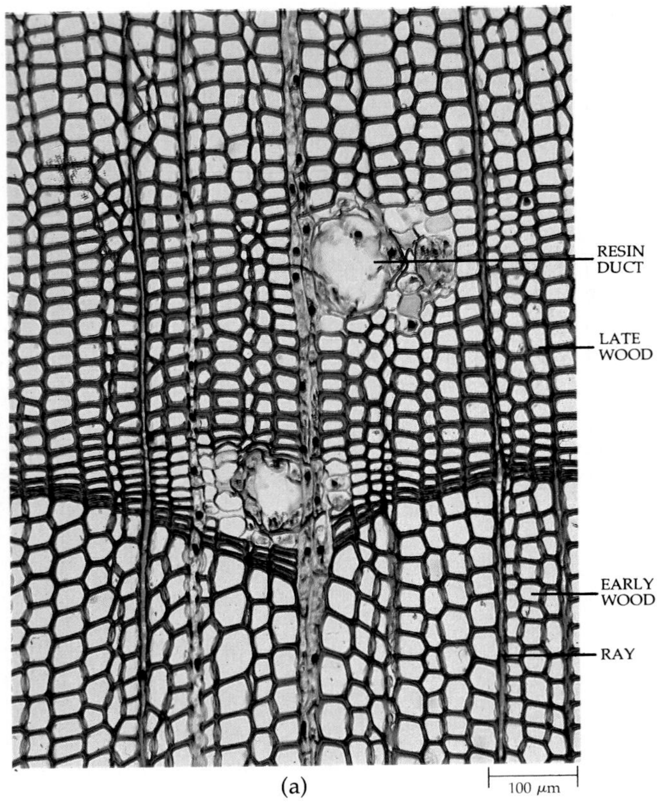

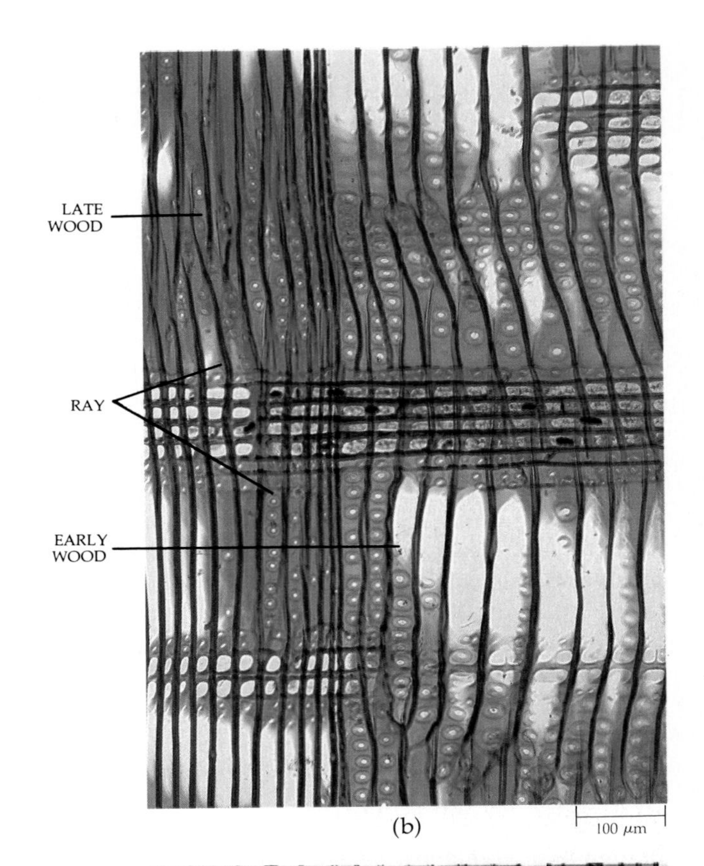

24–21
Wood of white pine (Pinus strobus), *a conifer, in* (a) *transverse,* (b) *radial, and* (c) *tangential sections.*

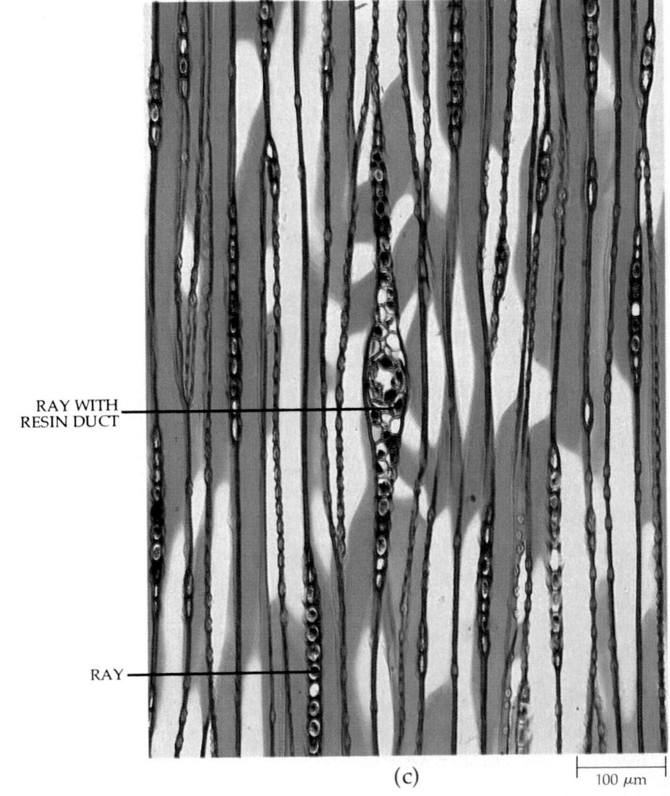

portion of the pit-membrane (Figure 24–23) and is slightly larger than the openings or apertures in the pit-borders (Figure 24–22). The pit-membrane is flexible, and under certain conditions, the torus may block one of the apertures and prevent the movement of water or gases through the pit-pair (Figure 24–22).

Figure 24–20 is a three-dimensional diagram of the wood of white pine *(Pinus strobus)* based on the three wood sections shown in Figure 24–21. In sections cut at right angles to the long axis of the root or stem—transverse or cross sections—the tracheids appear angular or squarish, and the rays can be seen on end traversing the wood (Figure 24–21a). There are two kinds of longitudinal sections—radial and tangential. Radial sections are cut parallel with the rays, and in such sections, the rays appear as sheets of cells oriented at right angles to the vertically elongated tracheids of the axial system (Figures 24–21b and 24–22d). Tangential sections are cut at right angles to the rays and reveal the width and height of the rays. In *Pinus* the rays are one cell wide, except for those containing resin ducts (Figure 24–21c). Details of white pine wood are shown in Figure 24–22.

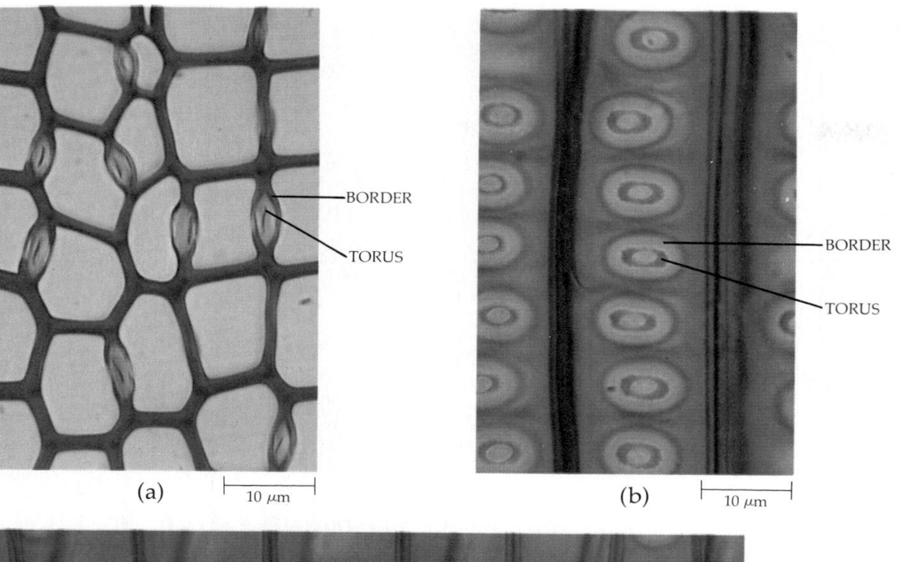

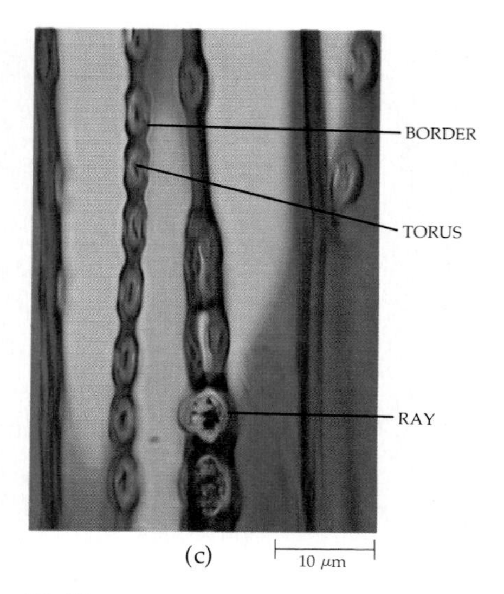

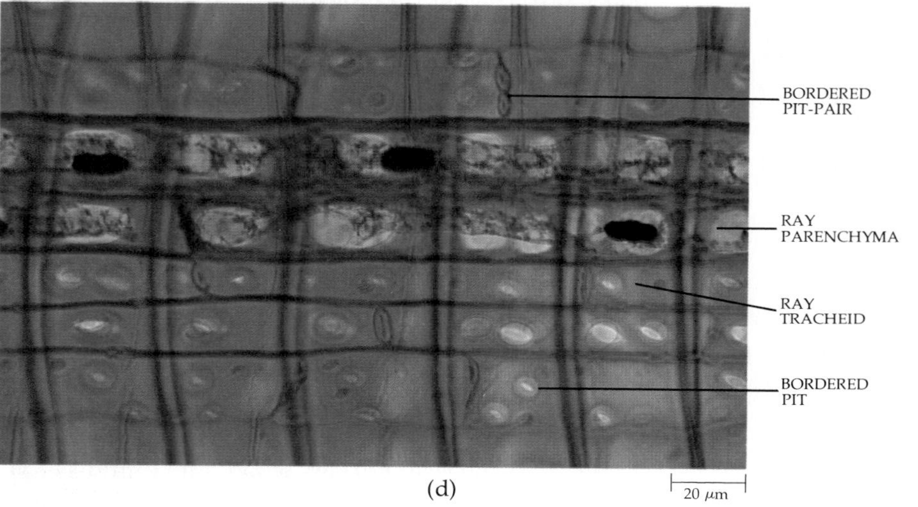

24–22
Details of white pine (Pinus strobus) *wood.* (a) *Transverse section, showing bordered pit-pairs of tracheids.* (b) *Radial section, showing the face-view of bordered pit-pairs in walls of tracheids.* (c) *Tangential section, showing bordered pit-pairs of tracheids.* (d) *Radial section, showing ray. The rays of pine and other conifers are composed of ray tracheids and ray parenchyma cells. Notice the bordered pits of ray tracheids.*

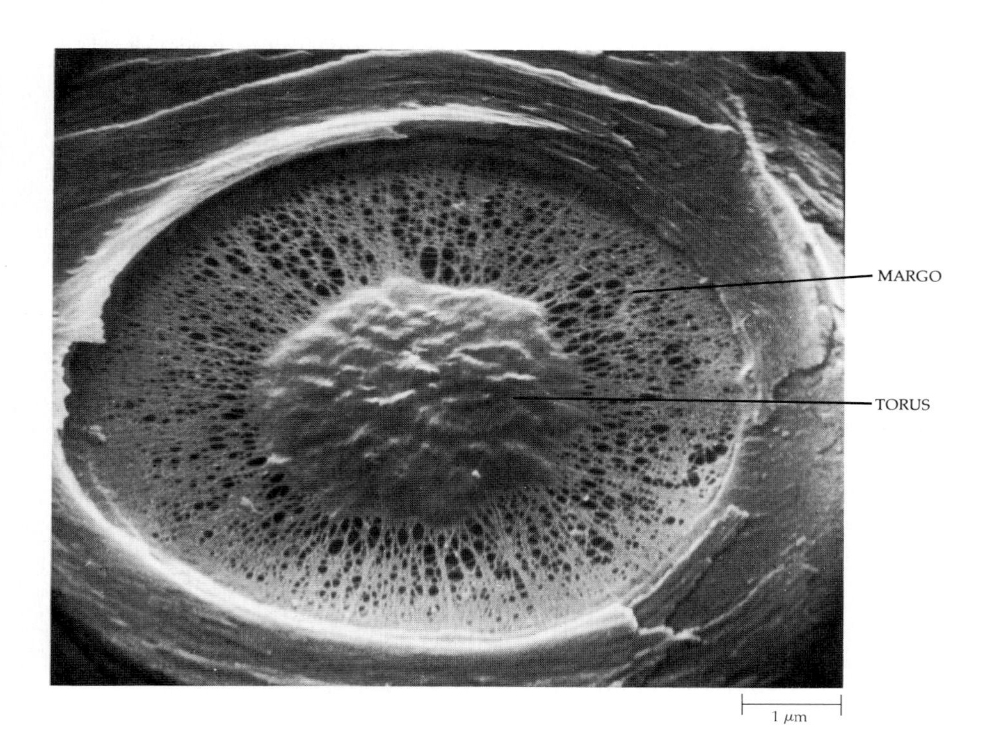

24–23
Scanning electron micrograph of pit membrane of bordered pit-pair in white pine (Pinus strobus) *tracheid. The thickened part of the membrane is the torus. The part of the membrane surrounding the torus is called the margo.*

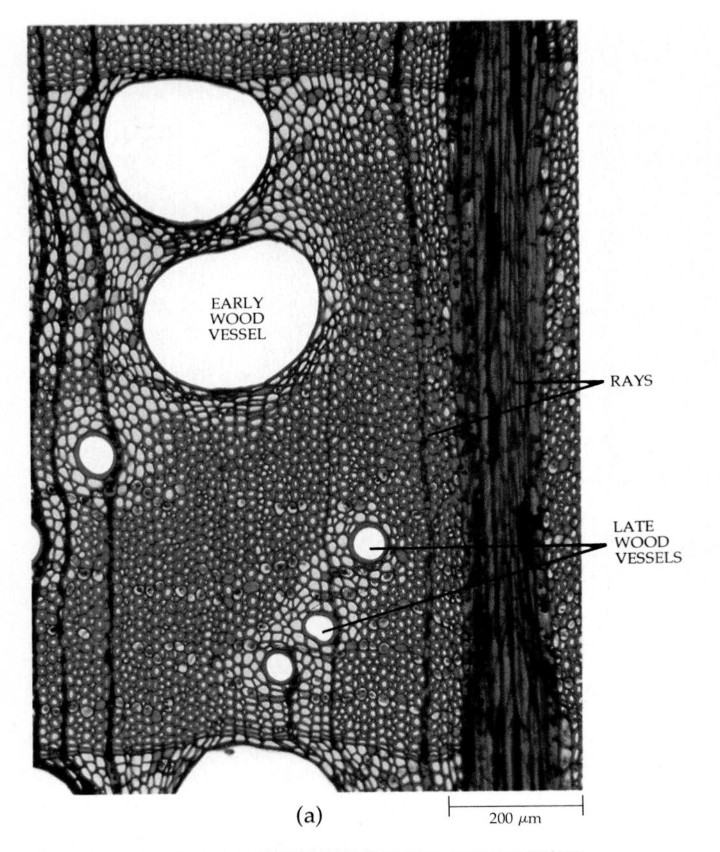

(a)

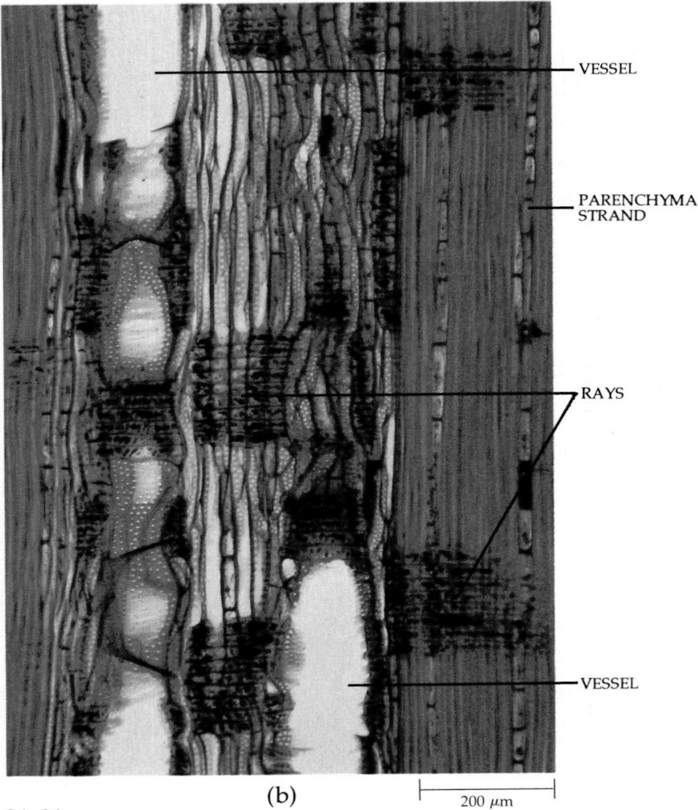

(b)

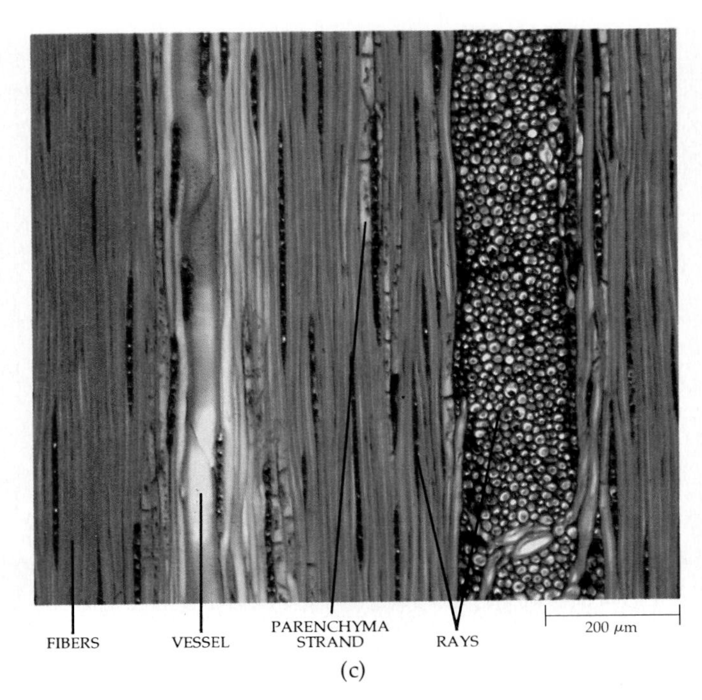

(c)

24–24
Wood of red oak (Quercus rubra), *in* (a) *transverse,* (b) *radial, and* (c) *tangential sections.*

Dicot Woods

Wood structure in dicotyledons is much more varied than in conifers, owing in part to a great number of cell types in the axial system, including vessel members, tracheids, several types of fibers, and parenchyma cells (Figure 24–24; see also Figure 21–11, page 421). The presence of vessel members, in particular, distinguishes dicot woods from conifer wood.

The rays of dicot woods are often considerably larger than those of conifer wood. In conifer wood, the rays are predominately one cell wide, and most range from 1 to 20 cells high. The rays of dicot woods range from one to many cells wide and from one to several hundred cells high. In some dicot woods, such as oak, the large rays can be seen with the unaided eye (Figure 24–13). The large rays of the red oak wood illustrated in Figure 24–24c are 12 to 30 cells wide and hundreds of cells high. Besides the large rays, oak wood has numerous rays that are only one cell wide. In red oak wood, the rays make up, on average, about 21 percent of the volume of the wood. Overall, the rays of hardwoods average about 17 percent of the volume of the wood; the average for conifer wood is about 8 percent.

As in conifer wood, transverse sections of dicot woods reveal radial files of cells of both the axial and radial systems derived from the cambial initials (Figures 24–24 and 24–25). The files may not be as orderly as in conifer wood, however, for enlargement of the vessels and elongation of fibers tend to push many of the cells out of position. The displacement of rays by vessel members is particularly conspicuous in the transverse section of red oak *(Quercus rubra)* wood shown in Figure 24–25a.

24–25
Transverse sections of wood, showing growth layers. (a) *Red oak* (Quercus rubra). *The large vessels of ring-porous wood such as red oak are found in the early wood. The dark vertical lines are rays.* (b) *Tulip tree* (Liriodendron tulipifera), *a diffuse-porous wood.*

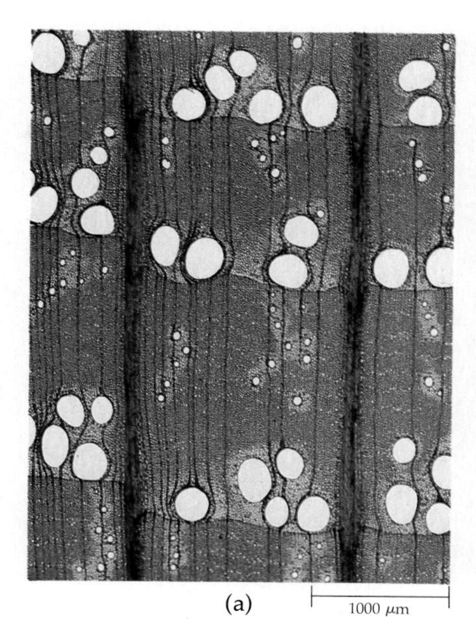

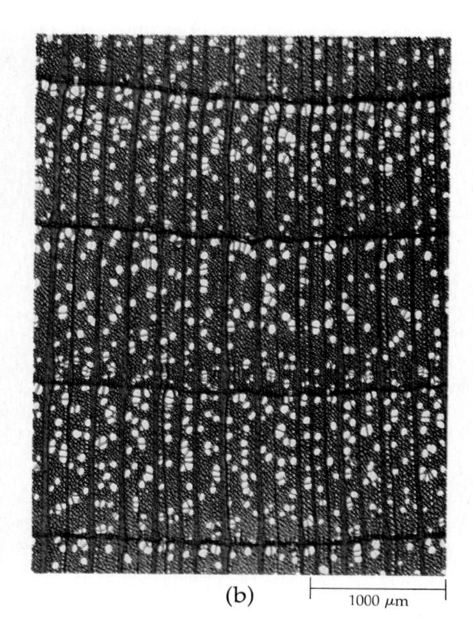

Growth Rings

The periodic activity of the vascular cambium, which is a seasonally related phenomenon in temperate zones, produces growth increments, or *growth rings*, in both secondary xylem and secondary phloem (in the phloem the increments are not always readily discernible). If a growth layer represents one season's growth, it is called an *annual ring*. Abrupt changes in available water and other environmental factors may be responsible for the production of more than one growth ring in a given year; such rings are called false annual rings. Thus the age of a given portion of the old woody stem can be estimated by counting the growth rings, but the estimates may be inaccurate if false annual rings are included.

The width of individual growth rings may vary greatly from year to year as a function of such environmental factors as light, temperature, rainfall, available soil water, and length of the growing season. The width of a growth ring is a fairly accurate index of the rainfall of a particular year. Under favorable conditions—that is, during periods of adequate or abundant rainfall—the growth rings are wide; under unfavorable conditions, they are narrow.

In semiarid regions, where there is very little rain, the tree is a sensitive rain gauge. An excellent example of this is the bristlecone pine *(Pinus longaeva)* of the western Great Basin (Figure 24–26). Each growth ring is different, and a study of the rings tells a story that dates back thousands of years. The oldest-known living specimen of bristlecone pine is 4900 years old. Dendrochronologists—scientists who conduct historical research through the growth rings of trees—have been able to match samples of wood from living and dead trees, and in this way they have built up a continuous series of rings dating back more than 8200 years. The widths of the growth rings of bristlecone pines at the higher elevations (the upper tree line) have also been found to be closely related to temperature changes, and a record of average ring width in these trees provides a valuable guide to past temperatures and climatic conditions. For example, in the White Mountains of California, the summers were relatively warm from 3500 B.C. to 1300 B.C., and the tree line was about 150 meters above its present level. Summers were cool from 1300 B.C. to 200 B.C.

The structural basis for the visibility of growth layers in the wood is the difference in density of the wood produced early in the growing season and that produced later (Figures 24–21, 24–24, and 24–25). The early wood is less dense (with larger cells and proportionally thinner walls) than the late wood (with narrower cells and proportionally thicker walls). In a given growth layer, the change from early to late wood may be very gradual and almost imperceptible. However, where the late wood of one growth layer abuts on the early wood of a following growth layer, the change is abrupt and thus clearly discernible.

In dicot woods, size differences of the vessels, or pores, in early and late woods are quite marked, the pores of the early wood being distinctly larger than in the late wood. (The term *pore* is used by the wood anatomist for a vessel seen in cross section.) Such woods are called ring-porous woods (Figures 24–24a and 24–25a). In other dicot woods, the pores are fairly uniform in distribution and size throughout the growth layer. These woods are called diffuse-porous woods (Figure 24–25b). In ring-porous woods, almost all the water is conducted in the outermost growth layer, at speeds about ten times greater than in diffuse-porous woods.

(a)

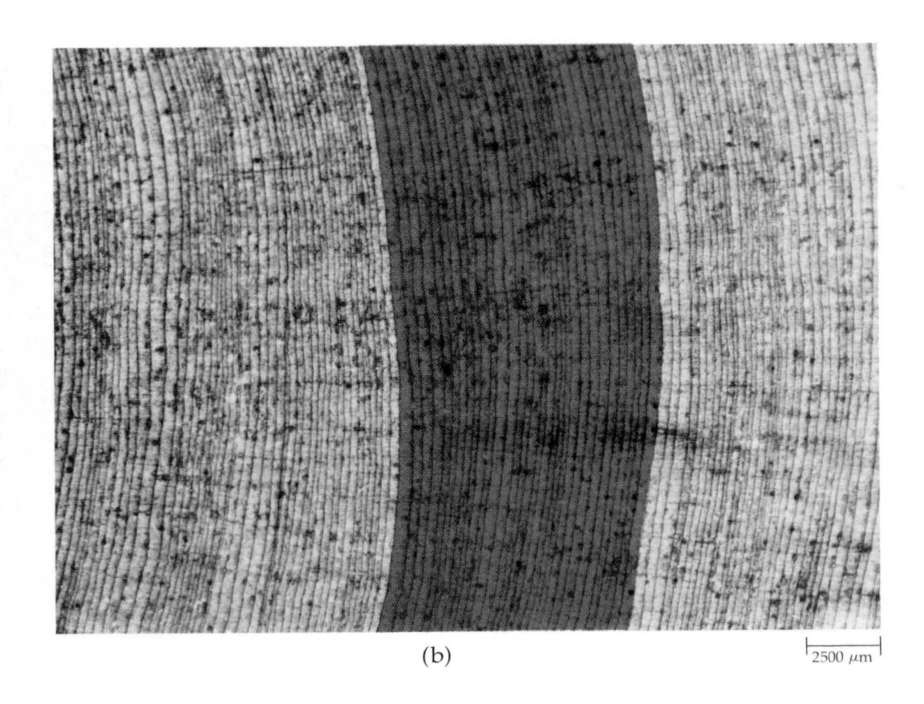

(b)

24–26
(a) *Bristlecone pine* (Pinus longaeva) *in the White Mountains of eastern California. These pines, which grow near the timberline, are the oldest living trees; one tree has reached an age of 4900 years.*
(b) *A transverse section of wood from a bristlecone pine, showing the variation in width of annual rings. This section begins approximately 6260 years ago; the band of rings in color represents the thirty years from 4240 B.C. to 4210 B.C. The overlapping patterns of rings in dead trees have made it possible to determine relative precipitation extending back some 8200 years.*

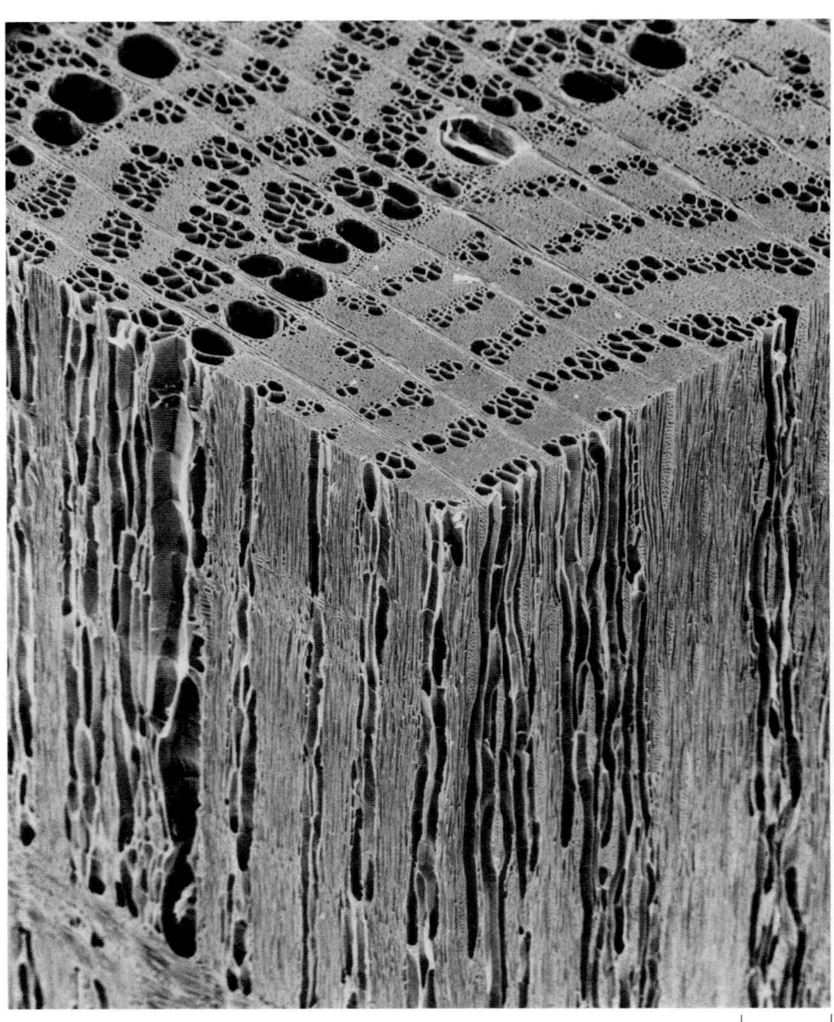

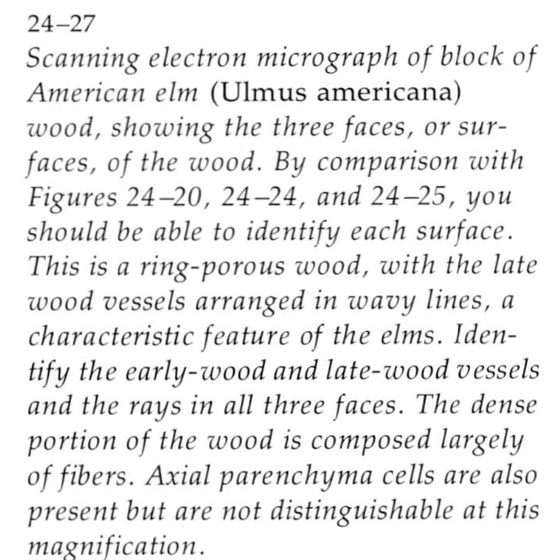

24–27
Scanning electron micrograph of block of American elm (Ulmus americana) *wood, showing the three faces, or surfaces, of the wood. By comparison with Figures 24–20, 24–24, and 24–25, you should be able to identify each surface. This is a ring-porous wood, with the late wood vessels arranged in wavy lines, a characteristic feature of the elms. Identify the early-wood and late-wood vessels and the rays in all three faces. The dense portion of the wood is composed largely of fibers. Axial parenchyma cells are also present but are not distinguishable at this magnification.*

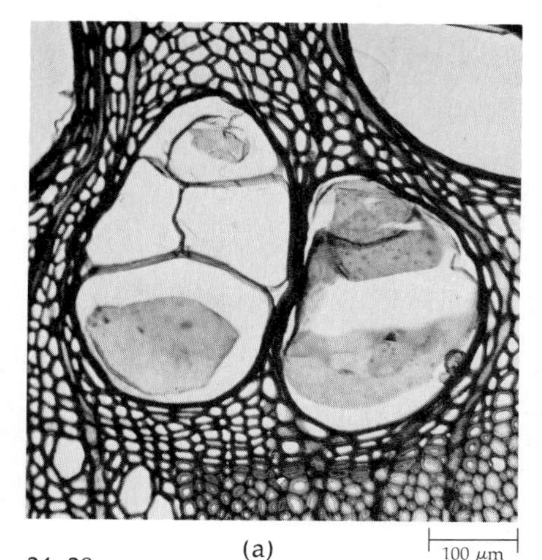

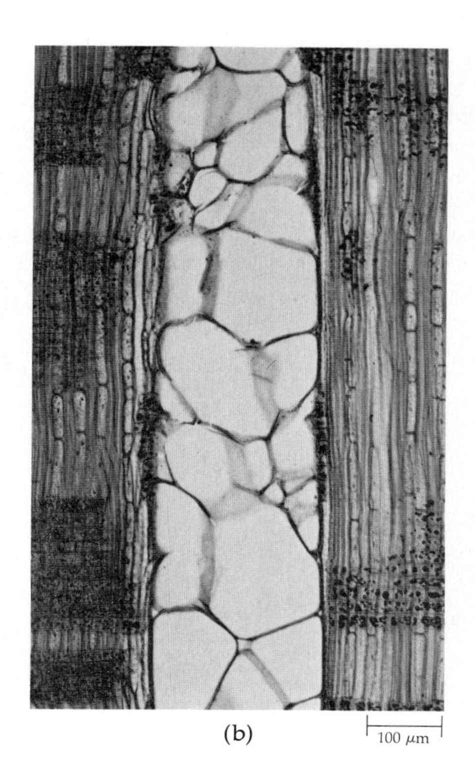

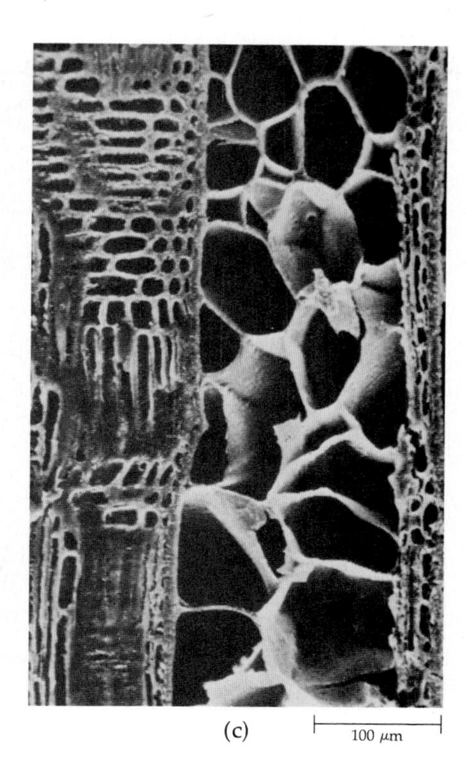

24–28
Tyloses, balloonlike outgrowths of parenchyma cells, which partially or completely block the lumen of the vessel. (a) *Transverse and* (b) *longitudinal sections showing tyloses in vessels of white oak* (Quercus alba), *as seen with a light microscope.* (c) *Tyloses in vessels of shagbark hickory* (Carya ovata), *as seen with a scanning electron microscope.*

Sapwood and Heartwood

As the wood grows older and no longer serves as a conducting tissue, its parenchyma cells eventually die. Before this happens, however, the wood often undergoes visible changes, which involve the loss of reserve food substances and the infiltration of the wood by various substances (such as oils, gums, resins, and tannin), which color and sometimes make it aromatic. This often darker, nonconducting wood is called *heartwood,* while the generally lighter conducting wood is called *sapwood* (see Figure 24–13, page 485). In many woods, tyloses are formed in the vessels when they become nonfunctional (Figure 24–28). *Tyloses* are balloonlike outgrowths from ray or axial parenchyma cells through pit cavities in the vessel wall. They may completely occlude the lumen (space bounded by the cell wall) of the vessel. Tyloses often are induced to form prematurely or unnaturally by plant pathogens and can result in death of the plant. Some of the so-called wilting diseases exert their effects in this way, by restricting the flow of water to the shoot.

The proportion of sapwood to heartwood and the degree of visible difference between them varies greatly from species to species. Some trees, such as maple *(Acer),* birch *(Betula),* and ash *(Fraxinus),* have thick sapwoods; whereas others, such as locust *(Robinia),* catalpa *(Catalpa),* and yew *(Taxus),* have thin sapwoods. Still other trees, such as the poplars *(Populus),* willows *(Salix),* and firs *(Abies),* have no clear distinction between sapwood and heartwood.

SUMMARY

Secondary growth (the increase in girth in regions that are no longer elongating) occurs in all gymnosperms and in most dicotyledons and involves the activity of the two lateral meristems—the vascular cambium and the cork cambium, or phellogen. Herbaceous plants may undergo little or no secondary growth, whereas woody plants—trees and shrubs—may continue to increase in thickness for many years. Figure 24–29 presents summaries of root and stem development of a woody plant, beginning with the apical meristem and ending with the secondary tissues produced during the first year's growth.

The vascular cambium contains two types of initials—fusiform initials and ray initials. Through periclinal divisions, the fusiform initials give rise to the components of the axial system, and ray initials produce ray cells, which form the vascular rays, or radial system. Increase in circumference of the cambium is accomplished by anticlinal division of the initials.

The first cork cambium in most stems originates in a layer of cells immediately below the epidermis. In the root, the first cork cambium arises in the pericy-

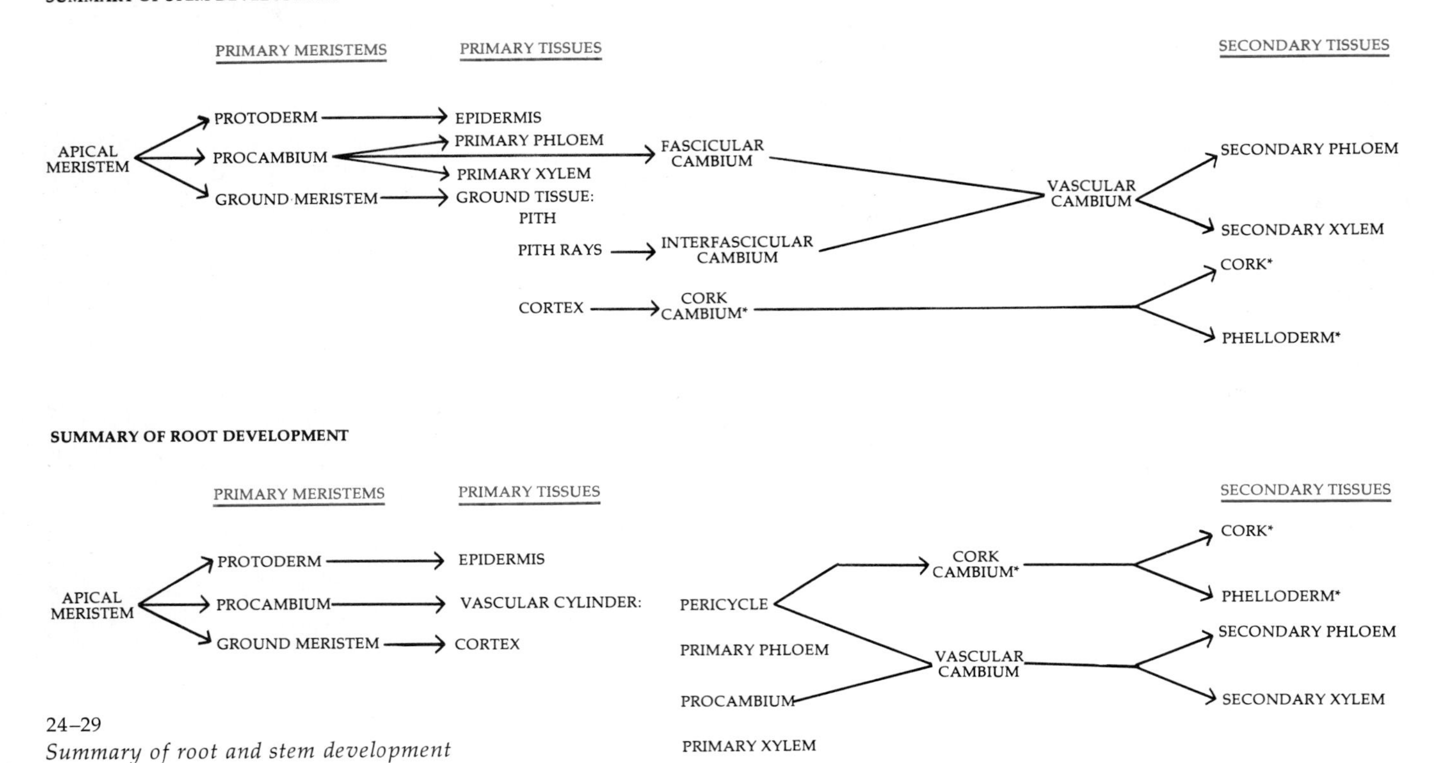

24–29
Summary of root and stem development in a woody dicot during the first year of growth.

cle. The cork cambium produces cork toward the outside, and phelloderm toward the inside. Together, the cork cambium, cork, and phelloderm comprise the periderm. Although most of the periderm consists of compactly arranged cells, isolated areas called lenticels have numerous intercellular spaces.

The bark consists of all tissues outside the vascular cambium. In old roots and stems, most of the phloem comprising the bark is nonfunctional. Sieve elements are short-lived, and generally only the present year's growth increment contains conducting, or functional, sieve elements. After the first periderm, subsequently formed periderms originate deeper and deeper in the bark from parenchyma cells of nonfunctional phloem.

Woods are classified as either softwoods or hardwoods. All so-called softwoods are conifers and all so-called hardwoods are dicotyledons. Compared with dicot woods, those of conifers are simple, consisting of tracheids and parenchyma cells. Some contain resin ducts. Dicot woods may contain a combination of all of the following cell types: vessel members, tracheids, several types of fibers, and parenchyma cells.

Growth layers that correspond to yearly increments of growth are called annual rings. The difference in density between the late wood of one growth increment and the early wood of the following increment makes it possible to distinguish the growth layers. In some kinds of plants, the nonconducting heartwood is visibly distinct from the actively conducting sapwood.

SUGGESTIONS FOR FURTHER READING

CORE, HAROLD A., WILFRED A. CÔTÉ, and ARNOLD C. DAY: *Wood Structure and Identification*, 2nd ed., Syracuse University Press, Syracuse, N.Y., 1979.*

A beautifully illustrated manual of wood structure, with a key for wood identification and an illustrated glossary.

CUTLER, DAVID F.: *Applied Plant Anatomy*, Longman Inc., New York, 1978.*

An interestingly written textbook on the fundamentals of plant anatomy showing some of the ways in which plant anatomy can be applied to solve many important everyday problems.

CUTTER, ELIZABETH G.: *Plant Anatomy,* Part I: *Cells and Tissues,* 2nd ed., Addison-Wesley Publishing Co., Inc., Reading, Mass., 1978.

An introduction to plant cells and tissues, with emphasis on experimental work.

———: *Plant Anatomy,* Part II: *Organs,* Addison-Wesley Publishing Co., Inc., Reading, Mass., 1971.*

An introduction to the organs of the plant and possible causes underlying their development.

EPSTEIN, EMANUEL: "Roots," *Scientific American* 228(5): 48–58, 1973.

A discussion of root structure and function, including a discussion of the mechanism of ion uptake by the root.

ESAU, KATHERINE: *Plant Anatomy,* 2nd ed., John Wiley & Sons, Inc., New York, 1965.

The standard work in the field; a well-illustrated book that considers all aspects of plant anatomy.

———: *Anatomy of Seed Plants,* 2nd ed., John Wiley & Sons, Inc., New York, 1977.

A shorter book than the preceding entry; an excellent textbook and reference.

FAHN, ABRAHAM: *Plant Anatomy,* 2nd ed., Pergamon Press, Inc., Elmsford, N.Y., 1974.*

A well-illustrated, up-to-date textbook considering all aspects of plant anatomy.

O'BRIEN, TERENCE P., and MARGARET E. McCULLY: *Plant Structure and Development,* The Macmillan Company, New York, 1969.

A pictorial and physiological approach to plant structure and development.

RAY, PETER M.: *The Living Plant,* 2nd ed., Holt, Rinehart & Winston, Inc., New York, 1972.*

A short, readable account of plant growth and development by one of the leading workers in the field.

STEEVES, TAYLOR A., and IAN M. SUSSEX: *Patterns in Plant Development,* Prentice-Hall, Inc., Englewood Cliffs, N.J., 1972.

A structural approach to plant development, with emphasis on experimental and analytical data.

TORREY, JOHN G., and DAVID T. CLARKSON (Eds.): *The Development and Function of Roots,* Academic Press, Inc., New York, 1975.

A collection of articles by recognized specialists in the fields of root structure and function.

WARDLAW, CLAUDE W.: *Morphogenesis in Plants: A Contemporary Study,* 2nd ed., Methuen & Co. Ltd., London, 1968.

A fascinating book by an acknowledged master of the field.

ZIMMERMANN, MARTIN H., and CLAUDE L. BROWN: *Trees: Structure and Function,* Springer-Verlag, New York, 1975.

An up-to-date discussion of how trees work, with emphasis on structure as it relates to function.

*Available in paperback.

SECTION 6 Growth Regulation and Growth Responses

25–1
Photoperiodism is the phenomenon that brings the plants of many species into flower simultaneously, year after year. Flowering is induced by the interaction of external (environmental) and internal (hormonal) factors. This is a cherry tree (Prunus) *blooming in the spring in Brooklyn, New York.*

CHAPTER 25

Regulating Growth and Development: The Plant Hormones

A plant, in order to grow, needs light from the sun, carbon dioxide from the air, and water and minerals, including nitrogen, from the soil. From these things, it makes more of its own substance, turning simple materials into the complex organic molecules of which living things are composed. As discussed in the previous section, the plant does far more than simply increase its mass and volume as it grows. It differentiates, develops, and takes shape, forming a variety of cells, tissues, and organs. How can a single cell, the fertilized egg, be the source of the myriad tissues and organs that make up the extraordinary individual known as a "normal" plant? Many of the details of how these processes are regulated are not known, but it has become clear that normal development depends on the interplay of a number of internal and external factors. The principal internal factors that regulate plant growth and development are chemical, and they are the subject of this chapter. Some of the external factors—light, temperature, day length, gravity, and so on—that affect plant growth are discussed in Chapter 26.

Plant hormones play a major role in regulating growth. The term *hormone* was coined by investigators studying animal physiology; it refers to organic substances that are produced in one tissue and transported to another tissue, where their presence results in physiological responses. Hormones are active in very small quantities. In the shoot of a pineapple plant *(Ananas comosus),* for example, only 6 micrograms of indoleacetic acid, a common plant hormone, are found per kilogram of plant material. One enterprising plant physiologist calculated that the weight of the hormone in relation to that of the shoot is comparable to the weight of a needle in 20 metric tons of hay.

The word *hormone* comes from the Greek word *hormaein,* meaning "to excite." It is now clear, however, that many hormones have inhibitory influences. Therefore, rather than thinking of hormones as stimulators, it is perhaps more useful to consider them as chemical regulators. But this term also needs qualification, for the response to the particular "regulator" depends not only on its content (chemical structure) but on how it is "read" by the recipient (tissue specificity).

The following discussion reflects the historical development of our knowledge of plant hormones. Accordingly, it begins with auxin, which made the concept of hormones in plants a valid one.

AUXIN

Some of the first recorded experiments on growth-regulating substances were performed by Charles Darwin and his son Francis and were reported in *The Power of Movement in Plants,* published in 1881. The Darwins first made systematic observations of the bending of plants toward light, which is known as phototropism (see page 518), using seedlings of canary grass *(Phalaris canariensis)* and oats *(Avena sativa).* They then showed that if they covered the upper portion of the coleoptile (the sheathlike, protective structure covering the shoot of grass seedlings) with a cylinder of metal foil or a hollow tube of glass blackened with India ink and exposed the plant to a lateral light (that is, light from the side), the characteristic bending did not occur (Figure 25–2). If, however, the tips were enclosed in transparent glass tubes, bending occurred normally. To the Darwins these experiments indicated the existence of a "communication" between the tip, that is, the tissue receiving the light stimulus, and the rest of the coleoptile, where the growth response occurs. They further stated, "We must therefore conclude that when seedlings are freely exposed to a lateral light some influence is transmitted from the upper to the lower part, causing the latter to bend."

In 1926, the Dutch plant physiologist Frits W. Went succeeded in isolating this "influence" from the coleoptile tips. Went cut off the coleoptile tips from a number of oat *(Avena)* seedlings and placed them for about an hour on a slice of agar with their cut surfaces in contact with the agar. (Agar is a gelatinous substance derived from certain red algae; it is commonly used as a neutral growth medium.) He then cut the agar into small blocks and placed the blocks on one side of the stumps of the decapitated plants, which were kept in the dark during the entire experiment. Within one hour, he observed a distinct bending *away* from the side on which the agar block was placed (Figure 25–3). Agar blocks that had not been exposed to a coleoptile tip produced either no bending or a slight bending toward the side on which the agar block had been placed. Agar blocks that had been exposed to a section of coleoptile from lower on the shoot also produced no physiological effect.

By these experiments, Went showed that the coleoptile tip exerted its effect by means of a chemical substance rather than a physical stimulus, such as an electrical one. Went named this chemical substance auxin, from the Greek word *auxein,* meaning "to increase."

The curvature of an *Avena* coleoptile away from the side bearing an auxin-containing agar block is caused by asymmetric distribution of auxin, which in turn causes asymmetric increase in cell size. The condi-

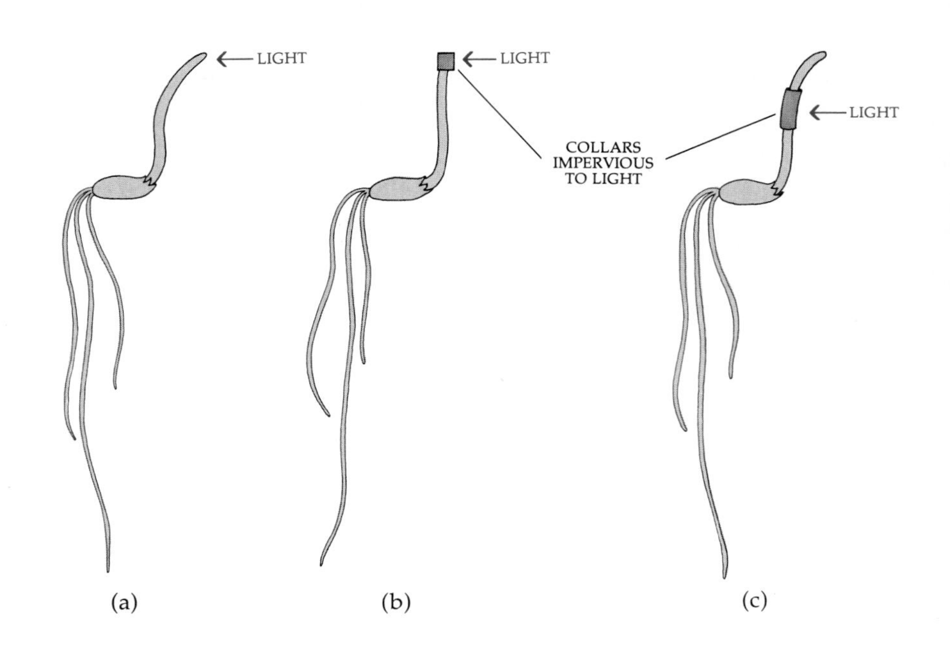

25–2
The Darwins's experiment. (a) *Seedlings normally bend toward the light.* (b) *When the tip of a seedling was covered by a light-proof collar, this bending did not occur. (Bending occurred normally when the tip of a seedling was covered with a transparent collar.)* (c) *When the collar was placed below the tip, the characteristic light response took place. From these experiments, the Darwins concluded that, in response to light, an "influence" that causes bending was transmitted from the tip of the seedling to the area below the tip, where bending normally occurs.*

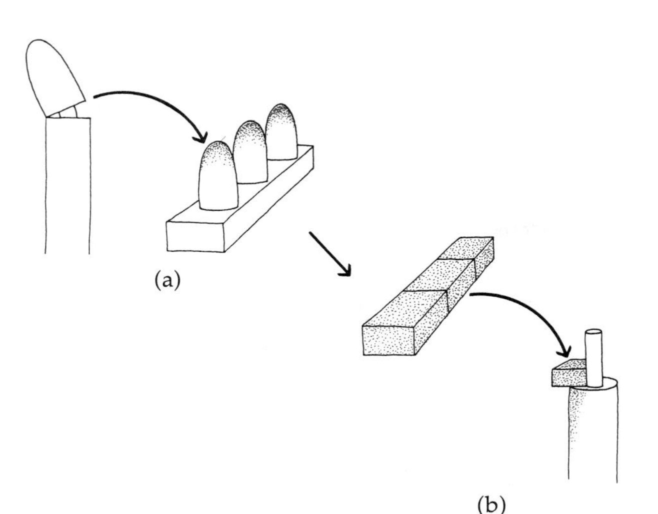

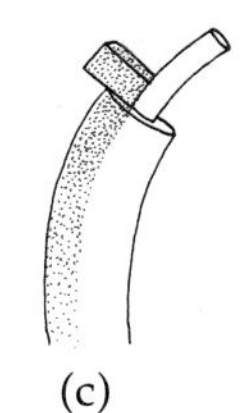

25–3
Went's experiment. (a) *Went removed the coleoptile tips from seedlings and placed the tips on agar for about an hour.* (b) *The agar was then cut into small blocks and placed on one side of the decapitated shoots of the seedlings.* (c) *The seedlings, which were kept in the dark during the entire experiment, subsequently were observed to bend away from the side on which the agar block was placed. From this result, Went concluded that the "influence" that caused the seedling to bend was chemical and that it accumulated on the side away from the light.*

tions for handling *Avena* seedlings and placing the agar block have been standardized to such an extent that the angle of curvature (measured with a protractor) can be used to determine the quantity of auxin in the agar block. Use of this technique, known as the *Avena* curvature bioassay, enabled investigators to isolate and identify the naturally occurring auxin called indoleacetic acid (abbreviated IAA). (A *bioassay* is a method of quantitatively determining the concentration of a substance by its effect on the growth of a suitable organism under controlled conditions.)

As can be seen in Figure 25–4, IAA closely resembles the amino acid tryptophan (see Figure 2–12, page 52). Tryptophan is probably the precursor from which IAA is formed in the living plant. Auxin is produced in the coleoptile tips of grasses and in shoot tips. Although IAA has been found in root tips, most evidence indicates that it is transported there via the vascular cylinder from the base of the root. It is probably abundant in embryos and is also found in young leaves and in flowers and fruits.

Shortly after the discovery of auxin and the recognition of its role in regulating cell elongation, an inhibitory effect of auxin was discovered in relation to the growth of lateral buds. For instance, if the apical meristem of a bean plant is removed, the lateral buds begin to grow. However, if auxin is immediately applied to the cut surface, the growth of the buds is inhibited. As with the phototropic response of *Avena* seedlings, this apical dominance demonstrates a form of "communication" between two plant tissues mediated by IAA. Furthermore, in both phenomena the "influence" moves from the growing tip to the base of the plant. This is because auxin is actively transported in plants from the tips of shoots to the base of the plant; that is, in a basipetal direction. The movement usually occurs in the tissue as a whole,

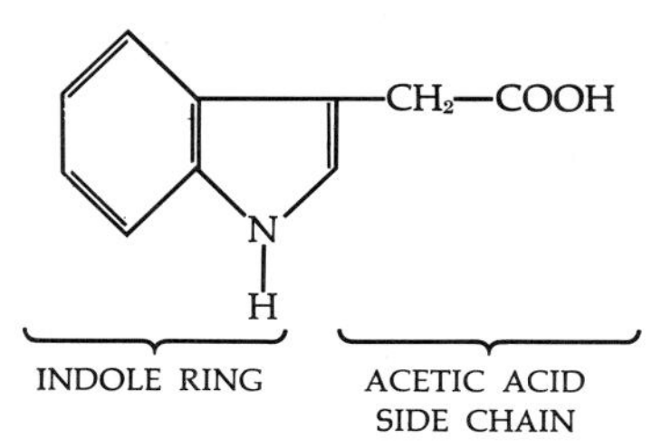

INDOLEACETIC ACID (IAA)

O—CH_2—COOH
Cl
Cl

2,4-DICHLOROPHENOXYACETIC ACID (2,4-D)

CH_2—COOH

α-NAPHTHALENACETIC ACID (NAA)

25–4
Indoleacetic acid (IAA) is the best-known naturally occurring auxin. Dichlorophenoxyacetic acid (2,4-D), a synthetic auxin, is widely used as an herbicide. Naphthalenacetic acid (NAA), another synthetic auxin, is commonly employed to induce the formation of adventitious roots in cuttings and to reduce fruit drop in commercial crops. The synthetic auxins, unlike IAA, are not readily broken down by natural plant enzymes and microbes and so are better suited for commercial purposes.

rather than in the "pipelines" of the xylem and phloem. The transport process is presumed to involve an interaction between IAA and the plasma membranes of the plant cells.

Application of auxin to plants causes a variety of effects, which differ from time to time, species to species, and most particularly, from tissue to tissue. Like many other physiologically active compounds, auxin is toxic at high concentrations. The weed killer 2,4-D is a synthetic auxin, one of many that have been manufactured commercially for a wide variety of applications (Figure 25–4).

Auxin and Cell Differentiation

Auxin influences the differentiation of the vascular tissue in the elongating shoot. If one cuts a wedge out of a stem of *Coleus* in such a way as to sever and remove portions of the vascular bundles, new vascular tissues will be formed from cells in the pith and will connect with the bundles in the uncut regions. If one takes away the leaves and buds above the excision, the formation of new cells is delayed. If one adds IAA to the petiole of the cut leaf just above the excision, formation of vascular tissue resumes. Auxin similarly plays an important role in the joining of vascular traces from developing leaves to the bundles in the stem.

Similar effects are seen in calluses. (A callus is a mass of undifferentiated cells that forms when a plant is wounded or when isolated cells are grown in tissue culture). If one takes a callus of lilac *(Syringa)* pith and grafts a bud onto it, vascular tissue is induced in the callus. Similarly, if the callus is grown in a medium containing auxin and sugar (the sugar is necessary because the callus does not contain photosynthetic cells), vascular tissues form. And, as discovered by R. H. Wetmore and his co-workers, by adjusting the amount of sugar in the medium, one can induce formation of xylem alone, xylem and phloem, or phloem alone. A low concentration of sucrose (1.5 to 2.5 percent) favors xylem, 4 percent favors phloem, and a percentage in between produces both. This illustrates the subtlety of growth regulator function and also draws attention to the critical factor that growth regulators *never* act alone; rather, they operate in concert with other internal factors such as sugar or other regulators of plant growth.

Auxin and the Vascular Cambium

In woody plants, auxin promotes the growth of the cambium. When the meristematic region of the shoot begins to grow in the spring, auxin moving down from the shoot tip stimulates cambial cells to divide, forming secondary phloem and secondary xylem. Again, these effects are modulated by other growth-regulating substances in the plant body.

Experiments with externally applied IAA and gibberellic acid indicate that, in the intact plant, interactions between auxins and gibberellins determine the relative rates of production of secondary phloem and secondary xylem. For example, IAA and gibberellic acid individually stimulate cambial activity when applied to a wide variety of woody plants. However, IAA in the absence of gibberellic acid stimulates only xylem development. Phloem development occurs with gibberellic acid alone, but both xylem and phloem development reach a maximum in the presence of both IAA and gibberellic acid.

25–5
The holly cuttings in the upper row were treated with auxin 21 days before the picture was taken. The cuttings in the lower row were not. Note the growth of adventitious roots in the treated plants.

Auxin and Root Growth

The first practical application of auxin involved its promoting effect on the formation of adventitious roots in cuttings (Figure 25–5). The practice of dipping cuttings in auxin solutions is commercially important, especially for the cultivation of woody plants that are produced on a large scale by vegetative propagation in nurseries. Application of large quantities of auxin to already growing roots, however, usually inhibits their growth.

Auxin and Fruit Growth

Auxin promotes the growth of fruit. Ordinarily, if the flower is not pollinated and fertilization does not take place, the fruit will not develop. In some plants, fertilization of one egg cell is sufficient for normal fruit development, but in others, such as apples or melons, which have many seeds, several must be fertilized for the ovary wall to mature and become fleshy. By treating the female flower parts of certain species with auxin, it is possible to produce parthenocarpic fruit (from *parthenos,* meaning "virgin"), which is fruit produced without fertilization, such as seedless tomatoes, cucumbers, and eggplants.

Apparently, developing seeds are a source of auxin. In the strawberry (*Fragaria* × *ananassa*), if the seeds enclosed in the fruits, which are achenes, are removed during the fruit's development, the fruit stops growing altogether. If a narrow ring of seeds is left, the fruit (actually the fleshy receptacle) forms a bulging girdle of growth in the area of the seeds. If auxin is applied to the denuded receptacle, growth proceeds normally (Figure 25–6).

Auxin and Abscission

Auxin is produced in young leaves, but it does not appear to have direct effects on the rate of leaf growth. However, auxin does affect abscission—the dropping of leaves or other plant parts. As leaves grow older, certain reusable ions and molecules are returned to the stem, among them magnesium ions, amino acids (derived from proteins), and sugars (some derived from starch). Then, in some plants at least, enzymes break down the cell walls in the separation layer of the abscission zone, across the base of the petiole (see

25–6
(a) *Normal strawberry fruit.* (b) *Strawberry from which all the seeds were removed.* (c) *Strawberry in which three horizontal rows of seeds were left.* (d) *Growth in strawberry induced by one developing seed.* (e) *Growth induced by three developing seeds. If a paste containing auxin is applied to the strawberry from which the seeds have been removed, the strawberry grows normally.*

PLANTS IN TEST TUBES

As long ago as the 1930s, scientists developed techniques for the growing of plant cells in test tubes. Tiny fragments of meristem are implanted under sterile conditions in a medium containing minerals and various combinations of organic compounds. Under these conditions, the meristematic cells proliferate to form clumps of undifferentiated cells. Subsequently, it was discovered that, by adjusting the hormone balance in the medium, it was possible to make these cells differentiate and grow into mature plants. Using these techniques, hundreds or even thousands of subcultures of meristematic tissue can be produced in a relatively short time and in a small space. By altering the hormone balance, each of these groups of plant cells can be turned into a small but perfect plant. This cloning technique has already been employed in the culture of orchids, and it is currently being extended to other plants.

Using similar techniques, it has become possible within the last decade to grow isolated protoplasts—plant cells without cell walls—in the laboratory. These protoplasts can be grown as isolated cells, like cultures of bacteria. Alternatively, if they are grown in a suitable medium, these protoplasts will regenerate cell walls, multiply, and differentiate into whole plants. The ability of nonembryonic plant cells to become embryonic cells and then—given proper conditions for growth and development—to develop into entire plants is known as totipotency; the cells themselves are said to be totipotent.

Cultured cells are being used in a variety of ways. For instance, it has been found that they can be used in rapid screening tests to determine resistance to infectious diseases or to detect nutritional requirements. In this way, a scientist not only can work much more rapidly but also can do assays with millions of cells growing in a very small space, as compared to the far fewer number of plants that can be grown in a field or in a greenhouse. Moreover, it has been found that it is possible to fuse the protoplasts of two different plant species, thus creating a hybrid.

The first plant created by this technique was a hybrid of tobacco (Nicotiana tabacum). *Interspecific hybrid plants have also been produced by protoplast fusion within the genera that include petunia* (Petunia), *carrot* (Daucus), *and potato* (Solanum). *More recently, intergeneric hybrid plants have been obtained by protoplast fusion between the potato* (Solanum tuberosum) *and the tomato* (Lycopersicon esculentum), *both of which are members of the nightshade family (Solanaceae), as well as between two members of the mustard family (Brassicaceae). This success has raised the hopes of producing entirely new "superplants"—organisms that might be capable, for instance, of simultaneously carrying out C_4 photosynthesis and fixing nitrogen. So far, fusion of protoplasts from cells as distantly related as corn and soybeans has been successful, but it has not been possible to induce these hybrid cells to develop into plants.*

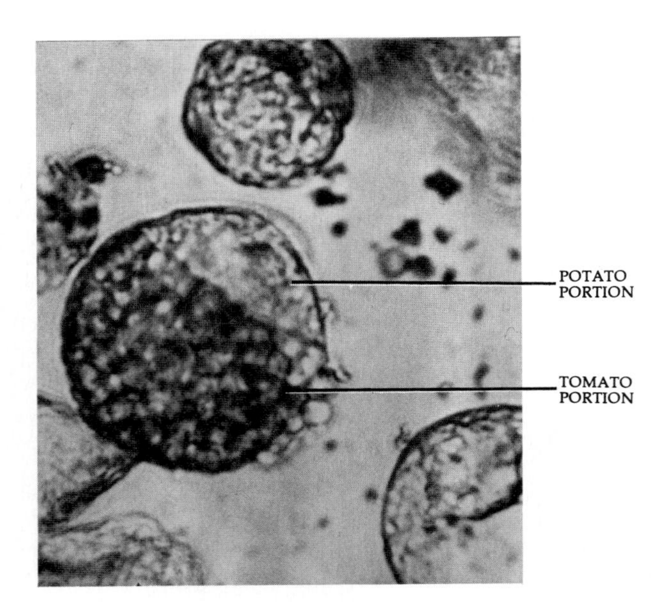

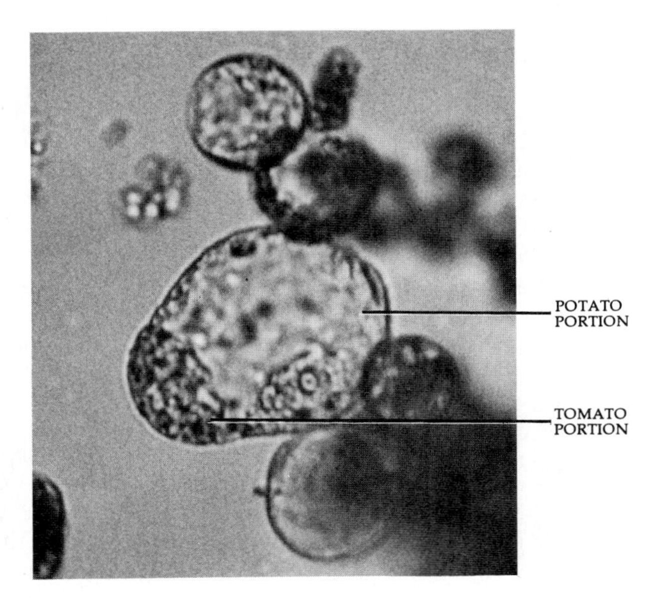

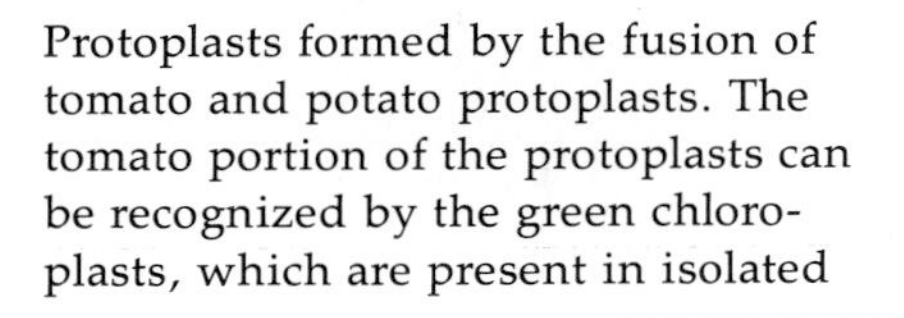

Protoplasts formed by the fusion of tomato and potato protoplasts. The tomato portion of the protoplasts can be recognized by the green chloroplasts, which are present in isolated tomato mesophyll protoplasts. The potato portion is colorless, because the isolated potato protoplasts contain only proplastids. (Research by Georg Melchers, Maria D. Sacristán, and Anthony A. Holder. Professor Melchers is at the Max-Planck-Institut für Biologie, Tübingen, West Germany.)

Figure 23–26, page 466). The cell wall changes may include weakening of the middle lamella and hydrolysis of the cellulosic walls themselves. Cell division may precede the actual separation. If cell division occurs, the newly formed cell walls are the ones largely affected by the degradation phenomena. Beneath the separation layer, a protective layer composed of heavily suberized cells is formed, further isolating the leaf from the main body of the plant before it drops. Eventually, the leaf is held to the plant only by a few strands of vascular tissue, which may be broken by the enlargement of parenchyma cells in the separation layer. Abscission has been correlated with a diminished production of auxin in the leaf, among other factors; under many circumstances, abscission can be prevented by the application of auxin.

The control of abscission of leaves, flowers, and fruits is extremely important in agriculture. Auxin and, more recently, ethylene have been used commercially for treatment of a number of plant species. For instance, auxin prevents leaf and berry drop from evergreen holly *(Ilex aquifolium)* and therefore minimizes losses during shipment. Auxin also prevents preharvest drop of citrus fruits. On the other hand, large amounts of auxin promote fruit drop, and auxin has been used for the thinning of fruit in the production of olives, apples, and other tree fruits.

Auxin and the Control of Weeds

Synthetic auxins have been used extensively for the control of weed pests in agricultural lands. In economic terms, this is the major practical use for plant growth regulators in the world today. Although a number of compounds can be employed, phenoxy auxins such as 2,4-D and its chemical derivatives are important weed control agents in that they represent approximately 20 percent of the total used. The continued use of these chemicals will depend on a number of factors, including cost-effectiveness and potential or real hazards to human health.

How Does Auxin Control Elongation?

Auxin increases the plasticity of the cell wall. When the cell wall softens, the cell enlarges because of the water pressure of its contents. As the water pressure is reduced by expansion of the cell, the plant cell takes up more water, and the cell thus continues to enlarge until it encounters sufficient resistance from the wall. (The factors involved in the uptake of water are discussed in Chapter 3.)

The softening of the cell wall is brought about by a complex series of interactions that are not yet fully understood. It is known that auxin causes increases in concentrations of RNAs and proteins. Furthermore, enzymes believed to be specifically involved in cell wall loosening and biosynthesis of new wall material have been shown to be affected. However, because auxin can, in some circumstances, stimulate elongation as quickly as three minutes after its application to the coleoptiles, it seems that its effect on protein biosynthesis may not be related to elongation. Unfortunately, the interpretation of the pertinent experiments is equivocal. One possibility is that the presence of auxin decreases the pH of the cell wall and, thereby, activates loosening enzymes that are present in a latent form. Rapid effects of auxin on the loss of hydrogen ions from coleoptile cells have been demonstrated, but more research needs to be done before the so-called pH hypothesis can be accepted as the biochemical mode of auxin action.

CYTOKININS

The discovery of auxin stimulated investigators to look for other types of chemicals regulating plant growth because, by analogy with animals, it seemed unlikely that growth and development in plants would be under the control of a single hormone. For instance, it was known that auxin often inhibited lateral bud growth when applied to decapitated plants. Were there natural hormones that counteracted the auxin effect?

Folke Skoog and his colleagues at the University of Wisconsin developed a method for studying hormonal regulation of bud growth in isolated plant tissues and organs in test tubes. When a stem segment of the tobacco plant *(Nicotiana tabacum)* was placed on a tissue culture medium containing sugar, vitamins, and various salts, cell division occurring at the cut surface resulted in the formation of an undifferentiated callus and, occasionally, an organized shoot which grew out as a leafy stem. Addition of auxin to the medium in sufficient amounts inhibited bud formation and growth, as expected. On the other hand, adenine promoted bud formation in the tissue culture and counteracted the inhibitory effects of auxin. The concentration of adenine required for bud formation was very high, however—too high for it to be considered a hormone—so Skoog and his colleagues set out to identify substances with greater biological activity in the test systems.

Initially, Skoog and his colleagues worked with the coconut *(Cocos nucifera)* because Johannes van Overbeek, and later F. C. Steward and his research group, had shown that coconut milk (which is a liquid endosperm) is a rich source of growth-promoting substances for tissue cultures. After many years of effort, Skoog and his co-workers succeeded in producing a thousandfold purification of a growth factor, but they could not isolate it. So, changing course, they tested a variety of purine-containing substances—largely nucleic acids—in the hope of finding a new source of the material.

Pursuing this new course, Carlos O. Miller, then a postdoctoral fellow in Skoog's laboratory, searched for bottles with a nucleic acid label. He tested one marked "Herring Sperm DNA" and found that it caused tobacco cells to divide. More herring sperm DNA was ordered. To the consternation of the investigators, these fresh preparations did not work. As a last resort, they tried another old sample. This one, too, was active. Apparently, the factor was to be found only as a breakdown product of DNA. Subsequently, they found that a variety of preparations of DNA that had aged, or had been "aged" artificially by heating in acid solution, contained material that produced the cell-division response.

Miller, Skoog, and their co-workers eventually succeeded in isolating the growth factor from one of these DNA preparations and identifying its chemical nature. They called this substance kinetin and named the class of growth regulators to which it belongs the *cytokinins,* because of its involvement in cytokinesis, or cell division. As shown in Figure 25–7, kinetin resembles the purine adenine, which was the clue that led to its discovery. Kinetin, which probably does not exist in plants in nature, has a relatively simple structure, and biochemists were soon able to synthesize a number of other, related compounds that behaved like cytokinins. Eventually, a natural cytokinin was isolated from kernels of corn *(Zea mays);* called zeatin, it is the most active of the naturally occurring cytokinins, although a few still more active synthetic cytokinins have now been produced.

Cytokinins have now been isolated from many different species of higher plants, where they are found primarily in actively dividing tissues, including seeds, fruits, and roots. These hormones have also been found in bleeding sap—the sap that drips out of pruning cuts, cracks, and other wounds in many types of plants.

Although applications for cytokinins are not as extensive as those for auxin, they have been important in plant development research and for plant propagation by tissue culture, as discussed below. Treatment of lateral buds with cytokinin will often cause the buds to grow, even in the presence of auxin, thus modifying apical dominance.

Cytokinins and Cell Division

Studies of interactions involving auxin and cytokinins are helping physiologists understand how plant hormones work to produce the total growth pattern of the plant. Apparently, the undifferentiated plant cell has two courses open to it: either it can enlarge, divide, enlarge, and divide again, or it can elongate without undergoing cell division. The cell that divides repeatedly remains essentially undifferentiated, or embryonic, whereas the elongating cell tends to differentiate, or become specialized. In studies of tobacco stem tissues, the addition of IAA to the tissue culture produced rapid cell expansion, so that giant cells were formed. Kinetin alone had little or no effect. IAA plus kinetin resulted in rapid cell division, so that large numbers of relatively small, undifferentiated cells were formed. In other words, the addition of the kinetin, together with IAA (although not kinetin alone), switched the cells to a meristematic course (Figure 25–8).

25–7
Note the similarities between the purine adenine and these four cytokinins. Kinetin and 6-benzylamino purine (BAP) are synthetic cytokinins which are commonly used. Zeatin and i⁶ Ade have been isolated from plant material.

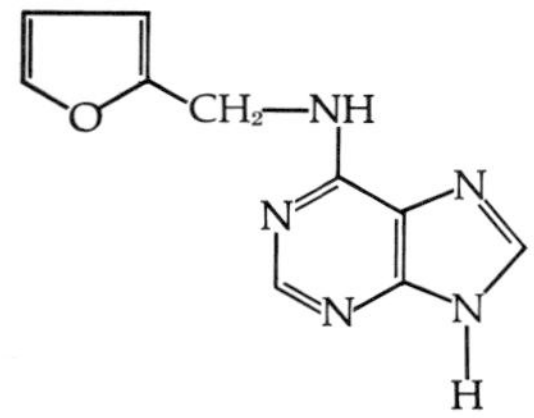

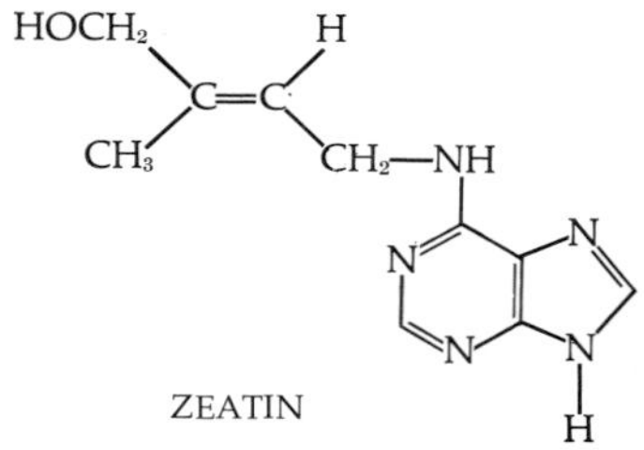

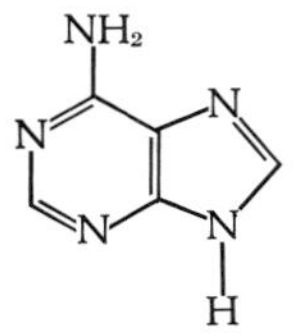

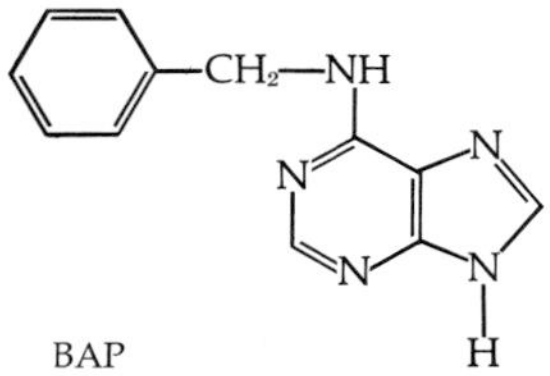

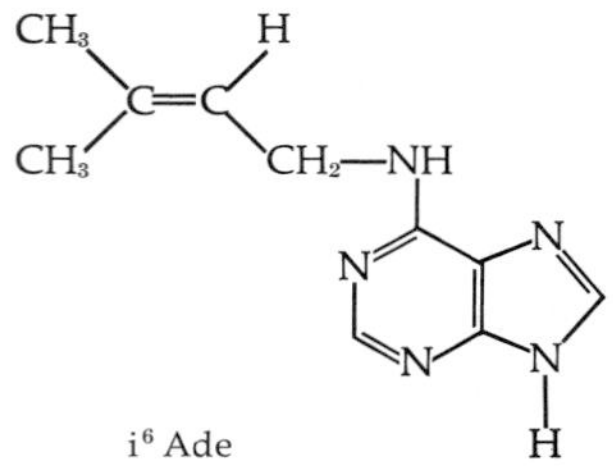

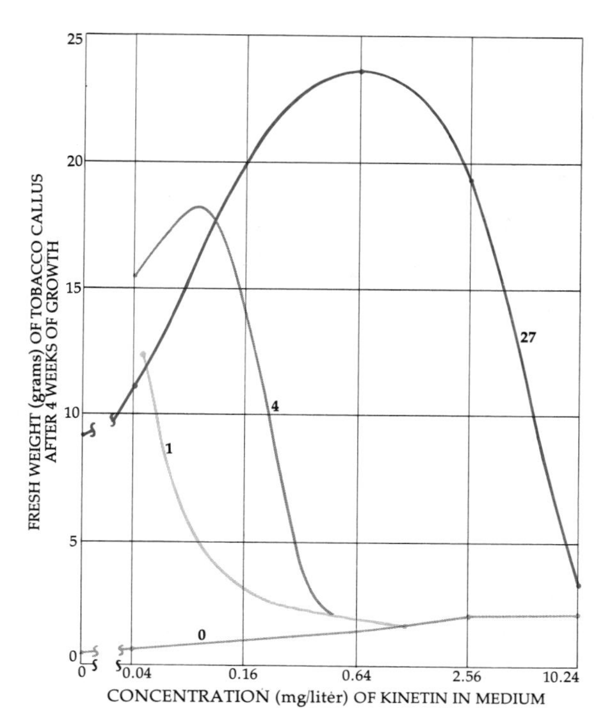

25–8
Auxin-kinetin interactions in the growth of tobacco callus. The concentrations of IAA supplied (mg per liter) are indicated on the curves. The higher the concentration of IAA, the greater the growth-promoting range of kinetin concentrations. In the absence of IAA, kinetin has only a slight growth-promoting effect. In the absence of kinetin, IAA promotes the growth of callus up to about 10 mg.

In another tissue culture study, using tuber tissue of the Jerusalem artichoke *(Helianthus tuberosus)*, it was shown that a third substance, the calcium ion, can modify the action of the auxin-cytokinin combination. In this study, IAA plus low concentrations of kinetin was shown to favor cell enlargement, but as Ca^{2+} was added to the culture, there was a steady shift in the growth pattern from cell enlargement to cell division. High concentrations of calcium prevent the cell wall from expanding, and at such concentrations the cell switches course and divides. Thus, not only do hormones modify the effects of hormones, but these combined effects are, in turn, modified by nonhormonal factors, such as calcium.

Cytokinins and Organ Formation in Tissue Cultures

In the presence of a high concentration of auxin, callus tissue frequently gives rise to organized roots. In tobacco pith callus, the relative concentrations of auxin and kinetin determine whether roots or buds form. With higher concentrations of auxin, roots are formed; with higher concentrations of kinetin, buds are formed; and when the two are present in roughly equal concentrations, the callus continues to produce undifferentiated cells (Figure 25–9). These same effects have been observed in tissues of a large number of plant types. Economically, probably the major practical application of cytokinins is in tissue culture propagation of some types of houseplants.

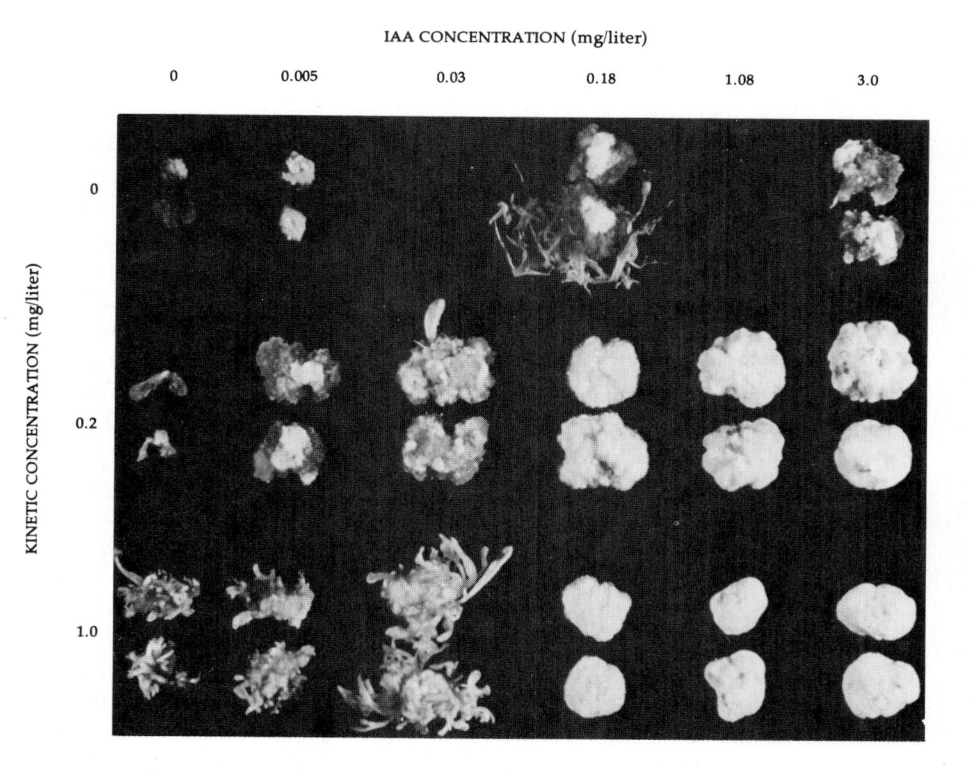

25–9
Effect of increasing IAA concentration at different kinetin levels on growth and organ formation of tobacco callus cultured on nutrient agar. Note that very little growth occurred without the addition of either IAA or kinetin. Higher levels of IAA alone promoted root formation, whereas they repressed bud formation when used alone or in combination with kinetin. The higher level of kinetin was more effective than the lower level in promoting bud development, but both kinetin levels were too high for promotion of root growth.

Cytokinins and Leaf Senescence

In most species of plants, the leaves begin to turn yellow as soon as they are removed from the plant. This yellowing, which is due to a loss of chlorophyll, can be prevented by cytokinins. Leaves of the cocklebur *(Xanthium strumarium)*, for example, when excised and floated on plain water, turn yellow in about ten days. If kinetin (10 mg per liter) is present in the water, much of the chlorophyll and the fresh appearance of the leaf are maintained. If excised leaves are spotted with kinetin-containing solutions, the spots remain green while the rest of the leaf yellows. Furthermore, if a cytokinin-spotted leaf contains radioactive amino acids, labeled with ^{14}C, it can be shown that the amino acids migrate from other parts of the leaf to the cytokinin-treated areas. Such studies, which have also been carried out on radishes and other plants, led to the hypothesis that senescence in leaves, and probably in other plant parts as well, results from the progressive "turning off" of segments of DNA, with a consequent loss of messenger RNA production and protein synthesis. These investigations have led to the proposal that cytokinins prevent the DNA from being turned off and so promote continued enzyme synthesis and continued production of such compounds as chlorophyll.

One interpretation of the prevention of senescence in detached leaves by cytokinins is that leaves normally do not synthesize sufficient cytokinin to meet their own requirements. Considerable evidence indicates that the root is the center of cytokinin production in a vegetatively growing plant and that cytokinins produced in the root move to the shoot and there influence shoot development. During reproductive growth the developing fruit also becomes a cytokinin-producing site.

How Do Cytokinins Work?

Ever since the isolation of the first cytokinins from preparations of nucleic acids, plant physiologists have suspected that these hormones might in some way be involved with the nucleic acids. Molecules of transfer RNA (tRNA) contain a number of unusual bases (see Figure 7–9, page 125). In some types of transfer RNA, the natural cytokinin i^6Ade (^{6}N-isopentenyladenine), itself an unusual base, is incorporated into the molecule. For example, i^6Ade is found in serine and tyrosine tRNA molecules, in which it is located immediately adjacent to the anticodon. It is still not known, however, if its presence or its position in the tRNA molecule is related to its activity in promoting cell division. Considerable research remains to be done before we know how cytokinins work.

ETHYLENE

Over a period of years, the discovery of auxin has led, more or less directly, to the isolation of kinetin and the recognition of cytokinin effects on plant growth and development. Ethylene, on the other hand, was known to have effects on plants long before its relationship to auxin was discovered; that is, long before it was considered to be a plant hormone.

The "botanical" history of *ethylene*, a simple hydrocarbon ($H_2C = CH_2$), goes back to the 1800s, when city streets were lighted with lamps that burned illuminating gas. In Germany, leaking illuminating gas from gas mains was found to cause defoliation of shade trees along the streets. As gas became more extensively used for street illumination, this phenomenon was reported by many investigators.

In 1901, Dimitry Neljubov, a graduate student at the Botanical Institute of St. Petersburg University in Russia, demonstrated that ethylene was the active component of illuminating gas. Neljubov noticed that exposure of pea seedlings to illuminating gas caused stems to grow in a horizontal direction. When the gaseous components of illuminating gas were individually tested for effects, all were inactive except ethylene, which caused horizontal growth at concentrations as low as 0.06 parts per million (ppm) in air. Neljubov's findings have since been confirmed by many other investigators, and it is now known that ethylene exerts a major influence on many, if not all, aspects of growth, development, and senescence in plants.

Ethylene, although a gas under physiological conditions of temperature and pressure, is considered to be a plant hormone because it is a natural product of metabolism and because it acts and interacts with other plant hormones in trace amounts. Its effects can be observed particularly during critical periods in the life cycle of the plant.

Ethylene currently is the easiest hormone to assay. Since it is a gas that is evolved by tissues, it requires no extraction or purification prior to analysis by gas chromatography. At the cellular level, the ethylene-synthesizing system is apparently localized on the surface of the plasma membrane.

Ethylene and Fruit Ripening

Ripening in fruit involves a number of changes. In fleshy fruits, the chlorophyll is degraded and other pigments may form, changing the fruit color. Simultaneously, the fleshy part of the fruit softens. This is a result of the enzymatic digestion of pectin, the principal component of the middle lamella of the cell wall. When the middle lamella is weakened, cells are able to slip past one another. During this same period, starches and organic acids or, as in the case of the avocado *(Persea americana),* oils are metabolized into sugars. As a consequence of these changes, fruits become conspicuous and palatable and thus attractive to animals that eat the fruit and so scatter the seed.

During the time of ripening of many fruits, there is a large increase in cellular respiration, evidenced by an increased uptake of oxygen. This phase is known as the *climacteric.* The relationship between the climacteric and the other events of fruit ripening is not known, but the ripening of fruits can be delayed by suppressing the intensity of the climacteric. For example, cold suppresses it, and in some fruits, exposure to low temperatures stops the climacteric permanently. Fruits can be stored for very long periods of time in a vacuum; under such conditions, the amount of available oxygen is minimal, thus suppressing cellular respiration; and ethylene, which speeds up the onset of the climacteric, is held at low levels. After the climacteric, senescence sets in, and the fruit becomes susceptible to invasions by fungi and other microorganisms.

In the early 1900s, many fruit growers made a practice of improving the color and increasing the sweetness of citrus fruits by "curing" them in a room with a kerosene stove. (Long before, however, the Chinese ripened fruits in rooms where incense was being burned.) It was believed that the heat of the stove ripened the fruits. Ambitious fruit growers, who installed more modern equipment, found to their sorrow that this was not the case. As subsequent experiments demonstrated, it is actually the incomplete combustion products of the kerosene that are responsible. The most active combustion product was a gas identified as ethylene. As little as 1 ppm of ethylene in the air will speed up the onset of the climacteric.

As early as 1910, it was reported that gases emanating from oranges hastened the ripening of bananas, but it was not until almost 25 years later that ethylene was identified as a natural product of numerous fruits and plant tissues. The amounts produced by plants are very small, and new and extremely sensitive assay methods had to be developed before it could be proved that ethylene production began before the climacteric, even though the largest amounts coincide with the climacteric. When this was established, ethylene became generally accepted as a natural plant-growth regulator. It has now been found in fruits (in all the types tested), flowers, leaves, leafy stems, and roots of many different plant species and also in some species of fungi.

The effect of ethylene on fruit ripening has agricultural importance. A major use of ethylene is for promotion of ripening of tomatoes that are picked green and stored in ethylene until marketed. It is also used to hasten ripening of walnuts and grapes.

Ethylene and Abscission

Ethylene promotes abscission of leaves, flowers, and fruits in a variety of plant species. In leaves, ethylene presumably triggers the enzymes that cause the cell wall dissolution associated with abscission. Ethylene is used commercially to promote fruit loosening in cherries, blackberries, grapes, and blueberries, thus making mechanical harvesting possible. It is also used as a fruit-thinning agent in commercial orchards of prunes and peaches.

Ethylene and Sex Expression

Ethylene appears to play a major role in determining the sex of flowers in monoecious plants (those plants having male and female flowers borne on the same individual). In cucurbits (family Cucurbitaceae), for example, high levels of gibberellins are associated with maleness, and treatment with ethylene changes the expression of sex to femaleness. In studies with cucumbers *(Cucumis sativus),* it was found that female buds evolved greater quantities of ethylene than male buds. In addition, cucumbers grown under light periods of short days, which promote femaleness, evolved more ethylene than those grown under long days. Hence, in cucurbits, ethylene apparently participates in the regulation of sex expression and is associated with the promotion of femaleness.

Ethylene and Auxin

In specific concentrations, auxin causes a burst of ethylene production in certain parts of some plants. It is believed that some of the effects on fruits and flowers once attributed to auxin are related to the effect of auxin on ethylene production.

Cell shape and size are controlled in part by the interaction of auxin and ethylene, which regulate cell expansion in opposing ways. In the garden pea *(Pisum sativum),* for example, concentrations of IAA that result in the deposition of transversely oriented cell-wall microfibrils enhance elongation of stem segments but restrict lateral growth. By contrast, higher concentrations of IAA, which induce much greater ethylene production and result in deposition of longitudinally oriented microfibrils, bring about considerable radial growth and swelling of the stem segments.

It is important to realize that the final shape and size of cells, as influenced by ethylene, are the result of its interaction not only with auxin but also with gibberellic acid and with cytokinin as well. Moreover, normal growth and development require complex hormonal interactions which involve cytokinins, gibberellins, and abscisic acid, as well as auxin and ethylene.

ABSCISIC ACID

At certain times, the survival of the plant depends on its ability to restrain its growth or its reproductive activities. Following the early discovery of the growth-promoting hormones, plant physiologists began to speculate that regulatory hormones with inhibitory actions might also be found. Finally, in 1949, it was discovered that the dormant buds of ash and potatoes contained large amounts of growth inhibitors. These inhibitors blocked the effects induced by IAA in the *Avena* coleoptile. When dormancy in the buds was broken, the inhibitor contents declined. These inhibitors became known as *dormins.*

During the 1960s, several investigators reported the discovery in leaves and fruits of a substance capable of accelerating abscission. One of these, called abscisin, was identified chemically. In 1965, one of the dormins was also identified chemically, and abscisin and this dormin were found to be identical. The compound is now known as *abscisic acid.*

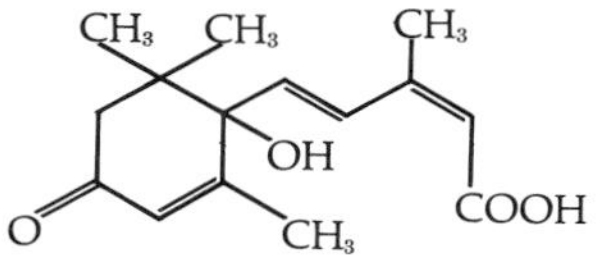

CH_3 CH_3 CH_3 OH COOH O CH_3

ABSCISIC ACID

Abscisic acid (abbreviated ABA) is collected largely from the ovary bases of fruits. The fruit of the cotton plant *(Gossypium)* has proved to be a particularly rich source. The largest amounts of abscisic acid are found at the time of fruit drop. Abscisic acid produced in core cells of the root cap has been implicated in the response of roots to gravity (see Chapter 26, page 519).

Application of abscisic acid to vegetative buds changes them to winter buds by converting the leaf primordia into bud scales. Its inhibitory effects on buds can be overcome by gibberellin. The appearance of alpha-amylase, an enzyme induced by gibberellin in the seeds of barley *(Hordeum vulgare),* is inhibited by abscisic acid, which seems to depress protein production in general. Auxin, on the other hand, seems to act both by interacting with the plasma membrane and by accelerating the production of specific proteins. Hence, auxin is apparently antagonistic to abscisic acid in its action.

If a drop of abscisic acid is spotted on a leaf, the treated areas yellow rapidly, even though the rest of the leaf stays green, an effect that is opposite to that of the cytokinins. Whether this is a direct or indirect action is not known at present.

There are currently no practical uses for abscisic acid, perhaps because of the limited knowledge about its physiology and biochemistry. Abscisic acid may become of tremendous importance in future agricultural practice, however, particularly in desert regions. There is reason to believe that the tolerance of some plants to stress conditions, such as drought, is directly related to their ability to produce abscisic acid. Furthermore, it is known that abscisic acid can cause the stomata of some plants to close, thus preventing water loss from leaves and lowering the plant's total water requirement. If plant geneticists can incorporate this characteristic related to abscisic acid into a wider range of plants, it may become possible to develop new crop plants that can be grown successfully in desert environments.

GIBBERELLINS

Unlike the other hormones, the discovery of the gibberellins is not closely related to the discovery of auxins. In fact, the early history of gibberellin research is quite independent of the work on auxin.

In 1926, the same year that Went first performed his experiments with blocks of agar, E. Kurosawa of Japan was studying a disease of rice *(Oryza sativa)* called "foolish seedling disease," in which the plants grew rapidly, were spindly, pale-colored, and sickly, and tended to fall over. The cause of these symptoms, Kurosawa discovered, was a chemical produced by a fungus, *Gibberella fujikuroi,* which was parasitic on seedlings. From the generic name of its source, this chemical substance was named *gibberellin.*

Gibberellin was isolated and identified chemically by biochemists in Japan in the 1930s, but for several decades it attracted little interest. Then, in 1956, gibberellin was successfully isolated from a plant (the seed of the bean *Phaseolus vulgaris*) rather than a fungus. Since that time, gibberellins have been isolated from many species of plants, and it is now generally believed that they probably occur in all plants. They are present in varying amounts in all parts of the plant, but the highest concentrations are found in immature seeds. More than 57 gibberellins now have been isolated from plant tissues and identified chemically. They vary slightly in structure (Figure 25–10), as well as in biological activity. The best studied of the group is GA_3 (known as gibberellic acid), which is also produced by the fungus *Gibberella fujikuroi.*

The gibberellins have dramatic effects on stem elongation in intact plants. A marked increase in the growth of the shoot is the most general response seen in higher plants; often the stems become long and thin and the leaves pale in color. The gibberellins stimulate both cell division and cell elongation and affect leaves as well as stems.

25–10
Three of the more than 57 gibberellins that have been isolated from natural sources. Gibberellic acid (GA_3) *is the most abundant in fungi and the most biologically active in many tests. The arrows indicate minor structural differences that distinguish the other two examples of gibberellins,* GA_7 *and* GA_4.

GIBBERELLIC ACID (GA_3)

GA_7

GA_4

Gibberellins and Dwarf Mutants

The most remarkable results are seen when gibberellins are applied to certain plants that are single-gene dwarf mutants (Figure 25–11). Under gibberellin treatment, such plants become indistinguishable from normally tall, nonmutant plants. This striking effect leads to the speculation that the result of the mutation, in biochemical terms, is a loss of the plant's ability to synthesize its own gibberellins. One bioassay for these hormones is this growth response in dwarf plants, particularly dwarf corn, an effect that cannot be duplicated by auxin or any of the other known hormones.

25–11
The plant on the right was treated with gibberellin; the one on the left served as a control. The plants are contender beans, a dwarf variety of the common bean (Phaseolus vulgaris).

Gibberellins and Seeds

The seeds of most plants require a period of dormancy before they will germinate. In certain plants, dormancy usually cannot be broken except by exposure to cold or to light. In many species, including lettuce, tobacco, and wild oats, gibberellins will substitute for the dormancy-breaking cold or light requirement and promote the growth of the embryo and the emergence of the seedling. Specifically, the gibberellins enhance cell elongation, making it possible for the root to penetrate the growth-restricting seed coat or fruit wall. This effect of gibberellin has at least one practical application. Gibberellic acid hastens seed germination and thus insures uniformity in the production of the barley malt used in brewing.

Gibberellins and Juvenility

The juvenile stages of some plants are different from their adult stages, and this is often reflected in the juvenile and adult forms of leaves. Among annual dicots, the bean plant *(Phaseolus),* whose juvenile leaves are simple and adult leaves are compound (trifoliate), provides an excellent example of such heterophylly (see Figure 20–2, page 406). Among the perennials, many species of *Eucalyptus* show striking differences between their juvenile and adult leaves (Figure 25–12).

(a)

(b)

25–12
(a) *Juvenile and* (b) *mature leaves of* Eucalyptus globulus, *showing the great differences that can occur within a given species. The juvenile leaves are softer and opposite to one another. Their only layer of palisade parenchyma is just below their upper epidermis. The mature leaves are hard, spirally arranged, and hang vertically. In the mature leaves, both surfaces are equally exposed to the light, and there are two layers of palisade parenchyma.*

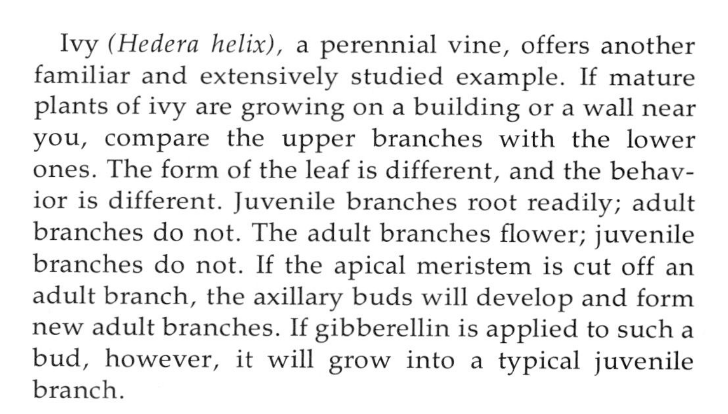

Ivy *(Hedera helix),* a perennial vine, offers another familiar and extensively studied example. If mature plants of ivy are growing on a building or a wall near you, compare the upper branches with the lower ones. The form of the leaf is different, and the behavior is different. Juvenile branches root readily; adult branches do not. The adult branches flower; juvenile branches do not. If the apical meristem is cut off an adult branch, the axillary buds will develop and form new adult branches. If gibberellin is applied to such a bud, however, it will grow into a typical juvenile branch.

Gibberellins and Flowering

Some plants, such as cabbages *(Brassica oleracea),* carrots *(Daucus carota),* and the biennial henbane *(Hyoscyamus niger),* form rosettes before flowering. (In a rosette, leaves develop but the internodes between them do not elongate.) In these plants, flowering can be induced by exposure to long days, to cold (as in the biennials), or both. Following the appropriate exposure, the stems elongate—a phenomenon known as bolting—and the plants flower. Application of gibberellin to such plants causes bolting and flowering without appropriate cold or long-day exposures (Figure 25–13). Elongation is brought about by an increase both in the number of cell divisions at certain locations and in the elongation of cells resulting from those divisions. Gibberellin can thus be used to produce early seed production of biennial plants. By treating a plant such as lettuce *(Lactuca sativa)* with gibberellic acid, seeds can be obtained after only one growing season.

Gibberellins and Pollen and Fruit Development

Gibberellins have been shown to stimulate pollen germination and the growth of pollen tubes in a number of plants, including lilies, lobelias, petunias, and peas. Like auxin, gibberellins can cause the development of parthenocarpic fruits, including apples, currants, cucumbers, and eggplants. In some fruits,

25–13
Bolting in cabbage (Brassica oleracea) *produced by gibberellin treatment. The plant on the right was treated once a week for eight weeks.*

25–14
The effect of gibberellic acid on growth of Thompson Seedless grapes, a cultivar of Vitus vinifera. *The slender bunch of grapes was untreated, whereas the fuller bunch was treated with gibberellic acid.*

such as mandarin oranges, almonds, and peaches, the gibberellins have been effective in the promotion of fruit development where auxin was not. The major application of gibberellin, however, is in the production of table grapes. In the United States, large amounts of gibberellic acid are applied annually to such grape varieties as the Thompson Seedless, a cultivar, or cultivated variety, of *Vitis vinifera*. Treatment causes larger berries and a much looser appearance of the clusters of grapes (Figure 25–14).

How Do Gibberellins Work?

The most important studies on the mechanism of gibberellin action were carried out simultaneously by investigators in Japan, Australia, and the United States. These studies, which trace the sequence of events in the germination of a barley seed and the early growth of its embryo, show the key role played by gibberellins in this sequence. In addition, they provide one of the best examples of how hormones integrate the biochemistry and physiology of the different tissues of a whole plant.

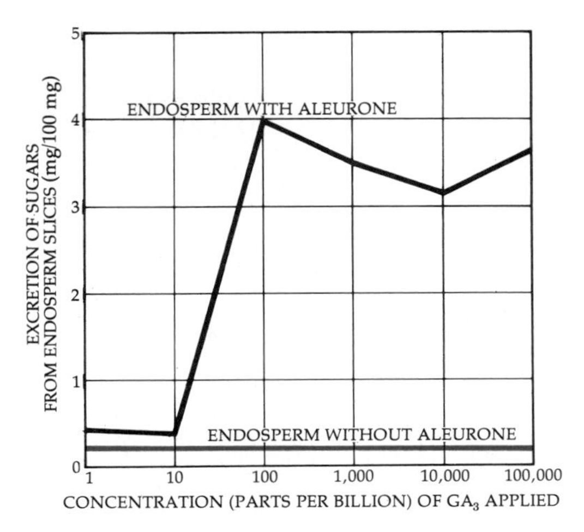

25–15
The release of sugar from endosperm can be induced by gibberellin (GA_3) *treatment. These data show that sugars are produced only when the aleurone layer is present. It is, in fact, the aleurone layer that is the source of the enzyme alpha-amylase, which digests the starches stored in the endosperm.*

In barley *(Hordeum vulgare)* and other grass seeds, there is a specialized layer of endosperm cells, called the aleurone layer (see Figure 20–4, page 408), just inside the seed coat. The cells in the aleurone layer are rich in protein. When the seeds begin to germinate, triggered by the imbibition of water, the embryo releases gibberellins. In response to these hormones, the aleurone cells synthesize hydrolytic enzymes, the principal one being alpha-amylase, the enzyme that breaks down starch into sugar (Figure 25–15). The enzymes digest the stored food reserves of the starchy endosperm. These food reserves are released in the form of sugars and amino acids, which are absorbed by the scutellum and are then transported to the growing regions of the embryo (Figure 25–16). In this way, the embryo calls forth the substances needed for its growth at the moment it requires them.

Investigators believe that gibberellins activate certain genes, causing the synthesis of specific messenger RNA molecules, which, in turn, direct the synthesis of the enzymes. It has not been proved, however, that gibberellins act directly on the gene, although it has been shown that both RNA and protein synthesis take place and are necessary for the appearance of the enzymes. Whatever the details of the mechanism of gibberellin action in aleurone cells, it is clear that the cells of the aleurone layer constitute a highly differentiated tissue poised to respond to the demands—mediated by gibberellins—of the growing embryo. It is not known whether the way in which gibberellins work in seeds is related to their effects on other plant organs.

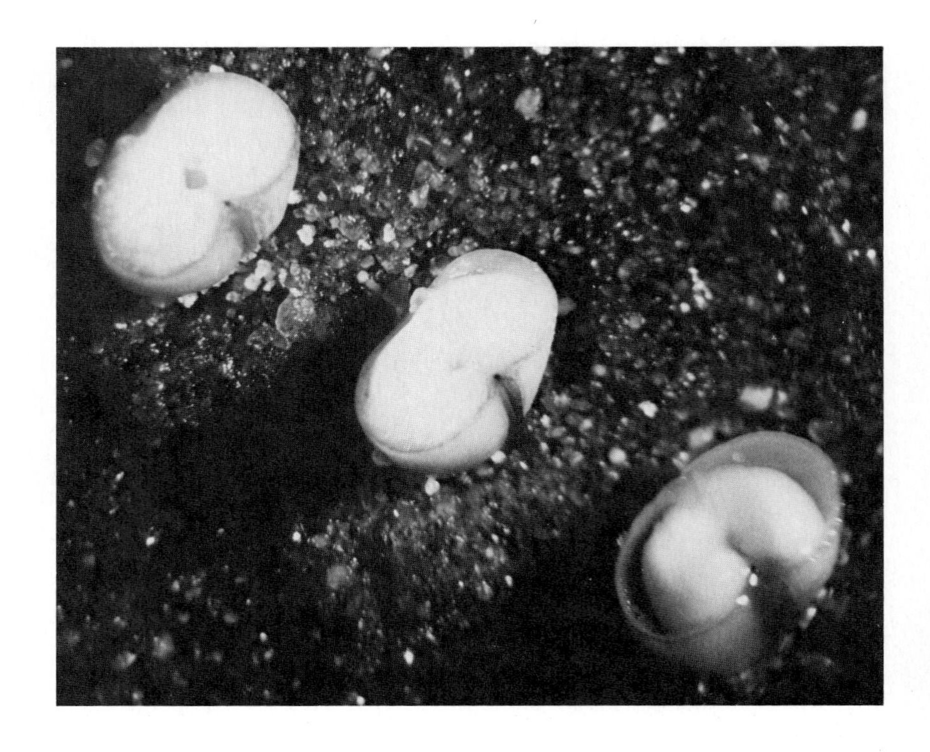

25–16
Action of gibberellin in barley seeds. Each of these three seeds has been cut in half and the embryo removed. Forty-eight hours before the picture was taken, the seed at the top left was treated with plain water. The seed in the center was treated with a solution of 1 part per billion of gibberellin, and the seed at the bottom right was treated with 100 parts per billion of gibberellin. Digestion of the starchy storage tissue has begun to take place in the treated seeds.

SUMMARY

Hormones are important chemical regulators of growth in both animals and plants. A hormone is a chemical produced in certain tissues of the organism and transported to other tissues, where it causes a physiological response. Hormones are biologically active in extremely small amounts.

Naturally occurring auxin is a hormone that is produced in the apical meristems of shoots and the tips of coleoptiles. Auxin travels unidirectionally toward the base of the plant, where it controls the lengthening of the shoot and the coleoptile, chiefly by promoting cell elongation. Studies indicate that its effect on cell elongation is achieved in some indirect way by a relaxation of the cellulose fibrils of the cell wall, permitting the cell to expand. Auxin also plays a role in differentiation of vascular tissue and initiates cell division in the vascular cambium. It often inhibits growth in lateral buds, thus maintaining apical dominance. The same quantity of auxin that promotes growth in the stem inhibits growth in the main root system. Auxin promotes the formation of adventitious roots in cuttings and retards abscission in leaves, flowers, and fruits. In fruits, auxin produced by seeds or the pollen tube stimulates growth of the ovary wall. Its capacity to produce such varied effects is believed to result from differential responses of the various target tissues.

The cytokinins, a second class of growth hormones, were first discovered as a consequence of their capacity to promote cell division and bud formation in cultures of plant tissues. They are chemically related to certain components of nucleic acids. Cytokinins can also act in concert with auxin to cause cell division in plant tissue culture. In tobacco pith cultures, a high concentration of auxin promotes root formation, while a high concentration of cytokinins promotes bud formation. In intact plants, cytokinins promote the growth of lateral buds, acting in opposition to the effects of auxin. Cytokinins prevent senescence in leaves by stimulating protein synthesis.

Ethylene is a gas produced by the incomplete combustion of hydrocarbons. It is also a natural growth regulator in plants, producing a number of distinct physiological responses, such as ripening in fruit and abscission.

Abscisic acid is a growth-inhibiting hormone that is found in dormant buds and in fruits, with a maximum amount present just before the fruit drops. Abscisic acid has been shown to oppose the effects of all three types of growth-stimulating hormones in various laboratory tests.

The gibberellins were first isolated from a parasitic fungus that causes abnormal growth in rice seedlings. They were subsequently found to be natural growth hormones present in many plants. The most dramatic effects of gibberellins are seen in dwarf plants, in which the application of gibberellins restores normal growth, and in plants with a rosette form of growth, in which gibberellins cause bolting. Gibberellins cause seed germination in grasses. In the barley seed, the embryo releases gibberellins that cause the aleurone layer of the endosperm to produce several enzymes, including alpha-amylase, which breaks down the starch stored in the endosperm, releasing sugar. The sugar nourishes the embryo and promotes the germination of the seed.

Exposing a plant tissue to a hormone can be com pared to putting a quarter in a vending machine. You may get the morning newspaper or a candy bar. The result depends not so much on the quarter as on the machine in which you put it. Similarly, the effects of plant hormones depend largely on the target tissues and the chemical environment in which these tissues are located.

(a)

(b)

26–1
Geotropic responses in the shoot of a young tomato (Lycopersicon esculentum) *plant. The plant in* (a) *was placed on its side and kept in a stationary position, whereas that in* (b) *was placed upside down and held in position in a ring stand. Although originally straight, the stems bent and grew upward. If a horizontally placed potted plant is slowly rotated around the stem on its axis, no curvature (geotropic response) will occur; that is, the plant will continue to grow horizontally. Can you explain the differences in the growth of a horizontally rotated plant and the plants shown here?*

CHAPTER 26

External Factors and Plant Growth

Living things must regulate their activities in accordance with the world around them. Animals, being mobile, can change their circumstances to some extent—foraging for food, courting a mate, and seeking or even making shelters in bad weather. A higher plant, by contrast, is immobilized once it sends down its first root. However, higher plants do have the ability both to respond and make adjustments to a wide range of changes in their external environment. This ability is manifested chiefly in changing patterns of growth.

THE TROPISMS

Perhaps the most familiar interaction between plants and the external world is the bending of the growing shoot tips of plants toward light (see Figure 25–2, page 502). This growth response, now known as *phototropism,* is caused by the elongation—under the influence of the plant hormone auxin—of the cells on the shaded side of the tip. What role does the light play in the phototropic response? Three possible answers to this question have been suggested: (1) light decreases the auxin sensitivity of the cells on the lighted side; (2) light destroys auxin; or (3) light drives auxin to the shaded side of the growing tip.

To choose among these hypotheses, Winslow Briggs and his co-workers carried out a series of experiments that were based on earlier work by Frits Went (see Figure 25–3, page 503). These investigators first showed that the same total amount of auxin is produced by the tip in the light as by the tip in the dark. Following the exposure to light, however, the amount diffused from the shaded side is greater than the amount diffused from the lighted side. If the tip is split and a barrier, such as a thin piece of glass, is placed between the two halves, this differential distribution of auxin diffusion no longer occurs. In other words, Briggs clearly demonstrated that auxin (or perhaps a precursor) migrates from the light side to the dark side and that the bending of the shoot is in response to this unequal distribution. Experiments using the auxin IAA (indoleacetic acid) labeled with ^{14}C have clearly shown that it is the auxin that migrates.

It has been determined that light of wavelengths from 400 to 500 nanometers is the most effective in inducing this hormone migration in a shoot tip. This observation indicates that a pigment absorbing blue light mediates the effect. The identity of the blue-light photoreceptor (which also mediates many other responses in plants and fungi) has not been definitively established, but present evidence indicates that it is a yellow-colored pigment called a flavin.

Another familiar tropism is *geotropism*, or gravitropism, a response to gravity. Seedlings conspicuously manifest geotropism. If a seedling is placed on its side, its root will grow downward and its shoot will grow upward. In shoots that for any reason are oriented horizontally, differences both in gibberellin and in auxin concentration develop between the upper and lower sides. Collectively, these cause the lower side of the shoot to elongate more than the upper side, and the shoot to grow upward. When it becomes vertical, the lateral asymmetries in hormone concentration disappear, and growth continues in the upright direction (Figure 26–1).

Asymmetries of growth-regulating agents are less well understood in roots, but it is known that they cause the upper side of the root to outgrow the lower side so that the root bends downward. There is overwhelming evidence that the site of perception of the gravitational stimulus in roots lies in the core cells of the root cap. In addition, there is convincing evidence for the theory that gradients of growth-controlling substances within the root cap are brought about by relatively small movements of starch-containing plastids (amyloplasts). When a root is placed in a horizontal position, the plastids, which were sedimented near the transverse walls of the vertically growing roots, slide downward and come to rest near what were previously vertically oriented walls (Figure 26–2). After several hours, the root curves downward and the plastids return to their previous position along the transverse walls. How the movement of these gravity sensors (statoliths) is translated into hormonal gradients has yet to be convincingly explained. Although the role of IAA in signal transmission in the shoot is well established, there is little evidence for a similar role for IAA in the root. In fact, some workers have failed to find IAA in the root cap at all. On the other hand, abscisic acid is found in the root cap; it has been shown to be redistributed (like IAA in the shoot tip), moving from the cap into the root proper and acting as an inhibitor on the cells in the region of elongation on the lower side of a horizontally oriented root. It now appears that abscisic acid may play an important role in the geotropic response of roots.

(a)

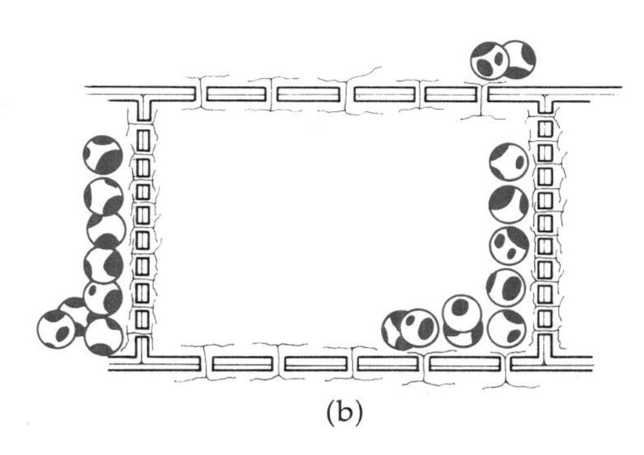

(b)

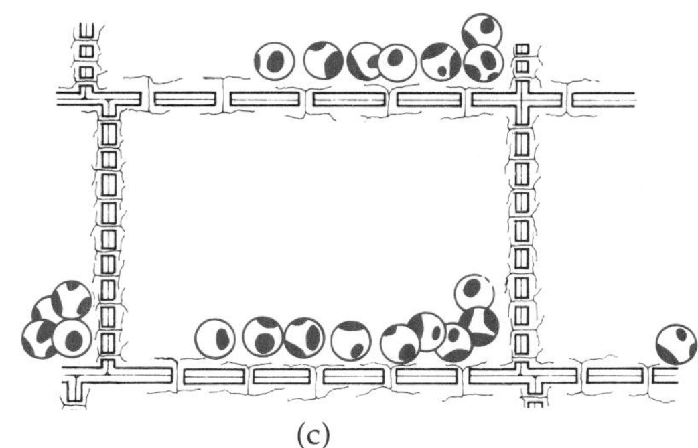

(c)

26–2
This diagram illustrates the response to gravity of amyloplasts (starch-containing plastids) in the core cells of a root cap. (a) *The amyloplasts are normally sedimented near the transverse walls in the root cap of a root growing vertically downward. When the same root is placed on its side,* (b) *and* (c)*, the amyloplasts slide toward the normally vertical walls that are now parallel with the soil surface. The evidence is convincing that this movement of particles plays an important role in the production of gradients of growth-controlling substances that lead to vertical root growth.*

26–3
Leaves of the wood sorrel (Oxalis), *during the day* (a) *and at night* (b). *One hypothesis concerning the function of such "sleep" movements is that they prevent the leaves from absorbing moonlight on bright nights, thus protecting the photoperiodic phenomena discussed later in this chapter. Another hypothesis, proposed by Darwin a century ago, is that the folding protects against heat loss from the leaves at night.*

(a)

(b)

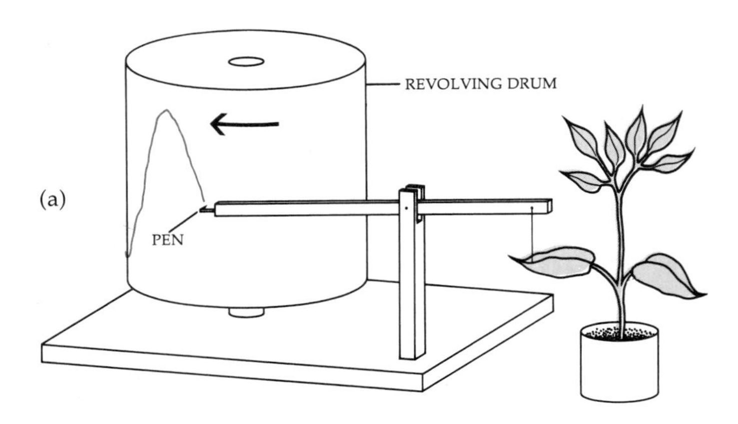

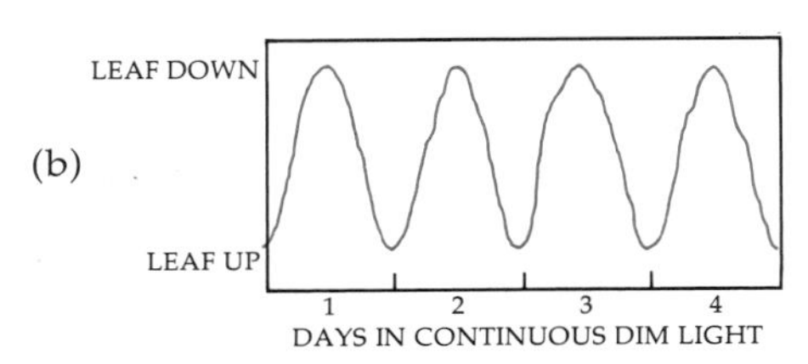

26–4
In many plants, the leaves move outward, to a position perpendicular to the stem and the sun's rays during the day and upward, toward the stem, at night. These "sleep movements" can be recorded on a revolving drum using a delicately balanced pen-and-lever system attached to a leaf by a fine thread (a). *Many plants, such as the garden bean* (Phaseolus vulgaris) *shown here, will continue to exhibit these movements for several days even when kept in continuous dim light.* (b) *A recording of this circadian rhythm, showing its persistence under a condition of constant dim light.*

CIRCADIAN RHYTHMS

It is a common observation that some plants open their flowers in the morning and close them at dusk or spread their leaves in the sunlight and fold them toward the stem at night (Figure 26–3). As long ago as 1729, the French scientist Jean-Jacques de Mairan noticed that these diurnal movements continue even when the plants are kept in dim light (Figure 26–4). More recent studies have shown that less evident activities, such as photosynthesis, auxin production, and the rate of cell division, also have regular daily rhythms, which continue even when all environmental conditions are kept constant. These regular, approximately 24-hour cycles are called *circadian rhythms*, from the Latin words *circa*, meaning "approximately," and *dies*, meaning "a day." Circadian rhythms, which have been found to exist almost universally throughout the plant and animal kingdoms, appear to be absent in the prokaryotes.

Are the Rhythms Endogenous?

Are these rhythms actually internal—that is, caused by factors entirely within the organism—or is the organism keeping itself in tune with some external stimulus? For years now, biologists have debated whether it might not be some environmental force, such as cosmic rays, the magnetic field of the earth, or the earth's rotation, that was setting the rhythms. Two important observations bearing on this question were that the rhythms studied were circadian (that is, the period of a rhythm is *about* 24 hours) and that there were slight individual differences among organisms.

Attempts to settle this recurrent controversy have led to countless experiments under an extraordinary variety of conditions. Organisms have been taken

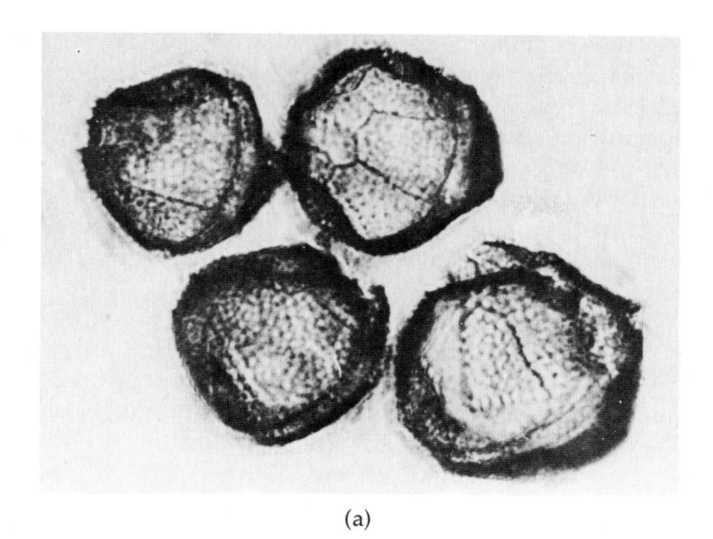

(a)

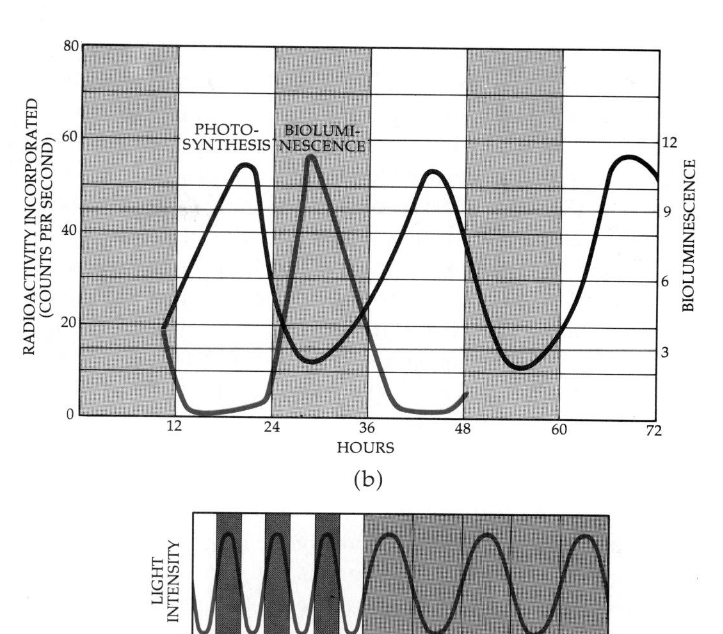

(b)

(c)

26–5
(a) *Photomicrograph of the dinoflagellate* Gonyaulax polyedra, *a single-celled marine alga.* (b) *In* G. polyedra, *three separate functions follow separate circadian rhythms: bioluminescence, which reaches a peak in the middle of the night (the curve in color); photosynthesis, which reaches a peak in the middle of the day (the black curve); and cell division (not shown), which is restricted to the hours just before dawn. If* Gonyaulax *is kept in continuous dim light, these three functions continue to occur with the same rhythm for days and even weeks, long after a number of cell divisions have taken place.* (c) *The rhythm of bioluminescence in* Gonyaulax, *like most circadian rhythms, can be altered by modifying the cycles of illumination. For example, if the investigators expose cultures of the algae to alternating light and dark periods, each 6 hours in duration, the rhythmic function will become entrained to this imposed cycle (left). If the cultures are then placed in continuous dim light, the organisms will return to their original rhythm of about 24 hours.*

down in salt mines, shipped to the South Pole, flown halfway around the world in airplanes, and, most recently, orbited in satellites. Although there is still a vocal minority that believes that circadian rhythms are under the influence of a subtle geophysical factor, most workers now agree that the rhythms are endogenous—that is, they are controlled internally. The internal timing mechanism is often referred to as the organism's *biological clock.*

Setting the Clock

Under constant environmental conditions, the period of the circadian rhythm is *free-running*—that is, its natural period (usually between 21 and 27 hours) does not have to be reset at each cycle. It behaves as a self-sustained oscillator. Although circadian rhythms originate within the organisms themselves, the environment acts as a synchronizing agent (or "Zeitgeber," German for "time giver"). In fact, the environment is responsible for keeping a circadian rhythm in step with the daily 24-hour light-dark cycle. If a circadian rhythm of a plant were greater or less than 24 hours, the rhythm would soon get out of step with the 24-hour light-dark cycle. Then, if only the circadian rhythm were followed, a phenomenon such as flowering, which generally takes place in the light period, would occur at a different time each day, including the dark period. It is therefore necessary for the plant to become resynchronized—that is, to become *entrained* to the 24-hour day.

Entrainment is the process by which a periodic repetition of light and dark, or some external cycle, causes a circadian rhythm to remain synchronized with the same cycle as the entraining factor. Light-dark cycles and temperature cycles are the principal factors in entrainment (Figure 26–5).

Another feature of interest is that these rhythms do not automatically speed up as the temperature rises. The reason one might expect them to speed up is that biochemical activities—an internal biological clock certainly must have a biochemical basis—take place more rapidly at high temperatures than at low ones. Some clocks run slightly faster as the temperature rises, but others go more slowly, and many are almost unchanged. Thus, the biological clock must contain within its workings a compensatory mechanism—a feedback system that allows it to adjust to temperature changes. Such a temperature-compensating feature would be very important in plants.

Some recent evidence suggests that cell membranes may be the key to the biological clock. Although the mechanism is unknown, the established membrane functions that might conceivably be involved include the regulation of ion movements into and out of cells and subcellular compartments, and the regulation of energy metabolism via the energy-transducing membranes of chloroplasts and mitochondria.

Some biological rhythms regulate the interaction between organisms. For example, some plants secrete nectar at specific times of the day. As a result, bees—which have their own biological clocks—become accustomed to visiting these flowers at these times, thereby ensuring maximum rewards for themselves and cross-pollination for the flowers.

For most organisms, however, the use of the biological clock for such special purposes is probably a secondary one. The primary usefulness of the clock is that it enables the plant or animal to respond to the changing seasons of the year by accurately measuring changing day length. In this way, changes in the environment trigger responses that result in adjustments of growth, reproduction, and other activities of the organism.

PHOTOPERIODISM

Fifty years ago a mutant appeared in a field of tobacco plants *(Nicotiana tabacum)* growing near Washington, D.C. The new variety had unusually large leaves and eventually grew to a height of more than 3 meters. As the season progressed, the regular plants flowered, but Maryland Mammoth, as the variety came to be known, just grew bigger and bigger. Two investigators from the U.S. Department of Agriculture, W. W. Garner and H. A. Allard, took cuttings from the Maryland Mammoth and placed them in a greenhouse, where they would be protected from frost. These cuttings flowered in December, although by then they were only 1.5 meters tall, half the size of their parent. New Maryland Mammoths grew from their seed, and these, too, did not flower until almost winter.

Coincidentally, Garner and Allard were carrying out experiments with the Biloxi variety of soybean *(Glycine max)*. Agriculturalists were interested in spacing out the soybean harvest by making successive sowings of seeds at two-week intervals from early May through June. But spacing out the planting had no effect; all the plants, no matter when the seeds were sown, came into flower at the same time—in September.

The investigators started growing these two kinds of plants—Maryland Mammoth tobacco and Biloxi soybeans—under a wide variety of controlled conditions of temperature, moisture, nutrition, and light. They eventually found that the critical factor in both species was the length of day. Neither plant would flower unless the day length was shorter than a critical number of hours. Consequently, soybeans, no matter when they were planted, all flowered as soon as the days became short enough, which was in September; the Maryland Mammoth, no matter how tall it grew, would not flower until December, when the days had become even shorter.

Garner and Allard called this phenomenon *photoperiodism*. Photoperiodism is a biological response to a change in the proportions of light and dark in a 24-hour daily cycle. Photoperiodism has now been shown to initiate mating and other activities in animals as diverse as codling moths, spruce budworms, aphids, and potato worms, as well as many species of fish, birds, and mammals.

Long-Day Plants and Short-Day Plants

Garner and Allard went on to test and confirm their discovery with many other species of plants. Following this single lead, they were able to answer a host of questions that had long troubled both professional botanists and amateur gardeners. Why, for example, is there no ragweed *(Ambrosia)* in northern Maine? *Answer:* Because ragweed starts producing flowers only when the duration of daylight is 14½ hours or less. The long summer days do not shorten to 14½ hours in northern Maine until August, and then there is not enough time for ragweed seed to develop before the frost. *Question:* Why does spinach *(Spinacia oleracea)* not grow in the tropics? *Answer:* Because spinach needs at least 14 hours of light per day for a period of at least two weeks in order to flower, and this combination of conditions never happens in the tropics.

The investigators found that plants are of three general types, which they called *short-day*, *long-day*, and *day-neutral*. Short-day plants flower in early spring or fall; they must have a light period shorter than a critical length. For instance, the common cocklebur *(Xanthium strumarium)* is induced to flower by 16 hours or less of light (Figures 26–6 and 26–7).

26–6
The relative length of day and night determines when plants flower. The four curves depict the annual change of day length in four North American cities at four different latitudes. The horizontal color lines indicate the effective photoperiod of three different short-day plants. The cocklebur, for instance, requires 16 hours or less of light. In Miami, it can flower as soon as it matures, but in Winnipeg, the buds do not appear until early August, so late that the frost will probably kill the plants before the seed is set.

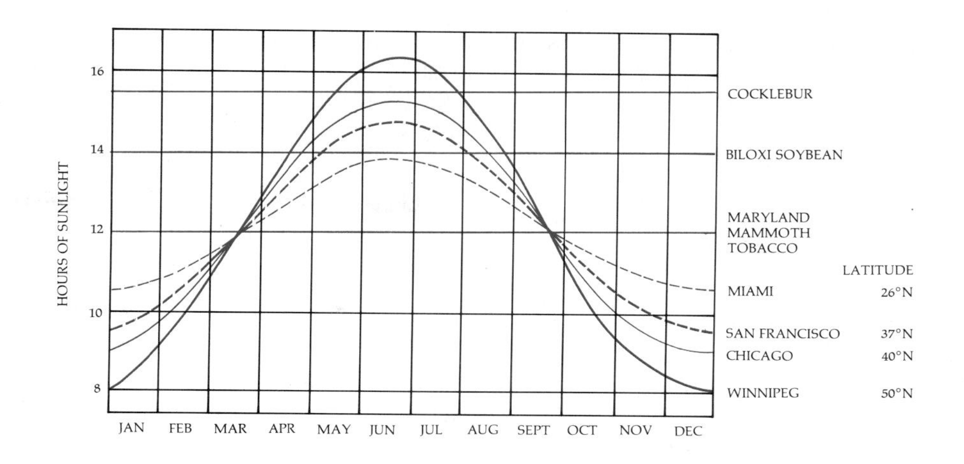

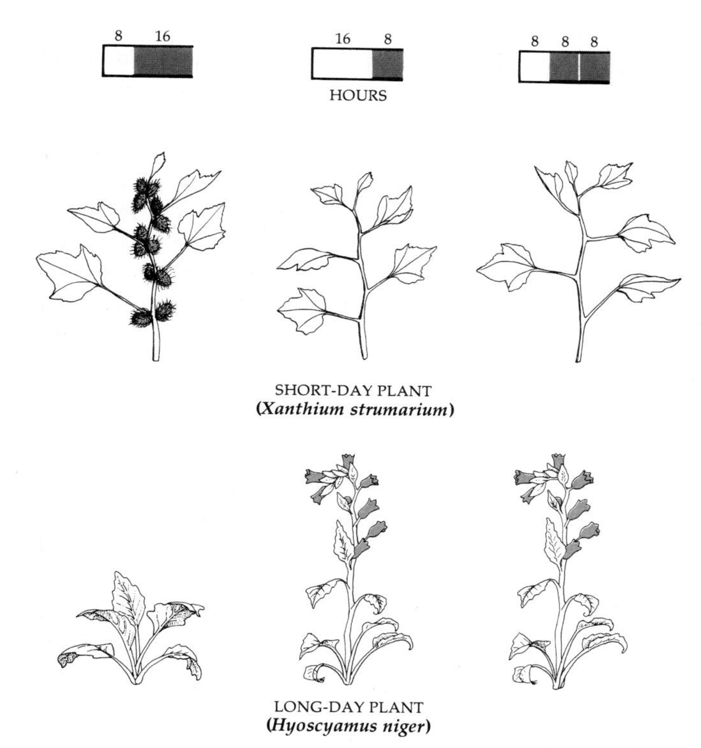

26–7
Short-day plants flower when the photoperiod is less *than some critical value. The common cocklebur* (Xanthium strumarium) *requires less than 16 hours of light to flower. Henbane* (Hyoscyamus niger) *requires about 10 hours (depending on temperature) or more to flower. However, if the dark period is interrupted by a flash of light,* Hyoscyamus *will flower even on a short-day period. A "pulse" of light during the dark period has the opposite effect in a short-day plant—it prevents flowering. The bars at the top indicate the duration of light and dark periods in a 24-hour day.*

Other short-day plants are some chrysanthemums, poinsettias, strawberries, and primroses.

Long-day plants, which flower chiefly in the summer, will flower only if the light periods are longer than a critical length. Spinach, some potatoes, some wheat varieties, lettuce, and henbane *(Hyoscyamus niger)* are examples of long-day plants (Figure 26–7).

Cocklebur and spinach will both bloom if exposed to 14 hours of daylight, yet only one is designated as a long-day plant. The important factor is not the absolute length of the photoperiod but whether it is longer or shorter than some critical interval. Day-neutral plants flower without respect to day length.

Within individual species of plants that cover a large north-south range, different photoperiodic ecotypes (locally adapted variants of an organism) have often been observed. Thus, in many prairie grasses, species that may occur from southern Canada to Texas, northern ecotypes flower before southern ones when they are grown together in a common environment. Different populations are precisely adjusted to the demands of the photoperiodic regime where they occur.

The photoperiodic response can be remarkably precise. At 22.5°C, the long-day plant *Hyoscyamus niger* (henbane) will flower when exposed to photoperiods of 10 hours and 20 minutes (Figure 26–7). But with a photoperiod of 10 hours, it will not flower at this temperature. Environmental conditions also affect photoperiodic behavior. For instance, at 28.5°C, henbane requires 11½ hours of light, whereas at 15.5°C, it requires only 8½ hours.

The response varies with different species. Some plants require only a single exposure to the critical day-night cycle, whereas others, such as spinach, require several weeks of exposure. In many plants, there is a correlation between the number of induction cycles and the rapidity of flowering or the number of flowers formed. Some plants have to reach a certain degree of maturity before they will flower, whereas others will respond to the appropriate photoperiod when they are seedlings. Some plants, when they get older, will eventually flower even if not exposed to the appropriate photoperiod, although they will flower much earlier with the proper exposure.

Measuring the Dark

In 1938, another pair of investigators, Karl C. Hamner and James Bonner, began a study of photoperiodism, using the cocklebur as their experimental tool. As mentioned previously, the cocklebur is a short-day plant, requiring 16 hours or less of light per 24-hour cycle to flower. It is particularly useful for experimental purposes because a single exposure under laboratory conditions to a short-day cycle will induce flowering two weeks later, even if the plant is immediately returned to long-day conditions. The cocklebur plant can withstand a great deal of rough treatment; for example, it can survive even if all its leaves are removed. Hamner and Bonner showed that it is the leaf blade of the cocklebur that perceives the photoperiod. A completely defoliated plant cannot be induced to flower. But if as little as one-eighth of a fully expanded leaf is left on the stem, the single short-day exposure induces flowering.

In the course of these studies, in which they tested a variety of experimental conditions, Hamner and Bonner made a crucial and totally unexpected discovery. If the period of darkness is interrupted by as little as a one-minute exposure to light from a 25-watt bulb, flowering does not occur. Interruption of the light period by darkness has absolutely no effect on flowering. Subsequent experiments with other short-day plants showed that they, too, required periods of uninterrupted darkness rather than uninterrupted light.

On the basis of the findings of Garner and Allard, commercial growers of chrysanthemums had found that they could hold back blooming in the short-day plants by extending the daylight with artificial light. On the basis of the new experiments by Hamner and Bonner, they were able to delay flowering simply by switching on lights in the middle of the night.

What about long-day plants? They also measure darkness. A long-day plant that will flower if it is kept in a laboratory in which there is light for 16 hours and dark for 8 hours will also flower on 8 hours of light and 16 hours of dark if the dark is interrupted by even a brief exposure of light.

CHEMICAL BASIS OF PHOTOPERIODISM

Hamner and Bonner had shown that when the dark period is interrupted by a single flash of light from an ordinary bulb, the cocklebur will not flower. The Beltsville group, following this lead, began to experiment with light of different wavelengths, varying the intensity and duration of the flash. They found that red light at about 660 nanometers (orange-red) was most effective in preventing flowering in the cocklebur and other short-day plants. It was also the most effective, they found, in promoting flowering in long-day plants.

The next important clue to the response of plants to the relative proportions of light and darkness came from a team of research workers at the U.S. Department of Agriculture Research Station in Beltsville, Maryland. The clue was found in the report of an earlier study performed with lettuce *(Lactuca sativa)* seeds. Lettuce seeds germinate only if they are exposed to light. This requirement is true of many small seeds, which need to germinate in loose soil and near the surface in order for the seedlings to be sure of breaking through. The earlier workers, in studying the light requirement of germinating lettuce seeds, had shown that red light stimulated germination and that light of a slightly longer wavelength (far-red) inhibited germination even more effectively than did no illumination at all.

The Beltsville group found that when red light was followed by far-red light, the seeds did not germinate. The red light most effective in inducing germination in seeds was light of the same wavelength as that involved in the flowering response—about 660 nanometers. Furthermore, they found that the light most effective in inhibiting the effect produced by red

(a)

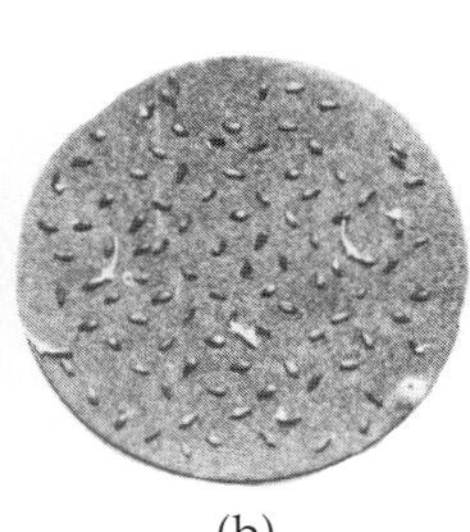

(b)

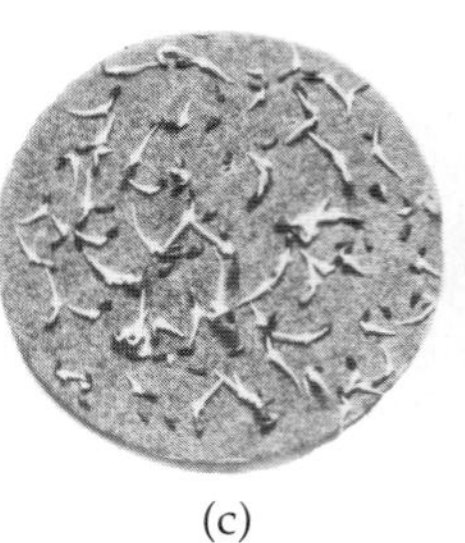

(c)

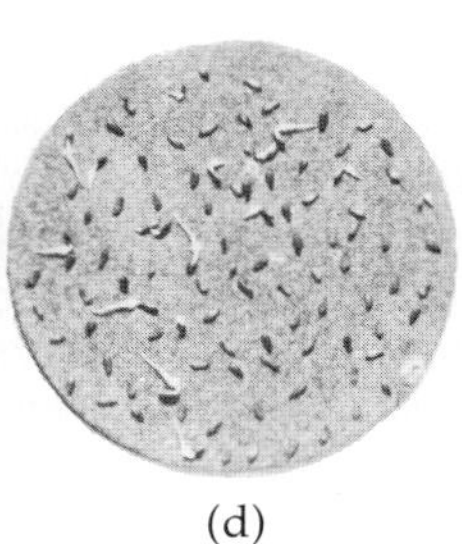

(d)

26–8
Light and the germination of lettuce seeds. (a) *Seeds exposed briefly to red light;* (b) *seeds exposed to red light, followed by far-red light;* (c) *seeds exposed to a sequence of red, far-red, red; and* (d) *seeds exposed to a red, far-red, red, far-red sequence. Whether or not the seeds germinated depended on the final wavelength in the series of exposures—red promoting and far-red inhibiting germination.*

light was light of a wavelength of 730 nanometers. The series of flashes could be repeated over and over; the number of flashes did not matter, but the nature of the final one did. If the series ended with a red flash, the majority of the seeds germinated. If it ended with a far-red flash, the majority did not germinate (Figure 26–8).

Far-red light was tried on short-day and long-day plants, with the same on-off effect. Far-red light alone, when given during the dark period, had no effect. But a flash of far-red light immediately following a flash of red light canceled the effects of the red light.

Discovery of Phytochrome

Plants contain a pigment that exists in two different interconvertible forms: P_r (a form that absorbs red light) and P_{fr} (a form that absorbs far-red light). When a molecule of P_r absorbs a photon of red light of 660 nanometers in wavelength, it is converted to P_{fr} in a matter of milliseconds; when a molecule of P_{fr} absorbs a photon of far-red light of 730 nanometers, it is very quickly reconverted to the P_r form. These are called *photoconversion reactions.* The P_{fr} form is biologically active (that is, it will trigger a response such as seed germination), whereas P_r is inactive. The pigment molecule can thus function as a biological switch, turning responses on or off depending on which form it is in.

The lettuce seed germination experiments are easily understood in these terms. Since P_r absorbs red light most efficiently (Figure 26–9), this wavelength will drive a high proportion of the molecules into the P_{fr} form, thereby inducing germination. Subsequent far-red light absorbed by the P_{fr} form (Figure 26–9) will drive essentially all of the molecules back to P_r, thereby canceling the effect of the prior red light.

What about flowering under natural day-night cycles? Since white light contains both red and far-red wavelengths, both forms of the pigment are exposed simultaneously to photons that are efficient in driving their photoconversion to the opposite form. After a few minutes in the light, then, a photoequilibrium is established whereby the forward ($P_r \rightarrow P_{fr}$) and backward ($P_{fr} \rightarrow P_r$) reactions are balanced. A constant proportion of the phytochrome population is in each form at any instant under these conditions (about 60 percent P_{fr} in noontime sunlight), and this proportion is maintained while the light remains on.

When plants are switched to darkness at the end of the light period, the level of P_{fr} steadily declines over a period of several hours. If a high level of P_{fr} is regenerated by pulse irradiation with red light in the middle of the dark period (Figure 26–7), it will inhibit flowering in short-day (that is, "long-night") plants that otherwise would have flowered; and it will promote flowering in long-day (that is, "short-night") plants that otherwise would not have flowered. In either case, the effect of the red pulse in regenerating high levels of P_{fr} can be canceled by an immediate subsequent far-red pulse which reconverts the P_{fr}.

In 1959, Harry A. Borthwick and his co-workers at Beltsville gave this pigment the name of *phytochrome* and presented conclusive physical evidence that it existed. The principal characteristics of the pigment, as they are presently understood, are summarized schematically in Figure 26–10. The molecule is synthesized continuously as P_r and accumulates in this form in dark-grown plants. Light causes photoconversion of P_r to P_{fr}, which induces a biological re-

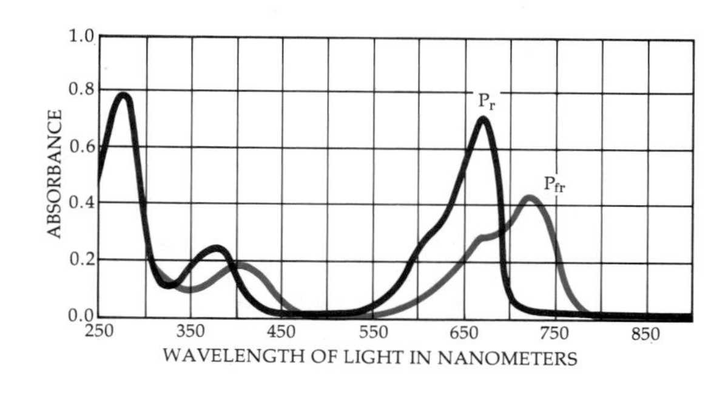

26–9
Absorption spectra of the two forms of phytochrome, P_r and P_{fr}. This difference in the absorption spectra made it possible to isolate the pigment.

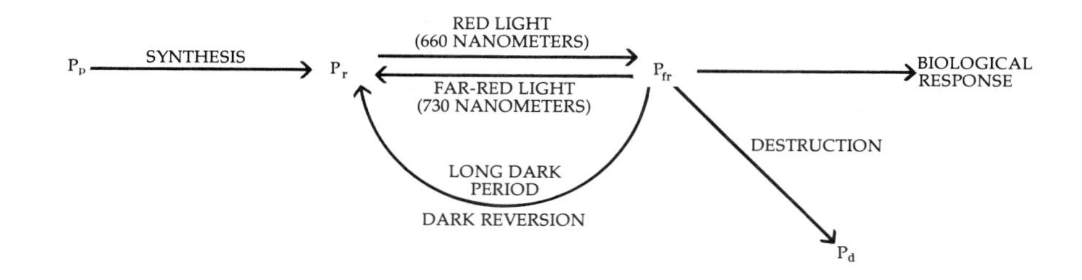

26–10
Phytochrome is first synthesized in the P_r form from amino acids (designated P_p for precursor). P_r changes to P_{fr} when exposed to red light. P_{fr} is the active form that induces a biological response. P_{fr} is converted back to P_r when exposed to far-red light. In darkness, P_{fr} reverts to P_r or is destroyed (designated P_d for destruction product).

sponse. P_{fr} can be converted to P_r by photoconversion with far-red light or reversion to P_r in the dark, a process termed "dark reversion" that occurs over a period ranging from several minutes to hours. P_{fr} can also be lost through irreversible denaturation, by a process termed "destruction" that occurs over several hours and probably involves hydrolysis by a protease. All three alternate pathways for P_{fr} removal provide the potential for reversing induced responses.

Isolation of Phytochrome

Phytochrome is present in plants in small amounts relative to such pigments as chlorophyll. To distinguish the phytochrome, one needs to use a spectrophotometer that is sensitive to extremely small changes in light absorbency. Such a spectrophotometer was not introduced until some seven years after the existence of phytochrome was proposed; in fact, the first use of the new instrument was to detect and isolate this pigment.

In order to avoid interference from chlorophyll which, like phytochrome, also absorbs light of about 660 nanometers, dark-grown seedlings (in which chlorophyll has not yet developed) were chosen as the source from which to isolate phytochrome. The pigment proved to be blue in color (why might you expect this color?), and it showed the characteristic red to far-red conversion in the test tube by reversibly changing color slightly in response to red or far-red light.

The phytochrome molecule was found to contain two distinct parts: a light-absorbing portion (the chromophore) and a large protein portion. The chromophore is much like the phycobilins that serve as accessory pigments in cyanobacteria and red algae.

The way in which phytochrome works has not yet been established, although several hypotheses have been proposed.

Other Phytochrome Responses

Phytochrome has now been shown to be involved in a number of other plant responses. The germination of many seeds, for instance, occurs in the dark. In the seedlings, the stem elongates rapidly, pushing the shoot (or, in most monocots, the cotyledon) up through the dark soil. During this early stage of growth, there is essentially no enlargement of the leaves, which would interfere with the passage of the shoot through the soil. Soil is not necessary for this growth pattern; any seedling grown in the dark will be elongated and spindly and will have small leaves. It will also be yellow to colorless, because plastids do not turn green until exposed to light. Such a seedling is said to be *etiolated* (Figure 26–11).

When the seedling tip emerges into the light, the etiolated growth gives way to normal plant growth. In dicots, the hook unbends, the stem growth rate may slow down somewhat, and leaf growth begins. In grasses, growth of the mesocotyl (the part of the embryo axis between scutellum and coleoptile) stops, the stem elongates, and the leaves open.

A dark-grown bean seedling that receives five minutes of red light a day, for instance, will show these light effects beginning on the fourth day. If the exposure to red light is followed by a five-minute exposure to far-red, none of the changes usually produced by the red light will appear (Figure 26–12). Similarly, in the seedlings of grains, termination of mesocotyl growth is triggered by exposure to red light, and the effect of red light is canceled by far-red.

A recent report from England suggests that an important function of phytochrome in plants growing in a natural environment is the detection of shading by other plants. Radiation below 700 nanometers is almost completely reflected or absorbed by vegetation, whereas that between 700 and 800 nanometers (in the far-red range) is largely transmitted. This causes a

26–11
Dark-grown seedlings, such as the ones on the left, are thin and pale; they have longer internodes and smaller leaves than normal seedlings, such as those on the right. This group of physical characteristics, known as etiolation, has survival value for the seedling because it increases its chances of reaching light before its stored energy supplies are used up.

26–12
All three bean plants received eight hours of light each day. The center plant was exposed to five minutes of elongation-promoting far-red light at the start of each dark period. The plant on the right received the same five-minute exposure to far-red light followed by a five-minute exposure to red light, which counteracted the effect of the far-red light. The plant on the left served as the control; its phytochrome is in the P_{fr} *form as it enters the dark period because of its exposure to daylight.*

dramatic upward shift in the equilibrium ratio of P_r to P_{fr} (that is, more P_{fr} driven to P_r) in shaded plants and results in a rapid increase in the rate of internodal elongation.

Reversible red to far-red reactions are also involved in anthocyanin formation in the apple, turnip, and cabbage; in the germination of seeds; in changes in the chloroplasts and other plastids; and in a tremendous variety of other plant responses in all phases of the plant cycle.

Phytochrome and Photoperiodism

When the existence of phytochrome was first demonstrated, its discoverers hypothesized that its behavior might explain the phenomenon of photoperiodism—that is, that the reversible red to far-red reactions might be involved in the time-measuring mechanism, or biological clock. According to this hypothesis, P_{fr} inhibits flowering in short-day plants but promotes flowering in long-day plants. In short-day plants, P_{fr} would accumulate in the light and be removed in the subsequent dark period by destruction or dark reversion. When the nights were long enough, all (or a critical amount) of P_{fr} would be removed, and flowering would no longer be inhibited. Long-day plants, on the other hand, would require short nights, during which the P_{fr} would not be completely destroyed; if the night were short enough, sufficient P_{fr} would remain at the end of it to promote flowering.

Experiments have shown, however, that P_{fr} will disappear in many plants within three or four hours of darkness. On the basis of these experiments and other observations, it is now generally agreed that the time-measuring phenomenon of photoperiodism is not controlled by the interconversion of P_r and P_{fr} alone. A more complex explanation must be sought.

HORMONAL CONTROL OF FLOWERING

Hamner and Bonner, in their early cocklebur experiments, showed that the leaf "perceived" the light, which caused the bud to flower. Apparently, some substance that has profound effects on growth and development is transmitted from leaf to bud. This hypothetical substance has been termed the flowering hormone, or floral stimulus.

The early experiments on the floral stimulus were carried out independently in several laboratories in the 1930s. Some of the first, those of the Soviet plant physiologist M. Kh. Chailakhyan, carried out just a few years before the first cocklebur studies, are representative. Using the short-day plant *Chrysanthemum indicum,* Chailakhyan showed that if the upper portion of the plant was defoliated and the leaves on the lower part were exposed to a short-day induction period, the plant would flower. If, however, the upper, defoliated part was kept on short days and the lower, leafy part on long days, no flowering occurred. He interpreted these results as indicating that the leaves formed a hormone that moved to the stem apex and initiated flowering. Chailakhyan named this hypothetical hormone *florigen,* the "flower maker."

Subsequent experiments showed that the flowering response will not take place if the leaf is removed immediately after photoinduction. But if the leaf is left on the plant for a few hours after the induction cycle is complete, it can then be removed without affecting flowering. The flowering hormone can pass through a graft from a photoinduced plant to a noninduced plant. Unlike auxin, however, which can pass through agar or other nonliving tissue, florigen can travel from one plant tissue to another only if there are anatomical connections of living tissue between them. If a branch is girdled, that is, if a circular strip of bark is removed, florigen movement ceases. On the basis of these data, it was concluded that florigen moves by means of the phloem system, the way most organic substances are transported in plants.

26–13
Grafts of Nicotiana silvestris *(a long-day plant) onto the* N. tabacum *cultivar Trabezond (a day-neutral plant). Under long days, flower formation in the day-neutral tobacco was accelerated by the graft (as in the plant on the left). Under short days, the day-neutral tobacco remained vegetative (flowering inhibited) for the duration (90 to 94 days) of the experiment (as in the plant on the right).*

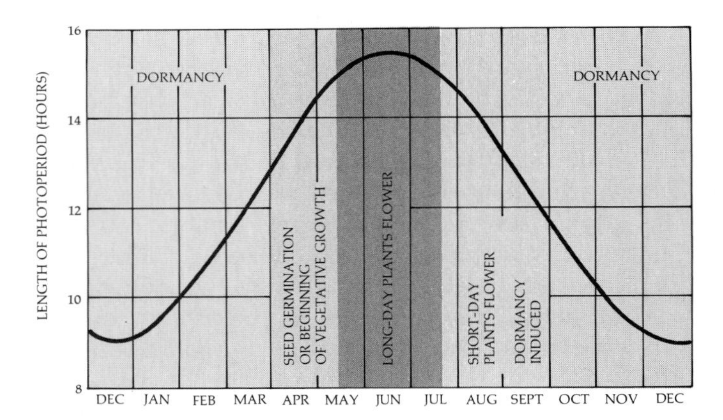

26–14
Relationship between day length and the developmental cycle of plants in the North Temperate Zone.

In some plants—the Biloxi soybean *(Glycine max)* is an example—leaves must be removed from the grafted receptor plant or it will not flower. This observation suggests that the leaves of noninduced plants may produce an inhibitor. In fact, some investigators have concluded on the basis of such evidence that there is no substance that initiates flowering but rather a substance that inhibits flowering, until it is removed. Strong evidence now suggests that, at least in certain plants, both inhibitors and promoters are involved in the control of flowering.

The most convincing evidence for the existence of both flower-inducing and flower-inhibiting substances in the same plant is provided by the recent experimental studies of Anton Lang of Michigan State University and M. Kh. Chailakhyan and I. A. Frolova of the K. A. Timiryazev Institute of Plant Physiology in Moscow. These researchers chose three kinds of tobacco plants for their studies: the day-neutral *Nicotiana tabacum* cultivar Trabezond, the short-day tobacco cultivar Maryland Mammoth, and the long-day *Nicotiana silvestris*. They found that flower formation in the day-neutral tobacco was accelerated by a graft union with the long-day plant when the grafts were kept on long days, as well as by a graft union with the short-day tobacco when the grafts were kept on short days, as compared with grafted day-neutral controls. When long-day plants grafted to day-neutral plants were exposed to short days, flowering in the day-neutral receptor was greatly inhibited (Figure 26–13). By contrast, when short-day plants grafted to day-neutral plants were exposed to long days, there was little or no delay in flowering.

These results indicate that the leaves of the long-day plants are capable of producing flower-inducing substances on long days and flower-inhibiting substances on short days, and that both substances can be transferred through a graft union, an indication that both substances are transported through the plant. In the case of the short-day tobacco, the leaves apparently produce little or no flower-inhibiting substance when exposed to long days. If such substances are produced by the short-day plant, they are much less effective in retarding flowering than those produced by the long-day plants.

Evidence for the existence of both flower-inducing and flower-inhibiting substances is very compelling, although attempts to isolate them have thus far been unsuccessful.

DORMANCY

Plants do not grow at the same rate all of the time. During unfavorable seasons, they limit their growth or cease to grow altogether. This ability enables plants to survive periods of water scarcity or low temperature.

Dormancy is a special condition of arrested growth. After periods of ordinary rest, growth resumes when the temperature becomes milder or when water or any other limiting factor becomes available again (Figure 26–14). A dormant bud or embryo, however, can be "activated" only by certain, often quite precise, environmental cues. This adaptation is of great survival importance to the plant. For example, the buds of plants expand, flowers are formed, and seeds germinate in the spring—but how do they recognize spring? If warm weather alone were enough, in many years all the plants would flower and all the seedlings would start to grow during Indian summer, only to be destroyed by the winter frost. The same could be said for any one of the warm spells that often punctuate the winter season. The dormant seed or bud does not respond to these apparently favorable conditions because of endogenous inhibitors which must first be removed or neutralized before the period of dormancy can be terminated. In contrast to this reluctance to grow too rapidly, commercial seeds are artificially selected for their readiness to germinate promptly when they are exposed to favorable conditions, a trait that would be a great hazard for wild seeds.

Dormancy in Seeds

Almost all seeds growing in areas with marked seasonal temperature variations require a period of cold prior to germination. Many seeds require drying before they germinate (although some may be nondormant before they dry out). This requirement prevents their germination within the moist fruit of the parent plant. Some seeds, as in the case of lettuce, require exposure to light, but others are inhibited by light. Some seeds will not germinate in nature until they have become abraded, as by soil action. Such abrasion wears away the seed coat, permitting water or oxygen to enter the seed and, in some cases, removing the source of inhibitors. Hard seed coats that interfere with water absorption and embryo enlargement are common among legumes.

The seeds of some desert species germinate only when sufficient rain has fallen to leach away inhibitory chemicals present in the seed coat. The amount of rainfall necessary to wash off these germination inhibitors is directly related to the supply of water that the desert plant needs to become established as a seedling.

Some seeds may survive a long time in the dormant condition, enabling them to exist for many years, decades, and even centuries under favorable conditions. In 1879, seeds of 20 species of common Michigan weeds were stored for an experiment designed to continue 160 years. At the last sampling, the seeds of three species were still viable. Although this demonstration of endurance is impressive, it does not approach the record for seeds of the sacred lotus *(Nelumbo nucifera)* found by a Japanese botanist in a peat deposit in Manchuria. Radiocarbon dating showed the seeds to be some 2000 years old, but when the seed coats were filed to permit water to enter, every seed germinated.

In 1967, even this record was broken with seeds of the arctic tundra lupine *(Lupinus arcticus)*. These seeds were found in a frozen lemming burrow in the Yukon with animal remains estimated by carbon dating to be at least 10,000 years old. Their cold requirement having been fulfilled, a sample of the seeds germinated within 48 hours (Figure 26–15).

During recent years, plant scientists have become increasingly interested in the factors involved in maintaining seed viability. Various enzyme systems may fail progressively in the stored seed, eventually leading to a complete loss of viability. Under what circumstances might viability best be prolonged? Such questions are relevant to the worldwide interest in developing seed banks with the goal of preserving the genetic characteristics of the original wild-type varieties of various crop plants for use in future

26–15
Lupinus arcticus *grown from a seed at least 10,000 years old. The seeds were found by a Yukon mining engineer in lemming burrows buried deeply in permafrost silt of Pleistocene age.*

breeding programs. The need for such seed banks arises from the progressive replacement of older varieties with newer ones and the elimination of the replaced varieties, chiefly through the destruction of their habitats. In addition, many wild species of plants are in danger of becoming extinct for similar reasons, and the seeds of these plants should also be preserved, if possible (see "The Conservation of Plants," page 609).

Dormancy in Buds

As with seeds, the buds of many plant species require cold to break dormancy. If branches of flowering trees and shrubs are cut and brought inside in autumn, they do not flower; but if the same branches are left outside until late winter or early spring, they will bloom in the warmer indoor temperatures. Deciduous fruit trees, such as the apple, chestnut, and peach, cannot be grown in climates where the winters are not cold. Similarly, bulbs such as those of tulips, hyacinths, and narcissus can be "forced," that is, made to bloom inside in the winter, but only if they have previously been outside or in a cold place. As discussed in Chapter 23, such bulbs are actually large buds in which the leaves are modified for storage.

Investigations carried out under controlled conditions confirm these suggestions that cold is required for the breaking of dormancy in many species. Most varieties of peach, for example, must remain for 600 to 900 hours at temperatures below 4°C before they will respond to the activating influence of warmer temperatures and longer days. Some plants will respond to a brief exposure to freezing temperatures; if one bud of a greenhouse-cultivated lilac bush is briefly exposed to freezing temperatures, that bud alone will break into bloom soon after. Cold is not required to break dormancy in all cases, however. In the potato, for instance, in which the "eyes" are dormant buds, at least two months of dry storage are the chief requirement; temperature is not a factor. In many plants, particularly trees, the photoperiodic response breaks winter dormancy, with the dormant buds being the receptor organs. Photoperiodism often regulates the onset of dormancy as well, presumably through some hormonal mechanism.

Application of gibberellins may sometimes break dormancy. For instance, gibberellin treatment of a peach bud may induce development after the bud has been kept for 164 hours below 8°C. Does this mean that under normal conditions, increase in gibberellin terminates the dormancy? Not necessarily. Dormancy may be a state of balance between growth inhibitors and growth stimulators. Addition of any growth stimulator (or removal of inhibitors, such as abscisic acid) may alter the balance so that growth begins.

There does not seem to be any common mechanism by which dormancy is induced or broken. This fact, although it considerably multiplies the problems of the plant physiologist, is certainly in keeping with our sense of evolutionary history. Dormancy became advantageous to plants only comparatively recently, when the seed plants began to spread into a variety of different ecological domains. Presumably, dormancy evolved independently in many groups of plants, each of which found its solution separately among the survivors, often in a different way.

COLD AND THE FLOWERING RESPONSE

Cold also may affect the flowering response. For example, if winter rye *(Secale cereale)* is planted in the autumn, it germinates during the winter and flowers the following summer, 7 weeks after growth begins. If it is planted in the spring, it does not flower for 14 weeks, and it remains vegetative throughout most of the growing season. In 1915, the German plant physiologist Gustav Gassner discovered that he could influence the flowering of winter rye and other cereal plants by controlling the temperature of the germinating seeds. He found that if the seeds of the winter strain are kept at near-freezing (1°C) temperatures during germination, the winter rye, even when planted in late spring, will flower the same summer it is planted. This procedure, which came to be known as *vernalization,* from the Latin *vernus,* meaning "spring," is now a common practice in agriculture.

Even after vernalization, the plant must be subjected to a suitable photoperiod, usually long days. The vernalized winter rye behaves like a typical long-day plant, flowering in response to the long days of summer. A similar example is seen in a biennial strain of henbane *(Hyoscyamus niger).* The vegetative rosette, which culminates the first year's growth, flowers only if it is exposed to cold. After cold exposure, it becomes a typical long-day plant, with the same photoperiodic response as that of the annual strain.

As the above examples indicate, in some plants the cold treatment affects the photoperiodic response. Spinach, ordinarily a long-day plant, does not usually flower until the days are 14 hours in length. If the spinach seeds are cold-treated, however, they will flower when the days are only 8 hours long. Similarly, cold treatment of the clover *Trifolium subterraneum* can completely remove its dependency on day length for flowering.

In the biennial henbane and most other biennial long-day plants that form rosettes, gibberellin treatment can substitute for cold exposure. If gibberellin is applied, such plants will elongate rapidly and then flower. Application of gibberellin to short-day plants or to nonrosette long-day plants has little effect on (or inhibits) the flowering response. However, if gibberellin synthesis is inhibited when the plant is being exposed to the appropriate inductive cycle, the plant will not flower unless its gibberellin is replaced.

TOUCH RESPONSES IN PLANTS

Plants also respond to touch. One of the most common examples is seen in tendrils, which are modified leaves in some species of plants and modified stems in others (see page 471). The tendrils wrap around any object with which they come in contact (Figure 26–16) and so enable the plant to cling and climb. The response can be rapid; a tendril may wrap around a support one or more times in less than an hour. Cells touching the support shorten slightly and those on the other side elongate. There is some evidence that auxin plays a role in this response.

Recent studies by M. J. Jaffe of Ohio University have revealed that the tendrils of young Alaska pea plants *(Pisum sativum)* can store sensory information and retrieve it and respond at a later time. For example, when tendrils were held in the dark for three days and then rubbed, they would not coil until they were subsequently illuminated. Moreover, the tendrils could be kept in the dark for as long as two hours after being rubbed, and still coil immediately upon being illuminated. Although the sensory information could be stored during the dark period, the motor

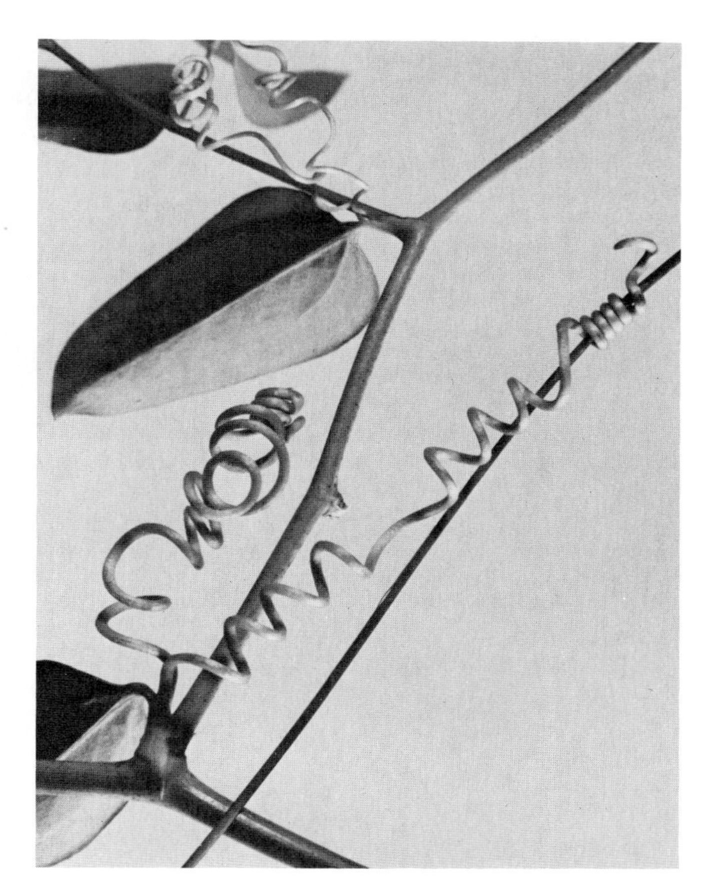

26–16
Tendrils of greenbrier (Smilax). *Twisting is caused by different growth rates on the inside and outside of the tendril.*

(a)

(b)

(c)

26–17
The sensitive plant (Mimosa pudica). (a) *The normal position of leaves and leaflets. Responses to touch with a needle are shown in* (b) *and* (c). *The responses result from changes in turgor pressure in certain cells of jointlike thickenings (pulvini) at the base of the leaflets. Only a single leaflet need be stimulated for the response in* (c) *to occur.*

26–18
Touch response in the Venus flytrap (Dionaea muscipula). *Here, an unwary fly, attracted by nectar secreted on the leaf surface, can be seen on a leaf, before and after its closure. Each leaf half has three sensitive hairs, which control the "trap." The fly triggered the traplike closing of the leaf halves when it brushed against one sensitive hair twice or two hairs in close succession.*

function was unable to proceed in the dark. Why the motor function could not proceed in the dark remains unexplained. It has been suggested, however, that ATP (which apparently is needed for coiling) may be used up during the dark period and is restored during illumination by photosynthesis; or that a coiling inhibitor accumulates in the dark and is quickly removed in the light.

A more spectacular touch response is seen in the sensitive plant *(Mimosa pudica),* in which the leaflets and sometimes entire leaves droop suddenly when touched (Figure 26–17). This response is a result of a sudden change in turgor pressure in certain cells of the jointlike thickenings called pulvini (singular: pulvinus) at the base of leaflets and leaves. The loss of water from these cells follows the migration of potassium ions from them. Only a single leaflet needs to be stimulated; the stimulus then moves to other parts of the leaf and throughout the plant.

Two distinct mechanisms, one electrical and the other chemical, appear to be involved with the spread of the stimulus in the sensitive plant. There is some controversy about the survival value of this touch response to the plant. *Mimosa pudica* often grows in arid, exposed areas where it may be subjected to drying winds; strong winds may shake the leaves enough to make them fold up, thus conserving water. Another suggestion is that the wilting response makes the plant unattractive to large herbivores. Finally, it is conceivable that the folding response could startle herbivorous insects; there have been claims that other "nonsensitive" species of *Mimosa* growing near *Mimosa pudica* show more evidence of attack by insects.

The triggering of turgor changes by touch is also involved in the capture of prey by the carnivorous Venus flytrap *(Dionaea muscipula).* Each leaf half is equipped with three sensitive hairs. When an insect walks on one of these leaves, attracted by the nectar on the leaf surface, it brushes against the hairs, triggering the traplike closing of the leaf. The toothed edges mesh, the leaf halves gradually squeeze shut, and the insect is pressed against digestive glands on the inner surface of the trap (Figure 26–18).

The trapping mechanism is so specialized that it can distinguish between living prey and inanimate objects, such as pebbles and small sticks, that fall on the leaf by chance: the leaf will not close unless two of its hairs are touched in succession or one hair is touched twice.

The connection between electrical activity and the movement of plant parts in various plant species has been studied by Barbara G. Pickard and her colleagues at Washington University in St. Louis. Most notable are the careful studies carried out by Pickard and Stephen E. Williams on how the sundew plant *(Drosera)* traps insects. The upper surfaces of the club-shaped leaves are covered with many large hairs—each about 3 millimeters long—which are enlarged at the tip (see Figure 27–10b, page 546). A sticky droplet surrounding the tip of each hair attracts and catches insects. When the insect is caught on the tips of the outer hairs, these hairs bend in, carrying the insect to the center of the leaf, where it is digested by enzymes secreted by the leaf. Studies in which microelectrodes were placed in the hairs have revealed that the rapid response of the outer hairs results from an abrupt change in the electrical potential across the plasma membrane. The stimulus is received in the enlarged tip of the hair and transmitted down the stalk, which responds by bending.

IN CONCLUSION

Each developmental event in the life of a plant is under the control not of any single factor but of a complex variety of factors. Internal and environmental factors interact. They may enhance, modify, or neutralize one another. In the words of William S.

Hillman of the Brookhaven National Laboratory, "After all, if plants were as simple as the physiologist might wish, there would be nothing left to do."

SUMMARY

Plants possess a variety of adaptations that enable them to detect and respond to alterations in their environment. Phototropism, or the bending of the growing shoot toward light, is an example of such an adaptation. The differential growth of the seedling is caused by lateral migration of the growth hormone auxin under the influence of light. The photoreceptor for this response is a blue-light absorbing pigment. Geotropism, or gravitropism, is the response to the pull of the earth's gravity by a shoot or a root. The movement of auxin to the lower surface of a horizontally oriented shoot causes the shoot to bend upward. The downward bending of a horizontally oriented root appears to result in part from the inhibition by abscisic acid on cell elongation in the lower surface of the root.

Circadian rhythms are cycles of activity in an organism that recur at intervals of about 24 hours under constant environmental conditions. They are probably endogenous, caused not by an external factor, such as alternating light and darkness or the earth's rotation, but by some internal timing mechanism within the organism. Such a timing mechanism, the chemical and physical nature of which is unknown, is called the biological clock. The possession of a biological clock makes it possible for an organism to perceive changes in external diurnal cycles, such as the lengthening and shortening of days as the seasons progress. In this way, activities such as dormancy, leaf abscission, and flowering can be brought into synchrony with the external environment.

Photoperiodism is the response of organisms to changing 24-hour cycles of light and darkness. Such responses control the onset of flowering in many plants. Some plants will flower only when the periods of light exceed a critical length. Such plants are known as long-day plants. Other plants, short-day plants, flower only when the periods of light are less than some critical period. Day-neutral plants flower regardless of photoperiods. Factors such as temperature and the age of the plant may affect the photoperiodic response. Interruption of the dark phase of the photoperiod, even by a brief flash of light, can serve to reverse the photoperiodic effects.

Phytochrome, a pigment commonly present in small amounts in the tissues of higher plants, is sensitive to the transitions between light and darkness. The pigment can exist in two forms, P_r and P_{fr}. The P_r form absorbs red light with a wavelength of 660 nanometers and is thereby converted to P_{fr}. The P_{fr} form absorbs far-red light (730 nanometers) and is converted to P_r. The P_{fr} form is also lost from the cell in the dark by reversion to P_r or by destruction. P_{fr} is the active form of the pigment; it promotes flowering in long-day plants and inhibits flowering in short-day plants. P_{fr} also is responsible for changes that take place in seedlings as they penetrate the soil to the light, for germination of seeds, and for development of anthocyanins.

In both long-day and short-day plants, the photoperiod is perceived in the leaves but the response takes place in the bud. Phytochrome is the photoreceptor that perceives the environmental stimulus and appears to interact with an endogenous circadian rhythm in an undetermined fashion to generate a floral stimulus. Although it has not yet been isolated or identified, this chemical substance—now termed florigen—moves from the leaves to the bud, where it induces flowering. Experiments have indicated that the chemical travels through the plant by way of the phloem system and that its structure and function are similar in long-day, short-day, and day-neutral plants. Strong evidence suggests that, at least in certain plants, both flower-inducing and flower-inhibiting substances are involved in flowering.

Alternation of periods of growth and cessation of growth permits the plant to survive water shortages and extremes of hot or cold. Dormancy is a special condition of arrested growth in which the entire plant, or such structures as seeds or buds, do not renew growth without special environmental cues. The requirement for such cues, which include cold exposure, dryness, and a suitable photoperiod, prevents the tissue from breaking dormancy during superficially favorable conditions, such as those found within the succulent fruit of the parent plant or during the warmth of Indian summer. There is apparently no uniform mechanism, common to all plant groups, for the induction and breaking of dormancy. Vernalization refers to the promotion of flowering in winter strains by keeping seeds at low temperatures. Hormones, cold, and light interact to modify plant responses.

Some species of plants respond to touch. Examples include the winding of tendrils, the collapse of the leaves of the sensitive plant, the triggered closing of the carnivorous Venus flytrap, and the enfoldings of the tentacles of the sundew. Some of these responses have been shown to be mediated by electrical signals.

SUGGESTIONS FOR FURTHER READING

BRADY, JOHN: *Biological Clocks, Studies in Biology, No. 104,* University Park Press, Baltimore, Md., 1979.*

An interesting and well-written introduction to the subject of biological clocks and their experimental study.

GALSTON, ARTHUR W., PETER J. DAVIES, and RUTH L. SATTER: *The Life of the Green Plant,* 3rd ed., Prentice-Hall, Inc., Englewood Cliffs, N.J., 1980.*

A comprehensive, up-to-date, and basic description of the functioning of the green plant, especially suited to persons without an advanced background in biology or chemistry.

KENDRICH, RICHARD E., and BARRY FRANKLAND: *Phytochrome and Plant Growth, Studies in Biology, No. 68,* Edward Arnold Publishers, London, 1976.*

A concise, well-illustrated introduction to the molecular properties and physiology of phytochromes.

LEOPOLD, A. CARL, and P. E. KRIEDEMAN: *Plant Growth and Development,* 2nd ed., McGraw-Hill Book Company, New York, 1975.

An important research work that provides much material of interest in connection with this and following chapters.

MOHR, HANS: *Lectures on Photomorphogenesis,* Springer-Verlag, New York, 1972.*

An in-depth treatment of the effects of light on plant development, primarily in seedlings of the wild mustard.

RAY, PETER M.: *The Living Plant,* 2nd ed., Holt, Rinehart & Winston, Inc., New York, 1972.*

A short, readable account of plant growth and development that has proven extemely useful for students.

SALISBURY, FRANK B.: *The Biology of Flowering,* Natural History Press, Garden City, N.Y., 1971.

A semipopular account of research in the field, concentrating mainly on the short-day plant Xanthium.

SALISBURY, FRANK B., and CLEON W. ROSS: *Plant Physiology,* 2nd ed., Wadsworth Publishing Co., Inc., Belmont, Calif., 1978.

A detailed and useful review of the entire subject.

SMITH, HARRY: *Phytochrome and Photomorphogenesis,* McGraw-Hill Book Company, New York, 1975.

A particularly well-written account of the role of the phytochrome pigment in photomorphogenesis.

STEEVES, TAYLOR A., and IAN M. SUSSEX: *Patterns in Plant Development,* Prentice-Hall, Inc., Englewood Cliffs, N.J., 1972.

An exceptionally well-written and well-illustrated text which should be read by all students interested in plant morphogenesis.

THIMANN, KENNETH V.: *Hormone Action in the Whole Life of Plants,* University of Massachusetts Press, Amherst, Mass., 1977.

An outstanding review of the ways in which hormones initiate and control the growth and development of plants; written by a scholar who has participated in many of the fundamental discoveries in his field.

VINCE-PRUE, DAPHNE: *Photoperiodism in Plants,* McGraw-Hill Book Company, New York, 1975.

A comprehensive treatment of the photoperiodic control of flowering, as well as other plant responses to light.

*Available in paperback.

SECTION 7 Uptake and Transport in Plants

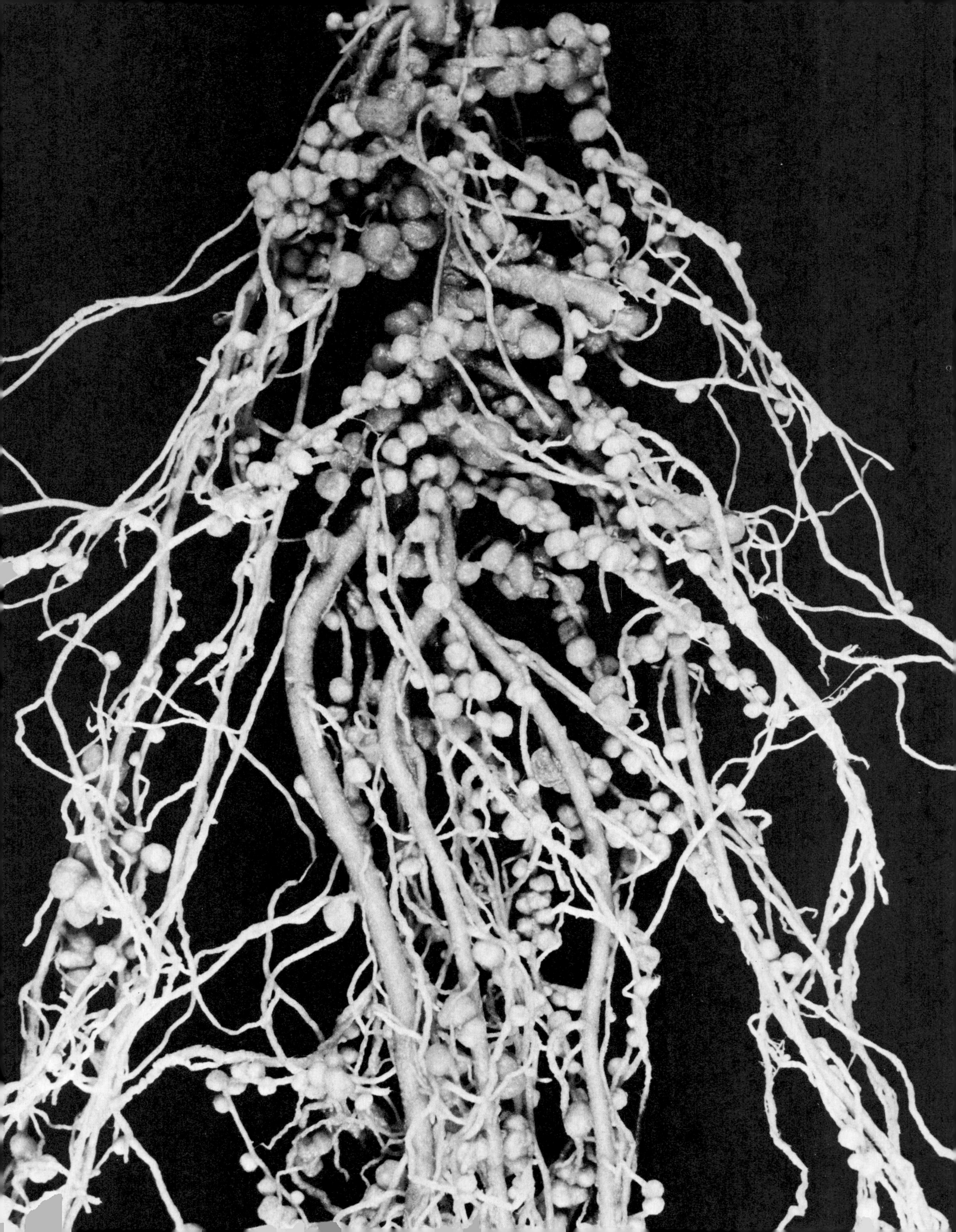

CHAPTER 27

Plant Nutrition and Soils

27–1
The symbiotic relationship between soil bacteria of the genus Rhizobium *and the cortical root cells of the bird's-foot trefoil* Lotus corniculatus, *a legume, leads to the development of conspicuous root nodules. The bacteria living within the nodules use organic compounds supplied by the plant as an energy source for their metabolic activities, which include nitrogen fixation. The legume, in return, obtains a ready supply of nitrogen in a form it can use to manufacture plant proteins.*

GENERAL NUTRITIONAL REQUIREMENTS

Plants must obtain from the environment the specific raw materials required in the complex biochemical reactions necessary for cell maintenance and plant growth. In addition to light, higher plants need water and certain chemical elements for metabolism and growth. Much of the evolutionary development of plants has involved the formation of specialized tissues and mechanisms that permit efficient acquisition of these raw materials and their distribution to living cells throughout the plant.

Like plants, animals need water and specific chemical elements for metabolism, but animals must also be able to obtain external supplies of sugars or other compounds that can serve as an energy source, as well as certain amino acids and vitamins. Thus, by comparison, the nutritional demands of plants are relatively simple. Under favorable environmental conditions, most green plants not only use light energy to transform CO_2 and H_2O into organic compounds for their energy source, but they also synthesize all their required amino acids and vitamins, using inorganic nutrients drawn from the environment.

The self-sufficiency of nitrogen-fixing cyanobacteria is particularly impressive. Through the process of nitrogen fixation, they convert atmospheric nitrogen into forms of nitrogen that can be utilized in the synthesis of amino acids and proteins.

Generally, plant nutrition involves the uptake from the environment of all the necessary raw materials for essential biochemical processes, the distribution of these materials within the plant, and their utilization in metabolism and growth.

Essential Inorganic Nutrients

As early as 1800, chemists and plant biologists had analyzed plants and demonstrated that certain chemical elements were absorbed from the environment.

However, opinions differed on whether the absorbed elements were impurities or constituents required for essential functions. By the mid-1880s it had been established that at least ten of the chemical elements present in plants were necessary for normal growth. In the absence of any one of these elements, plants displayed characteristic abnormalities of growth or symptoms of deficiencies, and often such plants did not reproduce normally. These ten elements—carbon, hydrogen, oxygen, potassium, calcium, magnesium, nitrogen, phosphorus, sulfur, and iron—were designated as essential chemical elements for plant growth. They are also referred to as essential minerals, or *essential inorganic nutrients.*

In the early 1900s, it was established that manganese was also an essential element. During the next 50 years, with the aid of improved techniques for removing impurities from nutrient cultures, five additional elements—zinc, copper, chlorine, boron, and molybdenum—were determined to be essential, with chlorine being recognized only in 1954. At present, these 16 elements are generally considered to be essential for most plants.

Nutrient Concentrations in Plants

Chemical analyses for essential inorganic nutrients are a useful means of indicating the relative amounts of various elements required for the normal growth of different species of plants. Typical results of the chemical analysis of three plant species for the essential inorganic nutrients (other than carbon, hydrogen, and oxygen), as well as two nonessential elements, are presented in Table 27–1. Plant analyses of this sort are particularly valuable in agriculture as a guide to the nutritional well-being of plants and the need for applications of fertilizer. Potential nutritional deficiencies in livestock that consume specific plants can also be predicted by inorganic analyses.

The concentrations of specific elements in plants are known to vary over a considerable range. On the basis of the usual concentrations in plants, the essential inorganic nutrients can be divided into two broad groups—macronutrients and micronutrients. In general, *macronutrients* are elements that are required in large amounts, and *micronutrients* are those required in very small, or trace, amounts (Table 27–2).

Potassium probably reaches higher concentrations in plants than any element other than carbon and oxygen. When the soil concentration is very high, as much as 10 percent of the oven-dried weight of plants may be due to accumulations of potassium.

Certain plant species and taxonomic groups are characterized by unusually high or low contents of specific elements (Figure 27–2). As a result, plants growing in the same nutrient medium may differ markedly in nutrient content. Compare the analyses in Table 27–1 for corn (a monocot) and alfalfa (a dicot). Dicots generally require greater amounts of calcium and boron than do monocots.

(a)

(b)

27–2
(a) *Plants of the mustard family, such as wintercress* (Barbarea vulgaris), *use sulfur in the synthesis of the mustard oils that give these plants both their name and their characteristic sharp taste.* (b) *Horsetails* (Equisetum) *incorporate silicon into their cell walls, making them indigestible to most herbivores but useful, at least in colonial North America, for scouring pots and pans.*

Table 27–1 *Examples of Inorganic Analysis of Plants*

	CONCENTRATION OF ELEMENT		
ELEMENT	ALFALFA	CORN*	WHITE OAK**
Potassium	2.77%	1.86%	0.85%
Calcium	1.70	0.40	0.82
Magnesium	0.41	0.27	0.36
Nitrogen	3.12	2.81	2.19
Phosphorus	0.35	0.28	0.19
Sulfur	0.29	0.18	0.13
Iron	190 ppm***	110 ppm	126 ppm
Manganese	62	80	572
Zinc	57	27	22
Copper	9	6	8
Chlorine	8800	3100	43
Boron	35	14	38
Molybdenum	1.40	1.03	6.21
Sodium	4300 ppm	127 ppm	210 ppm
Cobalt	0.21	0.16	—

* Shoot only; grain not included in analysis.
** Leaf and twig growth of current year.
*** Abbreviation for parts per million; ppm equals units of an element by weight per million units of oven-dried plant material; 1% equals 10,000 ppm.

Nutritional studies have established that a number of elements are essential only for limited groups of plants or for plants grown under specific environmental conditions. Legumes, such as alfalfa *(Medicago sativa)*, benefit from the addition of cobalt to the culture medium; it is not the alfalfa, however, that requires cobalt but the symbiotic nitrogen-fixing bacteria growing in association with the roots of the alfalfa. As indicated in Table 27–1, sodium is present in many plants in relatively high concentrations. It has been recognized for many years that in some species sodium can partially replace plant requirements for potassium. Relatively recent studies have demonstrated that sodium is an essential element for some species. For example, sodium seems to be required by certain halophytes (plants that grow in salty soils) and by at least some plant species that utilize C_4 photosynthesis.

Table 27–1 includes two elements—sodium and cobalt—that are not essential for all crop plants but are essential for herbivorous animals that consume these crop plants. For example, if the cobalt concentration in a forage crop is less than 0.1 ppm, animals eating these plants are likely to display symptoms of cobalt deficiency.

FUNCTIONS OF INORGANIC NUTRIENTS IN PLANTS

Inorganic nutrients are essential in many aspects of growth and metabolism. Table 27–2 lists the inorganic nutrients required by most plants, the forms in which they most commonly are absorbed from the environment, the usual range of concentrations in plants, and some of the functions of each nutrient.

Specific Versus Nonspecific Functions

Inorganic ions affect osmosis (Chapter 3) and thus help to regulate water balance. Because several inorganic ions can serve interchangeably in this role, in many plants this particular requirement is described as *nonspecific*. On the other hand, an inorganic nutrient may function as part of an essential biological molecule; in this case the requirement is highly *specific*. An example of a specific function is the presence of magnesium in the chlorophyll molecule (see Figure 6–8, page 103). Some inorganic nutrients are essential constituents of cellular membranes, whereas others control the permeability of such membranes. Other inorganic nutrients are indispensable components of a variety of enzyme systems that catalyze biological reactions in the cell. Still others provide a proper ionic environment in which biological reactions can occur.

Because inorganic nutrients fill such basic needs and are involved in such fundamental processes, the effects of deficiencies are typically quite widespread, affecting a wide variety of structures and functions in the plant body.

Catalysts

A key role of the inorganic nutrients is their participation as catalysts in some of the enzymatic reactions of the plant cell. In some cases, they are an essential structural part (a "prosthetic group") of the enzyme. In other cases, they serve as activators or regulators of certain enzymes. Potassium, for instance, which probably affects 50 to 60 enzymes, is believed to regulate the conformation of some proteins. Changing the shape of an enzyme could, for example, expose or obstruct reaction sites (see Figures 2–16, page 55).

Electron Transport

Many of the biochemical activities of cells, including photosynthesis and respiration, are oxidation-reduction reactions. In such reactions, electrons are often transferred to or from a molecule that functions as an electron acceptor. Among the important electron acceptors are the cytochromes, which contain iron (see Figure 5–9, page 92).

Table 27–2 *A Summary of the Functions of Inorganic Nutrients in Plants*

ELEMENT	PRINCIPAL FORM IN WHICH ELEMENT IS ABSORBED	USUAL CONCENTRATION IN HEALTHY PLANTS (% OF DRY WEIGHT)	IMPORTANT FUNCTIONS
Macronutrients			
Carbon	CO_2	~44%	Component of organic compounds.
Oxygen	H_2O or O_2	~44%	Component of organic compounds.
Hydrogen	H_2O	~6%	Component of organic compounds.
Nitrogen	NO_3^- or NH_4^+	1–4%	Amino acids, proteins, nucleotides, nucleic acids, chlorophyll, and coenzymes.
Potassium	K^+	0.5–6%	Enzymes, amino acids, and protein synthesis. Activator of many enzymes. Opening and closing of stomata.
Calcium	Ca^{2+}	0.2–3.5%	Calcium of cell walls. Enzyme cofactor. Cell permeability.
Phosphorus	$H_2PO_4^-$ or HPO_4^{2-}	0.1–0.8%	Formation of "high-energy" phosphate compounds (ATP and ADP). Nucleic acids. Phosphorylation of sugars. Several essential coenzymes. Phospholipids.
Magnesium	Mg^{2+}	0.1–0.8%	Part of the chlorophyll molecule. Activator of many enzymes.
Sulfur	SO_4^{2-}	0.05–1%	Some amino acids and proteins. Coenzyme A.
Micronutrients			
Iron	Fe^{2+} or Fe^{3+}	25–300 ppm	Chlorophyll synthesis, cytochromes, and nitrogenase.
Chlorine	Cl^-	100–10,000 ppm	Osmosis and ionic balance; probably essential in photosynthetic reactions that produce oxygen.
Copper	Cu^{2+}	4–30 ppm	Activator of certain enzymes.
Manganese	Mn^{2+}	15–800 ppm	Activator of certain enzymes.
Zinc	Zn^{2+}	15–100 ppm	Activator of many enzymes.
Molybdenum	MoO_4^{2-}	0.1–5.0 ppm	Nitrogen fixation. Nitrate reduction.
Boron	BO_3^- or $B_4O_7^{2-}$	5–75 ppm	Influences Ca^{2+} utilization. Functions unknown.
Elements Essential to Some Plants or Organisms			
Cobalt	Co^{2+}	Trace	Required by nitrogen-fixing microorganisms.
Sodium	Na^+	Trace	Osmotic and ionic balance, probably not essential for many plants. Required by some desert and salt-marsh species. May be required by all plants that utilize C_4 photosynthesis.

Structural Components

Some mineral elements serve as structural components of cells, either as part of a physical structure (see Figure 27–2b) or as part of the chemical structures involved in cellular metabolism. Calcium combines with pectic acid in the middle lamella of the plant cell wall. Phosphorus occurs in the sugar phosphate backbone of the DNA helix, in RNA, and in the phospholipids of the cellular membranes. Nitrogen is an essential component of amino acids, chlorophylls, and nucleotides. Sulfur is found in two amino acids, thus forming an important structural element in proteins.

Osmosis

The movement of water into and out of plant cells, as discussed in Chapter 3, is largely dependent on the concentration of solute in the cells and in the surrounding medium. The uptake of ions by the plant cells thus may result in the entry of water into the cell. The resultant hydrostatic pressure from within the cell produces expansion of the immature cell, which is the chief cause of cellular growth, and is responsible for turgor in the mature cell (see page 63). This is an example of conversion of energy from one form to another by a living system; the chemical (ATP) energy expended in the active uptake of ions by the plant cell is translated into the physical energy of water movement.

Effects on Cell Permeability

Calcium has a direct effect on the physical properties of the cellular membranes. When there is a calcium deficiency, membranes seem to lose their integrity, and solutes within the membranes or cells leak out.

THE SOIL

Soil is the primary nutrient medium for plants. Soils must provide plants not only with physical support but also adequate inorganic nutrients at all times, as well as with adequate water and a suitable gaseous environment for the root systems. Understanding the origins of soils and their chemical and physical properties in relation to plant growth requirements is critical in planning for the nutrition of field crops.

Weathering of the Earth's Crust

The inorganic nutrients utilized by plants are derived from the atmosphere and from the weathering of rocks in the crust of the earth. The earth is composed of about 92 naturally occurring elements, which are often found in the form of minerals. *Minerals* are naturally occurring inorganic compounds that are usually composed of two or more elements in definite proportions by weight (see Appendix A). Quartz (SiO_2) and calcite ($CaCO_3$) are examples of mineral compounds.

Most rocks consist of several different minerals and are divided into three groups based on origin and formation. *Igneous* rocks such as granite are derived directly from molten material and, for the most part, were originally formed when the earth cooled and solidified. As a result of weathering, igneous rocks and other kinds of rocks may be broken down into soluble and insoluble component parts; after being transported by water, wind, or glaciers, these components form new deposits—usually in water—that in time become cemented and solidified into *sedimentary* rocks, such as shale, sandstone, and limestone. Although sedimentary rocks form only about 5 percent of the earth's crust, they are of great importance because they occur extensively at or near the surface. Under the extreme heat and pressure deep within the earth, sedimentary and igneous rock may be transformed into a third type of rock—*metamorphic*. In this way, quartzite is formed from sandstone, slate from shale, and marble from limestone.

Weathering processes, involving the physical disintegration and chemical decomposition of minerals and rocks at or near the earth's surface, produce the inorganic materials from which soils are formed. Weathering involves freezing and thawing or heating and cooling, which cause substances in the rocks to expand and contract, splitting rocks apart. Water and wind often carry the rock fragments great distances, exerting a scouring action that breaks the fragmented rock into even smaller particles. Water enters between the particles, and soluble materials dissolve in the water. Water, in combination with carbon dioxide and air impurities such as sulfur dioxide and oxides of nitrogen, forms dilute acids that help to dissolve materials that are less soluble in pure water. Soil formation may occur at the site of weathering, or the parent materials may be transported elsewhere by gravity, wind, water, or glaciers.

Soils also contain organic materials. If light and temperature conditions permit, bacteria, fungi, algae, lichens, and bryophytes and small vascular plants gain a foothold on or among the weathered rocks and minerals. Growing roots also split rocks, and their disintegrating bodies and those of the animals associated with them add to the accumulating organic material. Finally, larger plants move in, anchoring the soil in place with their root systems, and a new community begins.

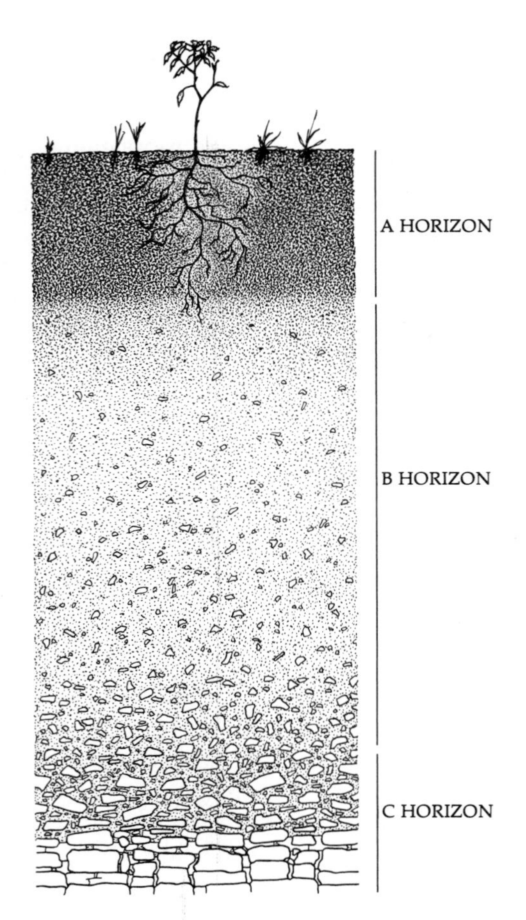

27–3
The three primary horizons, or soil layers, recognized in a typical soil.

27–4
The fibrous root systems of grasses bind and anchor prairie soil in place.

Examining a vertical section of soil (Figure 27–3), one can see variations in the color, the amount of living and dead organic matter, the porosity, the structure, and the extent of weathering. These variations generally result in a succession of rather distinct layers that soil scientists refer to as *horizons*. A minimum of three horizons is recognized.

The A horizon (sometimes called the "topsoil") is the upper region, that of the greatest physical, chemical, and biological activity. The A horizon contains the greatest portion of the soil's organic material (Figure 27–4), both living and dead, such as large amounts of dead and decaying leaves and other plant parts, insects and other small arthropods, earthworms, protozoa, nematodes, and decomposer organisms (Figure 27–5).

The B horizon is a region of deposition. Iron oxide, clay particles, and small amounts of organic matter are among the materials carried from the A horizon to the B horizon by water percolating, or moving down, through the soil. The B horizon contains much less organic material and is less weathered than the horizon above it.

The C horizon is composed of the broken-down and weathered rocks and minerals from which the true soil in the two upper horizons is formed.

Soil Composition

Soils consist of solid matter and pore space (the space around the soil particles). Different proportions of air and water occupy the pore space, depending on prevailing moisture conditions. The soil water is present primarily as a film on the surfaces of soil particles.

The fragments of rocks and minerals in the soil vary in size from sand grains, which can be seen easily with the naked eye, to clay particles too small to be seen even under the low power of a light microscope. The following classification is one scheme for categorizing soil particles (also known as soil separates) according to size:

Separate	Diameter (in micrometers)
Coarse sand	200–2000
Fine sand	20–200
Silt	2–20
Clay	Less than 2

Soils contain a mixture of particles of different sizes and are divided into textural classes according to the proportions of different particles present in the mixture. For example, soils that contain 35 percent or more clay and 45 percent or more sand are sandy clays; those containing 40 percent or more clay and 40 percent or more silt are silty clays. Loam soils contain sand, silt, and clay in proportions that result in ideal agricultural soils (Figure 27–6).

The solid matter of soils consists of both inorganic and organic materials, with the proportions varying greatly in different soils. The organic component includes the remains of organisms in various stages of decomposition, as well as a wide range of living plants and animals. Structures as large as tree roots may be included, but the living phase is dominated by fungi, bacteria, and other microorganisms.

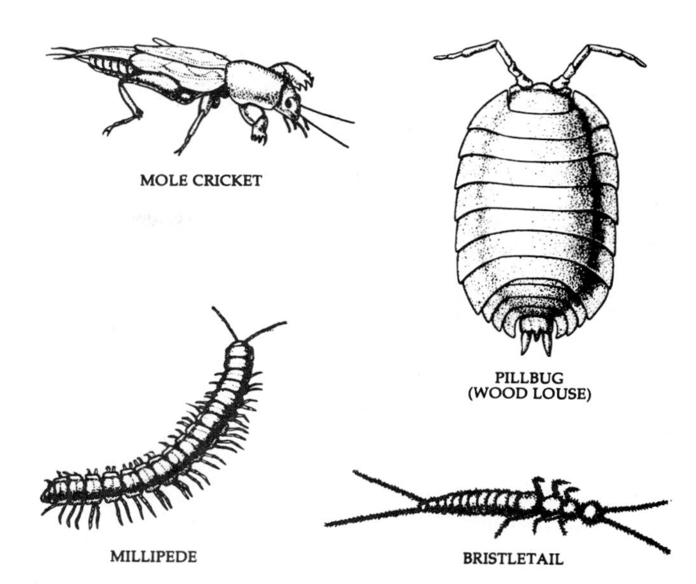

27–5
Plants share the soil with a vast number of other living organisms. The four soil animals shown here are common in the soils of temperate deciduous forests. They play an important role in breaking down dead organic matter—both plants and other animals.

Cation Exchange

The inorganic nutrients taken in through the roots of plants are present in the soil solution as ions. Most metals form positively charged ions, that is, cations, such as Ca^{2+}, K^+, and Na^+. Clay particles provide a reservoir of such cations for the plant—at various points on their crystalline lattice there is an excess of negative charge, where cations can be bound and thus held against the leaching action of percolating soil water.

The cations bound in this way to the clay particles can be replaced by other cations (in a process called *cation exchange*) and then released into the soil solution, where they become available for plant growth. This is one reason clay particles are an essential component of productive soils.

The principal negatively charged ions, or anions, found in soil are NO_3^-, SO_4^{2-}, HCO_3^-, and OH^-. Anions are leached out of the soil more rapidly than cations because they do not attach to clay particles. An exception is the phosphate ion (PO_4^{3-}), which is retained against leaching because it forms insoluble precipitates and is specifically adsorbed, or held on the surface, of compounds containing iron, aluminum, and calcium.

The acidity or alkalinity of soil is related to the availability of inorganic nutrients for plant growth. Soils vary widely in pH, and many plants have a narrow range of tolerance on this scale. In alkaline soils, some cations are precipitated, and such elements as iron, manganese, copper, and zinc may thereby become unavailable to plants.

27–6
A soil sample from McClain County, Illinois. The black loam of the corn belt is among the richest and most fertile soils in the world.

Soils and Water

Approximately 50 percent of the total soil volume is represented by pore space, which is occupied by varying proportions of air and water, depending on moisture conditions. When no more than half the pore space is occupied by water, adequate oxygen is available for root growth and other biological activity.

Following a heavy rain or irrigation, soils retain a certain amount of water and remain moist even after gravity has removed loosely bound water. If the

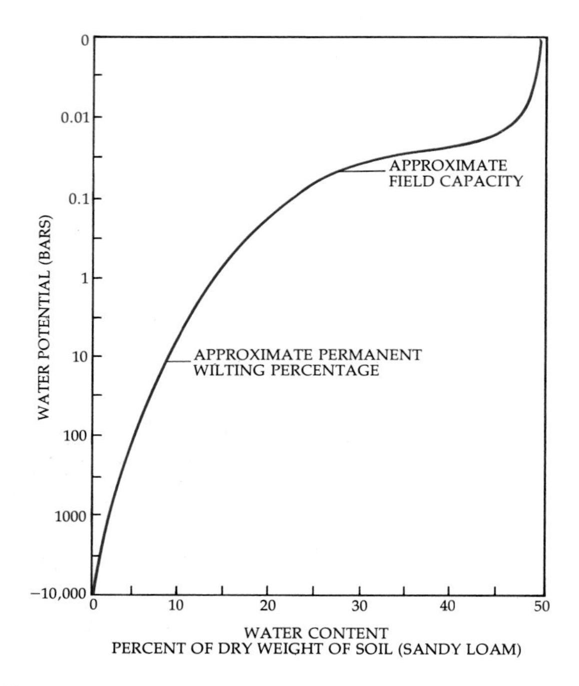

27–7
Relation between the soil water potential and the soil water content in a sandy loam soil.

fragments that make up a soil are large, the pores and spaces between them will be large; water will drain through the soil rapidly, and relatively little will be available for plant growth in the A and B horizons. Because of their finer pores, clay soils are able to hold a much greater amount of water against the action of gravity. Thus, clay soils may retain three to six times more water than a comparable volume of sand. The percentage of water that a soil can hold against the action of gravity is called its *field moisture capacity*.

If a plant is allowed to grow indefinitely in a sample of soil and no water is added, the plant will eventually not be able to absorb water rapidly enough to meet its needs, and it will droop and wilt. When wilting is severe, plants fail to recover even when placed in a humid chamber. The percentage of water remaining in a soil when such irreversible wilting occurs is called the *permanent wilting percentage* of that soil.

Figure 27–7 shows the relationship between the soil water content and the potential with which that water is held by a sandy loam soil. The forces that retain water in the soil can be expressed in the same terms (in this case, water potential) as the forces for water uptake that develop in cells and tissues (see Chapter 3, pages 59–60). The potential of the soil water decreases gradually with a decrease in the soil

THE WATER CYCLE

The earth's supply of water is stable and is used over and over again. Most of the water (98 percent) is present in oceans, lakes, and streams. Of the remaining 2 percent, some is frozen in polar ice and glaciers, some is found in the soil, some is in the atmosphere as water vapor, and some is in the bodies of living organisms.

Sunshine evaporates water from the oceans, lakes, and streams, from the moist soil surfaces, and from the bodies of living organisms, drawing the water back up into the atmosphere, from which it falls again as rain. This constant movement of water from the earth into the atmosphere and back again is known as the water cycle. The water cycle is driven by solar energy.

Some of the water that falls on the land percolates down through the soil until it reaches a zone of saturation. In the zone of saturation, all holes and cracks in the rock are filled with water. Below the zone of saturation is solid rock through which the water cannot penetrate. The upper surface of this zone of saturation is known as the water table.

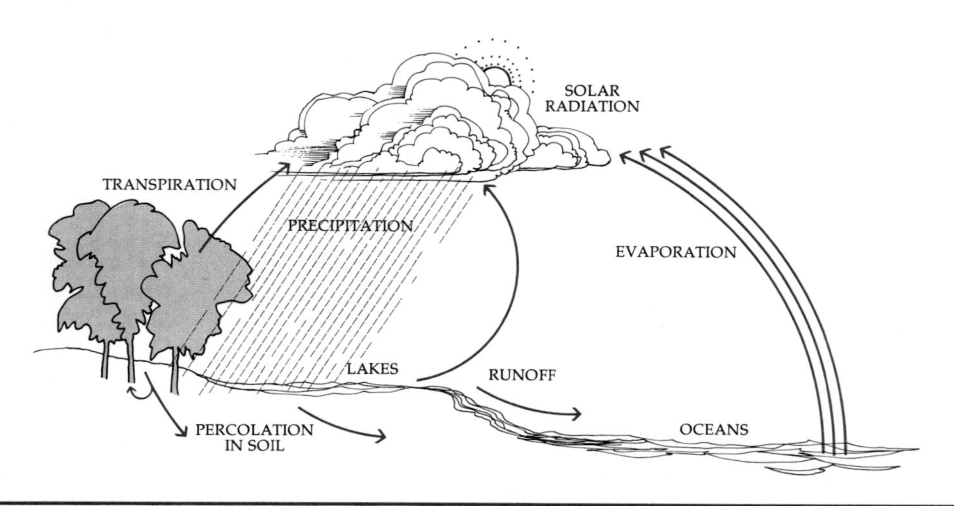

moisture below field moisture capacity. The soil water potential then decreases sharply as the moisture content approaches the permanent wilting percentage of approximately −15 bars.

NUTRIENT CYCLES

The bulk of the organic matter of soil is made up of dead leaves and other plant material, together with the decomposing bodies of animals. This organic debris is mixed with the inorganic particles of the soil, and in this mixture live astonishing numbers of small organisms that spend all or a part of their lives beneath the soil surface. A single teaspoon of soil may contain 5 billion bacteria, 20 million small filamentous fungi, and 1 million protozoa. The soil animals and microorganisms break down the organic matter, releasing its inorganic nutrients so that they can be reutilized by the plants (Figure 27–8). Thus, except for the nutrients that leach out of the soil, are carried away by streams and rivers, and eventually precipitate in the ocean, substances that are taken from the soil are constantly returned to it. Both macronutrients and micronutrients are recycled constantly through plant and animal bodies, returned to the soil, broken down, and taken up into plants again. Each element has a different cycle, involving many different organisms and different enzyme systems. The end results are the same, however—a significant amount of the element is constantly returned to the soil and is available for plant use.

NITROGEN AND THE NITROGEN CYCLE

The nitrogen in soil is derived from the earth's atmosphere. Although the atmosphere is 78 percent nitrogen, most living things cannot use atmospheric nitrogen to make proteins and other organic substances. Unlike carbon and oxygen, nitrogen is chemically unreactive. The highly specialized capacity for converting atmospheric nitrogen into a form that can be used by cells is limited to a few prokaryotes. This unique process, called nitrogen fixation, will be discussed shortly.

On a world-wide basis, usable nitrogen is the major limiting nutrient in crop plant growth. The processes by which nitrogen is circulated from the atmosphere through plants and the soil by the action of living organisms are illustrated in the nitrogen cycle shown in Figure 27–9.

27–8
Among the largest of the soil organisms are the earthworms. Earthworms make burrows in the earth by passing soil and organic matter through their digestive tracts and depositing them in the form of castings. In a single year, their combined activities may produce as much as 500 metric tons of castings per hectare. The castings are very fertile, containing 5 times the nitrogen content of the surrounding soil, 7 times the phosphorus, 11 times the potassium, 3 times the magnesium, and 2 times the calcium.

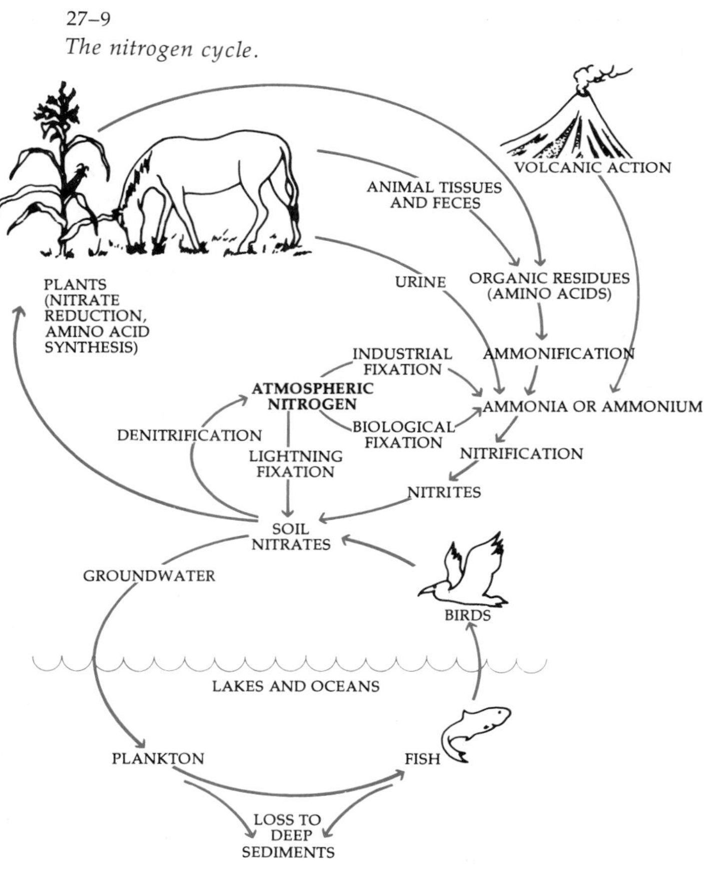

27–9
The nitrogen cycle.

Ammonification

Much of the soil nitrogen is derived from dead organic materials in the form of complex organic compounds such as proteins, amino acids, nucleic acids, and nucleotides. These nitrogenous compounds are usually rapidly decomposed into simple compounds by soil-dwelling saprobic bacteria and various fungi. These microorganisms use the protein and amino acids for the formation of their own proteins and release excess nitrogen in the form of ammonium ions (NH_4^+) by a process known as *ammonification*. The nitrogen may be given off as ammonia gas (NH_3), but this usually occurs only during the decomposition of large amounts of nitrogen-rich material, as in a manure pile or a compost heap. Usually the ammonia produced by ammonification is dissolved in the soil water, where it combines with protons to form the ammonium ion.

Nitrification

Several species of bacteria common in soils are able to oxidize ammonia or ammonium ions. The oxidation of ammonia, or *nitrification*, is an energy-yielding process, and the energy released in the process is used by these bacteria to reduce carbon dioxide in much the same way the photosynthetic autotrophs use light energy in the reduction of carbon dioxide. Such organisms are known as chemosynthetic autotrophs (as distinct from photosynthetic autotrophs). The chemosynthetic nitrifying bacterium *Nitrosomonas* is primarily responsible for oxidation of ammonia to nitrite ions (NO_2^-).

$$2NH_3 + 3O_2 \longrightarrow 2NO_2^- + 2H^+ + 2H_2O$$

Nitrite is toxic to higher plants, but it rarely accumulates in the soil. *Nitrobacter*, another genus of bacteria, oxidizes the nitrite to form nitrate ions (NO_3^-), again with a release of energy:

$$2NO_2^- + O_2 \longrightarrow 2NO_3^-$$

Because of the rapidity with which nitrite is oxidized to nitrate under most soil conditions, nitrate is the form in which almost all nitrogen is absorbed by plants.

A few species of plants are able to use animal proteins directly as a nitrogen source. These carnivorous

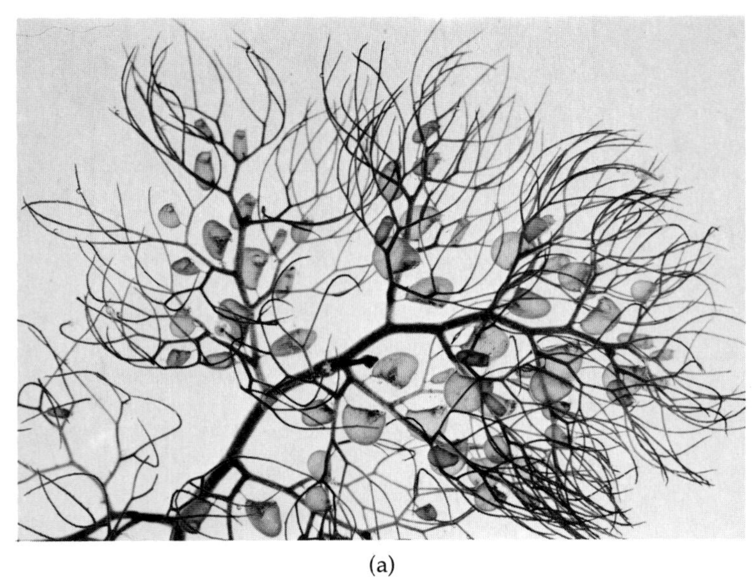

(a)

(b)

27–10
Carnivorous plants can obtain minerals, including nitrogen, from the bodies of their prey. (a) *The common bladderwort* (Utricularia vulgaris) *is a free-floating aquatic plant. The traps are tiny, flattened, pear-shaped bladders. Each bladder has a mouth guarded by a hanging door. The tripping mechanism consists of four stiff bristles near the lower free edge of the door. When a small animal brushes against these bristles, the hairs distort the lower edge of the door, causing it to spring open. Water then rushes into the bladder, carrying the animal inside, and the door snaps shut behind it. The animal decays by bacterial action, and released minerals and organic compounds are taken up by the cellular walls of the trap. The undigested exoskeletons remain within the bladders.* (b) *The sundew* (Drosera intermedia) *is a tiny plant, often only a few centimeters across, with club-shaped hairs on the upper surface of its leaf. The tips of these hairs secrete a clear, sticky liquid that attracts insects. When an insect is caught in the secretion, the hairs bend inward until the leaf finally curves around the insect. The hairs then secrete digestive enzymes.*

CARNIVOROUS PLANTS

Carnivorous plants obtain minerals, including fixed nitrogen, from animal prey. In some of these plants, flies and other insects are caught on the leaf surface, as in (a) *the butterwort* (Pinguicula grandiflora). *The capture of small insects is accomplished by numerous stalked glands scattered over the leaf. Each of these glands, as seen in the scanning electron micrograph* (b), *bears a globular droplet of mucilaginous secretion, so that the leaf is sticky, like flypaper. Insects coming in contact with the secretion pull it out into strands which set to form strong cables, some of which are visible holding down the ant in* (c). *The more the insect struggles, the more glands it touches, and the more tightly it is held onto the leaf. Scattered among the stalked glands are sessile glands* (b), *which bear no surface material until they are stimulated by the struggling of the captured prey. The sessile glands then pour out an enzyme-containing secretion which quickly forms a pool around the insect. These enzymes digest the prey, and the products accumulate in the secretion pool. After the completion of the digestion, the pool is absorbed into the leaf, and the digestion products are distributed to the growing parts of the plant.*

The digestive enzymes are synthesized in the head cells of the sessile glands. Until stimulation these enzymes are stored in enlarged vacuoles and in the wall, where they can be detected by suitable cytochemical techniques. As can be seen in (d), *for example, the localization of the enzyme acid phosphatase in the walls of the gland head is revealed by the distribution of the dark reaction product. The discharge of the enzymes is brought about by the rapid passage of water through the gland head. This process is driven by the pumping of chloride ions from an underlying reservoir cell. The "ion pumps" are present in the membranes of an intervening endodermislike cell; they are quiescent in the inactive gland but are quickly activated by the stimulus of the prey.*

Similar secretion mechanisms are found in other carnivorous plants, including the Venus flytrap (Dionaea muscipula). *In the sundews* (Drosera), *the glands are more complex and are borne on stalks, forming the so-called tentacles, or elongated hairs. The cells of the gland head secrete both a sticky mucilage and various digestive enzymes, so that these cells function both in the capture and the digestion of the prey.*

(a) 1 cm

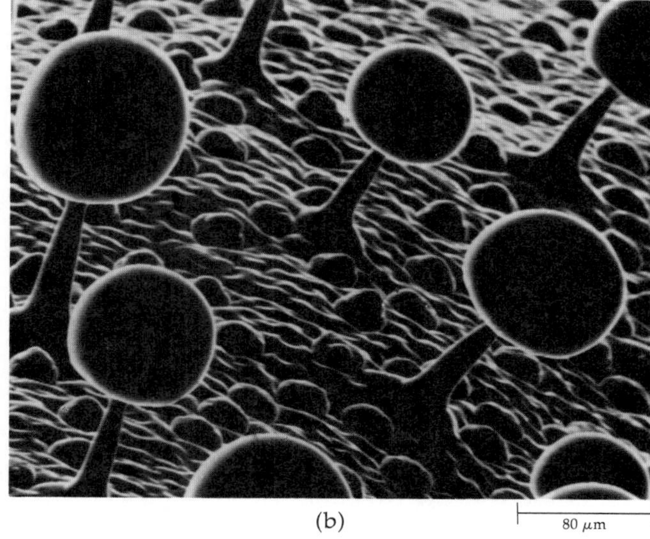

(b) 80 μm

(d) 10 μm

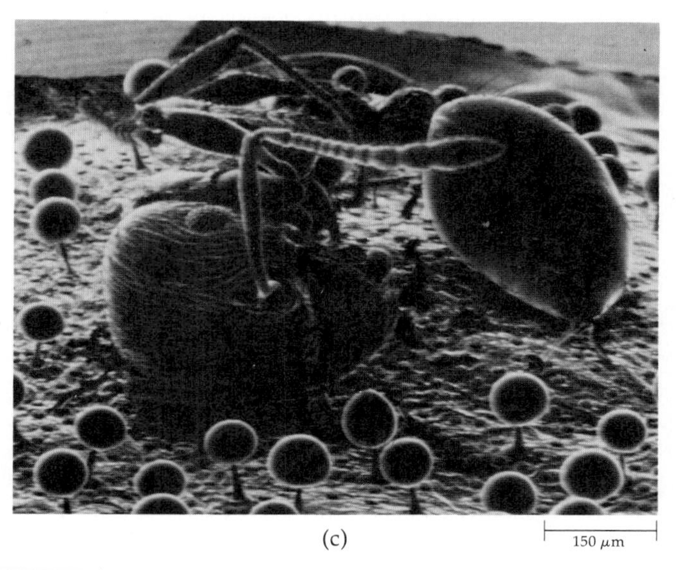

(c) 150 μm

plants (Figure 27–10) have special adaptations that are used to lure and trap insects and other very small animals. The plants digest the trapped organisms, absorbing the nitrogenous compounds they contain, as well as other organic compounds and minerals, such as potassium and phosphate. Most of the carnivores of the plant world are found in bogs, a habitat that is usually quite acidic and thus not favorable for the growth of nitrifying bacteria.

Assimilation of Nitrogen

Once the nitrate ions enter the plant cell, they are reduced back to ammonium ions. This reduction process requires energy, in contrast to nitrification, which involves the oxidation of NH_4^+ and releases energy. The ammonium ions formed by reduction are transferred to carbon-containing compounds to produce amino acids and other nitrogen-containing organic compounds. This process is known as *amination*. The incorporation of nitrogen into organic compounds takes place largely in young, growing root cells. The initial stages in the metabolism of nitrogen also appear to occur in the root; almost all the nitrogen ascending the stem in the xylem is already in the form of organic substances, largely molecules of various amino acids.

Formation of Amino Acids

Amino acids are formed from ammonium ions and keto acids. The latter are usually products of the metabolic breakdown of sugars. The major amino acid that is formed in this fashion is glutamic acid—the chief carrier of nitrogen through the plant body. From the amino acid produced by the amination of a keto acid, other amino acids are formed by the process of *transamination*—the transfer of the amino group ($—NH_2$) on one amino acid to a keto acid, producing a second amino acid.

Plants, either by amination or transamination, can make all the amino acids they require, starting from inorganic nitrogen. Animals can make only about 8 of the 20 amino acids they require and must obtain the others in their diet. Thus, the animal world is completely dependent on the plant world for its proteins, just as it is for its carbohydrates.

Other Nitrogen-Containing Compounds

Other important organic nitrogen-containing compounds include the nucleotides, such as ATP, ADP, NAD, and NADP; chlorophyll and other similar organic molecules with porphyrin ring structures; and the nucleic acids, DNA and RNA. Many of the vitamins, such as the vitamin B group, contain nitrogen. These, like the amino acids, can be synthesized from inorganic nitrogen by plants, but animals must obtain them from plants.

Nitrogen Loss

As discussed, the nitrogen-containing compounds of green plants are returned to the soil with the death of plants (or animals that have eaten the plants) and are reprocessed by soil organisms. Nitrate dissolved in the soil water is then taken up by plant roots and reconverted to organic compounds. In the course of this cycle, a certain amount of nitrogen is always "lost," in the sense that it becomes unavailable to the plants in specific ecosystems.

A main source of nitrogen loss from specific ecosystems is the removal of plants from the soil. Soils under cultivation often show a steady decline of nitrogen content. Nitrogen may also be lost when topsoil is carried off by soil erosion or when ground cover is destroyed by fire. Nitrogen is also removed by leaching; nitrates and nitrites, both of which are anions, are particularly susceptible to being washed away by water percolating through the soil.

Under anaerobic conditions, nitrate is often reduced to volatile forms of nitrogen, such as nitrogen gas (N_2) and nitrous oxide (N_2O), which then return to the atmosphere. This process of reduction, called *denitrification*, is carried out by numerous microorganisms. The low oxygen conditions necessary for denitrification are characteristic of waterlogged soils and such habitats as swamps and marshes. A fresh supply of readily decomposable organic matter provides the energy source required by denitrifying bacteria and promotes denitrification if other conditions are appropriate.

Sometimes a high proportion of the nitrogen present in the soil is unavailable to plants. This immobilization comes about when there is an excess of carbon present. When organic substances rich in carbon but poor in nitrogen—such as straw—are abundant in the soil, the microorganisms attacking these substances will need more nitrogen than they contain in order to make full use of the carbon present. As a consequence, they use not only the limited nitrogen present in the straw or similar material but also all the soluble nitrogen salts available in the soil. Eventually, this imbalance tends to right itself as the carbon is given off as carbon dioxide by microbial respiration and the ratio of nitrogen to carbon in the soil increases.

Nitrogen Fixation

If the nitrogen that is removed from the soil were not steadily replaced, virtually all life on this planet

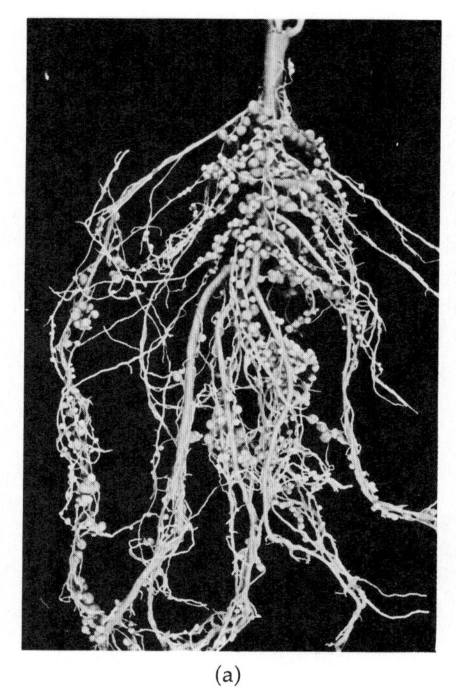

(a)

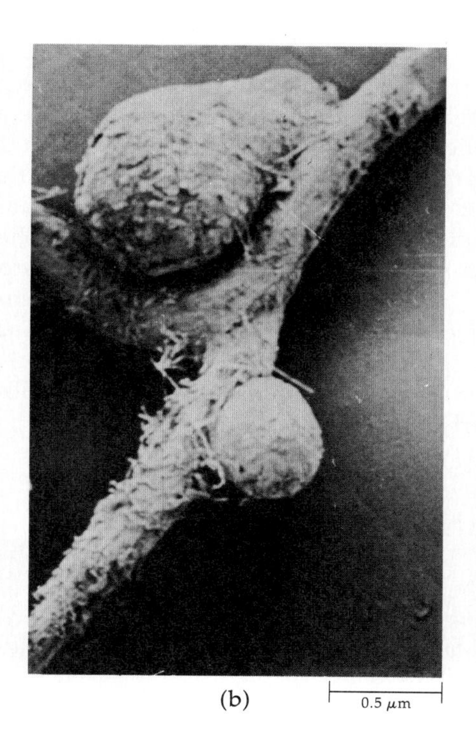

(b)

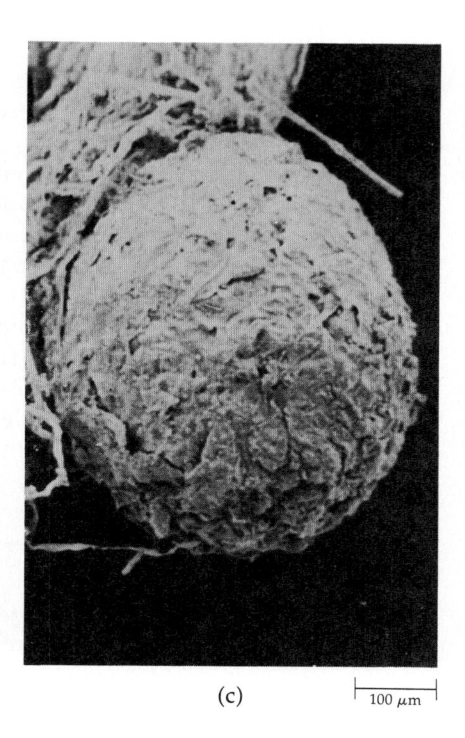

(c)

27–11
(a) *Nitrogen-fixing nodules on the roots of a soybean* (Glycine max) *plant. These nodules are the result of a symbiotic relationship between the root cells of this leguminous plant and the bacterium* Rhizobium. (b) *and* (c) *Scanning electron micrographs of nitrogen-fixing nodules on the clover* Trifolium repens.

would slowly disappear. Nitrogen is replenished in the soil by nitrogen fixation.

Nitrogen fixation is the process by which gaseous nitrogen from the air is incorporated into organic compounds and, thereby, is brought back into circulation in the nitrogen cycle. Nitrogen fixation, which can be carried out to a significant extent only by certain bacteria, is a process on which all living organisms are now dependent, just as all organisms are ultimately dependent on photosynthesis as the source of their energy.

Of the various classes of nitrogen-fixing organisms, the symbiotic bacteria are by far the most important in terms of total amounts of nitrogen fixed. The most common of the nitrogen-fixing bacteria is *Rhizobium*, a bacterium that invades the roots of leguminous plants (angiosperms of the family Fabaceae), such as alfalfa, clover, peas, and beans (see Figures 27–1 and 27–11).

The beneficial effects on the soil that are derived from growing leguminous plants have been recognized for centuries. Theophrastus, who lived in the third century B.C., wrote that the Greeks used crops of broad beans to enrich the soils. Where leguminous plants are grown, some of the "extra" nitrogen may be released into the soil, becoming available for other plants. In modern agriculture, it is common practice to rotate a nonleguminous crop, such as corn, with a leguminous crop, such as alfalfa. The leguminous plants are then either harvested for hay, leaving behind the nitrogen-rich roots, or, better still, they are simply plowed under. A crop of alfalfa that is plowed back into the soil may add as much as 300 to 350 kilograms of nitrogen per hectare of soil. As a conservative estimate, 150 to 200 million metric tons of fixed nitrogen are added to the earth's surface each year by such biological systems.

An industrial process for the chemical fixation of atmospheric nitrogen was developed in 1914. Since that time, the commercial production of fixed nitrogen has increased steadily to the current level of approximately 50 million metric tons per year. Most of this nitrogen is used in agricultural fertilizers. Industrial fixation, unfortunately, is accomplished at a high energy cost in terms of fossil fuels.

Small amounts of nitrogen can also be fixed by lightning and brought to the earth in rainfall. Rainwater sometimes also brings down ammonia that has escaped into the atmosphere. Measurements at an experimental station in England over a five-year period showed that the rainwater brought down 7.1 kilograms of nitrogen per hectare each year.

Nitrogen-Fixing Symbioses

In the symbiotic associations between *Rhizobium* and legumes, the leguminous plant supplies the bacteria with carbon compounds as an energy source for nitrogen-fixation and other metabolic activities and also provides a protective environment. The legume, in return, obtains nitrogen in a form usable for the production of plant proteins.

The *Rhizobium* bacteria (rhizobia) enter the root hairs of leguminous plants when the plants are still seedlings (see Figures 27–12 and 27–13), although the details of this process are not clearly understood. Once inside the root hair, the enlarged, invading bacteria, which are called bacteroids, induce the plant cell to produce a cellulose tube, or infection thread, through which the bacteria move from the root hair to cells of the cortical region of the root (Figure 27–13c). Infection threads within the cortical cells eventually become branched and populated with multiplying bacteria. Bacteroids enclosed in vesicles are released from the infection thread into the cytoplasm of the host cells. Infection by the bacteria stimulates repeated division of the cortical cells of the legume root to form tumorlike growths known as nodules (Figure 27–11).

LECTINS

Lectins are carbohydrate-binding proteins that are generally isolated from plant sources. They were first discovered in 1888, when extracts of castor beans were found to cause red blood cells to agglutinate (adhere to one another). This agglutination results from the multivalent nature of lectin proteins, which allows them to bind to the surface carbohydrates of more than one cell. The agglutination of red blood cells by plant proteins gave rise to the name "phytohemagglutinins." More recently, this term has been replaced by "lectin" (derived from the Latin legere, *meaning "to choose") because of the specific nature of lectin binding. Lectins from various plants display different sugar-binding specificity. For example, the lectin from the lima bean,* Phaseolus lunatus, *binds to the sugars in human blood group A antigen, causing specific agglutination of type A red blood cells, whereas the lectin of another legume,* Bandeiraea simplicifolia, *causes a similarly specific reaction of type B red blood cells. In addition to the clinical applications that arise from this specificity, lectins have been used in numerous studies of cell-surface carbohydrates during the last 20 years.*

Lectins have also been isolated from bacteria, fungi, lichens, and animals that range from invertebrates to mammals. Among the higher plants, lectins are widespread, having been identified in more than 800 species representing some 80 families. Lectins are most common in the legume family, where they often occur in high levels in the seeds. Lectins also have been identified in many other plant tissues, although few of the nonseed lectins have been characterized or studied in detail.

In the past few years, a great deal of interest has focused on the physiological role of plant lectins, and a number of interesting roles have been proposed. For example, some scientists have suggested that legume lectins may mediate the specific recognition between legume roots and bacteria of the genus Rhizobium *during the infection process leading to the nitrogen-fixing symbiosis. Other scientists have suggested that lectins play a role in defense mechanisms, protecting the plant against bacteria, fungi, insects, and herbivores.*

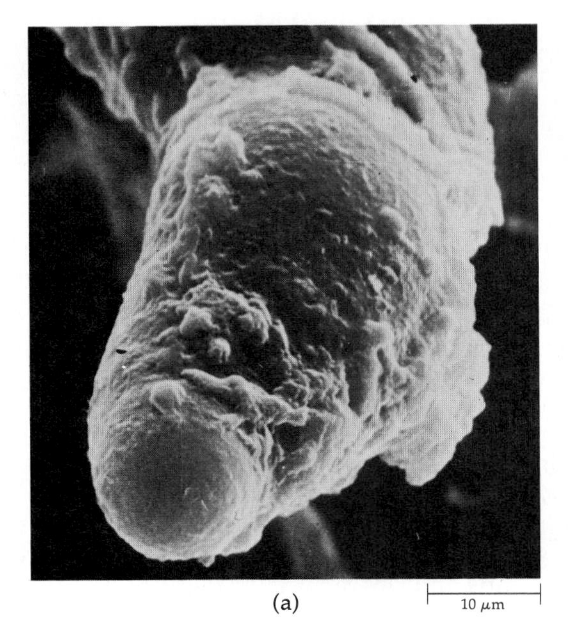

(a)

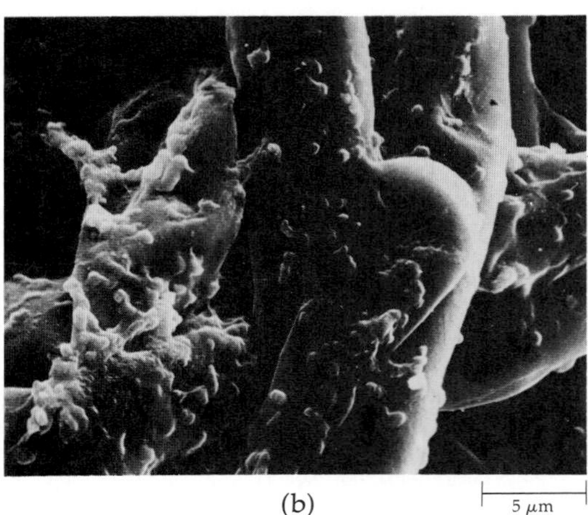

(b)

27–12
(a) *Tip of root hair of a seedling of cluster clover* (Trifolium glomeratum), *showing several rhizobia and some soil particles.* (b) *Root hairs, showing rhizobia and other soil bacteria adhering to the surfaces.*

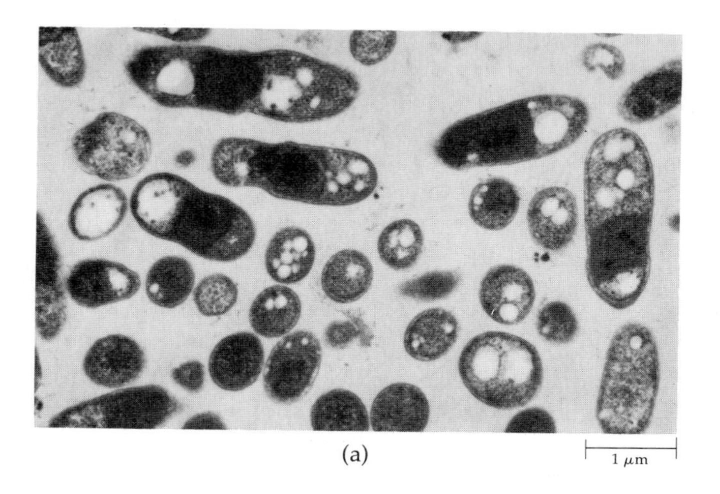

(a)

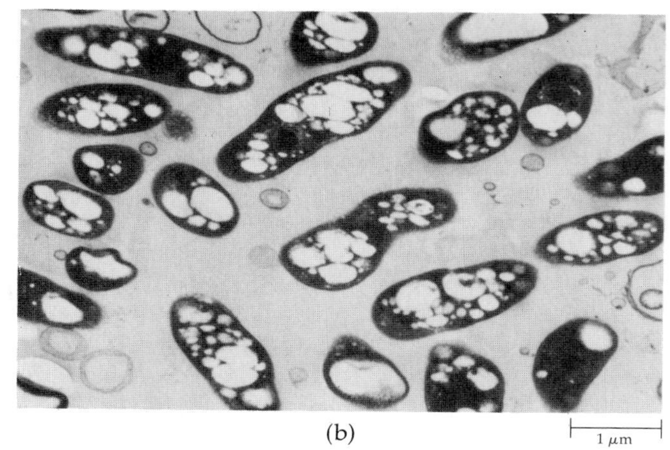

(b)

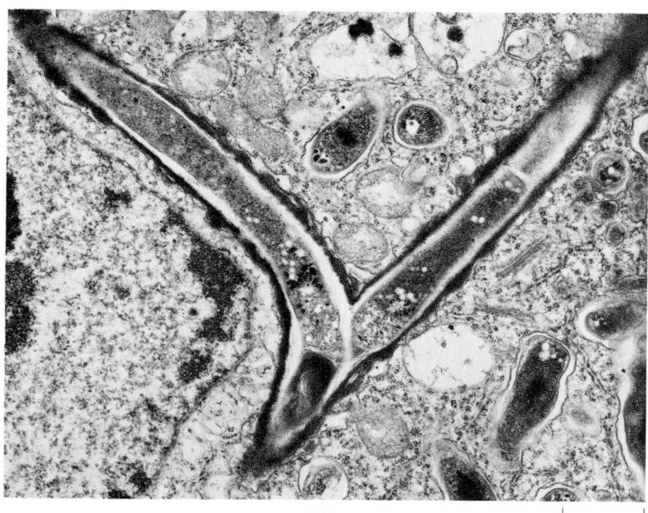

(c)

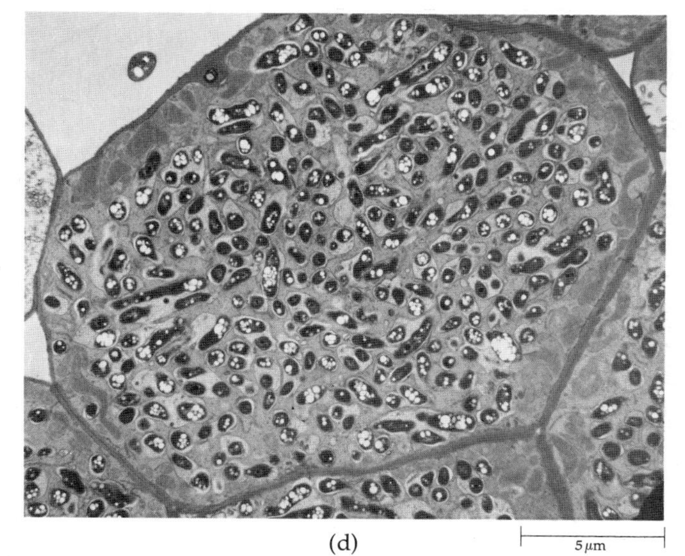

(d)

27–13
Rhizobium. (a) *Bacteria in their free-living form.* (b) *The bacteroid form that the rhizobia assume when they enter the root cells.* (c) *An infection thread.* (d) *Cross section of an infected nodule showing a cell with numerous bacteria.*

The symbiosis between a species of *Rhizobium* and a legume is quite specific; for example, bacteria that invade and induce nodule formation in clover *(Trifolium)* roots will not induce nodules on the roots of soybeans *(Glycine)*. Specific laboratory-grown strains (inocula) of *Rhizobium* are now commercially available. By mixing particular bacteria with the legume seed at the time of planting, farmers can insure that the appropriate bacteria are available in the soil to form effective associations for nitrogen fixation in the legume crop. The mechanism by which the *Rhizobium* bacteria interact with only the specific legume root is currently being investigated. The bacteria and legume may recognize each other via a plant protein (lectin) found on the root surface. Lectin binds to a polysaccharide on the surface of an effective *Rhizobium* cell but not to the surface polysaccharides of ineffective bacteria.

The legumes are by far the largest group of plants that enter into a nitrogen-fixing partnership with symbiotic bacteria. There are, however, numerous nitrogen-fixing symbioses that involve plants other than legumes. The alder tree *(Alnus)*, for example, forms nodules that are induced by and contain nitrogen-fixing actinomycetes (moldlike bacteria), rather than *Rhizobium*. Sweet gale *(Myrica gale)*, sweet fern *(Comptonia)*, and various species of *Ceanothus* also form symbiotic associations with actinomycetes.

Another symbiotic relationship is of considerable practical interest in certain parts of the world. *Azolla* is a small floating water fern, and *Anabaena* is a nitrogen-fixing cyanobacterium which lives in the pores of the *Azolla* fronds (Figure 27–14). *Azolla* infected with *Anabaena* may contain as much as 50 kilograms of nitrogen per hectare. In the Far East, for example, heavy growths of *Azolla-Anabaena* are permitted to develop on rice paddies. The rice plants eventually shade out the *Azolla*, and as the fern dies, nitrogen is released for use by the rice plants.

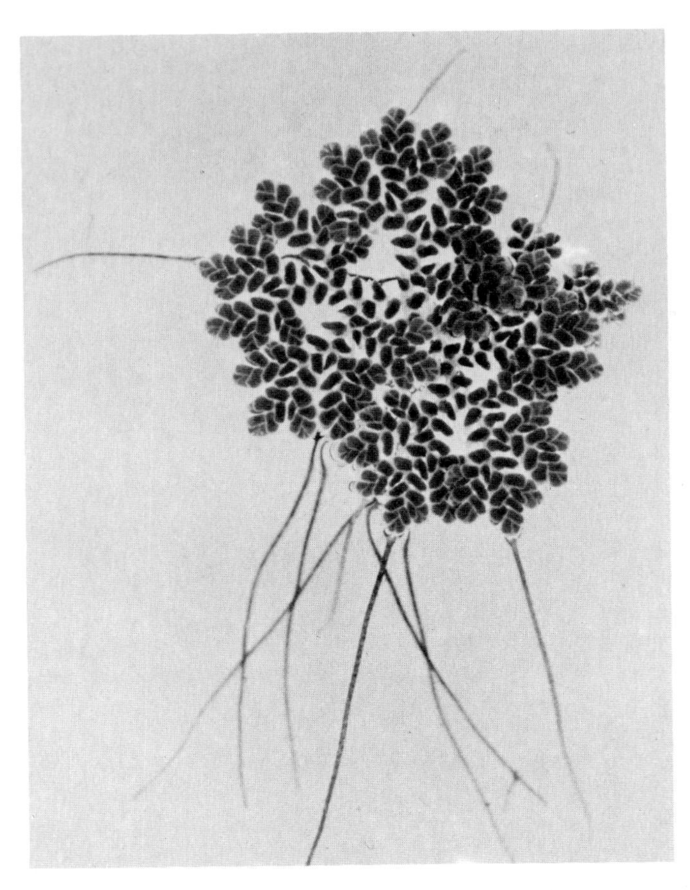

27–14
Azolla caroliniana, *a water fern that grows in symbiotic association with the cyanobacterium* Anabaena *(see Figure 11–10, page 191).*

Free-Living Nitrogen-Fixing Microorganisms

Nonsymbiotic bacteria of the genera *Azotobacter* and *Clostridium* are both able to fix nitrogen. *Azotobacter* is aerobic, and *Clostridium* is anaerobic; both are common saprophytic soil bacteria. It is estimated that they probably add about 7 kilograms of nitrogen to a hectare of soil per year. Another important group includes many photosynthetic bacteria, such as cyanobacteria.

The differences between nitrogen-fixation by free-living bacteria and nitrogen-fixation by bacteria living in symbiotic associations are not always distinct. For example, the aerobic nitrogen-fixing bacteria *Azotobacter* regularly grow around the roots of certain grasses, such as sugarcane *(Saccharum officinarum)*, where they play a significant role in providing nitrogen to the plants. The grasses (and other plants) presumably exude organic compounds that serve as an energy source for these "free-living" bacteria. Thus an association that approximates symbiosis results.

THE PHOSPHORUS CYCLE

When compared with the nitrogen cycle, the phosphorus cycle (Figure 27–15) seems relatively simple, primarily because several steps are not limited to specific groups of microorganisms. The phosphorus cycle also differs from the nitrogen cycle in that the earth's crust, rather than the atmosphere, is the primary reservoir of phosphorus. As previously discussed, the weathering of rocks and minerals over long periods of time is the source of most of the phosphorus in the soil solution.

Compared to nitrogen, the amount of phosphorus required by plants is relatively small (see Tables 27–1 and 27–2). Nevertheless, of all the elements for which the earth's crust is the primary reservoir, phosphorus is the most likely to limit plant growth. In Australia, for example, where the soils are extremely weathered and deficient in phosphorus, the distribution and limits of native plant communities are often determined by the available soil phosphate.

Phosphorus circulates from plants to animals and is returned to the soil in organic forms in residues and wastes; these organic forms of phosphorus are converted to inorganic phosphate and thus again become available to plants (Figure 27–15).

Through erosion, pollution, and loss in drainage water, large amounts of phosphorus are discharged into rivers and streams; this phosphorus eventually reaches the oceans, where it is deposited in sediments as precipitates and in the remains of organisms. In the past, the use of guano (deposits of seabird feces) as agricultural fertilizer returned some of the ocean phosphorus to terrestrial ecosystems. However, most of the phosphorus in deep-sea sediments will become available only as a result of major geological uplifts. To counter this loss, deposits of phosphate rock are being mined on a large scale for use as agricultural fertilizer.

HUMAN IMPACT ON NUTRIENT CYCLES

Normal functioning of the phosphorus, nitrogen, and other nutrient cycles requires the orderly transfer of elements between steps in a cycle in order to prevent buildup or depletion of nutrients at any one stage. Over millions of years, organisms have been provided with needed quantities of essential inorganic nutrients as a result of the normal functioning of such cycles. However, in recent years, the activities of humans have had drastic effects on some cycles, and in time this may lead to harmful accumulations and depletions of nutrients at specific stages of one or more nutrient cycles. For example, increased soil erosion has accelerated phosphorus loss from soils. Also,

27–15
The phosphorus cycle.

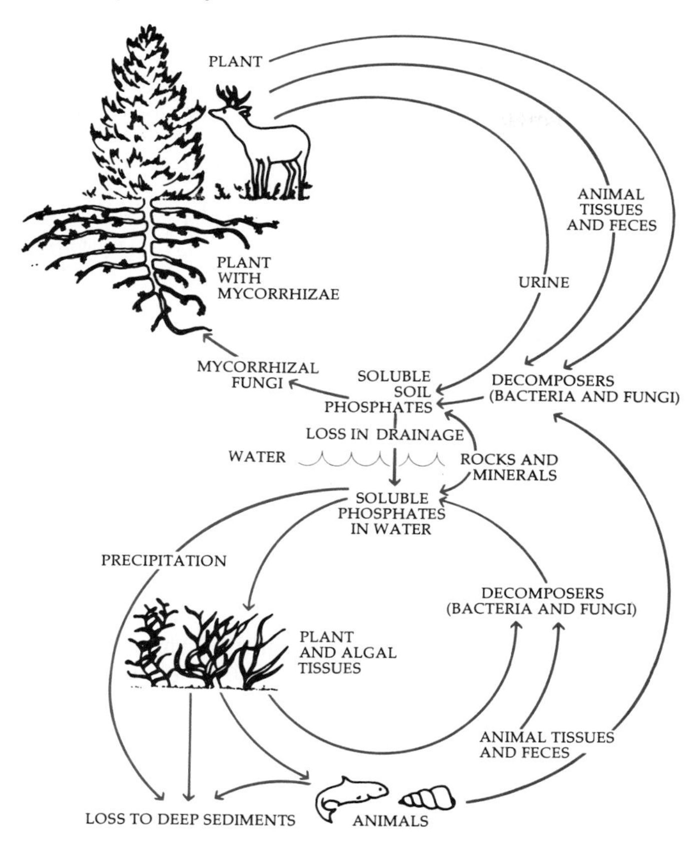

rather than recycling the phosphorus present in sewage effluents, the effluents have often been discharged into streams, and the phosphorus has thus been transported (and lost) to the oceans.

Normal functioning of the nitrogen cycle involves a balance between fixation processes that remove nitrogen from the atmospheric reservoir and denitrification reactions that return nitrogen to the atmosphere. Recently there has been massive introduction of fixed nitrogen (nitrates) into the environment through the extensive use of commercial fertilizers. Because nitrogen in the form of nitrates is readily leached from soil, increased nitrogen pollution of groundwater and lakes and streams has resulted. To add to this problem, the marshes and wetlands that are primary sites of denitrification are being destroyed at an alarming rate through their conversion to building sites, agricultural land, and dump sites.

To reduce the loss of soil nitrates by leaching, another manipulation of the nitrogen cycle has been suggested and could become common agricultural practice. This involves the application of an organic chemical which selectively inhibits the nitrifying bacteria *Nitrosomonas* for a limited period of time and then is broken down in the soil. This practice would help retain most of the fertilizer nitrogen in the ammonia form in which it is applied until the nitrogen can be incorporated by plants. As is often true of such an environmental manipulation, this possibility involves trade-offs. For example, high levels of ammonia can be quite toxic to some crop plants.

Although the quantitative data that would permit a conclusive evaluation are not yet available, there is concern that nitrous oxide (N_2O) produced during nitrification of ammonia and denitrification of nitrate may be contributing to the partial destruction of the stratospheric ozone layer which protects humans from damage by ultraviolet radiation.

SOILS AND AGRICULTURE

In natural situations, the elements present in the soil recirculate and so become available again for plant growth. As discussed previously, negatively charged clay particles are able to bind such positively charged ions as Ca^{2+}, Mg^{2+}, and K^{+}. The ions are removed from the particles by the roots of the plant, either directly or after they pass into the soil solution. In general, the cations that are required by plants are present in large amounts in fertile soils, and the amounts removed by a single crop are small. However, when a series of crops is grown on a particular field and the nutrients are continuously removed from the cycle as the crops are harvested, some of these cations (most commonly potassium) may become depleted to such an extent that fertilizers containing the missing element must be added.

Programs for supplementing nutrient supplies for agricultural and horticultural crops should be based on a bookkeeping approach that equates amounts of nutrients required to produce a specific crop and the nutrients available to the plants from all sources. Often the soil and plant residues cannot supply the required nutrients, and supplemental amounts must be provided in applications of commercial fertilizers, organic residues such as compost, or a combination of the two.

Nitrogen, phosphorus, and potassium are the three elements that are commonly included in commercial fertilizers. Fertilizers are usually labeled with a formula that indicates the percentage of each of these elements. A 10-5-5 fertilizer, for example, is one that contains 10 percent nitrogen (N), 5 percent phosphorus pentoxide (P_2O_5), and 5 percent potassium oxide (K_2O).

Other essential inorganic nutrients, although required in very small amounts, can sometimes become limiting factors in soils on which crops are grown.

COMPOST

Composting, a practice as old as agriculture itself, has recently attracted increased interest as a means of utilizing organic wastes by converting them to fertilizer. The starting product is any collection of organic matter—leaves, kitchen garbage, animal manure, straw, lawn clippings, sewage sludge, sawdust—and the population of bacteria and other microorganisms normally present. The only other requirements are oxygen and moisture. Grinding of the organic matter is not essential, but it provides greater surface area for microbial attack and thus speeds the process.

In a compost heap, microbial growth accelerates quite rapidly, generating heat, much of which is retained because the outer layers of organic matter act as insulation. In a large heap (2 meters × 2 meters × 1.5 meters, for instance), the interior temperature rises to about 70°C; in small heaps, it usually reaches 40°C. As the temperature rises, the population of decomposers changes, with thermophilic and thermotolerant forms replacing the organisms previously present. As the original forms die, their organic matter also becomes part of the product. A useful side effect of the temperature increase is that most of the common pathogenic bacteria that may have been present, for example, in sewage sludge, are destroyed, as are cysts, eggs, and other immature forms of plant and animal parasites.

With the passage of time, changes in pH *also occur in a compost heap. The initial* pH *value is usually slightly acidic (about 6), which is comparable to the liquid portion of most plant material. During the early stages of decomposition, the production of organic acids causes a further acidification, with the* pH *decreasing to a value of about 4.5 to 5.0. However, as the temperature rises, the* pH *also increases; the composted material eventually levels off at slightly alkaline values (7.5 to 8.5).*

An important factor in composting (as in any biological growth process) is the ratio of carbon to nitrogen. A ratio of about 30 to 1 (by weight) is optimal. If the carbon ratio is higher, microbial growth slows. If the nitrogen ratio is higher, some escapes as ammonia. If the compost materials are quite acidic, limestone (calcium carbonate) may be added to balance the pH*; however, if too much of this buffer is added, it will increase the nitrogen loss.*

Studies involving municipal compost piles in Berkeley, California, demonstrated that if large piles were kept moist and aerated, composting could be completed in as little as two weeks. Generally, though, three months or more during the winter is needed to complete the process. If compost is added to the soil before the composting process is complete, it may temporarily deplete the soil of soluble nitrogen.

Because it greatly reduces the bulk of plant wastes, composting can be a very useful means of waste disposal. In Scarsdale, New York, for example, leaves composted in a municipal site were reduced to one-fifth their original volume. At the same time, they formed a useful soil conditioner, improving both the aeration and water-holding capacity. Chemical analyses indicate, however, that a rich compost commonly contains, in dry weight, only about 1.5 to 3.5 percent nitrogen, 0.5 to 1.0 percent phosphorus, and 1.0 to 2.0 percent potassium, far less than commercial fertilizers. However, unlike such fertilizers, compost can be a source of nearly all the elements known to be needed by plants. Compost provides a continuous balance of nutrients, releasing them gradually as it continues to decompose in the soil.

As commercial fertilizers are becoming more expensive and less available and our waters are becoming increasingly polluted with fertilizer runoff and organic wastes, composting is an increasingly attractive alternative.

PLANT NUTRITION RESEARCH

Research on the inorganic nutrients essential for crop plants—particularly on the quantities of those nutrients required for optimal crop yields and on the capacities of various soils to provide the nutrients—has been of great practical value in agriculture and horticulture. Because of the steady increase in worldwide food needs, this type of research undoubtedly will continue to be essential.

Soil Deficiencies and Toxicities

Modification and manipulation of soils by adding nutrients in fertilizers, by raising the pH with lime, or by removing excess salts by leaching with water may not be the only means of improving and maintaining crop production in below-optimum soils. By using the knowledge and techniques of plant breeding and plant nutrition, it may be possible to select and develop cultivars of crop species that are better adapted for growth in nutrient-deficient environments than are present commercial cultivars. The validity of this research approach is confirmed by the occurrence of wild plants in nutritional environments that are very different from the average soil environments in which crop plants are grown; examples are acidic sphagnum bogs, in which the pH may be less than 4.0, and mine tailings, which often contain high concentrations of potentially toxic metals, such as zinc and nickel.

Research to develop bean *(Phaseolus vulgaris)* strains that are tolerant of potassium-deficiency stress has recently been carried out. Strains of beans were col-

27–16
Comparison of potassium-deficiency symptoms in the bean (Phaseolus vulgaris). *In the bean strain on the left, the plant is tolerant of a deficiency level of potassium, whereas the plant on the right is clearly a nontolerant strain.*

lected from around the world and grown in a nutrient solution containing less potassium than is required for optimal bean growth. All other nutrients were at optimal concentrations. The extremes in bean growth under the imposed potassium-deficiency stress are shown in Figure 27–16. Strain 58 (on the left) is nearly normal; strain 63 (on the right) shows the symptoms of severe potassium-deficiency. In another aspect of this relatively new area of research, Emanuel Epstein and his associates at the University of California at Davis have had extraordinary success in isolating barley *(Hordeum vulgare)* strains tolerant of high salt concentrations. The most tolerant strains grew reasonably well even when irrigated with sea water rather than fresh water.

Efficiency of Nitrogen Fixation

Manipulation of biological nitrogen fixation also offers tremendous potential for improved efficiency in nitrogen utilization. One aspect of research in this area is concerned with improving the efficiency of the *Rhizobium*-legume association, for example, through the genetic screening of both legumes and bacteria to identify combinations that would result in increased fixation in specific environments. This could result from a greater photosynthetic efficiency in legumes, so that more carbohydrate is available for bacterial nitrogen fixation and growth. In the pursuit of this possibility, however, it must be recognized that nitrogen fixation requires considerable energy and that any increase in fixation might be at the expense of shoot productivity.

A second research approach is to develop additional and more effective associations of free-living nitrogen-fixing bacteria and higher plants. In Brazil in the early 1970s, several types of nitrogen-fixing bacteria were found growing in association with the roots of certain tropical grasses; for example, the grass *Digitaria* was found to support populations of the bacterium *Spirillum lipoferum*. Similar associations with some of the world's major food crops, such as corn and sugarcane, have since been reported. While the practical benefits to agriculture could be tremendous, the effectiveness of associations of nitrogen-fixing bacteria and grasses such as corn is still uncertain.

Using the most sophisticated techniques of molecular biology, probably the most exciting research approach is that of genetic modifications and transfer of the genes necessary for nitrogen-fixation from one organism to another. Transfer of the appropriate genes between organisms has already been accomplished. The cluster of genes that are responsible for nitrogen-fixation was transferred from the bacterium *Klebsiella pneumoniae* to *Escherichia coli*. The transfer was accomplished by incorporating the genes in a plasmid (see page 194) and introducing the plasmid into *E. coli* cells (Figure 27–17). The genetically modified *E. coli* fixed nitrogen under some conditions.

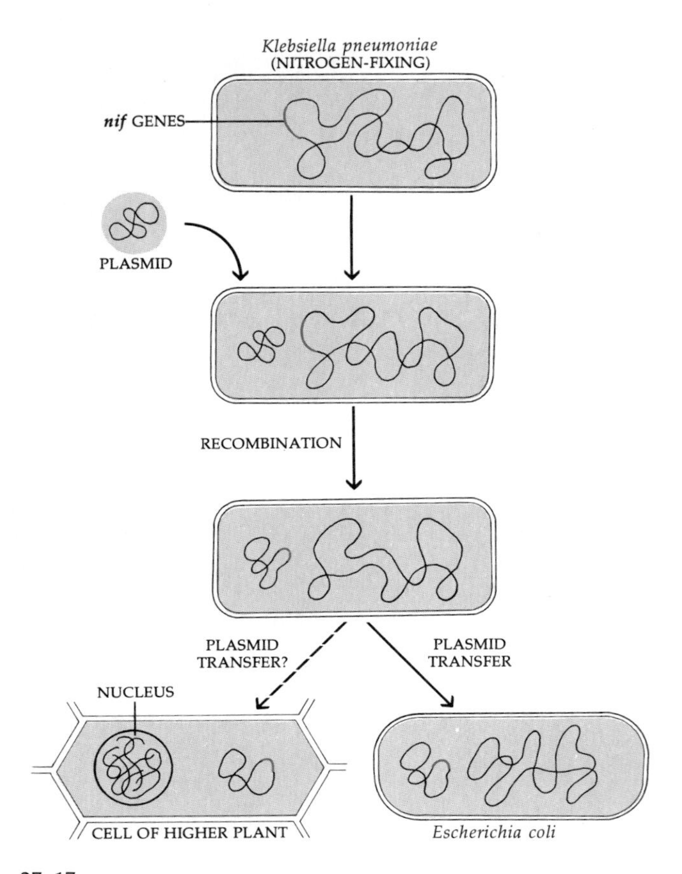

27–17
The transfer of nitrogen-fixing (nif) *genes from a nitrogen-fixing bacterium to an organism incapable of nitrogen fixation has been accomplished. The* nif *genes from* Klebsiella pneumoniae *were incorporated into a segment of extrachromosomal DNA (a plasmid) and then were transferred to* Escherichia coli, *a bacterium that cannot fix nitrogen. Although an essential initial step, gene transfer does not ensure that the recipient organism will fix nitrogen. The genetically modified* E. coli, *for example, produced the enzyme nitrogenase, which is required for nitrogen fixation, but was able to fix nitrogen only when the enzyme was experimentally protected from oxygen. That is, the "new" nitrogen-fixing abilities of the bacteria did not include the ability to shield the critical enzyme from oxidation.*

It is conceivable that nitrogen-fixing genes could be transferred to nonfixing species such as corn, but this involves a number of problems beyond transfer of the genes, such as the protection of the nitrogen-fixing enzyme from oxygen which inhibits fixation. Consequently, research workers in this field are not optimistic that an effective transfer of this type will be easily accomplished.

Effects of Pollution

The toxic effects of various inorganic agents discharged into the environment as pollutants also are of current research interest. Crop plants, for example, may be adversely affected by heavy metals such as copper and cadmium. Aquatic ecosystems, in particular, have been damaged because they have become common disposal sites for industrial and municipal wastes. The introduction of nitrogen and phosphorus—the primary eutrophication nutrients—into freshwater ecosystems has resulted in massive growths of algae and aquatic flowering plants, thus seriously reducing the recreational value of affected lakes and streams.

Both terrestrial and, particularly, aquatic ecosystems are subject to widespread environmental damage because of acid rain. Acid rain results from the interaction of sulfur dioxide and oxides of nitrogen—derived principally from the combustion of fossil fuels—with atmospheric moisture to form sulfuric and nitric acids. These acids impart a high degree of acidity to the rainfall. In parts of the Scandinavian countries, in the northeastern and upper midwestern United States and in southeastern Canada, the pH of rainwater is commonly in the range of 4.0 to 4.5 and on occasion is below 4.0. In contrast, the pH of rainwater in equilibrium with carbon dioxide in an unpolluted atmosphere is approximately 5.6. Acid rain can have adverse effects on plants, the weathering of rocks and minerals, the solubilities of potentially toxic metals in the environment, and even human health. Hundreds of soft-water lakes in the United States, Canada, and Scandinavia are being affected. These lakes frequently are found at high altitudes, such as in the Adirondack Mountains of northern New York, and they lack the buffering provided by bicarbonates and carbonates that neutralize acid rain in hard-water lakes. The increased acidities now developing in hundreds of such lakes are severely affecting the reproduction of game fish.

SUMMARY

A total of 16 inorganic nutrients are required by most plants for normal growth. Of these, carbon, hydrogen, and oxygen are derived from air and water. The rest are absorbed by roots in the form of ions. These 13 elements are categorized as macronutrients or as micronutrients. The macronutrients are nitrogen, potassium, calcium, phosphorus, magnesium, and sulfur. The micronutrients are iron, chlorine, copper, manganese, zinc, molybdenum, and boron. Some inorganic nutrients, such as sodium and cobalt, are essential only for specific organisms (Table 27–1).

Inorganic nutrients perform a number of important roles in cells. They regulate osmosis and affect cell permeability. Some also serve as structural components of cells, as components of critical metabolic compounds, and as activators and components of enzymes (Table 27–2).

The chemical and physical properties of soils are critical in determining their capabilities to provide the inorganic nutrients, water, and the conditions necessary for maximum crop plant production. The weathering of rocks and minerals provides fragments that represent the inorganic component of soils. All of the inorganic nutrients except nitrogen are derived from weathering processes. In addition, soils contain organic matter and pore space occupied by varying proportions of water and gases. Under agricultural conditions, nitrogen, phosphorus, and potassium are the nutrients most often limiting to plant growth and most frequently added to soils in fertilizers.

Each essential inorganic nutrient is circulated in a complex cycle among organisms within ecosystems and between those organisms and environmental reservoirs of the nutrient. The circulation of nitrogen through the soil, through the bodies of plants and animals, and back to the soil again is known as the nitrogen cycle. Nitrogen reaches the soil in the form of organic material of plant and animal origin. These substances are decomposed by soil organisms. Ammonification—the release of ammonium (NH_4^+) ions from nitrogen-containing compounds—is carried out by soil bacteria and fungi. Nitrification is the oxidation of ammonium ions to form nitrites and nitrates; these two processes are carried out by two different types of bacteria. Nitrogen enters plants almost entirely in the form of nitrates. Within the plants, nitrates are reduced to ammonium ions. Amino acids are formed from reactions that produce glutamic acid (amination), or by the transfer of an amino group from one amino acid to a keto acid producing another amino acid (transamination). These organic compounds are eventually returned to the soil, completing the nitrogen cycle.

Nitrogen is lost from the soil by crop removal, erosion, fire, leaching, and the action of denitrifying bacteria. Nitrogen is added to the soil by nitrogen fixation, which is the incorporation of elemental nitrogen into organic components. Biological nitrogen fixation is carried out entirely by prokaryotic organisms. These include bacteria *(Rhizobium)*, which are symbionts of leguminous plants, free-living bacteria, and actinomycetes in symbiotic relationship with a number of plants other than legumes. In agriculture, plants are removed from the soil. As a consequence, nitrogen and other elements are not recycled, as they are in nature; therefore, they must be replenished in either an organic form or an inorganic form.

CHAPTER 28

The Movement of Water and Solutes in Plants

28–1
Most of the water taken up by plants is transpired; that is, it is lost by evaporation from the plant body. Only small fractions are retained. For example, corn (Zea mays) *transpires more than 98 percent of the water it absorbs.*

The ability of the plant to transport both organic and inorganic nutrients, including water, throughout the plant body is critical in determining the ultimate structure and function of its component parts, as well as the development and form of the plant as a whole. In the first part of this chapter, we examine the movement of water and solutes through the plant body from the soil to the aerial plant parts. Then, toward the end of the chapter, we consider the movement of solutes and water from the sites of photosynthesis to the nonphotosynthetic parts of the plant body. We begin with a description of transpiration because this process is a major determining factor in the movement of water through the plant body.

MOVEMENT OF WATER THROUGH THE PLANT BODY

Transpiration

In the early eighteenth century, Stephen Hales noted that plants "imbibe" a much greater amount of water than animals. He calculated that one sunflower plant, bulk for bulk, "imbibes" and "perspires" 17 times more water than a human every 24 hours. Indeed, the total quantity of water absorbed by any plant is enormous—far greater than that used by any animal of comparable weight. An animal uses less water because much of its water is recirculated through its body over and over again, in the form (in vertebrates) of blood plasma and other fluids. In plants, more than 90 percent of the water taken in by the roots is released by the plant into the air as water vapor (Table 28–1). This process is known as *transpiration*, which is defined as the loss of water vapor by any part of the plant body, although leaves are by far the principal organs of transpiration.

Why do plants lose such large quantities of water to transpiration? This question can be answered by re-

Table 28–1 *Water Loss by Transpiration in One Plant in a Single Growing Season*

PLANT	WATER LOSS (LITERS)
Cowpea *(Vigna sinensis)*	49
Potato *(Solanum tuberosum)*	95
Wheat *(Triticum aestivum)*	95
Tomato *(Lycopersicon esculentum)*	125
Corn *(Zea mays)*	206

After J. F. Ferry, *Fundamentals of Plant Physiology,* The Macmillan Company, New York, 1959.

viewing the structure of the leaf. The chief function of the leaf is photosynthesis, which is the source of all the food for the entire plant body. The necessary energy for photosynthesis comes from sunlight. Therefore, for maximum photosynthesis, a plant must spread a maximum surface to the sunlight. But sunlight is only one of the requirements for photosynthesis; the chloroplast also needs carbon dioxide. Under most circumstances, carbon dioxide is readily available in the air surrounding the plant, but in order for carbon dioxide to enter the plant cell, which it does by diffusion, it must go into solution, because the plasma membrane is nearly impervious to the gaseous form of carbon dioxide. Hence, the gas must come into contact with a moist cell surface. However, wherever water is exposed to air, evaporation occurs. Plants have developed a number of special adaptations that limit evaporation, but all of these cut down the supply of carbon dioxide. In other words, the uptake of carbon dioxide for photosynthesis and the loss of water by transpiration are inextricably bound together in the life of the green plant.

Absorption of Water by Roots

The root system serves to anchor the plant in the soil and, above all, to meet the tremendous water requirements of the leaves. Almost all the water that a plant takes from the soil enters through the younger parts of the root. Absorption takes place directly through the epidermis of the root. The root hairs, located several millimeters above the root tip, provide an enormous area for absorption (Figure 28–2; Table 28–2). From the root hairs, the water moves through the cortex, the endodermis (the inner layer of cortical cells), and the pericycle, and into the primary xylem. Once in the conducting elements of the xylem, the water moves upward through the root and stem and into the leaves.

Table 28–2 *Number of Root Hairs per Square Centimeter of Root Surface in Three Plant Species*

PLANT	ROOT HAIR DENSITY
Loblolly pine *(Pinus taeda)*	217
Black locust *(Robinia pseudo-acacia)*	520
Rye *(Secale cereale)*	2500

After J. F. Ferry, *Fundamentals of Plant Physiology,* The Macmillan Company, New York, 1959.

(a)

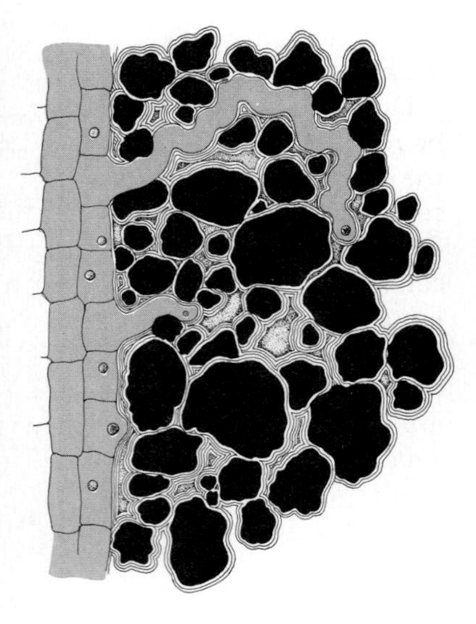

(b)

28–2
(a) *A primary root of radish* (Raphanus sativus) *seedling, showing root hairs.* (b) *Root hairs surrounded by soil particles with water adhering to them.*

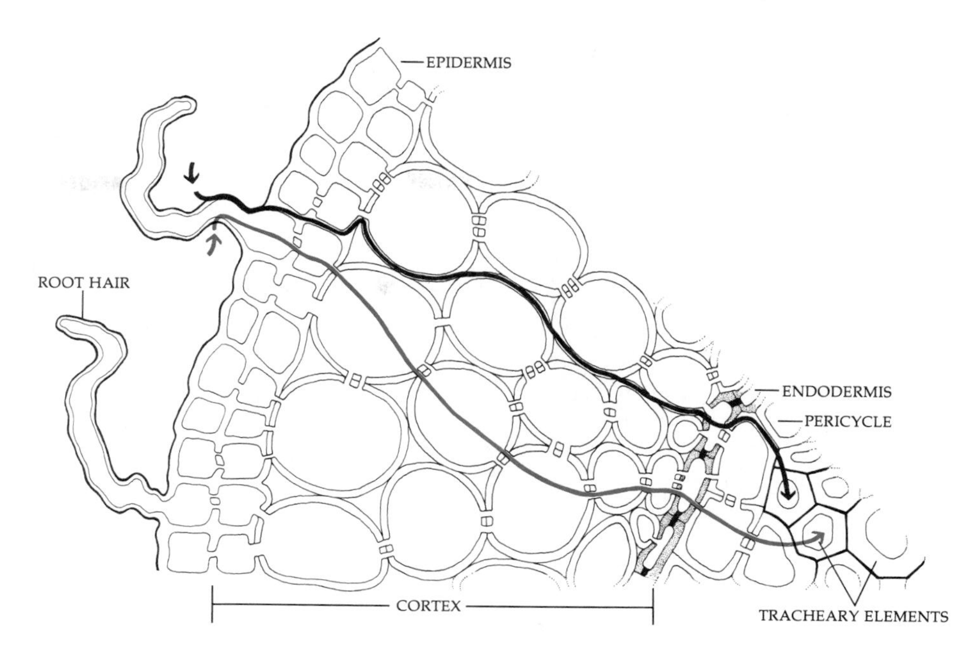

28–3
Principal pathways for the movement of water (black line) and inorganic ions (colored line) from the soil, across the epidermis and cortex, and into the tracheary elements, or water-conducting elements, of the root. The water follows a mostly apoplastic pathway until it reaches the endodermis, where apoplastic movement is blocked by the Casparian strips. The Casparian strips force the water to cross the plasma membranes and protoplasts of the endodermal cells on its way to the xylem. Having crossed the plasma membrane on the inner surface of the endodermis, the water may once again enter the apoplastic pathway and make its way into the lumina of the tracheary elements. The inorganic ions are actively absorbed by the epidermal cells and then follow a symplastic pathway across the cortex and into parenchyma cells from which they are secreted into the tracheary elements.

The pathway followed by water and solutes in the plant may be *apoplastic* (that is, via the cell walls) or *symplastic* (from protoplast to protoplast via plasmodesmata) or a combination of the two. The main pathway for water across the epidermis and cortex of the root is apoplastic (Figure 28–3). At the endodermis, however, the water is forced to traverse the plasma membranes and protoplasts of the tightly packed endodermal cells because of the presence of the water-impermeable Casparian strip in their radial and transverse walls (see page 438). The endodermis, therefore, forms an osmotic barrier between the cortex and the vascular cylinder of the root.

During periods of rapid transpiration, water may be removed from around the root hairs so quickly that the soil becomes depleted; water will then move from some distance away toward the root hairs through fine pores in the soil. In general, however, the roots come into contact with additional water by growing. For example, under normal conditions, roots of apple trees grow an average of about 3 to 9 millimeters a day; roots of prairie grasses may grow more than 13 millimeters a day; and the main roots of corn plants average 52 to 63 millimeters a day. The results of such rapid growth can be remarkable: a four-month-old rye plant has over 10,000 kilometers of roots and many billions of root hairs.

The *transpiration stream*, in addition to keeping the shoot provided with water, distributes inorganic ions to the shoot as well (Figure 28–4). After ions are absorbed by the outer cells of the root, they are transferred across the cortex and finally are secreted into the xylem. When transpiration is occurring, the ions are carried rapidly throughout the plant.

Root Pressure and Guttation

When transpiration is very slow or absent, as it is at night, the root cells may still secrete ions into the xylem. Because the vascular tissue of the root is surrounded by the endodermis, ions do not tend to leak back out of the xylem. Therefore, the water potential (page 59) of the xylem becomes more negative, and water moves into the xylem by osmosis through the surrounding cells. In this manner a positive pressure, called *root pressure*, is created, and it forces both water and dissolved ions up the xylem (Figure 28–5a).

Dewlike droplets of water at the tips of grass leaves in the early morning demonstrate the effects of root pressure. These droplets are not dew—which is water that has condensed from the air—but come from within the leaf by a process known as *guttation*. They exude not through the stomata but through special openings called hydathodes, which occur at the tips and margins of leaves (Figure 28–5b). The water of guttation is literally forced out of the leaves by root pressure.

Root pressure is least effective during the day, when the movement of water through the plant is the fastest, and the pressure never becomes high enough to force water to the top of a tall tree. Moreover, many plants, such as pines, develop no root pressure at all.

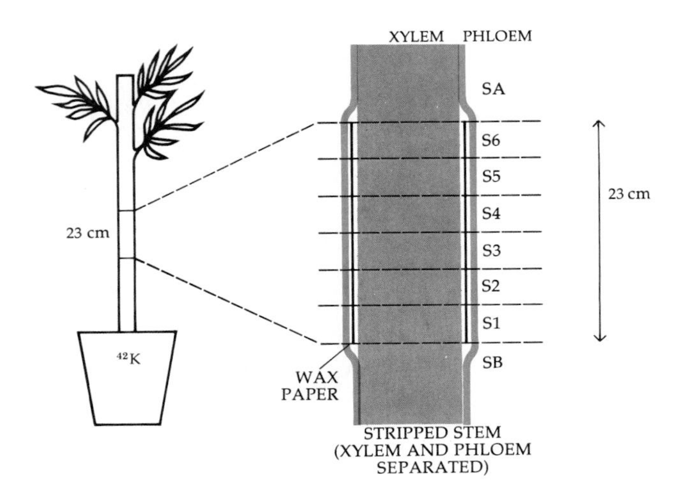

		ppm ^{42}K IN PHLOEM	ppm ^{42}K IN XYLEM
ABOVE STRIP	SA	53	47
STRIPPED SECTION	S6	11.6	119
	S5	0.9	122
	S4	0.7	112
	S3	0.5	98
	S2	0.3	108
	S1	20	113
BELOW STRIP	SB	84	58

28–4
This demonstration, using radioactive potassium (^{42}K) added to the soil water, shows that the xylem is the channel for the upward movement of both water and inorganic ions. Wax paper was inserted between the xylem and the phloem to prevent lateral transport of the isotope. A comparison of the relative amounts of radioactive potassium detected in each segment of the stem is given in the table at the right.

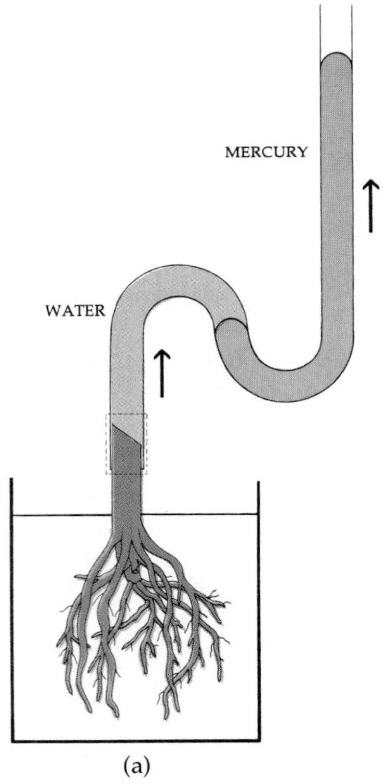

(a)

(b)

28–5
(a) *Demonstration of root pressure in the cut stump of a plant. Uptake of water by the plant roots causes the mercury to rise in the column. Pressures of 3 to 5 bars have been demonstrated by this method.*

(b) *Guttation droplets at the leaf tips of barley* (Hordeum vulgare) *also demonstrate the presence of root pressure. These droplets are not condensation from water vapor in the surrounding air; rather, they are forced out of the leaf through special openings, called hydathodes, at the leaf tips.*

Thus, root pressure can perhaps be regarded as a by-product of the mechanism of pumping ions into the xylem and as a subsidiary means of moving water into the shoot under special conditions.

Passive Water Absorption

During periods of high transpiration rates, ions accumulated in the xylem of the root are swept away in the transpiration stream, and the amount of osmotic movement across the endodermis decreases. At such times, the roots become passive absorbing surfaces through which water is pulled by bulk flow generated in the transpiring shoots. Some investigators believe that practically all of the absorption of water by the roots of transpiring plants occurs in this passive manner.

Water Transport

Water enters the plant by the roots and is given off, in large quantities, by the leaf. How does the water get from one place to another, often over large vertical distances? This question has intrigued many generations of botanists.

The general pathway that the water follows in its ascent has been clearly identified. You can trace this pathway in a simple experiment. Place a cut stem in water that is colored with any harmless dye (preferably, cut the stem under the water to prevent air from entering the conducting elements of the xylem) and then trace the path of the liquid into the leaves. The stain quite clearly delineates the conducting elements of the xylem. Recent experiments using radioactive isotopes confirm that the isotope and, presumably, the water do indeed travel by way of vessel elements (or tracheids) in the xylem. In the experiment shown in Figure 28–4, care had to be taken to separate the xylem from the phloem. Earlier experiments in which this separation was not made produced ambiguous results, because there is a great deal of lateral movement from the xylem into the phloem. This lateral movement, however, as the experiment shows, is not necessary for the overall movement of water and minerals from soil to leaf.

Hence, this is the path that water takes, but how does the water move? Logic suggests two possibilities: it can be pushed from the bottom or pulled from the top. (A third possibility, involving active pumps, or "hearts," along the way has been proposed from time to time but is no longer seriously considered by botanists.) The first of these possibilities has already been eliminated, however. Root pressure, as noted previously, does not exist in all plants, and in those plants in which it is present, it is not sufficient to push water to the top of a tall tree. Moreover, the simple experiment just described (that involving the cut stem) rules out root pressure as a crucial factor. So we are left with the hypothesis that water is pulled up through the plant body, and this hypothesis is correct according to all present evidence.

The Cohesion-Adhesion-Tension Mechanism

When water evaporates from the wall surfaces bordering the intercellular spaces in the interior of a leaf during transpiration, it is replaced by water from within the cell. This water diffuses across the plasma membrane, which is freely permeable to water but not to the solutes of the cell. As a result, the concentration of solutes within the cell increases, and the water potential of the cell decreases. A gradient of water potential then becomes established between this cell and adjacent, more saturated cells. These cells, in turn, gain water from other cells until eventually this chain of events reaches a vein and exerts a "pull" or tension on the water of the xylem. Because of the extraordinary cohesiveness of water, this tension is transmitted all the way down the stem to the roots, so that water is withdrawn from the roots, pulled up the xylem, and distributed to the cells that are losing water to the atmosphere (Figure 28–6). However, this loss makes the water potential of the roots more negative, thus increasing their ability to extract water from the soil.

This theory of water movement is known as the cohesion-tension theory, because it depends on the cohesiveness of water, which permits it to withstand tension (Figure 28–7). However, the theory might

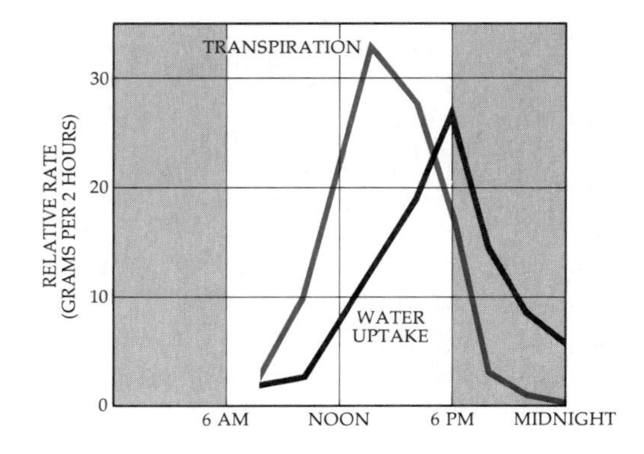

28–6
Measurements of water movement in ash (Fraxinus) *trees shows that a rise in water uptake follows a rise in transpiration. These data suggest that the loss of water generates the force needed for its uptake.*

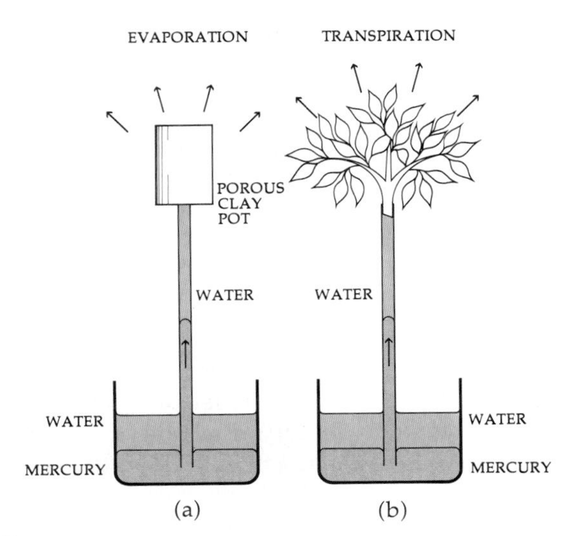

28–7
(a) *A simple physical system that demonstrates the cohesion-adhesion-tension theory. A porous clay pot is filled with water and attached to the end of a long, narrow glass tube that is also filled with water. The water-filled tube is placed with its lower end below the surface of a volume of mercury contained in a beaker. Water evaporates from the pores in the pot and is replaced by water "pulled up" through the tube in a continuous column. As the water rises, mercury is pulled up into the tube to replace it.* (b) *Transpiration from leaves results in sufficient water loss to create a similar negative pressure. Consequently, the cohesion-adhesion-tension theory is often called the "transpiration-pull theory."*

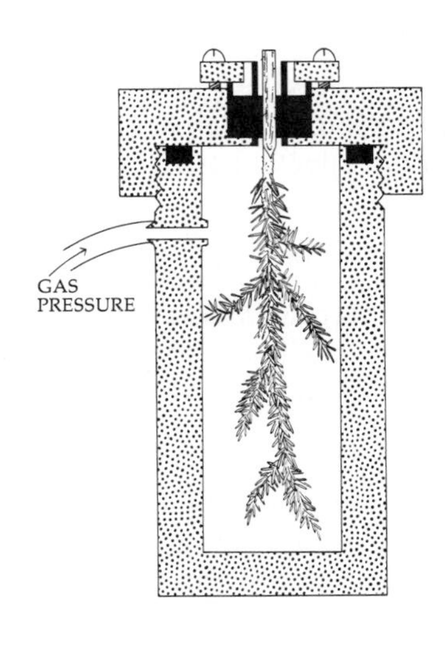

28–8
Measuring the water tension in the xylem. A branch whose xylem tension the investigator wishes to measure is cut off and placed in a pressure "bomb." When the branch is cut, some of the xylem sap—which was under tension before the branch was cut—recedes into the xylem below the cut surface. Pressure is raised in the bomb until sap emerges from the cut end of the stem. Presuming that equal pressure is required to force the sap in either direction, the positive pressure needed to force out the sap is ideally equal to the tension that existed in the branch before it was severed.

better be known as the *cohesion-adhesion-tension theory*, because adhesion of the water molecules to the walls of the tracheids and vessels of the xylem and to the cell walls of the leaf and root cells is just as important for the rise of water as cohesion and tension. The cell walls along which the water moves have evolved as a very effective water-attracting surface, taking maximal advantage of water's adhesiveness and thus providing a situation in which cohesiveness is readily expressed. Moreover, the small diameter of the xylem conduits—vessels and tracheids—through which the water moves contributes to the reliability of the system.

There is no doubt that the tensile strength of water is great enough to prevent the pulling apart of water molecules under the tension required to move water up the xylem of tall trees. For example, it has been demonstrated that a column of water in a fine capillary tube is capable of withstanding a tension of −264 bars; the estimated tension required to move water to the top of a giant redwood *(Sequoia sempervirens)* is only about −20 bars.

How can the cohesion-adhesion-tension theory be tested? One way to test this theory directly is by measuring the tension of water within the xylem. When a twig is cut from a transpiring tree, the columns of water in the vessels abruptly recede below the cut surfaces. By mounting the twig in a pressure chamber, as shown in Figure 28–8, it is possible to apply pressure to the leaves until the curved upper surface of the water columns appear (when magnified) at the cut surface of the twig. The magnitude of the pressure required to return the water to the cut surface is equal to the magnitude of the tension under which the water existed in the twig before its excision. Results obtained by this method are entirely consistent with the predictions of the cohesion-adhesion-tension theory.

A second set of data that is in accord with the cohesion-adhesion-tension theory indicates that the movement of water begins at the top of the tree. The velocity of sap flow in various parts of the tree has been measured by an ingenious method involving a small heating element to warm the xylem contents for a few seconds and a sensitive thermocouple to detect the moment at which the heated xylem sap moves past a specific point (see Figure 28–9). As shown in the graph, in the morning the sap begins to flow first in the twigs, as tension arises close to the leaves, and later in the trunk. In the evening, the flow diminishes first in the twigs, as water loss from the leaves diminishes, and later in the trunk.

A third set of supporting data comes from measurements of minute changes in the diameter of the tree trunk (Figure 28–10). The shrinking of the trunk occurs because of the negative pressures in the water passages of the xylem. The water molecules clinging to the sides of the vessels pull them inward. When transpiration begins in the morning, first the upper part of the stem shrinks, as water is pulled out of the xylem before it can be replenished from the roots; then the lower part shrinks. Later in the day, as the transpiration rate decreases, the upper trunk expands before the lower trunk does.

Note that the energy for the evaporation of water molecules—and thus for the movement of water and inorganic nutrients through the plant body—is supplied not by the plant but directly by the sun. Note also that the movement is possible because of the extraordinary cohesive and adhesive properties of water to which the plant is so exquisitely adapted.

Regulation of Transpiration

As carbon dioxide—which is absolutely essential for photosynthesis—enters a leaf, water vapor is lost to the atmosphere by transpiration, a phenomenon that can be extremely injurious to the plant. These two processes are not separable, at least not by any evolutionary solution yet devised by a plant. However, a number of special adaptations exist that minimize water loss while optimizing carbon dioxide gain.

Cuticle and Stomata

Leaves are covered by a cuticle that makes the surface of the leaf largely impervious both to water and to carbon dioxide. Only a small fraction of the water transpired by plants is lost through this protective outer coating, and another small fraction is lost through the lenticels in the bark. By far the largest amount of water transpired by a higher plant is lost through the stomata (Figure 28–11). Stomatal transpiration involves two steps: (1) evaporation of water from cell wall surfaces bordering the intercellular spaces, or air spaces, of the leaf, and (2) diffusion of the resultant water vapor from the intercellular spaces into the atmosphere by way of the stomata (see Figure 28–14).

Stomata are small openings in the epidermis surrounded by two guard cells, which can change their shape to bring about the opening and closing of the pores (Figure 28–12). Stomata are also found on

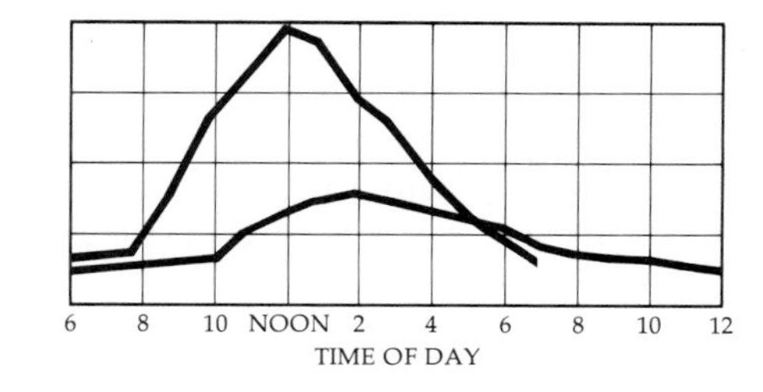

28–9
A method for measuring velocity of sap flow. A small heating element inserted into the xylem heats the ascending sap for a few seconds. A thermocouple above the heating element records the passing wave of heat. The experimenter times the interval between these two events. As shown in the graph, in the morning the sap begins to increase its velocity of flow first in the twigs (upper curve) and then in the trunk (lower curve). In the evening, velocity diminishes first in the twigs and then in the trunk.

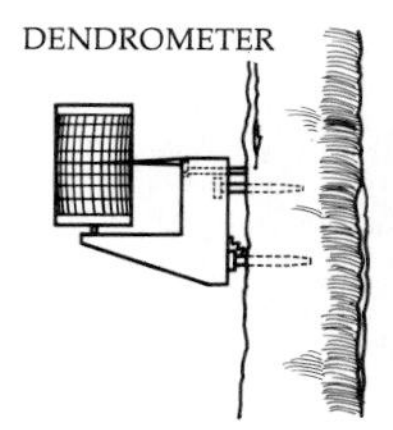

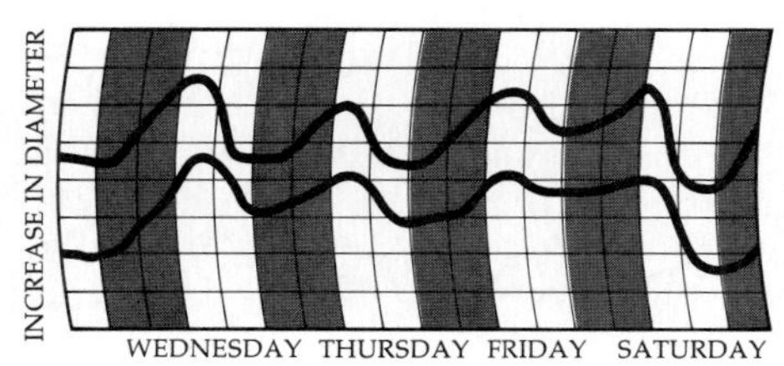

28–10
A dendrometer (left) records small daily fluctuations in the diameter of a tree trunk. Simultaneous measurements at two different heights indicate, as shown in the graph, that in the morning shrinkage occurs in the upper trunk slightly before it occurs in the lower. These data suggest that transpiration from the leaves "pulls" water out of the trunk before it can be replenished from the roots. The shaded strips signify nighttime.

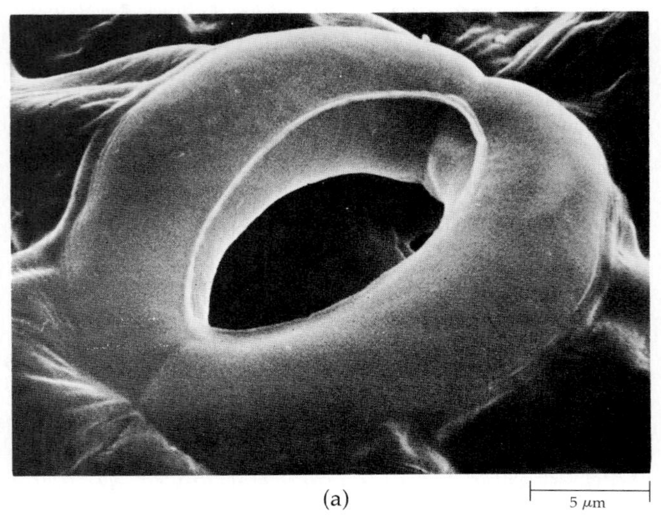

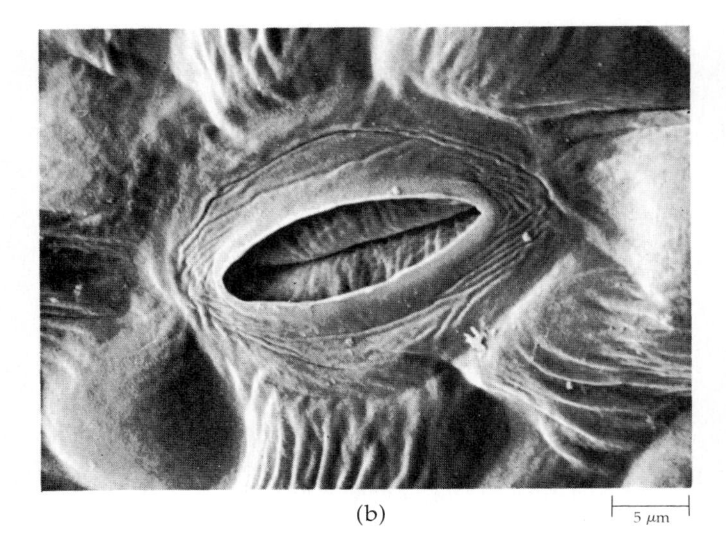

28–11
Scanning electron micrographs showing (a) *open stoma in epidermis of cucumber* (Cucumis sativus) *leaf, and* (b) *closed stoma in epidermis of parsley* (Apium petroselinum) *leaf. The stomata lead into a honeycomb of air spaces that surround the thin-walled, photosynthetic mesophyll cells within the leaf. The air spaces are saturated with water vapor that has evaporated from the surfaces of the mesophyll cells.*

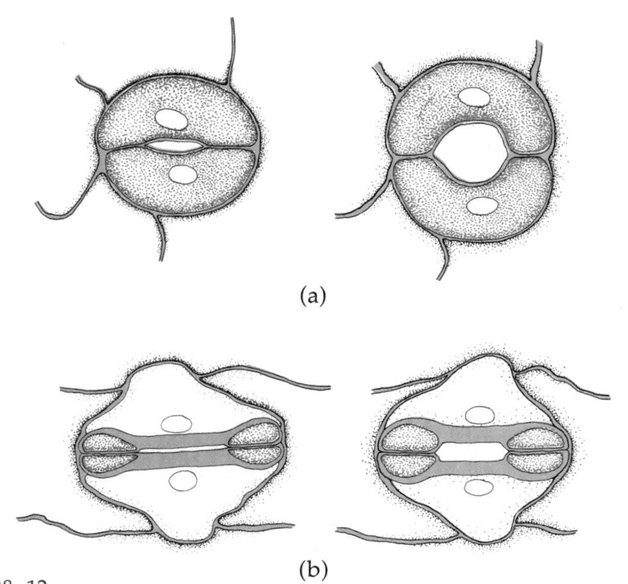

28–12
The mechanisms of stomatal movements. (a) *In many species, the guard cells are kidney-shaped. When the guard cells are turgid, the inflated ends of the cells push them apart and the stoma opens. The walls appear to expand along the curved axes of the cells. The guard cells come together when they lose turgor.* (b) *In the grasses, the guard cells are dumbbell-shaped. Here again, inflation of the ends of the guard cells pushes the cells apart and the stoma opens. The inflatable ends of the dumbbell-shaped guard cells are separated by a rigid middle portion. Both* (a) *and* (b) *show two stages of opening.*

young stems, but they are far more abundant on leaves. The number of stomata may be quite large; for example, there are approximately 12,000 stomata per square centimeter of leaf surface in tobacco leaves. The stomata lead into a honeycomb of air spaces within the leaf that surround the thin-walled mesophyll cells. The air in these spaces—which make up 15 to 40 percent of the total volume of the leaf—is saturated with water vapor that has evaporated from the damp surfaces of the mesophyll cells. Although the stomatal openings take up only about 1 percent of the total leaf surface, more than 90 percent of the water transpired by the plant is lost through the stomata. The rest is lost through the cuticle.

Closing of stomata not only prevents the loss of water vapor from the leaf but, as mentioned, also prevents the entry of carbon dioxide into the leaf. A certain amount of carbon dioxide, however, is produced by the plant during respiration, and as long as light is available, this carbon dioxide can be used to sustain a very low level of photosynthesis even when the stomata are closed.

The Mechanism of Stomatal Movements

Stomatal movements result from changes in turgor pressure within the guard cells. Opening occurs when solutes are actively accumulated in the guard cells. The accumulation of solutes results in a movement of water into the guard cells and a buildup of turgor pressure in excess of that in the surrounding epidermal cells. Stomatal closing is brought about by the reverse process: with a decline in guard-cell solutes, water moves out of the guard cells and the turgor pressure decreases. Thus, turgor is maintained or lost due to the passive osmotic movement of water into or out of the cells along a gradient of water potential.

The major solute responsible for these gradients in water potential is the potassium ion (K^+). This ion has been found in the guard cells of open stomata of more than 50 species, including CAM plants whose stomata open at night. Techniques for estimating potassium levels within a single guard cell show that potassium levels rise when the stomata open and drop when the stomata close. The surrounding cells provide the required reservoir of potassium ions. The gradient of potassium between the guard cells and surrounding cells changes significantly, accompanied by the osmotic flow of water and resultant turgor changes.

With the positively charged K^+ transported in such large amounts, negatively charged ions (anions) are needed to counter the charge. Two anions in particular have been implicated in this regard—chloride and malate.

Factors Affecting Stomatal Movements

A number of environmental factors affect stomatal opening and closing, water loss being the major influence. When the turgor of a leaf drops below a certain critical point, which varies with different species, the stomatal opening becomes smaller. The effect of water loss overrides other factors affecting the stomata, but stomatal changes can occur independently of overall water gain or loss by the plant. The most conspicuous example is found in the many species in which the stomata open regularly in the morning and close in the evening, even though there may be no changes in water available to the plant.

During periods of water stress in many plants there is a marked increase in the level of abscisic acid (ABA). When "fed" or applied to leaves, ABA causes stomatal closure within a few minutes; moreover, the effect of ABA on stomatal movement is readily reversible. Experimental evidence suggests that solute loss from guard cells begins when ABA of mesophyll origin arrives at the stomata, signaling the stomata that the mesophyll cells are experiencing water stress. How ABA acts on guard cells remains to be determined.

Other factors that affect stomatal movement include carbon dioxide concentration, light, and temperature. In most species, an increase in CO_2 concentration causes the stomata to close. The magnitude of this response to CO_2 varies greatly from species to species and with the degree of water stress a given plant has suffered or is suffering. In corn *(Zea mays)*, the stomata may respond to changes in CO_2 in a matter of seconds. The site for sensing the level of CO_2 has been demonstrated to be located within the guard cells.

In most species, the stomata open in the light and close in the dark. This can be explained in part by the photosynthetic utilization of CO_2 which brings about a reduction in the CO_2 level within the leaf. Light may, however, have a more direct effect on stomata. Blue light has long been known to stimulate stomatal opening independently of CO_2. For example, guard-cell protoplasts of onion *(Allium cepa)* swell in the presence of K^+ when illuminated with blue light. The blue-absorbing pigment (a flavin or a flavoprotein located in the tonoplast and possibly the plasma membrane) promotes K^+ uptake by the guard cells.

Within normal ranges of temperature (10 to 25°C), changes in temperature have little effect on stomatal behavior, but temperatures higher than 30 to 35°C can lead to stomatal closure. The closing can be prevented, however, by holding the plant in air that contains no carbon dioxide, which suggests that temperature changes work primarily by affecting the concentration of carbon dioxide in the leaf. An increase in temperature results in an increase in respiration and a concomitant increase in the concentration of intercellular carbon dioxide, which may actually be the cause of stomatal closure in response to heat. Many plants in hot climates close their stomata regularly at midday, apparently because of the effect of temperature on carbon dioxide accumulation and because of dehydration of leaves as the water loss of transpiration exceeds the water uptake by absorption.

Although the stomata of most plants are open during the day and closed at night, this is not true of all plants. A wide variety of succulents—including cacti, the pineapple *(Ananas comosus)*, and members of the stonecrop family (Crassulaceae) among others—open their stomata at night, when conditions are least favorable to transpiration. The Crassulacean acid metabolism (CAM) characteristic of such plants has a pathway for carbon flow not substantially different from that of C_4 plants, as discussed in Chapter 6. At night, when their stomata are open, the CAM plants take in carbon dioxide and convert it into organic acids. During the day, when their stomata are closed, the carbon dioxide is released from these organic acids for use in photosynthesis.

Factors Affecting the Rate of Transpiration

Although stomatal opening and closing are the major factors affecting the rate of transpiration, there are a number of other factors both in the environment and in the plant itself. One of the most important of these is temperature. The rate of water evaporation doubles for every temperature rise of about 10°C. However, because evaporation cools the leaf surface, its temperature does not rise as rapidly as that of the surrounding air. As noted previously, stomata close when temperatures exceed 30 to 35°C.

Humidity is also important. Water is lost much more slowly into air already laden with water vapor. Leaves of plants growing in shady forests, where the humidity is generally high, typically spread large luxuriant leaf surfaces, because their main problem is getting enough light, not losing water. In contrast, plants of grasslands or other exposed areas often have narrow leaves, with relatively little leaf surface. They get all the light they can use but are constantly in danger of excess water loss.

Air currents also affect the rate of transpiration. A breeze cools your skin on a hot day because it blows away the water vapor that has accumulated near the skin surface and so accelerates the rate of evaporation of water from your body. Similarly, wind blows away the water vapor from leaf surfaces. Sometimes, if the air is very humid, wind may actually decrease the transpiration rate by cooling the leaf, but a dry breeze will greatly increase evaporation. Leaves of plants that grow in exposed, windy areas are often hairy; these hairs are believed to protect the leaf surface from wind action and so slow the rate of transpiration by stabilizing the boundary layer of air over the leaf surface.

THE MOVEMENT OF INORGANIC NUTRIENTS THROUGHOUT THE PLANT BODY

Uptake of Inorganic Nutrients

The uptake, or absorption, of inorganic ions takes place through the epidermis of the root. Ion movement from the epidermis across the cortex to the endodermis of the root may be by an apoplastic pathway, a symplastic pathway, or a combination of the two. Current evidence suggests that the major pathway is symplastic. Ion uptake by the symplastic route begins at the plasma membrane of the epidermal cells. The ions then move from the epidermal cell protoplasts to the first layer of cortical cells through the plasmodesmata in the epidermal-cortical cell walls (see Figure 28–3). Radial movement of the ions continues in the cortical symplast—from protoplast to protoplast via plasmodesmata—through the endodermis and into the parenchyma cells of the vascular cylinder by diffusion, aided by cytoplasmic streaming within the individual cells.

How the ions enter the vessels (or tracheids) of the xylem from the parenchyma cells of the vascular cylinder has been the subject of considerable debate. It was suggested that the ions leak passively from the parenchyma cells into the vessels, but there is now substantial evidence that the ions are secreted into the vessels from the parenchyma cells by an active, carrier-mediated membrane transport (see page 66).

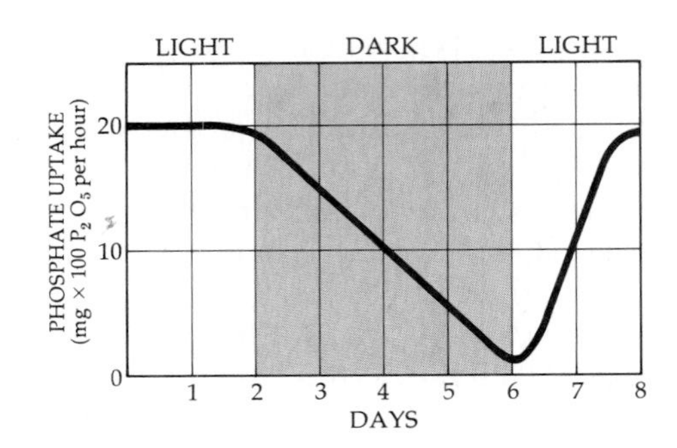

28–13
The rate of phosphate absorption by corn (Zea mays) *plants fell to near zero after four days of continuous darkness. It began to rise again when the plants were reilluminated. These and other data indicate that salt uptake in plants is a process that requires energy.*

Active Solute Absorption

The mineral composition of root cells is far different from that of the medium in which the plant grows. For example, in one study, cells of pea *(Pisum sativum)* roots were found to have a concentration of potassium ions 75 times greater than that of the nutrient solution. Similarly, in another study, the vacuoles of rutabaga (*Brassica napus* var. *napobrassica)* cells were shown to contain 10,000 times more potassium than the external solution.

Since substances do not diffuse against a concentration gradient, it is clear that minerals are absorbed by *active transport*. Support for this hypothesis comes from observations indicating that the uptake of minerals is an energy-requiring process. For instance, if roots are deprived of oxygen, or poisoned so that respiration is curtailed, mineral uptake is drastically decreased. Also, if a plant is deprived of light, it will cease to absorb salts after carbohydrate reserves are exhausted and will finally release them back into the soil solution (Figure 28–13). Hence, ion transport from the soil to the vessels of the xylem requires two active, carrier-mediated membrane events: (1) uptake at the plasma membrane of the epidermal cells and (2) secretion into the vessels at the plasma membrane of the vascular parenchyma cells.

Transmembrane Potential

The active transport of ions across the plasma membrane may result in a difference of electrical charge on the two sides of the membrane. When this occurs, a voltage difference, called the *transmembrane potential,* develops across the membrane. The hydrogen ion, H^+, is one of the principal cations involved with the generation of the transmembrane potential; with the pumping of H^+ to the outside, a negative potential develops on the inside of the cell. Once generated, the transmembrane potential can have a marked effect on further ion movements. For example, a negative interior electrical potential will attract positively charged ions such as K^+ but will repel negatively charged ions such as Cl^-.

Transport of Inorganic Nutrients

Once the inorganic ions have been secreted into the vessels of the xylem, they are rapidly distributed upward and throughout the plant in the transpiration stream. Some ions move laterally from the xylem into surrounding tissues of the stems, while others are transported into the leaves.

Much less is known about the pathways followed by the ions in leaves than in roots. Within the leaf, the ions are transported along with the water in the leaf apoplast, that is, in the cell walls. Some ions may remain in the transpiring water and reach the main regions of water loss—the stomata and other epidermal cells. Most eventually enter the protoplasts of the leaf cells, probably by carrier-mediated transport mechanisms similar to those demonstrated in roots. The ions may then move symplastically to other parts of the leaf, including the phloem. Inorganic ions can also be absorbed in small amounts through the leaves; consequently, fertilization of some crop plants that involves the direct application of micronutrients to the foliage has become a standard agricultural practice.

Substantial amounts of the inorganic ions imported into the leaves through the xylem are exchanged with the phloem of the leaf veins and exported from the leaf with sucrose in the assimilate stream (see the discussion of translocation that follows). For instance, in the annual white lupine *(Lupinus albus)*, transport in the phloem accounted for more than 80 percent of the fruit's vascular intake of nitrogen and sulfur and 70 to 80 percent of its phosphorus, potassium, magnesium, and zinc. The uptake of such inorganic ions by developing fruits is undoubtedly coupled to the flow of sucrose in the phloem.

Recycling may occur in the plant as nutrients reaching the roots in the descending assimilate stream are transferred to the ascending transpiration stream of the xylem. Only those ions that can move in the phloem, and which are said to be *phloem-mobile*, can be exported from the leaves to any great extent. For example, K^+, Cl^-, and $H_2PO_4^{2-}$ are readily exported from leaves, whereas Ca^{2+} is not. Solutes such as calcium are said to be *phloem-immobile*.

TRANSLOCATION: THE MOVEMENT OF SUBSTANCES IN THE PHLOEM

As discussed in Chapters 21 through 23, the xylem and phloem together form a continuous vascular system that penetrates practically every part of the plant. Whereas the water and inorganic solutes ascend the plant in the xylem, or the transpiration stream, the sugars manufactured during photosynthesis move out of the leaf in the phloem—which is also known as the *assimilate stream*—to sites where they are utilized, such as growing shoot and root tips, and to sites of storage, such as fruits, seeds, and the storage parenchyma of stems and roots (Figure 28–14).

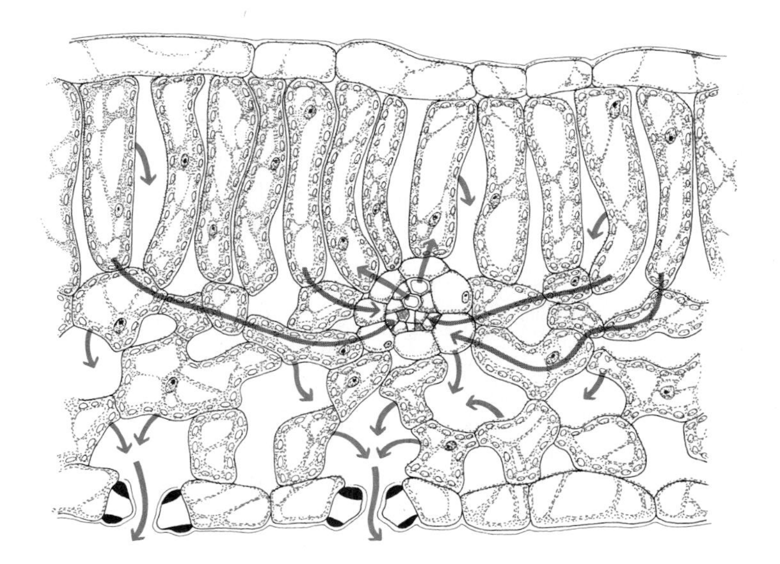

28–14
Diagram of leaf showing the pathways followed by water molecules of the transpiration stream as they move from the xylem of a small, or minor, vein to the mesophyll cells, evaporate from the surface of the mesophyll cell walls, and then diffuse out of the leaf through an open stoma (colored lines).

Also shown are the pathways followed by sugar molecules manufactured during photosynthesis as they move from mesophyll cells to the phloem of the same vein and enter the assimilate stream. The sugar molecules manufactured in the palisade parenchyma cells are believed to move to the spongy parenchyma cells and then laterally to the phloem via the spongy cells (gray lines).

Assimilate movement is said to follow a source-to-sink pattern. The principal *sources* of assimilate solutes are the photosynthesizing leaves, but storage tissues may also serve as important sources. All plant parts unable to meet their own nutritional needs may act as *sinks*, that is, as importers of assimilates. Thus, storage tissues act as sinks when they are importing assimilates and as sources when they are exporting assimilates.

Source-sink relations may be relatively simple and direct, as in young seedlings, where cotyledons containing reserve food often represent the major source and growing roots represent the major sink. In older plants, the upper, most recently formed (mature) leaves commonly export assimilates primarily toward the shoot tip; the lower leaves export assimilates primarily to the roots; and those in between export assimilates in both directions (Figure 28–15a). This pattern of assimilate distribution is markedly altered during the change from vegetative to reproductive growth. Developing fruits are highly competitive sinks that monopolize assimilates from the nearest leaves, and frequently from distant ones as well, often causing a decline or virtual cessation of vegetative growth (Figure 28–15b).

Evidence of Sugar Transport in Phloem

Early evidence supporting the role of the phloem in assimilate transport came from observations of trees from which a complete ring of bark had been removed. As noted in Chapter 24, the bark in older stems is composed largely of phloem and contains no xylem. When a photosynthesizing tree is "girdled" in this manner, the bark above the ring becomes swollen, indicating an accumulation of assimilates moving downward in the phloem from the photosynthesizing leaves (Figure 28–16).

28–15
Diagrams of a plant in (a) *the vegetative stage and* (b) *the fruiting stage. The arrows indicate the direction of assimilate transport in each stage.*

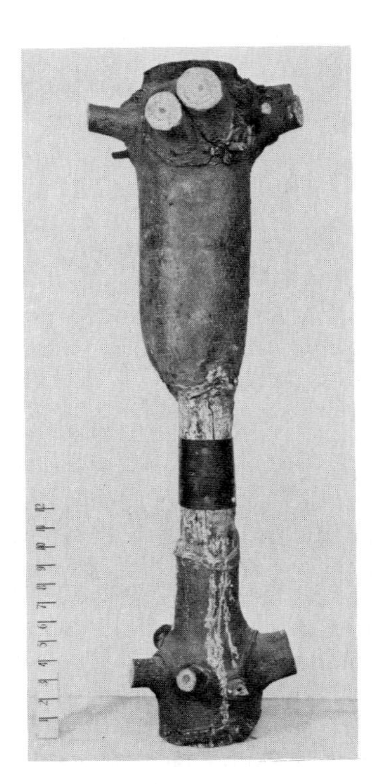

28–16
Portion of white pine (Pinus strobus) *stem that had been "girdled." (A metal band has been placed around the girdled portion of the stem to prevent it from breaking.) Notice the differences in development of the stem immediately above and below the area where the ring of bark is missing. Food manufactured by leaves of the whorl of branches below the girdled area stimulated growth in thickness of the stem below the girdle and prevented it from dying.*

Much more convincing evidence of the role of the phloem in assimilate transport was obtained with radioactive tracers. Before such tracers were available, it was necessary to cut into the intact plant to introduce dyes and other substances in an attempt to study certain transport phenomena. However, when the high hydrostatic pressures of the sieve tubes are released at the time the sieve tubes are severed, the contents of the sieve elements surge toward the cut surfaces, greatly disturbing the system. As discussed in Chapter 21, this phenomenon is responsible for the formation of slime (P-protein) plugs in injured sieve elements. With the use of radioactive tracers, it is now possible to experiment with entire plants and thus to obtain a fairly clear understanding of normal transport phenomena. The results of experiments with radioactive assimilates (such as ^{14}C-labeled sucrose) confirmed the movement of such substances in the phloem. More recently, such studies have shown conclusively that sugars are transported in the sieve tubes of the phloem (see "Radioactive Tracers and Autoradiography in Plant Research," page 572).

The Aphid in Phloem Research

Much valuable information on movement of substances in the phloem has come from studies utilizing aphids—small insects that suck the juices of plants. Most species of aphids are phloem feeders. When these aphids insert their modified mouth parts, or stylets, into a stem or leaf, they extend them until the tips of the stylets puncture a conducting sieve tube (Figure 28–17). The turgor pressure of the sieve tube then forces the sieve-tube sap through the aphid's digestive tract and out its posterior end as droplets of "honeydew." If feeding aphids are anesthetized and severed from their stylets, exudation often continues from the cut stylets for many hours. The sieve-tube exudate can then be collected from the cut ends of the stylets with a micropipette. Analyses of exudates obtained in this manner reveal that sieve-tube sap contains 10 to 25 percent dry matter, 90 percent or more of which is sugar—mainly sucrose in most plants. Low concentrations (less than 1 percent) of amino acids and other nitrogen-containing substances are also present.

Data obtained from studies utilizing aphids and radioactive tracers indicate that the rates of longitudinal movement of assimilates in the phloem are remarkably fast. For example, in one series of experiments utilizing severed aphid stylets, it was estimated that the sieve-tube sap was moving at a rate of about 100 centimeters per hour at the sites of the stylet tips.

(a)

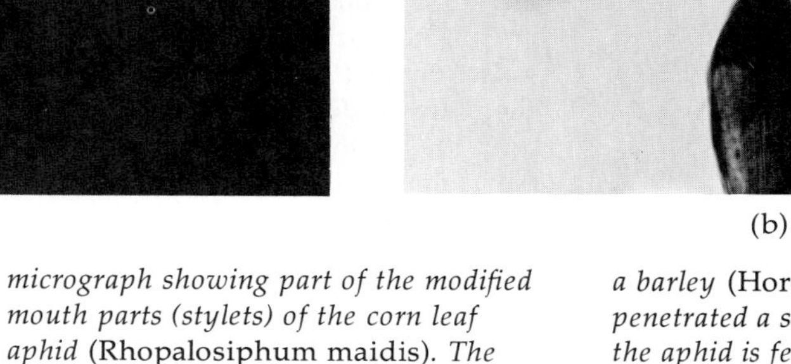

(b)

28–17
(a) *Aphid* (Longistigma caryae) *feeding on a linden* (Tilia americana) *stem. A droplet of "honeydew" can be seen emerging from the aphid.* (b) *A photomicrograph showing part of the modified mouth parts (stylets) of the corn leaf aphid* (Rhopalosiphum maidis). *The stylets have traversed the lower part of a barley* (Hordeum vulgare) *leaf and penetrated a sieve element, upon which the aphid is feeding.*

RADIOACTIVE TRACERS AND AUTORADIOGRAPHY IN PLANT RESEARCH

Radioactive tracers can be used in a number of ways to study the synthesis, transport, and use of materials within plants. Initially, in any tracer study, the radioactive isotope must be incorporated into the plant. Radioactive carbon, for example, will be taken up by a plant if its leaves are exposed to carbon dioxide that contains carbon 14, or radioactive phosphorus will be taken up by a plant if the roots are exposed to a solution containing ions of phosphorus 32.

The length of time the plant must be exposed to the radioactive material is determined by the information that the investigators hope to obtain. For example, in studies designed to determine the time required for carbon dioxide to be incorporated into the various products of photosynthesis, a sequence of exposure times would be used. In studies focusing on the location of a particular product formed by the metabolic processes that involve the radioactive substance, the length of exposure would depend on the time required for the chemical reactions under study to occur.

In whole-plant autoradiography, the plant is quick-frozen and freeze-dried after exposure to the radioactive substance. The plant is then flattened and pressed against a sheet of X-ray film. Radiation given off by the isotope exposes the film adjacent to the portions of the plant in which the tracer is located. By comparing the flattened plant with the developed film, investigators can determine the location of the radioactive substance within the plant.

In tissue autoradiography (histoautoradiography), the freeze-dried plant tissues are embedded in paraffin, resin, or a similar material. Next, they are sliced into very thin sections, which are mounted on microscope slides. The tissue sections are then placed in contact with a photographic emulsion or film. The radiation from the isotope exposes the film in contact with the portions of the tissue section containing the radioactive material. After an appropriate interval of time, the film is developed. Comparison, under a microscope, of the developed film and the underlying tissue section reveals the exact location of the radioactive substance in the plant tissues.

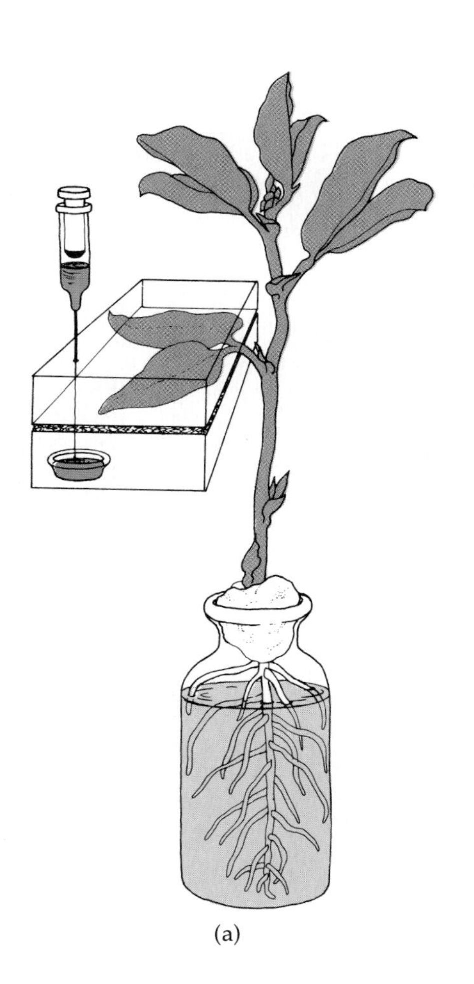
(a)

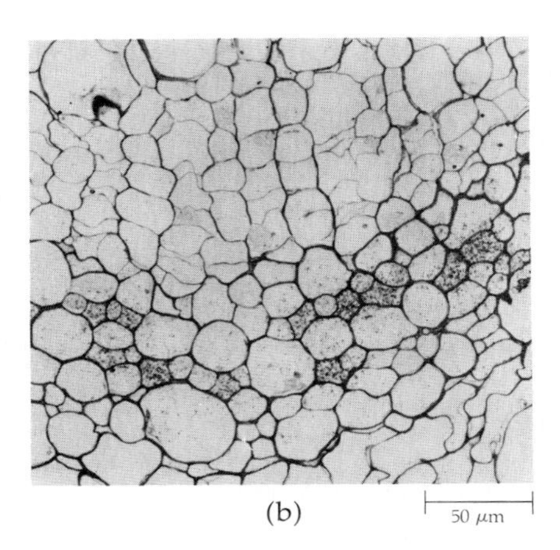

(b)

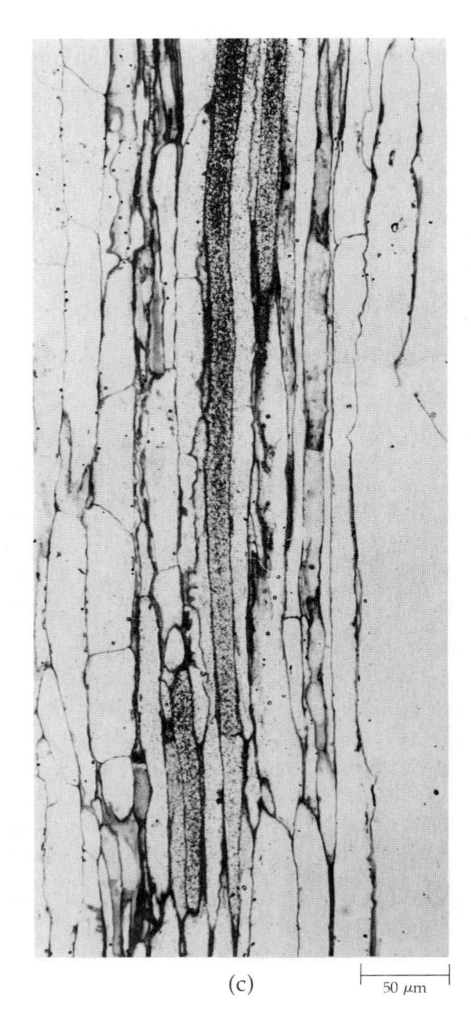

(c)

(a) Two leaflets of a broad bean plant *(Vicia faba)* were enclosed in a Plexiglas container in which $^{14}CO_2$ was generated and were exposed to radioactive carbon dioxide and light for 35 minutes. During that time, the $^{14}CO_2$ was incorporated into sugars, which were then being transported to other parts of the plant. A cross section (b) and a longitudinal section (c) from the stem were placed in contact with autoradiographic film for 32 days. When the film was developed and compared with the underlying tissue sections, it was apparent that the radioactivity (visible as dark grains on the film) was confined almost entirely to the sieve tubes. (See Figure 22–5, page 435, for another example of this technique.)

The Mechanism of Phloem Transport: Pressure Flow

Several mechanisms have been proposed over the years to explain assimilate transport in the sieve tubes of the phloem. Probably the earliest was that of diffusion, followed by that of cytoplasmic streaming. Normal diffusion and cytoplasmic streaming of the kind found in higher plant cells were largely abandoned as possible translocation mechanisms when it became known that the velocities of assimilate transport (typically 50 to 100 centimeters per hour) were far too great for either of these two phenomena to account for long-distance transport via sieve tubes.

Alternate hypotheses have been advanced to explain the mechanism of phloem transport, but only one, the pressure-flow hypothesis, satisfactorily accounts for practically all of the data obtained in experimental and structural studies on phloem. All of the other hypotheses have serious deficiencies.

Originally proposed in 1927 by the German plant physiologist Ernst Münch, the pressure-flow hypothesis is clearly the simplest and, at present, the most widely accepted explanation for long-distance assimilate transport in sieve tubes. It is the simplest explanation because it depends only on osmosis as the driving force for assimilate transport.

Briefly stated, the *pressure-flow hypothesis* asserts that assimilates are transported from source to sink along a gradient of turgor (hydrostatic) pressure developed osmotically. The principle underlying this hypothesis can be illustrated by a simple physical model consisting of bulbs permeable only to water and connected by glass tubes (Figure 28–18). Initially, the first bulb (A) contains a sugar solution, and the second bulb (B) contains only water. When these interconnected bulbs are placed in distilled water, water will enter the first bulb by osmosis, thereby increasing the turgor pressure within that bulb. This pressure will be transmitted through the tube to the second bulb, causing the sugar solution to move by bulk, or mass, flow to the second bulb and forcing water out of it. If the second bulb is connected with a third one containing water or a sucrose concentration lower than that now in the second one, the solution will flow from the second to the third bulb by the same process, and so on indefinitely down the turgor gradient.

In the plant, sucrose produced by photosynthesis in a leaf is actively secreted into the sieve tubes of the small veins (Figure 28–19). This active process, called *phloem loading*, decreases the water potential in the sieve tube and causes water entering the leaf in the transpiration stream to move into the sieve tube by osmosis. With the movement of water into the sieve tube at this source, the sucrose is carried passively by the water to a sink, such as a storage root, where the sucrose is actively removed (unloaded) from the sieve tube. The removal of sucrose results in an increased water potential in the sieve tube at the sink and the subsequent movement of water from the sieve tube there. The sucrose may be either utilized or stored at the sink, but most of the water returns to the xylem and is recirculated in the transpiration stream.

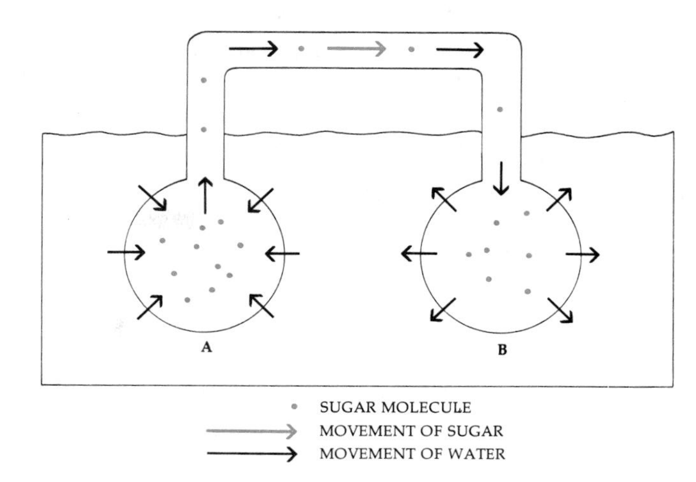

28–18
A model of the pressure-flow hypothesis. Bulbs A *and* B, *which are interconnected and are permeable to water, are placed in a bath of distilled water. Bulb* A *contains a higher concentration of sucrose than bulb* B. *Water enters bulb* A *from the medium by osmosis, thus increasing the hydrostatic pressure and pushing the solution to bulb* B. *If bulb* B *were connected to a third bulb with a still lower concentration of sucrose (just as sieve-tube elements are connected in a series), hydrostatic pressure building up in* B *would push the solution into the third bulb.*

Note that the pressure-flow hypothesis casts the sieve tubes in a *passive* role in the movement of the sugar solution through them. Active transport is also involved in the pressure-flow mechanism; however, it is not directly involved with the long-distance transport through the sieve tubes, but rather with the loading and unloading of sugars and other substances into and out of the sieve tubes at the sources and sinks. The metabolic energy required for the loading and unloading is expended by companion cells or parenchyma cells bordering the sieve tubes, rather than by the sieve tubes.

Phloem loading is a selective process. As mentioned previously, sucrose is by far the most common sugar transported; in addition, all of the sugars found in sieve-tube sap are nonreducing sugars. Certain

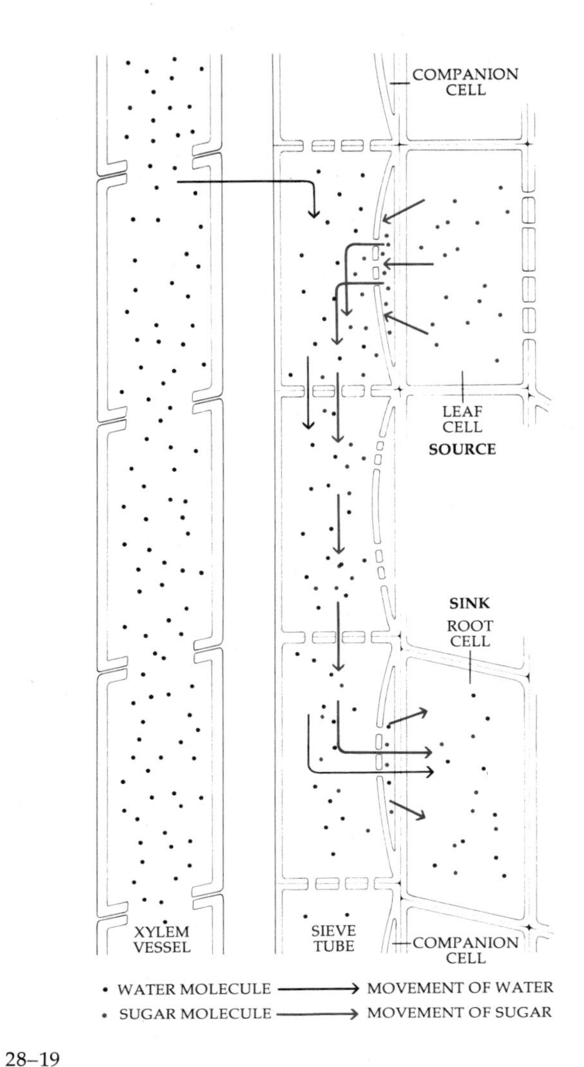

28–19
The pressure-flow mechanism as it is thought to occur in the plant body. Sugar enters a companion cell by active transport and then moves into the sieve tube via the many cytoplasmic connections in the common cell wall of the sieve tube and companion cell. As a consequence of the increased concentration of sugar, the water potential is decreased, and water enters the sieve tube by osmosis. Sugar is removed (unloaded) in the root by an active transport process, and the sugar concentration falls; as a result, the water potential is increased, and water leaves the sieve tube. Because of the active secretion (loading) of sugar into the sieve tube-companion cell complex at the source and its active removal from the sieve tube-companion cell complex at the sink, a flow of sugar solution takes place, along a gradient of turgor pressure, between source and sink.

amino acids and ions also are selectively loaded into the phloem. Moreover, it now seems clear that phloem loading involves active transport across the plasma membrane of the companion cells or parenchyma cells bordering the sieve tubes. From the companion cells and parenchyma cells, the substances enter the sieve tubes through the many plasmodesmatal connections found in their common walls.

SUMMARY

In plants, more than 90 percent of the water taken in by the roots is given off into the air as water vapor. This loss of water vapor by plant parts is called transpiration, and most of the water transpired by higher plants is lost through the stomata in the leaves.

Absorption of water takes place largely through the root hairs, which provide an enormous surface area for water uptake. In some plants, when the roots absorb water from the soil and transport it into the xylem, the water within the xylem builds up positive pressure, or root pressure. This osmotic uptake depends on the transport of inorganic ions from the soil into the xylem by the living cells of the root and can result in a phenomenon called guttation, in which liquid water is forced out through special openings in the tips or margins of the leaves. The water follows a largely apoplastic pathway across the epidermis and cortex until it reaches the endodermis, where further movement is blocked by the Casparian strips. The water must pass through the plasma membrane and protoplast of the endodermal cells on its way to the xylem.

From the roots, water moves to the leaves by means of the xylem. The current and widely accepted theory of water movement to the top of tall plants through the xylem is the cohesion-adhesion-tension theory. According to this theory, water within the vessels is under tension because the water molecules cling together in continuous columns pulled by evaporation from above. It has been demonstrated that water has sufficient tensile strength to withstand such tension when contained in small-diameter conduits. Other supporting evidence includes observations that water in the xylem is under tension, that water movement in trees begins in the topmost branches, and that the trunk of a tree shrinks slightly at the beginning of water movement.

The rate of transpiration is affected by such factors as carbon dioxide concentration in the intercellular spaces (and, conversely, the exposure of the leaf to carbon dioxide from the air), light, temperature, atmospheric humidity, air currents, and availability of soil water. Most of these factors have an effect on the behavior of stomata. Stomatal opening and closing are controlled by changes in turgor of the guard cells, which are closely correlated with changes in the potassium ion level within the guard cells. Abscisic acid and blue light also play roles in stomatal movement. The stomata open when the guard cells become turgid and close when they become flaccid.

Inorganic nutrients become available to plants in soil solution in the form of ions. Plants employ metabolic energy to concentrate the ions they require. Most ions are taken up by active transport processes, whereas others flow in passively due to the water potential across the plasma membrane that is created by the actively moving ions and their pumps. Inorganic ions follow a mostly symplastic pathway from the epidermis to the xylem.

Research on the movement of substances in the phloem has been greatly aided by the use of aphids and radioactive tracers. Analyses of sieve-tube sap reveal that it contains sugar—mainly sucrose—and small quantities of nitrogenous substances. Rates of longitudinal movement of substances in the phloem are far in excess of the normal rate of diffusion of sucrose in water—the rates typically range from 50 to 100 centimeters per hour.

According to the pressure-flow hypothesis, assimilates move from source to sink along a turgor pressure gradient which is developed osmotically. Sugars are actively secreted into (loaded) and absorbed from (unloaded) the sieve tube by companion cells or parenchyma cells at source and sink, respectively, resulting in mass flow of solution in the sieve tube. The role of the sieve tube is a passive one.

SUGGESTIONS FOR FURTHER READING

BEM, ROBYN: *Everyone's Guide to Home Composting,* Van Nostrand Reinhold Company, New York, 1978.*

A comprehensive, easy-to-read manual that discusses the construction and monitoring of a compost heap, as well as the uses for compost. Included are such topics as the nutrient content of various compost materials, green manure crops, and essential plant nutrients.

BRADY, NYLE C.: *The Nature and Properties of Soils,* 8th ed., Macmillan Publishing Company, Inc., New York, 1974.

A comprehensive elementary text on soil science.

BRILL, WINSTON J.: "Biological Nitrogen Fixation," *Scientific American* 236:68–81, 1977.

_______: "Nitrogen Fixation: Basic to Applied," *American Scientist* 67:458–66, 1979.

The above two articles are excellent reviews of a complex subject presented in an easy-to-understand and readable style.

BUNTING, BRIAN T.: *The Geography of Soils,* Aldine Publishing Co., Chicago, 1965.*

A brief description of the main soil groups of the world and their influence on plant distribution.

CRAFTS, ALDEN. S., and CARL E. CRISP: *Phloem Transport in Plants,* W. H. Freeman and Co., San Francisco, 1971.

A definitive text on phloem structure and function. Experimental methods employed in phloem research and the influence of environmental factors on translocation are discussed. Five hypotheses for the phloem-transport mechanism are reviewed, and strong evidence in support of pressure flow is presented.

EPSTEIN, EMANUEL: *Mineral Nutrition of Plants: Principles and Perspectives,* John Wiley & Sons, Inc., New York, 1972.

A thorough coverage of the whole field of mineral nutrition, well illustrated and written by one of the leading students of the subject.

FARB, PETER: *Living Earth,* Pyramid Publications, Inc., New York, 1969.*

An eloquent description of the teeming life that exists below the surface of the ground. A book to read for pleasure.

International Symposium on Nitrogen Fixation, Washington State University Press, Pullman, Wash., 1975.

A collection of outstanding papers on all aspects of this very active field of research.

LUETTGE, ULRICH, and NOE HIGINBOTHAM: *Transport in Plants,* Springer-Verlag, New York, 1979.

A comprehensive, up-to-date review of the various processes by which inorganic and organic substances are transported in plants, from the level of cellular organelles to long-distance movements in trees.

MENGEL, K., and E. A. KIRKBY: *Principles of Plant Nutrition,* Der Bund AG, Bern, Switzerland, 1978.

The physiological and agricultural aspects of plant nutrition are effectively integrated in this presentation.

NOGGLE, FRITZ: *Introductory Plant Physiology,* Prentice-Hall, Inc., Englewood Cliffs, N.J., 1976.

A text that can be useful to the student with training only in general biology and general chemistry.

PEEL, A. J.: *Transport of Nutrients in Plants,* John Wiley & Sons, Inc., New York, 1974.

A concise, up-to-date account of the physiology of long-distance transport in plants. Presents a balanced view of transport in phloem and xylem. A good text for the student who is unfamiliar with transport processes.

*Available in paperback.

QUISPEL, A. (Ed.): *Biology of Nitrogen Fixation,* Frontiers of Biology Ser., vol. 33, Elsevier-North Holland Publishing Co., New York, 1974.

Specialists in various aspects of research on nitrogen fixation contributed chapters to this volume.

SALISBURY, FRANK B., and CLEON W. ROSS (Eds.): *Plant Physiology,* 2nd ed., Wadsworth Publishing Co., Inc., Belmont, Calif., 1978.

A detailed and useful review of the entire subject.

SUTCLIFFE, JAMES: *Plants and Water,* St. Martin's Press, Inc., New York, 1968.*

An excellent short account of water relationships, including a review of experimental work in this area.

VIORST, JUDITH: *The Changing Earth,* Bantam Books, Inc., New York, 1967.*

A well-written introduction for the layman to the science of modern geology.

WARDLAW, IAN F., and J. B. PASSIOURA (Eds.): *Transport and Transfer Processes in Plants,* Academic Press, Inc., New York, 1976.

The proceedings of a symposium designed to examine how the various forms of long-distance and short-distance transport operate and interact in the whole plant.

ZIMMERMANN, MARTIN H., and CLAUD L. BROWN: *Trees: Structure and Function,* Springer-Verlag, New York, 1975.

This book is devoted to those aspects of tree physiology that are peculiar to tall woody plants. The emphasis is on function and includes material not found in general physiology texts.

ZIMMERMANN, MARTIN H., and JOHN A. MILBURN (Eds.): *Transport in Plants I, Phloem Transport,* Encyclopedia of Plant Physiology, vol. I, Springer-Verlag, New York, 1976.

A collection of articles by leading researchers in the field of phloem structure and function.

*Available in paperback.

SECTION 8 Ecology

CHAPTER 29

The Dynamics of Ecosystems

29–1
A redwood (Sequoia sempervirens) *forest along the coast of northern California.*

Ecology is the study of the interactions of organisms with one another and with their environment. As a science, it attempts to explain why particular plants and animals can be found living in one area and not in others, why there are so many organisms of one sort and so few of another, and what changes one might expect the interactions among them to produce in a particular area.

In following these discussions, a few definitions will be helpful. A *community* consists of all the plants, animals, and other organisms that live in a particular area. An *ecosystem* includes not only the aggregation of living organisms (biotic factors) but also the nonliving (abiotic) elements of the environment with which they are interacting. *Biomes* are large complexes of communities of living organisms that are characterized by distinctive vegetation and climate, such as deserts or grasslands. The principal terrestrial biomes are described in Chapter 30.

INTERACTIONS BETWEEN ORGANISMS

None of the organisms living in a community—whether a patch of woodland, a pasture, a pond, or a coral reef—exists in isolation. Each organism is involved in a number of relationships, both with other organisms and with factors in the nonliving environment. These interactions are discussed under three headings: mutualism, competition, and plant-herbivore interactions.

Mutualism

Mutualism is a form of biological interaction in which the growth and survival of both interacting populations are enhanced. In nature, neither population can survive without the other. The formation of lichens is a familiar example. Another is the relationship between legumes and the nitrogen-fixing bacteria that occur on their roots. Also, some of the closely linked pollination relationships discussed in Chapter 19, such as that between the yucca moth and the *Yucca* plant, can be described as mutualism.

One of the most interesting and significant examples of mutualism in the plant world concerns the interaction between fungi and seed plants. As discussed in Chapter 12, the roots of most vascular plants are associated with fungi to form compound structures known as mycorrhizae, without which the normal growth of the plants is impossible. Mycorrhizal associations appear to have played a crucial role in the first invasion of the land by plants.

The fungi that form mycorrhizal associations in most plants are zygomycetes; the associations formed are called endomycorrhizae. In some groups of conifers and dicotyledons—mainly trees—the associations are mostly with basidiomycetes and certain ascomycetes; these associations are called ectomycorrhizae. Some of these latter associations, which are particularly characteristic of relatively pure stands of trees growing at high latitudes in the Northern Hemisphere or at high elevations, are highly specific. The basidiomycete *Boletus elegans,* for example, is known to associate only with the larch *(Larix),* a conifer. Other fungi, such as *Cenococcum geophilum,* have been discovered living in association with more than a dozen genera of forest trees.

As more is learned about mycorrhizal relationships, their importance to the higher plants becomes more evident. In many higher plants, uninfected individuals are rarely encountered under natural conditions, even though growth may be possible without fungi if other conditions are narrowly regulated. Most vascular plants are dual organisms in the same sense that lichens are dual organisms, although the relationship may not be obvious above ground. As University of Wisconsin soil scientist S. A. Wilde has stated, "A tree removed from the soil is only a part of the whole plant, a part surgically separated from its . . . absorptive and digestive organ." For most plants, the mycorrhizal fungi play a vital role in absorption and digestion.

Competition

Under experimental conditions, when two kinds of plants are grown together for a long enough time in a simple environment, one is always eliminated. Similarly, under natural conditions, two species cannot coexist indefinitely in the same habitat. This is a simplified statement of what Garrett Hardin has called the competitive exclusion principle. If they are growing together and utilizing any of the same essential resources, the individuals of one or both species will be smaller or fewer in number than they would be if they were growing alone. Ecologists group interactions of this sort under the general heading of *competition*. If the environment is complex, as in nature, various organisms may use it in very different ways, in effect, subdividing the habitat. They may then continue to coexist indefinitely.

In a bog, for example, mosses of the genus *Sphagnum* often appear to form a continuous cover, and several species are usually involved. How can these species continue to coexist? When the situation is examined in more detail, it is found that there are semiaquatic species growing along the bottoms of the wettest hollows; other species grow in drier places on the sides of the hummocks, which they help to form; and still other species grow only in the driest conditions on the tops of the hummocks, where they are eventually succeeded by one or more species of flowering plants. Therefore, although all the species of *Sphagnum* coexist, in the sense that they are all present in the same bog, they actually occupy different *microhabitats* and continually replace one another as the characteristics of each microhabitat change.

If the populations of coexisting species are kept at low levels, they may not eliminate one another. This effect was seen in England during this century, when a severe epidemic of myxomatosis, a disease caused by a virus, drastically reduced the population of rabbits. Formerly, the grasses growing on the chalky soils were kept closely cropped by the rabbits, and many different kinds of flowering plants were able to grow in this habitat. After the decline in the number of rabbits, the grass cover of the chalk soils became deeper and more dense and many of the formerly abundant species of flowering plants became rare (Figure 29–2). Similar effects are often seen when comparing grazed and ungrazed pastures or grasslands.

Unlike animals, green plants are dependent upon a single process for the conversion of energy, namely, photosynthesis. Competition in plants is manifested largely in terms of the "struggle for light," and plants that grow in the shade of others have evolved mechanisms for carrying on photosynthesis at low light intensities. Variations in plant height, arrangement of leaves, and shape of crown seem to be significant factors in the subdivision of the community's environment, whether it is low grassland or tall forest.

Most competitive situations are quite complicated. There are a number of different ways of expressing the relative success of two species growing together. Figure 29–3, for example, shows the variations in performance (dry weight of yield per hectare) that occurred with changes in density of corn *(Zea mays)* plants. With increasing density, the dry weight of the corn plants (shoots plus ears) increased within the limits of the experiment. When there were more than seven plants per square meter, the dry weight of the shoots less ears increased more rapidly than that of

(a)

(b)

29–2
Kingsley Vale, Sussex, England, an area of chalk grassland (a) *before and* (b) *after the elimination of rabbits by the viral disease myxomatosis. The first photograph was taken in April 1955, the second in July 1961.*

the shoots alone, but the dry weight of the ears decreased markedly.

Population ecologists consider that two types of selection operate in natural situations. In what is called *r* selection, rapid breeding and the production of many seeds is called for; in *K* selection, maximum adjustment to the environment and a relatively lower reproductive potential is needed. Many of the plants in which *r* selection predominates are annuals or short-lived perennials that occur in fluctuating, open communities, such as disturbed ground; many of those in which *K* selection predominates are perennials of relatively closed communities, such as most trees and shrubs.

Other aspects may be important in competitive interactions. Many organisms produce chemical substances that either inhibit their own growth, resulting in increased spacing of like individuals, or the growth of other species. For example, the fungus *Penicillium chrysogenum,* which grows on organic substrates such as seeds, produces significant quantities of penicillin in nature. Often, however, the fungus is replaced by bacteria, such as *Bacillus cereus,* that produce penicillinases (enzymes that break down penicillin).

More complex relationships are evident in the higher plants. In coastal California, for example, a bare zone normally occurs between shrub and grass communities. The feeding activity of rodents, rabbits, and birds, which find shelter in the shrubs, is concentrated in this zone; if this activity is prevented by means of wire-mesh exclosures, annual herbs grow vigorously within what normally would be a bare zone. In addition, plants such as purple sage *(Salvia leucophylla)* produce volatile terpenes, which inhibit the establishment of seedlings of many species of plants in their vicinity. Such plants may not grow in the bare zone even if exclosures are constructed but often grow only at the border between the bare zone and the fully developed grassland,

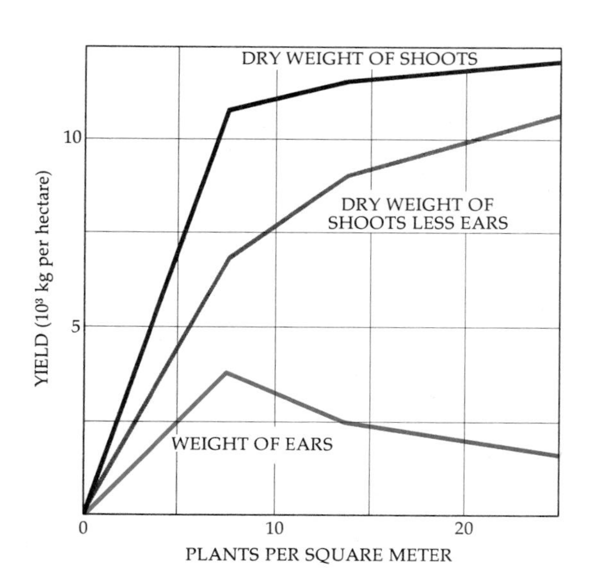

29–3
The influence of plant density on yield in corn (Zea mays) *as measured in three different ways. If ears of corn of the maximum dry weight is the farmer's goal, an optimum density is between seven and eight plants per square meter.*

29–4
Purple sage (Salvia leucophylla) *shrubs produce terpenes that evaporate and spread through the air to inhibit the growth of other plants in their vicinity. This clump of* Salvia *shrubs, growing near Santa Barbara, California, is surrounded first by a completely bare zone and then by a zone of inhibited grassland inhabited by stunted annual herbs.*

where toxic substances are less abundant and the grassland is still somewhat open (Figure 29–4).

Chemical inhibition of one species of plant by another is called allelopathy. Toxic chemicals often play an important role in structuring communities.

Plant-Herbivore Interactions

Vast areas of Australia were at one time covered with spiny clumps of prickly pear cactus *(Opuntia)*, a plant that was introduced to the continent from Latin America. Fertile lands became useless for grazing, and the economy of great stretches of the interior was severely threatened. Today, the cactus has been nearly eliminated by a cactus moth *(Cactoblastis cactorum)* discovered in South America and deliberately introduced into Australia. The larvae of this moth destroy the cactus plants by eating them. The moth, once abundant, now can scarcely be found, even by a careful inspection of the remaining clumps of cactus; yet there is no doubt that it continues to exert a controlling influence over the populations of the plant in Australia (Figure 29–5).

In general, the effects of herbivores on plants are profound, both in the short and the long term. As discussed in Chapter 19, these interactions have led to the production by plants of a wide variety of chemical defenses—in the form of biochemicals once regarded as "secondary plant substances." The ability of plants to produce such toxic chemicals and retain them in their tissues gives them a tremendous competitive advantage. This advantage is analogous to that which is achieved by the production of thorns or tough, leathery leaves.

The protection of plants by their production of toxic molecules has economic implications. Normally, plants of the squash family (Cucurbitaceae) produce bitter terpenes in their fruits and leaves, and these substances protect them from the attacks of most herbivores. Under cultivation, however, these substances have been bred out in order to improve the flavor of the fruits, and the plants have again become palatable to herbivores. Special steps, such as spraying with insecticides, are now taken to protect them. The cultivated watermelon *(Citrullus vulgaris)* also can be attacked by a much wider range of insects than its wild relatives, and insects must be discouraged if the crop is to be grown successfully.

Pollination relationships are a specialized form of plant-herbivore interaction in which a particular portion of the plant (often nectar) is eaten and pollination takes place. Here, the emphasis is on attracting the herbivore, and as described in Chapter 19, this has produced a great diversity of angiosperm flowers. In this sort of selective race, both plants and animals become better adapted to one another, and increasingly specialized pollination systems arise.

The secondary plant substances that animals ingest may in turn play a role in the ecological relationships of the animals. Some insects store these poisons within their tissues and are then protected from their own predators by them (Figure 29–6). Some sex attractants in insects are derived from the plants on which they feed. The insects concentrate these substances and then use them to attract the opposite sex of their own species.

The most intricate coevolutionary systems involving plant-herbivore interactions occur in the tropics, where many more kinds of organisms are found than in temperate regions. Trees and shrubs of the genus *Acacia* occur widely in the tropical and subtropical regions of the world. The interaction that occurs between certain species of *Acacia* in the lowlands of Mexico and Central America and the ants that inhabit their thorns provides a remarkable example of the complexity that can be involved in plant-animal interactions. The particular relationship between the bull's-horn acacias *(Acacia)* and the inhabitants of their thorns, ants of the genus *Pseudomyrmex,* is a good example (see Figure 29–7).

(a)

(b)

29–5
(a) *Dense prickly pear cactus* (Opuntia inermis), *growing in a mixed scrub forest in Queensland, Australia, in October 1926;* (b) *the same forest in October 1929, after the cacti were destroyed by the deliberately introduced South American moth,* Cactoblastis cactorum. *First introduced in May 1925, the larvae of this moth destroyed the cacti on more than 120 million hectares of rangeland.*

(a)

(b)

29–6
(a) *The monarch butterfly* (Danaus plexippus) *obtains cardiac glycosides derived from the plants of the milkweed family (Asclepiadaceae) upon which its larvae* (b) *feed. As a result, the monarch is unpalatable to birds and other vertebrates. It "advertises" this fact by bright orange-and-black adult coloration and by larvae that are conspicuously banded with white, yellow, and black. Even the eggs of the monarch, which are bright yellow and conspicuous, contain enough cardiac glycosides to be protected.*

(a)

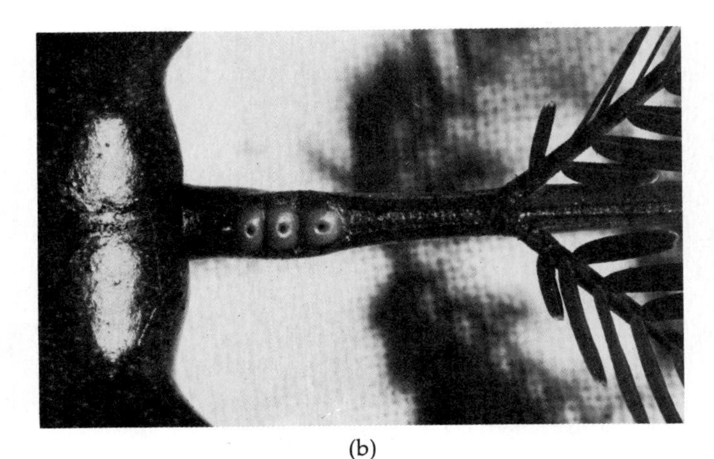

(b)

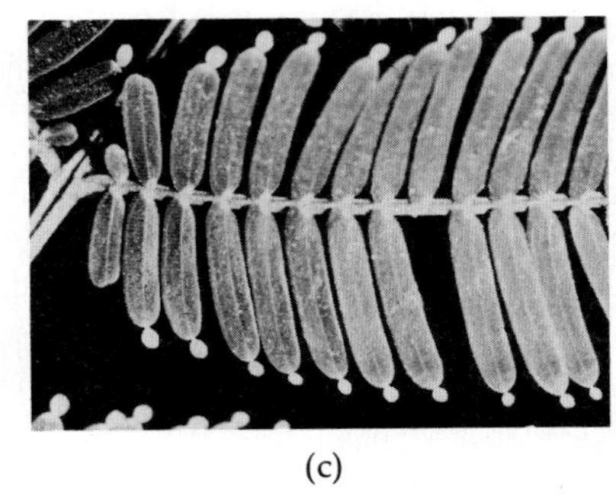

(c)

(d)

(e)

(f)

29–7
Ants and acacias. (a) *Sucker shoots regenerating from cut stumps of bull's-horn acacia* (Acacia cornigera); *the stump on the right is occupied by a healthy ant colony. The ants live in the hollow thorns (being modified stipules, the acacia "thorns" are technically spines).* (b) Acacia collinsii, *showing the nectaries at the base of the petioles of the compound leaves.* (c) *Beltian bodies at the end of the leaflets of* Acacia collinsii. (d) *Worker ants* (Pseudomyrmex ferruginea) *attacking the tendrils of a vine growing on* Acacia cornigera. *Obtaining all of their food from the plant, the ants in turn girdle all plants that come into contact with it and kill most other insects that attempt to feed on the plant.* (e) *Regeneration and growth of* Acacia cornigera *in the lowlands of Veracruz, Mexico. The stumps on the left are occupied by ants.* (f) *A single individual* Acacia cornigera *overtopping the dense second-growth vegetation in the tropical lowlands of Mexico. If this plant were not occupied by ants, it would probably have died as a small seedling, overtopped by other vegetation and devoured by insects.*

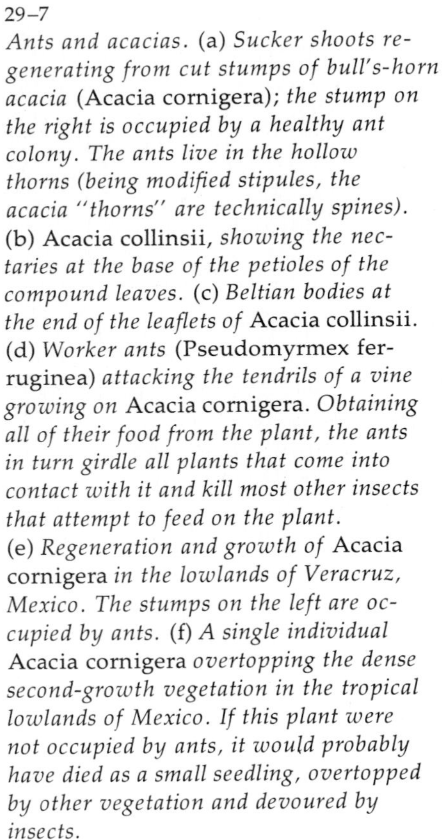

The bull's-horn acacias have a pair of greatly swollen thorns more than 2 centimeters long at the base of each leaf. Nectaries are borne on the petioles, and small nutritive organs known as Beltian bodies are borne at the tip of each leaflet. The ants live in the thorns and obtain sugars from the nectaries and oils and proteins by eating the Beltian bodies. The acacias grow extremely rapidly and are particularly common in disturbed areas, where the competition for light is intense.

Thomas Belt first described the relationship between *Pseudomyrmex* and the swollen-thorn acacias in his book, *The Naturalist in Nicaragua* (1874). Since then, there has been controversy about whether the presence of the ants actually benefited the acacia plants. This question was finally and definitively solved in 1964 by Daniel Janzen, who was then at the University of California. He found that the worker ants, which swarm over the surface of the plant, bite and sting animals of all sizes that contact the plant, thus protecting it from the activities of herbivores and ensuring a home for themselves and their fellow ants. Moreover, whenever the branches of other plants touch an inhabited acacia tree, the ants girdle the bark, thus destroying the invading branches and producing a tunnel to the light through the rapidly growing tropical vegetation.

When Janzen removed the ants from the plants artificially—by poisoning them or clipping off the portions of the plant that contained ants—or when an acacia was naturally unoccupied, growth was extremely slow, and the plant usually died after a few months as a result of insect damage and shading by other species. Plants inhabited by ants grew very rapidly, soon reaching 6 meters or more in height and overtopping the other second-growth vegetation. These ants occur nowhere else and are completely dependent upon the nectaries and Beltian bodies of the acacias for food. Therefore, it is clear that the ant-acacia system is as much a dual biological entity as, for example, a lichen. One element cannot survive without the other in the community of which it is a part.

There are many other kinds of relationships that link organisms that occur together. For example, trees in a forest, as well as lower herbs, are often joined by their roots in a phenomenon known as root grafting. Nutrients present in one plant can thus be transferred to another in complex and unexpected patterns; the survival of one species in an area may literally depend on the presence of another species with which it forms root grafts. Stumps of trees may live indefinitely, even though they have no photosynthetic surface of their own, because their roots are linked with those of other individuals in the forest from which they obtain nutrients.

Viewed as a whole, the relationships within a community are incredibly complex. Plants that occur together affect one another in an endless variety of ways, a few of which are just beginning to be understood. It may not be the individual in a species that is important in survival but that the linked group of individuals and the kinds of interactions that they have with one another determine the success of a particular species growing in a particular place.

CYCLING OF NUTRIENTS

In terms of its nutrient supply, an ecosystem is more or less self-sustaining. One of the most important reasons for this autonomy is the continuous cycling of chemical elements between organisms and the environment. The pathways of some of these essential elements, known as nutrient cycles, were discussed in Chapter 27. Ideally, nothing is removed from the ecosystem and nothing is used up, so that the pool of nutrients is continually renewed and continually available for the growth of organisms. The rate of flow from the nonliving pool to the organisms and back again, the amount of available material in the nonliving pool, and the form of this pool differ from element to element.

Recycling in a Forest Ecosystem

Studies of a deciduous forest ecosystem have shown that the plants of this community play a major role in the retention of nutrient elements. The studies were made in the Hubbard Brook Experimental Forest in the White Mountain National Forest of New Hampshire. The investigators first established a procedure for determining the mineral budget—input and output, or "profit" and "loss"—of different areas in the forest. By analyzing the content of rain and snow, they were able to estimate input, and by constructing concrete weirs that channeled the water flowing out of selected areas, they were able to calculate output (Figure 29–8). A particular advantage of the site is that bedrock is present just below the soil surface, so that very little material leaches downward; that is, the soil water percolates only a short distance.

The investigators discovered that the natural forest was extremely efficient in conserving its mineral elements. For example, the annual net loss of calcium from the ecosystem was 9.2 kilograms per hectare. This represents only about 0.3 percent of the calcium in the system. In the case of nitrogen, the ecosystem was actually accumulating this element at a rate of about 2 kilograms per hectare per year. There was a similar, though somewhat smaller, net gain of potassium.

PLANT DEFENSE IN SOLANACEAE

Many of the plants of the Solanaceae family—which contains such economically important crops as potatoes (Solanum), *tomatoes* (Lycopersicon), *and tobacco* (Nicotiana)—*have evolved glandular hairs as a major source of protection against insect pests. These hairs release a sticky and, in certain instances, toxic exudate. The cultivated potato* (Solanum tuberosum) *does not have such hairs, but a few related wild potato species, such as* S. berthaultii, *possess abundant glandular hairs on their leaves and stems. These hairs release a sticky substance when touched by insects, and the insects become trapped and eventually starve to death. This defense mechanism is effective against many insects, including such major potato pests as aphids, flea beetles, and leafhoppers. Attempts are currently under way in several countries to breed this characteristic into high-yielding varieties of the cultivated potato, which readily hybridizes with* S. berthaultii.

Scanning electron micrograph (a) *shows an insect-trapping hair on the foliage of* Solanum berthaultii. *A sticky exudate is released when the four-lobed head of the hair ruptures on contact with an insect. Electron micrographs* (b), (c), *and* (d) *show an aphid* (Myzus persicae) *glued to the stem of* S. berthaultii *by the exudate from the hairs, each at successively higher magnifications.*

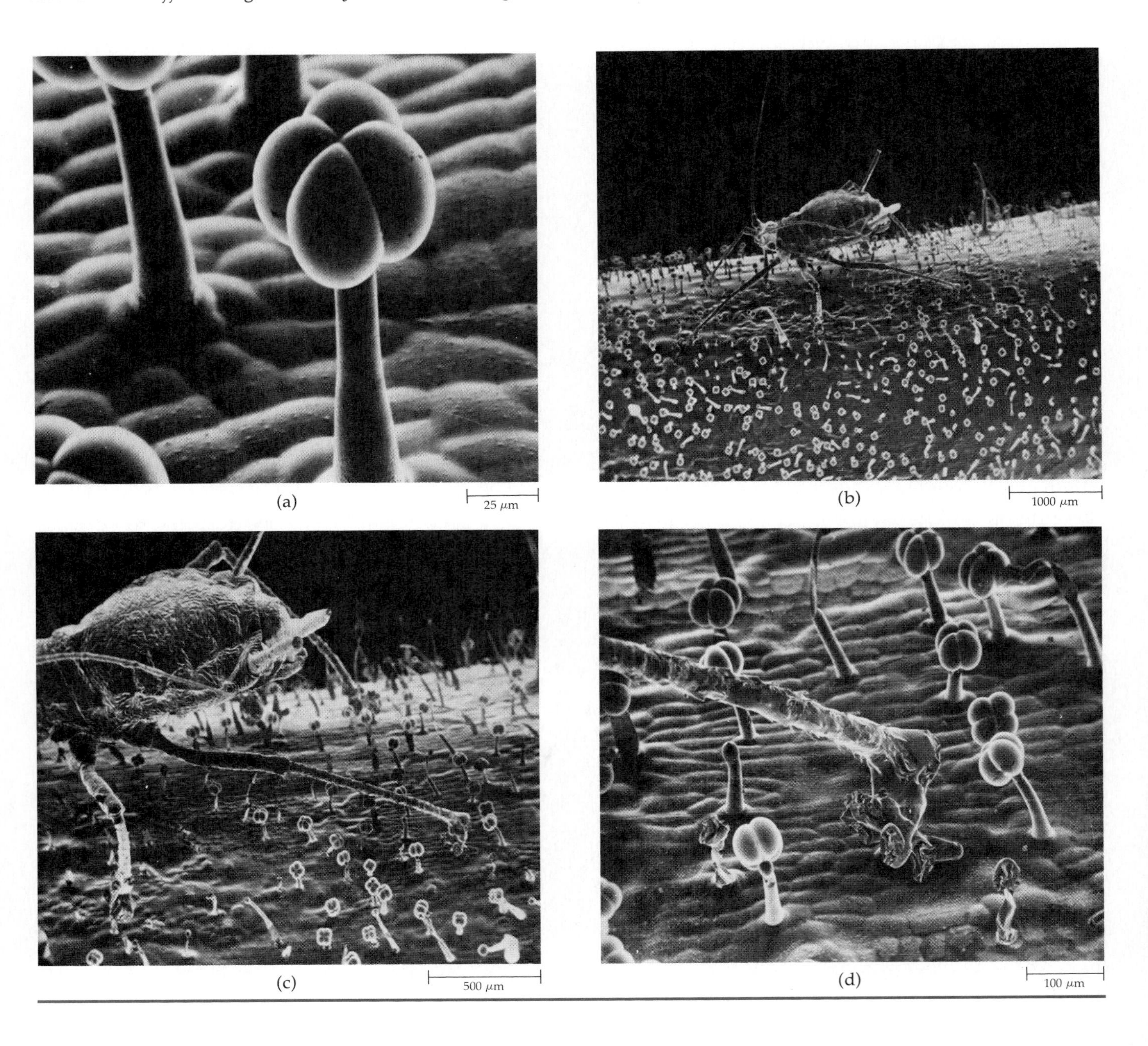

In the winter of 1965–66, all of the trees, saplings, and shrubs in one 15.6-hectare area of the forest were cut down. No organic materials were removed, however, and the soil was undisturbed. During the following spring, the area was sprayed with an herbicide to inhibit regrowth. During the four months from June to September 1966, the runoff of water from the area was 4 times higher than in previous years. Net loss of calcium was 10 times higher than in the undisturbed forest, and potassium loss was 21 times higher. The most severe disturbance was seen in the nitrogen cycle. The tissues of dead plants and animals continued to be decomposed to ammonia or ammonium ions, which then were acted upon by nitrifying bacteria to produce nitrates, the form in which nitrogen is usually assimilated by higher plants. However, no higher plants were present, and the nitrate, a negatively charged ion, was not held in the soil. The net loss of nitrogen averaged 120 kilograms per hectare per year from 1966 to 1968. As a side effect, the stream that drained the area became polluted with algal blooms, and its nitrate concentration increased to levels above those established by the U.S. Public Health Service for drinking water.

TROPHIC LEVELS

In addition to its abiotic, or nonliving, components, each ecosystem includes two biotic elements—autotrophs and heterotrophs. Autotrophs are mainly photosynthesizing organisms which are able to use light energy to manufacture their own food. Heterotrophs cannot manufacture their own food and use the organic molecules made by the autotrophs as a food source. Several feeding levels, or *trophic levels*, are recognized among the heterotrophs. First, there are the *primary consumers*, or herbivores. Second, there are the *secondary consumers*, or carnivores and parasites, which feed on the primary consumers. Finally, there are the *decomposers*. All these levels are represented in any fairly complicated ecosystem.

In a given ecosystem, organisms from each of the trophic levels make up what is called a *food chain* (Figure 29–9). The relationships between the organisms involved in a food chain regulate the flow of energy through the ecosystem. The length and complexity of such food chains vary a great deal. Usually an organism has more than one source of food and is itself preyed upon by more than one kind of or-

29–8
Weir in the Hubbard Brook Experimental Forest in New Hampshire. Water from each of six experimental ecosystems was channeled through a weir, such as this one built where the water leaves the watershed, and was analyzed for chemical elements. The trees and shrubs in the watershed behind this weir have been cut down. The experiments showed that such deforestation greatly increased the loss of nutrient elements from the system.

29–9
A food chain. A three-toed box turtle (Terrapene carolina triunguis) *feeds on a snail that has fed on the mushrooms that are decomposing organic matter in the soil.*

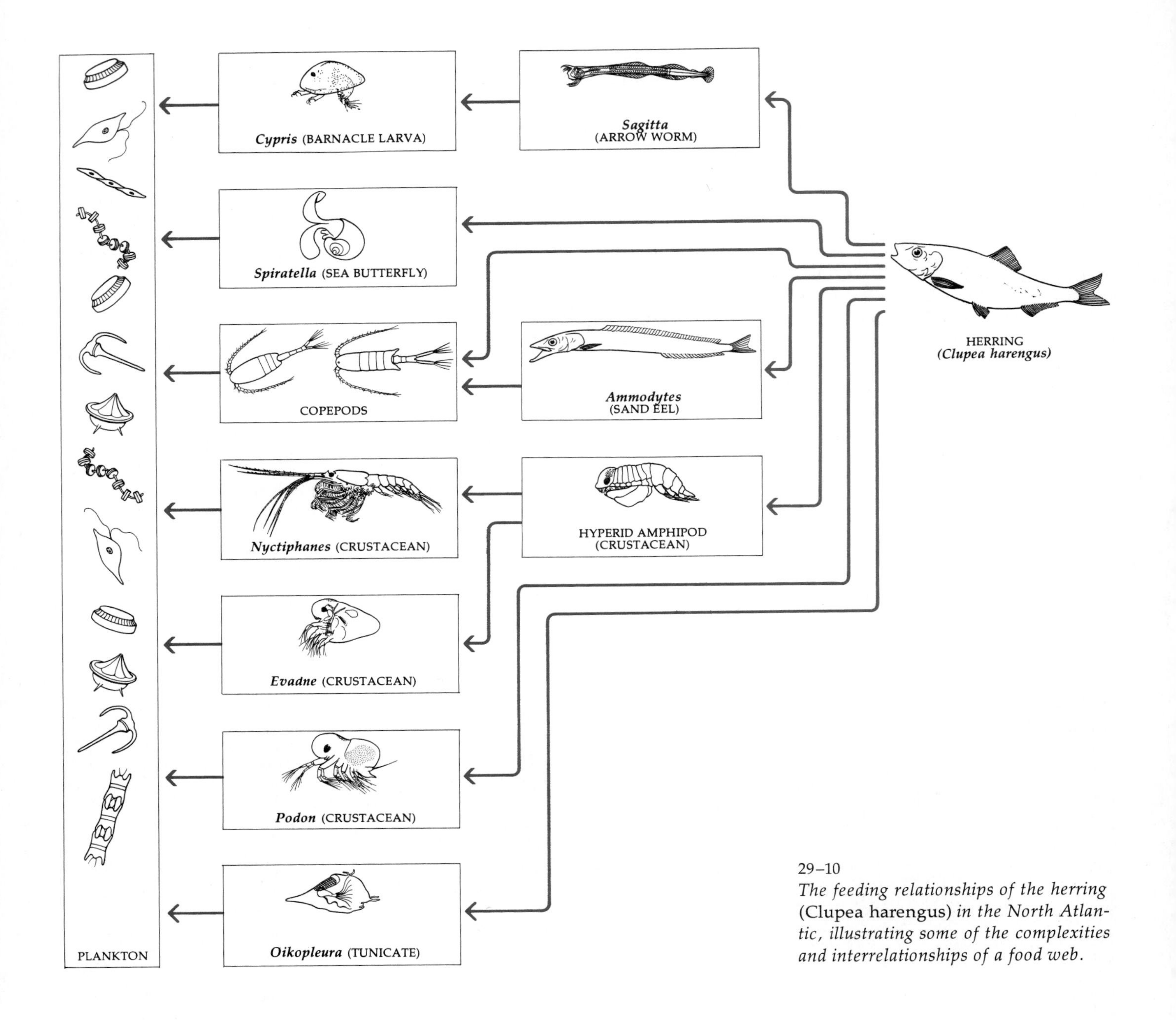

29–10
The feeding relationships of the herring (Clupea harengus) *in the North Atlantic, illustrating some of the complexities and interrelationships of a food web.*

ganism. Under these circumstances, it is perhaps more correct to speak of a *food web* (Figure 29–10). The complexity of trophic relationships has a number of important implications for the overall properties of the ecosystem.

The Flow of Energy

In an ecosystem, the flow of energy begins with photosynthesis and the production of carbohydrate molecules. Energy does not cycle in the ecosystem; it enters as energy captured by photosynthesizing organisms and is then gradually dissipated. A large proportion is lost at each step of a food chain.

To begin with, a very large amount of *biomass* is produced each year. (Biomass is a convenient shorthand term for organic matter; it includes woody parts of trees, stored food, bones, and so on.) It is currently estimated that about 200 billion metric tons of organic material are produced each year. Despite this enormous figure, however, photosynthesizing organisms are not really very efficient in converting the sun's energy into organic compounds. Generally, only about 1 percent of the light that falls on a plant is actually utilized (see Chapter 4). However, particularly productive stands of vegetation may convert up to 3 percent of the annual incident radiation into chemical energy.

When the organic material produced by plants is consumed by an herbivore, energy is released. Most of this energy is lost as heat, and a fraction is converted to animal tissue. In general, only about 10 percent of the usable energy of the plant is added to the mass of the herbivore; the remainder is lost in respiration. A similar gain-loss relationship is found at each succeeding level. Thus, if an average of 1500 calories of light energy per square meter of land surface is utilized by plants per day, about 15 calories are converted to plant material. Of these, about 1.5 calories are incorporated into the bodies of the herbivores that eat the plants, and about 0.15 calorie is incorporated into the bodies of the carnivores that prey on the herbivores.

To give a concrete example, Lamont Cole of Cornell University, in his studies of Cayuga Lake, has calculated that for every 1000 calories of light energy utilized by algae in the lake, about 150 calories are reconstituted as small aquatic animals and 30 calories as smelt. If we were to eat these smelt, we would gain about 6 calories from the original 1000 calories used by the algae. But if trout eat the smelt and we then eat the trout, we gain only about 1.2 calories from the original 1000 calories. From this example it is clear that there is more energy available to us if we eat smelt rather than the trout that feed on the smelt; yet trout are considered a delicacy, smelt a coarse fish. Under conditions of starvation, humans must turn to an all-plant diet, not being able to afford the tenfold loss in energy that occurs when these plants are fed to animals. For humans to make the maximum use of the solar energy trapped by plants, we must become mainly herbivores.

Food chains are generally limited to three or four links; the amount of food remaining at the end of a long food chain is so small that few organisms can be supported by it. Body size also plays a role in the structure of food chains. For example, an animal constituting one link generally has to be large enough to capture prey from the previous, lower link on the food chain.

Owing to the relationships just discussed, the total biomass at successive trophic levels in an ecosystem generally decreases sharply, setting up the sort of relationship described by the expression "pyramid of mass" (Figure 29–11). If energy is measured, it follows the same rapid decrease characteristic of mass; there is a "pyramid of energy," with far less energy in the bodies of all the predators present in a given community, for example, than in all the plants. In general, there are also far more individuals at the lower levels than at the higher levels, which leads to a "pyramid of numbers." It also follows that if all the organisms in an ecosystem are divided into size classes, the small animals will be far more numerous than the large ones.

DEVELOPMENT OF ECOSYSTEMS

Succession

Some plant communities remain the same year after year, whereas others change rapidly. A cleared woodlot is rapidly colonized by the remaining trees of the vicinity; likewise, a pasture eventually gives way to a forest. A similar series of events occurs in naturally open areas, such as lakes, meadows, or rocky hillsides. All of these are examples of *succession*, a process that is continuous and worldwide in scope.

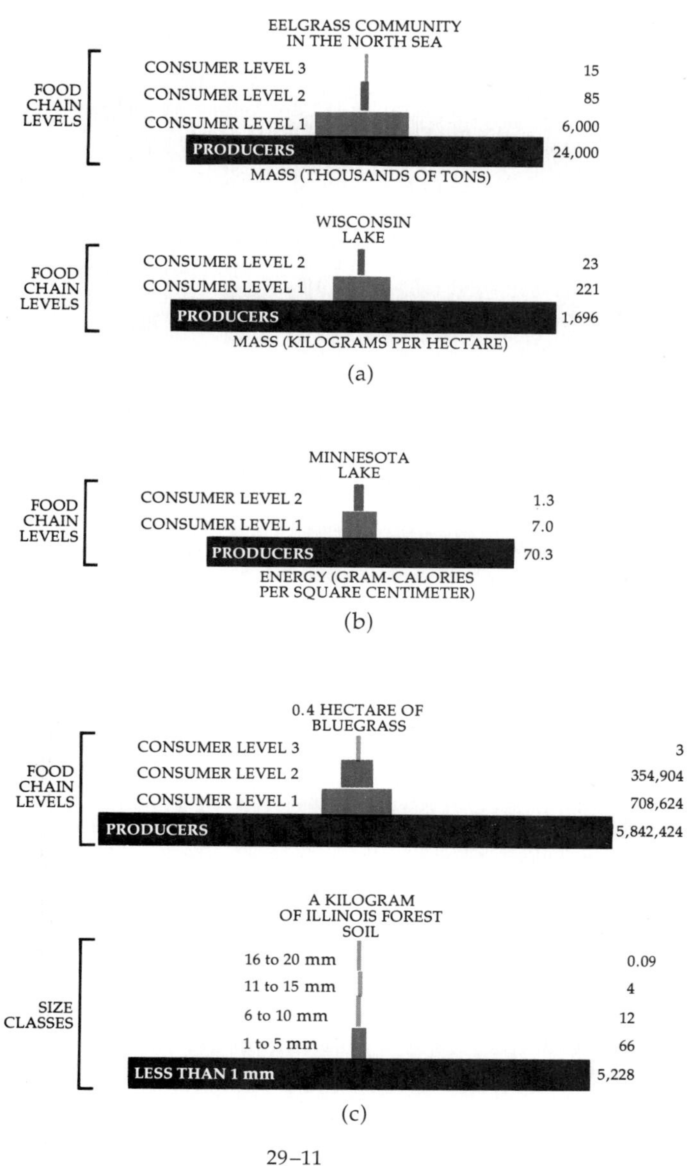

29–11
Pyramids of (a) *mass,* (b) *energy, and* (c) *numbers of organisms in various communities. A relatively small amount of mass or energy is transferred to each successively higher level.*

29–12
(a) *Emerging vegetation grows along the edge of a pond.* (b) *Aquatic plants with floating leaves, such as water lily* (Nymphaea odorata), *grow across the surface of a pond and eventually choke out bottom-dwelling plants.* (c) *Water hyacinths play a similar role in warmer climates.* (d) *Marsh grasses, sedges, and cattails growing on an old pond bed complete the process of succession.*

Succession occurs at a variable rate in all temporarily open areas. Some ponds, for example, fill with aquatic plant remains and debris; emergent vegetation builds soil; the site is taken over by meadow; moisture-loving shrubs may come in; and finally the forest characteristic of the region develops in the meadow that was formerly a pond (Figure 29–12). In another example, lichens growing on dry rocks break down the rocks directly because of the chemicals they secrete; soil accumulates around the bases of the lichens; they then give way to mosses and finally to flowering plants (Figure 29–13). The roots of the flowering plants probe cracks, breaking the rocks down further. Eventually, perhaps after many centuries, the rock is reduced to soil, and the soil is occupied by the forest or other vegetation types characteristic of the region. Other examples of succession are shown in Figure 29–14.

The Climax Community

An ecosystem undergoes various stages of succession until, finally, a mature ecosystem, or more commonly, a *climax community*, is produced. Although the nature of the particular climax community varies according to the climate of the area, such a community is relatively self-sustaining. The organisms in the community have achieved a most intricate set of interrelationships. Each climax community constitutes part of a self-contained ecosystem driven by solar energy.

29–13
An early stage of succession on a rocky slope. Lichens have begun to accumulate soil, and a small fern has sprung up in a small crevice.

(a)

(b)

29–14
(a) *Seedling trees of balsam fir* (Abies balsamea) *growing under and replacing quaking aspen* (Populus tremuloides) *in northern Minnesota—a stage in forest succession leading to a climax community of white spruce and balsam fir.* (b) *Seedling of a red maple* (Acer rubrum) *rising above needles of white pine* (Pinus strobus). *Mature white pines filter the light so that their own seedlings cannot survive and only those tolerant of shade, such as those of maples and oaks, can gain a foothold.*

Some aspects of succession and climax have great significance for humans. For example, when western European settlers first came to California in large numbers, they found a magnificent forest of sugar pine *(Pinus lambertiana)* along much of the length of the Sierra Nevada. Later, although conservationists tried to preserve some of these forests in national parks and forests, many of the stands of pines were eventually replaced by other trees, such as white fir *(Abies concolor)* and incense cedar *(Calocedrus decurrens)*. Why did this change take place?

The answer is that the sugar pine was a member of a climax community before the advent of a large human population. Thereafter, the forest that contained it was no longer a climax community for the

(a)

29–15
(a) *When fire sweeps through a forest, secondary succession—with regeneration from nearby unburned stands of vegetation—is initiated. Some plants produce sprouts from the base, others seed abundantly on the burn. In one group of pines, the closed-cone pines, the cones do not open to release their seeds until they have been exposed to fire.* (b) *Sugar pines* (Pinus lambertiana) *in the southern Sierra Nevada of California. With the control of forest fires by humans, sugar pines are being replaced by other trees such as the incense cedar* (Calocedrus decurrens), *the large tree on the right.*

(b)

region. The variable was fire, which was greatly reduced after the influx of human inhabitants. Without periodic, lightning-set fires of low intensity racing through the groves, a thick growth of brush and smaller trees arose and prevented reproduction of the sugar pines. Only a system of controlled burning can preserve the remaining groves of sugar pine in their original form (Figure 29–15).

Recolonization

When humans alter a landscape, changes are made in the community structure. Given sufficient time, however, successional processes may gradually restore the original vegetation to the area. For example, in the northern hardwood forests of North America and Eurasia, it is estimated that 60 to 80 years may be required to replace the plant biomass and nutrients lost in harvesting the trees. In other communities, the process may be faster or slower.

In abandoned fields, denuded sand dunes, and on the streets of the ghost towns of the American West, succession is taking place, and ecosystems that more and more closely resemble those of adjacent, less-disturbed areas are being produced (Figure 29–16).

Following natural disasters, recolonization produces similar changes. For example, in August 1883, a violent explosion destroyed half of the island of Krakatau, in the Java Straits about 40 kilometers from Java; the remaining half was covered by a layer of pumice and ash more than 30 meters thick. The

29–16
An old, abandoned field in Wake County, North Carolina, with a young stand of loblolly pine (Pinus taeda). *The plow furrows of the abandoned field are still visible.*

neighboring islands of Verlaten and Lang were also buried, and the entire assemblage of plants and animals on these islands was wiped out. Soon afterward, however, the recolonization of the island began, and the expected number (based on the number originally occupying the area) of about 30 species of land and freshwater birds was reached in about 30 years. Recolonization by plants also proceeded rapidly, with a total of over 270 species being recorded for the island of Krakatau by 1934.

In the Mexican state of Michoacán, Paricutín Volcano erupted violently in February 1943. During the period of the eruption, all vegetation was destroyed over an area of about 13,000 hectares. Within three years after the eruption ceased in March 1952, lichens, algae, and mosses were growing on the lava flows. Ferns appeared a year later, and flowering plants appeared on the rim of the crater within five years. It is likely that complete weathering of the volcanic rock and reforestation may take several centuries. Figure 29–17 shows another example of the results of a volcanic eruption.

29–17
Yapoah Crater, a volcanic cinder cone east of the Cascade Mountains in central Oregon. Succession leading to the establishment of a climax forest on such a cone may take centuries and is often interrupted by further volcanic activity before it is complete.

Structural Changes During Succession

A number of important changes in the structure of an ecosystem occur during the course of succession.

First, the kinds of plants and animals change continuously. Under dry conditions, as on a bare, rocky hillside, the first organisms to become established are those that are able to endure all environmental conditions. These include lichens and mosses, and such animals as mites, arachnids, and small insects. Appropriately, these are called "pioneer organisms," and they are said to form a pioneer community. In general, they are relatively simple organisms, and they form short food chains, that is, ones with few links. They depend on disturbance for their continued existence in the area. Often the plants, animals, and other organisms characteristic of this particular successional stage are present only at this time. They are not present in the intermediate or final stages of succession.

Second, the amount of organic matter in the community that is incorporated into living organisms (or that is in the remains of once-living organisms) increases during the course of succession. Larger and more complex organisms appear, and food chains lengthen. Some of the new organisms persist for a long time in the community, and others are relatively short-lived. The kinds of interrelationships between organisms become more and more complex.

Third, the overall diversity of species, particularly heterotrophic groups of organisms, tends to increase with succession. The relationships between these organisms are more and more difficult to discern, and the feeding range of each species is apt to become narrower and narrower as the community approaches its mature condition.

Finally, autotrophic organisms such as plants occupy each site more rapidly than heterotrophic organisms. Therefore, in any successional series, the maximum production for the site is reached relatively early. Afterward, respiration for the entire ecosystem continues to increase, but the net production for the ecosystem decreases as the proportion of the heterotrophs and the complexity of the food chains increase.

SUMMARY

The ecosystem is the highest level of biological integration; it is a self-sustaining system driven by energy from the sun, in which the regulated cycling of essential materials takes place.

Some of the relationships that occur in ecosystems can be grouped under three headings: mutualism, competition, and plant-herbivore relationships. In mutualism, both interacting populations benefit. Examples include the formation of lichens, the growth of nitrogen-fixing bacteria in nodules on the roots of legumes, mycorrhizal associations between fungi and the roots of higher plants, and closely linked pollinator-flower relationships.

Competitive interactions are found between most plants that grow together. One of the most important kinds of interaction concerns competition for light. Plants have also evolved chemicals that limit or enhance the success of other plants in their vicinity, and such allelopathic relationships are of great importance in determining the structure of communities.

Herbivores control the reproductive potential of plants by destroying their photosynthetic surface or by directly eating their reproductive structures. Plants counter these attacks through the evolution of spines, tough leaves, or most importantly, chemical defenses. When an insect has overcome these chemical defenses, it not only has a new and often largely untapped food resource at its disposal but may also utilize the toxic substances to gain a degree of protection from its own predators. Pollination relationships are a special case of plant-herbivore interactions in which the emphasis is on attraction, not repulsion. Some plant-herbivore interactions involve a high degree of mutualism; the obligate interaction between the bull's-horn acacias and their associated ants in eastern Mexico is such a system.

An ecosystem consists of nonliving elements and two different kinds of living elements—autotrophs and heterotrophs. Among the heterotrophs are the primary consumers, or herbivores; the secondary and tertiary consumers, or carnivores and parasites; and finally, the decomposers. The organisms found at these levels are members of food chains or food webs.

Energy flows through an ecosystem, with about 1 percent of the incident light being converted into chemical energy by green plants. When these plants are consumed, about 10 percent of their potential energy is stored at the next trophic level; a similar degree of efficiency characterizes transfers further up the food chain. The amounts of energy remaining after several transfers are so small that food chains rarely exceed three or four links in length.

Succession occurs when an area is denuded by artificial or natural means, or when a new area appears, such as a lava bed. The kinds of plants and animals change continuously, some being characteristic only of early stages of succession and others only of middle stages. The amount of organic material that is incorporated into organisms increases during succession, for this is a time when the biomass of the community grows rapidly. The diversity of species in the ecosystem increases greatly, and the relationships between them become more and more complex. Production (related to the number of photosynthetic organisms) reaches a peak early, and then respiration increases rapidly in the later stages of succession. Eventually, succession results in the production of a climax community, which reproduces itself indefinitely unless there are major environmental changes.

Chapter 30

Terrestrial Biomes

30–1
Three of the organisms that occur together in the redwood (Sequoia sempervirens) *community along the coast of California and Oregon are bracket, or shelf, fungus, a basidiomycete; sword fern* (Polystichum munitum); *and redwood sorrel* (Oxalis oregona). *Redwoods are shown with some of their other associated plants in Figure 29–1.*

A *biome* is a climatically controlled group of plants and animals that has a characteristic appearance and that is distributed over a wide geographical area. Terrestrial biomes can be classified in a number of different ways, but the categories that are discussed in this book provide a basis for an overall account of the vegetation of the world.

The distribution of biomes (see Figure 30–5, page 600) results from three kinds of physical factors: (1) global patterns of air circulation, particularly the directions in which the prevailing moisture-bearing winds blow; (2) the distribution of heat from the sun and the relative seasonality of different portions of the earth; and (3) such geological factors as the distribution of mountains and their height and orientation. In very general terms, the biomes that are geologically oldest are those found nearest the equator and at the lowest elevations.

LIFE ON THE LAND

Land plants face a variety of problems. They are subjected to periodic drought and to rapid diurnal and seasonal changes in temperature. They must survive during unfavorable seasons, they must often grow on substrates of varying mineral composition, and they are subject to the action of gravity (which affects land plants much more directly than it does water plants). Except at high elevations, however, oxygen is more uniformly distributed than in the sea, and carbon dioxide is readily available.

The earth's land areas are discontinuous, and this separation of land masses has an important effect on the distribution of organisms. In addition, the radical changes that often occur from place to place on the land mean that the distribution of any particular kind of land organism is apt to be much more limited than would be the distribution of an organism of similar size and motility in the sea.

Local Modifications in Climate

Plants and plant communities vary as the land varies. For instance, the mean atmospheric temperature decreases about 0.5°C for each degree of latitude increase. Increases in elevation produce a similar effect, and in general, a change in mean atmospheric temperature corresponding to a one degree increase in latitude occurs with each rise of about 100 meters in elevation. This relationship has important consequences for the distribution of land organisms. For example, plants and animals characteristic of arctic regions may approach or even reach the equator, particularly in mountain ranges running from north to south (Figure 30–2).

There are, however, important differences between high-latitude habitats and high-altitude habitats. In the mountains the air is clearer and the solar radiation is more intense. Most of the water vapor in the atmosphere—which plays a major role in preventing heat from radiating away from the earth at night—occurs below 2000 meters. Consequently, nights in the mountains are often much cooler than those at lower elevations in the same latitude (Figure 30–3). Moreover, there is pronounced seasonal variation both in day length and in temperature near the poles and relatively little variation near the equator. Those "arctic" organisms that do range toward the equator in the mountains must make physiological adjustments for such differences between high-latitude (arctic) and high-altitude (tropical) environments.

There are often pronounced temperature variations from slope to slope on a particular mountain. In the middle latitudes of the Northern Hemisphere, for example, sunlight reaches only the southern and western slopes of mountains in the winter, and such slopes are often drier than the northern and eastern slopes of the same mountains. Furthermore, the afternoon sun in the summer is much hotter than the morning sun because of a decrease in the water vapor in the atmosphere. Because of the interaction of these two factors, the driest slope of a mountain in the Northern Hemisphere is usually that oriented toward the southwest. (Which would you expect to be the driest slope of a mountain in the Southern Hemisphere?) On low hills in an otherwise barren arctic tundra, clusters of small trees and shrubs may occur on the southern slope. For the same reasons, "moss" grows, or is better developed, on the northeast side of trees in the Northern Hemisphere.

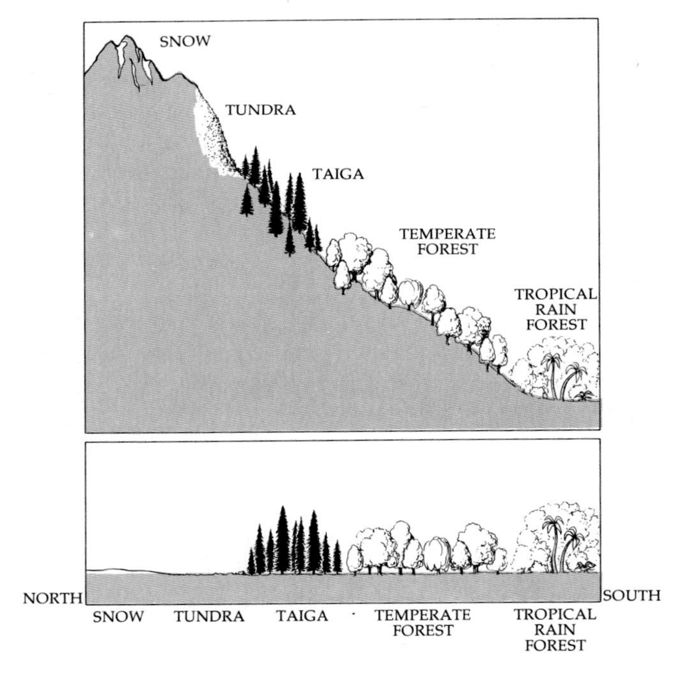

30–2
We can experience a similar series of plant communities whether we travel north for hundreds of kilometers or ascend a mountain. This particular kind of relationship between altitude and latitude was first pointed out by Alexander von Humboldt.

30–3
In high mountains in the tropics, the temperatures do not change seasonally, but freezing may occur every night of the year. These unusual conditions have resulted in the evolution of plants with bizarre growth forms, such as these giant senecios (family Asteraceae) on Mt. Kenya in East Africa. In such plants, the apical meristem is protected from freezing by the tightly infolded leaves, thick fleshy construction, and a matted covering of reflective silvery hairs.

ALEXANDER VON HUMBOLDT

Alexander von Humboldt (1769–1859) was perhaps the greatest scientific traveler who ever lived. Humboldt ranged widely across the trackless interior of Latin America around the start of the nineteenth century and climbed some of its highest mountains. Exploring the region between Ecuador and central Mexico, Humboldt was the first to recognize the incredible diversity of tropical life and, consequently, the first to realize just how many species of plants and animals there must be in the world.

In his travels, Humboldt was impressed with the fact that plants tended to occur in repeatable groups, or communities, and that whenever there were similar conditions—relating to climate, soil, or biological conditions—similar groupings of plants appeared. He also discovered a second major ecological principle—the relationship between altitude and latitude. He found that climbing a mountain in the tropics was analogous to traveling farther north (or south) from the equator. Humboldt illustrated this point with his well-known diagram of the zones of vegetation on Mt. Chimborazo in Ecuador.

On his return from Latin America in 1804, Humboldt visited the United States for eight weeks. He spent three of these weeks as Thomas Jefferson's guest at Monticello, talking over many matters of mutual interest, and it is thought that Humboldt's enthusiasm for exploring America encouraged Jefferson's own great scheme for the exploration of the western United States. Thus, it is fitting that Humboldt's name is commemorated in the names of several counties, mountain ranges, and rivers in the American West. After returning to Berlin, Humboldt lived for more than half a century, dying in his ninetieth year. He was one of the greatest writers and scientists of his era.

As a result of the complex interactions between varying local conditions, the distribution of biomes is not what we would expect if the earth were a perfectly round sphere. In the following pages, some examples are provided of the ways in which climate, topography, and soil type interact to control the distributions of the major vegetation types.

TROPICAL RAIN FOREST

More species of plants and animals live in the tropical rain forest than in all the rest of the biomes of the world combined. Neither water nor temperature are limiting factors during any part of the year. Although there are many species, there are few individuals per species; a species may be represented only once per hectare. Not only is there a large number of different organisms in the tropical rain forest but their interrelationships are more complex than those of plants and animals in any other biome (Figure 30–4).

Little light penetrates to the floor of the tropical rain forest, and the rainfall is generally between 200 and 400 centimeters per year. There is little accumulation of organic debris; decomposers rapidly break down leaves and stems and the bodies of animals that fall from the canopy. Nutrients released from this breakdown are quickly absorbed by mycorrhizal roots or are leached from the soil by the rain. Although there may be notable variation in precipitation from month to month, there is no pronounced dry season.

Generally, plants in the tropical rain forest have not evolved particular mechanisms that would permit them to survive unfavorable seasons of drought or cold. Nearly all the plants are woody, and woody vines are abundant. There is a large flora of *epiphytes*, which grow on the branches of other plants in the illuminated zone far above the forest floor. Epiphytes, which have no direct contact with the forest floor, obtain water and minerals from the humid air of the canopy. Along with the epiphytes and climbing vines, many kinds of animals have moved into the treetops; this is the area of the tropical rain forest in which animal life is most abundant and diverse.

RODENT POLLINATION OF TROPICAL PLANTS

New and complex interrelationships between plants and animals continue to be discovered in the tropics. In 1979, for example, Cecile Lumer of the New York Botanical Garden discovered the first New World example of pollination of flowers by rodents. In this nighttime photograph, a rice rat (Oryzomus devius) *is shown drinking nectar from a flower of the epiphytic plant* Blakea chlorantha *in a Costa Rican cloud forest. The flowers of this plant exhibit several characteristics that can be related to the attraction of nonflying mammals—inconspicuous green blossoms that open at night and a well-hidden source of copious, sucrose-rich nectar. When the rodent grasps the flower and inserts its tongue to probe for nectar, the pollen is released explosively from the anthers, dusting the rodent's face. The pollen is then transferred to the stigma of the next flower visited. The pollination of flowers by mammals other than bats has also been discovered recently among marsupials in Australia and South America, lemurs in Madagascar, monkeys in South America, and rodents in Hawaii and South Africa.*

Because so little light reaches the forest floor, there are very few herbaceous plants, and those that do occur are mostly epiphytes or grow in clearings. Many plants in the tropical rain forest are trees; they are usually larger than those found in temperate forests, often reaching 40 to 60 meters in height. Furthermore, the trees are diverse; there are seldom fewer than 40 species per hectare. This is in sharp contrast with temperate forests, where there are rarely more than a few species per hectare. The trees of the tropical rain forest are remarkably homogeneous in appearance. Generally, they branch only near the crown. Because their roots are usually shallow, they often have buttresses at the base of the trunk to provide a firm, broad anchorage. Their leaves are medium-sized, leathery, and dark green; their bark is thin and smooth; and their flowers are generally inconspicuous and greenish or whitish in color. Such forests often form several layers of foliage, the lower layers consisting of seedlings of the taller species, plus a few lower-growing species.

There are three major areas of the world in which the tropical rain forest is well developed. The largest is in the Amazon basin of South America, with extensions into coastal Brazil, Central America, and eastern Mexico, as well as some of the islands in the West Indies. In Africa, there is a large area of rain forest in the basin of the Zaïre, with an extension along the west coast of Liberia. The third area of rain forest extends from Ceylon and eastern India to Thailand, the Philippines, the large islands of Malaysia, and a narrow strip along the northeast coast of Queensland, Australia (see Figure 30–5).

The tropical rain forest now forms about half the forested area of the earth, but it is in the process of being systematically destroyed by human activities. The rapidly expanding human population in the tropics, coupled with traditional patterns of land ownership, have made the usual practices of tropical agriculture—clear-cutting, followed by short-term cultivation—immensely destructive when carried out on such a wide scale. Many kinds of plants and animals will die in the course of this destruction.

One reason for this particular level of destruction is the nature of tropical soils. Many of these soils are conditioned by high and constant temperatures and abundant rainfall and are relatively infertile. Some of them are known as latosols—red clays largely leached of their nutrients. Tree roots do not reach deeply into tropical soils, although the processes leading to soil formation may extend to depths greater than 15 meters. When clearing is carried out, the leaching process accelerates greatly, and the soils either erode rapidly or form thick, impenetrable crusts on which cultivation is not possible. Despite this, the tropical forests of the world are being cut and burned at an ever-increasing rate, mainly to produce fields which become completely useless to agriculture within a few years. It is estimated that by the end of the present century, most of the tropical rain forests will have disappeared, with the exception of those in the western Amazon Basin and central Africa.

(a)

(b)

(c)

(d)

(e)

30–4
Rain forest. (a) *Dense growth at the edge of a Venezuelan forest.* (b) *A huge buttressed tree in a lowland forest of West Africa.* (c) *Stinkhorn fungus* (Dictyophora duplicata). (d) *Epiphytic bromeliads on a tree trunk.* (e) *Fruit of the tropical American tree* Theobroma cacao, *the seeds of which are the source of cacao, or chocolate.*

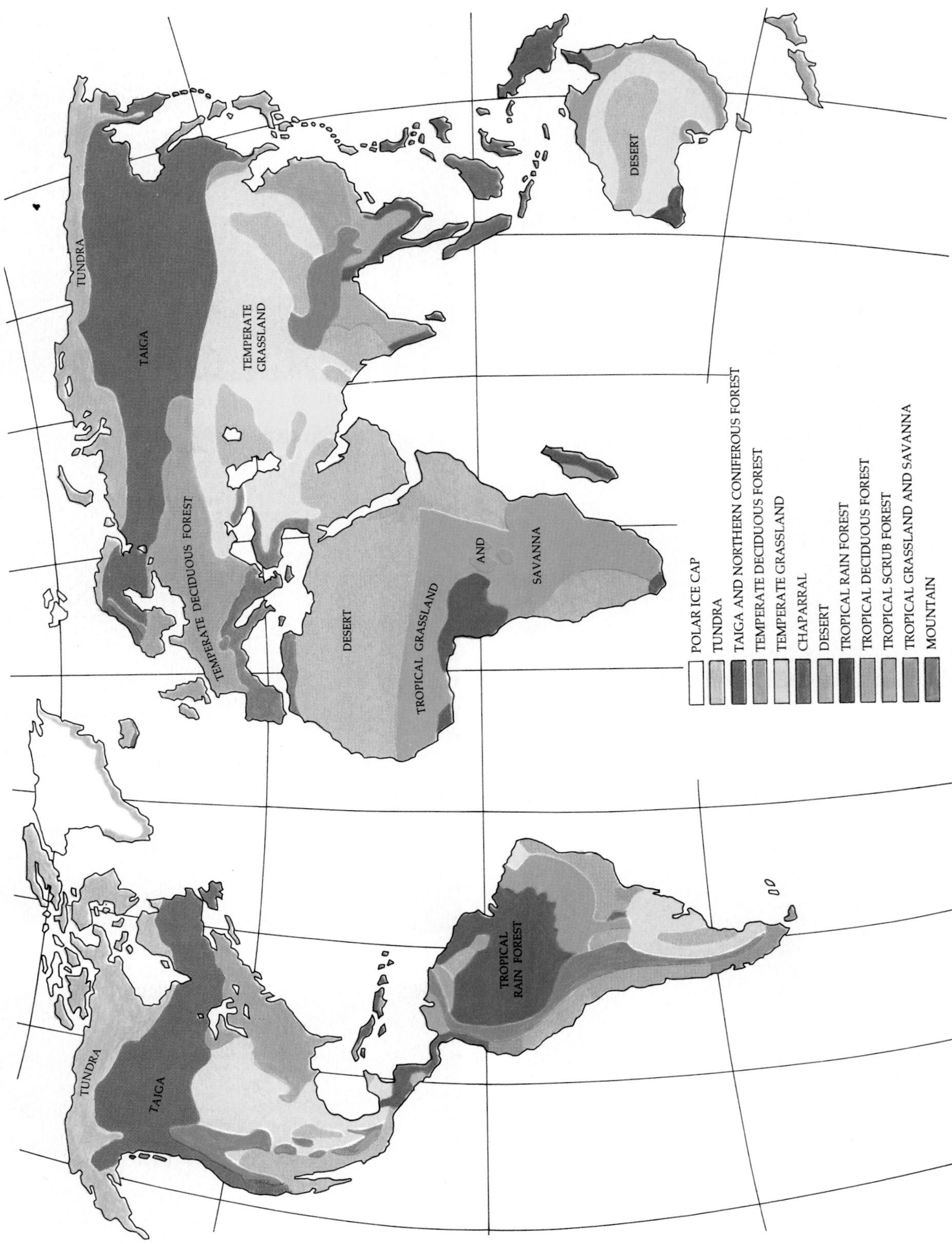

TUNDRA
TAIGA
TROPICAL RAIN FOREST
TUNDRA
TAIGA
TEMPERATE DECIDUOUS FOREST
TEMPERATE GRASSLAND
DESERT
TROPICAL GRASSLAND
AND
SAVANNA
DESERT

POLAR ICE CAP
TUNDRA
TAIGA AND NORTHERN CONIFEROUS FOREST
TEMPERATE DECIDUOUS FOREST
TEMPERATE GRASSLAND
CHAPARRAL
DESERT
TROPICAL RAIN FOREST
TROPICAL DECIDUOUS FOREST
TROPICAL SCRUB FOREST
TROPICAL GRASSLAND AND SAVANNA
MOUNTAIN

30-5
Biomes of the world.

30–6
A savanna in Kenya, with zebras, a giraffe, and an impala. The transitional nature of this biome, relative to the characteristic vegetation of the tropical rain forest biome and the desert biome, is evident in the grasses, shrubs, and short trees seen here. The trees in the background are acacias.

SAVANNA

The savanna is a transitional area between the evergreen tropical rain forest biome and the desert biome (Figure 30–6). Savannas usually have much lower annual rainfall than the tropical forest—frequently in the range of 90 to 150 centimeters a year. There is also a wider fluctuation in average monthly temperatures, owing to the seasonal drought and sparse covering of vegetation.

Many plant communities in the tropics are characterized by a marked period of drought each year. One of the most widely distributed plant communities is the tropical grassland, which has widely scattered trees that are generally deciduous, losing their leaves in the dry season. Such regions of tropical grassland cover large areas of East Africa (see Figure 30–5), but they are also found on the margins of rain forests everywhere.

In Southeast Asia, there are extensive areas of monsoon forest. Here there is high precipitation during portions of the year, when the moisture-bearing monsoon is blowing off the ocean, but there also exists a well-defined dry season during which the trees lose their leaves. Thorn forests occur under drier conditions found in other parts of the world; they are characterized by dense, spiny, low trees. In drier regions, the thorn forests grade into the desert.

The existence of the thorn forests and open grassy plains of central Africa depends to a large extent on periodic burning. The winter is long and hot, and the natives frequently burn the plain to produce young grass for the game that they hunt, as well as for their domestic herds of grazing animals.

In savanna communities, because of the scattered distribution of trees, the forest floor is generally well illuminated, and perennial herbs (mostly grasses) are common. Bulbous plants, which are able to withstand periodic burning, are abundant. Because of the dense cover of perennial herbs made possible by the abundant, seasonal rainfall, there are few annual herbs. Epiphytes are also rare.

The trees found in savanna communities often have thick bark. They are well branched but are seldom more than 15 meters tall. Almost all of them are deciduous; they lose their leaves at the start of the dry season and flower in a leafless condition. Their leaves are smaller than those characteristic of the evergreen trees of the rain forest and so lose less water by transpiration.

30–7
North American deserts. The Sonoran Desert stretches from southern California to western Arizona and south into Mexico. Some of its characteristic plants are shown in Figures 30–8a and 30–8b. North of the Sonoran Desert is the Mojave Desert; one of its characteristic plants is the Joshua tree (Yucca brevifolia). *The Mojave Desert includes Death Valley, the lowest point on the continent (90 meters below sea level), which is only 130 kilometers from Mt. Whitney, the highest point in the continental United States (at an elevation of more than 4000 meters). The Mojave Desert blends into the Great Basin Desert, a cold desert bounded by the Sierra Nevada to the west and the Rocky Mountains to the east. Vast stretches of the Great Basin Desert are dominated by sagebrush* (Artemisia tridentata) *and rabbitbush* (Chrysothamnus). *To the east of the Sonoran Desert is the Chihuahuan Desert.*

(a)

30–8
Desert plants. (a) *Washington palms* (Washingtonia filifera) *in the Sonoran Desert of Baja California.* (b) *Community of saguaro* (Carnegiea gigantea) *and barrel cactus* (Ferocactus) *in the Sonoran Desert of southern Arizona.* (c) *Carpet of annual plants flowering in the Mojave Desert of California.*

DESERT

The great deserts of the world are all located in the zones of atmospheric high pressure that flank the tropics at about 30° north latitude and 30° south latitude, and they extend poleward in the interior of the large continents (see Figure 30–5). Many deserts are characterized by less than 10 centimeters of rain per year. In the Atacama Desert of coastal Peru and northern Chile, the average rainfall is less than 2 centimeters per year. Extensive deserts are located in North Africa and in the southern part of the African continent, in the Near East, in western North and South America, and in Australia. The Sahara Desert, which extends all the way from the Atlantic coast of Africa to Arabia, is the largest in the world. Australia's desert, which covers some 44 percent of the continent, is the next largest. Less than 5 percent of North America is desert (Figure 30–7).

Desert regions are characterized by very high temperatures; summer temperatures of over 36°C are common. On the other hand, because of the sparse plant cover, there is little water vapor in the desert atmosphere, and the consequent rapid radiation of heat at night creates large differences in temperature in the course of a single 24-hour period.

(b)

(c)

The annual distribution of rainfall in desert areas generally reflects that of the bordering areas. On the equatorial side, it rains in the summer; on the pole-ward side, it rains in the winter. Between the two, as in the lowlands of Arizona, there may be two annual peaks of precipitation. As a result, in such an area there are generally two periods of active plant growth—one in the winter and one in the summer—and different plants are active in each period. In general, the patterns of activity of desert plants reflect their respective origins—the present deserts of the world are of relatively recent origin, and their plants and animals have been selected from those found in the bordering areas.

Annual plants are most important, both in number and kind, in the desert and semiarid regions of the world (Figure 30–8c). Because of the erratic supply of water, perennial herbs do not succeed well in such regions, and there is no dense covering of perennials which would normally inhibit the growth of annuals. Because of their very rapid growth, annuals can spring up in the open areas during the limited periods when water is available. The seeds of these plants can survive in the soil during the long periods of drought, which sometimes extend over many years. Then, when sufficient water is available, they can germinate rapidly.

The relatively few perennial herbs that do occur in the desert are generally bulbous and dormant for much of the year. Most of the taller plants are either succulents—such as the cacti, euphorbias, and other characteristic desert plants (Figure 30–8b)—or have small leaves that are either leathery or are shed during unfavorable seasons. Usually the leaves have a thicker cuticle and fewer stomata than those of plants in less arid regions. During dry periods, photosynthesis often occurs in the stems as well as in the leaves. Many of the succulent plants have adopted CAM photosynthesis (see pages 111 and 473) and absorb carbon dioxide only at night. Temperatures for maximum photosynthesis are often much higher for desert plants than for other species, and in many, the orientation of the leaves minimizes heat absorption. Woody plants either have wide-ranging roots that effectively absorb the periodic rainfall (the majority of the desert shrubs) or are restricted to washes and arroyos where the moisture is concentrated (most desert trees).

GRASSLANDS

The grasslands include a wide variety of plant communities; some are related to savannas, others to deserts, and still others to temperate deciduous forests. As the amount of precipitation decreases, the major grasslands of the world often grade into deserts toward the equator. The richer grasslands have as much as 100 centimeters of precipitation per year; they grade into temperate deciduous forests where the moisture supply is more abundant.

Grasslands form the zone that lies between deserts and temperate woodlands, occurring where the amount of rainfall is intermediate between that characteristic of these two biomes. When disturbed, grasslands have often been changed into either woods or deserts. The failure of humans to recognize the existence of the delicate ecological balance in grasslands was the major cause of the "dust bowl" disasters of the central United States in the 1930s (Figure 30–9).

30–9
The prairie soils were once so bound together with the roots of grasses that they could not be cultivated until adequate plows were developed. But once the plants were removed by overgrazing or careless cultivating, prairie soil rapidly deteriorated and was carried away by the wind. This photograph, taken in Texas County, Oklahoma, in 1937, vividly recalls the "dust bowl" conditions that led to migrations away from the central United States. John Steinbeck's novel, The Grapes of Wrath, *was based on the experiences of these migrants.*

The grasslands have traditionally been heavily exploited for agriculture, both as pastureland and, when cleared, for cultivation. Much effort has been made in learning how to convert other biomes into grassland and, once the conversion has been made, learning how to maintain them in this condition. The most productive soils for agriculture occur in areas formerly occupied by tall-grass prairies.

Grasslands generally occur over large areas in the interior portions of the continents (see Figure 30–5). In North America, there is a transition from the more desertlike, western, short-grass prairie (the Great Plains), through the moister, richer, tall-grass prairie (the Corn Belt), to the eastern temperate deciduous woodland (Figures 30–10 and 30–11). Grasslands become drier and drier at increasing distances from the Atlantic Ocean and the Gulf of Mexico, which are the major sources of moisture-bearing winds in the eastern half of the North American continent.

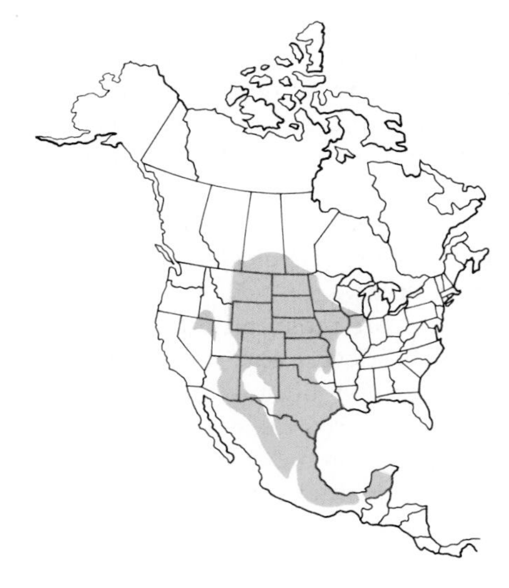

30–10
Distribution of the grasslands of North America.

Perennial bunchgrasses and sod-forming grasses are dominant, but other perennial herbs are common. Although the growth of grassland plants may be somewhat seasonal, there is little room for the development of annual herbs, and these are essentially absent in this biome. Occasionally, annuals and weeds from other areas become established in disturbed areas, such as around the burrows of animals, near buildings, and along roadsides.

The great grasslands of the world are inhabited by herds of grazing mammals that are associated with a complex of large predators. Many of these, such as the American bison, have been hunted almost to the point of extinction and survive mainly in refuges, having given way to herds of domestic animals and cultivated fields.

30–11
(a) *Tall-grass prairie dominated by big bluestem* (Andropogon gerardi) *in eastern Kansas. The plants shown here are just over two meters tall.* (b) *Short-grass prairie in northeastern Colorado, with a young pronghorn antelope* (Antilocapra americana). *The bluish clumps of vegetation are fringed sagebrush* (Artemisia frigida). (c) *Overgrazed short-grass prairie in northeastern Colorado.*

(a)

(b)

(c)

TEMPERATE DECIDUOUS FOREST

Temperate deciduous forest is almost absent in the Southern Hemisphere but is represented in all the major land masses of the north (see Figure 30–5). This type of forest reaches its best development in areas with warm summers and relatively cold winters (Figures 30–12 and 30–13). Annual precipitation generally ranges from about 75 to 250 centimeters. The deciduous (leaf-dropping) habit may be related to the unavailability of water during much of the winter.

The ecology of the temperate deciduous forest is somewhat different from that of evergreen forests. In winter, the trees are leafless and tree activity is greatly reduced. In spring, a variety of herbaceous plants bursts forth in profusion on the well-illuminated forest floor (Figure 30–14a and b). Before the appearance of leaves, which dramatically reduce the light reaching the forest floor, most of the early spring herbs have set their seeds. Very few annual plants occur in the deciduous forest, probably because they lack the storage organs that enable the perennials to grow rapidly during the relatively short time that conditions are favorable. The sequence of seasonal change in one of these forests is shown in Figure I–5 on page 6.

One of the outstanding characteristics of the temperate deciduous forest is the similarity of the plants found in its three main regions in the Northern Hemisphere. Deserts found in various regions of the world, for example, are generally inhabited by very different groups of plants that have converged in some of their ecological characteristics (see "Convergent Evolution," page 473). In contrast, the plants of the temperate deciduous forest are a much more uni-

(a)

(b)

30–12
The deciduous forests of the southern Appalachians are among the richest temperate-zone forests in the world, containing many kinds of hardwoods. (a) *Fog over the Great Smokies.* (b) *Autumn in the mountains of North Carolina.*

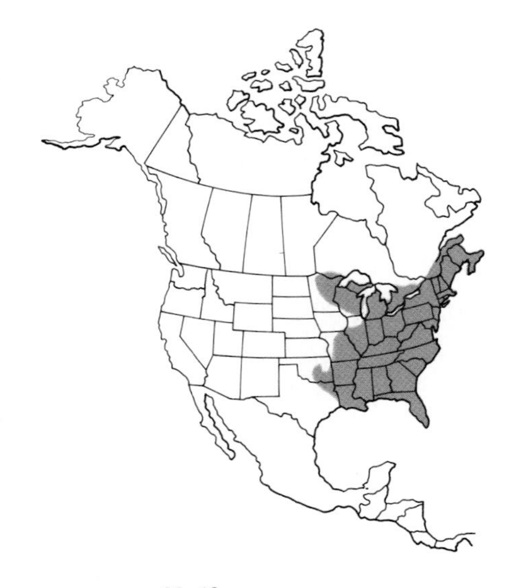

30–13
Distribution of the deciduous forests of North America.

(a)

(b)

30–14
Examples of perennial herbs on the floor of a temperate forest. (a) *Among the common harbingers of spring is the white-flowered bloodroot* (Sanguinaria canadensis), *which derives its name from the color of its sap (latex).* (b) *The marsh marigold* (Caltha palustris) *forms colonies in wet woods throughout temperate North America and Eurasia.*

Temperate forests may include (c) *slash pine* (Pinus elliottii), *such as this example from Florida.* (d) *Western coniferous forest in the Cascade Mountains of Washington, dominated by Douglas fir* (Pseudotsuga menziesii).

(c)

(d)

form group. For example, the herbaceous plants of the forests of Japan resemble those of eastern North America much more closely than either resembles those of western North America. Most of the deciduous trees and their associated herbs were eliminated in the western United States during the latter half of the Tertiary and during the Pleistocene era as the amount of summer rainfall was greatly reduced. In their place now is the magnificent coniferous forest of the western mountains (Figure 30–14d), which contain such trees as the coastal redwood *(Sequoia sempervirens),* the big tree *(Sequoiadendron giganteum),* the Douglas fir *(Pseudotsuga menziesii),* and the sugar pine *(Pinus lambertiana).*

A local but distinctive plant association has evolved from mixed deciduous-evergreen forests in climatic areas that are characterized by cool, moist winters and hot, dry summers. Such climates are found along the shores of the Mediterranean, over a large part of California and into some adjacent states, in the coastal portions of central Chile, in southwestern Africa, and along portions of the coast of southern and southwestern Australia. The plants in these areas—often evergreen or summer deciduous trees and shrubs—have relatively short growing seasons in fall and spring, being limited by low temperatures in winter and by drought in summer (Figure 30–15). In such a climate, precipitation and the need for water

30–15
Chaparral in the mountains of San Diego County, California. This sort of plant association, consisting of broad-leaved, drought-resistant, and, often, spiny evergreen shrubs, occurs only in limited areas, which like the deserts, originated independently in different parts of the world. The shrubs found in each area were derived from completely different ancestral stocks and have become very similar in appearance as they responded over millions of years to comparable mediterranean climates—hot, dry summers, followed by moist, cool winters during which most of the growth occurs. Similar plant associations occur in central Chile (where they are known as matorral), around the Mediterranean (where they are often called maquis), in South Africa, and western Australia.

are not correlated, for the luxuriant growth of spring is followed by the drought of summer (Figure 30–16).

Each of the areas of mediterranean climate is isolated from the others, and each has evolved its own distinctive assemblage of plants and animals. The degree of ecological convergence has been high, however, and the chaparral of California is closely similar in appearance to the matorral of Chile or the maquis of the Mediterranean, even though the plants are essentially unrelated. Seasonal drought enhances the importance of edaphic (soil-related) and biotic variation, and small differences in precipitation often have profound effects on the vegetation and animal life present in the area. Hence, these areas often have high proportions of extremely local species of plants and animals.

30–16
Open woodland dominated by blue oaks (Quercus douglasii) *and digger pines* (Pinus sabiniana) *near Murphys, California, in May. In summer-dry areas, the trees are often widely spaced and frequently evergreen.*

There are about 235,000 species of flowering plants in existence today. Approximately a third (85,000) of these are native to temperate regions, and the balance are found in the tropics. A vast number of tropical plants are in danger of extinction in the wild within the next hundred years because the human populations of most tropical countries continue to double every 20 to 25 years and the forests are rapidly being cleared for wood and marginal cultivation. About half of the world's tropical forests have already been altered, and at least another 250,000 square kilometers are being disrupted each year.

So little is known of the plants of the tropics that many have not even been given scientific names. The samples of these plants that are preserved may well be all that are passed on to our descendents in the twenty-first century and beyond. The useful properties these plants possess can certainly be determined better today, when species are still in existence, than at any point in the future.

In temperate regions, about 5 percent of the 85,000 native species are in current danger of extinction. Habitat destruction is only one problem. Overgrazing by domestic animals, use of fertilizers and herbicides, introduction of foreign plants without their natural controls, and destruction of pollinators all can endanger plant species.

Of the approximately 20,000 species, subspecies, and varieties of native higher plants of the continental United States, at least 10 percent warrant concern, according to a study completed in 1978 by the Smithsonian Institution. About 90 species are recently (within the last 200 years) extinct or presumed extinct; about 850 are endangered (currently in danger of extinction throughout all or part of their range); and more than 1200 are threatened (likely to become endangered within the foreseeable future).

The Tiburon mariposa lily *(Calochortus tiburonensis)* is restricted to a single hilltop on the Tiburon Peninsula on the northern shore of San Francisco Bay. Discovered for the first time about 1970, this lily provides an excellent example of the extreme restriction of plant species characteristic of regions with a mediterranean climate.

Hawaii's *Rollandia angustifolia,* a plant of the lobelia family, is pollinated by birds whose curved bills match the curve of the flowers.

More than half of the approximately 2200 species of native plants in the Hawaiian Islands are endangered or recently extinct. The plants of islands all over the world are in particular danger of extinction, having evolved in isolation and having few natural defenses against alien plants or animals.

Extinct in the wild by 1806, *Franklinia alatamaha,* a handsome tree belonging to the camellia family, was discovered in 1765 on a small plot of land near the Alatamaha River in Georgia and was named in honor of Benjamin Franklin.

30–17
The taiga, which covers hundreds of thousands of square kilometers in the cooler part of the North Temperate Zone, is dominated by white spruce (Picea glauca) *and tamarack* (Larix laricina), *which become smaller as they extend northward. This photograph was taken in northern Manitoba, Canada.*

TAIGA

This northern coniferous forest, or boreal forest, is characterized by severe winters, with a persistent cover of snow. The climates that produce such biomes develop only in the interior of large continental masses at the appropriate latitudes and are thus virtually absent from the Southern Hemisphere (see Figure 30–5). Owing to the influence of the prevailing westerlies between 40° and 50° north latitude, the western portions of North America and Eurasia are characterized by milder climates than the eastern portions. Consequently, taiga is found somewhat farther north towards the Pacific coast than it is along the Atlantic coast (Figure 30–18). The northern limits of taiga are ultimately limited by the rigorous arctic conditions.

In the taiga, most of the precipitation falls in the summer; the cold winter air in these regions has a very low moisture content. The rate of evaporation is low, and lakes, bogs, and marshes are common (Figure 30–17).

"Taiga" is the Russian word for vegetation of the sort found in this biome, which extends over vast areas of the Soviet Union and also in North America (Figure 30–18). The taiga biome is flanked on the south by either deciduous forest or grassland, depending on the precipitation. Similar plant communities extend southward in the mountains (see Figure 30–2, page 596). A number of genera of coniferous trees are common, such as spruce *(Picea)*, hemlock *(Tsuga)*, fir *(Abies)*, and, in relatively warm, dry areas, pine *(Pinus)*, with a lesser representation of willows *(Salix)* and birches *(Betula)*, particularly in moist places. Perennial herbs are common, and there are a few shrubs. Annual plants are virtually absent.

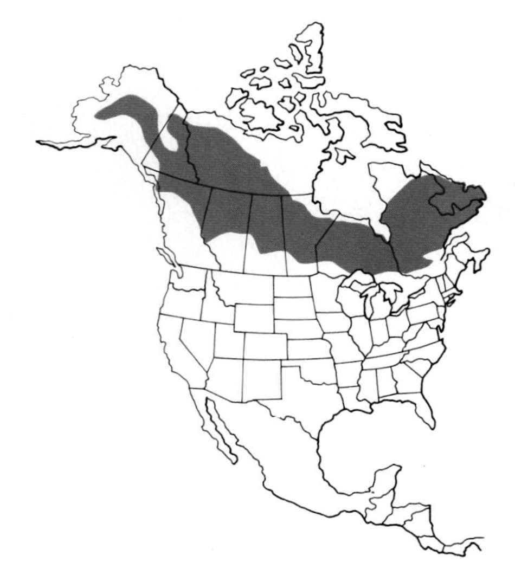

30–18
Taiga of North America.

(a)

(b)

30–19
(a) *Tundra in Mt. McKinley National Park during the Alaskan autumn. Although precipitation is limited, tundra soils are usually moist. This is because the permanent ice that lies just below the surface traps the water at the surface and because the evaporation rates in the cool northern climates are low.* (b) *Mosses and crustose lichens on rocks in the tundra of coastal Alaska.*

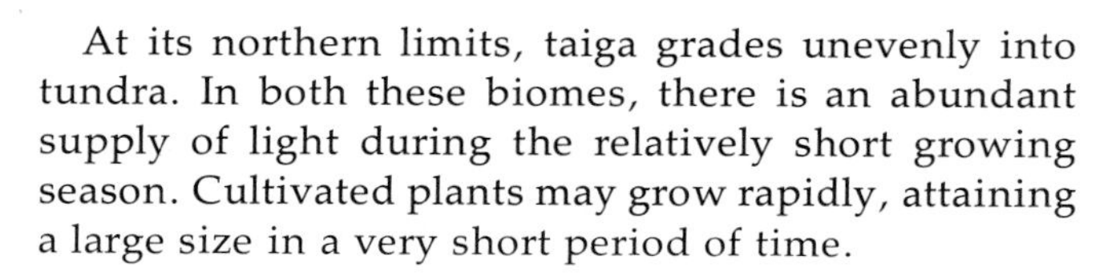

At its northern limits, taiga grades unevenly into tundra. In both these biomes, there is an abundant supply of light during the relatively short growing season. Cultivated plants may grow rapidly, attaining a large size in a very short period of time.

TUNDRA

Tundra, a treeless region extending to the farthest northern limits of plant growth, is an enormous biome occupying one-fifth of the earth's land surface (see Figure 30–5). It is best developed in the Northern Hemisphere and is mostly found north of the Arctic Circle, although it extends farther south along the eastern sides of the continents (Figures 30–19 and 30–20). The tundra essentially comprises one huge band across Eurasia and North America, with outposts of similar vegetation extending southward in the mountains (see Figures 30–2 and 30–21). Some species of plants that occur in this biome have wide circumpolar ranges.

In general, permanent ice, or permafrost, exists within less than a meter of the surface, and ground conditions in the tundra are usually moist because the water cannot percolate through the soil because of the ice. The precipitation is usually less than 25 centimeters per year, but much of this water is held in the surface layers of the soil and accumulated vegetation. Nitrogen is generally present in very short supply. Evaporation is also slow because of the relatively low moisture-holding capacity of the air at low temperatures; some areas are so dry as to constitute true polar deserts.

For plant growth to occur, the mean temperature must be above freezing for at least one month of the

30–20
Tundra of North America.

30–21
Tundra on the Olympic Peninsula in Washington. Such mountain tundra is comparable in many respects with that found hundreds of miles to the north in the arctic. Here, however, forested slopes are found within a hundred meters or so of such alpine meadows. There is much more snow than in arctic areas, and the relatively long, cold nights of the alpine summer do not occur in the arctic, where there is continuous daylight during the growing season.

year. The growing season in many areas of tundra is less than two months. A wide variety of perennials occurs, but there are almost no annuals and few woody plants. Vegetative propagation is characteristic of many of the perennials, and this may be correlated with the uncertainties of setting seed in the brief arctic summer. Large woody plants are absent, probably because of the low temperature and permafrost, which permits only shallow root penetration; most of the volume of tundra vegetation—as much as 96 to 98 percent in some plant communities—is underground.

SUMMARY

The tropical rain forest, where neither water nor low temperatures are limiting factors, is the richest terrestrial biome. Here the trees are evergreen and characterized by large leathery leaves. There is a poorly developed layer of herbs on the forest floor, but there are many vines and epiphytes at higher levels. Tropical soils are often very poor in nutrients, and they may erode or promptly become rocklike when the forest is cleared.

The tropical communities characterized by a seasonal drought are termed savannas. Away from the equator, savannas grade into desert, which is characterized by very low precipitation and high daytime temperatures.

The major grasslands of the world lie between deserts and deciduous woodland. They are a fertile region for agriculture, and in the past they often supported large herds of grazing mammals. The numbers of grazing herbivores have been seriously depleted by the activities of humans.

In the temperate deciduous forest, most of the trees lose their leaves during the cold, often snowy winters when moisture may be unavailable for growth. Communities of similar origin, now less rich in their variety of woody plants, are the coniferous forests of western North America and the evergreen and summer deciduous scrub communities found in the widely scattered areas with a mediterranean climate.

The taiga is a vast northern coniferous forest that extends in unbroken bands across Eurasia and North America. At its northern limits, taiga grades into tundra, a treeless region that also extends around the Northern Hemisphere, mostly above the Arctic Circle, in a band broken only by bodies of water.

SUGGESTIONS FOR FURTHER READING

BILLINGS, W. D.: *Plants, Man, and the Ecosystem,* 3rd ed., Wadsworth Publishing Co., Inc., Belmont, Calif., 1978.*

An excellent summary of the field, with an emphasis on the physiology of individual organisms.

BORMANN, F. H., and G. E. LIKENS: *Pattern and Process in a Forested Ecosystem,* Springer-Verlag, New York, 1979.

An integrated description of the structure, function, and development of the hardwood ecosystem in northern New England, based largely on the Hubbard Brook Ecosystem Study.

BRAUN, E. LUCY: *The Deciduous Forests of Eastern North America,* Macmillan Publishing Co., Inc., New York, 1967.

A detailed and well-documented study of one of the richest temperate biomes.

COLINVAUX, PAUL A.: *Introduction to Ecology,* John Wiley & Sons, Inc., New York, 1973.

A modern synthesis of ecology, emphasizing the interrelationships between ecology and evolution.

CURTIS, JOHN T.: *The Vegetation of Wisconsin: An Ordination of Plant Communities,* University of Wisconsin Press, Madison, Wis., 1959.

An outstanding study of the vegetation of the state by an ecologist whose influence on the field was extensive.

DAUBENMIRE, REXFORD: *Plants and Environment: A Textbook of Plant Autecology,* 3rd ed., John Wiley & Sons, Inc., New York, 1974.

A standard text, written by a community-oriented ecologist.

EMLEN, J. MERRITT: *Ecology: An Evolutionary Approach,* Addison-Wesley Publishing Co., Menlo Park, Calif., 1973.

A theoretical consideration of ecology, including the construction of models, which are applied imaginatively in a variety of field situations.

EMSLEY, MICHAEL and KJELL SANDVED: *Rain Forests and Cloud Forests,* Harry N. Abrams, Inc., New York, 1979.

A magnificently illustrated account of two rapidly disappearing but extraordinarily rich ecosystems.

GLEASON, HENRY A., and ARTHUR CRONQUIST: *The Natural Geography of Plants,* Columbia University Press, New York, 1964.

This handsome, nontechnical volume presents an accurate survey of the principles of plant ecology and plant geography.

KREBS, CHARLES J.: *Ecology: The Experimental Analysis of Distribution and Abundance,* 2nd ed., Harper & Row, Publishers, Inc., New York, 1978.

Aptly described by its subtitle, this outstanding book provides a sound basis for understanding ecology.

ODUM, EUGENE P.: *Fundamentals of Ecology,* 3rd ed., W. B. Saunders Co., Philadelphia, 1971.

The standard text in the field, written by one of the leaders of American ecology; many animal examples are used.

RICHARDS, PAUL W.: *The Tropical Rain Forest,* Cambridge University Press, New York, 1952.

A standard reference on a complex, rapidly disappearing biome.

WHITTAKER, ROBERT H.: *Communities and Ecosystems,* 2nd ed., Macmillan Publishing Co., Inc., New York, 1975.*

A concise introductory text that outlines the characteristics of interacting systems of organisms.

*Available in paperback.

APPENDIX A

Fundamentals of Chemistry

ATOMS

All matter is composed of *atoms* (from the Greek *atomos,* meaning "indivisible"), which are the smallest complete units of *elements*. There are 92 naturally occurring elements, and each is unique in the structure of its atoms. Each type of atom has a characteristic number of *protons*—positively charged particles—in its nucleus (center). The number of protons ranges from the lightest element—hydrogen—which has 1 proton, to the heaviest—uranium—which has 92. The *atomic number* of an element represents the number of protons in the nucleus of one atom (Table A–1). Outside the nucleus of the atom are *electrons*—negatively charged particles—that are attracted by the positive charges of the protons. The way electrons are arranged in an atom determines the chemical properties of that atom, and chemical reactions involve changes in the number and the distribution of an atom's electrons.

Atoms also contain neutrons, which are uncharged particles of about the same mass as protons. The *atomic mass* of an element is essentially equal to the

Table A–1 *Atomic Number and Mass of an Atom of Some Common Elements*

ELEMENT	ATOMIC NUMBER	ATOMIC MASS*
Hydrogen (H)	1	1
Helium (He)	2	4
Carbon (C)	6	12
Nitrogen (N)	7	14
Oxygen (O)	8	16
Sodium (Na)	11	23
Phosphorus (P)	15	30
Sulfur (S)	16	32
Chlorine (Cl)	17	35
Calcium (Ca)	20	40

*For the most common isotope

number of protons and neutrons in the nucleus of one atom. By comparison, electrons are so light that their weight is usually disregarded. For instance, when you weigh yourself, only about 30 grams, or about 1 ounce, of your total weight is made up of electrons.

Isotopes

Not all atoms of the same element have the same atomic mass. These different kinds of atoms are known as *isotopes*. They have the same number of protons—and hence the same atomic number—but different numbers of neutrons. For example, the common form of hydrogen, with its one proton, has an atomic mass of 1 and is symbolized as ^{1}H, or simply H. Deuterium, ^{2}H, is an isotope of hydrogen that contains one proton and one neutron, and so it has an atomic mass of 2. Tritium, ^{3}H, a third isotope of hydrogen, has one proton and two neutrons; it has an atomic mass of 3 (Figure A–1). Like many (but not all) of the less common isotopes, tritium is *radioactive*, which means that its nucleus is unstable and emits energy when it changes into a more stable form. Both deuterium and tritium have nearly the same chemical properties as the more common isotope of hydrogen (^{1}H), and either can substitute for it in chemical reactions. However, if an atom gains a proton along with two neutrons, it is no longer hydrogen but helium. It now has an atomic number of 2 and an atomic mass of 4 (Figure A–1). The fusion of hydrogen nuclei (protons) to form helium is the source of energy at the heart of the sun and also provides the terrible destructive force of the hydrogen bomb.

Many naturally occurring isotopes are radioactive. All the heavier elements—atoms that have 84 or more protons in their nucleus—are unstable and, therefore, radioactive. All radioactive isotopes emit nuclear particles at a rate proportional to the number of atoms present; they are said to undergo radioactive "decay" as they change to another element. The rate of decay is measured in terms of half-life: the half-life of a radioactive isotope is defined as the time in which half the atoms in a sample have changed into another isotope or into a stable element. Because the half-life of an element is constant, it is possible to calculate the fraction of decay that will occur for a given isotope over a given period of time.

Half-lives vary widely, depending on the isotope. The radioactive nitrogen isotope ^{13}N has a half-life of only 10 minutes; tritium has a half-life of 12.25 years. The most common isotope of uranium (^{238}U) has a half-life of 4.5 billion years. The uranium atom decays through a series of isotopes and is eventually transformed to an isotope of lead (^{206}Pb).

Isotopes play a number of important roles in biological research. One use is in dating the age of fossils or the rocks in which fossils are found. For example, the proportion of ^{238}U to ^{206}Pb in a given rock sample is a good indication of how long ago that rock was formed. (The lead formed as a result of the decay of uranium is not the same as the lead commonly present in the original rock, ^{204}Pb.) Isotopes are also used as radioactive tracers. The use of radioactive carbon dioxide ($^{14}CO_2$) has played an important role in enabling plant physiologists to trace the path of carbon in photosynthesis, as described in Chapter 6 (see also "Radiocarbon Dating," page 47). A third use of isotopes is in autoradiography, a technique in which a sample of material containing a radioactive isotope is placed on a sheet of photographic film. Energy emitted from the isotope leaves traces on the film and so reveals the exact location of the isotope within the specimen; Figure 22–5, page 435, is an example of an autoradiograph (see also "Radioactive Tracers and Autoradiography in Plant Research," page 572).

Electrons and Orbitals

As early as 400 to 500 B.C., Greek philosophers suggested that matter cannot be forever divided into smaller and smaller parts. However, the modern concept of the atom as the fundamental unit of the chemical elements is less than 200 years old, and our ideas about its structure have undergone many changes in that time. These ideas, or hypotheses, have usually been presented in the form of models, as are many scientific hypotheses.

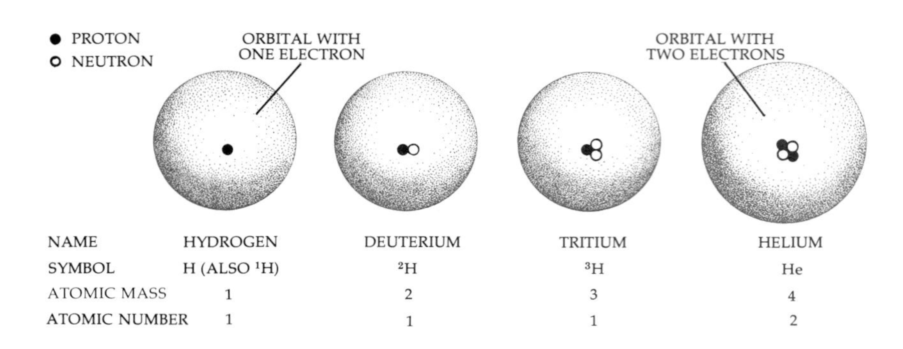

A–1
Diagrams of atoms of hydrogen, deuterium, tritium, and helium. Note that hydrogen, deuterium, and tritium each have only a single proton and a single electron, so they are chemically similar even though they differ in the number of neutrons in their nuclei. Helium, which has one more proton and one more electron, is chemically very different from hydrogen and its isotopes.

A–2
The Böhr planetary model of a carbon atom, as seen in (a) *two dimensions and* (b) *three dimensions.*

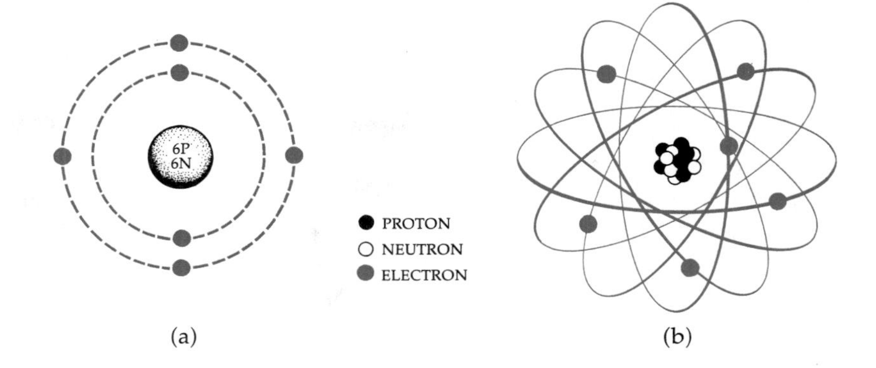

The earliest model was of an indivisible atom, resembling a billiard ball. When it was realized that electrons could be removed from the atom, the billiard-ball model gave way to the plum-pudding model, in which the atom was represented as a solid, positively charged mass with negatively charged particles—electrons—embedded in it. Subsequently, however, physicists found that an atom is in fact mostly open space. The distance from electron to nucleus, experiments indicated, is about 100,000 times the diameter of the nucleus; the electrons are so exceedingly small that the space is almost entirely empty. The physicist Niels Böhr proposed a type of planetary model in which the electrons in the atom were depicted as moving in definite orbits around the nucleus, with a specific energy associated with each orbit (Figure A–2).

The current model of electron configurations is quite different from all previous ones; it reflects our increased knowledge of the behavior of electrons. According to this model, the electron moves unpredictably around the nucleus, and its position at any given moment cannot be known with certainty. For convenience, its pattern of motion is defined as the volume of space in which the electron can be found 90 percent of the time. This volume is known as the electron's orbital. Each orbital can hold a maximum of two electrons. In this model, however, the electrons in the atom have definite energies, or energy levels.

The energy levels in the atom are roughly arranged into *shells*, each of which is composed of one or more *subshells*. Each subshell contains one or more variously shaped *orbitals*. The first two electrons occupy a single spherical orbital. Thus, 90 percent of the time, hydrogen's single electron moves about the nucleus within a single spherical orbital, as do the two electrons of helium. This single spherical orbital, with its maximum of two electrons, makes up the first shell.

The second shell is composed of two subshells and four orbitals, each of which can hold two electrons. The first subshell has a single spherical orbital, and the second subshell has three dumbbell-shaped orbitals. The axes of these three orbitals are perpendicular to one another (Figure A–3). The spherical orbital is filled first, followed by the dumbbell-shaped ones. The second shell can hold a total of eight electrons. Atoms with electrons in as many as seven shells are known.

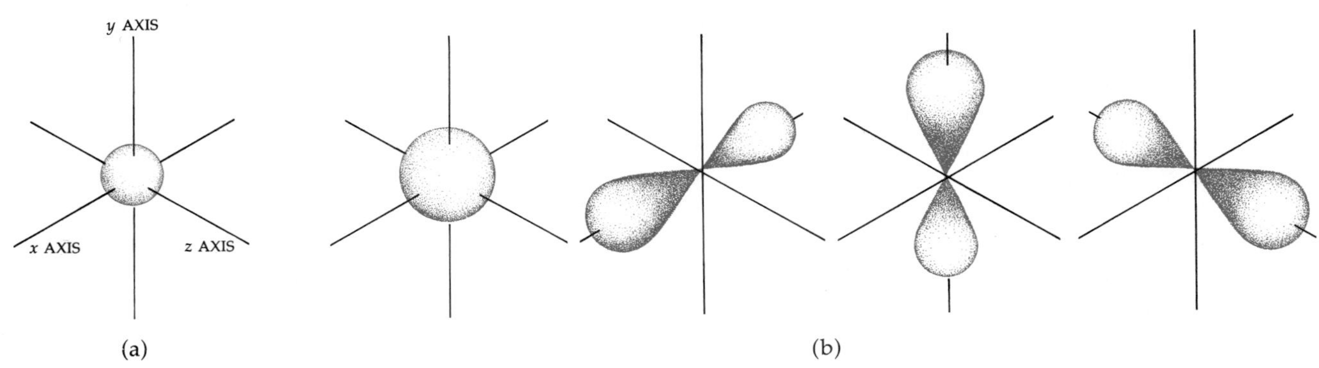

A–3
Orbital models. (a) *The two electrons in the first shell of an atom occupy a single spherical orbital.* (b) *The second shell has four orbitals, each containing two electrons. One of these orbitals is spherical and the other three are dumbbell-shaped. The axes of the dumbbell-shaped orbitals (indicated by the lines) are perpendicular to one another. The nucleus is at the intersection of the axes.*

Atoms tend to stabilize, or complete their energy levels, and the chemical behavior of atoms is governed by this tendency. For instance, helium (atomic number 2), neon (atomic number 10), and argon (atomic number 18) have completely filled outer energy levels and so tend to be unreactive; they are called the "noble" gases because of this apparent "disdain" for reacting with other elements. Atoms of hydrogen (atomic number 1), lithium (atomic number 3), sodium (atomic number 11), and potassium (atomic number 19) have a single electron in their outermost energy level, and they tend to lose this electron. As a consequence of such a loss, each has one more proton than electron and therefore acquires a positive charge: H^+, Li^+, Na^+, and K^+. By contrast, fluorine and chlorine, with atomic numbers of 9 and 17, respectively, tend to gain an electron in order to complete an outer energy level and so become negatively charged: F^- and Cl^-. Similarly, an atom with two electrons in its outer energy level may lose both of them, thus acquiring a double positive charge. For example, magnesium (atomic number 12) and calcium (atomic number 20) become Mg^{2+} and Ca^{2+}. Such charged atoms are known as *ions*. Positively charged ions are called *cations*, and negatively charged ions are called *anions*.

Ions make up less than 1 percent of the weight of most living matter, but they play particularly crucial roles. For instance, K^+ is the principal positively charged ion in most cells, and many essential biological reactions cannot proceed in its absence. Both Na^+ and K^+ are involved in the active transport of sugar and amino acids across the plasma membrane in many plant and animal cells. The calcium cation Ca^{2+} has a direct effect on the physical properties of the membrane; Mg^{2+} forms a part of chlorophyll—the molecule that traps the radiant energy from the sun.

Electrons and Energy

As was noted previously, electrons, which are negatively charged, are attracted to the atomic nucleus because of the positive charge of the protons. The orbital occupied by an electron is determined by the amount of energy of the electron; this energy is in the form of potential energy. The following analogy may be useful: A boulder on flat ground may be said to have no energy. If you push it up a hill, you give it energy—potential energy. So long as it sits on the top of the hill, it neither gains nor loses energy. If it rolls down the hill, however, it loses its potential energy as it rolls back toward its original position on level ground.

The electron is like the boulder in that an input of energy can raise it to a higher energy level—to a position farther away from the nucleus. As long as it remains at this higher level, it possesses this added energy. Also, the electron tends to go to its lowest possible energy level, just as the boulder rolls downhill (Figure A–4).

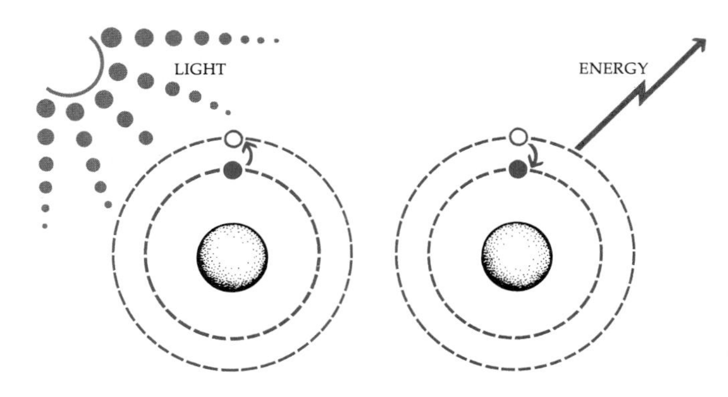

A–4
When an input of energy—such as light energy—boosts an electron to a high energy level, the electron possesses additional potential energy which is released when the electron returns to its previous energy level.

In a given atom, the first spherical orbital is the lowest energy level. The four orbitals of the second level are occupied by electrons with more energy, and so on. Energy is needed to move a negatively charged electron farther away from the positively charged nucleus, just as energy is needed to push a boulder to the top of a hill. However, unlike the boulder on the hill, the electron cannot be pushed part way up. With an input of energy, electrons may move from a lower energy level to any one of several higher ones, but they cannot move to someplace in between. For an electron to move from one level to another, the atom must absorb a discrete packet of energy, known as a quantum, which contains precisely the amount of energy needed for the transition.

Electronegativity

The atomic nuclei of different elements have various degrees of attraction for electrons. The strength of the attraction depends on the number of protons in the nucleus, the number of electrons, and their proximity to the nucleus. The affinity of an atom for electrons is called *electronegativity*. Electronegativity is expressed on a scale of 0 to 4. Helium and the other unreactive noble gases have electronegativities of 0. At the other end of the scale is fluorine, which has an electronegativity of 4. The value for oxygen, the next most electronegative element, is 3.5. Electronegativity values for certain elements are given in Table A–2.

When an electron moves from an atom that is less electronegative to one that is more electronegative, it moves "downhill"—energetically speaking—and en-

Table A–2 *Electronegativity for Atoms of Some Common Elements*

Oxygen (O)	3.5
Nitrogen (N)	3.0
Chlorine (Cl)	3.0
Carbon (C)	2.5
Sulfur (S)	2.5
Hydrogen (H)	2.1
Phosphorus (P)	2.1
Sodium (Na)	0.9

ergy is released, as potential energy is released when a boulder rolls downhill.

In the cells of photosynthesizing green plants and algae, the radiant energy of sunlight raises electrons to a higher energy level. In the course of a series of electron-transfer reactions (which are described in Section 2), the electrons are passed downhill, and the radiant energy of sunlight is changed to the chemical energy on which nearly all life on earth depends.

BONDS AND MOLECULES

Atoms may be held together by forces known as *chemical bonds*; an assembly of atoms held together by chemical bonds is called a molecule. When atoms interact to form a bond, only electrons in their outer shells (termed the valence shells) are involved. There are two general types of chemical bonds: ionic bonds and covalent bonds.

Ionic Bonds

Positive and negative ions attract one another. Bonds involving the mutual attraction of ions of opposite charge are known as *ionic bonds*. The ionic bond is a very common type of bond in inorganic molecules. Thus, the sodium ion (Na^+)—with its single positive charge—is attracted to the chloride ion (Cl^-)—with its single negative charge (Figure A–5). The calcium ion (Ca^{2+}), with its double positive charge, can attract and hold two Cl^- ions, forming $CaCl_2$—the subscript 2 indicates that two atoms of chlorine are present for each atom of calcium.

The combining capacity of an element is called its *valence*. A valence is the number of positive or negative charges on an ion that consists of a single atom (a monatomic ion) in an ionically bonded substance, or the number of pairs of electrons shared by an atom in a covalently bonded substance. The valences of Na^+ and Cl^- are 1, and the valence of Ca^{2+} is 2. Thus, Na^+ combines with Cl^- in a ratio of 1 to 1, and Ca^{2+} combines with Cl^- in a ratio of 1 to 2.

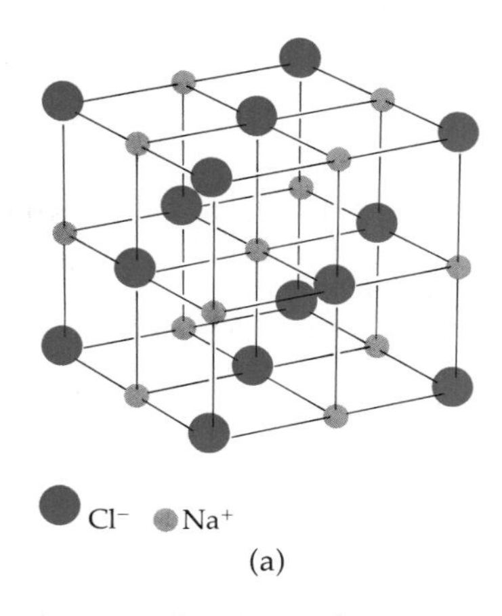

(a)

(b)

A–5
Sodium (Na), *which has only one electron in its outer energy level, becomes more stable if it loses the electron. Chlorine* (Cl), *which has seven electrons in its outer energy level, becomes more stable if it gains one. When sodium and chlorine interact, sodium loses its single electron and chlorine gains it; following this transaction, sodium has a positive charge* (Na^+) *and chlorine has a negative one* (Cl^-). *Such charged atoms are called ions.* (a) *Oppositely charged ions attract one another. Table salt is crystalline* NaCl, *a latticework of alternating* Na^+ *and* Cl^- *ions held together by their opposite charges. Such bonds between oppositely charged ions are known as ionic bonds.* (b) *The regularity of the latticework is reflected in the structure of salt crystals.*

A–6
In a molecule of hydrogen, each atom shares its single first orbital electron with the other atom. As a result, both atoms effectively have a filled first orbital, containing two electrons—a highly stable configuration. The orbital shown here—a molecular orbital—indicates the volume in which the two electrons can be found 90 percent of the time.

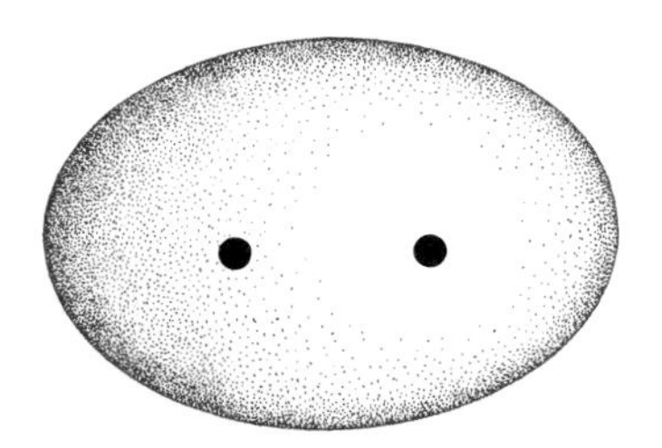

Covalent Bonds

Another way for an atom to complete its outer energy level is by *sharing* electrons with another atom. Bonds formed by shared pairs of electrons are known as *covalent bonds*. Covalent bonds figure prominently in organic chemistry, which is essentially the chemistry of carbon-containing compounds. The simplest covalent bond is that found between hydrogen atoms in the hydrogen molecule (Figure A–6). In a covalent bond, the shared pair of electrons forms a new orbital that encompasses both atoms. In such a bond, each electron spends part of its time around one nucleus and part of its time around the other. Thus, the electron sharing both completes the outer energy level and keeps the nuclear charge neutral.

Carbon, which has an atomic number of 6, needs to share four electrons to achieve a filled and therefore stable outer energy level; and so carbon is able to form covalent bonds with as many as four other atoms. When these bonds are formed, the electron pairs assume new orbitals. These orbitals are distributed symmetrically in space, forming a regular tetrahedron (Figure A–7). The capacity of carbon to form four covalent bonds is crucial to its central role in the chemistry of living systems.

Double and Triple Bonds

There are various ways in which atoms can participate in covalent bonds and satisfy their valence requirements. Oxygen, with two unpaired electrons in its outer electron orbitals, has a valence of 2. Carbon and oxygen can thus form a simple compound, car-

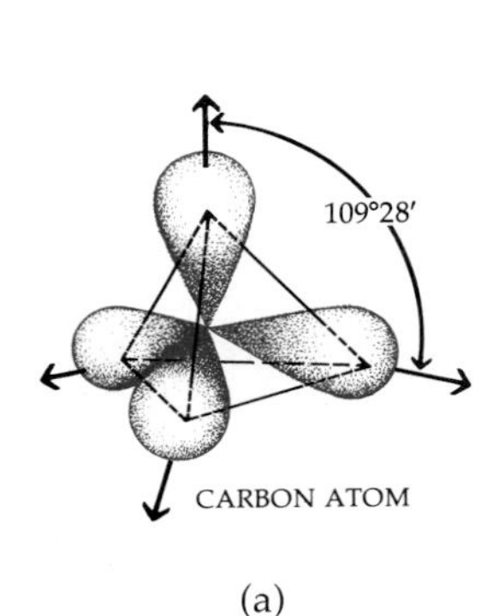

(a)

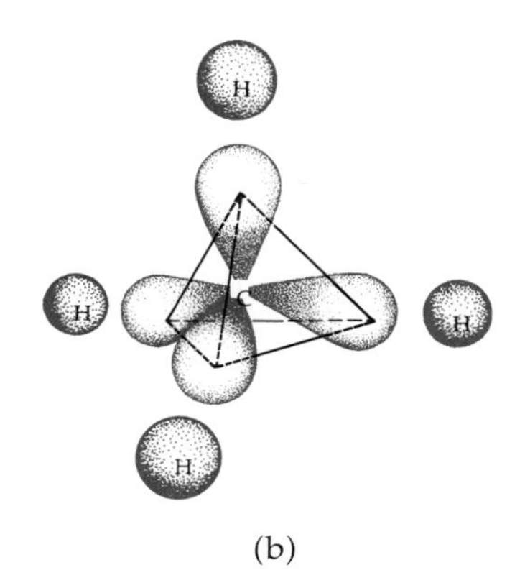

(b)

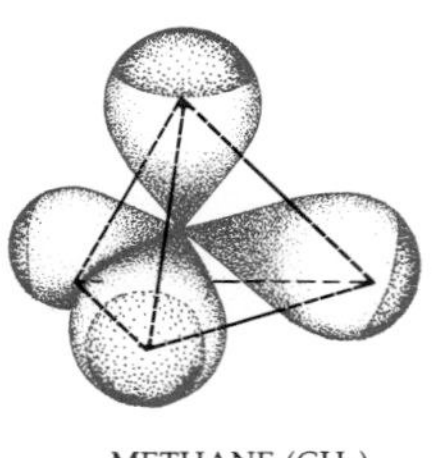

(c)

A–7
(a) *When carbon forms covalent bonds with four other atoms, new orbitals are formed. These new orbitals, which are all the same shape, are oriented toward the four corners of a tetrahedron. Thus, the four orbitals are spatially separated as much as possible.* (b) *When a carbon atom reacts with hydrogen to form a methane molecule, the four unpaired electrons of the carbon atom form pairs with each of the unpaired electrons in four hydrogen atoms.* (c) *Each pair of electrons moves in a new molecular orbital. The molecule has the shape of a regular tetrahedron.*

bon dioxide (CO_2), in which each oxygen atom shares two pairs of electrons with a central carbon atom. Such bonds in which two atoms are held together by two pairs of electrons (four electrons) are called double bonds. They are symbolized in a structural formula by two lines connecting the atomic symbols: O=C=O. Carbon atoms can form double or even triple bonds (in which three pairs of electrons are shared) with each other as well as with other atoms.

Electrons shared in double and triple bonds form orbitals that differ in shape from the orbitals filled by single electron pairs. For instance, when four bonds satisfy the electron requirements of carbon, they are directed toward the four corners of a tetrahedron that has the carbon atom at its center, as in Figure A–7. When two bonds are replaced by a double bond, the remaining single bonds form the arms of a Y with the double bond as its leg (Figure A–8). When two double bonds are made by a single carbon atom, as in carbon dioxide, the three bonded atoms lie in a straight line.

Single bonds are flexible, allowing atoms to rotate in relation to one another. Double and triple bonds hold the atoms relatively rigid in relation to one another. The presence of double bonds in a molecule can make a significant difference in its properties. For example, both fats and oils are composed of carbon and hydrogen atoms that are covalently bonded to one another, but in oils some of the bonds are double, and in fats the bonds are all single.

Polar Covalent Bonds

The electrons in covalent bonds are not always shared equally between the two atoms involved. Some atoms have a greater attractive force for the electrons than others—that is, they are more electronegative. Because of this differential attraction, the shared electrons tend to spend more time around the more electronegative atom. As a result, the more electronegative atom in the molecule will have a slightly negative charge, and the less electronegative atom will have a slightly positive one, since its nuclear charge is not entirely neutralized (Figure A–9). The polar properties of some covalent bonds have very important consequences for living things. For example, many of the special properties of water, upon which life depends, derive largely from its polar nature, as is explained below.

Ionic, covalent, and polar covalent bonds may be considered to be different versions of the same type of bond. The differences depend on the differences in electronegativity among the combining atoms. In a wholly nonpolar covalent bond, the electrons are shared equally; such bonds can exist only between identical atoms, as in H_2, Cl_2, O_2, and N_2. In polar covalent bonds, electrons are shared unequally. In ionic bonds, there is an electrostatic attraction between the negatively charged and positively charged ions as a result of their having gained or lost electrons.

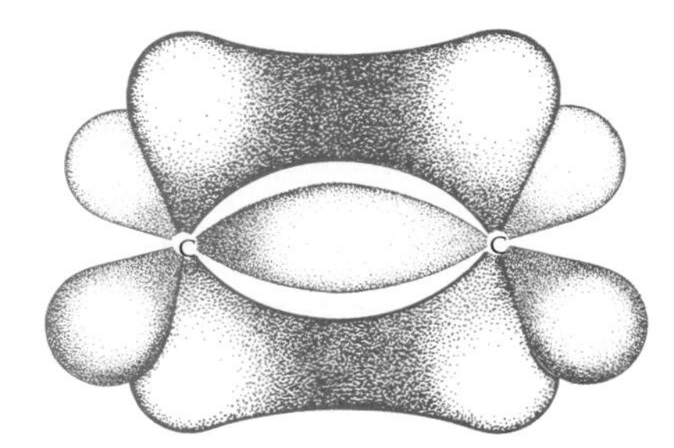

A–8
An orbital model showing carbon-carbon double bond of ethylene. One pair of electrons occupies the inner orbital between the two carbon atoms. The other pair of electrons occupies the outer orbital, which has two phases—one above the plane of the two nuclei and one below. This creates a rigid bond, about which the atoms cannot rotate. The two smaller orbitals extending from each carbon atom contain one electron each and can form covalent bonds with other atoms.

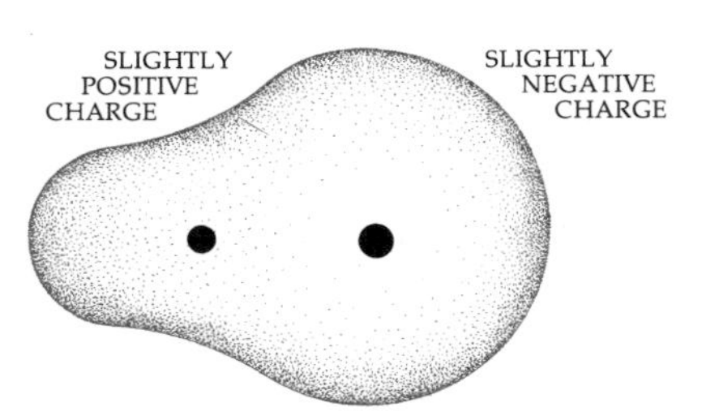

A–9
In a polar molecule, such as hydrogen chloride (HCl), the shared electrons tend to spend more time around the more electronegative atom—in this case, the chlorine atom. As a result, chlorine has a slightly negative charge, and the less electronegative atom (hydrogen) has a slightly positive charge.

Atomic and Molecular Weights

The *atomic weight* of any element is an average value for the naturally occurring mixture of isotopes of that element relative to the common isotope of carbon (^{12}C) which has an atomic mass of 12. Hypothetically, an element exactly twice as heavy as carbon would have an atomic weight of 24; an element one-half as heavy would have an atomic weight of 6. Because atoms are much too small to be weighed individually, they are measured in amounts called *gram atoms*. One gram atom of an element is that amount of the element in grams that equals its atomic weight. For example, 1 gram atom of carbon is that amount of carbon that weighs 12 grams, and 1 gram atom of hydrogen weighs about 1 gram (actually 1.008 grams because of the presence of isotopes in any naturally occurring sample). The *molecular weight* of a substance is the sum of the atomic weights of all the atoms in one molecule. A *gram molecule* is the amount of a substance, in grams, that equals its molecular weight. One gram molecule of oxygen gas (O_2) weighs about 32 grams; 1 gram molecule of hydrogen gas (H_2) weighs about 2 grams; and 1 gram molecule of CO_2 weighs about 44 grams.

The gram molecule is called a *mole*. The number of particles in 1 mole of any substance (whether atoms, ions, or molecules) is always the same: 6.022×10^{23}. For example, 1 mole of water contains 6.022×10^{23} water molecules, and 1 mole of glucose ($C_6H_{12}O_6$) contains 6.022×10^{23} glucose molecules. This numerical constant is known as *Avogadro's number*. The mole is useful for defining quantities of substances involved in chemical reactions.

In order to make water, for instance, one would combine 2 moles of hydrogen (4 grams) with 1 mole of oxygen (32 grams); in other words, four hydrogen atoms for every two oxygen atoms. Two moles of water would be produced, each weighing 18 grams. Similarly, to make table salt (NaCl), one would combine 1 gram atom of sodium (about 23 grams) and 1 gram atom of chlorine (about 35.5 grams).

Functional Groups

Sometimes clusters of atoms joined by covalent bonds tend to react together as a group. One such class of atomic clusters is represented by organic radicals, or *functional groups*. The specific chemical properties of organic molecules are determined primarily by their functional groups. The $-OH$ group is an example ($-OH$, the functional group, is called hydroxyl; OH^-, the ion, is called hydroxide). When one hydrogen and one oxygen are bonded covalently, one electron on the oxygen is left over for sharing. A compound formed when a hydroxyl group replaces one or more of the hydrogens in a hydrocarbon (a compound consisting only of carbon and hydrogen) is known as an alcohol.

Thus, when one hydrogen atom in methane (CH_4) is replaced by a hydroxyl group, the compound becomes methanol (CH_3OH), a pleasant-smelling, poisonous alcohol noted for its ability to cause blindness and death. Ethane similarly becomes ethanol (C_2H_5OH), which is present in all alcoholic beverages. Glycerol, $C_3H_5(OH)_3$, is an alcohol that contains three hydroxyl groups.

The carboxyl group ($-COOH$) is a functional group that gives a compound the properties of an acid. Table A–3 shows some functional groups of considerable biological importance.

WATER AND THE HYDROGEN BOND

Water is composed of a number of very small molecules held together by the mutual attraction of positively and negatively charged atoms. Each water molecule is made up of two atoms of hydrogen and one atom of oxygen, held together by two covalent bonds.

The water molecule as a whole is neutral in charge, having an equal number of electrons and protons. However, the molecule is electrically asymmetric—that is, it is polar. This polarity occurs because oxygen is more electronegative than hydrogen. The paired electrons in the outer orbitals spend more time around the oxygen nucleus than they do around the hydrogen nuclei. Consequently, the region near the oxygen nucleus has two weakly negative zones, and each of the regions near the hydrogen nuclei has a

Table A–3 *Some Functional Groups that Play Important Roles in Organic Compounds*

GROUP	NAME
$-OH$	Hydroxyl
$-NH_2$	Amino
$-C(=O)-CH_3$	Acetyl
$-C-C(=O)-C-$	Keto
$-C(=O)-OH$	Carboxyl
$-O-P(=O)(OH)-OH$	Phosphate

A–10
(a) *An orbital diagram of a single water molecule. Note the four-cornered distribution of positive and negative charges.*
(b) *As a result of these slight positive and negative charges, each water molecule can form hydrogen bonds (indicated by dashed lines) with four other water molecules.*

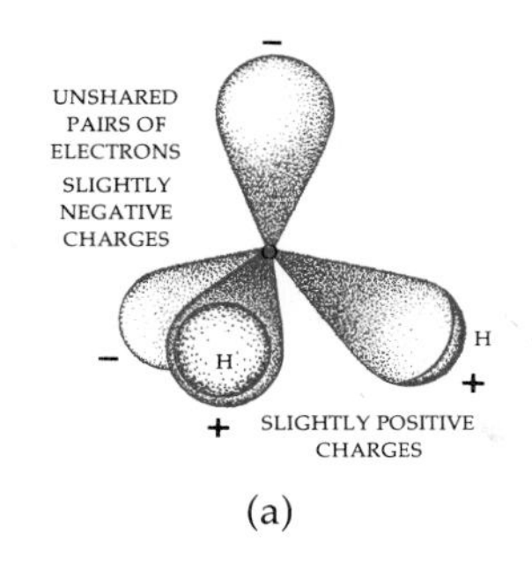

(a)

(b)

weakly positive zone. Thus, the water molecule—in terms of its polarity—is four-cornered, with two positively charged corners and two negatively charged ones (Figure A–10a).

When any of these charged regions comes close to the oppositely charged region of another water molecule, the force of the attraction forms a bond between them—a *hydrogen bond*. Hydrogen bonds are found not only in water but can form between any hydrogen atom that is covalently bonded to an electronegative atom—usually oxygen or nitrogen—and the electronegative atom of another molecule. In liquid water, hydrogen bonds form between the negative "corners" of one water molecule and the positive "corners" of another. Thus, every water molecule can establish hydrogen bonds with four other water molecules. Liquid water is made up of water molecules bound together in this way, as shown in Figure A–10b.

Any single hydrogen bond is relatively weak and has an exceedingly short lifetime; on an average, such bonds last about 1/100,000,000,000th of a second. But, as one is broken, another is formed. As a group, however, hydrogen bonds have considerable strength, making water both liquid and stable under ordinary conditions of pressure and temperature.

WATER AS A SOLVENT

Many substances within living systems are found in solution. A *solution* is a uniform mixture of the molecules of two or more substances. The substance present in the greatest amount—usually a liquid—is called the *solvent*; the substances present in lesser amounts are called *solutes*. The polarity of the water molecules is responsible for the capacity of water to act as a solvent. The polar water molecules tend to separate substances such as NaCl into their constituent ions. Then, as shown in Figure A–11, the water molecules cluster around and segregate the charged ions.

Many of the molecules important in living systems,

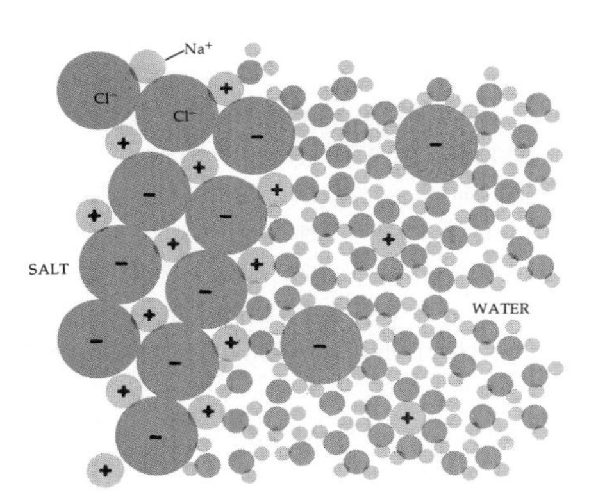

A–11
Because of the polarity of water molecules, water can serve as a solvent for polar atoms or molecules. This diagram shows sodium chloride (NaCl) dissolving in water as the water molecules cluster around the individual ions, separating them from each other. Substances that dissociate into ions in solution produce solutions that can conduct electricity. Such solutions are called electrolytes.

such as glucose, also have polar areas. Such regions of partial positive or negative charges arise in the neighborhood of covalently bonded atoms of unequal electronegativity. Thus, these molecules attract water molecules and are dissolved in the water.

Molecules that readily dissolve in water are called *hydrophilic* ("water-loving"). Such molecules slip into aqueous solution easily because their partially charged regions attract water molecules and thus compete with the attraction between the water molecules themselves.

Molecules that lack polar regions, such as fats, tend to be insoluble in water. The hydrogen bonding between the water molecules acts as a force to exclude the nonpolar molecules. As a result of this exclusion, nonpolar molecules tend to cluster together in water, just as droplets of fats tend to coalesce on the surface of chicken soup, for example. Such molecules are said to be *hydrophobic* ("water-fearing").

ACIDS AND BASES

Acids taste sour, like sour milk, citrus fruits, and vinegar. Bases taste flat, like milk of magnesia, and feel slippery and soapy in solution.

To define "acid" and "base" in chemical terms, it is easiest to begin by looking at water. Water consists of two atoms of hydrogen and one of oxygen held together by covalent bonds. Water molecules also have a slight tendency to ionize—separate into H^+ and OH^- ions. (The H^+ ions tend to combine with other H_2O molecules to produce the hydronium ion, H_3O^+, but we shall omit the extra ion from our consideration in this discussion.) In any given volume of pure water, a very small but constant number of water molecules will be dissociated into ions. The number is constant because the tendency of water to dissociate is exactly offset by the tendency of the ions to reunite; thus, even as some are ionizing, an equal number of others are re-forming, a state known as dynamic equilibrium.

In pure water, the number of H^+ ions exactly equals the number of OH^- ions. This is necessarily the case, because neither ion can be formed without the other when only H_2O molecules are present. A solution acquires the properties we recognize as acidic when the number of H^+ ions exceeds the number of OH^- ions; conversely, a solution is basic when the concentration of OH^- ions exceeds that of the H^+ ions. There is always an inverse relationship between the concentration of H^+ and OH^- ions; when H^+ is high, OH^- is low, and vice versa. This is because the product of their concentrations is a constant. Thus, at 25°C,

$$[H^+][OH^-] = 1 \times 10^{-14} \text{ mole}$$

We now can define our terms chemically:

1. An *acid* is a substance that donates H^+ ions to a solution. Because an H^+ ion is really a proton, acids can also be defined as *proton donors*.
2. A *base* is a substance that decreases the number of H^+ ions, or protons. More specifically, a base is a *proton acceptor*. The OH^- ion is a base because it can accept a proton and thus be neutralized:

$$H^+ + OH^- \longrightarrow H_2O$$

3. In any acid-base reaction, there is always a proton donor and a proton acceptor.

Strong and Weak Acids and Bases

Hydrochloric acid (HCl) is an example of a common acid. It is a strong acid, meaning that it tends to be almost completely ionized in an aqueous solution into H^+ and Cl^- ions. Sodium hydroxide (NaOH) is a common strong base; in an aqueous solution, it exists entirely as Na^+ and OH^- ions. Weak acids and weak bases are those that ionize only slightly. Compounds that contain the carboxyl group (−COOH) are often weak acids because the hydrogen atom may partially dissociate from the carboxyl group to yield a proton:

$$R{-}COOH \rightleftharpoons R{-}COO^- + H^+$$

In this reaction, R represents any chemical structure that may be attached to a carboxyl group.

Compounds that contain the amino group ($-NH_2$) act as weak bases, because the amino group has a weak tendency to accept hydrogen ions, thereby forming NH_3^+:

$$R{-}NH_2 + H^+ \rightleftharpoons R{-}NH_3^+$$

The pH Scale

Chemists define degrees of acidity by means of the pH scale. In the expression "pH," the "p" stands for "power" and the "H" stands for the concentration of hydrogen ions.

In a liter of pure water, 0.0000001 mole of hydrogen ions can be detected. For convenience, this is written in terms of a power of ten, 10^{-7}, and in terms of the pH scale, it is simply referred to as pH 7 (see Table A–4). At pH 7, the concentrations of free H^+ and OH^- ions are exactly the same, and pure water is thus "neutral." Any pH below 7 is acidic, and any pH above 7 is basic. The lower the pH number, the higher the concentration of hydrogen ions. Thus, pH 2 means 10^{-2} mole of hydrogen ions per liter of water, or 0.01 mole per liter—a much higher concentration than 0.0000001. A difference of one pH unit represents a tenfold difference in the concentration of hydrogen ions.

Lemon juice has a pH of about 2, as do the stomach contents of humans and other animals. Orange juice has a pH of about 3; human blood has a pH of 7.4. The best soil pH for most plants is about 6.4, but alkaline soils have a pH between 7 and 9, and peat bogs may have a pH as low as 3.

CHEMICAL REACTIONS

All chemical reactions involve the breaking of some bonds and the formation of new bonds. According to present concepts, atoms or molecules react with one another only when they collide with sufficient force to overcome the initial forces of repulsion. The force required varies with the nature of the atoms or molecules; the more stable their initial state, the more forceful the collision must be for a reaction to occur. In any given group of atoms, it is likely that some proportion of them is moving with sufficient energy for a reaction to occur, but often this proportion is so small that the reaction, for all practical purposes, does not take place.

Reaction rates can be increased by increasing the likelihood of forceful collisions. One way to do this is to raise the temperature, thereby increasing the average velocity at which the atoms or molecules move and so increasing their likelihood of colliding with sufficient force. Sometimes, as in the case of methane (natural gas), a spark is all that is needed. Once the reaction begins, it liberates heat that is transferred to the other CH_4 molecules until all are moving rapidly enough to react almost simultaneously with explosive force. Driving a reaction by heat is a method commonly used in chemical laboratories, as well as in industry.

A second way to increase the rate of a reaction is to increase the number of reacting molecules. Chemists working in research laboratories or in industry usually work with pure chemicals in high concentrations. By this means, the chemical reactions of interest are not only speeded up, but they are also easier to control.

A third way to increase a reaction rate is to use catalysts. Catalysts lower the *energy of activation* of a reaction, that is, the minimum energy that must be available in a collision for a reaction to be initiated (Figure A–12). Metals, such as iron, nickel, and platinum, are commonly used as catalysts in industrial laboratories. Certain molecules apparently tend to cluster on the surfaces of such metals, thus increasing the likelihood of the necessary kinds of close encounters between the reactants. Although they participate in the reactions, catalysts are not used up, so they can be used over and over again (see Chapter 4, pages 77–81).

Table A–4 *The* pH *Scale*

	CONCENTRATION OF H^+ IONS (MOLES PER LITER)		pH	CONCENTRATION OF OH^- IONS (MOLES PER LITER)	
	1.0	10^{0}	0	10^{-14}	
	0.1	10^{-1}	1	10^{-13}	
	0.01	10^{-2}	2	10^{-12}	
Acidic	0.001	10^{-3}	3	10^{-11}	
	0.0001	10^{-4}	4	10^{-10}	
	0.00001	10^{-5}	5	10^{-9}	
	0.000001	10^{-6}	6	10^{-8}	
Neutral	0.0000001	10^{-7}	7	10^{-7}	
		10^{-8}	8	10^{-6}	0.000001
		10^{-9}	9	10^{-5}	0.00001
		10^{-10}	10	10^{-4}	0.0001
Basic		10^{-11}	11	10^{-3}	0.001
		10^{-12}	12	10^{-2}	0.01
		10^{-13}	13	10^{-1}	0.1
		10^{-14}	14	10^{0}	1.0

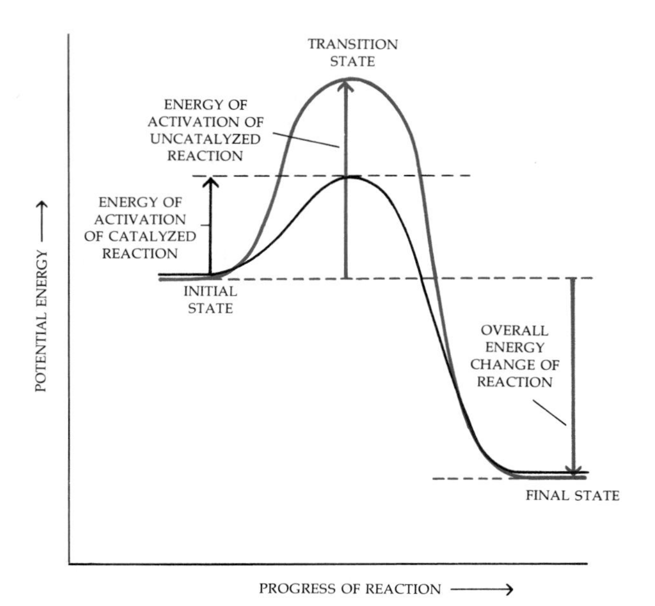

A–12
Chemical reactions require a continuous input of energy—called the activation energy—in order to take place. An uncatalyzed reaction requires more "input," or activation energy, than does a catalyzed one, such as an enzymatic reaction. The lower activation energy in the presence of the catalyst is often within the range of energy possessed by the molecules, and so the reaction can occur with little or no added energy. Note, however, that the overall energy change from the initial state to the final state is the same with or without the catalyst.

Types of Reactions

Chemical reactions can be classified into a few general types. One type can be represented by the expression:

$$A + B \longrightarrow AB$$

An example of this sort of reaction is the combination of hydrogen gas with oxygen gas to produce water:

$$2H_2 + O_2 \longrightarrow 2H_2O$$

A reaction may also take the form of a dissociation:

$$AB \longrightarrow A + B$$

For example, the equation for the formation of water can be reversed to show the breakdown of water into its component elements:

$$2H_2O \longrightarrow 2H_2 + O_2$$

This means that water molecules yield hydrogen and oxygen gases.

A reaction may also involve an exchange, taking the form:

$$AB + CD \longrightarrow AD + CB$$

An example of such a reaction is the combination of hydrochloric acid and sodium hydroxide to form table salt and water:

$$HCl + NaOH \longrightarrow NaCl + H_2O$$

Oxidation-Reduction

The passing of an electron from one molecule to another is known as an oxidation-reduction reaction. The loss of an electron is known as *oxidation*, and the compound that loses the electron is said to be oxidized. The reason electron loss is called oxidation is that oxygen, with its high electronegativity, is often the electron acceptor.

Reduction involves the gain of an electron. Oxidation and reduction occur simultaneously; the electron lost by the oxidized atom is accepted by another atom, which is thus reduced. (The fact that reduction means a gain of an electron seems paradoxical and may make it difficult to remember the distinction between the two terms. It may help to recall that the gain of an electron *reduces* the charge of an atom or molecule.)

Oxidation-reduction reactions may involve only a solitary electron, as when sodium loses an electron and becomes oxidized to Na^+, and chlorine gains an electron, thereby reducing its charge (Cl^-). Often the electron travels with a proton, so oxidation involves the removal of hydrogen atoms, and reduction involves the gain of hydrogen atoms. When glucose is burned (oxidized), hydrogen atoms are lost by the glucose molecule and gained by oxygen:

$$C_6H_{12}O_6 + 6O_2 \longrightarrow 6CO_2 + 6H_2O$$

Chemical Equilibrium

Some chemical reactions can go in either direction, as discussed previously. When net change ceases, the reaction is said to be at equilibrium. In the reaction

$$A + B \rightleftharpoons C + D$$

the point of equilibrium is reached when as many molecules of C and D are being converted to molecules of A and B as molecules of A and B are being converted to molecules of C and D.

The concentration of reactants does *not* have to equal the concentration of products in order for equi-

A–13
A graph of the changes in concentration of products and reactants in a reversible reaction. At first, only molecules of A and B are present. The reaction begins when A and B start to yield products C and D. At the end of two minutes, the concentrations of A + B and C + D are equal. As the reaction proceeds, the concentration of C + D will continue to increase to the point of chemical equilibrium (at about six minutes) and will thereafter remain greater than the concentration of A + B. This is the proportion at which the rates of forward and reverse reactions are the same.

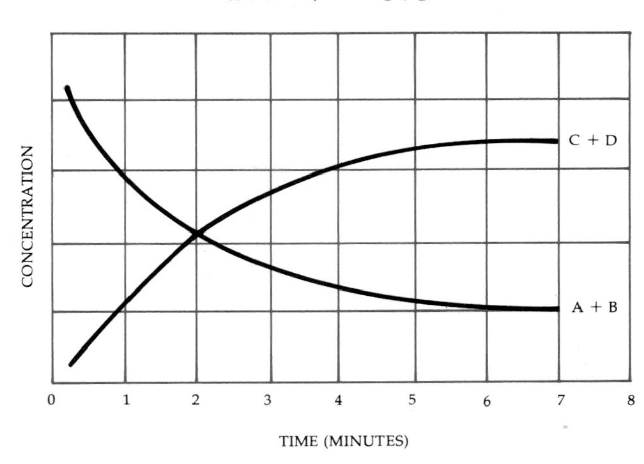

librium to be established; only the *rates* of the forward and reverse reactions must be the same. Consider the reaction above. The different lengths of the arrows indicate that there is more C + D present at equilibrium than A + B. If only A and B molecules are present initially, the reaction occurs at first to the right, with A and B molecules converting into C and D molecules. Figure A–13 shows the relative changes in concentration as the reaction continues. As C and D accumulate, the rate of the reverse reaction increases, and at the same time, the rate of the forward reaction decreases because of the decreasing concentrations of A and B. After about six minutes, the rates of the forward and reverse reactions equalize, and no further changes in concentration take place. The proportions of A + B and C + D will remain the same, but there will always be more C and D molecules.

The Energy Factor

Although a chemical equation must be balanced in terms of its chemical components, one crucial factor is often missing—an indication of the energy changes that accompany the reaction. As discussed earlier, forces of electrical attraction between atoms produce the chemical bonds that hold a compound together. Depending on the strength of these forces, each compound has a particular energy content, or *bond energy* (Figure A–14). The total bond energy of a compound can be defined as the amount of energy required to break that compound into its constituent atoms. All chemical reactions involve the rearrangement of bonds; hence, they are accompanied by changes in energy. Most chemical reactions result in the loss of energy to the surroundings.

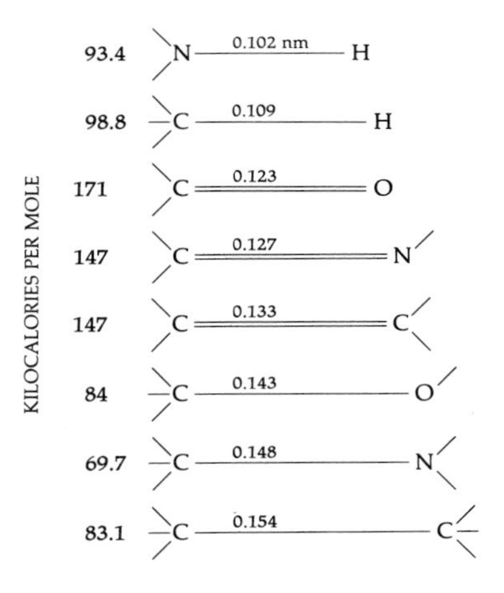

A–14
A chemical bond is a force holding atoms together. The strength of a bond is measured in terms of the energy required to break it. The figures at the left indicate the number of kilocalories per mole needed to break the bonds between the pairs of atoms shown. The lines connecting the atoms represent the bonds, and the figures above them represent the characteristic center-to-center distances between the atoms involved in the bond. Double lines indicate double bonds, which hold the atoms closer together and thus are stronger.

A–15
A calorimeter. A known quantity of glucose or some other material is ignited electrically. As it burns, the rise in the temperature of the water is measured. Based on the specific heat of water, one can then calculate the number of calories released by the burning of the sample.

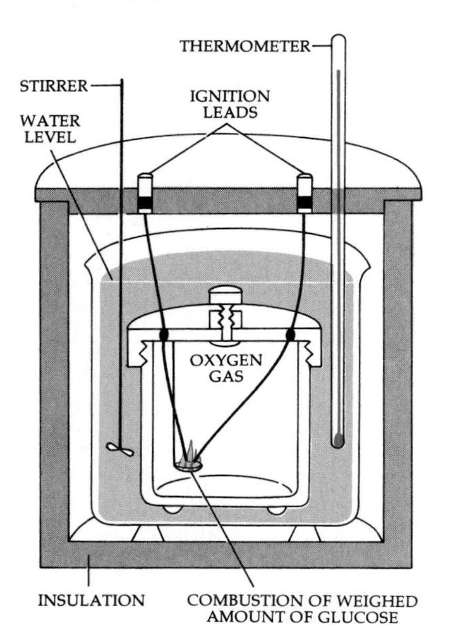

Consider, for example, the burning of methane, represented by the following equation:

$$CH_4 + 2O_2 \longrightarrow CO_2 + 2H_2O$$

This reaction can be set in motion by a spark (which is what causes explosions in coal mines), and then energy is released in the form of heat. The amount of energy released (that is, 213 kilocalories per mole of methane) can be measured quite precisely (Figure A–15). This release of energy can be expressed by a simple equation: $\Delta H = -213$ kcal. The Greek letter delta (Δ) represents "change" and H represents "heat content." In general, the change in heat content is approximately equal to the change in potential energy. The minus sign indicates that energy has been released. (A calorie is defined as the amount of heat necessary to raise the temperature of 1 gram of water by 1°C; 1000 calories = 1 kilocalorie. In expressing the heat- or energy-producing value in a food that is oxidized, the calorie unit ordinarily used is actually a "large Calorie"—that is, a kilocalorie.)

Similarly, changes in energy occur in the chemical reactions that take place in living systems. However, living systems have evolved ways to minimize the activation energy required for molecules to react, as well as ways to convert a portion of the energy released in some reactions into more usable forms than heat (see Chapter 4).

Endergonic and Exergonic Reactions

A reaction that requires an input of energy is said to be *endergonic* (from the Greek *endon*, meaning "within," and *ergon*, meaning "work"), because energy must be put into it. A reaction that liberates energy is an *exergonic* reaction. An endergonic reaction can be thought of as an "uphill" reaction and an exergonic reaction as a "downhill" reaction. Only exergonic (downhill) reactions can proceed spontaneously.

Endergonic reactions do not occur by themselves; they must be coupled to some downhill process in such a way that the energy released by the downhill process can be used for the uphill process. Thus, the energy released in the downhill process must be at least as great as that required for the uphill process. In living systems, the uphill and downhill movements are often accomplished in very small stages, so that large amounts of energy are not required or released all at once.

APPENDIX B

Metric Table

	FUNDAMENTAL UNIT	QUANTITY	NUMERICAL VALUE	SYMBOL	ENGLISH EQUIVALENT
Area		hectare	10,000 m^2	ha	2.471 acres
Length	meter			m	39.37 inches
		kilometer	1000 (10^3) m	km	0.62137 mile
		centimeter	.01 (10^{-2}) m	cm	0.3937 inch
		millimeter	.001 (10^{-3}) m	mm	
		micrometer	.000001 (10^{-6}) m	μm	
		nanometer (millimicron)	.000000001 (10^{-9}) m	nm (mμ)	
		angstrom	.0000000001 (10^{-10}) m	Å	
Mass	gram			g	0.03527 ounce
		kilogram	1000 g	kg	2.2 pounds
		milligram	.001 g	mg	
		microgram	.000001 g	μg	
Time	second			sec	
		millisecond	.001 sec	msec	
		microsecond	.000001 sec	μsec	
Volume (solids)	cubic meter			m^3	35.314 cubic feet
		cubic centimeter	.000001 m^3	cm^3	0.061 cubic inch
		cubic millimeter	.000000001 m^3	mm^3	
Volume (liquids)	liter			l.	1.06 quarts
		milliliter	.001 liter	ml	
		microliter	.000001 liter	μl	

Temperature Conversion Scale

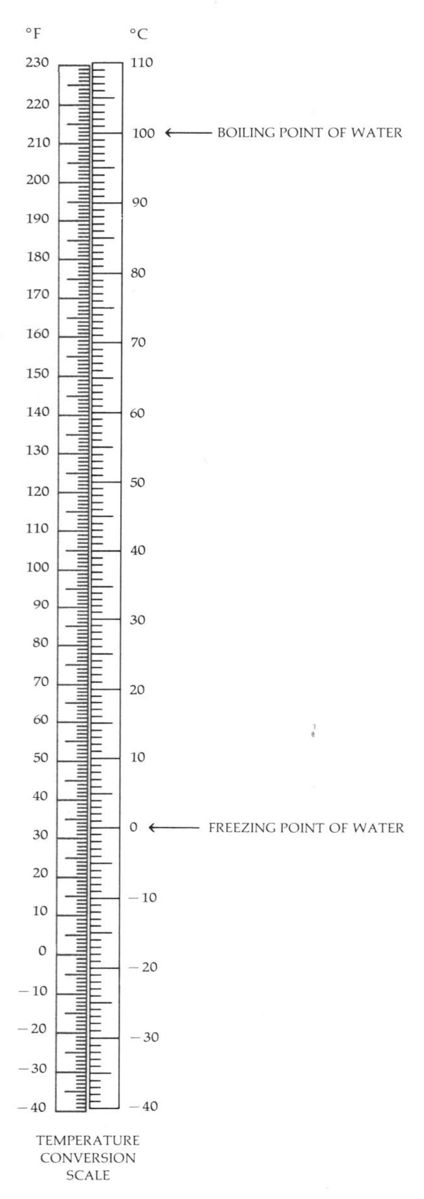

TEMPERATURE CONVERSION SCALE

FOR CONVERSION OF FAHRENHEIT TO CENTRIGRADE, THE FOLLOWING FORMULA CAN BE USED:

$$°C = \frac{5}{9}(°F - 32)$$

FOR CONVERSION OF CENTIGRADE TO FAHRENHEIT, THE FOLLOWING FORMULA CAN BE USED:

$$°F = \frac{9}{5}°C + 32$$

APPENDIX C

Classification of Organisms

There are several alternative ways to classify organisms. The one presented here follows the overall scheme described at the end of Chapter 10, in which organisms are divided into five major groups, or kingdoms: Monera, Protista, Animalia, Fungi, and Plantae. The chief taxonomic categories are kingdom, division (phylum), class, order, family, genus, species.

The classification outlined below includes the divisions of Protista, except those considered Protozoa, as well as the Fungi and Plantae. Certain classes given prominence in this book are also included, but the listings are far from complete. The number of species given for each group is the estimated number of living species that have been described and named. Only groups that include living species are described.

KINGDOM MONERA

Prokaryotic cells that lack a nuclear envelope, plastids and mitochondria, and 9-plus-2 flagella. Monera are unicellular but sometimes aggregate into filaments or other superficially multicellular bodies. Their predominant mode of nutrition is absorption, but some groups are photosynthetic or chemosynthetic. Reproduction is primarily asexual, by fission or budding, but portions of DNA molecules may also be exchanged between cells under certain circumstances. They are motile by simple flagella, by gliding, or they may be nonmotile.

About 2500 species of bacteria and bacterialike organisms are recognized at present. The recognition of species is not comparable with that in eukaryotes and is based largely upon metabolic features. One group, the class Rickettsiae—very small bacterialike organisms—occurs widely as parasites in arthropods and may contain tens of thousands of species, depending upon the classification criteria used; they have not been included in the estimate given here. No satisfactory classification of the Monera has yet been proposed. One of the groups included is the cyanobacteria, or blue-green algae, whose photosynthesis is based on chlorophyll *a*. Although some 7500 species of cyanobacteria have been described, a more reasonable estimate puts the total number of these specialized bacteria at about 200 distinct nonsymbiotic species.

KINGDOM FUNGI

Eukaryotic multinucleate or rarely unicellular organisms in which the nuclei occur in a basically continuous mycelium; this mycelium becomes septate in certain groups and at certain stages of the life cycle. Fungi are heterotrophic; they obtain their nutrition by absorption. Reproductive cycles typically include both sexual and asexual phases. There are about 100,000 valid species of fungi to which names have been given, and many more will eventually be found. Some have been named two or more times; this is particularly so for fungi that may be classified both as ascomycetes and as members of the Fungi Imperfecti.

DIVISION

DIVISION ZYGOMYCOTA: Terrestrial fungi with the hyphae septate only during the formation of reproductive bodies; chitin predominant in the cell walls. The class includes about 600 described species, of which about 25 occur as components of the endomycorrhizae that are found in about 80 percent of all vascular plants.

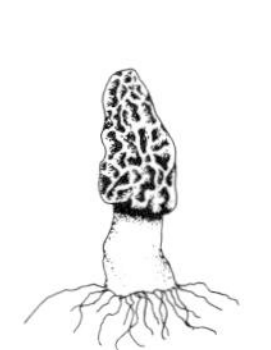

DIVISION ASCOMYCOTA: Terrestrial and aquatic fungi with the hyphae septate but the septa perforated; complete septa cut off the reproductive bodies, such as spores or gametangia. Chitin is predominant in the cell walls. Sexual reproduction involves the formation of a characteristic cell—the ascus—in which meiosis takes place and within which ascospores are formed. The hyphae in many ascomycetes are packed together into complex "fruiting bodies" known as ascocarps. Yeasts are unicellular ascomycetes that reproduce asexually by budding. There are about 30,000 species, in addition to some 25,000 described species of Fungi Imperfecti, in which sexual stages do not occur or are not known.

Lichens. The lichens are ascomycetes that have obligate symbiotic relationships with cyanobacteria or unicellular green algae that multiply within their densely packed hyphae. There are about 25,000 described species.

DIVISION BASIDIOMYCOTA: Terrestrial fungi with the hyphae septate but the septa perforated; complete septa cut off reproductive bodies, such as spores. Chitin is predominant in the cell walls. Sexual reproduction involves formation of basidia, in which meiosis takes place and on which the basidiospores are borne. Basidiomycetes are dikaryotic during most of their life cycle, and there is often complex differentiation of "tissues" within their basidiocarps. They are components of most ectomycorrhizae. There are some 25,000 described species.

KINGDOM PROTISTA

Eukaryotic unicellular or multicellular organisms. Their modes of nutrition include ingestion, photosynthesis, and absorption. True sexuality is present in most divisions. They move by means of 9-plus-2 flagella or are nonmotile. Fungi, plants, and animals are specialized multicellular groups derived from Protista. The divisions treated in this book are categorized as heterotrophic protista (water molds and slime molds; the first four divisions) and autotrophic protista (the algae).

DIVISION

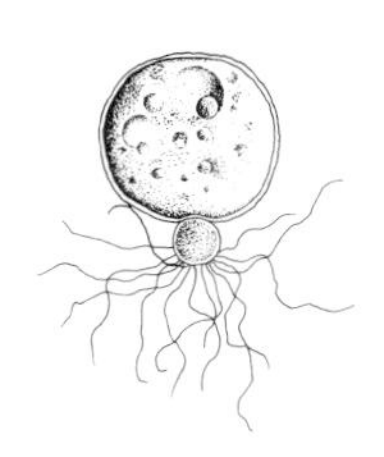

DIVISION OOMYCOTA: Water molds and related organisms. Aquatic or terrestrial organisms with motile cells characteristic at certain stages of their life cycle. The flagella are two in number—one tinsel and one whiplash. Their cell walls are composed of glucose polymers, including, in some cases, cellulose. There are about 475 species.

DIVISION CHYTRIDIOMYCOTA: Chytrids. Aquatic heterotrophic organisms with motile cells characteristic at certain stages in their life cycle. These motile cells have a single, posterior, whiplash flagellum. Their cell walls are composed of chitin, but other polymers may also be present. There are about 750 species.

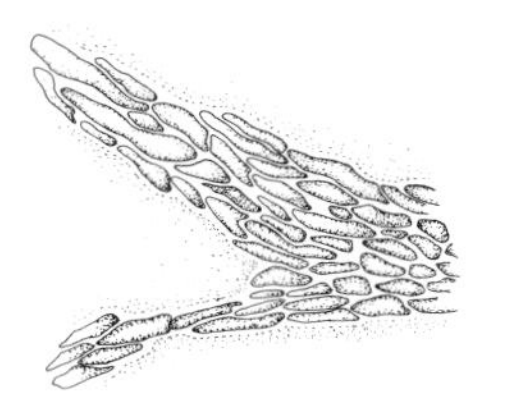

DIVISION ACRASIOMYCOTA: Cellular slime molds. Heterotrophic organisms which exist as separate amoebae (called myxamoebae). Eventually, the myxamoebae swarm together to form a pseudoplasmodium, within which they retain their individual identity. Ultimately, the pseudoplasmodium differentiates into a compound sporangium. Sexual reproduction is unknown. The principal mode of nutrition is by ingestion. There are 7 genera, with about 26 species.

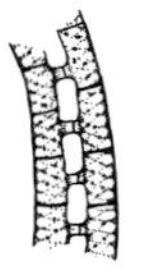

DIVISION MYXOMYCOTA: Plasmodial slime molds. Heterotrophic amoeboid organisms that form a multinucleate plasmodium that creeps along as a mass and eventually differentiates into sporangia, each of which is multinucleate and eventually gives rise to many spores. Sexual reproduction is occasionally observed. The predominant mode of nutrition is by ingestion. There are about 450 species.

DIVISION CHLOROPHYTA: Green algae. Unicellular or multicellular photosynthetic organisms characterized by the presence of chlorophyll *a*, chlorophyll *b*, and various carotenoids. The carbohydrate food reserve is starch. Motile cells have two lateral or apical whiplash flagella. True multicellular genera do not exhibit complex patterns of differentiation. Multicellularity has arisen at least three times, and quite possibly more often. There are about 7000 known species.

Class Chlorophyceae. Those green algae in which cell division involves a phycoplast—a system of microtubules parallel to the plane of cell division. The nuclear envelope persists throughout mitosis and chromosome division occurs within it. Motile cells, if present, possess flagella at the anterior end.

Class Charophyceae. Green algae in which cell division often involves a phragmoplast—a system of microtubules perpendicular to the plane of cell division. The nuclear envelope breaks down during the course of mitosis. Motile cells, if present, possess flagella, which are often lateral.

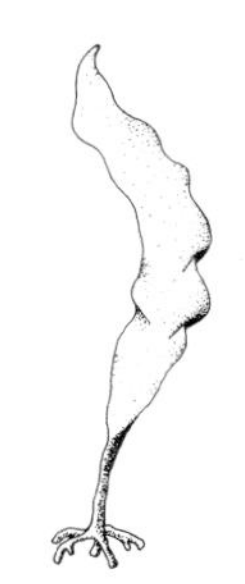

DIVISION PHAEOPHYTA: Brown algae. Multicellular marine algae characterized by the presence of chlorophyll *a*, chlorophyll *c*, and fucoxanthin. The carbohydrate food reserve is laminarin. Motile cells are biflagellated, with one forward flagellum of the tinsel type and one trailing flagellum of the whiplash type. A considerable amount of differentiation is found in some of the kelps, with specialized conducting cells for transporting the products of photosynthesis to the dimly lighted regions of the body. There is, however, no differentiation into roots, leaves, and stems, as in the vascular plants. There are about 1500 species.

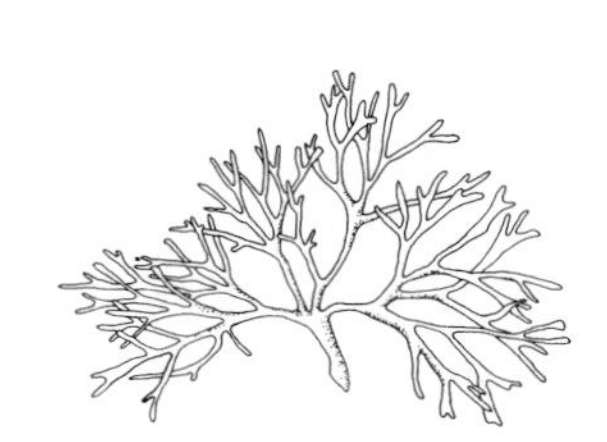

DIVISION RHODOPHYTA: Red algae. Primarily marine algae characterized by the presence of chlorophyll *a* and phycobilins. Their carbohydrate food reserve is floridean starch. No motile cells are present at any stage in the complex life cycle. The vegetative body is built up of closely packed filaments in a gelatinous matrix and is not differentiated into roots, leaves, and stems. It lacks specialized conducting cells. There are some 4000 species.

DIVISION CHRYSOPHYTA: Diatoms and golden-brown algae. Autotrophic organisms that possess chlorophyll *a*, chlorophyll *c*, and fucoxanthin. Food is stored as the carbohydrate leucosin or as large oil droplets. Cell walls consisting mainly of pectic compounds, sometimes heavily impregnated with siliceous materials. There may be as many as 11,500 living species.

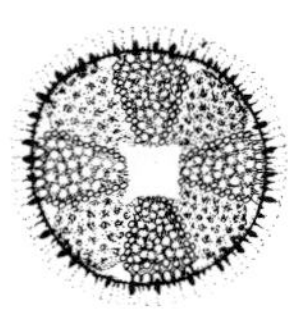

Class Bacillariophyceae. Diatoms. Chrysophyta with double siliceous shells, the two halves of which fit together like a pillbox. They are sometimes motile by the secretion of mucilage fibrils along a specialized groove, called the raphe. There are nearly 10,000 living species, plus an extremely large number of extinct species.

Class Chrysophyceae. Golden-brown algae. A diverse group of organisms including flagellate, amoeboid, and nonmotile forms, some naked and others with a cell wall that may be ornamented with siliceous scales. There are about 1500 species.

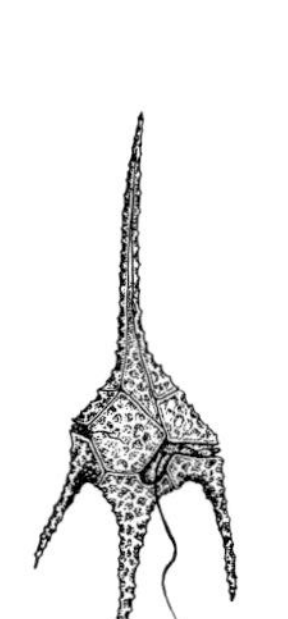

DIVISION PYRROPHYTA: Autotrophic organisms possessing chlorophylls *a* and *c*. Food is stored as starch. Cell walls contain cellulose. This division contains some 1100 species, mostly biflagellate organisms, of which the great majority are dinoflagellates. These all have lateral flagella, one of which beats in a groove that encircles the organism. Sexual reproduction is unknown. Their mitosis is unique.

DIVISION EUGLENOPHYTA: The euglenoids. Autotrophic (or sometimes derived heterotrophic) organisms with chlorophylls *a* and *b*. They store food as paramylon, an unusual carbohydrate. Euglenoids usually have a single apical flagellum of the tinsel variety and a contractile vacuole. The flexible pellicle is rich in proteins. Sexual reproduction is unknown. Euglenoids occur mostly in fresh water. There are some 800 species.

KINGDOM PLANTAE

Autotrophic green plants possessing advanced tissue differentiation; diploid phase (sporophyte) includes an embryo; haploid phase (gametophyte) produces gametes by mitosis; primarily terrestrial.

DIVISION

DIVISION BRYOPHYTA: Liverworts, hornworts, and mosses. Multicellular plants with photosynthetic pigments and food reserves similar to those of the green algae. They have multicellular gametangia with a sterile jacket layer. The sperm are biflagellated. Gametophytes and sporophytes both exhibit complex multicellular patterns of development, but conducting tissues are usually absent and, when present, are not well differentiated. Most photosynthesis in these primarily terrestrial plants is carried out by the gametophyte, upon which the sporophyte is dependent, at least initially. There are some 16,000 species.

Class Hepaticopsida. The liverworts. The gametophytes are thallose or leafy, the rhizoids are single-celled, and the sporophytes, which lack stomata, are relatively simple structures. There are about 6000 species.

Class Anthocerotopsida. The hornworts. The gametophytes are thallose. The sporophyte grows from a basal intercalary meristem for as long as conditions are favorable. Stomata are present on the sporophyte. There are about 100 species.

Class Muscopsida. The mosses. The gametophytes are leafy. Sporophytes have complex patterns of dehiscence. Rhizoids are multicellular. Stomata are present on the sporophyte. There are about 9500 species.

DIVISION PSILOPHYTA: Whisk ferns. Homosporous vascular plants with or without microphylls and with extremely simple sporophytes; no differentiation between root and shoot. Motile sperm. Two genera and several species.

DIVISION LYCOPHYTA. The lycopods. Homosporous and heterosporous vascular plants with microphylls; extremely diverse in appearance. All lycophytes have motile sperm. There are five genera, with about 1000 species.

DIVISION SPHENOPHYTA: The horsetails. Homosporous vascular plants with jointed stems marked by conspicuous nodes and elevated siliceous ribs. Sporangia are borne in a strobilus at the apex of the stem. Leaves are scalelike. Sperm are motile. Although they are thought to have evolved from a megaphyll, the leaves of the horsetails are structurally indistinguishable from microphylls. There is one genus, *Equisetum,* with 15 living species.

DIVISION PTEROPHYTA: The ferns. Mostly homosporous, although some are heterosporous. All possess a megaphyll. The gametophyte is more or less free-living and usually photosynthetic. Multicellular gametangia and free-swimming sperm are present. There are about 12,000 species.

DIVISION CONIFEROPHYTA: The conifers. Seed plants with active cambial growth and simple leaves; ovules and seeds exposed; sperm are not flagellated. The most familiar group of gymnosperms. There are some 50 genera, with about 550 species.

DIVISION CYCADOPHYTA: The cycads. Seed plants with sluggish cambial growth and pinnately compound, palmlike or fernlike leaves; ovules and seeds exposed. The sperm are flagellated and motile but are carried to the vicinity of the ovule in a pollen tube. Cycads are gymnosperms. There are 10 genera, with about 100 species.

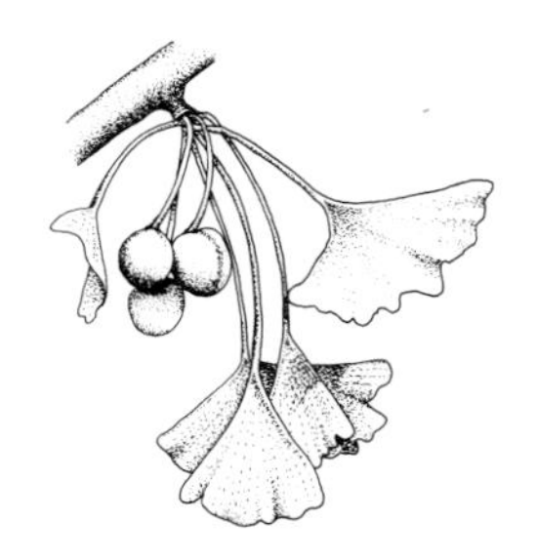

DIVISION GINKGOPHYTA: Ginkgo. Seed plants with active cambial growth and fan-shaped leaves with open dichotomous venation; ovules and seeds exposed; seed coats fleshy. Sperm are carried to the vicinity of the ovule in a pollen tube but are flagellated and motile. They are gymnosperms. There is only one species.

DIVISION GNETOPHYTA: Seed plants with many angiospermlike features. The only gymnosperms in which vessels occur. Motile sperm are absent. There are 3 very distinctive genera, with about 70 species.

DIVISION ANTHOPHYTA: The flowering plants. Seed plants in which ovules are enclosed in a carpel and seeds are borne within fruits. They are extremely diverse vegetatively but are characterized by the flower, which is basically insect-pollinated. Other modes of pollination, such as wind pollination, have been derived in a number of different lines. The gametophytes are much reduced, with the female gametophyte often consisting of only seven cells at maturity. Double fertilization involving the two sperm of the mature microgametophyte gives rise to the zygote (sperm and egg) and to the primary endosperm nucleus (sperm and polar nuclei); the former becomes the embryo and the latter becomes a special nutritive tissue, called the endosperm. There are about 235,000 species.

Class Monocotyledones. The monocots. Flower parts are usually in threes; leaf venation is usually parallel; primary vascular bundles in the stem are scattered; true secondary growth is not present; there is one cotyledon. There are about 65,000 species.

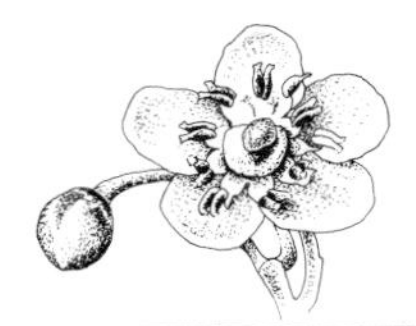

Class Dicotyledones. The dicots. Flower parts are usually in fours or fives; leaf venation is usually netlike; primary vascular bundles in the stem are in a ring; there is true secondary growth with vascular cambium commonly present; there are two cotyledons. There are about 170,000 species.

Some other common terms used to describe major groups of plants deserve mention here. In systems in which the algae and fungi are regarded as plants, they are often grouped as a subkingdom, Thallophyta, the thallophytes: organisms with no highly differentiated tissues, such as root, stem, or leaf, and no vascular tissues (xylem and phloem). The bryophytes and vascular plants are then grouped into a second subkingdom, Embryophyta, in which the zygote develops into a multicellular embryo still encased in an archegonium or an embryo sac. All embryophytes are marked by a heteromorphic alternation of generations.

Although they are no longer used in formal schemes of classification, such terms as "algae," "thallophytes," "vascular plants," and "gymnosperms" are still sometimes useful in an informal sense. An even earlier scheme divided all plants into "phanerogams," for those with flowers, and "cryptogams," for those lacking flowers; these terms are occasionally seen today as well.

Viruses have not been included in the preceding system of classification; their naming is haphazard, the first worker who discovers a particular one giving it an English name or some other designation that may persist if it is accepted by other workers. It would be absurd to try to estimate the number of kinds, or "species" of virus, since new forms are discovered in virtually every insect or plant that is examined critically for their presence.

APPENDIX D

Geologic Eras

ERAS AND DURATION	PERIOD	EPOCHS	LIFE FORMS	CLIMATES AND MAJOR PHYSICAL EVENTS
CENOZOIC 65 million years	Quaternary	Recent Pleistocene	Age of humans. Extinction of many large mammals, including woolly mammoths. Deserts on large scale.	Fluctuating cold to mild. Four glacial advances and retreats (Ice Age); uplift of Sierra Nevada.
	Tertiary	Pliocene	Herbaceous plants abundant as climates diversify. Large carnivores. First known appearance of human-apes.	Cooler. Continued uplift and mountain-building, with widespread extinction. Uplift of Panama results in joining of North and South America.
		Miocene	Spread of grasslands as forests contract. Grazing animals, apes, whales.	Moderate. Extensive glaciation begins again in Southern Hemisphere.
		Oligocene	Origin of Asteraceae and many other modern families of flowering plants. Large, browsing mammals. Apes appear. Madro-Tertiary Geoflora expands.	Rise of Alps and Himalayas. Lands generally low. Volcanoes in Rocky Mountains.
		Eocene	Grasslands begin to expand. Primitive horses, tiny camels, modern and giant types of birds.	Mild to very tropical. Many lakes in western North America. Australia separates from Antarctica; India collides with Asia.
		Paleocene	First known primitive primates and carnivores.	Mild to cool. Wide, shallow continental seas largely disappeared.
MESOZOIC 160 million years	Cretaceous		Angiosperms appear and become abundant. Age of reptiles. Extinction of dinosaurs at end of period. Marsupials, insectivores.	Lands low and extensive. Elevation of Rocky Mountains cuts off rain. Africa and South America separate.
	Jurassic		Gymnosperms, especially cycads; ferns. Dinosaurs' zenith. Flying reptiles, small mammals. Birds appear.	Mild. Continents low. Large areas in Europe covered by seas.
	Triassic		Forests of gymnosperms and ferns. First dinosaurs. Primitive mammals appear.	Continents mountainous. Large areas arid. Eruptions in eastern North America. Continents joined in one mass.
PALEOZOIC 400 million years	Permian		Origin of conifers, cycads, and ginkgos; earlier forest types wane. Reptiles evolve.	Extensive glaciation in Southern Hemisphere. Appalachians formed by end of Paleozoic; most of seas drained from North America.
	Carboniferous Pennsylvanian Mississippian		Forests, ferns, lycophytes, sphenophytes, gymnosperms. Major groups of fungi exist. Age of amphibians. First reptiles. Variety of insects. Sharks abundant.	Warm. Lands low, covered by shallow seas or great coal swamps. Mountain-building in eastern United States, Texas, Colorado. Moist, equable climate; conditions like those in temperate or subtropical zones—little seasonal variation, water plentiful.
	Devonian		Rise of land plants. Extinction of primitive vascular plants. Origin of modern subclasses of vascular plants. Age of fishes. Amphibians appear. Mollusks abundant. Lung fishes.	Europe mountainous with arid basins. Mountains and volcanoes in eastern United States and Canada. Rest of North America low and flat. Sea covered most of land.
	Silurian		Earliest vascular plants. Modern groups of algae and fungi. Rise of fishes, reef-building corals, and shell-forming sea animals. Invasion of land by arthropods.	Mild. Continents generally flat. Mountain-building in Europe. Again flooded.
	Ordovician		Possible invasion of land by plants. First fungi. Shell-forming sea animals.	Mild. Shallow seas, continents low; sea covered United States. Limestone deposits; microscopic plant life thriving.
	Cambrian		First primitive fishes. Age of invertebrates. Trilobites, brachiopods, other animals.	Mild. Extensive seas. Seas spilled over continents.
PRECAMBRIAN 4 billion years			Origin of life: Prokaryotes; eukaryotic cells and multicellularity by close of period.	Dry and cold to warm and moist. Formation of earth's crust. Extensive mountain-building. Shallow seas. Glaciation in eastern Canada. Planet cooled. Components of different densities separated under influence of gravity.

Glossary

Å: *See* angstrom.

a- [Gk. *a-*, not, without]: Prefix, negates the succeeding part of the word; "an-" before a vowel.

abscisic acid [L. *abscissus*, to cut off]: A plant hormone involved in abscission and dormancy.

abscission (ăb•sizh'ŭn): The dropping off of leaves, flowers, fruits, or other plant parts, usually following the formation of an abscission zone.

abscission zone: The area at the base of a leaf, flower, or fruit, or other plant part containing tissues that play a role in the separation of a plant part from the plant body.

absorption spectrum: The spectrum of light waves absorbed by a particular pigment.

accessory bud: A bud generally located above or on either side of the main axillary bud.

accessory cell: *See* subsidiary cell.

accessory pigment: A pigment that captures light energy and transfers it to chlorophyll *a*.

achene: A simple, dry, one-seeded indehiscent fruit; the seed coat is not adherent to the pericarp.

acid: A substance that dissociates, releasing hydrogen ions (H^+) but not hydroxyl ions (OH^-), having a pH in solution of less than 7; a proton donor; *see* base.

actinomorphic [Gk. *aktis*, ray, + *morphe*, form]: Pertaining to a type of flower that can be divided into two equal halves in more than one longitudinal plane; also called radially symmetrical or regular; *see* zygomorphic.

action spectrum: The spectrum of light waves that elicits a particular reaction.

active transport: The pumping of a substance across a cellular membrane from a point of lower concentration to one of higher concentration (against a concentration gradient); an energy-requiring process.

ad- [L. *ad-*, toward, to]: Prefix meaning "toward" or "to."

adaptation [L. *adaptare*, to fit]: A peculiarity of structure, physiology, or behavior that aids in fitting an organism to its environment.

adaptive radiation: The evolution from one kind of organism to several divergent forms, each specialized to fit a distinct and diverse way of life.

adenine (ăd'e•nēn): A purine base present in DNA, RNA, and nucleotide derivatives, such as ADP and ATP.

adenosine triphosphate: Called ATP; the major source of usable chemical energy in metabolism. On hydrolysis, ATP loses one phosphate to become adenosine diphosphate (ADP), releasing usable energy.

adhesion [L. *adhaerere*, to stick to]: The sticking together of unlike objects or materials.

adnate [L. *adnasci*, to grow to or on]: Said of fused unlike parts, as stamens and petals; *see also* connate.

ADP: *See* adenosine triphosphate.

adsorption [L. *ad-*, to, + *sorbere*, to suck in]: The adhesion of a liquid, gaseous, or dissolved substance to a solid, resulting in a higher concentration of the substance.

adventitious [L. *adventicius*, not properly belonging to]: Referring to a structure arising from an unusual place, such as buds at other places than leaf axils, roots growing from stems or leaves.

aeciospore (ē'sĭ•o•spor) [Gk. *aikia*, injury, + *spora*, seed]: A binucleate spore of rust fungi; produced in an aecium.

aecium, *pl.* **aecia:** In rust fungi, a cuplike structure in which aeciospores are produced.

aerobic [Gk. *aer*, air, + *bios*, life]: Requiring free oxygen for respiration.

aerobic respiration: *See* respiration.

after-ripening: Term applied to the metabolic changes that must occur in some dormant seeds before germination can occur.

agar: A gelatinous substance derived from certain red algae; used as a solidifying agent in the preparation of nutrient media for growing microorganisms.

aggregate fruit: A fruit developing from the several separate carpels of a single flower.

akinete: A vegetative cell transformed into a thick-walled, resistant spore in cyanobacteria.

albuminous cell: Certain ray and axial parenchyma cells in gymnosperm phloem that are closely associated with the sieve cells both morphologically and physiologically.

aleurone [Gk. *aleuron*, flour]: Proteinaceous material, usually in the form of small granules, occurring in the outermost cell layer of the endosperm of wheat and other grains.

alga, *pl.* **algae** (ăl′ga, al′je): A photosynthetic eukaryotic organism lacking multicellular sex organs (except for the charophytes); the blue-green algae, or cyanobacteria, are photosynthetic bacteria.

alkali [Arabic *alqili,* the ashes of the plant saltwort]: A substance with marked basic properties.

alkaline: Pertaining to substances that release hydroxyl ions (OH^-) in water, having a pH greater than 7.

alkaloids: Nitrogen-containing ring compounds produced by plants; physiologically active in vertebrates. Many have a bitter taste and some are poisonous; examples are nicotine, morphine, quinine, caffeine, and strychnine.

allele (ă·lēl′) [Gk. *allelon,* of one another, + *morphe,* form]: One of the two or more alternative states of a gene that occupy the same position (locus) on homologous chromosomes. Alleles are separated from each other at meiosis.

allelopathy [Gk. *allelon,* of each other, + *pathos,* suffering]: The inhibition of one species of plant by chemicals produced by another plant.

allopolyploid [Gk. *allos,* different, + *poly,* many, + *ploos,* fold]: A polyploid in which the different sets of chromosomes come from different species or widely different strains.

alternate: Referring to bud or leaf arrangement in which there is one bud or one leaf at a node.

alternation of generations: A reproductive cycle in which a haploid ($1n$) phase, the gametophyte, gives rise to gametes which, after fusion to form a zygote, germinate to produce a diploid ($2n$) phase, the sporophyte. Spores produced by meiotic division from the sporophyte give rise to new gametophytes, completing the cycle.

amino acids [Gk. *Ammon,* referring to the Egyptian sun god, near whose temple ammonium salts were first prepared from camel dung]: Nitrogen-containing organic acids, the units, or "building blocks," from which protein molecules are built.

ammonification: Decomposition of amino acids and other nitrogen-containing organic compounds, resulting in the production of ammonia (NH_3) and ammonium ions (NH_4^+).

amoeboid [Gk. *amoibe,* change]: Moving or eating by means of pseudopodia (temporary cytoplasmic protrusions from the cell body).

amphi- [Gk. *amphi-,* on both sides]. Prefix meaning "on both sides," "both," or "of both kinds."

amylase (ăm′ĭ·las): An enzyme that hydrolyzes starch.

amyloplast: A leucoplast (colorless plastid) that forms starch grains.

an- [Gk. *an-,* not, without]: Prefix, equivalent to "a-," meaning "not" or "without"; used before vowels and "h."

anabolism [Gk. *ana-,* up, + *-bolism* (as in metabolism)]: The constructive part of metabolism; the total chemical reactions involved in biosynthesis.

anaerobic [Gk. *an-,* without, + *aer,* air, + *bios,* life]: Applied to cells that can live without oxygen; strict anaerobes cannot live in the presence of oxygen.

analogous [Gk. *analogos,* proportionate]: Applied to structures similar in function but different in evolutionary origin.

anaphase [Gk. *ana,* away, + *phasis,* form]: A stage in mitosis in which the chromatids of each chromosome separate and move to opposite poles; similar stages in meiosis in which chromatids or paired chromosomes move apart.

anatomy: In botany, the area that deals with the internal structure of organisms.

andro- [Gk. *andros,* man]: Prefix meaning "male."

androecium [Gk. *andros,* man, + *oikos,* house]: (1) Stamens of a flower, collectively; (2) in leafy liverworts, a tubular sheath surrounding the antheridia.

angiosperm [Gk. *angion,* a vessel, + *sperma,* a seed]: Literally, a seed borne in a vessel (carpel); thus one of a group of plants whose seeds are borne within a mature ovary (fruit).

angstrom [after A. J. Ångstrom, a Swedish physicist, 1814–74]: A unit of length equal to 0.0001 micrometer; abbreviated Å.

anion [Gk. *anienai,* to go up]: A negatively charged ion.

anisogamy [Gk. *aniso,* unequal, + *gamos,* marriage]: The condition of having dissimilar motile gametes.

annual [L. *annulus,* year]: A plant in which the entire life cycle is completed in a single growing season.

annual ring: In wood, the growth layer formed during a single year; *see also* growth layer.

annulus [L. *anus,* ring]: In ferns, a row of specialized cells in a sporangium; in gill fungi, the remnant of the inner veil forming a ring on the stalk.

anterior: Situated before or toward the front; in the case of flagella, the end that moves forward.

anther [Gk. *anthos,* flower]: The pollen-bearing portion of a stamen.

antheridiophore [Gk. *anthos,* flower, + *phoros,* bearing]: In some liverworts, a stalk that bears antheridia.

antheridium: Multicellular male gametangium of plants other than seed plants; unicellular male gametangium in some algae and fungi.

anthocyanin [Gk. *anthos,* flower, + *kyanos,* dark blue]: A water-soluble blue or red pigment found in the cell sap.

antibiotic [Gk. *anti,* against or opposite, + *biotikos,* pertaining to life]: Natural organic substances that retard or prevent the growth of organisms; generally used to designate substances formed by microorganisms that prevent growth of other microorganisms.

anticlinal: Perpendicular to the surface.

antipodals: Three (sometimes more) cells of the mature embryo sac located at the end opposite the micropyle.

apical dominance: Influence exerted by a terminal bud in suppressing the growth of lateral buds.

apical meristem: The growing point, composed of meristematic tissue, at the tip of the root or shoot in a vascular plant.

apomixis [Gk. *apo,* separate, away from, + *mixis,* mingling]: Reproduction without meiosis or syngamy; vegetative reproduction.

apoplast [Gk. *apo,* away from, + *plastos,* molded]: The cell wall continuum of a plant or organ; the movement of substances via the cell walls is called apoplastic movement or transport.

apothecium [Gk. *apotheke,* storehouse]: A cup-shaped or saucer-shaped open ascocarp.

arch-, archeo- [Gk. *arche, archos,* beginning]: Prefix meaning "first," "main," or "earliest."

archegoniophore [Gk. *archegonos,* the first of a race, + *phoros,* bearing]: In some liverworts, a stalk that bears archegonia.

archegonium, *pl.* **archegonia:** Multicellular female sex organ, containing an egg; found in the Bryophyta and some vascular plants.

aril (ăr'ĭl) [L. *arillus,* grape, seed]: An accessory seed covering, often formed by an outgrowth at the base of the ovule; often brightly colored, which may aid in dispersal by attracting animals that eat it and, in the process, carry the seed away from the parent plant.

artifact [L. *ars,* art, + *facere,* to make]: A product that exists because of an extraneous, especially human, agency and does not occur in nature.

ascocarp [Gk. *askos,* bladder, + *karpos,* fruit]: A fruiting body of the ascomycetes, generally either an open cup, a vessel, or a closed sphere lined with specialized cells called asci, in which nuclear fusion and meiosis occur.

ascogenous hyphae [Gk. *askos,* bladder, + *genous,* producing]: Hyphae containing paired haploid male and female nuclei; develop from an ascogonium and eventually give rise to asci.

ascogonium: The oogonium or female gametangium of the ascomycetes.

ascospore: A fungal spore produced within an ascus.

ascus, *pl.* **asci:** A specialized cell, characteristic of the ascomycetes, in which two haploid nuclei fuse to produce a zygote that immediately divides by meiosis; at maturity, an ascus contains ascospores.

aseptate [Gk. *an-,* not, + L. *septum,* fence]: A term used to describe algal or fungal filaments lacking cross walls; an equivalent term is nonseptate.

asexual reproduction: Any reproductive process, such as fission or budding, that does not involve the union of gametes.

atom [Gk. *atomos,* indivisible]: The smallest unit into which a chemical element can be divided and still retain its characteristic properties.

atomic nucleus: The central core of an atom, containing protons and neutrons, around which electrons orbit.

atomic number: The number of protons in the nucleus of an atom.

atomic weight: The weight of a representative atom of an element relative to the weight of an atom of carbon ^{12}C, which has been assigned the integral value of 12.

ATP: *See* adenosine triphosphate.

auto- [Gk. *autos,* self, same]: Prefix, meaning "same" or "self-same."

autoecious [Gk. *auto,* self, + *oikia,* dwelling]: In some rust fungi, completing the life cycle on a single species of host plant.

autopolyploid: A polyploid in which all the chromosomes come from the same species, usually resulting from the doubling of chromosomes in a single individual.

autoradiograph: A photographic print made by a radioactive substance acting upon a sensitive photographic film.

autotroph [Gk. *autos,* self, + *trophos,* feeder]: An organism that is able to synthesize the nutritive substances it requires from inorganic substances in its environment; *see also* heterotroph.

auxin [Gk. *auxein,* to increase]: A plant growth-regulating substance; controls cell elongation, among other effects.

axial system: In secondary xylem and secondary phloem, the term applied collectively to cells derived from fusiform cambial initials. The long axes of these cells are oriented parallel with the main axis of the root or stem. Also called longitudinal system and vertical system.

axil [Gk. *axilla,* armpit]: The upper angle between a twig or leaf and the stem from which it grows.

axillary: Term applied to buds or branches occurring in the axil of a leaf.

bacillus, *pl.* **bacilli** (ba•sĭl'ŭs) [L. *baculum,* rod]: A rod-shaped bacterium.

backcross: The crossing of a hybrid with one of its parents or with a genetically equivalent organism; a cross between an individual whose genes are to be tested and one that is homozygous for all recessive genes involved in the experiment.

bacteriophage [Gk. *bakterion,* little rod, + *phagein,* to eat]: A virus that parasitizes bacterial cells.

bacterium, *pl.* **bacteria:** A unicellular prokaryotic organism, including cyanobacteria, or blue-green algae.

bark: A nontechnical term applied to all tissues outside the vascular cambium; *see also* inner bark and outer bark.

basal body: Cylinder-shaped cytoplasmic structure that forms the basal portion of the cilium or flagellum; identical in form and structure to the centriole.

base: A substance that, upon dissociation, releases hydroxyl ions (OH^-), but not hydrogen ions (H^+), having a pH in solution of more than 7; *see* acid.

basidiocarp [L. *basidium,* a little pedestal, + *carpus,* a fruit]: A fruiting body of the basidiomycetes.

basidiospore: A spore of the basidiomycetes; produced within and borne upon a basidium following nuclear fusion and meiosis.

basidium, *pl.* **basidia:** A specialized reproductive cell of the basidiomycetes, often club-shaped, in which nuclear fusion and meiosis occur.

berry: A simple fleshy fruit, which includes a fleshy ovary wall and one or more carpels and seeds; examples are grapes, tomatoes, and bananas.

bi- [L. *bis,* double, two]: Prefix meaning "two," "twice," or "having two points."

biennial: A plant that normally requires two growing seasons to complete its life cycle. Only vegetative growth occurs the first year, often resulting in the formation of an over-wintering rosette; the flowering and fruiting occur in the second year.

bilaterally symmetrical: *See* zygomorphic.

biological clock [Gk. *bios,* life, + *logos,* discourse]: The internal timing mechanism that governs what seem to be the innate biological rhythms of some organisms.

biomass: Total dry weight of all organisms in a particular habitat or area.

biome: A complex of communities of very wide extent, characterized by distinctive vegetation and climate; for example, all grassland areas collectively form the grassland biome.

biosphere: The zone of air, land, and water at the surface of the earth occupied by living things.

biotic: Relating to life.

bisexual flower: A flower that has at least one functional stamen and one functional carpel.

bivalent [L. *bis*, twice, + *valere*, to be strong]: A pair of synapsed chromosomes.

blade: The broad, expanded part of a leaf; also called the lamina.

body cell: The cell of the male gametophyte, or pollen grain, of gymnosperms, which divides mitotically to form two sperm.

bordered pit: A pit in which the secondary wall arches over the pit membrane.

bract: A modified, usually reduced leaflike structure.

branch root: *See* lateral root.

bryophytes (brī′o•fīts): Nonvascular, terrestrial green plants, such as mosses, hornworts, and liverworts.

bud: (1) An embryonic shoot of a plant; (2) a vegetative outgrowth of yeasts and some bacteria as a means of asexual reproduction.

bulb: A short underground stem covered by enlarged and fleshy leaf bases containing stored food.

bulk flow: The overall movement of water or some other liquid induced by gravity, pressure, or an interplay of both.

bundle scar: Scar or mark left on leaf scar by vascular bundles broken at the time of leaf fall, or abscission.

bundle sheath: Layer or layers of cells surrounding a vascular bundle; may consist of parenchyma or sclerenchyma cells, or both.

bundle-sheath extension: A group of cells extending from a bundle sheath of a vein in the leaf mesophyll to either or both upper and lower epidermis; may consist of parenchyma, collenchyma, or sclerenchyma.

C_3 pathway: *See* Calvin cycle.

C_4 pathway: The set of reactions through which carbon dioxide is fixed to a compound known as phosphoenolpyruvate (PEP) to yield oxaloacetate, a four-carbon compound.

callose: A complex branched carbohydrate, which is a common wall constituent associated with the sieve areas of sieve elements; may develop in reaction to injury in sieve elements and parenchyma cells.

callus [L. *callos*, hard skin]: Undifferentiated tissue; a term used in tissue culture, grafting, and wound healing.

calorie [L. *calor*, heat]: The amount of energy in the form of heat required to raise the temperature of one gram of water 1°C; in making metabolic measurements, the kilocalorie (kcal) is generally used. A kilocalorie, or Calorie, is the amount of heat required to raise the temperature of one kilogram of water 1°C.

Calvin cycle: The series of enzymatically mediated photosynthetic reactions during which carbon dioxide is reduced to 3-phosphoglyceraldehyde and the carbon dioxide acceptor, ribulose 1,5-bisphosphate, is regenerated. For every six molecules of carbon dioxide entering the cycle, a net gain of two molecules of glyceraldehyde 3-phosphate results.

calyptra [Gk. *kalyptra*, covering for the head]: The hood or cap, which partially or entirely covers the capsule of some species of mosses; it is formed from the expanded archegonial wall.

calyx (kā′lĭks) [Gk. *kalyx*, a husk, cup]: Sepals collectively; the outermost flower whorl.

CAM: *See* Crassulacean acid metabolism.

cambial zone: A region of thin-walled, undifferentiated meristematic cells between the secondary xylem and secondary phloem; consists of cambial initials and their recent derivatives.

cambium [L. *cambiare*, to exchange]: A meristem that gives rise to parallel rows of cells; commonly applied to the vascular cambium and the cork cambium, or phellogen.

capsule: (1) A dehiscent, dry fruit that develops from two or more carpels; (2) a slimy layer around the cells of certain bacteria; (3) the sporangium of Bryophyta.

carbohydrate [L. *carbo*, ember, + *hydro*, water]: An organic compound consisting of a chain of carbon atoms to which hydrogen and oxygen are attached in a 2:1 ratio; includes sugars, starch, glycogen, and cellulose.

carbon cycle: Worldwide circulation and utilization of carbon atoms.

carbon fixation: The conversion of CO_2 into organic compounds during photosynthesis.

carnivorous: Feeding upon animals, as opposed to plants (herbivorous); also refers to plants that are able to utilize proteins obtained from trapped animals, chiefly insects.

carotene (kăr′o•tēn) [L. *carota*, carrot]: A yellow or orange pigment found in chloroplasts and chromoplasts of plants.

carotenoids (kă•rŏt′e•noids): A class of fat-soluble pigments that includes the carotenes (yellow and orange pigments) and the xanthophylls (yellow pigments); found in chloroplasts and chromoplasts of plants.

carpel [Gk. *karpos*, fruit]: A leaflike organ in angiosperms that encloses one or more ovules; one of the members of the gynoecium.

carpellate: *See* pistillate.

carpogonium [Gk. *karpos*, fruit, + *gonos*, offspring]: In red algae, the female gametangium.

carposporangium [Gk. *karpos*, fruit, + *spora*, seed, + *angeion*, vessel]: In red algae, a carpospore-containing cell.

carpospore: In red algae, the single diploid protoplast found within a carposporangium.

carrier molecule: *See* permease.

caryopsis [Gk. *karyon*, a nut, + *opsis*, appearance]: Simple, dry, one-seeded indehiscent fruit, with the pericarp firmly united all around the seed coat; a grain characteristic of the grasses (family Poaceae).

Casparian strip [Robert Caspary, a German botanist]: A bandlike region of primary wall containing suberin and lignin; found in anticlinal—radial and transverse—walls of endodermal cells.

catabolism [Gk. *katabole,* throwing down]: Collectively, the chemical reactions resulting in the breakdown of complex materials and involving the release of energy.

catalyst [Gk. *katalysis,* dissolution]: A substance that accelerates the rate of a chemical reaction but is not used up in the reaction; enzymes are catalysts.

cation [Gk. *katienai,* to go down]: A positively charged ion.

catkin: A spikelike inflorescence of unisexual flowers; found only in woody plants.

cell [L. *cella,* small room]: The structural unit of living organisms; in plants, consists of cell wall and protoplast.

cell division: The division of the cytoplasm into two equal parts; brought about in higher plants by the formation of a cell plate, which grows outward to the margins of the cell; also referred to as cytokinesis.

cell plate: The structure that forms at the equator of the spindle in dividing cells of plants and a few green algae during early telophase; the ontogenetic predecessor of the middle lamella.

cell sap: The fluid contents of the vacuole.

cellulase: An enzyme that hydrolyzes cellulose.

cellulose: A carbohydrate; the chief component of the cell wall in most plants; an insoluble complex carbohydrate formed of glucose molecules attached end to end.

cell wall: The rigid outermost layer of the cells found in plants, some protista, and most prokaryotes.

central mother cells: Relatively large vacuolate cells in a subsurface position in apical meristems of shoots.

centric diatom [Gk. *kentron,* center]: A radially symmetrical diatom.

centriole [Gk. *kentron,* center, + L. *-olus,* little one]: A cytoplasmic structure generally found in animal cells and in flagellate cells of other groups; usually outside of the nuclear membrane, which doubles before mitosis; the two centrioles then move apart and organize the spindle apparatus.

centromere [Gk. *kentron,* center, + *meros,* a part]: That portion of the chromosome to which the spindle fibers are attached; also called the kinetochore.

chalaza [Gk. *chalaza,* small tubercle]: The region of an ovule or seed where the funiculus unites with the integuments and the nucellus.

chemical potential: The activity or free energy of a substance; it is dependent upon the rate of motion of the average molecule and the concentration of the molecules.

chemoautotrophic: Refers to organisms (bacteria) that are able to manufacture their own basic foods with chemical energy; *see* autotroph.

chiasma (kī•ăz'ma) [Gk. *chiasma,* a cross]: The X-shaped figure formed by the meeting of two nonsister chromatids of homologous chromosomes; probable site of crossing-over.

chitin (kī'tĭn) [Gk. *chiton,* tunic]: A tough, resistant polysaccharide forming the cell walls of certain fungi, the exoskeleton of arthropods, and the epidermal cuticle or other surface structures of many other invertebrates.

chlor- [Gk. *chloros,* green]: Prefix meaning "green."

chlorenchyma: General term applied to parenchyma cells that contain chloroplasts.

chlorophyll [Gk. *chloros,* green, + *phyllon,* leaf]: The green pigment of plant cells, necessary for photosynthesis; also found in some protista and some prokaryotes.

chloroplast: A plastid in algal and green plant cells in which chlorophylls are contained; site of photosynthesis.

chlorosis: Loss or reduced development of chlorophyll.

chroma- [Gk. *chroma,* color]: Prefix meaning "color."

chromatid [Gk. *chroma,* color, + L. *-id,* daughters of]: One of the two daughter strands of a duplicated chromosome which are joined by a single centromere.

chromatin: The deeply staining nucleoprotein complex of chromosomes.

chromatophore [Gk. *chroma,* color, + *phorus,* a bearer]: In some bacteria, a discrete vesicle delimited by a single membrane and containing photosynthetic pigments.

chromoplast: A plastid containing pigments other than chlorophyll, usually yellow and orange carotenoid pigments.

chromosome [Gk. *chroma,* color, + *soma,* body]: One of the bodies in the cell nucleus, along which the genes are located; visualized as threads or rods of chromatin which appear in contracted form during mitosis and meiosis.

cilium, *pl.* **cilia** (sĭl'ĭ•ŭm) [L. *cilium,* eyelash]: A short, hairlike structure present on the surface of certain specialized cell types; usually numerous and arranged in rows.

circadian rhythms [L. *circa,* about, + *dies,* a day]: Regular rhythms of growth and activity, which occur on an approximately 24-hour basis.

circinate vernation [L. *circinare,* to make round, + *vernare,* to flourish]: As in ferns, the coiled arrangement of leaves and leaflets in the bud; it uncoils gradually as the leaf develops further.

cisterna, *pl.* **cisternae** [L. *cistern,* a reservoir]: A flattened or saclike portion of the endoplasmic reticulum or a dictyosome (Golgi body).

cladophyll [Gk. *klados,* shoot, + *phyllon,* leaf]: A branch resembling a foliage leaf.

clamp connection: In the basidiomycetes, a lateral connection between adjacent cells of a dikaryotic hypha; ensures that each cell of the hypha will contain two dissimilar nuclei.

class: A taxonomic category between division and order.

cleistothecium [Gk. *kleistos,* closed, + *thekion,* small receptacle]: A closed, spherical ascocarp.

climax community: Final or stable community in successional series that is more or less in equilibrium with existing environmental conditions.

cline: A graded series of changes in some characteristics within a species, correlated with some gradual change in climate or another geographical factor.

clone [Gk. *klon,* twig]: A population of individuals descended by mitotic division from a single ancestor.

closed bundle: A vascular bundle in which a cambium does not develop.

coalescence [L. *coalescere,* to grow together]: The union of floral parts of the same whorl, as petals to petals.

coccus, *pl.* **cocci** (kŏk'ŭs) [Gk. *kokkos,* a berry]: A spherical bacterium.

codon (kō'dŏn): Sequence of three adjacent nucleotides that form the code for a single amino acid.

coenocytic (se•nō•sĭ'tic) [Gk. *koinos,* shared in common, + *kytos,* a hollow vessel]: A term used to describe an organism or part of an organism that is multinucleate, the nuclei not separated by walls; siphonaceous; siphonous.

coenzyme: An organic molecule, or nonprotein organic cofactor, which plays an accessory role in enzyme-catalyzed processes, often by acting as a donor or acceptor of a substance involved in the reaction; ATP and NAD are common coenzymes.

cofactor: One or more nonprotein components required by enzymes in order to function; some cofactors are ions, whereas others are nonprotein organic substances called coenzymes.

cohesion [L. *cohaerere,* to stick together]: Union or holding together of parts of the same materials.

coleoptile (kō'le•op'till) [Gk. *koleos,* sheath, + *ptilon,* feather]: The sheath enclosing the apical meristem and leaf primordia of the grass embryo; often interpreted as the first leaf.

coleorhiza (kō'le•o•rī'ză) [Gk. *koleos,* sheath, + *rhiza,* root]: The sheath enclosing the radicle in the grass embryo.

collenchyma [Gk. *kolla,* glue]: A supporting tissue composed of collenchyma cells; often found in regions of primary growth in stems and in some leaves.

collenchyma cell: Elongated living cell with irregularly thickened primary cell wall.

colloid (kŏl'oid): A permanent suspension of fine particles.

community: All the organisms inhabiting a common environment and interacting with one another.

companion cell: A specialized parenchyma cell associated with a sieve-tube member in angiosperm phloem and arising from the same mother cell as the sieve-tube member.

competition: The effect of a common demand by two or more organisms on a limited supply of food, water, light, minerals, etc.

complete flower: A flower having four whorls of floral parts—sepals, petals, stamens, and carpels.

compound: A combination of atoms in a definite ratio, held together by chemical bonds.

compound leaf: A leaf whose blade is divided into several distinct leaflets.

concentration gradient: The concentration difference of a substance per unit distance.

cone: *See* strobilus.

conidiophore: Hypha on which one or more conidia are produced.

conidium, *pl.* **conidia** [Gk. *konis,* dust]: An asexual fungal spore not contained within a sporangium; it may be produced singly or in chains; most conidia are multinucleate.

conifer: A cone-bearing tree.

conjugation: The process in bacteria, protozoa, and certain algae and fungi by which two individuals transfer nuclear material.

conjugation tube: As in the green alga *Spirogyra,* a tube through which a gamete or gametes may move to unite with other gametes.

connate (kŏn'āt): Said of similar parts that are united or fused, as petals fused in a corolla tube; *see also* adnate.

convergent evolution [L. *convergere,* to turn together]: The independent development of similar structures in forms of life that are unrelated or only distantly related; often found in organisms living in similar environments.

cork, or phellem: A secondary tissue produced by a cork cambium; made up of polygonal cells, nonliving at maturity, with walls infiltrated with suberin, a waxy or fatty material resistant to the passage of gases and water vapor; the outer part of the periderm.

cork cambium, or phellogen: The lateral meristem that forms the periderm, producing cork (phellem) toward the surface (outside) of the plant and phelloderm toward the inside; common in stems and roots of gymnosperms and dicots.

corm: A thickened underground stem, upright in position, in which food is accumulated, usually in the form of starch.

corolla [L. *corona,* crown]: Petals, collectively; usually the conspicuously colored flower whorl.

corolla tube: A tubelike structure resulting from the fusion of the petals along their edges.

cortex: Ground-tissue region of a stem or root bounded externally by the epidermis and internally by the vascular system; a primary-tissue region.

cotyledon (kŏt'ĭ•lē'dŭn) [Gk. *kotyledon,* cup-shaped hollow]: Seed leaf; generally stores food in dicotyledons and absorbs food in monocotyledons.

coupled reactions: Reactions in which energy-requiring chemical reactions are linked to energy-releasing reactions.

covalent bond: A chemical bond formed between atoms as a result of the sharing of two electrons.

Crassulacean acid metabolism: A variant of the C_4 pathway; phosphoenolpyruvate fixes CO_2 in C_4 compounds at night and then, during the daytime, the fixed CO_2 is transferred to the ribulose bisphosphate of the Calvin cycle within the same cell.

cristae, *sing.* **crista:** The enfoldings of the inner mitochondrial membrane which form a series of crests or ridges.

crop rotation: The practice of growing different crops in regular succession to aid in the control of insects and diseases, to increase soil fertility, and to decrease erosion.

crossing over: The exchange of corresponding segments of genetic material between chromatids of homologous chromosomes at meiosis.

cross-pollination: The transfer of pollen from the anther of one plant to the stigma of a flower of another plant.

cross section: *See* transverse section.

cryptogam: An archaic term for all organisms except the flowering plants (phanerogams), animals, and heterotrophic protista.

cultivar: A variety of plant found only under cultivation.

cuticle: Waxy or fatty layer on outer wall of epidermal cells.

cutin [L. *cutis,* skin]: Fatty substance deposited in many plant cell walls and on outer surface of epidermal cell walls, where it forms a layer known as the cuticle.

cyclosis (sī•klō'sis) [Gk. *kyklosis,* circulation]: The streaming of cytoplasm within a cell.

-cyte, cyto- [Gk. *kytos,* hollow vessel, container]: Suffix or prefix meaning "pertaining to the cell."

cytokinesis: *See* cell division.

cytokinin [Gk. *kytos,* hollow vessel, + *kinesis,* motion]: Class of plant growth substances that promote cell division, among other effects.

cytology: The study of the cell.

cytoplasm: Term commonly used to refer to the protoplasm of the cell exclusive of the nucleus.

cytoplasmic ground substance: The least differentiated part of the cytoplasm, as seen with the electron microscope, and the part surrounding the nucleus and various organelles; also called hyaloplasm.

cytosine: A pyrimidine base found in the nucleic acids DNA and RNA.

day-neutral plants: Plants that flower without regard to day length.

de- [L. *de-,* away from, down, off]: Prefix meaning "away from," "down," or "off"; for example, dehydration means removal of water.

deciduous [L. *decidere,* to fall off]: Shedding leaves at a certain season.

decomposers: Organisms (bacteria, fungi, heterotrophic protistans) in an ecosystem that break down organic material into smaller molecules that then are recirculated.

dehiscence [L. *de,* down, + *hiscere,* split open]: The opening of an anther, fruit, or other structure, permitting the escape of reproductive bodies contained within.

denitrification: The process by which nitrogen is released from the soil by the action of denitrifying bacteria.

deoxyribonucleic acid (DNA): Carrier of genetic information in cells; composed of chains of phosphate, sugar molecules (deoxyribose), and purines and pyrimidines; capable of self-replication as well as determining RNA synthesis.

dermal tissue system: The epidermis or the periderm.

desmid: A freshwater green alga whose cell wall is in two sections united by a narrow constriction, or isthmus; sexual reproduction is by the union of amoeboid gametes.

desmotubule [Gk. *desmos,* to bind, + L. *tubulus,* small tube]: The tubule traversing a plasmodesmatal canal and uniting the endoplasmic reticulum of the two adjacent cells.

determinate growth: Growth of limited duration, as is characteristic of floral meristems and of leaves.

deuterium: Heavy hydrogen; a hydrogen atom, the nucleus of which contains one proton and one neutron. (The common nucleus of hydrogen consists of only one proton.)

dichotomy: The division or forking of an axis into two branches.

dicotyledon: A plant whose embryo has two cotyledons; one of the two classes of angiosperms, Dicotyledones; often abbreviated as dicot.

dictyosome: *See* Golgi body.

differentially permeable membrane: A membrane through which substances diffuse at different rates.

differentiation: A process by which a relatively unspecialized cell undergoes a progressive change to a more specialized cell; the specialization of cells and tissues for particular functions during development.

diffuse-porous wood: A wood in which the pores, or vessels, are fairly uniformly distributed throughout the growth layers or in which the size of pores changes only slightly from early wood to late wood.

diffusion [L. *diffundere,* to pour out]: The movement of suspended or dissolved particles from a more concentrated region to a less concentrated region as a result of the random movement of individual molecules; the process tends to distribute such particles uniformly throughout a medium.

digestion: The conversion of complex, usually insoluble foods into simple, usually soluble forms by means of enzymatic action.

dikaryon [Gk. *di,* two, + *karyon,* a nut]: In fungi, mycelium with paired nuclei, each usually derived from a different parent.

dikaryotic: In fungi, having pairs of nuclei within cells or compartments.

dimorphism [Gk. *di,* two, + *morphe,* form]: The condition of having two distinct forms, such as sterile and fertile leaves in ferns, or sterile and fertile shoots in horsetails.

dioecious [Gk. *di,* two, + *oikos,* house]: Unisexual; having the male and female (or staminate and ovulate) elements on different individuals of the same species.

diploid: Having two sets of chromosomes; the $2n$ number characteristic of the sporophyte generation.

disk flowers: The actinomorphic, tubular flowers that compose the central part of a head of flowers in most Asteraceae; contrasted with flattened, zygomorphic ray-shaped flowers (ray flowers) on the margins of the head.

distal: Situated away from or far from point of reference (usually the main part of body); opposite of proximal.

division: One of the major kinds of groups used by botanists in classifying organisms; the taxonomic category between kingdom and class.

DNA: *See* deoxyribonucleic acid.

dominant: Applied to a gene that exerts its full phenotypic effect regardless of its allelic partner; a gene that masks the effect of its allele.

dormancy: A special condition of arrested growth in which the plant and such plant parts as buds and seeds do not begin to grow without special environmental cues. The requirement for such cues, which include cold exposure and a suitable photoperiod, prevents the breaking of dormancy during superficially favorable growing conditions.

double fertilization: The fusion of the egg and sperm (resulting in a $2n$ fertilized egg, the zygote) and the simultaneous fusion of the second male gamete with the polar nuclei (resulting in a $3n$ primary endosperm nucleus); a unique characteristic of all angiosperms.

doubling rate: The length of time required for a population of a given size to double in number.

drupe [Gk. *dryppa*, overripe olive]: A simple, fleshy fruit, derived from a single carpel, usually one-seeded, in which the inner fruit coat is hard and may be adherent to the seed.

druse: A compound, more or less spherical crystal with many component crystals projecting from its surface; composed of calcium oxalate.

early wood: The first-formed wood of a growth increment; it contains larger cells and is less dense than the subsequently formed late wood; replaces the term "spring wood."

eco- [Gk. *oikos*, house]: Prefix meaning "house" or "home."

ecology: The study of the interactions of organisms with their physical environment and with each other; the results of such interactions.

ecosystem: A system of interactions between living organisms and their nonliving environment.

ecotype [Gk. *oikos*, house, + L. *typus*, image]: A locally adapted variant of an organism, differing genetically from other ecotypes.

ectoplast: *See* plasma membrane.

edaphic [Gk. *edaphos*, ground, soil]: Pertaining to soil conditions that influence plant growth.

egg: A nonmotile female gamete.

egg apparatus: The egg cell and synergids located at the micropylar end of the female gametophyte, or embryo sac, of angiosperms.

elater [Gk. *elater*, driver]: (1) An elongated, spindle-shaped, sterile cell in the sporangium of a liverwort sporophyte (aids in spore dispersal); (2) clubbed, hygroscopic bands attached to the spores of the horsetails.

electrolyte: A substance that dissociates into ions in aqueous solution and so makes possible the conduction of an electric current through the solution.

electron: A subatomic particle with a negative electric charge equal in magnitude to the positive charge of the proton, but with a mass of 1/1837 of that of the proton. Electrons surround the atom's positively charged nucleus and determine the atom's chemical properties.

electron-dense: In electron microscopy, not permitting the passage of electrons and so appearing dark.

element: A substance composed of only one kind of atom; one of about 100 distinct natural or synthetic types of matter that, singly or in combination, compose virtually all materials of the universe.

embryo [Gk. *en*, in, + *bryein*, to swell]: A young sporophytic plant, before the start of a period of rapid growth (germination in seed plants).

embryogeny: The formation of the embryo.

embryo sac: The female gametophyte of angiosperms, generally an eight-nucleate, seven-celled structure; the seven cells are the egg cell, two synergids and three antipodals (each with a single nucleus), and the central cell (with two nuclei).

endergonic: Energy-requiring, as in a chemical reaction; applied to an "uphill" process.

endo- [Gk. *endon*, within]: Prefix meaning "within."

endocarp [Gk. *endon*, within, + *karpos*, fruit]: The innermost layer of the mature ovary wall, or pericarp.

endocytosis [Gk. *endon*, within, + *kytos*, hollow vessel]: The uptake of material into cells by means of invagination of the plasma membrane; if solid material, called phagocytosis; if dissolved material, called pinocytosis.

endodermis [Gk. *endon*, within, + *derma*, skin]: A single layer of cells forming a sheath around the vascular region in roots and some stems; characterized by a Casparian strip within radial and transverse walls. In roots and stems of seed plants, the endodermis is the innermost layer of the cortex.

endogenous [Gk. *endon*, within, + *genos*, race, kind]: Arising from deep-seated tissues, as in the case of lateral roots.

endomembrane system: Collectively, the cellular membranes that form a continuum (plasma membrane, tonoplast, endoplasmic reticulum, Golgi bodies, and nuclear envelope).

endoplasmic reticulum: A complex, three-dimensional membrane system of indefinite extent present in eukaryotic cells, dividing the cytoplasm into compartments and channels; often coated with ribosomes (rough endoplasmic reticulum).

endosperm [Gk. *endon*, within, + *sperma*, seed]: A tissue, containing stored food, that develops from the union of a male nucleus and the polar nuclei of the central cell; it is digested by the growing sporophyte either before or after the maturation of the seed; found only in angiosperms.

entrainment: The process by which a periodic repetition of light and dark, or some other external cycle, causes a circadian rhythm to remain synchronized with the same cycle as the modifying, or entraining, factor.

entropy: A measure of the randomness or disorder of a system.

enzyme: A complex protein produced in living cells that, even in very low concentration, speeds up (catalyzes) the rate of a chemical reaction.

epi- [Gk. *epi*, upon]: Prefix meaning "upon" or "above."

epicotyl: The upper portion of the axis of an embryo or seedling, above the cotyledons (seed leaves) and below the next leaf or leaves.

epidermis: The outermost layer of cells of the leaf and of young stems and roots; primary in origin.

epigyny [Gk. *epi*, upon, + *gyne*, woman]: Floral organization in which the sepals, petals, and stamens apparently grow from the top of the ovary; as opposed to hypogyny.

epiphyte (ĕp′ĭ•fit): An organism that grows upon another organism but is not parasitic on it.

epistatic [Gk. *epistasis*, a stopping]: Term used to describe a gene the action of which determines whether or not the effects of another gene will occur.

ethylene: A simple hydrocarbon that is a plant growth substance involved in the ripening of fruit; $H_2C = CH_2$.

etiolation (e′tĭ•o•lā′shŭn) [Fr. *étioler*, to blanch): A condition involving increased stem elongation, poor leaf development, and lack of chlorophyll; found in plants growing in the dark or with a greatly reduced amount of light.

eukaryote [Gk. *eu,* good, + *karyon,* kernel]: A cell that has a membrane-bound nucleus, membrane-bound organelles, and chromosomes in which the DNA is associated with proteins; an organism composed of such cells.

eustele [Gk. *eu-,* good, + *stele,* pillar]: A stele in which the primary vascular tissues are arranged in discrete strands around a pith; typical of gymnosperms and angiosperms.

evolution: The derivation of progressively more complex forms of life from simple ancestors; Darwin proposed natural selection to explain the manner in which evolution takes place.

exergonic [L. *ex,* out, + Gk. *ergon,* work]: Energy-yielding, as in a chemical reaction; applied to a "downhill" process.

exine: The outer wall layer of a spore or pollen grain.

exocarp [Gk. *exo,* without, + *karpos,* fruit]: The outermost layer of the mature ovary wall, or pericarp.

eyespot: Also stigma; a small, pigmented structure in flagellate unicellular organisms that may be sensitive to light.

F_1: First filial generation in a cross between any two parents; F_2 and F_3 are the second and third generations.

facilitated diffusion: Carrier-assisted transport of solutes driven by a concentration gradient.

family: A taxonomic group between order and genus; the ending on the family name in animals and heterotrophic protista is *-idae;* in all other organisms it is *-aceae.*

fascicle (făs'ĭ•k'l) [L. *fasciculus,* a small bundle]: A bundle of pine leaves or other needlelike leaves of gymnosperms; a now obsolete meaning is a vascular bundle.

fascicular cambium: The vascular cambium originating within a vascular bundle, or fascicle.

fats: Organic compounds containing carbon, hydrogen, and oxygen, as in carbohydrates; the proportion of oxygen to carbon is much less in fats than it is in carbohydrates; fats in the liquid state are called oils.

fermentation: The extraction of energy from organic compounds without the involvement of oxygen.

ferredoxin: Electron-transferring proteins of high iron content; some are involved in photosynthesis.

fertilization: The fusion of two gametes, especially their nuclei, to form a diploid zygote.

fiber: An elongated, tapering, generally thick-walled sclerenchyma cell of vascular plants; its walls may or may not be lignified; it may or may not have a living protoplast at maturity.

fibril: Submicroscopic threads composed of cellulose molecules, which constitute the form in which cellulose occurs in the cell wall.

field moisture capacity: The percentage of water a particular soil will hold against the action of gravity.

filament: (1) The stalk of a stamen; (2) a term used to describe the threadlike bodies of certain algae or fungi.

fission: Asexual reproduction involving the division of a single-celled individual into two new single-celled individuals of equal size.

fitness: The reproductive capacity of an organism.

flagellum, *pl.* **flagella** [L. *flagellum,* whip]: A fine, long thread, composed of protoplasm, protruding from the surface of a cell; longer than a cilium, but having the same internal structure; capable of a vibratory motion; used in locomotion and feeding; common in algae and motile gametes.

flavoprotein: A dehydrogenase that contains a flavin and often a metal and plays a major role in oxidation; abbreviated FP.

floral tube: A cup or tube formed by the fusion of the basal parts of the sepals, petals, and stamens, often in plants that have an inferior ovary.

floret: One of the small flowers that make up the composite inflorescence or the spike of the grasses.

floridean starch: The reserve or storage carbohydrate of the red algae.

florigen [L. *flor-,* flower, + Gk. *-genes,* producer]: A plant hormone that promotes flowering.

flower: The reproductive structure of angiosperms; a complete flower includes calyx, corolla, androecium (stamens), and gynoecium (carpels), but all contain at least one stamen or one carpel.

follicle [L. *folliculus,* small ball]: A dry, dehiscent fruit derived from a single carpel and opening along one side.

food chain, food web: A chain of organisms existing in any natural community such that each link in the chain feeds on the one below and is eaten by the one above; there are seldom more than six links in a chain, with autotrophs on the bottom and the largest carnivores at the top.

fossil [L. *fossilis,* dug up]: The remains, impressions, or traces of an organism that has been preserved in rocks found in the earth's crust.

FP: *See* flavoprotein.

free energy: Energy available to do work.

frond: The leaf of a fern; any large, divided leaf.

fruit: In angiosperms, a mature, ripened ovary (or group of ovaries), containing the seeds; also applied informally, as in "fruiting body," to reproductive structures of other groups of organisms, together with any adjacent parts that may be fused with it at maturity.

frustule [L. *frustulum,* a small piece]: The siliceous, two-valved cell wall, or shell, of a diatom.

fucoxanthin (fu'ko•zăn'thĭn) [Gk. *phykos,* seaweed, + *xanthos,* yellowish-brown]: A brownish pigment found in brown algae and certain other groups of protista.

fundamental tissue system: *See* ground tissue system.

funiculus [L. *funiculus,* small rope or cord]: The stalk of the ovule.

fusiform initials [L. *fusus,* spindle]: The vertically elongated cells in the vascular cambium that give rise to the cells of the axial system in the secondary xylem and secondary phloem.

gametangium, *pl.* **gametangia** [Gk. *gamein,* to marry, + L. *tangere,* to touch]: General term applied to any cell or organ in which gametes are formed.

gamete [Gk. *gamete,* wife]: The mature functional haploid reproductive cell whose nucleus fuses with that of another gamete of opposite sex (fertilization) resulting in formation of a diploid cell, or zygote.

gametophore [Gk. *gamein,* to marry, + *phoros,* bearing]: In the bryophytes, a fertile stalk that bears gametangia.

gametophyte: In plants having alternation of generations, the haploid (1*n*), gamete-producing phase.

gel: A mixture of substances having a semisolid or solid constitution.

gemma, *pl.* **gemmae** (jĕm'ă) [L. *gemma,* bud]: A small mass of vegetative tissue; an outgrowth of the thallus, for example, in liverworts or certain fungi; it can develop into an entire new plant.

gene: A unit of heredity in the chromosome; a sequence of nucleotides in a DNA molecule that codes for a polypeptide.

gene flow: The exchange of genes between different populations.

gene frequency: The incidence (relative occurrence) of a particular allele in a population.

gene pool: All the alleles of all the genes in a population.

generative cell: (1) In many gymnosperms, the cell of the male gametophyte that divides to form the stalk and body cells; (2) in angiosperms, the cell of the male gametophyte that divides to form two sperm.

genetic code: The system of nucleotide triplets in DNA and RNA that carries genetic information; referred to as a code because it determines the amino acid sequence in the enzymes and other protein components synthesized by the organism.

genetic recombination: The occurrence of gene combinations in the progeny that are different from the combinations present in the parents.

genotype: The genetic constitution, latent or expressed, of an organism, as contrasted with the phenotype; the sum total of all the genes present in an individual.

genus, *pl.* **genera:** The taxonomic group between family and species; includes one or more species that have certain characteristics in common.

geotropism [Gk. *ge,* earth, + *tropes,* turning]: Direction of growth induced by the force of gravity; also called gravitropism.

germination [L. *germinare,* to sprout]: The beginning or resumption of growth by a spore, seed, bud, or other structure.

gibberellins (jĭb•ĕ•rĕ'lĭns) [*Gibberella,* a genus of fungi]: A group of growth hormones, the best known effect of which is to increase the elongation of plant stems.

gill: The plates on the underside of the cap in the class of gill fungi (basidiomycetes).

girdling: The removal from a woody stem of a ring of bark extending inward to the cambium; also called ringing.

glucose: A common six-carbon sugar ($C_6H_{12}O_6$); the most common monosaccharide in most organisms.

glycogen [Gk. *glykys,* sweet, + *gen,* of a kind]: A carbohydrate similar to starch, the reserve food of bacteria, including cyanobacteria; also known as cyanophycean starch.

glycolysis: A process in which sugar is changed anaerobically to pyruvic acid, with the liberation of a small amount of usable energy.

glyoxysome: A microbody containing enzymes necessary for the conversion of fats into carbohydrates during the germination of many seeds.

Golgi body (gôl'je): In eukaryotes, a group of flat, disk-shaped sacs that are often branched into tubules at their margins; they seem to function as collecting and packaging centers for substances the cells manufacture; also called dictyosomes in plants. The term "Golgi apparatus" is used to refer collectively to all of the Golgi bodies of a given cell.

grafting: A union of different individuals in which a portion, called the scion, of one individual is inserted into a root or stem, called the stock, of the other individual.

grain: *See* caryopsis.

grana, *sing.* **granum:** Structures within chloroplasts, seen as green granules with a light microscope and as a series of stacked thylakoids with an electron microscope; the grana contain the chlorophylls and carotenoids and are the actual site of the light reactions of photosynthesis.

gravitropism: *See* geotropism.

ground meristem [Gk. *meristos,* divisible]: The primary meristem, or meristematic tissue, that gives rise to the ground tissues.

ground tissue system: All tissues other than the epidermis (or periderm) and the vascular tissues; also called fundamental tissue system.

growth layer: A layer of growth in the secondary xylem or secondary phloem; *see also* annual ring.

growth ring: A growth layer in the secondary xylem or secondary phloem, as seen in transverse section; may be called a growth increment, especially where seen in other than transverse section.

guanine: A purine base found in DNA and RNA.

guard cells: Pairs of specialized epidermal cells surrounding a pore, or stoma; changes in turgor of a pair of guard cells causes opening and closing of the pore.

guttation [L. *gutta,* a drop]: The exudation of liquid water from leaves due to root pressure.

gymnosperm [Gk. *gymnos,* naked, + *sperma,* seed]: A seed plant with seeds not enclosed in an ovary; the conifers are the most familiar group.

gynoecium [Gk. *gyne,* woman, + *oikos,* house]: The aggregate of carpels in the flower of a seed plant.

habit [L. *habitus,* condition, character]: Characteristic form or bodily appearance of an organism.

habitat [L. *habitare,* to inhabit]: The natural environment of an organism; the place where it is usually found.

haploid [Gk. *haploos,* single]: The state in which each chromosome is represented only once (1*n*).

hardwood: A name commonly applied to the wood of a dicot tree.

Hardy-Weinberg law: The mathematical expression of the relationship between relative frequencies of two or more alleles in a population; it demonstrates that the frequencies of dominant and recessive genes tend to remain constant.

haustorium, *pl.* **haustoria** [L. *haustus,* from *haurire,* to drink, draw]: A projection of fungal hypha, stem, or plant part that acts as a penetrating and absorbing organ.

heartwood: Nonliving and commonly dark-colored wood in

which no water transport occurs; it is surrounded by sapwood.

hemicellulose (hĕm'ĭ•sĕl'u•lōs): Polysaccharide resembling cellulose but more soluble and less ordered; found particularly in cell walls.

herb [L. *herba,* grass]: A nonwoody seed plant with a relatively short-lived aerial portion.

herbaceous: A general term referring to any nonwoody plant.

herbarium: A collection of dried and pressed plant specimens.

herbivorous: Feeding upon plants.

heredity [L. *heredis,* heir]: The transmission of characteristics from parent to offspring through the gametes.

hermaphrodite [Gk. for Hermes and Aphrodite]: An organism possessing both male and female reproductive organs.

hetero- [Gk. *heteros,* different]: Prefix meaning "other" or "different."

heterocyst [Gk. *heteros,* different, + *cystis,* a bag]: A large, transparent, thick-walled cell in the filaments of certain cyanobacteria.

heteroecious (hĕt'er•ē'shŭs) [Gk. *heteros,* different, + *oikos,* house]: As in some rust fungi, requiring two different host species to complete the life cycle.

heterogamy [Gk. *heteros,* different, + *gamos,* union or reproduction]: Reproduction involving two types of gametes.

heterokaryotic [Gk. *heteros,* other, + *karyon,* kernel]: In fungi, having two or more genetically distinct types of nuclei within the same mycelium.

heteromorphic [Gk. *heteros,* different, + *morphe,* form]: A term used to describe a life history in which the haploid and diploid generations are dissimilar in form.

heterosis [Gk. *heterosis,* alteration]: Hybrid vigor, the superiority of the hybrid over either parent in any measurable character.

heterosporous: Having spores of two kinds, usually designated as microspores and megaspores.

heterothallic [Gk. *heteros,* different, + *thallus,* sprout]: A term used to describe a species, the individuals of which are self-sterile or self-incompatible; two compatible strains or individuals are required for sexual reproduction to take place.

heterotroph [Gk. *heteros,* other, + *trophos,* feeder]: An organism that cannot manufacture organic compounds and so must feed on organic materials that have originated in other plants and animals; *see also* autotroph.

heterozygous: Having two different alleles at the same locus on homologous chromosomes.

hilum [L. *hilum,* a trifle]: (1) Scar left on seed after separation of seed from funiculus; (2) the part of a starch grain around which the starch is laid down in more or less concentric layers.

holdfast: (1) Basal part of an algal thallus that attaches it to a solid object; may be unicellular or composed of a mass of tissue; (2) cuplike structures at the tips of some tendrils, by means of which they become attached.

homeo-, homo- [Gk. *homos,* same, similar]: Prefix meaning "similar" or "same."

homeostasis (hō'me•ō•stā'sĭs) [Gk. *homos,* similar, + *stasis,* standing]: The maintaining of a relatively stable internal physiological environment or equilibrium in an organism, population, or ecosystem.

homokaryotic [Gk. *homos,* same, + *karyon,* kernel]: In fungi, having nuclei with the same genetic makeup within a mycelium.

homologous chromosomes: Chromosomes that associate in pairs in the first stage of meiosis; each member of the pair is derived from a different parent; also called homologues.

homology [Gk. *homologia,* agreement]: A condition indicative of the same phylogenetic, or evolutionary, origin, but not necessarily the same in present structure and/or function.

homosporous: Having but one kind of spore.

homothallic [Gk. *homos,* same, + *thallus,* sprout]: A term used to describe a species in which the individuals are self-fertile.

homozygous: Having identical alleles at the same locus on homologous chromosomes.

hormogonium, *pl.* **hormogonia:** A portion of a filament of a cyanobacterium that becomes detached and grows into a new filament.

hormone [Gk. *hormaein,* to excite]: A chemical substance produced usually in minute amounts in one part of an organism, from which it is transported to another part of that organism on which it has a specific effect.

host: An organism on or in which a parasite lives.

humus: Decomposing organic matter in the soil.

hyaloplasm: *See* cytoplasmic ground substance.

hybrid: Offspring of two parents that differ in one or more heritable characteristics; offspring of two different varieties or of two different species.

hybridization: The formation of offspring between unlike parents.

hybrid vigor: *See* heterosis.

hydrogen bond: A weak bond between a hydrogen atom attached to one oxygen or nitrogen atom and another oxygen or nitrogen atom.

hydrolysis [Gk. *hydro,* water, + *lysis,* loosening]: Splitting of one molecule into two by addition of the H^+ and OH^- ions of water.

hydrophyte [Gk. *hydro,* water, + *phyton,* a plant]: A plant that depends on an abundant supply of moisture or that grows wholly or partly submerged in water.

hymenium [Gk. *hymen,* a membrane]: The layer of asci on an ascocarp, or of basidia on a basidiocarp, plus any associated sterile hyphae.

hyper- [Gk. *hyper,* above, over]: Prefix meaning "above" or "over."

hypertonic: Having a concentration high enough to gain water across a permeable membrane from another solution.

hypha, *pl.* **hyphae** [Gk. *hyphe,* web]: A single tubular filament of a fungus or oomycete; the hyphae together comprise the mycelium.

hypo- [Gk. *hypo,* less than]: Prefix meaning "under" or "less."

hypocotyl: The portion of an embryo or seedling situated between the cotyledons and the radicle.

hypocotyl-root axis: The embryo axis below the cotyledon or cotyledons, consisting of the hypocotyl and the apical meristem of the root or the radicle.

hypodermis [Gk. *hypo,* under, + *derma,* skin]: A layer or layers of cells beneath the epidermis, which are distinct from the underlying cortical or mesophyll cells.

hypogyny [Gk. *hypo,* under, + *gyne,* woman]: Floral organization in which the sepals, petals, and stamens are attached to the receptacle below the ovary; as opposed to epigyny.

hypothesis [Gk. *hypo,* under, + *tithenai,* to put]: A temporary working explanation or supposition based on accumulated facts and suggesting some general principle or relation of cause and effect; a postulated solution to a scientific problem that must be tested by experimentation and, if disproved or shown to be unlikely, is discarded.

hypotonic: Having a concentration low enough to lose water across a permeable membrane to another solution.

IAA: *See* indoleacetic acid.

imbibition (ĭm·bĭ·bĭsh′ŭn): Adsorption of water and swelling of colloidal materials because of the adsorption of water molecules onto the internal surfaces of the materials.

imperfect flower: A flower lacking either stamens or carpels.

imperfect fungi: Fungi reproducing only by asexual means, or in which the sexual cycle has not been observed; most Fungi Imperfecti are ascomycetes.

inbreeding: The breeding of closely related plants or animals; in plants, it is usually brought about by repeated self-pollination.

incomplete dominance: The condition that results when two different alleles together produce an effect intermediate between the effects of these same genes in the homozygous condition.

incomplete flower: A flower lacking one or more of the four kinds of floral parts, that is, lacking sepals, petals, stamens, or carpels.

indehiscent (ĭn′de·hĭs′ĕnt): Remaining closed at maturity, as are many fruits (samaras, for example).

independent assortment: *See* Mendel's second law.

indeterminate growth: Unrestricted or unlimited growth, as with a vegetative apical meristem that produces an unrestricted number of lateral organs indefinitely.

indoleacetic acid, or IAA: A naturally occurring auxin; a growth regulator.

indusium, *pl.* **indusia** (ĭn·dū′zi·ŭm) [L. *indusium,* a woman's undergarment]: Membranous growth of the epidermis of a fern leaf that covers a sorus.

inferior ovary: An ovary that is completely or partially attached to the calyx; the other floral whorls appear to arise from its top.

inflorescence: A flower cluster, with a definite arrangement of flowers.

initial: (1) In the meristem, a cell that remains within the meristem indefinitely and at the same time, by division, adds cells to the plant body; (2) a meristematic cell that eventually differentiates into a mature, more specialized cell or element.

inner bark: In older trees, the living part of the bark; the bark inside the innermost periderm.

integument: The outermost layer or layers of tissue enveloping the nucellus of the ovule; develops into the seed coat.

inter- [L. *inter,* between]: Prefix meaning "between," or "in the midst of."

intercalary [L. *intercalare,* to insert]: Descriptive of meristematic tissue or growth not restricted to the apex of an organ, that is, growth in the region of the nodes.

interfascicular cambium: The vascular cambium arising between the fascicles, or vascular bundles, from interfascicular parenchyma.

interfascicular region: Tissue region between vascular bundles in stem; also called pith ray.

internode: The region of a stem between two successive nodes.

interphase: The state between two mitotic or meiotic cycles.

intine: The inner wall layer of a spore or pollen grain.

intra- [L. *intra,* within]: Prefix meaning "within."

ion: An atom or molecule that has lost or gained one or more electrons. By this process, known as ionization, it becomes electrically charged.

irregular flower: A flower in which one or more members of at least one whorl differ in form from other members of the same whorl.

iso- [Gk. *isos,* equal]: Prefix meaning "equal"; like "homeo-" or "homo-."

isogamy: A type of sexual reproduction in algae and fungi in which the gametes (or gametangia) are alike in size.

isomer [Gk. *isos,* equal, + *meros,* part]: One of a group of compounds identical in atomic composition but differing in structural arrangement, for example, glucose and fructose.

isomorphic [Gk. *isos,* equal, + *morphe,* form]: A term used to describe a life history in which the haploid and diploid generations are similar in form.

isotonic: Having the same osmotic concentration.

isotope: One of several possible forms of a chemical element, differing from other forms in the number of neutrons in the atomic nucleus, but not in chemical properties.

karyogamy [Gk. *karyon,* kernel, + *gamos,* marriage]: The union of two nuclei following plasmogamy.

kelp: A common name for any of the large brown algae.

kinetin [Gk. *kinetikos,* causing motion]: A purine that probably does not occur in nature but that acts as a cytokinin in plants.

kinetochore: *See* centromere.

kingdom: The chief taxonomic category, for example, Monera or Plantae.

Kranz anatomy [Ger. *Kranz,* wreath]: The wreathlike arrangement of mesophyll cells around a layer of large bundle-sheath cells, forming two concentric layers around the vascular bundle, as is typically found in the leaves of C_4 plants.

Krebs cycle: The series of reactions that results in the oxidation of pyruvic acid to hydrogen atoms, electrons, and carbon dioxide. The electrons, passed along electron-carrier molecules, then go through the oxidative phosphorylation and terminal oxidation processes. Also called the tricarboxylic acid cycle or TCA cycle.

lamella (*la*•mĕl'*a*) [L. *lamella,* thin metal plate]: Layer of cellular membranes, particularly photosynthetic, chlorophyll-containing membranes; *see also* middle lamella.)

lamina: *See* blade.

laminarin: The principal storage product of the brown algae; a polymer of glucose.

lateral meristems: Meristems that give rise to secondary tissue; the vascular cambium and cork cambium.

lateral root: A root that arises from another, older root; also called a branch root, or secondary root, if the older root is the primary root.

late wood: The last part of the growth increment formed in the growing season; it contains smaller cells and is more dense than the early wood; replaces the term "summer wood."

latosol [L. *later,* brick]: A red, iron-containing soil found in the tropics.

leaching: The downward movement and drainage of minerals, or inorganic ions, from the soil by percolating water.

leaf buttress: A lateral protrusion below the apical meristem; represents the initial stage in the development of a leaf primordium.

leaf gap: Region of parenchyma tissue in the primary vascular cylinder above the point of departure of the leaf trace or traces.

leaflet: One of the parts of a compound leaf.

leaf primordium [L. *primordium,* beginning]: A lateral outgrowth from the apical meristem that will eventually become a leaf.

leaf scar: A scar left on the twig when the leaf falls.

leaf trace: That part of the vascular bundle extending from the base of the leaf to its connection with a vascular bundle in the stem.

legume [L. *legumem,* leguminous plant]: (1) A member of the Fabaceae, the pea or bean family; (2) a type of dry fruit developed from one carpel and opening along two sides.

lenticels (len'tĭ•sĕls) [L. *lenticella,* a small window]: Spongy areas in the cork surfaces of stem, roots, and other plant parts that allow interchange of gases between internal tissues and the atmosphere through the periderm.

lepidodendrids: Dominant lycopod trees of the late Carboniferous period.

leucoplast (lū'ko•plăst) [Gk. *leuko,* white, + *plasein,* to form]: A colorless plastid, commonly the center of starch formation.

leucosin: The principal storage product of the golden-brown algae and diatoms.

liana [Fr. *liane,* from *lier,* to bind]: A large, woody vine that climbs upon other plants.

lichen [Gk. *leichen,* thallus plants growing on rocks and trees]: An ascomycete fungus which characteristically incorporates living algal cells, forming a symbiotic relationship that results in a composite organism.

life cycle: The entire sequence of phases in the growth and development of any organism from time of zygote formation to gamete formation.

lignin: One of the most important constituents of the secondary wall, although not all secondary walls contain lignin; after cellulose, lignin is the most abundant plant polymer.

ligule [L. *ligula,* small tongue]: A minute outgrowth or appendage at the base of grass leaves and leaves of certain Lycophyta.

linkage: The tendency for certain genes to be inherited together owing to the fact that they are located on the same chromosome.

lipid [Gk. *lipos,* fat]: A large variety of organic fat or fatlike compounds, including fats, oils, steroids, phospholipids, and carotenes.

locule (lŏk'ūl) [L. *loculus,* small chamber]: A cavity within a sporangium or a cavity of the ovary in which ovules occur.

locus, *pl.* **loci:** The position on a chromosome occupied by a particular gene.

long-day plants: Plants that must be exposed to light periods longer than some critical length for flowering to occur; they flower primarily in summer.

lumen [L. *lumen,* light, an opening for light]: The space bounded by the plant cell wall.

lysis [Gk. *lysis,* a loosening]: A process of disintegration or cell destruction.

lysogenic bacteria: Bacteria-carrying viruses (phages) that eventually break loose from the bacterial chromosome and set up an active cycle of infection, producing lysis in their bacterial hosts.

lysosome [Gk. *lysis,* loosening, + *soma,* body]: An organelle, bounded by a single membrane, and containing hydrolytic enzymes capable of breaking down proteins and other complex macromolecules.

macrofibril: An aggregation of microfibrils, visible with the light microscope.

macromolecule [Gk. *makros,* large]: A molecule of very high molecular weight; refers specifically to proteins, nucleic acids, polysaccharides, and complexes of these.

macronutrients [Gk. *makros,* large, + L. *nutrire,* to nourish]: Inorganic chemical elements required in large amounts for plant growth, such as nitrogen, potassium, calcium, phosphorus, magnesium, and sulfur.

maltase: An enzyme that hydrolyzes maltose to glucose.

marginal meristem: The meristem located along the margin of a leaf primordium and forming the blade.

mating type: A strain of organisms incapable of sexual reproduction with one another but capable of such reproduction with members of other strains of the same organism.

mega- [Gk. *megas,* great, large]: Prefix meaning "large."

megagametophyte [Gk. *megas,* large, + *gamos,* marriage, + *phyton,* plant]: In heterosporous plants, the female gametophyte; located within the ovule of seed plants.

megaphyll [Gk. *megas,* great, + *phyllon,* leaf]: A generally large leaf with several to many veins; its leaf trace is associated with a leaf gap; in contrast to microphyll.

megasporangium, *pl.* **megasporangia:** A sporangium in which megaspores are produced; *see* nucellus.

megaspore: In heterosporous plants, a haploid ($1n$) spore that develops into a female gametophyte; in some groups, megaspores are larger than microspores.

megaspore mother cell: A diploid cell in which meiosis will occur, resulting in the production of four megaspores; also called a megasporocyte.

megasporocyte: *See* megaspore mother cell.

megasporophyll: A leaf or leaflike structure bearing megasporangia.

meiosis (mī•ō'sĭs) [Gk. *meioun,* to make smaller]: The two successive nuclear divisions in which the chromosome number is reduced from diploid ($2n$) to haploid ($1n$) and segregation of the genes occurs; gametes or, as in higher plants, spores may be produced as a result of meiosis.

Mendel's first law: The factors for a pair of alternate characteristics are separate and only one may be carried in a particular gamete (genetic segregation).

Mendel's second law: The inheritance of one pair of characteristics is independent of the simultaneous inheritance of other traits, such characteristics "assorting independently" as though there were no others present (later modified by the discovery of linkage).

meristem [Gk. *merizein,* to divide]: The undifferentiated plant tissue from which new cells arise.

meso- [Gk. *mesos,* middle]: Prefix meaning "middle."

mesocarp [Gk. *mesos,* middle, + *karpos,* fruit]: The middle layer of the mature ovary wall, or pericarp, between the exocarp and endocarp.

mesophyll: The photosynthetic ground tissue (parenchyma) of a leaf, located between the layers of epidermis.

mesophyte [Gk. *mesos,* middle, + *phyton,* a plant]: A plant that requires abundant soil water and a relatively humid atmosphere.

messenger RNA (mRNA): The RNA that carries genetic information from the gene to the ribosome, where it determines the order of the amino acids in a polypeptide.

metabolism [Gk. *metabole,* change]: The sum of all chemical processes occurring within a living cell or organism.

metaphase: Stage of mitosis or meiosis during which the chromosomes lie in the central plane of the spindle.

metaxylem [Gk. *meta,* after]: The part of the primary xylem that differentiates after the protoxylem; it reaches maturity after the portion of the plant part in which it is located has finished elongating.

micro- [Gk. *mikros,* small]: Prefix meaning "small."

microbody: An organelle bounded by a single membrane and containing a variety of enzymes; generally associated with one or two cisternae of endoplasmic reticulum. Peroxisomes and glyoxysomes are kinds of microbodies.

microfibril: A threadlike component of the cell wall, composed of cellulose molecules, visible only with the electron microscope.

microgametophyte [Gk. *mikros,* small, + *gamos,* marriage, + *phyton,* plant]: In heterosporous plants, the male gametophyte.

micrometer: A unit of microscopic measurement convenient for describing cellular dimensions; 1/1000 of a millimeter; its symbol is μm.

micronutrients [Gk. *mikros,* small, + L. *nutrire,* to nourish]: Inorganic chemical elements required only in very small, or trace, amounts for plant growth, such as iron, chlorine, copper, manganese, zinc, molybdenum, and boron.

microphyll [Gk. *mikros,* small, + *phyllon,* leaf]: A small leaf with one vein and one leaf trace not associated with a leaf gap; in contrast to megaphyll.

micropyle: In the ovules of seed plants, the opening in the integuments through which the pollen tube usually enters.

microsporangium: A sporangium within which microspores are formed.

microspore: A spore that develops into a male gametophyte.

microspore mother cell: A cell in which meiosis will occur, resulting in four microspores; in seed plants, often called a pollen mother cell; also called a microsporocyte.

microsporocyte: *See* microspore mother cell.

microsporophyll: A leaflike organ bearing one or more microsporangia.

microtubule [Gk. *mikros,* small, + L. *tubulus,* little pipe]: Narrow (about 25 nanometers in diameter), elongate, nonmembranous tubule of indefinite length occurring in the cytoplasm of many eukaryotic cells and flagella.

middle lamella: The layer of intercellular material, rich in pectic compounds, cementing together the primary walls of adjacent cells.

mimicry [Gk. *mimos,* mime]: The superficial resemblance in form, color, or behavior of certain organisms (mimics) to other more powerful or more protected ones (models), resulting in protection, concealment, or some other advantage for the mimic.

mineral: A chemical element or compound occurring naturally as a result of inorganic processes.

mitochondrion, *pl.* **mitochondria:** A double-membrane-bound organelle found in eukaryotic cells; contains the enzymes of the Krebs cycle and the electron-transport chain; the major source of ATP in nonphotosynthetic cells.

mitosis (mɪtō'sĭs) [Gk. *mitos,* thread]: A process during which the duplicated chromosomes divide longitudinally and the daughter chromosomes then separate to form two genetically identical daughter nuclei; usually accompanied by cytokinesis.

mole: The name for a gram molecule. The number of particles in 1 mole of any substance; always equal to Avogadro's number: 6.022×10^{23}.

molecular weight: The relative weight of a molecule when the weight of the common carbon atom is taken as 12; the sum of the relative weights of the atoms in a molecule.

molecule: Smallest possible unit of a compound, consisting of two or more atoms.

mono- [Gk. *monos,* single]: Prefix meaning "one" or "single."

monocotyledon: A plant whose embryo has one cotyledon; one of the two great classes of angiosperms, Monocotyledones; often abbreviated as monocot.

monoecious (mo•nē'shŭs) [Gk. *monos,* single, + *oikos,* house]: Having the anthers and carpels produced in separate flowers but borne on the same individual.

monokaryotic [Gk. *monos,* one, + *karyon,* kernel]: In fungi, having a single haploid nucleus within one cell or compartment.

monosaccharide [Gk. *monos,* single, + *sakcharon,* sugar]: A simple sugar, such as five-carbon and six-carbon sugars.

-morph, morph- [Gk. *morphe,* form]: Suffix or prefix meaning "form."

morphogenesis: The development of form.

morphology [Gk. *morphe,* form, + *logos,* discourse]: The study of form and its development.

mRNA: *See* messenger RNA.

mucigel: A slime-sheath covering the surface of many roots.

multiple epidermis: A tissue composed of several cell layers derived from the protoderm; only the outer layer assumes characteristics of a typical epidermis.

multiple fruit: A cluster of matured ovaries produced by a cluster of flowers, as in the pineapple.

mutagen [L. *mutare,* to change, + Gk. *genaio,* to produce]: An agent that increases the mutation rate.

mutant: A mutated gene or an organism carrying a gene that has undergone a mutation.

mutation: An inheritable change of a gene from one allelic form to another.

mutualism: The living together of two or more organisms in an association that is mutually advantageous.

myc-, myco- [Gk. *mykes,* fungus]: Prefix meaning "pertaining to fungi."

mycelium [Gk. *mykes,* fungus] The mass of hyphae forming the body of a fungus or oomycete.

mycology: The study of fungi.

mycoplasmas [Gk. *mykes,* fungus, + *plasma,* something molded]: The smallest of the known prokaryotic organisms.

mycorrhiza, *pl.* **mycorrhizae:** The symbiotic combination of the hyphae of certain fungi with the root of a plant.

NAD: *See* nicotinamide adenine dinucleotide.

NADP: *See* nicotinamide adenine dinucleotide phosphate.

nannoplankton (năn'o•plăngk'tŏn) [Gk. *nanos,* dwarf, + *planktos,* wandering]: Plankton with dimensions of less than 70 to 75 micrometers.

natural selection: The differential reproduction of genotypes resulting from interactions among a variety of phenotypes and the environment.

nectary [Gk. *nektar,* the drink of the gods]: In angiosperms, a gland that secretes a sugary fluid which pollinators utilize as food.

netted venation: The arrangement of veins in the leaf blade resembles a net; characteristic of dicot leaves; also called reticulate venation.

neutron [L. *neuter,* neither]: An uncharged particle with a mass slightly greater than that of a proton, found in the atomic nucleus of all elements except hydrogen, in which the nucleus consists of a single proton.

niche: The role played by a particular species in its ecosystem.

nicotinamide adenine dinucleotide (NAD): A coenzyme which functions as an electron acceptor.

nicotinamide adenine dinucleotide phosphate (NADP): A coenzyme which functions as an electron acceptor; similar in structure to NAD except that it contains an extra phosphate group.

nitrification: The conversion of ammonium ions or ammonia to nitrate, a process carried out by specific bacteria and fungi.

nitrogen base: A nitrogen-containing molecule having basic properties (tendency to acquire an H atom); a purine or pyrimidine; one of the building blocks of nucleic acids.

nitrogen cycle: Worldwide circulation of nitrogen atoms in which certain microorganisms take up atmospheric nitrogen and convert it into other forms which may be assimilated into the bodies of other organisms; excretion, burning, and bacterial and fungal action in dead organisms return nitrogen atoms to the atmosphere.

nitrogen fixation: Incorporation of atmospheric nitrogen into nitrogen compounds available to green plants, a process that can be carried out only by certain microorganisms, or by higher plants in symbiotic association with microorganisms.

nitrogen-fixing bacteria: Soil bacteria that convert atmospheric nitrogen into nitrogen compounds.

node [L. *nodus,* knot]: The part of a stem where one or more leaves are attached; *see* internode.

nodules: Enlargements or swellings on the roots of legumes and certain other plants inhabited by symbiotic nitrogen-fixing bacteria.

nonseptate: *See* aseptate.

nucellus (nu•sĕl'ŭs) [L. *nucella,* a small nut]: Tissue composing the chief part of the young ovule, in which the embryo sac develops; equivalent to a megasporangium.

nuclear envelope: The double membrane surrounding the nucleus of a cell.

nucleic acid: An organic acid consisting of joined nucleotide complexes; the two types are deoxyribonucleic acid (DNA) and ribonucleic acid (RNA).

nucleolar organizer: A special area on certain chromosomes associated with the formation of the nucleolus.

nucleolus (nu•klē'o•lŭs) [L. *nucleolus,* a small nucleus]: A spherical body composed chiefly of RNA and protein found in the nucleus of eukaryotic cells; site of production of ribosomes.

nucleoplasm: The ground substance of a nucleus.

nucleotide: A single unit of nucleic acid, composed of a phosphate, a five-carbon sugar (either ribose or deoxyribose), and a purine or a pyrimidine.

nucleus: (1) A specialized body within the eukaryotic cell bounded by a double membrane and containing the chromosomes; (2) the central part of an atom of a chemical element.

nut: A dry, indehiscent, hard, one-seeded fruit, generally produced from a gynoecium of more than one fused carpel.

obligate anaerobe: An organism that is metabolically active only in the absence of oxygen.

-oid [Gk. *oid,* like, resembling]: Suffix meaning "like" or "similar to."

ontogeny [Gk. *on,* being, + *genesis,* origin]: The development, or life history, of all or part of an individual organism.

oo- [Gk. *oion,* egg]: Prefix meaning "egg."

oogamy: Sexual reproduction in which one of the gametes (the egg) is large and nonmotile, and the other gamete (the sperm) is smaller and motile.

oogonium (ō′o•gō′nĭ•*ŭ*m): In oomycetes and certain algae, a unicellular female sex organ that contains one or several eggs.

oospore: The thick-walled zygote characteristic of the oomycetes.

open bundle: A vascular bundle in which a vascular cambium develops.

operator gene: The site at which the protein product of a regulator gene works.

operculum (o•pûr′ku•l*ŭ*m) [L. *operculum,* lid]: In mosses, the lid of the sporangium.

operon: A group of adjacent genes that are under the control of a single operator gene.

opposite: Term applied to buds or leaves occurring in pairs at a node.

order: A category of classification above family and below class; composed of one or more families.

organ: A structure composed of different tissues, such as root, stem, leaf, or flower parts.

organelle (ôr′găn•ĕl′): A formed body in the cytoplasm of a cell.

organic: Pertaining to living organisms in general, to compounds formed by living organisms, and to the chemistry of compounds containing carbon.

organism: Any individual living creature, either unicellular or multicellular.

osmosis (ŏs•mō′sĭs) [Gk. *osmos,* impulse or thrust]: The diffusion of water, or any solvent, across a differentially permeable membrane; in the absence of other forces, movement of water during osmosis will always be from a region of greater water potential to one of lesser water potential.

osmotic potential: *See* solute potential.

osmotic pressure: The potential pressure that can be developed by a solution separated from pure water by a differentially permeable or semipermeable membrane; it is an index of the solute concentration of the solution.

outcrossing: Cross-pollination between individuals of the same species.

outer bark: In older trees, the dead part of the bark; the innermost periderm and all tissues outside it; also called rhytidome.

ovary [L. *ovum,* an egg]: An enlarged basal portion of a carpel or of a gynoecium composed of fused carpels; the ovary becomes the fruit.

ovule (ō′vūl) [L. *ovulum,* a little egg]: A structure in seed plants containing the female gametophyte with egg cell, all being surrounded by the nucellus and one or two integuments; when mature, the ovule becomes a seed.

ovuliferous scale: In certain conifers, the appendage or scalelike shoot to which the ovule is attached.

oxidation: Loss of an electron by an atom. Oxidation and reduction (gain of an electron) take place simultaneously, because an electron that is lost by one atom is accepted by another. Oxidation-reduction reactions are an important means of energy transfer within living systems.

oxidative phosphorylation: The formation of ATP from ADP and inorganic phosphate that takes place in the electron-transport chain of the mitochondrion.

pairing of chromosomes: Side-by-side association of homologous chromosomes.

paleobotany [Gk. *palaios,* old]: The study of fossil plants.

palisade parenchyma: A leaf tissue composed of columnar chloroplast-bearing parenchyma cells with their long axes at right angles to the leaf surface.

panicle (păn′ĭk′l) [L. *panicula,* tuft]: An inflorescence, the main axis of which is branched, and whose branches bear loose flower clusters.

para- [Gk. *para,* beside]: Prefix, meaning "beside."

paradermal section [Gk. *para,* beside, + *derma,* skin]: Section cut parallel with the surface of a flat structure, such as a leaf.

parallel venation: The pattern of venation in which the principal veins of the leaf are parallel or nearly so; characteristic of monocots.

paraphysis, *pl.* **paraphyses** [Gk. *para,* beside, + *physis,* growth]: As in certain fungi and brown algae, a sterile hypha growing among reproductive cells in the fruiting body.

parasexual cycle: The fusion and segregation of heterokaryotic haploid nuclei in certain fungi to produce recombinant nuclei.

parasite: An organism that lives on or in an organism of a different species and derives nutrients from it.

parenchyma (p*a*•rĕng′kĭ•m*a*) [Gk. *para,* beside, + *en,* in, + *chein,* to pour]: A tissue composed of parenchyma cells.

parenchyma cell: Living, generally thin-walled cell, concerned with one or more of the many physiological activities in plants; varies in size and form; the most common cell type in the plant.

parthenocarpy [Gk. *parthenos,* virgin, + *karpos,* fruit]: The development of fruit without fertilization; parthenocarpic fruits are usually seedless.

passage cell: Endodermal cell of root that retains thin wall and Casparian strip when other associated endodermal cells develop thick secondary walls.

pathogen [Gk. *pathos,* suffering, + *genesis,* beginning]: An organism that causes a disease.

pathology: The study of plant or animal diseases, their effects on the organism, and their treatment.

pectin: A complex organic compound present in the intercellular layer and primary wall of plant cell walls; the basis of fruit jellies.

pedicel (pĕd'ĭ•sĕl): The stalk of an individual flower in an inflorescence.

peduncle (pe•dŭng'k'l): The stalk of an inflorescence or of a solitary flower.

pennate diatom [L. *pennate,* feathered]: A bilaterally symmetrical diatom.

peptide: Two or more amino acids linked by peptide bonds.

peptide bond: The type of bond formed when two amino acid units are joined end to end by the removal of a molecule of water; the bonds always form between the carboxyl (—COOH) group of one amino acid and the amino ($—NH_2$) group of the next amino acid.

perennial [L. *per,* through, + *annus,* a year]: A plant that persists from year to year and usually produces reproductive structures in two or more different years.

perfect flower: A flower having both stamens and carpels; hermaphroditic flower.

perfect stage: Phase of the life history of a fungus that includes sexual fusion and the spores associated with such fusions.

perforation plate: Part of the wall of a vessel member that is perforated.

peri- [Gk. *peri,* around]: Prefix meaning "around" or "about."

perianth (pĕr'ĭ•ănth) [Gk. *peri,* around, + *anthos,* flower]: (1) The petals and sepals taken together; (2) in leafy liverworts, a tubular sheath surrounding an archegonium, and later, the developing sporophyte.

pericarp [Gk. *peri,* around, + *karpos,* fruit]: Fruit wall which develops from the mature ovary wall.

periclinal: Parallel with the surface.

pericycle [Gk. *peri,* around, + *kykos,* circle]: Tissue, generally of root, bounded externally by the endodermis and internally by the phloem.

periderm [Gk. *peri,* around, + *derma,* skin]: Outer protective tissue that replaces epidermis when it is destroyed during secondary growth; includes cork, cork cambium, and phelloderm.

perigyny [Gk. *peri,* about, + *gyne,* female]: Floral organization in which the sepals, petals, and stamens are attached to the margin of a cup-shaped extension of the receptacle; superficially, the sepals, petals, and stamens appear to be attached to the ovary.

perisperm [Gk. *peri,* around, + *sperma,* seed]: Food-storing tissue derived from the nucellus in some seeds.

peristome (pĕr'ĭ•stōm) [Gk. *peri,* around, + *stoma,* a mouth]: In mosses, a fringe of teeth around the opening of the sporangium.

perithecium: A spherical or flask-shaped ascocarp.

permanent wilting percentage: The percentage of water remaining in a soil when a plant fails to recover from wilting even if placed in a humid chamber.

permeable [L. *permeare,* to pass through]: Usually applied to membranes through which liquid substances may diffuse.

permease: Transport protein, or carrier molecule, that assists in the transport of substances across cellular membranes; not permanently altered in the process.

peroxisome: A microbody that plays an important role in glycolic acid metabolism associated with photosynthesis.

petal: A flower part, usually conspicuously colored; one of the units of the corolla.

petiole: The stalk of the leaf.

pH: A symbol denoting the relative concentration of hydrogen ions in a solution; pH values run from 0 to 14, and the lower the value the more acidic a solution, that is, the more hydrogen ions it contains; pH 7 is neutral, less than 7 is acidic, more than 7 is alkaline.

phage: *See* bacteriophage.

phagocytosis: *See* endocytosis.

phellem: *See* cork.

phelloderm (fĕl'o•dûrm) [Gk. *phellos,* cork, + *derma,* skin]: A tissue formed inwardly by the cork cambium, opposite the cork; inner part of the periderm.

phellogen (fĕl'lo•jĕn): *See* cork cambium.

phenotype: The physical appearance of an organism resulting from interaction between its genetic constitution (genotype) and the environment.

phloem (flō'ĕm) [Gk. *phloos,* bark]: Food-conducting tissue basically composed of sieve elements, various kinds of parenchyma cells, fibers, and sclereids.

phloem loading: The process by which substances (primarily sugars) are actively secreted into the sieve tubes.

phosphate: A compound formed from phosphoric acid by replacement of one or more hydrogen atoms.

phospholipids: Compounds closely related to fats; glycerol is attached to only two fatty acids, with the third space occupied by a phosphorus-containing molecule; important components of cellular membranes.

phosphorylation (fŏs'fo•rĭl•ā'shŭn) [Gk. *phosphoros,* bringing light]: A reaction in which phosphate is added to a compound, e.g., the formation of ATP from ADP and inorganic phosphate.

photo-, -photic [Gk. *photos,* light]: Prefix or suffix meaning "light."

photolysis: The light-dependent oxidative splitting of water molecules that takes place in Photosystem II of the light reactions of photosynthesis.

photon [Gk. *photos,* light]: The elementary particle of light.

photoperiodism: Response to duration and timing of day and night; a mechanism evolved by organisms for measuring seasonal time.

photophosphorylation [Gk. *photos,* light, + *phosphoros,* bringing light]: The formation of ATP in the chloroplast during photosynthesis.

photorespiration: The light-dependent production of glycolic acid in chloroplasts and its subsequent oxidation in peroxisomes.

photosynthesis [Gk. *photos,* light, + *syn,* together, + *tithenai,* to place]: The conversion of light energy to chemical energy; the production of carbohydrates from carbon dioxide and water in the presence of chlorophyll by using light energy.

photosystem: A discrete unit of organization of chlorophyll and other pigment molecules embedded in the thylakoids of chloroplasts and involved with the light-requiring reactions of photosynthesis.

phototropism [Gk. *photos,* light, + *trope,* turning]: Growth in which the direction of the light is the determining factor, as the growth of a plant toward a light source; turning or bending in response to light.

phragmoplast: A spindle-shaped system of fibrils, which arises between two daughter nuclei at telophase and within which the cell plate is formed during cell division, or cytokinesis. The fibrils of the phragmoplast are composed of microtubules.

phycobilins: A group of water-soluble accessory pigments, including phycocyanins and phycoerythrins, which occur in the red algae and cyanobacteria.

phycology [Gk. *phykos,* seaweed]: The study of algae.

phycoplast: A system of microtubules, in the cells of chlorophycean green algae, that develops between the two daughter nuclei parallel to the plane of cell division.

phyllo-, phyll- [Gk. *phyllon,* leaf]: Prefix meaning "leaf."

phyllode (fĭl'ōd): A flat, expanded petiole replacing the blade of a leaf in photosynthetic function.

phylogeny [Gk. *phylon,* race, tribe]: Evolutionary relationships among organisms; developmental history of a group of organisms.

physiology: The study of the activities and processes of living organisms.

phyto-, -phyte [Gk. *phyton,* plant]: Prefix or suffix meaning "plant."

phytochrome: A phycobilinlike pigment found in cytoplasm of green plants that is associated with the absorption of light; photoreceptor for red to far-end light; involved in a number of timing processes, such as flowering, dormancy, leaf formation, and seed germination.

phytoplankton: Autotrophic plankton.

pigment: Substance that absorbs light, often selectively.

pileus [L. *pileus,* a cap]: The caplike part of the mushroom basidiocarp and of certain ascocarps.

pinna, *pl.* **pinnae** (pĭn'a) [L. *pinna,* feather]: A primary division, or leaflet, of a compound leaf or frond; may be divided into pinnules.

pinocytosis: *See* endocytosis.

pistil [L. *pistillum,* pestle]: Central organ of flowers typically consisting of ovary, style, and stigma; a pistil may consist of one or more fused carpels.

pistillate: Pertaining to a flower with one or more carpels but no functional stamens; also called carpellate.

pit: A recess or cavity in a cell wall where the secondary wall does not form.

pit membrane: The middle lamella and two primary cell walls between two pits.

pit-pair: Two opposite pits plus the pit membrane.

pith: The ground tissue occupying the center of the stem or root within the vascular cylinder; usually consists of parenchyma.

pith ray: *See* interfascicular region.

placenta, *pl.* **placentae** [L. *placenta,* a cake]: The part of the ovary wall to which the ovules or seeds are attached.

placentation: The manner of ovule attachment within the ovary.

plankton [Gk. *planktos,* wandering]: Free-floating, mostly microscopic, aquatic organisms.

plaque: Clear area in a sheet of cells resulting from the killing or lysis of contiguous cells by viruses.

-plasma, plasmo-, -plast [Gk. *plasma,* form, mold]: Prefix or suffix meaning "formed," or "molded"; examples are protoplasm, "first-molded" (living matter), and chloroplast, "green-formed" (body).

plasma membrane or plasmalemma: Outer boundary of the protoplast, next to the cell wall; consists of a single membrane unit; also called cell membrane and ectoplast.

plasmid: A relatively small fragment of DNA that can exist free in the cytoplasm of a bacterium and can be integrated into and then replicated with the bacterial chromosome by a process known as genetic recombination.

plasmodesma, *pl.* **plasmodesmata:** The minute cytoplasmic threads that extend through openings in cell walls and connect the protoplasts of adjacent living cells.

plasmodium (plăz•mo'dĭ•*ŭ*m): Stage in life cycle of myxomycetes (plasmodial slime molds); a multinucleate mass of protoplasm surrounded by a membrane.

plasmogamy [Gk. *plasma,* form, + *gamos,* marriage]: Union of the protoplasts of sex cells, or gametes, not accompanied by union of their nuclei.

plasmolysis (plăz•mŏl'ĭ•sĭs) [Gk. *plasma,* form, + *lysis,* a loosening]: The separation of the protoplast from the cell wall due to removal of water from the protoplast by osmosis.

plastid (plăs'tĭd): Organelle in the cells of certain groups of eukaryotes that is the site of such activities as food manufacture and storage; plastids are bounded by a double membrane.

pleiotropy [Gk. *pleros,* more, + *trope,* a turning]: The capacity of a gene to affect a number of different characteristics.

plumule [L. *plumula,* a small feather]: The first bud of an embryo; the portion of the young shoot above the cotyledons.

pneumatophores [Gk. *pneuma,* breath, + *-phoros,* carrying]: Negatively geotropic extensions of the root systems of some trees growing in swampy habitats; they grow upward and out of the water to assure adequate aeration.

polar nuclei: Two nuclei (usually), one derived from each end (pole) of the embryo sac, which become centrally located; they fuse with a male nucleus to form the primary ($3n$) endosperm nucleus.

pollen [L. *pollen,* fine dust]: A collective term for pollen grains.

pollen grain: A microspore containing a mature or immature microgametophyte (male gametophyte).

pollen mother cell: *See* microspore mother cell.

pollen sac: A cavity in the anther that contains the pollen grains.

pollen tube: A tube formed after germination of the pollen grain; carries the male gametes into the ovule.

pollination: The transfer of pollen from where it was formed (e.g., the anther) to a receptive surface (e.g., the stigma).

poly- [Gk. *polys,* many]: Prefix meaning "many."

polyembryony: Having more than one embryo within the developing seed.

polygenic inheritance: The inheritance of quantitative characteristics which is determined by the combined effects of multiple genes.

polymer (pŏl'ĭ•mĕr): A large molecule composed of many of the same molecular subunits.

polymerization: The chemical union of monomers such as glucose or nucleotides to form polymers such as starch or nucleic acid.

polynucleotides: Long-chain molecules composed of units (monomers) called nucleotides; DNA is a polynucleotide.

polypeptide: Numerous amino acids linked together by peptide bonds.

polyploid (pŏl'ĭ•ploid): Referring to a plant, tissue, or cell with more than two complete sets of chromosomes.

polysaccharide: A carbohydrate composed of many monosaccharide units joined in a long chain, such as glycogen, starch, and cellulose.

polysome or polyribosome: An aggregation of ribosomes; ribosomes actively involved in protein synthesis.

pome (pōm) [Fr. *pomme,* apple]: A simple fleshy fruit, the outer portion of which is formed by the floral parts that surround the ovary and expand with the growing fruit; found only in one subfamily of the Rosaceae (apples, pears, quince, and so on).

population: Any group of individuals, usually of a single species.

P-protein: Phloem-protein; a proteinaceous substance found in cells of angiosperm phloem, especially in sieve-tube members; also called slime.

preprophase band: A ringlike band of microtubules that outlines the equatorial plane of the future mitotic spindle in many cells; may play a role in positioning of the nucleus before mitosis begins.

primary endosperm nucleus: The result of the fusion of a sperm nucleus and the two polar nuclei.

primary growth: In plants, growth originating in the apical meristems of shoots and roots, as contrasted with secondary growth.

primary meristem or meristematic tissue (mĕr'ĭ•ste•măt'ĭk): A tissue derived from the apical meristem; of three kinds: protoderm, procambium, and ground meristem.

primary pit-field: Thin area in a primary cell wall through which plasmodesmata pass, although plasmodesmata may occur elsewhere in the wall as well.

primary plant body: The part of the plant body arising from the apical meristems and their derivative meristematic tissues; composed entirely of primary tissues.

primary root: The first root of the plant, developing in continuation of the root tip or radicle of the embryo; in gymnosperms and dicots it becomes the taproot.

primary tissues: Cells derived from the apical meristems and primary meristematic tissues of root and shoot; as opposed to secondary tissues derived from a cambium; primary growth results in an increase in length.

primary wall: The wall layer deposited during the period of cell expansion.

primordium, *pl.* **primordia** [L. *primus,* first, + *ordiri,* to begin to weave]: A cell or organ in its earliest stage of differentiation.

pro- [Gk. *pro,* before]: Prefix meaning "before" or "prior to."

procambium (pro•kăm'bĭ•*ŭ*m) [L. *pro,* before, + *cambiare,* to exchange]: A primary meristematic tissue that gives rise to primary vascular tissues.

proembryo: Embryo in early stages of development, before the embryo proper and suspensor become distinct.

progymnosperms: An extinct group of Paleozoic seedless vascular plants; believed to be the likely progenitors of gymnosperms.

prokaryote [Gk. *pro,* before, + *karyon,* kernel]: A cell lacking a membrane-bound nucleus or membrane-bound organelles; a bacterium.

promoter: A specific site at the beginning of an operon at which a polymerase can start its synthesis.

prophage (prō'fāj'): Noninfectious phage units linked with the bacterial chromosome which multiply with the growing and dividing bacteria but do not bring about lysis of the bacteria. Prophage is a stage in the life cycle of a temperate phage.

prophase [Gk. *pro,* before, + *phasis,* form]: An early stage in nuclear division, characterized by the shortening and thickening of the chromosomes and their movement to the metaphase plate.

proplastid: A minute self-reproducing body in the cytoplasm from which a plastid develops.

prop roots: Adventitious roots arising from the stem above soil level and helping to support the plant; common in many monocots, for example, corn *(Zea mays).*

protease (prō'te•ās): An enzyme that digests protein by hydrolysis of peptide bonds.

protein [Gk. *proteios,* primary]: A complex organic compound composed of many (100 or more) amino acids joined by peptide bonds.

prothallial cell [Gk. *pro,* before, + *thallos,* sprout]: The sterile cell or cells found in the male gametophytes, or microgametophytes, of heterosporous plants; believed to be remnants of the vegetative tissue of the male gametophyte.

prothallus (pro•thăl'ŭs): In ferns and some other relatively unspecialized vascular plants, the more or less independent gametophyte; also called prothallium.

proto- [Gk. *protos,* first]: Prefix meaning "first," for example, Protozoa, "first animals."

protoderm [Gk. *protos,* first, + *derma,* skin]: Primary meristematic tissue that gives rise to epidermis.

proton: A subatomic, or elementary, particle, with a single positive charge equal in magnitude to the charge of an electron and a mass of 1; the basic component of every atomic nucleus; a proton can also be thought of as the nucleus of the lightest and most abundant hydrogen isotope.

protonema, *pl.* **protonemata** [Gk. *protos,* first, + *nema,* a thread]: Filamentous growth, an early stage in development of the gametophyte of mosses and certain liverworts.

protoplasm: A general term for the living substance of all cells.

protoplast: The protoplasm of an individual cell; in the case of a plant cell, the unit of protoplasm inside the cell wall.

protostele [Gk. *protos,* first, + *stele,* pillar]: The simplest type of stele, containing a solid column of vascular tissue.

protoxylem: The first part of the primary xylem, which matures during elongation of the plant part in which it is found.

proximal (prŏk'sĭ•măl) [L. *proximus,* near]: Situated near the point of reference, usually the main part of a body or the point of attachment; opposite of distal.

pseudo- [Gk. *pseudes,* false]: Prefix meaning "false."

pseudoplasmodium: A multicellular mass of individual amoeboid cells, representing the aggregate phase in the cellular slime molds.

purine (pū'rēn): A nitrogenous base with a double-ring structure, such as adenine or guanine; one of the components of nucleic acids.

pyramid of energy: Energy relationships among various feeding levels involved in a particular food chain; autotrophs (at the base of the pyramid) represent the greatest amount of available energy; herbivores are next; then primary carnivores; secondary carnivores; etc.

pyrenoid [Gk. *pyren,* the stone of a fruit, + L. *oides,* like]: A body found in the chloroplasts of certain algae and hornworts that seems to be associated with starch deposition.

pyrimidine: A nitrogenous base with a single-ring structure, such as cytosine, thymine, or uracil; one of the components of nucleic acids.

quantasome [L. *quantus,* how much, + Gk. *soma,* body]: Granules located on the inner surfaces of chloroplast lamellae; believed to be involved in the light-requiring reactions of photosynthesis.

quiescent center: The relatively inactive initial region in the apical meristem of a root.

raceme [L. *racemus,* bunch of grapes]: An indeterminate inflorescence in which the main axis is elongated but the flowers are borne on pedicels that are about equal in length.

rachis (rā'kĭs) [Gk. *rhachis,* a backbone]: Main axis of spike; axis of fern leaf (frond) from which pinnae arise; in compound leaves, the extension of the petiole corresponding to the midrib of an entire leaf.

radially symmetrical: *See* actinomorphic.

radial section: A longitudinal section cut parallel to the radius of a cylindrical body, such as a root or stem; in the case of secondary xylem, or wood, and secondary phloem, parallel to the rays.

radial system: In secondary xylem and secondary phloem, term applied to all the rays, the cells of which are derived from ray initials; also called horizontal system and ray system.

radicle [L. *radix,* root]: The embryonic root.

radioisotope: An unstable isotope of an element that decays or disintegrates spontaneously, emitting radiation; also called a radioactive isotope.

raphe [Gk. *raphe,* seam]: (1) Ridge on seeds, formed by the stalk of the ovule, in those seeds in which the stalk is sharply bent at the base of the ovule; (2) groove on the shell, or frustule, of a diatom.

raphides (răf'ĭ•dēz) [Gk. *rhaphis,* a needle]: Fine, sharp, needlelike crystals of calcium oxalate found in the vacuoles of many plant cells.

ray flowers: *See* disk flowers.

ray initial: An initial in the vascular cambium that gives rise to ray cells of secondary xylem and secondary phloem.

reaction center: The chlorophyll molecule of a photosystem capable of using energy in the photochemical reaction.

receptacle: That part of the axis of a flower stalk that bears the floral organs.

recessive: Describing a gene whose phenotypic expression is masked by a dominant allele; heterozygotes are phenotypically indistinguishable from dominant homozygotes.

reduction [L. *reductio,* a bringing back; originally "bringing back" a metal from its oxide]: Gain of an electron by an atom that takes place simultaneously with oxidation (loss of an electron by an atom), because an electron that is lost by one atom is accepted by another.

regular: *See* actinomorphic.

regulator gene: A gene that prevents or represses the activity of the structural genes in an operon.

replicate: Produce a facsimile or a very close copy; used to indicate the production of a second molecule of DNA exactly like the first molecule or of a sister chromatid.

resin duct: A tubelike intercellular space lined with resin-secreting cells (epithelial cells) and containing resin.

respiration: An intracellular process in which food is oxidized with the release of energy. The complete breakdown of sugar or other organic compounds to carbon dioxide and water is termed aerobic respiration, although the earlier steps are anaerobic.

reticulate venation: *See* netted venation.

rhizobia [Gk. *rhiza,* root, + *bios,* life]: Bacteria of the genus *Rhizobium* which may be involved with leguminous plants in a symbiotic relationship that results in nitrogen fixation.

rhizoids [Gk. *rhiza,* root]: (1) Branched rootlike extensions of fungi and algae that absorb water, food, and nutrients; (2) root-hairlike structures in liverworts, mosses, and some vascular plants, usually borne by the gametophyte generation.

rhizome: A more or less horizontal underground stem.

ribonucleic acid (RNA): Type of nucleic acid formed on chromosomal DNA and involved in protein synthesis; composed of chains of phosphate, sugar molecules (ribose), and purines and pyrimidines; the genetic material of many viruses.

ribose: A five-carbon sugar; a component of RNA.

ribosome: A small particle composed of protein and RNA; the site of protein synthesis.

ring-porous wood: A wood in which the pores, or vessels, of the early wood are distinctly larger than those of the late wood, forming a well-defined ring in cross sections of the wood.

RNA: *See* ribonucleic acid.

root: The usually descending axis of a plant, normally below ground, and serving to anchor the plant and to absorb and conduct water and minerals.

root cap: A thimblelike mass of cells covering and protecting the growing tip of a root.

root hairs: Tubular outgrowths of epidermal cells of the root in the zone of maturation.

root pressure: The pressure developed in roots as the result of osmosis which causes guttation of water from leaves and exudation from cut stumps.

runner: *See* stolon.

samara: Simple, dry, one-seeded or two-seeded indehiscent fruit with pericarp-bearing, winglike outgrowths.

sap: (1) A name applied to the fluid contents of the xylem or the sieve elements of the phloem; (2) the fluid contents of the vacuole are called cell sap.

saprobe [Gk. *sapros,* rotten, + *bios,* life]: An organism that secures its food directly from nonliving organic matter.

sapwood: Outer part of the wood of stem or trunk, usually distinguished from the heartwood by its lighter color, in which active conduction of water takes place.

savanna: A tropical or subtropical grassland containing scattered trees; transitional areas between the evergreen tropical rain forest and the deserts.

schizo- [Gk. *schizein,* to split]: Prefix meaning "split."

schizocarp (skĭz'o•kärp): Dry fruit with two or more united carpels which split apart at maturity.

sclereid [Gk. *skleros,* hard]: A sclerenchyma cell with a thick, lignified secondary wall having many pits. Variable in form but typically not very long; may or may not be living at maturity.

sclerenchyma (skle•rĕng'kĭ•má) [Gk. *skleros,* hard, + L. *enchyma,* infusion]: A supporting tissue composed of sclerenchyma cells, including fibers and sclereids.

sclerenchyma cell: Cell of variable form and size with more or less thick, often lignified, secondary walls; may or may not be living at maturity; includes fibers and sclereids.

scutellum (sku•tĕl'ŭm) [L. *scutella,* a small shield]: Single cotyledon of grass embryo, specialized for absorption of the endosperm.

secondary growth: In plants, growth derived from secondary or lateral meristem, the vascular and cork cambiums; secondary growth results in an increase in girth; as contrasted with primary growth.

secondary plant body: The part of the plant body produced by the vascular cambium and the cork cambium; consists of secondary xylem, secondary phloem, and periderm.

secondary root: *See* lateral root.

secondary tissues: Tissues produced by the vascular cambium and cork cambium.

secondary wall: Innermost layer of the cell wall, formed in certain cells after cell elongation has ceased; secondary walls have a highly organized microfibrillar structure.

seed: A structure formed by the maturation of the ovule of seed plants following fertilization.

seed coat: The outer layer of the seed, developed from the integuments of the ovule.

seedling: A young sporophyte developing from a germinating seed.

segregation: The separation of the chromosomes (and genes) from different parents at meiosis.

semipermeable membrane: A membrane that is permeable to water but not to solutes.

sepal (se'păl) [L. *sepalum,* a covering]: One of the outermost flower structures which usually encloses the other flower parts in the bud; a unit of the calyx.

septate [L. *septum,* fence]: Divided by cross walls into cells or compartments.

sessile (sĕs'ĭl) [L. *sessilis,* of or fit for sitting, low, dwarfed]: Attached directly by the base; referring to a leaf lacking a petiole or to a flower or fruit lacking a pedicel.

seta, *pl.* **setae** (sē'ta) [L. **seta,** bristle]: In bryophytes, the stalk that supports the capsule, if present; part of the sporophyte.

sexual reproduction: The fusion of gametes followed by meiosis and recombination at some point in the life cycle.

sheath: (1) The base of a leaf that wraps around the stem, as in grasses; (2) a tissue layer surrounding another tissue, such as a bundle sheath.

shoot: The above-ground portions, such as the stem and leaves, of a vascular plant.

short-day plants: Plants that must be exposed to light periods shorter than some critical length for flowering to occur; they flower in early spring or fall.

shrub: A perennial woody plant of relatively low stature, typically with several stems arising from or near the ground.

sieve area: A portion of sieve-element wall containing clusters of pores through which the protoplasts of adjacent sieve elements are interconnected.

sieve cell: A long, slender sieve element with relatively unspecialized sieve areas and with tapering end walls that lack sieve plates; found in the phloem of gymnosperms and lower vascular plants.

sieve element: The cell of the phloem that is involved in the long-distance transport of food substances; further classified into sieve cells and sieve-tube members.

sieve plate: The part of the wall of sieve-tube members bearing one or more highly differentiated sieve areas.

sieve tube: A series of sieve-tube members arranged end-to-end and interconnected by sieve plates.

sieve-tube member: One of the component cells of a sieve tube; found primarily in flowering plants and typically associated with a companion cell; also called sieve-tube element.

silique [L. *siliqua,* pod]: The fruit characteristic of the mustard family; two-celled, the valves splitting from the bottom and leaving the placentae with the false partition stretched between.

simple fruit: A fruit derived from one carpel or several united carpels.

simple leaf: An undivided leaf; as opposed to a compound leaf.

simple pit: Pit not surrounded by an overarching border of secondary wall; as opposed to a bordered pit.

siphonaceous, siphonous [Gk. *siphon,* a tube, pipe]: In algae, multinucleate cells without cross walls; coenocytic.

siphonostele [Gk. *siphon,* pipe, + *stele,* pillar]: A type of stele containing a hollow cylinder of vascular tissue surrounding a pith.

slime: *See* P-protein.

softwood: A name commonly applied to the wood of a conifer.

solute: A dissolved substance.

solute potential: The change in free energy or chemical potential of water produced by solutes; carries a negative (minus) sign; also called osmotic potential.

solution: Usually liquid, in which the molecules of the dissolved substance, the solute (e.g., sugar), are dispersed between the molecules of the solvent (e.g., water).

somatic cells [Gk. *soma,* body]: The differentiated, usually diploid ($2n$) cells composing body tissues of multicellular plants and animals.

soredium, *pl.* **soredia** [Gk. *soros,* heap]: A specialized reproductive unit of lichens, consisting of a few cyanobacterial or green algal cells surrounded by fungal hyphae.

sorus, *pl.* **sori** (sō'rŭs) [Gk. *soros,* heap]: A group or cluster of sporangia.

specialized: (1) Of organisms, having special adaptations to a particular habitat or mode of life; (2) of cells, having particular functions.

species, *pl.* **species** [L. kind, sort]: A kind of organism; species are designated by binomial names written in italics.

specificity: Uniqueness, as in proteins in given organisms or enzymes in given reactions.

sperm: A mature male sex cell or gamete, usually motile and smaller than the female gamete.

spermagonium, *pl.* **spermagonia** (spûr'ma•gō'nĭ•ŭm) [Gk. *sperma,* sperm, + *gonos,* offspring]: In the rust fungi, the structure that produces spermatia.

spermatangium, *pl.* **spermatangia** (spûr'ma•tăn'jĭ•ŭm) [Gk. *sperma,* sperm, + L. *tangere,* to touch]: In the red algae, the cell that produces spermatia.

spermatium, *pl.* **spermatia** [Gk. *sperma,* sperm]: In the red algae and some fungi, a minute, nonmotile male gamete.

spermatophyte [Gk. *sperma,* seed, + *phyton,* plant]: A seed plant.

spherosome: Single, membrane-bound spherical structures in the cytoplasm of plant cells, many of which contain mostly lipids and apparently are centers of lipid synthesis and accumulation.

spike [L. *spica,* head of grain]: An indeterminate inflorescence in which the main axis is elongated and the flowers are sessile.

spikelet: The unit of inflorescence in grasses; a small group of grass flowers.

spindle fibers: A group of microtubules that extend from the centromeres of the chromosomes to the poles of the spindle or from pole to pole in a dividing cell.

spine: A hard, sharp-pointed structure; usually a modified leaf, or part of a leaf.

spirillum, *pl.* **spirilli** [L. *spira,* coil]: A long coiled or spiral bacterium.

spiroplasma: Long, thin, helical mycoplasma.

spongy parenchyma: A leaf tissue composed of loosely arranged, chloroplast-bearing cells.

sporangiophore (spo•răn'jĭ•o•fōr') [Gk. *spora,* seed, + *pherein,* to carry]: A branch bearing one or more sporangia.

sporangium, *pl.* **sporangia** (spo•răn'jĭ•ŭm) [Gk. *spora,* seed, + *angeion,* a vessel]: A hollow unicellular or multicellular structure in which spores are produced.

spore: A reproductive cell, usually unicellular, capable of developing into an adult without fusion with another cell.

spore mother cell: A diploid ($2n$) cell that undergoes meiosis and produces (usually) four haploid cells (spores) or four haploid nuclei.

sporophyll (spō'ro•fĭl): A modified leaf or leaflike organ that bears sporangia; applied to the stamens and carpels of angiosperms, fertile fronds of ferns, etc.

sporophyte (spō'ro•fīt): The spore-producing, diploid ($2n$) phase in the life cycle of a plant having alternation of generations.

sporopollenin: The tough substance of which the exine, or outer wall, of spores and pollen grains is composed.

stalk cell: One of two cells produced by division of the generative cell in developing pollen grains of gymnosperms; it is a sterile cell and eventually degenerates.

stamen (stā'mĕn) [L. *stamen,* thread]: The part of the flower producing the pollen, composed (usually) of anther and filament; collectively, the stamens make up the androecium.

staminate (stăm'ĭ•nat): Pertaining to a flower having stamens but no functional carpels.

starch [M.E. *sterchen,* to stiffen]: A complex insoluble carbohydrate, the chief food storage substance of plants; composed of several hundred glucose units ($C_6H_{10}O_5$) and readily broken down enzymatically into these components.

statoliths [Gk. *statos,* stationary, + *lithos,* stone]: Gravity sensors; starch grains or other bodies in the cytoplasm.

stele (stē'le) [Gk. *stele,* a pillar]: The central cylinder, inside the cortex, of roots and stems of vascular plants.

stem: The part of the axis of vascular plants that is above ground, as well as anatomically similar portions below ground (rhizomes, corms, etc.)

stem bundle: Vascular bundle belonging to the stem.

sterigma, *pl.* **sterigmata** [Gk. *sterigma,* a prop]: A small, slender protuberance that bears a basidiospore; it is borne by a basidium.

stigma: (1) The region of a carpel serving as a receptive surface for pollen grains and on which they germinate; (2) light-sensitive eyespot of some algae.

stipe: A supporting stalk, such as the stalk of a gill fungus or the leaf stalk of a fern.

stipule (stĭp'ūl): A leaflike appendage on either side of the basal part of a leaf of some groups of plants.

stolon (stō'lŏn) [L. *stolo,* shoot]: A stem that grows horizontally along the ground surface and may form adventitious roots, such as runners of the strawberry plant.

stoma, *pl.* **stomata** (stō'ma) [Gk. *stoma,* mouth]: A minute opening bordered by guard cells in the epidermis of leaves and stems through which gases pass; also used to refer to the entire stomatal apparatus—the guard cells plus their included pore.

strobilus, *pl.* **strobili** (strŏb'ĭ•lŭs) [Gk. *strobilos,* a cone]: A reproductive structure consisting of a number of modified leaves (sporophylls) or ovule-bearing scales grouped terminally on a stem; a cone.

stroma [Gk. *stroma,* anything spread out]: The ground substance of plastids.

structural gene: Any gene that produces a protein, in distinction to the other genes of an operon (operator and regulator genes).

style [Gk. *stylos,* a column]: Slender column of tissue which arises from the top of the ovary and through which the pollen tube grows.

sub- [L. *sub,* under, below]: Prefix meaning "under" or "below"; for example, subepidermal, "underneath the epidermis."

suberin (sū'bẽr•ĭn) [L. *suber,* the cork oak]: Fatty material found in the cell walls of cork tissue and in the Casparian strip of the endodermis.

subsidiary cell: An epidermal cell morphologically distinct from other epidermal cells and associated with a pair of guard cells; also called an accessory cell.

subspecies: A subdivision of a species.

substrate [L. *substratus,* strewn under]: The foundation to which an organism is attached; substance acted on by an enzyme.

substrate phosphorylation: Phosphorylation—the formation of ATP from ADP and inorganic phosphate—that takes place during glycolysis.

succession: In ecology, the slow, orderly progression of changes in community composition during development of vegetation in any area, from initial colonization to the attainment of the climax typical of a particular geographic area.

succulent: A plant with fleshy, water-storing stems or leaves.

sucker: A sprout produced by the roots of some plants and that gives rise to a new plant.

sucrase (sū'krās): An enzyme that hydrolyzes sucrose into glucose and fructose; also called invertase.

sucrose (sū'krōs): A disaccharide (glucose plus fructose) found in many plants; the primary form in which sugar produced by photosynthesis is translocated.

superior ovary: An ovary free and separate from the calyx.

suspension: A heterogeneous dispersion in which the dispersed phase consists of solid particles sufficiently large that they will settle out of the fluid dispersion medium under the influence of gravity.

suspensor: A structure in the embryo of many vascular plants that pushes the terminal part of the embryo into the endosperm.

symbiosis (sĭm'bı•ō'sĭs) [Gk. *syn,* together with, + *bios,* life]: The living together in close association of two or more dissimilar organisms; includes parasitism (in which the association is harmful to one of the organisms) and mutualism (in which the association is advantageous to both).

symplast [Gk. *syn,* together with, + *plastos,* molded]: The interconnected protoplasts and their plasmodesmata; the movement of substances in the symplast is called symplastic movement, or symplastic transport.

sympodium, *pl.* **sympodia:** A stem bundle and its associated leaf traces.

syn-, sym- [Gk. *syn,* together with]: Prefix, meaning "together."

synapsis: The pairing of homologous chromosomes that occurs prior to the first meiotic division; crossing over occurs during synapsis.

synergids (sĭ•nûr'jĭds): Two short-lived cells lying close to the egg in the mature embryo sac of the ovule of flowering plants.

syngamy [Gk. *syn,* together with, + *gamos,* marriage]: The process by which two haploid cells fuse to form a diploid zygote; fertilization.

synthesis: The formation of a more complex substance from simpler ones.

systematics: Scientific study of the kinds and diversity of organisms and of the relationships between them.

taiga: The northern coniferous forest, or boreal forest, characterized by severe winters and persistent winter snow cover.

tangential section: A longitudinal section cut at right angles to the radius of a cylindrical structure, such as a root or stem; in the case of secondary xylem, or wood, and secondary phloem, at right angles to the rays.

tapetum (ta•pē'tŭm) [Gk. *tapes,* a carpet]: Nutritive tissue in the sporangium, particularly an anther.

taproot: The primary root of a plant formed in direct continuation with the root tip or radicle of the embryo; forms a stout, tapering main root from which arise smaller, lateral branches.

taxon: General term for any one of the taxonomic categories such as species, class, order, or division, into which living organisms are classified.

taxonomy [Gk. *taxis,* arrangement, + *nomos,* law]: The science of the classification of organisms.

teliospore (tē′lĭ•ō•spōr′): In the rust fungi, a thick-walled spore in which karyogamy and meiosis occur and from which basidia develop.

telium, *pl.* **telia:** In the rust fungi, the structure that produces teliospores.

telophase: The last stage in mitosis and meiosis, during which the chromosomes become reorganized into two new nuclei.

temperate phage: A bacterial virus that may remain latent in its host bacterial cell; in this latent (prophage) state, it is associated with the bacterial chromosome and is replicated with it.

template: A pattern or mold guiding the formation of a negative or complement; a term applied especially to DNA duplication, which is explained in terms of a template hypothesis.

tendril [L. *tendere,* to extend]: A slender coiling structure, usually a modified leaf or part of a leaf, that aids in support of the stems.

tepal: Unit structure of a perianth that is not differentiated into sepals and petals.

test cross: A cross of a dominant with a homozygous recessive; used to determine whether the dominant is homozygous or heterozygous.

tetrad (tĕt′răd): A group of four spores formed from a spore mother cell by meiosis.

tetraploid (tĕt′rə•ploid) [Gk. *tetra,* four, + *ploos,* fold]: Twice the usual, or diploid ($2n$), number of chromosomes (that is, $4n$).

tetrasporangium, *pl.* **tetrasporangia** [Gk. *tetra,* four, + *spora,* seed, + *angeion,* vessel]: In certain red algae, a sporangium in which meiosis occurs, resulting in the production of tetraspores.

tetraspore [Gk. *tetra,* four, + *spora,* seed]: In certain red algae, the four spores formed by meiotic division in the tetrasporangium of a spore mother cell.

tetrasporophyte [Gk. *tetra,* four, + *spora,* seed, + *phyton,* plant]: In certain red algae, a diploid individual that produces tetrasporangia.

thallophyte: A term previously used to designate fungi and algae collectively, now largely abandoned.

thallus (thăl′ŭs) [Gk. *thallos,* a sprout]: A type of plant body that is undifferentiated into root, stem, or leaf.

theory [Gk. *theorein,* to look at]: A generalization based on observation and experiments conducted to test the validity of a hypothesis and found to support the hypothesis.

thermodynamics [Gk. *therme,* heat, + *dynamis,* power]: The study of energy exchanges, using heat as the most convenient form of measurement of energy. The first law of thermodynamics states that in all processes, the total energy of the universe remains constant. The second law of thermodynamics states that the entropy, or degree of randomness, tends to increase.

thorn: A hard, woody, pointed branch.

thylakoid [Gk. *thylakos,* sac, + *oides,* like]: A saclike membranous structure in cyanobacteria and the chloroplasts of eukaryotic organisms; in chloroplasts, stacks of thylakoids form the grana; chlorophylls are found within the thylakoids.

thymine: A pyrimidine occurring in DNA but not in RNA; *see also* uracil.

tissue: A group of similar cells organized into a structural and functional unit.

tissue culture: A technique for maintaining fragments of plant or animal tissue alive in a medium after removal from the organism.

tissue system: A tissue or group of tissues organized into a structural and functional unit in a plant or plant organ. There are three tissue systems: dermal, vascular, and ground, or fundamental.

tonoplast [Gk. *tonos,* stretching tension, + *plastos,* formed, molded]: The cytoplasmic membrane surrounding the vacuole in plant cells; also called vacuolar membrane.

torus, *pl.* **tori:** The central thickened part of the pit-membrane in the bordered pits of conifers and some other gymnosperms.

tracheary element: The general term for a water-conducting cell in vascular plants; tracheids and vessel members.

tracheid (trā′ke•ĭd): An elongated, thick-walled conducting and supporting cell of xylem. It has tapering ends and pitted walls without perforations, as contrasted with a vessel member. Found in nearly all vascular plants.

transduction: The transfer of genetic material (DNA) from one bacterium to another by a temperate phage.

transfer cell: Specialized parenchyma cell with wall ingrowths that increase the surface area of the plasma membrane; apparently functions in the short-distance transfer of solutes.

transfer RNA (tRNA): Low-molecular-weight RNA that becomes attached to an amino acid and guides it to the correct position on the ribosome for protein synthesis; there is at least one tRNA molecule for each amino acid.

transformation: A genetic change produced by the incorporation into a cell of DNA from another cell.

transition region: The region in the primary plant body showing transitional characteristics between structures of root and shoot.

translocation: (1) In plants, the long-distance transport of water, minerals, or food; most often used to refer to food transport; (2) in genetics, the interchange of chromosome segments between nonhomologous chromosomes.

transpiration [Fr. *transpirer,* to perspire]: The loss of water vapor by plant parts; most transpiration occurs through stomata.

transverse section: A section cut perpendicular, or at right angles, to the longitudinal axis of a plant part.

tree: A perennial woody plant generally with a single stem (trunk).

tricarboxylic acid cycle or TCA cycle: *See* Krebs cycle.

trichogyne [Gk. *trichos,* a hair, + *gyne,* female]: In the red algae and certain ascomycetes and basidiomycetes, a receptive protuberance of the female gametangium for the conveyance of spermatia.

trichome [Gk. *trichos,* hair]: An outgrowth of the epidermis, such as a hair, scale, and water vesicle.

triose [Gk. *tries,* three, + *ose,* suffix indicating a carbohydrate]: Any three-carbon sugar.

triple fusion: In angiosperms, the fusion of the second male gamete, or sperm, with the polar nuclei, resulting in formation of a primary endosperm nucleus, which is most often triploid ($3n$).

triploid [Gk. *triploos,* triple]: Having three complete chromosome sets per cell ($3n$).

tritium: Radioactive isotope of hydrogen, ^3H. The nucleus of a tritium atom contains one proton and two neutrons, whereas the more common hydrogen nucleus consists only of a proton.

tRNA: See transfer RNA.

-troph, tropho- [Gk. *trophos,* feeder]: Suffix or prefix meaning "feeder," "feeding," or "nourishing"; for example, autotrophic, "self-nourishing."

trophic level: A step in the movement of energy through an ecosystem.

tropism [Gk. *trope,* a turning]: A response to an external stimulus in which the direction of the movement is usually determined by the direction from which the most intense stimulus comes.

tube cell: In male gametophytes, or pollen grains, of seed plants, the cell that develops into the pollen tube.

tuber [L. *tuber,* swelling]: An enlarged, short, fleshy underground stem, such as that of the potato.

tundra: A treeless circumpolar region, best developed in the Northern Hemisphere and mostly found north of the Arctic Circle.

tunica-corpus: The organization of the shoot apex of most angiosperms and a few gymnosperms, consisting of one or more peripheral layers of cells (the tunica layers) and an interior (the corpus). The tunica layers undergo surface growth (by anticlinal divisions), and the corpus undergoes volume growth (by divisions in all planes).

turgid (tûr'jĭd) [L. *turgidus,* swollen]: Swollen, distended, referring to a cell that is firm due to water uptake.

turgor pressure [L. *turgor,* a swelling]: The pressure within the cell resulting from the movement of water into the cell.

tylose [Gk. *tylos,* a lump]: A balloonlike outgrowth from a ray or axial parenchyma cell through the pit in a vessel wall and into the lumen of the vessel.

umbel (ŭm'bĕl) [L. *umbella,* sunshade]: An inflorescence, the individual pedicels of which all arise from the apex of the peduncle.

unicellular: Composed of a single cell.

unisexual: Usually applied to a flower lacking either stamens or carpels; a perianth may be present or absent.

unit membrane: A visually definable, three-layered membrane, consisting of two dark layers separated by a lighter layer.

uracil (ū'ra•sĭl): A pyrimidine found in RNA but not in DNA; *see also* thymine.

uredinium, *pl.* **uredinia** [L. *uredo,* a blight]: In rust fungi, the structure that produces uredospores.

uredospore [L. *uredo,* a blight, + *spora,* spore]: In rust fungi, a reddish, binucleate spore produced in summer.

vacuolar membrane: *See* tonoplast.

vacuole [L. *vacuus,* empty]: A space or cavity within the cytoplasm filled with a watery fluid, the cell sap; part of the lysosomal compartment of the cell.

valve: One of the two halves of the diatom cell wall; a diatom shell, or frustule.

variation: The differences that occur within the offspring of a particular species.

variety: A group of plants or animals of less than species rank.

vascular [L. *vasculum,* a small vessel]: Pertains to any plant tissue or region consisting of or giving rise to conducting tissue; e.g., xylem, phloem, vascular cambium.

vascular bundle: A strand of tissue containing primary xylem and primary phloem (and procambium if still present) and frequently enclosed by a bundle sheath of parenchyma or fibers.

vascular cambium: A cylindrical sheath of meristematic cells, the division of which produces secondary phloem and secondary xylem.

vascular rays: Ribbonlike sheets of parenchyma that extend radially through the wood, across the cambium, and into the secondary phloem; they are always produced by the vascular cambium.

vascular tissue system: All the vascular tissues in a plant or plant organ.

vegetative: Of, relating to, or involving propagation by asexual processes; also referring to nonreproductive plant parts.

vegetative reproduction: (1) In seed plants, reproduction by means other than by seeds; apomixis; (2) in other organisms, reproduction by vegetative spores, fragmentation, or division of the plant body. Unless a mutation occurs, each daughter cell individual is genetically identical with its parent.

vein: A vascular bundle forming a part of the framework of the conducting and supporting tissue of a leaf or other expanded organ.

velamen [L. *velumen,* fleece]: A multiple epidermis covering the aerial roots of some orchids and aroids; also occurs on some terrestrial roots.

venation: Arrangement of veins in leaf blade.

venter [L. *venter,* belly]: The enlarged basal portion of an archegonium containing the egg.

vernalization [L. *vernalis,* spring]: The induction of flowering by cold treatment.

vessel [L. *vasculum,* a small vessel]: A tubelike structure of the xylem composed of elongate cells (vessel members) placed end to end and connected by perforations. Its function is to conduct water and minerals through the plant body. Found in nearly all angiosperms and a few other vascular plants (e.g., the Gnetophyta).

vessel member: One of the cells composing a vessel; also called vessel element.

viroids: The smallest known agents of infectious disease.

virus: A submicroscopic, noncellular particle; composed of a nucleic acid core and a protein shell; viruses reproduce only within host cells.

volva [L. *volva,* a wrapper]: A cuplike structure at the base of the stalk of certain mushrooms.

wall pressure: The pressure of the cell wall exerted against the turgid protoplast; opposite and equal to the turgor pressure.

water potential: The algebraic sum of the solute potential and the pressure potential, or wall pressure; the potential energy of water.

water vesicle: An enlarged epidermal cell in which water is stored; a type of trichome.

weed [O.E. *weod,* used at least since the year 888 in its present meaning]: Generally an herbaceous plant not valued for use or beauty, growing wild, and regarded as using ground or hindering the growth of useful vegetation.

whorl: A circle of leaves or of flower parts.

wild type: In genetics, the phenotype or genotype that is characteristic of the majority of individuals of a species in a natural environment.

wood: Secondary xylem.

xanthophyll (zăn'tho•fĭl) [Gk. *xanthos,* yellowish-brown, + *phyllon,* leaf]: A yellow chloroplast pigment; a member of the carotenoid group.

xerophyte [Gk. *xeros,* dry, + *phyton,* a plant]: A plant that has adapted to arid habitats.

xylem [Gk. *xylon,* wood]: A complex vascular tissue through which most of the water and minerals of a plant are conducted; characterized by the prescence of tracheary elements.

zeatin (zē'ā•tĭn): Plant hormone; a natural cytokinin isolated from corn.

zoosporangium: A sporangium bearing zoospores.

zoospore (zō'o•spōr): A motile spore, found among algae, oomycetes, and chytrids.

zygomorphic [Gk. *zygo,* pair, + *morphe,* form]: A type of flower capable of being divided into two symmetrical halves only by a single longitudinal plane passing through the axis; also called bilaterally symmetrical.

zygospore: A thick-walled, resistant spore that develops from a zygote, resulting from the fusion of isogametes.

zygote (zı'gōt) [Gk. *zygotos,* paired together]: The diploid ($2n$) cell resulting from the fusion of male and female gametes.

Illustration Acknowledgments

All photographs not credited herein are by Ray F. Evert.

I–1 M. Kage, Peter Arnold, Inc.

I–2 *(a)* G. J. Breckon; *(b)* L. West, Bruce Coleman, Inc.

I–3 Grant Heilman

I–4 After W. Troll, 1937, *Vergleichende Morphologie der Höheren Pflanzen*, vol. 1, pt. 1, Verlag von Gebrüder Borntraeger, Berlin

I–5 J. H. Gerard

I–6 L. M. Stone, Bruce Coleman, Inc.

I–8 E. S. Ross

I–9 *(a)* Wide World Photos; *(b)* International Rice Research Institute

I–10 Grant Heilman

1–1 M. A. Walsh

1–2 A. Ryter

Essay, page 18, Rare Book Division, The New York Public Library

1–5 *(a)* D. Branton

1–6 After W. Braune, A. Leman, and H. Taubert, 1967, *Pflanzenanatomisches Praktikum*, VEB Gustav Fischer Verlag, Jena

1–7 After D. Branton and R. B. Park, 1967, *Journal of Ultrastructural Research*, **19**:283–303

1–9 R. B. Park

1–12 After E. G. Cutter, 1978, *Plant Anatomy*, pt. 1, Cells and Tissues, 2nd ed., Addison-Wesley Publishing Company, Reading, Mass.

1–13 S. E. Eichhorn

1–14 J. W. Perry

Essay, page 25, P. Echlin

1–17 S. E. Eichhorn

1–19 R. R. Dute

1–21 M. Kruatrachue and R. F. Evert, 1977, *The American Journal of Botany*, **64**:310–25

1–22 After D. J. Morré and H. H. Mollenhauer, 1974, in *Dynamic Aspects of Plant Ultrastructure*, A. W. Robards (Ed.), McGraw-Hill Book Company, New York

1–24 R. R. Powers

1–26 H. J. Hoops and G. L. Floyd

1–27 R. D. Preston

1–29 After K. Esau, 1977, *Anatomy of Seed Plants*, 2nd ed., John Wiley and Sons, Inc., New York

1–31 After R. D. Preston, 1974, in Robards, *op. cit.*

1–37 A. S. Bajer, University of Oregon

1–39 R. R. Dute

1–40 R. R. Dute

1–41 J. Cronshaw

1–42 R. F. Evert and B. P. Deshpande, 1970, *The American Journal of Botany*, **57**:942–61.

1–43 P. K. Hepler

1–44 After Esau, *op. cit.*

2–4 *(e)* L. M. Beidler

2–9 B. E. Juniper

3–1 W. G. Whaley

3–2 From H. Curtis, 1979, *Biology*, 3rd ed., Worth Publishers, Inc., New York

3–3 R. Winch, Photo Researchers

3–4 From Curtis, *op. cit.*

3–5 After A. L. Lehninger, 1975, *Biochemistry*, 2nd ed., Worth Publishers, Inc., New York

3–8 After S. J. Singer and G. L. Nicolson, 1972, *Science*, **175**:720–31; copyright 1972 by the American Association for the Advancement of Science

3–9 After J. L. Hall and D. A. Baker, 1977, *Cell Membranes and Ion Transport*, Longman, Inc., New York

3–10 From Curtis, *op. cit.*

3–11 K. B. Raper

4–1 Hale Observatories

4–2 After Curtis, *op. cit.*

4–3 After Lehninger, *op. cit.*

Essay, page 76, L. Jacobi

4–4 H. Wright, Freelance Photographers Guild

4–5 From Curtis, *op. cit.*

4–6 After Lehninger, *op. cit.*

4–7 From Curtis, *op. cit.*

4–9 From Curtis, *op. cit.*

5–1 R. D. Warmbrodt

5–6 After A. J. Vander, J. N. Sherman, and D. S. Luciano, 1969, *Human Physiology*, McGraw-Hill Book Company, New York

5–7 After Lehninger, *op. cit.*

Essay, page 91, after P. C. Hinkle and R. E. McCarty, 1978, *Scientific American*, **238**:104–23

Essay, page 93, Y. Haneda

5–12 *(b)* Grant Heilman

6–2 R. L. Gherna, American Type Culture Collection

Essay, page 99, Cambridge University Library

6–3 H. Towner

6–4 From Curtis, *op. cit.*

6–7 Prepared by Govindjee

6–10 D. Branton

6–13 S. G. Pallardy and T. T. Kozlowski, 1980, *New Phytologist*, **85:**363–68

7–1 A. Sparrow

7–4 *(a)* From J. D. Watson, 1968, *The Double Helix*, Atheneum, New York

7–5 J. Cairns

7–6 From Curtis, *op. cit.*

7–7 From Curtis, *op. cit.*

7–9 S. H. Kim, 1973, *Science*, **179:**285–88; copyright 1973 by the American Association for the Advancement of Science

7–10 O. L. Miller, Jr., B. A. Hamkalo, and C. A. Thomas, Jr., 1970, *Science*, **169:**392–95; copyright 1970 by the American Association for the Advancement of Science

7–14 *(a)* through *(c)* K. B. Raper; *(d)* J. T. Bonner

8–1 V. Orel, The Moravian Museum, Brno, Czechoslovakia

8–2 *(a)* C. F. Woodcock; *(b)* S. Brown, from U. Goodenough, 1978, *Genetics*, 2nd ed., Holt, Rinehart and Winston, Inc., New York

8–3 P. B. Moens

8–4 B. John

8–5 G. Östergren

8–6 Photographs by W. Tai

8–11 H. Towner

8–13 C. G. G. J. van Steenis

9–1 Radio Times Hulton Picture Library

9–5 After K. Mather and B. J. Harrison, 1959, *Heredity*, **3:**1–52

9–6 After S. Carlquist, 1965, *Island Life*, The Natural History Press, Garden City, N.Y.

9–7 A. P. Nelson

Essay, page 155, *(right)* W. H. Hodge, Peter Arnold, Inc.

9–9 B. Crandall, Carnegie Institution of Washington, Publication 540

9–10 From Carlquist, *op. cit.*

9–12 M. A. Nobs, Carnegie Institution of Washington, Publication 623

9–13 D. Myrick, Jepson Herbarium, University of California, Berkeley

9–14 Color photographs by H. Angel; *(a)* through *(c)* C. J. Marchant

10–1 G. J. Breckon

10–2 Burndy Library, Norwalk, Conn.

10–3 *(a)* L. West; *(b)* R. Carr, Bruce Coleman, Inc; *(c)* E. S. Ross

10–4 *(a)*, *(c)* E. V. Gravé; *(b)* H. Forest

10–5 *(a)* C. Blacklock; *(b)* M. P. Godomski, National Audubon Society Collection/Photo Researchers; *(c)* E. R. Degginger, Bruce Coleman, Inc.; *(d)* L. Blacklock; *(e)* L. West

10–6 *(a)* E. S. Ross; *(b)* D. S. Neuberger; *(f)* S. M. Carpenter; *(c)* W. H. Amos, Bruce Coleman, Inc.; *(d)* L. E. Graham; *(e)* Oxford Scientific Films, Animals Animals; *(g)* Runk/Schoenberger, Grant Heilman; *(h)*, *(i)* E. V. Gravé; *(j)* D. P. Wilson/Eric and David Hosking

10–7 *(a)*, *(d)* R. Carr, Bruce Coleman, Inc.; *(b)*, *(i)* J. W. Perry; *(c)* E. S. Ross, Bruce Coleman, Inc.; *(e)* J. Shaw, Bruce Coleman, Inc.; *(f)* through *(h)* J. Dermid; *(j)* E. Beals

10–8 *(a)* M. Jost; *(b)* L. A. Staehelin

Essay, page 179, *(top)* J. Dermid; *(bottom)* Grant Heilman

10–9 After G. L. Stebbins, 1960, in *The Evolution of Life*, vol. 1, S. Tax (Ed.), The University of Chicago Press, Chicago, Ill.

11–1 L. D. Simon

11–2 J. W. Schopf

11–3 *(a)* P. L. Grilione and J. Pangborn, 1975, *Journal of Bacteriology*, **124:**1558; *(b)* National Medical Audiovisual Center; *(c)* H. Lechevalier; *(d)* R. S. Wolfe

11–4 *(a)*, *(b)* D. Greenwood; *(c)* J. L. Pate

11–5 *(a)* E. V. Gravé; *(c)* Turtox/Cambosco; *(d)* E. J. Ordal

11–6 D. A. Cuppels and A. Kelman, 1980, *Phytopathology*, **70:**1110–15

11–7 H. Stolp and M. P. Starr, 1963, *Antoine van Leeuwenhoek*, **29:**217–48

11–8 C. C. Brinton, Jr., and J. Carnahan

11–9 C. Robinow

11–10 R. D. Warmbrodt

11–11 R. D. Warmbrodt

11–12 A. Berkaloff, J. Bourguet, P. Favard, and M. Guinnebault, 1967, *Introduction à la Biologie: Biologie et Physiologie Cellulaire*, Hermann Collection, Paris

11–14 J. Lederberg and E. M. Lederberg

Essay, page 194, R. Crane, *Life Magazine*, © 1980 Time, Inc.

Essay, page 195, Department of Plant Pathology, Cornell University

11–15 E. H. Newcomb

11–16 Walter Reed Army Institute of Research

11–17 After M. C. Pelczar and R. D. Reid, 1965, *Microbiology*, McGraw-Hill Book Company, New York

11–18 *(a)* Parke, Davis & Co.; *(b)* after R. Y. Stanier, M. Doudoroff, and E. A. Adelberg, 1963, *The Microbial World*, 2nd ed., Prentice-Hall, Inc., Englewood Cliffs, N.J.

11–19 National Medical Audiovisual Center

11–20 W. A. Niering

11–21 After G. N. Agrios, 1978, *Plant Pathology*, 2nd ed., Academic Press, Inc., New York

Essay, pages 202 and 203, *(a)* J. F. Worley, U. S. Department of Agriculture; *(b)* M. V. Parthasarathy; *(c)*, *(d)* K. Maramorosch

11–22 From Lehninger, *op. cit.*

11–23 J. D. Almeida and A. F. Howatson, 1963, *Journal of Cell Biology*, **16:**616–20

Essay, page 206, National Communicable Disease Center

11–24 *(a)* L. D. Simon; *(b)* K. Allen

11–25 *(a)* G. Gaard and R. W. Fulton; *(b)*, *(c)* G. Gaard and G. A. deZoeten

11–26 *(a)* K. Maramorosch; *(b)* E. Shikata; *(c)* E. Shikata and K. Maramorosch

Essay, page 209, Th. Koller and J. M. Sogo, Swiss Federal Institute of Technology, Zürich

12–1 J. Dermid

12–2 R. E. Hutchins

12–3 E. V. Gravé

12–5 M. D. Coffey, B. A. Palevitz, and P. J. Allen, 1972, *The Canadian Journal of Botany*, **50:**231–40

12–6 H. C. Hoch and D. P. Maxwell

12–9 After T. Delevoryas, 1966, *Plant Diversification*, Holt, Rinehart and Winston, Inc., New York

Essay, page 218, drawing from A. H. R. Buller, *Researches on Fungi*, vol. 6, Longman, Inc., New York; *(a)*, *(b)* E. V. Gravé; *(c)* R. M. Page

12–10 *(a)* C. Bracker; *(c)* J. Cooke, 1969, *The American Journal of Botany*, **56:**335–40

12–12 After Delevoryas, *op. cit.*

Essay, page 222, U. S. Department of Agriculture

12–14 Bevilaqua/Prato, Agenzia Fotografica, Luisa Ricciarini, Milan

12–15 *(b)* W. A. Russin

Essay, page 224, D. Pramer, 1964, *Science*, **144:**382–88; copyright 1964 by the American Association for the Advancement of Science

12–16 L. West

12–17 *(a)* A–Z Collection, Photo Researchers; *(b)* E. S. Ross

12–18 *(a)* J. R. Clawson, Photo Researchers; *(b)* E. S. Ross; *(c)* L. West

12–19 *(b)* V. Ahmadjian

12–20 *(a)* A. E. Staffan; *(b)* J. Shaw; *(c)* G. J. Breckon

12–21 *(a)*, *(b)* J. Burton, Bruce Coleman, Inc.

12–22 *(a)* J. Burton, Bruce Coleman, Inc.; *(b)* J. Markham, Bruce Coleman, Inc.; *(c)*, *(d)* J. W. Perry

12–23 E. S. Ross

12–24 V. Ahmadjian

12–25 J. Keller

12–26 H. C. Hoch and R. J. Howard (in press) *Experimental Mycology*

12–27 *(b)* C. Heintz and D. Niederpruem

12–29 U. S. Department of Agriculture

12–30 R. G. Wasson

12–32 E. S. Ross

12–35 S. A. Wilde

12–36 F. W. Went

12–37 R. J. Molina, J. M. Trappe, and G. S. Strickler, 1978, *The Canadian Journal of Botany,* **56:**1691–95

12–38 B. Zak, U. S. Forest Service

12–39 R. D. Warmbrodt

13–1 Runk/Schoenberger, Grant Heilman

13–2 *(a)* A. W. Barksdale; *(b)* A. W. Barksdale, 1963, *Mycologia,* **55:**493–501

Essay, page 224 *(a), (b)* A. W. Barksdale; *(c)* A. W. Barksdale, 1963, *Mycologia,* **55:**627–32

13–4 After J. H. Niederhauser and W. C. Cobb, 1959, *Scientific American,* **200:**100–112

13–5 L. P. Gauriloff, 1978, in *Lower Fungi in the Laboratory,* M. S. Fuller (Ed.); copyright 1978 by Department of Botany, University of Georgia

13–7 K. B. Raper

13–8 *(c)* R. E. Hutchins; *(d)* Runk/Schoenberger, Grant Heilman

14–1 J. Dermid

14–2 D. P. Wilson/Eric and David Hosking

Essay, page 254, *(top)* L. E. Graham; *(bottom)* L. E. Graham and J. M. Graham, 1980, *Transactions of the American Microscopical Society,* **99:**160–66

14–3 G. L. Floyd

14–4 G. E. Palade

Essay, page 258, after J. L. Salisbury and G. L. Floyd, 1978, *Science,* **202:**975–77; copyright 1978 by the American Association for the Advancement of Science

14–8 L. E. Graham

14–9 D. L. Kirk

14–10 H. J. Marchant and J. D. Pickett-Heaps, 1972, *Australian Journal of Biological Science,* **25:**1199–213

14–11 *(a)* J. Dermid; *(b)* R. E. Hutchins

Essay, page 263, from R. F. Skagel *et al.*, 1966, *An Evolutionary Survey of the Plant Kingdom,* Wadsworth Publishing Company, Inc., Belmont, Calif.

14–13 G. L. Floyd, K. D. Stewart, and K. R. Mattox, 1971, *Journal of Phycology,* **7:**306–9

14–17 D. P. Wilson/Eric and David Hosking

14–18 Skagel *et al., op. cit.*

14–19 J. D. Pickett-Heaps

14–20 Carolina Biological Supply

14–21 E. V. Gravé

14–22 *(a)* L. S. Radford, from R. D. Wood and K. Imahori, 1964, *A Revision of the Characeae,* vol. 2, J. Cramer, Weinheim, Germany

14–23 D. P. Wilson/Eric and David Hosking

14–28 Skagel *et al., op. cit.*

14–29 *(a)* D. Longanecker; *(b)* W. R. Taylor, 1950, *Plants of Bikini and Other Northern Marshall Islands,* University of Michigan Press, Ann Arbor, Mich; *(c)* D. P. Wilson/Eric and David Hosking; *(d)* G. M. Smith, 1955, *Cryptogamic Botany,* vol. 2, *Bryophytes and Pteridophytes,* 2nd ed., McGraw-Hill Book Company, New York

Essay, page 275, A. Miura, Tokyo University of Fisheries

14–31 *(a)* G. D. Hanna, California Academy of Sciences; *(b)* F. Rossi; *(c)* G. A. Fryxell

14–33 G. A. Fryxell

14–36 *(a), (b)* D. P. Wilson/Eric and David Hosking; *(c)* A. R. Loeblich, III

14–37 D. Longanecker

Essay, page 282, L. E. Graham

Essay, page 283, D. F. Kubai and H. Ris

Essay, page 284, L. E. Graham

15–1 W. H. Hodge, Peter Arnold, Inc.

15–2 R. E. Hutchins

15–3 C. Hébant, 1975, *Journal of the Hattori Botanical Laboratory,* **39:**235–54

15–6 *(a), (b)* Smith, *op. cit.*; *(c)* R. E. Magill, Botanical Research Institute, Pretoria

15–8 A. R. Grove

Essay, page 294, after C. T. Ingold, 1939, *Spore Discharge in Land Plants,* Clarendon Press, Oxford

15–10 D. S. Neuberger

15–11 *(a), (b)* J. J. Engel, 1980, *Fieldiana, Botany* (New Series), **3:**1–229; *(c)* J. J. Engel

15–14 Turtox/Cambosco

15–15 D. S. Neuberger

15–16 *(a)* H. A. Thornhill, National Audubon Society Collection/Photo Researchers; *(b)* R. E. Hutchins

15–18 *(a)* L. E. Anderson; *(b)* L. West; *(c)* R. E. Hutchins; *(d)* Runk/Schoenberger, Grant Heilman

15–19 R. E. Magill, Botanical Research Institute, Pretoria

15–20 After Ingold, *op. cit.*

16–1 Field Museum of Natural History

16–2 *(a)* After D. Edwards, 1970, *Palaeontology,* **13:**451–61; *(b)* after H. C. Bold, 1973, *Morphology of Plants,* 3rd ed., Harper & Row, Publishers, New York

16–3 *(a)* After J. Walton, 1940, *Fossil Plants,* Macmillan, Inc., New York; *(b)* after J. Walton, 1964, *Phytomorphology,* **14:**155–60; *(c)* after F. M. Heuber, 1968, in *International Symposium on the Devonian System,* vol. 2, D. H. Oswald (Ed.)

16–4 After H. P. Banks, 1970, *Evolution and Plants of the Past,* Wadsworth Publishing Company, Inc., Belmont, Calif.

16–5 After Troll, *op. cit.*

16–7 After K. K. Namboodiri and C. B. Beck, 1968, *The American Journal of Botany,* **55:**464–72

16–8 After Smith, *op. cit.*

16–10 After Smith, *op. cit.*

16–11 From H. N. Andrews, 1963, *Science,* **142:**925–31; copyright 1963 by the American Association for the Advancement of Science; after A. G. Long, 1960, *Transactions of the Royal Society of Edinburgh,* **64:**29–44, 201–15, and 261–80

16–12 After Long, *op. cit.*

16–13 J. M. Pettitt and C. B. Beck, 1968, *Contributions from the Museum of Paleontology, University of Michigan,* **22:**139–54; *(c)* J. M. Pettitt and C. B. Beck, 1967, *Science,* **156:**1727–29; copyright 1967 by the American Association for the Advancement of Science

17–1 After P. Kukuk, 1938, *Geologie des Niederrheinisch-Westfälischen Steinkohlengebietes,* Verlag von Julius Springer, Berlin

17–6 *(a)* D. Cameron; *(b)* R. Schmid

17–10 *(a)* D. S. Neuberger; *(b)* R. Carr, Bruce Coleman, Inc.

17–13 *(b)* D. S. Neuberger

17–19 After A. J. Eames, 1936, *Morphology of Vascular Plants: Lower Groups,* McGraw-Hill Book Company, New York

17–20 *(a)* D. S. Neuberger

17–23 *(a)* R. E. Hutchins

17–24 After A. S. Foster and E. M. Gifford, Jr., 1974, *Comparative Morphology of Vascular Plants,* 2nd ed., W. H. Freeman and Company, San Francisco

Essay, pages 330 and 331, *(a)* after M. Hirmer, 1927, *Handbuch der Paläobotanik,* vol. 1, Druck and Verlag von R. Oldenbourg, Munich and Berlin; *(b)* W. N. Stewart and T. Delevoryas, 1956, *Botanical Review,* **22:**45–80; *(c)* after Banks, *op. cit.*

17–26 *(a)* C. Schoepf; *(b)* W. H. Hodge, Peter Arnold, Inc.; *(c)* American Fern Society, Wherry Collection; *(e)* E. S. Ross

17–29 J. Dermid

17–30 C. Niedorf

17–33 I. Mikukami and J. Gall, 1966, *Journal of Cell Biology*, **29:**97–111

18–1 After C. B. Beck, 1962, *The American Journal of Botany*, **49:**373–82

18–2 C. B. Beck, 1970, *Biological Reviews*, **45:**379–400

18–3 After S. E. Scheckler, 1975, *The American Journal of Botany*, **62:**923–34

18–4 After C. B. Beck, 1971, *The American Journal of Botany*, **58:**758–84

18–5 J. Dermid

18–6 *(a)* J. Dermid

18–7 *(a)* R. Spurr and J. Spurr, Bruce Coleman, Inc.

18–8 J. Kummerow

18–11 W. M. Harlow

18–12 W. A. Russin

18–13 *(c)* G. J. Breckon

18–15 *(a)* W. A. Russin

18–20 Photograph by J. Dermid

18–21 W. H. Hodge, Peter Arnold, Inc.

18–22 J. W. Perry

18–23 H. H. Iltis

18–24 Grant Heilman

18–25 *(a)* L. West; *(b)* J. Burton, Bruce Coleman, Inc.

18–26 G. Ahrens, Bruce Coleman, Inc.

18–27 Sichuan Institute of Biology

18–28 Photograph, Carolina Biological Supply

18–29 D. A. Steingraeber

18–30 *(b)* J. W. Perry

18–31 *(a)* J. W. Perry; *(b)* Runk/Schoenberger, Grant Heilman

18–33 *(a)* J. Dermid; *(b)* W. H. Hodge, Peter Arnold, Inc.

18–34 C. H. Bornman

18–35 E. S. Ross

18–36 From P. Maheshwari, 1950, *An Introduction to the Embryology of the Angiosperms*, McGraw-Hill Book Company, New York

18–37 *(a)* J. Dermid; *(b)* D. S. Neuberger; *(c)* R. Carr

18–38 *(a)*, *(b)* G. J. Breckon; *(c)* E. S. Ross

18–39 *(a)* D. S. Neuberger; *(b)* E. S. Ross; *(c)* E. R. Degginger, Earth Scenes

18–41 *(a)*, *(c)* L. West; *(b)*, *(e)* J. H. Gerard; *(d)* Grant Heilman

18–44 E. S. Ross

18–45 *(a)* Runk/Schoenberger, Grant Heilman; *(c)* L. West

18–47 After Foster and Gifford, *op. cit.*

18–48 *(a)* P. Echlin; *(b)*, *(c)* J. Heslop-Harrison and Y. Heslop-Harrison; *(d)* J. Mais

19–1 *(a)* E. S. Ross; *(b)* D. L. Dilcher

19–2 E. Dorf

19–4 *(a)* J. H. G. Johns, New Zealand Forest Service; *(b)* W. H. Hodge, Peter Arnold, Inc.

Essay, page 376, D. L. Dilcher, Paleobotanical Laboratory, Indiana University

19–5 After I. W. Bailey and B. G. L. Swamy, 1954, in *Contributions to Plant Anatomy*, I. W. Bailey (Ed.), Ronald Press, New York

19–6 After G. H. M. Lawrence, 1951, *Taxonomy of Vascular Plants*, The Macmillan Company, New York

19–8 *(a)*, *(b)* Populi and Paesi, Agenzia Fotografica, Lùisa Ricciarini, Milan; *(c)* G. J. Breckon

19–10 *(a)*, *(c)* E. S. Ross; *(b)* J. W. Perry

19–11 *(b)* A. Sabarese; *(c)* E. S. Ross; *(d)* J. Dermid

19–12 *(a)* E. S. Ross

Essay, pages 382 and 383, J. Heslop-Harrison

19–13 L. West

19–14 *(a)* J. Dermid; *(b)* L. West; *(c)* E. S. Ross

19–15 E. S. Ross

19–16 L. West

19–17 T. Eisner

19–18 E. S. Ross

19–19 *(a)*, *(c)* E. S. Ross; *(b)* L. B. Thien

19–20 *(a)* L. West; *(b)* E. S. Ross

19–21 M. P. L. Fogden, Bruce Coleman

19–22 R. A. Tyrell

19–23 Oxford Scientific Films

19–24 D. J. Howell

19–25 *(a)*, *(b)* T. Hovland, Grant Heilman; *(c)* E. S. Ross; *(d)* J. F. Skvarla, University of Oklahoma

19–27 D. S. Neuberger

19–29 After Skagel *et al.*, *op. cit.*

19–30 *(a)* Runk/Schoenberger, Grant Heilman; *(b)* W. Harlow, Photo Researchers

19–31 U. S. Forest Service

19–32 After Skagel *et al.*, *op. cit.*

19–34 J. Dermid

19–35 K. H. Maslowski, National Audubon Society Collection/Photo Researchers

19–36 After Skagel *et al.*, *op. cit.*

19–38 H. Harrison, Grant Heilman

19–39 *(a)*, *(c)* W. H. Hodge, Peter Arnold, Inc.; *(b)* Peter Arnold

20–1 J. Dermid

Essay, page 409, Grant Heilman

20–6 After Foster and Gifford, *op. cit.*

21–1 R. D. Warmbrodt and R. F. Evert, 1979, *The American Journal of Botany*, **66:**412–40

21–5 M. C. Ledbetter and K. R. Porter, 1970, *Introduction to the Fine Structure of Plant Cells*, Springer-Verlag, New York

21–12 H. A. Core, W. A. Côté, and A. C. Day, 1979, *Wood: Structure and Identification*, 2nd ed., Syracuse University Press, Syracuse, N. Y.

21–13 *(a)* I. B. Sachs, U. S. Forest Products Laboratory; *(b)* J. Heslop-Harrison

21–17 *(a)*, *(b)* R. F. Evert, W. Eschrich, and S. E. Eichhorn, 1973, *Planta*, **109:**193–210

21–18 J. S. Pereira

21–19 *(a)* M. A. Walsh

21–20 *(c)* After K. Esau, 1965, *Plant Anatomy*, 2nd ed., John Wiley and Sons, Inc., New York

22–1 After Braune, Leman, and Taubert, *op. cit.*

22–2 *(a)* L. M. Chace, National Audubon Society Collection/Photo Researchers; *(b)* J. Dermid

22–5 F. A. L. Clowes

22–11 H. T. Bonnett, Jr., 1968, *Journal of Cell Biology*, **37:**199–205

22–12 After Braune, Leman, and Taubert, *op. cit.*

22–14 J. Dermid

22–15 D. H. Franck

22–17 D. H. Franck

23–3 M. E. Gerloff

23–6 *(c)* W. Eschrich

23–10 After A. Fahn, 1974, *Plant Anatomy*, 2nd ed., Pergamon Press, New York

23–13 After Esau, *op. cit.*, and G. S. Avery, Jr., 1933, *The American Journal of Botany*, **20:**565–92

23–15 R. D. Meicenheimer

23–24 T. Pray, 1954, *The American Journal of Botany*, **41:**659–70

23–25 *(a)* J. Dermid; *(b)* T. Pray

Essay, page 465, *(top)* H. C. Jones, Tennessee Valley Authority; *(bottom)* J. S. Jacobson and A. C. Hill (Eds.), 1970, *Recognition of Air Pollution Injury to Vegetation: A Pictorial Atlas*, Air Pollution Control Association, Pittsburgh, Pa.

23–28 After J. B. Payer, 1857, *Traité D'Organogénie Comparée de la Fleur*, Librairie de Victor Masson, Paris

23–33 D. H. Franck

23–34 *(b)* J. Dermid

23–33 D. H. Franck

24–23 I. B. Sachs, Forest Products Laboratory, U. S. Department of Agriculture

24–26 *(a)* U. S. Forest Service; *(b)* C. W.

Ferguson, Laboratory of Tree-Ring Research, University of Arizona

24–27 Core, Côté, and Day, *op. cit.*

24–28 *(c)* I. B. Sachs, Forest Products Laboratory, U.S. Department of Agriculture

25–1 M.B. Winter

25–5 U. S. Department of Agriculture

25–6 J. P. Nitsch, 1950, *The American Journal of Botany* **37**:211–15

Essay, page 506, G. Melchers, M. D. Sacristán, and A. A. Holder, 1978, *Carlsberg Research Communications*, **43**:203–18

25–8 After T. Murashige and F. Skoog, 1962, *Physiologia Plantarum*, **15**:473–97

25–9 F. Skoog and C. O. Miller, 1957, *Symposia of the Society for Experimental Biology*, **11**:118–31

25–11 S. W. Wittwer

25–12 H. Towner

25–13 S. W. Wittwer

25–14 Abbott Laboratories

25–15 J. van Overbeek, 1966, *Science*, **152**:721–31; copyright 1966 by the American Association for the Advancement of Science

25–16 J. E. Varner

26–2 After B. E. Juniper, 1976, *Annual Review of Plant Physiology*, **27**:385–406

26–3 R. F. Trump, Photo Researchers

26–4 After A. W. Galston, 1968, *The Green Plant*, Prentice-Hall, Inc., Englewood Cliffs, N. J.

26–5 *(a)* A. Loeblich, III; *(b)*, *(c)* after B. Sweeney, 1969, *Rhythmic Phenomena in Plants*, Academic Press, New York

26–6 After A. W. Naylor, 1952, *Scientific American*, **186**:49–56

26–7 After P. M. Ray, 1963, *The Living Plant*, Holt, Rinehart and Winston, Inc., New York

26–8 U. S. Department of Agriculture

26–12 U. S. Department of Agriculture

26–13 A. Lang, M. Kh. Chailakhyan, and I. A. Frolova, 1977, *Proceedings of the National Academy of Sciences*, **74**:2412–16

26–14 After Naylor, *op. cit.*

26–15 A. E. Porsild, C. R. Harrington, and G. A. Mulligan, 1967, *Science*, **158**:113–14; copyright 1967 by the American Association for the Advancement of Science

26–16 J. Dermid

26–17 J. Dermid

26–18 Grant Heilman

27–1 The Nitragin Company, Inc.

27–2 *(a)* J. H. Gerard, National Audubon Society Collection/Photo Researchers; *(b)* R. E. Hutchins

27–3 From J. Bonner and A. W. Glaston, 1952, *Principles of Plant Physiology*, W. H. Freeman and Company, San Francisco

27–4 R. H. Wright, National Audubon Society Collection/Photo Researchers

27–6 Standard Oil Company

27–7 After B. S. Meyer *et al.*, 1973, *Introduction to Plant Physiology*, D. Van Nostrand Company, New York

27–8 U. S. Department of Agriculture

27–10 *(a)* Carolina Biological Supply; *(b)* R. E. Hutchins

Essay, page 547, Y. Heslop-Harrison

27–11 *(a)* The Nitragin Company, Inc.; *(b)*, *(c)* Oxford University Botany School

27–12 P. J. Dart

27–13 R. R. Herbert, R. D. Holsten, and R. W. F. Hardy, E. I. duPont de Nemours Co.

27–16 G. C. Gerloff and W. H. Gabelman (in press), "Inorganic Plant Nutrition," in *Encyclopedia of Plant Physiology* (New Series), vol. 12, A. Läuchli and R. L. Bieleski (Eds.), Springer-Verlag, New York

27–17 After W. J. Brill, 1977, *Scientific American*, **236**:68–81

28–1 Grant Heilman

28–2 *(a)* Jack Dermid

28–4 After M. Richardson, 1968, *Translocation in Plants*, Edward Arnold Publishers, Ltd., London

28–5 *(a)* After Richardson, *op. cit.*

28–6 After A. C. Leopold, 1964, *Plant Growth and Development*, McGraw-Hill Book Company, New York

28–7 After Richardson, *op. cit.*

28–8 After P. F. Scholander, H. T. Hammel, E. D. Bradstreet, and E. A. Hemmingsen, 1965, *Science*, **148**:339–46; copyright 1965 by the American Association for the Advancement of Science

28–9 After M. H. Zimmerman, *Scientific American*, **208**:132–42, March 1963

28–10 After Zimmerman, *op. cit.*

28–11 *(a)* L. M. Beidler; *(b)* J. H. Troughton

28–12 After K. Raschke, 1979, in *Encyclopedia of Plant Physiology* (New Series), vol. 7, W. Haupt and M. E. Feinleib (Eds.), Springer-Verlag, New York

28–17 *(a)* M. H. Zimmerman

Essay, page 572, *(b)*, *(c)* E. Fritz

29–1 U. S. Forest Service

29–2 A. S. Thomas

29–3 After J. L. Harper, 1961, *Symposia of the Society for Experimental Biology*, **15**:1–39

29–4 C. H. Muller, 1966, *Bulletin of the Torrey Botanical Club*, **93**:332–51

29–5 J. Mann, Australian Department of Lands

29–6 H. Harrison, Grant Heilman

29–7 D. H. Janzen

Essay, page 586, R. W. Gibson, Rothamsted Experimental Station, England

29–8 G. E. Likens

29–9 J. H. Gerard, National Audubon Society Collection/Photo Researchers

29–10 After J. Phillipson, 1966, *Ecological Energetics*, Edward Arnold Publishers, Ltd., London

29–11 After G. G. Simpson and W. S. Beck, 1965, *Life: An Introduction to Biology*, 2nd ed., Harcourt, Brace, Jovanovich, Inc., New York

29–12 *(a)*, *(c)*, *(d)* L. West; *(b)* J. Dermid

29–13 L. West

29–14 *(a)* U. S. Forest Service; *(b)* J. Dermid

29–15 *(a)* J. Dermid; *(b)* U. S. Forest Service

29–16 J. Dermid

29–17 U. S. Forest Service

30–1 D. Brokaw

30–2 From Curtis, *op. cit.*

30–3 S. Carlquist

Essay, page 597, The Bettmann Archive

Essay, page 598, R. Schoer

30–4 *(a)*–*(c)* E. S. Ross; *(d)*, *(e)* E. Beals

30–5 From Curtis, *op. cit.*

30–6 J. Van Wormer, Bruce Coleman, Inc.

30–7 From Curtis, *op. cit.*

30–8 E. S. Ross

30–9 U. S. Department of Agriculture

30–10 From Curtis, *op. cit.*

30–11 *(a)* L. C. Hurlbert, Kansas State University; *(b)* M. Travis; *(c)* J. L. Dodd

30–12 *(a)* L. West; *(b)* J. Dermid

30–13 From Curtis, *op. cit.*

30–14 *(a)* J. Dermid; *(b)* L. Blacklock; *(c)* J. Dermid; *(d)* K. Gunnar, Bruce Coleman, Inc.

30–15 and 30–16 E. Beals

Essay, page 609, *(top)* E. S. Ross; *(bottom, left)* K. M. Nagata; *(bottom, right)* E. S. Ayensu, National Museum of Natural History, Smithsonian Institution

30–17 J. Bartlett and D. Bartlett, Bruce Coleman, Inc.

30–19 *(a)* C. Blacklock; *(b)* E. Beals

30–20 From Curtis, *op. cit.*

30–21 E. Beals

A–1 through A–4 After Curtis, *op. cit.*

A–5 *(a)* From D. M. Callewaert and J. Genyea, 1980, *Basic Chemistry*, Worth Publishers, Inc., New York; *(b)* D. Scharf, Peter Arnold, Inc.

A–6 through A–10 After Curtis, *op. cit.*

A–12 After Lehninger, *op. cit.*

A–13 through A–15 From Curtis, *op. cit.*

Index*

ABA (abscisic acid), 512
Abies, 350, 495, 610 (*see also* firs)
  balsamea, **350, 591**
  concolor, 591
abiotic factors, 579, 587, 594
abscisic acid (ABA), 512, 517, 519, 530, 533
  and stomatal movements, 567
abscisin, 512
abscission, 466, 475, 505, 507, 511, 533
  and auxin, 505, 507, 517
  and ethylene, 511, 517
  zone, 466, 475, 507
absolute temperature, 77, 83
absorption,
  nutritive mode, 177–78
  by roots, of inorganic nutrients, 568–69
  by roots, of water, 560–63
  spectra, 101–2, **525**
Acacia (acacia), 582, **601**
  -ant symbiosis, 582, **584,** 585, 594
  bull's-horn, 582, **584,** 585, 594
  collinsii, **584**
  cornigera, 582, **584**
Acaena, **397**
accessory cell, **427,** 428
accessory pigments
  carotenoids, 103–4
  chlorophylls, 103
  phycobilins, 103–4
Acer, **395, 466,** 495
  negundo, 487
  rubrum, **591**
  saccharinum, **457**
  saccharum, **457**
Acetabularia (alga), 262
acetaldehyde, 94
acetyl CoA, 88–90, 92
N-acetylglucosamine, 214
acetyl group (CH_3CO), 88–89, **92**
achene, 394, **395,** 505
Achlya, 242
  ambisexualis, **242, 244**
acid phosphatase, 547
acid rain, 464–65, 556
acids, 624
  fatty, 46, 50–51
  hydrochloric (HCl), 624
acorn, 395, **404, 405**
Acrasiomycota (acrasiomycetes; *see* slime molds)
actinomycetes, **186,** 187, 551–52, 557
actinomycin, 190
action spectrum, 101–2
activation energy, 74
active membrane transport, 64, 66–67, 69, 568–69, 574
active site, 78–79
active solute absorption, 568
i^6 Ade, 508, 510
adenine, 56, **57,** 118–19, **120,** 121–22, 507–8
  nucleotide of, 56
  structure of, 80, 508
adenosine
  diphosphate (*see* ADP)
  monophosphate (*see* AMP)
  triphosphate (*see* ATP)
adenovirus, **205**
adhesion (adhesiveness), 564–65
Adiantum, **333, 334**
Adler, Julius, 190
adnation, 362, 380, 467
ADP, 56, 58, 82–95, 105, 548
aecia, aeciospores, 236
Aedes, **387**
aerobes, 552
aerobic conditions, 84
Aesculus
  hippocastum, **365, 447,** 477, **487**
  pavia, **458**
aethalium, 248
after-ripening, 413
Agallia constricta, 208
agar, 275, 502–3
Agaricus
  bisporus, 232
  campestris, 175, 179, 232
Agave, 475
agglutination by lectins, 550
Agraulis vanillae, **168, 169**
agricultural revolution, 9
agriculture, 549, 553–55, 557, 598, 604, 612
  primitive, **9**
Agrobacterium tumefaciens, 195
Agropyron, **391**
  cristatum, **137**
Agrostis tenuis, 111, 153, **438, 562**
A horizon (topsoil), 542, 544
air circulation and biomes, 595
air currents and transpiration, 568
air pollution, 288
air (intercellular) spaces, 438, 450, 452, 462–63, 475, 484–85, 489, 565
akinetes, 192
alanine, **52**
albuminous cell, **424,** 426, 431
alcohol, 196
  ethanol (ethyl), 94–95
  fermentation of, 94, 214, 221
  in lichens, 230
alder (*Alnus*), 198, 551
aldolase, 86
Aleurodiscus amorphus, 231
aleurone layer, **408,** 409, 516–17
Alexopoulos, C. I., 242
alfalfa, 433, **448,** 451, 538–39, 549
algae, 251–86, 589, 593 (*see also by name of group*)
  abundance of in soil, 187
  and bioluminescence, 521
  blooms of, 556
  brown, 130–31, **172,** 268–72, 287 (*see also* brown algae)
  cell walls of, 252–53
  chlorophylls of, 253
  and cyanobacteria compared, 283
  derivation of, 283
  economic uses of, 275
  food reserves of, 253 (*see also* laminarin, leucosin, paramylon)
  as food source, 259, 268, 275
  freshwater, 255–68
  general characteristics of, 252–53
  green, 1, 172, 225, **230,** 255–68, 287, 296–97 (*see also* green algae)
  as ancestors of plants, 282
  marine, 259, 262, **263,** 265, 268–74, **521**
  phytoplanktonic, 251–52
  red, 173, 272–74, 526 (*see also* red algae)
  symbiotic, 254
algin (alginic acid), 253, 270, 275
alkaloids, 397–98, 400
Allard, H. A., 522, 524
allele, 139–40, 146, 149
  dominant/recessive, 139–42, 146, 149
  homozygous/heterozygous, 139–40, 146, 150
allelopathy, 582, 594
Allium cepa, **19, 39, 407,** 414, **434,** 567
Allomyces, 180, 245, 265
  macrogynus, **246**
allosteric
  effector, 128
  interactions, 128
almonds, 515
Alnus (alder), 198, 551
  rubra, 422
Alopecurus pratensis, 383
alpha-amylase, 512, 516–17
alpha-carbon atom, 53
alpha-galacturonic acid, **49**
alpha-glucose monomers, **48, 49**
alpha helix, 54
alternation of generations, 180–81
  in ancestor of plants, 287
  in bryophytes, 289, 303
  in Chytridiomycota, 245–47
  heteromorphic, 181, 270, **271,** 273, 288, 303, 371
  isomorphic, 181, 245–46, 262, 265, **266,** 270, 273, **274**
  in vascular plants, 287–89, **311**
altitude-latitude relationship, 596
aluminum (Al), 543
Amanita, 232
  muscaria, **228**
  phalloides, **232**
Ambrosia, 522
  psilostachya, **365**
Amaryllis (amaryllis),
  belladonna, 388
  pink-flowered, 388
American elm, **42**
amination, 548, 557
amino acids, 51, **52,** 53, 58 (*see also by name*)
  in biochemical pathways, 79, 89, 548, 557
  and cytokinins, 510
  in DNA, 121
  and enzymes, 79
  formation of, 548, 557
  and genetic code, 123–26
  group, **52,** 53, 79
  and IAA, 503
  and phytochrome, **525**
  as protein components, 46, 51, 53–56, 58, 126, 541
  in protein synthesis, 46, 51, 53–56, 58, 123–26, 131
  residues, 53, **54,** 55
  sequence, 53–54, 58
  structure, **52,** 53
  synthesis, 124, **125, 127,** 128, 131, 537
amino-acyl-tRNA, 126
amino group ($-NH_2$), 548
ammonia (NH_3), 197, 546, 549, 553, 587
  conversion to nitrates, 197, 546, 553, 557
  oxidation of, 197, 546
ammonification, 197, 546, 557
ammonium ions (NH_4^+), 197, 540, 546, 548, 557, 587
amoebas, 67, **128,** 129
  and cellular slime molds, 247
  and cyanobacteria, 204
amoeboid organisms, **68,** 128–30, 247–49, 633
Amorphophallus titanum, 143
AMP, 82, 128–30
amylase, 56
amylopectin, **48**
amyloplasts, **45, 48,** 519
amylose, **48**
Anabaena, **192,** 551, **552**
  azollae, **191, 192**
  cylindrica, **176**
anabolism, 77–78, 83
anaerobes, 185, 210, 552
anaerobic conditions, 84–85
Ananas comosus, 112, 501, 567
anaphase
  meiotic, **116,** 117, 135, **136, 137**
  mitotic, 36, **37, 38,** 40–41
Andreaea, **301,** 303
Andreaeidae, 296
Andreaeobryum, 303
androecium, 295, 359, 378
Andropogon gerardi, **605**
Aneurophyton, 339
angiosperms (flowering plants), 312–13, 356–401, 590, 593, 595
  adaptations of, 373–74
  aquatic, 590
  evolution of, 372–401
  and gymnosperms compared, 356, 371
  life cycle of, 363–69, **370,** 371
  nongreen, **357**
  origin of, 372–76
  reproductive structures of, 139
  specialized families of, 379–82
  in the Tertiary period, 384, 607
Animal Kingdom, 178, 182
animals, 4, 585
  aquatic, small, 589, 593, 595, 597
  dependency on plants, 548
  in food chains, webs, 589
  grazing, 601, 604, 609, 612
  migratory, 8
  reproduction of, 180–81
  as seed distributors, 396–97, 400
  soil, 542–43, 545
anions (negatively charged ions), 540, 543, 548, 567, 569

* Numbers in **boldface** indicate illustrations

anisogamy, 258, **259**
annual rings, 493, 496
annuals, 4, 476, 513, 581, 601, **602,** 603–4, 606, 610
annular scars, 265
annular wall thickening, 423
annulus, 335
Antarctica, 185
Antarcto-Tertiary Geoflora, **375**
antelope,
pronghorn, **605**
antenna pigments, 104
anther, 359, 363, **364, 378, 379,** 382
antheridia
of algae, 265, 268, **272, 279**
of bryophytes, 288–89, **290, 293,** 295–96, **297,** 298, **299, 302,** 303
of *Equisetum,* 329
of ferns, **312,** 335, **336**
of fungi, 220, **221**
of *Lycopodium,* 321, **322**
of oomycetes, 242, **243,** 244
of *Psilotum,* **319**
of *Selaginella,* 324, **325, 326**
of vascular plants, 312
antheridiol, 244
antheridiophore, 292
Anthoceros, 296–97
Anthocerotopsida, 291, 296–97, 303, 634 (*see also* bryophytes; hornworts)
anthocyanins, 26, 302, 356–71, 392–93, 527, 533
Anthophyta, 339, 356–71 (*see also* angiosperms)
anthrax, 187
Anthreptes collarii, **389**
antibiotics
and bacteria, **186,** 187, 190, 200
fungal origin of, 224, 239
resistance to, 190
and viruses, 204
anticlinal divisions, 478, 495
anticodon, 124, **125,** 126, **127,** 131, 510
antigens, 550
Antilocapra americana, **605**
antipodals, 368
ants, 582, **584,** 585, 594
apex, vegetative to floral, 467, 475
aphids, 522, 586
-stylet method, 571
Apiaceae, **379,** 395
apical cell, 291, 298
apical dominance, 445, 503, 508
apical meristems, 4, 307, 412, 415–16, 445, 503, 514, 517
of embryo, 405, 415
of liverworts, 291
organization in shoot, 446
reproductive (floral), 414
of root, 307–8, 348, 405, 412, 415, 434–35
root and shoot, comparison of, 446
of shoot, 307–8, 342, 348, 405, 412, 415, 446–47, 475
vegetative, 414, 446
Apis mellifera, **385**
Apium
graveolens, **456,** 474
petroselinum, **566**
apomictic races, species, 162
apomixis, 154, 162–63, 165
apoplast, 561, 569
apoplastic transport (movement, pathway), 478
apothecia, 219, **228**
apple, 155, **363,** 393, 477, 485, 503, 514, 524, 527, 530
aquatic plants, 332
ferns, **332,** 335, 552
arabinose, **49**
Araceae, **384**
Arachis hypogaea, 471
arachnids, 594
Araucaria
araucana, 350
excelsa, 350
Araucarites longifolia, **373**
Arceuthobium, **395**
Archaeopteris, **338**
macilenta, **340**
Archebacteria, 197
archegonia, 288
of bryophytes, 288–89, **290, 293, 295,** 296, 298, **299, 302,** 303
of *Equisetum,* 329
of ferns, 312, 335, **336**
of gymnosperms, 340, 371
of *Lycopodium,* 321, **322**
of pine, 346, **347**
of *Psilotum,* **319**
of *Selaginella,* 324, **325, 326**
of vascular plants, 312
archegoniophore, **290,** 292
Archeosperma arnoldii, 314
arctic, alpine populations, 157, 162
arctic environments, 596
Arctium minus, **397**
Arctostaphylos, 159
Arcyria, **248**
arginine, **52,** 133
aril, 350, **351,** 396
Aristolochia, **484**
Aristotle, 96
arrowroot, 412
arrow-woods, **373**
arroyos, 603
Artemisia
biennis, 175
frigida, **605**
tridentata, **602**
Arthrobotrys dactyloides, 224
arthropods,
and chitin, 214
as soil components, 542
artichoke, Jerusalem, 509
ascidians, 196
Asclepiadaceae, 398, 473, **583** (*see also* milkweed)
Asclepias, 396
curassavica, 397
Asclera ruficollis, **384**
ascocarp, ascospores, 219, 221, 226, **228,** 239
Ascodesmis nigricans, **219**
ascogonium, 220–21
Ascomycota (ascomycetes), 219–22, 239, 580, 632
evolution of, 215
life cycle of, 221
ascus, 219–21, 239
asexual reproduction (*see* reproduction *and under specific organisms*)
ash, 394, **395,** 495, 512, 563
green, **458, 459, 487**
asparagine, **52**
asparagus, 472
Asparagus officinalis, 472
aspartate (aspartic acid), 52, 108–9, 112
aspen, quaking, **591**
Aspergillus, **213, 223,** 224
nidulans, 216
niger, **223**
oryzae, 222, 224
soyae, 222
Asplenium rhizophyllum, 155
assimilate transport, 569–74
Asteraceae (Compositae, composite), **365, 372,** 379–81, 394–95, 400
aster yellows, 203
atmosphere, 1, 2, 541, 545
atom (*see also specific elements*)
Böhr (planetary) model of, **617**
electrical charge of, 617–21
electronic structure of, 617–21
naturally occurring, 45, 615
oxidation of, 626
in oxidation-reduction, 626
structure of, 617–21
atomic composition of representative organisms, **45**
atomic number, 615–16
ATP, 22, 25, 56, 58, 67, 74, 81, 83–95, 105–6, 112, **128,** 532, 541, 548
Atriplex
patula, 111
rosea, 111
aureomycin, 187
Austrobaileya, **378**
auteocious state, 235
autoradiography, 121, 572
autotrophs, 2, 546, 587, 594
chemosynthetic, 546
auxin, 502–7, 517, 527 (*see also* IAA; 2,4-D; NAA)
and abscisic acid, 512
and abscission, 505, 507, 517
and apical dominance, 503
and cell differentiation, 504
and cell elongation, 507
and cell turgor, 63
chemical structure of, 513
and circadian rhythms, 520
-cytokinin interactions, 508–9
and ethylene, 511
and fruit growth, 505, 517
and geotropism, 519, 533
-gibberellin interactions, 504
inhibitory effects of, 503, 507, 517
and light response, 502, 518
naturally occurring, 503, 517
and phototropism, 518–19, 533
and plasma membrane interaction, 512
and root growth, 505
synthetic, **503,** 504, 507
and touch responses, 531
and vascular cambium, 479, 504
auxospores, 277
Avena, 502–3, 512
sativa, 11, 502
Avicennia germinans, 441, **442**
avocado, 511
Axelrod, Daniel, 373–74
axial core, 134
axial system, 477, 485, 489, 495
axil, leaf, 414
Azolla, 203, 551–52
caroliniana, **552**
Azotobacter, 552

baby blue eyes, **380**
Bacillariophyceae, 634 (*see also* diatoms)
bacilli, 187, 210
Bacillus, 202
cereus, 581
megaterium, **191**
bacitracin, 187
backcrossing, 140
bacteria, 1, 2, 16, **17,** 43, **68,** 102, 118, **119,** 184–204, 210, 631 (*see also by name*)
capsule of, 191
cell wall of, 190–91, 210
chemoautotrophic, 197, 210
chemosynthetic autotrophs, 546
chlorophyll in, 103, 196
chromosomes of, **119, 120,** 191–92, 210
colonies of (*see* colonies)
conjugation of, 189, 210
as decomposers, 187, 196, 210, **545,** 546, 548, **553**
ecological role of, 187, 197–204
evolution of, 185, 197
and fermentation, 94
filamentous, **186,** 188, 210
free-living, 552, 556–57
freshwater, 203
function of pigments in, 196
genetic recombination in, 178, 193
gliding, **186**
green sulfur, 103, 196
heterotrophic, 196, 199, 210
and human diseases, 199–200
lectins in, 550
lipopolysaccharides in, 190
methanogenic, 197
moldlike, 551
myxobacteria, 186
nitrifying, denitrifying (soil), 197, 546, 548, 553, 587
nitrogen-fixing (soil), 198, **536, 537,** 539, 548–52, 555, **556,** 557, 579, 594
pathogenic, 187, 199–201, 554
penicillinases, producers of, 581
photosynthetic, 97, 103, 196, 210, 552
phototaxis in, 190
and plant diseases, 200–201
purple, 103
purple nonsulfur, 196
purple sulfur, 97, 196
reproductive techniques of, 193
saprobic, 196
saprophytic (soil), 197, 546, 552 (*see also by name*)
sensory responses in, 190
structure of, 184, 187–91
symbiotic, 196, 199, 548–51, **552,** 555, 557
tuberculosis and, 199
virus infection of, 118, **119, 184,** 193
bacterialike organisms, 199
bacteriochlorophyll, 103
bacteriophage, 118, **119,** 193, 204, 206, **207,** 208
temperate, 206
bacteroids, 550, **551**
balsa, 489
bananas, 155, **357,** 388, 511
Bandeiraea simplicifolia, 550
banyan tree, 441
BAP, 508
bar, 545
defined, 60
Barbarea vulgaris, **538**
barberry, 235–36
bark, **343,** 484–89, 496, 585
of savanna trees, 601
of tropical rain forest trees, 598
Barksdale, Alma, 244
barley, 235, 512, 555, **571**
foxtail, **174**
seeds, 515–17
wild, 9
barley malt, 513
basal bodies, 30, **31,** 41, 258
basidiocarp, 231–32
Basidiomycota (basidiomycetes), 228–36, 240, 580, **595,** 632
evolution of, 215
life cycles of, 234–35
basidiospores, basidia, 229–32, 234, 240
basswood, **422** (*see also* linden)
Bateson, William, 144
bats as pollinators, 390, 400, 598
Bdellovibrio bacteriovorus, 189
Beagle, H.M.S., 147
bean, **43, 406,** 408, 414, 512–13, **520,** 526–27, 549, 554, **555** (*see also Phaseolus*)
broad, 572
castor, **406,** 414, 550
contender, 513
lima, **141,** 550
beeches, 433
silver, **375**
bees, 384–86, **387,** 393, 400, 522
beetles, 152, **372,** 382, 384, 398, 400, 586
beets, 369, 393
beggar-ticks, 380
Beggiatoa, 186
Begonia, **24**
bell-shaped curve, 145
Belt, Thomas, 585
Beltian bodies, **584,** 585
6-benzylamino purine (BAP), 508
Berberis vulgaris, 236
berry, 393
beta-carotene, 103, **104**
betacyanins, 393
beta-galactosidase, 126, **127**
betalins, **393**
Beta vulgaris, 369, 429, 443, **444**
Betula, 495, 610
papyrifera, **392,** 488
B horizon (subsoil), 542, 544
bicarbonates, 556
Bidens, 380
biennials, 432, 476, 514, 530
big tree, 350, 607
binomial system, 170, 175, 182
bioassay
for auxin, 503
for ethylene, 510
for gibberellin, 513
biochemical coevolution, 397–400

biological clocks, 521–22, 527, 533 (*see also* circadian rhythms)
  and cellular membranes, 522
biological control, 204, 245
biological revolution, 2
biological system of classification, 169–78, 182
biological time, 8
bioluminescence, 93, **521**
bioluminescent algae, **280, 521**
biomass, 588–89, 592, 594
biomes, 6, 595–612 (*see also under specific type*)
  defined, 579, 595
  distributions of, 597–98, **600, 602, 604, 606, 610, 611**
biosphere, 2, 8, 96, 169, 197, 251
biosynthesis of sucrose, 47
biosynthetic pathways, 77–83
biotic factors, 579, 587, 608
birches, 495, 610
  paper, **392,** 488
bird of paradise, **389**
birds, 581, 593
  as fruit-bearers, 396–97, 400
  as pollinators, **388,** 390, 400
bird's-foot trefoil, **536, 537**
bird's-nest fungi, **229**
bisexual flower, 382–83
bisexual plants, 292, 296, 298
bison, American, 604
bittersweet, 396
bivalents (chromosomes), 134–35, 146
blackberries, 155, 163, 465, 511
black-eyed Susan, 380
Blackman, F. F., 101–2
Blackmore, Richard, 190
bladder (seaweed), **269**
bladderwort, **546**
Blakea chlorantha, 598
bleeding sap, 508
blending inheritance theory, 149
blights, 187, 201
  chestnut, 219
  fire, 187
  potato, 244–45, 250
bloodroot, white-flowered, 607
bloom(s)
  of cyanobacteria, 191, 203
  on grapes, 94
  of red dinoflagellates, 280
bluebell, **360**
blueberries, 511
blue-green algae (*see* cyanobacteria)
blue light and stomatal movements, 567
body (somatic) cell, 346
bogs, 303, 548, 554, 580, 610
Böhr, Niels, 98, 617
bolete, **229**
Boletus elegans, 580
bolting, 514, **515,** 517
bonds
  carbon, 50, 83, 621
  carbon-hydrogen, 50, 83
  chemical, 47, 102, 619–22
  conjugated, 103
  covalent, 82, 620–22
  disulfide, 53, **54,** 79
  double, 620
  high-energy, 82–83, 85, 87, 95
  hydrogen (*see* hydrogen bonds)
  ionic, 619
  peptide, 53, 126, **127**
  polar covalent, 621
  phosphate, 82–85, 87, 105
  triple, 620
Bonner, James, 524, 527
Bordeaux mixture, 244
bordered pit, **34,** 35
Borlaug, Norman, 11
boron (B), 538–40, 557
  ions (BO_3^-; $B_4O_7^{2-}$), 540
Borthwick, Harry A., 525
Botanical Congress, International, 170
Botrychium, 333, **334**
  virginianum, **29, 334**
botulism, 200
Bougainvillea, 393
boxelder, 487
Boyer, Paul D., 91
Brachythecium, 301
bracts, **392**
bran, 409
branching
  dichotomous, 292, **293,** 305–6, 314
branch traces, 454
Brassica
  napus var. *napobrassica,* 568
  nigra, **397**
  oleracea, 163, 474, 514, **515**
Brassicaceae, 237, 394, 398, 506
Braun, Armin C., 195
Briggs, Winslow, 518
bristlecone pine, **494**
British soldiers, 226
Bromeliaceae, 112
bromeliads, 112
brown algae, 268–72 (*see also* algae)
  conducting tissue in, 268, 270
  evolution of, 270
  life cycles of, 270, **271, 272**
  tissue differentiation in, 268, **270**
Brucella, 199
Bruchidae, bruchids, 152
Bryidae, 296–302
Bryophyta (bryophytes), 178, 287–303, 634 (*see also* Anthocerotopsida; bryophytes; Hepaticopsida; Muscopsida)
bryophytes, 287–303, 634 (*see also* hornworts; liverworts; mosses)
  fossil, 287
  and green algae compared, 287, 296–97, 303
  and vascular plants compared, 287–88, 303
Bryum, **301**
buckeye, red, **458**
buckwheat, 394
bud(s), 5, 445, **447, 487,** 512
  adventitious, 155
  and auxin, 503, 507, 517
  and cytokinins, 507, 509, 517
  dormant, 529–30
  mixed, **447**
  primordia, 445–46
  terminal, 445, **447**
  winter, 512
budding, 220, 239
bugs, true, 398
bulbous plants, 601, 603
bulbs, 155, 474–75, 530
bulk flow, 59–60, 69
bulliform cell, **462**
bundle scar, **487**
bundle sheath,
  in leaf, **460, 461, 462,** 463, 475
  in stem, **451,** 452, **453**
bundle-sheath cells, **14, 15,** 108–9, **110,** 112
bundle-sheath extension, **462,** 463
burdock, **397**
Burgeff, H., 215
butter-and-eggs, 396
buttercup, 394, **436, 437, 439,** 451
  Bermuda, **370**
butterflies, **168,** 169, 384, 388, 398, 400, **583**
  painted-lady, **388**
  monarch, **583**
butternut, **487**
butterwort, 547
buttressed trees, 598, **599**
butyric acid, 354
cabbage, 20, 26, 163, 398, 474, 514, **515,** 527
cacao, **599**
Cactaceae, 111, 473
Cactoblastis cactorum, 582, **583**
cactus, 111, 356, 388, 393, **472, 473,** 567, 603 (*see also by name*)
  barrel, **602, 603**
  cereus, **473**
  Christmas, 207
  fishhook, **174**
  giant saguaro, 200, **358**
  moth, 582, **583**
  organ-pipe, **390**
  peyote, **399**
  prickly-pear, 582, **583**
cadmium, 556
caffeine, **397**
calactin, **397**
Calamites, **316, 317,** 327, 330
calcite ($CaCO_3$), 541
calcium (Ca), **49,** 538–41, 543, 557, 585
  carbonate, 272
  loss, 587
  oxalate, 24, **25,** 398
calcium ion (Ca^{2+}), 540, 543, 553, 569
  and auxin-cytokinin action, 509
callase, 364
Callistephus chinensis, 203
Callixylon newberryi, **339**
callose, 364, 383, **424,** 426
callus, 504, 507, 509
Calocedrus decurrens, **174,** 591, **592**
Calochortus tiburonensis, 609
caloric, 74
calories, 589
  in ATP, 82
  defined, 628
  in glucose, 76
  in pyruvate, 88
  released in glycolysis, 88
calorimeter, **628**
Caltha palustris, **386,** 393, **607**
Calvin cycle, 107–8, 110–12
Calvin, Melvin, 107
Calypte anna, 389
calyptra, **288,** 290–91, **302**
calyx, 359, **395, 468, 469, 470, 471**
cambial activity, 504
cambial initials, 477–78, 495
cambium
  cork, 308, 416, 428, 440, 476, 482–83, 485, 488, 495–96
  fascicular, **448,** 480
  interfascicular, **448,** 451, 480
  reactivation of, 479
  supernumerary, 443, **444**
  vascular, 4, 308, 440, 449–51, 476–80, 482, 484, 495–96, 504
camellia family, 609
CAM photosynthesis, 111–12, 473, 603
CAM plants, 111–12
  and stomatal behavior, 111, 567
cancer, 208–9 (*see also* tumors, plant)
Candida, 222
Cannabis sativa, 398, **399**
Cantharellus aurantiacus, **171**
Capsella bursa-pastoris, **410, 411**
Capsicum frutescens, 207
capsules, 394, **395** (*see also* sporangia)
  bacterial, 184, 191
  of bryophytes, **288,** 290, **291,** 294–95, 298, **300, 301,** 302–3
  gelatinous or mucilaginous, 184
carbohydrates, 46–50, 84, 550, 555, 588 (*see also by name*)
  formation of, 96
  in membrane structure, 64
  oxidation of, 84–95
  in seeds, 369
carbon (C), 1, 3, 45–46, 538, 540, 557, 615–16, 620–21, 626–27
  atom combinations, 620–21
  -containing compounds, 45, **620–21, 627**
  fixation of, 112
  14 (^{14}C), 47, 510, 615
  radioactive (^{14}C), 47, 615
  reduction of, 101, 106–8
carbonates, 556
carbon cycle, 212
carbon dioxide (CO_2), 4, 97, 100, 540, 556, 595
  in CAM plants, 111–12, 567, 603
  concentration and stomatal movement, 567
  in dark reactions, 106–8
  fixation of, 101, 107, 112
  in 4-carbon pathway, 108–9
  and Krebs cycle, 88–90
  radioactive, 616
  reduction of, 112
  in 3-carbon pathway, 107
Carboniferous period, 215, 304, 310, 316, 320, 327, 330–31, 333, 337, 341, 372
Carboniferous swamp forest, 304, 316–17
carboxyl group (COOH), 50, 79, 622
cardiac glycosides, **397,** 398, **583**
Carnegiea gigantea, 200, **358, 602, 603**
carnivores, 587, 589, 594
carnivorous plants, 472, 532–33, **546,** 547–48
carotenes, 103
carotenoids, 21–22, 44, **102,** 103, **104,** 112, 196, 203, 253, 287, 392–93, **365**
carpels, 359, 361, 376–78, **379,** 380–81, 393, 400, 467, **468, 469, 470, 471**
carpogonium, carposporangium, carpospores, 274
carrageenan, **273,** 275
carrier-assisted (mediated) transport, 66, 69, 568–69
carrier hypothesis, 66
carrier molecules, 66, 69
carrot, 130, 432, 442, 506, 514
Carya ovata, **458, 476, 488, 495**
caryopsis, 394
Casparian strip, 438–40, 444, 547, 561, 574
castor bean, **406,** 414, **423,** 550
catabolic reactions, 83
catabolism, 78
Catalpa, 495
catalysts, 56, 539
categories, taxonomic, 175
cation (positively charged ions; divalent and monovalent), 540, 543, 553, 567, 569, 619
  exchange, 543
catkins, **361, 392**
cattail, 358, **590**
Cattleya, 442
Ceanothus, 198, 551
  hybrid populations, **162**
cedar, incense, **174,** 591, **592**
Celastrus scandens, 396
celery, **456,** 474
cell(s), 14–44
  chemical components of, 45–58, 538–40, 557
  cycle, 36–44
  differentiation, 417, 504, 506
  and diffusion, 61–62
  division, 36, 41–42, 44, 417, 508, 512, 520, **521** (*see also* cytokinesis; mitosis)
  egg, 312, 409
  elongation, 503, 507–8, 511–12, 517
  eukaryotic, 14–44, 94
  eukaryotic and prokaryotic compared, 16, 43, 177
  expansion (growth), 541
  nutritive, 363
  plate, **37, 38,** 41–43
  pneumococcal, 118
  prokaryotic, 16, **17,** 25, 43, 175–76
  sap, 24, 26, 44, 392
  sperm, 30, 312
  sporogenous, 363
  structure, 15–44, **59,** 541, 543, 557 (*see also specific types*)
  theory, 15, 18
  three-dimensional diagram, **20**
  types, summary, 430–31
cell differentiation (*see also* cell)
  and auxins, 504, 506
  and hormonal interactions, 507, 509
cell division, 36, 41–43, 184, 508, 510, 512, 517 (*see also* cell; meiosis; mitosis)
  in algae, 255–56, 259, 264–66, 268, 276, **277,** 281, 285
  in bacteria, 192
  in bryophytes and vascular plants, 287
  and circadian rhythms, 520
  by furrowing, 255–56
cell elongation (*see also* cell)
  and hormones, 507, 512–13
cell membrane, 15, 522, 539, 541 (*see also* plasma membrane)

cell permeability, 539, 541, 557
cell-plate formation, 41–43
  in algae, 255–56, **264, 266,** 287
  in bryophytes and vascular plants, 287
cellular membranes, 16, 64, **65,** 69
cellulose, 32–35, 44–45, **49,** 58, 366
  in cellular slime molds, 247
  in cell wall structure, 32–34, 242, 247, 287
  crystalline and optical properties of, 32
  enzymatic digestion of, 47
  tube, 550
cell wall, **17,** 18, 32–35
  algal, 252–53, 255, 268, 270, 272, 276, 279
  bacterial, 190–91, 210
  of cellular slime molds, 247
  of chytrids, 245
  components of, 32–33
  of *Escherichia coli,* **17**
  fungal, 214
  growth of, 35
  of horsetails, 538
  layers of, 33–35, 44
  middle lamella of, 33, **34,** 44, 507, 510, 541, 543
  movement across/along, 478, 561, 574
  mucilaginous compounds in, 270, 272
  of parenchyma cell, **34,** 418
  and plant hormones, 507, 517
  primary, 33–35, 44
  of prokaryotes, 17, 190–91, 210
  secondary, 33–35, 44, 419–22, 440
  siliceous compounds in, 276
  of water molds, 242
Cenococcum geophilum, 580
central cell, 368
central mother cells, 446
central strand, 288, 298
centriole, 27, 41, 215, 242, 247, 252
centromere, **37,** 39–41, 44, 135, **136, 137**
century plant, 475
Cephalozia, 294
Cerambycidae, **372**
cerambycid beetles, 398
Ceratium, **279**
  tripos, **280**
Ceratocystis ulmi, 219
cereals/grains, 526, 530
C_4 photosynthesis, 108–9, 112, 506
C_4 plants, 108–9, 111, 539
Chaetomium, 213
  erraticum, **219**
Chaetomorpha, **32**
Chailakhyan, M. Kh., 527–28
chalazal end, 368
chalcone, 393
Chaney, Ralph, 350
chanterelle fungus, **171**
chaparral, 608
Chara, **172, 268**
Charales, 267–68
Chargaff, Erwin, 118
charge, electrical, 616–19
Charophyceae, 255–56, 266–68, 633
  cell division in, 255–56
charophytes, 266–68
cheese, 187, 222
chemical bond energy, 83, 85
chemical coupling hypothesis, 91
chemical defenses, 582, 594
chemical messengers (*see* hormones)
chemical reactions, 56, 58
chemiosmotic coupling hypothesis, 91
chemoautotroph, 197, 210
chemoreceptors, 190
chemosynthetic autotrophs, 546
chemotaxis, 190
Chenopodium glaucum, **453**
cherries, 26, 155, **362,** 393, 396–97, 511
chestnut, 530
  horse, **365**
chiasma, chiasmata, 134, **135, 136,** 143, 146
Chilton, Mary-Dell, 195
Chimaphila umbellata, 380
Chinese aster, 203
chitin, 50, 214, 245
Chlamydomonas, 175, 180–81, 253, 256–59, 285
  life cycle of, 257
Chlorella, 254, 259
chloride ions (Cl^-), 540, 547
chlorine (Cl), 538–40, 557, 567, 569, 615, **619**
Chlorophyceae, 255–65, 633
  cell division in, 255–56
chlorophyll(s), 1, 21, 44, 96, 101–3
  a, 102–5, 112, 196, 253, 287, 548
  absorption of light by, 101–2
  absorption spectra of, 101–2
  b, 102–3, 196, 253, 287
  bacterio-, 103
  c, 103, 253
  chlorobium, 103
  and cytokinins, 510
  structure of, 103, 539, 541
Chlorophyta, 172–73, 255–68, 633 (*see also* green algae)
chloroplast, **14, 20,** 21–22, 44
  in algae, 255, 257, 259, 262, 264, 267, 281, 283
  in bryophytes, 292, 296
  in C_3, C_4 plants, 109, **110**
  development of, **23**
  division of, 22
  evolutionary origin of, 25, 255
  function of, 96
  origin of, **176**
  in photosynthesis, 96, 104, 107, 109, **110,** 112
  and phytochrome, 527
chocolate, **599**
cholera, 187
Chondromyces crocatus, **186**
Chondrus crispus, **273**
C horizon (soil), 542
Chorthippus parallelus, **135**
chromatid
  in meiosis, 134–35, **136, 137,** 146
  in mitosis, **37,** 39
chromatin, 20, 44, 133
chromatophores, 184
chromomeres, 133
chromophore, 526
chromoplasts, 22, 44
chromosome(s), 16, 20, **37,** 38, **116,** 117–21
  bacterial, **119, 120,** 184, 192–93
  bivalents, 134–35, 146
  crossing-over, 134, **137,** 146
  daughter, 37, **38,** 44
  in dinoflagellates, 284
  and DNA, 16, 20, 25, 36, 44, 117–21, 176–77
  eukaryotic, 117, 133–37, 146, 184
  and evolution of eukaryotes, 178, 181
  fungal, 215–16
  homologous, 132–39, 146
  human, 121
  and meiosis, 134–39, 146
  and mutation, 143–44
  number, 20, 132–33, **136,** 138, 163–65
  and polyploidy, 163–65
  and proteins, 16, 20, 44, 133
Chrysanthemum indicum, 527
chrysanthemums, 523–24
chrysomelid beetles, 398
Chrysophyceae (*see* golden-brown algae)
Chrysophyta, 276–79, 285, 683 (*see also* diatoms; golden-brown algae; yellow-green algae)
Chrysothamnus, **602**
chymotrypsin, 79
Chytridium confervae, **245**
chytrids (Chytridiomycota), **180,** 181, 245, 250
Cicuta maculata, **360**
cilia, 30
Cinchona, **399**
circadian rhythms, 520–22, 533 (*see also* biological clocks)
  endogenous, 520–21, 533
  entrainment, 521
  free-running period, 521
  Zeitgeber, 521
circinate vernation, 334
Cirsium pastoris, **381**
cis-aconitate (*cis*-aconitic acid), 89, 92
cisternae, 26–27
citrate (citric acid), 88–89, 92, 224
Citrullus vulgaris, 582
citrus, 155, 507, 511
Citrus aurantium, **468, 469**
Cladonia
  cristatella, **226, 227, 230**
  subtenuis, **226**
Cladophora, 262, **263,** 265, 270
cladophylls, 472
Cladosporium herbarum, 213
clamp connections, 222, 231
Clasmatocolea
  humilis, 295
  puccioana, 295
class, classification by, 175, 182
classification, binomial system of, 170, 175
  of living things, 169–83
  of major groups of organisms, 175–76, 631–36
Clausen, Jens, 156
Clavariaceae, **171**
Claviceps purpurea, 222
clay, 542, 553
  red, 598, 612
  as soil, 544, 598, 612
clear-cutting, 598
cleistothecia, 219
climacteric, 511
climate, 374, 400, 596–97
clinal variation, 157
clines, 157, 165
cloning, 506
Clostridium, 208, 552
  botulinum, 200
  tetani, **187**
clover, 153, 208, 530, 549, 551 (*see also Melilotus; Trifolium*)
clover leafhopper, 208
club moss(es), **174,** 296
Clupea harengus, **588**
cluster, **550**
Cnidoscolus aconitifolius, **168, 169**
CoA, 88–90, 92
coal age plants, 330–31, 337
coal deposits, 330
coalescence, 362, 380, 467
cobalt (Co), 539–40, **557**
  ions, (Co^{2+}), 540
cocci, 187
Cochleanthus, **464**
cocklebur, 510, 522–24, 527
cockroaches, 50
cocoa, **397**
coconut, **357,** 393, **394,** 396, 507
  milk, **394,** 507
  palm, 203
Cocos nucifera, 203, **357, 375, 394,** 507
Code of Botanical Nomenclature, International, 170
Codium, **263**
  magnum, 255, 262
codon, 124, **125,** 126, **127,** 133
Coelomomyces, 245
coenocytic pattern, 214, 216, 242, 262, 278, **279**
coenzyme, 80, 83, 90, 105, 112
coenzyme A, 88–90
coevolution,
  biochemical, 397–400
  plant-herbivore, 383–92, 396, 582
cofactors, 79, 83
  ions as, 80, 83
  nonprotein organic, 80, 83
Coffea arabica, **397**
coffee, **397**
cohesion (cohesiveness), 65, 563–65
cohesion-adhesion-tension mechanism, 563–65, 574
Colacium, 281
colchicine, 163
cold, resistance to, 375
colds, 204
cold treatment
  and dormancy, 529–30
  and flowering response, 530–31
  and germination, 529
  tolerance, 597
Cole, Lamont, 589
Coleochaete, 267, 282
coleoptile, 409, 414, 502–3, 507, 512, 517, 526
coleorhiza, 409, 414
Coleus, 504
  blumei, **445, 446, 455**
coliphage T_7, 209
collenchyma, 417–18, **419,** 429–30
collenchyma cell (*see* collenchyma)
colonies
  algal, 260–61
  bacterial, 193
  plant, 159
columbines, 394
  red, 388
column of orchid flower, 382
communities
  climax, 590–92, 594
  defined, 579
  evolution of, 7
  forest, 591
  grass, 581
  pioneer, 594
  plant, 7, 375, 552, 580–82, 585, 589, 596–612
  redwood, **595**
  shrub, 581
companion cell, **424, 425,** 426, 431, 574
competition, 2, 579, 580–82, 594
competitive exclusion principle, 580
composites, **372,** 379, 381, 383, **388, 395,** 400
compost, composting, 553–54
compounds, chemical, 541
Comptonia, 551
concentration gradient, 61
conceptacles, **272**
Conchocelis, 275
conducting tissue, 268–69 (*see* central strand; hydroids; leptoids)
cones (strobili), 382
  of *Abies,* **350**
  of cycads, 353
  of *Cycas,* **353**
  of *Encephalartos altensteinii,* **353**
  of *Ephedra,* **355**
  of *Equisetum,* **327, 328,** 329
  of *Juniperus,* **351**
  of *Larix,* **350**
  of *Pinus,* 342, 344, **345,** 346, 348
  of *Selaginella,* 323, **324, 325**
  of *Taxus,* **351**
  of *Welwitschia,* **355**
  of *Zamia,* **353**
conformational coupling hypothesis, 91
conidium, conidiophore, 219–20, **221,** 222, **223,** 239
conifers (Coniferophyta), 338–52, 371–72, 382, 390, 580, 635
conjugation tubes
  in algae, 257, 267
Conopholis americana, **357**
conservation, 609
  mass, law of, 76
consumers
  primary, 587, 594 (*see also* herbivores)
  secondary, 587, 594 (*see also* carnivores; parasites)
continental drift and angiosperm evolution, 374–75
continents, super-, 374, 400
convergent evolution, 309, 473
Convolulaceae, **357**
Convoluta roscoffensis, 254
Cooksonia, 305
  caledonica, 305
  hemisphaerica, 305
copper (Cu), 538–40, 543, 556–57
  ions (Cu^{2+}), 540
Coprinus, **171, 232**
  atramentaris, **3**
  lagopus, 231
coral fungus, **171, 229**
coralline algae, 272–73

coral reef(s), 272, 579
-building, 254
fungus, **171**
Cordaitales, 330
Cordaites (cordaites), **316, 317,** 331, 338
cork, 4, 18, 308, 428, 482–84, 496
commerical, 488–89
corms, 155, 326, 474–75
corn, 361, **391,** 506, 508, 538–39, 549, 555–56, 560, 580 *(see Zea mays)*
belt, **543,** 604
dwarf, 513
earworm, 204
leaf blight, 12
smut, **228**
cornflowers, 26
corolla, 359, 378, 380, **468, 469, 470, 471**
cortex, 5, 305–6, **308**
of root, 437–40, 444, 550, 560–61
of stem, 447, **448,** 449–50
corymb, **361**
Corynebacterium, 201
diphtheriae, 199
cotton, **380,** 512
cottonwoods, 391
cotyledons, 348, 369, 405, 408–11, 414, 526
coupled reactions, 82
cowpea, 560
Crassulaceae, 111, 567
Crassulacean acid metabolism *(see* CAM photosynthesis)
Crataegus, 163
flava, **473**
Cretaceous period, 278, 313, 350, 372–73, **374,** 376, 399
Crick, Francis, 118–19, **120,** 122, 131, 205
cristae, 22, **23,** 88, 91
crops, agricultural, 9, 164–65, 539, 541, 545, 549, 551, 553–57, 586
crossing-over, 134, **137,** 146
in fungi, 216, 239
cross walls *(see* septum)
crown gall(s), 195
crozier *(see* hook, fungal)
crucifer (Brassicaceae), 383
cryptococcosis, 222
Cryptococcus, 222
Cryptophyta, 284
Cryptothecodinium (Gyrodinium) cohnii, 283
crystalline bodies in diatoms, 276–77
crystals, 24, **25, 420**
crystal violet, 190
C-terminal, 53
cucumber, 566
Cucumis sativus, 511, 566
Cucurbitaceae, 511, 582
Cucurbita maxima, **418, 421, 422, 425**
cucurbits, 511
culture techniques, 506
Culver's-root, **457**
Cumegloia, **273**
Cupressus, 350
goveniana, **351**
cupules, seed, **314**
currants, 514
Cuscuta cuspidata, **357**
cuticle, 4, 288, 296, 303–4, 342, 383, 428, 438, 565, 603
cutin, 33, 50, 304, 428
Cyanea, 158
cyanidin, 392, **393**
Cyanidium caldarium, 176
cyanobacteria (blue-green algae), 2, 16, **25,** 43, 184, 202–4, 296, 526, 537, 551, **552** *(see also* algae)
accessory pigments in, 103–4, 196, 203
and chloroplast origin, 25, 255
ecology of, 202–3
evolution of, 197, 204
as functional chloroplasts, 197
locomotion of, 188
nitrogen-fixing, 203–4, 552
and photosynthesis, 196
reproduction in, 192
structure of, 188
symbiotic, 203–4, 280
cyanogenic substances, 397
cyanophycean starch, 196
Cyathea, 333
Cyathus striatus, **229**
Cycadophyta, 339–40, 353, 371
cycadophytes, 372
cycads (Cycadophyta), 340, 353, 635
Cycas revoluta, **353**
cycles
nitrogen, 545–49, 552–53, 557, 587
nutrient, 545–46, 548–53, 557
phosphorus, 552–53
seasonal growth, 476–77, 604, 611–12
water, 544
cyclic electron flow, 186
cyclic photophosphorylation, 106
cyclosis *(see* cytoplasmic streaming)
Cymbidium, **174**
Cynthia annabella, **388**
Cyperaceae, 237
cypress, 350
bald, 350, **352,** 441
Gowen, **351**
Cypripedium acaule, **464**
cypsela, 395
Cyrtomium falcatum, **335**
cysteine, **52,** 53
cytochromes, 91–92, 105, 539
cytokinesis, 36, **38,** 41–44, **364,** 369, 508
cytokinin(s), 508–10
-auxin interactions, 508–9, 517
and cell division, 508, 510, 517
mechanism of action, 510
natural, 508, 510
and organ formation, 509, 517
in seeds, 508
structure of, 508
synthetic, 508
cytoplasm, 15, 17–20, 43, 184, 191 *(see also specific components of cytoplasm)*
cytoplasmic ground substance, 18, 85, 95
cytoplasmic strands in *Volvox,* 260
cytoplasmic streaming (cyclosis), 18, 61, 69, 568
cytosine, 56, **57,** 118–19, **120,** 121–22

2, 4–D, **503,** 504, 507
Dactylella drechsleri, 224
Dactylis glomerata, 153, 383
Danaus plexippus, **583**
dandelion, 155, **174,** 380, 395, **433**
dark reactions, 100–101, 106–12
Darwin, Charles, 15, 132, 372, 375, 502, 520
Darwin, Francis, 502
dates, 393
dating methods, 47, 185
Daucus carota, 130, 442, 506, 514
Dawes, Clinton, 275
dawn redwood, 350, **352**
Dawsonia superba, 298
daylength and flowering response, 522–25, 527–28, 530, 533
day-neutral plants, 522–23, 528, 533
decarboxylation, 94
deciduous forests, 585, 587, 606–8, 612
deciduous trees, **350,** 354, 374, 601, 607
decomposers, 187, 196, 210, 542, **545,** 546, 553–54, 557, 587, 594, 597 *(see also under* bacteria; fungi)
fungal, 212, 239
defense mechanisms, 397–98, 400, 550, 582, 586, 594
deforestation, 587
Degeneria, **378**
dehiscence *(see also* spore dispersal)
of anther, **364,** 366, 368
delphinidin, **393**
delphiniums, 26
denaturation, 55–56
dendrochronology, 493
dendrometer, **565**
denitrification, 197, 546, 548, 553, 557
deoxyribonucleic acid, 16, 56 *(see also* DNA)
deoxyribonucleotides, **57**
deoxyribose sugar, 56, **57,** 58, **120,** 121–22, 124
derivatives
of cambial initials, 477, **478**
of meristematic cells, 416
dermal tissue system, 417
dermatophytes, 224
deserts, 601–4, **608,** 612
desmids, 267
desmotubule, 35, **36,** 68
destroying angel, 232
Desulfovibrio, 198
Deuteromycetes *(see* Fungi Imperfecti)
development
control of, 128–31
of embryo, 409–12
of flower, 467–71
of leaf, 454–56
of plant, 130–31
regulation of, by hormones, 501–17
of root, 433–36, 496
Devonian period, 287, 305–7, 314–15, 319, 327, 337–40, 371, 637
deVries, Hugo, 143
diatomaceous earth, 278
diatoms, 173, 276–78, 634 *(see also* algae)
cell wall of, 276
centric, **173, 276,** 277
life cycle of, 277
motility in, 277–78
pennate, **173, 276,** 277
dichlorophenoxyacetic acid *(see* 2,4-D)
Dicksonia, **333**
dicot(s), 309, 356–58, 364, 379, 476, 538, 580, 636
herbaceous, 476
and monocots compared, 358, 369
Dicotyledones, 358, 371
Dictyophora duplicata, **599**
dictyosomes *(see* Golgi bodies)
Dictyostelium
aureum, **68**
discoideum, 128–31, 247, 249
differentiation
in algae, 261, 263, 268, 270
of bacterial spores, 192
cellular, 130–31, 417
control of, 128–31
in leaf tissue, 454–63
physiological, 157
in stem tissues, 446–52
diffusion, 60–62, 69, 565, 568, **569,** 573
Digitalis purpurea, **386**
Digitaria, 555
sanguinalis, 111
dihydroxyacetone phosphate, 86
dikaryotic (cells), dikaryosis, 220–21, 231, 234, **235,** 236
dilated ray, 485
Dilcher, David L., 376
dinoflagellates (Pyrrophyta), 279–81, **521** *(see also* algae)
cell wall of, 253, 279
and cyanobacteria, 280
dioecious plants, 362, 391
Dionaea muscipula, 532, 547
diosgenin, 399
Diospyros, 35
diphtheria, 187, 199
Diplococcus pneumoniae, **187,** 193, 199
diploid
cells 132–34, 138
number, 20, 132–33, 146
organisms, 133–34, 138–42, 146, 162–63
state, 163, 178, 180–81
diploidy, evolution of, 178, 181–82
Diplosoma virens, 196
disaccharides, 46–47, 58
Dischidia rafflesiana, 442, **443**
disease *(see also by name)*
bacterial and viral, 118, 187, 193, 199–201, 205–6, **207,** 208–10
of fish, 242
fungal, 219, 224, 230, 234–36
infectious, 209
lethal, 199
organisms, 187, 193, 199–201, 242, 244–45
disk flower, 380
Distephanus speculum, **278**
disulfide bonds, 53, **54**
diurnal movements, 520
division, classification by, 175, 182
DNA, 16, 20, 56, **57,** 58, 117–28, 510, 548
in bacteria, **17,** 118, **119, 121,** 205–6, 208
base pairing, **120,** 121–22
building blocks of, 56, **57,** 58
in chloroplasts, 22, 25
in chromosomes, 117
composition of, 118
and cytokinins, 510
diffraction studies of, 119
double helix, 119–21, 541
of eukaryotes, 20, 132–33, 176–77
extrachromosomal, **556**
herring sperm, 508
in mitochondria, 24–25
of prokaryotes, 16, 20, 132–33, 175–76
in protein synthesis, 123–24, 126
replication of, 36, **37,** 122
structure, 78, 118–22, **120, 122, 123**
T-, 195
viral, 205–6, **207,** 208–10
Watson-Crick model, 119–21
Doctrine of Signatures, 291
dodder, **357**
Dodecatheon pauciflorum, **360**
dogwood, 384
dominance, incomplete, 140–41, 146
dominant/recessive characteristics (phenotypes), 139–42, 146
dormancy, 528–31, 533, 603
breaking of, 529–30, 533
in buds, 529–30, 533
cambium, 479
in lichens, 226–27, 240
in seeds, 413, 529, 533
in sporangia, 246
dormins, 512
double fertilization, 368, **369,** 371
double helix, 119–21
Drosera, 532, 547
intermedia, **546**
Drosophila melanogaster, 143
cross-over studies of, 143
drought, 373, 375, 400, 512, 595, 597, 601, 603, 607–8, 612
drugs, 222, 224
drupe, **357,** 393, **394**
druses, **24**
Dryopteris marginalis, **334**
duckweeds, 356
dust bowl, 604
Dutch elm disease, 219
dwarf mistletoe, **395**
dwarf plants, 513, 517
dynamic equilibrium, 85
dysentery, 199

early wood, 489, **490,** 493, **494,** 496
earth, 1
crust, 541–42, 552
earthstars, 229
earthworm, as soil components, 542, **545**
eastern skunk cabbage, 85
Echinocereus, **387**
ecology, 577–613
defined, 579
ecosystem, 7, 557, 579–94
aquatic, 556
cycling of nutrients in, 552, 557, 585, 587
development of, 589–94
interactions between organisms, 579–86, 594
productivity in, 594
trophic levels, 587–89
ecocline, 157
ecotype(s), 156–57, 165
differentiation, 157
and photoperiod, 523
studies in California, 156–57
variation, 156–57
Ectocarpus, 270

ectomycorrhizae, 237–38, 580
edaphic factors, 608
egg apparatus, 368, 371
egg cell, 368
eggplant, 505, 514
Einstein, Albert, 76, 98
elaters, 290, 294, 329
elder, 384
elderberry, **418, 450, 451, 481, 483**
electrical
impulses and touch responses, 532–33
signals in plants, 532–33
electric current, 98
electromagnetic spectrum, 98–99
electron(s), 93, 615–22
acceptor/donor, 80, 100, 105–6, 112, 539
in atoms, 615–22
carriers, 90–91, 95, 101, 105–6
flow, cyclic, 106
flow, noncyclic, **105,** 106
of hydrogen atoms, 615–19, **620**
microscope, 16
in photosynthetic cycle, 105–6, 112
transfer, 90–91, 94
transfer molecules, 105–6
electron transport chain, 84, 90–93, 95, 105, 112, 539
electronegativity, 618, 621
elements *(see also specific types)*
cycling of, 7
essential, 537–41, 556–57
of living matter, 45
elm, 394, **395**
American, **42,** 487, **494**
Elodea, 18, **19,** 63, **64, 100**
embryo (young sporophyte), 313, 315, 466–67
angiosperm, 368, 405–9, 414
in bryophytes, 288, 290
corn, **414**
development in arrowroot, **412**
development in shepherd's purse, **410, 411**
dicot, 369, 405, **406,** 408
dormancy of, 413
and evolution of seed, 313, 338
of ferns, **336,** 337
formation in angiosperms, 409–12
of *Ginkgo,* 354
grass, **408,** 409, 414
growth of, 516
of gymnosperms, 340
of *Lycopodium,* 321, **322,** 410
monocot, 369, 405, **407, 408,** 409
of pine, 346, 348, 410
in plants, 288
of *Psilotum,* 318
sac, **367,** 368, 371
of seed plants, 338
of *Selaginella,* 324, 410
wheat, **408**
embryogeny
in angiosperms, 369, 371, 405
in lower vascular plants, 369
in pine, 346, 348, **349,** 369, 371
$E = mc^2$, 76
Encephalartos altensteinii, **353**
Encyclia, **382**
endangered species, 609
endergonic reaction, 75, 78, 82, 628
endocarp, 369, 396
endocytosis, 67, 69
endodermis, **328,** 560–61, 563, 574
in pine leaf, 343
in root, 438–40, 444
endogenous rhythms, 520–21, 533
endomembrane system, 28–29
endomycorrhizae, 237–40, 580
endomycorrhizal relationship, 304
endoplasmic reticulum (ER), **19,** 27–28, 44
endosperm, 368–69, 371, **394,** 408–9, 412, 414–15, 507, 516–17
endothelium, **410**
Endothia parasitica, 219
energy (*see also* phosphorylation; photosynthesis; respiration)
activation, 74
biological flow of, 73–83
carrier, 83–84
cellular, 88, 546
chemical, 1, 47, 73, 84, 96, 101, 112, 588, 594, 617–19, 625–28
chemical bond, 47, 102
conversion, 1, 75, 83, 96, 101–2, 541
in ecosystems, 588
electrical, 2, 75, 98
for evaporation, 565
-exchange system, 83
flow in an ecosystem, 587–89, 594
heat, 74–77, 83, 85, 589
kinetic, 74
levels, 90–92, 102, 105–6, 112
light, 1, 74–75, 96, 98, 101–2, 112, 588–89
"loss," 74–75, 83
mechanical, **60,** 73
potential, **60,** 74–75, 83, 594
release of in respiration, 84, 93–94, 589
solar (radiant), 47, 73, 83, 96, 544, 589–90, 616
storage, 7, 84–85
transfer, 7, 81–83, 86–87, 94, 103
transformation of, 56
yield in photosynthesis, 106, 112
Engelmann, T. W., 102
enolase, 87
Entogonia, **276**
entrainment, 521
entropy, 75–76, 83
environment, 579–80
influence on selection, 148, 151–54
modification/adaptation, 383, 151–57, 165
enzymatic pathways, 80–81
enzyme(s), 56, 58, 77
and abscission, 511
activating, 124
activity, regulation of, 81, 83
in amino acid synthesis, 123–26
catalytic, 56, 58, 78, 83, 539
conformation, 79
digestive, 47, 414, 440, 472, 532, **546,** 547
in electron transport chain, 95
fungal, 212, 214, **216**
glycolytic, 85–87
in glyoxosomes, 24
in Golgi bodies, 27
hydrolytic, 516
in Krebs cycle, 95
and living systems, 77–78
mineral function in, 539
in mitochondria, 22, 84, 88
nitrogen-fixing, 556
in photosynthesis, 100, 107
in prokaryotes, 184
restriction, 194
in seeds, 529
specificity, 78
synthesis, 510, 516
systems, 539
viral, 206
Eocene epoch, **375**
Ephedra, 354, **355**
california, **355**
epicotyl, 405, 414
epidemics, 206
epidermis, 305, **308,** 417, 427–29
of *Eucalyptus cloeziana* leaf, **51**
of leaf, 342–43, 459, **460, 461, 462**
multiple, 427, **429,** 442, **461**
of root, 436–38, 444, 475
of stem, 447, **448,** 449
epigyny, **362,** 363, **379,** 380
Epilobium, 396
Epiphyllum, **472**
epiphytes, 319, 356, 442, **443,** 597–98, **599,** 601, 612
epistasis, 144–46
Epstein, Emanuel, 555
equatorial plane (in meiotic/mitotic division), 37–40, 135, **136, 137**
equilibrium, state of, 77
Equisetum, 309, 327, **328,** 329–30, 337, 354, **538**
arvense, **327, 328**
× *ferrissii,* **165**
hybrids, **165**
hyemale, **27, 39, 40,** 165, **327**
laevigatum, **165**
sylvaticum, **174**
telmateia, **329**
Eremosphaera, 259
ergot, ergotism, 222
Ericaceae, 238–39
mycorrhizae of, 238–39
Erigeron, **470, 471**
erosion, soil, 548, 552, 557, 598, 612
Erwinia, 201
amylovora, **187,** 201
carnegieana, **200**
tracheiphila, 201
Erysiphe aggregata, **219**
erythrocyte, **133**
Escherichia coli, **17,** 51, 68, **187, 189,** 190, 193–94, 555, **556**
and conjugation, **189**
and DNA studies, 121–22
and gene transcription studies, 126, **127,** 128
sensory responses in, 190
and viral infection, **184,** 204, 206, **207,** 208
essential oils, 397
ethanol (ethyl alcohol), 94–95
ethylene, 396, 507, 510–11
and abscission, 511, 517
and auxin, 511
and fruits, 510–11, 517
and gibberellin, 511
etiolation of seedlings, 526
Eucalyptus, 356, 388, 513
cloeziana, 51
globulus, **427,** 514
jacksonii, **356**
Eucheuma isiforme, 275
euchromatin, 133
Eudorina, 260–61
Euglena, 63, 281–82
euglenoids (Euglenophyta), 281–82, 285, 634 (*see also* algae)
nonphotosynthetic, 281
pellicle of, 281
and plants compared, 283
eukaryotes, 16, 85, 96, 132–33, 176–77
DNA of, 16, 20, 44
evolution of, 178, 180, 181
and prokaryotes compared, 17, 126, 132–33, 175–78
reproductive cycles of, 132–33
eukaryotic cell, 2, 15–44, 59, 94, 283
Euonymus americanus, 396
Euphorbia, 475
Euphorbiaceae, 473
euphorbias, 603
Eurystoma angulare, **313**
eustele, **309,** 310, 314
eutrophication, 556
evaporation, 544, 563, 565, **569**
lifting power of, 564
evergreens, 606–7, 610, 612
evolution, 3, 15, 147–66
agents of, 382, 384–92
of androecium, 378
angiosperm, 372–401
of bacteria, 185, 197
biochemical coevolution, 397–400
of bryophytes, 287, 302
of carpel, 377
change, rapid, 153
of chloroplast, 25
of communities, 7
convergent, 606, 608
of diploidy, 178, 180–81
of eukaryotic algae, 283
the eustele, 310
of the flower, 377–82, 384–93
of flowering plants, 372–401
of fruits, 393–97
of fungi, 215
of gametophytes of vascular plants, 312
genetic systems, **180**
of glycolytic sequence, 85
of green algae, 255–56, 261–62
of gymnosperms, 339–41
of gynoecium, 377
mechanisms, 159–65
of mitochondrion, 25
of mycorrhizae, 239
of photosynthesis, 106, 197
of plants, 3–4, 12, 287–88, 303
of the seed, 313
of tracheary elements, 309
of vascular plants, 287–88, 304–15, 338, 399–400
of virus, 209–10
evolutionary relationships among organisms, 182
exergonic reaction, 47, 75, 77–78, 82–84, 87–88, 628
exine, **365,** 366
exocarp, 369
exocytosis, 68–69
exoskeletons, 50, **214**
extraterrestrial life, 186–87
Exuviaella, **279**
eyespot (stigma), 257, 260, 281, 283

Fabaceae, 152, 549 (*see also* legumes)
facilitated diffusion, 66, 69
facultative anaerobes, 185
FAD, 90–92, 95
$FADH_2$, 90–93, 95
Fagaceae, 238, **362**
Fagara
ailanthoides, **152**
dipetala, **152**
fairy rings, 232, **233**
fall coloring, 393
fallout, radioactive, 227
family, 175, 182
fats, 46, 50, 58, 89
fatty acids, 46, 50–51, 89, 196
saturated/unsaturated, 50
feces, 552
feedback inhibition, 128
fermentation, 84, 95
alcoholic, 94, 221
energy yield in, 93–95
lactate, 94
of wine and/or beer, 94, 221–22
ferns, 309, 333–37, 372–73, **591,** 593, 634–35, 637
bracken, **334**
cinnamon, **332, 334**
diversity of, 333
grape, 333
Hart's-tongue, **332**
life cycle of, **336**
marsh, 174
Paleozoic, 339
rattlesnake, **334**
seed, **307,** 330–31
sword, **595**
tree, 331, **332,** 333
walking, 155
water, **332,** 335, 551–52
whisk, **318,** 634
wood, **334**
fertilization, 132, 313, 315 (*see also* syngamy)
in angiosperms, 359, 368, **369,** 371
in bryophytes, 289
cross-, 382
and fruit development, 505
in gymnosperms, 340, 353, 371
in mosses, **302**
in pine, 346, **347**
in seedless vascular plants, 340
Fertile Crescent, 9
fertilization tubes, **242, 243**
fertilizers, 11, 538, 549, 552–54, 557, 609
Festuca, 238
fibers, 419, **420**
bast, 419
libriform, **421**
in phloem, **448,** 449–50, 480–81
in xylem, **420**
fiber-tracheid, **421**
fibrils, cellulose, 32, **33,** 34
Ficus
bengalensis, 441
elastica, **429**
fiddleheads, 334
field moisture capacity, 544–45

fields
cultivated, 604
old, 592, **593**
filament, 359, 378
fire, 548, 557, 592, 601
fireflies, 93
fireweed, 396
firs, 341, 350, 610
balsam, **350, 591**
Douglas, 350, 607
white, 591
fish, 522, 556
fission, 192, 220, 239
and division of proplastids, 22
five-kingdom system (scheme), 177–78
flagella, 30, **31,** 44
of algae, 246, 256, 260–61, 279, 281
of chytrids, 241, 245
of prokaryotes, 189
structure of, 30, **31,** 44
tinsel, 30, 242, 270, 281
of water molds, 241–42, 249–50
whiplash, 30, 242, 245, 256, 270
flatworm, marine, 254
flavin, 519
flavin adenine dinucleotide (*see* FAD)
flavin mononucleotide (*see* FMN)
flavonoids, 392, 397
flavonols, 393
flax, **214,** 419
fleabane, **470, 471**
Fleming, Sir Alexander, 224
flies, **384,** 386, **387,** 400, **532,** 547
florets, **392**
Florey, Howard, 224
florigen, 527, 533
flour, 409
flower(s), 359–63, 371, 400 (*see also* angiosperms)
aquatic, 556
bisexual, 382–83
carpellate (ovulate), 361, **362,** 380, 383, **391**
colors of, 359, 384–86, 388, 390, 392–93, 598
complete, 362
definition of, 359
development of 467–71
evolution of, 377–82, 384–93
imperfect, 361
incomplete, 362
irregular (zygomorphic), 363
odors of, 384, **387,** 388
parts of, 359, 377–78
perfect, 361
placentation of, 361
primitive, 377–79, 400
regular (actinomorphic), 363
staminate, 361, **362,** 383, **391**
terminology, 359, 361–63
trends in evolution of, 378–79, 400
of tropical rain forest trees, 598
flowering, 102, 527–28, 533
and cold treatment, 530
and gibberellins, 514
hormonal control of, 527–28, 533
induction of, 527–28, 533
inhibitors to, 528, 533
and light, 525
and photoperiod, 522–24, 527
and phytochrome, 527
plants (*see* angiosperms)
flower pot plant, 442, **443**
fluid mosaic model, 64, **65**
fluorescence, 102
fluorides, 464
fly agaric, **228**
FMN, **90,** 91
follicle, 394
Fontinalis, 298
food chain, web, 587–89, 594
food-conducting cells (*see also* leptoids)
food-conducting tissue, 423–25
food poisoning, 199 (*see also* botulism)
food reserves (*see* glycogen; laminarin; leucosin; paramylon; starch)
food (assimilate) transport, 569–74
foolish seedling disease, 512
foot, 290, **291,** 294, 296–98, **302,** 303, 308, 318, **336,** 337
Foraminifera (marine protozoa), 277
forest(s), 590 (*see also* reforestation)
boreal, 610
Carboniferous swamp, **304, 316, 317**
climax, 591, **593**
cloud, 598
coniferous, 607, 610
deciduous, 3, 6, 585, 587, 606–8, 610, 612
evergreen, 606–7, 610, 612
fires, 592
monsoon, 601
redwood, **578, 579**
scrub, mixed, **583**
temperate deciduous, 585, 604, 606–8, 612
temperate evergreen, **607**
thorn, 601
tropical rain, 356, 597–98, **599,** 612
fossil fuels, 330, 549
fossils, 9, 372
of algae, 255, 268, 272, 278, 282
of angiosperms, 372–76, 399
of beetles, 372
of bryophytes, 287
coal age plants, **304,** 330–31
distribution of redwoods, 352
earliest known, 1, 185
fuels, 549, 556
of fungi, 215
of gymnosperms, 310, 313, **314, 373**
living, 350
of *Metasequoia,* 350, **352**
pollen, 373
of progymnosperms, **338,** 339
record of bacteria, 185
of vascular plants, 239, 287, 305–6, 316, 330–31, **338,** 339–40
wood, **339**
four-carbon (C_4) photosynthetic pathway, 108–9, 112, 539
four-carbon (C_4) plants, 108–9, 111
efficiency of, 110–11
evolution of, 111
foxglove, **386**
Fragaria, 157
× *ananassa,* 154, 505
fragmentation (*see* reproduction)
Franklin, Rosalind, 119
Franklinia alatamaha, 609
Fraxinus, **395,** 495, 563
pennsylvanica, **458, 459, 487**
free-energy change, 76, 83
freeze-fracture, -etch preparation, **19, 22, 104**
Frisch, Karl von, 385
fritillary butterfly, **168, 169**
Fritschiella, 263
Frolova, I.A., 528
fronds, 333
fructose, **46,** 47, 56, 82, 86
1,6-bisphosphate, 86, 108
6-phosphate, 86
fruit(s), 363, 369, 393–97, 400, 581, **599**
accessory, 393, 400
aggregate, 393, 400
animal-borne, 396–97, 400
color, 396–97
cytokinins in, 508
dehiscent, 394, **395,** 400
dispersal of, 395–97, 400
evolution of, 393–97
growth and development of, 368–69, 505
indehiscent, 394, **395,** 400
multiple, 393, 400
parthenocarpic, 505
and plant hormones, 505, 512, 517
ripening, 510–11
simple, 393–94, 400
thinning, 511
water-borne, 396, 400
wind-borne, 395–96, 400
fruit drop and auxin, **503,** 507, 512
fruiting bodies, 231, 240
Frullania, 294–95
frustules, 276
fuchsia, 388
fucoxanthin, 253, 270, 276 (*see also* carotenoids; xanthophylls)
Fucus, 130, **172,** 270
life cycle of, 272
vesiculosus, 269
Fuller, R. Buckminster, 205
fumarate (fumaric acid), 89, 92
functional groups (chemical), 622
fundamental tissue system, 417
fungal mantle, 238
fungi, 3, 4, 178, 212–40, 631–32 (*see also* by *name*)
anaerobic, 214
bracket, **595**
as decomposers, 212, 239, 543, 545–46
destructiveness of, 212–13, 235
economic importance of, 212–13, 219, 221–22, 224, 234–35
and fermentation, 94
jelly, 234, **235**
lectins in, 550
life cycles of, **217, 221, 234, 235**
mycorrhizal, 237–39, 312, 317, **357,** 580
parasitic, 214, 219, 222, 224, 234
pathogenic, 219, 224, 230, 234–36 (*see also* rusts; smuts)
predaceous, 224
and relationship to environment, 214
reproductive techniques of, 215–16
saprobic, 214, 235, 239, 546
shelf, **595**
soil, 187, 197, 212, 224, 237, 545, 557
spores of, 215 (*see also* spores *and specific spore types*)
stinkhorn, **599**
symbiotic, 237–40 (*see also* lichens)
wood-rotting, 229, 231
Fungi Imperfecti, 222–25, 632
Fungi, kingdom, 178, 182
fungus roots (*see* mycorrhizae)
funiculus, 336, 394, 409

GA_3, GA_4, GA_7, 513
Gaillardia pulchella, **381**
galactose, **49,** 126, **127**
Galapagos Islands, 147–48
tortoise, 148
gametangia, 312 (*see also* antheridia; archegonia)
of algae, 265–66, 268
of ancestors of plants, 287
of bryophytes, 287–88, **290,** 298, **299**
of fungi, 215–16, **217,** 220, **221**
of vascular plants, 287, 312
gamete(s), 20, 30, **31,** 178, 180–81
of algae, 258, 261, 264–65, 267–68, **271, 274,** 277
of angiosperms, 366, 368
defined, 134
of fungi, 215
in genetics, 134, 150
of gymnosperms, 346
iso-, **257,** 258, 264, 267
gametophores, 292 (*see also* antheridiophores; archegoniophores)
gametophyte(s), 141, 181, 183, 311–12
of algae, **266, 271, 274,** 287
of ancestors of plants, 287
of angiosperms, 363, **365,** 371
bisexual, 311, 314, 318, **319,** 321, 329, 335, **336**
of bryophytes, 288–89, 303
of *Equisetum,* 329
of ferns, 335, **336**
of gymnosperms, 340, 371
of hornworts, 296–97, 303
of liverworts, 291, 293, 295,303
of *Lycopodium,* 321, **322, 323**
of mosses, 296–98, 302–3
of pines, 344, **345,** 346, **347,** 348
of *Psilotum,* 318, **319**
of *Selaginella,* 324, **325, 326**
unisexual, 311–12, 314, 324, 329
of vascular plants, 304, 312
gangrene, 208
Ganoderma tsugae, **229**
Garner, W. W., 522, 524
Gassner, Gustav, 530
Geastrum triplex, **229**
gemma cup, 295
gemmae, 295, 302
gene, 139, 383
activation of, 124, 516
bacterial, 125, 195
behavior in populations, 149–51
chemistry of, 117–18
chromosomal location of, 139
and DNA, discovery of, 118–21, 131 (*see also* DNA)
dominant/recessive (alleles), 139–42, 146
epistatic, 144
eukaryotic, 194
flow, 154
frequency, 149–51
interactions, 144–45, 151
linkage, 142–43, 146
nitrogen-fixing (nif), 555–56
pool, 150
regulator, 126–28
transcription, 126–28, 133
transfer, 555–56
generation
filial (F_1, F_2), 139–44
generative cell, 344, **345,** 346, 366, 368, 371
genetic(s)
balance, internal, 151
changes and mutations, 143–44, 146, 193, 206, 210
code, 16, 124
crosses, 139–42, 144–45, 149–50
diversity and the Green Revolution, 11–12
drift, 160–61
engineering, 194–95
of eukaryotic organisms, 132–46
experiments, 132, 139–44
factors, response to, 151–53
homeostasis, 151
information, 18, 56, 117, 121 (*see also* hereditary material)
isolation (*see also* reproductive isolation), **159, 160, 161**
material, 118–21, 144
recombination, 159, 178, 383
segregation, 140–42
self-incompatibility, 383
systems, evolution of, 178, 181
tools of, 219
Genomosperma
kidstonii, **313**
latens, **313**
genotype, 139–40, 146, 149–51, 383
genus, classification, 169–70, 175, 182
geological factors and biomes, 595
geologic eras, 637
geotropism, **518,** 519, 533
Gephyrocapsa, **278**
Geranium (geranium), **388**
maculatum, **388**
geraniums, 26
germination
and cold treatment, 529
epigeous, **406,** 413
hypogeous, **406,** 414
inhibition of, 524
and light quality, 524
and phytochrome, 526–27
of pollen, 514
requirements for, 413
seed, 152, **406, 407,** 413–15, 445, 516–17, 524–25, 529, 533, 603
germ tube, **245**
Gibberella fujikuroi, 512
gibberellic acid (GA), 512–13
gibberellins, 512–17 (*see also* gibberellic acid)
and abscisic acid, 512
and cell division, 512
and cell elongation, 512–13
and dormancy, 513, 530–31
and dwarf mutants, 513, 517
and flowering, 514, 530–31
and geotropism, 519
and IAA, 504
and juvenility, 513–14
mechanism of action of, 514–16
and pollen and fruit development, 514–15

  and seeds, 513–14, 517
  structure of, 513
Gibbs, Josiah Willard, 76
gill fungi, 232
gills, 232, 234
Ginkgo, 635
  biloba, 353–54
Ginkgophyta, 339–40, 354, 371, 635
ginkgos, 372
giraffe, **601**
girdling, 570, 585
glands, 547 (*see also* nectaries)
glandular tissue (*see* stigmatic tissue)
***Glaucoystis nostochinearum*, 25**
Gleditsia triacanthes, **458, 459**
Gleichenia dichotoma, 312
Gloeotrichia, **188**
glucose, **32,** 33, **48, 49,** 56, 58, 82, 84–87, 108, **127,** 622, 626
  facilitated diffusion of, 66
  ^{14}C-labeled, 230
  6-phosphate, 86
glumes, **392**
glutamic acid, 52, 548, 557
glutamine, 52
glyceraldehyde, **46**
  3-phosphate, 86, **107,** 108, 112
  3-phosphate dehydrogenase, 87
glycerate 1,3-bisphosphate, 87
glycerate 2-phosphate, 87
glycerate 3-phosphate, 87
glycerate 3-phosphate kinase, 87
glycerol, 46, 50–51, 89
glycine, 52, **127**
Glycine, 551
  max, 111, 522, 527, **549**
glycocalyx, 191
glycogen, 48, 196, 214, 239 (*see also* starch, cyanophycean)
glycolate oxidase, 266
glycolysis, 84–88, 94–95
  anaerobic pathway, 94
  overall equation, 87
glycolytic pathway, 85–88
glycoprotein, 33–34
glycosides, 397
  cardiac, **397,** 398, **583**
  mustard oil, 398
glyoxysomes, 24
Gnetophyta, 309, 339–40, 354–56, 371, 635
  and angiosperms compared, 356
Gnetum, 354, **355**
golden-brown algae, **30,** 253, 278, 634 (*see also* algae)
  and brown algae compared, 276
  reproduction in, 278
goldenrod, 157, **360**
goldeye lichen, **226**
Golgi apparatus, bodies, 26–27, **29,** 42, 44, 434
  and cell wall formation, 27
  Golgi (secretory) vesicle, 27, **29,** 68
Gondwanaland, 374, 400
Gonium, 260–61
gonorrhea, 187, 198
Gonyaulax
  catanella, 280
  excavata, 280
  polyedra, **521**
goosefoot family, 393
Gossypium, **380,** 512
G phases of cell cycle, 36–37, 44
gradient, environmental, 596–97
grafting, 585
gram
  atom, molecule, 622
  negative microorganisms, 190–91
  positive microorganisms, 190
  staining technique, 190–91
Gram, Hans Christian, 190
grana (granum), 21, 44, 96, 109, **110**
granite, 541
grape, 26, 94, 242, 244, 250, 393, 396, 471, **472,** 511, 515
grasses, 154, 235, **357,** 358, 383, **391, 392,** 394, 552, 580, 601
  annual blue, **462**
  bent, **438, 562**
  big bluestem, **605**
  bunch-, 604
  canary, 502
  crab-, 111
  creeping bent, 111
  ecotypes of, 523
  foxtail, 383
  Kentucky blue, 111, 153, 162
  marsh, **590**
  orchard, 153, 383
  prairie, 561, 604, **605**
  rye, 383, 530
  saltmarsh, British, 164
  sod-forming, 604
  sugarcane, 552
  tropical, 555
  Wales, 153
  wheat, **135, 137**
grasshopper chromosomes, **135**
grasslands, 581–82, 601, 604–5, 610, 612
gravitropism, 519, 533 (*see also* geotropism)
gravity, 544, 595
  and geotropism, 434, 519, 533
grazing, 580, 582, 604, 609, 612
green algae, 253, 255–68, 285, 633 (*see also* algae)
  alternation of generations in, 262, 265
  as ancestors of plants, 255, 283, 304
  cell division in, 255–56
  life cycles of, **257, 266**
  and plants compared, 255, 283
  reproduction in, 257–59, 261–62, 264–68
  siphonous, 262
  siphonous line of, 262
  volvocine line of, 260–61
greenbrier, **531**
Green, David E., 91
Green Revolution, 11–12
Griffith, F., 193
ground meristem, 410, 436, 447, **450,** 475
ground substance, 18
growth, 417
  abnormal, 538
  bud, **447,** 479
  deficiency symptoms of, 538, 555
  determinate, 342, 359, 454, 475
  embryonic, 409–12
  "forced," 530
  increments, layers, 493, 496
  indeterminate, 342, 344, 454, 475
  inhibitors, 512, 517 (*see also* hormones, plants)
  intercalary, 454
  normal, 538
  primary, 4, 307, 416
  regulators, 501–17 (*see also by name*)
  rings, 493, 496
  in roots, 433–36
  of root tip, 413
  secondary, 4, 308, 476–97
  of shoot tip, 415
  in stems, 446–47
growth regulators, 518–20, 527, 530 (*see also by name*)
Gruneberg, Hans, 144
guanine, 56, **57,** 118–19, **120,** 121–22
guano, 552
guard cells, 427–28, 565–67
guttation, 516–62, 574
Gymnodinium
  costatum, **279**
  neglectum, **279**
gymnosperms, 309–10, 312, 338–56, 371–73, 382 (*see also* conifers; Coniferophyta; cycads; *Ephedra; Ginkgo; Gnetum; Welwitschia*)
  and angiosperms compared, 356, 371
  life cycle, **349,** 371
  origin of, 339–40
  vessel-containing, 340
gynoecium, 359, 377, **380,** 393, 395

Habenaria elegans, **387**
habitat(s), 580, 595
  adaptation to, 153, 156–57, 159, 163–64
  microhabitats, 580
Haemanthus katherinae, **38**
hairs, 396, **596** (*see also* trichomes)
  glandular, 586
  root, 550
  sensitive, 532
  stigmatic, 377
Hales, Stephen, 559
Halictidae, **387**
hallucinogens, **223,** 224, 399
Halobacterium halobium, 190
halophytes, 66, 539
Hamamelis, 396
Hammer, Karl C., 524, 527
haploid
  cells, 134, **137,** 146
  number, 20, 132, **136,** 146
  organisms, 134, 146, 178, 181
Haplopapppus gracilis, 20
Hardin, Garrett, 580
hardwoods, 489, 496 (*see also by name*)
Hardy, C. H., 149
Hardy-Weinberg law, 149–51, 165
  exceptions to, 150–51
Hatch-Slack pathway, 108
haustoria, 214, 230, 239, 340
hawthorns, 163, **473**
hay, 544
hay fever, **365**
hazelnuts, 395
head, inflorescence, **361,** 380–81, **397,** 400
heartwood, 485, 495–96
heat, 2, 85, 102
  and biomes, 595
heat energy, 74–77, 83–85
Hedera helix, 441, 514
Helianthus
  annuus, **381**
  tuberosus, 509
helical wall thickenings, 423
Heliothis zea, 204
helium, 76
helix
  alpha, 54
  double, 119–21
Helminthosporium maydis, 12
Helmont, Jan Baptista van, 96
hemicellulose, 32–33, 44, 255
Hemitrichia serpula, **248**
hemlock, 238, 350, 489, 610
hemp, 419
henbane, 514, 523, 530
hepatica
  round-lobed, **358, 384**
Hepatica americana, **358, 384**
Hepaticopsida, 291–95, 303, 634 (*see also* bryophytes; liverworts)
herbicide, **503,** 587, 609
herbivores, 8, 398, 400, 550, 582, 585, 587, 589, 594, 612 (*see also* animals; mammals; plant-herbivore interactions)
herbs, 6, 476, 495, 598, 607
  annual, 476, 581, 601, **602,** 603–4, 606, 610
  biennial, 476, 514, 530
  perennial, 477, 581, 601, 603–4, 606–7, 610, 612
hereditary information, 118, 121–22
heredity, 117
  chemistry of, 117–31
Hericium coralloides, **229**
herring, **588**
herring sperm (*see* DNA)
Heterobasidiomycetes, 234–36
  life cycle of, **235**
heterochromatin, 133
heterocysts, **192,** 204
heteroecious state (condition), 235
heterokaryosis, heterokaryon, 215–16, 231, 234, 239
heterophylly, 513
heterospory, 311–12, 314
  in angiosperms, 363, 371
  in ferns, 335
  in gymnosperms, 344, 363, 371
  in *Isoetes*, 326
  in progymnosperms, 340
  in *Selaginella*, 323, 325
heterothallic individuals, 219, 242
heterotrophic protista, 177–78, 241–50
heterotrophs, 2–3, 196, 199, 210, 212, 241–50, 587, 594
hexokinase, 86
hexoses, 46
hibiscus, 388
Hiesey, William, 156
Hillman, William S., 532–33
hilum, in seeds, **406,** 409
Himantandra, **378**
Himantothallus, 272
histidine, 52
histones, 133, 215, 283
historadiography, 572
holdfast, 264–66, 268, **269, 271**
holly, **396, 504**
  American, 361
  evergreen, 507
homeostasis, genetic, 151
Homobasidiomycetes, **229,** 232–34
  life cycle of, 234
homokaryosis, 215–16, 231
homologous (chromosomes), 132–39, 146
homospory, 311–12, 314
  in *Equisetum*, 329
  in ferns, 335
  in *Lycopodium*, 321
  in progymnosperms, 340
  in *Psilotum*, 318
homothallic individuals, 219, 242
homozygous, heterozygous, 139–41, 151, 165
honeydew, 571
honey guides, 386
honey locust, **458, 459**
honeysuckle, 488
hook (and seed germination), 406, 414
hook, fungal, 220
Hooke, Robert, 18
Hordeum
  jubatum, **174**
  vulgare, 512, 516, 555, 571
horizons, soil, 542, 544
hormogonium, 192
hormone(s), 501–17 (*see also by name*)
  of chytrid, **246**
  defined, 501–2
  fungal, **217**
  influence on differentiation, 130–31
  and mycorrhizae, **238**
  of water mold, 244
hornet, **213**
hornworts, 288, 291, 296–97, 303, 634 (*see also* bryophytes)
horseradish, 398
horsetails, **27, 39, 165, 174,** 327–29, **538**
horticulture, 553–54
Hoya carnosa, 112
humans, 8, 20, 553, 556
human impact on nutrient cycles, 552–53
humans and ecosystems, 589, 591–92, 598, 604, 609
Humboldt, Alexander von, **596,** 597
humidity and transpiration, 568
hummingbirds, 388, **389,** 390, 400
hummocks, 530
hunter-gatherers, 8
hyacinths, 530
hydathodes, 561
hybridization, 159–65
  barriers to, 157–58, 165
hybrids, 157, 586
  fertile, 157–59, 163–65
  and genetic recombination, 159
  polyploid, 163–65
  populations, stable, 162, 164
  produced by protoplast fusion, 506
  protoplasts, 506
  sterile, 157, 162, 164
hydra, 254
hydration, 65
hydrocarbon chains, 50
Hydrodictyon, 262
  reticulatum, 262
hydrogen (H_2), 1, 3, 45–46, 76, 538, 540, 556, 615–16, 620–25

bonds, 49, 54–55, 99, **120**, 121, **125**, 622–23
donors, 196, 624
ion (H^+) and transmembrane potential, 568–69
ions, **49**, 90, 92, 624
isotope(s), 121
phosphates ($H_2PO_4^-$, HPO_4^{2-}), 540
sulfide (H_2S), 97, **186**, 198
in water molecule, 622–23
hydrogen carbonate ion (HCO_3^-), 543
hydroids, 288–89
hydrolysis, 47, 50, 56, 58, 82, 84, 86, 89
hydronium ions (H_3O^+), 624–25
hydrophilic/hydrophobic interactions, 50–51, 64, **65**
hydrophytes, 459, **461**
hydrostatic pressure, 60, 63, 541, 571, 573
hydroxyl ions (OH^-), 543, 624–25
groups, **49**
hygroscopic wall thickenings, 294
hymenium, hymenial layer, 219, **220**, **232**
Hyoscyamus niger, 514, 523, 530
hypanthium, 362
hypertonic solution, 63, 69
hyphae, 213–14, 241 (*see also* haustoria; rhizoids)
ascogenous, 220, **221**
aseptate, 216
coenocytic, 215–16, 239, 242
receptive, 236
septate, 215, 219, 222, 230
hypocotyl, 408, 414
hypocotyl-root axis, 348, 408
hypodermis, 342
hypogyny, 362
hypotonic solution, 63, 69

IAA, 501, 503–4, 511, 519
i⁶ Ade, 508, 510
ice plant, **429**, 475
icosahedron, 205, 210
igneous rocks, 541
Ilex aquifolium, 507
montana, **396**
imbibition, 65, 227, 516
immune reactions, 224
impala, **601**
Impatiens, 396
incense cedar, 174
incompatibility response, 383
independent assortment, 141–42, 146
Indian pipe, **337**
indoleacetic acid (*see* IAA)
induced-fit hypothesis, 79
induction
in enzyme synthesis, 126–28
floral, 467
indusia, 334, **335**
infection(s), (*see* diseases)
process and symbiosis, 549–50, **551**
infection thread, bacterial, 549–50
inflorescences, 359, **360**, **361**, 391
influenza, 206
Ingenhousz, Jan, 97
ingestion, 248
nutritive mode, 178
inhibitor(s), 512, 517, 530–32, 581–82, 594
endogenous, 529
initials, 416
apical, 434–35, 444, 446
cambial, 477
fusiform, 477–78
ray, 477–78
inky cap mushrooms, **171**
inorganic molecules, oxidation of, 198
insecticides, 582
insects, 547, 582, 586, 594 (*see also under* pollination)
colors of, 398, 583
and lectins, 550
as plant feeders, 382, 384–88, **584**
as soil components, 542
integuments, 313, 315, 338, 371
of angiosperms, 366, **367**, 369
of pines, 344, 348
intercalary meristem, 446–47
interfascicular region, 447, 450–51
interferon, 194
internodal elongation, 446, **447**
internodes, 359, 414, 445–46, **526**, 527
interphase
meiotic, 134
mitotic, 36, **37**, 44
intine, 366
inversions, chromosomal, 144
invertebrates, lectins in, 550
ion pumps, 547
ions, 24, 539–40, 557, **619**, 623 (*see also* anions; cations; specific ions)
and active transport, 568–69
concentrations in green alga, **60**
functions of, 540
and inorganic nutrient uptake, 541, 557, 568–69
phloem-immobile, 569
phloem-mobile, 569
transport in plant, 569
Ipomoea batatas, 443
irises, 155, 358
iron (Fe), 91, **92**, 538–40, 543, 557
ions (Fe^{2+}; Fe^{3+}), 540
oxide, 542
island species, oceanic, 147–48, 152, 158, 160–61
isocitrate (isocitric acid), 89
Isoetes, 310, 326–27
muricata, **28**, **326**
isogamy, 215, 258, **259**
isolation and differentiation, 154–55, 157–58
isoleucine, 52
isomerase, 86
⁶N-isopentenyladenine (i⁶Ade), 508, 510
isotonic solutions, 62, 63
isotopes (*see also* dating methods)
atomic, 616
radioactive (as tracers), 47, 100, **562**, **572**
ivy, 441, 514

jacket layer (sterile)
of gametangia, 288–89, 303
of sporangia, 288
Jacob, François, 128
Jaffe, M. I., 531
Janzen, Daniel, 152, 585
Jefferson, Thomas, 597
jelly fungus, 234, **235**
Joshua tree, **602**
Juglans cinerea, **487**
Juniperus (junipers), 350
communis, **351**
Jupiter, 186
jute, 419
juvenility, 513–14

Kalanchoë, 156
daigremontiana, 112, 155
karyogamy, 220, **221**, 231, **234**, **235**, 249
Keck, David, 156
kelp, 252, 268–71, 275 (*see also* brown algae)
Kentucky bluegrass, 111
kernel, **408**, **414**
kerosene, 511
keto acids, 196, 548, 557
α-ketoglutarate (α-ketoglutaric acid), 89, 92
kilocalories, 75–76, 82
kinases, 82
kinetic energy, 74
kinetin, 508–10
kinetochore, 39, **40**
kingdom
classification by, 175, 177–78, 182
schemes of, 175
Klebsiella pneumoniae, 555, **556**
Klebsormidium, 266, 267
subtillissimum, **266**
kohlrabi, 474
Kornberg, Arthur, 122
Kranz anatomy, 109
Krebs cycle, 84, 88–90, **92**, 93, 95, 107
K selection, 581
Kummerow, Jochen, **342**
Kurosawa, E., 512

labellum, **387**
lactate (lactic acid), 94
Lactobacillus, 202
acidophilus, **171**
lactose, 47, 126, **127**, 131
(lac) system, 126, 131
Lactuca
biennis, 175
sativa, 514, 524
lakes, 553, 556, 589, 610
Lamarck, Jean Baptiste de, 147
Laminaria, 270
digitata, **269**
life cycle of, **271**
ochroleuca, **269**
saccharina, **269**
laminarin, 253
land plants, 382, 595
Lang, Anton, 528
"language of life," 118, 124
lanthanum, 439
larch, 580
European, **350**
Larix, 580
decidua, **350**
laricina, **610**
late wood, 489, **490**, 493, **494**, 496
latex, **607**
Lathyrus odoratus, 144
Latin names, 169
latosols, 598
lava, 593–94
leaching, 543, 548, 553–54, 557, 585, 597–98
lead (Pb), 616
leaf, 4–5
adaptations for light, 580
adult, 513–14
of angiosperms, 374
arrangement of, 457
arrangements, 580
association with stem, 453–54
axil, 5, 414
blade (lamina), 308, 311, 457
of bryophytes, 288, 295, 298, 302
buttress, 454
of conifers, 341
of cycads, 353
of desert plants, 603
development of, 454–56
of dicots, 457, 459, 462–63, 475
of *Ephedra*, 354, **355**
epidermis of, 459, **460**, **461**
of *Equisetum*, 310–11, 327
evolution of, 310–11
of ferns, 333–34
floating, 590
folding and unfolding, 462
function, 4, 307–8, 445
gap, **309**, 310, **333**, 453, **454**
of *Ginkgo*, 353–54
of *Gnetum*, 354, **355**
ground tissue of (*see* mesophyll)
and heat absorption, 603
of hydrophyte, 459, 461
of *Isoetes*, 310, 326
juvenile, **341**, 513–14
leathery, 582, 598, 603, 612
of *Lycopodium*, 320
mesophyll of, 108–9, **110**, 112, **308**, 342–43, 462, 475, **569**
of mesophyte, 459, 462
modifications, 471–75
of monocots, 457, **458**, 459, **462**, 463–64
morphology of, 457–59
origin (evolutionary) of, 310–11
photoperiodic responses in, 524
of pine, 341–43
primordium (primordia), 445–46, 454, **455**, 456
of savanna trees, 601
scar, 466, **487**
of *Selaginella*, 323
sheath, **458**
stalk (*see* petiole)
structure, 459–64
succulent, 475
traces, 310, 453, **454**
of tropical rain forest trees, 598, 612
of vascular plants, 307–11, 314
veins (venation), 454, **455**, 462–63, **464**
of *Welwitschia*, **355**
of xerophyte, 459, **461**
leafhopper, 203, 208, 586
leaflets, 456–59
leaves
as soil components, 542, 545
tough, 594
Lecidea albocaerulescens, **230**
lectins, 550–51
Legionella pneumophila, 200
Legionnaire's disease, 200
legumes, 383, 394, **395**, 529, **536**, **537**, 539, 550–51, 555–56, 579, 594
leguminous plants, 152, 197–98, 549, 557
Lemaireocereus, **390**
lemma, **392**
lenticels, **483**, 484, **487**, 496
lepidodendrids, 320, 330
Lepidodendron, **316**, 330
Lepiota procera, **233**
leptoids, 288–89
Leptonycteris, **390**
Lerner, I. Michael, 151
lettuce, 513–14, 523–25, 529
leucine, **52**, **127**
leucoplasts, 22, **23**
leucosin, 253, 276
Lewin, Ralph A., 196
lichens, 225–27, 230, 239–40, 579–80, 585, 590, **591**, 593–94, 632
algal or cyanobacterial component of, 225–26, **227**, 230
biology of, 226–27
cortex of, **227**
crustose, **225**, **611**
foliose, **225**
fruticose, **226**
fungal component of, 225, **227**, 230
goldeye, **226**
imbibition by, 227
lectins in, 550
organization of, 227
reproduction in, 226
structure of, **227**, 228
lid (*see* operculum)
life cycle(s) of angiosperms and gymnosperms compared, 317, 371 (*see also specific group or organism*)
principal types of, **180**, **181**
of a vascular plant, **311**
ligase, 194
light, 1, 2 (*see also* energy; solar radiation)
absorption of, 101–2
intensity in biomes, 596, 598, 601, 606
particle theory of, 98
and phototropism, 518
reactions, 100–6, 112
red, far-red, 524–27, 533
and seed germination, 524
and stomatal movements, 567
wavelengths of, 101–2, 519, 524–26
wave theory of, 98
lignin, 32–34, 44, 47, 197, 308, 438, **440**
ligule of *Selaginella* leaf, 323–24
lilac, 161, **162**, **460**, **461**, 504, 530
lilies, 358, **359**, **364**, **365**, **366**, **367**, **369**, 384, **387**, 514
lotus, **380**
Tiburon mariposa, 609
water, **358**, **420**, **461**
Lilium, **134**, **364**, **366**, **367**, **369**
henryi, **359**
longiflorum, **365**
lime, 554
limestone, 541
Linaria, 396
linden, **420**, **422**, **424**, 426, **448**, 449, 482, **485**, **487**

Linnaeus, Carolus, **169,** 170, 175
Linum usitatissimum, 214
lip cells, 335
lipid(s), 45–46, 50–51, 58, 270
droplets, 21–22, 28
as membrane component, 50–51, 64, **65**
in seeds, 369
lipopolysaccharide, 190
Liriodendron tulipifera, 463
Liriophyllum, **372,** 376
Lithocarpus densiflora, **362**
liverworts, 173, 288, 290–95, 303 (*see also* bryophytes)
asexual reproduction in, 294–95
sexual reproduction in, 292–94
Ljungdahl, Lars, 185
loam, 542, **543**
Lobaria verrucosa, **227**
lobelia, 514
family, 609
lock-and-key hypothesis, 78
lockjaw, **186**
locule, 359, 380
locus (on chromosome), 139
incompatibility, 383
locust
black, **458, 459,** 477, **486, 487,** 560
honey-, **458, 459**
lodicules, **392**
long-day plants, 522–25, 527–28, 530, 533
Longistigma caryae, **571**
Lonicera, 488
Lophophora williamsii, **399**
Lotus corniculatus, **536, 537**
lotus, **380,** 529
Lumer, Cecile, 598
luminescent organisms, 93
lupine, **360**
arctic tundra, 529
white, 569
Lupinus
albus, 569
arcticus, 529
diffusus, **360**
Lycogala, **248**
Lycoperdon ericeterum, **228**
Lycopersicon esculentum, **429,** 506, **518,** 560, 586
Lycophyta, 309–12, 317, 319–27, 330, 337, 635 (*see also Isoetes; Lycopodium; Selaginella*)
Lycopodium, 304, 312, 320–24, 330, 337
clavatum, **321**
complanatum, **174**
life cycle of, 322
lucidulum, **321**
obscurum, **320**
lycopods, **307,** 320, 323
lycopod trees (*see* lepidodendrids)
Lyell, Charles, 148
Lysichiton americanum, **384**
lysine, 52, 133
lysis, 184, 206
lysosomes, lysosomal activity, 26, 32
lysozyme, 54–55, 206

Macrocystis, 268, **269,** 270, 275
pyrifera, **270**
macrofibrils, cellulose, 32, **33**
macronutrients, 538–40, 545, **557**
Macrosteles fascinfrons, 203
madder, **397**
magnesium (Mg), **49, 103,** 538–39, 557, 618
in chlorophyll molecule, 539
ion (Mg^{2+}), 540, 553, 618
Magnolia (magnolia), 378, 384, 393
family, 378
grandiflora, **379**
virginiana, **357**
maidenhair tree, 340, 353–54
Mairan, Jean-Jacques de, 520
malaria, **399**
malate (malic acid), 88, 92, 108–9, 111–12
Malloch, D. W., 239
Malpighi, Marcello, 422
Malthus, Reverend Thomas, 148
Malus, 362, **363,** 477, 485
Malvaceae, **380**
mammals (*see also by name*)
grazing, 604
heat energy in, 85
lectins in, 550
as pollinators, 598
Mammillaria wildii, **174**
manganese (Mn), 538–40, 543, 557
ions (Mn^{2+}), 540
mangrove
black, 441, **442**
red, 441
Manila hemp, 419
mannitol, 253, 270
mannose, 49, 253
mantle, fungal or hyphal, **238**
manzanitas, 159
maple, 395, 466
red, 26, **591**
silver, **457**
sugar, **457**
maquis, 608
marble, 541
Marchant, Harvey, 255
Marchantia, **173, 290, 291,** 292, **293,** 294–95
polymorpha, **292**
margo, **491**
marijuana, 398, **399**
marshes, 548, 553, 610
marsh marigold, **607**
Marsilea, **332, 335**
mass, 76
Mastigocladus, 203
maternity plant (mother-of-thousands), 111–12
mating
random/nonrandom, 149–50, 165
strains, 217
mattorral, 608
Maxwell, James Clerk, 98
meadows, 589–90
measurement, units of, 17, 629
Medicago, 451
sativa, 433, **448,** 451, 539
medicinal compounds, 399
mediterranean-type climates, 608–9, 612
medulla (lichen), **227**
Medullosa noei, 331
megagametogenesis, 366, 368
megagametophytes, 312–14, 339, 363, 366, 368, 371
megaphylls, **309,** 310–11, 314, 333–34, 337, 339,371
evolution of, 310–11, 314
megasporangia, 311, 313, 323, **324,** 338, 340, 344, 346, 366, 371, 400
megaspores, 311–14, 324, 338, 344, 346, 366, 368, 371
megasporocytes (megaspore mother cells), 324, **325,** 344, **346,** 366, **367**
megasporogenesis, 366, 368
megasporophylls, **309,** 310–11, 314, 323, **353,** 359
meiosis, 132, 134–39 (*see also* life cycles)
in angiosperms, 363, 366, **367**
consequences of, 138
and evolution, 163
first division, 134–35, 146
gametic, **180,** 270, **272**
in gymnosperms, 344, 346
and mitosis contrasted, 138–39
in mosses, **302**
prophase, 134–35, **136, 137**
second division, **116,** 117, 146
sporic, **180,** 270
and syngamy, 132, 139, 146
zygotic, **180,** 215, 258, 262, 264–65, 267
meiotic spindle (apparatus), 135, **136**
Melampsora lini, **214**
Melilotus alba, 208
melon family, 378
membrane(s), 16
and biological clocks, 522
cellular, **17**
components of, 50–51
differentially permeable, 62
dynamics, 28–29
and electrical impulses, 532
functions of, 15, 59, 68
and membrane systems, 18, 27–28, 44
plasma, 2, 15, **17,** 18, 43
structure of, 16, 50–51, 64, **65**
tonoplast (vacuolar membrane), 18, **23,** 24, 44
transport across, 64, 66–67, 69
unit, 16, **17,** 18, 20, 22, 43, 51
Mendel, Gregor, 137, 139–42, 149
Mendelian genetics, 132
Mendelian ratio, 139–42, 149
meristem, 4, 416 (*see also* apical, ground, and primary meristems; procambium; vascular cambium)
in hornworts, 296, 303
lateral, 416
marginal, 454, **455**
peripheral, 446
pith, 446
meristematic cells, tissues, 417, 477–79
Merrill, Elmer D., 350
Mertensia virginica, **360**
mescaline, **399**
Mesembryanthemum crystallinum, **429,** 475
mesocarp, 369
mesocotyl, growth of, 526
mesophyll (*see* leaf mesophyll)
mesophytes, 459, 462, 475
Mesozoic era, 272, 313, 331, 353–54, 372
metabolism, 77–78, 83
metamorphic rocks, 541
metaphase
meiotic, 135, **136, 137,** 138
mitotic, 36, **37, 38,** 40
metaphloem, 452, **453**
Metasequoia, 350, **352**
glyptostroboides, **352**
metaxylem, 423, **436, 437,** 440, 452, **453**
methionine, **52**
metric table, 629
micelles, 32, **33**
Micrasterias, 267
microbodies, **23,** 24, 110
Microcoleus vaginatus, 202
microcyst, 249
microfibrils, 32–34, 423, 495, 511 (*see also* cell wall)
microgametogenesis, 363–64, 366
microgametophytes, 311–12, 314, 324, **325,** 363, **365, 366,** 368
microhabitats, 580
micronutrients, 538–39, 545, 557
microorganisms, 543, 552
microphyll, **309,** 310–11, 314, 320, 323, 326–27, 337
micropyle, 338, 344, 346, 366, 368, 382, 406, 410
microspectral apparatus, 102
microscope,
electron, 16–17
light, 16
scanning electron, 16–17
transmission electron, 16
microsporangia, 311, 323, **324,** 344, **351,** 354, 359, 363
microspores, 311, 314, 324, **325,** 344, 363, **364**
microsporocytes (microspore mother cells), 324, **325,** 344, 363, **364**
microsporogenesis, 363–64, 366
microsporophylls, 323, 344, **351,** 359, **378**
microtubules, 29–31, 44, 284 (*see also* spindle fibers, apparatus)
and cell-plate formation, 41–43, 163, 255–56
and cell wall formation, 29, 35
in flagella and cilia, 30, **31,** 44, 177
and positioning of nucleus, 39
and spindle fibers, 40
middle lamella, 507, 510, 541
midrib, midvein, 454, **460, 461,** 462
migration, differential, 150, 165
Miki, Shigeru, 350
mildew (*see also* mold)
downy, 242, 244, 250
powdery, 219
milkweed, 394, 396, **397,** 398, **583**
family, 473
foul-smelling, **384, 385**
Miller, Carlos O., 508
mimicry, 386, 398
Mimosa, 532
pudica, **531,** 532
Mimulus cardinalis, **389**
minerals, 3, 541, (*see also* soils; nutrients)
cycles, 545–53
in enzyme catalyzation, 539, 557
mint family, 378
Miocene period, 350
Mississippian period, 340, 637
Mitchell, Peter, 91
mites, 594
mitochondrion, **15,** 22, **23,** 24–25, **28,** 44, **84,** 110
and energy conversion, 88
evolutionary origin of, 25
function, 22, 23, **84,** 88, 91
and phosphorylation, 91
structure of, 22, **23,** 91
mitosis, 36–41, 44, 366, 368
in algae, 255–56
duration of, 41
in fungi, 215
and meiosis contrasted, 138–39
in mosses, **302**
Mnium, **299**
molds, 219, 239, 241–50 (*see also* water molds)
bread, 216–17, 219
cheese, 222
mole (gram molecule), 622
molecular biology, **119**
molecular weight, 615
molecules, 615
and energy-exchange, 81–83
energy storage in, 84–85
excited, 99, 105
organic, 2, 45–58
rate of motion in, 625
molybdenum (Mo), 538–40, 557
ions (MoO_4^{2-}), 540
Monera, kingdom, 177, 182, 184, 631
monkeyflower, scarlet, **389**
monkey-puzzle tree, 350
monocots, 309, 356, **357,** 358, 364, 379, 381, **382,** 526, 635
and dicots compared, 358, 369
Monocotyledones, 358, 371, 635
Monod, Jacques, 128
monoecious plants, 361, **362,** 511
monokaryotic hyphae, monokaryosis, 231, **234**
monomers, 46, 51
mononucleosis, 208
monosaccharides, 46–47, 58
Monotropa uniflora, 357
Morchella, **220**
esculenta, **228**
morel, 219
Morel, George, 195
Morgan, Thomas Hunt, 143
Moricandia, **458**
morning glory family, **357**
morphogenesis, 417
Morus alba, **457**
mosquitoes, 386, **387**
mosses, 173, 288, 291, 296–303, 580, 590, 593–94, **611** (*see also* bryophytes)
aquatic, 298
asexual reproduction in, 302
club, 296
cushiony, 298
feathery, 298, **300**
granite, 296, 303
hairy, **300**
Irish, 266, 273, 296
life cycle of, 302
peat, 296, 302–3
reindeer, **226,** 227, 296
sea, 296
sexual reproduction in, 297–303
Spanish, 296

true, 296–302
urn, 288
moths, 384, 388, 398, 400, 582, **583**
hawk-, 388
skipper, **388**
mountain lilacs, **162,** 198
mRNA (*see also* RNA)
and cytokinins, 510
and gibberellins, 516
mucilage, 188, 202, 449
-containing cavities, 296
layer, 383
mucilaginous compounds, 270, 272, 383, 547
mulberry, **457**
multicellularity, 177
in prokaryotes, 184
multilayered structure (MLS), 282
multinucleate cells (*see* fungi, coenocytic)
Münch, E., 573
muramic acid, 176
Musa × *paradisiaca*, **357**
muscles, 94, 179
Muscopsida, 291, 296–303, 634 (*see also* bryophytes; mosses)
mushroom, **171, 228, 229,** 230, 232, **233,** 234, 240, **587**
edible, 179
life cycle of, **234**
poisonous, **228,** 232
mustard, 394, **397,** 398, 506, **538**
mutant, 143, 193
mutation
and adaptation, 382
early studies of, 143
effects of single-gene changes, 143
and fungal hyphae, 216
and the operon, 128
in prokaryotes and viruses, 193, 206, 210
rate, 144, 150, 206
and selection, 119, 144
as source of variation, 143–44, 146, 150
in *Volvox*, 261
mutualism, 579–80, 594
mycelium, 213–14, 216, 219, 230–31
Mycena lux-coeli, 93
Mycobacterium tuberculosis, 188, 199
mycologists, 213
Mycoplasma (mycoplasmas), 202–3
mycoplasmalike organisms (MLOs), 202–3
mycorrhizae, 237–40, 287, 580, 594, 597
vesicular-arbuscular, 237
mycorrhizal fungus, 357, 580
of angiosperms, **357**
of *Lycopodium*, 321
of *Psilotum*, 317
Mydaea urbana, **229**
myketos, **213**
Myrica, 198
gale, 551
myxamoeba, 130
myxobacteria, 186
myxomatosis, 580, **581**
Myxomycota (myxomycetes; *see* slime molds)
Myzus persicae, 586

NAA, **503**
NAD, 56, 80–81, 85–95, 548
NADH + H^+, 80, 85
$NADH_2$, 85–95
NADP, 105–6, 112, 548
$NADPH_2$, 105–6, 112
nannoplankton, 251
naphthalenacetic acid (NAA), **503**
narcissus, 530
natural selection, 119, 144, 150–51, **160, 161,** 382
neck,
canal cells, 289, **302**
nectar, 378, 382, 384, 386, **387,** 388, 390, 400, 522, 532, 582, 598
nectaries, 378, 382, 386, **387,** 388, **584,** 585
nectarine, 175
negative pressure in xylem, 564
Neisseria gonorrhoeae, 198
Neljubov, Dimitry, 510
Nelumbo
lutea, **380**
nucifera, 529
nematode, 224
as soil components, 542
Nemophila menziesii, **380**
neomycin, **186,** 187
Nepeta cataria, 170
Nereocystis, 268
Nerium oleander, 461
Neurospora, 144, **216,** 219
crassa, **215**
neutron, 76
Newton, Sir Isaac, 98
n-formyl methionine, 126
nickel, 554
Nicotiana, 388, **465,** 586
silvestris, 528
tabacum, **397,** 506–7, 522, 528
nicotinamide, 80
adenine dinucleotide (NAD), 80, 85
ring, 80
nicotine, **397**
Niel, C. B. van, 97, 100, 196
Nigella, **160, 161**
nightshade, 399, 506
Nilson-Ehle, H., 145
Nitella
hyalina, **268**
ion concentration in cytoplasm, **60**
nitrate ions (NO_3^-), nitrite ions (NO_2^-), 197, 204, 210, 540, 543, 546, 548, 553, 557, 587
nitric acid, 556
pollutant, 465
nitrification, 197, 546, 548, 553, 557
Nitrobacter, 197, 546
nitrogen (N), 1, 6, 538–41, 545–49, 556–57, 611
assimilation of, 548
base, 56–58
-containing ring (*see* porphyrin ring)
-containing ring compounds, 397
cycle, 197–98, 545–53, 557
dioxide, 465
fixation
biological, 197–98, 296, 537, 545, 548–53, 555, **556,** 557
efficiency of, 555
industrial, 549
by lightning, 549
in rainwater, 549
-fixing organisms, 187, 198, 537, 539, 549–52, 557
atmospheric, 537, 545, 549, 553
gas (N_2), 548
ions (NO_3^-; NO_2^-; NH_4^+), 540
loss, 548, 553, 587
oxides, 541, 556
nitrogenase, **556**
nitrogenous compounds, 546
Nitrosococcus nitrosus, **187**
Nitrosomonas, 197, 546, 553
nitrous oxide (N_2O), 548, 553
Noctiluca scintillans, **280**
nodal structure, **449,** 453, **454**
node, 5, 359, 445
nodules,
root, **536, 537, 549,** 550–51, 594
noncyclic electron transport flow, **105,** 106
noncyclic photophosphorylation, 106
nonpolar molecules (heads), 51, 62, 64, 66
nori, 275
Nostoc, **171,** 204, 225, 230, 296
commune, **188**
Nothofagus menziesii, **375**
N-terminal, 53, 126
nucellus, 313, 338, 340, 344, 346, 366, 368–69
nuclear division, comparison of in kingdoms, 215
nuclear envelope, 18, **19,** 27, 39–40, **41, 43,** 44, 134–35
nuclear pores, 18, **19**
nucleic acids, 45–46, 56, 508, 510, 548 (*see also specific compounds*)
and cytokinins, 508, 510, 517
and viral, 206, 208
nucleolus, **14,** 20, **37,** 39–40, 124, 134–35
nucleoplasm, 20
nucleoprotein, 205
nucleosome, 133
nucleotides, 80, 541, 548
in ATP, 56, 58
in genetic code, 123–26, 131
in Krebs cycle, 88
and mutation, 144
in nucleic acids, 56, **57,** 58, 119, 121–25
nucleus, **15,** 16, 18, **19,** 20, 44, 176–77, 284
atomic, 615–18
diploid and haploid, 20
fungal, 215–16
interphase, 36, **37,** 44
nudibranch, 255
nutrients (inorganic nutrients) (*see also* soils)
concentrations in plants, 538–39
cycles, 545–46, 548–53, 557, 585, 587, 594, 597
cycles and humans, 552–53
essential inorganic, 537–38, 552–54, 556–57
functions of inorganic, 539–41, 557
movement and recirculation in plants, 545, 557, 568–69 (*see also* minerals/nutrients; vascular system; xylem)
source of inorganic, 541
nutrition, plant,
general requirements of, 537
and plant breeding, 554
research, 554–56
and soils, 536–57
nutritive organs, 585
nuts, 395
Nymphaea, **358**
odorata, **358, 420, 461, 590**
Nypa fruticans, **375**

oak, 169, 361, **362,** 391, 421, 433, **591**
black, **488**
blue, **608**
cork, 488–89
red, **422, 457, 485, 492**
tan-bark, **362**
white, **495,** 539
oak-leaved goosefoot, **453**
oats, 111, 502, 513
obligate anaerobes, 185
Ochroma lagopus, 489
Oedermeridae, **384**
Oedogonium, 264–65
Oenothera, 128, 170, 388
biennis, 170, 175
glazioviana, 143
muelleri, 170
oil droplets (*see also* lipid droplets), 276
oils, 50, 585
essential, 397
mustard, **538**
old man's beard, **225**
Olea europaea, **429**
oleander, **461**
oleic acid, 50
olive, 393, **429,** 507
Oliver, F. W., 331
Onagraceae, **379**
onion, **19, 39, 407,** 414, 434, 567
oogamy, 258, **259,** 261, 265, 267, 288
oogonium (-ia)
in algae, 265, 267, **268, 272, 279**
in oomycetes, **242, 243**
Oomycota (oomycetes), 241–45, 249–50, 632
life cycle of, **243**
oospore, 242, **243**
operator gene, 128
operculum, **288,** 298, 302
operon, 128
Ophioglossum, 20
Ophrys, 386
opines, 195
Opuntia (*see also* cactus, prickly-pear)
inermis, 582, **583**
orange(s), 511, 515
orbitals, electron, 617
orchard grass, 153
orchid, 238–39, 358, 381, **382,** 386, **387,** 395, 400, 442, 506 (*see also by name*)
bee, **387**
Cymbidium, **174**
epiphytic, 112
lady's slipper, 381, **464**
mycorrhizae of, 238–39
propagation of, 506
Orchidaceae, 238, 379, 381, 400
order, classification by, 175, 182
Ordovician period, 215
organelle, 16, 18, 43 (*see also specific types*)
organic compounds, 45–58, 540, 588 (*see also by name*)
recycling of, 26, 198, 553
organic matter, 542–43, 545, 557, 588–89, 594
organs, 307, **308**
storage, 606
Origin of Species, On the, 132
Oryza sativa, 111, **357,** 512
Oryzomus devius, 598
Oscillatoria, **171, 188**
osmosis, 62–63, 69, 539, 541, 557, 561, 567, 573–75
osmotic potential, pressure, 62
Osmunda, **335**
cinnamomea, **332, 334**
outcrossing, 390–91
ovary, 359, 361, 383, 393, 400, **468, 469**
development into fruit, 363, 369, 371, 505, 517
inferior, **357,** 363, 379–82, 393–94, **470**
superior, 362, 379, 394
Overbeek, Johannes van, 507
ovipositor, **389**
ovule(s), 313, 315, 338
of angiosperms, 358–59, 366, **367,** 377, 380–81, 392, 400
of *Ginkgo*, 354
of gymnosperms, 340, 344, 382
of pines, **347**
of yews, 350
ovuliferous scale, 344, 349
Oxalis, **520**
oregana, **595**
pes-caprae, **370**
oxaloacetate (oxaloacetic acid), 89–90, 108–9, 112
oxidation, 626–28 (*see also* oxygen, reduction of)
of carbon, 90
of glucose, 84–95, **628**
of pyruvate, 91
-reduction reaction, 626
oxidative phosphorylation, 90–92
oxygen (O_2), 1–4, 538, 540, 557, 595 (*see also* oxidation)
as atmospheric element, 557
atom (O_2), 45–46, 84
discovery of, 97
in electron transport, 90, 92, 95
evolution of, 197
heavy isotope of, 100
as an inhibitor, 556
and Krebs cycle, 90
and photosynthesis, 100, 105–106
and respiration, 84, 95
in water molecule, 622–23
oxygen debt, 94
Oxyria digyna, 157
oyster mushroom, **171**
ozone, 2, 99, 553
injury, 465

Paeonia californica, **133**
pairing (chromosomal), 134, **136, 137,** 146, 163

palea, **392**
Paleozoic era, 252, 313, 327, 339, 354, 371–72, 637
palmettos, **373**
palmitic acid, 50, 89
palms, 358, **375,** 441, 476
  coconut, **357, 375**
  fan, 373
  sago, 353
  Washington, **602, 603**
PAN, 465
Pandorina, 260–61
panicle, **361**
pansy, 170
pantothenic acid, 88
Papaver (*see also* poppy), **395**
Papaveraceae, 394
papillae of stigma, 383
pappus, 380, 395
paramylon, 253, 281, 295
paraphyses, 299
parasexuality, 215–16, 222, 239
parasites, 587, 594
  angiosperm, **357,** 358, **395**
  bacterial, 199
  fungal, 214, 219, 222, 224, 234
paratyphoid, 199
parenchyma, 417–18, 429–30
  axial, 489
  palisade, **460, 461,** 462, 475, **514**
  in phloem, **424,** 426
  ray, 423, **491**
  spongy, **460, 461,** 462, 475
  storage, 418, 442–43
  wood (xylem), 423, 489
parenchyma cells (*see* parenchyma)
parental species and hybridization, 159–65
Parmelia perforata, **225**
parsley, 395, 566
parthenocarpic fruit, 505, 514
Parthenocissus quinquefolia, 471
passage cells, 440
passion flower, 388
pasteurization, 199
pasture plants, 153
pastures, 579–80, 589, 604
pathogen(s), 12, 187, 199–201, 204–5 (*see also* bacteria; rusts; smuts)
  animal, 199–200, 208–9
  fungal, 219, 224, 230, 234–36
  plant, 200–201, 205
pea, 139–42, **406,** 408, **458,** 510–11, 514, 531, 549
  family, 152, 378, 394
  Mendel's experiments on, 139–42
peach, 175, 393, 511, 515, 530
peacock, 152
peanut, 471
pear, 393, 420, 485
  industry, 187
peat bogs, pH of, 625
peat mosses, 296, 302–3 (*see also Sphagnum)*
pectic acid, **49,** 541
pectic compounds (substances), 32–33, 47, **49,** 253, 276, 383, 396, 434
pectic layer, 383
pectin, 44, 49, 366, 510
pedicel, 359
peduncle, 359
pelargonidin, **393**
Pellaea glabella, 334
pellicle, 281-82, 285
penicillin, 198, 200, 224, 239, 581
penicillinase(s), 198, 200, 581
Penicillium, 222, 223, 224
  camembertii, 222
  chrysogenum, 239, 581
  dupontii, **223**
  roquefortii, 222
Pennsylvanian, 330, 637
Penstemon hybrids, **163**
3-pentadecanedienyl catechol, **398**
pentoses, 46, 56
peony, 26, **133**
PEP, carboxylase, 108, 111–12
Peperomia, 475
pepper plant, 207
peptide bond, 53, 126, **127**
peptidoglycan layer, 190
perennials, 477, 513, 581, 601, 603–4, 606, **607,** 610, 612
perforation plate, 422
perforations, in vessel members, 421–22
perianth, 359, 378, 393, 395
  in liverworts, 295
pericarp, 369, 394, 396, **408,** 409, 414
periclinal division, 477–78, 495
pericycle, **423, 424, 439,** 440, **441,** 444, 479–80, 495–96
periderm, 308, 343, 417, 428, 429, 474, 482–88, 495–96
perigyny, 362, **379**
perisperm, 369, 408, 412
peristome, 298, 301
perithecium, 219
permafrost, 611–12
permanent wilting percentage, 544–45
permeases, 66
Permian period, 310, 320, 331, 338, 341
Peronosporales, 242
peroxisomes, 24, 110
peroxyacetyl nitrate, 465
Persea americana, 511
persimmon, 35
petals, 359, 378, 380, 382, 400
petioles, 457, 507
  of *Acacia* leaves, **584,** 585
Petunia (petunia), 506, 514
PGA, 107, 112
pH, 553–54, 556, 624–25
Phaeophyta, 268–72, 285, 633 (*see also* brown algae)
phage (*see* bacteriophage)
phagocytosis, 67–69
Phalaris canariensis, 502
Phallus impudicus, **229**
Phaseolus
  lunatus, 141, 550
  vulgaris (bean), **43, 406,** 413, 512–13, **520,** 554, **555**
phellem, 482 (*see also* cork)
phelloderm, 428, 482–84, 496
phellogen, 482, 484, 495–96
phenotype, 139–40, 142, 145, 151, 153, 156
phenylalanine, **52, 125**
Philodendron, 85
phloem, 4–5, 304, 307, 309, 423–26, 429, 440
  disease in, 202–3
  and florigen, 527, 533
  functional, **485, 486,** 489, 496
  loading, 573
  nonfunctional, **485, 486,** 489, 496
  and plant hormones, 504
  primary, 307, 440, 449–50
  secondary, 308, 477–81, 484–85, 489, **496**
  secondary, of pines, 343
  transport (translocation), 569–74
phlox family, 378
phosphate (PO_4^{2-}), 50–51, 56, **57,** 58, **120,** 121, **123, 125,** 540, 543, 548
  high-energy bond, 82–88, 108
  inorganic (P_i), 87, 91, 105, 552
phosphoenolpyruvate (PEP), 87, 108–9, 112
  carboxylase, 108, 111–12
phosphofructokinase, 86
phosphoglucoisomerase, 86
3-phosphoglycerate (PGA), 107, 112
phosphoglyceromutase, 87
phosphoglycolic acid, 111
phospholipid bilayer, 64, **65,** 69
phospholipids, 50–51, 58, 541
phosphorescence, 102
phosphorus (P), 6, 50, 538–41, 553, 556–57
  cycle, 552–53
  ions ($H_2PO_4^-$; HPO_4^{2-}), 540
  loss from soil, 552–53
  pentoxide, (P_2O_5), 553
phosphorylation, 82
  oxidative, 90, 92
  photo-, 105–6, 112
  substrate, 87
photoelectric cell, 101
photoelectric effect, 98
photolysis, 105
photons, 98, 104
photoperiodism, 98–99, 522–27, 533
  chemical basis of, 524
  and dormancy, 530
  effects on animals, 522
  and flowering, 522–24, 527
  and measuring the dark, 524
  and phytochrome, 527, 533
  and temperature, 523, 533
  and vernalization, 530, 533
photoreceptor, 218, 519, 533
photorespiration, 110–12
photosynthesis, 1–2, 4, 21, 47, 83, 96–113, 580
  and algae, 253, 255
  in bacteria, 196, 210
  balanced equation, 100
  in bryophytes, 290, 298
  CAM, 111–12, 603
  and chloroplasts, 21
  and circadian rhythm, 520, **521**
  in desert plants, 603
  in an ecosystem, 558
  efficiency, 555
    in C_4 plants, 110–11
  evolution of, 197
  4-carbon pathway, 108–9, 112, 539
  generalized equation, 100
  in lichens, 226–27
  pigments involved in, 21, 103–4
  and shade plants, 580
  and temperature, 603
  3-carbon pathway, 107–8, 112
photosynthetic pigments (*see* accessory pigments; carotenoids; chlorophylls; phycobilins)
photosystem(s), 104, 112
  I, 105–6, 112
  II, 105–6, 112
phototaxis, 190, 218
phototropism, 98, 102, 502, 518–19, 530, 533
phragmoplast, 41, **43,** 256, 266, 282, 285, 287
phycobilins, 103–4, 196, 203, 253, 272, 285, 287, 526
phycocyanin, 196
phycoerythrin, 196
phycoplast, 255–56, 264, 285
Phyllitis scolopendrium, **332**
Phylloglossum, 319
phylogenetic relationships, 182, 307
phylum, 175
Physarum, **172,** 248
Physcomitrium turbinatum, 288
phytochrome, 525–26, 533
  dark reversion of, 526–27, 533
  photoconversion, reactions of, 525–26
  and photoperiodism, 527
phytohemagglutinins, 550
Phytophthora infestans, 244–45, 250
phytoplankton, 251, 276, 279, 285
Picea, 1, 350, 610 (*see also* spruces)
  glauca, **610**
Pickard, Barbara G., 532
Pickett-Heaps, J. D., 255
Pierinae, 398
pigment(s), 103–4 (*see also by name)*
  absorption spectrum of, 101–2, 525
  accessory, 101–2, 196, 287
  antenna, 104
  in chloroplasts, 21
  in chromoplasts, 22
  and genetics, 141
  light absorption by, 101–2
  and photoperiodism, 525
  photosynthetic, 103–4, 112, 181, 196, 210
  P_r, P_{fr}, 525–27, 533
  P_{680}, P_{700}, 105–6, 112
  in vacuole, 26
pileus, 232
pili, 189, 210
Pilobolus, **218**
Pinaceae, 238
pine(s), 238, 341–49, 443, 488–89, 610 (*see also Pinus)*
  bristlecone, **341,** 493, **494**
  closed-cone, 348, **592**
  digger, **345, 608**
  eastern white, **345**
  jack, **344,** 348
  life cycle of, **349**
  loblolly, 560, **593**
  longleaf, **341**
  Norfolk Island, 350
  pinyon, **341, 345**
  ponderosa, **4, 345**
  red, **341, 345**
  reproduction in, 344, 349
  slash, **607**
  sugar, **174,** 345, 591–91, 607
  western yellow, **4, 345**
  white, **237,** 477, **489,** 490, **491, 570, 591**
pineapple, 112, 393, 501, 567
Pinguicula grandiflora, 547
pinnae, 334
Pinnularia, **276**
pinocytosis, 67–69
Pinus, **238,** 341, **343, 344, 345, 346,** 489, 610
  banksiana, **344,** 348
  edulis, **341, 345**
  elliottii, **607**
  fascicles (needle bundles), **341, 342**
  lambertiana, **174, 345,** 591, **592, 604**
  longaeva, **341,** 493, **494**
  palustris, **341**
  ponderosa, **4, 345**
  radiata, **341**
  resinosa, **341, 345**
  sabiniana, **345, 608**
  seedling, **341**
  strobus, **237, 345,** 477, **489,** 490, **491, 570, 591**
  taeda, 560, **593**
pioneer organisms, 594
Pirozynzki, K. A., 239
pistil, 359
Pisum sativum, **406,** 414, **458,** 511, 531
pit, 34–35, 421, **422,** 489
pit cavity, 34–35
pit-fields, primary, 33, **34**
pit membranes, **34,** 35, 423, 490, **491**
pit-pairs, 35, 423, 489, **491**
pith, 5, **308,** 309–10 314, 340, 466, 504
  in roots, **437,** 440
  in stems, 447, **448,** 449–50
placenta, 361, 366
placentation, 361
Placobranchus ocellatus, 254
plagues, 206
plane tree, **429, 488**
plankton, 251, 276, 279, 285 (*see also* nannoplankton; phytoplankton; zooplankton)
plant(s)
  and air pollution, 464–65
  body, 5, 305, 307–11, 466–67
  classification, chemical compounds in, 397
  Devonian, 305–6, 314
  -herbivore interactions, 579, 582–86, 594, 598
  lectins in, 550
  and humans, 8
  nutrition and soils, 536–57
  Paleozoic, 313
  seed, 304, 313–14
  succulent, 111, 475
  tissues, 417–31
  water-storing, 475
Plantae, kingdom, 178, 182, 634 (*see also* bryophytes; vascular plants)
plant breeding, 235, 554, 581
plasma membrane(s) (plasmalemma), 15, 17–18, 29, 43, 59, 64, **137,** 184, 191, 434
  and active transport, 66, 568, 574
  and auxin, 504, 512
  and carrier molecules, 66
  and electrical impulses, 532–33
plasmid, 193–95, 555, **556**
plasmodesmata, 28, 35–36, 44, 68, 177, 263, 266–67, 364
plasmodiocarp, 248
plasmodium, **172,** 247–50

plasmogamy, 220, **221**, 231, **234**, **235**, 249
plasmolysis, 63, **64**
Plasmopara viticola, 242, 250
plastids, 20–22, 130, 526–27
Platanus, **159**
occidentalis, **488**
orientalis, **429**
Platycerium bifurcatum, **416**
Platymonas convolutae, 254
pleiotropy, 144–45, 151, 165
Pleistocene era, 8, 10, 160, **529**, 607
Pleodorina (see Eudorina)
Pleonosporium dasyoides, **273**
pleuro-pneumonialike organism (PPLO), 202 (*see also* mycoplasma)
Pleurotus ostreatus, **171**
Pliny, 235
plum, 26
plumule, 405–407, 413, 414, 445
pneumatophores, 441–42
pneumonia, 187, 193, 224
pneumonococci, 118
Poa
annua, **462**
pratensis, 111, 153, 162
Poaceae, 394 (*see also* grasses)
Poanes hobomok, **388**
Pogonatum brachyphyllum, **300**
poinsettias, 523
poison ivy, **398**
polarity, 409–10
polar molecules (heads), 51, 62, 64, 66
polar nuclei, **367**, 368
polio, 204
pollen (grains), 16
of angiosperms, 359, 363–64, **365**, 366, 368, 373, **379**, 383, **385**, **389**, 390, **391**, 392, 400, 514
fossil, 363, 373
of gymnosperms, 340, 373, 382, 400
of pine, 344, **345**, 346
wall, 364, **365**
pollen sacs, 359, 363, **364**
pollen tubes, 312, 383, 514
of angiosperms, 363, 366, 368
of gymnosperms, 340, **345**, 346
pollination, 312
of angiosperms, 363, **366**, 368, 371, **372**, 382, 384–92, 400, 594, 609
by bats, 390, 400, 598
by birds, 388, **389**, 390, 400, 609
cross-, 382–83, 390
of gymnosperms, 340, 371, 382
by herbivores, 582, 594
by insects, **372**, 382, 384–86, **387**, 388, 390, 400
by lemurs, 598
by mammals, 598
by marsupials, 598
mechanisms of, 383
by monkeys, 598
of orchids, 386
of pines, 344, 346
by rodents, 598
self-, 383
by wind, 382, 390–92, 400
pollinium, 382, 386, **387**
pollutants, 464–65
pollution, 464–65, 552–53, 556, 587
and cyanobacteria, 203
polyembryony, 340, 346
polygenic inheritance, 144–46, 151, 165
Polygonatum, **368**
polymers, 46, 51
polymixin B, 187
polynomials, 169–70
polypeptide, 53–55, 58, 79, 126, **127**, 190
polyploidy number, 163–65
polyploidy, 163–65
Polypodium virginianum, **334**
Polyporus sulphureus, **171**
polysaccharides, 32, 47, **48**, 58, 190, 551 (*see also by name*)
in fungi, 214, 239
in cell walls, 32, 383
Polysiphonia life cycle, **274**
polysomes (polyribosomes), **19**, 26, 44
Polystichum munitum, **595**
Polytrichum, **289**, 298
juniperinum, **298**
piliferum, **297**
pome, 393
pond(s), 579, 590
ion concentration in, **60**
Pontecorvo, G., 216
poplars, 396, 433
poppy, 394, **395**
California, 384
opium, 398
population(s), 383, 579
arctic, alpine, 157, 162
diploid, 149, 165
explosion, 10, 12
and Hardy-Weinberg equilibrium, 150, 165
haploid, evolutionary change in, 153
hybrid, 157–63, 165
mean value, 150
of microorganisms, 153
and mutation rate in bacteria, 193
natural, 383
natural and environmental pressures, 153, 156
and rapid evolutionary change, 153
and selective genetic change, 151, 165
Populus, 106, **394**, 495
tremuloides, **591**
pores
in *Azolla*, 551
in bryophytes, 292, 302
in pollen grains, 366, 368
Poria latemarginata, **231**
Porolithon craspedium, **273**
Porphyra, 275
porphyrin ring, 91, **92**, **103**, 548
portulaca family, 393
Postelsia palmiformis, **172**
potassium (K), 66–67, 538–40, 553, 557, 585
deficiency, 554, **555**
ion (K^+), 540, 543, 548, 553
and stomatal movements, 567
and touch response, 532
loss, 587
oxide (K_2O), 553
potato, **23**, **45**, **48**, 155, 398–99, 506, 512, 523, 530, 586
beetle, 398
blight, 245, 250, 344
eye (bud), 474, 512, 530
famine, Ireland, 244–45
propagation, 155
scab, **186**
spindle tuber disease (PSTV), 209
sweet, 155, 443
white (Irish), 244–45, 445, 474, 530, 560
wild, 586
Potentilla glandulosa, ecotypes of, 156
Power of Movement in Plants, The, 502
pox viruses,
chicken-, 204–5
small-, 204
PPLO (*see* Mycoplasma)
P-protein (slime), **425**, 426, 431
P-protein (slime) bodies, **425**, 426
prairies, 604, **605**
Precambrian period, 272, 637
precipitation, 544, 598, 607
in deserts, 602–3
in grasslands, 604
in mediterranean climates, 607–8
in monsoon forests, 601
in savannas, 601, 612
in the taiga, 610
in temperate deciduous forests, 606
in tropical rain forest, 597, 612
in the tundra, 611
predators, 582, 589, 594, 604
pressure bomb, 564
pressure-flow mechanism (hypothesis), 573–75
prey, 589
prickles, 472
Priestley, Joseph, 97
primary endosperm nucleus, 368–69
primary growth, 307, 416
in roots, 433–36
in stems, 446–52
primary meristems, 410–11, 416, 436, 444, 447, 475
primary pit-fields, 33, **34**
primary plant body, 307, 416, 466–67
primary structure
comparison in root and stem, 479
of root, 437–40
of stem, 447–52
primary tissues, 307–8, 416
of root, 433–40
of stem, 446–52
primary vascular tissues, 437, 440, 444, 447, 466–67
primordium
bud, 445–46
leaf, 445–46, 454–56
root, 440–41
primrose, evening, 143, 170, 388
primroses, 523
prism, 98
procambium, 410, 416, 421, 436, 444, **445**, 447, 449–54, **455**
Prochloron, **196**, 284
productivity, ecosystem, 594
proembryo, 410
progymnosperm, **307**, **338**, 339–40, 371
Aneurophyton-type, 339
Archaeopteris-type, 340
prokaryotes, 85, 124, 126, **171**, 175–76, 184–204, 210, 520, 557 (*see also* bacteria; cyanobacteria)
and anaerobic respiration, 94
and eukaryotes compared, 16, 126, 175–78, 189
genetic recombination in, 178, 193
genetics of, 192–93
locomotion in, 188–89
photosynthetic, 97, 103, 196
proline, **52**
prophage, 206
prophase
meiotic, 134–35, **136**, **137**, 146
mitotic, 36, **37**, **38**, 39
proplastids, 22, **23**
proteases, 526
prosthetic groups, 539
protective layer, 466 (*see* abscission zone)
protein, 45–46, 51–56, 58, 585 (*see also by name*)
biosynthesis, 123–26, 131
as cell-tissue components, 51, 53, 55
as cell-wall components, 33, 34
chloroplast, 107
coat in virus, 205–6, **207**, 208
and DNA/RNA interactions, 123–26, **127**
and enzymes, 55–56, 117, 123–26
and genetic changes, 144, 151
globular, 53, 55–56
histone, 133, 215, 283
layer, 383
lectin (carbohydrate-binding), 550–51
as membrane component, 64, **65**
molecule, 51, 58
and phytochrome, 526
plant, 550–51
in seeds, 369, 516
self-incompatibility, 383
specificity of, 550
structure of, 51–56, 58, 121
synthesis, 507, 510, 516–17, 537
viral, 205–6
proteinase inhibitors, 398
prothallial cells, 324, 344, **345**
prothallus, 335
Protista, kingdom, 177–78, 182, 630, 632–34 (*see also* algae; slime molds; water molds)
autotrophic, 251–86
heterotrophic, 241–50
protoderm, 410, 436, 444, **445**, 447, **450**, 475
proton, 76, 90, 92, 615–18
protonema, 296–98, 302–3
protopectin, **49**
protophloem, 423, **436**, 452
protoplasm, 18
protoplast, 18
bacterial, 191
fusion, 506
isolation of, 506
protostele, 309–10, 314
of *Lycopodium*, 320
of progymnosperms, 339
of *Psilotum*, 317, **318**
protoxylem, 423, **436**, **437**, **439**, 440, 450, 452, **453**
protozoa, 47, 177, 187, 277–78
as soil components, 542, 545
Prunella vulgaris, **153**
prunes, 511
Prunus, **362**, 397
persica var. *nectarina*, 175
persica var. *persica*, 175
Psaronius, 331
Pseudomonas, 201
marginalis, **189**
solanacearum, **187**
Pseudomyrmex ferruginea, 582, **584**, 585
pseudoplasmodia, 129–30
pseudopodium
in peat moss, 302
Pseudotsuga, 350
menziesii, 607
P_{700}, 105–6, 112
Psilocybe mexicana, **233**
psilocybin, **233**
Psilophyta, 310, 317–19, 337, 634 (*see also Psilotum; Tmesipteris*)
Psilophyton, **306**, 309–11
Psilotum, 312, 317–19, 337
nudum, **318**, **319**
P_{680}, 105–6, 112
psychedelic drugs, **233**
Pteridium aquilinum, **334**
Pteridospermales, 330–31
Pterophyta, 317, 332–37, 635
Ptilium crista-castrensis, 300
Puccinia graminis, 235–36
puffballs, **229**, 230, 232, 234
pulvinus, **531**, 532
purines, 56, **57**, 121–22, 508
Puya raimondii, 477
pyramids of energy, mass, numbers, **589**
pyrenoid, 257, 259, 262, 264, 267, 281, 296
pyrimidines, 56, **57**, 121–22, 124
Pyrrophyta, 279–81, 285, 634 (*see also* dinoflagellates)
Pyrus communis, **32**, **420**, 488
pyruvate (pyruvic acid), 85, 87–88, 91–92, 94–95, 108–9
kinase, 87

quantasomes, 22
quantum, 98
quartz (SiO_2), 541
quartzite, 541
Quercus, 169, **362**, **421**
alba, **495**
coccinea, 158
douglasii, **608**
rubra, **422**, **457**, **485**, **492**
suber, 488
velutina, 158, **488**
quiescent center, 434–35
quillworts, **28**, 326
quinine, 399
quinones, 397

rabbitbush, **602**
rabbits, 580, **581**
raceme, **361**
rachilla, **392**
rachis, 9, 334, 457
radial section (surface), **485**, **489**, 490, **491**, **492**
radial system, 478
radiation, 98–99, 588 (*see also* energy; light; spectrum)
infrared, 99
ionizing, 99
ultraviolet, 2
radicle, 408, 413–14
radioactive dating method, 47, 616

radioactive tracers, **562,** 571–72
radioautography (*see* autoradiography)
radish, 163, 510, **560**
ragweed, **365,** 522
rain, 544, 549
  acid, 556
rain forest, **599** (*see also* forest, tropical rain)
Ranunculus, **436, 437, 439,** 451, 476
Raper, John R., 244
Raper, Kenneth B., 247
Raphanus sativus, 163, **560**
raphe, 278
raphides, **25,** 398
raspberry, 155, 393, 396
rat, rice, 598
ray(s), 423, 478, 480, **481,** 482, **486, 489,** 490, 495
  of conifer wood, 492
  of dicot wood, 492
  dilated, 485
  pith, 447, **450,** 480
ray flowers, 380
ray parenchyma, **491**
ray tracheid, **491**
reactants, 78–79, 83
reaction center, 104, 112
  Pigment System I, 105–6, 112
  Pigment System II, 105–6, 112
Reboulia hemisphaerica, **292**
receptacle (flower), 359, 362, 378, **379,** 505
recombinant bacteria, 194
recombinant plasmids, 194
recombination (*see also* heterokaryosis; parasexuality)
  genetic, 158, 178, 193, 210, 239
recycling
  in forest ecosystem, 598
  of nutrients in plants, 569
red alder, **422**
red algae, 104, 272–74, 633 (*see also* algae)
  economic uses of, 272, 275
  life cycle of, **274**
red blood cells, **133,** 550
red/far-red reactions, 524–27, 533
Red Sea, 203
red tides, 280
reduction, 626
redwood, 340, 564, **578, 579, 595,** 607 (*see also Sequoia sempervirens*)
  dawn, 350, **352**
  family, **352**
reforestation, **350,** 593
region
  of cell division, 435, **436,** 475
  of elongation, **432,** 435, **436,** 475
  internodal and nodal, in corn stem, **449**
  of maturation, 436, 475
Regnellidium diphyllum, **23, 26, 84**
replica plating, 193
repression in enzyme synthesis, 128
reproduction
  asexual, 154–55, 219–20, 222, 239, 257, 259, 261–62, 264–65, 267, 276, **277,** 278, 280–81, 294–95, 302
  in bryophytes, 288–90, 292–94, 296–99
  by cell division, 177–78, 192
  by fission, 192, 220
  by fragmentation, 192, 267, 294, 302
  by gemmae, 294–95, 302
  in isolation, 157–58
  in pines, 344–49
  sexual, 132, 178, 180–82, 193, 215–16, 239, 257–59, 261–62, 264–68, 270, **271,** 273, **274, 277, 279,** 281, 288–90, 292–94, 296–99, 344–49, 383
  vegetative, 154–55, 162, 294–95, 302, 612
reproductive potential, 581
reproductive systems, 178, 180–81
reservoir in *Euglena,* 281
resin, 382, 489
resin duct, 342, 489, **490,** 496
respiration 2, 84–95, 511, 594
  aerobic pathway, 88–93
  anaerobic pathway, 94–95
  summary, **92,** 95
resurrection plant, 323
retinene, **104**
reverse transcriptase, 194
R group, **52,** 53
Rheum rhaponticum, **419,** 474
rhizobia, 550, **551**
Rhizobium, 198–99, **536, 537,** 549–51, 555–57
rhizoids, 214, 216–17, 245, **246,** 262–63, 289, 291–92, 296, 298
rhizomes, 154–55, 305
  of *Equisetum,* 327
  of ferns, 333
  of *Lycopodium,* 320
  of *Psilotum,* 317, **318**
rhizomorph, **238**
Rhizophora mangle, 441
rhizoplasts, 258
Rhizopus stolonifer, 216, **217**
Rhodophyta, 272–74, 283, 285, 633 (*see also* red algae)
Rhopalosiphum maidis, **571**
rhubarb, **419,** 474
Rhynia, 305, 309
  major, **306**
Rhyniophyta, 305–6, **307,** 314
D-ribitol, 230
riboflavin, **90**
ribonucleic acid (*see* RNA)
ribonucleotide, 57
ribose sugar, **46,** 56, **57,** 58, 124
ribosomes, **17,** 22, 24–26, **28,** 44, 124, **125, 127,** 131, 191
ribulose 1,5-bisphosphate (RuBP), 107, 111–12
  carboxylase, 107, 111
Riccia, 292, 294–95
Ricciocarpus, 292, 294–95
  natans, **393**
rice, 9, 11, 111, 203, **357,** 512, 551
Ricinus communis, **406,** 413
Rickettsia prowazekii, 199
ring compounds, 397
ringing experiments, 570
ringworm, 224
RNA, 24–26, 56, **57,** 58, 122, 541, 548
  and auxin, 507
  and cytokinins, 510
  and gibberellins, 516
  mRNA (messenger RNA), 124, **125,** 126, **127,** 128, 131
  polymerase, 124, **125**
  rRNA (ribosomal RNA), 124, **125, 127,** 131
  synthesis, 123–26
  tRNA (transfer RNA), 124, **125,** 126, **127**
  viral, 205, **207,** 208, 210
Robinia pseudo-acacia, **458, 459,** 477, **486, 487,** 560
rocks,
  types of, 541
rockweed, **172, 269**
rocky slopes, 589, **591,** 592
rodents, 581
Rollandia angustifolia, 609
root(s), 4–5, 307, 314 (*see also* systems; vascular plants)
  adaptations, 441–44
  adventitious, 154, 413, 441, 474, **503, 504,** 505
  aerial, 441, **442**
  of angiosperms, 432–44, 590
  apical meristem of, 307–8, 405, 412, 415, 434–36, 444
  and auxins, 503–5, 517
  cap, 348, 408, **432,** 434, 519
  comparison with stem, 437, 446, 479, 496
  and cytokinins, 508–9, 517
  development of, 433–40
  development, summary of, 496
  of dicots, **432, 436, 437,** 438
  embryonic, 408, 413
  of *Equisetum,* 327
  feeder, 433, 438
  of ferns, 333
  function of, 4, 432, 444, 560
  geotropism, 512, 519
  grafting, 585
  of grasses, 433, **542,** 555, **604**
  growth regions of, 434–36, 475
  of gymnosperms, 438
  hairs, 4, **432,** 436–38, 444, 550, 560–61, 574
  initiation of, 440, 509
  lateral (branch, secondary), 413, 440–41
  and lectins, 550–51
  of legumes, 550–51
  of *Lycopodium,* 320
  of monocots, **436, 437,** 438, 440
  and nitrogen assimilation, 548
  nodules, 198, **536, 537, 549,** 551, 594
  and nutrient uptake, 543, 548, 557
  origin (evolutionary), 310, 314
  phloem of, 440, 444
  of pines, **341**
  pressure, 561–63
  primary, 413, 432
  primary structure of, 437–40
  primordium, 440, **441**
  prop, 441
  secondary growth of, 479–80
  secondary structure of, 479–80
  of *Selaginella,* 323
  storage, 442–43, **444**
  structure, 437–40
  systems, 5, 307, 413, 432–33, 541, **542,** 603
  systems, extent of, 433, 561
  tip, 405
  vascular cambium of, 440, 443, **444,** 479–80
  xylem of, 440
root-fungi associations, 237–39
root-hair zone, **432,** 436
Roper, J. A., 216
Rosa, 169
Rosaceae, **379**
rose, 26, 393
  wild, 384
rosette, 514, 517, 530
roundworm (*see* nematode)
r selection, 581
rubber plant, **429**
Rubus, 163, 465
RUBP, 107, 111–12
  carboxylase, 107, 111
ruminants, 50
runner, 154
rusts, 230, 234–36
  wheat, life cycle of, 235
rutabaga, 568
rye, 111, 222, 235, 383, 433, 437
  winter, 530

Sabalites montana, **373**
Saccharomyces cerevisiae, **220,** 222
Saccharum officinarum, 552
sacred lotus, 529
sagebrush, **602, 605**
sage, purple, 581, **582** (*see also Salvia leucophylla*)
saguaro, **602, 603**
St. Anthony's fire, 222
Salicaceae, 238, 396
Salix, 169, **441, 480,** 495, 610 (*see also* willows)
Salmonella, 199
Salsola kali, 396 (*see also* tumbleweeds)
salt (NaCl)
  crystalline (ionic), **619, 623**
  -marsh plants, 164
  table (*see* sodium chloride)
saltbushes, 111
salt, excess, 554–55
Salvia, **385**
  leucophylla, 581, **582**
Salvinia natans, **332**
samaras, 394, **395**
Sambucus, 450
  canadensis, **418, 450, 451,** 481, **483**
sand
  coarse, 542
  dunes, 592
  fine, 542
  as soil, 544
sandstone, 541
sandy clay, 542
sandy loam, 544
Sanguinaria canadensis, **607**
Sansevieria, **25,** 112
sap (*see also* cytoplasm)
  cell, 24, 26, 44
  exudation from aphids, 571
  flow rate in phloem, 571
  micropylar, 382
  sieve-tube, 571
saponins, 397
saprobes, 242, 245 (*see also* bacteria; fungi)
Saprolegnia, 242, **243**
saprophytes
  angiosperm, **357,** 358
sapwood, **485,** 495–96
Sarcoscypha coccinea, **228**
Sargassum, 268, **269**
saturated/unsaturated fatty acids, 50
saturation, zone of, 544
savannas, 601, 604, 612
scales
  of gymnosperm cones, **373**
  of liverworts, 292
scarlet cup, 228
schizocarp, 395
Schizosaccharomyces octosporus, **220**
Schleiden, Matthias, 18
Schof, J. William, 185
Schwann, Theodor, 18
sclereids, **32,** 419–20, 430
sclerenchyma, 419, **420,** 429, **451,** 452
sclerenchyma cells (*see* sclerenchyma)
Scott, D. H., 331
scouring rushes, 327
scutellum, 409, 516, 526
sea lettuce, 265
sea moss, 296
sea otters, 252
sea palms, **172**
seasons, 595–96, 601, 606–7, 610
  and growth, 604, 611–12
seaweed, 268, 272, 275 (*see also* brown *and* red algae)
Sebdenia polydactla, **173**
Secale, 164
  cereale, 111, 222, 383, 433, 530
secondary growth, 4, 307–8, 476–97
  in *Botrychium,* 333
  effect on primary body of root, 479–80
  effect on primary body of stem, 480–82
  in *Isoetes,* 326
secondary plant body, 308, 416–17
secondary plant substances, 397–98, 400, 582
secretion, 27
secretory vesicles, 27, 29
sedges, **590**
sedimentary rocks, 541
Sedum, 475
seed(s), 9, 313–14, 405–9
  of angiosperms, 338, 359, 368–69, 371
  and auxin, 505
  -bearing structures of *Ephedra procera,* **355**
  coat, 313–15, 338, 348, 369, 371, 409, 516
  cupules, 314
  cytokinins in, 508
  development of, 346, 348, 368–69, 410–12
  dispersal, 348, **351,** 380, 395, 397
  dormancy of, 338, 513, 529
  of *Encephalartos altensteinii,* **353**
  evolution of, 313, 338
  ferns, 331, 338–39, 372
  germination, **406, 407,** 413–15, 515, 517, 524–27, 529, 533, 603
  germination and gibberellins, 513
  of *Ginkgo,* 354
  of grasses, 516
  of gymnosperms, 338, 340, 346, 348, 371
  habit, prerequisites of, 371
  lectins in, 550
  of *Magnolia,* **379**

of Orchidaceae, 381
of pines, 346, 348
plants, 304, 313–14, 330–31, 338–71, 635–37
production, 581
-scale complexes, 344
stored food in, 338–39, 348, 369, 371
viability of, 529
seedless vascular plants, 316–37, 339–40, 634–35
seedling, 413
dark-grown, 526
establishment of, 414
etiolated, 526
and geotropism, 519
light-grown, 526
Selaginella, 304, 309, 323–25, 330, 337
kraussiana, **19, 323**
life cycle of, **325**
rupestris, **323**
selection, 150, 165, 582 (*see also* natural selection)
artificial, 148
in diploid organisms, 144
in *Drosophila melanogaster*, 151
r and *K*, 581
response to, 151–53
semiarid environments, 603
senecios, **596**
senescence
and cytokinins, 510, 517
and ethylene, 510–11
sensitive plant, 531–32
sensory responses in bacteria, 190
sepals, 359, 378, **380**, 382, 400, 467, **468, 469**
of Asteraceae, 380
of Orchidaceae, 382
separation layer, 466 (*see also* abscission zone)
septum, septate (pattern), 215, 219, 222, 230
Sequoiadendron giganteum, 350, 607
Sequoia sempervirens, 341, 350, 564, **578, 579, 595**, 607
serine, **52, 127**
in tRNA, **127**, 510
seta (stalk), 288, **289**, 290, **291**, 294–95, 298, **302**, 303
sex attractants, 582
sex cells, 30
sex expression and ethylene, 511
sex organs (*see* gametangia)
shade tolerance, 580, 591
shagbark hickory, **458, 476, 488, 495**
shale, 541
shaman, Mexican, 233
sheath
bundle, **451**, 452, **453, 462**, 463
gelatinous or mucilaginous, 184, 202
shelf fungi, **229**, 230
shepherd's purse, 410–11
Shigella dysentarizae, 199
shoot
apex, 405, 415, 445–46
apical meristem of, 342, 405, 412, 414–15, 445–46
axillary (lateral), 415
emergence during germination, 414–15
reproductive (floral), 414
system, 5, 307, 415, 445
tip, 405, **445**
vegetative, 414, 446, 467
shooting-star, **360**
shoot-root ratio, 433
short-day plants, 522–25, 527–28, 530, 533
shrubs, 476–77, 495, 581–82, 587, 589, 596, **601**, 603, 607, **608**, 610, 612
primitive gymnosperm, 372
sieve area, 425
sieve cell, 343, **424**, 425–26, 431
sieve element, 309, 425–26
sieve plate, 425
in algae, 270
sieve tube, 426
in algae, 268, 270
exudate contents of, 571
sieve-tube member, **424**, 425–26, 431
Sigillaria, **316**
siliceous compounds, 253, 276
silicon (Si), **538**
silique, 394
Silphium, **366**
silt, 542
silty clay, 542
Silurian period, 287, 305, 309, 314, 637
silverfish, 47
simple pit, **34**, 35
sinigrin, **397**
siphonostele, 309–10, 314
of ferns, 333
of *Psilotum*, 317
siphonous algae, 262
sirenin, 246
16-dehydropregnenolone (16D), 399
Skoog, Folke, 507–8
skunk cabbage, **384, 385**
eastern, 85
western, **384**
slate, 541
sleep movements in leaves, **520**
slime (P-protein), **425**, 426, 431
slime (P-protein) bodies, **425**, 426
slime mold(s), 67, 128–30, 241, 247–50, 632–33 (*see also* cellular slime molds; plasmodial slime molds)
AMP and, 128–29
cellular, 67, **68**, 128–31
and fungi compared, 241
life cycle of cellular, 128–30
life cycle of plasmodial, 249
plasmodial, 247–50
pseudoplasmodium of, 129–30
slime (P-protein) plug, **425**, 426, 571
slime sheath, 247
slopes and environments, 596
slug (*see* pseudoplasmodium)
smelt, 589
Smilax, 531
smog, 464
smuts, **228**, 230, 234
snails, **587**
snake plant, **25**, 112
snapdragon family, 141, 378
snow, 610
sodium (Na), 539–40, 557, 615–19, **623**
chloride (NaCl), 616, **619, 623**, 626
hydroxide (NaOH), 626
ions (Na^+), 540, 543, **619, 623**
soft rot, 201
softwoods, 489, 496
soil(s) (*see also* clay), 541–45
acid, alkaline condition of, 543
and biomes, 597
and agriculture, 553–54
and cation exchange, 543
chalk, 580
classification of, 542
composition of, 542–43, 557
deficiences and toxicities, 554
erosion, 552
formation, 541–43, 557, 590
grassland, 604
horizons, 542, 544
latosols, 598, 612
living components of, 542–43, 545
microorganisms, organisms, 187, 197–98, 542–43, 552
mineral content of, 541
organic components, 541–42, 545, 557, **587**
particles, 542, 545, **550**
and plant growth, 541
and plant nutrition, 536–57
pore space of, 542–43, 557
prairie, 604
recycling of nutrients in, 545
separates, 542
textural classes of, 542
toxic conditions of, 553–54, 557
tropical, 598, 612
tundra, 611
and water, 543–45
water potential of, 545
Solanaceae, 506, 586
Solanum, 399, 506
berthaultii, 586
tuberosum, **23**, 506 560, 586
solar
energy, 544, 589–90
radiation, 596
solasodine, 399
Solidago
altissima, **360**
virgaurea, 157
Solomon's seal, **368**
solute
concentration, 62–63
potential, 62
solution, 62
solvent, 62
D-sorbitol, 230
soredia, 226, **227**
sorghum, 111
Sorghum vulgare, 111
sori, 334, **335**
fungal, 236
sorrel
redwood, **595**
wood, **520**
sources and sinks, 570, 573, 575
source-sink relations, 570
soybean, 111, 222, 506, **549**, 551
Biloxi, 522, 527
Spanish "moss," 112
sparrow tree, **396**
Spartina, **164**
alterniflora, 164
maritima, 164
× *townsendii*, 164
spasms, 222
special creation theory, 147
species, 157
classification by, 170, 175, 182
definition of, 175
diversity of, 594, 597
Species Plantarum, 170
spectrophotometer, 101, 526
spectrum, 98–99, 101
electromagnetic, 98–99
of visible (white) light, 98–99
sperm, 30
of algae, 259, 261, 265, **271, 277**
of angiosperms, 366, 368, 371
biflagellate, 289, 303
of bryophytes, 289, **302**, 303
of chytrids, 245
of conifers, 340
of cycads, 340
of *Equisetum*, 329
of ferns, 30, 335
of *Ginkgo*, 340
of Gnetophyta, 340
of gymnosperms, 30, 346, **347**, 371
of liverworts, 30
of *Lycopodium*, 321, **322**
of mosses, 30, **302**
of pine, 346, **347**
of *Psilotum*, 318
of seedless vascular plants, 340
of *Selaginella*, 324, **325**
of water molds, 242, **243**
spermagonia, 236
spermatangia, 274
spermatia, 236, 274
spermatozoid, 282
Sphaeropteris glauca, **332**
Sphagnidae, 296, 303
Sphagnum, **173, 300, 301**, 302–3, 580
Sphenophyta, 309, 311, 317, 327–30, 337, 635
spheroids, *Volvox*, 261
spherosomes, 28–29
spike, **361**
spinach, 522–23, 530
Spinacia oleracea, 522
spindle fibers, apparatus, 39–41, 135, **136**, 215, 220
spines, 472, **473, 584**, 594
spiraea, 384
spirilli, 187, 210
Spirillum lipoferum, 555
spirochaetes, 186
Spirogyra, 267
Spiroplasma citri, 203
spiroplasmas, 202–3
Spizella arborea, **396**
sporangia (*see also* capsule; megasporangia; microsporangia)
of algae, **266, 271, 274**
of *Allomyces*, 246
of angiosperms, 338, 359, 363, 366, 371, 378
of bryophytes, 288, 290, 294, 296–97, **301**, 303
of cellular slime molds, 247
of cycads, 353
of *Equisetum*, **328**, 329
of ferns, 334–35
of fungi, 214, 216, **217**, 218, 239
of gymnosperms, 338, 340, 371
of *Isoetes*, 326, **327**
of Lycopodium, 321, **322**
of pines, 344, 346, 354
of plasmodial slime molds, **172**, 248, **249**
of Psilotum, 318
of *Selaginella*, 323, **324**
of water molds, **242, 243**
sporangiophore, 216, **217**, 218, **328**, 329
spore dispersal, 291, 294, 296, 339
in *Equisetum*, 329
in hornworts, 296
in liverworts, **290**, 292, 294
in mosses, 298, 301–3
sporophylls,
of angiosperms, 359
of gymnosperms, 340
of *Lycopodium*, 320–321
of *Selaginella*, 323
sporophyte, 181, 183
of algae, **266, 271, 274**
of ancestor of plants, 287, 304
of angiosperms, 370–71, 403–97
of bryophytes, 288–89, 303
of *Equisetum*, 327, 329
of ferns, 332–34
of granite mosses, 303
of hornworts, 296–97, 303
of *Isoetes*, 326, **327**
of liverworts, **290, 291**, 292, 294–95
of *Lycopodium*, 320
of peat mosses, **301**, 302–3
of pines, 348–49, 371
of *Psilotum*, 317, **318**
of seed plants, 338
of *Selaginella*, 323
of true mosses, 298, **300, 301, 302**, 303
of vascular plants, 289, 304, 307–11, 314
sporopollenin, **365, 366**
sprouts, 155
spruce, 341, 350, 433, 610
white, **591, 610**
spurge family, **168, 169**, 473
spur of flower, 388
squash, **418, 421, 422, 425**
squaw-root, **357**
stain, fluorescent, 383
stalk, 232 (*see also* seta; stipe)
stalk cell, 346
stamens, 359, 361, 378, **379**, 380–81, **392**, 400, 467, **468, 469, 470, 471**
Stamnostoma huttonense, **313**
Stanley, Wendell, 205
Stapelia schinzii, 384, **385**
Staphylinidae, **384**
Staphylococcus, 200, 224
epidermidis, 198
star, birth of, **72**, 73
starch, **45**, 46–47, **48**, 84, 109, 253
cyanophycean, 196
digestion of, 516
floridean, 253, 272
as food reserve, **45**, 47, **48**, 287
in plastids, 22, **45, 48**
statoliths, 519
stearic acid, 50
Steinbeck, John, **604**
steles, 309–10, 314
stem, 4–5, 307–8, 314
apical meristem of, 445–46
association with leaf, 453, **454**
of bryophytes, 288
bundle, 453
comparison with root, 437, 446, 475
of conifer, 447
development, summary of, 496

of dicot, 447
elongation, 512
of *Equisetum*, 327, **328,** 329
fleshy, 474–75
function of, 445
of herbaceous dicot, **448,** 451
-leaf relation, 453–54
of *Lycopodium*, 320
modifications, 471–75
modified, 474, 531 (*see also* tendril)
of monocot, **449, 452, 453**
periderm of, 482–84
phloem of, 449
of *Pinus*, **343**
primary structure of, 447–52
of *Psilotum*, 317, **318**
secondary growth of, **479,** 480–82
secondary structure of, **479,** 480–82
of *Selaginella*, 323, **324**
storage, 474–75
succulent, 475
vascular cambium of, 449–50, **479,** 480–82
woody, external features of, **487**
of woody dicot, 449–50
xylem of, **448,** 449–52
Stemonitus, **172**
sterigmata, **231,** 232
sterile bract, 344
sterility, 157, 162, 164
Steward, F. C., 130, 507
stigma (eyespot), 257, 281, 283
of flower, 359, 368, 377, **379, 380,** 382–83, **385, 389, 391,** 392, 400
stigmatic surfaces, 377, 383
stigmatic tissue, 368
stimulus
bacterial response to, 190
in cellular slime molds, 128–29
in euglenoids, 282
response to in diatoms, 277–78
stinkhorns, **229,** 230, 232, **599**
stipe, 232 (*see also* stalk)
in algae, 268, **269, 271**
stipules, 457, **458, 584**
stolon, 154, 216, **217**
stomata, 4–6, 106, 305, **308,** 427–28, 475
in CAM plants, 111–12
of desert perennials, 603
development of, 428
factors affecting movement of, 567
of hornworts, 288, 296, 303
in leaf epidermis, 342, 459, **460, 461,** 475, 566–67
mechanism of opening and closing of, 566–67
of mosses, 288, 298, 303
in stem epidermis, 449, 484, 566–67
structure and function of, 428
and transpiration, 565–67
stomatal crypt, **461**
stone cells, **32, 420**
stonecrops, 111, 475, 567
stoneworts, 267–68
storage
food, 338–39, 348, 369, 442–43, **444,** 474
leaf, 474–75
parenchyma, 418, 442–43
root, 432, 442–43, **444**
in seed, 348, 369
stem, 474–75
water, 474–75
straw, 548
strawberry, 154, 393, 505, 523
strawberry bush, 396
streams, 553, 556
Strelitzia reginae, **389**
Streptococcus, 199
lactis, **187**
Streptomyces
fradiae, **186**
scabies, **186**
streptomycin, 187
Strid, Arne, 160–61
strobili (*see also* cones)
of *Equisetum*, **327, 328,** 329
of *Lycopodium*, 320–21, **322**
of pine, 342, 344, **345,** 346, 348
of *Selaginella*, 323
stroma, 20, **21, 96,** 107, 112
style, 359, 368, 377, **379,** 382–83, 400
Stylites, 319
suberin, 33, 50, 438, 483, 507
subsidiary cell, 427–28
subsoil (*see* B horizon)
subsporangial swelling, 218
substrate, 78, 83
subtilin, 187
succession, 580, 589–94
succinate (succinic acid), 89, 92
succulency, 475
succulents, 111–12, 475
suckers, 155
sucrase, 56
sucrose, 47, 56, 58, 82, 84
sugar(s), 46–47, **48, 49,** 58 (*see also by name*)
beet, 432, 443, **444**
5-carbon, 46
and lectins, 550
in lichens, 230
maple, 457
molecules, **46, 47, 49**
in nucleic acids, 56, **57,** 58
phosphate, 121
and plant hormones, 504
simple, 46
6-carbon, 46
transport in phloem, 569–74
sugarcane, 111, 552, 555
Suillus bovinus, **229**
sulfate (SO_4^{2-}), 198, 540
sulfur (S), 97, **119,** 196, 198, 538–41, 619
in chemoautotrophs, **186,** 197
cycle, 198, 210
dioxide, oxides, 227, 288, 464, 541, 556
fungus, **171**
ions (SO_4^{2-}), 540, 543
in photosynthetic bacteria, 97
sulfuric acid, 556
as pollutant, 464
sun, 1, 76 (*see also* energy; solar)
sunbird, collared, **389**
sundew, **546,** 547
sunflower, 20, **381,** 408, 532–33
family of, 378
supernumerary cambia, 443, **444**
survival of the fittest, 148
suspended animation, 186, 226
suspensor, 324, 346, 410–11
swamps, 548
sweet,
fern, 551
gale, 551
sweet potato, 155, 432, 443
symbiosis (*see* mutualism)
symbiotic associations, 198, 203–4, 237–39, 254, 272, 277, 280, **536, 537,** 539, 549–51, **552** (*see also* lichens; bacteria)
symbiotic fungus
of *Lycopodium*, 321
of *Psilotum*, 317
symplast, 68, 568
symplastic transport (movement, pathway), 68, 478, **561,** 568-69
Symplocarpus foetidus, 85, **384, 385**
sympodium, 453
synapsis, 134
synaptonemal complex, 134
Synechococcus, 203
synergids, 368
syngamy, 132, 139, 180–81, 368, 371
Synura petersenii, **30**
syphilis, 186, 224
Syringa, 504
vulgaris, **460, 461**
Szent-Györgi, Albert, 1

Tagetes, **22**
taiga, 610–12
tamarack, **610**
tangential section (surface), **485, 489,** 490–91, **492**
tapetum, 363, **364,** 366
taproot, **341,** 432, **433**
Taraxacum, **395, 433**
officinale, **174**
taxa, 175
Taxaceae, 350, **351**
Taxodiaceae, 350, **352**
Taxodium, 350
distichum, *352*
taxonomic categories, 175, 631
and inorganic nutrients, 538
taxonomist, 175
Taxus, **351,** 495
canadensis, **424**
T-DNA, 195
tea, **397**
Tegeticula yucasella, 389
telia, 236
teliospores, 236
telophase
meiotic, 135, **136, 137**
mitotic, 36, **37, 38,** 40–41, 44
Teloschistes chrysophthalmus, **226**
temperate deciduous forests, **543,** 585, 598, 604, 606–8, 612
temperate evergreen forests, 607
temperate regions, 609
temperature
and altitude, 596
and biological clocks, 522
diurnal, 595–96, 612
and evaporation, 567
and latitude, 596
seasonal, 595
and stomatal movement, 567
temperature conversion scale, 630
tendrils, 471, **472, 584**
and touch responses, 531–33
tension (negative pressure), 564–65
tepals, **359, 380**
terminal-bud-scale scars, **447, 487**
termites, 50
terpenes, 581–82
terpenoids, 397
terramycin, 187
Terrapene carolina triunguis, **587**
Tertiary period, 341, 350, 384, 607
testcross, backcross, 140
tetanus, 187 (*see also Clostridium tetani*)
tetrads, 137
of spores, 363–64, 366
tetrahydrocannabinol (THC), **399**
Tetraselmis, 258
tetrasporangia, tetraspores, 274
tetrasporine line (*see* green algae)
tetrasporophyte, 274
T4 bacteriophage, **207**
Thalassioria nordenskioeldii, 276
thallus, 265, 273, 296
in algae, 265, 273
of liverworts, 291–92
Thea sinensis, **397**
theca, 279
Thelypteris palustris, **174**
Theobroma cacao, **397**
theobromine, **397**
Theophrastus, 549
therapeutics (*see* antibiotics)
thermocouple, 565
thermodynamics, 73, 83
first law of, 74–75, 83
second law of, 75–77, 83, 85, 96
thermonuclear reaction, 73
thermophilic organisms, 554
thermotolerant organisms, 554
Thiobacillus, 198
Thiothrix, 188
thorns, 472, **473,** 582, **584**
of *Acacia*, 582, 585
three-carbon (C_3) photosynthetic pathway, 107–8
three-carbon (C_3) plants, 108
threonine, **52**
thrush, 223
Thuidium delicatulum, **298**
thylakoids, 21, 44, 96, 102, 104, 107, **191**
thymine, 56, **57,** 118, **120,** 121–22, 124
Tiburon mariposa lily, 609
Tilia, 449, 485
americana, **420, 422, 424,** 426, **448,** 449, **482, 485, 487**
Tillandsia usneoides, 112
time's arrow, 75
tissue culture, 507–9
tissues, 307–9, 314, 417 (*see also by name*)
complex, 417
concept of, 417
simple, 417
tissue systems, 307, 410, 417, 437
Tmesipteris, 317, 319, 337
lanceolata, **319**
parva, **319**
TMV (tobacco mosaic virus), 205, **207**
toadstools, 230, 232, 240
tobacco, 388, **397, 454, 455, 456,** 506–9, 513, 522, 528, 586 (*see also Nicotiana tabacum*)
Maryland Mammoth, 522, 528
mosaic virus (TMV), 205, **207**
Trabezond, 528
tomato(es), 393, 398, **429,** 505–6, 511, **518,** 560, 586
tonoplast, 18, **23,** 24, 44, 64, 67
topsoil, 542
torus (tori), 489–90, **491**
totipotency, 506
touch-me-not, 396
touch responses, 531–33
Toxicodendron radicans, **398**
toxins, 199, **200,** 204, 280, 553–54, 556, 582, 586 (*see also* secondary plant substances)
tracers, radioactive, **562,** 571–72
tracheary elements, 309, 421–22
evolution of, 309
tracheids, 305, 309, 343, 421–23, 430, 489–90, **491,** 492
traits (dominant/recessive), 139–42
transamination, 548, 557
transcription, 131
regulating gene, 126–28, 133
of RNA from DNA, 124–26
schematic view of, **127**
transduction, 193, 210
transfer cells, **416,** 418
transformation (bacterial), 118, 193, 210
transfusion tissue, 343
transition region, 466, **467**
translation, 131
of mRNA into protein, 126
schematic view of, **127**
translocation, 270, 569–74
transmembrane potential, 568–69
transmitting tissue, 368
transpiration, 559–60, 565–68, 574–75, 601
transpiration-pull theory, 564
transpiration stream, 561
transverse section (surface), **485, 489,** 490–91, **492, 493**
Trebouxia, 225, **230**
tree(s), 476–77, 485, 495, 580–82, 586, 589, 596, **599,** 603 (*see also by name;* forests; gymnosperms)
water movement in, 563–65
treeline, 493
Trentepohlia, **225**
Treponema pallidum, **186**
tricarboxylic acid (TCA) cycle (*see* Krebs cycle)
Trichodesmium, 203
trichogyne, 220, **221,** 274
trichomes, 400, 427, 429 (*see also* hairs)
Tridachnidae, 254
Trifolium, 550–51
glomeratum, **550**
repens, 153, **549**
subterraneum, 530
Trillium grandiflorum, **3**
Triloboxylon ashlandicum, **339**
Trimerophyta, 305–6, 339
trimerophytes, 339
Trimerophyton, **306**
triple fusion, 368, **369**
Triticale, 11, 164–65
Triticum, 111, 164
aestivum, 135, **408,** 409, **419,** 560
erectum, 116, 117
tritium (^3H), **121**
tRNA (*see also* RNA)
and cytokinins, 510

trophic levels, 587–89, 594
tropical environments, 596, 609
tropical rain forests, 356, 597–99, 601, 612
tropism(s), 518–19 (*see also* geotropism; phototropism)
trout, 589
truffles, 219, **228**
tryptophan, **52,** 503
Tsang Wang, 350
Tsuga, 350, 489, 610 (*see also* hemlocks)
  heterophylla, **238**
T2 bacteriophage, 184
tube cell, 344, **345,** 366
tube nucleus, 368
tuber(s), 155, 474–75
tuberculosis, 187
Tuber melanosporum, 228
tubulin, 29
Tulasnellales, 239
tulips, 530
tulip tree, **463**
tumbleweeds, 396
tumors, plant, 195, 208, 550
tundra, 596, 611–12
tunica-corpus, 446
Turesson, Göte, 156
turgor, 63, 69, 541, **566**
turgor pressure, 63, 566–67
  and touch responses, **531,** 532
turnips, 527
turtle, three-toed box, **587**
twigs, **487**
2,4-D, **503,** 504, 507
tyloses, 495
typhoid fever, 199
typhus, 199
tyrosine, **52**
tyrosine tRNA, 510
tyrothricin, 187

Ulmus, **395**
  americana, **42,** 487, **494**
Ulothrix, 264–66
ultraviolet portion of spectrum, 385–86, 393
ultraviolet radiation, 392
ultraviolet rays, 2
Ulva, 265, 270
  lactuca, **265**
  life cycle of, **266**
Ulvaria, **31**
umbel, **361**
undulant fever, 199
unisexual plants, 292, 296, 298
uracil, 56, **57,** 124, 126
uredinia, uredospores, 236
urn moss, 288
Urtica, **429**
Usnea, **225,** 287
Ustilago maydis, **228**
Utricularia vulgaris, **546**

vacuole, 18, **23,** 24, 26, **28, 42,** 44, 67
  contractile, 129, 257, 281
  gas, 191
  and solutes, 63, 67
vacuum, 98
valine, **52**
Valonia, 262, **263**
valves, of diatoms, 276, **277**
variety, classification by, 175
vascular
  rays, 477–78
  systems, 4, 339, 447, **449,** 453, **454**
  transition, 466–67
vascular bundle(s), 447, **448, 449,** 450–52
  closed, open, 451
  differentiation of, 450-52
vascular cambium, 4, 308, 353, 416, **421,** 477–79, 495
  and auxin, 504, 517
  of *Botrychium,* 333
  of *Isoetes,* 326
  in root, 440, 443, 479–80
  in stem, **343,** 449–51, **479,** 480–82
vascular cylinder
  of roots, 440, 444
  of stem, 447, **448,** 449
vascular plants, 4, 5, 178, 304–15, 317–401
  and bryophytes compared, 287–88, 303
  characteristics of, 311
  early, 305–6
  evolution of, 287, 304–15, 372–401
  generalized life cycle of, **311**
  groups of, 634–36
  introduction to, 304–15
  reproductive systems of, 311–14
vascular strand
  of *Equisetum,* 328
  of flowers, 378
vascular tissue(s) (*see by name*)
  induction of, by hormones, 504, 517
Vaucheria, 278–79
vectors, 195, 199, 203, 208
vegetation types (*see* biomes)
vegetative reproduction (*see* reproduction)
velamen, 442
venation, 462–63, **464**
venter, 289–90, **302**
Venus flytrap, 532–33, 547
vernalization, 530, 533
Veronicastrum virginicum, **457**
Verrucaria serpuloides, 225
vertebrates
  heat energy in, 85
vesicles, 550
vessel members (elements), 309, 390, 421–23, 430, 492, 496
vessels, 309, 422, **492,** 493, **494,** 495
  in Gnetophyta, 309, 356
Viburnum marginatum, **373**
Vicia faba, 5, 572
Vigna sinensis, 560
vine(s), 354, **355,** 356, 477, 514, **584,** 612
  woody, 597
Viola, **170**
  papilionacea, **170**
  rostrata, **170**
  tricolor var. *hortensis,* **170**
violets
  common, **170**
  long-spurred, **170**
  reproduction of, 154
viral diseases, 204–6, 208–9
Virchow, Rudolf, 18
Virginia creeper, 471
viroids, 209
virus, 118–19, 204–10, 580
  as bacteriophage, 118, **119,** 206–8
  and cancer, 208–9
  contractile sheath of, **207**
  Epstein-Barr, 208–9
  and evolution, 209–10
  human cold, 204
  influenza, 206
  plant, 206–7
  polio, 204, 206, 208
  replication of, 206
  rhabdo-, 206–7
  RNA, 194
  structure of, 205
  Tulare apple mosaic, **207**
  wound tumor, 208
vision, 99
vitamin, 548
  A, 103, **104**
  B-complex, 88
  B_2, **90**
  B_{12}, 279
  synthesis of, 537
Vitis, 471–72, 488
  vinifera, 515
Vittaria guineensis, **17, 28**
volcanoes, 592–93
volva, 232
volvocine line, 260
Volvox, **172,** 260–61
  carteri, 261
Vorticella, 254
Voyage of the Beagle, The, 147–48

wake-robin, **116,** 117
Wald, George, 99
Wallace, Alfred Russel, 375
Wallace's line, 374
wall pressure, 63
walnuts, 408, 511
washes, 603
Washingtonia filifera, **602, 603**
wasps, 386, 400
water
  absorption of, 560–63, 574
  balance, 539
  characteristics of, 65
  as component of cell sap, 24
  as component of living matter, 45
  conducting cells, 288, 298 (*see also* hydroids)
  cycle, 544
  drainage, 552
  formation of in respiration, 84, 92
  of guttation, 561, **562**
  molecule (H_2O), 65, 540
  photolysis, 105
  potential, 59–60, 63, 69
    of soil, 544–45
  pressure, in cells, 62, 507
  and respiratoin, 87, 92
  table, 544
  tensile strength of, 564, 574
  transport, 563–65
  uptake of, 507, 560–63, 574
  vapor, 559, 565, 568, 574, 596, 602
water felt, **278**
water hemlock, **360**
water hyacinth, **590**
water lily family, 378, **590**
watermelon, 582
water molds, 241–45, 249–50
watershed, 585–86
waterwheel, 60, 74
Watson, James D., 118–19, **120,** 122, 131, 205
wavelengths (*see also* light)
  energy of, 98–99
  radio, 98
  X-rays, 98
wax deposits, **427**
waxes, 33, 50, **51,** 58
wax plant, 112
weathering of rocks, 541–42, 552, 556–57, 593
weed control, 507
weeds, 9, 154, 604
Weinberg, G., 149
weir, **587**
Weissman, Charles, 194
Welwitschia, 354, 356
  mirabilis, 112, **355**
Went, Frits W., 502, **503,** 512, 518
wetlands, 553
Wetmore, R. H., 504
wheat, 9, 20, 111, 145, 164, **408,** 409, **419,** 523, 560
  bran, 409
  germ, 409
  production, 11–12
whirling whips, 279
White Mountains, 493, **494**
whooping cough, 199
Wilde, S. A., 580
Wilkins, Maurice, 119, 121
Williams, Stephen E., 532
willow(s), 96, 238, 362, 396, **441, 480,** 495, 610
wilt (disease), 201
wilting, 64, 544
wine, 94, 222
winterberry, **396**
wintercress, **538**
wintergreen, 380
witch hazel, 396
Woese, Carl R., 197, 202
Wolffia, 356
  microscopica, **356**
wood, 489–95
  of conifer, 489–93, **494,** 495
  of dicot, 489, 492–93, **494,** 495
  diffuse-porous, 493
  early, **489,** 493, 496
  late, **489,** 493, 496
  of progymnosperms, **339**
  ring-porous, 493
woodlands, 579, 608
woody plants, 603, 612
work, 74
wort, 291
wound tumor virus, 208

Xanthomonas, 201
  campestris, 201
Xanthium strumarium, 510, 522, **523**
xanthophylls, 103, 253, 270
Xanthophyceae, 278–79, (*see* yellow-green algae)
xerophytes, 459
X-ray diffraction studies, 119
xylan, **49**
xylem, 4–5, 15, 304, 314, 417, 421–23, 429
  and plant hormones, 504
  primary, 307, 421, 423, 440, 444, 449–52
  of progymnosperms, 339
  secondary, 308, 421, 423, 478, 489–95
  secondary, of conifer, 489–91
  secondary, of dicot, 492
  secondary, of pines, 343
  and water transport, 563–65, 574
xylose (xylans), **49**

yams, wild, 399
Yapoah Crater, **593**
yeast, 94, 220–22, 239
yellow fever, 204
yellow-green algae, 278–79 (*see also* algae)
yew(s), **424,** 495
yucca
  flower, **389**
  moth, **389,** 580
Yucca, 580
  brevifolia, **602**

Zamia pumila, 353
Zea mays (corn), **15, 21, 36, 59, 68, 108, 109, 110,** 111, 128, **145,** 179, 391, **407, 414,** 415, **427, 434, 435, 436, 437, 441, 449, 452, 453,** 508, **559,** 560, 567, **568,** 580, **581**
zeatin, 508
zeaxanthin, **104**
zebra, **601**
Zebrina, **23**
Zigadenus fremontii, **387**
zinc (Zn), 538–40, 543, 554, 557
  ions (Zn^{2+}), 540
zone of saturation, 544
zooplankton, 251
zoospores
  of algae, 262, 264–66, **271,** 280
  of chytrids, 241, 245, **246,** 250
  of water molds, 241–42, **243,** 249–50
Zooxanthellae, 280
Zosterophyllophyta, 305–6, **307,** 314
Zosterophyllum, 305, **306,** 309, 320
Zygocactus truncatus, 207
Zygomycota (zygomycetes), 216–18, 239, 580, 632
  evolution of, 215
  life cycle of, **217**
zygospore(s)
  of algae, **257,** 258, **267**
  of fungi, 216, **217**
zygotes, 132, 178
  of algae, **257,** 258, 264, 266, **271, 272, 277,** 282
  of angiosperms, 368–69
  of bryophytes, 288, 290, **302**
  of chytrids, **246**
  division of, 178, 181–82, 409–10, 412
  of fungi, 220
  of gymnosperms, **349**
  of plasmodial slime molds, **249**
  of water molds, 242, **243**